Epigenetic Regulations in Plants

This book provides a detailed overview of the significance of epigenetic regulation as an important mechanism underlying the regulation of gene expression in plants. It highlights the role played by epigenetic mechanisms in the overall physiology of plants and how these mechanisms can improve quality traits by conferring tolerance and resistance against abiotic and biotic stresses, respectively.

The book is organized into four distinct sections, beginning with an introduction to the subject and a generalized discussion on the different epigenetic markers. Chapters then go on to discuss epigenetic responses in model plants, including *Arabidopsis*, rice and cereal crops, and the role epigenetics plays in controlling physiological processes and providing protection against various forms of stress. The volume introduces readers to the application of epigenetics, providing them with a clear road map to undergo qualitative and quantitative improvement of plant species to enhance productivity and performance. This can be achieved through a better understanding of the different components of epigenetic machinery, chromatin accessibility and transcription factors that regulate the expression of concerned genes.

This book is an essential read for researchers and scientists working in the fields of plant physiology, molecular biology and biotechnology, as well as in related disciplines such as biochemistry, agricultural science and botany. It will benefit university libraries and research institutes dealing with epigenetics and its applications in improving plant growth, development and quality, and in overcoming different stresses.

Epigenetic Regulations in Plants

Mechanisms, Plant Development and Stress Responses

Edited by
Aryadeep Roychoudhury

CRC Press
Taylor & Francis Group
Boca Raton London New York

CRC Press is an imprint of the
Taylor & Francis Group, an **informa** business

Designed cover image: Shutterstock

First edition published 2026
by CRC Press
2385 NW Executive Center Drive, Suite 320, Boca Raton FL 33431

and by CRC Press
4 Park Square, Milton Park, Abingdon, Oxon, OX14 4RN

CRC Press is an imprint of Taylor & Francis Group, LLC

ISBN: 978-1-032-94313-8 (hbk)
ISBN: 978-1-032-96565-9 (pbk)
ISBN: 978-1-003-58991-4 (ebk)

DOI: 10.1201/9781003589914

Typeset in Times
by Deanta Global Publishing Services, Chennai, India

Contents

Preface

Waddington first coined the term "epigenetics", which combines the terms "epigenesis" and "genetics", and defined it as "the study of the interaction between a gene and its product that produces the phenotype". The transcription of a gene is not only influenced by DNA sequence but also depends on the manner of gene organization in the chromosome. Transcription factors and RNA polymerase bind to the promoter region, only upon opening up of the chromatin architecture, thereby switching on gene expression. Epigenetic regulation, consisting of DNA and histone modifications and chromatin remodeling, is an important mechanism underlying the regulation of gene expression in plants. Three vital epigenetic markers, namely small non-coding RNAs, DNA methylation and posttranslational histone modifications, together with the level of chromatin condensation, are important for gene regulation and influence chromatin accessibility to the transcriptional machinery. The propagation of epigenetic markers during gametophyte development determines the mechanisms of transgenerational epigenetic inheritance.

The spatiotemporal regulation of gene expression and epigenetic regulators are the determining factors for plant growth and development. Epigenetic regulation is the determining factor in controlling vital physiological activities in plants, like photosynthesis-related carbon fixation pathways, photoperiod-mediated flowering, sulfur homeostasis and plant response to sulfur deficiency, nitrogen signaling and improving nitrogen use efficiency and the selection of hybrids with better performance. Epigenetics provide essential biological information that could be directly used to enhance crop tolerance and adaptability. Plant epigenetic regulation in response to stress holds immense importance in agriculture and can revolutionize crop breeding and genetic engineering strategies. Plants employ different types of stress memories, such as somatic, intergenerational and transgenerational, regulated by epigenetic changes. Such epigenetic memory helps plants to respond better during recurrent stress and prepare their offspring to face future assaults.

Despite such tremendous potential of epigenetic mechanisms in controlling gene expression in diverse physiological processes in plants and in the tolerance/resistance mechanism against various abiotic and biotic stresses, there is no comprehensive latest book focusing exclusively on epigenetic regulation in plants, vividly describing its significance in the overall physiology and improving quality traits of plants. This book provides a detailed overview of the significance of different epigenetic mechanisms in model plants including cereals, rice and *Arabidopsis* and encompasses several chapters that highlight how such epigenetic regulation governs the physiological aspects and defense against different abiotic and biotic stresses. The book is exhaustive with 30 chapters, which are systematically organized under different sections, with the first section containing chapters that provide a generalized discussion on the different epigenetic markers, especially for beginners. The latter sections are more specialized in comprehending such regulation in the plant system.

This book introduces the readers primarily to the application part of epigenetics in controlling vital physiological activities and protection against stress episodes. It is unique of its kind since, to my knowledge, there is no such latest exhaustive book available on the market with the same title. This book definitely provides the reader a clear roadmap to achieve qualitative and quantitative improvement of plant species and improve productivity by better understanding the different components of epigenetic machinery and chromatin accessibility to transcription factors for upregulating or downregulating the expression of concerned genes. I firmly believe that the book will be beneficial to different universities and research institutes dealing with epigenetics and their applications to improve plant growth and development, quality and overcome the effect of different stresses.

Aryadeep Roychoudhury
Professor
Discipline of Life Sciences, School of Sciences
Indira Gandhi National Open University, New Delhi

About the Editor

Aryadeep Roychoudhury is currently working as Professor in the Discipline of Life Sciences, Indira Gandhi National Open University, New Delhi, India. Earlier, he served as Assistant Professor at the Postgraduate Department of Biotechnology, St. Xavier's College (Autonomous), Kolkata, West Bengal, India. He earned his BSc (Hons) in Botany from Presidency College, Kolkata, and MSc in Biophysics and Molecular Biology from the University of Calcutta, West Bengal, India. He pursued his doctoral research at Bose Institute, Kolkata, and earned his PhD from Jadavpur University, Kolkata, in the area of stress biology in plants. Following his PhD work, he joined as Research Associate (post-doctorate), pursuing translational research on transgenic rice. He is presently involved in active research in the field of abiotic stress responses in plants with perspectives to physiology, molecular biology and cell signaling under diverse stress conditions. He has 26 years of research experience in the concerned discipline. He has handled several government-funded projects as a principal investigator and supervised five PhD students. Professor Roychoudhury has published over 300 articles in peer-reviewed journals and chapters in books of international and national repute. He has edited 18 books with publishers such as Wiley, Elsevier, Springer, Taylor & Francis Group and Nova and has also handled special issues as Guest Editor for several renowned international journals. He is a regular reviewer of articles in high-impact, international journals, an editorial board member of several journals and Life Member of different scientific associations and societies. He is the recipient of the Young Scientist Award 2019, conferred upon him by the International Foundation for Environment and Ecology, at the University of Allahabad, Prayagraj, Uttar Pradesh, India. He also received the Distinguished Scientist Award – 2024 conferred upon him by the Institute of Ecotoxicology and Environmental Sciences, India. His name is included in the Stanford University's list of world's top 2% most influential scientists.

Contributors

Dilawar Ali
Department of Plant Breeding and Genetics
Bahauddin Zakariya University
Multan, Pakistan

Qurat ul ain Ali
Department of Agricultural Sciences and Technology
Atta-ur-Rahman School of Applied Biosciences (ASAB)
National University of Sciences and Technology (NUST)
Islamabad, Pakistan

Sami Ul-Allah
Department of Plant Breeding and Genetics
Bahauddin Zakariya University
Multan, Pakistan

M. S. Amritha
Plant Physiology and Biochemistry Division
Department of Botany
University of Calicut
Malappuram, India

Irshad Arshad
New Mexico Environment Department
Santa Fe, New Mexico, USA

Asad Azeem
Department of Plant Breeding and Genetics
University of Layyah
Layyah, Pakistan

Abhinaba Banerjee
Department of Biotechnology
St. Xavier's College (Autonomous)
Kolkata, India

Ritesh Banerjee
School of Biological and Environmental Sciences
Shoolini University
Solan, India

Susmita Behara
Department of Life Science
Guru Nanak Institute of Pharmaceutical Science and Technology
Kolkata, India

Savita Bhardwaj
Department of Life Sciences
MCM DAV College
Kangra, India

Mousumi Bhowal
Department of Botany
Hislop College
Nagpur, India

Nikha Borah
Microbiology Laboratory
Department of Life Sciences
Dibrugarh University
Dibrugarh, India

Bornita Bose
Department of Life Science and Biotechnology
Jadavpur University
Kolkata, India

Sayak Chakraborty
Department of Life Science
Guru Nanak Institute of Pharmaceutical Science and Technology
Kolkata, India

Ashish Chandran
Department of Botany
Feroze Gandhi College
Raebareli, India

Channakeshaviah Chikkaputtaiah
Biological Sciences and Technology Division
CSIR-North East Institute of Science and Technology (CSIR-NEIST)
Jorhat, India

Gourav Chouhan
Department of Botany
JNV University
Jodhpur, India

Ritika Dadhwal
School of Biological and Environmental Science
Shoolini University
Solan, India

Yanchen Dolma
Department of Life Sciences
MCM DAV College
Kangra, India

Suman Ganger
Department of Microbiology
Vivekanand Education Society's College of Arts, Science and Commerce
Mumbai, India

Nimra Ghafoor
Department of Agricultural Sciences and Technology
Atta-ur-Rahman School of Applied Biosciences (ASAB)
National University of Sciences and Technology (NUST)
Islamabad, Pakistan

Dibakar Ghosh
Plant Biochemistry Laboratory
Department of Botany
University of North Bengal
Siliguri, India

Abhik Gogoi
Department of Agricultural Biotechnology
Assam Agricultural University
Jorhat, India

Ankita Gogoi
Microbiology Laboratory
Department of Life Sciences
Dibrugarh University
Dibrugarh, India

Titir Guha
Department of Biology
IISER Tirupati
Panuguru, India

Heeya Gupta
Department of Biotechnology
St. Xavier's College (Autonomous)
Kolkata, India

Ashmaul Husna
Department of Life Science
Guru Nanak Institute of Pharmaceutical Science and Technology
Kolkata, India

Poonam Jodha
Department of Botany
JNV University
Jodhpur, India

Vishal Kamble
Department of Microbiology
Shahid Veerpatni Laxmi Mahavidyalaya
Titave, India

Bharat Kapoor
Department of Hotel Management and Tourism
Guru Nanak Dev University
Amritsar, India

Dhriti Kapoor
School of Biological and Environmental Sciences
Shoolini University
Solan, India

Subhasis Karmakar
ICAR-National Rice Research Institute
Cuttack, India

Sana Khalid
Department of Environmental Sciences
COMSATS University Islamabad
Vehari, Pakistan

Ayesha Khan
Department of Molecular Biology and Genetics
Faculty of Engineering and Natural Sciences
Istinye University
Istanbul, Turkey

Bushra Hafeez Kiani
Department of Biological Sciences
International Islamic University
Islamabad, Pakistan

Eyerusalem Goitom Kidane
Department of Biotechnology
Vel Tech Rangarajan Dr. Sagunthala R&D Institute of Science and Technology
Chennai, India

Ajay Kumar
Department of Life Sciences
MCM DAV College
Kangra, India

Prajita Kundu
Department of Chemistry
University of Calcutta
Kolkata, India

Rita Kundu
Centre of Advanced Study
Department of Botany
University of Calcutta
Kolkata, India

Priya Kushwaha
Plant Nutrition and Stress Physiology Laboratory
Department of Botany
University of Lucknow
Lucknow, India

Konkona Lahiri
Department of Biotechnology
St. Xavier's College (Autonomous)
Kolkata, India

Sivarama Krishna Lakkaboyana
Department of Chemistry
Vel Tech Rangarajan Dr. Sagunthala R&D Institute of Science and Technology
Chennai, India

Payal Lodha
Department of Botany
University of Rajasthan
Jaipur, India

Parama Majumdar
Department of Botany
Hislop College
Nagpur, India

Snehalata Majumdar
Department of Life Science
Guru Nanak Institute of Pharmaceutical Science and Technology
Kolkata, India

Swayambhik Mukherjee
Department of Biotechnology
St. Xavier's College (Autonomous)
Kolkata, India

Faiza Munir
Department of Agricultural Sciences and Technology,
Atta-ur-Rahman School of Applied Biosciences (ASAB)
National University of Sciences and Technology (NUST)
Islamabad, Pakistan

Behzad Murtaza
Department of Environmental Sciences
COMSATS University Islamabad
Vehari, Pakistan

Hafiz Muhammad Nadeem
Department of Environmental Sciences
COMSATS University Islamabad
Vehari, Pakistan

Ratul Nath
Microbiology Laboratory
Department of Life Sciences
Dibrugarh University
Dibrugarh, India

Aditi Nayak
Department of Life Science
Guru Nanak Institute of Pharmaceutical Science and Technology
Kolkata, India

Jayanta K. Nayak
Department of Anthropology
Central University of Odisha
Koraput, India

Nisha Sam Nirmala
PG Department of Biotechnology
Dwaraka Doss Goverdhan Doss Vaishnav College
Chennai, India

Louis Noble
Plant Physiology and Biochemistry Division
Department of Botany
University of Calicut
Malappuram, India

Suraj K. Padhi
Department of Biodiversity and Conservation of Natural Resources
Central University of Odisha
Koraput, India

Harshata Pal
Amity Institute of Biotechnology
Amity University Kolkata
Kolkata, India

Prashant Palkar
Department of Microbiology
Shahid Veerpatni Laxmi Mahavidyalaya
Titave, India

Debabrata Panda
Department of Biodiversity and Conservation of Natural Resources
Central University of Odisha
Koraput, India

Apexa Pareek
Department of Botany
University of Rajasthan
Jaipur, India

Suman Parihar
Department of Botany
JNV University
Jodhpur, India

Jagannath Patil
NAAC Adviser
Bangalore, India

Sachin Patil
Department of Botany
Dr. Ghali College
Gadhinglaj, India

Jos T. Puthur
Plant Physiology and Biochemistry Division
Department of Botany
University of Calicut
Malappuram, India

Swastika Rakshit
Department of Life Science
Guru Nanak Institute of Pharmaceutical Science and Technology
Kolkata, India

Ashwini Ravi
PG Department of Biotechnology
Dwaraka Doss Goverdhan Doss Vaishnav College
Chennai, India

Chandan Roy
Department of Biology
IISER Tirupati
Panguru, India

Swarnendu Roy
Plant Biochemistry Laboratory, Department of Botany
University of North Bengal
Siliguri, India

Aryadeep Roychoudhury
Discipline of Life Sciences, School of Sciences
Indira Gandhi National Open University
New Delhi, India

Shelvy S.
Division of Plant Physiology
ICAR-Indian Agricultural Research Institute
New Delhi, India

Uma S.
PG Department of Biotechnology
Dwaraka Doss Goverdhan Doss Vaishnav College
Chennai, India

Archana Sahani
Division of Plant Physiology
ICAR-Indian Agricultural Research Institute
New Delhi, India

Lekshmy Sathee
Division of Plant Physiology
ICAR-Indian Agricultural Research Institute
New Delhi, India

Pamarthi Satyanarayan
School of Law
Vel Tech Rangarajan Dr. Sagunthala R&D Institute of Science and Technology
Chennai, India

Sabarinathan Selvaraj
Department of Seed Science and Technology, College of Agriculture
Odisha University of Agriculture and Technology
Bhubaneswar, India

Shahab
Department of Agricultural Sciences and Technology, Atta-ur-Rahman School of Applied Biosciences (ASAB)
National University of Sciences and Technology (NUST)
Islamabad, Pakistan

Muhammad Shahid
Department of Environmental Sciences
COMSATS University Islamabad
Vehari, Pakistan

Noorie Shama
Department of Life Science
Guru Nanak Institute of Pharmaceutical Science and Technology
Kolkata, India

Dhriti Sharma
Siddarth Govt. College
Nadaun, India

Hanuman Sharma
Department of Botany
University of Rajasthan
Jaipur, India

Lata Sharma
Department of Botany
JNV University
Jodhpur, India

Renu Sharma
Department of Botany
University of Rajasthan
Jaipur, India

Sakshi Sharma
School of Biological and Environmental Sciences
Shoolini University
Solan, India

Gyan Singh Shekhawat
Department of Botany
JNV University
Jodhpur, India

Haritha Si
Division of Plant Physiology
ICAR-Indian Agricultural Research Institute
New Delhi, India

Aman Singh
Division of Plant Physiology
ICAR-Indian Agricultural Research Institute
New Delhi, India

Amit Kumar Singh
Department of Botany
University of Lucknow
Lucknow, India

Ankur Singh
KIIT-Technology Business Incubator (KIIT-TBI)
KIIT University
Bhubaneswar, India

Bijendra Kumar Singh
Department of Botany
Feroze Gandhi College
Raebareli, India

Satyajit Singh
Department of Botany
Feroze Gandhi College
Raebareli, India

Dhanawantari L. Singha
Department of Botany
Rabindranath Tagore University
Bhopal, India

Sumali Sucharita
KIIT-Technology Business Incubator (KIIT-TBI)
KIIT University
Bhubaneswar, India

Arpita Talukdar
Department of Botany
Rabindranath Tagore University
Bhopal, India

Ramsha Tariq
Department of Agricultural Sciences and Technology, Atta-ur-Rahman School of Applied Biosciences (ASAB)
National University of Sciences and Technology (NUST)
Islamabad, Pakistan

Tasveer Zahra Tariq
Department of Botany
Bahauddin Zakariya University Multan
Vehari, Pakistan

Mayank Varun
Department of Botany
Hislop College
Nagpur, India

Rahul Verma
Plant Nutrition and Stress Physiology Laboratory
Department of Botany
University of Lucknow
Uttar Pradesh, India

Jianxu Wang
State Key Laboratory of Environmental Geochemistry
Institute of Geochemistry
Chinese Academy of Sciences
Guiyang, P.R. China
and
Institute of Foundation Engineering, Water and Waste Management
School of Architecture and Civil Engineering, University of Wuppertal
Wuppertal, Germany

Muhammad Younis
Department of Environmental Sciences
COMSATS University Islamabad
Vehari, Pakistan

Section I

Introduction to Epigenetics

1 Molecular Components of Chromatin Serving as Epigenetic Modifiers in Plants

Susmita Behara, Swastika Rakshit, Titir Guha, Prajita Kundu, Rita Kundu, Snehalata Majumdar and Aditi Nayak

1.1 INTRODUCTION

1.1.1 Definition of Chromatin and Its Role in Gene Regulation

Chromatin is a complex of DNA and proteins and is considered the building block of chromosomes. It plays a key role in transcription (expression of genes) and replication (duplication of DNA) (Bonato et al., 2024). These stainable thread-like structures visible within the nucleus during mitosis were first identified as chromatin by Walther Flemming in the 1880s. Researchers soon found out that chromatin was composed of both DNA and proteins, while interest in its structure only developed after it was realized that DNA is the "transforming principle". In 1974, Kornberg proposed that chromatin consists of repeating units made of two copies of each of the four proteins (H2A, H2B, H3 and H4) along with 200 base pairs of DNA. Around the same time, Olins and Olins visualized this chromatin structure as "beads on a string", which later came to be known as nucleosomes. It can be categorized in terms of nucleosome organization, the existence of histone variants and the post-translational modifications (PTMs) of histones. PTMs involve the addition or removal of small chemical groups, like methyl or acetyl groups, on the tail of the histone. Histone variants are specialized proteins that replace standard histones in nucleosomes at specific genomic regions, thereby changing chromatin structure and function. For example, H3.3 replaces H3 in active genes, while centromeric protein A (CENP-A) is specific to centromeres (Morrison & Thakur, 2021).

The chromatin structure is dynamic, which facilitates the response of the cell to environmental changes and developmental stimuli. Epigenetic memory in plants ensures that specific gene expression patterns are consistently passed down over generations. The stress responses are also governed by chromatin through the activation or repression of genes related to adaptation and survival. The chromatin is arranged into units named topologically associating domains (TADs), which are typically self-interacting, thereby physically bringing the regulatory components, like enhancer and promoter nearby (Han et al., 2024). This spatial arrangement controls gene expression by protecting genes from unfavorable regulatory signals.

1.1.2 Overview of Epigenetics and Its Significance in Plant Biology

Plants respond to environmental conditions and stimuli through epigenetic regulation. Epigenetic regulation critically controls plant growth and development, reproduction, disease resistance and resilience to abiotic stress caused by variations in temperature, salt concentration and nutrient levels. The epigenetic mechanisms regulate the key traits in crops like yield and productivity through a variety of modifications like DNA methylation, histone modification and small RNA-induced alterations. Gene expression is also modified, thereby affecting various processes such as growth, germination and fruit formation (Abdulraheem et al. 2024). Long-term epigenetic regulation plays a significant role in evolution, which allows phenotypic flexibility. Short-term changes are required for survival under stress conditions. A few enzymes are implicated in stress-induced changes in epigenetic regulation. DRM2, a methyltransferase, is responsible for stress-triggered changes in DNA methylation. Stress-induced histone modifications include histone acetyl transferases and histone deacetylases (HDACs).

DNA methylation regulates and modifies gene expression and genomic stability in plants. The addition of a methyl group to the cytosine base results in the formation of 5-methylcytosine or 5-mc, which governs many processes like maintaining genome stability, regulating gene activity, guiding development and responding to environmental stress. Both methylation and demethylation interchangeably dominate the gene expression patterns.

Histone modifications in plants have been generally associated with both abiotic and biotic stress responses. Recent studies report that the treatment of *Arabidopsis thaliana* with the histone deacetylase inhibitor trichostatin A enhanced the resistance against the neurotropic

DOI: 10.1201/9781003589914-2

fungus *Botrytis cinerea*. Drought stress increases the acetylation of histone H3, which influences stress-responsive genes in *Arabidopsis thaliana*. Cold stress has been reported to enhance chromatin accessibility in potato (*Solanum tuberosum*) through histone modifications (H3K4me3 and H3K27me3) related to the activated genes (Xiong et al., 2024). Several epigenetic processes and activated mechanisms in response to external stressors are listed in Table 1.1.

1.1.3 Chromatin Modifications in Response to Environmental Stress

Plants are exposed to a diverse range of environmental stress conditions, which involve either a response to stress stimuli or the consequences of that response. The underlying mechanism of the response is highly significant for the growth, development and survival of the plant. These modifications greatly affect the DNA structure and its interaction with different proteins. Several changes have been reported in the levels of 5-methylcytosine, 5-hydroxymethylcytosine and 8-oxo-7,8-dihydroguanine in response to artificially induced accelerated aging (Vo et al., 2024). DNA methylation, microRNAs, histone acetylation and miRNAs have modulatory effects on gene regulation. Epigenetic changes are relatively stable over a lifetime and are strongly associated with many biological processes, development and diseases. These modifications regulate gene expression and are recognized as the critical molecular interface for environmental factors and the genome (Rossnerova et al., 2020). Epigenetic changes are indeed more frequent than genetic mutations, and most of them are adaptive responses to environmental stimuli.

1.2 STRUCTURE OF CHROMATIN

1.2.1 DNA and Histone Protein: Basic Units of Chromatin

DNA, the double-helical molecule that represents the genetic blueprint of life, carries instructions for cell division, growth and function. The nearly 2 m of DNA contained within a single eukaryotic cell must fit within a nucleus just a few micrometers in diameter. The basic repeating unit of chromatin is known as a nucleosome, formed by DNA loops around the histone proteins to reach a compact structure. The 147 base pairs of DNA in a nucleosome are coiled above an octamer of histone proteins approximately 1.65 times (Li et al., 2024). The four core histones, H2A, H2B, H3 and H4, are repeated twice in this octamer, which contributes to the highly ordered chromatin folding, thus producing the first level of chromatin organization that characterizes an organism's "beads on a string" structure, as represented in Figure 1.1 (Faroozani et al., 2022).

Histone proteins are rich in positively charged amino acids such as arginine and lysine. These positively charged amino acids allow DNA to bind to histones and form strong ionic interactions with the negatively charged phosphate backbone of DNA. The interaction is crucial for both chromatin structural stability and nucleosome assembly. Histones play a significant role in the modulation of gene expression in addition to their structural function (Chen et al., 2024). Several PTMs, such as acetylation, methylation,

TABLE 1.1
Epigenetic regulation and responses to different environmental stress conditions

Plant species	Epigenetic process	Mechanism	Citation
Populus	DNA methylation, histone modifications	Changes in methylation and histone acetylation during stress	Liu et al. (2024)
Zea mays	Histone modification	Histone acetylation enhance drought stress response	Zheng et al. (2021)
Oryza sativa	DNA methylation	Methylation of promoters of genes involved in salt tolerance	Wang et al. (2023); Banerjee et al. (2018)
Solanum tuberosum	Histone modification	Histone modifications (H3K4me3 and H3K27me3) during cold stress	Kondhare et al. (2021)
Arabidopsis thaliana	DNA methylation	Salt stress alters DNA methylation levels affecting gene expression	Hou et al. (2022)
Brassica napus	RdDM or RNA directed DNA methylation	Targeting transposons using siRNA	Li et al. (2021)
Triticum aestivum	Histone methylation	H3K4me3 is implicated in cold response and vernalization	Ucar et al. (2024)
Solanum lycopersicum	Histone acetylation	Fruit ripening gene expression is increased by H3 acetylation	Ming et al. (2023)
Glycine max	Chromatin remodeling	Genes responding to drought are modulated by the SWI/SNF complex	Lima et al. (2022)
Medicago truncatula	Long non-coding RNA (IncRNA)	IncRNA regulate nitrogen fixation genes	Zanetti et al. (2024)

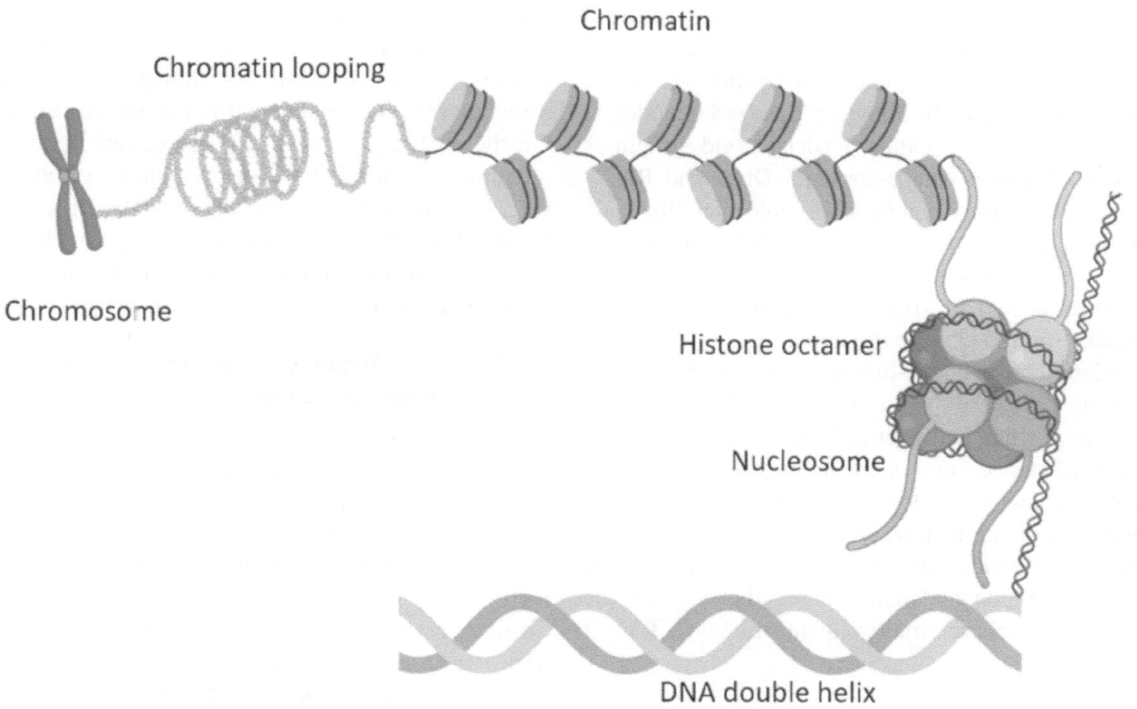

FIGURE 1.1 Structure of chromatin.

phosphorylation and ubiquitination, have been recognized to modify the N-terminal tails of histones protruding from the nucleosome core These modifications regulate both the chromatin accessibility to transcription factors and other regulatory proteins, in addition to their contribution to chromatin compaction.

In accordance with the highly dynamic and non-static nucleosome architecture, chromatin switches between several compaction states. Euchromatin is the less compact type of chromatin that is transcriptionally active, allowing the gene regulatory machinery to access the DNA (Li et al., 2024). Heterochromatin, the more compact type, is transcriptionally silent and more commonly found in regions with fewer genes and repetitive sequences. Chromatin remodelers are ATP-dependent enzymes that move, expel or recognize nucleosomes to reveal or obstruct particular DNA sequences, thus regulating the transition between these states. Histones provide a scaffold for epigenetic regulation, which allows stable gene expression patterns to be transmitted from one generation of cells to the next without altering the underlying DNA sequence; for example, histone deacetylases (HDACs) remove acetyl groups, leading to chromatin compaction and gene repression. The histone acetylation, regulated by histone acetyltransferases (HATs), relaxes chromatin structure and promotes gene activation. This is also dependent on the type of lysine or arginine residues being modified as well as the number of methyl groups being added. The alteration determines the fate of the histone methylation, whether it will activate or repress the transcription. Histone H3 lysine 4 trimethylation (H3K4me3) indicates active promoters, whereas histone H3 lysine 27 trimethylation (H3K27me3) is generally associated with gene repression (Chen et al., 2024).

In addition to core histones, higher-order chromatin organization is critically dependent on linker histone H1. The 30-nm fiber structure represents the advanced stage of chromatin compaction and is stabilized when H1 attaches to DNA between nucleosomes. This higher-order arrangement adds another level of control over gene expression and further governs DNA accessibility (Czajka et al., 2021).

1.2.1.1 Role of Non-Histone Proteins and RNA in Chromatin Structure

Non-histone proteins and RNA molecules regulate the dynamic and organized structure of chromatin. The controlled coordination governs the functional architecture of the genome. A wide array of structural and regulatory non-histone proteins is categorized into scaffold proteins, transcription factors and chromatin remodelers, which support the organization of higher-order compaction of chromatin and nucleosome positioning. Highly conserved complexes like cohesin and condensin are involved in the structural maintenance of chromosomes (Jin et al., 2024), which ensures the development of chromatin loops and domains for interactions between promoters and distant regulatory elements. To retain the chromatin flexibility and sensitivity to transcriptional or replicative demands, chromatin remodelers like the SWI/SNF complex slide or reposition nucleosomes using ATP hydrolysis (Chen et al., 2023).

RNA is a key regulator of chromatin structure. Specifically, non-coding RNAs (ncRNAs) play pivotal roles in gene regulation and chromatin architecture (Wierzbicki et al., 2021). Long non-coding RNAs (lncRNAs), such as

XIST and HOTAIR, recruit histone-modifying enzymes and other factors to specific genomic loci, thereby acting as scaffolds or guides for chromatin-modifying complexes, targeting the DNA region by allowing localized remodeling of chromatin. The three-standard nucleic acid structures consisting of displaced single-stranded DNA and RNA–DNA hybrids are also formed by RNA molecules (Jin et al., 2024). R-loops are a form of chromatin slicing machinery directed to specific genomic areas by small RNAs like siRNAs and piRNAs, and they frequently cause heterochromatin formation through DNA methylation.

Interactions between non-histone proteins, RNAs and chromatin are essential to preserve the integrity of the genome structure that facilitates the alterations between the closed heterochromatic and open euchromatic states upon any environmental or developmental stimulus. Therefore, the contribution of non-histone proteins and RNAs to chromatin structure provides crucial insights into the genome structure organization and hence paves the way for the development of novel crop improvement strategies (Tang et al., 2023).

1.2.2 Types of Chromatins

1.2.2.1 Euchromatin vs. Heterochromatin: Functional Implications

Euchromatin and heterochromatin have functional and structural distinctions in gene expression, cellular activity and in the arrangement of the genome. Both chromatins are characterized by distinctive histological features, epigenetic indicators and functional roles in controlling genetic activity.

The sparsely packed chromatin type has an open structure that allows the machinery of transcription and transcription factors to access DNA sequences, thereby dynamically regulating gene expression in response to environmental and developmental signals. Histone modifications that are associated with transcriptional activities, such as histone H3 lysine 4 trimethylation (H3K4me3) and histone acetylation, are generally found in higher concentrations. Chromatin, in open configuration, facilitates transcription, DNA repair and replication. The gene-rich euchromatic regions are crucial for cellular processes involving signal transduction, cell differentiation and stress responses that require rapid and flexible gene expression (Morrison et al., 2021). In comparison, heterochromatin is transcriptionally silent and is highly packed. It is divided into two major categories: constitutive and facultative heterochromatin. Constitutive heterochromatin plays structural roles in chromosome stability and segregation and is typically localized around centromeres and telomeres. Facultative heterochromatin is mostly associated with repressed gene activity during development or differentiation and is subject to switching on or off (Morrison et al., 2021).

Functionally, both cellular differentiation and genomic stability require interaction between euchromatin and heterochromatin. Heterochromatin areas ensure that chromosome architecture during mitosis and meiosis is properly restricted, preventing transposable elements and arresting repetitive elements from erroneous transcription. On the contrary, euchromatin makes rapid and reversible alterations in the expression of genes, which assist the cells in responding to environmental stimulus (Table 1.2). Chromatin remodelers, histone modifying enzymes and non-coding RNAs altogether closely maintain this dynamic balance between switching on and off genes.

1.2.2.2 The Dynamic Nature of Chromatin Structure in Plants

The dynamics of plant chromatin structure are essential for the regulation of gene expression and allow the plants to respond to developmental cues and changes in the environment. The chromatin architecture is continually being remodeled through constant changes in DNA methylation, histone modification and the integration of histone variation by remodeling complexes. The regulation of transcriptional activities of genes in response to internal and external stimuli occurs through these alterations that affect the accessibility of chromatin. Chromatin dynamics play a critical role in stress responses as they are essential for the rapid activation or repression of genes related to stress. Histone acetylation is associated with the transcriptional activation of genes that respond to histone methylation, based on a particular methylation mark. This can eventually activate or repress gene expression, much like the case above (Cheng et al., 2024). Environmental stressors induce changes in DNA methylation patterns, but these ensure stable gene expression through generations due to epigenetic memory. Chromatin remodelers such as the SWI/SNF and ISWI complexes actively move nucleosomes to promote or limit access to transcriptional machinery. This helps regulate genes in pathways linked to development and stress. Non-coding RNAs also contribute significantly to chromatin regulation by targeting particular genomic loci for histone modifications and DNA methylation (Cheng et al., 2024). These dynamic chromatin alterations interact to allow plants to respond rapidly to changes in the environment while maintaining the integrity of their genomes. Crop improvement and climate change stress resilience can be achieved by using knowledge of epigenetic control mechanisms by studying chromatin dynamics in plants.

Scientists have uncovered that traits can sometimes be passed from one generation to the next, not just through DNA but also through chemical changes that can affect how genes are processed. These changes, known as epimutations, can influence traits without altering the genetic code. They include modifications like DNA methylation, changes to histone proteins (which help package DNA) and small RNA molecules that regulate gene expression. While this type of inheritance occurs in many organisms, such as yeast, plants and animals, it appears in mammals such as

TABLE 1.2
A summary of distinctive features of euchromatin and heterochromatin

Feature	Euchromatin	Heterochromatin	Citation
Definition	Open chromatin structure	Highly condensed chromatin structure	Morrison et al. (2021)
Gene activity	Active transcription associated with the expression of genes	Silenced transcription associated with gene repression	Morrison et al. (2021)
DNA sequence content	Gene-rich areas that contain coding sequence	Includes repetitive DNA sequence, such as transposable elements and satellite DNA	Morrison et al. (2021)
Histone modification	Enhanced with histone acetylation and active histone markers such as H3K4me	Enriched in histone markers that suppress, such as H3K9me3 and H3K27me3	Morrison et al. (2021)
DNA methylation	Transcription activation due to low DNA methylation levels	Elevated DNA methylation, which encourages the suppression of transcription	Morrison et al. (2021)
Structural role	Adaptable structure that promotes DNA replication, transcription and repair	Structural function in preserving repetitive DNA, chromosome segregation, and genomic stability	Morrison et al. (2021)
Localization	Found in regulatory components including enhancers and promoters	Present in pericentromeric, telomere and centromere region	Morrison et al. (2021)
Dynamism	Extremely dynamic and able to switch between active and dormant modes quickly	Although facultative heterochromatin is generally stable, it can change into euchromatin in some circumstances	Morrison et al. (2021)
Key proteins involved	RNA polymerase ll, transcription factors and chromatin remodelers (such as the SWI/SNF complex)	Histone deacetylases (HDACs), DNA methyltransferases (e.g., DNMTs) and heterochromatin protein1 (HP1)	Morrison et al. (2021)
Biological role	Promotes cell differentiation, gene expression and reaction to external stimuli	Preserves genomic stability and inhibit transposable elements	Morrison et al. (2021)

humans. However, the underlying mechanisms for common patterns in transmitting epimutations are yet to be unfolded. In some cases, they are copied directly when cells divide, while in others, the changes are gone but can be reconstructed using molecular signals in the next generation. Understanding these processes might indicate how genetic and epigenetic adaptations shape traits over time, potentially impacting evolution (Fitz-James & Cavalli, 2022).

1.3 EPIGENETIC MECHANISMS IN PLANTS

1.3.1 DNA Methylation

Plants show a distinct pattern of DNA methylation. Methylation develops at CG sites and CHG and CHH sites (where H can be A, T or C). This makes plant DNA methylation more detailed than in mammals. The methylation steps in plants are about 24% at CG sites, 6.7% at CHG sites and 1.7% at CHH sites. Unlike in mammals, where methylation mostly affects genes, plants primarily use methylation to silence transposons, also known as jumping genes, and successive DNA sequences to keep them inactive. Certain enzymes help in controlling DNA methylation, such as DRM2 (similar to Dnmt3), which adds new methylation; MET1 (similar to Dnmtl), which maintains CG methylation; and CMT3, which maintains CHG methylation; DRM2 also contributes to CHH methylation, but it has to keep adding methyl groups continuously because CHH sites do not have a symmetric pattern like CG or CHG. There is still ongoing research, as DNA methylation plays a crucial role in regulating genes, development and adaptation in both plants and animals (Law & Jacobsen, 2010).

There is an exceptional process where small RNA molecules guide for matching of DNA sequences. Once they detect the target DNA, they help by adding cytosine methylation (a chemical mark) to silence that area. This occurs in all sequence contexts (CG, CHG and CHH sites). In this mechanism, plants use RNA interference machinery, which involves enzymes called Dicer that cut RNA into small pieces. Argonaute helps to guide these small RNAs to their DNA targets. DRM2 is the main enzyme that mediates methylation of the targeted DNA. A summary of enzymes involved in the methylation process is listed in Table 1.3. Plant-specific RNA polymerases such as Pol IV and Pol V perform vital but different roles. Pol IV helps in forming the small RNAs required for RdDM. Pol V helps in guiding these small RNAs to their matching DNA sequences. Scientists have formed various tools to control DNA methylation at particular locations in plant genomes. These tools use two main approaches (Gallego-Bartolomé, 2020):

a. Small interfering RNAs: These small RNA molecules guide the natural RNA-directed DNA methylation machinery of the plants to the specific DNA sequences. Once there, methylation enzymes like DRM2 merge methyl groups to mute the genes.

TABLE 1.3
Enzymes and methylation process with their functions

Methylation process	Enzyme	Target site	Functions	References
De novo methylation	DRM2	CG, CHG, CHH	Establishes new methylation	Fang et al. (2021)
CG maintenance	MET1	CG	Copies methylation after DNA replication	Zubko et al. (2012)
CHG maintenance	CMT3	CHG	Reinforces CHG methylation	Wendte et al. (2019)
CHH maintenance	CMT2, DRM2	CHH	Maintains methylation at transposons	Fang et al. (2021)
Demethylation	ROS1, DME	CG, CHG, CHH	Removes methylation to reactivate genes	Fang et al. (2021)

b. Programmable DNA-binding proteins: This process directly bonds the DNA methylation machinery to a particular location in the genome. Scientists apply special proteins that can be programmed to adhere to a selected DNA sequence. These proteins then deliver DNA methylation enzymes to that region, modifying the DNA accurately. Examples of such proteins include CRISPR-dCas9, an altered version of CRISPR-Cas9, and zinc finger proteins. Both approaches control gene expression by adding or removing DNA methylation, which facilitates the study of plant genetics and crop improvement.

1.3.2 Histone Modifications

Plants alter their responses to repeated stress exposure through the utilization of their epigenetic memory, which helps them in modifying gene activity without transforming their DNA sequence. This process occurs through histone protein modification, which helps to pack DNA tightly in the nucleus (Yamaguchi, 2021). The histones have small outgrowths called tails, where chemical alterations occur because of the addition or removal of methyl or acetyl groups. These changes act like "switches" that can turn genes on or off, allowing plants to react to stress swiftly. Recent studies in the model plant *Arabidopsis* have offered maps with regard to where these histone alterations occur across the genome. Pfluger and Wagner (2007) have disclosed their biological necessity, showing how they support plants adapting to different environments. Histone methylation can trigger or silence genes, depending on where the methyl group is added. Histone acetylation induces gene activation. These modifications support plants in adjusting their gene expression, impacting the DNA compaction and improving their stress adaptation (Pfluger & Wagner, 2007),

1.3.2.1 Types of Histone Modifications in Plants

Histone methylation: It can either trigger or repress gene expression. The enzymes such as histone methyltransferases (HMTs) and methyl groups; for example, H3K4me3 triggers gene expression, H3K9me2 helps silence genes and H3K27me3 represses stress-responsive genes until required (Fang et al., 2021).

Histone acetylation: It loosens the chromatin, thus making genes more available for activation. The enzymes involved are histone acetyltransferases, which help in adding acetyl groups, and histone deacetylases, which remove them; for example, H3K9ac triggers gene expression, while H4K16ac is associated with stress responses (Le et al., 2025).

Histone phosphorylation: It helps in signal transduction during stress responses. The enzymes involved are kinases, which add phosphate groups and H3S10ph, associated with stress responses and cell division.

Histone ubiquitination: It marks histones for deletion or changes in gene activation. The enzymes involved are ubiquitin ligases, which add ubiquitin, and deubiquitinases, which remove it (Le et al., 2025).

1.3.2.2 Histone Modifications under Stress Responses

When a plant is stressed due to drought or cold, the signals are sent inside the cell. Specific enzymes modify histones by the addition or removal of chemical groups. If the modification leads to loosened chromatin, the genes are activated, helping the plant to survive. If the modification tightens chromatin, the genes are repressed, blocking unnecessary activity. Some of these alterations remain even after withdrawal of the stress responses, allowing the plant to react faster if the same stress occurs again (epigenetic memory). For example, prior to drought exposure, the stress-responsive genes that help in controlling water retention are repressed by H3K27me3. During drought, the H3K27me3 is removed, and histones are acetylated, thus activating stress-related genes. The role of histone modifications has been represented diagrammatically in Figure 1.2. Some modifications remain, so that the plant reacts more quickly if drought occurs again after recovery.

1.3.3 Role of Non-coding RNA in Epigenetic Regulation

Long non-coding RNAs are RNA molecules that are longer than 200 nucleotides but do not encode proteins. Despite this, they play a vital role in supporting plants to survive under various stress conditions. Long RNAs interact with DNA, RNA and proteins, influencing important processes such as chromatin remodeling, RNA metabolism and

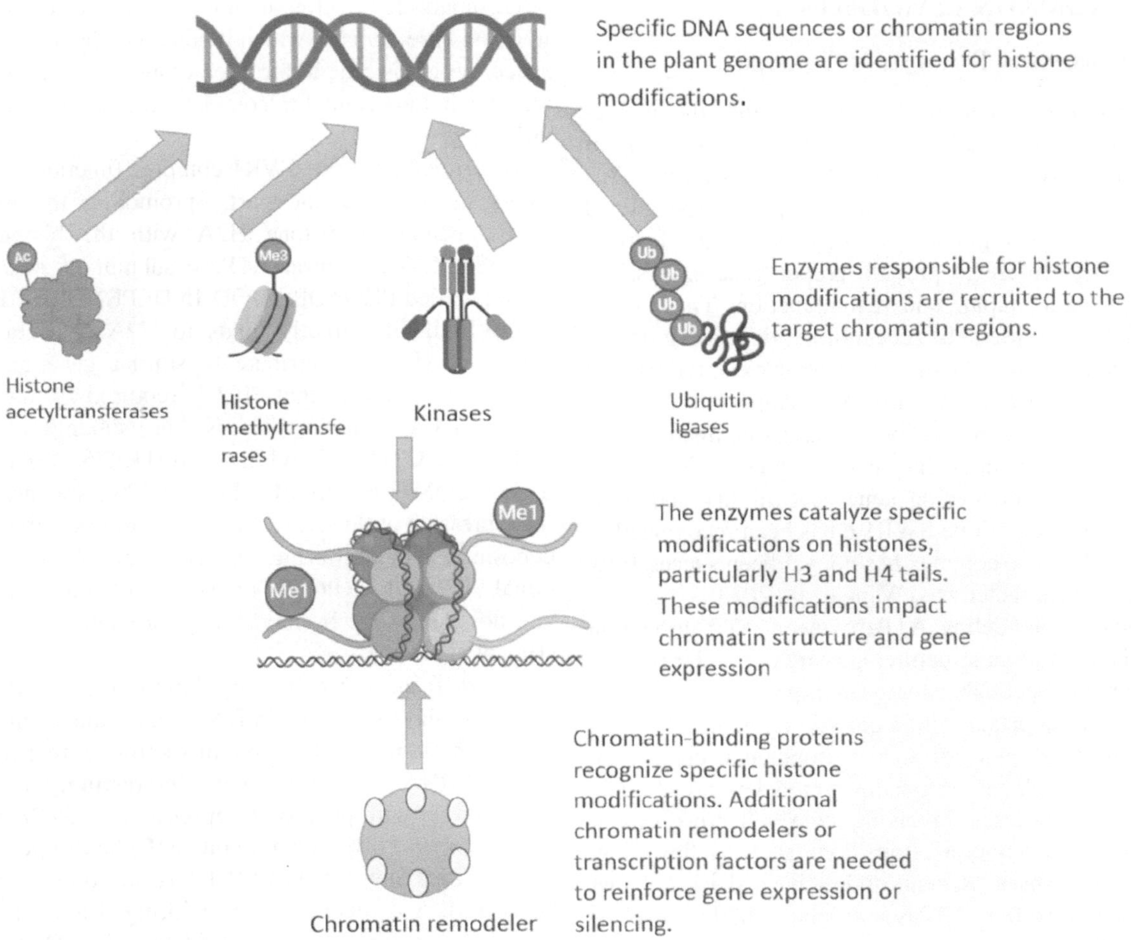

FIGURE 1.2 Various histone modifications targeting specific DNA sequences, impacting chromatin structure and gene expression.

protein alterations. Hence, they operate the expression of other genes through epigenetic modifications as well as at the transcriptional and post-transcriptional levels. Recent studies on computational biology and functional genomics have disclosed that long RNAs are included in various biological processes, including flowering time, reproductive organ growth and responses to biotic and abiotic stress. Long RNA has also targeted plant protection against pathogens, summarizing the known molecular mechanisms by which they support plants to resist diseases, although in-depth knowledge is limited and research is still ongoing. Many lncRNAs have been recognized in plants, especially in species like *Arabidopsis thaliana*, wheat (*Triticum aestivum*), rice (*Oryza sativa*), maize (*Zea mays*), *Medicago truncatula* and grapevine (*Vitis vinifera*). Most stated lncRNAs are stress-responsive, supporting plants to adapt to environmental conditions and protect against pathogens. However, they are also involved in plant growth processes, including flowering phase regulation, root nodule development, polar auxin transport and stress response; for example, in *Arabidopsis*, the *FLOWERING LOCUS C* (*FLC*) gene, which regulates flowering time, is epigenetically regulated by lncRNAs COOLAIR and COLDAIR. These lncRNAs mobilize the PHD-PRC2 complex, which modifies histones and silences *FLC*, ensuring proper flowering control. Other lncRNAs, like EARLY NODULIN 40 (END40) and AUXIN REGULATED PROMOTER LOOP (APOLO), perform important roles in nodule growth and polar auxin transport, respectively. ENOD40 supports legume nodule development by transferring RNA-binding proteins from the nucleus to the cytoplasm. APOLO, on the other hand, influences polar auxin transport by regulating chromatin loop dynamics, influencing the expression of the *PID* gene, which controls auxin flow (Sharma et al., 2022).

1.3.3.1 MicroRNAs

The genetic reprogramming in plants is regulated by epigenetic factors, which fine-tune gene expression without changing the DNA sequence. One major group of regulators in this whole process is microRNAs (miRNAs). These small RNA molecules regulate gene activity at the post-transcriptional level, meaning that they control the balance and translation of mRNAs. Reports suggest that the *DICER-LIKE1* (*DCL1*) gene is required for miRNA production (Banerjee et al., 2016). Beyond *Arabidopsis*, many gymnosperms, monocots and dicots, involving economically important crops, have differential miRNA expression in embryogenic cultures (Wójcikowska et al., 2020).

1.4 CHROMATIN REMODELING

1.4.1 Chromatin Remodeling Complexes

These are enzymes that non-covalently change the interaction between DNA and histone octamers, repositioning or evicting nucleosomes to alter the DNA accessibility. This allows genes to be turned on or off as required. Specialized chromatin-remodeling and modified complexes help to control this structure to provide proper gene expression, DNA replication, repair and recombination. These complexes alter the chromatin accessibility, depending on the cellular requirements. Among these complexes, INO80 and SWR1 are ATP-dependent chromatin remodelers that play a pivotal role in gene regulation by regulating histone variant H2A.Z incorporation into nucleosomes. This method is important for influencing gene accessibility and transcriptional activity. While SWR1-c has been recognized in plants, the INO80-c complex has not yet been successfully separated and characterized (Wang et al., 2023).

Chromatin remodeling ATPases need ATP utilization to alter the chromatin structure, making particular regions of the genome accessible for gene expression. This process plays an important role in controlling gene activity in response to developmental and environmental conditions. In animals and plants, chromatin remodelers are the key to biological processes, involving stem cell preservation, differentiation, phenotypic transitions such as flowering, and stress responses. Recent studies have disclosed that chromatin remodeling ATPases exhibit a higher level of selectivity than previously thought, which means that their activity is finely tuned to particular genes and the environment. Scientists have recognized different mechanisms that lead to this selectivity, emphasizing the difficulty of chromatin regulation in plants. The ATPase is the major subunit of chromatin remodeling complexes and has a rare feature in its ATPase domain, which is separated by a spacer. This spacer region is crucial for interacting with other subunits, making it important for complex formation (Han et al., 2015).

In *Arabidopsis*, the SWR1 complex functions similarly to its metazoan counterparts, promoting the exchange of the canonical histone H2A with the histone variant H2A.Z. The central ATPase subunit of *Arabidopsis* SWR1, called PHOTOPERIOD-INDEPENDENT EARLY FLOWERING1, directly binds to H2A.Z. Mutations in PIE1 and H2A.Z contribute to similar gene expression defects, emphasizing that PIE1 is required for integrating H2A.Z into chromatin. Various other subunits of SWR1, such as ACTIN-RELATED PROTEIN6 (ARP6) and SWR1 COMPLEX SUBUNIT6 (SWC6), are included in the repression of flowering. Studies have shown that H2A.Z deposition at the flowering repressor gene *FLC* needs functional PIE1 and ARP6. This means that SWR1 helps regulate flowering time by modifying chromatin at key genes (Han et al., 2015).

Apart from flowering regulation, the SWR1 complex also plays key roles in DNA repair and homologous recombination, as *SWR1* mutants show increased sensitivity to DNA-damaging agents. Interestingly, PIE1 may also interact with polycomb repression, as *PIE1* mutants can suppress Polycomb mutant leaf phenotypes. These observations suggest that PIE1 has additional purposes beyond H2A.Z incorporation, making it a vital regulator of chromatin dynamics in plants. The IN080 complex supports chromatin organization and genomic stability during cell differentiation (Han et al., 2015). Various triggers of chromatin remodeling are presented in Figure 1.3.

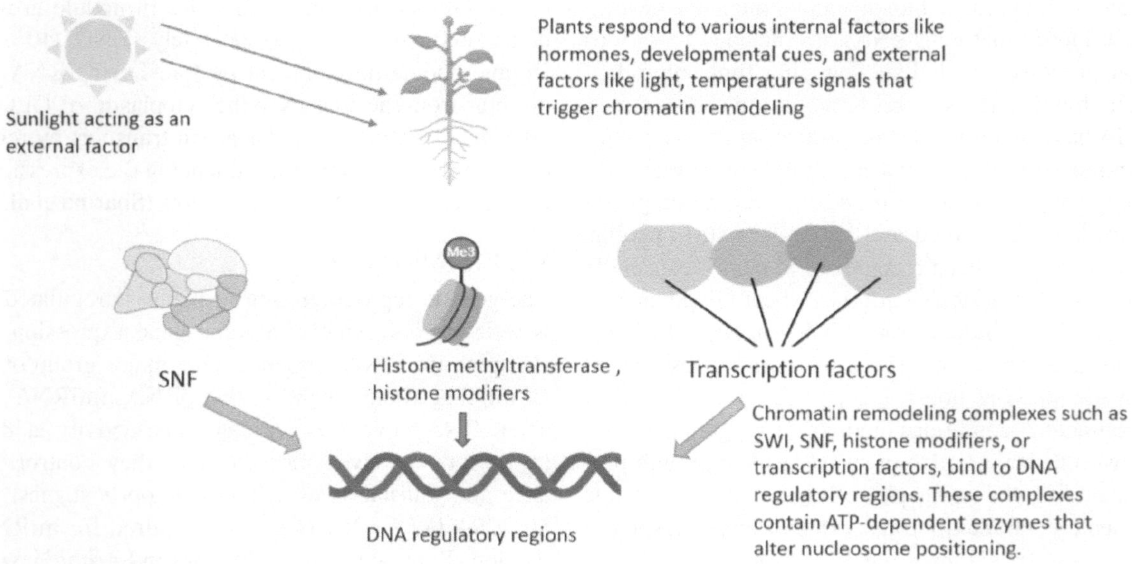

FIGURE 1.3 Different internal factors like hormones, developmental cues and external factors like light, temperature and pressure trigger the chromatin remodeling complexes, which help in altering nucleosome positioning.

1.4.2 Chromatin Remodeling Process

Chromatin remodeling plays a crucial role in controlling plant senescence by changing the availability of senescence-associated genes to transcription factors and other regulatory proteins. Since senescence includes extensive genetic reprogramming, chromatin remodeling allows the necessary genes to be either activated or suppressed in response to internal growth cues and external stress stimuli. This method is carried out by chromatin remodelers, which are ATP-dependent enzyme complexes that shift, eject or restore nucleosomes, thereby altering the chromatin structure. During senescence, chromatin remodeling controls the expression of key transcription factors (TFs) that influence senescence-related genes; for example, the SWR1 chromatin remodeling complex regulates the incorporation of the histone variant H2A.Z, which is connected to gene activation and repression in different growth processes, including leaf senescence. Similarly, Polycomb Group (PcG) proteins and their related chromatin remodeling complexes are included in repressing genes that slow down senescence, provided that the method progresses at the appropriate level. Conversely, ATP-dependent chromatin remodelers, such as SWI/SNF complexes, can execute gene activation by repositioning nucleosomes and making the key regulatory regions more accessible. Additionally, chromatin remodeling engages with other epigenetic mechanisms, such as DNA methylation and histone alterations; for instance, histone acetylation by histone acetyltransferases generally contributes to chromatin relaxation and active gene expression, executing senescence-related processes. On the contrary, histone deacetylases (HDACs) can eject these acetyl marks, which leads to chromatin compaction and gene repression, thereby slowing down the senescence. Moreover, chromatin remodeling is also controlled by small RNA-mediated operations, such as microRNAs (miRNAs) and long non-coding RNAs (lncRNAs), which support and modulate the recruitment of chromatin remodelers to particular senescence-associated genes. This dynamic and extremely coordinated interaction between chromatin remodeling, histone modifications, DNA methylation and RNA-mediated regulation allows plant senescence to occur in a regulated and timely manner. Understanding these methods can support the development of strategies to manipulate senescence for agricultural outcomes, such as broadening leaf lifespan and increasing crop productivity (Miryeganeh, 2022).

Nitric oxide (NO) plays a crucial role in chromatin remodeling by regulating the activity of main chromatin-modifying enzymes, thereby controlling gene expression in response to developmental and environmental conditions. One of the primary ways by which NO leads to chromatin remodeling is through S-nitrosation, which is a post-translational modification that changes the function of proteins, including histone acetyltransferases (HATs), histone deacetylases (HDACs), methyltransferases and demethylases. By inhibiting HDACs, NO executes histone acetylation, forming a relaxed chromatin structure that increases transcriptional accessibility. Additionally, NO can control histone methylation by targeting histone methyltransferases and demethylases, which control the activation or repression of particular genes. Furthermore, NO can affect DNA methylation by regulating DNA methyltransferases (DNMTs) and demethylases, thereby performing a role in long-term gene control. Through these mechanisms, NO acts as a dynamic signaling molecule that fine-tunes chromatin structure, allowing plants to adjust their gene expression profiles in response to biotic and abiotic stress, growth transitions and ecological shifts. This emphasizes NO as a vital player in connecting redox signaling with epigenetic regulation, ultimately guiding plant development and stress flexibility (Ageeva-Kieferle et al., 2019). The basic principle of chromatin remodeling is diagrammatically presented in Figure 1.4.

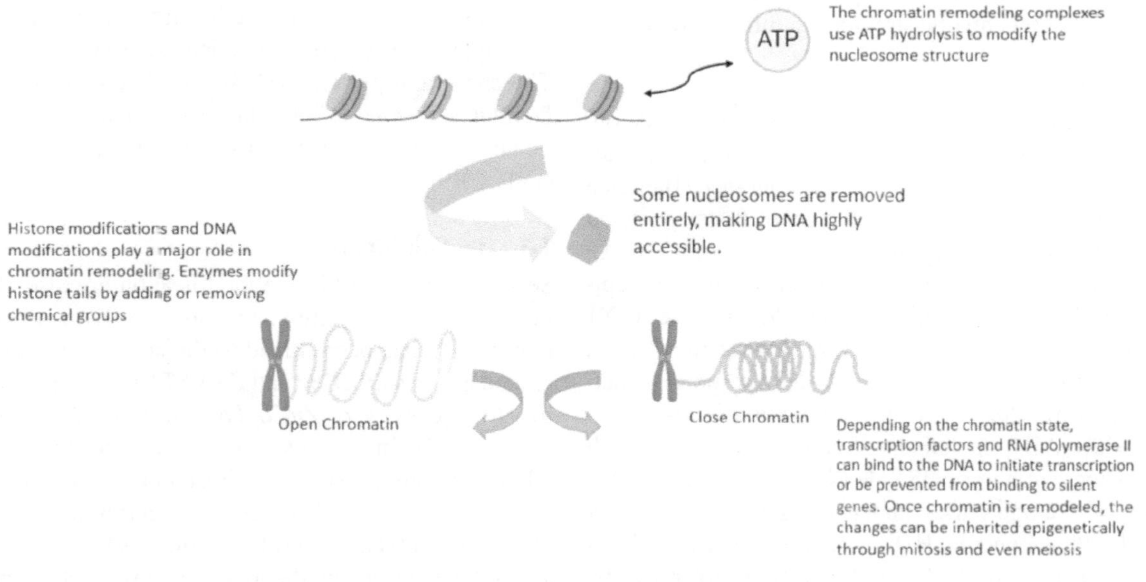

FIGURE 1.4 Chromatin remodeling complexes use ATP hydrolysis to modify the nucleosome structure for remodeling of the chromatin so that the changes can be inherited epigenetically through mitosis and meiosis.

1.5 EPIGENETIC MODIFIERS AND PLANT STRESS RESPONSES

1.5.1 Stress-Induced Chromatin Changes

Chromatin regulation plays crucial roles in controlling gene expression and the overall stress response mechanism of the plant. In plants, modifications in histones and changes in DNA methylation are closely related to the expression of stress-responsive genes, enabling adaptation to environmental challenges. Various chromatin regulators have been identified as key players in modulating gene networks under different abiotic stress conditions. Plants can sense environmental stimuli and also rely on complex gene regulatory networks to modulate metabolite accumulation, enabling survival and adaptation to different stress conditions (Ashapkin et al., 2020). Recent research has shown that histone modifications like histone methylation and acetylation, accompanied by DNA methylation, are associated with gene expression changes in response to different abiotic stress, and these changes often occur rapidly with environmental cues or they may also progress gradually along with the transcriptional changes to enhance physiological balance and stress adaptability (Kim et al., 2015). However, it is not yet well understood whether the changes in chromatin directly lead to changes in gene expression, or if gene expression changes occur first and then result in chromatin alterations. Plants can also retain epigenetic "memory" of past stress, enabling their offspring to inherit various stress resilience features. Additionally, miRNAs regulate gene expression and interact with epigenetic pathways, influencing chromatin remodeling and stress response genes (Chaffai et al., 2024). Together, these processes form a complex network that can help plants adapt to challenging conditions, and ongoing research is needed to better understand these mechanisms and their potential for developing stress-resistant crops. In this section, the recent findings that are associated with epigenetic modifications in response to different abiotic stress are summarized below.

1.5.1.1 Drought Stress

Histone modifications, such as methylation or acetylation, affect chromatin structure and thus regulate expression patterns of genes related to drought stress tolerance (Banerjee & Roychoudhury, 2017). Mostly, the genes that are related to hormone signaling pathways, like abscisic acid (ABA), or drought-responsive genes (DRG) undergo different epigenetic modifications. The histone methyltransferase ATX1 plays a key role in activating *NCED3*, a gene involved in ABA biosynthesis, and regulates drought-responsive networks (Le et al., 2025). Histone acetylation is also associated with ABA-mediated drought responses, with increased acetylation of H3K9, H3K18, H3K27 and H4K5 observed in rice and barley under drought stress (Seem et al., 2024). Plant-specific HD2-type histone deacetylases (HDACs), such as AtHD2C, modulate stress tolerance by altering ABA sensitivity and transpiration rates, indicating their broader role in abiotic stress adaptation (Han et al., 2016). Drought-responsive genes, including Response to Desiccation (*RD*), viz., *RD20* and *RD29A*, exhibit increased histone modifications such as H3K4me3 and H3K9ac under severe drought conditions, correlating with nucleosome loss and transcriptional activation (Kim et al., 2021). Upon rehydration, histone deacetylation and nucleosome replacement contribute to resetting chromatin status and repressing stress-induced gene expression. Additionally, histone modifications are linked to drought stress memory, as H3K4me3 marks persist on genes like *RD29B* and *RAB18* following repeated drought exposure, facilitating faster transcriptional responses (Godwin & Farrona, 2020). These findings highlight the conserved epigenetic regulation of drought stress responses across plant species, emphasizing histone modifications as key players in stress adaptation.

1.5.1.2 Heat Stress

Heat acclimation is also governed by the alteration of specific gene expression patterns regulated through histone modifications in response to heat stress. Certain histone modifications undergo rapid changes, while others adjust gradually. The histone variant H2A.Z, including subtypes like H2A, H2A.Bbd, H2A.X and H2A.Z, is known to play crucial roles in temperature sensing by modulating nucleosome occupancy (Talbert & Henikoff, 2014). In *Arabidopsis*, the chromatin remodeling complex SWR1 facilitates the ATP-dependent replacement of H2A with H2A.Z, a process regulated by the *ACTIN-RELATED PROTEIN6* (*ARP6*) gene, which is essential for temperature-responsive gene regulation (Aslam et al., 2019). In *Brachypodium*, mutations in H2A.Z impair heat acclimation (Jin et al., 2024). Additionally, histone acetylation supports H2A.Z deposition, suggesting a broader role for histone variants in transcriptional modulation under temperature stress (Cortijo et al., 2017). Heat stress enhances histone acetylation (H3K9ac and H4K5ac) in maize tissues, while its effect on histone methylation varies, highlighting their role in chromatin decondensation (He et al., 2022). The chromatin protein BRUSHY1 (BRU1) in *Arabidopsis* facilitates epigenetic inheritance and maintains transcriptional activation of heat stress memory genes (Brzezinka et al., 2019).

1.5.1.3 Salt Stress

Several patterns of histone modification are also observed in response to salt stress. Increased H3K9 acetylation in maize roots has been linked to the activation of cell wall-related genes *ZmEXPB2* and *ZmXET1*, potentially mediated by *HAT* genes, viz., *ZmHATB* and *ZmGCN5* (Tersenidis et al., 2024). In *Arabidopsis*, the transcriptional adaptor ADA2b modulates HAT activity, and its mutation results in hypersensitivity to salt stress, highlighting the importance of histone acetylation in salinity tolerance (Vlachonasios et al., 2021). Conversely, the loss of HDAC protein-encoding genes such as *HD2C*, *HDA6* and *HDA19* also increases salt sensitivity (Luo et al., 2012). The role of SUMOylation in stress responses is also emerging, as SUMOylation of chromatin modifiers like GCN5 and HDA19 has been implicated in transcriptional regulation under stress (Sharma et al., 2023). These findings suggest that histone modifications and epigenetic regulators fine-tune gene expression to enhance plant resilience to salt stress.

1.5.1.4 Cold Stress

Histone modifications are also essential for plant responses to cold stress, including vernalization and cold acclimation, by regulating gene expression (Banerjee et al., 2017). Long-term cold exposure induces vernalization, while short-term non-freezing cold exposure enhances freezing tolerance, with both processes involving distinct pathways. Cold stress triggers widespread transcriptional changes, affecting up to 20% of genes, with epigenetic regulators like *HDA6* playing a role in stress adaptation (Ramakrishnan et al., 2022). In maize, increased *HDAC* expression during cold acclimation leads to global histone deacetylation, while in *Arabidopsis*, histone modifications such as reduced H3K27me3 levels at cold-responsive genes and increased histone acetylation in stress-associated genes contribute to tolerance (Hereme et al., 2021). Key regulators such as ADA2b, HOS1 and HOS15 influence histone modifications, impacting chromatin structure and gene expression (Chung et al., 2022). ADA2b and GCN5 regulate histone acetylation, affecting cold-responsive gene activation, while HOS1 modulates chromatin status at *FLC* and represses the key cold tolerance genes (Ding et al., 2024). HOS15, a homolog of human TBL1, is linked to histone H4 deacetylation and stress tolerance (Kim et al., 2024). Although vernalization mechanisms are well studied, further research is needed to fully understand how histone modifications regulate cold acclimation and enhance freezing tolerance in plants.

1.5.2 Mechanisms Underlying Stress Memory through Epigenetic Modifications

Chromatin acts as a crucial interface between environmental signals and gene regulation, which leads to different stress responses through histone modifications, nucleosome remodeling and DNA methylation (Banerjee & Roychoudhury, 2017). Stress-induced transcription factors recruit histone-modifying complexes, such as the H3K4 methyltransferase complex, and methyltransferases like SDG8 and SDG25 to influence plant immunity through H3K4 and H3K36 methylation (Le et al., 2025). The chromatin remodeler BRAHMA (BRM) typically represses ABA-responsive genes under normal conditions but is regulated by ABA-dependent phosphorylation during stress (Bhadouriya et al., 2021). Other chromatin remodelers and DNA methylation pathways also contribute to stress defense, though their precise roles are still being explored. Determining whether chromatin modifications directly regulate stress responses or result from indirect developmental and metabolic changes remains a complex challenge.

Plants can also retain adaptive responses beyond initial stress exposure through stress memory. While many chromatin modifications reset once stress subsides, some, such as those involved in vernalization, create long-term epigenetic changes. In *Arabidopsis thaliana*, vernalization accelerates flowering by epigenetically silencing *FLOWERING LOCUS C (FLC)* through H3K27 trimethylation (Schmitz et al., 2008). This modification, orchestrated by a polycomb complex activated by cold, remains stable for weeks or months, allowing plants to respond effectively to seasonal cues. Histone modifications are also pivotal in stress memory, particularly in systemic acquired resistance (SAR) (Zhao et al., 2019). Exposure to bacterial pathogens or chemical agents like acibenzolar S-methyl induces sustained histone modifications, primarily H3K4me2 and H3K4me3 enrichment (Zhu et al., 2016). These modifications persist for several days and contribute to transcriptional memory in primed tissues, including previously untreated leaves. While H3K4me3 is strongly associated with gene expression, the role of H3K4me2 in transcription remains less understood (Howe et al., 2017).

Another unique example of stress-induced memory comes from the report of Nguyen et al. 2020, which states that plants exposed to high temperature develop heat stress memory, which enhances their ability to tolerate recurring heat waves. This involves prolonged activation of specific memory genes and sustained H3K4 hypermethylation. Unlike transient heat shock responses, these epigenetic marks persist for several days. The transcription factor HSFA2 is essential for maintaining H3K4me2/3 levels, although its exact mechanism is yet to be fully understood. Genome-wide profiling of plants subjected to salt stress revealed a reduction in H3K27me3 levels at specific genomic loci after mild salt treatment (Sun et al., 2019). This epigenetic shift correlated with improved drought tolerance but did not directly regulate stress-responsive genes. These findings suggest that chromatin remodeling plays a role in long-term stress adaptation, though the precise mechanisms remain to be elucidated.

Beyond histone modifications, nucleosome positioning significantly impacts stress memory. The *FORGETTER1* (*FGT1*) gene encodes a chromatin remodeler necessary for maintaining heat stress memory genes. Similarly, the CHROMATIN ASSEMBLY FACTOR-1 (CAF-1) complex modulates nucleosome dynamics, and its mutants exhibit constitutive priming responses (Kovalchuk, 2023). These studies highlight chromatin accessibility as a critical factor in sustaining stress memory. Plants repeatedly exposed to desiccation stress develop distinct transcriptional responses. Some genes exhibit enhanced re-induction, linked to prolonged H3K4me3 hypermethylation and RNA polymerase II accumulation (Yu et al., 2025). While ABA signaling is crucial for transcriptional memory, additional yet-to-be-identified factors contribute to this regulatory process.

Histone H3K4 methylation is a conserved marker of somatic stress memory across various organisms (Fabrizio et al., 2019). Reduced nucleosome occupancy during stress priming also influences chromatin-based memory, though the precise interplay between these factors remains unclear. Typically, somatic stress memory lasts between five to seven days under different abiotic stress conditions (Sharma et al., 2022). The factors limiting its duration are still unknown but may relate to energy allocation. As researchers identify regulators of stress memory with minimal impact on plant growth, future studies will assess their potential benefits under natural environmental conditions. Understanding these mechanisms could help develop crops with improved resilience to climate variability.

1.5.3 Transgenerational Epigenetic Inheritance

1.5.3.1 Evidence for the Inheritance of Epigenetic Changes across Generations

To establish transgenerational stress memory, certain criteria must be met, including meiotic stability and persistence for at least one stress-free generation without underlying DNA sequence changes. Additionally, since offspring develop on the parent plant, the maternal or paternal influences must also be taken into account. Transgenerational memory describes changes that persist beyond the first stress-free generation, whereas intergenerational memory is observed only in the immediate progeny and may not always have an epigenetic origin. Stable epialleles, such as the peloria mutant in *Linaria vulgaris*, exemplify long-term heritable changes in plants (Lämke & Bäurle, 2017). This mutation modifies flower symmetry through promoter methylation rather than alterations in DNA sequence, occasionally reverting to the wild-type phenotype. The extent to which such epigenetic modifications function as adaptive responses to stress remains an open question.

Few reports highlight that transgenerational priming enhances stress tolerance and adaptation in crop plants. Environmental priming agents can induce heritable stress resilience, as seen in *Triticum aestivum*, where drought and hydro-halo priming improved drought tolerance and yield across generations (Luis et al., 2023). Similarly, UV-B priming in *Oryza sativa* boosted drought resistance in successive generations (Sen et al., 2022). Beyond stress tolerance, priming enhances DNA repair and genome flexibility, as observed in *Arabidopsis thaliana* (Luis et al., 2023). However, the stability of priming-induced epigenetic modifications remains uncertain. Cold and osmotic priming in *A. thaliana* maintained H3K27me3 marks for days post-stress, suggesting short-term persistence (Kwon et al., 2009; Sani et al., 2013). In *T. aestivum*, thermopriming upregulated *LSD1* and RNA methyltransferases, enhancing thermotolerance and grain yield (Wang et al., 2016). While evidence supports the inheritance of priming effects, understanding the mechanisms of regulating epigenetic stability across generations remains crucial for optimizing stress resilience in crops.

1.5.3.2 Mechanisms Underlying Transgenerational Epigenetic Inheritance

The inheritance of epigenetic modifications is constrained by the resetting of the epigenome during gametogenesis and early embryonic development. Extensive DNA demethylation occurs in pollen and endosperm, while histone H3 modifications in gametes and the zygote shift independently of DNA replication (Han et al., 2019). Despite these erasure mechanisms, some epigenetic marks may persist or be replaced by alternative modifications. Small noncoding RNAs (siRNAs) play a pivotal role in maintaining epigenetic memory across generations in plants (Junaid et al., 2024). In mammals, specific nucleosomes with epigenetic marks remain in sperm, while in *Arabidopsis*, mutants lacking CG methylation demonstrate stable transmission of hypomethylated loci over multiple generations (Stassen et al., 2018). However, RNAi-mediated pathways eventually restore DNA methylation, resembling the progressive silencing of transposable elements and transgenes in plants (Martienssen & Chandler, 2013). Evidence suggests that siRNAs contribute to transgenerational inheritance by facilitating de novo methylation during gametogenesis (Cao et al., 2024).

1.6 CONCLUSION

The dynamic features of chromatin structure are determined by the regulated interplay among the diverse molecular components, like epigenetic modifiers and chromatin remodelers. This ensures the genomic stability and plays a pivotal role in alterations of the genomic responses to environmental stressors. The epigenetic modification plays a crucial role in the mediation of environmental adaptability and the regulation of development. The major components of epigenetic regulators include the enzymes associated with DNA methylation, the complexes involved in histone modification, the chromatin remodelers and the ncRNAs. The silencing of transposable elements and heterochromatin condensation are highly important for the repression of the chromatin structure. Enzymes like DRM, CMTs and MET1 are associated with methylation. Demethylation is induced to initiate chromatin decompaction and improve chromatin accessibility (Bhadouriya et al., 2021). HATs and HDACs play a pivotal role in histone modification and regulate chromatin function by adjusting histone acetylation levels. Cellular identity is maintained by the polycomb repressive complexes and is the key player during the developmental process (Luo et al., 2020). The SWI/SNF complexes are the major family of chromatin remodelers and play an instrumental role in stress management and developmental processes (Candela Ferre et al., 2024). In addition to the epigenetic regulation by histone modifications, the ncRNAs also guide the DNA methylation process. Reports suggest that they are pivotal in the DNA damage response and stress management (Luo et al., 2020). Nevertheless, stress-induced epigenetic changes are not always consistently inherited, with studies indicating that maternal transmission is often the dominant route. While plants exhibit heritable stress responses, it remains unclear whether these responses arise from random epigenomic instability or represent adaptive regulatory shifts that confer a selective advantage under environmental stress. The long-term impact of altered epigenetic stability requires further investigation, as the accumulation of stable or metastable epialleles could influence plant fitness. Mechanisms that limit stress-induced epigenetic modifications may function during seed aging, but whether this regulation is mediated by specific pathways or results from random loss of stress-induced changes remains uncertain.

REFERENCES

Abdulraheem, M. I., Xiong, Y., Moshood, A. Y., Cadenas-Pliego, G., Zhang, H., & Hu, J. (2024). Mechanisms of plant epigenetic regulation in response to plant stress: Recent discoveries and implications. *Plants*, 13(2), 163.

Ageeva-Kieferle, A., Rudolf, E. E., & Lindermayr, C. (2019). Redox-dependent chromatin remodeling: A new function of nitric oxide as architect of chromatin structure in plants. *Frontiers in Plant Science*, 10, 625.

Ashapkin, V. V., Kutueva, L. I., Aleksandrushkina, N. I., & Vanyushin, B. F. (2020). Epigenetic mechanisms of plant adaptation to biotic and abiotic stresses. *International Journal of Molecular Sciences*, 21(20), 7457.

Aslam, M., Fakher, B., Jakada, B. H., Cao, S., & Qin, Y. (2019). SWR1 chromatin remodeling complex: A key transcriptional regulator in plants. *Cells*, 8(12), 1621.

Banerjee, A., & Roychoudhury, A. (2017). Epigenetic regulation during salinity and drought stress in plants: Histone modifications and DNA methylation. *Plant Gene*, 11, 199–204.

Banerjee, A., & Roychoudhury, A. (2018). The gymnastics of epigenomics in rice. *Plant Cell Reports*, 37, 25–49.

Banerjee, A., Roychoudhury, A., & Krishnamoorthi, S. (2016). Emerging techniques to decipher microRNAs (miRNAs) and their regulatory role in conferring abiotic stress tolerance of plants. *Plant Biotechnology Reports*, 10(4), 185–205.

Banerjee, A., Wani, S. H., & Roychoudhury, A. (2017). Epigenetic control of plant cold responses. *Frontiers in Plant Science*, 8, 1643.

Bhadouriya, S. L., Mehrotra, S., Basantani, M. K., Loake, G. J., & Mehrotra, R. (2021). Role of chromatin architecture in plant stress responses: An update. *Frontiers in Plant Science*, 11, 603380.

Bonato, A., Chiang, M., Corbett, D., Kitaev, S., Marenduzzo, D., Morozov, A., & Orlandini, E. (2024). Combinatorics and topological weights of chromatin loop networks. *Physical Review E*, 109(6), 064405.

Brzezinka, K., Altmann, S., & Bäurle, I. (2019). BRUSHY1/TONSOKU/MGOUN3 is required for heat stress memory. *Plant, Cell & Environment*, 42(3), 771–781.

Cao, S., & Chen, Z. J. (2024). Transgenerational epigenetic inheritance during plant evolution and breeding. *Trends in Plant Science*, 29(11), 1203–1223.

Candela-Ferre, J., Diego-Martin, B., Pérez-Alemany, J., & Gallego-Bartolomé, J. (2024). Mind the gap: Epigenetic regulation of chromatin accessibility in plants. *Plant Physiology*, 194(4) 1998–2016.

Chaffai, R., Ganesan, M., & Cherif, A. (2024). Noncoding RNA regulation in plant responses to abiotic stress. In *Plant Adaptation to Abiotic Stress: From Signaling Pathways and Microbiomes to Molecular Mechanisms* (pp. 417–455). Singapore: Springer Nature Singapore.

Chen, K., Yuan, J., Sia, Y., & Chen, Z. (2023). Mechanism of action of the SWI/SNF family complexes. *Nucleus*, 14(1), 2165604.

Cheng, Y. J., Wang, J. W., & Ye, R. (2024). Histone dynamics responding to internal and external cues underlying plant development. *Plant Physiology*, 194(4), 1980–1997.

Chung, S., Kwon, C., & Lee, J. H. (2022). Epigenetic control of abiotic stress signaling in plants. *Genes & Genomics*, 44(3), 267–278.

Cortijo, S., Charoensawan, V., Brestovitsky, A., Buning, R., Ravarani, C., Rhodes, D., John van Noort, Jaeger, K. E. & Wigge, P. A. (2017). Transcriptional regulation of the ambient temperature response by H2A. Z nucleosomes and HSF1 transcription factors in Arabidopsis. *Molecular Plant*, 10(10), 1258–1273.

Czajka, K., Mehes-Smith, M., & Nkongolo, K. (2022). DNA methylation and histone modifications induced by abiotic stressors in plants. *Genes & Genomics*, 44(3), 279-297.

de Lima, F., Boldt, A. B. W., Kava, V. M., Galli-Terasawa, L. V., & Adamoski, D. (2022). Epigenetics' role in the common bean (*Phaseolus vulgaris* L.) and soybean (*Glycine max* (L.) Merr.) nodulation: A review. *Plant Molecular Biology Reporter*, 40(3), 471–481.

Ding, Y., Shi, Y., & Yang, S. (2024). Regulatory networks underlying plant responses and adaptation to cold stress. *Annual Review of Genetics*, 58, 43–65.

Fabrizio, P., Garvis, S., & Palladino, F. (2019). Histone methylation and memory of environmental stress. *Cells*, 8(4), 339.

Fang, J., Leichter, S. M., Jiang, J., Biswal, M., Lu, J., Zhang, Z. M., ... & Song, J. (2021). Substrate deformation regulates DRM2-mediated DNA methylation in plants. *Science Advances*, 7(23), eabd9224.

Fitz-James, M. H., & Cavalli, G. (2022). Molecular mechanisms of transgenerational epigenetic inheritance. *Nature Reviews Genetics*, 23(6), 325–341.

Foroozani, M., Holder, D. H., & Deal, R. B. (2022). Histone variants in the specialization of plant chromatin. *Annual Review of Plant Biology*, 73(1), 149–172.

Gallego-Bartolomé, J. (2020). DNA methylation in plants: Mechanisms and tools for targeted manipulation. *New Phytologist*, 227(1), 38–44.

Godwin, J., & Farrona, S. (2020). Plant epigenetic stress memory induced by drought: A physiological and molecular perspective. In Spillane C. and McKeown C. P. (Eds) *Plant Epigenetics and Epigenomics: Methods and Protocols*, 243–259. Publisher: Springer (Humana Press / Springer Science).

Han, M. H., Park, J., & Park, M. (2024). Advances in the multimodal analysis of the 3D chromatin structure and gene regulation. *Experimental & Molecular Medicine*. 25;56(4):763–771.

Han, Q., Bartels, A., Cheng, X., Meyer, A., An, Y. Q. C., Hsieh, T. F., & Xiao, W. (2019). Epigenetics regulates reproductive development in plants. *Plants*, 8(12), 564.

Han, S. K., Wu, M. F., Cui, S., & Wagner, D. (2015). Roles and activities of chromatin remodeling ATPases in plants. *The Plant Journal*, 83(1), 62–77.

Han, Z., Yu, H., Zhao, Z., Hunter, D., Luo, X., Duan, J., & Tian, L. (2016). *AtHD2D* gene plays a role in plant growth, development, and response to abiotic stresses in *Arabidopsis thaliana*. *Frontiers in Plant Science*, 7, 310.

He, S., Zhang, Y., Wang, J., Wang, Y., Ji, F., Sun, L., ... & Hao, F. (2022). H3K4me2, H4K5ac and DNA methylation function in short-and long-term heat stress responses through affecting the expression of the stress-related genes in *G. hirsutum*. *Environmental and Experimental Botany*, 194, 104699.

Hereme, R., Galleguillos, C., Morales-Navarro, S., & Molina-Montenegro, M. A. (2021). What if the cold days return? Epigenetic mechanisms in plants to cold tolerance. *Planta*, 254(3), 46.

Hou, Y., Yan, Y., & Cao, X. (2022). Epigenetic regulation of thermomorphogenesis in *Arabidopsis thaliana*. *Abiotech*, 3(1), 12–24.

Howe, F. S., Fischl, H., Murray, S. C., & Mellor, J. (2017). Is H3K4me3 instructive for transcription activation? *Bioessays*, 39(1), 1–12.

Jin, Q., Chachar, M., Ali, A., Chachar, Z., Zhang, P., Riaz, A., ... & Chachar, S. (2024). Epigenetic regulation for heat stress adaptation in plants: New horizons for crop improvement under climate change. *Agronomy*, 14(9), 2105.

Jin, X., Li, X., Teixeira da Silva, J. A., & Liu, X. (2024). Functions and mechanisms of non-histone protein acetylation in plants. *Journal of Integrative Plant Biology*, 66(10), 2087–2101.

Junaid, M. D., Chaudhry, U. K., Şanlı, B. A., Gökçe, A. F., & Öztürk, Z. N. (2024). A review of the potential involvement of small RNAs in transgenerational abiotic stress memory in plants. *Functional & Integrative Genomics*, *24*(2), 74.

Kim, J. H. (2021). Multifaceted chromatin structure and transcription changes in plant stress response. *International Journal of Molecular Sciences*, 22(4), 2013.

Kim, J. M., Sasaki, T., Ueda, M., Sako, K., & Seki, M. (2015). Chromatin changes in response to drought, salinity, heat, and cold stresses in plants. *Frontiers in Plant Science*, 6, 114.

Kim, Y. J., Kim, W. Y., & Somers, D. E. (2024). HOS 15-mediated turnover of PRR 7 enhances freezing tolerance. *New Phytologist*, 244(3), 798–810.

Kondhare, K. R., Kumar, A., Patil, N. S., Malankar, N. N., Saha, K., & Banerjee, A. K. (2021). Development of aerial and belowground tubers in potato is governed by photoperiod and epigenetic mechanism. *Plant Physiology*, 187(3), 1071–1086.

Kovalchuk, I. (2023). Role of epigenetic factors in response to stress and establishment of somatic memory of stress exposure in plants. *Plants*, 12(21), 3667.

Kwon, C. S., Lee, D., Choi, G., & Chung, W. I. (2009). Histone occupancy-dependent and -independent removal of H3K27 trimethylation at cold-responsive genes in Arabidopsis. *The Plant Journal*, 60(1), 112–121.

Lämke, J., & Bäurle, I. (2017). Epigenetic and chromatin-based mechanisms in environmental stress adaptation and stress memory in plants. *Genome Biology*, 18, 1–11.

Law, J. A., & Jacobsen, S. E. (2010). Establishing, maintaining and modifying DNA methylation patterns in plants and animals. *Nature Reviews Genetics*, 11(3), 204–220.

Le, H., Simmons, C. H., & Zhong, X. (2025). Functions and mechanisms of histone modifications in plants. *Annual Review of Plant Biology*, 76, 551–578.

Li, M., Sun, W., Wang, F., Wu, X., & Wang, J. (2021). Asymmetric epigenetic modification and homoeolog expression bias in the establishment and evolution of allopolyploid *Brassica napus*. *New Phytologist*, 232(2), 898–913.

Li, W., Hu, J., Song, F., Yu, J., Peng, X., Zhang, S., ... & Zhu, P. (2024). Structural basis for linker histone H5–nucleosome binding and chromatin fiber compaction. *Cell Research*, 34(10), 707–724.

Louis, N., Dhankher, O. P., & Puthur, J. T. (2023). Seed priming can enhance and retain stress tolerance in ensuing generations by inducing epigenetic changes and trans-generational memory. *Physiologia Plantarum*, 175(2), e13881.

Luo, M., Ríos, G., Sarnowski, T. J., Zhang, S., Mantri, N., Charron, J. B., & Libault, M. (2020). New insights into mechanisms of epigenetic modifiers in plant growth and development. *Frontiers in Plant Science*, 10, 1661.

Luo, M., Wang, Y. Y., Liu, X., Yang, S., Lu, Q., Cui, Y., & Wu, K. (2012). HD2C interacts with HDA6 and is involved in ABA and salt stress response in *Arabidopsis*. *Journal of Experimental Botany*, 63(8), 3297–3306.

Martienssen, R. A., & Chandler, V. L. (2013). Molecular mechanisms of transposon epigenetic regulation. In Fedoroff, N.V. (ed.) *Plant Transposons and Genome Dynamics in Evolution* (pp. 71–92). USA: John Wiley & Sons.

Ming, Y., Jiang, L., & Ji, D. (2023). Epigenetic regulation in tomato fruit ripening. *Frontiers in Plant Science*, 14, 1269090.

Miryeganeh, M. (2022). Epigenetic mechanisms of senescence in plants. *Cells*, 11(2), 251.

Morrison, O., & Thakur, J. (2021). Molecular complexes at euchromatin, heterochromatin and centromeric chromatin. *International Journal of Molecular Sciences*, 22(13), 6922.

Nguyen, H. M., Kim, M., Ralph, P. J., Marín-Guirao, L., Pernice, M., & Procaccini, G. (2020). Stress memory in seagrasses: First insight into the effects of thermal priming and the role of epigenetic modifications. *Frontiers in Plant Science*, 11, 494.

Pfluger, J., & Wagner, D. (2007). Histone modifications and dynamic regulation of genome accessibility in plants. *Current Opinion in Plant Biology*, 10(6), 645–652.

Ramakrishnan, M., Zhang, Z., Mullasseri, S., Kalendar, R., Ahmad, Z., Sharma, A., ... & Wei, Q. (2022). Epigenetic stress memory: A new approach to study cold and heat stress responses in plants. *Frontiers in Plant Science*, 13, 1075279.

Rossnerova, A., Izzotti, A., Pulliero, A., Bast, A., Rattan, S. I. S., & Rossner, P. (2020). The molecular mechanisms of adaptive response related to environmental stress. *International Journal of Molecular Sciences*, 21(19), 7053.

Sani, E., Herzyk, P., Perrella, G., Colot, V., & Amtmann, A. (2013). Hyperosmotic priming of *Arabidopsis* seedlings establishes a long-term somatic memory accompanied by specific changes of the epigenome. *Genome Biology*, 14, 1–24.

Schmitz, R. J., Sung, S., & Amasino, R. M. (2008). Histone arginine methylation is required for vernalization-induced epigenetic silencing of *FLC* in winter-annual *Arabidopsis thaliana*. *Proceedings of the National Academy of Sciences*, 105(2), 411–416.

Seem, K., Kaur, S., Selvan, T., & Kumar, S. (2024). Plant response to drought stress: Epigenomic perspective. In Bhatt D., Nath M., Badoni S., Joshi R. (eds.) *Current Omics Advancement in Plant Abiotic Stress Biology* (pp. 323–341). Academic Press. https://doi.org/10.1016/B978-0-443-21625-1.00022-1.

Sen, A., Puthur, J. T., Challabathula, D., & Brestič, M. (2022). Transgenerational effect of UV-B priming on photochemistry and associated metabolism in rice seedlings subjected to PEG-induced osmotic stress. *Photosynthetica*, 60(2), 219.

Sharma, M., Kumar, P., Verma, V., Sharma, R., Bhargava, B., & Irfan, M. (2022). Understanding plant stress memory response for abiotic stress resilience: Molecular insights and prospects. *Plant Physiology and Biochemistry*, 179, 10–24.

Sharma, M., Sidhu, A. K., Samota, M. K., Gupta, M., Koli, P., & Choudhary, M. (2023). Post-translational modifications in histones and their role in abiotic stress tolerance in plants. *Proteomes*, 11(4), 38.

Stassen, J. H., López, A., Jain, R., Pascual-Pardo, D., Luna, E., Smith, L. M., & Ton, J. (2018). The relationship between transgenerational acquired resistance and global DNA methylation in Arabidopsis. *Scientific reports*, *8*(1), 14761.

Sun, L., Song, G., Guo, W., Wang, W., Zhao, H., Gao, T., Lv, Q., et al. (2019). Dynamic changes in genome-wide histone3 lysine27 trimethylation and gene expression of soybean roots in response to salt stress. *Frontiers in Plant Science*, 10, 1031.

Talbert, P. B., & Henikoff, S. (2014). Environmental responses mediated by histone variants. *Trends in Cell Biology*, 24(11), 642–650.

Tang, J., Wang, X., Xiao, D., Liu, S., & Tao, Y. (2023). The chromatin-associated RNAs in gene regulation and cancer. *Molecular Cancer*, 22(1), 27.

Tersenidis, C., Poulios, S., Komis, G., Panteris, E., & Vlachonasios, K. (2024). Roles of histone acetylation and deacetylation in root development. *Plants*, 13(19), 2760.

Ucar, S., & Aydin, M. (2025). Effects of Epigenetic Inhibitors on Somatic Embryogenesis in Wheat (*Triticum aestivum L.*). *Iranian Journal of Science*, *49*(1), 1–10.

Vlachonasios, K., Poulios, S., & Mougiou, N. (2021). The histone acetyltransferase GCN5 and the associated coactivators ADA2: From evolution of the SAGA complex to the biological roles in plants. *Plants*, 10(2), 308.

Vo, B. T., Mas, P., & Johannes, F. (2024). Time's up: Epigenetic clocks in plants. *Current Opinion in Plant Biology*, 81, 102602.

Wang, H., Wang, L., Yang, M., Zhang, N., Li, J., Wang, Y., ... & Xu, S. (2023). Growth and DNA methylation alteration in rice (*Oryza sativa* L.) in response to ozone stress. *Genes*, 14(10), 1888.

Wang, X., Xin, C., Cai, J., Zhou, Q., Dai, T., Cao, W., & Jiang, D. (2016). Heat priming induces trans-generational tolerance to high temperature stress in wheat. *Frontiers in Plant Science*, 7, 501.

Wendte, J. M., Zhang, Y., Ji, L., Shi, X., Hazarika, R. R., Shahryary, Y., ... & Schmitz, R. J. (2019). Epimutations are associated with CHROMOMETHYLASE 3-induced de novo DNA methylation. *elife*, 8, e47891.

Wierzbicki, A. T., Blevins, T., & Swiezewski, S. (2021). Long noncoding RNAs in plants. *Annual Review of Plant Biology*, 72(1), 245–271.

Wójcikowska, B., Wójcik, A. M., & Gaj, M. D. (2020). Epigenetic regulation of auxin-induced somatic embryogenesis in plants. *International Journal of Molecular Sciences*, 21(7), 2307.

Yamaguchi, N. (2021). Heat memory in plants: Histone modifications, nucleosome positioning and miRNA accumulation alter heat memory gene expression. *Genes & Genetic Systems*, 96(5), 229–235.

Yu, M. H., Liao, W. C., & Wu, K. (2025). Histone methylation in plant responses to abiotic stresses. *Journal of Experimental Botany*, eraf058.

Zanetti, M. E., Blanco, F., Ferrari, M., Ariel, F., Benoit, M., Niebel, A., & Crespi, M. (2024). Epigenetic control during root development and symbiosis. *Plant Physiology*, 196(2), 697–710.

Zhao, T., Zhan, Z., & Jiang, D. (2019). Histone modifications and their regulatory roles in plant development and environmental memory. *Journal of Genetics and Genomics*, 46(10), 467–476.

Zheng, L., Ma, S., Shen, D., Fu, H., Wang, Y., Liu, Y., ... & Huang, J. (2021). Genome-wide identification of Gramineae histone modification genes and their potential roles in regulating wheat and maize growth and stress responses. *BMC Plant Biology*, *21*(1), 543.

Zhu, Q. H., Shan, W. X., Ayliffe, M. A., & Wang, M. B. (2016). Epigenetic mechanisms: An emerging player in plant-microbe interactions. *Molecular Plant-microbe Interactions*, 29(3), 187–196.

Zubko, E., Gentry, M., Kunova, A., & Meyer, P. (2012). De novo DNA methylation activity of methyltransferase 1 (MET1) partially restores body methylation in *Arabidopsis thaliana*. *The Plant Journal*, 71(6), 1029–1037.

2 Role of DNA Methylation in Plant Growth Regulation

Sachin Patil, Prashant Palkar, Vishal Kamble and Jaganath Patil

2.1 INTRODUCTION

Genomic expression is regulated by chromatin structure, which is governed by various processes such as histone variants, histone post-translational modifications and DNA methylation, all associated with epigenetic regulation. In eukaryotic cells, gene expression is greatly influenced by the dynamic chromatin environment associated with epigenetic mechanisms. Such mechanisms create various chromatin states for stress-responsive gene expression, which is important for adaptation to harsh environmental conditions. Recent studies have revealed that many epigenetic factors participate in abiotic stress responses, and various chromatin modifications are changed when plants are exposed to stressful environment (Chinnusamy & Zhu, 2009). For understanding the genomic expression during growth, development and environmental stress, one can know that information on genome sequence and its expression is required to modify the genotype (Chang et al., 2020). The developmental activities or environmental stress can induce epigenetic modifications in the genome, and a single genome in a plant cell gives rise to multiple epigenomes in response to growth, development and environmental stress (Lu & Hyun, 2021).

Waddington (1942) introduced the term *epigenetics* as the branch of biology that deals with the study of causal interactions between genes and their products which create the phenotype in an organism. However, in the original sense, epigenetics referred to all the molecular pathways modulating the expression of a genotype into a particular phenotype. Today, epigenetics is defined as the study of changes in gene function that are mitotically and/or meiotically heritable and that do not entail a change in DNA sequence. Epigenetic modification is maintained during a period of development, but reset between generations (i.e., intra-individual) or inherited across generations in plants (i.e., inter-generational). It must be maintained through cell divisions in tissues from which a germline will be established. In this way, epigenetic information can be sustained over developmental and generational timelines (Hemenway & Gehring, 2023). This information is also dynamic because modifications occur in specific cells or tissues during developmental stages as well as in response to certain environmental conditions.

Plants live in changing and challenging environment that are usually unfavorable or stressful for growth and development. The stressful environmental factors can be either biotic (diseases or pathogens, insects, pests, allelopathic effects) or abiotic (drought, salinity, extreme temperature, nutrient deficiencies, heavy metal stress and ultraviolet radiation). The adverse effects of environmental factors may threaten agricultural productivity (Sahu et al., 2013; Kim et al, 2015). Therefore, plant stress resistance is critical for agriculture, and this can be achieved through epigenetics. Recent studies showed the participation of epigenetic mechanisms in the response of plants to abiotic stresses. Hewezi et al. (2017) carried out a comprehensive examination of plant methylome changes prompted by various phytopathogens, underscoring the importance of DNA demethylation as a common feature of plant response during susceptible interactions. Li et al. (2018) surveyed the role of active DNA demethylation in plant developmental activities including pollen tube formation, embryogenesis, fruit ripening, stomatal development and nodule formation. Plants can remember and learn from their experiences, making them highly adaptable to their environment (Baurle et al., 2020; Bhadouriya et al., 2020). Inherited stress tolerance mechanisms vary among plant species based on genetic makeup, intensity and stress duration. Mutations in DNA sequences cause trait variations, which plant breeders use to improve plant populations due to alterations in chromatin states (Sudan et al., 2018). During stress responses, epigenetic modifications are known to significantly ensure the reprogramming of the plant transcriptome. Alterations to the chromatin structure mediate these changes. These modifications can be stable and passed on to subsequent generations, allowing plants to transmit stress memories across generations (Banerjee & Roychoudhury, 2018).

2.2 EPIGENETICS REGULATIONS IN PLANTS

Phenotypic characteristics of an individual, passed through either mitosis or meiosis, undergo various changes, variations and altered cell attributes that should be maintained after each cell division. The epigenetic alteration in plants, generally, was comprehensively studied by DNA methylation, and the process includes the addition of a methyl group to the fifth carbon position of a cytosine ring. Histone modification may be achieved by alterations in transposable elements that are often methylated and small RNAs capable

 DOI: 10.1201/9781003589914-3

TABLE 2.1
Enzymes required during DNA methylation

Sr. No.	Name of the enzyme	Role of enzyme in DNA methylation
	Met1 DNA methyltransferase	It plays a vital role in maintaining DNA methylation patterns, precisely at CG sites, by adding methyl groups to cytosine bases during DNA replication. However, a mutation in Met1 may lead to significant changes in gene expression and developmental abnormalities (Zangi et al., 2019)
	Chromomethylase 3 (CMT3)	Another important protein that is responsible primarily for adding methyl groups at the "CHG" site of DNA. This upholds genome stability by silencing transposons and regulating gene expression. Genomic stability is important during heat stress. In heat stress, the enzyme acts as a regulator of DNA methylation at non-CG sites and prevents excessive transposon activation. Thus, genomic stability is achieved (Nozawa et al., 2021)
	Chromomethylase 2 (CMT2)	It is a primary DNA methyltransferase responsible for CHH methylation at the ONSEN locus. It plays a role in chromatin assembly and disassembly and is involved in gene silencing (Moore et al., 2013)
	Small interfering RNAs (siRNAs)	It identifies the DNA sequences that need to be methylated, allowing the DNA methyltransferase enzymes to modify those sites and repress gene expression at those locations (Moore et al., 2013). DNA methylation is mediated by siRNAs known as RNA-directed DNA methylation (RdDM)

of directing DNA methylation and chromatin remodeling at their target sites (Banerjee & Roychoudhury, 2017). Chromatin structure then alters the availability of DNA to transcription factors and affects whether genes can be expressed. Thus, DNA methylation significantly influences histone modifications by creating a molecular environment that facilitates the recruitment of proteins that can then modify histones.

2.2.1 DNA Methylation Impacts Histone Modifications

Protein binding: Once DNA is methylated, specific binding proteins bind to the methylated DNA, which then allows them to bind to other proteins and form protein complexes that modify histones. This often leads to chromatin compaction and gene silencing.

Histone deacetylases (HDACs): The methylated DNA-binding proteins recruit the HDACs, which remove acetyl groups from histone tails, resulting in a more condensed chromatin structure and reduced gene expression.

Specific histone modifications: DNA methylation is often associated with specific histone modifications like H3K9 methylation, which is generally linked to transcriptional repression.

Feedback loop: The relationship can be a feedback loop, where histone modifications can also influence the pattern of DNA methylation, creating a complex regulatory network (Table 2.1).

2.3 MECHANISM OF DNA METHYLATION IN PLANT GROWTH REGULATION

DNA methylation is a complex process that involves various enzymes and cofactors. At the beginning of the process, the DNA methyltransferase enzyme recognizes the CpG dinucleotide. CpG dinucleotides are short sections of DNA that play a crucial role in gene regulation. Unlike most of the genome, these sequences are GC-rich. They contain large numbers of CpG dinucleotides that are predominantly unmethylated. The modification in CpG alters the chromatin structure, leading to the suppression of gene transcription to regulate gene expression during plant development and stress response. In *Arabidopsis thaliana*, researchers have effectively delayed the flowering process achieved through DNA methylation that suppresses the expression of specific genes involved in flower development. In addition, DNA methylation also plays an important role in controlling leaf morphology, where it suppresses the expression of genes responsible for shaping and sizing leaves, resulting in the formation of smaller-sized leaves. Through this mechanism, *Arabidopsis* has been shown to enhance its resistance capacity against the pathogen *Hyaloperonospora arabidopsidis* (Hpa) for protection from downy mildew disease. Additionally, DNA hypermethylation plays a crucial role in improving plant immunity against two fungal pathogens, *Plectosphaerella cucumerina* and *Alternaria brassicicola* in *Arabidopsis*. Research has also found that *Pseudomonas syringae* pv. *tomato* (Pst), a type of bacterial pathogen, can trigger defense and hormone pathways achieved through DNA methylation.

2.3.1 DNA Methylation and Demethylation

Epigenetic modifications are heritable chemical variations that do not modify the DNA nucleotide sequence but can influence the phenotype (Henderson & Jacobsen, 2007). A heritable epigenetic modification in which the transfer of a methyl moiety from S-adenosyl methionine to the fifth position of the cytosine residue of DNA, catalyzed by enzymes, is known as DNA methylation. It is one of the fascinating subjects in plant biology because the dynamism and plasticity in the epigenome affect plant growth, development and evolution in response to the environment. In prokaryotes, DNA methylation acts as a defense mechanism in the host, whereas in eukaryotes, it controls the transposable element in the genome. The level of methylation varies extremely between organisms that display this modification (Bittar & Bitter, 1994; Steward et al., 2002). The suggested availability of methylated cytosines (5-mC) in plants ranges from 4 to 37%, depending upon the species, which promotes transcriptional repression by preventing the activators from binding to their target sites (Zhai et al., 2008). In plant genomes, 5-mC content varies due to the presence of methylated symmetric cytosines in CpNpG or random asymmetric C nucleotides that are often associated with DNA outside the nuclear genes. There are various sites, viz., CG sites, CHG sites (H indicates A, C or T) and CHH sites (asymmetrical site) where DNA methylation occurs and is catalyzed by enzymes such as DNA methyltransferase (MTase). In plants, tissue-specific, developmental stage-specific and stress-induced variations in DNA methylation have been reported.

DNA demethylation occurs through either an active or passive state. In the active mechanism, specific proteins demethylate the DNA sequence. During this process, DNA methylation involves a base excision repair pathway mediated by various DNA methylases that possess either only monofunctional DNA glycosylase acitivity or both glycosylase and lyase activities (bifunctional DNA glycosylase). 5-mC is removed as a free base, and a single base gap left behind is filled with a non-methylated cytosine by DNA polymerase and ligase (Vairapandi & Duker, 1993), for example, in *Arabidopsis,* four DNA methylations occur: one interferes with the maternal expression of an imprinted gene, while the others induce the transcriptional gene silencing (TGS) of transgenes and endogenous homologous loci related to the genes *DEMETER (DME1)* and *ROS1,* respectively, which code DNA glycosylases and remove 5-mC from DNA. These DNA demethylases also prevent the formation of stable hyper-methylated epialleles in plant genomes and thus help maintain equilibrium between methylated and unmethylated DNA. Passive DNA methylation occurs in dividing cells when DNA methylation maintenance activity is absent. In this process, methylated cytosines are replaced with unmodified cytosines during DNA replication. Incorporation of unmodified cytosines may be due to the loss of activity of maintenance of DNA methylases, i.e., MET1 and CMT3, during DNA replication. As a result of enzyme inactivity, there is subsequently a loss of DNA methylation in the subsequent generation (Zhu, 2009; Ibarra et al., 2012).

The importance of DNA methylation in plant growth and development is to provide a heritable epigenetic mark that could direct the developmental program of plants. It also provides a means of genome defense against the activity of parasitic mobile elements to reduce background transcriptional noise and memorize the pattern of gene activity by stabilizing gene silencing brought about by other mechanisms (Table 2.2).

2.3.2 DNA Methylation for Plant Breeding/Selection

An epigenetic modification through DNA methylation is increasingly being explored in plant breeding. Nowadays it is one of the potential tools to introduce new variations and improve crop traits by altering gene expression without changing the underlying DNA sequence. The DNA methylation mechanism allows breeders to select for desirable phenotypes related to stress tolerance, yield, flowering time and more, by manipulating the methylation patterns of specific genes of interest. Epigenetic modifications allow for quantitative changes in gene activity and have a

TABLE 2.2
Modification of genes, involved in abiotic and biotic stresses in different plant varieties, through DNA methylation

Plant	Stress	Genes	DNA methylation
Tomato	Drought	*Asr1*	CG hypermethylation and CHH hypomethylation (Gonzalez et al., 2011; Dominguez & Carrari, 2015)
Tomato	Drought	*Asr2*	CHH hypomethylation in the regulatory region (Gonzalez et al., 2013; Dominguez & Carrari, 2015)
Tobacco	Cold	*NtGPDL*	Hypomethylation (Choi & Sano, 2007)
Maize	Cold	*ZmMll*	Root-specific hypomethylation (Steward et al., 2002)
Soybean	Salinity	*Glyma11g02400*	Demethylation (Song et al., 2012)
Soybean	Salinity	*Glyma6g27950*	Hypomethylation (Song et al., 2012)
Soybean	Salinity	*Glyma20g30840*	Hypomethylation (Song et al., 2012)
Maize	Cold	Genome-wide	Global methylation shift (Tan, 2010)
Rice	Drought	Genome-wide	Genotype-dependent differential methylation (Wang et al., 2011)
Tomato	*Pseudomonas syringae*	*Pto*	CHH methylation (Mulaudzi et al., 2023)
Potato	*Phytophthora infestans*	*Rpi-blb1*	CHH methylation (DiBiase et al., 2024)

correlation with increasing hybrid vigor and performance of F_1 hybrid maize either at the individual protein level, enzymatic activity level or RNA level. It acts as a generator of epimutations and plays an important role in phenotypic consequences, leading to the formation of more permanent mutant alleles at a locus, while mutations only rarely lead to new epialleles.

2.3.3 Various Techniques for Analyzing and Manipulating DNA Methylation in Plants

Bisulfite sequencing: A widely used method to determine the methylation status of individual cytosine sites across the genome. It is considered a gold-standard technique for sensing DNA methylation, particularly useful for analyzing CpG islands with high methylation density. Bisulfite sequencing is a technique in which unmethylated cytosines (C) in the genome distinguish methylated from unmethylated cytosines. Coupled with high-throughput sequencing technology, it determines methylation status at CpG/CHG/CHH sites. It has been successfully applied in methylome analysis across various branches of eukaryotic phylogeny.

Methylation-sensitive amplified polymorphism (MSAP): A cost-effective method used to identify differentially methylated regions in the plant genome to analyze epigenetic changes related to environmental stress or development stages. In this technique, DNA methylation patterns are analyzed by comparing the digestion profiles of genomic DNA with different restriction enzymes, which are sensitive to methylation, allowing researchers to detect sites of the genome where methylation levels differ between samples.

Genome-wide association studies (GWAS): A technique used to identify genomic regions associated with phenotypic variation linked to DNA methylation patterns. This technique is used to examine the entire genome of a large group of individuals to identify genetic variations, usually single nucleotide polymorphisms (SNPs), that are statistically associated with a particular disease or trait, allowing researchers to pinpoint genes potentially involved in disease development and progression.

2.3.4 Applications of DNA Methylation in Plant Breeding

Crop improvement: Identifying and targeting genes associated with yield components like seed size or number of grains could lead to increased crop productivity.

Control the flowering period or time: By using this mechanism, modifying methylation patterns in genes regulating flowering time can allow breeders to modify crop varieties that grow in diverse environment.

Disease resistance varieties: Methylation can stimulate plant immune responses, possibly leading to enhanced resistance against the pathogens.

Boost stress tolerance capacity: By modifying methylation patterns in genes related to stress response, breeders can develop crops that are more resilient to abiotic stresses like drought, cold, flood and salinity.

2.3.5 Challenges in DNA Methylation

The complexity of regulation: DNA methylation is a complex mechanism, and it plays a crucial role in regulatory mechanisms for targeted manipulation. Therefore, maintaining regulation in the complex mechanism remains a challenge.

Tissue-specific patterns: Another important challenge is to maintain tissue-specific patterns. Because methylation patterns can vary across different plant tissues, careful consideration is required when selecting samples for analysis.

Epigenetic stability: The most important challenge is to maintain stability over generations. Ensuring that induced methylation changes remain stable over generations is a key challenge for successful application in breeding programs.

2.4 CONCLUSION

In the modern era, epigenetics is one of the important tools of molecular biology that generates wide prospects for plant improvement. DNA methylation has become the mechanism of preference for scientists investigating gene function and manipulating plants to create novel features, capable of stability in different environmental conditions. Therefore, epigenetics by using the DNA methylation mechanism can be enhanced, and transgene silencing could be mitigated with improved knowledge of the mechanisms.

ACKNOWLEDGMENT

Dr Sachin Patil is very much thankful to the principal, Dr. Ghali College, Gadhinglaj, for providing the facilities and constant support to write the book chapter. Mr Prashant Palkar and Mr Vishal Kambale are thankful to the management body of Shahid Veerpatni Laxmi Mahavidyalaya, Titve, for their constant support and encouragement to write the book chapter.

REFERENCES

Banerjee, A., & Roychoudhury, A. (2017). Epigenetic regulation during salinity and drought stress in plants: Histone modifications and DNA methylation. *Plant Gene*, 11, 199–204.

Banerjee, A., & Roychoudhury, A. (2018). The gymnastics of epigenomics in rice. *Plant Cell Reports*, 37, 25–49.

Baurle, I., & Trindade, I. (2020). Chromatin regulation of somatic abiotic stress memory. *Journal of Experimental Botany*, 71, 5269–5279.

Bhadouriya, S. L., Mehrotra, S., Basantani, M. K., Loake, G. J., & Mehrotra, R. (2020). Role of chromatin architecture in plant stress responses: An update. *Frontiers in Plant Science*, 11, 603380.

Bittar, E. E., & Bittar, N. (1994). *Principles of Medical Biology*. Jai Press.

Chang, Y.-N., Zhu, C., Jiang, J., Zhang, H., Zhu, J.-K., & Duan, C.-G. (2020). Epigenetic regulation in plant abiotic stress responses. *Journal of Integrative Plant Biology*, 62(5), 563–580.

Chinnusamy, V. and Zhu, J-K. 2009. Epigenetic regulation of stress responses in plants. *Current Opinion in Plant Biology*. 12 (2). 133–139.

Choi, C.S. and Sano, H. (2007). Abiotic-stress induces demethylation and transcriptional activation of a gene encoding a glycerophosphodiesterase-like protein in tobacco plants. *Molecular Genetics and Genomics*. 277(5): 589–600. doi: 10.1007/s00438-007-0209-1

DiBiase, C. N., Cheng, X., Lee, G., Moore, R. C., McCoy, A. G., Chilvers, M. I., Sun, L., Wang, D., Lin, F., & Zhao, M. (2024). DNA methylation analysis reveals local changes in resistant and susceptible soybean lines in response to *Phytophthora sansomeana*. *Plant Genetics and Genomics*, 14(10), jkae191. https://doi.org/10.1093/g3journal/jkae191.

Dominguez, D. P., & Carrari, F. (2015). ASR1 transcription factor and its role in metabolism. *Plant Signaling & Behavior*, 10(4), e992751. https://doi.org/10.4161/15592324.2014.992751.

González, A.I., Sáiz, A., Acedo, A., Ruiz M. L. and Polanco, C. (2013). Analysis of genomic DNA methylation patterns in regenerated and control plants of rye (*Secale cereale* L.). *Plant Growth Regulation*, 70, 227–236. https://doi.org/10.1007/s10725-013-9794-7

Hemenway, E. A., & Gehring, M. (2023). Epigenetic regulation during plant development and the capacity for epigenetic memory. *Annual Review of Plant Biology*, 74, 89–109.

Henderson, I. R., & Jacobsen, S. E. (2007). Epigenetic inheritance in plants. *Nature*, 447, 418–424.

Henderson IR and Jacobsen SE. (2009). Tandem repeats upstream of the Arabidopsis endogene SDC recruit non-CG DNA methylation and initiate siRNA spreading. *Genes & Development*. 15;22(12):1597–606. doi: 10.1101/gad.1667808.

Hewezi, T., Pantalone, V., Bennett, M., Stewart, C. N., & Burch-Smith, T. M. (2017) Phytopathogen-induced changes to plant methylomes. *Plant Cell Reports*, 37(1), 17–23. https://dx.doi.org/10.1007/s00299-017-2188-y.

Ibarra, C. A., Feng, X., Schoft, V. K., Hsieh, T. F., Uzawa, R., Rodrigues, J. A., et al. (2012). Active DNA demethylation in plant companion cells reinforces transposon methylation in gametes. *Science*, 337, 1360–1364.

Kim, J. M., Sasaki, T., Ueda, M., Sako, K., Seki, M. (2015). Chromatin changes in response to drought, salinity, heat, and cold stresses in plants. *Frontiers in Plant Science*. 6:114. https://doi.org/10.3389/fpls.2015.00114.

Li Y, Kumar, S. and Qian, W. (2018). Active DNA demethylation: mechanism and role in plant development. *Plant Cell Reports*. 37(1):77–85. doi: 10.1007/s00299-017-2215-z.

Lu, X. & Hyun, T. K. (2021). Histone deacetylase inhibitors improve MeJA-induced ginsenoside production in ginseng adventitious roots. *Industrial Crops and Products*, 171: 113909. https://doi.org/10.1016/j.indcrop. 2021.113909

Moore, L. D., Le, T., & Fan, G. (2013). DNA methylation and its basic function. *Neuropsychopharmacology*, 38, 23–38.

Mulaudzi, P. E., Koorsen, G., Mwaba, I., Mahomed, N. B., & Allie, F. (2023). The identification of the methylation patterns of tomato curly stunt virus in resistant and susceptible tomato lines. *Frontiers in Plant Science*, 14, 1135442. https://doi.org/10.3389/fpls.2023.1135442.

Nozawa, K., Chen, J., Jiang, J., Leichter, S. M., Yamada, M., Suzuki, T., Liu, F., Ito, H., & Zhong, X. (2021). DNA methyltransferase *CHROMOMETHYLASE3* prevents *ONSEN* transposon silencing under heat stress. *PLoS Genetics*, 17(8), e1009710. https://doi.org/10.1371/journal.pgen.1009710.

Sahu, P. P., Pandey, G., Sharma, N., Puranik, S., Muthamilarasan, M., Prasad, M. (2013). Epigenetic mechanisms of plant stress responses and adaptation. *Plant Cell Reports*. 32(8):1151–9. doi: 10.1007/s00299-013-1462-x. Epub 2013 May 30. PMID: 23719757.

Song, Y., Ji, D., Li, S., Wang, P., Li Q., Xiang, F. (2012). The dynamic changes of DNA methylation and histone modifications of salt responsive transcription factor genes in soybean. *PLoS ONE* 7(7): e41274. https://doi.org/10.1371/journal.pone.0041274

Steward, N., Yamaguchi, Y., Koizumi, N., & Sano, H. (2002). Periodic DNA methylation in maize nucleosomes and demethylation by environmental stress. *Journal of Biological Chemistry*, 277, 37741–37746.

Sudan, J., Raina, M., & Singh, R. (2018). Plant epigenetic mechanisms: Role in abiotic stress and their generational heritability. *Biotech*, 8, 172.

Tan MP (2010) Analysis of DNA methylation of maize in response to osmotic and salt stress based on methylation-sensitive amplified polymorphism. *Plant Physiology and Biochemistry*. 48:21–26

Vairapandi, M., & Duker, N. J. (1993). Enzymic removal of 5-methylcytosine from DNA by a human DNA-glycosylase. *Nucleic Acids Research*, 21, 5323–5327.

Waddington, C. H. (1942). The epigenotype. *Endeavour*, 1, 18–20.

Wang, W. S., Pan, Y. J., Zhao, X. Q., Dwivedi, D., Zhu, L. H., Ali, J., Fu, B. Y., & Li, Z. K. (2011). Drought-induced site-specific DNA methylation and its association with drought tolerance in rice (*Oryza sativa* L.). *Journal of Experimental Botany*, *62*(6), 1951–1960. https://doi.org/10.1093/jxb/erq391

Zangi, M., Najjar, M. B. B., Golalipour, M., & Ghdasi, M. (2019). *met1* DNA methyltransferase controls *TERT* gene expression: A new insight to the role of telomerase in development. *Cell Journal*, 22(1), 71–74. https://doi.org/10.22074/cellj.2020.6290.

Zhai, J., Liu, J., Liu, B., Li, P., Meyers, B. C., Chen, X., et al. (2008). Small RNA-directed epigenetic natural variation in *Arabidopsis thaliana*. *PLoS Genetics*, 4, e1000056.

Zhu, J. K. (2009). Active DNA demethylation mediated by DNA glycosylases. *Annual Review of Genetics*, 43, 143–166.

3 Methylation and Histone Editing in Epigenetic Regulation of Plants

Aryadeep Roychoudhury, Heeya Gupta and Konkona Lahiri

3.1 INTRODUCTION

Conrad Waddington first proposed the term "epigenetics" in 1942. It originally referred to the interconnections between genes and their specific products which eventually resulted in phenotypic expression. In recent times, it is widely known to be all the changes in the phenotype that are essentially heritable and occur without hindering the underlying sequence of DNA. Epigenetic processes are now understood to be heritable through the mechanism of meiosis and not mitotic cell division, as proposed originally. This was exhibited by plants and then in yeast, mice, *Drosophila* and humans.

The epigenetic processes coordinate essential genomic functions by the removal or addition of various modifications on histones or DNA, forming a complicated mesh of molecular signals that either cause the activation of transcription of certain genes or their repression. Nowadays, robust sequencing techniques allow for the accurate mapping of all these modifications and the analysis of their function in gene regulation. As compared to genetic changes that occur uniformly across the entire population of cells, epigenetic alterations are specific to a subset of cells and triggered by external factors such as the environment as well as internal factors such as aging or hormones (Banerjee & Roychoudhury, 2017a, b). These alterations are stable, long lasting and cause genetically identical cells to have phenotypes that are distinct but can be erased or reprogrammed during modifications in the programming of cells. In case of humans, large-scale reprogramming takes place in embryos and gametes, thus limiting the inheritance of acquired epigenetic modifications that are transgenerational (Sweatt et al., 2013; Walter & Hümpel, 2017). The normal epigenetic mechanisms are disrupted in cancer, causing the activation of transcription of pro-proliferative and anti-apoptotic genes like *CDKN2A* (Al Aboud et al., 2024).

The primary epigenetic mechanisms are histone modification, non-coding RNA (ncRNA)-associated gene regulation and DNA methylation.

This process governs the transfer of methyl groups to cytosines in CpG sequences, often in promoter regions, through the action of methyltransferases. Methylation in these regions of the DNA causes silencing of genes by the recruitment of proteins that suppress their transcription, reducing DNA accessibility of positively regulating transcription factors and assisting the formation of heterochromatin. DNA methylation is important for regulating genes that are tissue specific, for genomic imprinting and for inactivating the X-chromosome. In cancer, tumor suppressor genes are hypermethylated, whereas protooncogenes are hypomethylated.

Histone proteins, viz., H2A, H2B, H3 and H4, create an octameric structure for the DNA to wrap itself around. Post-translational alterations to histone proteins comprise acetylation, methylation, phosphorylation, ubiquitylation and sumoylation, which essentially alter DNA–histone interactions. Histone acetylation, as that of H3K9ac and H3K27ac, causes easy accessibility of transcription factors by weakening DNA–histone binding. Histone methylation, on the other hand, is more composite and is extremely context-dependent, based on the nature of residue and the type of histone protein being methylated. Phosphorylation primarily occurs during DNA damage repair, while ubiquitylation and sumoylation seem to have broad consequences on the expression of the gene.

3.2 DNA METHYLATION

The diverse profiles of gene expression established across multiple organs and cells in multicellular organisms are mediated by epigenetic mechanisms. DNA methylation is a significant epigenetic mechanism that involves a direct chemical change of the DNA (Tost, 2008). In 1948, Rollin Hotchkiss made a groundbreaking discovery of modified cytosine while analyzing a calf thymus preparation using paper chromatography. He proposed that this modified cytosine was 5-methylcytosine (5-mC), based on its separation pattern, which mirrored the behavior of thymine (methyluracil) when compared to uracil. Hotchkiss further suggested that this modified form of cytosine naturally occurs in DNA (Hotchkiss, 1948). A significant portion of experimental evidence points to the connection between DNA methylation and gene activity in addition to the well-established role of DNA methylation in restriction modification systems in bacteria and eukaryotic systems (Razin & Szyf, 1984). This epigenetic process is related to the dinucleotide sequence located at the fifth position

DOI: 10.1201/9781003589914-4

of the cytosine pyrimidine ring. Through its role in chromatin structure, CpG functions as one of the several key components of the epigenetic machinery that regulates as well as controls the expression of the gene. To create the genomic chromatin landscape, it works closely with chromatin-remodeling complexes and histone modifications (Tost, 2008). A greater proportion of CpG is found in CpG islands compared to the rest of the genome. These islands are around 1000 base pair long segments of DNA that are frequently unmethylated. Around 70% of gene promoters are found within CpG islands. Specifically, CpG islands frequently contain the promoters for housekeeping genes (Moore et al., 2012). Since its discovery in 1948, the fifth base of human DNA, 5-methylcytosine (5-mC), has generated considerable interest and controversy during attempts to understand its significance (Singal & Ginder, 1999).

This fundamental mechanism in epigenetics influences genomic expression without modifying the DNA sequence itself. In prokaryotes, methylation of DNA plays diverse roles, including regulating transcription, directing the removal of mismatches after replication, controlling replication of the DNA, coordinating the cell cycle, determining the pathogenicity of bacteria and distinguishing its own DNA from foreign DNA. It is essential for vital physiological functions like genomic imprinting, silencing of repetitive sequences of DNA and germline-specific genes and X-chromosome inactivation in mammals. Furthermore, DNA methylation has been identified in actively transcribed gene bodies, where it likely prevents cryptic transcription initiation, thereby ensuring fidelity in gene expression (Kadayifci et al., 2018).

3.2.1 Role and Importance of DNA Methylation in Epigenetics

Mammals have harnessed cytosine methylation as a heritable mechanism to modulate DNA-protein interactions, thereby facilitating long-term gene silencing (Jones & Takai, 2001). A key factor in controlling gene activity is the constant suppression of gene expression caused by methylation of CpG islands. Such regulation takes place either via attracting proteins linked to gene repression or by blocking the ability of transcription factors to bind to DNA. The dynamic process of DNA methylation involves both de novo methylation and demethylation, which allows the methylation patterns of the genome to evolve throughout development. Differentiated cells establish unique, enduring methylation patterns, which govern the expression of tissue-specific genes. When disruptions are spurred on by genetic mutations or environmental factors, including drug exposure, leading to mental deficits and neuropsychiatric illnesses, precise management of DNA methylation is essential for optimal cognitive function and brain activity. This makes DNA methylation a promising therapeutic target for such conditions (Moore et al., 2012). These modifications also serve essential purposes in various biological phenomena, such as genomic imprinting, regulating development, tissue-specific expression of the gene and inactivation of X-chromosome. For instance, it is indispensable for imprinting genes such as *H19* (silencing) and *Igf2* and *Igf2r* (activation). Similar to this, it also serves a critical function in maintaining and regulating inactivation of X-chromosome, especially in regulating the *Xist* gene expression. Although the role of methylation in genome stability remains less clearly defined, it is also thought to serve as a means of defense for the host against intragenomic parasites, including viruses and retroelements, many of which are methylation-inactivated. Over time, additional functions of methylation, such as genomic imprinting, likely evolved from this defensive role (Reik & Dean, 2001).

In cancer biology, DNA methylation emerges as a key player, influencing various stages of oncogenesis through diverse mechanisms. It not only contributes to genome stability but also impacts cellular processes integral to cancer progression (Laird & Jaenisch, 1996). Thus, DNA methylation is intricately related to epigenetics, driving numerous biological processes and underscoring its profound importance in both health and disease.

3.3 GENERAL OVERVIEW OF DNA METHYLATION MECHANISM

Enzymes involved: DNA methyltransferases (DNMTs) such as DNMT1, DNMT3a and DNMT3b.

DNA methyltransferases (DNMTs) are a family of enzymes that are responsible for catalyzing DNA methylation by employing a methyl donor like S-adenosyl-L-methionine to transfer a methyl group to the fifth position of the pyrimidine ring of cytosine residues. This process primarily occurs at CpG dinucleotides, with DNMTs playing distinct yet cooperative roles in establishing and maintaining DNA methylation patterns (Singal & Ginder, 1999).

3.3.1 DNMT1: The Maintenance Methyltransferase

The primary enzyme responsible for maintaining pre-existing methylation patterns is commonly known as DNMT1. During DNA replication, it accurately copies methylation marks from the original DNA strand to the newly synthesized daughter strand, ensuring the integrity of genome-wide information. Structurally, DNMT1 contains an N-terminal and a C-terminal domain, which are the regulatory and the catalytic domains, respectively. It also includes subdomains like the DMAP1-binding domain (connecting histone acetylation to the DNMT1 enzyme), the replication foci targeting sequence (RFTS, promoting post-replicative methylation) and the CXXC domain (binding unmethylated CpG dinucleotides). Although the function of its bromo-adjacent homology (BAH) domain remains unclear, these subdomains collectively mediate the interaction of DNMT1 with DNA and other proteins, highlighting its complexity and precision (Kadayifci et al., 2018).

3.3.2 DNMT3A and DNMT3B: The De Novo Methyltransferases

DNMT3A and DNMT3B are the two methyltransferases which establishes new methylation patterns on unmodified DNA during early development. These enzymes are crucial for creating methylation landscapes that are inherited and remodeled in differentiated cells. The C-terminal catalytic domain of the DNMT3 enzyme is structurally similar to that of the DNMT1 enzyme but possesses unique N-terminal regions and subdomains, including the Pro-Trp-Trp-Pro (PWWP) domain and the ATRX-DNMT3-DNMT3L (ADD) domain, which facilitate interactions with chromatin. Beyond their de novo functions, DNMT3A and DNMT3B also contribute to the stable maintenance and active remodeling of methylation patterns in mature cells (Moore et al., 2012).

3.3.3 Other DNMTs: DNMT3L and DNMT2

DNMT3L, although catalytically inactive, increases the activity of DNMT3A and DNMT3B. However, DNMT2 is not involved in DNA methylation but functions as a tRNA methyltransferase. The coordination of these enzymes is essential for regulating DNA methylation patterns and ensuring epigenetic fidelity (Kadayifci et al., 2018).

3.3.4 Dynamic Interplay of DNMTs in Methylation

DNA methylation involves two processes: de novo methylation and maintenance methylation. De novo methylation performed by DNMT3A and DNMT3B introduces methylation marks early in development, while maintenance methylation executed during replication by DNMT1 preserves these marks at hemimethylated CpG sites across cell divisions (Jeltsch & Jurkowska, 2014). Passive demethylation, on the other hand, occurs during replication in the absence of DNMT1. Together, DNMTs, along with active demethylation intermediates such as 5-hydroxymethylcytosine (5-hmC), create a dynamic system that enables precise regulation of gene expression and epigenetic memory. This intricate enzymatic machinery underscores the fundamental role of DNMTs in shaping and maintaining epigenetic landscapes, which are critical for development, tissue-specific expression of gene, and genome stability.

3.3.5 CpG Sites and Non-CpG Methylation

Sites that are highly concentrated with 5-methyl cytosines (5-mCs) and where 70–80% of cytosines are methylated are called CpG sites. The majority of 5-mCs are located in intergenic regions, gene bodies and repetitive sequences. Additionally, certain intergenic regions situated between genes influence nearby gene expression. Conversely, when CpG sites in promoter or enhancer regions are hypermethylated, these areas become heterochromatic, preventing transcription activators from binding and resulting in transcriptional silencing. Unlike the coding sections, the CpG islands in the promoter regions of active genes are usually unmethylated (Jang et al., 2017). Apart from CpG methylation, methylation of DNA has also been observed at non-CpG sites, like CpT, CpA and CpC. While methylation of non-CpG is now recognized as a universal phenomenon, its functions and mechanisms remain poorly understood and controversial. Some scholars suggest that widespread activation of de novo methylation machinery that targets CpG sites causes non-CpG methylation. Others postulate a relationship between tissue specificity, gene expression and non-CpG methylation. Thus, both CpG and non-CpG methylation play important roles in controlling the expression of genes and maintaining tissue-specific epigenetic landscapes. However, further research is required to fully elucidate the functional significance of non-CpG methylation (Jang et al., 2017).

3.3.6 Demethylation Processes: Passive (Cell Division) vs. Active (Enzymatic, e.g., TET Proteins)

DNA methylation can be removed through two distinct processes: passive demethylation, which is replication-dependent, and active demethylation, which involves enzymatic mechanisms (Zhu, 2009). Passive DNA demethylation occurs during DNA replication when DNA methyltransferases fail to methylate the newly synthesized strand. This results in a progressive dilution of methylation marks with each replication cycle, effectively reducing the methylation levels over time (Agius et al., 2006). In contrast, active demethylation is an enzymatically driven process that does not depend on DNA replication. In plants, this process is mediated by a family of plant-specific enzymes, including ROS1 (Repressor of Silencing 1) and DME (DEMETER). These enzymes are guided to specific loci by complexes such as MBD7-IDM, SWR1 or the Facilitates Chromatin Transactions (FACT) complex. The 5-methylcytosine (5-mC) gets eliminated from the DNA backbone by ROS1 or DME after its recruitment. The resultant single-nucleotide gap is repaired through the base-excision repair pathway, involving enzymes like apurinic/apyrimidinic site lyase (ARP, APE1L), zinc finger DNA 3′-phosphatase (ZDP) and DNA ligase I (LIG1) (Liu & Lang, 2019). The dynamic control of DNA methylation in the genome is highlighted by the interaction between passive and active demethylation, which ensures that appropriate epigenetic changes are preserved or restored as needed.

3.4 DNA METHYLATION IN PLANTS

3.4.1 Unique Aspects of DNA Methylation in Plants as Compared to Animals

In plant genomes, DNA methylation is distributed non-randomly and is enriched in repetitive regions such as transposable elements (TEs), centromeric repeats and rDNA (Slotkin & Martienssen, 2007; Feng et al., 2010; Zemach & Zilberman, 2010). Unlike animals, where methylation predominantly occurs in symmetric CG contexts, plant DNA methylation occurs in CG, CHG (symmetric) and CHH (asymmetric) contexts, where H represents A, T or C (He et al., 2011). In *Arabidopsis thaliana,* heterochromatic regions with transposons and repetitive DNA exhibit an elevated level of methylation. Additionally, interspersed transposon-associated methylation exists in euchromatic chromosome arms (Zhang et al., 2018). Three DNA methyltransferase families regulate methylation in plants: (i) DNA METHYLTRANSFERASE 1 (MET1): Homologous to DNMT1 in mammals, MET1 maintains symmetric methylation of CG; (ii) CHROMOMETHYLASES (CMT2 and CMT3): CMT3 is essential for maintaining CHG methylation, while CMT2 is involved in CHH methylation in heterochromatic regions; and (iii) DOMAINS REARRANGED METHYLTRANSFERASES (DRM1 and DRM2): These enzymes regulate de novo methylation with respect to sequences (Figure 3.1). DRM2, in particular, contributes significantly to methylation of CHH, as this context relies entirely on de novo methylation. DRM2 is guided to target regions via 24-nucleotide siRNAs in a pathway called RNA-directed DNA methylation (RdDM) (Bartels et al., 2018; He et al., 2011; Greenberg et al., 2011).

DNA methylation in plants encompasses two primary processes: maintenance methylation, which replicates existing patterns during DNA replication, and de novo methylation, which introduces new marks in response to cues, guided by small RNAs (Table 3.1).

3.4.2 Role of RNA-Directed DNA Methylation (RdDM)

An important mechanism in plant biology is RNA-directed DNA methylation (RdDM), wherein specific regions of DNA methylation are added under the influence of non-coding RNA molecules. Although RdDM is unique to plants, similar RNA-directed chromatin modification mechanisms are observed in fungi as well as in mammals. Angiosperms like *Arabidopsis thaliana* have the best understanding of the RdDM pathway; however, conserved components and small RNAs have also been identified in plants like ferns and gymnosperms. RdDM shares similarities with the RNAi pathway, as both produce small RNAs and involve proteins such as Dicer, Argonaute and

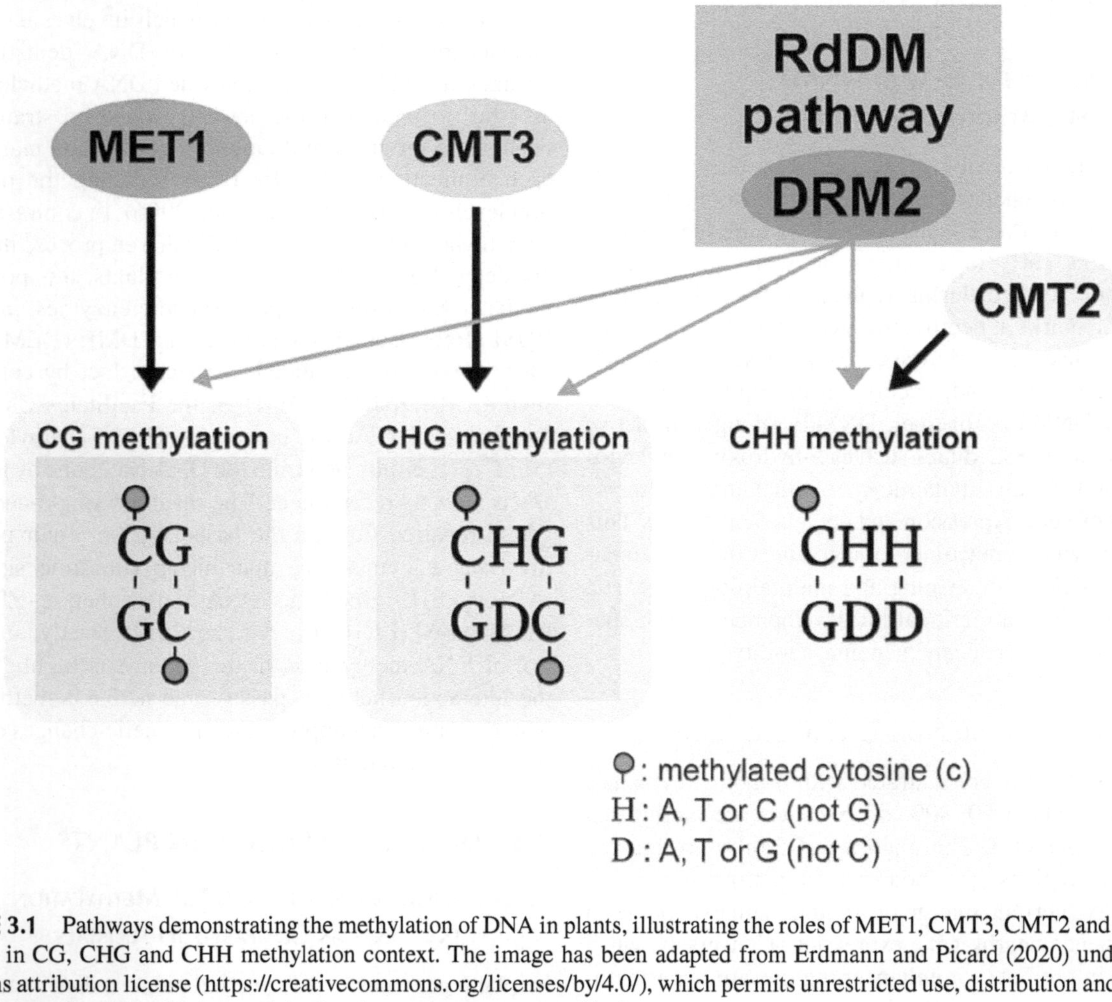

FIGURE 3.1 Pathways demonstrating the methylation of DNA in plants, illustrating the roles of MET1, CMT3, CMT2 and the RdDM pathway in CG, CHG and CHH methylation context. The image has been adapted from Erdmann and Picard (2020) under creative commons attribution license (https://creativecommons.org/licenses/by/4.0/), which permits unrestricted use, distribution and reproduction in any medium, provided the original author and source are credited.

TABLE 3.1
Difference between maintenance methylation and de novo methylation

Aspect	Maintenance methylation	De novo methylation
Definition	Preserves existing methylation patterns during DNA replication	Establishes new methylation marks on previously unmethylated DNA
Enzymes involved	DNA METHYLTRANSFERASE 1 (MET1) and CHROMOMETHYLASEs (CMTs)	DOMAINS REARRANGED METHYLTRANSFERASEs (DRM1/2)
Substrate preference	Targets hemimethylated DNA at symmetric sites (CG and CHG)	Targets unmethylated DNA, often guided by small RNAs
Methylation context	Symmetric sequences: CG and CHG	Symmetric (CG, CHG) and asymmetric (CHH) instances
Role	Maintains epigenetic stability across cell divisions	Introduces novel epigenetic changes in accordance to environmental or developmental stimuli
Mechanism	Identifies and replicates the pattern of methylation cytosines on the parent strand	Identifies and replicates the pattern of methylation cytosines on the parent strand
Importance	Ensures inheritance of epigenetic states	Allows plants to adapt to new environmental or developmental signals
Trigger	DNA replication	RNA signals or external stimuli

Source: Aufsatz et al. (2004), Li et al. (2020b) and Greenberg et al. (2011).

RNA-dependent RNA polymerase. RdDM is crucial for processes such as stress response, cell communication and maintaining genome stability by silencing transposable elements (Erdmann & Ficard, 2020).

3.4.3 Mechanism of RdDM

The RNA-directed DNA methylation (RdDM) pathway is a complex process that regulates gene expression and maintains genome stability by silencing transposons and repetitive sequences (He et al., 2011). It involves a series of coordinated steps:

1. **Initiation of RNA transcripts:** DNA-dependent RNA Polymerase IV (Pol IV) transcribes single-stranded RNA (ssRNA) from regions of DNA containing repeats and transposons. Pol IV transcription is facilitated by the chromatin remodeling protein (CLSY1) and helps produce small interfering RNAs (siRNAs), thereby expanding transcriptional gene silencing (TGS) regions.
2. **Formation of double-stranded RNA (dsRNA):** The single-stranded RNA (ssRNA) generated by Pol IV is changed into double-stranded RNA (dsRNA) by RNA-dependent RNA polymerase 2 (RDR2).
3. **Generation of siRNAs:** The DICER-like proteins, DCL2, DCL3 and DCL4 cleave the dsRNA to produce 24-nucleotide siRNAs, a process dependent on DCL proteins (Kumar & Mohapatra, 2021). These siRNAs are then stabilized by HUA ENHANCER 1 (HEN1), which methylates their 3′ ends.
4. **siRNA loading onto ARGONAUTE proteins:** The siRNAs bind to ARGONAUTE proteins (such as AGO4, AGO6 or AGO9), forming a functional effector complex.
5. **Scaffold RNA production by RNA polymerase V (Pol V):** The DDR complex, consisting of DRD1, DMS3 and RDM1, facilitates the recruitment of Pol V to the chromatin, which then transcribes scaffold non-coding RNAs (ncRNAs) at intergenic and repetitive loci, a process requiring transcription factors like RDM4/DMS4, DRD1 and DMS3.
6. **Formation of effector complex:** KTF1, a scaffold protein, binds to both Pol V-produced scaffold RNAs and AGO4-bound siRNAs. This interaction helps align the siRNAs with the scaffold RNAs, facilitating the recruitment of DNA methyltransferase DRM2.
7. **Recruitment of DNA methyltransferase DRM2:** The effector complex directs DRM2 (a de novo methyltransferase) to specific loci, where it catalyzes methylation of CG, CHG and CHH.
8. **Reinforcement of methylation:** siRNA amplification occurs through the cleavage of scaffold RNAs by AGO4, followed by the RDR2-mediated production of secondary siRNAs. This creates feedback loops between siRNA generation and DNA methylation, ensuring the maintenance of transcriptional gene silencing (TGS).

3.4.4 Outcomes of RdDM

1. Transposons and repetitive DNA sequences are silenced.
2. Gene expression is regulated through transcriptional silencing.
3. Stable epigenetic modifications are established and inherited to maintain genome stability.

3.4.5 Role of Demethylation in Plants

Demethylation of DNA in plants occurs through two mechanisms: active and passive. During the occurrence of replication when DNA methylation is not actively maintained, passive demethylation occurs, leading to a dilution of methylation in the newly synthesized strand. Active demethylation, however, is a more regulated process, facilitated via the plant-specific ROS1/DME family of enzymes. ROS1 and DME are assigned to track down the loci by complexes such as MBD7-IDM and SWR1, where they cleave the 5-methylcytosine (5-mC) from the backbone of DNA. This creates a single-nucleotide gap that is subsequently filled with an unmethylated cytosine through a base-excision repair pathway involving enzymes like ARP, APE1L, ZDP and DNA ligase I (Liu & Lang, 2019).

ROS1 was initially identified by screening for repressors of silencing in *Arabidopsis* plants, where it was found to prevent the silencing of transgenes caused by RNA-directed DNA methylation. ROS1 also alleviates over-methylation, helping to regulate the silencing of endogenous genes and transposons. Additionally, ROS3, a protein that binds to RNA, may assist in directing ROS1 to specific genomic targets. DME, which is primarily expressed in endosperm, is crucial for genome-wide demethylation and gene imprinting (He et al., 2011).

Active demethylation facilitates the activation of transposons as well as repetitive sequences, promoting the production of small interfering RNAs (siRNAs) in endosperms, which can be laced into embryos, contributing to DNA hypermethylation, which helps maintain genome stability in the developing embryo. The action of demethylation is depicted as a process of switching between methylated and unmethylated cytosines (He et al., 2011).

In *Arabidopsis,* the key demethylases identified are ROS1, DME and DEMETER-LIKE PROTEINS 2 and 3 (DML2, DML3). In mammals, ten-eleven translocation (TET)-like enzymes for demethylation are employed; however, in plants, these enzymes have not yet been identified. These TET enzymes catalyze the methylated cytosine oxidation. However, the oxidative products during active demethylation suggest the potential involvement of TET-like proteins in this process, as overexpression of a human TET protein in *Arabidopsis* induces significant genome-wide demethylation, supporting the existence of TET-like activity in plants (Lucibelli et al., 2022).

3.5 DNA Methylation under Different Conditions in Plants

3.5.1 Developmental Stages: Regulation during Seed Development, Germination and Flowering

By now, it is clear that methylation is one of the significant regulating processes important for plant development and dynamically modulates gene expression during key roles like seed development, flowering and germination. For instance, during seed development in soybean, CHH methylation increases from 6% in early stages to 11% in later stages, demonstrating its role in cellular and epigenetic regulation (An et al., 2017; Lin et al., 2017). Similarly, methylome analyses in *Arabidopsis* and rice have highlighted specific genes whose methylation states influence seed maturation and germination dynamics. Active demethylation, mediated by DNA demethylases like DML2, is pivotal during fruit ripening, as seen in tomato, where it activates ripening-associated genes while suppressing the inhibitors (Lang et al., 2017; Liu et al., 2015). Hence, a proper balance between methylation and demethylation controls the timing of blooming and embryogenesis, which is essential for the change from vegetative to reproductive stages.

Methylation also plays an essential role in the maintenance of genomic stability during the development of plants; for instance, methylation at CpG islands in promoter regions influences transcription factor binding, modulating gene expression to enable precise control over growth and differentiation (Kumar & Mohapatra, 2021). While hypermethylation often correlates with gene silencing, hypomethylation enables transcriptional activation, especially in euchromatic regions (Valledor et al., 2007). Such dynamic changes underscore the regulatory importance of DNA methylation during key developmental transitions in plants.

3.5.2 Environmental Stress Responses: Role in Heat, Drought, Salt and Cold Tolerance

Epigenetic changes are caused by environmental stressors like heat, drought, cold and salt, with methylation of DNA playing a significant role. Under conditions pertaining to stress, DNA methylation patterns change, enabling plants to modulate gene expression for stress tolerance. For example, methylation of transposable elements (TEs) near stress-responsive genes often silences these elements while enhancing gene expression under specific environmental stimuli (Le et al., 2014). In *Arabidopsis,* promoter methylation of the *ROS1* gene enhances its transcription under stress, highlighting a nuanced regulatory mechanism (Williams et al., 2015). Methylation also contributes to stress priming, where exposure to a stressor induces epigenetic changes that prepare plants for subsequent stress events. Hypomethylation, for instance, facilitates transcriptional activation of genes associated with endurance to salt and drought. Similarly, hypermethylation of certain genomic regions under cold stress helps maintain chromatin integrity, minimizing damage and preserving genome stability (Kumar & Mohapatra, 2021). These adaptive responses underscore the importance of methylation in DNA in ensuring resilience against environmental challenges.

3.5.3 Biotic Stress Response

RNA-directed DNA methylation (RdDM) was first identified to represent the reaction of plants to infections by viroids. Alongside RNA interference (RNAi), RdDM defends plants against viroids and viruses. Viral RNAs are detected and converted into small RNAs (sRNAs), which serve distinct purposes: RNAi utilizes these sRNAs to degrade viral RNA, while RdDM employs them to silence viral RNA. However, the mechanisms that enable RdDM and RNAi to differentiate between own transcripts and viral RNAs remain largely unclear. Mutants impaired in RdDM or other methylation-related processes often exhibit heightened sensitivity to viral infections. Beyond viral defense, RdDM also contributes to protecting plants from various biotic stresses, such as bacterial, fungal and herbivore attacks. Interestingly, depending on the pathogen, the loss of RdDM function might affect resistance in different ways; for instance, RdDM-deficient mutants display heightened susceptibility to the bacterium *Agrobacterium tumefaciens*, yet exhibit reduced vulnerability to *Pseudomonas syringae* infections. This duality underscores the intricate interactions between RdDM-mediated pathways and the diverse defense mechanisms of plants (Erdmann & Picard, 2020).

3.5.4 Epigenetic Memory in Plants under Stress

Plants exhibit a form of epigenetic memory that enables them to "remember" environmental conditions and adapt accordingly. This memory, mediated by DNA methylation, involves both short-term and heritable long-term changes; for instance, stress-induced DNA methylation changes persistently through mitosis, allowing plants to "store" information about prior stress exposure (Gallusci et al., 2022). Evidence from *Arabidopsis* indicates that approximately one-third of its genes are methylated in the gene body, facilitating transcriptional regulation and enabling memory retention (Zhang et al., 2006). Epigenetic memory is particularly evident in TEs, where methylation cycles regulate their activity. In maize, changes in cytosine methylation have been linked to the activation and silencing of transposons, influencing phenotypic plasticity and stress adaptation (Valledor et al., 2007). Furthermore, plants can transmit epigenetic information to subsequent generations, as seen in *Brassica rapa,* where methylation dynamics in reproductive tissues influence seed development and stress responses (Grover et al., 2018). Such memory mechanisms emphasize how crucial DNA methylation is with regard to improving plant fitness and survival in fluctuating environment.

3.6 MECHANISM OF DNA METHYLATION ACCORDING TO CONDITIONS

3.6.1 Under Normal Growth Conditions

Sustaining the integrity of the genome and coordinating the transcription depend on methylation of DNA. It ensures stable epigenetic marks by methylating cytosine residues, primarily in CpG islands, to suppress TEs and maintain heterochromatic regions. Acting as an "ON-OFF" switch for transcription, it silences specific genes when necessary. This balance of methylation and demethylation stabilizes chromatin structure, fostering consistent cellular function and development (Moore et al., 2012).

3.6.2 Mechanisms Induced by Stress

When exposed to environmental stress, DNA methylation demonstrates exceptional adaptability. Initially thought to function as a prokaryotic host defense mechanism, it has since been identified as a critical regulator of genome integrity in eukaryotes. This defense mechanism suppresses TEs to ensure that the genome remains intact over generations (Zhang et al., 2011). Additionally, epigenetic changes have been extensively documented during developmental processes and environmental stress in plants (Bartels et al., 2018), for example, methylation of cytosine residues in promoter regions represses transcription by affecting transcription factor binding and forming repressive chromatin structures through interactions with methylated DNA-binding proteins (Bird, 2002). DNA methylation also regulates gene/genome imprinting, silencing the genes via DNA and histone modifications. In *Arabidopsis*, rice and maize, demethylation of the maternal genome activates specific genes in the endosperm (Hsieh et al., 2009; Luo et al., 2011; Rodrigues et al., 2013; Waters et al., 2011). In rice, knockout of ALKBH1 (a 6-mA demethylase) increased 6-mA levels, leading to early flowering, suggesting the significant role of 6-mA in reproductive development (Zhou et al., 2018). Similarly, nucleosome remodelers like DDM1 influence 5-mC content in plants (Tan et al., 2016). Decreased 6-mA content and unchanged 5-mC levels in CRISPR/Cas9-induced *ddm1a/ddm1b* double mutants caused dwarfism and decreased seed-setting, highlighting the role of 6-mA in vegetative and reproductive development (Zhang et al., 2018). These findings underscore the interplay of DNA methylation and histone modifications in genetic regulation.

3.6.3 The Source of Diversity and Heritable Variations

Heritable epialleles are frequently produced by epigenetic modifications that last longer than environmental factors changing. These epialleles represent a valuable source of genetic diversity, particularly for crop breeding in species with limited genetic variation. Unlike natural mutations, epialleles form more rapidly, although their reversal rates are higher. Nevertheless, heritable epialleles contribute significantly to evolution. Epigenetic modifications brought on by the environment can be meiotically inherited and mitotically stable; for instance, the transcriptional activation of the Tos17 retrotransposon during rice tissue culture is repressed upon plant regeneration (Liu et al., 2004). Activation and transposition of Tos17 are mediated by DNA

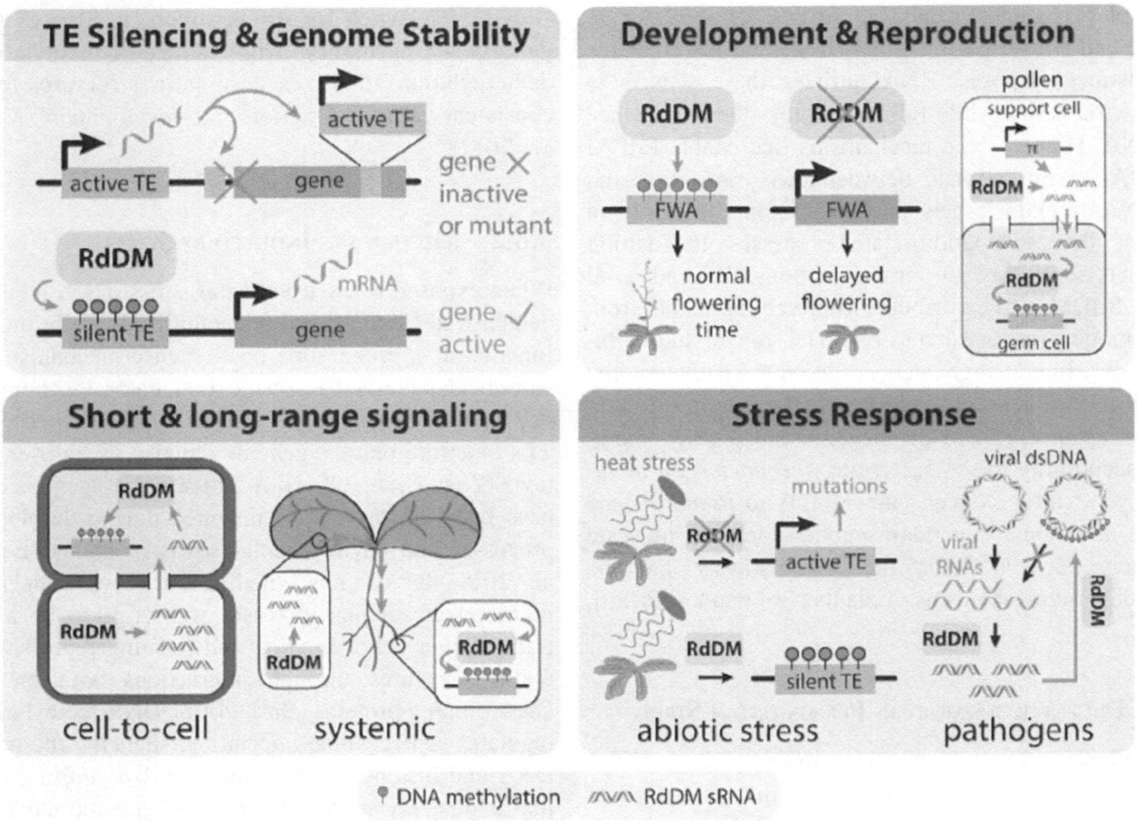

FIGURE 3.2 DNA methylation and RdDM sRNA pathways: the RdDM system plays functional roles in TE silencing and genomic stability, development, reproduction, short- and long-range signaling and stress responses. The image has been adapted from Erdmann and Picard (2020) under creative commons attribution license (https://creativecommons.org/licenses/by/4.0/), which permits unrestricted use, distribution and reproduction in any medium, provided the original author and source are cited.

hypomethylation during tissue culture, facilitated by DNA glycosylase/lyase (Cheng et al., 2006; Ding et al., 2007; La et al., 2011). The RNA-directed DNA methylation (RdDM) pathway responds to environmental stimuli, inducing epigenetic changes at specific loci to generate heritable epialleles (Manning et al., 2006;). Understanding the variance in epigenetic marks, their phenotypic impacts and the heritability of variables associated with epimarks is necessary to effectively use epialleles in crop breeding (Figure 3.2).

Genomic stability and regulatory flexibility are largely dependent on DNA methylation, which maintains gene expression balance under both stress and normal circumstances. While its role in generating epialleles offers immense potential for crop improvement, the challenges of estimating and utilizing epigenetic variability persist. As our knowledge of epigenomics deepens, so will our ability to harness these mechanisms to address global challenges in agriculture and food security.

3.7 HISTONE MODIFICATIONS

Histone modifications occur via two primary mechanisms. The first one concerns changes that impact the organization of the chromatin directly, over either long or short distances. The second one coordinates the binding of the effector molecules either positively or negatively (Bannister & Kouzarides, 2011). The dynamic regulation of histone modifications is carried out by "writers", for example, methyltransferases, ubiquitin-ligases, acetyltransferases, etc., and "erasers", which remove the said modifications like deacetylases, demethylases and deubiquitinases (Banerjee & Roychoudhury, 2018). The transcription factors or chromatin remodeling factors are the "readers" that recognize these modifications and further alter the landscape of the chromatin so as to change with respect to the regulation taking place (Zhao et al., 2019).

Acetyl CoA is utilized by histone acetyltransferases (HATs) in the form of a cofactor to catalyze the addition of acetyl groups (CH_3 COO^-) to lysine residues, specifically their positively charged ε-amino groups. Histone deacetyltransferases (HDACs) remove these acetyl groups (Boycheva et al., 2024). Methylation marks (tri-, di- and mono-marks) on lysines of histone proteins are deposited by histone methyltransferases (HMTs). In the case of *Arabidopsis*, the primary sites of methylation are H3K4, H3K9, H3K27 and H3K36. There have been found additional sites of methylation in mammals, i.e., H3K79 and H4K20 that are important in mammals but are controversial or not reported in plants (Hu & Du, 2022). Histone demethylases remove these methylation marks. Histone phosphorylation takes place due to the transfer of phosphate groups to histone residues such as serine, threonine or tyrosine from adenosine triphosphate. There are a number of kinases and phosphatases that add or remove phosphate groups from the histones (Raghubeer, 2024). Ubiquitin and ubiquitin-like, for example, small ubiquitin-like modifier (SUMO) proteins can be linked covalently to target histones. The modification is known to contribute to both proteolysis and proteolysis independent functions. The first histone ubiquitination discovered was in H2A and since then, various studies have been conducted on this modification (Feng & Shen, 2014). The following portions of this chapter delve deeper into the various kinds of histone modifications that take place in plants and their role in the intricate transcriptional regulation. Plants, being sessile and a crucial part of the environment, experience a dynamic range of stresses, which have been discussed first, after which a brief note on their normal development has been stated.

3.8 HISTONE MODIFICATIONS IN BIOTIC STRESS

With respect to the chromatin landscape, extensive transcriptional reprogramming has been observed as an important part of plant defense mechanisms (Birkenbihl et al., 2017). Defense genes get triggered due to secondary stress when plants are treated with salicylic acid (SA) or its analogue-benzothiadizole (BTH), influencing H3/H4 acetylation (ac) and H3K4 dimethylation/trimethylation (me2/3) in at particularly those regions of the chromatin that comprise promoters of genes encoding various salicylic acid (SA) pathway-based WRKY transcription factors. Expression of the aforementioned genes was prohibited in *npr1* mutants that were priming-deficient. These modifications were not directly activated by BTH, but a stronger and extensive expression was seen as a result of the above changes when the plants were infected with *Pseudomonas syringae* pv maculicola (Jaskiewicz et al., 2011). Systemic acquired resistance (SAR) in *Arabidopsis* has been seen to be connected to histone post-translational modifications that are related to the activation of gene expression and the priming of genes that are SA dependent (Conrath et al., 2015). β-aminobutyric acid (BABA) is a non-protein amino acid that functions as a synthetic priming activator, protecting *Arabidopsis* from the infection of the oomycete pathogen *Peronospora parasitica*. It triggers innate plant defense processes like hypersensitive response (HR) and callose deposition. This activator is capable of providing this protection to plants that lack pathways of salicylic acid, ethylene and jasmonic acid. Noted as similar to the already mentioned BTH, BABA does not cause the accumulation of mRNA of specific relevant genes but confers stronger resistance (Kang et al., 2022). In addition to *Peronospora parasitica*, BABA increases the resistance of plants to *Pectobacterium carotovorum* ssp. carotovorum (*Pcc*), a necrotrophic bacterium, and *Pseudomonas syringae* pv. tomato *(Pst)* DC3000, which is a hemibiotrophic bacterium, via similar mechanisms of priming. The microbe-associated molecular pattern (MAMP) and pattern-triggered immunity (PTI) specific gene expression levels in response to attack by the latter were increased by the acetylation of histones H3K9 and H3K14 along with the di-methylation of H3K4 after BABA treatment, as were the levels of immunity to *Pcc* and callose deposition (Po-Wen et al., 2013). Treatments with a functional synthetic analog of SA, viz., 2,6-dichloroisonicotinic acid (INA), and BABA to *Phaseolus vulgaris*, or the common bean, cause resistance to occur in the case of subsequent attack by *Pseudomonas syringae* pv. *phaseolicola* and modify trimethylation markers in H3K4 and H3K36 with respect to promoter-exon regions (Martínez-Aguilar et al., 2016). Interestingly, the ever-changing natural environment plays a critical role in forming the varied environmental histories of plants (Singh et al., 2014). *Arabidopsis* plants that were repeatedly exposed to heat, salt stress or cold were found to be more resistant than those grown in an environment that was comparatively more stable. Transcriptional activation and thereby epigenetic marks responsible for the same were found to be enhanced by a considerable margin.

It is typical of all histone methyltransferases (HKMTs) present in plants to contain a (Su (var)3-9, Enhancer-of-Zeste and Trithorax) domain or SET domain that is conserved evolutionarily. Distinct SET domains along with their neighborhood sequences have specificity to different lysine residues (H3K4, H3K36, H3K27 and H3K9). There are two families of plant histone demethylases (HKDMs), viz., Jumonji-C (JmjC) domain proteins and lysine-specific demethylase (LSD1) homologues. So far, a significant magnitude of HKDMs and HKMTs has been found responsible for plant responses to pathogen attacks. WRKY70 functions as a transcription factor gene at the JA and SA pathways interconnected with each other and is dependent on the trimethylation of H3K4 by the ARABIDOPSIS HOMOLOG OF TRITHORAX 1 (ATX1) for its transcription. The initiation of transcription of genes that are essentially defense responsive, such as PR1 and THI2.1, is also dependent on ATX1, which holds it under an "actively" modified state of chromatin so that the response is fast and efficient whenever necessary. This suggests that WRKY70 is a target of interest in the context of the activity of histone methylase, whereas PR1, regulated by the levels of SA, and THI2.1, which in turn is regulated by JA levels, are thus secondary targets. Plants that are essentially *atx1* mutants exhibit decreased resistance to hemibiotrophic bacteria such as *Pst* DC3000 due to decreased levels of PR1 (Alvarez-Venegas et al., 2007). It still remains elusive as to how exactly ATX1 works in order to target the levels of WRKY70, PR1 and THI2.1. However, there have been studies to comprehend the functioning of ATX-RELATED 7 (ATX7), which is one of its homologues. The binding of ATXR7 to plant-specific proteins such as MODIFIER OF SNC1 9 (MOS9) attunes the levels of transcription of *RECOGNITION OF PERONOSPORA PARASITICA 4 (RPP4), CONSTITUTIVE1 (SNC1) and SUPPRESSOR OF NPR1-1* via the methylation status of

H3K4. All of these are utilized in encoding plant-specific NLR proteins that play the role of intracellular sensors in order to recognize pathogen effectors and begin a strong as well as efficient immune response, as has been observed in the case of the immune response of *Arabidopsis* against the oomycete pathogen *Hyaloperonospora arabidopsis* Emwa1 (Xia et al., 2013). The intrinsic accumulation of SA and heightened resistance is caused when H3K4me1/2 mark is erased by an LSD1-family HKDM that is encoded by Arabidopsis *FLOWERING LOCUS D (FLD)* (Singh et al., 2014). Infections by several other strains of *Pst* are reported to trigger plant immunity by the removal of the same marker by *LSD1-LIKE 1 (LDL1)* and *LDL2*. As opposed to the wild-type plants, H3K4me1 levels increase in the double mutants *ldl1 ldl2* at the promoters of various defense genes, denoting that *LDL1* and *LDL2* demethylases promote defense. Further examination by the use of chromatin immunoprecipitation and *in silico* comparative analysis of RNA sequencing profiles strongly suggested that various WRKY transcription factors such as WRKY22/40/70 were also responsible for the enhanced immune responses of *ldl1* and *ldl2* (Noh et al., 2021). The accumulation of the H3K4me3 mark due to pathogen attack is also removed by demethylases like JMJ14 in order to similarly control the levels of transcription of defense-related genes, particularly those that are directly influenced by salicylic acid and pipecolic acid (Li et al., 2020a). Almost 15 *JmjC* genes are reported to be positively transcribed when rice plants are under attack by the pathogen *Xanthomonas oryzae* pv. oryzae (Xoo) causing the disease of bacterial blight. Of these 15, JMJ704 prevents the transcription of Os-11N3, NRR and OsWRKY62, all of which negatively regulate defense-related responses by mitigating H3K4me2/3 levels (Hou et al., 2015).

H3K36 methylation also regulates the transcription of *R* genes. The primary HKMT carrying out H3K36me2/3 in *Arabidopsis* is the SET DOMAIN GROUP 8 (SDG8), which precisely causes the deposition of the H3K36me3 mark at *LAZARUS 5* (*LAZ5*), encoding an RPS-4-like R protein. Moreover, SDG8 is crucial in plant defense against pathogenic fungi because it regulates a broad spectrum of genes in the ET or JA signaling pathways. Reduced resistance to fungal pathogens *Alternaria brassicola* and *Botrytis cinerea*, along with the bacterial pathogen *Pst* DC3000, is observed in mutant plants with loss of function of SDG8 (Berr et al., 2010; Zhang et al., 2020). It has thus been proposed that this H3K36 methylation, which is brought about by SDG8, might be acting as a memory with respect to a number of genes in the ET/JA signaling pathways, which aids in carrying out rapid transcription if and when necessary (Berr et al., 2010). The modifier also interacts with the carboxyl terminal domain (CTD) of RNA polymerase II (RNAPII), which suggests a direct correlation between RNAPII loading and H3K36me3 status when induced by SA levels (Zhang et al., 2020). *ECERIFERUM 3* (*CER3*) and *CAROTENOID ISOMERASE 2* (*CCR2*) levels are heavily influenced by SDG8 and encode proteins that activate cuticular wax and carotenoid biosynthesis. Loss of SDG8 in these cases resulted in reduced accumulation levels of lipid and loss of cuticle integrity, which interfered with plant immunity (Lee et al., 2016). SDG8 orthologues, SlSDG33 and SlSDG34, are necessary in tomato (*Solanum lycopersicum, Sl*) for the accumulation of H3K36me3 and H3K4me3 at specific locations. When loss of these markers occurs, it might increase resistance to *Botrytis cinerea.* It is thought that mutations occurring in SlSDG34 and SlSDG33 might be impeding a chromatin environment that is conducive to transcription by aiding the transcription of negative regulatory factors. This does not compromise growth but increases the tolerance of the plant (Bvindi et al., 2022).

H3K27me2/3 marks denote silenced regions of the chromatin. Arabidopsis LIKE HETEROCHROMATIN PROTEIN 1 (LHP1), via its chromodomain, interacts with H3K27me3 and participates in the spreading of the same. This in turn represses the MYC2 branch of the JA/ET pathway and plant immunity is vastly compromised (Ramirez-Prado et al., 2019). *Pst* DC3000 infection causes the Polycomb Repressive Complex 2 (PRC2) to deposit H3K27me2/3 and its SET domain containing components have somewhat partial non-redundant functions in the progression of HR (Dvořák Tomaštíková et al., 2021a). JMJ705 has a different target specificity as compared to that of JMJ704 (H3K27me2/3) but the response is similar in both cases, viz., increased resistance to *Xoo* in rice plants (Hou et al., 2015; Li et al., 2013).

Along with other methylation marks, H3K9me2, which is conferred by HKMT SU(VAR)3-9 HOMOLOG 4 (SUVH4)/KRYPTONITE (KYP) and its homologues, is also responsible for decreasing the levels of basal expression of *NLR/PRR* genes, whose activation, when triggered, inhibits normal development as well as growth of plants (Cambiagno et al., 2021). This means that the methylation state of H3K9 is crucial to determining proper immunity of plants as and when necessary. Evidence of this comes from the fact that the removal of H3K9me2 by JMJ25/JMJ27 is required for defense genes such as *PR1/2* to be correctly induced (Chan & Zimmerli, 2019; Dutta et al., 2017).

There are epigenetic alterations that take place in pathogens during the attack which regulate interactions between pathogens and plants. TrAP protein, which is geminivirus encoded, interacts with the catalytic domain of SUVH4/KYP and blocks its methylase activity, thus forming a regulatory connection between antiviral defense and viral counter defense and finally counteracting the phenomenon of plant defense (Castillo-González et al., 2015). *Phytophthora sojae* escapes the activity of the soybean *R* gene *Rps1b* by transcriptional polymorphism in the *Avr1b* gene with the help of PRC2 PsSu(z) 12, causing no alteration in its sequence. This PsSu(z)12 mutation results in the loss of H3K27 methylation in the region of *Avr1b*, causing more expression of the same, which in turn ends in failed immune recognition by the plant (Wang et al., 2020).

Plant response to pathogens also comprises histone acetylation, which is HAT-mediated, and histone deacetylation, which is HDAC-mediated. This leads to various changes in the chromatin landscape and thus the transcriptional status of various genes, as has been seen extensively in the case of methylation. As noted already, plants exposed to sub-lethal levels of heat, cold and salt have increased resistance to infection and already primed *PTI* genes. These *PTI* genes exhibit an accumulation of histone acetylation, which confers an active transcription status. Mutants of HAC1 encode a HAT that belongs to the class of HAC/CBP and are essentially incapable of carrying out proper priming of genes and fail to protect the plant body from attack (Singh et al., 2014). A coactivator complex is formed due to the association of HAC1/5 with NPR1 in the signaling pathway of SA and activates transcription of various PRs via the acetylation of histones (Jin et al., 2018). HAG1/GCN5 of *Arabidopsis* is a part of the SAGA transcriptional coactivator complex and acetylates various sites of histone H3. Mutations in *hag1/gcn5* lead to lower levels of H3K14ac at the downregulated target regions of the 5′ end of HAG1/GCN5 and simultaneously higher levels of the same modification at the upregulated target regions of the 3′ end of HAG1/GCN5. This established that HAG1/GCN5 actually carries out a dual role in defining the acetylation status of the histones, thereby also modifying the underlying genetic transcription (Wu et al., 2021). The expression of the defense genes is also inhibited by pathogen effectors that deliberately affect the functioning of HAT complexes. The effector of *Phytophthora sojae* PsAvh23 associates with the SAGA complex ADA2 subunit and does not allow it to interact with GCN5, causing a systemic dysregulation of defense genes (Kong et al., 2017). Catalase is known to function in order to protect cells from extensive oxidative damage due to increased reactive oxygen species (ROS) levels by catalyzing the degradation of H_2O_2 into oxygen and water. NcGCN5 regulates CATALASE-3 transcription positively in the filamentous fungus *Neurospora crassa* by the deposition of H3ac (Qi et al., 2018).

Arabidopsis HISTONE DEACETYLASE 6 (HDAC6) and HDAC19 are a class of RPD3-like HDACs and play essential roles in plant-pathogen interactions. Wound formation, pathogens such as *Alternaria brassicicola* and *Pst* induce expression of HDAC19. Loss of this histone modification causes the increased accumulation of SA and expression of *PR* genes, thereby enhancing resistance to *Pst* (Kang et al., 2022). The *PR1/2* promoter is a region where HDAC19 binds to carry out deacetylation of histones and regulate low basal levels of the same under normal conditions and when overstimulated expression is to be prevented during pathogen interactions (Choi et al., 2012).

HDAC6 represses the expression of *PR1/2* genes that are pathogen responsive by binding to them during pathogen infection as well as under normal conditions (Wang et al., 2017). SA biosynthesis is also dependent on the activity of HDAC6 and is inhibited by the regulation of expression levels of CBP60g and SARD1 by HDAC (Wu et al., 2021).

Ustilaginoidea virens is an effector in UvSec117 which targets the RPD3-like class HDAC OsHA701 in rice to the nucleus to lower H3K9ac levels and inhibit the activation of H3K9-mediated defense gene transcription, promoting pathogen infection (Chen et al., 2022). A significant drop in H3 and H4 acetylation levels is caused by the overexpression of BcRPD3 in *Botrytis cinerea*, leading to slower vegetative growth of the fungus and weakened virulence (Zhang et al., 2020).

HDAC AtHD2B of the HD2-class is a target of the MAPK-type kinase MPK. It re-localizes to the nucleoplasm from the nucleolus to inhibit the transcription levels of biotic responsive and flg-22 regulated genes (Latrasse et al., 2017). A second HD2-class HDAC which is AtHD2C has been observed to positively regulate the resistance of plants to infection by cauliflower mosaic virus (CaMV). It removes histone acetylation marks on the viral minichromosome and silences the viral genome. In response, the P6 protein of the virus interacts with the HDAC physically and affects the association of the latter with HDA6, promoting infection. CaMV infection levels increase drastically in *hd2c* mutants (Li et al., 2021).

There exist enzymatic inhibitors of HDACs that are also important in terms of the survival of the plant. Various HDACs along with their enzymatic functions are inhibited in vitro by the *Cochliobolus*-based cyclic tetrapeptide HC-toxin (HCT). When maize was treated with HCT, H3 and H4 were observed to be hyperacetylated, which suggests that the toxin establishes compatibility between the pathogen and maize by modifying the status of reversible acetylation of histones (Kang et al., 2022). NO can inhibit the functioning of HDACs; for example, a study in Arabidopsis found that when treated with a physiological NO donor GSNO, universal levels of various histone acetylation markers being examined increased. A comparison of genome-wide H3K9/14 acetylation profiles showed that histone acetylation and thereby HDAC inhibition regulated by NO is biased towards hyperacetylation and transcription of plant defense-specific genes (Mengel et al., 2017).

RING E3 ligase HISTONE MONOUBIQUITINATION1 (HUB1) and HUB2 in *Arabidopsis* control the levels of histone H2B monoubiquitination (H2Bub1). *Arabidopsis* mutants that are deficient in HUB1 show enhanced sensitivity to infections by *Botrytis cinerea* and *Alternaria brassicicola*, while its overexpression results in resistance to these infections (Dhawan et al., 2009). The H2Bub1 status is directly regulated by HUB1/HUB2 at the *R* gene, *SNC1*, to induce its expression, which is actually associated with autoimmune phenotypes and enhanced disease resistance (Zou et al., 2014).

A more recent discovery of histone modification is the lysine 2-hydroxyisobutyrylation (Khib), which is widely distributed and acts as a marker of transcriptional activation. In plants like rice, Khib carries out the regulation of the *R* genes and their transcription. *Ustilaginoidea virens* is capable of controlling Khib levels, where there is downregulation of a majority of Khib sites in H3 during any infection. In rice, HDAC705 is responsible for removing Khib, further lowering the resistance of plants to a range

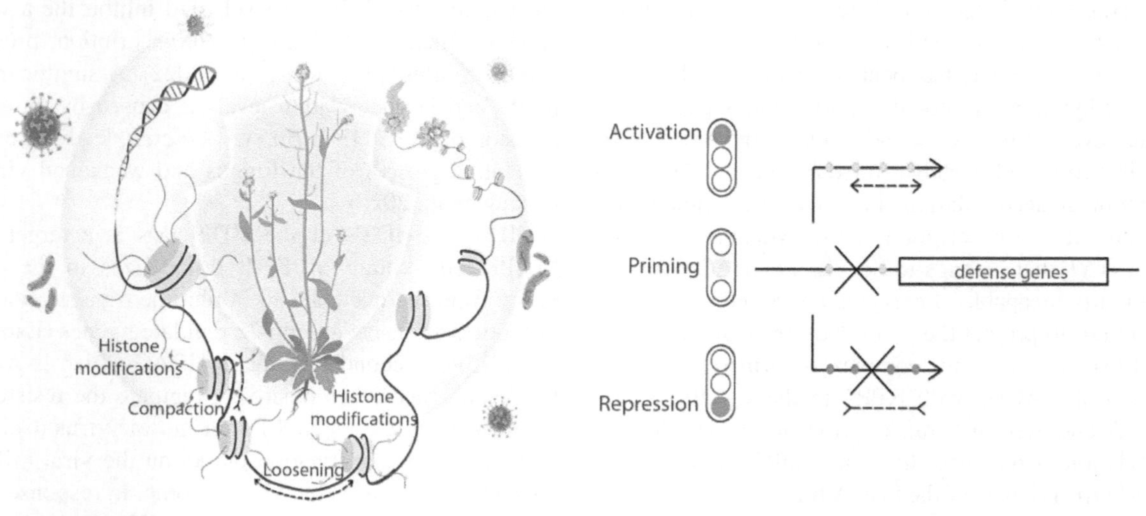

FIGURE 3.3 Plant pathogens and their effect on plant defense genes via histone modifications. This image has been adapted from Kang et al. (2022) under the creative commons attribution license (https://creativecommons.org/licenses/by/4.0/).

TABLE 3.2
Examples of key histone modifications that occur on certain histones and their influence on the plant defense response

Modification on	Effect on transcription regulation	Modified by	Effect	References
1. H3K4	Activation	ATX1, ATXR7	Affects the SA/JA pathway genes and the *NLR* genes	Alvarez-Venegas et al. (2007), Xia et al. (2013)
2. H3K4	Repression	FLD, LDL1/LDL2, AtJMJ14, OsJMJ704	Negatively affects plant immunity response; affects *WRKY* genes as well as SNI1	Hou et al. (2015), Li et al. (2020b), Noh et al. (2021), Singh et al. (2014)
3. H3K9	Activation	AtIBM1, AtJMJ27	Affects PR and PTI marker genes	Chan & Zimmerli (2019), Dutta et al. (2017)
4. H3K9	Repression	AtKYP, AtSUVH5	Affects various *PRR/NLR* genes and is responsible for induction of pathogen induced PCD	Castillo-González et al. (2015), Dvořák Tomaštíková et al. (2021b)
5. H3K27	Activation	AtREF6, OsJMJ705	Affects mainly *PRR/NLR* and other genes in response to biotic stress	Cambiagno et al. (2021) Li et al. (2013)
6. H3K27	Repression	AtCLF, AtSWN	Pathogen-induced programmed cell death (PCD) regulated	Dvořák Tomaštíková et al. (2021a)
7. H3K36	Activation	AtSDG8, SISDG33	Affects SA, defense and MAP-kinase genes along with resistance genes	Bvindi et al. (2022), Dvořák Tomaštíková et al. (2021a), Lee et al. (2016), X. Zhang et al. (2020)

of infections. This HDAC knockout enhances disease resistance of plants (Chen et al., 2021; Dai et al., 2014; Huang et al., 2018) (Figure 3.3 and Table 3.2).

3.9 HISTONE MODIFICATIONS IN ABIOTIC STRESS

3.9.1 Modifications during Heat Stress

Plants that are sessile are essentially sensitive to heat stress, which is one of the most impacting abiotic stresses that significantly restrict growth and provides lower yields (Ahuja et al., 2010). In order to aid plants in the endurance mechanisms in the event of such a stress as well as provide counter measures, there is a specific subset of heat responsive genes whose transcription is regulated by various histone modifications (Von Koskull-Döring et al., 2007). GCN5 is an important modulator that increases the acetylation of H3K14/K9 in the promoters of *HEAT SHOCK TRANSCRIPTION FACTOR A3* and *ULTRAVIOLET HYPERSENSITIVE6* genes, thus triggering thermotolerance. Plants that are *gcn5* mutants have profoundly low thermotolerance levels when exposed to heat stress (Ueda & Seki, 2020). The crosstalk between H3K9 acetylation and H3K4 demethylation causes an upregulation in the levels of Hsf (heat stress transcription factor), allowing for enhanced ROS levels that help in acclimatizing to stress (Hou et al., 2019). *ATX1* and *SDG25* are the two genes that are heat stress responsive, expressed in *Arabidopsis* during heat stress and later during the phase of recovery from the said stress. The RING finger E3 ligases, like those of OsHTAS, OsHIRP1 or OsHCI1, are overexpressed during heat stress and confer heat stress tolerance to rice plants (Han et al., 2021; J. Liu et al., 2015; Wei et al., 2020). Overexpression of SUMO E3 ligase OsSIZ1 of rice also increased the rate of photosynthesis in plants under heat stress. Post-fertilization, when rice plants were subjected to moderate levels of heat stress at 48 hours and 72 hours, there were decreased levels of H3K9 methylation associated with OsFIE1 (fertilization-independent endosperm 1), suggesting that epigenetic regulation is required to influence thermosensitivity during the development of endosperm (Folsom et al., 2014). The histone chaperones ASF1A and ASF1B are responsible for acquired as well as innate thermotolerance that takes place via the eviction of nucleosomes and the acetylation of H3K56 (Weng et al., 2014). Some other studies have noted that the activation of SUMO1/2 also causes the acclimation to rising temperature through HSFA1 family, which is the primary mediator of the response to heat stress. Plant mutants that lack SUMO1/2 exhibit thermosensitivity at a temperature of 28°C (Hammoudi et al., 2021). The presence of an additional E3 ligase, the XB3 ortholog (XBAT31), has been identified in *Arabidopsis* and is involved in the coordination of stabilizing the protein ELF3, influencing subsequently the growth of hypocotyl under high temperature (Zhang et al., 2021).

3.9.2 Some Modifications during Cold Stress

When temperature is extremely low (freezing and chilling), plants are dramatically affected and experience dysregulation in photosynthesis, wilting, chlorosis and uncontrolled apoptosis (Wang et al., 2016). As with heat stress, plants have developed mechanisms that are epigenetically regulated in order to cope with cold stress (Verma et al., 2022). The coping mechanism involves increasing the transcription of active genes and forming bivalent promoters. This is evident from the fact that the bivalent configuration of H3K4me3-H3K27me3 allows for the positive regulation of cold stress-related genes in tubers that are cold stored (Chang et al., 2020). Cold signaling takes place due to the cold-responsive element-binding factors (CBFs), viz., CBF1/CBF2/CBF3. These are crucial for the induction of the expression of certain genes that confer resistance to freezing. Transgenic rice plants showed ectopic overexpression in the case of CaPUB1, which is a U-box E3 ubiquitin ligase (Min et al., 2016). There is also an important function of *HOS15 (HIGH EXPRESSION OF OSMOTICALLY RESPONSIVE GENE15)* in regulating the degradation of HD2C when the plant experiences severe cold stress, which changes the inhibitory regions of the chromatin to an active state (Park et al., 2018).

3.9.3 Some Modifications during Salt Stress

Salt stress also causes a low growth and development rate in plants. An elevation in H3K9ac and H4K5 levels due to the working of two *HAT* genes (*ZmGCN5* and *ZmHATB*) is connected with the increase in genes that are cell-wall related in the roots of maize, required for a proper response to high salinity. A number of these aforementioned genes are observed in the coding and promoter regions of *ZmEXPB2* and *ZmXET1* (Li et al., 2014). GCN5 mRNA expression is triggered in salt stress, and *gcn5* mutants show an increase in the disintegration of the cell wall and sensitivity to salt. The acetylation of H3K9 and H3K14 carried out by GCN5 causes the expression of CTL1, a protein that is chitinase-like and important for the biosynthesis of the cell wall as well as resilience to salt stress (Zheng et al., 2019). There are also reports of histone deacetylases, functioning in a negative manner during exposure to salt stress. *OsHDA1* negatively affects the transcription of *OsSOS1* in rice (Cheng et al., 2018). Various other mechanisms by which histone modifications occur that aid in the adaptation of the plant to salt stress are yet to be worked upon and studied in detail (Singroha et al., 2022).

3.9.4 Some Modifications during Other Stress Conditions

Prevalent in case of UV radiations is methylation and acetylation of different residues on H3. When exposed to UV-B, the levels of H3K9/K14 ac are enhanced at the promoter region of *ELIP1* (early light inducible protein 1) in *Arabidopsis* and wheat plants. During DNA damage repair, H3K4me3 was seen at the loci which was gamma ray inducible, and H3K9 acetylation was observed at the locus of *AtAgo2*. The simultaneous modifications at H3K4 and H3K9 play a more crucial role than the changes in DNA methylation (Sharma et al., 2023). Drought stress triggers the working of various *HAT* genes, particularly 12 *OsHAT* genes in rice and 15 *BraHAT* genes in Chinese cabbage, in order to increase acetylation levels of residues of H3 (Eom & Hyun, 2018; Fang et al., 2014).

3.10 HISTONE MODIFICATIONS DURING NORMAL DEVELOPMENT

3.10.1 Some Modifications during Seed Dormancy and Development

H2B monoubiquitination influences the transcription of various genes that are dormancy related, such as *DELAY OF GERMINATION 1* (*DOG1*), *ATS2*, *NINE-CIS-EPOXYCAROTENOID DIOXYGENASE 9* (*NCED9*) and *ONE-CYSTEINE PEROXIREDOXIN 1* (*PER1*) to promote dormancy. Thus, this modification, in coordination with the transcription elongation factor TRANSCRIPT ELONGATION FACTOR IIS (TFIIS), carries out and establishes common targets for dormancy. Reduced dormancy is seen in plants with loss of *ARABIDOPSIS TRITHORAXRELATED 7* (*ATXR7*), which is a gene that encodes for an H3K4 methyltransferase, suggesting that this modification positively regulates seed development (Zhao et al., 2019). Contrary to this, KYP is an H3K9 methyltransferase that inhibits the expression of ABA INSENSITIVE 3 (ABI3) and DOG1, thus repressing dormancy (Zheng et al., 2012). The occurrence of this modification hence requires the demethylation of H3K4 by the LSD LIKE 1 (LDL1) and LDL2. The *ldl1 ldl2* double mutants show increased ABI3 and DOG1 expressions, which cause seed dormancy. This is also evident from the fact that these mutants show enhanced defense responses as well, as discussed earlier in this chapter. The histone deacetylase *HISTONE DEACETYLASE 9 (HDA9)* suppresses the expression of regulatory genes of photosynthesis such as *RUBISCO SMALL SUBUNIT 2B (RBCS2B)* and *RUBISCO ACTIVASE (RCA)* in dry seeds to prevent seedling traits. Moreover, the association formed by SWI-INDEPENDENT 3 (SIN3)-LIKE 1 (SNL1) with the histone acetylase HDA19 also represses histone acetylation and seed germination in tandem with its close homolog, SNL2. When stratification occurs and a temperature treatment (4°C for *Arabidopsis*) is performed, expression of SNL1 and SNL2 is seen to decline and acetylation of histones in the genes of the auxin pathway is restored, causing seeds to germinate and develop normally (Wang et al., 2013, 2016).

3.10.2 Some Modifications during Flowering

The length of the day is critical in determining flower development and providing flowering cues. Long days induce flowering in *Arabidopsis* by expressing *FLOWERING LOCUS T* (*FT*), which encodes a mobile protein translocated to the shoot apical meristem (SAM) from the vasculature of leaves. This expression of *FT* is repressed by the trimethylation of H3K27, carried out by PRC2, which comprises CLF (CURLY LEAF), SWN (SWINGER), EMF2 (EMBRYONIC FLOWER 2) and FIE (FERTILISATION-INDEPENDENT ENDOSPERM) (Zhao et al., 2019). The Nuclear Factor-Y subunit C (NF-YC) proteins work in contradiction to the deposition of H3K27me3 on the FT chromatin by interacting with CLF for a temporary period of time, thus activating the transcription of FT (Liu et al., 2018). As the trimethylation levels of H3K27 decrease, there is instead an increase seen in the trimethylation of H3K4. The balance in *FT* expression is thus regulated by the antagonizing functions of H3K27me3 and H3K4me3 (Zhao et al., 2019).

During a long day photoperiod, MUT9P-LIKE-KINASE (MLK4) is a protein that supports proper flowering time. It causes phosphorylation of H2A at serine residue 95 which only exists in plants and has not been found in yeast, *Drosophila* or humans. The interaction of *CIRCADIAN CLOCK ASSOCIATED 1* (*CCA1*) with MLK4 allows its binding to the promoter of *GIGANTEA* (*GI*), which in turn promotes flowering. Plants that are *mlk4* mutants show delayed flowering. It is interesting to note that MLK4 has no contribution to the regulation of flowering time under a short-day photoperiod. Su et al. (2017) experimentally analyzed these functions of MLK4 and hypothesized that the protein in question might have varied levels of mRNA or its activity is different under the influence of long-day and short-day photoperiods.

3.11 INTERPLAY OF HISTONE MODIFICATIONS AND DNA METHYLATION

Epigenetic mechanisms are shaped by chromatin remodeling, histone modifications and DNA methylation – all of which are crucial for regulating gene expression and maintaining genomic integrity. Tissue-dependent differentially methylated regions (T-DMRs) define the unique epigenetic identity of each cell type, enabling precise control of the function of genes without changing the DNA sequence. Dynamism between DNA methylation and histone modifications drives chromatin organization and gene regulation. Non-coding RNAs and chromatin remodeling factors also have an impact on this coordination, which guarantees epigenetic stability. Genome-wide profiling has revealed the interdependence of these systems, highlighting their role in cellular differentiation, development and disease pathogenesis. This understanding offers a foundation for therapeutic strategies targeting the epigenome in regenerative medicine and disease treatment (Ikegami et al., 2009).

3.12 CONCLUSION

DNA methylation and histone editing represent the foundational pillars of plant epigenetics, being essential in controlling development, gene expression and environmental adaptability. These intricate mechanisms weave a delicate balance between stability and flexibility, enabling plants to preserve heritable epigenetic marks while dynamically adjusting to ever-changing environmental conditions. By modulating genetic activity without altering the DNA sequence, these processes unlock transformative potential for agricultural innovation. These modifications not only influence the growth of plants and stress responses but also additionally function as invaluable tools for enhancing environmental resilience. Their ability to govern genomic expression equips plants to withstand stresses, both biotic and abiotic, positioning them as key drivers of sustainable farming practices. As our understanding of these mechanisms deepens, the potential to manipulate them offers unprecedented opportunities for advancing agriculture. From improving crop yields to developing climate-resilient plant varieties, these epigenetic tools address pressing global challenges such as food security and environmental degradation. Harnessing these capabilities could redefine modern agriculture, paving the way for innovative strategies that reduce ecological footprints while optimizing productivity. In this era of climate change and resource scarcity, the integration of epigenetic research into agricultural biotechnology is not merely a scientific pursuit, but a necessity for attaining sustainability and global food safety.

REFERENCES

Agius, F., Kapoor, A., & Zhu, J. (2006). Role of the *Arabidopsis* DNA glycosylase/lyase ROS1 in active DNA demethylation. *Proceedings of the National Academy of Sciences*, 103(31), 11796–11801. https://doi.org/10.1073/pnas.0603563103.

Ahuja, I., De Vos, R. C. H., Bones, A. M., & Hall, R. D. (2010). Plant molecular stress responses face climate change. *Trends in Plant Science*, 15(12), 664–674. https://doi.org/10.1016/j.tplants.2010.08.002.

Al Aboud, N. M., Tupper, C., & Jialal, I. (2024). Genetics, epigenetic mechanism. In StatPearls. StatPearls Publishing. http://www.ncbi.nlm.nih.gov/books/NBK532999/.

Alvarez-Venegas, R., Abdallat, A. A., Guo, M., Alfano, J. R., & Avramova, Z. (2007). Epigenetic control of a transcription factor at the cross section of two antagonistic pathways. *Epigenetics*, 2(2), 106–113. https://doi.org/10.4161/epi.2.2.4404.

An, Y. C., Goettel, W., Han, Q., Bartels, A., Liu, Z., & Xiao, W. (2017). Dynamic changes of genome-wide DNA methylation during soybean seed development. *Scientific Reports*, 7(1). https://doi.org/10.1038/s41598-017-12510-4.

Aufsatz, W., Mette, M., Matzke, A., & Matzke, M. (2004). The role of MET1 in RNA-directed de novo and maintenance methylation of CG dinucleotides. *Plant Molecular Biology*, 54(6), 793–804. https://doi.org/10.1007/s11103-004-0179-1.

Banerjee, A., & Roychoudhury, A. (2017a). Epigenetic regulation during salinity and drought stress in plants: Histone modifications and DNA methylation. *Plant Gene*, 11, 199–204.

Banerjee, A., & Roychoudhury, A. (2018). The gymnastics of epigenomics in rice. *Plant Cell Reports*, 37, 25–49.

Banerjee, A., Wani, S. H., & Roychoudhury, A. (2017b). Epigenetic control of plant cold responses. *Frontiers in Plant Science*, 8, 1643.

Bannister, A. J., & Kouzarides, T. (2011). Regulation of chromatin by histone modifications. *Cell Research*, 21(3), 381–395. https://doi.org/10.1038/cr.2011.22.

Bartels, A., Han, Q., Nair, P., Stacey, L., Gaynier, H., Mosley, M., Huang, Q. Q., Pearson, J. K., Hsieh, T., An, Y. C., & Xiao, W. (2018). Dynamic DNA methylation in plant growth and development. *International Journal of Molecular Sciences*, 19(7), 2144. https://doi.org/10.3390/ijms19072144.

Berr, A., McCallum, E. J., Alioua, A., Heintz, D., Heitz, T., & Shen, W.-H. (2010). *Arabidopsis* histone methyltransferase SET DOMAIN GROUP8 mediates induction of the jasmonate/ethylene pathway genes in plant defense response to necrotrophic fungi. *Plant Physiology*, 154(3), 1403–1414. https://doi.org/10.1104/pp.110.161497.

Bird, A. (2002). DNA methylation patterns and epigenetic memory. *Genes & Development*, 16(1), 6–21. https://doi.org/10.1101/gad.947102.

Birkenbihl, R. P., Liu, S., & Somssich, I. E. (2017). Transcriptional events defining plant immune responses. *Current Opinion in Plant Biology*, 38, 1–9. https://doi.org/10.1016/j.pbi.2017.04.004.

Boycheva, I., Bonchev, G., Manova, V., Stoilov, L., & Vassileva, V. (2024). How histone acetyltransferases shape plant photomorphogenesis and UV response. *International Journal of Molecular Sciences*, 25(14), 7851. https://doi.org/10.3390/ijms25147851.

Bvindi, C., Lee, S., Tang, L., Mickelbart, M. V., Li, Y., & Mengiste, T. (2022). Improved pathogen and stress tolerance in tomato mutants of SET domain histone 3 lysine methyltransferases. *New Phytologist*, 235(5), 1957–1976. https://doi.org/10.1111/nph.18277.

Cambiagno, D. A., Torres, J. R., & Alvarez, M. E. (2021). Convergent epigenetic mechanisms avoid constitutive expression of immune receptor gene subsets. *Frontiers in Plant Science*, 12, 703667. https://doi.org/10.3389/fpls.2021.703667.

Castillo-González, C., Liu, X., Huang, C., Zhao, C., Ma, Z., Hu, T., Sun, F., Zhou, Y., Zhou, X., Wang, X.-J., & Zhang, X. (2015). Geminivirus-encoded TrAP suppressor inhibits the histone methyltransferase SUVH4/KYP to counter host defense. *eLife*, 4, e06671. https://doi.org/10.7554/eLife.06671.

Chan, C., & Zimmerli, L. (2019). The histone demethylase IBM1 positively regulates *Arabidopsis* immunity by control of defense gene expression. *Frontiers in Plant Science*, 10, 1587. https://doi.org/10.3389/fpls.2019.01587.

Chang, Y., Zhu, C., Jiang, J., Zhang, H., Zhu, J., & Duan, C. (2020). Epigenetic regulation in plant abiotic stress responses. *Journal of Integrative Plant Biology*, 62(5), 563–580. https://doi.org/10.1111/jipb.12901.

Chen, X., Duan, Y., Qiao, F., Liu, H., Huang, J., Luo, C., Chen, X., Li, G., Xie, K., Hsiang, T., & Zheng, L. (2022). A secreted fungal effector suppresses rice immunity through host histone hypoacetylation. *New Phytologist*, 235(5), 1977–1994. https://doi.org/10.1111/nph.18265.

Chen, X., Xu, Q., Duan, Y., Liu, H., Chen, X., Huang, J., Luo, C., Zhou, D., & Zheng, L. (2021). *Ustilaginoidea virens* modulates lysine 2-hydroxyisobutyrylation in rice flowers during infection. *Journal of Integrative Plant Biology*, 63(10), 1801–1814. https://doi.org/10.1111/jipb.13149.

Cheng, C., Daigen, M., & Hirochika, H. (2006). Epigenetic regulation of the rice retrotransposon Tos17. *Molecular Genetics and Genomics*, 276(4), 378–390. https://doi.org/10.1007/s00438-006-0141-9.

Cheng, X., Zhang, S., Tao, W., Zhang, X., Liu, J., Sun, J., Zhang, H., Pu, L., Huang, R., & Chen, T. (2018). INDETERMINATE SPIKELET1 recruits histone deacetylase and a transcriptional repression complex to regulate rice salt tolerance. *Plant Physiology*, 178(2), 824–837. https://doi.org/10.1104/pp.18.00324.

Choi, S., Song, H., Han, S., Han, M., Kim, C., Park, J., Lee, Y., Jeon, J., Noh, Y., & Noh, B. (2012). HDA19 is required for the repression of salicylic acid biosynthesis and salicylic acid-mediated defense responses in *Arabidopsis*. *The Plant Journal*, 71(1), 135–146. https://doi.org/10.1111/j.1365-313X.2012.04977.x.

Conrath, U., Beckers, G. J. M., Langenbach, C. J. G., & Jaskiewicz, M. R. (2015). Priming for enhanced defense. *Annual Review of Phytopathology*, 53(1), 97–119. https://doi.org/10.1146/annurev-phyto-080614-120132.

Dai, L., Peng, C., Montellier, E., Lu, Z., Chen, Y., Ishii, H., Debernardi, A., Buchou, T., Rousseaux, S., Jin, F., Sabari, B. R., Deng, Z., Allis, C. D., Ren, B., Khochbin, S., & Zhao, Y. (2014). Lysine 2-hydroxyisobutyrylation is a widely distributed active histone mark. *Nature Chemical Biology*, 10(5), 365–370. https://doi.org/10.1038/nchembio.1497.

Dhawan, R., Luo, H., Foerster, A. M., AbuQamar, S., Du, H.-N., Briggs, S. D., Scheid, O. M., & Mengiste, T. (2009). HISTONE MONOUBIQUITINATION1 interacts with a subunit of the mediator complex and regulates defense against necrotrophic fungal pathogens in *Arabidopsis*. *The Plant Cell*, 21(3), 1000–1019. https://doi.org/10.1105/tpc.108.062364.

Ding, Y., Wang, X., Su, L., Zhai, J., Cao, S., Zhang, D., Liu, C., Bi, Y., Qian, Q., Cheng, Z., Chu, C., & Cao, X. (2007). SDG714, a histone H3K9 methyltransferase, is involved in Tos17 DNA methylation and transposition in rice. *The Plant Cell*, 19(1), 9–22. https://doi.org/10.1105/tpc.106.048124.

Dutta, A., Choudhary, P., Caruana, J., & Raina, R. (2017). JMJ 27, an *Arabidopsis* H3K9 histone demethylase, modulates defense against *Pseudomonas syringae* and flowering time. *The Plant Journal*, 91(6), 1015–1028. https://doi.org/10.1111/tpj.13623.

Dvořák Tomaštíková, E., Hafrén, A., Trejo-Arellano, M. S., Rasmussen, S. R., Sato, H., Santos-González, J., Köhler, C., Hennig, L., & Hofius, D. (2021a). Polycomb Repressive Complex 2 and KRYPTONITE regulate pathogen-induced programmed cell death in *Arabidopsis*. *Plant Physiology*, 185(4), 2003–2021. https://doi.org/10.1093/plphys/kiab035.

Eom, S. H., & Hyun, T. K. (2018). Histone Acetyltransferases (HATs) in Chinese cabbage: Insights from histone H3 acetylation and expression profiling of HATs in response to abiotic stresses. *Journal of the American Society for Horticultural Science*, 143(4), 296–303. https://doi.org/10.21273/JASHS04436-18.

Erdmann, R. M., & Picard, C. L. (2020). RNA-directed DNA methylation. *PLoS Genetics*, 16(10), e1009034. https://doi.org/10.1371/journal.pgen.1009034.

Fang, H., Liu, X., Thorn, G., Duan, J., & Tian, L. (2014). Expression analysis of histone acetyltransferases in rice under drought stress. *Biochemical and Biophysical Research Communications*, 443(2), 400–405. https://doi.org/10.1016/j.bbrc.2013.11.102.

Feng, J., & Shen, W.-H. (2014). Dynamic regulation and function of histone monoubiquitination in plants. *Frontiers in Plant Science*, 5. https://doi.org/10.3389/fpls.2014.00083.

Feng, S., Cokus, S. J., Zhang, X., Chen, P., Bostick, M., Goll, M. G., Hetzel, J., Jain, J., Strauss, S. H., Halpern, M. E., Ukomadu, C., Sadler, K. C., Pradhan, S., Pellegrini, M., & Jacobsen, S. E. (2010). Conservation and divergence of methylation patterning in plants and animals. *Proceedings of the National Academy of Sciences*, 107(19), 8689–8694. https://doi.org/10.1073/pnas.1002720107.

Folsom, J. J., Begcy, K., Hao, X., Wang, D., & Walia, H. (2014). Rice fertilization-independent endosperm1 regulates seed size under heat stress by controlling early endosperm development. *Plant Physiology*, 165(1), 238–248. https://doi.org/10.1104/pp.113.232413.

Gallusci, P., Agius, D. R., Moschou, P. N., Dobránszki, J., Kaiserli, E., & Martinelli, F. (2022). Deep inside the epigenetic memories of stressed plants. *Trends in Plant Science*, 28(2), 142–153. https://doi.org/10.1016/j.tplants.2022.09.004.

Greenberg, M. V., Ausin, I., Chan, S. W., Cokus, S. J., Cuperus, J. T., Feng, S., Law, J. A., Chu, C., Pellegrini, M., Carrington, J. C., & Jacobsen, S. E. (2011). Identification of genes required for de novo DNA methylation in *Arabidopsis*. *Epigenetics*, 6(3), 344–354. https://doi.org/10.4161/epi.6.3.14242.

Grover, J. W., Kendall, T., Baten, A., Burgess, D., Freeling, M., King, G. J., & Mosher, R. A. (2018). Maternal components of RNA-directed DNA methylation are required for seed development in *Brassica rapa*. *The Plant Journal*, 94(4), 575–582. https://doi.org/10.1111/tpj.13910.

Hammoudi, V., Beerens, B., Jonker, M. J., Helderman, T. A., Vlachakis, G., Giesbers, M., Kwaaitaal, M., & Van Den Burg, H. A. (2021). The protein modifier SUMO is critical for integrity of the *Arabidopsis* shoot apex at warm ambient temperatures. *Journal of Experimental Botany*, erab262. https://doi.org/10.1093/jxb/erab262.

Han, D., Chen, C., Xia, S., Liu, J., Shu, J., Nguyen, V., Lai, J., Cui, Y., & Yang, C. (2021). Chromatin-associated SUMOylation controls the transcriptional switch between plant development and heat stress responses. *Plant Communications*, 2(1), 100091. https://doi.org/10.1016/j.xplc.2020.100091.

He, X., Chen, T., & Zhu, J. (2011). Regulation and function of DNA methylation in plants and animals. *Cell Research*, 21(3), 442–465. https://doi.org/10.1038/cr.2011.23.

Hotchkiss, R. D. (1948). The quantitative separation of purines, pyrimidines, and nucleosides by paper chromatography. *Journal of Biological Chemistry*, 175(1), 315–332. https://doi.org/10.1016/s0021-9258(18)57261-6.

Hou, H., Zhao, L., Zheng, X., Gautam, M., Yue, M., Hou, J., Chen, Z., Wang, P., & Li, L. (2019). Dynamic changes in histone modification are associated with upregulation of Hsf and rRNA genes during heat stress in maize seedlings. *Protoplasma*, 256(5), 1245–1256. https://doi.org/10.1007/s00709-019-01364-4.

Hou, Y., Wang, L., Wang, L., Liu, L., Li, L., Sun, L., Rao, Q., Zhang, J., & Huang, S. (2015). JMJ704 positively regulates rice defense response against *Xanthomonas oryzae* pv. oryzae infection via reducing H3K4me2/3 associated with negative disease resistance regulators. *BMC Plant Biology*, 15(1), 286. https://doi.org/10.1186/s12870-015-0674-3.

Hsieh, T., Ibarra, C. A., Silva, P., Zemach, A., Eshed-Williams, L., Fischer, R. L., & Zilberman, D. (2009). Genome-wide demethylation of *Arabidopsis* endosperm. *Science*, 324(5933), 1451–1454. https://doi.org/10.1126/science.1172417.

Hu, H., & Du, J. (2022). Structure and mechanism of histone methylation dynamics in *Arabidopsis*. *Current Opinion in Plant Biology*, 67, 102211. https://doi.org/10.1016/j.pbi.2022.102211.

Huang, H., Luo, Z., Qi, S., Huang, J., Xu, P., Wang, X., Gao, L., Li, F., Wang, J., Zhao, W., Gu, W., Chen, Z., Dai, L., Dai, J., & Zhao, Y. (2018). Landscape of the regulatory elements for lysine 2-hydroxyisobutyrylation pathway. *Cell Research*, 28(1), 111–125. https://doi.org/10.1038/cr.2017.149.

Ikegami, K., Ohgane, J., Tanaka, S., Yagi, S., & Shiota, K. (2009). Interplay between DNA methylation, histone modification and chromatin remodeling in stem cells and during development. *The International Journal of Developmental Biology*, 53(2–3), 203–214. https://doi.org/10.1387/ijdb.082741ki.

Jang, H., Shin, W., Lee, J., & Do, J. (2017). CpG and Non-CpG methylation in epigenetic gene regulation and brain function. *Genes*, 8(6), 148. https://doi.org/10.3390/genes8060148.

Jaskiewicz, M., Conrath, U., & Peterhänsel, C. (2011). Chromatin modification acts as a memory for systemic acquired resistance in the plant stress response. *EMBO Reports*, 12(1), 50–55. https://doi.org/10.1038/embor.2010.186.

Jeltsch, A., & Jurkowska, R. Z. (2014). New concepts in DNA methylation. *Trends in Biochemical Sciences*, 39(7), 310–318. https://doi.org/10.1016/j.tibs.2014.05.002.

Jin, H., Choi, S.-M., Kang, M.-J., Yun, S.-H., Kwon, D.-J., Noh, Y.-S., & Noh, B. (2018). Salicylic acid-induced transcriptional reprogramming by the HAC–NPR1–TGA histone acetyltransferase complex in *Arabidopsis*. *Nucleic Acids Research*. https://doi.org/10.1093/nar/gky847.

Jones, P. A., & Takai, D. (2001). The role of DNA methylation in mammalian epigenetics. *Science*, 293(5532), 1068–1070. https://doi.org/10.1126/science.1063852.

Kadayifci, F. Z., Zheng, S., & Pan, Y. (2018). Molecular mechanisms underlying the link between diet and DNA methylation. *International Journal of Molecular Sciences*, 19(12), 4055. https://doi.org/10.3390/ijms19124055.

Kang, H., Fan, T., Wu, J., Zhu, Y., & Shen, W.-H. (2022). Histone modification and chromatin remodeling in plant response to pathogens. *Frontiers in Plant Science*, 13, 986940. https://doi.org/10.3389/fpls.2022.986940.

Kong, L., Qiu, X., Kang, J., Wang, Y., Chen, H., Huang, J., Qiu, M., Zhao, Y., Kong, G., Ma, Z., Wang, Y., Ye, W., Dong, S., Ma, W., & Wang, Y. (2017). A *Phytophthora* effector manipulates host histone acetylation and reprograms defense gene expression to promote infection. *Current Biology*, 27(7), 981–991. https://doi.org/10.1016/j.cub.2017.02.044.

Kumar, S., & Mohapatra, T. (2021). Dynamics of DNA methylation and its functions in plant growth and development. *Frontiers in Plant Science*, 12. https://doi.org/10.3389/fpls.2021.596236.

La, H., Ding, B., Mishra, G. P., Zhou, B., Yang, H., Del Rosario Bellizzi, M., Chen, S., Meyers, B. C., Peng, Z., Zhu, J., & Wang, G. (2011). A 5-methylcytosine DNA glycosylase/lyase demethylates the retrotransposon Tos17 and promotes its transposition in rice. *Proceedings of the National Academy of Sciences*, 108(37), 15498–15503. https://doi.org/10.1073/pnas.1112704108.

Laird, P. W., & Jaenisch, R. (1996). The role of DNA methylation in cancer genetics and epigenetics. *Annual Review of Genetics*, 30(1), 441–464. https://doi.org/10.1146/annurev.genet.30.1.441.

Lang, Z., Wang, Y., Tang, K., Tang, D., Datsenka, T., Cheng, J., Zhang, Y., Handa, A. K., & Zhu, J. (2017). Critical roles of DNA demethylation in the activation of ripening-induced genes and inhibition of ripening-repressed genes in tomato fruit. *Proceedings of the National Academy of Sciences*, 114(22). https://doi.org/10.1073/pnas.1705233114.

Latrasse, D., Jégu, T., Li, H., De Zelicourt, A., Raynaud, C., Legras, S., Gust, A., Samajova, O., Veluchamy, A., Rayapuram, N., Ramirez-Prado, J. S., Kulikova, O., Colcombet, J., Bigeard, J., Genot, B., Bisseling, T., Benhamed, M., & Hirt, H. (2017). MAPK-triggered chromatin reprogramming by histone deacetylase in plant innate immunity. *Genome Biology*, 18(1), 131. https://doi.org/10.1186/s13059-017-1261-8.

Le, T., Schumann, U., Smith, N. A., Tiwari, S., Au, P. C. K., Zhu, Q., Taylor, J. M., Kazan, K., Llewellyn, D. J., Zhang, R., Dennis, E. S., & Wang, M. (2014). DNA demethylases target promoter transposable elements to positively regulate stress responsive genes in *Arabidopsis*. *Genome Biology*, 15(9). https://doi.org/10.1186/s13059-014-0458-3.

Lee, S., Fu, F., Xu, S. Lee, S. Y., Yun, D.-J., & Mengiste, T. (2016). Global regulation of plant immunity by histone lysine methyltransferases. *The Plant Cell*, tpc.00012.2016. https://doi.org/10.1105/tpc.16.00012.

Li, D., Liu, R., Singh, D., Yuan, X., Kachroo, P., & Raina, R. (2020a). JMJ14 encoded H3K4 demethylase modulates immune responses by regulating defence gene expression and pipecolic acid levels. *New Phytologist*, 225(5), 2108–2121. https://doi.org/10.1111/nph.16270.

Li, H., Yan, S., Zhao, L., Tan, J., Zhang, Q., Gao, F., Wang, P., Hou, H., & Li, L. (2014). Histone acetylation associated up-regulation of the cell wall related genes is involved in salt stress induced maize root swelling. *BMC Plant Biology*, 14(1), 105. https://doi.org/10.1186/1471-2229-14-105.

Li, J., Li, J., Yang, D., Huang, H., Zhang, G., He, L., Pang, J., Lozano-Durán, R., Lang, Z., & Zhu, J. (2020b). Epigenetic memory marks determine epiallele stability at loci targeted by de novo DNA methylation. *Nature Plants*, 6(6), 661–674. https://doi.org/10.1038/s41477-020-0671-x.

Li, S., Lyu, S., Liu, Y., Luo, M., Shi, S., & Deng, S. (2021). Cauliflower mosaic virus P6 dysfunctions histone deacetylase HD2C to promote virus infection. *Cells*, 10(9), 2278. https://doi.org/10.3390/cells10092278.

Li, T., Chen, X., Zhong, X., Zhao, Y., Liu, X., Zhou, S., Cheng, S., & Zhou, D.-X. (2013). Jumonji C domain protein JMJ705-mediated removal of histone H3 lysine 27 trimethylation is involved in defense-related gene activation in rice. *The Plant Cell*, 25(11), 4725–4736. https://doi.org/10.1105/tpc.113.118802.

Lin, J., Le, B. H., Chen, M., Henry, K. F., Hur, J., Hsieh, T., Chen, P., Pelletier, J. M., Pellegrini, M., Fischer, R. L., Harada, J. J., & Goldberg, R. B. (2017). Similarity between soybean and *Arabidopsis* seed methylomes and loss of non-CG methylation does not affect seed development. *Proceedings of the National Academy of Sciences*, 114(45). https://doi.org/10.1073/pnas.1716758114.

Liu, J., Zhang, C., Wei, C., Liu, X., Wang, M., Yu, F., Xie, Q., & Tu, J. (2015b). The RING finger ubiquitin E3 ligase OsHTAS enhances heat tolerance by promoting H2O2-induced stomatal closure in rice. *Plant Physiology*, 170(1), 429–443. https://doi.org/10.1104/pp.15.00879.

Liu, R., & Lang, Z. (2019). The mechanism and function of active DNA demethylation in plants. *Journal of Integrative Plant Biology*, 62(1), 148–159. https://doi.org/10.1111/jipb.12879.

Liu, R., How-Kit, A., Stammitti, L., Teyssier, E., Rolin, D., Mortain-Bertrand, A., Halle, S., Liu, M., Kong, J., Wu, C., Degraeve-Guibault, C., Chapman, N. H., Maucourt, M., Hodgman, T. C., Tost, J., Bouzayen, M., Hong, Y., Seymour, G. B., Giovannoni, J. J., & Gallusci, P. (2015a). A DEMETER-like DNA demethylase governs tomato fruit ripening. *Proceedings of the National Academy of Sciences*, 112(34), 10804–10809. https://doi.org/10.1073/pnas.1503362112.

Liu, X., Yang, Y., Hu, Y., Zhou, L., Li, Y., & Hou, X. (2018). Temporal-specific interaction of NF-YC and CURLY LEAF during the floral transition regulates flowering. *Plant Physiology*, 177(1), 105–114. https://doi.org/10.1104/pp.18.00296.

Liu, Z. L., Han, F. P., Tan, M., Shan, X. H., Dong, Y. Z., Wang, X. Z., Fedak, G., Hao, S., & Liu, B. (2004). Activation of a rice endogenous retrotransposon Tos17 in tissue culture is accompanied by cytosine demethylation and causes heritable alteration in methylation pattern of flanking genomic regions. *Theoretical and Applied Genetics*, 109(1), 200–209. https://doi.org/10.1007/s00122-004-1618-8.

Lucibelli, F., Valoroso, M. C., & Aceto, S. (2022). Plant DNA methylation: An epigenetic mark in development, environmental interactions, and evolution. *International Journal of Molecular Sciences*, 23(15), 8299. https://doi.org/10.3390/ijms23158299.

Luo, M., Taylor, J. M., Spriggs, A., Zhang, H., Wu, X., Russell, S., Singh, M., & Koltunow, A. (2011). A genome-wide survey of imprinted genes in rice seeds reveals imprinting primarily occurs in the endosperm. *PLoS Genetics*, 7(6), e1002125. https://doi.org/10.1371/journal.pgen.1002125.

Manning, K., Tör, M., Poole, M., Hong, Y., Thompson, A. J., King, G. J., Giovannoni, J. J., & Seymour, G. B. (2006). A naturally occurring epigenetic mutation in a gene encoding an SBP-box transcription factor inhibits tomato fruit ripening. *Nature Genetics*, 38(8), 948–952. https://doi.org/10.1038/ng1841.

Martínez-Aguilar, K., Ramírez-Carrasco, G., Hernández-Chávez, J. L., Barraza, A., & Alvarez-Venegas, R. (2016). Use of BABA and INA as activators of a primed state in the common bean (*Phaseolus vulgaris L.*). *Frontiers in Plant Science*, 7. https://doi.org/10.3389/fpls.2016.00653.

Mengel, A., Ageeva, A., Georgii, E., Bernhardt, J., Wu, K., Durner, J., & Lindermayr, C. (2017). Nitric oxide modulates histone acetylation at stress genes by inhibition of histone deacetylases. *Plant Physiology*, 173(2), 1434–1452. https://doi.org/10.1104/pp.16.01734.

Min, H. J., Jung, Y. J., Kang, B. G., & Kim, W. T. (2016). CaPUB1, a hot pepper U-box E3 ubiquitin ligase, confers enhanced cold stress tolerance and decreased drought stress tolerance in transgenic rice (*Oryza sativa L.*). *Molecules and Cells*, 39(3), 250–257. https://doi.org/10.14348/molcells.2016.2290.

Moore, L. D., Le, T., & Fan, G. (2012). DNA methylation and its basic function. *Neuropsychopharmacology*, 38(1), 23–38. https://doi.org/10.1038/npp.2012.112.

Noh, S. W., Seo, R.-R., Park, H. J., & Jung, H. W. (2021). Two *Arabidopsis* homologs of human lysine-specific demethylase function in epigenetic regulation of plant defense responses. *Frontiers in Plant Science*, 12, 688003. https://doi.org/10.3389/fpls.2021.688003.

Park, J., Lim, C. J., Shen, M., Park, H. J., Cha, J.-Y., Iniesto, E., Rubio, V., Mengiste, T., Zhu, J.-K., Bressan, R. A., Lee, S. Y., Lee, B., Jin, J. B., Pardo, J. M., Kim, W.-Y., & Yun, D.-J. (2018). Epigenetic switch from repressive to permissive chromatin in response to cold stress. *Proceedings of the National Academy of Sciences*, 115(23). https://doi.org/10.1073/pnas.1721241115.

Po-Wen, C., Singh, P., & Zimmerli, L. (2013). Priming of the *Arabidopsis* pattern-triggered immunity response upon infection by necrotrophic *Pectobacterium carotovorum* bacteria. *Molecular Plant Pathology*, 14(1), 58–70. https://doi.org/10.1111/j.1364-3703.2012.00827.x.

Qi, S., He, L., Zhang, Q., Dong, Q., Wang, Y., Yang, Q., Tian, C., He, Q., & Wang, Y. (2018). Cross-pathway control gene CPC1/GCN4 coordinates with histone acetyltransferase GCN5 to regulate catalase-3 expression under oxidative stress in *Neurospora crassa. Free Radical Biology and Medicine*, 117, 218–227. https://doi.org/10.1016/j.freeradbiomed.2018.02.003.

Raghubeer, S. (2024). The influence of epigenetics and inflammation on cardiometabolic risks. *Seminars in Cell & Developmental Biology*, 154, 175–184. https://doi.org/10.1016/j.semcdb.2023.02.006.

Ramirez-Prado, J. S., Latrasse, D., Rodriguez-Granados, N. Y., Huang, Y., Manza-Mianza, D., Brik-Chaouche, R., Jaouannet, M., Citerne, S., Bendahmane, A., Hirt, H., Raynaud, C., & Benhamed, M. (2019). The polycomb protein LHP 1 regulates *Arabidopsis thaliana* stress responses through the repression of the MYC 2-dependent branch of immunity. *The Plant Journal*, 100(6), 1118–1131. https://doi.org/10.1111/tpj.14502.

Razin, A., & Szyf, M. (1984). DNA methylation patterns formation and function. *Biochimica Et Biophysica Acta (BBA) – Gene Structure and Expression*, 782(4), 331–342. https://doi.org/10.1016/0167-4781(84)90043-5.

Reik, W., & Dean, W. (2001). DNA methylation and mammalian epigenetics. *Electrophoresis*, 22(14), 2838–2843.

Rodrigues, J. A., Ruan, R., Nishimura, T., Sharma, M. K., Sharma, R., Ronald, P. C., Fischer, R. L., & Zilberman, D. (2013). Imprinted expression of genes and small RNA is associated with localized hypomethylation of the maternal genome in rice endosperm. *Proceedings of the National Academy of Sciences*, 110(19), 7934–7939. https://doi.org/10.1073/pnas.1306164110.

Sharma, M., Sidhu, A. K., Samota, M. K., Gupta, M., Koli, P., & Choudhary, M. (2023). Post-translational modifications in histones and their role in abiotic stress tolerance in plants. *Proteomes*, 11(4), 38. https://doi.org/10.3390/proteomes11040038.

Singal, R., & Ginder, G. D. (1999). DNA methylation. *Blood*, 93(12), 4059–4070. https://doi.org/10.1182/blood.v93.12.4059.

Singh, P., Yekondi, S., Chen, P.-W., Tsai, C.-H., Yu, C.-W., Wu, K., & Zimmerli, L. (2014a). Environmental history modulates *Arabidopsis* pattern-triggered immunity in a HISTONE ACETYLTRANSFERASE1–dependent manner. *The Plant Cell*, 26(6), 2676–2688. https://doi.org/10.1105/tpc.114.123356.

Singroha, G., Kumar, S., Gupta, O. P., Singh, G. P., & Sharma, P. (2022). Uncovering the epigenetic marks involved in mediating salt stress tolerance in plants. *Frontiers in Genetics*, 13, 811732. https://doi.org/10.3389/fgene.2022.811732.

Slotkin, R. K., & Martienssen, R. (2007). Transposable elements and the epigenetic regulation of the genome. *Nature Reviews Genetics*, 8(4), 272–285.

Su, Y., Wang, S., Zhang, F., Zheng, H., Liu, Y., Huang, T., & Ding, Y. (2017). Phosphorylation of histone H2A at serine 95: A plant-specific mark involved in flowering time regulation and H2A.Z deposition. *The Plant Cell*, 29(9), 2197–2213. https://doi.org/10.1105/tpc.17.00266.

Sweatt, J. D., Nestler, E. J., Meaney, M. J., & Akbarian, S. (2013). An overview of the molecular basis of epigenetics. In *Epigenetic Regulation in the Nervous System* (pp. 3–33). Elsevier. https://doi.org/10.1016/B978-0-12-391494-1.00001-X.

Tan, F., Zhou, C., Zhou, Q., Zhou, S., Yang, W., Zhao, Y., Li, G., & Zhou, D. (2016). Analysis of chromatin regulators reveals specific features of rice DNA methylation pathways. *Plant Physiology*, 171(3), 2041–2054. https://doi.org/10.1104/pp.16.00393.

Tost, J. (2008). DNA methylation: An introduction to the biology and the disease-associated changes of a promising biomarker. *Methods in Molecular Biology*, 3–20. https://doi.org/10.1007/978-1-59745-522-0_1.

Ueda, M., & Seki, M. (2020). Histone modifications form epigenetic regulatory networks to regulate abiotic stress response. *Plant Physiology*, 182(1), 15–26. https://doi.org/10.1104/pp.19.00988.

Valledor, L., Hasbún, R., Meijón, M., Rodríguez, J. L., Santamaría, E., Viejo, M., Berdasco, M., Feito, I., Fraga, M. F., Cañal, M. J., & Rodríguez, R. (2007). Involvement of

DNA methylation in tree development and micropropagation. *Plant Cell Tissue and Organ Culture (PCTOC)*, 91(2), 75–86. https://doi.org/10.1007/s11240-007-9262-z.

Verhoeven, K. J. F., Jansen, J. J., Van Dijk, P. J., & Biere, A. (2009). Stress-induced DNA methylation changes and their heritability in asexual dandelions. *New Phytologist*, 185(4), 1108–1118. https://doi.org/10.1111/j.1469-8137.2009.03121.x.

Verma, N., Giri, S. K., Singh, G., Gill, R., & Kumar, A. (2022). Epigenetic regulation of heat and cold stress responses in crop plants. *Plant Gene*, 29, 100351. https://doi.org/10.1016/j.plgene.2022.100351.

Von Koskull-Döring, P., Scharf, K.-D., & Nover, L. (2007). The diversity of plant heat stress transcription factors. *Trends in Plant Science*, 12(10), 452–457. https://doi.org/10.1016/j.tplants.2007.08.014.

Walter, J., & Hümpel, A. (2017). Introduction to epigenetics. In R. Heil, S. B. Seitz, H. König, & J. Robienski (Eds.), *Epigenetics* (pp. 11–29). Springer Fachmedien Wiesbaden. https://doi.org/10.1007/978-3-658-14460-9_2.

Wang, J., Meng, X., Yuan, C., Harrison, A. P., & Chen, M. (2016a). The roles of cross-talk epigenetic patterns in *Arabidopsis thaliana*. *Briefings in Functional Genomics*, 15(4), 278–287. https://doi.org/10.1093/bfgp/elv025.

Wang, L., Chen, H., Li, J., Shu, H., Zhang, X., Wang, Y., Tyler, B. M., & Dong, S. (2020). Effector gene silencing mediated by histone methylation underpins host adaptation in an oomycete plant pathogen. *Nucleic Acids Research*, 48(4), 1790–1799. https://doi.org/10.1093/nar/gkz1160.

Wang, Y., Hu, Q., Wu, Z., Wang, H., Han, S., Jin, Y., Zhou, J., Zhang, Z., Jiang, J., Shen, Y., Shi, H., & Yang, W. (2017). HISTONE DEACETYLASE 6 represses pathogen defence responses in *Arabidopsis thaliana*. *Plant, Cell & Environment*, 40(12), 2972–2986. https://doi.org/10.1111/pce.13047.

Wang, Z., Cao, H., Sun, Y., Li, X., Chen, F., Carles, A., Li, Y., Ding, M., Zhang, C., Deng, X., Soppe, W. J. J., & Liu, Y.-X. (2013). *Arabidopsis* paired amphipathic helix proteins SNL1 and SNL2 redundantly regulate primary seed dormancy via abscisic acid–ethylene antagonism mediated by histone deacetylation. *The Plant Cell*, 25(1), 149–166. https://doi.org/10.1105/tpc.112.108191.

Waters, A. J., Makarevitch, I., Eichten, S. R., Swanson-Wagner, R. A., Yeh, C., Xu, W., Schnable, P. S., Vaughn, M. W., Gehring, M., & Springer, N. M. (2011). Parent-of-origin effects on gene expression and DNA methylation in the maize endosperm. *The Plant Cell*, 23(12), 4221–4233. https://doi.org/10.1105/tpc.111.092668.

Wei, W., Chen, J., Zeng, Z., Kuang, J., Lu, W., & Shan, W. (2020). The Ubiquitin E3 ligase MaLUL2 is involved in high temperature-induced green ripening in banana fruit. *International Journal of Molecular Sciences*, 21(24), 9386. https://doi.org/10.3390/ijms21249386.

Weng, M., Yang, Y., Feng, H., Pan, Z., Shen, W., Zhu, Y., & Dong, A. (2014). Histone chaperone ASF1 is involved in gene transcription activation in response to heat stress in *Arabidopsis thaliana*. *Plant, Cell & Environment*, 37(9), 2128–2138. https://doi.org/10.1111/pce.12299.

Williams, B. P., Pignatta, D., Henikoff, S., & Gehring, M. (2015). Methylation-sensitive expression of a DNA demethylase gene serves as an epigenetic rheostat. *PLoS Genetics*, 11, e1005142. https://doi.org/10.1371/journal.pgen.1005142.

Wu, C.-J., Liu, Z.-Z., Wei, L., Zhou, J.-X., Cai, X.-W., Su, Y.-N., Li, L., Chen, S., & He, X.-J. (2021). Three functionally redundant plant-specific paralogs are core subunits of the SAGA histone acetyltransferase complex in *Arabidopsis*. *Molecular Plant*, 14(7), 1071–1087. https://doi.org/10.1016/j.molp.2021.03.014.

Wu, Z., He, L., Jin, Y., Chen, J., Shi, H., Wang, Y., & Yang, W. (2021). HISTONE DEACETYLASE 6 suppresses salicylic acid biosynthesis to repress autoimmunity. *Plant Physiology*, 187(4), 2592–2607. https://doi.org/10.1093/plphys/kiab408.

Xia, S., Cheng, Y. T., Huang, S., Win, J., Soards, A., Jinn, T.-L., Jones, J. D. G., Kamoun, S., Chen, S., Zhang, Y., & Li, X. (2013). Regulation of transcription of nucleotide-binding leucine-rich repeat-encoding genes SNC1 and RPP4 via H3K4 trimethylation. *Plant Physiology*, 162(3), 1694–1705. https://doi.org/10.1104/pp.113.214551.

Zemach, A., & Zilberman, D. (2010). Evolution of eukaryotic DNA methylation and the pursuit of safer sex. *Current Biology*, 20(17), R780–R785. https://doi.org/10.1016/j.cub.2010.07.007.

Zhang, H., Lang, Z., & Zhu, J. (2018). Dynamics and function of DNA methylation in plants. *Nature Reviews Molecular Cell Biology*, 19(8), 489–506. https://doi.org/10.1038/s41580-018-0016-z.

Zhang, L., Luo, P., Bai, J., Wu, L., Di, D.-W., Liu, H.-Q., Li, J.-J., Liu, Y.-L., Khaskheli, A. J., Zhao, C.-M., & Guo, G.-Q. (2021). Function of histone H2B monoubiquitination in transcriptional regulation of auxin biosynthesis in *Arabidopsis*. *Communications Biology*, 4(1), 206. https://doi.org/10.1038/s42003-021-01733-x.

Zhang, M., Xu, C., Von Wettstein, D., & Liu, B. (2011). Tissue-specific differences in cytosine methylation and their association with differential gene expression in sorghum. *Plant Physiology*, 156(4), 1955–1966. https://doi.org/10.1104/pp.111.176842.

Zhang, N., Yang, Z., Zhang, Z., & Liang, W. (2020a). BcRPD3-mediated histone deacetylation is involved in growth and pathogenicity of *Botrytis cinerea*. *Frontiers in Microbiology*, 11, 1832. https://doi.org/10.3389/fmicb.2020.01832.

Zhang, X., Yazaki, J., Sundaresan, A., Cokus, S., Chan, S. W., Chen, H., Henderson, I. R., Shinn, P., Pellegrini, M., Jacobsen, S. E., & Ecker, J. R. (2006). Genome-wide high-resolution mapping and functional analysis of DNA methylation in *Arabidopsis*. *Cell*, 126(6), 1189–1201. https://doi.org/10.1016/j.cell.2006.08.003.

Zhang, X., Ménard, R., Li, Y., Coruzzi, G. M., Heitz, T., Shen, W.-H., & Berr, A. (2020b). *Arabidopsis* SDG8 potentiates the sustainable transcriptional induction of the pathogenesis-related genes PR1 and PR2 during plant defense response. *Frontiers in Plant Science*, 11, 277. https://doi.org/10.3389/fpls.2020.00277.

Zhao, T., Zhan, Z., & Jiang, D. (2019). Histone modifications and their regulatory roles in plant development and environmental memory. *Journal of Genetics and Genomics*, 46(10), 467–476. https://doi.org/10.1016/j.jgg.2019.09.005.

Zheng, J., Chen, F., Wang, Z., Cao, H., Li, X., Deng, X., Soppe, W. J. J., Li, Y., & Liu, Y. (2012). A novel role for histone methyltransferase KYP/SUVH4 in the control of *Arabidopsis* primary seed dormancy. *New Phytologist*, 193(3), 605–616. https://doi.org/10.1111/j.1469-8137.2011.03969.x.

Zheng, M., Liu, X., Lin, J., Liu, X., Wang, Z., Xin, M., Yao, Y., Peng, H., Zhou, D., Ni, Z., Sun, Q., & Hu, Z. (2019). Histone acetyltransferase GCN 5 contributes to cell wall integrity and salt stress tolerance by altering the expression of cellulose synthesis genes. *The Plant Journal*, 97(3), 587–602. https://doi.org/10.1111/tpj.14144.

Zhou, C., Wang, C., Liu, H., Zhou, Q., Liu, Q., Guo, Y., Peng, T., Song, J., Zhang, J., Chen, L., Zhao, Y., Zeng, Z., & Zhou, D. (2018). Identification and analysis of adenine N6-methylation sites in the rice genome. *Nature Plants*, 4(8), 554–563. https://doi.org/10.1038/s41477-018-0214-x.

Zhu, J. (2009). Active DNA demethylation mediated by DNA glycosylases. *Annual Review of Genetics*, 43(1), 143–166. https://doi.org/10.1146/annurev-genet-102108-134205.

Zou, B., Yang, D.-L., Shi, Z., Dong, H., & Hua, J. (2014). Monoubiquitination of histone 2B at the disease resistance gene locus regulates its expression and impacts immune responses in *Arabidopsis*. *Plant Physiology*, 165(1), 309–318. https://doi.org/10.1104/pp.113.227801.

4 Non-coding RNAs in Epigenetic Regulation

Ashmaul Husna, Noorie Shama, Sayak Chakraborty, Chandan Roy, Rita Kundu, Snehalata Majumdar and Aditi Nayak

4.1 INTRODUCTION

Epigenetics refers to reversible, heritable changes in gene expression or phenotype that occur through mechanisms other than alterations in the DNA sequence. The changes influence the expressions and patterns in which the genes are read. They also determine the variety of cellular phenotypic characters (Mattick et al., 2023). They are important for development, cell differentiation, and environmental responses. The mechanisms of epigenetics include DNA methylation (addition of a methyl group to the cytosine base, associated with gene silencing), histone modification (acetylation, methylation and phosphorylation modification of histone proteins around which DNA is wrapped), chromatin remodeling (alteration of chromatin structure by ATP-dependent chromatin remodeling complex) and non-coding RNAs (Zhang et al., 2019). These non-coding RNAs (ncRNA) are transcribed to RNA but not translated into proteins. They are involved in regulatory functions like gene expression regulation, maintaining genome stability and influencing cellular responses to environmental changes (Verhoeff et al., 2023).

In eukaryotes, non-coding RNAs are up to 1000 nucleotides long and perform various regulatory functions during transcription, chromatin remodeling and cell differentiation. They target RNA polymerase II (enzyme for transcription) or transcription factors (TFs). They are divided into small, intermediate, and long non-coding RNA. The long non-coding RNAs (lncRNAs) interact with TF and chromatin-modifying complexes to influence transcription. Some recruit histone methyltransferases to specific chromatin for modifying it and its expression (Bure et al., 2022a). Small non-coding RNAs (sncRNA) and miRNAs bind to target mRNA during post-transcriptional regulation to repress translation or promote degradation. The ncRNAs recruit chromatin remodeling complexes to alter chromatin architecture by histone modification and DNA accessibility. They determine gene expression patterns during development, proliferation, and environmental stimuli. ncRNAs also help in maintaining genome stability and integrity by silencing transposable elements, significant in germline cells where transposons disrupt genome stability if not monitored and regulated (Yang et al., 2022a). ncRNAs act as sensors during environmental changes and stress and regulate stress-related genes or apoptosis to modulate stress responses. Abnormal expression or mutations of ncRNAs have been associated with various diseases, including types of cancers, cardiovascular diseases and neurological disorders; thus, their role is crucial in uncovering novel therapeutic strategies targeting specific ncRNA pathways.

Epigenetic regulation plays a vital role in plant adaptability by adjusting to changing environmental conditions and enhancing stress tolerance by enabling it to combat extreme drought, heat and pathogenic conditions (Yao et al., 2022). By identifying and enhancing epigenetic markers, researchers can develop crops that are more resistant to stressed conditions, ensuring food security and sustainability in agriculture. Plants utilize epigenetic mechanisms as environmental memory cells. This helps them synchronize their growth and division according to the environmental conditions, improving their survival rate. Phenotypic plasticity involves the formation of adaptive traits specific to varying stimuli. Some epigenetic changes are inherited by offspring, generation after generation, in response to environmental stress and enhance the stability and adaptability of the population. These are crucial to understand as they are vital for breeding programs that aim to develop stress-tolerant varieties of plants (Burenina et al., 2017).

4.2 TYPES OF NON-CODING RNAS

4.2.1 Small Non-coding RNAs (19–24 bp)

Small non-coding RNAs (snRNA) are of two types.

4.2.1.1 MicroRNAs (miRNAs)

miRNA has a similar mechanism of action to siRNA in mediating transcriptional gene silencing. miRNA regulates the direct cleavage and degradation of mRNA even though they are partially complementary (about six to seven nucleotides in the seed region). miRNA has the potential to regulate hundreds of different genes (Banerjee et al., 2016; Chakraborty & Roychoudhury, 2023). miRNA can restrict chromatin remodeling enzyme activity to change the whole DNA or the state of the chromatin. They can also induce it through histone modification; for example, histone deacetylase 4 (HDAC4) is a specific target for miR-140 in mouse embryonic cartilage tissue, suggesting that miRNA may be

DOI: 10.1201/9781003589914-5

involved in transcriptional gene silencing via the modification of histone proteins (Burenina et al., 2017). miR-320 can recruit Argonaute-1 (AGO1) to the POLR3D promoter, which facilitates the recruitment of EZH2. Further, EZH2 trimethylates at H3K27, leading to the silencing of the *POLR3D* gene (Kim et al., 2008). Few studies have demonstrated the existence of DNA methylation mediated by miRNA in plants.

4.2.1.2 Small Interfering RNAs (siRNAs)

Small interfering RNAs (siRNAs) silence genes at the post-transcriptional level to stop translation by cleaving specific targets on mRNA molecules. siRNAs are loaded onto Argonaute (AGO) proteins; the antisense single-stranded component of siRNA then guides the RNA-induced silencing complex (RISC) complex to the target mRNA (Pisignano & Ladomery, 2021). The catalytic property of RISC cleaves the mRNA. siRNA can silence EZH2, a histone methyltransferase that causes methylation of H2K27 and serves as an anchor point for CpG methylation and formation of silent chromatin. This DNA methylation and histone modification in cells lead to transcriptional gene silencing. A specific argonaute family protein (AGO4) in *Arabidopsis* causes the accumulation of DNA and histone methylation, mediating epigenetic regulation. This involves RNA interference (RNAi) where genes are regulated by RNA-induced silencing complex (RISC) which are guided by siRNA for efficient gene silencing regulation (Tassinari et al., 2021; Singh & Roychoudhury, 2023).

4.2.2 Intermediate-Sized RNAs (25–250 bp)

4.2.2.1 Small Nucleolar RNAs (snoRNAs)

Small nucleolar RNA (snoRNA) is an intermediate ncRNA with a recognized role in tuning ribosomal and spliceosomal function. It is responsible for guiding ribose methylation and pseudouridylation at targeted nucleotide residues of ribosomal and small nuclear RNAs. snoRNAs are increasingly being implemented in rRNA acetylation, modulation of splicing patterns, control of mRNA abundance and translational efficiency (D. Li et al., 2021). They are responsible for the synthesis of 200 modified nucleotides in rRNA as well as modifying spliceosomal snRNAs. The involvement of snoRNPs in the modification of snRNAs was first demonstrated with box C/D snoRNAs that direct the 2-O-methylation of U6 snRNA. They have a classical role in ribosome biogenesis. They are also involved in splicing and transcription regulation. Box C/D and H/ACA snoRNAs differ in their maturation pathways, with 5′ loop caps involved in the transcription of C/D snoRNAs (Bratkovič et al., 2020).

4.2.2.2 Small Nuclear RNAs (snRNAs)

Small nuclear RNA (snRNA) is an essential component of spliceosomes, responsible for pre-mRNA splicing by forming small nucleoproteins (snRNPs) and facilitating the removal of introns from pre-mRNA (Huang et al., 2022). The five major snRNPs are U1, U2, U4, U5 and U6. These are involved in the identification of the splice site and catalyzing the splicing reaction. snoRNPs guide the necessary chemical changes necessary for snRNA molecules, creating specific modifications needed for their effective function during splicing. Their involvement in both RNA modification and processing underscores their importance in maintaining proper gene regulation and cellular function (Lin et al., 2021).

4.2.3 Long Non-coding RNAs (lncRNAs) (>200 bp)

Long non-coding RNAs (lncRNAs) are of 200 nucleotides long and are found in the nucleus and cytoplasm. They have five category divisions: sense, antisense, bidirectional, intronic, and intergenic lncRNAs (Statello et al., 2021; Bhattacharya & Roychoudhury, 2024). They regulate gene expression through archetypal mechanisms such as signals (indicating specific cellular or environmental conditions), decoys (binding to TFs, preventing their binding to the target DNA sequence), guides (guiding regulatory proteins like chromatin-modifying complexes to specific genomic locations), and scaffolds (binding different proteins together to form functional complexes). lncRNAs have a major function in chromatin remodeling. H19 RNA is a genome imprinting ncRNA that can regulate gene expression through miRNA action mechanism as it contains a precursor of miR-675 in human and rat cells (Dey et al. 2014). They are spliced and polyadenylated for high cytosolic concentration. lncRNA is necessary for silencing Kcnq1ot1 and lgf2r/Air imprinted genes (Zhou et al., 2021). Xist is a 17-kb long ncRNA which is crucial for X chromosome inactivation by physically interacting with it. RepA (a 1.6 kb portion of Xist RNA) can induce H3K27 trimethylation by recruiting the polycomb repressive complex 2 (PRC2) to the inactivated X. HOTAIR RNA is a lncRNA that originates from the HOXC cluster. It can silence gene transcription in a region of up to 40 kb by recruiting PRC2 through HOTAIR. This then activates the formation of heterochromatin and gene silencing via trimethylation of H3K9 (Shaath et al., 2022).

4.3 MECHANISMS OF ACTION: NON-CODING RNAs IN EPIGENETIC REGULATION

Non-coding RNAs (ncRNAs) have appeared as main players in the complex landscape of epigenetic regulation, arranging gene expression patterns without being translated into proteins. miRNAs and long non-coding RNAs have gathered momentous attention for their roles in mediating epigenetic alterations and gene silencing. These multipurpose molecules exert their influence through numerous mechanisms affecting DNA methylation, histone modification and chromatin remodeling, ultimately shaping the transcriptional yield of cells (Kumar et al., 2020).

4.3.1 Epigenetic Modifications

ncRNAs energetically contribute to establishing and upholding DNA methylation patterns, a fundamental epigenetic mark that affects gene expression. DNA methylation, mainly occurring at cytosine residues in CpG dinucleotides, is catalyzed by DNA methyltransferases (DNMTs). ncRNA can directly or indirectly modulate the action of dNTP, thereby swaying DNA methylation patterns and subsequent gene expression. miR-148 has been shown to target DNMT1 and DNMT3B, leading to translation suppression or mRNA degradation (Ilieva, 2024). The interaction of miR-148 with DNMT3B mRNA leads to post-transcriptional gene silencing (PTGS), which could lead to translational suppression or mRNA deprivation. In addition to directly targeting DNMT, ncRNAs can indirectly affect DNA methylation by modulating the appearance of other epigenetic regulators or signaling pathways involved in DNA methylation.

Histone alterations, another crucial deposit of epigenetic regulation, involve the covalent modification of histone proteins that bundle DNA into chromatin. These modifications, including acetylation, methylation, and phosphorylation, along with ubiquitylation, can change chromatin structure and accessibility, swaying gene transcription. ncRNAs are implicated in regulating histone modifications by interrelating with histone-modifying enzymes such as histone deacetylase and histone methyltransferase. miRNAs can directly target histone-regulating enzymes, modulating their expression, and activity; for example, miR-449a targets HDAC1. Many miRNAs have been reported to interact and replace the expression of the HTML, EZH2 (enhancer of zest homolog 2), which is part of the Polycomb group repressive complex 2 (PRC 2). EZH2 is targeted by several miRNAs, including miR-101, miR-137, miR-26a, miR-98, miR-124, miR-214 and let-7 (Arif et al., 2020). Reduced levels of miRNAs capable of defeating DNMT1, DNMT3A3B and relieved edge 2 can lead to irregular DNA methylation patterns, silencing specific gene marks. Long non-coding RNAs contribute to the formation of distinct chromatin domains, swaying gene transcription, and cellular identity (Kumar et al., 2020).

4.3.2 Gene Silencing

Non-coding RNAs play a central role in facilitating transcriptional suppression, a process by which gene expression is silenced. Transcriptional repression is crucial for maintaining cellular individuality, regulating development and preventing aberrant gene expression. Non-coding RNA employs diverse mechanisms to achieve transcriptional repression, including DNA methylation, histone modification and chromatin remodeling.

miRNAs mainly mediate gene silencing by directly binding to target mRNAs characteristically within the 3′ UTR, leading to mRNA degradation or translational suppression (Kumar et al., 2023). This direct communication between miRNAs and target mRNAs results in the downregulation of protein expression.

lncRNAs employ numerous mechanisms to mediate transcriptional repression. Long non-coding RNAs can cooperate with chromatin-modifying developments such as the polycomb repressive complex and direct them to histone modification and transcriptional repression (Hung et al., 2020). Long non-coding RNAs also target the transcription factors, avoiding them from binding to their target DNA sequences. This prevents gene transcription. Furthermore, these can bring together the transcription factors, RNA polymerase and other regulatory proteins to specific genomic sites. They can also adjust the activity of transcription factors by manipulating gene expression.

In conclusion, non-coding RNAs, mainly miRNAs, and long non-coding RNAs, play a multifaceted role in the epigenetic guidelines by modulating DNA methylation. Based on modifications and chromatin remodeling, these versatile components contribute to gene silencing, swaying cellular processes.

4.3.3 Influence on Chromatin Architecture

Long non-coding RNAs (lncRNAs) exhibit a predominantly nuclear localization compared to messenger RNAs (mRNAs), although a significant portion is also found in the cytoplasm. This nuclear localization suggests that ncRNAs play a crucial role in assembling and recruiting various proteins to specific genomic regions, thereby contributing to the formation of three-dimensional nuclear structures that influence gene expression. Additionally, newly transcribed RNAs in the nucleus undergo various processing steps, such as splicing, polyadenylation, 5′ capping, and methylation. During these processes, ncRNAs impact RNA splicing and mRNA stability. Apart from their nuclear roles, ncRNAs found in the cytoplasm interact with other RNA molecules, such as competitive endogenous RNAs (ceRNAs), acting as modulators of gene expression (Tripathi et al., 2010).

Furthermore, lncRNAs regulate gene expression by serving as scaffolds for chromatin-modifying proteins, including methyltransferases, demethylases, acetyltransferases, and deacetylases. These proteins are recruited to specific genomic sites through either cis-regulation, which affects the transcription of nearby genes, or trans-regulation, which influences genes located at distant genomic regions. By modifying nuclear architecture and chromatin accessibility, lncRNAs play a significant role in regulating transcriptional activity and overall gene expression (Engreitz et al., 2016).

One of the most extensively studied long non-coding RNAs (lncRNAs) is the X-inactive specific transcript (Xist), which plays a crucial role in X chromosome inactivation (XCI) during the early stages of female embryonic development. Xist is expressed from one of the two X chromosomes and facilitates chromatin binding through scaffold attachment factor A (also known as hnRNPU). It also recruits the SMRT/histone deacetylase

1 (HDAC1)-associated repressor protein (SHARP), which in turn interacts with HDAC3, polycomb repressive complex 1 (PRC1) and polycomb repressive complex 2 (PRC2) (Splinter et al., 2011). Interestingly, while Xist expression and the SHARP-HDAC3 complex are essential for initiating XCI, Xist is not required for maintaining transcriptional silencing. Additionally, PRC2 is dispensable for initiating XCI, as studies have shown that its genetic deletion does not affect the initiation of transcriptional silencing, though it is necessary for maintaining gene inactivation (Kalantry et al., 2006).

Similar to Xist, several other lncRNAs are believed to interact with PRC2 to regulate gene silencing at specific genomic loci. One such lncRNA is HOTAIR, which is transcribed from the Homeobox (HOX) C gene cluster on chromosome 12 and co-expressed with *HOXC* genes. HOTAIR influences the expression of *HOXD* genes on chromosome 2 through transregulation. Additionally, it interacts with lysine-specific demethylase 1 and PRC2, promoting histone modifications such as H3K27 methylation and H3K4 demethylation, which contribute to gene repression (Tsai et al., 2010).

A recent study found that while HOTAIR strongly interacts with PRC2 to silence target genes through H3K27me3 deposition, the deletion of PRC2 components did not disrupt the silencing function of HOTAIR (Kuo et al., 2022). This suggests that PRC2 is not essential for initiating HOTAIR-mediated gene silencing. A similar mechanism has been observed in Xist-mediated gene silencing, further supporting the idea that lncRNAs like HOTAIR and Xist interact with PRC2 in ways that have broader implications for gene regulation. These findings highlight the complex role of lncRNAs in guiding chromatin modifications, involving multiple regulatory proteins.

Long non-coding RNAs (lncRNAs) serve as molecular scaffolds by interacting with multiple transcription factors (TFs) or other regulatory proteins simultaneously. These interactions lead to the formation of lncRNA-protein complexes that facilitate coordinated gene regulation. By acting as a structural framework, these lncRNAs enable transcriptional modulators to influence gene expression efficiently. One well-known example is the lncRNA HOTAIR, which plays a key role in recruiting chromatin-modifying complexes to regulate target genes (Tsai et al., 2010).

Non-coding RNAs (ncRNAs) are crucial for gene regulation and chromatin organization, primarily through their interactions with chromatin-associated proteins. Research available highlights the importance of these interactions in various cellular processes; for instance, lncRNAs like XIST and HOTAIR associate with chromatin-modifying complexes such as polycomb repressive complex 2 (PRC2) to facilitate gene silencing and chromatin remodeling . XIST, a widely studied lncRNA, coats the inactive X chromosome and recruits chromatin-modifying proteins to establish X-chromosome inactivation, ensuring proper gene dosage in females. Similarly, HOTAIR interacts with PRC2 and other chromatin regulators to modulate gene expression, playing a role in both development and disease progression (Tsai et al., 2010).

Apart from long non-coding RNAs (lncRNAs), other classes of non-coding RNAs (ncRNAs), such as small nuclear RNAs (snRNAs) and microRNAs (miRNAs), play a significant role in chromatin regulation. snRNAs assist in directing chromatin-remodeling complexes to specific genomic sites, influencing chromatin accessibility and RNA processing. Similarly, miRNAs can recruit chromatin-modifying enzymes to target genes, allowing for precise regulation of gene expression. Additionally, circular RNAs (circRNAs), a newly identified group of ncRNAs, have been shown to interact with chromatin-associated proteins, contributing to transcriptional regulation.

These interactions typically involve ncRNAs acting as molecular scaffolds or guides, ensuring that chromatin-modifying enzymes are accurately positioned at specific genomic regions. For instance, certain ncRNAs facilitate the recruitment of DNA demethylases such as TET2, which alters DNA methylation patterns and subsequently impacts gene expression. Recent technological advancements, including chromatin isolation by RNA purification followed by mass spectrometry (ChIRP-MS) and RNA antisense purification sequencing (RAP-seq), have greatly enhanced the ability to identify ncRNA-protein interactions with high precision. Ongoing research in this field continues to reveal the extensive role of ncRNAs in chromatin organization and gene regulation, offering deeper insights into their influence on cellular functions and disease development (Thouly et al., 2020).

4.4 NON-CODING RNAs: ORCHESTRATING STRESS TOLERANCE SIGNALING PLANTS

Plants are endlessly exposed to a myriad of ecological stresses, including drought, salinity, temperature and pathogen attack. These trigger a cascade of molecular and biochemical responses intended to mitigate injury and promote survival. Non-coding RNAs, mainly micro-RNAs and long non-coding RNAs, have arisen as critical supervisors of these stress response pathways. They regulate gene expression by initiating signaling cascades that improve plant resilience by working at the epigenetic and post-transcriptional stages. Non-coding RNAs enable plants to adapt and thrive in a challenging environment (Swaminathan et al., 2023).

4.4.1 Non-coding RNAs as Initiators of Stress Response Signaling

During stress, plants rapidly activate a network of signaling pathways that eventually lead to changes in gene expression and biological responses. Non-coding RNA plays a pivotal role in introducing and modulating these signaling pathways by substituting as both upstream regulators and downstream targets.

miRNAs typically control gene expression by binding to the target mRNAs, leading to mRNA deprivation or translational repression in the background of a stress response. miRNAs can target genes encoding transcription factors, signaling proteins, and metabolic enzymes involved in stress tolerance (Fahad et al., 2024). By downregulating negative controllers of stress response pathways, miRNAs can strengthen the signaling cascade and encourage the expression of downstream stress-responsive genes. For example, during drought conditions, specific miRNAs mark genes encoding proteins that prevent abscisic acid (ABA) signaling by repressive mechanisms. These inhibitory miRNAs improve ABA signaling, leading to reduced water loss and can amplify drought resistance (Zhakypbek et al., 2025).

Long non-coding RNAs also play a diverse role in stress response signaling. They can act as a molecular scaffold bringing organized proteins involved in signaling pathways and enabling their interactions. In stress responses, long non-coding RNAs can moderate the expression of genes encoding signaling receptor kinases and phosphatases (Heydari et al., 2024).

4.4.2 Specific Examples of Non-coding RNA-Mediated Regulation and Plant Resilience

Numerous studies have established the critical role of non-coding RNA in improving plant flexibility under various stress conditions (Table 4.1).

4.4.2.1 Drought Tolerance

Several miRNAs have been recognized as key regulators of drought tolerance in plants; for instance, miR-169 is upregulated during drought stress in *Arabidopsis thaliana* and targets a gene encoding a nuclear factor YA (NF-YA) (Li et al., 2008). Similarly, in rice, miR-393 is downregulated under drought stress, leading to amplified expression of its target genes encoding auxin receptors. This modulation of auxin signaling contributes to progress in root development and enhanced water uptake under drought conditions (Ji et al., 2023).

4.4.2.2 Salt Tolerance

Salinity is a major ecological stress that affects plant growth and efficiency. Non-coding RNAs have been shown to play a role in modifying salt tolerance mechanisms; for example, miR-399 is induced by salt stress in *Arabidopsis* and targets a gene encoding a phosphate transporter (Pegler et al., 2020) (Liu et al., 2016). Downregulating miR-399 helps uphold iron homeostasis and prevent salt-induced toxicity. Overexpression of ***miR-399*** in *Arabidopsis* improves salt tolerance, indicating its role in regulating plant response to salinity. Furthermore, long non-coding RNAs have also been concerned with salt tolerance. The long non-coding RNAs have been shown to interact with the chromatin-modifying component and promote the expression of genes involved in ion transport and osmoprotection, leading to increased salt tolerance in rice (Hasanuzzaman & Fujita, 2022).

4.4.2.3 Cold Tolerance

Exposure to low temperature can cause significant damage to the plants. Non-coding RNAs have been found to regulate cold tolerance, enhancing plant survival under freezing conditions; for instance, miR-163 is downregulated during cold stress in *Arabidopsis*, leading to increased expression of its target gene encoding a transcription factor involved in cold tolerance. This transcription factor activates the expression of downstream cold-responsive genes, enhancing plant freezing tolerance. Overexpression of miR-163 reduces freezing tolerance, confirming its role in regulating cold tolerance. Additionally, long non-coding RNAs have also been shown to regulate cold tolerance. The long non-coding RNA interacts with the PRC2 complex and promotes the silencing of *EFLC* genes (a major repressor of flowering; embryonic fibroblast Lineage Commitment). This silencing of *EFLC* allows plants to flower earlier, avoiding

TABLE 4.1
ncRNAs in epigenetic regulation in plants

ncRNA	Mechanism of action	Function	Stress context	Source
miR-393	Downregulated during drought, leading to rise in the expression of auxin receptor genes	Drought tolerance	Drought	Jiang et al. (2022)
miR-169	Target *NF-YA* transcription of mRNA, leading to decreased expression	Drought tolerance	Drought	Li et al. (2021)
miR-399	Induced by salt stress, targeting a phosphate transporter gene leading to downregulated phosphate uptake	Salt tolerance	Salt	Pegler et al. (2020)
miR-163	Downregulated during cold stress leading to increased expression of a transcription factor, involved in cold acclimation	Cold tolerance	Cold	Tiwari et al. (2020)
SID1-AS1	Interacts with chromatin changing complexes and promotes expression of ion transport and osmoprotection genes	Salt tolerance	Salt	Zhang et al. (2024)
CILnc1	Interacts with the PRC2 complex promoting silencing of *FLC* genes	Cold tolerance	Cold	Nielsen et al. (2024)

the damaging effects of late-season frost (Hasanuzzaman & Fujita, 2022).

4.4.2.4 Pathogen Resistance

Plants are constantly challenged by pathogens, and non-coding RNAs play a critical role in regulating plant immune responses. miRNA can target genes encoding components of the plant immune system. miR-393 targets genes encoding auxin receptors, and its overexpression leads to amplified disease resistance against certain bacterial pathogens. Long RNAs have also been shown to regulate plant immunity. The long non-coding RNA is induced by bacterial infection, interacts with the transcription factor and promotes the expression of genes involved in pathogen resistance, enhancing plant confrontation to bacterial infections (Waheed et al., 2021).

Finally, non-coding RNAs, particularly miRNAs and long-known coding RNAs, play a central role in controlling plant stress tolerance by initiating and modulating signaling pathways that regulate gene expression and biological responses (Figure 4.1).

4.4.3 Implications of Non-coding RNA for Plant Adaptability

4.4.3.1 Contribution to Phenotypic Plasticity

By altering gene expression in response to different stressors, non-coding RNAs (ncRNAs) play a critical role in regulating how plants adapt to changes in their environment. These ncRNAs, which include long non-coding RNAs (lncRNAs) and microRNAs (miRNAs), act as molecular mediators that affect pathways linked to stress. ncRNAs control mRNA stability, translation, and epigenetic alterations in response to harsh environmental factors like drought, salinity, high temperature or disease attacks, allowing plants to adjust their physiological responses. To conserve resources and improve tolerance, for example, some miRNAs are activated during drought stress to repress genes related to growth and development. Similarly, lncRNAs alter the expression of important genes necessary for long-term adaptation through their interactions with chromatin-modifying complexes. The importance of ncRNAs in supporting plant life and resilience in shifting environments is highlighted by their dynamic regulation (Chand Jha et al., 2021; Yadav et al., 2024).

As crucial regulators of environmental adaptability, ncRNA expression patterns are closely associated with plant stress responses. Through encouraging mRNA breakdown, inhibiting translation or causing epigenetic changes, these molecules such as miRNAs and lncRNAs manage the expression of genes. Under heat stress, certain miRNAs are activated to suppress genes that could make plants less able to withstand high temperatures. On the other hand, lncRNAs affect the transcription of stress-responsive genes by interacting with chromatin-modifying enzymes. When faced with environmental obstacles, these mechanisms enable plants to quickly modify their physiological and molecular processes. The crucial function of ncRNAs

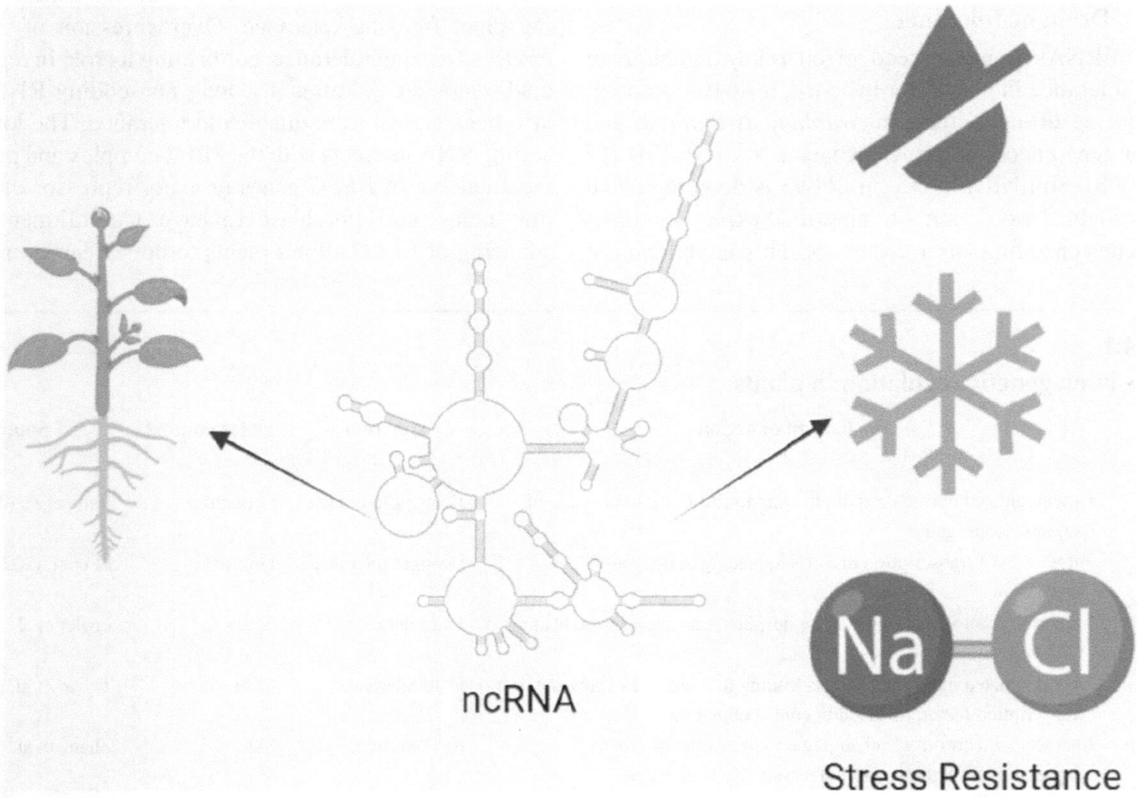

FIGURE 4.1 ncRNA conferring resistance to various abiotic stress factors.

in improving plant resilience and adaptability to changing environments is highlighted by their capacity to precisely control gene expression (Matsui & Seki, 2019a).

Variations in ncRNA expression are essential for stress tolerance; plants rely on ncRNAs to negotiate environmental stresses. By directing mRNAs for destruction, preventing translation or altering chromatin structure, miRNAs and lncRNAs both aid in the control of genes. Under drought, certain miRNAs are increased to inhibit genes linked to growth, putting survival ahead of growth. In a similar manner, lncRNAs enable long-term adaptation by modifying the expression of genes linked to stress through interactions with epigenetic regulators. Plants can adapt to problems including drought, salt, high temperature, and pathogen infestations due to the dynamic control of ncRNAs. The crucial function of ncRNAs in controlling stress responses emphasizes how crucial they are to maintaining plant productivity and survival in dynamic situations (Matsui & Seki, 2019b).

Genome-wide association studies (GWASs) have provided valuable insights into how methylation of non-coding RNAs (ncRNAs) influences various traits in living organisms. These studies have identified genetic variations that affect DNA methylation patterns in ncRNAs, ultimately impacting gene expression. As a result, these modifications contribute to differences in physical traits and susceptibility to certain diseases. By examining the interaction between genetic and epigenetic factors, researchers can gain a deeper understanding of inheritance patterns and how environmental factors shape gene activity.

One significant finding from GWASs is the association between single-nucleotide polymorphisms (SNPs) and changes in ncRNA methylation (Maurano et al., 2012). Some SNPs have been shown to influence traits like body mass index (BMI) and metabolic health by altering the regulatory roles of ncRNAs. Variations in SNPs can modify the methylation of microRNAs (miRNAs) and long non-coding RNAs (lncRNAs), which are crucial for metabolic and immune system regulation. This suggests that ncRNA methylation plays an essential role in maintaining biological stability and enabling organisms to adapt to environmental challenges (Wang et al., 2022).

Another emerging concept in this field is epiphenotyping, which involves using DNA methylation patterns to predict traits and disease risks. By analyzing ncRNA methylation profiles, scientists can identify individuals who may be at higher risk for conditions like cardiovascular diseases, neurodegenerative disorders, or autoimmune diseases. This approach holds promise for early disease detection and the development of personalized treatment strategies based on an individual's unique epigenetic profile. As research continues, ncRNA methylation may become a key tool in precision medicine, offering new opportunities for disease prevention and targeted therapies (Ludwig & Dwivedi, 2016).

According to GWASs, ncRNA methylation is essential for determining characteristics and illness risks. The intricate relationships among gene expression, epigenetic changes, and genetic differences have been made clear by these investigations. Gaining knowledge of these processes can help with illness risk prediction and the creation of novel therapies that focus on epigenetic modifications. Further studies in this area may provide additional understanding of how genes and the environment interact to influence plant development and health.

4.5 POTENTIAL USE OF ncRNAs IN EPIGENETIC REGULATION

The production of crop varieties that are both more nutrient-dense and more resistant to environmental challenges is being made possible by developments in non-coding RNA (ncRNA) research. ncRNAs, such as long non-coding RNAs (lncRNAs) and microRNAs (miRNAs), are essential for controlling gene expression, which enables plants to adjust to shifting environmental conditions. By modifying genes that respond to stress, these ncRNAs not chemicals affect a variety of biological pathways, assisting plants in surviving drought, salt, high temperature, and disease attacks. While certain miRNAs repress pathways that result in excessive water loss, others are known to regulate genes linked to water retention, improving drought tolerance. The activation of genes that increase plant resistance to salt and temperature extremes is another way that lncRNAs aid in chromatin remodeling.

Non-coding RNAs (ncRNAs) play a crucial role in enhancing crop nutrition by regulating metabolic pathways involved in nutrient uptake and storage. Certain miRNAs specifically influence the absorption and accumulation of essential minerals like iron and zinc, which are vital for human health. Increasing the expression of these ncRNAs in crops can lead to biofortified plant varieties with improved nutrient content, addressing micronutrient deficiencies, especially in populations that rely primarily on plant-based diets. The incorporation of ncRNA research into agricultural biotechnology offers great potential for crop enhancement. Advanced genetic tools such as CRISPR-Cas and RNA interference (RNAi) allow precise regulation of ncRNA activity, enabling the development of crops with enhanced stress tolerance and higher nutritional value. These advancements not only improve agricultural yield but also promote food security and sustainable farming by reducing the dependence on chemical fertilizers and pesticides.

Researchers can develop crops that have superior nutritional value and are more resilient to environmental stresses. Global agriculture continues to be impacted by climate change. Therefore, ncRNA-based inventions provide a practical means of guaranteeing food security. Farmers and consumers around the world will eventually gain from novel breeding techniques for high-yielding, nutrient-rich and stress-resistant crops made possible by ongoing research of ncRNA roles in plants.

4.6 RESEARCH GAPS AND FUTURE DIRECTIONS

Non-coding RNAs (ncRNAs) are important in epigenetic modifications and regulating gene expression; yet, several aspects of their functions remain unclear. Further studies are necessary in this particular field to get a proper understanding of their roles (Wei et al., 2017).

- Molecular mechanisms: More research is required to fully understand how ncRNAs interact with epigenetic regulators including DNA methyltransferases and histone-modifying enzymes.
- Tissue-specific roles: Although ncRNA expression differs among many cell types and tissues, a more thorough examination of the mechanisms underlying this specificity is necessary. Technology for advanced sequencing can yield more insightful information.
- Finding out how they interact with other ncRNAs and proteins as well as regulate the genes is crucial to understand (Mangiavacchi et al., 2023).
- Transgenerational epigenetic inheritance: Although the precise processes are still mostly unclear, some ncRNAs may help pass on epigenetic features from one generation to the next.
- Role in diseases: Although ncRNAs have been linked to diseases and neurological illnesses, further investigation is required to determine how they affect the epigenetic modifications that occur over the course of disease.
- Therapeutic applications: While ncRNA-based therapies have promise, further research is needed to address issues like stability, delivery strategies, and side effects.
- Environmental effects: Although environmental factors including pollution, stress, and food affect ncRNA activity, more research is needed to determine how they affect epigenetic regulation over the long run (Yao et al., 2019).

4.7 CONCLUSION

Non-coding RNA has emerged as a significant topic in contemporary genetics research, particularly as a novel method of epigenetic regulation. Nevertheless, up to now, researchers have possessed only a limited comprehension of how non-coding RNAs influence gene expression. Initially, miRNAs and siRNAs were perceived to operate independently, a belief supported by the clear distinctions among them. However, recent findings indicate that the functions of these pathways are becoming increasingly interconnected, forming a complex regulatory network. This raises the question of how living organisms have adapted to these diverse regulatory mechanisms. To address this issue, it is essential to first comprehend the specifics of the non-coding RNAs themselves. Currently, a significant challenge lies in understanding the distinct functions and interactions within this intricate RNA regulatory network in relation to protein-based regulatory mechanisms. Examining this network of non-coding RNAs is a particularly challenging task, necessitating a systematic approach to discover and analyze these RNAs. Achieving this will depend on continuous advancements and innovative methods in gene and genome scanning technologies to uncover the complete range of information and functions associated with non-coding RNAs. The ultimate objective is to elucidate the intricate regulatory mechanisms employed by non-coding RNAs and their interactions with both healthy cells and disease conditions. An excellent link between the RNA and chromatin world is provided by the capacity of the RNA-binding proteins (RBPs) to attach to non-coding regulatory RNAs and direct them to certain nascent transcripts. One special characteristic of RBPs is that they can identify particular RNA motifs in emerging RNAs, often known as RNA codes. This might help to explain why lncRNA complex targeting requires a basic level of transcription. An array of distinct RBPs may have the capacity to bind a lengthy stretch of RNA sequences that contain all of the appropriate RNA codes because lncRNAs frequently connect with many RNA-binding proteins, each of which has a preferred RNA sequence to bind. This is probably the main factor that links particular genomic locations to lncRNA complexes. Nuclear compartments and macromolecular machinery may be arranged using lncRNAs, in particular, as molecular scaffolds to carry out complex and accurate control of a variety of biological processes. New paradigms for epigenetic regulation will be aided by the knowledge of the location of certain loci and the organization of lncRNA complexes that target them. In this chapter, the various ways in which both long and short non-coding RNAs affect gene expression have been discussed. Many of these RNA transcripts are essential for the accurate localization of histone modification complexes or contribute to the process of DNA methylation. A modern definition of epigenetics might also encompass the gene silencing or upregulation initiated by non-coding RNAs, which occurs in regular biological processes and is influenced by epigenetic complexes. This expanded view could integrate harmful environmental factors that are transmitted through cell divisions, as observed in cancer cases. It would also account for significant variations in gene expression identified between cells with the same DNA sequence during differentiation or in developing tissues. Notably, targeting ncRNAs for either repression or overexpression can serve as an effective method for amplifying changes in the levels of subsequent effectors. Future research that elucidates the connection between environmental factors and alterations in ncRNA functionality is essential for a deeper comprehension of the role played by these transcripts. This understanding would enhance the modified interpretation of Francis Crick's "central dogma", illustrating that RNA is not merely a phase in the transition from DNA to protein but also plays a crucial role in providing regulatory feedback that affects both transcription and translation. Thus alongside the involvement of ncRNA in traditional epigenetics, we suggest that transcripts influenced by these mechanisms act as epigenetic intermediaries, relaying the message across an extensive network of dynamic gene expression regulation.

REFERENCES

Arif, K. M. T., Elliott, E. K., Haupt, L. M., & Griffiths, L. R. (2020). Regulatory mechanisms of epigenetic miRNA relationships in human cancer and potential as therapeutic targets. *Cancers*, 12(10), 2922. https://doi.org/10.3390/cancers12102922.

Banerjee, A., Roychoudhury, A., & Krishnamoorthi, S. (2016). Emerging techniques to decipher microRNAs (miRNAs) and their regulatory role in conferring abiotic stress tolerance of plants. *Plant Biotechnology Reports*, 10(4), 185–205.

Benetti R., Gonzalo S., Jaco I., Muñoz P., Gonzalez S., Schoeftner S., Murchison E., Andl T., Chen T., Klatt P., Li E., Serrano M., Millar S., Hannon G., Blasco M.A. (2008). A mammalian microRNA cluster controls DNA methylation and telomere recombination via Rbl2-dependent regulation of DNA methyltransferases. *Nature Structural & Molecular Biology*. 2008 Mar;15(3):268-79. doi: 10.1038/nsmb.1399. Epub 2008 Mar 2. Erratum in: Nat Struct Mol Biol. 15(9):998. PMID: 18311151; PMCID: PMC2990406.

Bhattacharya, A., & Roychoudhury, A. (2024). Plant non-coding RNAs and their regulation in salinity and heavy metal stress. In S. Penna, V. Kumar, & A. K. Srivastava (Eds.), *Non-Coding RNAs Molecular Tools for Crop Improvement* (pp. 128–147). CRC Press (Taylor & Francis group; Milton Park, Abingdon-on-Thames, United Kingdom).

Bratkovič, T., Božič, J., & Rogelj, B. (2020). Functional diversity of small nucleolar RNAs. *Nucleic Acids Research*, 48(4), 1627–1651. https://doi.org/10.1093/nar/gkz1140.

Bure, I. V., Nemtsova, M. V., & Kuznetsova, E. B. (2022a). Histone modifications and non-coding RNAs: Mutual epigenetic regulation and role in pathogenesis. *International Journal of Molecular Sciences*, 23(10). https://doi.org/10.3390/ijms23105801.

Burenina, O. Y., Oretskaya, T. S., & Kubareva, E. A. (2017). Non-coding RNAs as transcriptional regulators in eukaryotes. *Acta Naturae*, 9(4), 13–25.

Chakraborty, S., & Roychoudhury, A. (2023). MicroRNA-mediated regulation of heavy metal stress in plants. In D. Pandita & A. Pandita (Eds.), *Plant MicroRNAs and Stress Response* (pp. 183–197). CRC Press (Taylor and Francis group).

Chand Jha, U., Nayyar, H., Mantri, N., Siddique, K. H. M., & Figlerowicz, M. (2021). Non-coding RNAs in legumes: Their emerging roles in regulating biotic/abiotic stress responses and plant growth and development. *Mdpi.Com*. https://doi.org/10.3390/cells10071674.

Dey B. K., Pfeifer K., Dutta A. (2014). The H19 long noncoding RNA gives rise to microRNAs miR-675-3p and miR-675-5p to promote skeletal muscle differentiation and regeneration. *Genes & Development*, 2014 Mar 1;28(5):491–501. doi: 10.1101/gad.234419.113.

Engreitz, J., Ollikainen, N., Guttman, M. (2016). Long non-coding RNAs: Spatial amplifiers that control nuclear structure and gene expression. *Nature Reviews Molecular Cell Biology*, 17, 756–770.

Fahad, M., Tariq, L., Muhammad, S., & Wu, L. (2024). Underground communication: Long non-coding RNA signaling in the plant rhizosphere. *Plant Communications*, 5(7), 100927. https://doi.org/10.1016/j.xplc.2024.100927.

Hasanuzzaman, M., & Fujita, M. (2022). Plant responses and tolerance to salt stress: Physiological and molecular interventions. *International Journal of Molecular Sciences*, 23(9), 4810. https://doi.org/10.3390/ijms23094810.

Heydari, R., Karimi, P., & Meyfour, A. (2024). Long non-coding RNAs as pathophysiological regulators, therapeutic targets and novel extracellular vesicle biomarkers for the diagnosis of inflammatory bowel disease. *Biomedicine & Pharmacotherapy*, 176, 116868. https://doi.org/10.1016/j.biopha.2024.116868.

Huang, Z., Du, Y., Wen, J., Lu, B., & Zhao, Y. (2022). snoRNAs: Functions and mechanisms in biological processes, and roles in tumor pathophysiology. *Cell Death Discovery*, 8(1), 259. https://doi.org/10.1038/s41420-022-01056-8.

Hung, F.-Y., Chen, C., Yen, M.-R., Hsieh, J.-W. A., Li, C., Shih, Y.-H., Chen, F.-F., Chen, P.-Y., Cui, Y., & Wu, K. (2020). The expression of long non-coding RNAs is associated with H3Ac and H3K4me2 changes regulated by the HDA6-LDL1/2 histone modification complex in *Arabidopsis*. *NAR Genomics and Bioinformatics*, 2(3). https://doi.org/10.1093/nargab/lqaa066.

Ilieva, M. S. (2024). Non-Coding RNAs in neurological and neuropsychiatric disorders: Unraveling the hidden players in disease pathogenesis. *Cells*, 13(12). https://doi.org/10.3390/cells13121063.

Ji, J., Zeng, Y., Zhang, S., Chen, F., Hou, X., & Li, Q. (2023). The miR169b/NFYA1 module from the halophyte Halostachys caspica endows salt and drought tolerance in *Arabidopsis* through multi-pathways. *Frontiers in Plant Science*, 13. https://doi.org/10.3389/fpls.2022.1026421.

Jiang, J., Zhu, H., Li, N., Batley, J., & Wang, Y. (2022). The miR393-target module regulates plant development and responses to biotic and abiotic stresses. *International Journal of Molecular Sciences*, 23(16), 9477. https://doi.org/10.3390/ijms23169477.

Kalantry, S., Magnuson, T., & 2006, undefined. (2006). The Polycomb group protein EED is dispensable for the initiation of random X-chromosome inactivation. *Journals.Plos.Org*, 2(5), 656–664. https://doi.org/10.1371/journal.pgen.0020066.

Kim, D. H., Saetrom, P., Snøve, O Jr. and Rossi, J. J. (2008). MicroRNA-directed transcriptional gene silencing in mammalian cells. *Proceedings of the National Academy of Sciences USA*, 105: 16230–16235

Kim, K., Jutooru, I., Chadalapaka, G., Johnson, G., Oncogene, J. F. (2013). HOTAIR is a negative prognostic factor and exhibits pro-oncogenic activity in pancreatic cancer. *Oncogene*, 32, 1616–1625.

Kumar, S., Gonzalez, E. A., Rameshwar, P., & Etchegaray, J.-P. (2020). Non-coding RNAs as mediators of epigenetic changes in malignancies. *Cancers*, 12(12), 3657. https://doi.org/10.3390/cancers12123657.

Kumar, D., Sahoo, S. S., Chauss, D., Kazemian, M., & Afzali, B. (2023). Non-coding RNAs in immunoregulation and autoimmunity: Technological advances and critical limitations.

Journal of Autoimmunity, 134, 102982. https://doi.org/10.1016/j.jaut.2022.102982.

Kuo F. C., Neville M. J., Sabaratnam R., Wesolowska-Andersen A., Phillips D., Wittemans L. B. L., van Dam A. D., Loh N. Y., Todorčević M., Denton N., Kentistou K. A., Joshi P. K., Christodoulides C., Langenberg C., Collas P., Karpe F., Pinnick K. E. (2022). HOTAIR interacts with PRC2 complex regulating the regional preadipocyte transcriptome and human fat distribution. *Cell Reports*. 40(4):111136. doi: 10.1016/j.celrep.2022.111136.

Li, D., Yang, Y., Li, Y., Zhu, X., & Li, Z. (2021a). Epigenetic regulation of gene expression in response to environmental exposures: From bench to model. *Science of The Total Environment*, 776, 145998. https://doi.org/10.1016/j.scitotenv.2021.145998.

Li, J., Duan, Y., Sun, N., Wang, L., Feng, S., Fang, Y., & Wang, Y. (2021b). The miR169n-NF-YA8 regulation module involved in drought resistance in *Brassica napus* L. *Plant Science*, 313, 111062. https://doi.org/10.1016/j.plantsci.2021.111062.

Li W. X., Oono Y., Zhu J., He X. J., Wu J.M., Iida K., Lu X. Y., Cui X., Jin H., Zhu J. K (2008). The *Arabidopsis* NFYA5 transcription factor is regulated transcriptionally and post-transcriptionally to promote drought resistance. Plant Cell. (8):2238-51. doi: 10.1105/tpc.108.059444. Epub 2008 Aug 5. PMID: 1868–2547; PMCID: PMC2553615.

Lin, L.-M., Pan, Q., Sun, Y.-M., & Wang, W.-T. (2021). Small nucleolar RNA is potential as a novel player in leukemogenesis and clinical application. *Blood Science*, 3(4), 122–131. https://doi.org/10.1097/BS9.0000000000000091.

Liu, T., Zhang, H., Li, Y., Cheng, L., Ouyang, Y., Li, S., ... & Zhang, J. R. (2016). A Role for *Arabidopsis* miR399f in Salt, Drought, and ABA Signaling. *Frontiers in Plant Science*, 7, 182.

Ludwig, B., & Dwivedi, Y. (2016). Dissecting bipolar disorder complexity through epigenomic approach. *Molecular Psychiatry*, 21(11), 1490–1498. https://doi.org/10.1038/mp.2016.123.

Mangiavacchi, A., Morelli, G., & Orlando, V. (2023). Behind the scenes: How RNA orchestrates the epigenetic regulation of gene expression. *Frontiers in Cell and Developmental Biology*, 11. https://doi.org/10.3389/fcell.2023.1123975.

Matsui, A., & Seki, M. (2019a). The involvement of long noncoding RNAs in response to plant stress. *Methods in Molecular Biology*, 1933, 151–171. https://doi.org/10.1007/978-1-4939-9045-0_8.

Mattick, J. S., Amaral, P. P., Carninci, P., Carpenter, S., Chang, H. Y., Chen, L.-L., Chen, R., Dean, C., Dinger, M. E., Fitzgerald, K. A., Gingeras, T. R., Guttman, M., Hirose, T., Huarte, M., Johnson, R., Kanduri, C., Kapranov, P., Lawrence, J. B., Lee, J. T., … Wu, M. (2023). Long non-coding RNAs: Definitions, functions, challenges and recommendations. *Nature Reviews Molecular Cell Biology*, 24(6), 430–447. https://doi.org/10.1038/s41580-022-00566-8.

Maurano, M. T., Humbert, R., Rynes, E., Thurman, R. E., Haugen, E., Shafer, L. P., ... & Stamatoyannopoulos, J. A. (2012). Systematic localization of common disease-associated variation in regulatory DNA. *Science*, 337(6099), 1190–1195.

Muhammad, I. I., Kong, S. L., Akmar Abdullah, S. N., & Munusamy, U. (2019). RNA-seq and ChIP-seq as complementary approaches for comprehension of plant transcriptional regulatory mechanism. *International Journal of Molecular Sciences*, *21*(1), 167.

Nielsen, M., Menon, G., Zhao, Y., Mateo-Bonmati, E., Wolff, P., Zhou, S., Howard, M., & Dean, C. (2024). *COOLAIR* and PRC2 function in parallel to silence *FLC* during vernalization. *Proceedings of the National Academy of Sciences*, 121(4). https://doi.org/10.1073/pnas.2311474121.

Pegler, J. L., Oultram, J. M. J., Grof, C. P. L., & Eamens, A. L. (2020). Molecular manipulation of the miR399/PHO2 expression module alters the salt stress response of *Arabidopsis thaliana*. *Plants*, 10(1), 73. https://doi.org/10.3390/plants10010073.

Pisignano, G., & Ladomery, M. (2021). Epigenetic regulation of alternative splicing: How LncRNAs tailor the message. *Non-Coding RNA*, 7(1), 21. https://doi.org/10.3390/ncrna7010021.

Samynathan, R., Venkidasamy, B., Shanmugam, A., Ramalingam, S., & Thiruvengadam, M. (2023). Functional role of microRNA in the regulation of biotic and abiotic stress in agronomic plants. *Frontiers in Genetics*, 14. https://doi.org/10.3389/fgene.2023.1272446.

Shaath, H., Vishnubalaji, R., Elango, R., Kardousha, A., Islam, Z., Qureshi, R., Alam, T., Kolatkar, P. R., & Alajez, N. M. (2022). Long non-coding RNA and RNA-binding protein interactions in cancer: Experimental and machine learning approaches. *Seminars in Cancer Biology*, 86, 325–345. https://doi.org/10.1016/j.semcancer.2022.05.013.

Singh, A., & Roychoudhury, A. (2023). RNA Interference (RNAi) Technology: An effective tool in plant breeding. In A. Raina, M. R. Wani, R. A. Laskar, N. Tomlekova, & S. Khan (Eds.), *Advanced Crop Improvement, Volume 1* (pp. 309–320). Springer Nature Switzerland AG.

Splinter, E., De Wit, E., Ge, E., Nora, P., Klous, P., Van De Werken, H. J. G., Zhu, Y., Kaaij, L. J. T., Van Ijcken, W., Gribnau, J., Heard, E., & De Laat, W. (2011). The inactive X chromosome adopts a unique three-dimensional conformation that is dependent on Xist RNA. *Genes and Development*, 25, 1371–1383.

Statello, L., Guo, C.-J., Chen, L.-L., & Huarte, M. (2021). Gene regulation by long non-coding RNAs and its biological functions. *Nature Reviews Molecular Cell Biology*, 22(2), 96–118. https://doi.org/10.1038/s41580-020-00315-9.

Tassinari, M., Richter, S. N., & Gandellini, P. (2021). Biological relevance and therapeutic potential of G-quadruplex structures in the human noncoding transcriptome. *Nucleic Acids Research*, 49(7), 3617–3633. https://doi.org/10.1093/nar/gkab127.

Thouly, C., Masson, M. Le, Lai, X., Carles, C., Genes, G. V. (2020). Unwinding BRAHMA functions in plants. *Genes*, 11(1), 90 https://doi.org/10.3390/genes11010090.

Tiwari, B., Habermann, K., Arif, M. A., Weil, H. L., Garcia-Molina, A., Kleine, T., Mühlhaus, T., & Frank, W. (2020). Identification of small RNAs during cold acclimation in *Arabidopsis thaliana*. *BMC Plant Biology*, 20(1), 298. https://doi.org/10.1186/s12870-020-02511-3.

Tripathi, V., Ellis, J. D., Shen, Z., Song, D. Y., Pan, Q., Watt, A. T., Freier, S. M., Bennett, C. F., Sharma, A., Bubulya, P. A., Blencowe, B. J., Prasanth, S. G., & Prasanth, K. V. (2010). The nuclear-retained noncoding RNA MALAT1 regulates alternative splicing by modulating SR splicing factor phosphorylation. *Molecular Cell*, 39(6), 925–938. https://doi.org/10.1016/J.MOLCEL.2010.08.011.

Tsai, M. C., Manor, O., Wan, Y., Mosammaparast, N., Wang, J. K., Lan, F., Shi, Y., Segal, E., & Chang, H. Y. (2010). Long noncoding RNA as modular scaffold of histone modification complexes. *Science*, 329(5992), 689–693. https://doi.org/10.1126/SCIENCE.1192002.

Verhoeff, T. J., Holloway, A. F., & Dickinson, J. L. (2023). Non-coding RNA regulation of integrins and their potential as therapeutic targets in cancer. *Cellular Oncology*, 46(2), 239–250. https://doi.org/10.1007/s13402-022-00752-y.

Waheed, S., Anwar, M., Saleem, M. A., Wu, J., Tayyab, M., & Hu, Z. (2021). The critical role of small RNAs in regulating plant innate immunity. *Biomolecules*, 11(2), 184. https://doi.org/10.3390/biom11020184.

Wang, D., Cao, W., Yang, W., Jin, W., Luo, H., Niu, X., & Gong, J. (2022). Pancan-MNVQTLdb: Systematic identification of multi-nucleotide variant quantitative trait loci in 33 cancer types. *NAR Cancer*, 4(4). https://doi.org/10.1093/narcan/zcac043.

Wei, J.-W., Huang, K., Yang, C., & Kang, C.-S. (2017). Non-coding RNAs as regulators in epigenetics. *Oncology Reports*, 37(1), 3–9. https://doi.org/10.3892/or.2016.5236.

Yadav, A., Mathan, J., Dubey, A., RNA, A. S.-N., & 2024, undefined. (2024). The emerging role of non-coding RNAs (ncRNAs) in plant growth, development, and stress response signaling. *Mdpi.Com*. https://doi.org/10.3390/ncrna10010013.

Yang, K., Zeng, L., Ge, A., Wang, S., Zeng, J., Yuan, X., Mei, Z., Wang, G., & Ge, J. (2022a). A systematic review of the research progress of non-coding RNA in neuroinflammation and immune regulation in cerebral infarction/ischemia-reperfusion injury. *Frontiers in Immunology*, 13, 930171. https://doi.org/10.3389/fimmu.2022.930171.

Yao, Q., Chen, Y., & Zhou, X. (2019). The roles of microRNAs in epigenetic regulation. *Current Opinion in Chemical Biology*, 51, 11–17. https://doi.org/10.1016/j.cbpa.2019.01.024.

Yao, Z., Yang, Y., Sun, M., He, Y., Liao, L., Chen, K., & Li, B. (2022). New insights into the interplay between long non-coding RNAs and RNA-binding proteins in cancer. *Cancer Communications*, 42(2), 117–140. https://doi.org/10.1002/cac2.12254.

Zhakypbek, Y., Belkozhayev, A. M., Kerimkulova, A., Kossalbayev, B. D., Murat, T., Tursbekov, S., Turysbekova, G., Tursunova, A., Tastambek, K. T., & Allakhverdiev, S. I. (2025). MicroRNAs in plant genetic regulation of drought tolerance and their function in enhancing stress adaptation. *Plants*, 14(3), 410. https://doi.org/10.3390/plants14030410.

Zhang, X., Wang, W., Zhu, W., Dong, J., Cheng, Y., Yin, Z., & Shen, F. (2019). Mechanisms and functions of long non-coding RNAs at multiple regulatory levels. *International Journal of Molecular Sciences*, 20(22), 5573. https://doi.org/10.3390/ijms20225573.

Zhang, D., Zhang, D., Zhang, Y., Li, G., Sun, D., Zhou, B., & Li, J. (2024). Insights into the epigenetic basis of plant salt tolerance. *International Journal of Molecular Sciences*, 25(21), 11698. https://doi.org/10.3390/ijms252111698.

Zhou, S., Yu, X., Wang, M., Meng, Y., Song, D., Yang, H., Wang, D., Bi, J., & Xu, S. (2021). Long non-coding RNAs in pathogenesis of neurodegenerative diseases. *Frontiers in Cell and Developmental Biology*, 9. https://doi.org/10.3389/fcell.2021.719247.

5 CRISPR/Cas-Based Epigenetic Regulation in Plants

Dhanawantari L. Singha, Abhik Gogoi, Ankita Gogoi, Nikha Borah, Arpita Talukdar, Ratul Nath and Channakeshaviah Chikkaputtaiah

5.1 INTRODUCTION TO EPIGENETIC REGULATION IN PLANTS

Plant epigenetics has a profound and well-established background. The term "epigenetics" was introduced in 1940 by Conrad Waddington (1905–1975), a British developmental biologist and embryologist. It refers to the study of heritable changes in gene expression that occur without alterations to the DNA sequence. This mechanism plays a crucial role in regulating gene activity, influencing plant development, stress responses and adaptability to environmental challenges (Banerjee & Roychoudhury, 2017).

Plant stress poses a considerable challenge that hinders growth and productivity, resulting in detrimental environmental conditions that alter normal physiological processes and inhibit plant survival. The mechanisms underlying epigenetic transgenerational stress adaptation and the potential advantages plants may gain from their retained stress memory can be elucidated through essential ecologically driven experimental designs that involve appropriate plant species and stress conditions (Hewezi, 2018). One of the most important ways that plants react and adjust to stress is through epigenetic control. Although DNA methylation (DM) and histone modification are identified as having a prominent effect on gene expression, epigenetic research frequently ignores their effects on the three-dimensional chromosome structure. Itabashi et al. (2018) also highlighted the significance of epigenetic modulation of some agriculturally significant traits such as self-incompatibility, hybrid vigor and vernalization. A single genome within a plant cell can produce several epigenomes (the chemical markers that alter cell DNA and DNA-associated proteins) in response to growth and environmental signals that may initiate epigenetic alterations in the genome. Studies have shown that several enzymes, such as the DNA methyltransferase DRM2, are involved in stress-triggered changes in DNA methylation and have been observed in stress-induced epigenetic modifications. DNA methylation, non-coding RNAs and histone modifications are key epigenetic mechanisms that significantly influence gene expression and the development of plants. Various molecular techniques can be employed to study and modify epigenetic regulations in the plant genome. Among these, CRISPR/Cas-mediated genome editing stands out as a leading-edge technology for precise and efficient modifications in crops.

5.2 CRISPR/Cas SYSTEMS

Genome editing is a method used to precisely and efficiently alter the genetic code within the cellular environment by modifying its DNA, including insertion, deletion or replacement of genetic sequences. Among the widely used genome editing techniques, the clustered regularly interspaced short palindromic repeats (CRISPR)/Cas (CRISPR-associated protein) is the most recently developed. It stands out for its higher efficiency, versatility, cost-effectiveness and time-saving advantages (Kumar & Jain, 2015; Banerjee & Roychoudhury, 2023). The technology comprises the Cas9 enzyme that, when paired with a guide RNA (gRNA) that is devised to match a particular DNA sequence in the plant genome, cuts the DNA at the precise position, enabling changes to be fixed through the innate repair systems of the cells (Liu et al., 2017). CRISPR/Cas systems are of two classes – Class 1 (types I, III, IV), which uses multiple Cas proteins, and Class 2 (types II, V, VI), which includes 6 types and 33 subtypes and utilizes only a single effector protein with multiple domains (Hryhorowicz et al., 2023). By altering the DNA sequences that are already present in the non-transgenic plants, CRISPR can be used to level up the immunological defenses of plants. Additionally, it can introduce new defense systems from other organisms into transgenic plants or even swap defense proteins across different plant species. Mutations due to alterations may help to improve resistance to abiotic stresses like drought, high temperature, salt and nutrient shortages or deficiencies. This advanced method can also help agricultural development initiatives by producing new, valuable features in crops (Cardi, 2016). Soybeans, potatoes, flax seed, tomatoes, rapeseed, cotton, chrysanthemums, etc. are some of the crops whose DNA has been altered so far using the CRISPR/Cas9 system to change consumer-preferred traits that are significant for commercial use (Tang et al., 2021).

5.3 MECHANISMS OF EPIGENETIC REGULATION IN PLANTS

Epigenetic regulation in plants controls gene expression without altering DNA sequences, ensuring adaptability to environmental cues. DNA methylation, primarily at cytosine residues, represses gene activity by recruiting repressor

DOI: 10.1201/9781003589914-6

proteins or inhibiting transcription factors. Noncoding RNAs, including microRNAs (miRNAs) and small interfering RNAs (siRNAs), guide gene silencing via RNA-directed DNA methylation (RdDM) or post-transcriptional repression. Chromatin remodeling, mediated by histone modifications (e.g., acetylation and methylation), alters chromatin structure to regulate gene accessibility. These mechanisms interplay dynamically, enabling plants to respond to stress, developmental signals and environmental changes while maintaining genome stability and transgenerational inheritance of epigenetic marks. Table 5.1 represents the role of epigenetic processes in plant response to abiotic stresses.

5.3.1 DNA Methylation

DNA methylation (DM) in plants serves as a key epigenetic modification that has been extensively studied. This process typically involves adding a methyl group to the cytosine residues, predominantly at CpG dinucleotides. Stress-induced shifts in DM patterns can lead to significant changes in gene expression and phenotypic adaptation. Recent studies underline the dynamic nature of DM, showing its responsiveness to a wide range of environmental challenges (Abdulraheem et al., 2024). In plants, extensive methylation occurs on cytosine residues, facilitated by enzymes called DNA methyltransferases. Conversely, demethylation of methylated DNA is carried out by demethylases. This finely regulated methylation landscape plays a crucial role in both plant development and evolution (Lucibelli et al., 2022). Developmental defects in plants can result from abnormal DM. This process take place at CHH, CG and CHG sequences (H representing adenine, thymine or cytosine) and involves methylation of cytosine (5-mC) and adenine (6-mA). DNA methylation, which is highly concentrated in transposable elements, is crucial for suppressing their expression via transcriptional gene silencing (TGS). The presence of DM in gene regulatory regions further enhances TGS. Moreover, the methylation of repetitive sequences and intronic TEs has been shown to affect mRNA splicing mechanisms (Jogam et al., 2022; Karanthamalai et al., 2020). DM in plants involves three key processes: maintenance DNA methylation, which ensures the methylation of partially methylated symmetrical sequences; de novo DM, which introduces methylation to cytosine that were previously unmethylated; and DNA demethylation, which reverses the state of methylation. The plant methylome is predominantly preserved during the replication of DNA and cell division through the action of DNA methyltransferases, encompassing both maintenance and de novo methylases (Huang et al., 2010; Sun et al., 2022). Research indicates that distinct enzymes are involved in maintaining methylation in different contexts: CHROMOMETHYLASE3 (CMT3) preserves CHG methylation, MET1 DNA methyltransferase maintains CG methylation and DOMAINS REARRANGED METHYLTRANSFERASES (DRM1/DRM2) or CMT2 are responsible for CHH methylation (Du et al., 2012; Thiebaut et al., 2019). Cytosine methylation, both symmetric (mCpG and mCpHpG) and asymmetric (mCpHpH), is linked to chromatin repression in gene promoters, suppressing gene transcription. De novo methylation is catalyzed by the enzymes DRM1 and DRM2, while symmetric methylation of CG and CHG is maintained by MET1 and CMT3, respectively. Recent findings suggest that de novo methylation may also be performed by MET1 and CMT3; DRM1 and 2 play roles in maintaining symmetric methylation (Chinnusamy & Zhu, 2009). In plants, siRNAs mediate the establishment of de novo methylation, targeting cytosines within CG, CHG or CHH motifs. Symmetrical sites, such as CG and CHG motifs, allow maintenance methylation to preserve existing patterns, as their 5′ to 3′ sequences match on both the DNA strands. This symmetry facilitates the transfer of methylation marks to daughter strands during DNA replication and ensures that newly synthesized DNA is methylated after repair, maintaining consistency in patterns of methylation (Pikaard & Mittelsten Scheid, 2014). Demethylation of DNA can take place passively through the dilution of methylation marks during replication or actively via chemical modifications of 5-methylcytosine (5-mC), such as oxidation or deamination. These changes are resolved through nucleotide replacement by base excision repair (BER) or alternative repair pathways. Recent discoveries emphasize the role of enzymes like base-editing glycosylases, which may either work alongside or replace thymine DNA glycosylase in this demethylation process (Akhter et al., 2021).

5.3.2 Histone Modifications

Histone modifications predominantly take place on the N-terminal tails of histones, though some are present within the nucleosome core regions. These modifications, denoted as the "histone code", play a crucial role in defining chromatin functions and regulating transcriptional activity. Common types of histone modifications include ubiquitination, methylation, acetylation and phosphorylation (Zhao et al., 2019). Research initially focused on identifying enzymes acting as "writers" and "erasers" of histone modifications. Chemical groups are added to histone tails or core domains by writers, such as acetyltransferases (HATs), methyltransferases, kinases and ubiquitinases. In contrast, these modifications are removed by erasers, including deacetylases (HDACs), phosphatases, demethylases (HDMs) and deubiquitinases. Arginine and lysine residues are the primary sites for histone methylation, leading to enhanced basicity and hydrophobicity of histone tails. This modification alters the binding affinity between transcription factors and DNA, resulting in either the activation or repression of gene expression (Xu et al., 2017). By neutralizing the positive charge of histones, the acetylation of H3 and H4 facilitates increased DNA accessibility. Histone post-translational modifications (PTMs) are dynamic and regulated by various enzymes that add or erase specific marks, influencing chromatin structure and gene regulation. PTMs can either facilitate transcription by

TABLE 5.1
Role of epigenetic processes in plant response to abiotic stresses

Epigenetic process	Abiotic stress	Plants	Mechanisms/responses	References
DNA methylation	Salt stress	*Oryza sativa*	Salt-tolerant rice (Pokkali) adjusts DNA methylation under salt stress via increased demethylase activity, unlike the less adaptive IR-29. Epigenetic regulator mutations affect salinity tolerance traits like root length and biomass.	Ferreira et al. (2015)
Histone modification	Drought stress	*Arabidopsis thaliana*	The histone demethylase JMJ17 regulated dehydration stress responses by demethylating H3K4me1/2/3 and modulating gene expression. Loss of JMJ17 enhanced ABA sensitivity, highlighting its role in stress adaptation.	Huang et al. (2019)
DNA methylation/ histone modification	Temperature stress (high/ low)	*Hordeum vulgare*	High temperature (HT) stress in barley seedlings caused greater growth damage, increased DNA demethylation and higher H3K9ac and H3K4me3 levels compared to mild low temperature (MLT), leading to chromatin loosening and gene activation. HT also induced more DEGs than MLT.	Li et al. (2022)
DNA methylation	Cd stress	*A. thaliana*	Cd stress suppressed DNA demethylase genes and increased genome-wide DNA methylation, enhancing Fe nutrition through the upregulation of genes involved in Fe uptake, thereby increasing Cd tolerance in the plant	Fan et al. (2020)
DNA methylation	Salt stress	*Oryza sativa*	IL177-103 shows improved salinity tolerance with better physiological traits and stable, root-specific DNA methylation changes, including genotype-specific epigenetic markers as compared to IR-64.	Wang et al. (2015)
Histone modification	Drought stress	*A. thaliana*	The *Arabidopsis* trithorax-like factor ATX1 trimethylates histone H3 at lysine 4 (H3K4me3). It regulates dehydration stress by promoting ABA biosynthesis. Loss of ATX1 function reduced germination, stomatal control and stress tolerance.	Ding et al. (2011)
Histone modification	Wounding, salt stress	*Brassica rapa*	Histone H3 acetylation levels increased under salt and wounding stresses, as shown by immunoblot analysis, and were linked to BraHAT expression and HAT-HDAC interactions.	Eom & Hyun (2018)
Histone modification	Drought stress	*Gossypium hirsutum*	GhHDT4D, a histone deacetylase in cotton, enhances drought tolerance by suppressing GhWRKY33 transcription via reduced H3K9 acetylation.	Zhang et al. (2020)
DNA methylation	Cd toxicity	*Oryza sativa*	Cd exposure triggered DNA and H3K9 demethylation at the *OsZIP1* locus, which activated the expression of *OsZIP1*. This gene encoded a metal-detoxification transporter that prevented the excessive accumulation of Zn, Cu and Cd in rice.	Liu et al. (2019)
lncRNA	Drought stress	*Oryza sativa*	A transcriptome analysis revealed 238 lncRNAs linked to drought response memory under water-deficit stress in rice.	Li et al. (2019)
Non-coding RNAs	Drought stress	*Oryza sativa*	In drought-stressed rice, 191 lncRNAs, 2115 mRNAs and 32 miRNAs were differentially expressed. Functional analysis showed their involvement in hormone signaling, chlorophyll synthesis, protein synthesis and more.	Yang et al. (2022)
siRNA	Heat, NaCl and dehydration stress	*Triticum aestivum*	Northern blot analysis revealed that the expressions of siRNAs 002061_0636_3054.1 and 005047_0654_1904.1 were strongly downregulated by heat, NaCl and dehydration stress.	Yao et al. (2010)
Histone acetylation	Salt stress	*Zea mays*	Upregulation of cell wall-related genes (e.g., *ZmEXPA1, ZmEXPB2* and *ZmXET1*) and increased histone acetylation levels (H3K9, H4K5) were observed. Epigenetic regulation of cell wall genes under salt stress promoted adaptive root swelling.	Li et al. (2014)
DNA methylation	Cold stress	*Cicer arietinum* L.	Under cold stress, tolerant genotypes of chickpea showed higher demethylation, activating stress-responsive genes, alongside enhanced antioxidant activity and reduced damage, aiding cold tolerance.	Rakei et al. (2016)
DNA methylation	Drought, salt stress	*Oryza sativa* (IR-64, Nagina 22, Pokkali)	Whole genome bisulfite sequencing in rice cultivars revealed DNA methylation differences, with differentially methylated regions (DMRs) linked to the expression of stress-related genes and transposon influence.	Garg et al. (2015)

promoting transcription factor recruitment and chromatin accessibility or be guided by transcription processes like histone exchange during elongation. However, the mechanisms underlying the inheritance of histone PTM patterns during DNA replication are unclear, and not all PTMs are consistently conveyed to daughter cells once the initiating signals are removed (Lauria & Rossi, 2011). Histone modifications, like acetylation and methylation at specific lysine residues on histone 3 (H3) like H3K4, H3K9 and H3K27, are critical for gene regulation. Acetylation marks, including H3K9ac and H3K27ac, are correlated with gene activation, whereas methylation marks like H3K9me3 and H3K27me3 generally lead to transcriptional repression. In contrast, H3K4me3 is linked to transcriptional activation. Recent studies have uncovered the role of modifications like H3K4me2, H3K4me3, H3K9ac and various H4 acetylation marks in improving plant resilience to environmental stresses (Kumar & Mohapatra, 2024). Histone modifications are core regulators of plant responses to biotic and abiotic stresses.

5.3.3 Non-coding RNAs (ncRNAs)

ncRNAs are RNAs that do not have protein-coding ability and were once believed to function solely in post-transcriptional gene regulation. Notably, an increasing body of evidence highlights their vital role in epigenetic regulation, underscoring the central role of RNA in gene expression control (Wei et al., 2017; Bhattacharya & Roychoudhury, 2024). ncRNAs are categorized into long ncRNAs (greater than 200 nucleotides) and short ncRNAs (less than 200 nucleotides), with short ncRNAs including miRNAs, siRNAs and piRNAs (Heo et al., 2013). Moreover, research in plants has revealed that certain ncRNAs can move systemically over long distances, facilitating actions that span from post-transcriptional silencing to epigenetic alterations (Ramirez-Prado et al., 2017). lncRNAs exhibit significant structural and functional diversity, with their defining traits being their length and lack of protein-coding ability. Although their sequence conservation is generally low, their conserved locations relative to protein-coding genes hint at functional importance. Processed similarly to mRNAs, lncRNAs are present in both the nucleus and cytosol, contributing to epigenomic regulation (Peschansky & Wahlestedt, 2014). ncRNAs are essential in regulating gene expression in complex organisms by guiding cytosine methylation and histone modifications. They play a key role in various epigenetic processes, such as silencing, X-chromosome inactivation, transposon activity and paramutation. Noncoding genes produce miRNAs via RNA polymerase II. Pri-miRNA is processed by the Drosha-DGCR8 complex into pre-miRNA, which is transported to the cytoplasm via the exportin-5/Ran-GTP complex. In the cytoplasm, Dicer and TRBP convert the pre-miRNA into a miRNA/miRNA duplex. There, Dicer and TRBP convert the pre-miRNA into a miRNA/miRNA duplex. The mature miRNA is loaded onto the RISC complex, guiding it to regulate target mRNAs through cleavage or translational repression, while the miRNA duplex is typically degraded. These miRNAs control gene expression (Ali & Tang, 2024). Long noncoding RNAs (lncRNAs) originate from intronic, exonic and intergenic regions, with their numbers varying widely among plant species; for example, *Arabidopsis* has 2986 lncRNAs, while maize has 18,031. Exon-derived lncRNAs evolve at a pace that is slower than intergenic lncRNAs but faster than protein-coding sequences and mRNA UTRs. Despite sequence divergence among species, lncRNAs exhibit positional conservation and share splicing signals like "GT-AG" with protein-coding genes (Hou et al., 2019). lncRNAs facilitate the inheritance and transmission of stress-induced epigenetic modifications in plants through their involvement in epigenetic memory. They are also involved in managing hormone production (e.g., ABA and ethylene), alternative splicing and signaling pathways, highlighting their central role in plant biology.

5.3.4 Chromatin Remodeling

Chromatin dynamics in plants plays a pivotal role in cell cycle progression and is meticulously controlled by developmental cues and environmental signals. The chromatin structure is fundamental to key processes like genome replication, DNA repair, recombination and transcriptional silencing. Furthermore, its regulation is necessary for accurate transcriptional activation (Kim, 2019; Vergara & Gutierrez, 2017). The development of plants relies on tightly regulated spatiotemporal gene expression patterns, where chromatin remodeling is pivotal in their establishment and in maintaining epigenetic transcriptional states through mitotic divisions. Evolutionarily conserved chromatin remodeling complexes (CRCs) control chromatin structure by modifying nucleosome composition and interactions. A distinguishing feature of all CRCs is the presence of an ATPase/helicase catalytic core from the SWI2/SNF2 family, part of the broader SF2 helicase family. Named after the first discovered SWI/SNF CRC, these complexes harness energy from ATP hydrolysis to influence nucleosome positioning, control DNA accessibility and determine the incorporation of variant histone subspecies (Narlikar et al., 2002; Ojolo et al., 2018).

Chromatin remodeling machines, such as SWI/SNF, CHRAC and NURF, disrupt and reorganize nucleosome-DNA interactions in organisms like Drosophila, yeast and mammals. Powered by DNA-dependent ATPases of the SWI2/SNF2 superfamily, these complexes often function as transcriptional coactivators, facilitating transcription factor access to nucleosome-packaged DNA. However, some complexes negatively regulate transcription or assist in other processes like chromatin assembly. Chromatin remodeling enhances chromatin fluidity, reducing the activation energy required for transitions between chromatin states (Verbsky & Richards, 2001).

5.4 CRISPR/Cas9 AND ITS ROLE IN GENOME EDITING

The clustered regularly interspaced short palindromic repeats/CRISPR-associated protein 9 (CRISPR/Cas9) system is the leading technique and has garnered significant attention in the field of genome editing, which has become a cutting-edge area of molecular biology (Rao et al., 2022). The remarkable precision and flexibility of CRISPR/Cas9 have sparked a revolution in genome editing. It has many uses in crop growth, plant science, human gene therapy and medical research. In the CRISPR/Cas9 system, the Cas9 protein and guide RNA (gRNA) are the primary components. The gRNA directs Cas9 to the target DNA sequence, where Cas9 then performs its role as a molecular scissor, cleaving the DNA at specific sites (Figure 5.1). There are multiple crucial steps in the CRISPR/Cas9 genome editing process. The target DNA sequence is the focus of the Cas9-gRNA complex, which causes a double-stranded DNA break. After the DNA is cut, cellular repair processes utilize either the homology-directed repair (HDR) pathway or the non-homologous end joining (NHEJ) pathway to introduce alterations or mutations at the target site. The precise and programmable nature of CRISPR/Cas9 allow for exact genome alterations, turning it into a useful tool for basic research, medicine and agricultural improvement (Li et al., 2023).

The CRISPR/Cas9 system is a groundbreaking advancement in genome editing, driven by the association of its key elements: the Cas9 endonuclease and the gRNA. The gRNA, a short 20-nucleotide sequence, ensures precise targeting by interacting with DNA through the protospacer adjacent motif (PAM) (Asmamaw & Zawdie, 2021). Three base pairs ahead of the PAM sequence, the gRNA strategically positions a precise double-strand cleavage at the target DNA in cooperation with the Cas9 endonuclease. Usually repaired by NHEJ, this break frequently causes insertion or deletion (indel) alterations that impair gene function (Xue & Greene, 2021). Researchers have successfully used CRISPR/Cas9 to precisely modify the genomes across multiple organisms, leading to significant progress and a better understanding of its functioning. Its uses span several fields, most notably agronomic trait enhancement (Hossain, 2021).

5.4.1 Variants of CRISPR/Cas Systems

CRISPR/Cas systems are essential genome-editing tools used across various organisms, from bacteria to humans. They are classified into two classes: Class 1, with multi-Cas protein effector complexes (Types I, III and IV), and Class 2, featuring a single effector protein (Types II, V and VI) (Mohanraju et al., 2016). Type I, II and V systems can detect and cleave DNA; RNA can be edited by Type VI, and both RNA and DNA can be edited by Type III (Wang et al., 2019). It is yet unclear how the Type IV system affects DNA or RNA. Notably, Type II Cas9 and Type V Cas12a, which are broadly utilized for genome editing in eukaryotic cells, are included in Class 2. Besides modifying DNA, CRISPR/Cas systems are efficient at editing RNA. Class 2 Type VI CRISPR/Cas13 systems precisely cleave target single-stranded RNA (ssRNA) using a single RNA guided Cas13 protein with ribonuclease activity (Koonin et al., 2017).

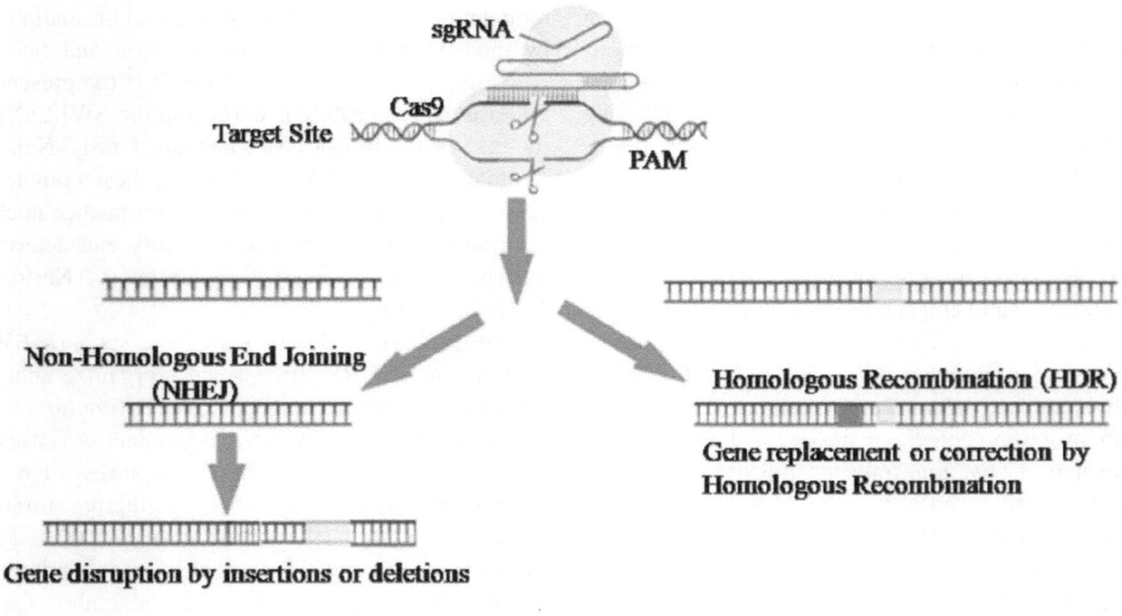

FIGURE 5.1 Schematic of the CRISPR/Cas9 genome editing system: CRISPR/Cas9-mediated genome editing involves a Cas9 protein guided by sgRNA to target a specific DNA sequence, introducing a double-strand break (DSB). Repair occurs via non-homologous end joining (NHEJ), leading to random insertions or deletions (indels) or homology-directed repair (HDR), using a donor template for precise modifications, enabling targeted genetic alterations.

5.4.2 CRISPR/Cas-Based Epigenetic Editing

The deliberate alteration of epigenetic markers is known as epigenetic editing (Voigt & Reinberg, 2013). One such use is the selective modification of epigenetic markers at a specific locus to investigate the dynamics through which targeted epigenetic modifications may impact the control of transcription and result in phenotypic changes. The main approach for attaining CRISPR/Cas9-assisted epigenome editing involves linking the Cas9 protein with a transcription repressor or activator domain, referred to as an epigenetic effector (epieffector) (Enriquez, 2016). Epigenome editing has been enabled by repurposing the CRISPR/Cas9 system, where a nuclease-null or "dead" Cas9 (dCas9) is used which allows aiming at specific DNA loci without cleavage (Xie et al., 2018). The catalytically deactivated Cas9 complex can be linked to epigenetic effectors, which enables it to localize the effectors to a particular locus and cause epigenetic changes. Simply altering the sgRNA sequence allows the dCas9 protein to be utilized again to target different loci (Nunez et al., 2021). As an engineering tool, CRISPR technology has many benefits, including the ability to target DNA segments of any length and modify them specifically and efficiently. With the advent of dCas9 fusion constructs, we now have a broad range of epigenome editing tools at our disposal to regulate the activation and inhibition of transcription of genes at the level of DM (Xiong et al, 2017), histone modifications (Chen et al., 2022) and even genes that code for noncoding RNAs (ncRNAs) (Phelan & Staudt, 2020).

5.4.3 CRISPR/Cas-Mediated DNA Methylation

5.4.3.1 Targeted DNA Methylation Using CRISPR/dCas9

Gene expression is influenced by DM, a significant epigenetic chemical modification. 5-Methylcytosine (5-mC), which is formed by attaching a methyl group to the fifth carbon position of cytosine, is the result of this modification. DNA methyltransferases (Dnmts), which are "writer" proteins, catalyze this process (Moore et al., 2013). The active removal of the methylation modification is managed by TET1-3 enzymes, and these enzymes, through their oxidase activity, add a hydroxyl group to the methyl group of 5-mC, resulting in the formation of a 5hmC modification, which can then be converted back to unmodified cytosine (Kohli & Zhang, 2013). When combined with Dnmts or TET domains, these "writer" and "eraser" proteins are used by several epigenome-editing technologies developed for DNA methylation control to add or remove chemical modifications to certain target genome loci using dCas9 systems. The fusion of the catalytic domains of Dnmt3a, Dnmt3L and TET1 to CRISPR/dCas9 to provide targeted DNA methylation control tools serves as evidence (Vojta et al., 2016) (Figure 5.2). Together with designed sgRNA molecules, this has been utilized to aim at and bind genomic loci, activate enzymatic DM activity and ultimately control the activation and suppression of the target gene. Previous research introduced a dCas9-SunTag DNMT3A dual system to study transcriptional changes related to localized epigenomic markers. The system uses a dCas9 protein linked to a SunTag peptide array and an antibody to the

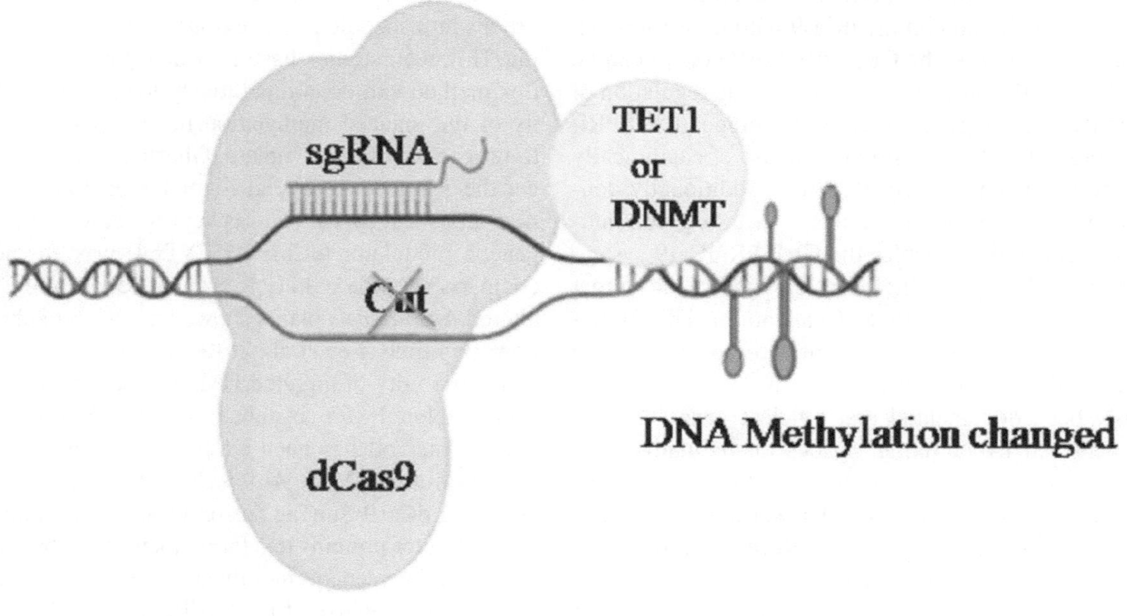

FIGURE 5.2 dCas9 fused to DNA methylation enzymes can be used to introduce site-specific epigenetic changes.

DNMT3A catalytic domain, addressing concerns about off-target effects and the need for specificity in epigenome-editing tools. The dual transfection of dCas9-SunTag with the DNMT3A single-chain antibody fusion (αGCN4-D3A) leads to more efficient and precise methylation at target sites, facilitated by the ability of SunTag peptide to recruit multiple αGCN4-D3A molecules. The dual aspect of this system also permits independent change in the regulation of the effector module (GCN4-D3A) and the DNA-targeting module (dC9Sun) (Fadul et al., 2023). A recent method employs dCas9-based systems devoid of epigenetic modifying enzymes to regulate DM through steric hindrance in a less confusing manner. The foundation of this epigenome editing technique is the observation that dCas9-gRNA attaching to a particular site inhibits linking and, as a result, DNA methyltransferase activity in dividing cells. When the enzymes are prevented from reaching the target site, the site will be passively demethylated by dilution as the cells divide. Off-target effects will be prevented because steric hindrance requires a strong, tight binding between the dCas9-gRNA and the target region (Sapozhnikov & Szyf, 2021).

5.4.4 Applications in Gene Silencing and Transcriptional Repression

Through sophisticated Cas9 protein modification, CRISPR/Cas9 can be employed to purposefully manage (activate or repress) a specific target gene in addition to genome editing. Scientists engineered dCas9 nuclease, an advanced modified variant of Cas9 endonuclease, by deactivating its HNH and RuvC domains (Jiang & Doudna, 2017). The dCas9 nuclease is incapable of DNA cleavage but retains the ability to bind DNA. By combining dCas9 with transcriptional activators or repressors, the CRISPR/dCas9 system can be tailored for specific purposes. This allows the regulation of expression of a target gene, either by silencing it (CRISPRi) or activating it (CRISPRa), through the use of catalytically inactive dCas9 (Dominguez et al., 2016). Additionally, coupling the dCas9 enzyme with markers such as green fluorescent proteins (GFP) enables the CRISPR/dCas9 system to visualize and pinpoint the exact location of the target gene within the cell (subcellular localization). This allows endogenous loci in living cells to be labeled and imaged site-specifically for future use (Anton et al., 2018). It has previously been documented that efficient gene silencing can be achieved by targeting various cis-regulatory elements or promoter regions with the DNA-binding domains of zinc finger proteins (ZFP) and Transcription Activator-Like Effector Nucleases (TALENs). Recently, a proof-of-concept for CRISPR-based epigenome editing to repress both proximal and distal regulatory elements has been established through effector-driven gene inhibition using CRISPR-dCas9 fusions. In one experiment, a nuclease-inactive dCas9 was combined with a repressor domain (KRAB), a method commonly seen in eukaryotic ZFPs. This type of gene silencing is facilitated by the KRAB-induced complex recruitment, primarily through protein-protein interactions between the KAP1 corepressor and other components that initiate heterochromatin formation at specific sites (Groner et al., 2010). Another study compared the suppressive effects of dCas9-KRAB with a dCas9 fusion to the histone demethylase lysine-specific demethylase 1 (LSD1). LSD1 catalyzes the removal of methyl marks on H3K4 and H3K9 residues and has been implicated in the repression of enhancer regions. The study showed that while dCas9-KRAB inhibition is probably caused by promoter silencing or heterochromatin spreading, dCas9-LSD1 induces repression by silencing target enhancers in an effector-driven approach without significantly altering local chromatin architecture (Kearns et al., 2015).

5.4.5 Case Studies of Targeted DNA Methylation (DM) in Plants Using CRISPR/dCas9

CRISPR/dCas9 technology for targeted DM in plants has surfaced as a potent method for accurate epigenetic alterations, providing scientists with the ability to control gene expression without altering the genetic sequence. This technique utilizes a catalytically inactive Cas9 (dCas9) protein, which, although devoid of endonuclease activity, maintains its ability to bind specific DNA sequences guided by single-guide RNAs (sgRNAs). By fusing dCas9 with DNA methyltransferase domains, it is possible to direct methylation to specific genomic loci, thereby influencing gene expression in a controlled manner. In plants, targeted DM has been accomplished by dCas9 fusion with the catalytic domain of the RNA-directed DM (RdDM) methyltransferase DRM2. This fusion allows for the incorporation of methyl groups into specific DNA regions, resulting in gene silencing. However, studies have indicated that the efficiency of this method can be comparatively low; also, the heritability of the induced methylation in the void of the effector transgene may be incomplete (Ghoshal et al., 2021). To target the *FT* gene, a crucial regulator of flowering time in *Arabidopsis*, a CRISPR/dCas9 construct with the MS2 epigenetic modulator (dCas9-MS2VP64) was engineered. In comparison to the wild type, around 65% of the CRISPRa-altered *Arabidopsis* plants showed a moderate alteration in flowering time (Lee et al., 2019). To enhance the efficiency and heritability of targeted DM, alternative strategies have been explored; for instance, the CRISPR/dCas9-based SunTag method has been adopted to design precise gene activation and DM in *Arabidopsis thaliana.* This system utilizes a dCas9-SunTag fusion to assemble multiple copies of effector proteins to a target locus, thereby amplifying the desired epigenetic modification (Papikian et al., 2019). Additionally, the use of the SunTag-dCpf1-TET1cd system has been developed to generate targeted DNA demethylation and activation of genes in *A. thaliana.* This system employs a deactivated Cpf1 (dCpf1) incorporated into the catalytic domain of the TET1 enzyme, which facilitates the removal of methyl groups from the specific DNA regions, resulting in gene activation (Zheng & He, 2023). Targeted demethylation of DNA and activation of the target gene, like *FWA* and *CACTA1* (AT2G12210) in *Arabidopsis* (Gallego-Bartolome et al., 2018) and *FIE1* and *Tos17* rice (Tang et al., 2021), have been accomplished recently using the CRISPR/dCas9-based SunTag-TET1cd system.

These advancements in CRISPR/dCas9-assisted targeted methylation and demethylation systems offer promising avenues for plant functional genomics and crop enhancement. By enabling precise epigenetic alterations, researchers can modulate gene expression to boost preferred traits, such as stress resilience and crop yield, without incorporating foreign DNA sequences into the plant genome. However, difficulties continue in improving the effectiveness, precision and heritability of these targeted epigenetic modifications. Additional studies are required to refine these tools and fully understand the mechanisms governing their function in different plant species and genomic contexts.

5.5 CRISPR/Cas-BASED NON-CODING RNA MODULATION

ncRNAs play integral roles in regulating epigenetic mechanisms, influencing chromatin dynamics, transcriptional activity and post-transcriptional modifications. These molecules, including lncRNAs, circular RNAs (circRNAs) and miRNAs, are critical in coping tolerance mechanisms against abiotic stresses and in developmental processes (Wei et al., 2017; Mangiavacchi et al., 2023), for instance, miR393 regulates the signaling pathways of auxin, enhancing bacterial resistance in *Arabidopsis* (Navarro et al., 2006). With CRISPR/Cas technologies, precise manipulation of ncRNAs has advanced our understanding of these regulatory networks, enabling targeted modulation for specific outcomes in plant biology.

CRISPR/Cas systems have proven versatile for studying and modulating ncRNAs. CRISPR/Cas9 facilitates targeted knockouts or activation of ncRNA genes by employing nuclease-dead Cas9 (dCas9) fused with transcriptional effectors (Hazan & Bester, 2021). In a study, researchers employed a dCas9 fused with transcriptional repressors to suppress specific lncRNAs. This method enabled them to explore the roles of these lncRNAs in gene regulation and plant development; for instance, directing dCas9 to the promoter regions of certain lncRNAs successfully reduced their expression, resulting in noticeable changes in stress response traits and growth patterns (Basak & Nithin, 2015). Similarly, CRISPR/Cas13 directly targets RNA molecules, offering the ability to deplete specific ncRNAs or introduce precise modifications. In rice, CRISPR/Cas13-mediated degradation of miR319, a key regulator of jasmonic acid pathways, conferred improved biotic stress resistance (Kavuri et al., 2022; Zegeye et al., 2022). Base editors and CRISPR activation/inhibition (CRISPRa/i) systems extend these capabilities by allowing targeted nucleotide changes or transcriptional modulation without introducing double-strand breaks (Z. Liu et al., 2022). Notably, CRISPRa using dCas9-VP64 was employed to upregulate lncRNA ENOD40 in *Medicago truncatula*, significantly enhancing nitrogen fixation under nutrient-deficient conditions (Gultyaev et al., 2023). In another notable example, CRISPR/Cas9-mediated editing of miR399 in barley improved tolerance to phosphate starvation by enhancing phosphate translocation and homeostasis (Abdullah, 2023). Furthermore, CRISPR/Cas9 was utilized to knockout miR528 in rice, leading to improved tolerance to oxidative stress by upregulating superoxide dismutase activity (Zhou et al., 2017). Such studies highlight the potential of CRISPR/Cas technologies to tailor ncRNA functions for enhanced stress responses and resource efficiency in crops.

In practical applications, CRISPR/Cas-mediated ncRNA modulation addresses challenges in agriculture by improving plant stress tolerance, development and pathogen resistance. For instance, in one study it was demonstrated that the CRISPR/Cas9-mediated knockout of miR827 enhances phosphorus uptake and utilization efficiency in rice, addressing nutrient limitations effectively (Bundó et al., 2024). Similarly, targeted disruption of lncRNA LAIR in rice accelerated flowering time, aiding adaptation to shorter growing seasons (Y. Wang et al., 2018). Enhancing pathogen resistance has also been achieved through ncRNA modulation; knocking down miR482 in tomato strengthened resistance to *Phytophthora infestans* by elevating NLR protein levels (Hong et al., 2021). Another groundbreaking case involved the modulation of miR444 in wheat to enhance salinity tolerance by regulating ethylene response factors and cytokinin pathways (Gruszka et al., 2021; Zhang et al., 2024). Such case studies underscore the transformative potential of CRISPR/Cas technologies in ncRNA research, paving the way for sustainable agricultural innovations.

5.6 ADVANCES IN CRISPR/Cas EPIGENETIC TOOLS

Advances in CRISPR/Cas epigenetic tools have enabled precise modulation of gene expression without altering DNA sequences. Modified Cas9 variants, such as dCas9, combined with epigenetic effectors like DNA methyltransferases or histone acetyltransferases, allow targeted epigenome editing. These tools enhance understanding of gene regulation, disease mechanisms and therapeutic potential in plants and humans. Several key areas of advancement in epigenetic research facilitated by CRISPR/Cas applications include the following.

5.6.1 dCas9 Fusions with Epigenetic Modifiers

The CRISPR/Cas9 system, along with its inactive variant dCas9, has transformed genetic manipulation by enabling precise and targeted changes essential for phenotype modification. dCas9 can regulate gene expression by binding transcriptional regulators to promoter regions near transcription start sites. Additionally, its fusion with epigenetic modulators like methylation or deacetylation enzymes allows for accurate epigenome editing. Researchers have also utilized dCas9 by attaching regulatory or reporter proteins, unlocking its potential for reversible gene activation or repression, and epigenetic modifications, showcasing its broad applicability in biological research (Hilton et al., 2015; Lowder et al., 2018; Moradpour & Abdulah, 2019). Transcriptional activation at specific loci using dCas9 is achieved by recruiting or fusing various activators. In plants, the VP64 domain, composed of four repeats of Herpes simplex viral protein 16, is the most commonly utilized. Other activators, such as VP64-p65-Rta (VPR), which combines VP64 with domains from RelA NF-kappaB (p65) and Epstein–Barr virus Rta protein, as well as trans-activation domains like HSF1 and the EDLL motif from ERF/EREBP transcription factors, have also proven effective. Direct fusions of dCas9 with VP64, EDLL or HSF1 generally lead to moderate activation of target genes, demonstrating the adaptability of the method (Dubois & Roudier, 2021). Recent research has emphasized dCas9-linked epigenetic enzymes for DNA methylation modification. Fusions of dCas9 with DNMT3A/B enable targeted deposition of 5-methylcytosine (5-mC), while dCas9-TET1/2 promotes the generation of 5-hydroxy-mC (5-hmC), a critical step in DNA demethylation (O'Geen et al., 2017). dCas9, guided by sgRNA, binds DNA with precision but lacks cleavage activity, allowing it to interfere with transcription and reversibly silence genes without altering the genome permanently. Unlike the standard CRISPR/Cas9 system, which creates double-strand breaks for gene editing, dCas9 regulates gene expression by recruiting transcriptional activators at its N and C terminal sites. This system is widely used for genetic screening, protein overexpression and fine-tuned regulation of target gene expression through various activators and repressors (Li et al., 2017; Moradpour & Abdulah, 2019; Piatek et al., 2015; Polstein & Gersbach, 2015).

5.6.2 CRISPRa and CRISPRi Technologies

CRISPR/Cas systems regulate gene expression through two approaches: CRISPR activation (CRISPRa) and interference (CRISPRi). In CRISPRa, dCas9 is linked to transcriptional activators, like VP64, p65 or Rta, and directed to specific promoters by gRNA to boost the transcription of target genes, simulating the role of natural transcription factors. On the other hand, CRISPRi involves dCas9 combined with repressors like the KRAB domain to suppress gene expression (Azeez et al., 2024; Casas-Mollano et al., 2020; Omachi & Miner, 2022). The CRISPRa system was employed in *Arabidopsis* to enhance the endogenous expression of the *PAP1* and *AVP1* genes. *PAP1* expression increased 2.0- to 3.0-fold, resulting in purple leaves resembling *PAP1* overexpressors. *AVP1* expression was enhanced 2.0- to 5.0-fold, leading to traits such as more leaves, larger leaf areas and improved drought tolerance, resembling *AVP1* overexpressors (Park et al., 2017).

CRISPRi was initially developed for targeted transcriptional repression in *Escherichia coli* and human cells. It enables precise genomic targeting without altering the DNA sequence. By fusing dCas9 to a transcriptional repressor domain (TRD), a chimeric dCas9-TRD protein is formed, enhancing transcriptional interference efficiency (Qi et al., 2023). The CRISPRi-based method for gene knockdown offers inducibility and reversibility, facilitating the dynamic and temporal regulation of specific genes (Zhang et al., 2021). CRISPRi provides a reversible and efficient method for RNA-guided suppression of target gene expression, relying solely on the sgRNA sequence for specificity. The recruitment of dCas9 to the sgRNA recognition complex blocks gene expression by interfering with RNA polymerase binding, transcription factor functions and transcriptional elongation (Sharma et al., 2024). Transgenic tomato plants with *SlLBD40* knocked out using CRISPRi were created to investigate its function. The knockout lines exhibited better drought tolerance, characterized by minimal wilting, reduced water loss rates and higher midday leaf water potential, compared to plants overexpressing *SlLBD40*, which showed severe wilting. These findings indicate that *SlLBD40* serves as a negative regulator of water stress resilience through jasmonic acid signaling, and its knockout significantly improves drought resilience in tomatoes (Liu et al., 2020).

5.6.3 Multiplexed Epigenome Editing

Epigenetic effectors are attached to dCas9, enabling precise targeting of specific loci to induce changes in gene expression. This system allows for reusability by simply altering the sgRNA sequence and can perform specific or multiplex modifications. It offers significant advantages for regulating gene expression via histone modifications, DNA methylation and noncoding RNA regulation, providing a powerful toolkit for epigenome control (Fadul et al., 2023). Multiplex genome editing can be accomplished by expressing Cas9 (or its derivatives) alongside multiple gRNAs targeting different sites. Traditional delivery approaches encompass microinjection or using plasmids with multiple sgRNA-expressing cassettes. A more advanced approach is to compact multiple gRNAs within a synthetic gene and use an RNA-processing system to extract individual gRNAs, such as the use of Csy4 endoribonuclease with Cas9 (Xie et al., 2015). A major advantage of CRISPR/Cas9 is its capacity for multiplex genome editing. While TALENs and ZFNs can achieve multiplex editing, it is limited to targeting conserved sequences shared between sites or requiring multiple distinct paired nucleases. In contrast, CRISPR/Cas9 enables efficient targeting of multiple sites by simply supplying multiple unique gRNAs alongside the Cas9 protein (Minkenberg et al., 2017). Rather than engineering new proteins to enhance specific traits in plants, a simpler approach is to add multiple gRNAs to the transformation construct. This method, using a single Cas protein with several gRNAs, enables the efficient generation of multiple gene mutations across different plant species. (Abdelrahman et al., 2021). In a particular study, successful mutagenesis was achieved for seven target sequences across five genes in *Nicotiana benthamiana* and *A. thaliana*. Using multiple sgRNAs:proCas9 improved gene knockout success, and protoplasts proved useful for rapid efficiency assessment (Li et al., 2013). Researchers developed a multiplex CRISPR/Cas9 system for *Arabidopsis* that co-expresses six sgRNAs in a single vector using three RNA polymerase III promoters. Targeting six *PYL* ABA receptor genes, they identified one T1 line with mutations in all six genes, achieving mutagenesis efficiencies of 13–93%. In T2 progeny, this line showed a 37% germination rate in the presence of ABA, with homozygous sextuple mutants confirmed in T3. This system demonstrates high effectiveness and suitability for complex genome editing of plants (Zhang et al., 2016). Multiplex genome editing has been efficiently utilized in plants to enhance traits like herbicide tolerance. Its scope is now broadening to include diverse areas of agricultural enhancement, such as metabolic engineering, molecular farming and hormone biosynthesis with the capability to achieve more than 100 simultaneous targeting events (Kumar et al., 2023). In an experiment, researchers used the CRISPR/Cas9-assisted multiplex genome editing system to mutate three key genes regulating rice grain weight at the same time. This efficient system enables the prompt generation and pyramiding of useful alleles, offering an advantage over conventional cross-breeding and quantitative trait loci (QTL) methods. The CRISPR/Cas9 approach lessens the incorporation of non-targeted genomic fragments, minimizing the chances of unintended traits (Xu et al., 2016).

5.6.4 Innovations in Delivery Methods

For successful genome editing, effective delivery of CRISPR-Cas9 components is vital. Nanomaterials such as polymeric nanoparticles, viral vectors and lipid-based nanoparticles play a key role in enhancing delivery by safeguarding CRISPR-Cas9 components from degradation, assisting in cellular uptake and facilitating endosomal escape (Aljabali et al., 2024). Viral vectors have emerged as promising tools for delivering CRISPR reagents into germline cells for DNA-free and heritable gene editing. Autonomously replicating RNA viruses like TRV and BMSV have been optimized for sgRNA delivery, with TRV targeting dicots and BMSV achieving heritable and multiplexed editing in wheat. Enhancements like fusing sgRNAs to mobile RNA (e.g., Flowering Locus T) improve systemic distribution. Additionally, SYNV, a negative-strand RNA virus with higher cargo potential, enables delivery of both Cas9 and sgRNAs for efficient DNA-free editing (Laforest & Nadakuduti, 2022).

5.7 APPLICATIONS OF CRISPR/Cas-BASED EPIGENETIC REGULATION IN PLANTS

CRISPR/Cas systems offer precise genetic modifications, enhancing crop yield and nutritional value, especially in staple crops like maize and rice (Chen et al., 2024). Epigenetic modifications, including DM, histone modifications and ncRNAs, are essential for regulating stress-responsive genes, influencing plant phenotypes and ensuring proper gene expression. By altering chromatin structure and epigenetic marks, these mechanisms play a crucial role in stress tolerance and crop improvement, making them a significant area of plant research (Jogam et al., 2022). CRISPR/Cas9 technology offers transformative potential for improving plant productivity and abiotic stress tolerance. It is not only valuable for studying gene functions but also for advancing plant improvement by boosting tolerance to both abiotic and biotic stresses, regulating secondary metabolites and facilitating cell imaging (Karlson et al., 2021). The CRISPR/Cas system has identified drought-responsive genes in cereals, particularly members of the SAPK family in the SnRK2 group, which are activated by abscisic acid. These genes boost plant growth under abiotic stresses such as salinity, osmotic stress and water stress, especially in the early phases (Maharajan et al., 2022). For instance, the abscisic acid-responsive element-binding protein AREB1/ABF2 is crucial for drought stress response and adaptation. Using CRISPRa with dCas9 fused to a histone acetyltransferase, researchers enhanced AREB1 expression, improving drought tolerance in transgenic plants. Overexpression increased chlorophyll content, antioxidant activity and soluble sugars while reducing reactive oxygen species. Additionally, several ABA-inducible genes were positively regulated, highlighting CRISPRa dCas9-HAT as a powerful resource for elevating drought tolerance (de Melo et al., 2020). In another study, CRISPR/Cas9 was employed to generate rice-rolled leaf mutants by targeting *SRL1* and *SRL2* (*semi-rolled leaf*) genes because leaf rolling reduces transpiration and stomatal permeability during drought stress. Cultivating semi-rolled leaf genotypes could help improve crop yield in drought-prone areas. These mutants showed improved drought tolerance with higher survival rates, antioxidant enzyme activities and ABA contents under drought stress. Proteomic analysis revealed the upregulation of stress-responsive proteins in the CRISPR mutants (Liao et al., 2019).

Viruses can significantly impact plant growth, causing major losses in agriculture by integrating their genetic material into plant genomes for replication. Plants defend against viral invasion through RNA interference (RNAi), but some viruses release suppressor proteins that inhibit this silencing process. CRISPR/dCas9, bypassing the RNAi pathway, offers a promising method to target and disrupt viral RNA. Recent studies have explored that using Cas protein variants like FnCas9 and LshCas13a can enhance plant immunity and inhibit viruses (Abudayyeh et al., 2016, 2017; Karlson et al., 2021). CRISPR/Cas13a (previously C2c2), a type VI-A ribonuclease, targets and cleaves single-stranded RNA (ssRNA) of phage genomes. A study demonstrated that CRISPR/Cas13a could interfere with the RNA virus turnip mosaic virus (TuMV) in plants. Using crRNAs addressing HC-Pro and GFP sequences, CRISPR/Cas13a showed stronger interference compared to those directed at other regions, like the coat protein (CP). Cas13a also processed pre-crRNAs into active crRNAs. These findings suggested that CRISPR/Cas13a is a promising tool for RNA virus immunity and RNA manipulation in plants (Aman et al., 2018). Weeds and insects are major biotic stress factors affecting cereal growth and yield, with weeds causing the most harm by competing for resources. Although herbicides are used to control weeds, repeated application can lead to resistance. Glufosinate, a commonly used herbicide, is often harmful to crops. In this study, OsSPL8, an SPL factor, was identified as a repressor of glufosinate resistance by preventing crucial genes implicated in the resistance mechanism. Knockout of *OsSPL8* using CRISPR/Cas9 improved resilience to salinity and drought stresses. Transcriptomic analysis revealed that OsSPL8 suppressed stress-responsive genes and promoted the genes that are related to growth (He et al., 2025). *Xanthomonas oryzae* pv. *oryzae* causes bacterial blight in rice, a significant disease in West Africa and South Asia. Studies have shown that sucrose transporters (SWEET) can elevate the growth and yield of rice under bacterial blight infection. CRISPR/Cas9-assisted knockout of the *OsSWEET13* gene improved rice growth and resistance to the disease (Zhou et al., 2015).

CRISPR/Cas technology can modulate the production of secondary metabolites, which plants use for survival, adaptation and communication with their environment. These metabolites are valuable for nutrition, medicine and agrochemicals. Advances in understanding biosynthetic pathways and identifying key targets have enabled the manipulation of plant secondary metabolism (Mipeshwaree Devi et al., 2023). By targeting specific genes, CRISPR/Cas9 enhances bioactive substances in secondary metabolism, optimizing the chemical composition of plants. Gene-editing strategies enhance phytometabolite pathways, improving the secondary metabolite synthesis for industrial and medicinal applications (Borah et al., 2024). In *Salvia miltiorrhiza*, phenolic acids are the key bioactive substances, and their production can be enhanced through ABA stimulation. Using CRISPR/Cas9 to knockout the bZIP2 factor-encoding gene led to an increase in phenolic acid levels, while overexpression of bZIP2 suppressed their production by downregulating the *PAL* gene. This finding highlighted the effectiveness of CRISPR/Cas9 in altering the biosynthetic pathway of secondary metabolites (Shi et al., 2021). A study used an upgraded CRISPR-Cas9 method to insert a 5.2 kb carotenoid biosynthesis cassette into two genomic-safe harbors in rice. The resulting marker-free rice plants had high carotenoid levels in seeds without impacting morphology or yield. Whole-genome sequencing validated no miss-targeted mutations, highlighting the potential of CRISPR-Cas9 for marker-free gene insertion and genetic advancement in rice and other crops (Dong et al., 2020).

5.8 FUTURE DIRECTIONS AND PERSPECTIVES

The future of CRISPR/Cas-based epigenetic regulation in plants has numerous applications in the area of agriculture and also for understanding plant biology in detail. One of the important areas is the advancement of more precise and specific CRISPR tools that can target epigenetic marks with fewer off-target effects. CRISPR/Cas could be utilized to modify DM or histone modifications, leading to control over gene expression in plants (Zhang et al., 2021; Jogam et al., 2022). The tool can be useful for regulating traits like drought tolerance and disease resistance, and even enhancing crop yield. This can be achieved by activating or silencing targeted genes through CRISPR/Cas. The researchers can potentially create plants that are more adaptable to climate disruption or have enhanced nutritional value. In the future, integration of CRISPR/Cas with omics technologies like transcriptomics, proteomics and metabolomics could bring more advances in epigenetics research. These combinations will allow scientists to investigate the impact of epigenetic editing at a systems level, helping them to better understand the regulatory networks in plants. Additionally, progress in single-cell sequencing methods will also serve a vital function in identifying cell-specific epigenetic changes, which can provide more precise insights into plant growth and stress adaptation mechanisms.

However, a crucial concern for the future is to improve the delivery mechanisms of CRISPR/Cas systems into plant cells (Sandhya et al., 2020; Leforest & Nadakuduti, 2022). Currently, approaches like *Agrobacterium*-assisted transformation and gene gun methods are commonly used, but they have constraints, especially for certain plant species or tissues. The development of more efficient delivery systems, such as nanotechnology-based approaches or virus-mediated delivery, could help overcome these barriers and make epigenetic editing more accessible for a diverse array of plants. Despite the immense potential of CRISPR/Cas in epigenetic research, gaps still remain in our understanding of how epigenetic changes are inherited in plants. Further research is needed to explore the stability of these changes across generations and their long-term impacts on plant populations.

5.9 CONCLUSION

CRISPR/Cas-based epigenetic regulation in plants plays an important role in advancing agricultural biotechnology. By enabling the precise modification of gene expression without modifying the underlying DNA sequence, this technology serves as a sound tool for improving crop traits. The ability to modify DNA methylation or histone modifications in a targeted manner allows researchers to regulate genes associated with flowering time, stress tolerance and nutrient efficiency. This approach will enhance understanding of epigenetic mechanisms and also shall offer solutions to concerns such as food accessibility and climate variability. However, the CRISPR/Cas technology still has some challenges, comprising off-target effects, delivery mechanisms and regulatory concerns. Despite these hurdles, CRISPR/Cas tools have the capability to revolutionize plant breeding by delivering precise, efficient and environmentally sustainable solutions. Moreover, the epigenetic changes induced using such tools can be reversible and heritable, providing a unique advantage over traditional genetic engineering methods. As research progresses, it is likely that CRISPR/Cas-based epigenetic editing will become an indispensable tool in plant sciences, facilitating progress for more resilient and productive crops.

REFERENCES

Abdelrahman, M., Wei, Z., Rohila, J. S., & Zhao, K. (2021). Multiplex genome-editing technologies for revolutionizing plant biology and crop improvement. *Frontiers in Plant Science*, 12. https://doi.org/10.3389/fpls.2021.721203.

Abdullah, S. N. A. (2023). Molecular regulation for maintenance of plant phosphate homeostasis in crop plants as basis for precision genome engineering—[V1] (2023060686). Preprints. https://doi.org/10.20944/preprints202306.0686.v1.

Abdulraheem, M. I., Xiong, Y., Moshood, A. Y., Cadenas-Pliego, G., Zhang, H., & Hu, J. (2024). Mechanisms of plant epigenetic regulation in response to plant stress: Recent discoveries and implications. *Plants*, 13(2), 163.

Abudayyeh, O. O., Gootenberg, J. S., Essletzbichler, P., Han, S., Joung, J., Belanto, J. J., Verdine, V., Cox, D. B. T., Kellner, M. J., Regev, A., Lander, E. S., Voytas, D. F., Ting, A. Y., & Zhang, F. (2017). RNA targeting with CRISPR–Cas13. *Nature*, 550(7675), 280–284. https://doi.org/10.1038/nature24049.

Abudayyeh, O. O., Gootenberg, J. S., Konermann, S., Joung, J., Slaymaker, I. M., Cox, D. B. T., Shmakov, S., Makarova, K. S., Semenova, E., Minakhin, L., Severinov, K., Regev, A., Lander, E. S., Koonin, E. V., & Zhang, F. (2016). C2c2 is a single-component programmable RNA-guided RNA-targeting CRISPR effector. *Science*, 353(6299), aaf5573. https://doi.org/10.1126/science.aaf5573.

Akhter, Z., Bi, Z., Ali, K., Sun, C., Fiaz, S., Haider, F. U., & Bai, J. (2021). In response to abiotic stress, DNA methylation confers epigenetic changes in plants. *Plants*, 10(6), 1096. https://doi.org/10.3390/plants10061096.

Ali, S., & Tang, Y. (2024). Noncoding RNA-mediated regulation of DNA methylation: Insights into plant epigenetic mechanisms. *Journal of Plant Growth Regulation*. https://doi.org/10.1007/s00344-024-11462-0.

Aljabali, A. A. A., El-Tanani, M., & Tambuwala, M. M. (2024). Principles of CRISPR-Cas9 technology: Advancements in genome editing and emerging trends in drug delivery. *Journal of Drug Delivery Science and Technology*, 92, 105338. https://doi.org/10.1016/j.jddst.2024.105338.

Aman, R., Ali, Z., Butt, H., Mahas, A., Aljedaani, F., Khan, M. Z., Ding, S., & Mahfouz, M. (2018). RNA virus interference via CRISPR/Cas13a system in plants. *Genome Biology*, 19(1), 1. https://doi.org/10.1186/s13059-017-1381-1.

Anton, T., Karg, E., & Bultmann, S. (2018). Applications of the CRISPR/Cas system beyond gene editing. *Biology Methods and Protocols*, 3(1), bpy002.

Asmamaw, M., & Zawdie, B. (2021). Mechanism and applications of CRISPR/Cas-9-mediated genome editing. *Biologics: Targets and Therapy*, 15, 353–361.

Azeez, S. S., Hamad, R. S., Hamad, B. K., Shekha, M. S., & Bergsten, P. (2024). Advances in CRISPR-Cas technology and its applications: Revolutionising precision medicine. *Frontiers in Genome Editing*, 6. https://doi.org/10.3389/fgeed.2024.1509924.

Banerjee, A., & Roychoudhury, A. (2017). Epigenetic regulation during salinity and drought stress in plants: Histone modifications and DNA methylation. *Plant Gene*, 11, 199–204.

Banerjee, A., & Roychoudhury, A. (2023). Genome editing using CRISPR/Cas9 system for crop improvement, nutrient enrichment, and abiotic stress tolerance in rice. In D. Pandita & A. Pandita (Eds.), *CRISPR/Cas-Mediated Genome Editing in Plants* (pp. 133–137). New York: Apple Academic Press, Inc. Co- published with CRC Press (Taylor & Francis).

Basak, J., & Nithin, C. (2015). Targeting non-coding RNAs in plants with the CRISPR-Cas technology is a challenge yet worth accepting. *Frontiers in Plant Science*, 6, 1001. https://doi.org/10.3389/fpls.2015.01001.

Bhattacharya, A., & Roychoudhury, A. (2024). Plant non-coding RNAs and their regulation in salinity and heavy metal stress. In S. Penna, V. Kumar, & A. K. Srivastava (Eds.), *Non-Coding RNAs Molecular Tools for Crop Improvement* (pp. 128–147). CRC Press (Taylor & Francis group).

Borah, A., Singh, S., Chattopadhyay, R., Kaur, J., & Bari, V. K. (2024). Integration of CRISPR/Cas9 with multi-omics technologies to engineer secondary metabolite productions in medicinal plant: Challenges and prospects. *Functional & Integrative Genomics*, 24(6), 207. https://doi.org/10.1007/s10142-024-01486-w.

Bundó, M., Val-Torregrosa, B., Martín-Cardoso, H., Ribaya, M., Campos-Soriano, L., Bach-Pages, M., Chiou, T.-J., & San Segundo, B. (2024). Silencing Osa-miR827 via CRISPR/Cas9 protects rice against the blast fungus *Magnaporthe oryzae*. *Plant Molecular Biology*, 114(5), 105. https://doi.org/10.1007/s11103-024-01496-z.

Cardi, T. (2016). Cisgenesis and genome editing: Combining concepts and efforts for a smarter use of genetic resources in crop breeding. *Plant Breed,* 135(2), 139–147.

Casas-Mollano, J. A., Zinselmeier, M. H., Erickson, S. E., & Smanski, M. J. (2020). CRISPR-Cas activators for engineering gene expression in higher eukaryotes. *The CRISPR Journal*, 3(5), 350–364. https://doi.org/10.1089/crispr.2020.0064.

Chen, E., Lin-Shiao, E., Trinidad, M., Saffari Doost, M., Colognori, D., & Doudna, J. A. (2022). Decorating chromatin for enhanced genome editing using CRISPR-Cas9. *Proceedings of the National Academy of Sciences*, 119(49), e2204259119.

Chen, F., Chen, L., Yan, Z., Xu, J., Feng, L., He, N., Guo, M., Zhao, J., Chen, Z., Chen, H., Yao, G., & Liu, C. (2024). Recent advances of CRISPR-based genome editing for enhancing staple crops. *Frontiers in Plant Science*, 15. https://doi.org/10.3389/fpls.2024.1478398.

Chinnusamy, V., & Zhu, J. K. (2009). Epigenetic regulation of stress responses in plants. *Current Opinion in Plant Biology,* 12(2), 133–139.

de Melo, B. P., Lourenço-Tessutti, I. T., Paixão, J. F. R., Noriega, D. D., Silva, M. C. M., de Almeida-Engler, J., Fontes, E. P. B., & Grossi-de-Sa, M. F. (2020). Transcriptional modulation of AREB-1 by CRISPRa improves plant physiological performance under severe water deficit. *Scientific Reports*, 10(1), 16231. https://doi.org/10.1038/s41598-020-72464-y.

Ding, Y., Avramova, Z., & Fromm, M. (2011). The Arabidopsis trithorax-like factor ATX1 functions in dehydration stress responses via ABA-dependent and ABA-independent pathways. *The Plant Journal*, 66(5), 735–744. https://doi.org/10.1111/j.1365-313X.2011.04534.x.

Dominguez, A. A., Lim, W. A., & Qi, L. S. (2016). Beyond editing: Repurposing CRISPR–Cas9 for precision genome regulation and interrogation. *Nature Reviews Molecular Cell Biology*, 17(1), 5–15.

Dong, O. X., Yu, S., Jain, R., Zhang, N., Duong, P. Q., Butler, C., Li, Y., Lipzen, A., Martin, J. A., Barry, K. W., Schmutz, J., Tian, L., & Ronald, P. C. (2020). Marker-free carotenoid-enriched rice generated through targeted gene insertion using CRISPR-Cas9. *Nature Communications*, 11(1), 1178. https://doi.org/10.1038/s41467-020-14981-y.

Du, J., Zhong, X., Bernatavichute, Y. V., Stroud, H., Feng, S., Caro, E., Vashisht, A. A., Terragni, J., Chin, H. G., Tu, A., Hetzel, J., Wohlschlegel, J. A., Pradhan, S., Patel, D. J., & Jacobsen, S. E. (2012). Dual binding of chromomethylase domains to H3K9me2-containing nucleosomes directs DNA methylation in plants. *Cell*, 151(1), 167–180. https://doi.org/10.1016/j.cell.2012.07.034.

Dubois, A., & Roudier, F. (2021). Deciphering plant chromatin regulation via CRISPR/dCas9-based epigenome engineering. *Epigenomes*, 5(3), Article 3. https://doi.org/10.3390/epigenomes5030017.

Enríquez, P. (2016). Focus: Epigenetics: CRISPR-mediated epigenome editing. *The Yale Journal of Biology and Medicine*, 89(4), 471.

Eom, S. H., & Hyun, T. K. (2018). *Histone Acetyltransferases (HATs) in Chinese Cabbage: Insights from Histone H3 Acetylation and Expression Profiling of HATs in Response to Abiotic Stresses*. https://doi.org/10.21273/JASHS04436-18.

Fadul, S. M., Arshad, A., & Mehmood, R. (2023). CRISPR-based epigenome editing: Mechanisms and applications. *Epigenomics*, 15(21), 1137–1155. https://doi.org/10.2217/epi-2023-0281.

Fan, S. K., Ye, J. Y., Zhang, L. L., Chen, H. S., Zhang, H. H., Zhu, Y. X., Liu, X. X., & Jin, C. W. (2020). Inhibition of DNA demethylation enhances plant tolerance to cadmium toxicity by improving iron nutrition. *Plant, Cell & Environment*, 43(1), 275–291. https://doi.org/10.1111/pce.13670.

Ferreira L. J., Azevedo V., Maroco J., Oliveira M. M., Santos A.P. (2015). Salt tolerant and sensitive rice varieties display differential methylome flexibility under salt stress. *PLoS ONE*, 10(5): e0124060.

Gallego-Bartolomé, J., Gardiner, J., Liu, W., Papikian, A., Ghoshal, B., Kuo, H. Y., Zhao, J. M., Segal, D. J., & Jacobsen, S. E. (2018). Targeted DNA demethylation of the Arabidopsis genome using the human TET1 catalytic domain. *Proceedings of the National Academy of Sciences of the United States of America*, 115(9), E2125–E2134.

Garg, R., Narayana Chevala, V. V. S., Shankar, R., & Jain, M. (2015). Divergent DNA methylation patterns associated with gene expression in rice cultivars with contrasting drought and salinity stress response. *Scientific Reports*, 5(1), 14922. https://doi.org/10.1038/srep14922.

Ghoshal, B., Picard, C. L., Vong, B., Feng, S., & Jacobsen, S. E. (2021). CRISPR-based targeting of DNA methylation in *Arabidopsis thaliana* by a bacterial CG-specific DNA methyltransferase. *Proceedings of the National Academy of Sciences*, 118(23), e2125016118.

Groner, A. C., Meylan, S., Ciuffi, A., Zangger, N., Ambrosini, G., Dénervaud, N., ... & Trono, D. (2010). KRAB–zinc finger proteins and KAP1 can mediate long-range transcriptional repression through heterochromatin spreading. *PLoS Genetics*, 6(3), e1000869.

Gruszka, D., Bajguz, A., Li, Q.-F., Hayat, S., Hansson, M., Wang, X., & Li, J. (Eds.). (2021). An update on brassinosteroids: Homeostasis, crosstalk, and adaptation to environmental stress. *Frontiers Media SA*. https://doi.org/10.3389/978-2-88966-906-6.

Gultyaev, A. P., Koster, C., van Batenburg, D. C., Sistermans, T., van Belle, N., Vijfvinkel, D., & Roussis, A. (2023). Conserved structured domains in plant non-coding RNA enod40, their evolution and recruitment of sequences from transposable elements. *NAR Genomics and Bioinformatics*, 5(4), lqad091. https://doi.org/10.1093/nargab/lqad091.

Hazan, J., & Bester, A. C. (2021). CRISPR-based approaches for the high-throughput characterization of long non-coding RNAs. *Non-Coding RNA*, 7(4), Article 4. https://doi.org/10.3390/ncrna7040079.

He, D.-Y., Liang, Q.-Y., Xiang, C.-B., & Xia, J.-Q. (2025). Loss of OsSPL8 function confers improved resistance to glufosinate and abiotic stresses in rice. *Plant, Cell & Environment*, 48(1), 682–698. https://doi.org/10.1111/pce.15168.

Heo, J. B., Lee, Y.-S., & Sung, S. (2013). Epigenetic regulation by long noncoding RNAs in plants. *Chromosome Research*, 21(6), 685–693. https://doi.org/10.1007/s10577-013-9392-6.

Hilton, I. B., D'Ippolito, A. M., Vockley, C. M., Thakore, P. I., Crawford, G. E., Reddy, T. E., & Gersbach, C. A. (2015). Epigenome editing by a CRISPR-Cas9-based acetyltransferase activates genes from promoters and enhancers. *Nature Biotechnology*, 33(5), 510–517. https://doi.org/10.1038/nbt.3199.

Hong, Y., Meng, J., He, X., Zhang, Y., Liu, Y., Zhang, C., Qi, H., & Luan, Y. (2021). Editing miR482b and miR482c simultaneously by CRISPR/Cas9 enhanced tomato resistance to phytophthora infestans. *Phytopathology®*, 111(6), 1008–1016. https://doi.org/10.1094/PHYTO-08-20-0360-R.

Hossain, M. A. (2021). CRISPR-Cas9: A fascinating journey from bacterial immune system to human gene editing. *Progress in Molecular Biology and Translational Science*, 178, 63–83.

Hou, J., Lu, D., Mason, A. S., Li, B., Xiao, M., An, S., & Fu, D. (2019). Non-coding RNAs and transposable elements in plant genomes: Emergence, regulatory mechanisms and roles in plant development and stress responses. *Planta*, 250(1), 23–40. https://doi.org/10.1007/s00425-019-03166-7.

Hryhorowicz, M., Lipiński, D., & Zeyland, J. (2023). Evolution of CRISPr/Cas systems for precise genome editing. *International Journal of Molecular Science*, 24(18):14233.

Huang, J., Wang, H., Xie, X., Zhang, D., Liu, Y., & Guo, G. (2010). Roles of DNA methyltransferases in Arabidopsis development. *African Journal of Biotechnology*, 9(50), Article 50.

Huang, S., Zhang, A., Jin, J. B., Zhao, B., Wang, T.-J., Wu, Y., Wang, S., Liu, Y., Wang, J., Guo, P., Ahmad, R., Liu, B., & Xu, Z.-Y. (2019). *Arabidopsis* histone H3K4 demethylase JMJ17 functions in dehydration stress response. *New Phytologist*, 223(3), 1372–1387. https://doi.org/10.1111/nph.15874.

Itabashi, E., Osabe, K., Fujimoto, R., & Kakizaki, T. (2018). Epigenetic regulation of agronomical traits in Brassicaceae. *Plant Cell Reports*, 37, 87–101.

Jiang, F., & Doudna, J. A. (2017). CRISPR–Cas9 structures and mechanisms. *Annual Review of Biophysics*, 46(1), 505–529.

Jogam, P., Sandhya, D., Alok, A., Peddaboina, V., Allini, V. R., & Zhang, B. (2022). A review on CRISPR/Cas-based epigenetic regulation in plants. *International Journal of Biological Macromolecules*, 219, 1261–1271. https://doi.org/10.1016/j.ijbiomac.2022.08.182.

Karanthamalai, J., Chodon, A., Chauhan, S., & Pandi, G. (2020). DNA N6-Methyladenine modification in plant genomes—A glimpse into emerging epigenetic code. *Plants*, 9(2), Article 2. https://doi.org/10.3390/plants9020247.

Karlson, C. K. S., Mohd-Noor, S. N., Nolte, N., & Tan, B. C. (2021). CRISPR/dCas9-based systems: Mechanisms and applications in plant sciences. *Plants*, 10(10), Article 10. https://doi.org/10.3390/plants10102055.

Kavuri, N. R., Ramasamy, M., Qi, Y., & Mandadi, K. (2022). Applications of CRISPR/Cas13-Based RNA editing in plants. *Cells*, 11(17), 2665. https://doi.org/10.3390/cells11172665.

Kearns, N. A., Pham, H., Tabak, B., Genga, R. M., Silverstein, N. J., Garber, M. & Maehr, R. (2015). Functional annotation of native enhancers with a Cas9–histone demethylase fusion. *Nature Methods*, 12(5), 401–403.

Kim, J.-H. (2019). Chromatin remodeling and epigenetic regulation in plant DNA damage repair. *International Journal of Molecular Sciences*, 20(17), Article 17. https://doi.org/10.3390/ijms20174093.

Kohli, R. M., & Zhang, Y. (2013). TET enzymes, TDG and the dynamics of DNA demethylation. *Nature*, 502(7472), 472–479.

Koonin, E. V., Makarova, K. S., & Zhang, F. (2017). Diversity, classification and evolution of CRISPR-Cas systems. *Current Opinion in Microbiology*, 37, 67–78.

Kumar, M., Prusty, M. R., Pandey, M. K., Singh, P. K., Bohra, A., Guo, B., & Varshney, R. K. (2023). Application of CRISPR/Cas9-mediated gene editing for abiotic stress management in crop plants. *Frontiers in Plant Science*, 14. https://doi.org/10.3389/fpls.2023.1157678.

Kumar, S., & Mohapatra, T. (2024). Epigenetic modifications in genome help remembering the stress tolerance strategy adopted by the plant. *Frontiers in Bioscience-Landmark*, 29(3), Article 3. https://doi.org/10.31083/j.fbl2903126.

Kumar V, Jain M. The CRISPR-Cas system for plant genome editing: advances and opportunities. *Journal of Experimental Botany*. 66(1):47–57.

Laforest, L. C., & Nadakuduti, S. S. (2022). Advances in delivery mechanisms of CRISPR gene-editing reagents in plants. *Frontiers in Genome Editing*, 4. https://doi.org/10.3389/fgeed.2022.830178.

Lauria, M., & Rossi, V. (2011). Epigenetic control of gene regulation in plants. *Biochimica et Biophysica Acta (BBA) - Gene Regulatory Mechanisms*, 1809(8), 369–378. https://doi.org/10.1016/j.bbagrm.2011.03.002.

Lee, J. E., Neumann, M., Duro, D. I., & Schmid, M. (2019). CRISPR-based tools for targeted transcriptional and epigenetic regulation in plants. *PLoS One*, 14(9), e0222778.

Li, B., Yang, C., An, B. Wang, H., Albaqami, M., Abou-Elwafa, S. F., Xu, L., & Xu, Y. (2022). Comparative transcriptomic and epigenetic analyses reveal conserved and divergent regulatory pathways in barley response to temperature stresses. *Physiologia Plantarum*, 174(4), e13727. https://doi.org/10.1111/ppl.13727.

Li, H., Yan, S., Zhao, L., Tan, J., Zhang, Q., Gao, F., Wang, P., Hou, H., & Li, L. (2014). Histone acetylation associated up-regulation of the cell wall related genes is involved in salt stress induced maize root swelling. *BMC Plant Biology*, *14*(1), 105. https://doi.org/10.1186/1471-2229-14-105.

Li, J.-F., Norville, J. E., Aach, J., McCormack, M., Zhang, D., Bush, J., Church, G. M., & Sheen, J. (2013). Multiplex and homologous recombination–mediated genome editing in *Arabidopsis* and *Nicotiana benthamiana* using guide RNA and Cas9. *Nature Biotechnology*, 31(8), 688–691. https://doi.org/10.1038/nbt.2654.

Li, P., Yang, H., Wang, L., Liu, H., Huo, H., Zhang, C., Liu, A., Zhu, A., Hu, J., Lin, Y., & Liu, L. (2019). Physiological and transcriptome analyses reveal short-term responses and formation of memory under drought stress in rice. *Frontiers in Genetics*, 10. https://doi.org/10.3389/fgene.2019.00055.

Li, T., Yang, Y., Qi, H., Cui, W., Zhang, L., Fu, X., ... & Yu, T. (2023). CRISPR/Cas9 therapeutics: Progress and prospects. *Signal Transduction and Targeted Therapy*, 8(1), 36.

Li, Z., Zhang, D., Xiong, X., Yan, B., Xie, W., Sheen, J., & Li, J.-F. (2017). A potent Cas9-derived gene activator for plant and mammalian cells. *Nature Plants*, 3(12), 930–936. https://doi.org/10.1038/s41477-017-0046-0.

Liao, S., Qin, X., Luo, L., Han, Y., Wang, X., Usman, B., Nawaz, G., Zhao, N., Liu, Y., & Li, R. (2019). CRISPR/Cas9-induced mutagenesis of semi-rolled Leaf1,2 confers curled leaf phenotype and drought tolerance by influencing protein expression patterns and ROS scavenging in rice (Oryza sativa L.). *Agronomy*, 9(11), Article 11. https://doi.org/10.3390/agronomy9110728.

Liu, L., Zhang, J., Xu, J., Li, Y., Guo, L., Wang, Z., Zhang, X., Zhao, B., Guo, Y.-D., & Zhang, N. (2020). CRISPR/Cas9 targeted mutagenesis of SlLBD40, a lateral organ boundaries domain transcription factor, enhances drought tolerance in tomato. *Plant Science*, 301, 110683. https://doi.org/10.1016/j.plantsci.2020.110683.

Liu, X. S., Feng, S. J., Zhang, B. Q., Wang, M. Q., Cao, H. W., Rono, J. K., Chen, X., & Yang, Z. M. (2019). OsZIP1 functions as a metal efflux transporter limiting excess zinc, copper and cadmium accumulation in rice. *BMC Plant Biology*, 19(1), 283. https://doi.org/10.1186/s12870-019-1899-3.

Liu, Z., Chen, S., Lai, L., & Li, Z. (2022). Inhibition of base editors with anti-deaminases derived from viruses. *Nature Communications*, 13(1), 597. https://doi.org/10.1038/s41467-022-28300-0.

Lowder, L. G., Zhou, J., Zhang, Y., Malzahn, A., Zhong, Z., Hsieh, T.-F., Voytas, D. F., Zhang, Y., & Qi, Y. (2018). Robust transcriptional activation in plants using multiplexed CRISPR-Act2.0 and mTALE-Act systems. *Molecular Plant*, 11(2), 245–256. https://doi.org/10.1016/j.molp.2017.11.010.

Lucibelli, F., Valoroso, M. C., & Aceto, S. (2022). Plant DNA methylation: An epigenetic mark in development, environmental interactions, and evolution. *International Journal of Molecular Sciences*, 23(15), 8299. https://doi.org/10.3390/ijms23158299.

Maharajan, T., Krishna, T. P. A., Rakkammal, K., Ceasar, S. A., & Ramesh, M. (2022). Application of CRISPR/Cas system in cereal improvement for biotic and abiotic stress tolerance. *Planta*, 256(6), 106. https://doi.org/10.1007/s00425-022-04023-w.

Mangiavacchi, A., Morelli, G., & Orlando, V. (2023). Behind the scenes: How RNA orchestrates the epigenetic regulation of gene expression. *Frontiers in Cell and Developmental Biology*, 11. https://doi.org/10.3389/fcell.2023.1123975.

Minkenberg, B., Wheatley, M., & Yang, Y. (2017). Chapter seven—CRISPR/Cas9-enabled multiplex genome editing and its application. In D. P. Weeks & B. Yang (Eds.), *Progress in Molecular Biology and Translational Science* (Vol. 149, pp. 111–132). Academic Press. https://doi.org/10.1016/bs.pmbts.2017.05.003.

Mipeshwaree Devi, A., Khedashwori Devi, K., Premi Devi, P., Lakshmipriyari Devi, M., & Das, S. (2023). Metabolic engineering of plant secondary metabolites: Prospects and its technological challenges. *Frontiers in Plant Science*, 14. https://doi.org/10.3389/fpls.2023.1171154.

Mohanraju, P., Makarova, K. S., Zetsche, B., Zhang, F., Koonin, E. V., & Van der Oost, J. (2016). Diverse evolutionary roots and mechanistic variations of the CRISPR-Cas systems. *Science*, 353(6299), aad5147.

Moore, L. D., Le, T., & Fan, G. (2013). DNA methylation and its basic function. *Neuropsychopharmacology*, 38(1), 23–38.

Moradpour, M., & Abdulah, S. N. A. (2019). CRISPR/dCas9 platforms in plants: Strategies and applications beyond genome editing. *Plant Biotechnology Journal*, 18(1), 32–44. https://doi.org/10.1111/pbi.13232.

Narlikar, G. J., Fan, H.-Y., & Kingston, R. E. (2002). Cooperation between complexes that regulate chromatin structure and transcription. *Cell*, 108(4), 475–487. https://doi.org/10.1016/S0092-8674(02)00654-2.

Navarro, L., Dunoyer, P., Jay, F., Arnold, B., Dharmasiri, N., Estelle, M., Voinnet, O., & Jones, J. D. G. (2006). A plant miRNA contributes to antibacterial resistance by repressing auxin signaling. *Science*, 312(5772), 436–439.

Nuñez, J. K., Chen, J., Pommier, G. C., Cogan, J. Z., Replogle, J. M., Adriaens, C., ... & Weissman, J. S. (2021). Genome-wide programmable transcriptional memory by CRISPR-based epigenome editing. *Cell*, 184(9), 2503–2519.

O'Geen, H., Ren, C., Nicolet, C. M., Perez, A. A., Halmai, J., Le, V. M., Mackay, J. P., Farnham, P. J., & Segal, D. J. (2017). dCas9-based epigenome editing suggests acquisition of histone methylation is not sufficient for target gene repression. *Nucleic Acids Research*, 45(17), 9901–9916. https://doi.org/10.1093/nar/gkx578.

Ojolo, S. P., Cao, S., Priyadarshani, S. V. G. N., Li, W., Yan, M., Aslam, M., Zhao, H., & Qin, Y. (2018). Regulation of plant growth and development: A review from a chromatin remodeling perspective. *Frontiers in Plant Science*, 9. https://doi.org/10.3389/fpls.2018.01232.

Omachi, K., & Miner, J. H. (2022). Comparative analysis of dCas9-VP64 variants and multiplexed guide RNAs mediating CRISPR activation. *PLOS ONE*, 17(6), e0270008. https://doi.org/10.1371/journal.pone.0270008.

Papikian, A., Liu, W., Gallego-Bartolomé, J., & Jacobsen, S. E. (2019). Site-specific manipulation of *Arabidopsis* loci using CRISPR-Cas9 SunTag systems. *Nature Communications*, 10(1), 729.

Park, J.-J., Dempewolf, E., Zhang, W., & Wang, Z.-Y. (2017). RNA-guided transcriptional activation via CRISPR/dCas9 mimics overexpression phenotypes in *Arabidopsis*. *PLOS ONE*, 12(6), e0179410. https://doi.org/10.1371/journal.pone.0179410.

Peschansky, V. J., & Wahlestedt, C. (2014). Non-coding RNAs as direct and indirect modulators of epigenetic regulation. *Epigenetics*, 9(1), 3–12. https://doi.org/10.4161/epi.27473.

Phelan, J. D., & Staudt, L. M. (2020). CRISPR-based technology to silence the expression of IncRNAs. *Proceedings of the National Academy of Sciences*, 117(15), 8225–8227.

Piatek, A., Ali, Z., Baazim, H., Li, L., Abulfaraj, A., Al-Shareef, S., Aouida, M., & Mahfouz, M. M. (2015). RNA-guided transcriptional regulation in planta via synthetic dCas9-based transcription factors. *Plant Biotechnology Journal*, 13(4), 578–589. https://doi.org/10.1111/pbi.12284.

Pikaard, C. S., & Mittelsten Scheid, O. (2014). Epigenetic regulation in plants. *Cold Spring Harbor Perspectives in Biology*, 6(12), a019315. https://doi.org/10.1101/cshperspect.a019315.

Polstein, L. R., & Gersbach, C. A. (2015). A light-inducible CRISPR-Cas9 system for control of endogenous gene activation. *Nature Chemical Biology*, 11(3), 198–200. https://doi.org/10.1038/nchembio.1753.

Qi, Q., Hu, B., Jiang, W., Wang, Y., Yan, J., Ma, F., Guan, Q., & Xu, J. (2023). Advances in plant epigenome editing research and its application in plants. *International Journal of Molecular Sciences*, 24(4), 3442. https://doi.org/10.3390/ijms24043442.

Rakei, A., Maali-Amiri, R., Zeinali, H., & Ranjbar, M. (2016). DNA methylation and physio-biochemical analysis of chickpea in response to cold stress. *Protoplasma*, 253(1), 61–76. https://doi.org/10.1007/s00709-015-0788-3.

Ramirez-Prado, J. S., Ariel, F., Benhamed, M., & Crespi, M. (2017). Plant epigenetics: Non-coding RNAs as emerging regulators. In N. Rajewsky, S. Jurga, & J. Barciszewski (Eds.), *Plant Epigenetics* (pp. 129–147). Springer International Publishing. https://doi.org/10.1007/978-3-319-55520-1_7.

Rao, Y., Yang, X., Pan, C., Wang, C., & Wang, K. (2022). Advance of clustered regularly interspaced short palindromic repeats-Cas9 system and its application in crop improvement. *Frontiers in Plant Science*, 13, 839001.

Sandhya, D., Jogam, P., Allini, V. R., Abbagani, S., & Alok, A. (2020). The present and potential future methods for delivering CRISPR/Cas9 components in plants. *Journal of Genetic Engineering and Biotechnology*, 18(1), 25. https://doi.org/10.1186/s43141-020-00036-8.

Sapozhnikov, D. M., & Szyf, M. (2021). Unraveling the functional role of DNA demethylation at specific promoters by targeted steric blockage of DNA methyltransferase with CRISPR/dCas9. *Nature Communications*, 12(1), 5711.

Sharma, M., Sidhu, A. K., Samota, M. K., Shah, P., Pandey, M. K., & Gangurde, S. S. (2024). Technological advancements in the CRISPR toolbox for improving plant salt tolerance. *Discover Agriculture*, 2(1), 102. https://doi.org/10.1007/s44279-024-00105-3.

Shi, M., Du, Z., Hua, Q., & Kai, G. (2021). CRISPR/Cas9-mediated targeted mutagenesis of bZIP2 in Salvia miltiorrhiza leads to promoted phenolic acid biosynthesis. *Industrial Crops and Products*, 167, 113560. https://doi.org/10.1016/j.indcrop.2021.113560.

Sun, M., Yang, Z., Liu, L., & Duan, L. (2022). DNA methylation in plant responses and adaption to abiotic stresses. *International Journal of Molecular Sciences*, 23(13), 6910. https://doi.org/10.3390/ijms23136910.

Tang, G., Teotia, S., Tang, X., & Singh, D. (Eds.). (2021). *RNA-Based Technologies for Functional Genomics in Plants*. Cham: Springer.

Thiebaut, F., Hemerly, A. S., & Ferreira, P. C. G. (2019). A role for epigenetic regulation in the adaptation and stress responses of non-model plants. *Frontiers in Plant Science*, 10. https://doi.org/10.3389/fpls.2019.00246.

Verbsky, M. L., & Richards, E. J. (2001). Chromatin remodeling in plants. *Current Opinion in Plant Biology*, 4(6), 494–500. https://doi.org/10.1016/S1369-5266(00)00206-5.

Vergara, Z., & Gutierrez, C. (2017). Emerging roles of chromatin in the maintenance of genome organization and function in plants. *Genome Biology*, 18(1), 96. https://doi.org/10.1186/s13059-017-1236-9.

Voigt, P., & Reinberg, D. (2013). Epigenome editing. *Nature Biotechnology*, 31(12), 1097–1099.

Vojta, A., Dobrinić, P., Tadić, V., Bočkor, L., Korać, P., Julg, B., ... & Zoldoš, V. (2016). Repurposing the CRISPR-Cas9 system for targeted DNA methylation. *Nucleic Acids Research*, 44(12), 5615–5628.

Wang, F., Wang, L., Zou, X., Duan, S., Li, Z., Deng, Z., ... & Chen, S. (2019). Advances in CRISPR-Cas systems for RNA targeting, tracking and editing. *Biotechnology Advances*, 37(5), 708–729.

Wang, W., Huang, F., Qin, Q., Zhao, X., Li, Z., & Fu, B. (2015). Comparative analysis of DNA methylation changes in two rice genotypes under salt stress and subsequent recovery. *Biochemical and Biophysical Research Communications*, 465(4), 790–796. https://doi.org/10.1016/j.bbrc.2015.08.089.

Wang, Y., Luo, X., Sun, F., Hu, J., Zha, X., Su, W., & Yang, J. (2018). Overexpressing lncRNA LAIR increases grain yield and regulates neighbouring gene cluster expression in rice. *Nature Communications*, 9(1), 3516. https://doi.org/10.1038/s41467-018-05829-7.

Wei, J.-W., Huang, K., Yang, C., & Kang, C.-S. (2017). Non-coding RNAs as regulators in epigenetics (Review). *Oncology Reports*, 37(1), 3–9. https://doi.org/10.3892/or.2016.5236.

Xie, K., Minkenberg, B., & Yang, Y. (2015). Boosting CRISPR/Cas9 multiplex editing capability with the endogenous tRNA-processing system. *Proceedings of the National Academy of Sciences*, 112(11), 3570–3575. https://doi.org/10.1073/pnas.1420294112.

Xie, N., Zhou, Y., Sun, Q., & Tang, B. (2018). Novel epigenetic techniques provided by the CRISPR/Cas9 system. *Stem Cells International*, 2018(1), 7834175.

Xiong, T., Meister, G. E., Workman, R. E., Kato, N. C., Spellberg, M. J., Turker, F., ... & Novina, C. D. (2017). Targeted DNA methylation in human cells using engineered dCas9-methyltransferases. *Scientific Reports*, 7(1), 6732.

Xu, R., Yang, Y., Qin, R., Li, H., Qiu, C., Li, L., Wei, P., & Yang, J. (2016). Rapid improvement of grain weight via highly efficient CRISPR/Cas9-mediated multiplex genome editing in rice. *Journal of Genetics and Genomics = Yi Chuan Xue Bao*, 43(8), 529–532. https://doi.org/10.1016/j.jgg.2016.07.003.

Xu, Y., Zhang, S., Lin, S., Guo, Y., Deng, W., Zhang, Y., & Xue, Y. (2017). WERAM: A database of writers, erasers and readers of histone acetylation and methylation in eukaryotes. *Nucleic Acids Research*, 45(D1), D264–D270. https://doi.org/10.1093/nar/gkw1011.

Xue, C., & Greene, E. C. (2021). DNA repair pathway choices in CRISPR-Cas9-mediated genome editing. *Trends in Genetics*, 37(7), 639–656.

Yang, X., Liu, C., Niu, X., Wang, L., Li, L., Yuan, Q., & Pei, X. (2022). Research on lncRNA related to drought resistance of Shanlan upland rice. *BMC Genomics*, 23(1), 336. https://doi.org/10.1186/s12864-022-08546-0.

Yao, Y., Ni, Z., Peng, H., Sun, F., Xin, M., Sunkar, R., Zhu, J.-K., & Sun, Q. (2010). Non-coding small RNAs responsive to abiotic stress in wheat (Triticum aestivum L.). *Functional & Integrative Genomics*, 10(2), 187–190. https://doi.org/10.1007/s10142-010-0163-6.

Zegeye, W. A., Tsegaw, M., Zhang, Y., & Cao, L. (2022). CRISPR-based genome editing: Advancements and opportunities for rice improvement. *International Journal of Molecular Sciences*, 23(8), 4454. https://doi.org/10.3390/ijms23084454.

Zhang, D., Zhang, Z., Unver, T., & Zhang, B. (2020). CRISPR/Cas: A powerful tool for gene function study and crop improvement. *Journal of Advanced Research*, 29, 207–221. https://doi.org/10.1016/j.jare.

Zhang, J.-B., He, S.-P., Luo, J.-W., Wang, X.-P., Li, D.-D., & Li, X.-B. (2020). A histone deacetylase, GhHDT4D, is positively involved in cotton response to drought stress. *Plant Molecular Biology*, 104(1), 67–79. https://doi.org/10.1007/s11103-020-01024-9.

Zhang, R., Xu, W., Shao, S., & Wang, Q. (2021). Gene silencing through CRISPR interference in bacteria: Current advances and future prospects. *Frontiers in Microbiology*, 12. https://doi.org/10.3389/fmicb.2021.635227.

Zhang, Z., Mao, Y., Ha, S., Liu, W., Botella, J. R., & Zhu, J.-K. (2016). A multiplex CRISPR/Cas9 platform for fast and efficient editing of multiple genes in *Arabidopsis*. *Plant Cell Reports*, 35(7), 1519–1533. https://doi.org/10.1007/s00299-015-1900-z.

Zhang, Z., Xia, Z., Zhou, C., Wang, G., Meng, X., & Yin, P. (2024). Insights into salinity tolerance in wheat. *Genes*, 15(5), Article 5. https://doi.org/10.3390/genes15050573.

Zhao, T., Zhan, Z., & Jiang, D. (2019). Histone modifications and their regulatory roles in plant development and environmental memory. *Journal of Genetics and Genomics*, 46(10), 467–476. https://doi.org/10.1016/j.jgg.2019.09.005.

Zheng, Y., & He, R. (2023). Targeted DNA demethylation of the *Arabidopsis* genome using the SunTag-dCpf1-TET1cd system. *Micropublication Biology*, 2023.

Zhou, J., Deng, K., Cheng, Y., Zhong, Z., Tian, L., Tang, X., Tang, A., Zheng, X., Zhang, T., Qi, Y., & Zhang, Y. (2017). CRISPR-Cas9 based genome editing reveals new insights into MicroRNA function and regulation in rice. *Frontiers in Plant Science*, 8, 1598. https://doi.org/10.3389/fpls.2017.01598.

Zhou, J., Peng, Z., Long, J., Sosso, D., Liu, B., Eom, J.-S., Huang, S., Liu, S., Vera Cruz, C., Frommer, W. B., White, F. F., & Yang, B. (2015). Gene targeting by the TAL effector PthXo2 reveals cryptic resistance gene for bacterial blight of rice. *The Plant Journal*, 82(4), 632–643. https://doi.org/10.1111/tpj.12838.

6 Emerging Technologies and Methodologies in Epigenetics Research

Parama Majumdar, Mayank Varun and Mousumi Bhowal

6.1 INTRODUCTION

Epigenetics pertains to the inheritance of changes that do not result from changes in DNA sequence. It examines the heritable traits that arise from the interaction between the genetic makeup and environmental factors, while leaving the core DNA unaltered. These traits can be passed down through cellular divisions and, in certain instances, across generations, showcasing an impressive level of adaptability. Molecular modifications of DNA that control gene expression without modifying the DNA sequence itself ensure the stability during cell division. This discipline has witnessed substantial expansion in recent years, characterized by pioneering discoveries and swift progress in comprehension and applications.

The concept of "epigenetics" was coined by C. H. Waddington in 1942 to describe how identical cells, once separated, follow distinct paths through the course of development. Further, the premise of epigenetics and the significant biological function of DNA methylation and demethylation were proposed by Griffith and Mahler in 1969, who suggested that this process might serve as a foundation for long-term memory within the brain (Griffith & Mahler, 1969). Further, in 1975, two pivotal papers were published that presented a molecular framework for the regulation of gene activity as well as the heritability of both active and inactive gene states. Enzymatic methylation of cytosine in DNA, commonly referred to as DNA modification, was proposed as the framework of this gene action, linking it to developmental processes and environmental response. Several concurrently developed ideas indicated that the pattern of methylation may be passed down through generations if a methylase enzyme is present, which identifies hemi-methylated DNA shortly after replication while refraining from acting on unmethylated DNA. The central idea was that DNA methylation could significantly influence gene expression, implying that variations in DNA methylation could elucidate the mechanisms behind the activation and deactivation of genes throughout development (Haig, 2004; Holliday, 2006).

The study of chromatin structure and gene expression has emerged as an extremely dynamic area of research. Chromatin exists in two forms: an open configuration that facilitates the access of transcriptional machinery, and a closed configuration that inhibits transcription. Modification of histones, especially through acetylation and methylation, is pivotal in mediating these structural and functional changes (Holliday, 1989, 1990). The significance of RNA in epigenetic processes has also gained considerable attention. Alternative splicing of gene transcripts is viewed as an epigenetic mechanism, resulting in the generation of multiple isoforms of a given protein, each exhibiting slightly different characteristics, and it is likely that distinct cell types express specific isoforms. However, the specificity of splicing events poses a challenge, which may be addressed by the presence of small RNA molecules that hybridize across splice junctions (Holliday & Murray, 1994).

Significant advancements have been achieved in recent years in the understanding of epigenetic regulation in response to environmental stimuli, particularly under stress conditions. The molecular mechanisms that govern chromatin dynamics and gene silencing have been elucidated, which have in turn enhanced the comprehension of novel phenotypes that emerge rapidly in reaction to environmental changes. It has also been demonstrated through multiple evidence that certain epigenetic modifications can be passed down to subsequent generations (Kakutani, 2002). There has been remarkable growth in the field of epigenetics in recent years, largely due to advancements in technology. Researchers can now map and analyze epigenetic modifications throughout the entire genome, in specific cell types, even at the level of individual cells (Mazan-Mamczarz et al., 2022).

In the recent years, major thrust has been laid upon the identification and characterization of the full spectrum of epigenetic modifications. This includes DNA methylation, histone alteration and chromatin accessibility across diverse cell types and tissues. Projects like ENCODE have generated comprehensive data regarding the human epigenome (Dunham et al., 2012a; Esteller, 2006; Kellis et al., 2014). Apart from mapping these epigenetic changes, research has also been focused on understanding the functional consequences of specific modifications and the mechanism by which they may influence gene expression and ultimately the phenotype. Exploring the molecular processes that regulate the establishment, maintenance and removal of epigenetic marks – and govern their roles in responding to environmental signals and developmental cues – is the chief backdrop of research that can further strengthen agroeconomy by combating the impact of climate change

DOI: 10.1201/9781003589914-7

on agricultural yield (Hauser et al., 2011; Paszkowski & Grossniklaus, 2011). Translating fundamental epigenetic findings into clinical applications has also been an area of prime focus in recent research, including the creation of new diagnostic biomarkers, therapeutic targets and strategies for personalized medicine (Y. Li, 2020).

In view of this cutting-edge research pace, familiarity with the innovative methods and technologies becomes crucial. Epigenomics stands out as one of the most promising areas of study, offering enhanced understanding of gene regulation by allowing the identification of quantitative changes, modifications and regulatory sequences that go beyond sequence information. Many new epigenomic techniques are being developed, using microarray platforms with different substrates, processing methods and data analysis tools to reveal extensive variations in DNA methylation and chromatin modifications. Projects like the Human Epigenome Project are actively functioning to leverage these technologies for a deeper understanding of normal human development and also chronic and systemic ailments. This chapter intends to concentrate on these new technologies and methods that are pushing the boundaries of epigenetic research, while also appreciating traditional techniques such as ChIP-seq and microarray-based methylation analysis. The chapter will delve into the foundational principles and diverse applications of emerging techniques within epigenetic research to enhance our understanding of gene regulation and cellular mechanisms.

6.2 EMERGING TECHNOLOGIES

6.2.1 Next-Generation Sequencing (NGS)

The advanced ability to sequence large volumes of DNA allows for precise and in-depth sequence analysis of entire genomes or targeted genomic segments from numerous samples. This results in the creation of a multitude of applications. Next-generation sequencing (NGS) studies provide in-depth and extensive insights into the epigenetic modifications present in the genomes of various species and cell types. In recent years, there has been a surge of interest in DNA methylation, particularly following the identification of 5-hydroxymethylcytosine and its significance in epigenetic reprogramming and the maintenance of pluripotency. This revitalized focus has coincided with advancements in methodologies that together facilitate high coverage and single-base resolution profiling of the mammalian methylome, even in limited cell quantities (Hu et al., 2021; E. Meaburn & Schulz, 2012). NGS has accelerated quantitative trait loci (QTL) mapping and gene discovery in crops. High-throughput NGS genotyping platforms have significantly enhanced the ability to create detailed datasets of polymorphisms, encompassing single nucleotide polymorphisms (SNPs), insertions and deletions (InDels), structural variations and genomic rearrangements, thereby enhancing the marker density and improving the resolution of QTL maps (Nguyen et al., 2019).

In eukaryotic organisms, epigenetic modifications at the nucleotide level are primarily restricted to cytosine, which is converted into 5-methylcytosine (5-mC) by DNA methyltransferases (Ehrlich & Wang, 1981). Additionally, plants possess enzymes capable of reversing this modification (He et al., 2011). The components involved in the demethylation pathways of metazoans have only been recently identified, and their functional implications are currently under active discussion. At the level of nucleosomes, the amino acids present in the N-terminal tails of histone proteins serve as sites for a diverse array of reversible biochemical modifications, which are catalyzed by a correspondingly extensive range of enzymes, each exerting different effects on chromatin structure (Ehrlich & Wang, 1981; Hajkova et al., 2010).

DNA methylation, histone modifications, study of the transcriptome and DNA–protein interactions, including RNA Pol II binding to DNA, on a genome-wide scale are examined through next-generation sequencing (NGS). Additionally, NGS techniques have been established to explore other facets of chromatin organization, such as DNA accessibility and long-range interactions. The analysis of the transcriptome via RNA sequencing and the study of RNA–protein interactions represent key components of epigenomics facilitated by NGS (Meaburn & Schulz, 2012), and some such analysis methods have been discussed here.

6.2.1.1 Whole-Genome Bisulfite Sequencing (WGBS) for DNA Methylation

Bisulfite (BS) sequencing of DNA has emerged as a distinct method for examining DNA methylation. The bisulfite treatment process takes advantage of the varying susceptibility of cytosine and 5-methylcytosine (5-mC) to deamination by bisulfite in acidic conditions. In this process, cytosine is converted to uracil, whereas 5-mC remains unaffected. The full potential of this technique became apparent when bisulfite-treated DNA was analyzed through shotgun sequencing (MethylC-seq), enabling the identification of 5-mC at single-base resolution (Urich et al., 2015). Whole-genome bisulfite sequencing (WGBS) facilitates the comprehensive detection of DNA methylation patterns across the genome at single-base resolution in all three sequence contexts (CG, CHG and CHH), which is particularly significant in plant studies. Three specific DNA methyltransferases are responsible for methylating cytosine in various sequence contexts: MET1 targets CG sites, CMT3 acts on CHG sites and DRM2 methylates CHH sites, each playing a unique role in the regulation of gene expression (Kumar et al., 2018). Whole-genome bisulfite sequencing analysis of *Arabidopsis thaliana* demonstrated that methylation is confined to the CG context, and methylation observed in CHG and CHH contexts was primarily within heterochromatin regions enriched in transposable elements and repeats. The levels of methylation exhibited considerable variation between CG and non-CG contexts. In *A. thaliana*, the methylation level in the CG context was noted to be 24%, whereas the methylation levels in the CHG and CHH contexts were recorded at 7% and 2%, respectively. Techniques termed BS-Seq are used to sensitively measure cytosine methylation on a genome-wide scale within specific sequence contexts (Cokus et al., 2008). Table 6.1 describes the methylation specificities of cytosine and its significance in plants. Figure 6.1 illustrates the flow of events while using 5-mC as an epigenetic marker.

Specific primers of polymerase chain reaction (PCR) are designed accordingly, leading to amplification and mapping at a single base resolution (Warnecke et al., 2002). Recently, BS treatment has been integrated with next-generation sequencing (NGS) techniques, producing reduced representation bisulfite sequencing (RRBS) or whole genome bisulfite sequencing (WGBS) data, which provides insights into the global genomic distribution of 5-mC (Grunau et al., 2001).

Assessment of the biases in PCR amplification, arising from BS-induced DNA degradation, has been used for certain pathbreaking studies. In mice, memory acquisition, consolidation and long-term storage have been found to involve DNA methylation in a manner that is specific to the context and varies in kinetics. Traditionally, DNA methylation has been viewed primarily as a regulator of transcription that facilitates gene silencing. Nevertheless, its regulation and role appear to be considerably more intricate within the adult brain, raising several pertinent questions. Beyond the proteins responsible for DNA methylation, various other epigenetic factors are implicated and interact with one another, such as histone modifications and chromatin remodeling. Recent findings suggest that non-CpG DNA methylation also plays a significant regulatory role (Vaissiére & Miller, 2011).

WGBS data has also been used to understand the concept of heterosis using the model plant *Arabidopsis*. Heterosis is a key biological occurrence defined by the enhanced performance of a hybrid compared to its parental lines across various characteristics. *Arabidopsis* hybrids were found to exhibit elevated levels of DNA methylation throughout their genomes, particularly within transposable elements. Notably, this increased methylation in hybrid genomes primarily occurred in regions that exhibited differential methylation between the two parental lines and were associated with small RNAs. This observation suggests that the RNA-directed DNA methylation (RdDM) pathway may influence DNA methylation patterns in hybrids. Furthermore, genes that are sensitive to alterations in the methylome were identified, and it was noted that these genes were transcriptionally repressed in both the reciprocal hybrids. Additionally, the growth vigor of F1 hybrids was adversely affected by treatment with a demethylating agent and by the disruption of functional small RNA production due to mutations in the *Arabidopsis* RNA methyltransferase HUA enhancer 1. Collectively, these findings indicate that the genome-wide reconfiguration of DNA methylation, guided by the RdDM pathway, may contribute to the phenomenon of heterosis (Shen et al., 2012).

Allele-specific epigenetic modifications are likely prevalent within the human genome. Observations indicate that allele-specific methylation (ASM) is quantitative rather than qualitative, exhibiting heterogeneity across various tissues and individuals, and is primarily influenced by genotype in cis (Zemach et al., 2010). Consequently, the comprehensive characterization of allelic methylation patterns in tissues relevant to disease represents a significant objective for next-generation sequencing (NGS) studies, as the integration of epiallelic data is crucial for the optimal detection and interpretation of genetic association signals. ASM may complicate the identification of variants associated with traits by creating hemizygous loci and obscuring the direct relationship between genotype and phenotype.

Such findings strongly imply that to comprehend and identify the molecular mechanisms underlying complex disorders, NGS-based methodologies that incorporate genome-wide, high-resolution, quantitative and allelic epigenetic, genetic

TABLE 6.1
Methylation of cytosine

Feature	Description	Significance in plants	Key enzymes involved	Proposed position as per BS-Seq
CG methylation	Methylation of cytosine followed by guanine	Often associated with transcriptional repression, particularly in gene promoters and transposable elements	MET1 (MAINTENANCE METHYLTRANSFERASE 1)	Euchromatin
CHG methylation	Methylation of cytosine followed by any base, then guanine	Maintained by CMT3 and often found in heterochromatin regions	CMT3 (CHROMOMETHYLTRANSFERASE 3)	Heterochromatin
CHH methylation	Methylation of cytosine followed by any two bases	Maintained by DRM2 and involved in RNA-directed DNA methylation (RdDM) pathway	DRM2 (DOMAIN REARRANGED METHYLTRANSFERASE 2)	Heterochromatin

Source: Cokus et al. (2008) and Kumar et al. (2018).

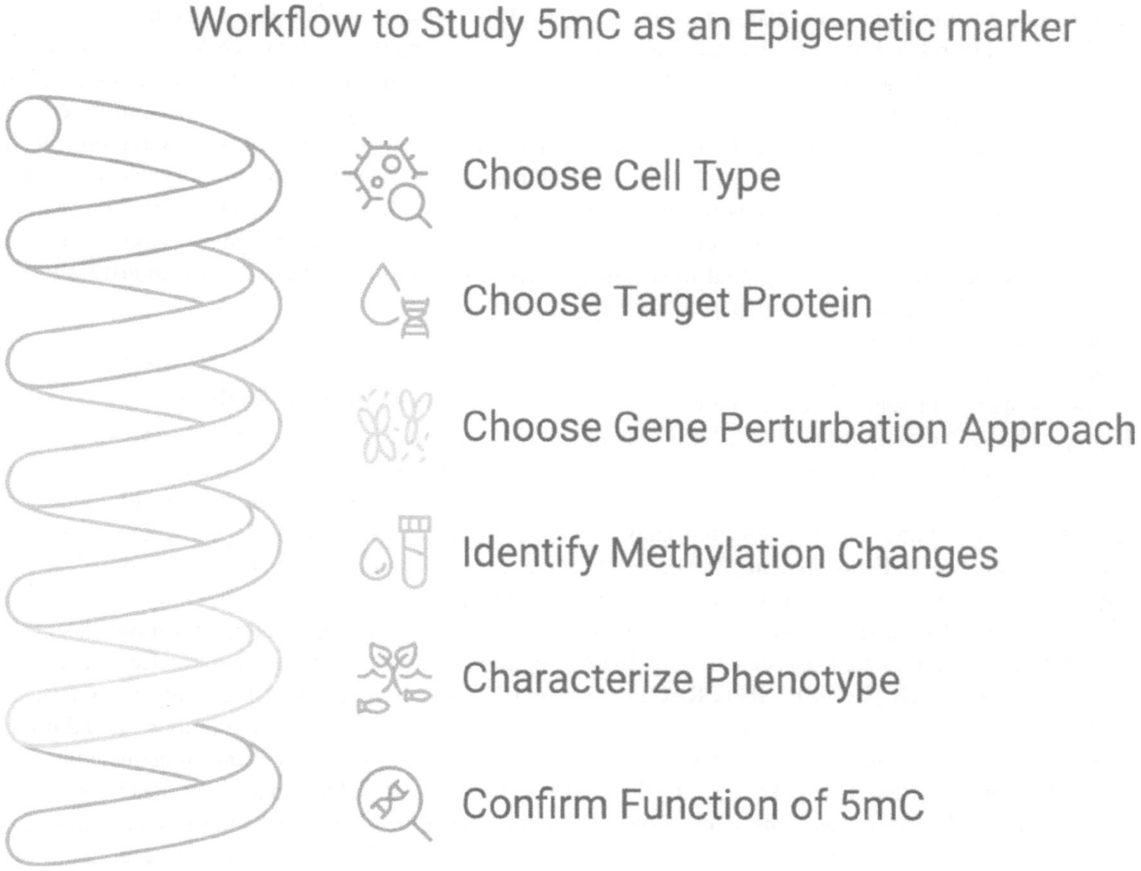

FIGURE 6.1 Workflow for use of 5-mC.

and transcriptomic data will be indispensable (Meaburn et al., 2010; Tycko, 2010).

6.2.1.2 Chromatin Immunoprecipitation Sequencing (ChIP-seq) for Histone Modifications

Chromatin immunoprecipitation technique, also known as ChIP, is a reliable method for investigating the complex dynamics of chromatin modifications and the interactions between proteins and DNA (Milne et al., 2009). This method enables scientists to identify specific DNA sequences and the proteins associated with them. The process requires binding of the proteins to DNA followed by fragmentation of the DNA and immunoprecipitation of the target protein along with its associated DNA fragments. Subsequently, the isolated DNA sequences are identified through techniques such as PCR. ChIP helps in identification of the functional role of the proteins of interest by detecting their localization within the genome. In the realm of epigenetic research, ChIP has been particularly relevant for mapping histone modifications (including methylation and acetylation), DNA methylation and the binding of regulatory proteins (such as transcription factors and chromatin remodeling complexes). All these elucidations have played an essential role in comprehending the epigenome (Gade & Kalvakolanu, 2012). Histone modifications can indicate distinct chromatin configurations and patterns of gene expression. This is observed in the modification of H3K4me3, which is commonly associated with active promoters, whereas H3K27me3 is linked to repressed genes. The techniques of ChIP and high-throughput sequencing integrated together form the ChIP-seq method, which has been used to facilitate comprehensive genome-wide mapping of these modifications. It helps to understand the diverse distribution of these modifications across various cell types and developmental stages.

ChIP-seq has been employed to map multiple histone modifications in human CD4+ T cells, illustrating the combinatorial characteristics of these marks and their relationship to gene expression (Bernstein et al., 2007a). This groundbreaking study has paved the way for further research on the significance of histone modifications in a range of biological processes, such as development, differentiation and the mechanisms of disease pathogenesis (Barski et al., 2007; Smith & Meissner, 2013).

DNA methylation, which is a crucial epigenetic marker, has also been analyzed by ChIP. Conventional methods like bisulfite sequencing provide direct assessments of DNA methylation levels. ChIP, however, is also capable of identifying the proteins that associate with methylated DNA, including MeCP2, which has a specific affinity for 5-methylcytosine and is essential in the process of transcriptional repression. ChIP assays focused on MeCP2 have been employed to investigate the involvement of DNA-binding proteins in neurological conditions such as Rett syndrome.

This has helped in the comprehension of the significance of DNA methylation and its related proteins in brain functionality (Amir et al., 1999; Shahbazian et al., 2002). ChIP can also be combined with bisulfite sequencing (ChIP-BS) to precisely assess the methylation status of DNA linked to a particular protein (Weber et al., 2005).

Another application of chromatin immunoprecipitation has been the study of regulatory proteins. Transcription factors are the key to gene regulation as they bind to specific DNA sequences, and ChIP-seq allows researchers to identify all the genomic sites where a particular transcription factor interacts. This helps in understanding the regulatory targets of the proteins and its role in gene expression networks. The ENCODE project has extensively utilized ChIP-seq to map the binding sites of various transcription factors throughout the human genome. This has further helped in creating a vital resource for understanding the regulatory mechanisms of gene (Dunham et al., 2012b; Feingold et al., 2004).

The identification of regulatory targets by ChIP has enabled the elucidation of regulatory networks that govern interactions of transcription factors associated with light signaling, such as PHYTOCHROME INTERACTING FACTORS (PIFs) (Casal & Fankhauser, 2023), as well as those involved in hormone responses, including AUXIN RESPONSE FACTORS (ARFs) Cherenkov et al., 2018). ChIP has been employed to investigate the interactions of transcription factors such as FLOWERING LOCUS C (FLC) and CONSTANS (CO) with their respective target genes, thereby elucidating their role in regulating flowering time in accordance with environmental signals, including light and temperature (Adrian et al., 2010; K. Liu et al., 2015).

While ChIP has its advantages, it also comes with some limitations. The method can be quite technically challenging, and the quality of the data obtained can be affected by various factors, including the specificity of the antibodies used, the efficiency of crosslinking and the conditions during sonication (Park, 2009; Kellis et al., 2014b). Additionally, ChIP captures a momentary view of protein–DNA interactions and may not reflect the temporal dynamics of these interactions. Consequently, it is often essential to integrate ChIP with other methodologies, such as RNA-seq for evaluating gene expression and live-cell imaging for investigating protein dynamics, to achieve a more thorough understanding of epigenetic regulation. ChIP has transformed the field of epigenetic research by facilitating the mapping of histone modifications, proteins associated with DNA methylation and the binding of transcription factors throughout the genome. The information obtained from ChIP-based methodologies has yielded critical insights into the intricate relationships between epigenetic markers and their effects on gene expression, development and various diseases. As sequencing technologies advance and innovative ChIP-based techniques emerge, this method is poised to continue as a fundamental element of epigenetic research, enhancing our comprehension of the complex mechanisms that regulate gene activity and cellular operations (Chen et al., 2018)

As a result, it is often necessary to combine ChIP data with other methods, like RNA-seq for assessing gene expression (Trapnell et al., 2012) and live-cell imaging for studying protein dynamics, to achieve a more comprehensive understanding of epigenetic regulation. Furthermore, improved ChIP-based techniques, such as CUT & RUN (Skene & Henikoff, 2017) and CUT & TAG (Kaya-Okur et al., 2019), which enhance signal-to-noise ratios and minimize background noise, are continuously advancing the field.

6.2.1.3 RNA Sequencing (RNA-seq) for Transcriptome Analysis

RNA sequencing (RNA-seq) is a technique for investigating the transcriptome to understand the functional consequences of epigenetic modifications and provides a quantitative evaluation of genes. Genes with differential expression have been identified using RNA sequencing (Wang et al., 2009). This technique is also used in the study of non-coding RNAs that play a significant role in the regulation of epigenetic processes. Recent advancements in RNA-seq have enabled the analysis of transcriptomic diversity within cellular populations. In addition, methodologies such as Global Run-On sequencing (GRO-seq) and Precision Run-On sequencing (PRO-seq) have yielded important information regarding nascent transcription and can be employed to investigate the immediate impacts of epigenetic changes on gene transcription (Mutz et al., 2013). Unlike conventional methods like microarrays, RNA-seq offers several advantages, such as increased sensitivity, a wider dynamic range and the ability to detect new transcripts and isoforms (Mortazavi et al., 2008).

To analyze the transcriptome, the RNA is isolated from a biological specimen, after which the RNA molecules are transformed into complementary DNA (cDNA) utilizing reverse transcriptase. Specific barcoded adaptors and sequencing primers are linked to the cDNA fragments. These fragments undergo sequencing through high-throughput next-generation sequencing (NGS) platforms. The obtained sequences are aligned with a reference genome or transcriptome, and the number of hits that correspond to each gene is quantified (CDNA (Copy DNA) | NHGRI.; Wang et al., 2009).

RNA sequencing (RNA-seq) ascertains the relationship between altered gene expression and the specific epigenetic markers. Examination of the transcriptome obtained through different levels of DNA methylation or histone modifications helps to identify genes whose expression is affected by these DNA modifications (Mortazavi et al., 2008). RNA-seq has also been employed to identify and characterize the non-coding RNAs, which play an enigmatic role in mechanisms such as gene silencing and chromatin remodeling (Hrdlickova et al., 2017).

In combination, RNA-seq may be used with other epigenomic methodologies, such as ChIP-seq and whole-genome bisulfite sequencing (WGBS), to gain a thorough comprehension of the epigenetic modifications that contribute to various biological processes (Mutz et al., 2013). Single-cell RNA sequencing allows researchers to study gene expression at the individual cell level, revealing expressional diversity at different developmental stages (Hwang et al., 2018). This is particularly relevant in epigenetics, where differences in epigenetic markers among cells can have significant functional consequences. To explore nascent transcription, techniques like Global Run-On sequencing (GRO-seq) and Precision Run-On sequencing (PRO-seq) are used to investigate nascent transcription, offering valuable information about the immediate effects of epigenetic changes on gene transcription (L. Liu et al., 2025).

6.2.2 Single-Cell Genomics

Most of the epigenomic analyses studies have provided an outline to the regulatory mechanisms of cellular functions. However, these studies have often been found to mask the inherent heterogeneity that exists within seemingly homogeneous cell populations. Single-cell epigenomics integrates precise epigenetic profiling, the isolation and barcoding of individual cells and high-throughput sequencing techniques to delineate the epigenetic landscapes present within groups of single cells. Populations of cells that appear uniform may exhibit significant variability in their epigenetic, transcriptional or phenotypic characteristics. Consequently, the application of bulk population epigenomics is restricted in certain contexts. In contrast, single-cell epigenomics offers the potential to overcome these constraints and pave the way for novel avenues in epigenetic research, particularly in critical areas that involve developing, dynamic and heterogeneous cellular compositions (Chang et al., 2008; Landan et al., 2012; Shalek et al., 2013).

Nucleosome positioning plays a vital role in gene expression and various DNA-related processes. The overall pattern of nucleosome positioning across the genome is shaped by a combination of factors, including the DNA sequence, ATP-dependent nucleosome remodeling enzymes and transcription factors such as activators, components of the preinitiation complex and elongating RNA polymerase II. These factors interact with one another, leading to variations in nucleosome positioning patterns among different genes and within cell populations, which in turn affects gene expression (Struhl & Segal, 2013). Histone modification maps, along with information regarding histone variants and nucleosome occupancy, have revealed typical nucleosomal arrangements surrounding transcription start sites, enhancers and insulators. Additionally, data on DNA methylation have demonstrated correlated patterns of CpG methylation and hydroxymethylation in these epigenetic hotspots. Nucleosome density patterns and DNA modifications to DNA accessibility are connected by these techniques like DNase hypersensitivity mapping and ATAC-seq. The complex and wide correlations observed between these local epigenetic characteristics and gene expression sharply show the potential involvement of each epigenetic pathway in the regulation of gene activity.

Remarkably changing aspects within gene regulatory regions have been analyzed by DNA methylation, principally emphasizing the role of TET proteins in active demethylation at transcription factor binding sites. As suggested by these findings, the local patterns of methylation were largely influenced by the underlying DNA sequence.

Thus, serving as a valuable indicator in the study of gene regulation in both healthy and diseased cells, the epigenetic modification even holds the potential of a biomarker. A significant breakdown of genomes into megabase-scale domains has resulted from the examination of epigenomic profiles. These domains encompass topologically associating domains (TADs), which are identified through techniques such as chromatin conformation capture (3C) and its derivatives, including Hi-C.

Along with this, there are DNA-replication domains, lamina-associated domains and domains that are inferred through computational methods based on multivariate analysis of histone modification maps. It is crucial to note that all these maps display a strong association with one another and with gene expression.

Certain techniques that have been particularly helpful in these studies are as follows:

1. *DNase-hypersensitivity mapping* is a technique employed to investigate active regulatory elements within the genome, focusing on areas characterized by a reduced density of nucleosomes. These regions exhibit increased susceptibility to cleavage by DNase I due to diminished nucleosome protection. The resulting DNA fragments from the cleaved chromatin are utilized to create a map of these sensitive regions throughout the genome. ATAC-seq (assay for transposase-accessible chromatin using high-throughput sequencing) performs a function similar to that of DNase-hypersensitivity mapping. This method identifies active regulatory sites that are marked by a lower nucleosome density. It employs the Tn5 transposase, which, due to steric hindrance, is capable of inserting sequencing adapter sequences exclusively into nucleosome-free regions.
2. *Chromatin conformation capture (3C)* is a technique utilized to investigate the three-dimensional structure of chromosomes through proximity ligation. This method involves the crosslinking of chromatin using a fixing agent, typically formaldehyde, followed by the digestion of DNA with a restriction enzyme that cuts at six or four base pairs. The final step is the ligation of the crosslinked chromatin. The resulting chimeric DNA template allows for the linear representation of regions that were previously in close spatial proximity.
3. *Hi-C* is an advanced method derived from chromatin conformation capture (3C) that facilitates genome-wide analysis of chromosome

conformation. This technique employs deep sequencing of chimeric 3C DNA templates, followed by a statistical examination of the distribution of ligation junctions across two-dimensional contact matrices. To improve the resolution of the assay, ultradeep sequencing or variations of Hi-C that focus on specific regions of interest can be implemented.

Single-cell RNA sequencing has swiftly become a prominent platform for genome-wide analysis of individual cells in recent years. Current methodologies enable the routine characterization of transcriptomes from hundreds or even thousands of single cells across various applications. The computational difficulties related to interpreting and modeling the intricate transcriptional states within mixed populations are progressively being resolved. The field of single-cell epigenomics stands to benefit significantly from the availability of transcriptional data. Initially, independent samples of cells can be analyzed at both transcriptional and epigenomic levels. This approach facilitates the computational integration of the data into a model that deduces epigenomic and transcriptional subpopulations with corresponding frequencies.

6.2.3 CRISPR-dCas9 Systems for Targeted Epigenetic Editing

The versatility of the CRISPR platform has opened new avenues for research: epigenetic editing. This promising domain, which utilizes catalytically inactive Cas9 fusions, facilitates accurate modifications to the epigenome without altering the basic DNA sequence (Azeez et al., 2024; Ganger et al., 2023).

The epigenome represents an avant-garde of inheritable biochemical alterations that overlap the DNA, playing an important role in the regulation of gene expression. Variations like modification of histones, DNA methylation, responses to environmental stimuli and a large variety of cellular-level changes are responsible for development and differentiation. A large number of diseases are connected with disruptions in the epigenetic framework.

CRISPR-dCas9 is a leading-edge progress in genomics, through which any specific targets within a genome can be accurately targeted by scientists. dCas9, a variant of the Cas9 enzyme, is used for this revolutionary technology. The technology does not possess catalytic activity, in conjunction with various effector domains, but enables the introduction of diverse epigenetic modifiers at selected sites, thereby allowing the precise addition or removal of specific epigenetic marks. Such capabilities make it an invaluable tool for exploring the functional implications of these marks (Qi et al., 2013), for example, when dCas9 is linked to DNA methyltransferases (DNMTs), it enables targeted DNA methylation, while its association with histone-modifying enzymes, such as histone acetyltransferases (HATs) or histone deacetylases (HDACs), permits accurate histone modifications (Ganger et al., 2023; Karlson et al., 2021).

The potential uses of CRISPR-dCas9 in the arena of epigenetic editing are huge and expanding rapidly. With regard to gene regulation, these systems have developed remarkable precision in activating or repressing gene expression. Through dCas9 fusions to promoter or enhancer regions, researchers can effectively change the epigenetic landscape, thereby influencing transcriptional outcomes (Huang et al., 2020). This will be a powerful approach for decoding the complicated regulatory networks governing gene expression and for identifying critical epigenetic marks associated with various biological processes. CRISPR-dCas9 epigenetic editing will be beyond simple gene regulation, presenting significant opportunities for disease modeling. By imitating the epigenetic changes linked to diseases in cellular models, valuable insights will be possible into the mechanisms behind various disorders, for instance, abnormal DNA methylation patterns associated with many cancers. Using CRISPR-dCas9-DNMT fusions, scientists can generate specific hypermethylation at the promoters of tumor suppressor genes, thereby effectively replicating the epigenetic modifications observed in cancer cells and uncovering new therapeutic targets (Pulecio et al., 2017; Tejedor et al., 2023). With regard to neurological disorders, where the dysregulation of histone modifications plays a crucial role, CRISPR-dCas9-HAT or HDAC fusions can be utilized to model relevant epigenetic changes and evaluate their impact on neuronal function (Allis & Jenuwein, 2016; Pulecio et al., 2017).

Despite being considerably potential, CRISPR-dCas9 faces many challenges of epigenetic editing, off-target effects, etc. Thus, to ensure the accuracy of planned modifications, careful design and authenticated experiments remain important. Further studies are required for the lasting stability and reversibility of these epigenetic changes.

The prompt progressions in CRISPR technology and along with a vast understanding of the epigenome are expanding this field of epigenetic editing. With its unparalleled precision and flexibility, it will revolutionize our understanding of gene regulation and disease modeling, potentially leading to revolutionary therapeutic strategies targeting the epigenome.

The expanding and advancing range of CRISPR-based epigenetic modifiers presents a promising future for this field, helping to discern the intricate relationships between genes and environmental factors, thereby deepening our comprehension of human health and disease (Huang et al., 2020).

6.3 CONFLUENCE OF EMERGING TECHNOLOGIES

Numerous groundbreaking technologies have arisen as the area of epigenetics advances. These epigenetic techniques can detect and examine chromatin state across different dimensions, from specific locus to genome-wide sequencing. The epigenetic techniques involve approaches, like utilizing high-quality antibodies, chromatin functional assays, imaging tools, high-throughput sequencing methods and extensive bioinformatics workflows.

DNA methylation represents a crucial epigenetic mechanism, involving modifications like 5-methylcytosine (5-mC), N6-methyladenine (6-mA) and 4-methylcytosine (4-mC). This is the most researched and comprehended DNA modification pattern. While 6-mA and 4-mC are predominantly observed in prokaryotic genomes, 5-mC is the most prevalent form of methylation in eukaryotes.

Recent developments in sequencing technologies enable researchers to read entire sequences, and this helps in exploring a greater scope of base modifications without the need for bisulfite treatment. Progress has also been made in methods capable of detecting oxidized versions of 5-mC, such as 5-hydroxymethylcytosine (Holliday, 1989; Zhang et al., 2015). Numerous technologies have been established to investigate the functions and dynamics of histone modifications, which represent a significant aspect of the epigenetic code. The majority of these technologies are built upon the framework of the chromatin immunoprecipitation (ChIP) assay discussed earlier in the chapter. The integration of ChIP-sequencing with 3C-based techniques, including Hi-C, enables the prediction of epigenetic landscapes within the three-dimensional architecture of chromatin. This approach elucidates the connections between patterns of histone modifications and the three-dimensional spatial organization of chromosomes, which play a crucial role in regulating biological functions in eukaryotic cells (Y. Li, 2021).

Understanding the changes in DNA bases, especially 5-mC, is crucial because they play a significant role in gene regulation, X chromosome inactivation, cellular differentiation and tumor development. The specific functions of 5-hmC in regulating gene expression are still not entirely clear, but it is linked to maintaining gene expression levels. Microbial genomes often show 6-mA modifications, which are key in controlling various biological functions. In *Caenorhabditis elegans*, 6-mA might carry heritable epigenetic information, while in *Drosophila*, it could help suppress transposons during embryonic development. The pathways for enzyme-mediated synthesis indicate that 5-hmU may play important roles in eukaryotic organisms, although its full significance is not completely understood. Different organisms exhibit 5-hmU and 7-mG modifications, each potentially influencing gene expression and other biological processes (Schübeler, 2015; Ziller et al., 2013). These modifications are essential for gene expression, growth, development and resilience to stress in organisms, thus impacting traits and aiding in the identification of diseases like cancer. The ability to dynamically detect epigenetic variations within cells or tissues is crucial for both scientific research and diagnostic purposes (Li, 2021; Zhang et al., 2015). Table 6.2 gives a brief summary of the modifications and their functions.

The initial detection of 5-methylcytosine (5-mC) relied on the methylation-sensitive amplified polymorphism (MSAP) method. The analysis of modified bases within animal and plant genomes was facilitated once their presence was confirmed in the overall base composition. Modification-sensitive restriction endonucleases (MSREs) like *Msp*I and *Hpa*II were employed for this process of mapping methylated cytosine and adenine. The *Mcr*BC enzyme identified methylated DNA through selective digestion (Xiong et al., 2019). Following this, a specific monoclonal antibody for 5-mC was employed to quantify methylcytosine across the entire genome, providing a more cost-effective and high-throughput alternative to high-performance liquid chromatography (HPLC)-based analyses. Initial methods such as methylation-sensitive PCR were also employed, succeeded by the creation of monoclonal antibodies aimed at quantifying methylcytosine. Subsequently, various efficient tools and techniques utilizing sodium-bisulfite conversion reactions have been developed to assess 5-mC levels with single-base resolution. The quantification of 5-hydroxymethylcytosine (5-hmC) has primarily utilized

TABLE 6.2

Cytosine modifications and their putative functions in model organisms

Modification	Organism	Known or proposed function	Associated processes	References
5-Hydroxymethylcytosine (5-hmC)	Mammals	Maintains gene expression levels, DNA demethylation intermediate	Gene regulation, DNA repair, development	Luo et al. (2016)
6-Methyladenine (6-mA)	Bacteria, *C. elegans*, *Drosophila*	Gene regulation, transposon suppression (*Drosophila*)	Various biological processes, heritable epigenetic information (*C. elegans*)	Greer et al. (2015)
5-Hydroxymethyluracil (5-hmU)	Eukaryotes	Potential role in DNA repair, stress response	Function not fully understood	Ito et al. (2011)
7-Methylguanosine (7-mG)	Various organisms	Potential role in gene expression, RNA processing	Various biological processes	Mitchell et al. (2010), J. Wang et al. (2019)

liquid chromatography and mass spectrometry techniques (LC-MS, HPLC-MS), although immunoassays are also being developed for this purpose. The initial quantification of 6-methyladenine (6-mA) was achieved through HPLC (Eick et al., 1983; Woodcock et al., 1984). Certain modified bases, such as 6-mA and 4-methylcytosine (4-mC), are present at low levels and can be detected using highly sensitive methodologies (Heyn & Esteller, 2015). Advanced techniques such as SMRT sequencing facilitate the reliable identification of frequently methylated DNA bases, including 5-mC, 6-mA and 4-mC. DA-6-mA-seq employs *Dpn*I to achieve precise detection of methylated adenine sites, necessitating only a small amount of input DNA and sequencing depth. Most commonly methylated DNA bases, including 5-mC, 6-mA and 4-mC, can be identified through SMRT sequencing, which requires only 25-fold coverage to ensure a higher confidence level in their detection (Davis et al., 2013) Table 6.3 lists techniques for the analysis of DNA modifications, and Figure 6.2 outlines the functions of these techniques.

A novel approach known as nanopore sequencing facilitates the direct profiling of the DNA methylome without the necessity for bisulfite conversion. This technique differentiates 5-mC from other nucleobases by employing electrical fields to maneuver DNA strands through minuscule openings. It detects bases, including their modified forms, by monitoring variations in ion current as DNA sequentially traverses the nanopores. Specialized nanopores, such as the MspA channel and solid-state nanopores (ssNPs), have been utilized to examine modified DNA bases through this sequencing method (Laszlo et al., 2013, 2014). A technique utilizing the MinION device from Oxford Nanopore Technology has been established for the direct identification of modified bases, including 5-mC, relying on electrical signals without the need for chemical alterations. This approach provides longer read lengths and facilitates rapid sequencing. While SMRT sequencing is associated with a higher error rate in base identification, typically ranging from 18% to 20%, nanopore sequencing demonstrates a lower error rate, between 6% and 10% (Simpson et al., 2017). Nevertheless, enhancements in throughput and sequencing depth are crucial for the routine analysis of methylomes in larger genomes using third-generation sequencing technologies.

6.4 EMERGING METHODOLOGIES

6.4.1 Computational Biology and Bioinformatics

Unprecedented advancements in technology have made it possible to conduct genome-scale analyses of epigenetic mechanisms, thereby facilitating comprehensive epigenome projects (Bernstein et al., 2007a). The epigenetic examination of embryonic stem (ES) cells has begun to reveal the fundamental processes underlying mammalian development (Surani et al., 2007). Furthermore, in the realm of cancer research, epigenetics introduces innovative strategies for early diagnosis and treatment (Jones & Baylin, 2007). Similarly in plants, an increasing number of studies are focusing on DNA methylation, which reveals methylation profiles and elucidates variations across different plant tissues. Moreover, while comparative analyses of DNA methylation among various plant tissue types and their dynamics during developmental processes are gradually emerging, they are essential for comprehending developmental and regulatory decisions. Applying this knowledge from plant model species to commercial crops could lead to the development of new varieties that exhibit enhanced stress resilience and improved yield (Omony et al., 2020). Disease resistance is one such trait, which is influenced by epigenetic modifications. A variety of techniques are available for the study of each of these epigenetic modifications, and each technique involves DNA sequencing followed by analysis of sequenced data using a variety of bioinformatics tools (Singh et al., 2024).

A variety of experimental methodologies have been established for the genome-wide mapping of epigenetic data. These methodologies typically adhere to a three-stage framework. Initially, the epigenetic data is biochemically transformed into genetic information, such as by enriching genomic regions that exhibit specific histone modifications within a DNA library. Subsequently, conventional DNA techniques, including tiling microarrays or sequencing, are employed. Finally, computational algorithms are utilized to deduce the epigenetic information from the data generated by the tiling arrays or sequencing outputs. All methods for epigenome mapping produce substantial volumes of data, necessitating effective strategies for low-level data processing and quality assurance (Bock & Lengauer, 2008).

ChIP-on-chip is a technique that integrates chromatin immunoprecipitation (ChIP) to enrich specific DNA fragments with the capabilities of tiling microarrays to identify variations between sample and control DNA (Buck & Lieb, 2004). Areas of the microarray that show significant over-representation in the immunoprecipitated DNA compared to control DNA are interpreted as being epigenetically modified or protein-bound, contingent upon the antibody utilized. A specific variant of ChIP-on-chip, known as methyl-DNA immunoprecipitation (MeDIP), involves the immunoprecipitation of purified DNA using an antibody that targets methylated cytosines, resulting in genomic maps that illustrate DNA methylation patterns. Although these methodologies are widely employed globally in numerous laboratories, the quality of antibodies remains a critical issue that necessitates careful monitoring. Additionally, it is essential to consider background noise caused by cross-hybridization and the variability in oligomer affinities during the data processing phase. To enhance data quality and ensure standardization, the Microarray and Gene Expression Data Society published a checklist outlining the minimum required information that should be reported for any ChIP-on-chip dataset (The MIAME Checklist-Update January, 2005).

TABLE 6.3
Techniques for analysis of DNA modifications

Technique	Target modification (s)	Resolution	Capabilities	Advantages	Limitations	References
Bisulfite Sequencing (BS-seq)	5-mC	Single base	Detects 5-mC in any DNA region	Gold standard for 5-mC mapping	Requires bisulfite conversion, can damage DNA	Chen et al. (2023)
Modification-Sensitive Restriction Endonucleases (MSREs)	5-mC, 6-mA	Restriction site	Simplifies mapping of 5-mC and 6-mA	Easy to use for targeted regions	Limited to restriction sites	Broday and Costa (2000)
Methylation-Sensitive PCR (MS-PCR)	5-mC	Amplicon	Detects methylation in specific amplicons	Simple and cost-effective	Limited to targeted regions	Khodadadi et al. (2021)
Monoclonal Antibodies	5-mC	Global	Quantifies global 5-mC levels	Provides overall methylation levels	Lacks single-base resolution	Kelkar and Deobagkar (2009)
Liquid Chromatography-Mass spectrometry (LC-MS)	5-mC, 5-hmC, 6-mA, etc.	Global/targeted	Quantifies various modified bases	Sensitive and accurate	Can be complex and expensive	Lai et al. (2019)
Single-Molecule Real-Time (SMRT) Sequencing	5-mC, 6-mA, 4-mC	Single base	Detects various modified bases directly	High confidence base calling	Can be expensive	Ardui et al. (2018)
DA-6-mA-seq	6-mA	Single base	Sensitive detection of 6-mA	Requires minimal DNA input and sequencing depth	Specific to 6-mA	Feng et al. (2024)
Nanopore	5-mC	Single base	Differentiates 5-mC from other nucleobases by application of electrical field	Low error rate (between 6% and 10%)	Can be expensive	Laszlo et al. (2013)

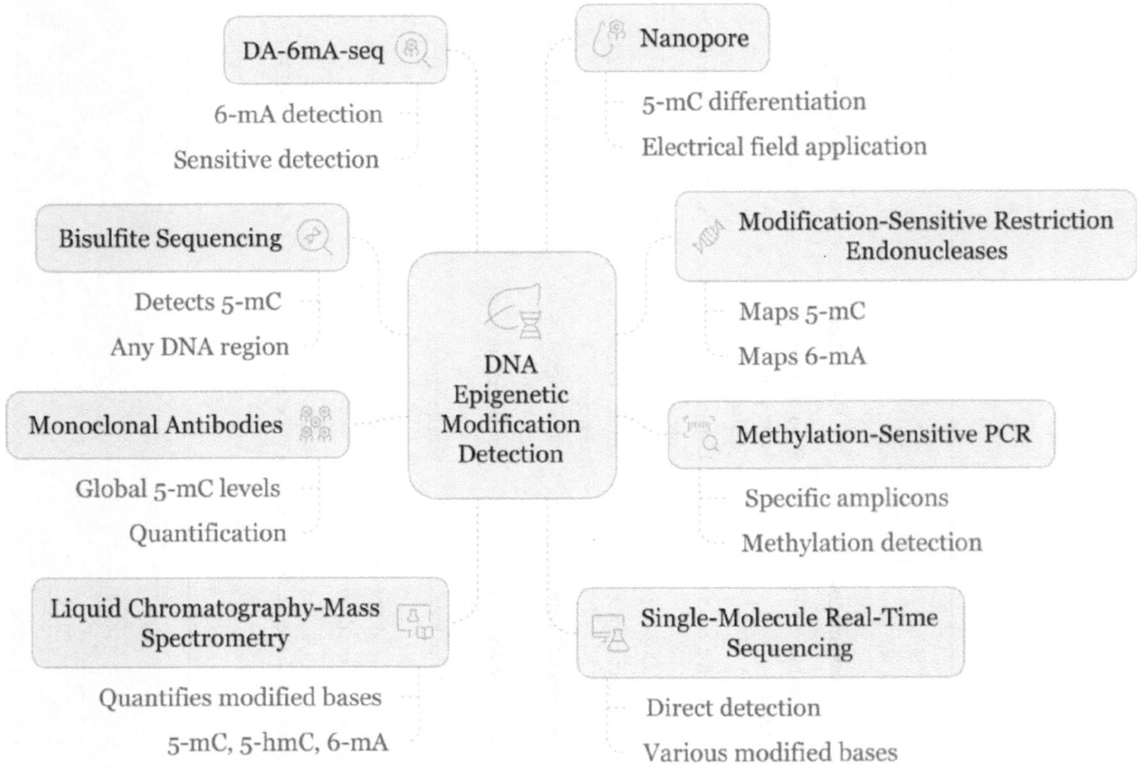

FIGURE 6.2 Analytical techniques for epigenetic modifications in DNA and their functions.

For ChIP-on-chip analysis, a significant bioinformatic challenge lies in generating a ranked list of overrepresented genomic regions based on raw probe intensities (Royce et al., 2005). While there are parallels to the analysis of tiling array data used for transcriptome mapping, most existing algorithms are specifically designed for peak detection in ChIP-on-chip datasets. The traditional and still prevalent approach consists of a three-step methodology. Initially, the microarrays undergo quantile normalization and are standardized to a uniform median intensity. Subsequently, a Wilcoxon rank sum test is conducted locally within a sliding window to assess differential hybridization, resulting in a *Z*-score for each probe. Finally, probes that are deemed significant are consolidated into regions of overrepresentation if they are in close proximity, and these regions are then ranked according to their aggregated *Z*-score. Recently, the introduction of hidden Markov models has enhanced the accuracy of detection (Cawley et al., 2004; Royce et al., 2005).

ChIP-seq represents an advancement over ChIP-on-chip, utilizing high-throughput DNA sequencing instead of tiling arrays to identify variations between sample and control DNA (Barski et al., 2007; Mikkelsen et al., 2007). This technique offers two primary benefits compared to ChIP-on-chip:

- the challenge of data normalization is minimized, as sequencing yields absolute read counts rather than relative hybridization scores, and
- recent advancements in sequencing-by-synthesis technologies, such as those developed by Roche/454 and Illumina/Solexa, enhance the cost-effectiveness of ChIP-seq.

The primary bioinformatics procedure in ChIP-seq involves the rapid and precise alignment of short sequence reads to the reference genome. In theory, any seed-based alignment tool, such as blastn, can be utilized for this purpose. However, it has been observed that seed alignment methods tailored for specific sequencing platforms can significantly enhance both speed and coverage. Currently, there are two commercial solutions available for short ChIP-seq reads: the ELAND tool, which is part of the Solexa analysis pipeline (http://www.solexa.com/), and the SXOligosearch software (http://www.synamatix.com/). Furthermore, a tailored alignment protocol has been established at the Broad Institute (Mikkelsen et al., 2007).

The National Human Genome Research Institute (NHGRI) established a public research consortium known as the Encyclopedia of DNA Elements (ENCODE) in September 2003, with the objective of identifying all the functional elements within the human genome sequence. This initiative followed the announcement by the International Human Genome Sequencing Consortium regarding the completion of the euchromatic sequence of the human genome. In the context of the ENCODE pilot project, specific segments of the human genome, amounting to 30 megabases and approximately 1% of the entire human genome, were chosen. These segments provided a basis for assessing the effectiveness and efficiency of a variety of methods and technologies aimed at identifying different functional elements within human DNA. The criteria for stratification included gene density and non-exonic conservation. The gene density score for a specific region was determined by calculating the percentage of bases that were either covered by genes listed in the ENSEMBL database or by the best BLAT alignments of human mRNA found in the UCSC browser database. The region was segmented into non-overlapping subwindows, each consisting of 125 bases. Subwindows that exhibited less than 75% alignment with the mouse sequence were excluded from further analysis. For the remaining subwindows, the non-exonic conservation score was derived from the percentage of bases that maintained at least 80% identity to the mouse sequence, provided that these bases did not align with ENSEMBL genes, GenBank mRNA BLASTZ alignments, Fgenesh++ gene predictions, TwinScan gene predictions, spliced EST alignments or repetitive sequences. Subsequently, ENCODE 2 and ENCODE 3 have also been completed, which include whole-genome analyses on the human and mouse genomes. The modENCODE Project also worked on whole-genome analyses of the genomes of *C. elegans* and *Drosophila melanogaster.* ENCODE has created a lot of data that can be accessed through the ENCODE Portal, the project's free database. This data comes from high-throughput experiments that map biochemical activities to find potential functional elements in the genomes that were studied (ENCODE Pilot Project, n.d.; Feingold et al., 2004).

6.4.2 Machine Learning and Artificial Intelligence

In recent decades, high-throughput experimental methodologies involving AI and machine learning have been on the rise. The laboratory techniques and biochemical processes conventionally used to study epigenetics tend to be both time-intensive and costly. To mitigate these issues, machine learning and artificial intelligence (AI) methodologies have been widely adopted to correlate epigenetic modifications with their phenotypic outcomes (Lecun et al., 2015).

Deep learning represents a distinct subset of machine learning that is capable of autonomously identifying features from unprocessed input data, eliminating the need for manually designed features. This approach is grounded in deep neural network (DNN) architectures, which have proven to achieve cutting-edge results across various fields, including image processing, natural language processing and speech recognition. The potential of deep learning in bioinformatics is substantial, as it aids in the examination of extensive high-throughput epigenomic datasets, forecasting gene expression, classifying diseases and analyzing enhancer–promoter interactions. Deep learning algorithms can be broadly classified into three categories: supervised learning, unsupervised learning and reinforcement learning (Ersavas et al., 2024; Zhang et al., 2020). In recent years AI and DL have been applied to epigenetic systems in order to predict gene expression, enhancer–promoter interactions, and, more specifically, as a disease marker to predict as well as diagnose diseases like cancer (Lecun et al., 2015).

6.4.2.1 Gene Expression

The expression of a specific gene is influenced by both the quantity and the rate of synthesis of its downstream functional products, including proteins or RNA. This process entails the production of functional RNA based on the genetic information encoded in genes as well as the identification of the genomic regions that undergo transcription. Numerous factors at various levels affect gene expression, including variations in the non-coding regions of DNA, the status of methylation and histone modifications, among others (Liu et al., 2020).

DeepChrome is aimed at analyzing gene expression by utilizing five histone modification (HM) marks across various human cell types. The study focused on correlating gene expression levels with signals from five specific HMs: H3K27me3, H3K9me3, H3K36me3, H3K4me3 and H3K4me1. The DeepChrome model was trained using data from 56 distinct cell types sourced from the REMC database. The input of the model consisted of a binned representation of histone marks, where the binning process covered a 10,000 base pair region centered on the transcription start site (TSS) of each gene, divided into bins of 100 base pair length (Singh et al., 2016).

AttentiveChrome is an additional program that employs the same benchmark datasets as DeepChrome. This program utilizes hierarchical attention-based models to explore the relationships among chromatin factors that influence gene regulation. Upon examining the five primary histone marks, the H3K36me3 mark was identified as indicative of gene body structure, while H3K4me3 and H3K4me1 were recognized as marks for promoters and enhancers, respectively. The H3K9me3 and H3K27me3 marks were classified as indicators of repressed genes. In the training phase, the AttentiveChrome approach implemented a dual-layer soft attention mechanism, one focusing on critical chromatin markers and the other on important regions within those markers, to forecast gene expression (Singh et al., n.d.). Other models like Transformer DeepLncLoc for prediction of post-transcriptional information and MRCNN and DeepCpG for DNA methylation analysis have also been extensively used (Tahir et al., 2024).

6.4.2.2 Enhancer–Promoter Interactions

Enhancer–promoter interactions (EPIs) are pivotal in the genome, facilitating transcriptional regulation that governs cell differentiation, gene expression and the mechanisms underlying various diseases. Enhancers influence the expression profiles of their associated genes by engaging directly with their promoter regions. The expression of a target gene is modulated by distal regulatory enhancer elements that interact with nearby promoter regions. Research has demonstrated that mutations within enhancer regions can disrupt these interactions, leading to dysregulation of the target gene. Conditions such as β-thalassemia and congenital heart disease have been linked to mutations in enhancers and promoters, resulting in altered EPIs. As a result, there is a considerable effort dedicated to developing methodologies for elucidating enhancer–promoter interactions based on one-dimensional genetic and epigenomic markers (Dong et al., 2012; Jing et al., 2020). The TargetFinder method utilized boosted trees alongside functional genomic signals to forecast EPIs. Following the publication of this foundational study, all research teams adopted the TargetFinder benchmark datasets for the development and evaluation of EPI prediction models. Additionally, a predictive approach known as EPI attention-based neural network (EPIANN) was introduced, which leveraged sequential features derived from DNA sequences to predict EPIs. The EPIANN method incorporates enhancer and promoter characteristics extracted from convolutional layers (Arora & Tollefsbol, 2021; Mao et al., n.d.; Whalen et al., 2016). EPI-DLMH, DeepDualEPI, StackEPI and EPI-Tran are some other models also working on the analysis of enhancer–promoter interactions.

6.4.2.3 Cancer Prediction and Diagnosis

DNA methylation patterns are known to undergo significant alterations across the genome during the initial phases of cancer development, thereby enabling the utilization of methylation status as a crucial indicator for the early identification of cancers. Consequently, the methylation states at specific CpG sites or within various sub-genomic regions can be integrated to create an algorithm that accurately leads to the construction and training of advanced cancer detection models (Kulis & Esteller, 2010). In this context, a deep learning-based approach known as Deep Integrating Sequence Individual Reads (DISMIR), aimed at the ultrasensitive detection of cancer, has been developed. The DISMIR methodology leverages methylation data alongside DNA sequences derived from plasma cell-free DNA whole genome bisulfite sequencing (WGBS). This approach introduces an innovative feature representation termed switching reads and switching regions, which facilitates the identification of cancer-specific differentially methylated regions (DMRs), thereby enhancing the resolution of cancer-related signals (Li et al., 2021). Deep learning methodologies typically necessitate a substantial quantity of samples for effective training. Nonetheless, the challenges associated with acquiring clinical samples, coupled with the high costs associated with whole genome bisulfite sequencing (WGBS) experiments, restrict the sample size of cfDNA WGBS data. Consequently, in this research, we treated each individual read from the switching regions as a separate sample rather than aggregating the statistics of all reads from a differentially methylated region (DMR). As a result, the deep learning model focused on learning the patterns of sequencing reads individually rather than as a collective. This approach ensured that the volume of individual reads was sufficiently large to satisfy the requirements of deep learning (Li et al., 2021; Tahir et al., 2024). Algorithms like MetaCancer, use DNA methylation and microRNA sequencing (RNA-Seq) for detection of cancer metastasis; the WGAN model for cancer diagnosis; and XAI-MethylMarker for the specific diagnosis of breast cancer (Albaradei et al., 2021; Rajpal et al., 2023; Xiao et al., 2021).

6.5 CHALLENGES AND FUTURE DIRECTIONS

Epigenetic modifications and their regulators serve as significant molecular components that influence essential physiological and pathological characteristics, thereby playing a crucial role in the progression of human diseases. These epigenetic modulators hold promise as disease biomarkers due to their numerous advantages and their ability to provide insights into gene function, which helps elucidate variations among patient endophenotypes. Epigenetic biomarkers can also reflect the impact of environmental factors and lifestyle choices on health and disease as well as track the effectiveness of therapeutic interventions (García-Giménez et al., 2017). Thus, the technologies employed for the analysis of these epigenetic biomarkers are continually advancing, becoming increasingly user-friendly. Laboratory professionals are gaining expertise more readily, and the techniques are becoming more cost-effective. A substantial number of epigenetic biomarker candidates are being proposed regularly, indicating integration of epigenetics into clinical laboratories and transformation of epigenetic markers into dependable biomarkers. Epigenetic biomarkers are increasingly recognized as valuable instruments for screening and early identification of a range of diseases, for prognostic assessment and treatment monitoring and for forecasting the potential risk of disease onset. In clinical settings, this advancement will facilitate the creation of in vitro diagnostic and prognostic tools, which may subsequently promote progress in the field of personalized medicine (García-Giménez et al., 2016, 2017).

A significant domain of application and advancement in epigenetics pertains to the study of plants that hold agronomic significance. Various agronomic strategies have been explored, including enhancing tolerance to abiotic and biotic stress, minimizing toxicity, improving nutritional quality, delaying ripening and optimizing post-harvest quality. Nevertheless, significant challenges to these approaches arise from the loss of agricultural land due to urban expansion in both developing and underdeveloped nations, severe climate fluctuations, plant diseases and reluctance of farmers to adopt modern agricultural practices. Plant epigenetic modifications are crucial for acclimatization, stress resilience, adaptation and evolutionary processes. The key agronomic traits in crops, such as flowering time, fruit development, environmental risk mitigation and crop immunity, are influenced by epigenetic changes, including DNA methylation, histone post-translational modifications and alterations in small non-coding RNAs (Krishnaswami et al., 2022). Recent developments have shed light on the epigenetic mechanisms governing plant responses to both abiotic and biotic stress, the epigenetic foundations of plant immunity, the evolution and roles of plant histones, the implications of epigenetic variation in plant breeding and the prospects of epigenome editing for crop enhancement. Future models in epigenomics aim to manipulate and selectively modulate proteins and metabolic pathways to combat pathogens, address critical diseases and boost agricultural productivity. The emergence of novel fields such as epigenome editing and RNA epigenetics is expected to deepen our comprehension of known epigenetic modifications and their implications in host-pathogen interactions, crop yield and responses to environmental changes (Alvarez-Venegas et al., 2019).

Recent advancements in epigenetics research include spatial epigenomics, long-read sequencing and CRISPR-based epigenome editing. Long-read sequencing provides in-depth insights into complex epigenetic modifications, while multi-omics approaches integrate various data types to offer a holistic view of the epigenome and its regulatory mechanisms. These methods enhance our understanding of gene regulation by revealing detailed information about the epigenome at the single-cell level and presenting a comprehensive overview through multi-omics strategies (C. Chen et al., 2023; Zhu et al., 2024). They also allow for the investigation of tumor heterogeneity and evolution, helping researchers to unravel complex epigenetic landscapes and pinpoint the mechanisms that silence tumor suppressor genes while promoting oncogenic potential, ultimately transforming biological insights from raw data (Sherif et al., 2024).

One of the primary challenges resides within the technical sphere. Although groundbreaking technologies such as single-cell epigenomics and CRISPR-based editing exhibit considerable potential, ensuring precision and reproducibility in these approaches remains a significant concern (Feingold et al., 2004). The epigenome functions in conjunction with various cellular elements, such as the transcriptome, proteome and metabolome. Consequently, the integration of epigenomics with other omics data is essential for achieving a comprehensive understanding of gene regulation and cellular activities. This requirement underscores the importance of creating multi-omics platforms and integrative analytical frameworks (Bersanelli et al., 2016). An initial strategy for multi-omics analysis involved the independent examination of various data types, followed by the integration of the findings to create a holistic representation of molecular interactions. As advancements in this domain have progressed, algorithmic meta-analysis frameworks and methodologies have emerged as the predominant approach for conducting thorough analyses of multi-omics data (Chen et al., 2023). The challenges of missing value management, variability across different omics, complexities in interpreting multi-omics models and issues related to data annotation, storage and computational resources represent significant yet often overlooked obstacles (Tarazona et al., 2021) .

Advancements in algorithms utilizing machine learning (ML) and deep learning (DL) techniques are enhancing the integration of multi-omics data, given the substantial benefits these methods offer in this domain. ML and DL excel in processing both linear and nonlinear datasets, employing supervised and unsupervised learning approaches for disease prediction and biomarker identification. Additionally, the emergence of open-source platforms and software is facilitating non-computational medical researchers in the integration and analysis of multi-omics data. Consequently, as the availability of open-source tools continues to expand for the integrated analysis of multi-omics data, the application of these methodologies will become increasingly accessible to a broader range of researchers. While these technologies significantly improve our understanding of gene regulation, there are still challenges to overcome, especially in improving the specificity and sensitivity of these methods. Additionally, combining epigenetic data with other biological datasets continues to be a complicated task. However, these new methodologies hold great promise for future research and therapeutic applications in the field of epigenetics.

REFERENCES

Adrian, J., Farrona, S., Reimer, J. J., Albani, M. C., Coupland, G., & Turck, F. (2010). cis-regulatory elements and chromatin state coordinately control temporal and spatial expression of FLOWERING LOCUS T in arabidopsis. *The Plant Cell*, 22(5), 1425. https://doi.org/10.1105/TPC.110.074682.

Albaradei, S., Napolitano, F., Thafar, M. A., Gojobori, T., Essack, M., & Gao, X. (2021). MetaCancer: A deep learning-based pan-cancer metastasis prediction model developed using multi-omics data. *Computational and Structural Biotechnology Journal*, 19, 4404–4411. https://doi.org/10.1016/J.CSBJ.2021.08.006.

Allis, C. D., & Jenuwein, T. (2016). The molecular hallmarks of epigenetic control. *Nature Reviews Genetics*, 17(8), 487–500. https://doi.org/10.1038/NRG.2016.59.

Alvarez-Venegas, R., De-la-Peña, C., & Casas-Mollano, J. A. (2019). Epigenetics in plants of agronomic importance: Fundamentals and applications: Transcriptional regulation and chromatin remodelling in plants: Second edition. *Epigenetics in Plants of Agronomic Importance: Fundamentals and Applications: Transcriptional Regulation and Chromatin Remodelling in Plants: Second Edition*, 1–415. https://doi.org/10.1007/978-3-030-14760-0/COVER.

Amir, R. E., Van Den Veyver, I. B., Wan, M., Tran, C. Q., Francke, U., & Zoghbi, H. Y. (1999). Rett syndrome is caused by mutations in X-linked MECP2, encoding methyl-CpG-binding protein 2. *Nature Genetics*, 23(2), 185–188. https://doi.org/10.1038/13810.

Ardui, S., Ameur, A., Vermeesch, J. R., & Hestand, M. S. (2018). Single molecule real-time (SMRT) sequencing comes of age: Applications and utilities for medical diagnostics. *Nucleic Acids Research*, 46(5), 2159–2168. https://doi.org/10.1093/NAR/GKY066.

Arora, I., & Tollefsbol, T. O. (2021). Computational methods and next-generation sequencing approaches to analyze epigenetics data: Profiling of methods and applications. *Methods*, 187, 92–103. https://doi.org/10.1016/J.YMETH.2020.09.008.

Azeez, S. S., Hamad, R. S., Hamad, B. K., Shekha, M. S., & Bergsten, P. (2024). Advances in CRISPR-Cas technology and its applications: revolutionising precision medicine. *Frontiers in Genome Editing*, 6, 1509924. https://doi.org/10.3389/FGEED.2024.1509924/FULL.

Barski, A., Cuddapah, S., Cui, K., Roh, T. Y., Schones, D. E., Wang, Z., Wei, G., Chepelev, I., & Zhao, K. (2007). High-resolution profiling of histone methylations in the human genome. *Cell*, 129(4), 823–837. https://doi.org/10.1016/J.CELL.2007.05.009/ATTACHMENT/F0709907-32C3-4AB2-BE76-039C1AE0CC07/MMC1.PDF.

Bernstein, B. E., Meissner, A., & Lander, E. S. (2007a). The mammalian epigenome. *Cell*, 128(4), 669–681. https://doi.org/10.1016/J.CELL.2007.01.033/ASSET/4713E2B2-47C1-44BE-AF63-DDF30BB690B9/MAIN.ASSETS/GR2.JPG.

Bersanelli, M., Mosca, E., Remondini, D., Giampieri, E., Sala, C., Castellani, G., & Milanesi, L. (2016). Methods for the integration of multi-omics data: Mathematical aspects. *BMC Bioinformatics*, 17(2), 167–177. https://doi.org/10.1186/S12859-015-0857-9/TABLES/1.

Bock, C., & Lengauer, T. (2008). Computational epigenetics. *Bioinformatics*, 24(1), 1–10. https://doi.org/10.1093/BIOINFORMATICS/BTM546.

Broday, L., & Costa, M. (2000). Detecting epigenetic changes: DNA methylation. *Current Protocols in Toxicology*, 3(1), 3.6.1–3.6.8. https://doi.org/10.1002/0471140856.TX0306S03.

Buck, M. J., & Lieb, J. D. (2004). ChIP-chip: Considerations for the design, analysis, and application of genome-wide chromatin immunoprecipitation experiments. *Genomics*, 83(3), 349–360. https://doi.org/10.1016/J.YGENO.2003.11.004.

Casal, J. J., & Fankhauser, C. (2023). Shade avoidance in the context of climate change. *Plant Physiology*, 191(3), 1475–1491. https://doi.org/10.1093/PLPHYS/KIAD004.

Cawley, S., Bekiranov, S., Ng, H. H., Kapranov, P., Sekinger, E. A., Kampa, D., Piccolboni, A., Sementchenko, V., Cheng, J., Williams, A. J., Wheeler, R., Wong, B., Drenkow, J., Yamanaka, M., Patel, S., Brubaker, S., Tammana, H., Helt, G., Struhl, K., & Gingeras, T. R. (2004). Unbiased mapping of transcription factor binding sites along human chromosomes 21 and 22 points to widespread regulation of noncoding RNAs. *Cell*, 116(4), 499–509. https://doi.org/10.1016/S0092-8674(04)00127-8/ATTACHMENT/2C2FBBD4-A1C2-4FD9-91D8-448B48CA7F25/MMC6.PDF.

cDNA (copy DNA) | NHGRI. (n.d.). Retrieved February 20, 2025, from https://www.genome.gov/genetics-glossary/copy-DNA-cDNA.

Chang, H. H., Hemberg, M., Barahona, M., Ingber, D. E., & Huang, S. (2008). Transcriptome-wide noise controls lineage choice in mammalian progenitor cells. *Nature*, 453(7194), 544–547. https://doi.org/10.1038/nature06965.

Chen, C., Wang, J., Pan, D., Wang, X., Xu, Y., Yan, J., Wang, L., Yang, X., Yang, M., & Liu, G. P. (2023). Applications of multi-omics analysis in human diseases. *MedComm*, 4(4), e315. https://doi.org/10.1002/MCO2.315.

Chen, X., Bhadauria, V., & Ma, B. (2018). ChIP-Seq: A powerful tool for studying Protein-DNA interactions in plants. *Current Issues in Molecular Biology*, 27, 171–180. https://doi.org/10.21775/CIMB.027.171.

Chen, X., Xu, H., Shu, X., & Song, C. X. (2023). Mapping epigenetic modifications by sequencing technologies. *Cell Death & Differentiation*, 32(1), 56–65. https://doi.org/10.1038/s41418-023-01213-1.

Cherenkov, P., Novikova, D., Omelyanchuk, N., Levitsky, V., Grosse, I., Weijers, D., & Mironova, V. (2018). Diversity of cis-regulatory elements associated with auxin response in *Arabidopsis thaliana*. *Journal of Experimental Botany*, 69(2), 329–339. https://doi.org/10.1093/JXB/ERX254.

Cokus, S. J., Feng, S., Zhang, X., Chen, Z., Merriman, B., Haudenschild, C. D., Pradhan, S., Nelson, S. F., Pellegrini, M., & Jacobsen, S. E. (2008). Shotgun bisulphite sequencing of the *Arabidopsis* genome reveals DNA methylation patterning. *Nature*, 452(7184), 215–219. https://doi.org/10.1038/nature06745.

Davis, B. M., Chao, M. C., & Waldor, M. K. (2013). Entering the era of bacterial epigenomics with single molecule real time DNA sequencing. *Current Opinion in Microbiology*, 16(2), 192. https://doi.org/10.1016/J.MIB.2013.01.011.

Dong, X., Greven, M. C., Kundaje, A., Djebali, S., Brown, J. B., Cheng, C., Gingeras, T. R., Gerstein, M., Guigó, R., Birney, E., & Weng, Z. (2012). Modeling gene expression using chromatin features in various cellular contexts. *Genome Biology*, 13(9), 1–17. https://doi.org/10.1186/GB-2012-13-9-R53/TABLES/1.

Dunham, I., Kundaje, A., Aldred, S. F., Collins, P. J., Davis, C. A., Doyle, F., Epstein, C. B., Frietze, S., Harrow, J., Kaul, R., Khatun, J., Lajoie, B. R., Landt, S. G., Lee, B. K., Pauli, F., Rosenbloom, K. R., Sabo, P., Safi, A., Sanyal, A., … Lochovsky, L. (2012a). An integrated encyclopedia of DNA elements in the human genome. *Nature*, 489(7414), 57–74. https://doi.org/10.1038/nature11247.

Ehrlich, M., & Wang, R. Y. H. (1981). 5-Methylcytosine in eukaryotic DNA. *Science (New York, N.Y.)*, 212(4501), 1350–1357. https://doi.org/10.1126/SCIENCE.6262918.

Eick, D., Fritz, H. J., & Doerfler, W. (1983). Quantitative determination of 5-methylcytosine in DNA by reverse-phase high-performance liquid chromatography. *Analytical Biochemistry*, 135(1), 165–171. https://doi.org/10.1016/0003-2697(83)90746-7.

ENCODE Pilot Project. (n.d.). Retrieved February 21, 2025, from https://www.genome.gov/Funded-Programs-Projects/ENCODE-Project-ENCyclopedia-Of-DNA-Elements/pilot.

Ernst, J., Kheradpour, P., Mikkelsen, T. S., Shoresh, N., Ward, L. D., Epstein, C. B., Zhang, X., Wang, L., Issner, R., Coyne, M., Ku, M., Durham, T., Kellis, M., & Bernstein, B. E. (2011). Mapping and analysis of chromatin state dynamics in nine human cell types. *Nature*, 473(7345), 43–49. https://doi.org/10.1038/nature09906.

Ersavas, T., Smith, M. A., & Mattick, J. S. (2024). Novel applications of convolutional neural networks in the age of transformers. *Scientific Reports*, 14(1). https://doi.org/10.1038/S41598-024-60709-Z.

Esteller, M. (2006). The necessity of a human epigenome project. *Carcinogenesis*, 27(6), 1121–1125. https://doi.org/10.1093/CARCIN/BGL033.

Feingold, E. A., Good, P. J., Guyer, M. S., Kamholz, S., Liefer, L., Wetterstrand, K., Collins, F. S., Gingeras, T. R., Kampa, D., Sekinger, E. A., Cheng, J., Hirsch, H., Ghosh, S., Zhu, Z., Patel, S., Piccolboni, A., Yang, A., Tammana, H., Bekiranov, S., … Harvey, S. C. (2004). The ENCODE

(ENCyclopedia Of DNA Elements) Project. *Science (New York, N.Y.)*, 306(5696), 636–640. https://doi.org/10.1126/SCIENCE.1105136.

Feng, X., Cui, X., Zhang, L. S., Ye, C., Wang, P., Zhong, Y., Wu, T., Zheng, Z., & He, C. (2024). Sequencing of N6-methyl-deoxyadenosine at single-base resolution across the mammalian genome. *Molecular Cell*, 84(3), 596–610.e6. https://doi.org/10.1016/J.MOLCEL.2023.12.021/ATTACHMENT/6F6EAE1B-E657-4989-9CD7-0EE84FC9EE2F/MMC12.PDF.

Gade, P., & Kalvakolanu, D. V. (2012). Chromatin immunoprecipitation assay as a tool for analyzing transcription factor activity. *Methods in Molecular Biology (Clifton, N.J.)*, 809, 85. https://doi.org/10.1007/978-1-61779-376-9_6.

Ganger, S., Harale, G., & Majumdar, P. (2023). Clustered regularly interspaced short palindromic repeats/crispr-associated (CRISPR/Cas) systems: Discovery structure, classification, and general mechanism. *CRISPR/Cas-Mediated Genome Editing in Plants*, 65–98. https://doi.org/10.1201/9781003331759-4/CLUSTERED-REGULARLY-INTERSPACED-SHORT-PALINDROMIC-REPEATS-CRISPR-ASSOCIATED-CRISPR-CAS-SYSTEMS-DISCOVERY-STRUCTURE-CLASSIFICATION-GENERAL-MECHANISM-SUMAN-GANGER-GEETANJALI-HARALE-PARAMA-MAJUMDAR.

García-Giménez, J. L., Seco-Cervera, M., Tollefsbol, T. O., Romá-Mateo, C., Peiró-Chova, L., Lapunzina, P., & Pallardó, F. V. (2017). Epigenetic biomarkers: Current strategies and future challenges for their use in the clinical laboratory. *Critical Reviews in Clinical Laboratory Sciences*, 54(7–8), 529. https://doi.org/10.1080/10408363.2017.1410520.

García-Giménez, J. L., Ushijima, T., & Tollefsbol, T. O. (2016). Epigenetic biomarkers: New findings, perspectives, and future directions in diagnostics. *Epigenetic Biomarkers and Diagnostics*, 1–18. https://doi.org/10.1016/B978-0-12-801899-6.00001-2.

Greer, E. L., Blanco, M. A., Gu, L., Sendinc, E., Liu, J., Aristizábal-Corrales, D., Hsu, C. H., Aravind, L., He, C., & Shi, Y. (2015). DNA methylation on N6-Adenine in C. elegans. *Cell*, 161(4), 868–878. https://doi.org/10.1016/J.CELL.2015.04.005.

Griffith, J. S., & Mahler, H. R. (1969). DNA ticketing theory of memory. *Nature*, 223(5206), 580. https://doi.org/10.1038/223580A0.

Grunau, C., Clark, S. J., & Rosenthal, A. (2001). Bisulfite genomic sequencing: Systematic investigation of critical experimental parameters. *Nucleic Acids Research*, 29(13), e65–e65. https://doi.org/10.1093/NAR/29.13.E65.

Haig, D. (2004). The (Dual) origin of epigenetics. *Cold Spring Harbor Symposia on Quantitative Biology*, 69, 67–70. https://doi.org/10.1101/SQB.2004.69.67.

Hajkova, P., Jeffries, S. J., Lee, C., Miller, N., Jackson, S. P., & Surani, M. A. (2010). Genome-wide reprogramming in the mouse germ line entails the base excision repair pathway. *Science (New York, N.Y.)*, 329(5987), 78–82. https://doi.org/10.1126/SCIENCE.1187945.

Hauser, M. T., Aufsatz, W., Jonak, C., & Luschnig, C. (2011). Transgenerational epigenetic inheritance in plants. *Biochimica et Biophysica Acta (BBA) – Gene Regulatory Mechanisms*, 1809(8), 459–468. https://doi.org/10.1016/J.BBAGRM.2011.03.007.

He, X. J., Chen, T., & Zhu, J. K. (2011). Regulation and function of DNA methylation in plants and animals. *Cell Research*, 21(3), 442–465. https://doi.org/10.1038/CR.2011.23.

Heyn, H., & Esteller, M. (2015). An adenine code for DNA: A second life for N6-methyladenine. *Cell*, 161(4), 710–713. https://doi.org/10.1016/j.cell.2015.04.021.

Holliday, R. (1989). DNA methylation and epigenetic mechanisms. *Cell Biophysics*, 15(1–2), 15–20. https://doi.org/10.1007/BF02991575.

Holliday, R. (1990). DNA methylation and epigenetic inheritance. *Philosophical Transactions of the Royal Society of London. Series B, Biological Sciences*, 326(1235), 329–338. https://doi.org/10.1098/rstb.1990.0015.

Holliday, R. (2006). Epigenetics: A historical overview. *Epigenetics*, 1(2), 76–80. https://doi.org/10.4161/EPI.1.2.2762.

Holliday, R., & Murray, V. (1994). Specificity in splicing. *BioEssays : News and Reviews in Molecular, Cellular and Developmental Biology*, 16(10), 771–774. https://doi.org/10.1002/BIES.950161013.

Hrdlickova, R., Toloue, M., & Tian, B. (2017). RNA-Seq methods for transcriptome analysis. *Wiley Interdisciplinary Reviews: RNA*, 8(1). https://doi.org/10.1002/WRNA.1364.

Hu, T., Chitnis, N., Monos, D., & Dinh, A. (2021). Next-generation sequencing technologies: An overview. *Human Immunology*, 82(11), 801–811. https://doi.org/10.1016/j.humimm.2021.02.012.

Huang, X., Chen, Z., & Liu, Y. (2020). RNAi-mediated control of CRISPR functions. *Theranostics*, 10(15), 6661. https://doi.org/10.7150/THNO.44880.

Hwang, B., Lee, J. H., & Bang, D. (2018). Single-cell RNA sequencing technologies and bioinformatics pipelines. *Experimental & Molecular Medicine*, 50(8), 1–14. https://doi.org/10.1038/s12276-018-0071-8.

Ito, S., Shen, L., Dai, Q., Wu, S. C., Collins, L. B., Swenberg, J. A., He, C., & Zhang, Y. (2011). Tet proteins can convert 5-methylcytosine to 5-formylcytosine and 5-carboxylcytosine. *Science (New York, N.Y.)*, 333(6047), 1300–1303. https://doi.org/10.1126/SCIENCE.1210597.

Jing, F., Zhang, S. W., & Zhang, S. (2020). Prediction of enhancer–promoter interactions using the cross-cell type information and domain adversarial neural network. *BMC Bioinformatics*, 21(1), 1–16. https://doi.org/10.1186/S12859-020-03844-4/FIGURES/4.

Jones, P. A., & Baylin, S. B. (2007). The epigenomics of cancer. *Cell*, 128(4), 683–692. https://doi.org/10.1016/J.CELL.2007.01.029/ASSET/05DE3C02-0A44-49A6-A56E-D9DF2B015B7B/MAIN.ASSETS/GR4.JPG.

Kakutani, T. (2002). Epi-Alleles in plants: Inheritance of epigenetic information over generations. *Plant and Cell Physiology*, 43(10), 1106–1111. https://doi.org/10.1093/PCP/PCF131.

Karlson, C. K. S., Mohd-noor, S. N., Nolte, N., & Tan, B. C. (2021). CRISPR/dCas9-based systems: Mechanisms and applications in plant sciences. *Plants*, 10(10), 2055. https://doi.org/10.3390/PLANTS10102055.

Kaya-Okur, H. S., Wu, S. J., Codomo, C. A., Pledger, E. S., Bryson, T. D., Henikoff, J. G., Ahmad, K., & Henikoff, S. (2019). CUT&Tag for efficient epigenomic profiling of small samples and single cells. *Nature Communications*, 10(1), 1–10. https://doi.org/10.1038/s41467-019-09982-5.

Kelkar, A., & Deobagkar, D. (2009). A novel method to assess the full genome methylation profile using monoclonal antibody combined with the high throughput based microarray approach. *Epigenetics*, 4(6), 415–420. https://doi.org/10.4161/EPI.4.6.9768.

Kellis, M., Wold, B., Snyder, M. P., Bernstein, B. E., Kundaje, A., Marinov, G. K., Ward, L. D., Birney, E., Crawford, G.

E., Dekker, J., Dunham, I., Elnitski, L. L., Farnham, P. J., Feingold, E. A., Gerstein, M., Giddings, M. C., Gilbert, D. M., Gingeras, T. R., Green, E. D., … Hardison, R. C. (2014b). Defining functional DNA elements in the human genome. *Proceedings of the National Academy of Sciences of the United States of America*, 111(17), 6131–6138. https://doi.org/10.1073/PNAS.1318948111.

Khodadadi, E., Fahmideh, L., Khodadadi, E., Dao, S., Yousefi, M., Taghizadeh, S., Asgharzadeh, M., Yousefi, B., & Kafil, H. S. (2021). Current advances in DNA methylation analysis methods. *BioMed Research International*, 2021(1), 8827516. https://doi.org/10.1155/2021/8827516.

Krishnaswami, V., Kumar, M., & Vijayaraghavalu, S. (2022). Advances in epigenetics for crop improvement and sustainable agriculture. *Plant Genomics for Sustainable Agriculture*, 351–370. https://doi.org/10.1007/978-981-16-6974-3_14.

Kulis, M., & Esteller, M. (2010). DNA methylation and cancer. *Advances in Genetics*, 70(C), 27–56. https://doi.org/10.1016/B978-0-12-380866-0.60002-2.

Kumar, S., Chinnusamy, V., & Mohapatra, T. (2018). Epigenetics of modified DNA bases: 5-methylcytosine and beyond. *Frontiers in Genetics*, 9, 429871. https://doi.org/10.3389/FGENE.2018.00640/BIBTEX.

Lai, W., Mo, J., Yin, J., Lyu, C., & Wang, H. (2019). Profiling of epigenetic DNA modifications by advanced liquid chromatography-mass spectrometry technologies. *TrAC Trends in Analytical Chemistry*, 110, 173–182. https://doi.org/10.1016/J.TRAC.2018.10.031.

Landan, G., Cohen, N. M., Mukamel, Z., Bar, A., Molchadsky, A., Brosh, R., Horn-Saban, S., Zalcenstein, D. A., Goldfinger, N., Zundelevich, A., Gal-Yam, E. N., Rotter, V., & Tanay, A. (2012). Epigenetic polymorphism and the stochastic formation of differentially methylated regions in normal and cancerous tissues. *Nature Genetics*, 44(11), 1207–1214. https://doi.org/10.1038/ng.2442.

Laszlo, A. H., Derrington, I. M., Brinkerhoff, H., Langford, K. W., Nova, I. C., Samson, J. M., Bartlett, J. J., Pavlenok, M., & Gundlach, J. H. (2013). Detection and mapping of 5-methylcytosine and 5-hydroxymethylcytosine with nanopore MspA. *Proceedings of the National Academy of Sciences of the United States of America*, 110(47), 18904–18909. https://doi.org/10.1073/PNAS.1310240110/SUPPL_FILE/SAPP.PDF.

Laszlo, A. H., Derrington, I. M., Ross, B. C., Brinkerhoff, H., Adey, A., Nova, I. C., Craig, J. M., Langford, K. W., Samson, J. M., Daza, R., Doering, K., Shendure, J., & Gundlach, J. H. (2014). Decoding long nanopore sequencing reads of natural DNA. *Nature Biotechnology*, 32(8), 829–833. https://doi.org/10.1038/NBT.2950.

Lecun, Y., Bengio, Y., & Hinton, G. (2015). Deep learning. *Nature*, 521(7553), 436–444. https://doi.org/10.1038/nature14539.

Li, J., Zhang, X., Zhang, W., Wang, H., Zhong, B., Xie, Z., Lv, H., & Wang, X. (2021). DISMIR: Deep learning-based noninvasive cancer detection by integrating DNA sequence and methylation information of individual cell-free DNA reads. *Briefings in Bioinformatics*, 22(6), 1–11. https://doi.org/10.1093/BIB/BBAB250.

Li, Y. (2020). Modern epigenetics methods in biological research. *Methods (San Diego, Calif.)*, 187, 104. https://doi.org/10.1016/J.YMETH.2020.06.022.

Li, Y. (2021). Modern epigenetics methods in biological research. *Methods*, 187, 104–113. https://doi.org/10.1016/J.YMETH.2020.06.022.

Liu, J., Li, J., Wang, H., & Yan, J. (2020). Application of deep learning in genomics. *Science China Life Sciences*, 63(12), 1860–1878. https://doi.org/10.1007/S11427-020-1804-5/METRICS.

Liu, K., Yuan, C., Li, H., Lin, W., Yang, Y., Shen, C., & Zheng, X. (2015). Genome-wide identification and characterization of auxin response factor (ARF) family genes related to flower and fruit development in papaya (*Carica papaya* L.). *BMC Genomics*, 16(1), 1–12. https://doi.org/10.1186/S12864-015-2182-0/FIGURES/7.

Liu, L., Zhao, Y., Hassett, R., Toneyan, S., Koo, P. K., & Siepel, A. (2025). Probabilistic and machine-learning methods for predicting local rates of transcription elongation from nascent RNA sequencing data. *Nucleic Acids Research*, 53(4), 13–14. https://doi.org/10.1093/NAR/GKAF092.

Luo, G. Z., Wang, F., Weng, X., Chen, K., Hao, Z., Yu, M., Deng, X., Liu, J., & He, C. (2016). Characterization of eukaryotic DNA N6-methyladenine by a highly sensitive restriction enzyme-assisted sequencing. *Nature Communications*, 7(1), 1–6. https://doi.org/10.1038/ncomms11301.

Mao, W., Kostka, D., & Chikina, M. (n.d.). *Modeling Enhancer-Promoter Interactions with Attention-Based Neural Networks*. https://doi.org/10.1101/219667.

Mazan-Mamczarz, K., Ha, J., De, S., & Sen, P. (2022). Single-cell analysis of the transcriptome and epigenome. *Methods in Molecular Biology (Clifton, N.J.)*, 2399, 21. https://doi.org/10.1007/978-1-0716-1831-8_3.

Meaburn, E. L., Schalkwyk, L. C., & Mill, J. (2010). Allele-specific methylation in the human genome: Implications for genetic studies of complex disease. *Epigenetics*, 5(7), 578. https://doi.org/10.4161/EPI.5.7.12960.

Meaburn, E., & Schulz, R. (2012). Next generation sequencing in epigenetics: Insights and challenges. *Seminars in Cell & Developmental Biology*, 23(2), 192–199. https://doi.org/10.1016/J.SEMCDB.2011.10.010.

Mikkelsen, T. S., Ku, M., Jaffe, D. B., Issac, B., Lieberman, E., Giannoukos, G., Alvarez, P., Brockman, W., Kim, T. K., Koche, R. P., Lee, W., Mendenhall, E., O'Donovan, A., Presser, A., Russ, C., Xie, X., Meissner, A., Wernig, M., Jaenisch, R., … Bernstein, B. E. (2007). Genome-wide maps of chromatin state in pluripotent and lineage-committed cells. *Nature*, 448(7153), 553–560. https://doi.org/10.1038/nature06008.

Milne, T. A., Zhao, K., & Hess, J. L. (2009). Chromatin Immunoprecipitation (ChIP) for analysis of histone modifications and chromatin-associated proteins. *Methods in Molecular Biology (Clifton, N.J.)*, 538, 409. https://doi.org/10.1007/978-1-59745-418-6_21.

Mitchell, S. F., Walker, S. E., Algire, M. A., Park, E. H., Hinnebusch, A. G., & Lorsch, J. R. (2010). The 5'-7-methylguanosine cap on eukaryotic mRNAs serves both to stimulate canonical translation initiation and to block an alternative pathway. *Molecular Cell*, 39(6), 950–962. https://doi.org/10.1016/J.MOLCEL.2010.08.021.

Mortazavi, A., Williams, B. A., McCue, K., Schaeffer, L., & Wold, B. (2008). Mapping and quantifying mammalian transcriptomes by RNA-Seq. *Nature Methods*, 5(7), 621–628. https://doi.org/10.1038/NMETH.1226.

Mutz, K. O., Heilkenbrinker, A., Lönne, M., Walter, J. G., & Stahl, F. (2013). Transcriptome analysis using next-generation sequencing. *Current Opinion in Biotechnology*, 24(1), 22–30. https://doi.org/10.1016/J.COPBIO.2012.09.004.

Nguyen, K. Le, Grondin, A., Courtois, B., & Gantet, P. (2019). Next-generation sequencing accelerates crop gene discovery. *Trends in Plant Science*, 24(3), 263–274. https://doi.org/10.1016/J.TPLANTS.2018.11.008.

Omony, J., Nussbaumer, T., Corresponding, R. G., Gutzat, R., & Mendel, G. (2020). DNA methylation analysis in plants: Review of computational tools and future perspectives. *Briefings in Bioinformatics*, 21(3), 906–918. https://doi.org/10.1093/BIB/BBZ039.

Park, P. J. (2009). ChIP-Seq: advantages and challenges of a maturing technology. Nature Reviews. *Genetics*, 10(10), 669. https://doi.org/10.1038/NRG2641.

Paszkowski, J., & Grossniklaus, U. (2011). Selected aspects of transgenerational epigenetic inheritance and resetting in plants. *Current Opinion in Plant Biology*, 14(2), 195–203. https://doi.org/10.1016/J.PBI.2011.01.002.

Pulecio, J., Verma, N., Mejía-Ramírez, E., Huangfu, D., & Raya, A. (2017). CRISPR/Cas9-based engineering of the epigenome. *Cell Stem Cell*, 21(4), 431. https://doi.org/10.1016/J.STEM.2017.09.006.

Qi, L. S., Larson, M. H., Gilbert, L. A., Doudna, J. A., Weissman, J. S., Arkin, A. P., & Lim, W. A. (2013). Repurposing CRISPR as an RNA-guided platform for sequence-specific control of gene expression. *Cell*, 152(5), 1173–1183. https://doi.org/10.1016/j.cell.2013.02.022.

Rajpal, S., Rajpal, A., Saggar, A., Vaid, A. K., Kumar, V., Agarwal, M., & Kumar, N. (2023). XAI-MethylMarker: Explainable AI approach for biomarker discovery for breast cancer subtype classification using methylation data. *Expert Systems with Applications*, 225, 120130. https://doi.org/10.1016/J.ESWA.2023.120130.

Royce, T. E., Rozowsky, J. S., Bertone, P., Samanta, M., Stolc, V., Weissman, S., Snyder, M., & Gerstein, M. (2005). Issues in the analysis of oligonucleotide tiling microarrays for transcript mapping. *Trends in Genetics*, 21(8), 466–475. https://doi.org/10.1016/j.tig.2005.06.007.

Schübeler, D. (2015). Function and information content of DNA methylation. *Nature*, 517(7534), 321–326. https://doi.org/10.1038/nature14192.

Shahbazian, M. D., Young, J. I., Yuva-Paylor, L. A., Spencer, C. M., Antalffy, B. A., Noebels, J. L., Armstrong, D. L., Paylor, R., & Zoghbi, H. Y. (2002). Mice with truncated MeCP2 recapitulate many Rett syndrome features and display hyperacetylation of histone H3. *Neuron*, 35(2), 243–254. https://doi.org/10.1016/S0896-6273(02)00768-7.

Shalek, A. K., Satija, R., Adiconis, X., Gertner, R. S., Gaublomme, J. T., Raychowdhury, R., Schwartz, S., Yosef, N., Malboeuf, C., Lu, D., Trombetta, J. J., Gennert, D., Gnirke, A., Goren, A., Hacohen, N., Levin, J. Z., Park, H., & Regev, A. (2013). Single-cell transcriptomics reveals bimodality in expression and splicing in immune cells. *Nature*, 498(7453), 236–240. https://doi.org/10.1038/nature12172.

Shen, H., He, H., Li, J., Chen, W., Wang, X., Guo, L., Peng, Z., He, G., Zhong, S., Qi, Y., Terzaghi, W., & Deng, X. W. (2012). Genome-wide analysis of DNA methylation and gene expression changes in two Arabidopsis ecotypes and their reciprocal hybrids. *The Plant Cell*, 24(3), 875–892. https://doi.org/10.1105/TPC.111.094870.

Sherif, Z. A., Ogunwobi, O. O., & Ressom, H. W. (2024). Mechanisms and technologies in cancer epigenetics. *Frontiers in Oncology*, 14, 1513654. https://doi.org/10.3389/FONC.2024.1513654/PDF.

Simpson, J. T., Workman, R. E., Zuzarte, P. C., David, M., Dursi, L. J., & Timp, W. (2017). Detecting DNA cytosine methylation using nanopore sequencing. *Nature Methods*, 14(4), 407–410. https://doi.org/10.1038/nmeth.4184.

Singh, K., Balyan, H. S., & Gupta, P. K. (2024). Bioinformatics for analyzing the role of epigenetics in plant disease resistance. *Bioinformatics for Plant Research and Crop Breeding*, 351–390. https://doi.org/10.1002/9781394209965.CH15.

Singh, R., Lanchantin, J., Robins, G., & Qi, Y. (2016). DeepChrome: Deep-learning for predicting gene expression from histone modifications. *Bioinformatics*, 32(17), i639–i648. https://doi.org/10.1093/BIOINFORMATICS/BTW427.

Singh, R., Lanchantin, J., Sekhon, A., & Qi, Y. (n.d.). *Attend and Predict: Understanding Gene Regulation by Selective Attention on Chromatin*. Retrieved February 21, 2025, from www.deepchrome.org.

Skene, P. J., & Henikoff, S. (2017). An efficient targeted nuclease strategy for high-resolution mapping of DNA binding sites. *ELife*, 6. https://doi.org/10.7554/ELIFE.21856.

Smith, Z. D., & Meissner, A. (2013). DNA methylation: Roles in mammalian development. *Nature Reviews. Genetics*, 14(3), 204–220. https://doi.org/10.1038/NRG3354.

Struhl, K., & Segal, E. (2013). Determinants of nucleosome positioning. *Nature Structural & Molecular Biology*, 20(3), 267–273. https://doi.org/10.1038/nsmb.2506.

Surani, M. A., Hayashi, K., & Hajkova, P. (2007). Genetic and epigenetic regulators of pluripotency. *Cell*, 128(4), 747–762. https://doi.org/10.1016/J.CELL.2007.02.010/ASSET/913D0067-7CEB-4AD5-9C8B-0BB30E6C08C8/MAIN.ASSETS/GR3.JPG.

Tahir, M., Norouzi, M., Khan, S. S., Davie, J. R., Yamanaka, S., & Ashraf, A. (2024). Artificial intelligence and deep learning algorithms for epigenetic sequence analysis: A review for epigeneticists and AI experts. *Computers in Biology and Medicine*, 183, 109302. https://doi.org/10.1016/J.COMPBIOMED.2024.109302.

Tarazona, S., Arzalluz-Luque, A., & Conesa, A. (2021). Undisclosed, unmet and neglected challenges in multi-omics studies. *Nature Computational Science*, 1(6), 395–402. https://doi.org/10.1038/s43588-021-00086-z.

Tejedor, J. R., Peñarroya, A., Gancedo-Verdejo, J., Santamarina-Ojeda, P., Pérez, R. F., López-Tamargo, S., Díez-Borge, A., Alba-Linares, J. J., González-del-Rey, N., Urdinguio, R. G., Mangas, C., Roberti, A., López, V., Morales-Ruiz, T., Ariza, R. R., Roldán-Arjona, T., Meijón, M., Valledor, L., Cañal, M. J., … Fraga, M. F. (2023). CRISPR/dCAS9-mediated DNA demethylation screen identifies functional epigenetic determinants of colorectal cancer. *Clinical Epigenetics*, 15(1), 133. https://doi.org/10.1186/S13148-023-01546-1.

The MIAME Checklist : Guidelines - GEO - NCBI. (2024). Retrieved November 10, 2025, from https://www.ncbi.nlm.nih.gov/geo/info/reviewer.html

Trapnell, C., Roberts, A., Goff, L., Pertea, G., Kim, D., Kelley, D. R., Pimentel, H., Salzberg, S. L., Rinn, J. L., & Pachter, L. (2012). Differential gene and transcript expression analysis of RNA-seq experiments with TopHat and Cufflinks. *Nature Protocols*, 7(3), 562–578. https://doi.org/10.1038/NPROT.2012.016.

Tycko, B. (2010). Allele-specific DNA methylation: Beyond imprinting. *Human Molecular Genetics*, 19(R2), R210. https://doi.org/10.1093/HMG/DDQ376.

Urich, M. A., Nery, J. R., Lister, R., Schmitz, R. J., & Ecker, J. R. (2015). MethylC-seq library preparation for base-resolution whole-genome bisulfite sequencing. *Nature Protocols*, 10(3), 475–483. https://doi.org/10.1038/nprot.2014.114.

Vaissiére, T., & Miller, C. A. (2011). DNA methylation: Dynamic and stable regulation of memory. *Biomolecular Concepts*, 2(6), 459. https://doi.org/10.1515/BMC.2011.046.

Wang, J., Alvin Chew, B. L., Lai, Y., Dong, H., Xu, L., Balamkundu, S., Cai, W. M., Cui, L., Liu, C. F., Fu, X. Y., Lin, Z., Shi, P. Y., Lu, T. K., Luo, D., Jaffrey, S. R., & Dedon, P. C. (2019). Quantifying the RNA cap epitranscriptome reveals novel caps in cellular and viral RNA. *Nucleic Acids Research*, 47(20), e130–e130. https://doi.org/10.1093/NAR/GKZ751.

Wang, Z., Gerstein, M., & Snyder, M. (2009). RNA-Seq: A revolutionary tool for transcriptomics. *Nature Reviews Genetics*, 10(1), 57–63. https://doi.org/10.1038/NRG2484.

Warnecke, P. M., Stirzaker, C., Song, J., Grunau, C., Melki, J. R., & Clark, S. J. (2002). Identification and resolution of artifacts in bisulfite sequencing. *Methods*, 27(2), 101–107. https://doi.org/10.1016/S1046-2023(02)00060-9.

Weber, M., Davies, J. J., Wittig, D., Oakeley, E. J., Haase, M., Lam, W. L., & Schübeler, D. (2005). Chromosome-wide and promoter-specific analyses identify sites of differential DNA methylation in normal and transformed human cells. *Nature Genetics*, 37(8), 853–862. https://doi.org/10.1038/NG1598.

Whalen, S., Truty, R. M., & Pollard, K. S. (2016). Enhancer–promoter interactions are encoded by complex genomic signatures on looping chromatin. *Nature Genetics*, 48(5), 488–496. https://doi.org/10.1038/ng.3539.

Woodcock, D. M., Crowther, P. J., Simmons, D. L., & Cooper, I. A. (1984). Sequence specificity of cytosine methylation in the DNA of the Chinese hamster ovary (CHO-K1) cell line. *Biochimica et Biophysica Acta*, 783(3), 227–233. https://doi.org/10.1016/0167-4781(84)90033-2.

Xiao, Y., Wu, J., & Lin, Z. (2021). Cancer diagnosis using generative adversarial networks based on deep learning from imbalanced data. *Computers in Biology and Medicine*, 135, 104540. https://doi.org/10.1016/J.COMPBIOMED.2021.104540.

Xiong, J., Ye, T. T., Ma, C. J., Cheng, Q. Y., Yuan, B. F., & Feng, Y. Q. (2019). N6-Hydroxymethyladenine: A hydroxylation derivative of N6-methyladenine in genomic DNA of mammals. *Nucleic Acids Research*, 47(3), 1268–1277. https://doi.org/10.1093/NAR/GKY1218.

Zemach, A., McDaniel, I. E., Silva, P., & Zilberman, D. (2010). Genome-wide evolutionary analysis of eukaryotic DNA methylation. *Science (New York, N.Y.)*, 328(5980), 916–919. https://doi.org/10.1126/SCIENCE.1186366.

Zhang, G., Huang, H., Liu, D., Cheng, Y., Liu, X., Zhang, W., Yin, R., Zhang, D., Zhang, P., Liu, J., Li, C., Liu, B., Luo, Y., Zhu, Y., Zhang, N., He, S., He, C., Wang, H., & Chen, D. (2015). N6-Methyladenine DNA modification in drosophila. *Cell*, 161(4), 893–906. https://doi.org/10.1016/J.CELL.2015.04.018.

Zhang, G., Liu, Y., & Jin, X. (2020). A survey of autoencoder-based recommender systems. *Frontiers of Computer Science*, 14(2), 430–450. https://doi.org/10.1007/S11704-018-8052-6/METRICS.

Zhu, C., Hao, Z., & Liu, D. (2024). Reshaping the landscape of the genome: Toolkits for precise DNA methylation manipulation and beyond. *JACS Au*, 4(1), 40–57. https://doi.org/10.1021/JACSAU.3C00671/ASSET/IMAGES/LARGE/AU3C00671_0005.JPEG.

Ziller, M. J., Gu, H., Müller, F., Donaghey, J., Tsai, L. T. Y., Kohlbacher, O., De Jager, P. L., Rosen, E. D., Bennett, D. A., Bernstein, B. E., Gnirke, A., & Meissner, A. (2013). Charting a dynamic DNA methylation landscape of the human genome. *Nature*, 500(7463), 477–481. https://doi.org/10.1038/nature12433.

Section II

Epigenetic Responses in Model Plants

7 Plant Epigenomes

Plasticity and Mapping to Understand Growth and Development

Savita Bhardwaj, Dhriti Sharma, Ajay Kumar, Yanchen Dolma, Bharat Kapoor and Dhriti Kapoor

7.1 INTRODUCTION

Eukaryotic organisms, including plants, contain their genomic DNA packaged into chromatin as stable nucleoprotein complexes that regulate both DNA accessibility and execute DNA-dependent functions (Ojolo et al., 2018). Epigenetics is defined as alterations in heritable gene regulation that bypass changes to DNA sequences during cell division through mitosis and meiosis (Kakoulidou et al., 2021). Thematically related changes include chromatin modifications that generate permanent alterations in transcriptional activity following removal of the initiating stimulus (Avramova, 2015). The epigenetic regulatory cascade works through multiple molecular agencies that involve both DNA methylation/demethylation mechanisms together with histone post-translational modifications (PTMs), chromatin remodeling elements and small and long noncoding RNAs (Pikaard & Scheid, 2014; Gallusci et al., 2016). Histone PTMs prove to be essential epigenetic markers because they accrue on core histone N-terminal tails (H2A, H2B, H3 and H4) during acetylation, methylation, phosphorylation and ubiquitination processes (Berr et al., 2011; Fortes & Gallusci, 2017).

In plants, dynamics and function of epigenetic marks have been improved by understanding the recent development of epigenome profiling for which various types of techniques have been developed (Schmitz & Zhang, 2011; Lee & Kim, 2014). Agronomic desired developmental or stress-related crop attributes can be improved by applying fundamental information among different aspects of plant resilience to environmental stresses which are regulated by heritable epigenetic variance. Plasticity greatly contributes to plants coping with and surviving in adverse environmental conditions due to their sessile nature (Sultan, 2000; Vannier et al., 2015). Alterations in growth, morphology, physiology or life history traits are some of the plastic responses of plants (Sultan, 2000). Hence, in response to these environmental constraints, a rapid response is provided by different developmental genetic pathways which support plasticity and those genes are selected which are responsible for these induced phenotypes (Martin & Pfennig, 2010; Pfennig et al., 2010). Directional or disruptive selection of new phenotypes can be caused in an organism when exposed to these environmental stresses if selection acts primarily on phenotype (Pfennig et al., 2010). Therefore, environmental induction followed by genetic adaptation of the changes, can produce unique phenotypes (West-Eberhard, 2005; Vannier et al., 2015).

Epigenetic changes are maintained via cell divisions in tissues in various plants so that these epigenetic changes can be inherited across generations, i.e., intergenerationally. Hence, maintenance of epigenetic changes can be done over developmental and generational aspects (Hemenway & Gehring, 2023). Addition and removal of modifications in various developmental aspects under varied ecological conditions in different cell and tissue types make epigenetic information dynamic. The distinction between stable epigenetic states, well preserved throughout cell divisions, as opposed to flexible epigenetic changes acquired or lost under diverse conditions or stages of development, has been a key contributor in this regard (Banerjee et al., 2017). Therefore, the present chapter discusses the role of epigenetics in crop improvement under normal physiological conditions as well as under stress conditions.

7.2 APPLICATION OF EPIGENOMICS IN CROP IMPROVEMENT

The epigenome comprises all epigenetic modifications affecting DNA bases (without changing the nucleotide sequence), histone proteins and small-RNA biogenesis within a cell (Banerjee & Roychoudhury, 2018; Kumar, 2019). In plants, epigenetic regulation during development and in response to environmental factors occurs through three primary mechanisms: DNA methylation, posttranslational histone modifications and the activity of noncoding RNA molecules. These noncoding RNAs include small RNAs, such as small interfering RNAs (siRNAs) and microRNAs (miRNAs), as well as long noncoding RNA. Several researchers have increasingly recognized the role of epigenetics in evolution, highlighting its function as an adaptive and responsive molecular mechanism in natural selection (Kumar, 2019). Extensive studies indicate that

DOI: 10.1201/9781003589914-9

climate change has already impacted agricultural crop production and food security (Frank et al., 2017; Ruane et al., 2018; Van Meijl et al., 2018). Moreover, biotic stressors, including pathogens, pests and weeds, continue to threaten global crop yield. Addressing these biotic and abiotic challenges through epigenetic research will be pivotal for enhancing crop adaptation and selecting superior genotypes, ultimately ensuring long-term agricultural sustainability (Kakoulidou et al., 2021).

7.3 EPIGENETIC MECHANISMS INVOLVED IN IMPROVING CROP TRAITS UNDER NORMAL PHYSIOLOGICAL CONDITIONS

Most agronomic traits are quantitative and governed by complex genetic regulation, posing challenges to crop improvement. Plant epibreeding, which harnesses epigenetic variation for crop development, offers extensive applications in crop enhancement (Gupta & Salgotra, 2022). Beyond genetic variation, a deeper understanding of how epigenetic modifications influence plant phenotypes has accelerated the crop improvement process (Merce et al., 2020).

7.3.1 DNA Methylation

A deeper understanding of epigenetic variations, such as DNA methylation, and their influence on plant phenotypes has created new opportunities to accelerate crop improvement. The regulation of cellular DNA methylation is precisely controlled by the coordinated activity of DNA methyltransferases and demethylases (Zhang & Zhu, 2024). Among various epigenetic modifications, DNA methylation remains the most extensively studied in plants and other eukaryotes (Shaikh et al., 2022). The heritability of DNA methylation patterns has been documented in plant species such as *Nicotiana tabacum* and *Arabidopsis thaliana* as well as in major crops like rice and wheat (Mercé et al., 2020; Molinier et al., 2006; Fieldes et al., 2005; Fieldes, 1994). Akimoto et al. (2007) exposed germinating rice seeds to the DNA methylation inhibitor, viz., 5-aza-deoxycytidine, and the surviving progeny were cultivated in the field for over a decade. Among these, a specific propagated line, designated as Line-2, exhibited acquired resistance to the bacterial pathogen *Xanthomonas oryzae* pv. *oryzae*. Similarly, a study on rapeseed epigenetic modulation revealed that exposing *Brassica napus* seedlings to 5-azacytidine led to the development of new *Brassica* lines with elevated seed protein levels and increased linoleic acid content (Amoah et al., 2012). The epiallele *epid1*, which encodes a GTP-binding protein, causes stunted growth. DNA methylation and histone acetylation together silence the *d1* genes, thereby regulating plant height in rice (Miura et al., 2009). Additionally, 6-phosphogluconolactonase 2 (PGL2), a key regulator of grain length, is positively influenced by *OsglHAT1*. The upregulation of *OsglHAT1* enhances agronomic traits such as grain length, grain weight, yield and overall biomass (Song et al., 2015).

Methylation of the *ZmMRP4* gene results in the *lpa1-241* phenotype in maize, characterized by high levels of inorganic phosphate in the seed (Pilu et al., 2009). The DEMETER-like DNA demethylase (DML) *SlDML2* is essential for regulating DNA methylation during tomato ripening. Inhibition of *SlDML2* via RNA interference led to ripening defects, which were associated with increased methylation in the promoter region, suppressing genes involved in fruit ripening and softening (Liu et al., 2015). Kundariya et al. (2020) utilized the MSH1 system in *Arabidopsis* and tomato to introduce rootstock epigenetic variation into grafting experiments. Introducing mutations in *dcl2*, *dcl3* and *dcl4* to the *msh1* rootstock disrupts siRNA production, uncovering RdDM targets involved in methylation re-patterning. Progeny from these grafting experiments exhibited improved growth vigor compared to controls. Notably, RdDM-induced changes can sometimes be retained and inherited across multiple generations without further intervention or manipulation, suggesting that RdDM could be a valuable tool for targeted epigenome editing (Shaikh et al., 2022). Additionally, the *CONSTANS* (*CO*) and *CO-LIKE* (*COL*) genes play a crucial role in regulating photoperiodicity in *Arabidopsis*. *COL2* is an epiallele found in allotetraploid cotton species. In wild cottons, COL2A is hypermethylated and silenced, while COL2D is repressed; however, in domesticated cotton varieties, COL2D is highly expressed due to the loss of methylation. Inhibiting DNA methylation results in the activation of *COL2* expression, and suppressing *COL2* expression in cultivated cotton delays flowering. Demethylation of COL2 enhances its expression, promoting photoperiodic flowering, which may have contributed to the global adaptation of cotton for cultivation (Song et al., 2017). Induced DNA methylation can also improve hybrid breeding and plant propagation through tissue culture, facilitating the creation of novel gene expression patterns in offspring (Surdonja et al., 2017; Temel et al., 2017), thereby aiding in the management of phenotypic traits. Furthermore, DNA methylation holds the potential to enhance plant resistance to viruses or even alter pathogen populations, making them more vulnerable to the defensive mechanisms of plants (Kantar et al., 2010).

7.3.2 Histone Modifications

Histones undergo a variety of post-translational modifications (PTMs), which are essential for regulating chromatin dynamics. The histone tails, comprising 15–38 amino acids, can be modified through processes such as methylation, acetylation, phosphorylation, ubiquitination, sumoylation, biotinylation, ribosylation and citrullination (Agarwal et al., 2020). In their study, Deng et al. (2022) showed that SlERF.F12 negatively regulates tomato fruit ripening by recruiting TPL2 and histone deacetylases (HDA1/HDA3) to repress the expression of ripening-related genes. This tripartite complex reduces H3K9Ac and H3K27Ac levels at the promoters of these genes, limiting histone acetylation and consequently delaying the ripening process.

In another study, histone H3K9 methylation levels were initially low during microspore reprogramming but increased as embryogenesis progressed, correlating with changes in gene expression. The compound BIX-01294 enhanced early reprogramming and totipotency by reducing both H3K9 methylation and DNA methylation. However, prolonged treatment hindered embryo differentiation, emphasizing the critical role of H3K9 methylation in embryogenesis (Berenguer et al., 2017). SlHDT3 positively regulates tomato fruit ripening by modulating ethylene biosynthesis, carotenoid accumulation and the expression of ripening-associated genes. RNAi lines of SlHDT3 showed delayed ripening, extended shelf life and reduced expression of genes involved in ethylene biosynthesis and cell wall metabolism, placing SlHDT3 upstream of SlMADS-RIN in the ripening regulatory network (Guo et al., 2017). In banana, MaERF11 inhibits fruit ripening by binding to GCC-box motifs in the promoters of *MaACO1* and expansin genes, suppressing their expression. It also recruits MaHDA1, a histone deacetylase, to mediate histone deacetylation, thereby linking ethylene signaling with chromatin modifications in the regulation of ripening (Han et al., 2016). Bolting is a significant agronomic trait in Chinese cabbage, but premature bolting can substantially decrease its commercial value, yield and quality. The early-bolting mutant *ebm1* in Chinese cabbage was associated with a 53 bp insertion in *Bra032169*, which encodes a histone methyltransferase homologous to *CLF* in *Arabidopsis*. This mutation likely disrupted the repression of flowering, leading to increased expression of the genes, viz., *FLC*, *FT*, *AG* and *SEP3*, which in turn accelerates bolting (Huang et al., 2020). The grain weight quantitative trait locus (QTL) *OsglHAT1* encodes a GNAT-like histone acetyltransferase, which enhances grain yield by promoting the growth of spikelet hulls and improving grain filling. The rare allele of *OsglHAT1*, associated with increased gene expression and elevated H4 acetylation levels, positively influences yield traits (Song et al., 2015). In the *lvp1* mutant, reduced levels of H3K36me2/3 at *MADS50* and *RFT1* loci suggest that variations in H3K36 methylation are a key mechanism driving the functional divergence between *RFT1*/Hd3a and *MADS50*/MADS51, with upregulated expression of *MADS50* and *RFT1* promoting flowering in rice (Sun et al., 2012). The maize histone deacetylase *HDA101* plays a crucial role in kernel development by regulating histone acetylation in genes specific to transfer cells, thereby influencing endosperm differentiation. It interacts with chromatin remodelers and represses a subset of inactive genes, underscoring its dual function in transcriptional regulation during seed development (Yang et al., 2016).

Histone modifications are crucial in regulating plant responses to abiotic stresses (Chinnusamy et al., 2008; Luo et al., 2012). In soybean, salinity stress affects gene expression by altering DNA methylation and histone modifications in key transcription factors. Increased levels of H3K4me3 and H3K9Ac, coupled with reduced H3K9 demethylation, were associated with the activation of stress-responsive genes, emphasizing the role of epigenetic regulation in salinity tolerance (Song et al., 2012). In rice seedlings, drought stress triggers acetylation of histones, including H3K9ac, H3K18ac, H3K27ac, H3ac and H4K5ac (Fang et al., 2014). Likewise, under submergence or nutrient starvation conditions, rice plants exhibit dynamic changes in genome-wide H3K9ac levels, which correlate with the activation of genes involved in stress responses (Lu et al., 2018). Downregulation of *OsSRT1* in rice results in increased histone H3K9ac and decreased H3K9me2, leading to the generation of H_2O_2, DNA fragmentation, cell death and lesion formation, which mimic hypersensitive responses during incompatible pathogen interactions (Huang et al., 2007). These findings highlight the significant role of histone modifications in plant responses to both biotic and abiotic stress, offering promising strategies to enhance crop development and productivity in an efficient and targeted manner.

7.3.3 Noncoding RNAs

Noncoding RNAs (ncRNAs) are integral to the regulation of plant growth, development and responses to stress at both the transcriptional and translational levels. This group includes long noncoding RNAs (lncRNAs), small RNAs and circular RNAs, all of which are recognized as key regulators of gene expression in plants. These ncRNAs have been shown to play pivotal roles in plant immunity and in the adaptation to both abiotic and biotic stresses (Summanwar et al., 2020; Bhattacharya & Roychoudhury, 2024). ncRNAs regulate the expression of endogenous or exogenous genes at the transcriptional, post-transcriptional or translational levels in a sequence-specific manner (Liu & Zhu, 2014). Numerous studies have highlighted the regulatory roles of miRNAs and lncRNAs in plant growth and development (Li & Zhang, 2016; Swarup & Denyer, 2018; Yu et al., 2019).

Many miRNAs and siRNAs regulate the transcription of defense-related genes by targeting transcription factors (Song et al., 2021; Chakraborty & Roychoudhury, 2023). In rice, *miR169* targets NF-YA transcription factors to negatively regulate immunity against *Magnaporthe oryzae* (Li et al., 2017). Additionally, *miR156* suppresses rice resistance to bacterial blight by targeting squamosa promoter-binding protein-like transcription factors *IPA1* and *SPL7* (Liu et al., 2019). Overexpression of the evolutionarily conserved *miR156* in tomato plants has been linked to fruit size (Zhang et al., 2016). Similarly, increased expression of *miR397* positively correlates with grain size and yield in rice by regulating its target gene, *LACCASE* (LAC) (Zhang et al., 2013). Numerous ncRNAs also play a role in regulating reactive oxygen species (ROS) production. In rice, overexpression of *miR398b* reduces the transcript levels of genes encoding superoxide dismutases (*CSD1*, *CSD2*, *SODX* and *CCSD*), resulting in elevated ROS production and enhanced resistance to *Magnaporthe oryzae* (Li et al., 2014, 2019). In tomato, *lncRNA33732* is activated by WRKY1, which in turn triggers the expression of *RBOH* and enhances ROS accumulation during the early stages of defense against *Phytophthora infestans* infection (Cui et al., 2019). Conversely, *miR528* suppresses viral resistance in rice by targeting and cleaving *L-ascorbate oxidase* (*AO*) mRNAs, leading to a reduction in ROS levels (Wu et al., 2017; Yao et al., 2019).

RNA silencing technology has been applied to enhance the nutritional value of crops (Dekeba 2021). RNA interference (RNAi) has been employed to develop tomato fruits with elevated levels of carotenoids and flavonoids, which are particularly beneficial for human health (Kamthan et al., 2015). Additionally, RNAi has been used to downregulate the starch branching enzyme in wheat, resulting in high-amylose wheat, which holds significant potential for improving human health (Man & Hong, 2013). RNA interference reduces the expression of *1-aminocyclopropane-1-carboxylate* (*ACC*) *oxidase*, a key gene in the ethylene biosynthesis pathway in tomato, effectively inhibiting ethylene production during ripening and allowing the fruit to have an extended shelf life (Dekeba, 2021). Virus control strategies rely on the ecological effect of increased carbon dioxide concentrations, which reduce the accumulation of cucumber mosaic virus in *Nicotiana tabacum* by interfering with the viral suppressor of RNA interference (VSR) 2b protein of CMV (Guo et al., 2021).

RNA interference has also been used to suppress two genes from *Agrobacterium tumefaciens* – *iaaM* and *ipt* – that are involved in crown gall tumor formation, significantly reducing tumor production in *Arabidopsis* (Kamthan et al., 2015). Insects that cause defoliation or extract sap from plants can severely damage them by stunting growth, weakening them and, in some cases, causing death (Bahadur et al., 2015). Transgenic tobacco lines expressing dsRNA targeting *vATPase* have shown a notable whitefly mortality rate, with target gene transcript levels in *Bemisia tabaci* reduced by more than 70% at a mortality rate of over 70% (Thakur et al., 2014; Raza et al., 2016; Shelby et al., 2020). RNA interference-mediated knockdown of *CsKrh1* significantly reduced the transcription of vitellogenesis (Vg) in *Chilo suppressalis*, playing a crucial role in controlling rice pests (Tang et al., 2020). Overexpression of *GmNFYA3* in *Arabidopsis* resulted in heightened sensitivity to salinity stress and exogenous ABA (Ni et al., 2013). Similarly, transgenic *Agrostis stolonifera* plants overexpressing the rice *miR319* gene (*OsamiR319*) exhibited enhanced tolerance to drought and salinity, which was associated with increased leaf wax content and improved water retention, though sodium uptake was reduced (Zhou & Luo, 2013). Overexpression of the *Osa miR319* gene increased cold stress tolerance (at 4°C) after plants underwent cold acclimatization (at 12°C) (Yang et al., 2013). In rice, overexpression of *OsPCF5* and *OsTCP21* resulted in the development of cold-resistant transgenic plants (Yang et al., 2013). The expression of circular RNAs (circRNAs) in response to bacterial pathogen infections, such as *Pseudomonas syringae* pv. *actinidiae* (PSA) in kiwi fruits and *Pectobacterium carotovorum* subsp. *Brasiliense* (PCB) in potato, highlights their involvement in biotic stress responses (Wang et al., 2017a; Zhou et al., 2018).

In rice, lncRNAs were found to regulate drought stress by targeting genes that encode zinc-finger proteins in the *qSDT2-1* region, a key drought QTL (Weidong et al., 2020). Similarly, in *Brassica rapa*, two heat-responsive lncRNAs were identified as endogenous target mimics of *miR164a*. The differential expression of both these lncRNAs and miRNA further underscores their pivotal roles in mediating the heat stress response (Wang et al., 2019b). Two long noncoding RNAs (lncRNAs), COOLAIR and COLD ASSISTED INTRONIC NON-CODING RNA (COLDAIR), play crucial roles in promoting flowering in plants under cold conditions (Whittaker & Dean, 2017). In tomato, during *Phytophthora infestans* infection, lncRNA16397 activates the expression of the *GLUTAREDOXIN 22* gene in cis, leading to enhanced resistance (Cui et al., 2017). These variations in ncRNAs provide potential avenues for improving desirable agronomic traits. Additionally, the *PHOTOPERIOD-SENSITIVE GENIC MALE STERILITY 1 (Pms1)* locus in rice encodes the lncRNA PMS1T, which is specifically expressed in the young panicles, further illustrating the role of lncRNAs in crop development (Fan et al., 2016).

7.4 ROLE OF EPIGENOMICS UNDER ABIOTIC STRESS CONDITIONS

Plants are remarkably potent in accentuating their signaling pathways and adapting themselves to the stressful changes in their ambient environment (salinity, drought, nutrient scarcity, cold, etc.) by undergoing epigenetic modifications. These epigenetic changes surpass the scope of genetically regulated mechanisms, thus encoding an additional level of endogenous stress control (Lu & Hyun, 2021; Hemenway & Gehming, 2023). Epigenomics significantly reprogram the genotypic and phenotypic expressions of stressed plants by modulating the transcriptome, and there are two major categories of epigenetic alterations: (a) long term and (b) short term. The first one serves as the molecular premise of phenotypic plasticity, which is consequently involved in the process of evolution, whereas the second type is essential in ensuring the survival of plants exposed to stressful conditions. Both of them form the basis for selecting the better-suited plants to an ever-changing environment, as documented in naturally occurring populations with the same type of genetic constitution, hinting at their epigenetic nature (Duarte-Aké et al., 2023).

Knowledge of epigenetic mechanisms in dealing with the abiotic stressors via modulating signal transduction systems could prove beneficial in making the plants more stress resistant with better agricultural production and quality (Akhter et al. 2021; Balazova et al., 2022). Altered nature of chromatin (DNA plus associated proteins) via methylation of DNA, modifications of histone proteins and some noncoding RNAs contribute to these stress-regulated epigenetic mechanisms. Epigenetic memory, where previous stress exposure induced an epigenetic imprint intact along with epigenetic control of hormone signaling and secondary metabolism, also substantiates the significance of epigenetic changes while combating different kinds of stress. Moreover, these changes are heritable and not only affect the genetic makeup of plants but also enable them to transmit stress-related experiences across the generations (Sudan et al., 2018; Jogam et al., 2022).

Methylation of DNA is a prerequisite for the regulation of gene expression and making the plant genome stable; for instance, the addition of methyl to the cytosine base of DNA gives rise to 5-methylcytosine (Chachar et al., 2022; Lee et al., 2023), which is vitally involved in the regulation of transcription, stability of the genome, pace of development and responses against abiotic stress inducers (Sadakierska-Chudy et al., 2015; Zhang et al., 2018). Methylation process could take place either actively or passively, and patterns of its manipulations (hypomethylation as well as hypermethylation) make plants behave as better performers in terms of being more productive, resistant to diseases and tolerant to diverse abiotic stress (Asensi-fabado et al., 2017; Zhang et al., 2023). Drought stress tolerance, for instance, is conferred on *Arabidopsis thaliana* due to DNA methylation by a specific gene *ATDM1* (Lee et al., 2023).

A research centered around salt-stressed sensitive and tolerant rice cultivars divulged the significant role of diverse DNA methylation levels which regulated the expression of genes dealing with high salinity stress (Ferreira et al., 2015). Additionally, the other outcomes of this study revealed the upregulation of the *DRM2* gene (DNA demethylase), which reduced the level of DNA methylation in salt-sensitive variants vis-a-vis tolerant ones, in response to salinity-induced stress. Research on apple under extremely low temperature and drought exposed *Populus trichocarpa* plants exhibited modifications in the extent of DNA methylation in order to incite the stress-responsive genes (Kumar et al., 2016; Saraswat et al., 2017).

Current research scenario has corroborated the underlying role played by the regulation of gene expression via dynamic alterations in the patterns of DNA methylation, which could enable the plants to withstand adverse abiotic conditions (Ferrari et al., 2020; Duarte-Aké et al., 2023).

Complex signaling mechanisms are triggered by different abiotic stressors like extremes of salt, water, temperature, etc., bringing about modifications of histone proteins. These changes regulate chromatin architecture plus gene expression, especially during stress-induced responses in plants. Research outcomes have demonstrated that alterations in the levels of histone acetylation during drought stress brought about upregulation of the expression of various stress-responsive genes as observed in *Arabidopsis thaliana* where drought elevates H3 histone acetylation levels (Wollmann et al., 2017; Nunez-Vazquez et al., 2022) or activates the enzyme *histone deacetylase* (*HDA6*) (Kim et al., 2015; Zhao et al., 2019). Likewise, salinity stress modifies the patterns of histone methylation, impacting the genes responsible for ionic homeostasis plus osmotic adjustments (Ueda & Seki, 2020), while cold stress enhances chromatin accessibility through bivalent histones, i.e., H3K4me3 and H3K27me3 alterations of the activated genes in *Solanum tuberosum* (Zeng et al., 2019). Extreme temperature also modulates heat-induced stress responses indicating the dynamic involvement of histone modifications during stressful conditions (Wu et al., 2019; Nunez-Vazquez et al., 2022). Histone modifications such as acetylation, methylation and phosphorylation mediated by protein kinases are also vital in alleviating abiotic stress, for example, extreme conditions of temperature, water availability or salt induce H3S10ph phosphorylation, which potentially leads to either silencing or upregulation of stress-related genes (Wang & Qiao, 2020).

The mitogen-activated protein kinase (MAPK) cascade is reported to take part in histone phosphorylation whereby it alters gene expression while dealing with stressful conditions. Other histone modifications, namely ADP-ribosylation, sumoylation and ubiquitination, also influence the structure of chromatin and the expression of stress-responsive genes; however, deeper insights into the stress-mitigating role of these genes are still needed to be worked out (Wang & Qiao, 2020). Response of plants to mitigate different abiotic stresses with the help of alterations in DNA methylation and histone modification are represented in Table 7.1.

Small noncoding RNAs (siRNAs and miRNAs) play a pivotal role in plant stress responses by modulating the expression of genes through gene silencing (Zheng et al., 2019; Begum, 2022). These RNA molecules navigate the RNA-induced silencing complex (RISC) to target mRNAs for degradation or repression during translation. Intrinsic hairpin loops form miRNAs, whereas precursors of either extrinsic or intrinsic double-stranded RNA give rise to siRNAs (Xie & Yu, 2015) and both of these RNAs base pair with target mRNA, causing its cleavage or inhibition at the translational level (Jagadhesan et al., 2022). miRNAs enhance adaptability to different abiotic stresses like heat and drought; for instance, miR159 modulates the genes of drought tolerance in *Arabidopsis thaliana* (Ferdous et al., 2015; Wang & Qiao, 2020).

Some of the research findings further suggest that stress-induced oxidative damage is regulated by some miRNAs, namely miR528 and miR398, in addition to the innate antioxidant defense system of plants (Qiu et al., 2018; Singroha et al., 2021; Yang et al., 2022), for example, in *Arabidopsis thaliana,* oxidative stress is observed to be mitigated by repressing the expression of the Cu-superoxide dismutase enzyme by miR398 (Singroha et al., 2021).

TABLE 7.1
Response of plants to mitigate different abiotic stresses

Epigenetic mechanism	Abiotic stress	Plant species	Effects	References
DNA Methylation	Drought	*Oryza sativa*	Deferred site specific methylation in genome	Ferreira et al. (2019), Kumar et al. (2023)
		Medicago sativa	Decline in DNA methylation	Ventouris et al. (2023)
		Zea mays	Enhancement in water retention	Min et al. (2016), Aravind et al. (2017)
	Cold	*Prunus simonii*	Methylation of cytosines	Kumar et al. (2023)
	Heat	*Brassica napus*	Betterment in DNA methylation	Shankar et al. (2016)
		Zea mays	Enhanced methylation levels with improved stress tolerance	Chen et al. (2019)
	Salinity	*Brassica napus*	Increase in methylation levels in sensitive cultivars contrary to tolerant ones	Rehman & Tanti (2020)
		Oryza sativa	Decrease in DNA methylation in tolerant varieties	Arora et al. (2022)
	Heavy metals	*Arabidopsis thaliana*	Elevation in DNA methylation levels	Sun et al. (2022)
		Trifolium repens	Long duration exposure causes decrease in methylation in the tolerant cultivars	Arora et al. (2022)
Histone Modification	Drought	*Arabidopsis thaliana*	Sensitivity to water scarcity is reduced by HDA9 through activation of H3K9 in fourteen genes	Zheng et al. (2016)
		Brassica rapa	Upregulation of nine *HAT* genes helping the stressed plants to cope with water deficit	Eom & Hyun (2018)
		Oryza sativa	Triggering of nine *HAT* genes for stress response	Hou et al. (2021)
	Heat	*Arabidopsis thaliana*	Interaction of HDA9 with PWR along with regulation of thermomorphogenesis through YUC8 plus phytochrome interacting factor 4	Shen et al. (2019)

Furthermore, these small RNAs function during transcription and even after transcription along with different chromatin remodeling systems whereby they permit or prohibit regulatory proteins from accessing DNA, leading to altered gene expression (Zhan & Meyers, 2023). Transposable elements (TEs) or different repetitive sequences also produce siRNA, which leads DRM2 (DNA methyltransferase) enzyme to particular genomic targets, where they perform gene silencing for controlling the expression of stress responsive genes as noted in heat stressed *Arabidopsis* (Gallego-Bartolomé, 2020). Additionally, these small noncoding RNAs are also involved intricately with histone modifiers or even with chromatin remodeling complexes to upgrade the stress responses of plants.

7.4.1 Chromatin Remodeling

Complexes formed by the remodeling of chromatin architecture have also been documented to epigenetically regulate gene expression, especially in cases of stress-induced responses (Pandey et al., 2016, Pecinka et al., 2020; Tonosaki et al., 2022). Such complexes not only alter the shape of chromatin and impact regulatory proteins but also change the accessibility of DNA to the transcriptional mechanisms (Pandey et al., 2016). Reprogrammed expression of stress-responsive genes brought about by dynamic modifications in structural parts of chromatin helps the plants to adjust and survive harsh conditions (Wang & Qiao, 2020). The energy generated due to ATP hydrolysis activates these complexes to slide, exclude or modify nucleosome composition for changing the expression of genes under stress (He et al., 2023). Histone proteins undergo posttranslational modifications with the help of histone-altering enzymes, which result in up- or down- regulating gene transcription. In highly condensed form, the chromatin complexes forbid the transcription factors, nuclear proteins and polymerases from gaining access to DNA, as some of the stress-induced architectural changes of chromatin otherwise make the DNA accessible (Bhadouriya et al., 2021). This vital regulation permits the plants to respond to diverse stressors without modifying the intrinsic genetic code. These remodeled complexes of chromatin complete the stress response of plants in collaboration with methylation of DNA, histone modifiers and silencing pathways mediated by small RNA, as depicted in Figure 7.1.

Their intricate crosstalk makes the stressed plants more resilient, adaptive and tolerant. Additionally, recent research has revealed that external signals could influence the action and specificity of these chromatin complexes, suggesting a direct link underlying between the environmental stimuli and epigenetic modifications which impact the stress-mediated gene expression (Paixão et al., 2019; Kim, 2021).

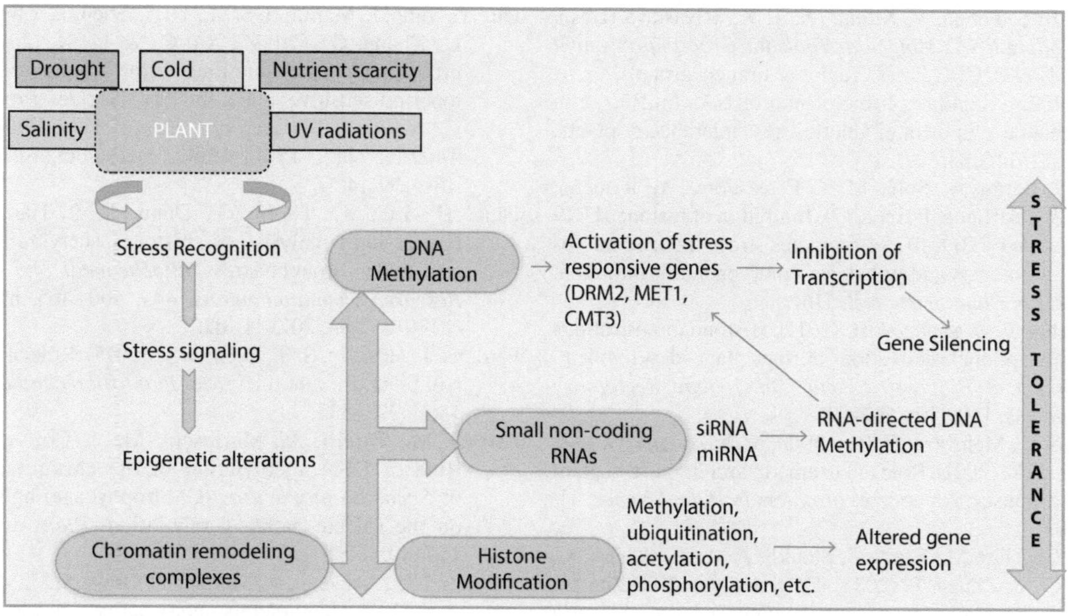

FIGURE 7.1 Summarized depiction of different epigenetic modifications vis-a-vis abiotic stresses.

7.5 CONCLUSION

Epigenetic processes, encompassing DNA methylation, histone modifications and noncoding RNAs, are fundamental in the regulation of plant growth, response to stress and the expression of agronomic traits. With plants increasingly exposed to both abiotic and biotic pressures driven by climate change and environmental factors, the manipulation and understanding of these epigenetic mechanisms offer significant potential for enhancing crop resilience and boosting agricultural yield. DNA methylation, for example, has demonstrated its potential in improving disease resistance and optimizing traits in crops such as rice, maize and rapeseed. Histone modifications are essential for regulating gene expression linked to stress adaptation and vital agronomic features, while noncoding RNAs, including microRNAs and long noncoding RNAs, play crucial roles in controlling stress-responsive genes, thereby aiding plant acclimatization to changing environmental conditions. Advancements in epigenetic research and the development of precise tools for modifying these mechanisms will be instrumental in expediting crop improvement and supporting sustainable agricultural practices amidst global challenges.

REFERENCES

Agarwal, G., Kudapa, H., Ramalingam, A., Choudhary, D., Sinha, P., Garg, V., & Varshney, R. K. (2020). Epigenetics and epigenomics: Underlying mechanisms, relevance, and implications in crop improvement. *Functional & Integrative Genomics*, 20, 739–761.

Akhter, Z., Bi, Z., Ali, K., Sun, C., Fiaz, S., Haider, F. U., & Bai, J. (2021). In response to abiotic stress, DNA methylation confers epigenetic changes in plants. *Plants*, 10(6), 1096.

Akimoto, K., Katakami, H., Kim, H. J., Ogawa, E., Sano, C. M., Wada, Y., et al. (2007). Epigenetic inheritance in rice plants. *Annals of Botany*, 100(2), 205–217.

Amoah, S., Kurup, S., Lopez, C. M. R., Welham, S. J., Powers, S. J., Hopkins, C. J., et al. (2012). A hypomethylated population of *Brassica rapa* for forward and reverse epigenetics. *BMC Plant Biology*, 12(1), 193.

Aravind, J., Rinku, S., Pooja, B., Shikha, M., Kaliyugam, S., Mallikarjuna, M. G., Kumar, A., Rao, A. R., & Nepolean, T. (2017). Identification, characterization, and functional validation of drought-responsive microRNAs in subtropical maize inbreds. *Frontiers in Plant Science*, 8, 941.

Arora, H., Singh, R. K., Sharma, S., Sharma, N., Panchal, A., Das, T., Prasad, A., & Prasad, M. (2022). DNA methylation dynamics in response to abiotic and pathogen stress in plants. *Plant Cell Reports*, 41(10), 1931–1944.

Asensi-Fabado, M. A., Amtmann, A., & Perrella, G. (2017). Plant responses to abiotic stress: The chromatin context of transcriptional regulation. *Biochimica et Biophysica Acta (BBA)-Gene Regulatory Mechanisms*, 1860(1), 106–122.

Avramova, Z. (2015). Transcriptional 'memory'of a stress: Transient chromatin and memory (epigenetic) marks at stress-response genes. *The Plant Journal*, 83(1), 149–159.

Bahadur, B., Rajam, M. V., Sahijram, L., & Krishnamurthy, K. V. (2015). Induced mutations and crop improvement. *Plant Biology and Biotechnology*, 1(23), 593–617.

Balazova, E., Balazova, A., & Oblozinsky, M. (2022). Epigenetic modifications in plants-impact on phospholipid signaling and secondary metabolism. *Chemické Listy*, 116(7), 416–422.

Banerjee, A., & Aryadeep Roychoudhury, A. (2017). Epigenetic regulation during salinity and drought stress in plants: Histone modifications and DNA methylation. *Plant Gene*, 11, 199–204.

Banerjee, A., & Roychoudhury, A. (2018). The gymnastics of epigenomics in rice. *Plant Cell Reports*, 37, 25–49.

Bhattacharya, A., & Roychoudhury, A. (2024). Plant non-coding RNAs and their regulation in salinity and heavy metal stress. In S. Penna, V. Kumar, & A. K. Srivastava (Eds.), *Non-Coding RNAs Molecular Tools for Crop Improvement* (pp. 128–147). CRC Press (Taylor & Francis group).

Begum, Y. (2022). Regulatory role of microRNAs (miRNAs) in the recent development of abiotic stress tolerance of plants. *Gene*, 821, 146283.

Berenguer, E., Bárány, I., Solís, M. T., Pérez-Pérez, Y., Risueño, M. C., & Testillano, P. S. (2017). Inhibition of histone H3K9 methylation by BIX-01294 promotes stress-induced microspore totipotency and enhances embryogenesis initiation. *Frontiers in Plant Science*, 8, 1161.

Berr, A., Shafiq, S., & Shen, W. H. (2011). Histone modifications in transcriptional activation during plant development. *Biochimica et Biophysica Acta (BBA)-Gene Regulatory Mechanisms*, 1809(10), 567–576.

Bhadouriya, S. L., Mehrotra, S., Basantani, M. K., Loake, G. J., & Mehrotra, R. (2021). Role of chromatin architecture in plant stress responses: An update. *Frontiers in Plant Science*, 11, 603380.

Chachar, S., Chachar, M., Riaz, A., Shaikh, A. A., Li, X., Li, X., Guan, C., & Zhang, P. (2022). Epigenetic modification for horticultural plant improvement comes of age. *Scientia Horticulturae*, 292, 110633.

Chakraborty, S., & Roychoudhury, A. (2023). MicroRNA-mediated regulation of heavy metal stress in plants. In D. Pandita & A. Pandita (Eds.), *Plant MicroRNAs and Stress Response* (pp. 183–197). CRC Press (Taylor and Francis group).

Chen, R., Li, M., Zhang, H., Duan, L., Sun, X., Jiang, Q., Zhang, H., & Hu, Z. (2019). Continuous salt stress-induced long non-coding RNAs and DNA methylation patterns in soybean roots. *BMC Genomics*, 20, 1–12.

Chinnusamy, V., Gong, Z., & Zhu, J. K. (2008). Abscisic acid-mediated epigenetic processes in plant development and stress responses. *Journal of Integrative Plant Biology*, 50, 1187–1195.

Cui, J., Jiang, N., Meng, J., Yang, G., Liu, W., Zhou, X., & Luan, Y. (2019). LncRNA33732-respiratory burst oxidase module associated with WRKY1 in tomato-*Phytophthora infestans* interactions. *The Plant Journal*, 97(5), 933–946.

Cui, J., Luan, Y., Jiang, N., Bao, H., & Meng, J. (2017). Comparative transcriptome analysis between resistant and susceptible tomato allows the identification of lncRNA16397 conferring resistance to *Phytophthora infestans* by co-expressing glutaredoxin. *Plant Journal*, 89, 577–589. https://doi.org/10.1111/tpj.13408.

Dekeba, D. G. (2021). Advances on the application of non-coding RNA in crop improvement. *African Journal of Biotechnology*, 20(11), 440–450.

Deng, H., Chen, Y., Liu, Z., Liu, Z., Shu, P., Wang, R., Hao, Y., Su, D., Pirrello, J., Liu, Y., Li, Z., Grierson, D., Giovannoni, J. J., Bouzayen, M., & Liu, M. (2022). SlERF.F12 modulates the transition to ripening in tomato fruit by recruiting the co-repressor TOPLESS and histone deacetylases to repress key ripening genes. *Plant Cell*, 34(4), 1250–1272. https://doi.org/10.1093/plcell/koac025.

Duarte-Aké, F., Us-Camas, R., & De-la-Peña, C. (2023). Epigenetic regulation in heterosis and environmental stress: The challenge of producing hybrid epigenomes to face climate change. *Epigenomes*, 7(3), 14.

Eom, S. H., & Hyun, T. K. (2018). Histone acetyltransferases (HATs) in Chinese cabbage: Insights from histone H3 acetylation and expression profiling of HATs in response to abiotic stresses. *Journal of the American Society for Horticultural Science*, 143(4), 296–303.

Fan, Y., Yang, J., Mathioni, S. M., Yu, J., Shen, J., Yang, X., Wang, L., Zhang, Q., Cai, Z., Xu, C., & Li, X. (2016). PMS1T, producing phased smallinterfering RNAs, regulates photoperiod-sensitive male sterility in rice. *Proceedings of the National Academy of Sciences of the United States of America*, 113, 15144–15149. https://doi.org/10.1073/pnas.1619159114.

Fang, H., Liu, X., Thorn, G., Duan, J., & Tian, L. (2014). Expression analysis of histone acetyltransferases in rice under drought stress. *Biochemical and Biophysical Research Communications*, 443, 400–405. https://doi.org/10.1016/j.bbrc.2013.11.102.

Ferdous, J., Hussain, S. S., & Shi, B. J. (2015). Role of micro RNA s in plant drought tolerance. *Plant Biotechnology Journal*, 13(3), 293–305.

Ferrari, M., Torelli, A., Marieschi, M., & Cozza, R. (2020). Role of DNA methylation in the chromium tolerance of *Scenedesmus acutus* (Chlorophyceae) and its impact on the sulfate pathway regulation. *Plant Science*, 301, 110680.

Ferreira, L. J., Azevedo, V., Maroco, J., Oliveira, M. M., & Santos, A. P. (2015). Salt tolerant and sensitive rice varieties display differential methylome flexibility under salt stress. *PLoS One*, 10(5), e0124060.

Ferreira, L. J., Donoghue, M. T., Barros, P., Saibo, N. J., Santos, A. P., & Oliveira, M. M. (2019). Uncovering differentially methylated regions (DMRs) in a salt-tolerant rice variety under stress: One step towards new regulatory regions for enhanced salt tolerance. *Epigenomes*, 3(1), 4.

Fieldes, M. A. (1994). Heritable effects of 5-Azacytidine treatments on the growth and development of flax (*Linum usitatissimum*) genotrophs and genotypes. *Genome*, 37, 1–11.

Fieldes, M. A., Schaeffer, S. M., Krech, M. J., & Brown, J. C. L. (2005). DNA hypomethylation in 5-azacytidine-induced early-flowering lines of flax. *Theoretical and Applied Genetics*, 111, 136–149.

Fortes, A. M., & Gallusci, P. (2017). Plant stress responses and phenotypic plasticity in the epigenomics era: Perspectives on the grapevine scenario, a model for perennial crop plants. *Frontiers in Plant Science*, 8, 82.

Frank S., Havlik P., Soussana J.-F., Levesque A., Valin H., Wollenberg E., Kleinwechter U., Fricko O., Gusti M., Herrero M., et al. (2017). Reducing greenhouse gas emissions in agriculture without compromising food security? *Environmental Research Letters*, 12, 105004. https://doi.org/10.1088/1748–9326/aa8c83.

Gallego-Bartolomé, J. (2020). DNA methylation in plants: Mechanisms and tools for targeted manipulation. *New Phytologist*, 227(1), 38–44.

Gallusci, P., Hodgman, C., Teyssier, E., & Seymour, G. B. (2016). DNA methylation and chromatin regulation during fleshy fruit development and ripening. *Frontiers in Plant Science*, 7, 807.

Guo, H., Ge, P., Tong, J., Zhang, Y., Peng, X., Zhao, Z., & Sun, Y. (2021). Elevated carbon dioxide levels decreases cucumber mosaic virus accumulation in correlation with greater accumulation of rgs-CaM, an inhibitor of a viral suppressor of RNAi. *Plants*, 10(1), 59.

Guo, J. E., Hu, Z., Li, F., Zhang, L., Yu, X., Tang, B., & Chen, G. (2017). Silencing of histone deacetylase SlHDT3 delays fruit ripening and suppresses carotenoid accumulation in tomato. *Plant Science*, 265, 29–38.

Gupta, C., & Salgotra, R. K. (2022). Epigenetics and its role in effecting agronomical traits. *Frontiers in Plant Science*, 13, 925688.

Han, Y. C., Kuang, J. F., Chen, J. Y., Liu, X. C., Xiao, Y. Y., Fu, C. C., & Lu, W. J. (2016). Banana transcription factor MaERF11 recruits histone deacetylase MaHDA1 and represses the expression of MaACO1 and expansins during fruit ripening. *Plant Physiology*, 171(2), 1070–1084.

He, M. Y., Ren, T. X., Jin, Z. D., Deng, L., Liu, H. J., Cheng, Y. Y., Li, Z. Y., Liu, X. X., Yang, Y., & Chang, H. (2023). Precise analysis of potassium isotopic composition in plant materials by multi-collector inductively coupled plasma mass spectrometry. *Spectrochimica Acta Part B: Atomic Spectroscopy*, 209, 106781.

Hemenway, E. A., & Gehring, M. (2023). Epigenetic regulation during plant development and the capacity for epigenetic memory. *Annual Review of Plant Biology*, 74(1), 87–109.

Hou, J., Ren, R., Xiao, H., Chen, Z., Yu, J., Zhang, H., Shi, Q., Hou, H., He, S., & Li, L. (2021). Characteristic and evolution of HAT and HDAC genes in Gramineae genomes and their expression analysis under diverse stress in Oryza sativa. *Planta*, 253, 1–22.

Huang, L., Sun, Q., Qin, F., Li, C., Zhao, Y., & Zhou, D. X. (2007). Down-regulation of a SILENT INFORMATION REGULATOR2-related histone deacetylase gene, OsSRT1, induces DNA fragmentation and cell death in rice. *Plant Physiology*, 144, 1508–1519.

Huang, S., Hou, L., Fu, W., Liu, Z., Li, C., Li, X., & Feng, H. (2020). An insertion mutation in Bra032169 encoding a histone methyltransferase is responsible for early bolting in Chinese cabbage (*Brassica rapa* L. ssp. *pekinensis*). *Frontiers in Plant Science*, 11, 547.

Jagadhesan, B., Das, S., Singh, D., Jha, S. K., Durgesh, K., & Sathee, L. (2022). Micro RNA mediated regulation of nutrient response in plants: The case of nitrogen. *Plant Physiology Reports*, 27(3), 345–357.

Jogam, P., Sandhya, D., Alok, A., Peddaboina, V., Allini, V. R., & Zhang, B. (2022). A review on CRISPR/Cas-based epigenetic regulation in plants. *International Journal of Biological Macromolecules*, 219, 1261–1271.

Kakoulidou, I., Avramidou, E. V., Baránek, M., Brunel-Muguet, S., Farrona, S., Johannes, F., Kaiserli, E., Lieberman-Lazarovich, M., Martinelli, F., Mladenov, V., & Maury, S. (2021). Epigenetics for crop improvement in times of global change. *Biology*, 10(8), 766.

Kamthan, A., Chaudhuri, A., Kamthan, M., & Datta, A. (2015). Small RNAs in plants: Recent development and application for crop improvement. *Frontiers in Plant Science*, 6, 208–210.

Kantar, M., Unver, T., & Budak, H. (2010). Regulation of barley MiRNAs upon dehydration stress correlated with target gene expression. *Functional & Integrative Genomics*, 10, 493–507.

Kim, J. H. (2021). Multifaceted chromatin structure and transcription changes in plant stress response. *International Journal of Molecular Sciences*, 22(4), 2013.

Kim, J. M., Sasaki, T., Ueda, M., Sako, K., & Seki, M. (2015). Chromatin changes in response to drought, salinity, heat, and cold stresses in plants. *Frontiers in Plant Science*, 6, 114.

Kumar, G., Rattan, U. K., & Singh, A. K. (2016). Chilling-mediated DNA methylation changes during dormancy and its release reveal the importance of epigenetic regulation during winter dormancy in apple (*Malus x domestica* Borkh.). *PLoS One*, 11(2), e0149934.

Kumar, S. (2019). Epigenetics and epigenomics for crop improvement: Current opinion. *Advances in Biotechnology and Microbiology*, 14, 555879.

Kumar, S. (2019). Epigenomics for crop improvement: Current status and future perspectives. *Journal of Genetics, Cellular and Biology*, 3, 128–134.

Kumar, S., Seem, K., & Mohapatra, T. (2023). Biochemical and epigenetic modulations under drought: Remembering the stress tolerance mechanism in rice. *Life*, 13(5), 1156.

Kundariya, H., Yang, X., Morton, K., Sanchez, R., Axtell, M. J., Hutton, S. F., & Mackenzie, S. A. (2020). MSH1-induced heritable enhanced growth vigor through grafting is associated with the RdDM pathway in plants. *Nature Communications*, 11(1), 5343.

Lee, S. I., & Kim, N. S. (2014). Transposable elements and genome size variations in plants. *Genomics & Informatics*, 12(3), 87.

Lee, S., Choi, J., Park, J., Hong, C. P., Choi, D., Han, S., Choi, K., Roh, T. Y., Hwang, D., & Hwang, I. (2023). DDM1-mediated gene body DNA methylation is associated with inducible activation of defense-related genes in *Arabidopsis*. *Genome Biology*, 24(1), 106.

Li, C., & Zhang, B. (2016). MicroRNAs in control of plant development. *Journal of Cellular Physiology*, 231, 303–313. https://doi.org/10.1002/jcp.25125.

Li, H., Yan, S., Zhao, L., Tan, J., Zhang, Q., Gao, F., et al. (2014). Histone acetylation associated up-regulation of the cell wall related genes is involved in salt stress induced maize root swelling. *BMC Plant Biology*, 14(10), 105. https://doi.org/10.1186/1471-2229-14-105.

Li, S., Lin, Y. J., Wang, P., Zhang, B., Li, M., Chen, S., et al (2019). The AREB1 transcription factor influences histone acetylation to regulate drought responses and tolerance in *Populus trichocarpa*. *Plant Cell*, 31, 663–686. https://doi.org/10.1105/tpc.18.00437.

Li, Y., Zhao, S. L., Li, J. L., Hu, X. H., Wang, H., Cao, X. L., Xu, Y. J., Zhao, Z. X., Xiao, Z. Y., Yang, N., et al. (2017). Osa-miR169 negatively regulates rice immunity against the blast fungus *Magnaporthe oryzae*. *Frontiers in Plant Science*, 8, 2.

Liu, M., Shi, Z., Zhang, X., Wang, M., Zhang, L., Zheng, K., Liu, J., Hu, X., Di, C., Qian, Q., et al. (2019). Inducible overexpression of Ideal Plant Architecture1 improves both yield and disease resistance in rice. *Nature Plants*, 5, 389–400.

Liu, R., & Zhu, J. K. (2014). Non-coding RNAs as potent tools for crop improvement. *National Science Review*, 1(2), 186–189.

Liu, R., How-Kit, A., Stammitti, L., Teyssier, E., & Gallusci, P. (2015). A DEMETER-like DNA demethylase governs tomato fruit ripening. *Proceedings of the National Academy of Sciences of the United States of America*, 112(34), 10804–10809.

Lu, X., & Hyun, T. K. (2021). The role of epigenetic modifications in plant responses to stress. *Botanica Serbica*, 45(1), 3–12.

Lu, Y., Xu, Q., Liu, Y., Yu, Y., Cheng, Z. Y., Zhao, Y., et al. (2018). Dynamics and functional interplay of histone lysine butyrylation, crotonylation, and acetylation in rice under starvation and submergence. *Genome Biology*, 19, 144. https://doi.org/10.1186/s13059-018-1533-y.

Luo, M., Liu, X., Singh, P., Cui, Y., Zimmerli, L., & Wu, K. (2012). Chromatin modifications and remodeling in plant abiotic stress responses. *Biochimica et Biophysica Acta-Gene Regulatory Mechanisms*, 1819, 129–136.

Man, Z., & Hong, L. (2013). MicroRNA-mediated gene regulation: Potential applications for plant genetic engineering. *Plant Molecular Biology*, 83(1–2), 59–75.

Martin, R. A., & Pfennig, D. W. (2010). Field and experimental evidence that competition and ecological opportunity promote resource polymorphism. *Biological Journal of the Linnean Society*, 100(1), 73–88.

Merce, C., Bayer, P. E., Fernandez, C. T., Batley, J., & Edwards, D. (2020). Induced methylation in plants as a crop improvement tool: Progress and perspectives. *Agronomy*, 10, 1484. https://doi.org/10.3390/agronomy10101484.

Min, H., Chen, C., Wei, S., Shang, X., Sun, M., Xia, R., Liu, X., Hao, D., Chen, H., & Xie, Q. (2016). Identification of drought tolerant mechanisms in maize seedlings based on transcriptome analysis of recombination inbred lines. *Frontiers in Plant Science*, 7, 1080.

Miura, K., Agetsuma, M., Kitano, H., Yoshimura, A., Matsuoka, M., Jacobsen, S. E., & Ashikari, M. (2009). A metastable DWARF1 epigenetic mutant affecting plant stature in rice. *Proceedings of the National Academy of Sciences*, 106(27), 11218–11223.

Molinier, J., Ries, G., Zipfel, C., & Hohn, B. (2006). Transgeneration memory of stress in plants. *Nature*, 442, 1046–1049.

Ni, Z., Hu, Z., Jiang, Q., & Zhang, H. (2013). GmNFYA3, a target gene of miR169, is a positive regulator of plant tolerance to drought stress. *Plant Molecular Biology*, 82(1–2), 113–129.

Nunez-Vazquez, R., Desvoyes, B., & Gutierrez, C. (2022). Histone variants and modifications during abiotic stress response. *Frontiers in Plant Science*, 13, 984702.

Ojolo, S. P., Cao, S., Priyadarshani, S. V. G. N., Li, W., Yan, M., Aslam, M., Zhao, H., & Qin, Y. (2018). Regulation of plant growth and development: A review from a chromatin remodeling perspective. *Frontiers in Plant Science*, 9, 1232.

Paixão, J. F. R., Gillet, F. X., Ribeiro, T. P., Bournaud, C., Lourenço-Tessutti, I. T., Noriega, D. D., Melo, B. P. D., de Almeida-Engler, J., & Grossi-de-Sa, M. F. (2019). Improved drought stress tolerance in *Arabidopsis* by CRISPR/dCas9 fusion with a histone acetyltransferase. *Scientific Reports*, 9, 1–9.

Pandey, G., Sharma, N., Pankaj Sahu, P., & Prasad, M. (2016). Chromatin-based epigenetic regulation of plant abiotic stress response. *Current Genomics*, 17(6), 490–498.

Pecinka, A., Chevalier, C., Colas, I., Kalantidis, K., Varotto, S., Krugman, T., Michailidis, C., Vallés, M. P., Muñoz, A., & Pradillo, M. (2020). Chromatin dynamics during interphase and cell division: Similarities and differences between model and crop plants. *Journal of Experimental Botany*, 71(17), 5205–5222.

Pfennig, D. W., Wund, M. A., Snell-Rood, E. C., Cruickshank, T., Schlichting, C. D., & Moczek, A. P. (2010). Phenotypic plasticity's impacts on diversification and speciation. *Trends in Ecology & Evolution*, 25(8), 459–467.

Pikaard, C. S., & Scheid, O. M. (2014). Epigenetic regulation in plants. *Cold Spring Harbor Perspectives in Biology*, 6(12), a019315.

Pilu, R., Panzeri, D., Cassani, E., Badone, F. C., Landoni, M., & Nielsen, E. (2009). A paramutation phenomenon is involved in the genetics of maize low phytic acid1-241 (lpa1-241) trait. *Heredity*, 102(3), 236–245.

Qiu, Z., He, Y., Zhang, Y., Guo, J., & Wang, L. (2018). Characterization of miRNAs and their target genes in He-Ne laser pretreated wheat seedlings exposed to drought stress. *Ecotoxicology and Environmental Safety*, 164, 611–617.

Raza, A., Malik, H. J., Shafiq, M., Amin, I., Scheffler, J. A., Scheffler, B. E., et al. (2016). RNA interference based approach to down regulate osmoregulators of whitefly (*Bemisia tabaci*): Potential technology for the control of whitefly. *PLoS One*, 11(4), e0153883.

Rehman, M., & Tanti, B. (2020). Understanding epigenetic modifications in response to abiotic stresses in plants. *Biocatalysis and Agricultural Biotechnology*, 27, 101673.

Ruane, A. C., Antle, J., Elliott, J., Folberth, C., Hoogenboom, G., Mason-D'Croz, D., Müller, C., Porter, C., Phillips, M. M., Raymundo, R. M., et al. (2018). Biophysical and economic implications for agriculture of +1.5° and +2.0 °C global warming using AgMIP coordinated global and regional assessments. *Climate Research*, 76, 17–39. https://doi.org/10.3354/cr01520.

Sadakierska-Chudy, A., Kostrzewa, R. M., & Filip, M. (2015). A comprehensive view of the epigenetic landscape part I: DNA methylation, passive and active DNA demethylation pathways and histone variants. *Neurotoxicity Research*, 27, 84–97.

Saraswat, S., Yadav, A. K., Sirohi, P., & Singh, N. K. (2017). Role of epigenetics in crop improvement: Water and heat stress. *Journal of Plant Biology*, 60, 231–240.

Schmitz, R. J., & Zhang, X. (2011). High-throughput approaches for plant epigenomic studies. *Current Opinion in Plant Biology*, 14(2), 130–136.

Shaikh, A. A., Chachar, S., Chachar, M., Ahmed, N., Guan, C., & Zhang, P. (2022). Recent advances in DNA methylation and their potential breeding applications in plants. *Horticulturae*, 8(7), 562. https://doi.org/10.3390/horticulturae8070562.

Shankar, R., Bhattacharjee, A., & Jain, M. (2016). Transcriptome analysis in different rice cultivars provides novel insights into desiccation and salinity stress responses. *Scientific Reports*, 6(1), 23719.

Shelby, E. A., Moss, J. B., Andreason, S. A., Simmons, A. M., Moore, A. J., & Moore, P. J. (2020). Debugging: Strategies and considerations for efficient RNAi-mediated control of the whitefly *Bemisia tabaci*. *Insects*, 11(11), 723.

Shen, Y., Lei, T., Cui, X., Liu, X., Zhou, S., Zheng, Y., Guérard, F., Issakidis-Bourguet, E., & Zhou, D. X. (2019). *Arabidopsis* histone deacetylase HDA 15 directly represses plant response to elevated ambient temperature. *The Plant Journal*, 100(5), 991–1006.

Singroha, G., Sharma, P., & Sunkur, R. (2021). Current status of microRNA-mediated regulation of drought stress responses in cereals. *Physiologia Plantarum*, 172(3), 1808–1821.

Song, L., Fang Y., Chen L., Wang J., & Chen X. (2021). Role of non-coding RNAs in plant immunity. *Plant Communications*, 2, 100180.

Song, Q., Zhang, T., Stelly, D. M., & Chen, Z. J. (2017). Epigenomic and functional analyses reveal roles of epialleles in the loss of photoperiod sensitivity during domestication of allotetraploid cottons. *Genome Biology*, 18, 1–14.

Song, X. J., Kuroha, T., Ayano, M., Furuta, T., Nagai, K., Komeda, N., & Ashikari, M. (2015). Rare allele of a previously unidentified histone H4 acetyltransferase enhances grain weight, yield, and plant biomass in rice. *Proceedings of the National Academy of Sciences*, 112(1), 76–81.

Song, Y., Ji, D., Li, S., Wang, P., Li, Q., & Xiang, F. (2012). The dynamic changes of DNA methylation and histone modifications of salt responsive transcription factor genes in soybean. *PLoS One*, 7(7), e41274.

Sudan, J., Raina, M., & Singh, R. (2018). Plant epigenetic mechanisms: Role in abiotic stress and their generational heritability. *3 Biotech*, 8, 1–12.

Sultan, S. E. (2000). Phenotypic plasticity for plant development, function and life history. *Trends in Plant Science*, 5(12), 537–542.

Summanwar, A., Basu, U., Rahman, H., & Kav, N. N. (2020). Non-coding RNAs as emerging targets for crop improvement. *Plant Science*, 297, 110521.

Sun, C., Fang, J., Zhao, T., Xu, B., Zhang, F., Liu, L., & Chu, C. (2012). The histone methyltransferase SDG724 mediates H3K36me2/3 deposition at MADS50 and RFT1 and promotes flowering in rice. *The Plant Cell*, 24(8), 3235–3247.

Sun, M., Yang, Z., Liu, L., & Duan, L. (2022). DNA methylation in plant responses and adaption to abiotic stresses. *International Journal of Molecular Sciences*, 23(13), 6910.

Surdonja, K., Eggert, K., Hajirezaei, M.-R., Harshavardhan, V. T., Seiler, C., von Wirén, N., Sreenivasulu, N., & Kuhlmann, M. (2017). Increase of DNA methylation at the HvCKX2. 1 promoter by terminal drought stress in barley. *Epigenomes*, 1, 9.

Swarup, R., & Denyer, T. (2018). miRNAs in plant development. In *Annual Plant Reviews Online*, 2, 689–712. https://doi.org/10.1002/9781119312 994.apr0649.

Tang, Y., He, H., Qu, X., Cai, Y., Ding, W., Qiu, L., & Li, Y. (2020). RNA interference-mediated knockdown of the transcription factor Krüppel homologue 1 suppresses vitellogenesis in *Chilo suppressalis*. *Insect Molecular Biology*, 29(2), 183–192.

Temel, A., Janack, B., & Humbeck, K. (2017). Drought stress-related physiological changes and histone modifications in barley primary leaves at *HSP17* gene. *Agronomy*, 7, 43.

Thakur, N., Upadhyay, S. K., Verma, P. C., Chandrashekar, K., Tuli, R., & Singh, P. K. (2014). Enhanced whitefly resistance in transgenic tobacco plants expressing double stranded RNA of v-ATPase A gene. *PLoS One*, 9(3), e87235.

Tonosaki, K., Fujimoto, R., Dennis, E. S., Raboy, V., & Osabe, K. (2022). Will epigenetics be a key player in crop breeding? *Frontiers in Plant Science*, 13, 958350.

Ueda, M., & Seki, M. (2020). Histone modifications form epigenetic regulatory networks to regulate abiotic stress response. *Plant Physiology*, 182(1), 15–26.

Van Meijl, H., Havlik, P., Lotze-Campen, H., Stehfest, E., Witzke, P., Domínguez, I. P., Bodirsky, B. L., Van Dijk, M., Doelman, J., Fellmann, T., et al. (2018). Comparing impacts of climate change and mitigation on global agriculture by 2050. *Environmental Research Letters*, 13, 064021. https://doi.org/10.1088/1748-9326/aabdc4.

Vannier, N., Mony, C., Bittebière, A. K., & Vandenkoornhuyse, P. (2015). Epigenetic mechanisms and microbiota as a toolbox for plant phenotypic adjustment to environment. *Frontiers in Plant Science*, 6, 1159.

Wang, A., Hu, J., Gao, C., Chen, G., Wang, B., Lin, C., Song, L., Ding, Y., & Zhou, G. (2019b). Genome-wide analysis of long non-coding RNAs unveils the regulatory roles in the heat tolerance of Chinese cabbage (*Brassica rapa* ssp. *chinensis*). *Scientific Reports*, 9, 5002. https://doi.org/10.1038/s41598-019-41428-2.

Wang, L., & Qiao, H. (2020). Chromatin regulation in plant hormone and plant stress responses. *Current Opinion in Plant Biology*, 57, 164–170.

Wang, Z., Liu, Y., Li, D., Li, L., Zhang, Q., Wang, S., & Huang, H. (2017a). Identification of circular RNAs in kiwifruit and their species specific response to bacterial canker pathogen invasion. *Frontiers in Plant Science*, 8, 413. https://doi.org/10.3389/fpls.2017.00413.

Weidong, Q. I., Hongping, C. H., Zuozhen, Y. A., Biaolin, H. U., Xiangdong, L., Bing, A., Yuan, L., Yu, H., Jiankun, X., & Fantao, Z. (2020). Systematic characterization of long non-coding RNAs and their responses to drought stress in Dongxiang wild rice. *Rice Science*, 27, 21–31.

West-Eberhard, M. J. (2005). Developmental plasticity and the origin of species differences. *Proceedings of the National Academy of Sciences*, 102(suppl_1), 6543–6549.

Whittaker, C., & Dean, C. (2017). The FLC locus: A platform for discoveries in epigenetics and adaptation. *Annual Review of Cell and Developmental Biology*, 33, 555–575. https://doi.org/10.1146/annurev-cellb io-100616-060546.

Wollmann, H., Stroud, H., Yelagandula, R., Tarutani, Y., Jiang, D., Jing, L., Jamge, B., Takeuchi, H., Holec, S., Nie, X., & Berger, F. (2017). The histone H3 variant H3. 3 regulates gene body DNA methylation in *Arabidopsis thaliana*. *Genome Biology*, 18, 1–10.

Wu, J., Yang, R., Yang, Z., Yao, S., Zhao, S., Wang, Y., Li, P., Song, X., Jin, L., Zhou, T., et al. (2017). ROS accumulation and antiviral defence control by microRNA528 in rice. *Nature Plants*, 3, 16203.

Wu, R., Wang, T., Richardson, A. C., Allan, A. C., Macknight, R. C., & Varkonyi-Gasic, E. (2019). Histone modification and activation by SOC1-like and drought stress-related transcription factors may regulate AcSVP2 expression during kiwifruit winter dormancy. *Plant Science*, 281, 242–250.

Xie, M., & Yu, B. (2015). siRNA-directed DNA methylation in plants. *Current Genomics*, 16(1), 23–31.

Yang, C., Li, D., Mao, D., Liu, X., Li, C., & Li, X. (2013). Over expression of microRNA319 impacts leaf morphogenesis and leads to enhanced cold tolerance in rice (*Oryza sativa* L.). *Plant, Cell & Environment*, 36(12), 2207–2218.

Yang, H., Liu, X., Xin, M., Du, J., Hu, Z., Peng, H., & Yao, Y. (2016). Genome-wide mapping of targets of maize histone deacetylase HDA101 reveals its function and regulatory mechanism during seed development. *The Plant Cell*, 28(3), 629–645.

Yang, Y., Huang, J., Sun, Q., Wang, J., Huang, L., Fu, S., ..., & He, Y. (2022). microRNAs: Key players in plant response to metal toxicity. *International Journal of Molecular Sciences*, 23(15), 8642.

Yao, S., Yang, Z., Yang, R., Huang, Y., Guo, G., Kong, X., Lan, Y., Zhou, T., Wang, H., Wang, W., et al. (2019). Transcriptional regulation of miR528 by OsSPL9 orchestrates antiviral response in rice. *Molecular Plant*, 12, 1114–1122.

Yu, Y., Zhang, Y., Chen, X., & Chen, Y. (2019). Plant noncoding RNAs: Hidden players in development and stress responses. *Annual Review of Cell and Developmental Biology*, 35, 407–431.

Zeng, Z., Zhang, W., Marand, A. P., Zhu, B., Buell, C. R., & Jiang, J. (2019). Cold stress induces enhanced chromatin accessibility and bivalent histone modifications H3K4me3 and H3K27me3 of active genes in potato. *Genome Biology*, 20, 1–17.

Zhan, J., & Meyers, B. C. (2023). Plant small RNAs: Their biogenesis, regulatory roles, and functions. *Annual Review of Plant Biology*, 74(1), 21–51.

Zhang, H., & Zhu, J. K. (2024). Epigenetic gene regulation in plants and its potential applications in crop improvement. *Nature Reviews Molecular Cell Biology*, 19, 1–17.

Zhang, H., Lang, Z., & Zhu, J. K. (2018). Dynamics and function of DNA methylation in plants. *Nature Reviews Molecular Cell Biology*, 19(8), 489–506.

Zhang, H., Tao, Z., Hong, H., Chen, Z., Wu, C., Li, X., Xiao, J., & Wang, S. (2016). Transposon-derived small RNA is responsible for modified function of WRKY45 locus. *Nature Plants*, 2, 16016.

Zhang, H., Zhang, X., & Xiao, J. (2023). Epigenetic regulation of nitrogen signaling and adaptation in plants. *Plants*, 12(14), 2725.

Zhang, Y. C., Yu, Y., Wang, C. Y., Li, Z. Y., Liu, Q., Xu, J., Liao, J. Y., Wang, X. J., Qu, L. H., Chen, F., & Xin, P. (2013). Overexpression of microRNA osmiR397 improves rice yield by increasing grain size and promoting panicle branching. *Nature Biotechnology*, 31, 848–852.

Zhao, T., Zhan, Z., & Jiang, D. (2019). Histone modifications and their regulatory roles in plant development and environmental memory. *Journal of Genetics and Genomics*, 46(10), 467–476.

Zheng, M., Liu, X., Lin, J., Liu, X., Wang, Z., Xin, M., Yao, Y., Peng, H., Zhou, D. X., Ni, Z., & Hu, Z. (2019). Histone acetyltransferase GCN 5 contributes to cell wall integrity and salt stress tolerance by altering the expression of cellulose synthesis genes. *The Plant Journal*, 97(3), 587–602.

Zheng, Y., Ding, Y., Sun, X., Xie, S., Wang, D., Liu, X., Su, L., Wei, W., Pan, L., & Zhou, D. X. (2016). Histone deacetylase HDA9 negatively regulates salt and drought stress responsiveness in Arabidopsis. *Journal of Experimental Botany*, 67(6), 1703–1713.

Zhou, M., & Luo, H. (2013). MicroRNA-mediated gene regulation: Potential applications for plant genetic engineering. *Plant Molecular Biology*, 83, 59–75.

Zhou, R., Zhu, Y., Zhao, J., Fang, Z., Wang, S., Yin, J., Chu, Z., & Ma, D. (2018). Transcriptome-wide identification and characterization of potato circular RNAs in response to Pectobacterium carotovorum Subspecies brasiliense Infection. *International Journal of Molecular Sciences*, 19, 71.

8 Mechanisms for Epigenetic Regulation in *Arabidopsis thaliana*

Bushra Hafeez Kiani, Irshad Arshad and Ayesha Khan

8.1 INTRODUCTION

Epigenetic regulation in *Arabidopsis* is essential for several biological processes, including flowering time control, stress responses, development and genome stability. Key epigenetic mechanisms in *Arabidopsis* include RNA-directed DNA methylation (RdDM), methylating DNA, modifying histones and the involvement of non-coding RNAs. Altogether, these processes establish and maintain gene expression patterns that govern plant response to developmental cues and surrounding stimuli (Whitehouse et al., 1999; Vanyushin, 2006).

Environmental factors and genetic variance influence diverse genotypic and phenotypic variance in organisms. Various external stimuli influence epigenetic and phenotypic variations; however, their pathways are still under investigation (Nordborg & Weigel, 2008). The stable inheritance of changes in genes that do not depend on DNA sequence is known as epigenetic state (Berger et al., 2009). Modifying histones and methylating DNA are some of the common and extensively investigated mechanisms involved in epigenetic states. Histones make up the nucleosome and can be modified in several ways either through ATP-dependent nucleosome remodeling and the removal and replacement of different types of histones (Whitehouse et al., 1999; Teif & Rippe, 2009; Banerjee & Roychoudhury, 2017) or through inducing changes covalently (Allfrey et al., 1964; Strahl & Allis, 2000). These changes can vary, depending on the stage of plant development or external stimuli, such as pathogenic interactions, temperature and water scarcity.

External stimuli can also methylate DNA through transposable elements (TEs), repeated sequences and gene silencing (Wagner, 2003; Vanyushin, 2006). RNA-dependent DNA methylation (RdDM) pathway is usually associated with the stability of DNA methylation (Law & Jacobsen, 2010), along with methyl transferases including DNA METHYLTRANSFERASE 1 (MET1) and methylases including CHROMOMETHYLASE 3 (CMT3). Since plants are stationary, these epigenetic modifications are needed to modulate genes, improve reproduction, increase survival rate and respond to external stimuli,

This chapter focuses on the model organism, *Arabidopsis thaliana,* and its epigenetic modifications that do not pass on to the next generation, i.e., mitotic/intra-generational and epigenetic modifications that pass on to the next generation, i.e., they are meiotic/transgenerational. This chapter also outlines *FLC* as a meiotic epigenetic state and *FWA* as a mitotic epigenetic state. Since both FLC and FWA are involved in the change from the vegetative to the reproductive stage; blooming serves as an indicator for examining the two types of epigenetic modifications discussed in this chapter.

8.2 MECHANISMS FOR EPIGENETIC REGULATION IN *ARABIDOPSIS THALIANA*

8.2.1 DNA Methylation Mechanism in *Arabidopsis thaliana*

DNA methylation is one of the most extensively studied epigenetic modifications in *Arabidopsis thaliana*, where it has a central part in maintaining genomic stability, gene expression regulation and silencing transposable elements (TEs). DNA methylation in plants typically takes place in three cytosine contexts: CXG, CXX and CG (where X represents any nucleotide except G). Each type of methylation serves distinct regulatory purposes and is maintained by specific enzymes, enabling a dynamic yet stable form of gene regulation (Wang et al., 2009) (Table 8.1).

8.2.1.1 Types of DNA Methylation and Their Roles

8.2.1.1.1 CG Methylation

This is the most stable and widespread form of DNA methylation in plants, often found within genes. In *Arabidopsis*, CG methylation is largely maintained through mitotic and meiotic cell divisions by the homologous gene of mammalian DNMT1, METHYLTRANSFERASE 1 (MET1). CG methylation primarily contributes to the stable repression of TEs, ensuring that they do not destabilize the genome (Albani et al., 2012).

8.2.1.1.2 CXG Methylation

CXG methylation is unique to plants and is typically associated with silencing repetitive DNA sequences and TEs. This type of methylation is maintained by the enzyme CHROMOMETHYLASE 3 (CMT3), which works in concert with histone modification proteins to reinforce silencing in heterochromatic regions. The dual regulation

DOI: 10.1201/9781003589914-10

TABLE 8.1
Mechanisms of epigenetic regulation in *Arabidopsis thaliana*

Mechanism	Key components	Function	Examples	Inheritance type
DNA methylation	MET1, CMT3, DRM2, DME, RdDM pathway (siRNAs, AGO4, RDR2, Pol IV, Pol V)	Silences transposons, repetitive sequences and some genes. Establishes stable gene expression patterns.	Silencing of the *FWA* gene in the sporophyte stage.	Meiotic (transgenerational)
Histone modifications	H3K9 methylation, H3K27 methylation, H3K4 methylation, acetylation, deacetylation	Alters chromatin structure to regulate gene expression.	H3K27me3-mediated silencing of the *FLC* gene during vernalization.	Mitotic (intra-generational)
RNA-directed DNA methylation (RdDM)	siRNAs, AGO4, RDR2, Pol IV, Pol V	Guides DNA methylation to specific sequences, typically transposons and repeats.	*FWA* gene regulation.	Meiotic (transgenerational)
Non-coding RNAs	miRNAs, siRNAs	Modulate chromatin structure and gene expression by interacting with chromatin modifiers or guiding RdDM.	miRNAs regulating flowering time genes, including *FLC*.	Mitotic (intra-generational)
Chromatin remodeling	ATP-dependent remodeling complexes (SWI/SNF, CHD)	Reorganizes nucleosomes to permit or restrict access to DNA for transcriptional machinery.	Remodeling during stress responses (e.g., heat or drought).	Both types, depending on context
Histone variants	H2A.Z, H3.3	Replace canonical histones to influence chromatin accessibility and transcription.	H2A.Z incorporation near promoters of environmentally responsive genes.	Mitotic (intra-generational)

modifying histones and methylating DNA ensure robust repression of TEs (Bartee et al., 2001; Mathieu et al., 2007).

8.2.1.1.3 CXX Methylation

CXX methylation is established in a sequence-independent manner and is especially dynamic. It is mainly found in the pericentromeric regions and is heavily influenced by the RdDM mechanism, which will be covered in the next section. DOMAINS REARRANGED METHYLTRANSFERASE 2 (DRM2) mediates CXX methylation and plays a crucial role in de novo methylation, especially in stress stimuli (Popova et al., 2013).

8.2.1.2 Key Enzymes in DNA Methylation

8.2.1.2.1 MET1 (METHYLTRANSFERASE 1)

MET1 maintains CG methylation during replication of DNA, ensuring that epigenetic information is faithfully transmitted across generations. This enzyme recognizes hemimethylated DNA strands after replication and adds methyl groups to restore symmetric CG methylation patterns (Zhang et al., 2006).

8.2.1.2.2 CMT3 (CHROMOMETHYLASE 3)

CMT3 maintains CXG methylation through a self-reinforcing mechanism involving histone H3K9 methylation. CMT3 interacts with histone methyltransferase proteins, like KRYPTONITE (KYP), to promote the formation of heterochromatin, providing an extra layer of repression for TEs and other repetitive sequences (Chan et al., 2006a).

8.2.1.2.3 DRM2 (DOMAINS REARRANGED METHYLTRANSFERASE 2)

DRM2 is essential for the de novo establishment of CXX methylation. In collaboration with components of the RdDM pathway, DRM2 adds methyl groups to previously unmethylated cytosines, effectively silencing sequences that may be activated by environmental changes or stress (Zilberman et al., 2004).

8.2.1.2.4 ROS1 and DME (Repressor of Silencing 1 and DEMETER)

These enzymes are responsible for active DNA demethylation in *Arabidopsis*. They play an essential role in counteracting excessive methylation, thereby reactivating genes needed for specific developmental processes or environmental responses.

8.2.1.3 Biological Functions of DNA Methylation in *Arabidopsis*

DNA methylation is critical for repressing TEs, which, if left unchecked, could disrupt gene function or genome integrity. By establishing methylation in TEs, particularly in the CXG and CG contexts, *Arabidopsis* limits the mutagenic potential of these elements. Interestingly, *Arabidopsis* exhibits a significant level of CG methylation within the bodies of actively expressed genes, unlike animals where methylation generally correlates with repression. Methylation in *Arabidopsis* is thought to play a role in preventing spurious transcription initiation within genes and maintaining efficient transcriptional output (Law & Jacobsen, 2010).

DNA methylation especially CXX methylation mediated by the RdDM pathway, is dynamically regulated under stress conditions. This epigenetic flexibility allows *Arabidopsis* to modify genetic expression in response to environmental cues, such as salt, water scarcity, etc., aiding in plant survival and adaptation. DNA methylation also contributes to plant development by regulating key developmental genes; for example, methylation patterns in certain promoter regions can dictate whether a gene remains active or silent, thereby influencing traits like flowering time or root development (Wierzbicki et al., 2009; Zheng et al., 2009).

8.2.2 RNA-Directed DNA Methylation (RdDM) Mechanism in *Arabidopsis thaliana*

The RdDM pathway is a unique and specialized epigenetics plant mechanism, where small RNAs direct DNA methylation to specific genomic regions, usually to regulate genes and silence transposons. In *Arabidopsis thaliana*, the RdDM pathway plays an essential role in defending against transposon activation and in fine-tuning responses to environmental stress, with a primary impact on CXX methylation sites, although it can also influence CG and CXG methylation under certain conditions (Liu et al., 2006).

8.2.2.1 Mechanism of RdDM

The RdDM pathway operates through a complex interplay between two key steps: small RNA biogenesis and DNA methylation targeting. Initially, double-stranded RNAs (dsRNAs) are synthesized, primarily from transposon sequences, by a plant RNA polymerase, RNA polymerase IV (Pol IV), unique to the RdDM pathway. The enzyme DICER-LIKE 3 (DCL3) processes dsRNAs into 24-nt siRNAs, which are loaded onto an ARGONAUTE protein (AGO4), forming a functional RNA-induced silencing complex (RISC) (Bender & Fink, 1995; Kakutani et al., 1996; Jacobsen & Meyerowitz, 1997; Soppe et al., 2000). This complex then interacts with RNA polymerase V (Pol V), which produces scaffold RNAs at the target loci. The DNA methylation by DRM2 is guided by the interactions between Pol V scaffold RNAs and AGO4-bonded siRNAs, specifically the methylation of cytosines in CXX targets (Liu et al., 2004; Rangwala et al., 2006).

8.2.2.2 Key Components of the RdDM Pathway

Several specialized enzymes and proteins work synergistically in the RdDM pathway to ensure precise targeting of methylation. Pol IV uses methylated DNA to transcribe single-stranded RNAs to initiate siRNA synthesis, which becomes substrates for DCL3, which then processes these RNAs into 24-nt siRNAs, essential for guiding the methylation machinery. RNA polymerase V (Pol V), on the other hand, generates non-coding RNAs at the methylation target sites, providing a scaffold for AGO4-siRNA complexes to bind and direct DRM2-mediated methylation. ARGONAUTE 4 (AGO4), a central player in the RdDM pathway, is the protein that binds siRNAs and targets Pol V transcripts for methylation (Saze & Kakutani, 2007; Johannes et al., 2009; Reinders et al., 2009; Foerster et al., 2011). Finally, DRM2, the enzyme that adds methyl groups to cytosines in the CXX targets, is driven by siRNA-guided interactions with the Pol V scaffold RNAs (Figure 8.1).

8.2.2.3 Biological Significance of RdDM in *Arabidopsis*

The RdDM pathway serves as a vital defense against TEs, which, if activated, can disrupt genome integrity by inserting themselves into functional genes. By silencing these elements via siRNA-mediated DNA methylation, *Arabidopsis* effectively limits transposon activity, preserving its genomic stability (Reinders et al., 2009). Beyond transposon silencing, RdDM contributes to gene regulation during stress responses. When *Arabidopsis* encounters

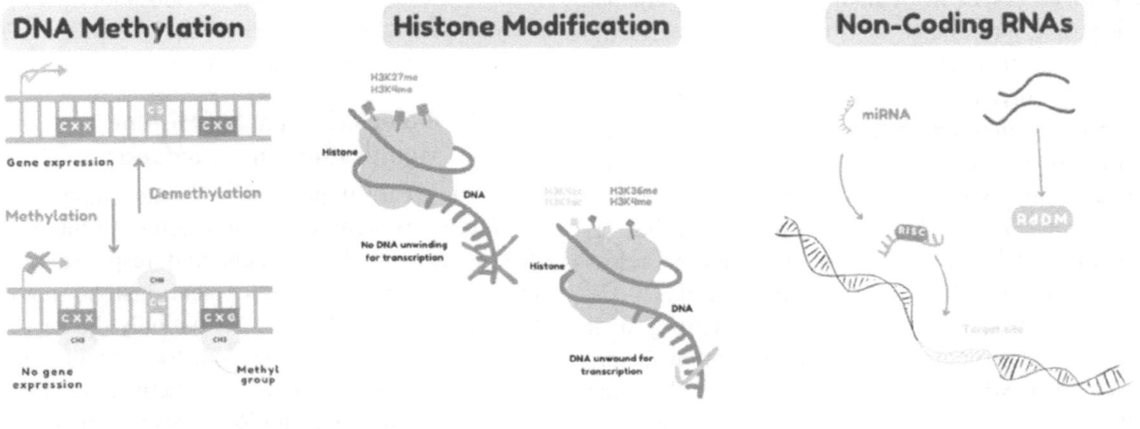

FIGURE 8.1 Overview of epigenic regulatory mechanisms.

environmental stresses, such as drought or pathogen attack, the RdDM pathway is dynamically activated to silence specific gene regions, allowing the plant to quickly adapt to the adverse conditions without changing its DNA sequence (Chan et al., 2006a, 2006b). This flexibility is particularly beneficial for plants, which are immobile and must adapt to their immediate environment for survival.

Additionally, the RdDM pathway plays a role in developmental regulation by silencing genes that should only be active at developmental stages; for instance, during seed development, RdDM can direct methylation to regulate gene expression patterns essential for proper embryo formation and maturation, ensuring that developmental genes are expressed in a precise temporal sequence. The RdDM pathway also has implications in epigenetic memory, enabling *Arabidopsis* to "remember" stress conditions and to be better prepared for recurrent stresses. This epigenetic memory, established through methylation marks, is sometimes heritable, allowing future generations to inherit the stress adaptation traits (Johnson et al., 2008; Woo et al., 2008; Greenberg et al., 2011).

8.2.3 Histone Modification Mechanism in *Arabidopsis thaliana*

Histone modifications have a key part in regulating genes in *Arabidopsis thaliana*, impacting chromatin structure and gene expression patterns. These modifications are chemical alterations to histone proteins, mainly on histones H3 and H4, that control how DNA is packaged and accessed within the nucleus. By adding or removing various chemical groups, histone modifications facilitate the ability of the plants to respond to both internal developmental cues and external environmental stimuli. The dynamic nature of these modifications, along with their reversibility, makes them essential for the flexible regulatory needs of *Arabidopsis* in diverse conditions (Xie et al., 2004).

8.2.3.1 Types of Histone Modifications and Their Functions

In *Arabidopsis*, histone modifications primarily include ubiquitination, phosphorylation, acetylation and methylation, each serving specific regulatory functions. Histone methylation is a prominent modification with versatile roles, as it can either upregulate or downregulate genes based on the methylation site. Methylation at lysine 4 on histone H3 (H3K4me), for instance, is commonly considered upregulation, whereas H3K9me and H3K27me are linked to gene repression. This regulation is particularly crucial for genes involved in developmental timing, as methylation can stabilize gene expression patterns across generations (Lindroth et al., 2004; Johnson et al., 2007). Acetylation of histones usually induces gene activity, as it loosens chromatin structures, letting transcription factors (TFs) bind to the DNA more readily. H3K9ac and H3K27ac modifications are essential in activating genes responsive to external stimuli like temperature and light, which are vital for plant adaptability.

Another key modification is histone phosphorylation, particularly phosphorylation at serine 10 on histone H3 (H3S10ph). This modification is often associated with gene activation during cell cycle progression and plays a vital role in facilitating DNA repair. Phosphorylation events help plants balance growth and stress responses, ensuring proper DNA maintenance under fluctuating conditions. Lastly, histone ubiquitination, particularly monoubiquitination at histone H2B (H2Bub1), has dual roles in both activating and repressing gene expression (Koornneef et al., 1991; Soppe et al., 2000). In *Arabidopsis*, H2Bub1 is particularly relevant for floral development and for genes involved in stress tolerance, highlighting its role in developmental and environmental adaptation.

8.2.3.2 Key Enzymes Involved in Histone Modifications

The addition and removal of histone modifications in *Arabidopsis* are governed by specific enzymes that recognize and alter histone tails. Histone methyltransferases (HMTs), like the SET DOMAIN GROUP (SDG) proteins, add methyl groups to specific histone residues. Different SDG proteins, including SDG4, SDG8 and SDG16, are involved in various processes from controlling flowering time to modulating stress responses; for example, SDG8 contributes to the activation of genes involved in flowering, while SDG16 influences growth regulation (Kinoshita et al., 2004). Histone demethylases (HDMs) counteract methylation by removing these marks, thus balancing gene activation and repression.

Histone acetyltransferases (HATs), such as GCN5, are responsible for adding acetyl groups to histones, particularly in response to environmental stimuli like light. GCN5 is integral to light-responsive gene activation, allowing *Arabidopsis* to adjust to day-night cycles. Conversely, histone deacetylases (HDACs), such as HDA6 and HDA19, play essential roles in repressing unnecessary genes during times of environmental stability. HDACs are also involved in transposon silencing and maintaining genome stability by limiting the expression of mobile genetic elements (Fujimoto et al., 2008). Kinases and phosphatases regulate histone phosphorylation and dephosphorylation, respectively, with enzymes like PROTEIN PHOSPHATASE 2A (PP2A) being pivotal in stress-responsive gene regulation.

8.2.3.3 Biological Impact of Histone Modifications in *Arabidopsis*

The biological impact of histone modifications in *Arabidopsis* is significant, influencing essential processes such as growth, development and response to environmental stress. During development, histone modifications are strictly regulated to ensure that only the key genes are expressed or silenced as needed; for example, the repressive histone mark H3 lysine 27 tri-methylation, H3K27me3, catalyzed by the Polycomb Repressive Complex 2 (PRC2), is crucial to silence genes that regulate flowering time (Lippman et al., 2004; Chan et al., 2006b). This ensures that flowering occurs only under favorable conditions. Similarly, acetylation marks, like H3K9ac, activate genes required for early seedling growth, highlighting how histone modifications control developmental phases.

In terms of environmental stress responses, histone modifications allow *Arabidopsis* to dynamically adjust its gene expression in response to external factors, such as temperature and water scarcity. Under water deficit conditions, histone acetylation levels increase at stress-responsive gene loci, promoting their activation and enabling the plant to adapt to limited water availability. Phosphorylation events also facilitate rapid activation of DNA repair genes, which helps maintain genomic integrity under stress, ensuring long-term survival (Kothapalli et al., 2005).

Finally, epigenetic memory of histone modifications in *Arabidopsis* allows the plant to "remember" previous environmental conditions. This is particularly important for plants in variable climates, as histone modifications retained after a stress event can prime the plant for faster responses if the stress reoccurs. This epigenetic memory, therefore, serves as an adaptive advantage, ensuring that *Arabidopsis* remains resilient in unpredictable environment (Shiio & Eisenman, 2003).

8.2.4 Small RNAs and RNA-Directed DNA Methylation (RdDM)

Small RNAs, particularly microRNAs (miRNAs) and siRNAs, are integral to epigenetic regulation in *Arabidopsis*, as they participate in directing DNA methylation and gene silencing. These RNAs function by guiding methyltransferases to specific genomic loci, facilitating DNA methylation in both gene and transposon regions and enabling the plant to dynamically respond to environmental stressors (Hepworth et al., 2002).

8.2.4.1 siRNAs and miRNAs in Gene Regulation

siRNAs are critical for the RdDM pathway, guiding DNA methylation to specified loci via ARGONAUTE protein interactions. These siRNAs are generated from precursors of double-stranded RNA and are vital for transposon silencing and gene regulation. In addition, microRNAs (miRNAs) play an essential regulatory role by targeting mRNAs for degradation or translational inhibition. miRNAs are involved in stress responses, growth regulation and developmental processes, acting as the main post-transcriptional gene regulators (Wood et al., 2006; De Lucia et al., 2008).

The RdDM pathway utilizes siRNAs to silence transposons and regulate genes in response to external conditions. The pathway involves nuclear RNA polymerase IV and V and AGO proteins, which process and direct siRNAs to complementary DNA regions for methylation. The RdDM pathway, therefore, allows *Arabidopsis* to modulate gene activity and suppress unwanted genomic elements, providing an additional layer of regulation for genome stability (Wang et al., 2009; Albani et al., 2012).

8.2.4.2 Biological Implications of RNA-Mediated Gene Silencing

The role of small RNAs in *Arabidopsis* extends to adaptation to environmental stress, as RNA-mediated silencing enables rapid responses to stress conditions. By gene regulation in response to high salinity, water deficit conditions and pathogen attack, siRNAs and miRNAs allow *Arabidopsis* to adapt by silencing or activating genes as needed. For example, under drought conditions, miRNAs can target specific mRNAs to modulate stress-responsive pathways, enhancing the plant's resilience (Banerjee et al., 2016). Small RNAs are also important for developmental processes, such as leaf morphogenesis, root growth and flowering, contributing to both structural and functional plasticity (Heo & Sung, 2011).

8.2.5 Chromatin Remodeling and Epigenetic Regulation

Chromatin remodeling is a fundamental process in epigenetic regulation, allowing *Arabidopsis* to control gene accessibility by altering chromatin structure. Chromatin remodelers use ATP to reposition, eject or restructure nucleosomes, thus modulating DNA accessibility for transcriptional machinery (Swiezewski et al., 2009).

8.2.5.1 ATP-Dependent Chromatin Remodeling Complexes

In *Arabidopsis*, ATP-dependent chromatin remodeling complexes, including ISWI, SWI/SNF and chromodomain-helicase-DNA-binding (CHD) complexes, play pivotal roles in regulating gene expression. The SWI/SNF complex, for instance, disrupts nucleosome structure, promoting transcriptional activation by exposing DNA-binding sites. This complex is essential in stress response genes, allowing *Arabidopsis* to swiftly activate defense genes under adverse conditions. The ISWI complex, conversely, promotes nucleosome positioning and is typically associated with gene repression, contributing to the maintenance of genome integrity by silencing unnecessary genes (Swiezewski et al., 2009).

Chromodomain-helicase DNA-binding (CHD) remodelers are crucial in developmental regulation, assisting in both the activation and repression of genes during growth phases. By altering chromatin accessibility, CHD remodelers ensure proper gene expression timing and spatial arrangement, which is vital for morphological development and stress adaptation (Finnegan et al., 2005).

8.2.5.2 Role in Development and Stress Responses

Chromatin remodeling complexes in *Arabidopsis* are integral to developmental transitions and stress adaptation. During early development, chromatin remodelers control gene expression by allowing or restricting access to transcription factors, ensuring accurate expression patterns. For instance, the regulation of flowering genes by chromatin remodelers ensures that flowering occurs in favorable conditions, optimizing reproductive success (Wierzbicki et al., 2009).

In response to biotic and abiotic stresses, chromatin remodeling complexes rapidly modify chromatin structure to either silence or activate stress-responsive genes. This flexibility in chromatin architecture enables *Arabidopsis* to conserve energy under non-stress conditions and quickly mount a defense response when stressed (Zheng et al., 2009). By balancing gene expression dynamically, chromatin remodeling contributes to *Arabidopsis* resilience and adaptability.

8.2.6 Epigenetic Memory and Transgenerational Inheritance

Epigenetic memory in *Arabidopsis* refers to the ability to retain epigenetic states across generations, allowing the offspring to inherit environmentally induced modifications. This transgenerational epigenetic inheritance is crucial for adaptation, as it enables plants to prepare future generations for conditions like those experienced by their predecessors (Saze & Kakutani, 2007; Johannes et al., 2009).

8.2.6.1 Mechanisms of Epigenetic Memory

Epigenetic memory is maintained through stable histone modifications and DNA methylation patterns that persist after environmental challenges subside. In *Arabidopsis*, environmental cues like drought and temperature fluctuations can induce lasting methylation or acetylation marks. These modifications are carried through mitotic divisions and, in some cases, through meiosis, allowing subsequent generations to retain adaptive advantages without changes in DNA sequence (Law & Jacobsen, 2010).

8.2.6.2 Role in Plant Adaptation and Evolution

The retention of epigenetic marks across generations enhances plant adaptation to fluctuating environmental conditions. By remembering prior stress exposure, *Arabidopsis* can prime gene expression patterns to respond more effectively to recurrent stress. This transgenerational inheritance mechanism offers a selective advantage, accelerating evolutionary responses in changing environment. It demonstrates how epigenetics can serve as a bridge between immediate physiological adaptation and long-term evolutionary change in *Arabidopsis* (Foerster et al., 2011).

8.3 MEIOTICALLY STABLE EPIGENETIC INHERITANCE

8.3.1 The Context of DNA Methylation in *Arabidopsis*

Highlighting the core processes behind methylating enzymes and RdDM-mediated silencing is crucial to understand how methylation of DNA may be inheritable in plants. The methylation of DNA takes place at gene loci where transcription does not typically occur, such as siRNA-producing sites, pericentric heterochromatin, repeated DNA sequences and transposons (Zhang et al., 2006). In *Arabidopsis*, cytosine sites are methylated at CG, CNG (where N is any nucleotide), and CXX (where X is either C, T or A) (Zhang et al., 2006; Popova et al., 2013). Multiple methylating enzymes are needed, depending on cytosine positions in the sequence. MET1, which is homologous to DNMT1 in mammals, is required for methylating CG sites. It uses pre-methylated parent strands as templates to deposit CH_3 groups on the target sequence. Methylating CG sites are crucial to silence constitutive heterochromatin and transposons by the mutants of MET1 increased pseudogene expression in pericentric heterochromatin (Zhang et al., 2006). Orthologs of mammalian DNMT3a/b, DOMAINS REARRANGED METHYLASE 1/2 (DRM1/2), and CMT3, a plant-specific methyltransferase, aid in methylating CNG, which is symmetric like methylating CG (Chan et al., 2006a); however, both methylations have distinct mechanisms (Gruenbaum et al., 1981; Chan et al., 2006a). CMT3 methylates targets by using the methylated template strand like MET1. CXX methylation, on the other hand, is asymmetric and does not readily transfer to a freshly duplicated DNA strand. Despite the different methylating sites, CMT3 and DRM1/2, the same methyltransferases involved in CNG methylation, also methylate in the CXX context (Zilberman et al., 2004; Law & Jacobsen, 2010). CXX methylation may additionally occur without the mechanism of RdDM and is regulated by the Snf2 family remodeler DDM1 (Lippman et al., 2004; Zemach et al., 2013).

Plants have CXX sites at every stage of their life cycle. Non-CG methylation, especially CXX methylation, is prevalent among human embryonic stem cell lines but diminishes when these cells undergo differentiation (Lister et al., 2009). The majority of euchromatic genes are upregulated in the *Arabidopsis* triple mutant *CMT3/DRM1/DRM2*, suggesting that non-CG methylation plays an essential part in the appropriate expression of the functioning genes in plants (Zhang et al., 2006). Multiple abnormalities in development are reported in these mutants, indicating that gene expression during development is regulated by non-CG methylation (Cao & Jacobsen, 2002).

8.3.2 RdDM: A Mechanism for DNA Methylation

RdDM is one means of methylating CXX or CNG sites. It works by producing 24-nt siRNAs that methylate DNA and silence chromatin regions with either transposons or repeated sequences, like those found in the centromere of the heterochromatin (Law & Jacobsen, 2010).

Given the possibility of double-stranded RNA (dsRNA) synthesis from repeated sequences or high expression levels, transgenes could potentially become RdDM targets. Pol IV and RNA-dependent RNA polymerase 2 (RDR2) work together to create dsRNA precursors. DCL3 can then convert those precursors into siRNA (Liu et al., 2006). These siRNAs may attach to transcripts generated by Pol V when paired with ARGONAUTE4 (Wierzbicki et al., 2009; Zheng et al., 2009). The RdDM machinery, which includes histone-modifying and DNA-methylating proteins, is recruited to the target loci by these scaffold-acting siRNAs (Wierzbicki et al., 2009).

The siRNA buildup for the corresponding targets is necessary for the ongoing functioning of this silencing process, indicating that these repeated sequences have basal transcriptional activity (Lister et al., 2008). Additionally, RdDM has been linked to several plant stresses, including tolerance to high heat (Popova et al., 2013). Furthermore, there are numerous well-researched epialleles in *Arabidopsis* that are characterized by distinct DNA methylating patterns with no differences in their DNA sequence (Bender & Fink, 1995; Kakutani et al., 1996; Jacobsen & Meyerowitz, 1997; Soppe et al., 2000; Liu et al., 2004; Rangwala et al., 2006; Mathieu et al., 2007; Saze & Kakutani, 2007; Johannes et al., 2009; Reinders et al., 2009; Foerster et al., 2011). RdDM is often needed to maintain these epialleles (Law & Jacobsen, 2010). Therefore, they are widely employed in genomic screenings to better understand the RdDM pathway mechanism and the proteins involved (Bender & Fink, 1995; Jacobsen & Meyerowitz, 1997; Saze & Kakutani, 2007; Reinders et al., 2009; Cao & Jacobsen, 2002; Jackson et al., 2002; Zilberman et al., 2003; Riddle & Richard, 2005; Chan et al., 2006a, 2006b; Johnson et al., 2008; Woo et al., 2008; Greenberg et al., 2011).

8.3.3 H3K9 Dimethylation Correlates with DNA Methylation

CNG methylation across the genome and H3K9 dimethylation (H3K9me2) are commonly associated, indicating that there may be an association between both epigenetic markers (Bernatavichute et al., 2008). RdDM gene silencing involves AGO4 recruiting histone H3K9 methyltransferase to the target region (Zilberman et al., 2003). AGO4 repression can lead to no methylation of H3K9 and DNA, which causes heterochromatin to reactivate, making this pathway essential to regulate DNA methylation (Zilberman et al., 2003; Xie et al., 2004).

Several histone methyltransferases deposit H3K9me2 in plants (Liu et al., 2007). In *Arabidopsis*, SUV6/5 and KRYPTONITE (KYP) are examples of H3K9 histone methyltransferases (Jackson et al., 2002). These enzymes may be associated in a loop of positive feedback because H3K9me2 and the methylation of DNA in CXG sites are frequently associated (Lindroth et al., 2004; Johnson et al., 2007). The SRA domain of KYP can identify methylated CXX or CNG cites (Lindroth et al., 2004). Likewise, the presence of a chromodomain in CMT3 that binds H3K9me2 may methylate DNA through CMT3, indicating that H3K9me2 and DNA methylation can reinforce one another.

8.3.4 FWA: A Transgenerationally Stable Epiallele

The *FWA* gene of *Arabidopsis* has epiallelic variants. Since significant levels of methylated DNA can be permanently transferred, *FWA* is latent in wild type (Koornneef et al., 1991; Soppe et al., 2000). DNA hypomethylation in the dominant epiallele *FWA-1* causes ectopic production of FWA, which in turn delays blooming (Soppe et al., 2000). The sequences produced by TEs at the promoter of the FWA locus are necessary to regulate DNA methylation. Additionally, it has been found that FWA chromatin was genetically connected to delayed blooming phenotypes seen in the *DDM1* mutant background, which results in DNA hypomethylation and a range of phenotypic abnormalities (Kakutani et al., 1996), indicating the importance of methylation in blooming. While the paternal allelic variant undergoes methylation, the *FWA* of the maternal allelic variant is demethylated following gamete development and subsequently imprints onto the endosperm (Kinoshita et al., 2004). Paternal and maternal allelic variants undergo methylation in the vegetative state, and MET1 keeps both of them silent through the life cycle (Kinoshita et al., 2004). In *Arabidopsis*, the shift from the vegetative stage to the flowering stage is influenced by ectopic expression of *FWA*.

Higher levels of FWA in vegetative cells lead to late blooming in a dominant *FWA-1* mutant (Soppe et al., 2000); wild-type *FWA* is exclusively produced in the endosperm and not in the vegetative tissues (Kinoshita et al., 2004) (Figure 8.2). The *FWA-1* mutated gene is transcribed due to a reduced level of DNA methylation in its promoter area

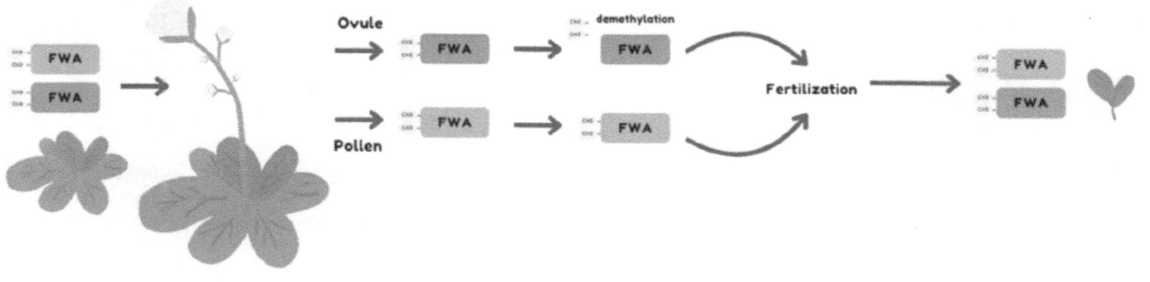

FIGURE 8.2 Schematic representation of *FWA* gene regulation through methylation and demethylation during ovule and pollen development. Fertilization results in the re-establishment of methylation patterns in the offspring, influencing developmental outcomes.

instead of an altered nucleotide sequence. A TE-like motif that contains tandem repeats causes *FWA* DNA methylation in the promoter area. Delayed blooming may be the consequence of ectopic FWA transcription in vegetative cells in the non-methylated state (Fujimoto et al., 2008). The RdDM mechanism controls the methylation of the repeated sequence (Lippman et al., 2004; Chan et al., 2006b). It is noteworthy that various *Arabidopsis* ecotypes have varied topologies for sequences like transposable elements (Fujimoto et al., 2008), which may indicate that there are differences in the epiallelic variant prevalence in the overall genome.

8.4 MITOTICALLY STABLE EPIGENETIC INHERITANCE

8.4.1 Vernalization

Reversible epigenetic changes such as modifying histones are frequently associated with shifts in the expression of genes caused by developmental and external factors. Based on the amino acids and type of modifications, these epigenetic marks – also referred to as the "Histone Code" – can exhibit either upregulatory or downregulatory activity (Jenuwein & Allis, 2001). Sumoylation (Shiio & Eisenman, 2003) and biotinylation (Kothapalli et al., 2005) are frequently linked to closely bounded chromatin and suppressed gene expression. Acetylating histones (Grunstein, 1997), ubiquitination (Pham & Sauer, 2000) and some phosphorylating activities (Wei et al., 1999) are frequently linked to loosely bounded chromatin and activated gene expression.

Depending on which amino acids are present, methylating histones could result in the suppression or activation of histones (Jenuwein & Allis, 2001). Long noncoding RNAs (lncRNAs) have, in recent times, become known as a source of gene transcription activity or repression by modifying histone residues (Weinberg & Morris, 2013). In the sections that follow, we explain how lncRNA and inhibitory histone modification might repress the floral repressor *FLC* intragenerationally, to speed up flower blooming.

8.4.2 The Role of Histone Modification in the Vernalization-Mediated Repression of FLC

Vernalization is an adaptation to extended low temperature and is arguably among the most extensively researched ecological epigenetic modifications in plants. Vernalization is necessary in several *Arabidopsis* ecological types to encourage floral transformation. Vernalization is an adaptation that increases the likelihood of reproduction and encourages blooming following extended exposure to low temperature in winters. Although temperature increases, plants retain the "recollections" of winter that vernalization offers them. FLC plays a key role as the main flowering suppressor during the vernalization mechanism of *Arabidopsis*.

FLOWERING LOCUS T (FT) and SUPPRESSOR OF CONSTANS 1 (SOC1) are floral integrators that are inhibited by FLC (Hepworth et al., 2002), which prevents the transitioning into the reproductive stage. Polycomb Repressive Complex 2 (PRC2) and high levels of VERNALIZATION INSENSITIVE 3 (VIN3) form complexes which lead to FLC chromatin H3K27me3 in low temperature (Wood et al., 2006; De Lucia et al., 2008). These cellular changes still persist in the plant memory even after the mild–moderate temperature returns. Interestingly, these epigenetic changes do not pass on to the next generation meiotically, and this allows FLC reactivation (Trevaskis et al., 2007) (Figure 8.3). Since *Arabidopsis* are annuals, FLC is reset when the seeds develop; however, perennials like *Arabis alpina* do not have a distinct cellular memory. *Arabidopsis* FLC and PERPETUAL FLOWERING 1 (PEP1) are orthologs, and they exhibit comparable expression (Wang et al., 2009; Albani et al., 2012).

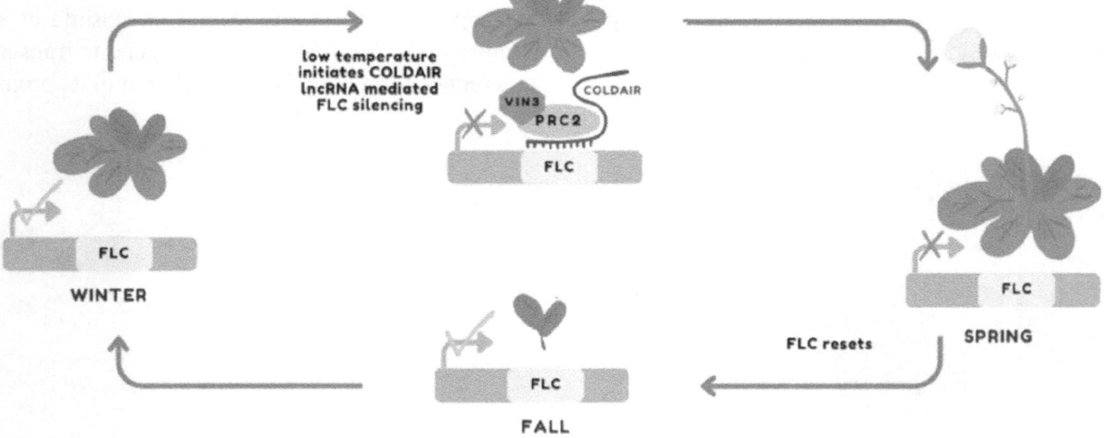

FIGURE 8.3 Seasonal regulation of *FLC* expression. During winter, low temperature initiates COLDAIR lncRNA-mediated silencing of *FLC* through the recruitment of PRC2 and VIN3, thereby suppressing flowering. In spring, *FLC* remains silenced, allowing flowering to occur. *FLC* expression resets, restoring its active state for the next cycle.

8.4.3 Long Noncoding RNAs

Apart from modifying histones, long noncoding RNAs (lncRNAs) like COOLAIR and COLDAIR mitotically suppress *FLC*. COLDAIR is produced via the initial intron of FLC from 5′ to 3′. When exposed to low temperature, COLDAIR forms and binds to PRC2. COLDAIR suppression results in fewer H3K27me3 as well as a reduction in epigenetic suppression, suggesting that it is a necessary element for PRC2 activation (Heo & Sung, 2011) (Figure 8.3). On the other hand, COOLAIR is a group of FLC transcripts coded from 3′ to 5′ that have alternative splicing and varying polyadenylations (Swiezewski et al., 2009). Although it does not directly attach to PRC2, COOLAIR is similarly produced in freezing temperature (Heo & Sung, 2011). According to certain theories, COOLAIR might play a part in early vernalization by transcriptional regulation of coding strands in a stage-dependent way (Swiezewski et al., 2009). When combined, *FLC* suppression through vernalization is an epigenomic occurrence where a cold environment acts as stimuli for lncRNAs and modifying histones. *FLC* can also be repressed by methylating its sequence (Finnegan et al., 2005), indicating that multiple mechanisms are involved in plant epigenetic regulation.

8.5 IMPORTANCE OF EPIGENETIC REGULATION IN *ARABIDOPSIS THALIANA*

Epigenetic regulation plays a crucial role in the development, evolution and adaptation of plants. In *Arabidopsis thaliana*, epigenetic processes such as DNA methylation, modifying histones and RNA interference are essential to regulate genes without making alterations in the DNA sequence. These regulatory mechanisms are dynamic and can be influenced by environmental stimuli, which makes epigenetics an essential factor in plant survival and stress resilience (Zilberman et al., 2004). Understanding the importance of epigenetic regulation in *Arabidopsis* offers insights into plant responses to various external challenges and the role of epigenetics in shaping plant evolution.

8.5.1 Role in Developmental Processes

Epigenetic regulation is integral to *Arabidopsis* development, ensuring that specific genes are activated or silenced at appropriate times. Processes such as seed germination, root and shoot development, and flowering rely heavily on epigenetic modifications to guide the expression of development-related genes; for instance, DNA methylation patterns established during seed development are essential for repressing genes that are not needed during early growth stages but may become necessary as the plant matures (Zemach et al., 2013). Epigenetic regulation also ensures that genes involved in flowering are activated at the right time, which is crucial for reproductive success. Without such precise control, *Arabidopsis* plants would face challenges in growth, resulting in lower fitness and adaptability (Lippman et al., 2004).

One well-studied example of epigenetic control in *Arabidopsis* is vernalization, the process by which long-term low-temperature exposure leads to spring flower blooming. The gene essential for this process, FLOWERING LOCUS C (FLC), is silenced epigenetically. During the epigenetic downregulation during winter, *FLC* is silenced by chromatin modifications, allowing flowering to occur when temperature rises, illustrating how epigenetics enable *Arabidopsis* to synchronize its life cycle with seasonal changes (Popova et al., 2013). This regulation not only conserves energy by ensuring that flowering occurs under favorable conditions but also increases reproductive success by aligning flowering times with optimal environmental cues.

8.5.2 Epigenetics in Stress Response and Adaptation

Arabidopsis is frequently exposed to external stimuli, including water scarcity, high saline environment and pathogen attacks, and its ability to respond to these stresses is partly governed by epigenetic regulation. Epigenetic mechanisms allow *Arabidopsis* to activate or silence specific genes in response to stress, providing flexibility and resilience. DNA methylation and histone modifications in response to external stimuli could modify genes in a way that enables rapid adaptation; for instance, under drought conditions, *Arabidopsis* can modulate genes involved in water conservation and root growth, allowing it to withstand periods of water scarcity. These epigenetic changes are often reversible, enabling the plant to return to its normal state once the stressor is removed (Lindroth et al., 2004).

Interestingly, certain epigenetic marks induced by environmental stresses can be inherited transgenerationally; for example, *Arabidopsis* plants exposed to stressful conditions, such as high salinity or pathogen attack, can pass on altered DNA methylation patterns to their offspring. These inherited epigenetic changes prime the next generation to better withstand similar stresses, providing an adaptive advantage. By remembering past environmental conditions, *Arabidopsis* can better prepare future generations for a fluctuating environment, enhancing its long-term survival and evolution (Kinoshita et al., 2004).

8.5.3 Genome Stability and Transposon Silencing

Epigenetic mechanisms are crucial for maintaining genome stability in *Arabidopsis*, particularly by silencing TEs that have the potential to disrupt gene function if allowed to move within the genome. Modifying histones and methylating DNA are the main players involved in keeping these transposons inactive. DNA methyltransferases, particularly in CpG and CXG contexts, silence transposons by adding methyl groups to cytosine bases, effectively preventing them from being transcribed (Lippman et al., 2004).

Transposon silencing is vital because active transposons can lead to mutations, gene disruptions and genomic instability. In *Arabidopsis*, the RdDM pathway utilizes small interfering RNAs (siRNAs) to target TEs for methylation, ensuring that they remain silenced. This regulation is crucial in maintaining the genomic integrity of the plant and preventing harmful genetic rearrangements (Chan et al., 2006b). Without such epigenetic control, transposon activity could increase, leading to reduced fitness and adaptability of *Arabidopsis* populations.

8.5.4 Contribution to Evolution and Genetic Diversity

Epigenetic regulation also contributes to the genetic diversity and evolutionary potential of *Arabidopsis*. While traditional genetic mutations alter DNA sequences, epigenetic modifications provide a reversible and dynamic means of generating phenotypic diversity. By creating alternative states of gene expression without permanent DNA changes, epigenetic modifications allow *Arabidopsis* to experiment with new traits that may be advantageous under certain environmental conditions. These epigenetically mediated traits can offer selective advantages, enabling *Arabidopsis* populations to adapt to diverse habitats and fluctuating environmental conditions over time (Jenuwein & Allis, 2001).

Moreover, epigenetic modifications can act as a source of phenotypic plasticity, allowing *Arabidopsis* plants to exhibit different traits in response to their environment. This plasticity is crucial in a rapidly changing environment, where flexibility in gene expression can determine survival. Over evolutionary time scales, epigenetic variations may become stabilized through genetic assimilation, where initially plastic responses become fixed, contributing to evolutionary change (De Lucia et al., 2008). Thus, epigenetic regulation enhances the adaptability of *Arabidopsis* and plays a fundamental role in its evolutionary success.

8.5.5 Insights for Crop Improvement and Agricultural Applications

Studying epigenetic regulation in *Arabidopsis thaliana* also has significant implications for crop improvement. *Arabidopsis* serves as a model for many crop plants, and understanding how epigenetic mechanisms control growth, stress responses and gene expression can inform strategies for enhancing crop resilience. For instance, if crops can be bred or genetically engineered to retain beneficial epigenetic marks in response to drought, they may exhibit improved tolerance to water-limited conditions (Trevaskis et al., 2007; Swiezewski et al., 2009).

Epigenetic research in *Arabidopsis* can also guide genome-editing strategies for crops, where targeted manipulation of epigenetic marks could enable more precise control of gene expression. By transferring insights gained from *Arabidopsis*, scientists could potentially develop crops that are better suited to withstand environmental stress, have higher yields and require fewer chemical inputs. This epigenetically informed approach aligns with sustainable agriculture goals, offering a path towards productive and resilient crops in the face of global climate issues (Jenuwein & Allis, 2001; Finnegan et al., 2005; Kothapalli et al., 2005).

8.6 CONCLUSION

Several cellular strategies for the maintenance of epigenetic modifications in both meiotic and mitotic states were highlighted in this chapter. Modification of histones (specifically H3K9me2), DNA methylation and the RdDM pathway are used by the meiotic or transgenerational state for silencing repeated sequences, transposons and heterochromatic areas. FWA is an excellent example of this kind of epigenetic alteration. On the contrary, intra-generational or mitotic states – particularly in the context of FLC – rely on lncRNAs and modifying histone modifications rather than methylating DNA.

External stimuli have an impact on FLC, suggesting a possible link between genetics and environmental factors. Furthermore, the epigenome opens the possibility of genome-independent modifications in gene expression, thereby enhancing the potential for adaptation and evolution.

REFERENCES

Albani, M. C., Castaings, L., Wo ̈tzel, S., Mateos, J. L., Wunder, J., Wang, R., & Coupland, G. (2012). PEP1 of *Arabis alpina* is encoded by two overlapping genes that contribute to natural genetic variation in perennial flowering. *PLoS Genetics*, 8, 1–14.

Allfrey, G., Faulkner, R., & Mirsky, A. E. (1964). Acetylation and methylation of histones and their possible role in the regulation of RNA synthesis. *Proceedings of the National Academy of Sciences*, 315, 786–794.

Banerjee, A., & Roychoudhury, A. (2017). Epigenetic regulation during salinity and drought stress in plants: Histone modifications and DNA methylation. *Plant Gene*, 11, 199–204.

Banerjee, A., Roychoudhury, A., & Krishnamoorthi, S. (2016). Emerging techniques to decipher microRNAs (miRNAs) and their regulatory role in conferring abiotic stress tolerance of plants. *Plant Biotechnology Reports*, 10(4), 185–205.

Bartee, L., Malagnac, F., & Bender, J. (2001). Arabidopsis cmt3 chromomethylase mutations block non-CG methylation and silencing of an endogenous gene. *Genes & Development*, 15, 1753–1758.

Bender, J., & Fink, G. R. (1995). Epigenetic control of an endogenous gene family is revealed by a novel blue, fluorescent mutant of *Arabidopsis*. *Cell*, 83, 725–734.

Berger, S. L., Kouzarides, T., Shiekhattar, R., & Shilatifard, A. (2009). An operational definition of epigenetics. *Genes & Development*, 23, 781–783.

Bernatavichute, Y., Zhang, X., Cokus, S., Pellegrini, M., & Jacobsen, S. (2008). Genome-wide association of histone H3 lysine nine methylation with CHG DNA methylation in *Arabidopsis thaliana*. *PLoS One*, 3, 1–11.

Cao, X., & Jacobsen, S. (2002). Role of the *Arabidopsis* DRM methyltransferases in de novo DNA methylation and gene silencing. *Current Biology*, 12, 1138–1144.

Chan, S., Henderson, I., Zhang, X., Shah, G., Chien, J., & Jacobsen, S. (2006a). RNAi, DRD1, and histone methylation actively target developmentally important non-CG DNA methylation in *Arabidopsis*. *PLoS Genetics*, 2, 791–797.

Chan, S., Zhang, X., Bernatavichute, Y., & Jacobsen, S. (2006b). Two-step recruitment of RNA-directed DNA methylation to tandem repeats. *PLoS Biology*, 4, 1923–1933.

De Lucia, F., Crevillen, P., Jones, A., Greb, T., & Dean, C. (2008). A PHD-polycomb repressive complex 2 triggers the epigenetic silencing of FLC during vernalization. *Proceedings of the National Academy of Sciences*, 105, 16831–16836.

Finnegan, E. J., Kovac, K. A., Jaligot, E., Sheldon, C. C., Peacock, W. J., & Dennis E. S. (2005). The downregulation of FLOWERING LOCUS C (FLC) expression in plants with low levels of DNA methylation and by vernalization occurs by distinct mechanisms. *Plant Journal*, 44, 420–432.

Foerster, A., Dinh, H., Sedman, L., Wohlrab, B., & Mittelsten Scheid, O. (2011). Genetic rearrangements can modify chromatin features at epialleles. *PLoS Genetics*, 7, 1–13.

Fujimoto, R., Kinoshita, Y., Kawabe, A., Kinoshita, T., Takashima, K., Nordborg, M., & Kakutani, T. (2008). Evolution and control of imprinted FWA genes in the genus *Arabidopsis*. *PLoS Genetics*, 4, 1–11.

Greenberg, M., Ausin, I., Chan, S., Cokus, S., Cuperus, J., Feng, S., & Jacobsen, S. (2011). Identification of genes required for de novo DNA methylation in *Arabidopsis*. *Epigenetics*, 6, 344–354.

Gruenbaum, Y., Naveh-Many, T., Cedar, H., & Razin, A. (1981). Sequence specificity of methylation in higher plant DNA. *Nature*, 292, 860–862.

Grunstein, M. (1997). Histone acetylation in chromatin structure and transcription. *Nature*, 389, 349–352.

Heo, J. B., & Sung, S. (2011). Vernalization-mediated epigenetic silencing by a long intronic noncoding RNA. *Science*, 331, 76–79.

Hepworth, S., Valverde F., Ravenscroft, D., Mouradov, A., & Coupland, G. (2002). Antagonistic regulation of flowering-time gene SOC1 by CONSTANS and FLC via separate promoter motifs. *EMBO Journal*, 21, 4327–4337.

Jackson, J. P., Lindroth, A. M., Cao, X., & Jacobsen, S. E. (2002). Control of CpNpG DNA methylation by the KRYPTONITE histone H3 methyltransferase. *Nature*, 416, 556–560.

Jacobsen, S. E., & Meyerowitz, E. M. (1997). Hypermethylated SUPERMAN epigenetic alleles in *Arabidopsis*. *Science*, 277, 1100–1103.

Jenuwein, T., & Allis, C. D. (2001). Translating the histone code. *Science*, 293, 1074–1080.

Johannes, F., Porcher, E , Teixeira, F. K., Saliba-Colombani, V., Simon, M., Agier, N., & Colot, V. (2009). Assessing the impact of transgenerational epigenetic variation on complex traits. *PLoS Genetics*, 5, 1–11.

Johnson, L. M., Bostick, M., Zhang, X., Kraft, E., Henderson, I., Callis, J., & Jacobsen, S. E. (2007). The SRA methyl-cytosine-binding domain links DNA and histone methylation. *Current Biology*, 17, 379–384.

Johnson, L. M., Law, J., Khattar, A., Henderson, I. R., & Jacobsen, S. E. (2008). SRA-domain proteins required for DRM2-mediated de novo DNA methylation. *PLoS Genetics*, 4, 1–13.

Kakutani, T., Jeddeloh, J., Flowers, S. K., Munakata, K., & Richards, E. J. (1996). Developmental abnormalities and epimutations associated with DNA hypomethylation mutations. *Proceedings of the National Academy of Sciences USA*, 93, 12406–12411.

Kinoshita, T., Miura, A., Choi, Y., Kinoshita, Y., Cao, X., Jacobsen, S. E., & Kakutani, T. (2004). One-way control of FWA imprinting in *Arabidopsis* endosperm by DNA methylation. *Science*, 303, 521–523.

Koornneef, M., Hanhart, C. J., & van der Veen, J. H. (1991). A genetic and physiological analysis of late flowering mutants in *Arabidopsis thaliana*. *Molecular Genetics and Genomics*, 229, 57–66.

Kothapalli, N., Camporeale, G., Kueh, A., Chew, Y. C., Oommen, A. M., Griffin, J. B., & Zempleni, J. (2005). Biological functions of biotinylated histones. *Journal of Nutritional Biochemistry*, 16, 446–448.

Law, J. A., & Jacobsen, S. E. (2010). Establish patterns in plants and animals. *Nature Reviews Genetics*, 11, 204–220.

Lindroth, A. M., Shultis, D., Jasencakova, Z., Fuchs, J., Johnson, L., Schubert, D., & Jacobsen, S. E. (2004). Dual histone H3 methylation marks at lysines 9 and 27 required for interaction with CHROMOMETHYLASE3. *EMBO Journal*, 23, 4286–4296.

Lippman, Z., et al. (2004). Role of transposable elements in heterochromatin and epigenetic control. *Nature*, 430, 471–476.

Lister, R., O'Malley, R. C., Tonti-Filippini, J., Gregory, B. D., Berry, C. C., Millar, H., & Ecker, J. R. (2008). Highly integrated single-base resolution maps of the epigenome in *Arabidopsis*. *Cell*, 133, 523–536.

Lister, R., Pelizzola, M., Dowen, R. H., Hawkins, R. D., Hon, G., Nery, J. R., & Ecker, J. R. (2009). Human DNA methylomes at A. L. Jones and S. Sung base resolution show widespread epigenomic differences. *Nature*, 462, 315–322.

Liu, J., He, Y., Amasino, R., & Chen, X. (2004). siRNAs targeting an intronic transposon in the regulation of natural flowering behavior in *Arabidopsis*. *Genes & Development*, 18, 2873–2878.

Liu, S., Yu, Y., Ruan, Y., Meyer, D., Wolff, M., Xu, L., & Shen, W. H. (2007). Plant SET- and RING-associated domain proteins in heterochromatinization. *Plant Journal*, 52, 914–926.

Liu, X., Jiang, F., Kalidas, S., Smith, D., & Liu, Q. (2006). Dicer-2 and R2D2 coordinately bind siRNA to promote assembly of the siRISC complexes. *RNA*, 12, 1514–1520.

Mathieu, O., Reinders, J., Caikovski, M., Smathajitt, C., & Paszkowski, J. (2007). Transgenerational stability of the *Arabidopsis* epigenome is coordinated by CG methylation. *Cell*, 130, 851–862.

Nordborg, M., & Weigel, D. (2008). Next-generation genetics in plants. *Nature*, 456, 720–723.

Pham, A., & Sauer, F. (2000). Ubiquitin-activating/conjugating activity of TAF250, a mediator of activation of gene expression in *Drosophila*. *Science*, 289, 2357–2360.

Popova, O. V., Dinh, H. Q., Aufsatz, W., & Jonak, C. (2013). The RdDM pathway is required for basal heat tolerance in *Arabidopsis*. *Molecular Plant*, 6, 396–410.

Rangwala, S. H., Elumalai, R., Vanier, C., Ozkan, H., Galbraith, D. W., & Richards, E. J. (2006). Meiotically stable natural epialleles of Sadhu, a novel *Arabidopsis* retroposon. *PLoS Genetics*, 2, 270–281.

Reinders, J., Wulff, B. B., Mirouze, M., & Marl, A. (2009). Compromised stability of DNA methylation and transposon immobilization in mosaic *Arabidopsis* epigenomes. *Genes & Development*, 23, 939–950.

Riddle, N. C., & Richards, E. J. (2005). Genetic variation in epigenetic inheritance of ribosomal RNA gene methylation in *Arabidopsis. Plant Journal*, 41, 524–532.

Saze, H., & Kakutani, T. (2007). Heritable epigenetic mutation of a transposon-flanked *Arabidopsis* gene due to lack of the chromatin-remodelling factor DDM1. *EMBO Journal*, 26, 3641–3652.

Shiio, Y., & Eisenman, R. N. (2003). Histone sumoylation is associated with transcriptional repression. *Proceedings of the National Academy of Sciences USA*, 100, 13225–13230.

Soppe, W. J., Jacobsen, S. E., Alonso-Blanco, C., Jackson, J. P., Kakutani, T., Koornneef, M., & Peeters, J. (2000). The late flowering phenotype of *fwa* mutants is caused by gain-of-function epigenetic alleles of a homeodomain gene. *Molecular Cell*, 6, 791–802.

Strahl, B. D., & Allis, C. D. (2000). The language of covalent histone medications. *Nature*, 403, 41–45.

Swiezewski, S., Liu, F., Magusin, A., & Dean, C. (2009). Cold-induced silencing by long antisense transcripts of an *Arabidopsis* Polycomb target. *Nature*, 462, 799–802.

Teif, V. B., & Rippe, K. (2009). Predicting nucleosome positions on the DNA: Combining intrinsic sequence preferences and remodeler activities. *Nucleic Acids Research*, 37, 5641–55.

Trevaskis, B., Hemming, M. N., Dennis, E. S., & Peacock, W. J. (2007). The molecular basis of vernalization-induced flowering in cereals. *Trends in Plant Science*, 12, 352–357.

Vanyushin, B. F. (2006). DNA methylation and epigenetics. *Russian Journal of Genetics*, 42, 985–997.

Wagner, D. (2003). Chromatin regulation of plant development. *Current Opinion in Plant Biology*, 6, 20–28.

Wang, R., Farrona, S., Vincent, C., Joecker, A., Schoof, H., Turck, F., & Albani, M. C. (2009). PEP1 regulates perennial flowering in Arabis alpina. *Nature*, 459, 423–427.

Wei, Y., Yu, L., Bowen, J., Gorovsky, M. A., & Allis, C. D. (1999). Phosphorylation of histone H3 is required for proper chromosome condensation and segregation. *Cell*, 97, 99–109.

Weinberg, M. S., & Morris, K. V. (2013). Long non-coding RNA targeting and transcriptional de-repression. *Nucleic Acid Therapeutics*, 23, 9–14.

Whitehouse, I., Flaus, A., Cairns, B. R., White, M. F., Workman, J. L., & Owen-Hughes, T. (1999). Nucleosome mobilization catalysed by the yeast SWI/SNF complex. *Nature*, 400, 784–787.

Wierzbicki, A., Ream, T., Jeremy, H., & Pikaard, C. (2009). RNA Polymerase V transcription guides ARGONAUTE4 to chromatin. *Nature Genetics*, 41, 630–634.

Woo, H. R., Dittmer, T., & Richards, E. J. (2008). Three SRA-domain methylcytosine-binding proteins cooperate to maintain global CpG methylation and epigenetic silencing in *Arabidopsis. PLoS Genetics*, 4, 1–13.

Wood, C. C., Robertson, M., Tanner, G., Peacock, W. J., Dennis, E. S., & Helliwell, C. (2006). The *Arabidopsis thaliana* vernalization response requires a polycomb-like protein complex that also includes VERNALIZATION INSENSITIVE 3. *Proceedings of the National Academy of Sciences USA*, 103, 14631–14636.

Xie, Z., Johansen, L. K., Gustafson, A. M., Kasschau, K. D., Lellis, A. D., Zilberman, D., & Carrington, J. C. (2004). Genetic and functional diversification of small RNA pathways in plants. *PLoS Biology*, 2, 642–652.

Zemach, A., Kim, M. Y., Hsieh, P. H., Coleman-Derr, D., Eshed-Williams, L., Thao, K., Harmer, S. L., & Zilberman, D. (2013). The Arabidopsis nucleosome remodeler DDM1 allows DNA methyltransferases to access H1-containing heterochromatin. *Cell*, 153, 193–205.

Zhang, X., Yazaki, J., Sundaresan, A., Cokus, S., Chan, S. W., Chen, H., & Ecker, J. R. (2006). Genome-wide high-resolution mapping and functional analysis of DNA methylation in arabidopsis. *Cell*, 126, 1189–1201.

Zheng, B., Wang, Z., Li, S., Yu, B., Liu, J. Y., & Chen, X. (2009). Intergenic transcription by RNA polymerase II coordinates Pol IV and Pol V in siRNA-directed transcriptional gene silencing in *Arabidopsis. Genes & Development*, 23, 2850–2860.

Zilberman, D., Cao, X., & Jacobsen, S. E. (2003). ARGONAUTE4 control of locus-specific siRNA accumulation and DNA and histone methylation. *Science*, 299, 716–719.

Zilberman, D., Cao, X., Johansen, L. K., Xie, Z., Carrington, J. C., & Jacobsen, S. E. (2004). Role of *Arabidopsis* ARGONAUTE4 in RNA-directed DNA methylation triggered by inverted re- peats. *Current Biology*, 14, 1214–1220.

9 Dynamics of Epigenetic Regulation in Physiological Processes, Gene Expression and Defense in Rice

Debabrata Panda, Suraj K. Padhi and Jayanta K. Nayak

9.1 INTRODUCTION

Epigenetic regulation refers to heritable alterations in gene expression that occur without modifications to the basic DNA sequence (Shi et al., 2015; Dinkar et al., 2024). These modifications are carried out via mechanisms such as histone alterations, DNA methylation and non-coding RNAs, which jointly influence chromatin structure and gene expression (Zhang et al., 2018). Epigenetic control is essential in plants for development, stress responses and maintaining genome stability (Law & Jacobsen, 2010). Rice (*Oryza sativa* L.), serving as a model organism for monocotyledonous plants, has been thoroughly investigated to comprehend the dynamics of epigenetic regulation in agricultural species (Feng et al., 2010; Shi et al., 2015). As per the report of the International Rice Genome Sequencing Project (IRGSP) (2005), the rice genome exhibits multiple epigenetic modifications, including methylation of DNA and alteration of histones, which are essential for controlling gene expression and maintaining genome integrity.

Epigenetic systems are essential for plants to adjust to environmental modifications, manage developmental processes and maintain genomic integrity (He et al., 2011). DNA methylation is recognized for its role in suppressing transposable elements (TEs) and preserving genomic integrity (Zemach et al., 2010). Histone changes, including acetylation and methylation, regulate chromatin accessibility, thus affecting transcriptional activity (Liu et al., 2010). Furthermore, short RNAs, such as siRNAs (small interfering RNAs) and miRNAs (micro RNAs), act as gene silencers after transcription and in protecting against viral infections (Vaucheret, 2006). Epigenetic mechanisms are crucial in rice, as they govern essential physiological processes including flowering time, stress responses and nutrient absorption (Chodavarapu et al., 2012; Zhu, 2016).

Rice offers a primary food source for more than half of the global population and acts as an ideal organism for genetic and epigenetic studies in cereals (Goff et al., 2002). The comparatively diminutive genome size of the organism, superior reference genome quality and the abundance of genetic resources render it an exemplary model for exploring epigenetic control (IRGSP, 2005). Moreover, rice demonstrates a diverse array of epigenetic phenomena, encompassing transgenerational epigenetic inheritance and stress-induced epigenetic alterations, which are of significant interest for both fundamental and applied study (Chodavarapu et al., 2012). Research indicates that epigenetic alterations in rice can be transmitted over generations, offering a way for plants to retain information about previous environmental circumstances (Liu et al., 2017; Mirouze & Paszkowski, 2011). Considering the pivotal role of epigenetics in rice biology, comprehending these mechanisms is crucial for formulating methods to enhance crop resilience and productivity amid global concerns like climate change and population expansion.

This chapter seeks to deliver a thorough examination of the dynamics of epigenetic regulation in rice, emphasizing its significance in physiological processes, gene expression and defense systems. This chapter highlights the fundamental mechanisms of epigenetic regulation, along with their impact on rice development and stress responses. Furthermore, this chapter describes the utilization of epigenetic methods and technologies in rice research and their prospects for enhancing crop quality. This chapter also aims to emphasize the significance of epigenetics in comprehending rice biology and its ramifications for sustainable agriculture by including recent advancements in the field. Additionally, epigenetic variation in rice germplasm might be utilized in breeding initiatives to enhance stress resilience and productivity.

9.2 PRINCIPLES OF EPIGENETIC MECHANISMS

Epigenetic activities are heritable alterations in the regulation of genes that occur without modifying the DNA sequence itself. The mechanisms comprise histone modification, methylation at DNA, non-coding RNAs and chromatin restructuring (Figure 9.1). They are essential for regulating gene expression, preserving genomic stability and reacting to environmental stimuli (Shi et al., 2015). We will examine these systems in depth, emphasizing their functions in rice (*Oryza sativa* L.), an ideal crop for investigating plant epigenetics.

DOI: 10.1201/9781003589914-11

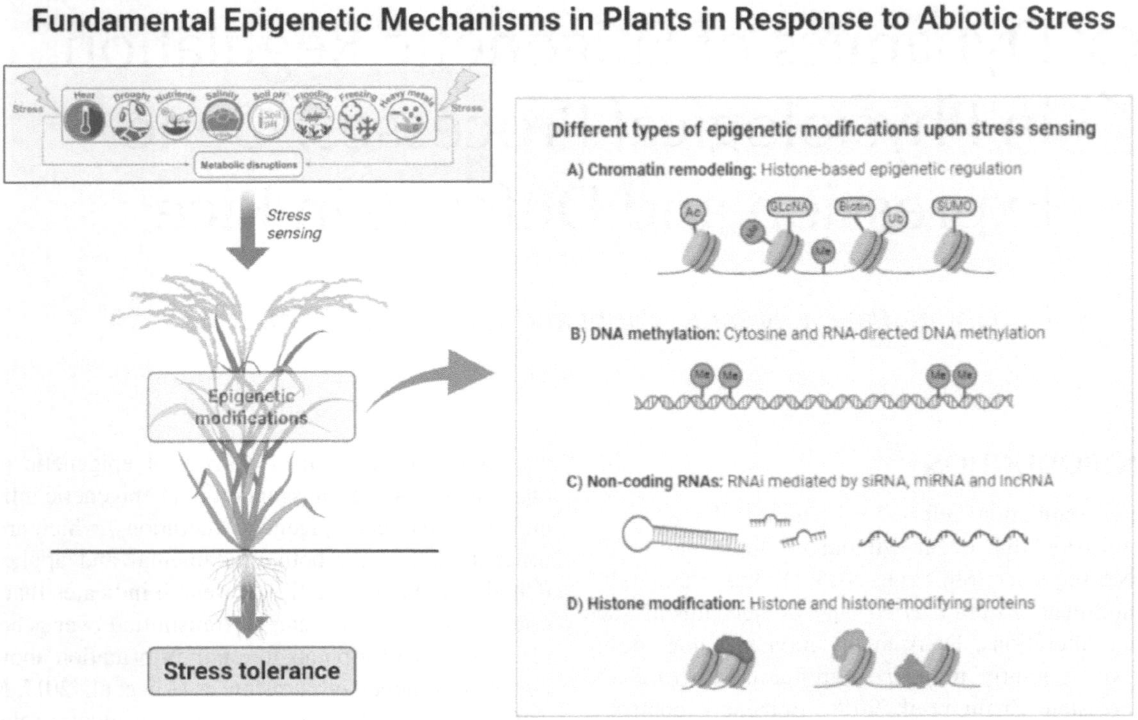

FIGURE 9.1 Epigenetic mechanisms such as methylation of DNA, histone alterations, non-coding RNAs and chromatin restructuring.

9.2.1 DNA Methylation

DNA methylation refers to the addition of a methyl group to the cytosine base within DNA, usually occurring at CpG dinucleotides. This alteration serves as a crucial epigenetic marker that governs gene expression, genomic imprinting and the silencing of transposable elements (Law & Jacobsen, 2010).

9.2.1.1 Function of DNA Methylation in Gene Suppression

The methylation of DNA is an established mechanism for gene silencing. It can inhibit gene expression by blocking the binding of transcription factors or by recruiting proteins that condense chromatin, making it less accessible for transcription (Law & Jacobsen, 2010; Shi et al., 2015). In plants, methylation in DNA is essential for the repression of TEs and the preservation of genomic integrity (Zhang et al., 2018; Li et al., 2012). In rice, hypermethylation of transposable elements inhibits their mobilization, which could otherwise result in genomic instability (Li et al., 2012; Shi et al., 2015). Moreover, DNA methylation in promoter regions frequently correlates with the downregulation of gene expression, particularly in stress-responsive genes subjected to abiotic stress situations.

9.2.1.2 DNA Methylation Patterns in Rice

Rice displays unique DNA methylation patterns that differ across tissues and developmental phases. Global DNA methylation levels are elevated in roots relative to leaves, indicating tissue-specific control (Yan et al., 2010). Moreover, environmental conditions like drought and salinity can modify DNA methylation patterns, resulting in alterations in gene expression that facilitate plant adaptation to stress (Zemach et al., 2010). Research indicates that DNA methylation in rice is dynamically controlled throughout seed development, influencing imprinting and seed viability (Xing et al., 2015; Shi et al., 2015).

9.2.2 Modifications of Histones in Rice

Modifications of histone proteins are covalent post-translational changes that affect chromatin architecture and gene expression. The changes include acetylation, methylation, phosphorylation and ubiquitination (Figure 9.2) (Bannister & Kouzarides, 2011).

9.2.2.1 Categories of Histone Modifications

1. **Acetylation:** An epigenetic change known as histone acetylation controls several genes important for numerous biological processes including physiological responses to stress and development (Kumar et al., 2021). The incorporation of acetyl groups onto lysine residues in histone tails often diminishes the positive charge of histones, resulting in a more flexible chromatin configuration and enhanced transcriptional activity. Histone acetylation in rice is associated with the induction of stress-sensitive genes in response to drought (Zhao et al., 2020; Shi et al., 2015).
2. **Methylation:** The methylation of histone can either boost or suppress gene expression, contingent upon the particular amino acid residue that is

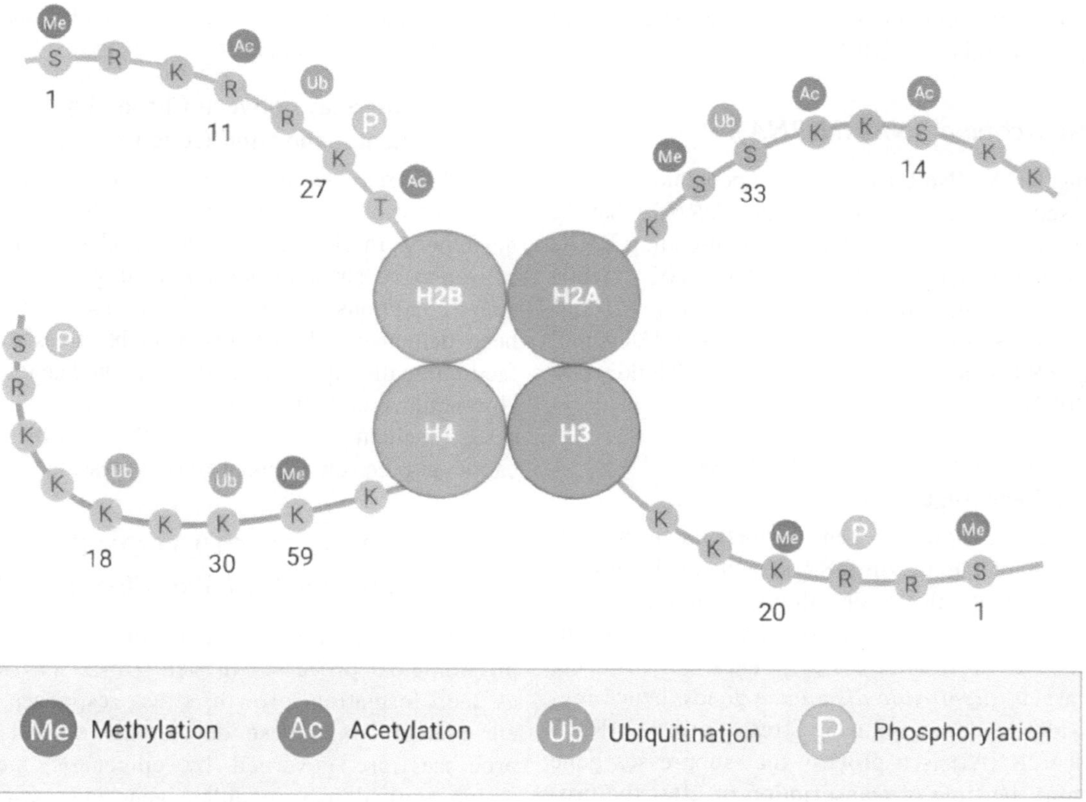

FIGURE 9.2 Schematic depiction of post-translational modifications in histones: acetylation, methylation, ubiquitination and phosphorylation sites on core histones (H2A, H2B, H3 and H4).

changed. It predominantly takes place on the side chains of lysine and arginine residues. Unlike acetylation and phosphorylation, histone methylation does not change the overall charge of the histone protein. Additionally, it is important to recognize the intricacy associated with this change; lysine can be mono-, di- or tri-methylated, while arginine undergoes mono- or symmetric/asymmetric di-methylation (Bannister & Kouzarides, 2011). Histone H3 methylation at lysine 4 (H3K4me) promotes gene activation, whereas methylation at lysine 27 (H3K27me) is associated with gene repression (Berr et al., 2011; Shi et al., 2015).

3. **Phosphorylation:** Similar to histone acetylation, histone phosphorylation is markedly dynamic. This occurs primarily on serine, threonine and tyrosine, especially in the N-terminal histone tails, however not exclusively. The modification levels are governed by phosphatases and kinases, which alternately add and remove the modification (Bannister & Kouzarides, 2011). Histone phosphorylation in rice is crucial in regulating meiosis and preserving genome stability (Nonomura et al., 2007; Shi et al., 2015).
4. **Ubiquitination:** According to Swatek and Komander (2016), the term "ubiquitination" or "ubiquitylation" refers to the enzymatic process of transferring one or more ubiquitin monomers to a protein substrate which regulates protein stability, localization and function. Ubiquitination is categorized into two types such as monoubiquitination and polyubiquitination (Oss-Ronen et al., 2022). This modification specifically takes place on residues of lysine. Drought stress tolerance cascade in rice and *Arabidopsis*, involving the ubiquitin ligase *OsPUB67* and its target proteins *OsDIS1* and *OsRZP34*, has been thoroughly investigated (Sharma et al., 2023). *OsPUB67* enhances drought tolerance by facilitating reactive oxygen species scavenging and stomatal closure during drought, while *OsDIS1* and *OsRZP34* function as detrimental regulators that promote stomatal opening under water deficit conditions. *OsPUB67* ubiquitinates the substrates such as *OsRZFP34* and *OsDIS1* for proteolytic destruction, resulting in enhanced stomatal closure (Sharma et al., 2023).

9.2.2.2 Influence of Histone Modifications on Chromatin Architecture

Histone modifications alter the interaction between histones and DNA, resulting in alterations to chromatin structure. Acetylation of histones H3 and H4 promotes an open chromatin configuration, hence facilitating passage for transcription factors and RNA polymerase II (Du et al., 2013). H3K9me is related to the development of heterochromatin and the repression of genes. In rice, these alterations are essential for the regulation of developmental genes and stress responses (Sui et al., 2012).

9.2.3 Non-coding RNAs (NcRNAs)

Non-coding RNAs (NcRNAs) are grouped into two categories based on their transcript size. NcRNAs having nucleotides (>200) are grouped as long non-coding RNAs (lncRNAs) (Bhattacharya & Roychoudhury, 2024), while NcRNAs having nucleotides (<200) are categorized as short NcRNAs. Short NcRNAs encompass siRNAs (small interfering RNAs) and miRNAs (microRNAs) (Ghildiyal & Zamore, 2009).

9.2.3.1 Function of miRNAs and siRNAs in Gene Regulation

The small RNAs modulate gene expression by hybridizing with complementary mRNA sequences, leading to the degradation of mRNA or inhibition of translation. In rice, miRNAs participate in regulating developmental processes as well as stress responses; for instance, miR398 targets *superoxide dismutase* genes, hence regulating oxidative damage during stress tolerance. RNA interference (RNAi) is a process that suppresses gene expression at the time of transcription or after the process of transcription (Singh & Roychoudhury, 2023). In rice, siRNAs are essential for combating viral infections and silencing transposable elements (Chodavarapu et al., 2012).

9.2.3.2 Epigenetic Regulation by Long Non-coding RNAs (lncRNAs)

LncRNAs are RNA molecules having nucleotides (>200) that modulate the expression of genes by several methods, like chromatin remodeling and transcriptional interference. In rice, lncRNAs have been demonstrated to regulate flowering timing and stress responses. The lncRNA COLDAIR engages with polycomb repressive complex 2 (PRC2) to inhibit genes related to flowering against cold stress (Shi et al., 2015).

9.2.4 Chromatin Remodeling

Chromatin remodeling involves the dynamic modification of chromatin architecture to control DNA accessibility. This process is facilitated by chromatin remodeling complexes that utilize ATP to reposition, evict or reconstruct nucleosomes (Shi et al., 2015).

9.2.4.1 Mechanisms of Chromatin Remodeling in Rice

In rice, chromatin remodeling is predominantly executed by SWI/SNF complexes, which remove nucleosomes to expose DNA for transcription. These complexes are crucial for regulating genes associated with development and stress responses. The SWI/SNF complex in rice regulates drought-related gene expression by modifying chromatin accessibility (Shi et al., 2015).

9.2.4.2 Interplay between Chromatin Remodeling and Gene Expression

Chromatin remodeling is crucial for regulating gene activity by controlling DNA accessibility to the transcription machinery. In rice, chromatin remodeling facilitates the activation of stress-response-related genes under abiotic stress conditions. The chromatin remodeler OsCHR4 has been demonstrated to boost drought stress tolerance by facilitating the expression of stress-related genes. Moreover, chromatin remodeling engages with other epigenetic processes, including methylation of DNA and histone modification, to precisely regulate gene expression.

9.3 EPIGENETIC REGULATION IN PHYSIOLOGICAL PROCESSES OF RICE

Epigenetic regulation is crucial in governing numerous physiological processes in rice (*Oryza sativa* L.), such as seed formation, growth, stress responses and nutrient metabolism. These developmental and metabolic processes are governed by epigenetic mechanisms, which meticulously modulate gene expression in reaction to developmental events and environmental changes (Banerjee & Roychoudhury, 2018). This section examines the function of epigenetics in these processes, emphasizing rice, and is substantiated by recent studies (Table 9.1).

9.3.1 Seed Development and Germination

9.3.1.1 Epigenetic Regulation of Seed Maturation

Seed maturity in rice entails the building of storage reserves and the attainment of desiccation tolerance, processes governed by epigenetic pathways. These epigenetic pathways are essential for regulating gene expression during seed maturation. The *OsFIE1* gene in rice, part of the PRC2, facilitates H3K27me3 modifications to inhibit genes associated with early germination and enhance seed maturity. Furthermore, DNA methylation at TEs maintains genomic integrity throughout seed formation (Zemach et al., 2010).

9.3.1.2 The Function of Epigenetics in Seed Dormancy and Germination

Epigenetic changes that respond to environmental stimuli affect seed dormancy and germination in rice. DNA demethylation, facilitated by rice-specific demethylases such as ROS1, is crucial for alleviating seed dormancy and initiating germination (Zhu et al., 1998). Histone alterations, including H3K27me3 and H3K4me3, regulate dormancy-associated genes such as *OsVP1* (Chen et al., 2021). These epigenetic modifications guarantee that rice seeds germinate optimally, improving crop establishment.

TABLE 9.1
Epigenetic regulation in physiological processes under stress conditions in some genes/enzymes of rice

Sl. no.	Name of gene/enzyme	Function	Type of regulation	Phenotypic effect	References
1	Histone methyltransferase – SDG708	Stimulates the expression of *OsNCED5* and *OsNCED3*	Epigenetic activation via H3K36 methylation	Overexpression improves grain yield and drought tolerance under water deficit condition	Chen et al., 2021
2	Histone demethylase – OsJMJ703	Regulates the level of H3K4me1/2/3, affecting the activity of '*Karma*' and *LINE1*	Epigenetic repression via H3K4 demethylation	Exhibit reduced stem elongation and growth. Overexpression causes drought sensitivity and panicle drooping, while *OsJMJ703* RNAi knockdown induces early flowering	Chen et al., 2013; Song et al., 2018
3	Histone demethylase – OsJMJ705	MeJA regulates H3K27me2/3 on biotic stress-responsive genes, while WOX11 recruits *OsJMJ705* to control shoot growth in rice	Epigenetic repression via H3K27 demethylation	The overexpression line improves resistance to blight disease caused by bacteria. Mutants have defective development in shoot	Li et al., 2013; Cheng et al., 2018
4	Histone demethylase – OsJMJ701 (Se14)	Eliminates the H3K4me3 modification of the *RFT1* gene. Se14 exerts a suppressive influence on blooming time under long-day circumstances	Epigenetic repression via H3K4 demethylation	The *Se14*-deficient mutant displays an early flowering phenotype	Yokoo et al., 2014
5	Histone demethylase – OsJMJ706	Modulates the *H3K9me2/3* genes involved in rice floral development	Epigenetic repression via H3K9 demethylation	*OsJMJ706* loss-of-function mutants exhibit abnormalities during the development of floral organs	Sun and Zhou, 2008
6	Inactive histone demethylase – OsJMJ715	Codes for a nuclear RING-type E3 ubiquitin ligase but is devoid of H3K9 demethylase activity	Ubiquitination-mediated regulation	OsJMJ715 negatively regulates rice resistance to the brown planthopper (BPH) by reducing BPH-induced ABA production and callose deposition	Lu et al., 2008; Zhang et al., 2021
7	Histone demethylase – OsJMJ710	Encodes the MYB48-1 transcription factor	Epigenetic repression via H3K36 demethylation	CRISPR knockout and RNAi knockdown mutants display increased drought tolerance	Zhao et al., 2022
8	HD2-type HDAC HDT701	Diminish the production of gibberellin – GA3ox1, GA3ox2, GA20ox2 and GA20ox3	Epigenetic repression via histone deacetylation	HDT701 overexpression postpones seed germination in the presence of ABA, salt and osmotic stress, although improves tolerance to salinity	Zhao et al., 2015
9	RPD3/HDA1-type HDAC – HDA704	Suppresses the expression of the transcription factors *DST* and *ABIL2*	Epigenetic repression via histone deacetylation	Augments tolerance to water deficit and salinity by facilitating stomatal closure, diminishing stomatal density, and lowering transpiration rates	Zhao et al., 2021
10	RPD3/HDA1 type HDAC – DSL1	Varied function in organogenesis	Epigenetic repression via histone deacetylation	CRISPR-Cas9 generated knockout *dsl1* gene exhibits pleiotropic developmental anomalies, including dwarfism, diminished vascular bundles, narrow leaves and reduced panicle size	Kubo et al., 2020
11	RPD3/HDA1 type – HDA710/ OsHDAC2	Modulate the acetylation of histones H3 and H4	Epigenetic repression via histone deacetylation	Genes responsive to ABA and salinity stress were increased in the CRISPR-Cas9-regulated mutant of the *hda710* gene, hence improving salinity stress tolerance	Ullah et al., 2021

(*Continued*)

TABLE 9.1 CONTINUED

Sl. no.	Name of gene/enzyme	Function	Type of regulation	Phenotypic effect	References
12	RPD3/HDA1 type – HDA705	Regulates and facilitates the expression of the genes responsible for gibberellin and abscisic acid biosynthesis	Epigenetic repression via histone deacetylation	HDA705 overexpression downregulates GA biosynthesis genes and upregulates ABA biosynthetic genes, consequently delaying seed germination, reducing root development, and enhancing osmotic stress tolerance	Zhao et al., 2016
13	OsHDAC1	OsHDAC1 directly deacetylates and inhibits OsGSK2, resulting in the accumulation of OsBZR1. OsBZR1 regulates auxin signaling genes	Epigenetic regulation via histone deacetylation	OsHDAC1 facilitates the creation of lateral root primordia, hence significantly influencing the development of lateral root	Hou et al., 2022
14	HDA703	HDA703 directly interacts with the GHD7 promoter and recruits OsBZR1, hence repressing the expression of *Ghd7* and significantly regulating the signaling of BR	Epigenetic repression via histone deacetylation	Lines with HDA703 over-expression exhibit accelerated growth, reduced plant height and leaf length, prompt emergence of new leaves and tillers, premature heading and seed maturation, and broader flag-leaf angles	Wang et al., 2020
15	Grain weight QTL–GW6a (OsglHAT1)	Codes proteins (GNAT-like) that regulate histone H4 acetylation	Epigenetic activation via histone acetylation	The overexpression of *OsglHAT1* enhances cell proliferation, resulting in enlarged spikelet hulls and expedited grain filling, ultimately enhancing yield	Song et al., 2015
16	$p34^{cdc2/CDC28}$-like histone H1 kinase	Gibberellin treatment boosts the activity of histone H1 kinase in the internode intercalary meristem	GA-responsive kinase activation	Enhanced cellular division facilitates lengthened internodes and aids in submergence tolerance	Sauter et al., 1995
17	Bub1-related kinase 1 (BRK1)	Attaches SHUGOSHIN1 to the centromere and catalyzes phosphorylation of histone H2A at Thr-134 and histone H3 at Ser-10 at the time of mitosis	Kinase-mediated histone modification	Ensures precise merotelic attachment of sister kinetochores during metaphase I and the equal segregation of homologous chromosomes	Wang et al., 2012
18	Ubiquitin E3 ligase – OsHUB2	OsHUB2 facilitates the monoubiquitination of histone H2B and interacts with the transcription factor OsbZIP46	Epigenetic activation via histone ubiquitination	Enhanced sensitivity to ABA and resilience to water deficit stress	Ma et al., 2019
19	Deubiquitinase – OsLMP1	Inhibits transcriptional activity and repressing the genes related to the production of SA	Epigenetic repression via histone deubiquitination	CRISPR-Cas9 induced deletion of OsLMP1 results in elevated *PR*-gene expression and enhanced resistance against blight diseases and blast	Sun et al., 2022
20	SUMO E3 Ligase – OsSIZ1	Involved in phosphorus and nitrogen uptake, transport and assimilation	SUMOylation-mediated regulation	Mutants exhibit reduced length of primary and adventitious roots	Wang et al., 2015
21	SWI/SNF2 protein – BRHIS1	Blocks further ubiquitination and suppresses the expression of target genes	Chromatin remodeling via histone ubiquitination	During rice blast attack, *BRHIS1* lifts repression in defense-related genes, enabling a swift immune response	Li et al., 2015

9.3.2 Expansion and Advancement

9.3.2.1 Epigenetic Modulation of Root and Shoot Development

Root and shoot development in rice is meticulously governed by epigenetic processes. DNA methylation and histone alterations influence gene expression during cell division and differentiation. The rice *OsWOX4* gene, which governs root formation, is epigenetically regulated by H3K27me3 modifications. Likewise, DNA methylation patterns affect shoot development, as seen by the modified phenotypes in rice *ddm1* (reduction in DNA methylation 1) mutants (Deng et al., 2016).

9.3.2.2 Regulation of Flowering Time via Epigenetic Mechanisms

The flowering time in rice is a crucial developmental change governed by epigenetic processes. The *OsFLC* gene, a homolog of *Arabidopsis* FLC, is repressed by H3K27me3 modifications in response to environmental stimuli. Moreover, DNA methylation at the promoter of the *OsMADS50* gene modulates blooming time in rice (Shi et al., 2015). These epigenetic alterations guarantee that flowering transpires under optimal environmental conditions, hence augmenting reproductive success.

9.3.3 Responses to Abiotic Stress

9.3.3.1 Epigenetic Alterations Induced by Drought Stress

Drought stress induces dynamic epigenetic modifications in rice that facilitate tolerance to water deficiency. Methylation of DNA and modification in histone govern the stress-responsive genes, including *OsDREB1A* and *OsLEA3*, and facilitate their expression. Hypermethylation of TEs during drought stress contributes to the preservation of genomic stability. Small RNAs, including miRNAs, participate in gene regulation and expression during drought stress conditions (Sunkar et al., 2012).

9.3.3.2 Epigenetic Regulation in Response to Salinity and Temperature Variations

Variations in salinity and temperature elicit epigenetic modifications in rice, hence augmenting stress resilience. Histone changes, including H3K4me3 and H3K9ac, correlate with the activation of salt-responsive genes such as *OsNHX1* (Rashid et al., 2022). Changes in DNA methylation have been noted in response to elevated salinity, influencing genes associated with ion homeostasis (Wibowo et al., 2016). Likewise, temperature variations modify histone modifications at genes that govern thermotolerance, such as *OsHSP70* (Yu et al., 2025).

9.3.4 Nutrient Absorption and Metabolism

9.3.4.1 Epigenetic Regulation of Nutrient Homeostasis

Nutrient homeostasis in rice is governed by epigenetic mechanisms that influence the expression of transporters and metabolic enzymes. DNA methylation modulates the expression of phosphate transporters (OsPT1) in phosphate-deficient environment (Verhoeven et al., 2010). Histone changes, including H3K27me3, affect the expression of genes related to iron homeostasis, such as *OsIRT1* (Xing et al., 2015).

9.3.4.2 Role of Epigenetics in Nitrogen and Phosphorus Metabolism

Epigenetic pathways are essential in the metabolism of nitrogen and phosphorus in rice. They govern the genes associated with nitrogen assimilation, including OsNR1 and OsNR2, and facilitate their expression (Yong-Villalobos et al., 2015). Phosphorus metabolism is similarly affected by epigenetic modifications in genes that encode phosphatases and transporters, including OsPAP10 (Wang et al., 2023). These epigenetic alterations guarantee effective nutrition utilization across diverse environmental situations.

9.4 EPIGENETIC MODULATION OF GENE EXPRESSION IN RICE

Epigenetic regulation is essential for modulating gene expression in rice (*Oryza sativa*), affecting developmental processes, stress responses and transgenerational inheritance. This section examines the processes and consequences of epigenetic control in rice, emphasizing tissue-specific gene expression, transgenerational epigenetic inheritance and stress-responsive gene regulation.

9.4.1 Gene Expression Specific to Tissues

Gene expression, specific to tissues in rice, is stringently governed by epigenetic processes. These systems guarantee the appropriate progress and functionality of roots, leaves and flowers. In rice, the epigenetic processes are crucial regulators of tissue-specific gene expression. Differential DNA methylation patterns have been identified between roots and shoots, affecting genes related to nutrient absorption and photosynthesis (Zemach et al., 2010). Histone changes, including H3K27me3, correlate with the suppression of floral development genes in vegetative tissues, hence regulating the time of flowering (He et al., 2010). Furthermore, ncRNAs perform a pivotal role in tissue-specific regulation by influencing mRNA stability and translation (Mirdar Mansuri et al., 2022). Epigenetic modifications in rice demonstrate dynamic alterations during development, indicating the necessity for meticulous gene regulation and their expression at various stages. DNA methylation levels in rice embryos diminish during seed germination but rise throughout vegetative growth, corresponding with the activation of growth-related genes (Zemach et al., 2010). Histone modifications, such as H3K4me3, are prevalent near the promoters of actively transcribed genes during early development, whereas H3K27me3 designates genes that are repressed in later developmental stages (He et al., 2010). These dynamic alterations underscore the adaptability of the rice epigenome in response to developmental signals.

9.4.2 Transgenerational Epigenetic Inheritance

Transgenerational epigenetic inheritance denotes the transfer of epigenetic modifications across generations, potentially affecting phenotypic characteristics in descendants. This phenomenon in rice is affected by genetic and environmental variables. Transgenerational epigenetic inheritance in rice is facilitated by the enduring transmission of DNA methylation patterns and histone changes during meiosis. Research indicates that DNA methylation at transposable elements (TEs) is preserved through generations, inhibiting their activation and safeguarding genome integrity (Zemach et al., 2010). Furthermore, short RNA molecules, including siRNAs, are crucial in directing DNA methylation and histone changes in offspring, hence facilitating the transfer of epigenetic states (Mirdar Mansuri et al., 2022). Environmental variables, including drought, salinity and thermal stress, can provoke epigenetic alterations that are inherited by later generations in rice. Exposure to drought stress induces alterations in DNA methylation patterns at stress-responsive genes, which are transmitted to progeny and augment their resilience to analogous stress conditions (Kumar et al., 2023). Heat stress has been demonstrated to modify histone alterations in rice, influencing gene expression and phenotypic characteristics in subsequent generations (Chen et al., 2016).

9.4.3 Epigenetic Modulation of Stress-Responsive Genes

Epigenetic control is essential for the commencement and suppression of genes related to stress response in rice, facilitating swift adaptation to biotic and abiotic challenges (Banerjee & Roychoudhury, 2017). Rice plants demonstrate epigenetic memory of stress responses, enabling them to react more efficiently to repeated stress circumstances. Exposure to drought stress generates alterations in epigenetic mechanisms at stress-responsive genes, such as *OsDREB1* and *OsNAC6*, which persist long after the stress is alleviated (Garg et al., 2015). Various type I MADS box genes, especially expressed in rice during the syncytial phase, exhibited differential regulation under moderate and severe heat stress. The repression and excessive accumulation of these genes correlate with premature and postponed cellularization under moderate and severe stress, respectively (Chen et al., 2016). This epigenetic memory improves the capacity of the plants to endure repeated stressors, illustrating the significance of epigenetics in stress adaptation.

Epigenetic modifications engage with stress signaling pathways to modulate gene expression in rice. Investigations in rice displayed the critical function of DNA methylation in regulating their adaptation to salinity and drought stress. The methylation of DNA specifically influences the rice plant response to different plant hormones, thereby regulating the expression of stress-response genes and improving overall resilience to stressful conditions (Ahmad et al., 2019). DNA methylation is vital for endurance to high temperature stress and for improving tolerance in rice (Wang et al., 2011). An examination of the mechanisms of DNA methylation in rice indicates that it typically influences the upregulation or downregulation of gene expression by targeting the promoter. The prevalence of small RNA is positively correlated with hypermethylated regions, and there exists an interplay between methylation of DNA, expression of genes and small RNA distribution. Regarding specific gene functions, *cld1* modulates curling of leaf in rice by influencing bulliform cell formation, but methylation of DNA could inhibit this process and diminish drought resilience (Yin et al., 2024).

9.5 EPIGENETIC REGULATION IN DEFENSE MECHANISMS IN RICE

Rice is a critical crop that faces significant biotic stresses, including fungal, bacterial and viral infections. Epigenetic regulation plays a vital function in modulating defense mechanisms in rice (Du et al., 2011). This section explores how these epigenetic mechanisms regulate immune responses, disease resistance genes and responses to biotic stress in rice.

9.5.1 Epigenetic Control of Immune Responses

The initial proof for the epigenetic regulation of plant resistance to biotic stimuli was the modulation of viral virulence through post-transcriptional gene silencing (PTGS), as reported by Ramirez-Prado et al. (2018). When plants are infected by RNA viruses, they detect the viral double-stranded RNA (dsRNA) molecules. Plant enzymes such as DCL2 and DCL4 chop the dsRNA into siRNAs to increase defense against the virus. Another defense mechanism, transcriptional gene silencing (TGS), offers long-term protection against DNA viruses through RNA directed DNA methylation (RdDM). Both PTGS and TGS not only help to defend against viral infections but also serve a crucial function in regulating the own genes of plants (Ashapkin et al., 2020).

9.5.1.1 Role of DNA Methylation in Pathogen Defense

Extensive research has explored the role of DNA methylation in plant immunity (Ramirez-Prado et al., 2018; Deleris et al., 2016). The first layer of active defense, known as pathogen-associated molecular pattern (PAMP)-triggered immunity (PTI), relies on the recognition of PAMPs or microbe-associated molecular patterns (MAMPs) by specialized pattern-recognition receptors (PRRs). Once these PAMPs are recognized, immune responses are activated, leading to a foundational level of protection. To weaken plant defense, pathogens release specialized effector molecules that suppress PTI. In response, plants counteract this attack by detecting these effectors through disease resistance proteins, triggering a stronger immune reaction known as effector-triggered immunity (ETI). Both PTI and ETI lead to significant changes in gene expression, which are controlled by epigenetic mechanisms (Ashapkin et al., 2020).

9.5.1.2 Histone Modifications in Immune Gene Activation

Histone changes, including acetylation and methylation, govern chromatin architecture and gene expression in immunological responses. In rice, the expression of defense-related genes is associated with specific histone modifications, including H3K9ac and H3K4me3; for instance, the allelic genes *OsWRKY45-1* and *OsWRKY45-2*, which produce proteins differing by only 10 amino acids, exhibit opposing roles in rice plant resistance to bacterial pathogens (Tao et al., 2009). Studies have shown that rice plants overexpressing *OsWRKY45-1* exhibit increased susceptibility to *Xoc* (*Xanthomonas oryzae* pv *oryzicola*) and *Xoo* (*Xanthomonas oryzae* pv *oryzae*), whereas *OsWRKY45-1* knockout plants demonstrate elevated resistance to these bacterial pathogens. Conversely, rice plants overexpressing *OsWRKY45-2* exhibited elevated resistance to *Xoc* and *Xoo*, whereas suppression of *OsWRKY45-2* led to enhanced susceptibility to these pathogens. Notably, overexpression of both *OsWRKY45-1* and *OsWRKY45-2* conferred increased resistance to *M. grisea* (Tao et al., 2009). Moreover, a distinct study identified the *OsWRKY45* transcription factor as crucial for benzothiadiazole-induced resistance to blast disease in rice (Shimono et al., 2007).

9.5.2 Epigenetic Regulation of Disease Resistance Genes

Epigenetic mechanisms regulate disease resistance genes in rice, including the silencing of susceptibility genes and the activation of *resistance* (*R*) genes.

9.5.2.1 Epigenetic Silencing of Susceptibility Genes

Susceptibility genes facilitate pathogen infection, and their epigenetic silencing can enhance disease resistance. In rice, the clade III subfamily of *OsSWEET* genes encodes transmembrane proteins that are essential for susceptibility to bacterial blight (BB). These genes are specially targeted by transcription activator-like effector (TALEs) from *Xoo*, facilitating sucrose transport out of plant cells, which supports bacterial growth and proliferation. The *OsSWEET* family of genes, which encodes sugar transporters, is targeted by bacterial pathogens to promote infection. Wang et al. (2024) reported that DNA methylation-mediated epigenetic silencing of *OsSWEET14* enhances resistance to *Xoo*, the pathogen responsible for bacterial blight disease in rice.

9.5.2.2 Activation of *Resistance* (*R*) Genes through Epigenetic Mechanisms

Resistance (*R*) genes are a crucial component of the plant defensome system, with both *R* genes and *defense response* (*DR*) genes playing a significant role in providing broad-spectrum resistance against rice blast disease (Devanna et al., 2022). While many defense-related genes show differential expression during host-pathogen interactions, this variation is primarily attributed to triggering defense signaling in the responses of resistant plants rather than differences in allelic expressions. This conclusion is based on gene expression studies (Zhou et al., 2002). So far, over 100 blast resistance genes have been mapped in rice, with 38 successfully cloned and characterized. The most effective strategy for managing the disease is identifying and utilizing *R* genes that provide broad-spectrum resistance against multiple pathogen races (Li et al., 2019; Devanna et al., 2022).

9.6 EPIGENETIC INSTRUMENTS AND TECHNOLOGIES IN RICE RESEARCH

Advancements in tools and technologies have transformed the study of epigenetics in rice, facilitating genome-wide profiling, accurate epigenome editing and intricate bioinformatics analysis. These instruments have markedly improved our comprehension of epigenetic control in rice, especially for stress responses, development and disease resistance. This section addresses essential technologies, such as genome-wide epigenetic profiling, CRISPR-mediated epigenome editing and bioinformatics tools for the interpretation of epigenetic data.

9.6.1 Comprehensive Epigenetic Profiling across the Genome

Genome-wide epigenetic profiling tools have facilitated the high-resolution mapping of DNA methylation and histone changes, yielding insights into the epigenetic landscape of rice.

9.6.1.1 Analysis of DNA Methylation by Bisulfite Sequencing

Bisulfite sequencing is a premier technique for examining methylation in DNA with single-base precision. Whole-genome bisulfite sequencing (WGBS) has been employed in rice to delineate DNA methylation patterns throughout the genome. WGBS demonstrated dynamic alterations in DNA methylation throughout rice growth and in reaction to biotic stress. This method has proved pivotal in discovering differentially methylated regions (DMRs) linked to stress responses and agronomic characteristics (Rajkumar et al., 2020).

9.6.1.2 ChIP-Seq for Mapping Histone Modifications

Chromatin immunoprecipitation coupled with sequencing (ChIP-seq) is essential for characterizing transcription factor and histone alteration interactions across the genome. Nonetheless, its implementation in economically significant plant organs (ESPOs) is problematic because of their robust cell walls and intricate components (Zhang et al., 2024). ChIP-seq has been employed in rice to investigate histone changes like H3K4me3, H3K27me3 and H3K9ac, which correlate with gene activation or repression. ChIP-seq research demonstrated that H3K27me3 marks are prevalent at repressed genes related to stress responses, whereas H3K4me3 marks correlate with actively transcribed genes (Zhao et al., 2020). These investigations have provided valuable insights into the role of histone alteration in regulating gene expression in rice.

9.6.2 CRISPR-Driven Epigenome Modification

CRISPR-based epigenome editing tools have emerged as potent technology for the investigation and alteration of epigenetic markers in rice.

9.6.2.1 Utilization of CRISPR/dCas9 in Epigenetic Research

CRISPR-Cas9 technology has revolutionized genome editing by offering precise and efficient methods for modifying genetic material. This powerful gene-editing system leverages the bacterial immune response to recognize and cut specific DNA sequences. It utilizes minute RNA molecule, sgRNA (single guide RNA), to guide Cas9 enzymes to a specific DNA location, causing a double-stranded break. This breakage can subsequently be rectified via cellular repair mechanisms, facilitating the incorporation of intended genetic alterations (Ansori et al., 2023). An intriguing use of this approach is the creation of a genome-wide mutant library for the identification of gene functions and genetic enhancement. Recently, CRISPR/Cas9-based gene editing has been used to enhance rice disease resistance by modifying disease-resistant genes to create new resistant varieties (Chen et al., 2024). Wang et al. (2016) used CRISPR/Cas9 to functionally knock down the *Pi21* gene in the Nanjing 9108 rice variety, generating mutants and paving the way for developing rice lines with improved resistance. Zhou et al. (2022) created hybrid rice germplasm with broad-spectrum resistance to rice blast by disabling the susceptibility genes.

9.6.2.2 Obstacles and Future Outlook

Notwithstanding its promise, CRISPR-based epigenome editing encounters obstacles like off-target effects, imperfect modifications and the intricacies of epigenetic regulation. Anticipated progress in CRISPR technology, such as enhanced specificity and the creation of novel epigenetic effectors, is projected to address these constraints and broaden the scope of epigenome editing in rice research (Hilton et al., 2015).

9.6.3 Bioinformatics Instruments for Epigenetic Data Examination

Bioinformatics serves as a proficient method for epigenomic profiling. A multitude of bioinformatics tools and software can analyze these epigenomic data. Numerous internet databases have extensive information regarding these adjustments. Contemporary strategies encompass several analytical and sequence-based techniques to get diverse forms of epigenetic data, which can be utilized to develop pharmaceuticals targeting disorders associated with epigenetic alterations (Ragini et al., 2023).

9.6.3.1 Databases and Resources for Rice Epigenetics

A multitude of databases and resources have been established to advance epigenetics research, including Methylome DB, MethDB, dbHiMo, EpimiR, MeInfoText database, Pubmeth, REBASE, MethPrimerDB, Histone Database and ChromDB (Ragini et al., 2023), for example, Gramene leverages annotated rice genes and genome assemblies. Besides Gramene, several other databases offer genetic variation information, such as indels and SNPs. These databases include RiceVarMap, the Rice SNP-Seek, PmiRKB and the SNP haplotype. The largest collection of rice genetic variants comes from the 3000 Rice Genome Sequencing Project, which is currently managed by IRIC. The project encompasses the Rice SNP-Seek Database, along with phenotypic data and variety-specific passport information (Garg & Jaiswal, 2016).

9.6.3.2 Computational Methods for Epigenetic Network Analysis

Algorithms and computational tools have been created to analyze epigenetic networks and discern regulatory links; for instance, software such as EpiExplorer, compEpiTools, CpGProD, MethBlAST and BiQ Analyser are employed to extract data and conduct mechanistic analyses of epigenetic alterations (Ragini et al., 2023). Machine learning methodologies have been employed to forecast patterns for DNA methylation and their functional implications on gene expression in rice (Hong et al., 2019). Network examination methods, like weighted gene co-expression network analysis (WGCNA), have been utilized to find epigenetic modules linked to stress responses (Langfelder & Horvath, 2008).

9.7 CONCLUSION

This chapter examined the complex role of epigenetic regulation in rice, highlighting its importance in defense mechanisms, stress responses and crop enhancement. Significant findings encompass the primary role of DNA methylation and histone alteration in modulating immunological responses, the epigenetic regulation of disease resistance genes and the dynamic epigenetic alterations prompted by biotic stress. Advanced methodologies, including bisulfite sequencing, ChIP-seq and CRISPR-based epigenome editing, have transformed our capacity to investigate and modify epigenetic markers in rice. Furthermore, bioinformatics technologies and multi-omics integration have yielded profound insights into the intricate regulatory networks that drive rice epigenetics. Epigenetic study yields significant consequences for rice biology and agriculture. Comprehending epigenetic mechanisms unveils novel pathways for the creation of stress-resistant and high-yielding rice cultivars. Epigenetic engineering, especially with CRISPR-based technologies, provides a promising method to accurately regulate gene expression and improve favorable features. Moreover, the integration of epigenetics with genomes and transcriptomics might expedite breeding programs and enhance resistance against climate change and new diseases. These innovations enhance food security and facilitate sustainable agriculture methods. Epigenetic control in rice is a swiftly advancing domain with significant potential to revolutionize agriculture. Despite substantial advancements, problems persist, including comprehending the intricacies of epigenetic connections, resolving ethical dilemmas and creating high-resolution instruments. Future studies must focus on elucidating the processes of epigenetic memory, investigating the enduring impacts of epigenetic alterations and incorporating epigenetics into breeding plans. By confronting these problems and utilizing epigenetic technology, we may fully harness the promise of rice epigenetics to satisfy the increasing demands of global agriculture.

REFERENCES

Ahmad, F., Farman, K., Waseem, M., Rana, R. M., Nawaz, M. A., Rehman, H. M., Abbas, T., Baloch, F. S., Akrem, A., Huang, J., & Zhang, H. (2019). Genome-wide identification, classification, expression profiling and DNA methylation (5mC) analysis of stress-responsive ZFP transcription factors in rice (*Oryza sativa* L.). *Gene*, 718, 144018. https://doi.org/10.1016/j.gene.2019.144018.

Ansori, A. N., Antonius, Y., Susilo, R. J., Hayaza, S., Kharisma, V. D., Parikesit, A. A., Zainul, R., Jakhmola, V., Saklani, T., Rebezov, M., Ullah, M. E., Maksimiuk, N., Derkho, M., & Burkov, P. (2023). Application of CRISPR-Cas9 genome editing technology in various fields: A review. *Narra Journal*, 3(2), e184. https://doi.org/10.52225/narra.v3i2.184.

Ashapkin, V. V., Kutueva, L. I., Aleksandrushkina, N. I., & Vanyushin, B. F. (2020). Epigenetic mechanisms of plant adaptation to biotic and abiotic stresses. *International Journal of Molecular Sciences*, 21(20), 7457. https://doi.org/10.3390/ijms21207457.

Banerjee, A., & Roychoudhury, A. (2017). Epigenetic regulation during salinity and drought stress in plants: Histone modifications and DNA methylation. *Plant Gene*, 11, 199–204.

Banerjee, A., & Roychoudhury, A. (2018). The gymnastics of epigenomics in rice. *Plant Cell Reports*, 7, 25–49.

Bannister, A., & Kouzarides, T. (2011). Regulation of chromatin by histone modifications. *Cell Research*, 21, 381–395. https://doi.org/10.1038/cr.2011.22.

Berr, A., Shafiq, S., & Shen, W. H. (2011). Histone modifications in transcriptional activation during plant development. *Biochimica et biophysica acta*, 1809(10), 567–576. https://doi.org/10.1016/j.bbagrm.2011.07.001.

Bhattacharya, A., & Roychoudhury, A. (2024). Plant non-coding RNAs and their regulation in salinity and heavy metal stress. In S. Penna, V. Kumar, & A. K. Srivastava (Eds.), *Non-Coding RNAs Molecular Tools for Crop Improvement* (pp. 128–147). CRC Press (Taylor & Francis group).

Chen, C., Begcy, K., Liu, K., Folsom, J. J., Wang, Z., Zhang, C., & Walia, H. (2016). Heat stress yields a unique MADS box transcription factor in determining seed size and thermal sensitivity. *Plant Physiology*, 171(1), 606–622. https://doi.org/10.1104/pp.15.01992.

Chen, J., Miao, Z., Kong, D., Zhang, A., Wang, F., Liu, G., Yu, X., Luo, L., & Liu, Y. (2024). Application of CRISPR/Cas9 technology in rice germplasm innovation and genetic improvement. *Genes*, 15(11), 1492. https://doi.org/10.3390/genes15111492.

Chen, Q., Chen, X., Wang, Q., Zhang, F., Lou, Z., Zhang, Q., & Zhou, D. X. (2013). Structural basis of a histone H3 lysine 4 demethylase required for stem elongation in rice. *PLoS Genetics*, 9(1), e1003239. https://doi.org/10.1371/journal.pgen.1003239.

Chen, W., Wang, W., Lyu, Y., Wu, Y., Huang, P., Hu, S., Wei, S., Jiao, G., Sheng, Z., Tang, S., Shao, G., & Luo, J. (2021). OsVP1 activates Sdr4 expression to control rice seed dormancy via the ABA signaling pathway. *The Crop Journal*, 9(1), 68–78.

Cheng, S., Tan, F., Lu, Y., Liu, X., Li, T., Yuan, W., Zhao, Y., & Zhou, D. X. (2018). WOX11 recruits a histone H3K27me3 demethylase to promote gene expression during shoot development in rice. *Nucleic Acids Research*, 46(5), 2356–2369. https://doi.org/10.1093/nar/gky017.

Deleris, A., Halter, T., & Navarro, L. (2016). DNA methylation and demethylation in plant immunity. *Annual Review of Phytopathology*, 54, 579–603. https://doi.org/10.1146/annurev-phyto-080615-100308.

Deng, X., Song, X., Wei, L., Liu, C., & Cao, X. (2016). Epigenetic regulation and epigenomic landscape in rice. *National Science Review*, 3(3), 309–327.

Devanna, B. N., Jain, P., Solanke, A. U., Das, A., Thakur, S., Singh, P. K., Kumari, M., Dubey, H., Jaswal, R., Pawar, D., Kapoor, R., Singh, J., Arora, K., Saklani, B. K., AnilKumar, C., Maganti, S. M., Sonah, H., Deshmukh, R., Rathour, R., & Sharma, T. R. (2022). Understanding the dynamics of blast resistance in rice-magnaporthe oryzae interactions. *Journal of Fungi (Basel, Switzerland)*, 8(6), 584. https://doi.org/10.3390/jof8060584.

Dinkar, V., Pandey, S., Kumar, A., Shiv, A., Lal, D., Bharati, A., ... & Chandra, A. K. (2024). Epigenetic regulations under plant stress: A cereals perspective. *Environmental and Experimental Botany*, 220, 105688. https://doi.org/10.1016/j.envexpbot.2024.105688.

Du, P., Wu, J., Zhang, J., Zhao, S., Zheng, H., Gao, G., Wei, L., & Li, Y. (2011). Viral infection induces expression of novel phased microRNAs from conserved cellular microRNA precursors. *PLoS Pathogens*, 7(8), e1002176. https://doi.org/10.1371/journal.ppat.1002176.

Du, Z., Li, H., Wei, Q., Zhao, X., Wang, C., Zhu, Q., Yi, X., Xu, W., Liu, X. S., Jin, W., & Su, Z. (2013). Genome-wide analysis of histone modifications: H3K4me2, H3K4me3, H3K9ac, and H3K27ac in *Oryza sativa* L. Japonica. *Molecular Plant*, 6(5), 1463–1472. https://doi.org/10.1093/mp/sst018.

Feng, S., Cokus, S. J., Zhang, X., Chen, P. Y., Bostick, M., Goll, M. G., ... & Jacobsen, S. E. (2010). Conservation and divergence of methylation patterning in plants and animals. *Proceedings of the National Academy of Sciences*, 107(19), 8689–8694. https://doi.org/10.1073/pnas.1002720107.

Garg, P., & Jaiswal, P. (2016). Databases and bioinformatics tools for rice research. *Current Plant Biology*, 7, 39–52.

Garg, R., Narayana Chevala, V., Shankar, R., & Jain, M. (2015). Divergent DNA methylation patterns associated with gene expression in rice cultivars with contrasting drought and salinity stress response. *Scientific Reports*, 5, 14922. https://doi.org/10.1038/srep14922.

Ghildiyal, M., & Zamore, P. D. (2009). Small silencing RNAs: An expanding universe. *Nature Reviews Genetics*, 10(2), 94–108. https://doi.org/10.1038/nrg2504.

Goff, S. A., Ricke, D., Lan, T. H., Presting, G., Wang, R., Dunn, M., ... & Briggs, S. (2002). A draft sequence of the rice genome (*Oryza sativa* L. ssp. japonica). *Science*, 296(5565), 92–100. https://doi.org/10.1126/science.1068275.

He, G., Zhu, X., Elling, A. A., Chen, L., Wang, X., Guo, L., Liang, M., He, H., Zhang, H., Chen, F., Qi, Y., Chen, R., & Deng, X. W. (2010). Global epigenetic and transcriptional trends among two rice subspecies and their reciprocal hybrids. *The Plant Cell*, 22(1), 17–33. https://doi.org/10.1105/tpc.109.072041.

Hilton, I. B., D'Ippolito, A. M., Vockley, C. M., Thakore, P. I., Crawford, G. E., Reddy, T. E., & Gersbach, C. A. (2015). Epigenome editing by a CRISPR-Cas9-based acetyltransferase activates genes from promoters and enhancers. *Nature Biotechnology*, 33(5), 510–517. https://doi.org/10.1038/nbt.3199.

Hong, W. J., Kim, Y. J., Chandran, A. K. N., & Jung, K. H. (2019). Infrastructures of systems biology that facilitate functional genomic study in rice. *Rice (New York, N.Y.)*, 12(1), 15. https://doi.org/10.1186/s12284-019-0276-z.

Hou, J., Zheng, X., Ren, R., Shi, Q., Xiao, H., Chen, Z., Yue, M., Wu, Y., Hou, H., & Li, L. (2022). The histone deacetylase 1/GSK3/SHAGGY-like kinase 2/BRASSINAZOLE-RESISTANT 1 module controls lateral root formation in rice. *Plant Physiology*, 189(2), 858–873. https://doi.org/10.1093/plphys/kiac015.

International Rice Genome Sequencing Project. (2005). The map-based sequence of the rice genome. *Nature*, 436(7052), 793–800. https://doi.org/10.1038/nature03895.

Kubo, F. C., Yasui, Y., Ohmori, Y., Kumamaru, T., Tanaka, W., & Hirano, H. Y. (2020). DWARF WITH SLENDER LEAF1 encoding a histone deacetylase plays diverse roles in rice development. *Plant and Cell Physiology*, 61(3), 457–469. https://doi.org/10.1093/pcp/pcz210.

Kumar, S., Seem, K., & Mohapatra, T. (2023). Biochemical and epigenetic modulations under drought: Remembering the stress tolerance mechanism in rice. *Life (Basel, Switzerland)*, 13(5), 1156. https://doi.org/10.3390/life13051156.

Kumar, V., Thakur, J. K., & Prasad, M. (2021). Histone acetylation dynamics regulating plant development and stress responses. *Cellular and Molecular Life Sciences*, 78, 4467–4486.

Langfelder, P., & Horvath, S. (2008). WGCNA: An R package for weighted correlation network analysis. *BMC Bioinformatics*, 9, 559. https://doi.org/10.1186/1471-2105-9-559.

Law, J. A., & Jacobsen, S. E. (2010). Establishing, maintaining, and modifying DNA methylation patterns in plants and animals. *Nature Reviews Genetics*, 11(3), 204–220. https://doi.org/10.1038/nrg2719.

Li, T., Chen, X., Zhong, X., Zhao, Y., Liu, X., Zhou, S., Cheng, S., & Zhou, D. X. (2013). Jumonji C domain protein JMJ705-mediated removal of histone H3 lysine 27 trimethylation is involved in defense-related gene activation in rice. *The Plant Cell*, 25(11), 4725–4736. https://doi.org/10.1105/tpc.113.118802.

Li, W., Chern, M., Yin, J., Wang, J., & Chen, X. (2019). Recent advances in broad-spectrum resistance to the rice blast disease. *Current Opinion in Plant Biology*, 50, 114–120. https://doi.org/10.1016/j.pbi.2019.03.015.

Li, X., Jiang, Y., Ji, Z., Liu, Y., & Zhang, Q. (2015). BRHIS1 suppresses rice innate immunity through binding to monoubiquitinated H2A and H2B variants. *EMBO Reports*, 16(9), 1192–1202. https://doi.org/10.15252/embr.201440000.

Li, X., Zhu, J., Hu, F., Ge, S., Ye, M., Xiang, H., Zhang, G., Zheng, X., Zhang, H., Zhang, S., Li, Q., Luo, R., Yu, C., Yu, J., Sun, J., Zou, X., Cao, X., Xie, X., Wang, J., & Wang, W. (2012). Single-base resolution maps of cultivated and wild rice methylomes and regulatory roles of DNA methylation in plant gene expression. *BMC Genomics*, 13, 300. https://doi.org/10.1186/1471-2164-13-300.

Liu, C., Lu, F., Cui, X., & Cao, X. (2010). Histone methylation in higher plants. *Annual Review of Plant Biology*, 61, 395–420. https://doi.org/10.1146/annurev.arplant.043008.091939.

Liu, H., Wu, Y., Cao, A., Mao, B., Zhao, B., & Wang, J. (2017). Genome-wide analysis of DNA methylation during ovule development of female-sterile rice fsv1. *G3 (Bethesda, Md.)*, 7(11), 3621–3635. https://doi.org/10.1534/g3.117.300243.

Lu, F., Li, G., Cui, X., Liu, C., Wang, X. J., & Cao, X. (2008). Comparative analysis of JmjC domain-containing proteins reveals the potential histone demethylases in *Arabidopsis* and rice. *Journal of Integrative Plant Biology*, 50(7), 886–896. https://doi.org/10.1111/j.1744-7909.2008.00692.x.

Ma, S., Tang, N., Li, X., Xie, Y., Xiang, D., Fu, J., Shen, J., Yang, J., Tu, H., Li, X., & Hu, H. (2019). Reversible histone H2B monoubiquitination fine-tunes abscisic acid signaling and drought response in rice. *Molecular Plant*, 12(2), 263–277. https://doi.org/10.1016/j.molp.2018.12.005.

Mirdar Mansuri, R., Azizi, A. H., Sadri, A. H., & Shobbar, Z. S. (2022). Long non-coding RNAs as the regulatory hubs in rice response to salt stress. *Scientific Reports*, 12(1), 21696. https://doi.org/10.1038/s41598-022-26133-x.

Mirouze, M., & Paszkowski, J. (2011). Epigenetic contribution to stress adaptation in plants. *Current Opinion in Plant Biology*, 14(3), 267–274. https://doi.org/10.1016/j.pbi.2011.03.004.

Nonomura, K., Morohoshi, A., Nakano, M., Eiguchi, M., Miyao, A., Hirochika, H., & Kurata, N. (2007). A germ cell specific gene of the ARGONAUTE family is essential for the progression of premeiotic mitosis and meiosis during sporogenesis in rice. *The Plant Cell*, 19(8), 2583–2594. https://doi.org/10.1105/tpc.107.053199.

Oss-Ronen, L., Sarusi, T., & Cohen, I. (2022). Histone monoubiquitination in transcriptional regulation and its mark on life: Emerging roles in tissue development and disease. *Cells*, 11(15), 2404. https://doi.org/10.3390/cells11152404.

Ragini, Sakshi, Mani, I., & Singh, V. (2023). Applications of bioinformatics in epigenetics. *Progress in Molecular Biology and Translational Science*, 198, 1–13. https://doi.org/10.1016/bs.pmbts.2023.03.023.

Rajkumar, M. S., Shankar, R., Garg, R., & Jain, M. (2020). Bisulphite sequencing reveals dynamic DNA methylation under desiccation and salinity stresses in rice cultivars. *Genomics*, 112(5), 3537–3548. https://doi.org/10.1016/j.ygeno.2020.04.005.

Ramirez-Prado, J. S., Abulfaraj, A. A., Rayapuram, N., Benhamed, M., & Hirt, H. (2018). Plant immunity: From signaling to epigenetic control of defense. *Trends in Plant Science*, 23(9), 833–844. https://doi.org/10.1016/j.tplants.2018.06.004.

Rashid, M. M., Vaishnav, A., Verma, R. K., Sharma, P., Suprasanna, P., & Gaur, R. K. (2022). Epigenetic regulation of salinity stress responses in cereals. *Molecular Biology Reports*, 49(1), 761–772. https://doi.org/10.1007/s11033-021-06922-9.

Sauter, M., Mekhedov, S. L., & Kende, H. (1995). Gibberellin promotes histone H1 kinase activity and the expression of cdc2 and cyclin genes during the induction of rapid growth in deepwater rice internodes. *The Plant Journal*, 7(4), 623–632. https://doi.org/10.1046/j.1365-313x.1995.7040623.x.

Sharma, M., Sidhu, A. K., Samota, M. K., Gupta, M., Koli, P., & Choudhary, M. (2023). Post-translational modifications in histones and their role in abiotic stress tolerance in plants. *Proteomes*, 11(4), 38. https://doi.org/10.3390/proteomes11040038.

Shi, J., Dong, A., & Shen, W. H. (2015). Epigenetic regulation of rice flowering and reproduction. *Frontiers in Plant Science*, 5, 803. https://doi.org/10.3389/fpls.2014.00803.

Shimono, M., Sugano, S., Nakayama, A., Jiang, C. J., Ono, K., Toki, S., & Takatsuji, H. (2007). Rice WRKY45 plays a crucial role in benzothiadiazole-inducible blast resistance. *The Plant Cell*, 19(6), 2064–2076. https://doi.org/10.1105/tpc.106.046250.

Singh, A., & Roychoudhury, A. (2023). RNA Interference (RNAi) technology: An effective tool in plant breeding. In A. Raina, M. R. Wani, R. A. Laskar, N. Tomlekova, & S. Khan (Eds.), *Advanced Crop Improvement, Volume 1* (pp. 309–320). Springer Nature Switzerland AG.

Song, T., Zhang, Q., Wang, H., Han, J., Xu, Z., Yan, S., & Zhu, Z. (2018). OsJMJ703, a rice histone demethylase gene, plays key roles in plant development and responds to drought stress. *Plant Physiology and Biochemistry*, 132, 183–188. https://doi.org/10.1016/j.plaphy.2018.09.007.

Song, X. J., Kuroha, T., Ayano, M., Furuta, T., Nagai, K., Komeda, N., Segami, S., Miura, K., Ogawa, D., Kamura, T., & Suzuki, T. (2015). Rare allele of a previously unidentified histone H4 acetyltransferase enhances grain weight, yield, and plant biomass in rice. *Proceedings of the National Academy of Sciences*, 112(1), 76–81. https://doi.org/10.1073/pnas.1421127112

Sui, P., Jin, J., Ye, S., Mu, C., Gao, J., Feng, H., Shen, W. H., Yu, Y., & Dong, A. (2012). H3K36 methylation is critical for brassinosteroid-regulated plant growth and development in rice. *The Plant Journal*, 70(2), 340–347. https://doi.org/10.1111/j.1365-313X.2011.04873.x.

Sun, J., Song, W., Chang, Y., Wang, Y., Lu, T., & Zhang, Z. (2022). OsLMP1, encoding a deubiquitinase, regulates the immune response in rice. *Frontiers in Plant Science*, 12, e814465. https://doi.org/10.3389/fpls.2021.814465.

Sun, Q., & Zhou, D. X. (2008). Rice jmjC domain-containing gene JMJ706 encodes H3K9 demethylase required for floral organ development. *Proceedings of the National Academy of Sciences of the United States of America*, 105(36), 13679–13684. https://doi.org/10.1073/pnas.0805901105.

Sunkar, R., Li, Y. F., & Jagadeeswaran, G. (2012). Functions of microRNAs in plant stress responses. *Trends in Plant Science*, 17(4), 196–203. https://doi.org/10.1016/j.tplants.2012.01.010.

Swatek, K. N., & Komander, D. (2016). Ubiquitin modifications. *Cell Research*, 26(4), 399–422. https://doi.org/10.1038/cr.2016.39.

Tao, Z., Liu, H., Qiu, D., Zhou, Y., Li, X., Xu, C., & Wang, S. (2009). A pair of allelic *WRKY* genes play opposite roles in rice-bacteria interactions. *Plant Physiology*, 151(2), 936–948. https://doi.org/10.1104/pp.109.145623.

Ullah, F., Xu, Q., Zhao, Y., & Zhou, D. X. (2021). Histone deacetylase HDA710 controls salt tolerance by regulating ABA signaling in rice. *Journal of Integrative Plant Biology*, 63(3), 451–467. https://doi.org/10.1111/jipb.13042.

Vaucheret, H. (2006). Post-transcriptional small RNA pathways in plants: Mechanisms and regulations. *Genes & Development*, 20(7), 759–771. https://doi.org/10.1101/gad.1410506.

Verhoeven, K. J., Jansen, J. J., van Dijk, P. J., & Biere, A. (2010). Stress-induced DNA methylation changes and their heritability in asexual dandelions. *The New Phytologist*, 185(4), 1108–1118. https://doi.org/10.1111/j.1469-8137.2009.03121.x.

Wang, F. Q., Fan, F. J., Li, W. Q., Zhu, J. Y., Wang, J., Zhong, W. G., & Yang, J. (2016). Knock-out efficiency analysis of Pi21 gene using CRISPR/Cas9 in rice. *Chinese Journal of Rice Science*, 30(5), 469. https://doi.org/10.16819/j.1001-7216.2016.6009.

Wang, H., Jiao, X., Kong, X., Liu, Y., Chen, X., Fang, R., & Yan, Y. (2020). The histone deacetylase HDA703 interacts with OsBZR1 to regulate rice brassinosteroid signaling, growth and heading date through repression of Ghd7 expression. *The Plant Journal*, 104(2), 447–459. https://doi.org/10.1111/tpj.14936.

Wang, H., Sun, R., Cao, Y., Pei, W., Sun, Y., Zhou, H., Wu, X., Zhang, F., Luo, L., Shen, Q., & Xu, G. (2015). *OsSIZ1*, a SUMO E3 ligase gene, is involved in the regulation of the responses to phosphate and nitrogen in rice. *Plant and Cell Physiology*, 56(12), 2381–2395. https://doi.org/10.1093/pcp/pcv162.

Wang, M., Tang, D., Luo, Q., Jin, Y., Shen, Y., Wang, K., & Cheng, Z. (2012). BRK1, a Bub1-related kinase, is essential for generating proper tension between homologous kinetochores at metaphase I of rice meiosis. *The Plant Cell*, 24(12), 4961–4973. https://doi.org/10.1105/tpc.112.105874.

Wang, W., Zhao, X., Pan, Y., Zhu, L., Fu, B., & Li, Z. (2011). DNA methylation changes detected by methylation-sensitive amplified polymorphism in two contrasting rice genotypes under salt stress. *Journal of Genetics and Genomics = Yi chuan xue bao*, 38(9), 419–424. https://doi.org/10.1016/j.jgg.2011.07.006.

Wang, X., Ju, Y., Wu, T., Kong, L., Yuan, M., Liu, H., Chen, X., & Chu, Z. (2024). The clade III subfamily of OsSWEETs directly suppresses rice immunity by interacting with OsHMGB1 and OsHsp20L. *Plant Biotechnology Journal*, 22(8), 2186–2200. https://doi.org/10.1111/pbi.14338.

Wang, Y., Wang, F., Lu, H., Lin, R., Liu, J., Liu, Y., Xu, J., Wu, Y., Wang, Z., Zhou, M., Mo, X., Wu, Z., Shou, H., Zheng, S., & Mao, C. (2023). Rice chromatin protein OsHMGB1 is involved in phosphate homeostasis and plant growth by affecting chromatin accessibility. *The New Phytologist*, 240(2), 727–743. https://doi.org/10.1111/nph.19189.

Wibowo, A., Becker, C., Marconi, G., Durr, J., Price, J., Hagmann, J., Papareddy, R., Putra, H., Kageyama, J., Becker, J., Weigel, D., & Gutierrez-Marcos, J. (2016). Hyperosmotic stress memory in Arabidopsis is mediated by distinct epigenetically labile sites in the genome and is restricted in the male germline by DNA glycosylase activity. *eLife*, 5, e13546. https://doi.org/10.7554/eLife.13546.

Xing, J., Wang, T., & Ni, Z. (2015). Epigenetic regulation of iron homeostasis in Arabidopsis. *Plant Signaling & Behavior*, 10(12), e1064574. https://doi.org/10.1080/15592324.2015.1064574.

Xing, M. Q., Zhang, Y. J., Zhou, S. R., Hu, W. Y., Wu, X. T., Ye, Y. J., Wu, X. X., Xiao, Y. P., Li, X., & Xue, H. W. (2015). Global analysis reveals the crucial roles of DNA methylation during rice seed development. *Plant Physiology*, 168(4), 1417–1432. https://doi.org/10.1104/pp.15.00414.

Yan, H., Kikuchi, S., Neumann, P., Zhang, W., Wu, Y., Chen, F., & Jiang, J. (2010). Genome-wide mapping of cytosine methylation revealed dynamic DNA methylation patterns associated with genes and centromeres in rice. *The Plant Journal*, 63(3), 353–365. https://doi.org/10.1111/j.1365-313X.2010.04246.x.

Yin, M., Wang, S., Wang, Y., Wei, R., Liang, Y., Zuo, L., Huo, M., Huang, Z., Lang, J., Zhao, X., Zhang, F., Xu, J., Fu, B., Li, Z., & Wang, W. (2024). Impact of abiotic stress on rice and the role of DNA methylation in stress response mechanisms. *Plants (Basel, Switzerland)*, 13(19), 2700. https://doi.org/10.3390/plants13192700.

Yokoo, T., Saito, H., Yoshitake, Y., Xu, Q., Asami, T., Tsukiyama, T., Teraishi, M., Okumoto, Y., & Tanisaka, T. (2014). Se14, encoding a JmjC domain-containing protein, plays key roles in long-day suppression of rice flowering through the demethylation of H3K4me3 of RFT1. *PLoS One*, 9(4), e96064. https://doi.org/10.1371/journal.pone.0096064.

Yong-Villalobos, L., González-Morales, S. I., Wrobel, K., Gutiérrez-Alanis, D., Cervantes-Peréz, S. A., Hayano-Kanashiro, C., Oropeza-Aburto, A., Cruz-Ramírez, A., Martínez, O., & Herrera-Estrella, L. (2015). Methylome analysis reveals an important role for epigenetic changes in the regulation of the *Arabidopsis* response to phosphate starvation. *Proceedings of the National Academy of Sciences of the United States of America*, 112(52), E7293–E7302. https://doi.org/10.1073/pnas.1522301112.

Yu, M. H., Liao, W. C., & Wu, K. (2025). Histone methylation in plant responses to abiotic stresses. *Journal of Experimental Botany*, eraf058. Advance online publication. https://doi.org/10.1093/jxb/eraf058.

Zemach, A., Kim, M. Y., Silva, P., Rodrigues, J. A., Dotson, B., Brooks, M. D., & Zilberman, D. (2010). Local DNA hypomethylation activates genes in rice endosperm. *Proceedings of the National Academy of Sciences of the United States of America*, 107(43), 18729–18734. https://doi.org/10.1073/pnas.1009695107.

Zemach, A., McDaniel, I. E., Silva, P., & Zilberman, D. (2010). Genome-wide evolutionary analysis of eukaryotic DNA methylation. *Science*, 328(5980), 916–919. https://doi.org/10.1126/science.1186366.

Zhang, H., Lang, Z., & Zhu, J. K. (2018). Dynamics and function of DNA methylation in plants. *Nature Reviews Molecular Cell Biology*, 19(8), 489–506. https://doi.org/10.1038/s41580-018-0016-z.

Zhang, Q., Zhong, W., Zhu, G., Cheng, L., Yin, C., Deng, L., Yang, Y., Zhang, Z., Shen, J., Fu, T., Zhu, J. K., & Zhao, L. (2024). aChIP is an efficient and sensitive ChIP-seq technique for economically important plant organs. *Nature Plants*, 10(9), 1317–1329. https://doi.org/10.1038/s41477-024-01743-7.

Zhang, Y., Chen, M., Zhou, S., Lou, Y., & Lu, J. (2021). Silencing an E3 ubiquitin ligase gene *OsJMJ715* enhances the resistance of rice to a piercing-sucking herbivore by activating ABA and JA signaling pathways. *International Journal of Molecular Sciences*, 22(23), e13020. https://doi.org/10.3390/ijms222313020.

Zhao, J., Li, M., Gu, D., Liu, X., Zhang, J., Wu, K., Zhang, X., da Silva, J. A. T., & Duan, J. (2016). Involvement of rice histone deacetylase HDA705 in seed germination and in response to ABA and abiotic stresses. *Biochemical and Biophysical Research Communications*, 470(2), 439–444. https://doi.org/10.1016/j.bbrc.2016.01.016.

Zhao, J., Zhang, J., Zhang, W., Wu, K., Zheng, F., Tian, L., Liu, X., & Duan, J. (2015). Expression and functional analysis of the plant-specific histone deacetylase HDT701 in rice. *Frontiers in Plant Science*, 5, e764. https://doi.org/10.3389/fpls.2014.00764.

Zhao, J., Zhang, W., da Silva, J. A. T., Liu, X., & Duan, J. (2021). Rice histone deacetylase HDA704 positively regulates drought and salt tolerance by controlling stomatal aperture and density. *Planta*, 254(4), 79. https://doi.org/10.1007/s00425-021-03729-7.

Zhao, L., Xie, L., Zhang, Q., et al. (2020). Integrative analysis of reference epigenomes in 20 rice varieties. *Nature Communications*, 11, 2658. https://doi.org/10.1038/s41467-020-16457-5.

Zhao, W., Wang, X., Zhang, Q., Zheng, Q., Yao, H., Gu, X., Liu, D., Tian, X., Wang, X., Li, Y., & Zhu, Z. (2022). H3K36 demethylase JMJ710 negatively regulates drought tolerance by suppressing MYB48-1 expression in rice. *Plant Physiology*, 189(2), 1050–1064. https://doi.org/10.1093/plphys/kiac095.

Zhou, B., Peng, K., Zhaohui, C., Wang, S., & Zhang, Q. (2002). The defense-responsive genes showing enhanced and repressed expression after pathogen infection in rice (*Oryza sativa* L.). *Science in China. Series C, Life Sciences*, 45(5), 449–467. https://doi.org/10.1360/02yc9050.

Zhou, Y., Xu, S., Jiang, N., Zhao, X., Bai, Z., Liu, J., Yao, W., Tang, Q., Xiao, G., Lv, C., Wang, K., Hu, X., Tan, J., & Yang, Y. (2022). Engineering of rice varieties with enhanced resistances to both blast and bacterial blight diseases via CRISPR/Cas9. *Plant Biotechnology Journal*, 20(5), 876–885. https://doi.org/10.1111/pbi.13766.

Zhu, J. K. (2016). Abiotic stress signaling and responses in plants. *Cell*, 167(2), 313–324. https://doi.org/10.1016/j.cell.2016.08.029.

Zhu, J. K., Liu, J., & Xiong, L. (1998). Genetic analysis of salt tolerance in arabidopsis. Evidence for a critical role of potassium nutrition. *The Plant Cell*, 10(7), 1181–1191. https://doi.org/10.1105/tpc.10.7.1181.

10 Insight into the Epigenetic Regulation of Cereal Crops in Controlling Physiology, Development and Stress Defense

Bornita Bose and Harshata Pal

10.1 INTRODUCTION

Cereals, as *Gramineae* family members, particularly sorghum, rice, barley, oat, rye, millet, wheat, maize and others, have been estimated to contribute 51% of calories along with a total of 47% protein as per reports published by the Food and Agricultural Organization (FAO). In addition, the FAO has reported an estimated global production value of cereals worth 2710.7 million tons for the 2019/2020 season (Adeboyejo et al., 2023). However, it is no surprise that the worldwide productivity and yield of cereal crops are being rapidly observed to decline due to global climatic changes, ultimately threatening global food security. These drastic climatic changes are majorly held accountable for imposing harsh environmental stress conditions such as salinity, drought, flood, cold, heat, pathogen infections, heavy metals and many others on food crops which demand crops to adapt morphological, physiological, biochemical and transcriptional pathways mostly governed by the major phytohormones (abscisic acid, gibberellins, ethylene, auxin, cytokinin, brassinosteroids, jasmonic acid and strigolactones) and triggering stress-induced genes regulated by transcription factors (TFs). In addition, recent research suggests that modifications in epigenetic patterns involving DNA methylation, modifications in histone proteins and non-coding RNAs elaborate the convoluted regulatory mechanisms governed by spatial and temporal repression, inactivation or activation of genes that are essential in defining varied developmental processes along with diverse phenotypic traits inherited over generations for defense against a multitude of biotic and abiotic stresses. These provide a scope to unveil the epigenetic plant adaptations to enhance stress tolerance and establish an effective strategy for developing and characterizing climate-resilient crops of agronomic importance (Choudhary & Muthamilarasan, 2022). This would allow us to overcome the stress-related obstacles largely faced by cereal crops.

The word epigenetics, a combination of the phrases "epigenesis" and "genetics", was invented by Waddington. According to him, epigenetics can be defined as "the study of the interaction between a gene and its product that produces the phenotype". The analysis, assessment and quantification study of epigenetic variations have meaningful impacts on key agronomical characteristics such as energy-use efficiency, yield components, respiration and seed quality, as well as developmental processes in plant systems and responses to environmental restrictions (Gupta & Salgotra, 2022). Sophisticated epigenetic strategies mitigated by different cereal crops as a coping mechanism to harsh and stressful environment differ at varied developmental stages throughout the lifetime of a crop; for instance, the vegetative state of a cereal crop is characterized by a reduction in light perception along with disturbances in carbon assimilation and transpiration-related processes, whereas malformation of female and male gametophytes, abortions and affected sterility mark the reproductive phase of plant development. In addition, the well-coordinated pathways involved in the transition of vegetative meristem to floral meristem are greatly disrupted in crops exposed to abiotic stresses (Begcy & Dresselhaus, 2018).

Chromatin state regulation of any given gene determines whether it is "off" or "on". A diverse range of biological processes, including DNA methylation, post-translational histone alterations, chromatin remodeling and non-coding RNAs, affect how accessible the chromatin is to the transcriptional machinery. Variations in phenotype and altered expression of genes are correlated with heritable epigenetic modifications influenced by environmental factors. These phenotypic differences improve different crop production qualities through epi-breeding. By altering the chromatin state, which supports plant growth and determines phenotypic response, epigenetic alterations affect gene expression (Kakoulidou et al., 2021). Furthermore, studies in this area highlight the integrated participation of chromatin modulations at specific loci such as DNA methylation, histone modifications and alterations in histone-DNA interactions, along with the regulation of chromatin modifiers during various stages of growth and development. Such epigenetic regulations mediate transcriptional gene control, meticulous chromatin state characterization and quantification,

DOI: 10.1201/9781003589914-12

and physiological phenotype of cereals, thereby naturally elevating agronomic traits, productivity and adaptability. These transcriptional regulations are fundamentally correlated with stress-responsive genes undergoing DNA methylation modifications which in turn determine plant responses to stress conditions.

By definition, DNA methylation or DNA-Me refers to the covalent methyl group addition at the fifth carbon and fourth nitrogen positions of cytosine (C) bases mediated by four DNA MTases, namely, DNA methyltransferase 1 (MET1), DNA methyltransferase 2 (DNMT2), domain-rearranged methyltransferase (DRM) and chromomethylase (CMT). In particular, plant-specific CMT has a distinctive N-terminal bromo-adjacent homology (BAH) domain and chromodomain (chromatin organization modifier) (Dinkar et al., 2024). Reports suggest that methylation in DNA regions at three different sequence sites, particularly, CG, CGH and CHH (where H indicates either A, T or C), is an important epigenetic mark investigated during vegetative development in cereals. While CG is predominantly found in genes and transposable elements (TEs), CGH and CHH modifications are abundant in TEs. Interestingly, exposure to abiotic stresses might cause inheritable expression patterns in crops leading to a phenomenon known as "epialleles", a condition where identical genes exhibit characteristic expression patterns because of epigenetic changes. Modification patterns in DNA methylation leading to the novel epiallele formation may assist in inducing plant adaptation to abiotic stress, development and improvement. Furthermore, environmental perturbations inversely alter genome-wide gene expression profiles by either hypermethylation or hypomethylation of DNA (Kumar & Singh, 2016). Research shows that the genome of rice (*Oryza sativa*) crops harbors four times higher total cytosine levels in comparison to *Arabidopsis* depending on the diverse abundance of TEs (Nobuta et al., 2007). Studies, which are prominently distinguishable from that of *Arabidopsis*, have also elucidated that preferential downstream DNA methylation, away from the transcription start site, is a significant signature in more than 50% of the protein-coding genes involving the promoter associated with transcriptional repression (Li et al., 2008; Banerjee & Roychoudhury, 2018).

Nucleosome, comprising DNA wound around a histone octamer (two copies of four histone proteins, viz., H2A, H2B, H3 and H4), forms the major fundamental unit. Histone proteins face post-translational histone N-terminal modifications (PTMs) including methylation, acetylation, ubiquitination and phosphorylation, playing vital roles in regulating epigenetic modifications. These modifications can alter the interface linking histones and DNA along with the interaction of regulatory proteins to DNA thereby influencing the dynamics of transcription of genes and chromatin shape. These mediate a diverse range of important biological mechanisms and processes that involve apoptosis, DNA replication, DNA repair and gene expression among many others. Most importantly, two important epigenetic marks, histone acetylation and methylation, are considered to be essentially involved in the genome compartmentalization of heterochromatin and euchromatin which determine developmental stages in plants. These alterations have been discussed with transcriptional activation and repression (Banerjee & Roychoudhury, 2017).

Histone modifications at mono-, di- and tri-methylated lysine and arginine have been well-studied. Histone H3 – 9th lysine residue (H3K9) methylation, along with H3K27 di-methylation, in particular, results in chromatin condensation, rendering DNA either transcriptionally inactive or transcriptionally active. On the other hand, H3K4 tri-methylation is noted to cause transcriptional activation. Such histone modifications, though said to be dynamic, have tendencies to enhance the expression of genes in stressed crops, reverting to their basal level of expression once they are relieved of stress. Maize RNA interference (RNAi) lines and knockout mutants of histone deacetylase, when triggered by stress, showed increased histone acetylation and prominent alteration in histone methylation patterns, which resulted in the activation of silenced genes (Kumar & Singh, 2016). Intriguingly, prominent epigenetic mechanisms associated with long-duration vernalization periods have been well documented where lysine 27 tri-methylation of histone 3 (H3K27me3) in association with a repressed state has been documented to be vital during vernalization in cereal crops and other plants (Diallo et al., 2012). Cross-talk between epigenetic patterns (DNA-histone ME) and gene expression is prominent when cytologically stained rice shoots and cultured cells portray densely methylated DNA TEs, devoid of any H3K4 methylation, whereas di- and/or tri-methylated H3K4 is observed along with methylated DNA in the majority of the protein-coding genes. This has been associated with the fact that suppressed transcription is not related to H3K4 methylation but to DNA-Me alone. Prominent H3K4me3 is associated with highly active transcription levels, while H3K4me2 results in moderate transcription; a high H3K4me3:H3K4me2 ratio has been seen to regulate transcriptional activity positively. On the contrary, the occurrence of H3K4 and DNA-Me together has been recorded to slightly reduce transcription (Li et al., 2008).

10.2 DNA METHYLATION AIDS IN CAUSING PHYSIOLOGICAL, BIOCHEMICAL, MORPHOLOGICAL AND TRANSCRIPTIONAL ALTERATIONS IN RESPONSE TO VARIOUS STRESS ENVIRONMENT

Scientific literature analysis indicates that the rice genome, in contrast to *Arabidopsis*, is vastly characterized by a high GC content and a greater heterochromatin proportion with an erratic distribution by TEs, which provides DNA methyltransferases with a higher potential for methylation of cytosine residues in a sequence-specific manner. This, in turn, emphasizes the important effects of rice methyltransferases on methylated repetitive sequences or TEs that actively contribute to developmental alterations by influencing gene expression (Figure 10.1).

In addition, distinct genomic environment during duplication events along with genome sizes in the two species also assist this phenomenon (Deng et al., 2016). Interestingly, drastic environmental changes, particularly temperature such as heat stress (HS), modify DNA methylation patterns, affecting seed development significantly, including delayed seed germination and modulating involved genes in phytohormone metabolism in rice grain filling. In the rice genome, the de novo RNA-directed DNA methylation (RdDM) pathway has been established to be governed by small RNAs (sRNAs) (Moritoh et al., 2012). With regard to the RdDM pathway, it aids in DNA-Me through either canonical or non-canonical pathways. The canonical RdDM pathway utilizes RNA-DEPENDENT RNA POLYMERASE 2 (RDR2) to transform single-stranded RNAs (ssRNAs), synthesized by RNA polymerase IV (Pol IV), into double-stranded RNAs (dsRNAs). These dsRNAs, which are cleaved by DICER-LIKE 3 (DCL3) into 24-nucleotide length small interfering RNAs (siRNAs), get incorporated into ARGONAUTE 6 (AGO6) and ARGONAUTE 4 (AGO4) (Zhang et al., 2018). These involved RNA Pol IV are in turn recruited by chromatin remodelers such as the CLASSY and SAWADEE HOMEODOMAIN HOMOLOG 1 (SHH1) family (Zhou et al., 2018). Similarly, the non-canonical RdDM pathways are guided by the small RNAs (sRNAs) (derived from viruses as well as Pol II transcripts) (Cuerda-Gil & Slotkin, 2016). Once the slicing of dsRNAs into 21–24 nucleotide sRNAs mediated by various DCL proteins has occurred, they are integrated with AGO proteins. These bind to complementary RNA, leading to cleavage or repression, causing post-transcriptional gene silencing (PTGS) (de Alba et al., 2013). After the immediate integration of these sRNAs takes place into AGO4/AGO6, complementary DNA sequences are allowed for methylation by DRM2 and Pol V, silencing the genes (Cuerda-Gil & Slotkin, 2016).

In alliance with DNA methyltransferase *OsDRM2* (a rice ortholog of *Arabidopsis* DOMAINS REARRANGED METHYLASE 2), which is responsible for methylation at the non-CG and CG regions, targeted interference of *OsDRM2* compromised vegetative as well as reproductive development, causing phenotypic differences in tiller number reduction, semi-dwarf stature, morphological abnormalities in spikelet and panicle, complete sterility and delayed or no heading, accompanied by an approximate methylation rate reduced to 14% in the cytosine residues (Arikit et al., 2013). In a comparative analysis to that of *Arabidopsis met1* mutants, impairment in seed development of rice crops has been found to be interrelated with seedling lethality, which could be precisely a critical consequence of a null mutation in one of the major CG methylases, *OsMET1b* (*osmet1-2*) (Hu et al., 2014). *OsDRM2*, in correlation with rice chromomethylase 3, *OsCMT3*, has been documented to be responsible for methylation in non-CG regions, which plays an indispensable role in rice development. The loss of function of *OsCMT3a* alone causes a wide spectrum of gene and TE suppression, which triggers pleiotropic phenotypes, especially during the reproductive stage (Deng

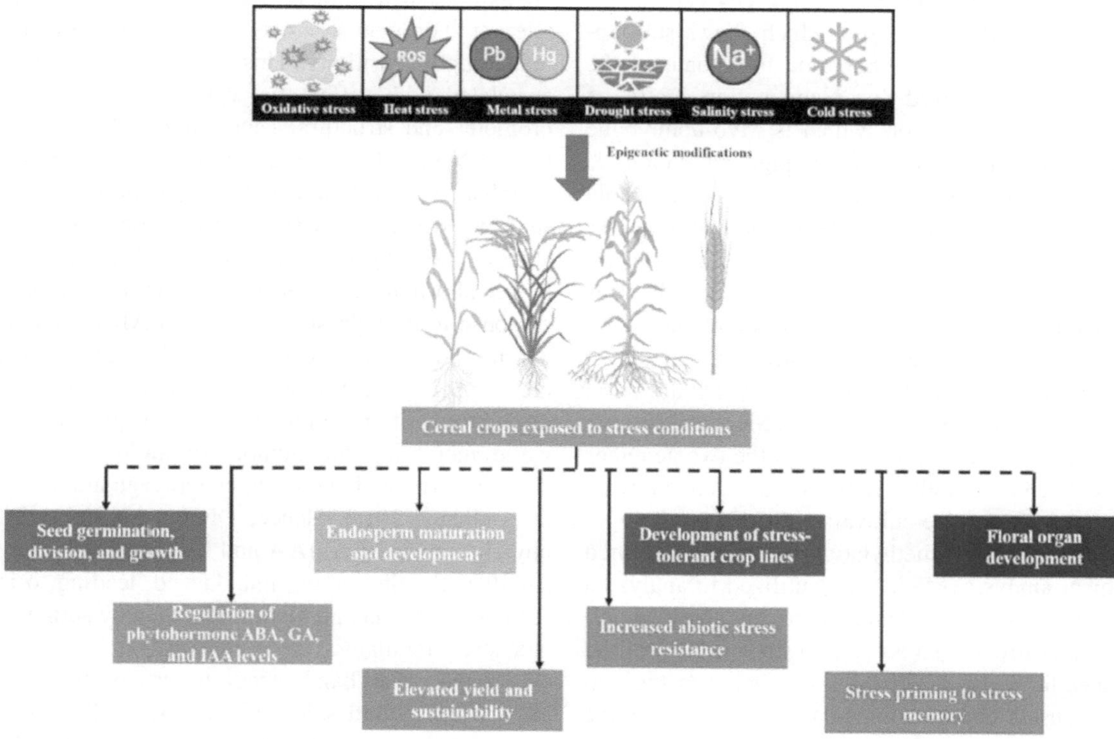

FIGURE 10.1 Overview of the epigenetic modifications mediating physiological, morphological, biochemical and transcriptional mechanisms in cereal crops on exposure to stress conditions.

et al., 2016). Besides, *CMT3* is recruited to the DNA by methylated histones in physical collaboration with heterochromatin protein 1 (*HP1*) to induce cytosine methylation. Studies suggest that the Cytokinin-Oxidase 2.1 (*CKX2.1*) gene promoter region of barley plants exhibits increased DNA-Me when compelled to terminal stress during the grain filling period (Surdonja et al., 2017). When exposed to water-deficient and increased temperature conditions during seed maturation, reduced/regulated seed dormancy has been observed to be correlated with a decline in sensitivity to phytohormone signaling molecules including auxin (IAA) and ABA (Singh et al., 2013). This decline, when confronted with stress, is attributed to the misregulation of numerous genes related to this hormone biosynthesis, particularly auxin response factors (*ARFs*), SNF1-related protein kinase 2 (*SNF1*), protein phosphatase 2C (*PP2C*), lipid phosphate phosphatase 2 (*LPP2*) and ABA insensitive 5 (*ABI5*). In conclusion, it can be said that the involved RdDM pathways are most likely affected initially, causing abnormal regulation of the aforementioned genes. However, further experimental data can highlight the fundamental reasons for the reduction in dormancy period during harsh stressful environment (Yu et al., 2016).

Barbara McClintock identified the epigenetic process of transposon silencing in maize. It lessens TE-induced genomic disruption and transposition. High levels of DNA methylation are predominantly found in silenced TEs, whereas active TEs often exhibit hypomethylation. Through post-translational histone modifications, DNA/RNA methylation and ncRNAs, epigenetic regulation affects diverse plant functions, including development regulation, response to external stimuli and adaptability. *ZmMRP4* gene methylation causes lpa1-241 phenotype, which shows substantially high concentrations of inorganic phosphate in the seed. Another study showed that pigmentation decline is linked to *p1* gene methylation, which is Myb-homologue and controls the manufacture of red pigment (Gupta & Salgotra, 2022). Early research on maize revealed that cold stress increased the expression of the retrotransposon-like gene *ZmMI1*, which was related to a decline in nucleosomal DNA methylation. Research conducted on F1 hybrids and their parents in maize showed that parents collected more DNA methylation sites under stressful conditions like intensive planting than their hybrids, which are more resistant to changes in DNA methylation. Significant variations in the methylation patterns between the two genomes were found in a latest study examining the genome-wide DNA-Me status of two rice cultivars with varying drought tolerance. Specifically, methylation-sensitive amplified polymorphism analysis (MSAP) was utilized to analyze a drought-tolerant line (DK151) and its drought-sensitive parent (IR64) under drought stress and no stress. 12.1% of the total site-specific methylation variations between the two lines were caused by developmental and tissue-specific alterations in DNA methylation and demethylation that were caused by drought. Remarkably, 29% of drought-induced methylation alterations remained unchanged after recovery, while 70% of them were reversed. These findings imply that response to dehydration conditions in rice is controlled by DNA methylation alterations, most likely through the stress-responsive gene activation or deactivation that causes drought adaptation (Wang et al., 2011).

Research study on heat-exposed barley (*Hordeum vulgare* L.) plants at 25°C during grain filling exhibited promoted seed germination compared to that of control (15°C). Based on theoretical study, gibberellic acid (GA)- and NADPH oxidase-generated reactive oxygen species (ROS) induce seed germination, whereas phytohormone abscisic acid (ABA) inhibits seed germination under normal conditions. On exposure to heat stress (HS), ABA biosynthesis (*HvNCED1* and *2*)-related genes are suppressed substantially, and genes encoding for ABA catabolism (*HvABA8'OH*) are upregulated significantly after imbibition. Genes related to NADPH oxidase (*HvRboh*) and GA biosynthesis (*HvHA20ox*, *HvGA3ox*) are upregulated at the same time. Further investigations showed that while hypomethylation occurs at the promoters of the *HvABA8'OH1*, *HvABA8'OH3*, *HvGA3ox2* and *HvRbohF2* genes of HS-treated barley seeds, the *HvNCED* gene promoter is marked with hypermethylation. In conclusion, it can be said that a diverse range of patterns can be observed in the DNA methylation of genes associated with promoting seed germination during HS-exposed barley grain filling (Sakai et al., 2022), as has been depicted in Table 10.1.

On the other hand, endosperm imprinting and maternal allele demethylation-related DNA glycosylase, close to Demeter (*DME*), have been found to influence stress response in barley. Elevated levels of *HvDME* methylation are associated with increased tolerance in barley varieties, tolerant to drought stress, as compared to the variety showing drought sensitivity. This variation has been relatively correlated with differential expression levels of specific promoter and structural genes of *HvDME* owing to varied DNA methylation (Kapazoglou et al., 2013). Moreover, established information on the regulation of gene expression by differential ABA concentration in ovaries during abiotic stress suggests that it is one of the main causes of flower abortion in stressed female crops; for instance, in response to drought stress, a rise in ABA levels in maize ovaries induces significantly delayed female organ development. ABA concentrations in the maize embryo sac during drought conditions inhibited style receptivity and growth while increasing the number of antipodal cells, which can be correlated with epigenetic regulations (Begcy & Dresselhaus, 2018). Hence, DNA methylation influences phytohormone ABA, IAA and GA concentrations that are significantly altered and imbalanced, leading to improper seed development and maturation, thereby acting as prominent stress regulators.

On the other hand, shoot tissues of the salt-tolerant (Kharchia-65) and salt-sensitive (HD-2329) genotypes of wheat (*Triticum aestivum*) have been reported to show considerably higher (approx. ~30–40%) levels of basal methylation than those of the root tissues. As roots primarily

TABLE 10.1
List of affected genes by DNA methylation under various stress conditions

Cereal crops	Affected genes	Stress condition	References
Rice	• MADS-box protein (*MADS87*) • MADS-box protein (*MADS69*) • MADS-box protein (*AGL36*)	Heat stress	Folsom et al. (2014)
Wheat	• Miniature inverted-repeat transposable elements (*MITEs*)	Heat stress	Li et al. (2014)
	• Protein phosphatase 2C (*PP2C*) • Auxin response factors (*ARFs*) • SNF1-related protein kinase 2 (*SNF1*) • Lipid phosphate phosphatase 2 (*LPP2*) • ABA insensitive 5 (*ABI5*)	Unfavorable stress environment	Yu et al. (2016)
Barley	• Demeter (*DME*)	Drought stress	Kapazoglou et al. (2013)
	• *CKX2.1*	Terminal stress	Surdonja et al. (2017)
	• ABA biosynthesis genes (*HvABA8'OH1*, *HvABA8'OH3*) • GA biosynthesis-related genes (*HvHA20ox*, *HvGA3ox*) • NADPH oxidase gene (*HvRbohF2*)	Heat stress	Sakai et al. (2022)
Maize	• Retrotransposon-like gene *ZmMI1*	Cold stress	Wang et al. (2011)

encounter waves of salt stress, it restricts Na^+ entry into the cells and tissues, which has been related to the prevalence of hypermethylation in the root tissues of the Kharchia-65 genotype. On the contrary, an upregulation in the levels of methylation in the salt-sensitive genotype could not have any effect on restricting Na^+ uptake into the tissues from the soil. Hence, it can be extrapolated that higher cytosine methylation levels induced by salt stress result in the establishment of the salt-tolerant genotype at the genomic level, distinct from that of the salt-sensitive phenotype through regulation of epigenetics and downregulating high-affinity potassium transporter (HKT)-encoding genes (Kumar & Singh, 2016). One of the most salt-tolerant genotypes, Kharchia-65, is demonstrated to employ physiological and biochemical features to endure stress and has gained genetic/epigenetic pathways to regulate *HKT* gene expression. To mediate structural and molecular changes, such as H_2O_2 accumulation, transcription factor induction and genetic and epigenetic control of gene expression through transcriptional and/or translational reprogramming for protective defense mechanisms, stress is recognized through the cell membrane and transmitted to diverse inducers. The salt-tolerant genotype can be applied to identify alleles or epialleles for the differentially expressed genes, followed by validation by EpiRILs or mapping populations for potential utilization in the stress breeding program (Kumar et al., 2017).

10.3 HISTONE MODIFICATIONS PLAY A CRUCIAL ROLE IN BRINGING ABOUT SIGNIFICANT CHANGES IN STRESS-RESPONSIVE GENES

A significant phenomenon termed the "epigenetic histone code" refers to the complex patterning of site-specific histone modifications at different amino acid sites which brings about an alteration in chromatin states as a survival response to external or environmental signals that regulate influence on transcriptional activity and impose their inheritance onto the offspring (Figure 10.2).

Among the various known modifications, the best-studied abiotic stress response-associated ones are the histone methylation/demethylation and histone acetylation/deacetylation reversible modifications at separate loci. Histone methylation/demethylation has been said to be controlled by certain histone methyltransferases (HMTs) and histone demethylases (HDMs), respectively. In contrast, histone acetylation, often correlated with gene activation, is catalyzed by histone acetyltransferases (HATs) and histone deacetylases (HDACs)-mediated histone deacetylation is, more often than not, associated with silencing of genes (Chinnusamy & Zhu, 2009). These enzymes that perform methylation mark addition, removal or reading out are termed writers, erasers and readers. The enzymes known as writers are dedicated to post-translational changes in proteins. The enzymes that remove post-translational modifications from proteins are known as erasers. The proteins called readers identify and attach to a substrate that has undergone post-translational modification (Xu et al., 2016). Histone methylation does not alter the electrostatic characteristics of histones; instead, it results in amino acid charging being retained (Ueda & Seki, 2020). Gene expression is altered, with both activating and repressive effects, primarily at the lysine and arginine residues of the H3 and H4 histones. Mono-methylation, di-methylation and tri-methylation of the lysine residues are all possible. Several methylation patterns influence gene expression in various ways (Zhao et al., 2019).

When confronted with abiotic stress such as heat, UV irradiation, drought, oxidative stress, high salinity and cold,

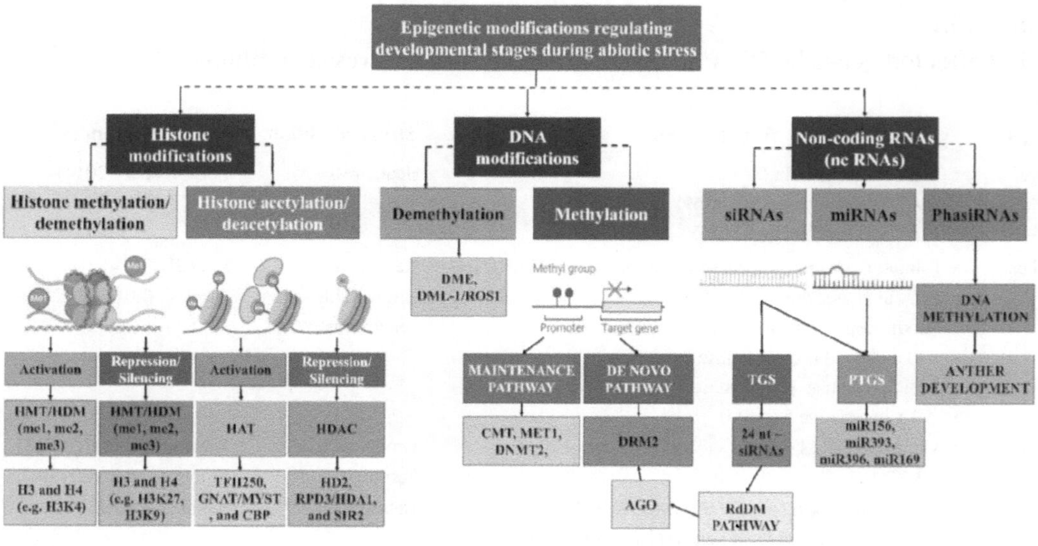

FIGURE 10.2 Epigenetic modifications regulating developmental stages during abiotic stress.

histone methylation/demethylation/acetylation/deacetylatio n patterns face alteration within the coding and promoter regions of the gene, resulting in either gene silencing or gene activation. For instance, H3-lysine 4 (H3K4me3) tri-methylation, mediated by histone methyltransferase from the Trithorax (TrxG) family, causes upregulated transcriptional levels in comparison to H3-lysine 27 (H3K27me3) tri-methylation catalyzed by a specific polycomb group (PcG) family member methyltransferase, resulting in gene repression. Additionally, various abiotic stress conditions also stimulate increased production of specific phytohormones, namely GA, ABA, salicylic acid (SA), ethylene and jasmonic acid (JA), which in turn directly or indirectly impact gene expression as an adaptive response to stress by crops. Through the study of expression patterns, HAT/HDAC-regulated histone acetylation/deacetylation along with HMT/HDM-mediated methylation/demethylation have been proposed to synergistically integrate with hormone signaling to epigenetically modulate several stress-responsive genes (Chinnusamy & Zhu, 2009).

10.3.1 Histone Methylation/Demethylation

The above-mentioned key multisubunit histone-modifying enzyme complex is polycomb repressive complex 2 (PRC2), which promotes epigenetic histone H3 lysine 27 (H3K27) tri-methylation primarily responsible for gene silencing (Liu & Liu, 2022). Several PRC2 members serve as major regulators of endosperm development, and the combination of their protein subunits contributes to seed formation, at the epigenetic level in cereals. The only PRC2 complex member in cereals with an expression pattern unique to endosperms is FIE1 (Tonosaki & Kinoshita, 2015). The sperm cell has methylation on the *FIE1* promoter, whereas the central cell does not. In cereals and other plants, the endosperm inherits an asymmetric pattern of parental DNA methylation following fertilization. Maternally derived alleles are expressed, but paternally derived *FIE1* is repressed. However, maternal and paternal *FIE1* alleles are severely suppressed during the early stages of seed formation following HS (Chen et al., 2016). *FIE1* inhibition in rice led to aberrant endosperm formation, providing compelling evidence that endosperm development in cereals is controlled by PRC2 through DNA methylation and histone modifications. The seed size of rice plants exposed to heat stress in the early stages of seed development was reduced coherently. Altered *FIE1* expression levels were associated with this decline. In heat-stressed rice seeds, several MADS-box genes, including *OsMADS87*, *OsMADS69* and *OsAGL36*, were improperly regulated in addition to *FIE1*, which reduced yield (Chen et al., 2016; Folsom et al., 2014). As opposed to H3K27me3, the H3K4me3 pattern is typically associated with transcriptional activation that has been reported to regulate stress-responsive genes in rice. By assisting PRC2 in recruiting to limit the expression of shoot branching-inhibitory genes via H3K27me3, NGR5 stimulates tillering in response to an increased nitrogen supply (Wu et al., 2020).

Similarly, FIE and enhancer of zeste [E(z)], two PRC2 complex components, have been demonstrated to influence seed size in barley by controlling endosperm development (Kapazoglou et al., 2013). Recently, *HvTX*, a gene encoding a trithorax-like H3K4 methyltransferase, has been recognized successfully followed by characterization in barley. When a drought-tolerant cultivar of barley was subjected to drought conditions, a significant increase in transcript levels of *HvTX* was observed (Shanker & Venkateswarlu, 2011). Heat stress in barley also caused a change in MADS-box gene expression, indicating that PRC2-mediated regulation might be susceptible to HS and might impact genes that are normally active in the early phases of endosperm development (Ejaz & von Korff, 2017). Furthermore, wheat varieties possessing heat sensitivity, when exposed to HS at the stage of seed maturation, remarkably altered germination rates and seed vigor by affecting hormones and storage products (Hasan et al., 2013). Compared to this, rice seeds exposed to moderate HS for 24 hours at the same stage showed higher and more rapid germination rates than non-stressed seeds. It has also been explicitly noted that the hormonal levels of ABA and GA strongly increase along with the amount of total starch in short-term HS-exposed seeds. These suggest that if exposed to post-zygotic 24-hour HS, embryos can potentially acquire the property of faster germination through priming by altering ABA sensitivity. However, HS for any period has been considered detrimental to seed germination, development and maturation. A detailed related study has revealed that ABA concentration-regulatory genes in crops are coordinated by stress-induced patterns of DNA methylation (Xing et al., 2015), for instance, H3K4me3 tri-methylation executed by the Trithorax protein ATX1, under water-deficit conditions, functions in both ABA-independent and ABA-dependent manners. In *atx1* plants, changes in physiological traits included increased sensitivity in response to dehydration stress, larger stomatal apertures and higher transpiration rates. Further studies revealed the fact that ATX1 is not only responsible for inducing an important gene, *NCED3*, that encodes one of the key ABA biosynthesis enzymes but also for triggering H3K4me3 as a dehydration stress response. This is followed by ATX1 exerting a direct effect on gene expression dependent on ABA biosynthesis and also indirect regulation of ABA-independent signaling pathways (Ding et al., 2011).

Notably, improper regulation of multiple gene subsets, synergistically aimed at maintaining hormonal pathways and cytoskeletal organization, has produced similar responses in drought stress-exposed wheat and HS rice seeds. In addition, stress also interfered with genes related to storage proteins in wheat such as glutenins, gliadins and avenins, resulting in strong repression of transcriptional levels (Chen et al., 2016). Intriguingly, the aforementioned mechanisms have been known to be regulated by PRC2, thereby illustrating the fact that PRC2 is most probably a crucial player in synchronizing seed maturation phases under drought-stress environments (Tonosaki & Kinoshita, 2015). Global methylation levels have been reported to be reduced in ryegrass (Tang et al., 2014) and rice (Karan et al., 2012) genotypes due to exposure to harsh environments. Phytohormones, auxin and ABA-regulatory genes at the time of seed formation have been witnessed to be mediated by global DNA methylation patterns in rice (Xing et al., 2015). Recent studies also indicate similar patterns in barley, which dictates the endosperm cellularization process to be, in particular, stress-sensitive (Surdonja et al., 2017). Therefore, it can be demonstrated that an integral cross-talk between DNA methylation, H3K27me3, H3K4me3 and ABA concentration, in combination with altered PRC2 regulation, is critically responsible for mediating endosperm cellularization, embryo patterning, cell/nuclear division and modified gene expression, thereby affecting phases of embryo development in environmental stress-exposed cereal crops. Likewise, members of the PRC2 complex are regulated by deleting the epigenetic marks on H3K9me3 under heat and drought stress conditions. MADS-box regulatory developmental genes, generally repressed during normal environmental conditions, are triggered by abiotic stress to reduce DNA methylation. This provokes a transcriptional shift, which brings about delayed and malformed seed development.

A GTP-binding protein that encodes for the rice epiallele epid1 results in stunted growth. DNA methylation and histone acetylation function synergistically to silence the *d1* genes, which affect the stature of rice plants. One epiallele correlated with the WEALTHY FARMERS PANICLE (WFP) is *OsSPL14*. Elevated *OsSPL14* expression is caused as a result of epigenetic alterations developing an increase in grain yield and panicle branching driven by epigenetic alterations. Additionally, *HMTase* genes that methylate histone H3K36 to induce flowering include *SET DOMAIN GENE 724* (*SDG724*), Class II member in the SET domain family. The SDG724 loss-of-function mutant *lvp1* exhibited delayed flowering under both long day (LD) and short day (SD), commonly linked to decreased *Heading date 3a* (*Hd3a*) expression and *RICE FLOWERING LOCUS T 1* (*RFT1*). The chromosomal region of *RFT1* alone reduced the transcriptional activation by H3K36me2/3 alteration, despite both the genes being close to 11.5 kb (Sun et al., 2012). In recent times, functional and molecular characterization of histone demethylases is of particular interest among active research work. Histone lysine methylation is removed by two different kinds of histone demethylases: proteins with the Jumonji C (JmjC) domain and proteins that resemble lysine-specific demethylase (LSD). Lysine-specific demethylases (LSDs) demethylate mono-methylated and di-methylated lysine, but proteins containing the Jumonji C (JmjC) domain selectively transfer methyl groups from di-methylated along with tri-methylated lysine (Tsukada et al., 2006). Studies on rice variants suggest that an H3K9 demethylase, JMJ706, jmjC domain-containing gene, is critically required for the development of floral organs. Besides, barley crops have been shown to retain a plant-specific histone demethylase, the PKDM7 subfamily member, which is recorded to be remarkably induced during drought conditions (Shanker & Venkateswarlu, 2011).

10.3.2 Histone Acetylation/Deacetylation

HAT-mediated histone acetylation involves the transfer of an acetyl moiety to the highly conserved lysine-rich ε-amino group located in the N-terminal extensions of core nucleosome histones, which in turn neutralizes the lysine bearing positive charges, thereby reducing their affinity towards the negatively charged DNA molecules. It increases hydrophobicity and counteracts the histone tail positive charges (Onufriev & Schiessel, 2019). This causes the chromatin state to change from a closed to an open state and reduces the histone protein affinity for negatively charged DNA (Li et al., 2019). This enhances chromatin relaxation followed by subsequent activation of transcription, causing transcription factor and RNA polymerase binding to the promoter region, initiating transcription. Type A and Type B HATs are distinguished by their subcellular localization. Acetylation of nuclear histones is the mechanism of Type A HATs, which in turn controls gene transcription and chromatin assembly. Conversely, the cytoplasm contains the Type B HAT proteins. Prior to histone integration into freshly duplicated chromatin, they catalyze the acetylation of histone H4 at lysine 5 and 12 in the cytoplasm. Type B HAT is a heterodimeric complex that is present in both the cytoplasm and the nucleus of maize, suggesting that it also has a role in the latter (Gupta & Salgotra, 2022).

Further, stress-exposed cereal crops have exhibited alterations in histone acetylation patterning; for instance, the *ALCOHOL DEHYDROGENASE 1* (*ADH1*) *and PYRUVATE DECARBOXYLASE* (*PDC1*) genes of rice seedlings, having 5′ and 3′ regulatory regions and coding regions, when subjected to submergence, were shown to possess H3K4me3 and H3 acetylation. This, in turn, upregulated the expression of *ADH1* and *PDC1* genes, which reduced back to their original basal level once the stress subsided, thereby indicating the underlying influence on the dynamic nature of histone modifications. Similarly, UV-radiation-exposed maize crops revealed elevated H3 and H4 acetylation in the UV-B-induced genes, specifically around the coding and promoter regions in a maize-UV-B-tolerant line; no such effect was observed in a UV-B-sensitive maize line (Casati et al., 2008). Numerous superfamilies including TFII250, GNAT/MYST and CBP, along with several other large protein complex subunits, are categorized under HATs. Significant HAT homologs have been reported in barley belonging to the GNAT/MYST family, namely *HvELP3*, *HvMYST* and *HvGCN5*. Their gene expression was extensively studied at different seed developmental stages accompanied by ABA treatment. Exogenous induction of barley seedlings with ABA caused the activation of all the *HAT* genes, out of which the majorly affected gene was *HvELP3*, which exhibited four times the expression levels in ABA-treated seedlings when compared to that of control. The other two genes, *HvMYST* and *HvGCN5*, were reported to increase by approximately two fold. These observations indicated that histone acetyltransferases function in an ABA-dependent manner as a response to abiotic stress and during seed development (Papaefthimiou et al., 2010). According to a different study on rice, acetylation of H3K9 close to the transcription start site can increase the expression of RFT1 in Hd3a-RNAi transgenic plants. Through H3K36me2/3, SDG725 – another member of the SET domain family class II – promotes flowering. Grain weight was linked to a quantitative trait locus (QTL) OsglHAT1, a new-type GNAT-like protein that was reported to have built-in H4 histone acetyltransferase activity. Grain length depends on PGL2 (6-PHOSPHOGLUCONOLACTONASE 2), which is favorably controlled by OsglHAT1. Grain length, grain weight, yield and total biomass are agronomic parameters that are positively impacted by increased expression of OsglHAT1 (Song et al., 2015).

Histone deacetylation can be demonstrated as a reversal process of acetylation, or in other words, the removal of an acetyl group from a lysine residue, which is mediated by HDACs in order to promote chromatin condensation and gene silencing.

i. Eukaryotes comprise three main families primarily homologous to yeast HDACs: a) the RPD3/HDA1 family; b) HD2, the plant-specific family; and c) the SIR2 family.
ii. Phylogenetic sequencing of the rice genome consists of three HDAC families:
    a) HDA1 family – further subdivided into four classes: Class I, Class II, Class III and Class IV;
    b) HD2 family – further classified into two classes HD2a and HD2b.
iii. Maize has been found to harbor 15 different types of HDACs: 10 HDA1 and many others, 4 HD2-like and 1 SIR2.
iv. Barley genome classified into:
    a) HDA1 family – subcategorized into Classes I, II, III and IV;
    b) HD2 family – subdivided into *HvHDAC2-1* and *HvHDAC2-2*.

Among the aforementioned families of HDACs, 18 HDAC genes belonging to the families of *HDA1, HD2* and *SIR2* of the rice genome were analyzed to scrutinize the differential expression pattern in collaboration with phytohormones pertaining to environmental stress. Upregulated expression of specific *HDA1* genes while downregulation of other gene families was reported in response to exogenous administration of ABA, JA and SA along with simultaneous osmotic, cold and salt stress conditions. In a particular external application of JA concentration, certain members of class I and class II HDA1 families, namely *HDA702*, *HDA705* and *HDA704*, respectively, were enhanced. On the contrary, similar conditions induced the repression of class IV (*HDA712*) members (Shanker & Venkateswarlu, 2011).

Explicit rice genome analysis implies that discriminating levels of histone and non-histone protein acetylation notably affect seed differentiation followed by maturation. The major distribution of the acetylated proteins facilitates carrying out carbohydrate metabolism, including synthesis, storage and breakdown along with the tricarboxylic acid (TCA) cycle, which has been marked in the starch biosynthesis pathway in rice, beginning five days after fertilization until seed maturation. OsSRT1 is an SIR2-type HDAC found in rice that influences the accumulation of starch in developing seeds while inhibiting the carbon metabolic flow of the glycolysis pathway (Huang et al., 2016). Identification and characterization of genes included in barley HDA1 and HD2 families during developmental phases, as well as ABA and JA-related stress response, were performed by induction of one member from each HDA1 class I, II, III, and IV with JA. *HvHDAC2-1* and *HvHDAC2-2* genes belonging to the barley HD2 family were exogenously administered with JA at time intervals of 6 h and 24 h. Findings indicated that transcript levels of *HvHDAC2-1* were substantially induced at a period of 24 h post-ABA treatment, whereas ABA-treated *HvHDAC2-2* expression markedly reduced after 6 h, but no major difference was observed post 24 h of treatment. Similarly, *HDT701* and *HDT702* rice homologs were induced by JA administration. However, barley homologs *HvHDAC2-1* and *HvHDAC2-2* exhibited varied ABA-regulated responses, unlike rice homologs (Shanker & Venkateswarlu, 2011). Hence, abiotic stress-triggered levels of histone acetylation can be regarded to have control over the transition phases of seed development in cereals, making it quite evident that only further studies can divulge the depths of transcriptional regulation associated with acetylation. Histone deacetylation, as well, involves non-permissive chromatin conformation, which leads to transcriptional suppression. The rice genome has notably exhibited suppression of ABA-related genes resulting from histone deacetylation on exposure to salt, cold and mannitol stress. Contrastingly, developing ovules of drought-stressed wheat and maize crops showed increased ABA concentration, which has been documented to potentially affect cell division, fertilization and hamper seed germination (Begcy & Dresselhaus, 2018).

One crucial histone deacetylase, HDA108, in maize, mediates plant height, leaf growth and fertility by lowering H3K9 di-methylation and raising nuclear H3 and H4 histone acetylation (Forestan et al., 2018). Overexpression of the *HDA101* (histone deacetylase) gene along with its decline brings about changes in the proportion of acetylated histones, causing morphological and developmental abnormalities (Rossi et al., 2007). HDA101 interacts with corepressors such as NFC103/MSI1 and SNL1/SIN3-like proteins, causing hypoacetylation of lysine 5 in histone H4 (H4K5Ac) on specific genes while having no effect on transcript levels. It regulates kernel size by regulating the transfer of cell-specific gene expression, and its removal causes hyperacetylation of histones in the target region while having little or no effect on transcript levels. HDA101 also targets dormant genes, which are linked with hyperacetylation and increased expression (Yang et al., 2016). Besides, stress conditions influence hyperacetylation at H3 and H4 lysine sites in the promoter region of cell cycle-related genes, which in turn induces prolonged cell cycle and, consequently, inhibits division and growth in maize (Zhao et al., 2014). A summary of major epigenetic modifications influencing morphological, physiological and developmental phases in stress-exposed cereal crops is listed in Table 10.2.

TABLE 10.2
Various epigenetic modifications influence morphological, physiological and developmental phases in stress-exposed cereal crops

Crop variety	Epigenetic modification	Stress condition	Effect on cereal crops	References
Maize	Histone deacetylation	–	*HDA108* mediates plant height, leaf growth and fertility	Forestan et al. (2018)
	Histone acetylation	Abiotic stress	Induces prolonged cell cycle and inhibits division and growth	Zhao et al. (2014)
	H3 and H4 acetylation	UV radiation	Developing maize-UV-B-tolerant lines	Casati et al. (2008)
	Histone acetylation	Drought	Increases ABA concentration, affects cell division and fertilization, and hampers seed germination	Begcy and Dresselhaus (2018)
Barley	Histone acetylation	Abiotic stress	Overexpressed *HvELP3* associated increased ABA levels during seed development	Papaefthimiou et al. (2010)
	DNA methylation	–	Dictates the endosperm cellularization process	Surdonja et al. (2017)
Wheat	DNA methylation	Heat stress	Alters germination rate and seed vigor, affects hormones and storage products	Hasan et al. (2013)
Rice	DNA methylation	Heavy metals, sublethal heat exposure and drought	Enhanced tolerance through heritable changes in gene expression	Ou et al. (2012)
	Histone deacetylation	Salt, cold and mannitol stress	Exhibits suppression of ABA-related genes	Begcy and Dresselhaus (2018)
	H3K4me3 and H3 acetylation	Submergence	Upregulated the expression of *ADH1* and *PDC1* genes	Casati et al. (2008)
	H3K9 demethylation	–	JMJ706 demethylase controls the development of floral organs	Shanker and Venkateswarlu (2011)
	H3K4me3/H3K27me3 methylation	Heat stress	Influences early endosperm development	Folsom et al. (2014)

10.4 ROLE OF NON-CODING RNAS (NcRNAs) IN COPING WITH STRESS

Chromatin modulators involving miRNAs and siRNAs, which follow small RNA-associated mechanisms, function by directly acting on chromatin and participating in the RdDM pathway. In particular, small RNAs or sRNAs can be grouped into major classes like small interfering RNAs (siRNAs), Piwi-interacting RNAs (piRNAs) and microRNAs (miRNAs), enclosing small nuclear RNA (U-RNAs), small nucleolar RNAs (snoRNAs), phased and secondary interfering RNAs (phasiRNAs), small rDNA-derived RNA (srRNAs) and tRNA-derived small RNAs (tsRNAs) in plants. Among cereals, the roles of siRNAs like phasiRNAs, derived from long dsRNAs, and miRNAs, originating from hairpin-shaped precursors of dsRNAs, are best investigated to understand their regulation in epigenetic studies under abiotic stress (Begcy & Dresselhaus, 2018; Bhattacharya & Roychoudhury, 2024).

Identification of numerous conserved miRNAs across species has been achievable through locus-specific research and extensive transcriptome investigations, which can also be used to explain their participation in stress response. The advancement of high-throughput sequencing technology over the past few years has made it possible to elucidate increasing miRNAs, including species-specific or extremely low-abundance miRNAs (Banerjee et al., 2016). This has led to the discovery of a rising number of short RNAs that react to biotic and abiotic stress in both monocots and dicots. In particular, cereal crops, i.e., wheat, rice, maize and the model plant for cereals, *Brachypodium*, have shown miRNA responsiveness to various abiotic stressors. For instance, a microarray study revealed that rice under high salinity and drought stress triggered several miRNAs. When rice roots and shoots were dehydrated, the upregulation of MiR169g was discovered. Remarkably, two dehydration-responsive elements (DREs) were reported to be present in the promoter of the *miR169g* gene. Similar to miR169g, it was discovered that high salinity conditions induced the rice *miR169n* gene. The promoter of rice *miR169n* comprises a cis-acting ABA-responsive element (ABRE), suggesting an ABA-mediated stress response (Zhao et al., 2007). Interestingly, both miRNAs target NF-YA, a transcription factor (TF) that has been demonstrated to decline in drought-prone environments. In rice, genome-wide profiling of miRNAs has identified 29 new miRNAs that were differentially expressed under drought (upregulation of 8 and downregulation of 11) (Zhou et al., 2010).

Dehydration can induce 28 novel miRNAs in barley, including Hvu-MIR156, Hvu-MIR166, Hvu-MIR171 and Hvu-MIR408. Among these, 34 miRNAs from 13 plant miRNA families showed significant alterations in expression following drought treatment of seedlings, according to microarray research in maize. It was discovered that under dehydration circumstances, miR474, which targets a gene-encoding proline dehydrogenase (PDH), an enzyme involved in the breakdown of proline, was enhanced. It is popularly known that plants store proline as a defense against drought stress. PDH accumulation decreased, while miR474 transcripts were enhanced in response to drought stress, indicating a miR474-dependent mechanism controlling proline levels in maize during drought. The expression of their target genes, MAPK (MITOGEN ACTIVATED PROTEIN KINASE), POD (PEROXIDASE) and PLD (PHOSPHOLIPASE D), was probably elevated as a consequence of the downregulation of other maize miRNAs, such as miR168, miR528 and miR167. It is interesting to note that these genes have an ABA-responsive element and are engaged in antioxidant defense of maize and ABA-induced stomatal movement (Wei et al., 2009).

PhasiRNA biogenesis in maize exhibits a high degree of spatiotemporal dynamism during anther development (Zhai et al., 2015), with 24-nt phasiRNAs accumulating preferentially during meiosis and 21-nt phasiRNAs accumulating preferentially during pre-meiosis. The latter continues to exist while pollen is being formed. Additionally, two new miRNAs, zma-MIR11969 and Zuma-MIR11970, as well as numerous 21- and 24-nt phasiRNAs were identified in a stage-specific investigation conducted in maize anthers throughout the meiosis stage, namely during zygotene (Dukowic-Schulze et al., 2016). Intriguingly, increased DNA methylation levels at phasiRNA loci were discovered, particularly in the setting of CHH, suggesting that phasiRNAs may play a part in cis-DNA methylation. During pollen development, 24-nt phasiRNA targets are still not identified. However, on the basis of its transcriptional profile, Argonaute (ZmAGO18b) in maize was recognized as a potential candidate (Fei et al., 2016). This discovery needs to be confirmed by more experimental data. Wheat phasiRNAs and sRNAs were shown to be significantly upregulated under heat and oxidative stress. It was proposed that pollen malformation results from the improper regulation of phasiRNAs and other sRNAs, which control the development of male gametophytes, during rising temperatures and other stressors like drought. This is frequently seen in abiotic stress situations and mutants of sRNA-regulated genes.

Several distinct miRNA expressions in cereals have also been shown to be significantly impacted by cold stress. Eighteen rice miRNAs that were differently expressed after cold treatment of rice seedlings were found using microarray analysis (Lv et al., 2010). During cold conditions, six miRNAs representing five families showed considerable upregulation, while twelve miRNAs representing ten families showed substantial decline. On the other hand, the known genes that produce TFs, such as scarecrow-like proteins, TCP family transcription factors, MADS-box proteins and homeodomain-leucine zipper proteins, are targeted by six conserved families (miR169, miR444, miR156, miR166, miR171 and miR319). Predictions show that the genes Os01g59660 and Os04g46860 are targeted by miR319a/b and miR171a in rice, respectively. Cognate miRNAs have been reported to be reduced under cold conditions, whereas Os01g59660 and Os04g46860 have been found to be activated under such environments. Hence, cold stress response can be inferred to share a negative relationship with miRNAs, which can be confirmed by 5′ RACE experiments. Further, rice miR444 has been noted to target MADS57 and MADS27, which in turn are upregulated instead of undergoing decline during cold stress (Shanker & Venkateswarlu, 2011). ABRE and gibberellin-responsive element (GARE), two of the cis-acting responsive elements related to phytohormones, have been discovered to be localized in the majority of cold-sensitive miRNAs. One ABRE element along with two GARE motifs harbored by the miR319 promoter serves as a record for ABA-regulated gene expression. Numerous other rice miRNAs are either increased or decreased by ABA and GA, and a recent study demonstrated that miR319 is downregulated by ABA and upregulated by GA levels (Liu et al., 2009).

SiRNAs have also been associated with the rice response to abiotic stress. The accumulation of rice siR441 and siR446 was inhibited by cold, drought, excessive salt and ABA treatment. The ability of rice to withstand abiotic stress may be influenced by siRNAs, as functional research revealed that siR441 and siR446 knockdown mutants were more susceptible to drought, cold or salt treatment than the wild type. MAIF1 (encoding an F-box protein), the verified target of siR441 and siR446, has been reported to have increased response to abiotic stress. Furthermore, the phenotype of transgenic rice plants with decreased accumulation of siR441 and siR446 was identical to that of plants overexpressing MAIF1. All of these findings dictate that rice siR441 and siR446 regulate MAIF1 to contribute to the abiotic stress response (Yan et al., 2011). The microRNA-regulated *OsSPL14* gene in rice has been documented to influence panicle branching and boost yield. Rice growth and development are impacted by a decline in OsARF18 expression mediated by miR160, which influences auxin signaling (Huang et al., 2016). Targeting AP2 (Apetala)-like TFs through miR172 influences panicle branching, floral development and blooming time (Wang et al., 2015a). Osa-miR397-mediated *OsLAC* gene downregulation causes laccase-like protein production that acts as a factor for grain size and yield (Zhang et al., 2013).

Differentially expressed miRNAs are involved in an adaptation process triggered by plants inhabiting less nitrogenous conditions (Zhao et al., 2013). While miRC10 and miRC68 may react to low soil nitrogen, miR169, miR399, miR408 and miR528 react to extremely low nitrogen stress. MiR528s, miR169s, miR166s and miR408/b alter root growth when there is a nitrate deficiency (Zhao et al., 2012; Zhao et al., 2013). When exposed to cadmium stress, miR169s, miR160f-5p, miR156s and miR171s target the transcription factors NFY, ARF, SBP and GRAS, respectively (Wang et al., 2019). miR156 plays a crucial role in developing meristems and the border of leaves by negatively regulating the SBP-box transcription factor *tasselsheath4* (*tsh4*), which permits the formation of lateral meristems and inhibits the commencement of leaves (Chuck et al., 2010). Negative regulation of *NAC1* expression by miR164 affects the formation of lateral roots (Li et al., 2012), as highlighted in Table 10.3. During the growth of maize seedlings, there can be notable variations in miRNA expression. Transcriptional control is the primary function of most miRNA target genes implicated in this pathway. By controlling growth and development, miR393, miR393, miR396 and miR156 target *TIR1*, *LAC*, *GRF* and *SPL*, respectively, and may be essential in governing grain filling (Jin et al., 2015). Throughout seed development, miR164 plays a regulatory role (Zheng et al., 2019). Nine miRNAs from the conserved zma-miR169 family control the development of the maize kernel (Xing et al., 2017).

TABLE 10.3
Role of several non-coding RNAs (ncRNAs) in crop development and establishing transcriptional stress memory

Crop variety	ncRNAs	Effect	References
Maize	miR393, miR393, miR396, miR156	Controls growth and development by targeting *TIR1*, *LAC*, *GRF* and *SPL*; Probable involvement in grain filling	Jin et al. (2015)
	miR164	Downregulation of *NAC1* expression that inhibits lateral root formation	Li et al. (2012)
	miR156	*tasselsheath4* (*tsh4*) downregulated, which induces lateral meristem development and inhibits leaf commencement	Chuck et al. (2010)
	PhasiRNA	Exhibits spatiotemporal dynamism during anther development	Zhai et al. (2015)
	miR474	Enhanced *PDH* levels under drought conditions to elevate proline as a defense mechanism	Wei et al. (2009)
Barley	miR172	Targets APETALA2 (AP2)-like TF and is associated with grain density and cleistogamous flowering	Houston et al. (2013)
Wheat	miR408	Influences timing of CAB Expression 1 TF and is associated with heading time	Zhao et al. (2016)
	miR159	It targets MYB TF and is associated with the development of anther	Wang et al. (2012)
Rice	siR441, siR446	Regulates MAIF1 (encoding an F-box protein) in response to abiotic stress	Yan et al. (2011)
	miR160	A decline in *OsARF18* expression influences auxin signaling	Huang et al. (2016)
	miR397	*OsLAC* gene downregulation serves for grain size and yield	Zhang et al. (2013)

10.5 ESTABLISHMENT OF TRANSCRIPTIONAL STRESS MEMORY THROUGH STRESS PRIMING

According to recent research, plants frequently remember certain ecological stress factors and environment they encounter over their lifetime. This phenomenon exhibited by plants, referred to as "plant memory" or "epigenetic memory", offers a strong foundation for a robust and prompt response to such future obstacles (Latzel et al., 2016). Plant memory, more especially, "stress memory", can be defined as a crucial characteristic that incorporates certain mechanisms, including persistent DNA methylation, which improves the plant's ability to withstand stress when they are previously acclimated and hardy to the same stress conditions (Crisp et al., 2016). These memories resulting from epigenetic modifications are important because they help plants withstand comparable environmental threats in the future (Ding et al., 2012). It has been reported that in order for plants to effectively handle such environmental conditions, they not only adapt to the current stress scenario but also retain this knowledge for future generations. Epigenetic modifications, which include modifications in histone and small RNA, DNA-Me, along with miRNA or siRNA expression for post-transcriptional gene silencing, lead to heritable alterations in the activities of genes and play a crucial role in regulating gene expression to cope with environmental challenges; for instance, in rice, variations in DNA methylation caused by nitrogen deprivation and a drop in the methylation level caused by treatment with 5-azacytidine can be inherited for at least three generations (Fieldes et al., 2005; Kou et al., 2011). Future generations may also inherit DNA methylation alterations induced by low-dose laser irradiation in sorghum (*Sorghum bicolor*) (Wang et al., 2010). Other research on rice has shown that stress memory, or an increase in stress tolerance, can be passed down to successive generations of offspring along with long-term stresses such as salinity, high temperature, nitrogen fertilizer and drought-induced alterations in DNA methylation. Furthermore, the enzymes responsible for epigenetic modifications are not directly involved; instead, proteins such as transcriptional factors act as mediators, helping these enzymes regulate gene expression to achieve their goals (Zheng et al., 2017).

Various studies have shown that certain changes triggered by environmental stress are permanent, retained by plants and passed on to the following generations as a stress memory. In other cases, the short-term solutions to deal with the stressful environment are unstable, and as soon as the stress subsides, the plants return to normal behavior. Stress memory can be generally categorized into two types that are induced in the early stages of plant growth: short-term stress memory and long-term stress memory. Plants with short-term stress memory may withstand a particular stress condition for up to ten days. The primary cause of this effect is transient morphological and biochemical metabolite alterations, after which the plants forget the stress event and return to their pre-stress growth state (Crisp et al., 2016). On the contrary, epigenetically regulated long-term stress memory may persist throughout the life of the stressed plant and can be passed on to its progeny. Through epigenetic changes in the progeny, this long-term stress memory can play a crucial role in the adaptation and evolution of plants to their surroundings (Zheng et al., 2017).

In addition to being able to withstand and survive a single kind of stress, plants have evolved to combat multiple environmental stressors at once. Stress priming is achieved through repeated exposure. Epigenetic features are largely dominant and pass on to the following generations, and initial stress exposure enables plants to respond better in the future, for instance, due to greater proline and relative water concentration in the leaves of pretreated plants during salt stress. Salt priming in sweet sorghum improved osmotic resistance. Root Na^+ efflux enhanced 8.5- and 3.9-fold in pretreated and non-pretreated plants, respectively, following seven days of salt stress. This suggests that salt priming decreased root Na^+ absorption, which in turn lowered Na^+ accumulation in roots and leaves of pretreated plants. By decreasing root Na^+ absorption and increasing osmotic resistance, salt priming often increases tolerance of sweet sorghum to salinity (Yan et al., 2015).

Drought treatment of winter wheat plants can improve crop resistance to heat and drought stress after flowering. Additionally, when the parental line was exposed to heat stress, the yield and photosynthetic and antioxidant capacity of the next generation were also enhanced (Wang et al., 2016). In winter wheat, heat stress during the jointing stage can significantly boost photosynthetic efficiency and antioxidant synthesis during the grain-filling stage. This improves heat resistance and lowers the yield loss of the plants that sprout from those seeds (Wang et al., 2014). Through stress memory, abiotic stress-exposed plants increase their resistance to stress by upregulating and downregulating sRNAs (miRNAs, siRNAs) that regulate transcription factors, plant hormones, reactive oxygen species (ROS), and negative and positive regulators (Banerjee et al., 2017; Zhang et al., 2019). The heat resistance exhibited by succeeding generations of wheat crops can be considerably increased by treating plants with semi-lethal high temperature during the blooming stage. In a different study on wheat, drought priming performed in the first, second and third generations led to higher grain yield, antioxidant capacity and leaf photosynthetic rate while lowering the O_2 release rate and H_2O_2 and MDA contents at the time of grain filling in the fourth generation. In addition, all subsequent generations of drought-primed plants had higher leaf water potential and higher levels of proline, glycine betaine, pyrroline-5-carboxylate synthetase and betaine aldehyde dehydrogenase in comparison to control non-primed plants (Wang et al., 2018). Moreover, drought-stressed wheat in the late growth stage can improve its resilience to low temperatures by adjusting photosynthesis and antioxidants as well as to heat and drought (Wang et al., 2015b). Similarly, drought treatment in oats, ryegrass and wheat (Wang et al., 2015b) favored frost resistance.

10.6 CONCLUSION AND FUTURE PERSPECTIVES

Flourishing documentation and advanced research indicate that stress-associated gene expression is significantly mediated by epigenetic modifications and genetic studies. Moreover, epigenetics influences phenotypic variance. Therefore, knowledge of epigenetics and epigenomics can explicitly describe the pathways by which plant phenotypes are affected by environmental factors (Agarwal et al., 2020). The discoveries regarding systems essential for maintaining yield and reproductive success in the face of abiotic stress now offer interesting opportunities; for instance, reproductive development is more susceptible to abiotic stress conditions than vegetative developmental stages, despite the fact that crops are subjected to stress throughout their life cycle. This is because stress can halt development and result in total sterility. Since various stresses often interact, comprehensive research on epigenetic regulatory mechanisms under abiotic stress circumstances during reproductive development needs to be systematically explored, studied and approached under field conditions. In the context of transgenerational memory, inheritable epigenetic modifications including DNA methylation, histone variants and tiny regulatory mechanisms are also not well studied in cereals, even though they may increase the offspring's resistance to stress. The transcriptional memory correlated with decisiveness regarding cell destiny, developmental shifts or stress reactions is reflected through epigenetic alterations. Reproduction frequently necessitates the deletion and reset of this memory. These alterations in plants are caused by "plant stress memory", which aids in instating stress management in future generations and improves plant response to stress in a challenging environment. Investigating the relationship between abiotic interactions and epigenetics holds considerable promise for addressing urgent challenges in crop improvement and phenotypic plasticity.

Pivotal information regarding developmental mechanisms, including embryo sac development and fertilization, remains implicit due to technical challenges related to their isolation and observation, even though new tissue- and cell-specific methodologies like fluorescence-activated cell sorting, ATAC sequencing and methylome sequencing make it easier to research stress-responsive reproductive development in spatiotemporal contexts. Furthermore, it is believed that with progressing sensitive techniques for single-cell genome sequencing, interactome, proteome and metabolome studies, as well as gene-editing technologies, systematic and detailed genome-wide studies will become possible along with identifying novel candidate genes that control the epigenetic responses of cereals to environmental stress (Farlik et al., 2015). By understanding the current epigenetic variability or epigenome alteration, breeders may be able to find cryptic variations and counteract the persistent problem of genetic erosion. In a single generation, epi-breeding programs may result in a vast variety of phenotypic heterogeneity, some of which may be inherited by subsequent generations (Dalakouras & Vlachostergios, 2021). Eventually, it ought to go beyond all restrictions and limits that lower the efficacy of crop breeding programs. Breeding populations with minimal genetic diversity would benefit greatly from novel epialleles. Comprehensive knowledge of the biochemical, physiological, genetic and epigenetic components of stress environments and the mechanisms of tolerance in plants will be beneficial for screening germplasm for stress breeding and for cloning the genes involved in stress tolerance, creating transgenics and improving breeding programs.

To meet the demands of a growing population, crop varieties of high yield must be state-of-the-art in the near future. Recent studies have improvised on the fact that epigenetic processes such as chromatin remodeling, histone modification and DNA methylation can greatly impact agronomic traits. Similarly, utilization of a diverse range of phenotypic variation can develop climate-resistant crops. Further, the advantage of the genome sequences of various crop species carrying epiallelic differences, which expands to among species too, if properly identified, can be quantified and reported. Epigenetic applications on rice, maize, wheat, barley, oats, sorghum and other cereal crops have shown significant changes in the agronomic features. Therefore, implementing epigenetic information opens the door to enormous possibilities for crop improvement. This can be achieved by applying several coming-of-age epigenetic approaches, including DNA methylation, histone modifications, regulation of transgenic expression and RNA interference (RNAi), which can generate new epialleles and crop varieties in terms of rapid plant adaptation, yield and other agronomic factors to meet modern food demands and progressive living standards.

Studies suggest the possibility of an integrated mechanism involving miRNAs and their targets that modulates abiotic stress responses throughout the formation of pollen and anther. Last but not least, further meticulous research and analysis are indispensable to explore the impact of additional sRNAs and chromatin remodeler genes, even though recent findings have emphasized the role of DNA methylation and sRNAs, particularly phasiRNAs and miRNAs, in male germline development during abiotic stress responses.

REFERENCES

Adeboyejo, F. O., Olatunde, S. J., Rustria, G. A., Azotea, A. N. B., Ostonal, J. M., Reyes, M. J. A., & Oyeyinka, S. A. (2023). Asian fermented cereal-based products. In *Indigenous Fermented Foods for the Tropics* (pp. 37–56). Elsevier.

Agarwal, G., Kudapa, H., Ramalingam, A., Choudhary, D., Sinha, P., Garg, V., Singh, V. K., Patil, G. B., Pandey, M. K., & Nguyen, H. T. (2020). Epigenetics and epigenomics: Underlying mechanisms, relevance, and implications in crop improvement. *Functional & Integrative Genomics*, 20, 739–761.

Arikit, S., Zhai, J., & Meyers, B. C. (2013). Biogenesis and function of rice small RNAs from non-coding RNA precursors. *Current Opinion in Plant Biology*, 16(2), 170–179.

Banerjee, A., & Roychoudhury, A. (2017). Epigenetic regulation during salinity and drought stress in plants: Histone modifications and DNA methylation. *Plant Gene*, 11, 199–204.

Banerjee, A., & Roychoudhury, A. (2018). The gymnastics of epigenomics in rice. *Plant Cell Reports*, 7, 25–49.

Banerjee, A., Aryadeep Roychoudhury, A., & Krishnamoorthi, S. (2016). Emerging techniques to decipher microRNAs (miRNAs) and their regulatory role in conferring abiotic stress tolerance of plants. *Plant Biotechnology Reports*, 10(4), 185–205.

Banerjee, S., Sirohi, A., Ansari, A. A., & Gill, S. S. (2017). Role of small RNAs in abiotic stress responses in plants. *Plant Gene*, 11, 180–189.

Begcy, K., & Dresselhaus, T. (2018). Epigenetic responses to abiotic stresses during reproductive development in cereals. *Plant Reproduction*, 31, 343–355.

Bhattacharya, A., & Roychoudhury, A. (2024). Plant non-coding RNAs and their regulation in salinity and heavy metal stress. In S. Penna, V. Kumar, & A. K. Srivastava (Eds.), *Non-Coding RNAs Molecular Tools for Crop Improvement* (pp. 128–147). CRC Press (Taylor & Francis group).

Casati, P., Campi, M., Chu, F., Suzuki, N., Maltby, D., Guan, S., Burlingame, A. L., & Walbot, V. (2008). Histone acetylation and chromatin remodeling are required for UV-B–dependent transcriptional activation of regulated genes in maize. *The Plant Cell*, 20(4), 827–842.

Chen, C., Begcy, K., Liu, K., Folsom, J. J., Wang, Z., Zhang, C., & Walia, H. (2016). Heat stress yields a unique MADS box transcription factor in determining seed size and thermal sensitivity. *Plant Physiology*, 171(1), 606–622.

Chinnusamy, V., & Zhu, J.-K. (2009). RNA-directed DNA methylation and demethylation in plants. *Science in China Series C: Life Sciences*, 52, 331–343.

Choudhary, P., & Muthamilarasan, M. (2022). Modulating physiological and transcriptional regulatory mechanisms for enhanced climate resilience in cereal crops. *Journal of Plant Physiology*, 278, 153815.

Chuck, G., Whipple, C., Jackson, D., & Hake, S. (2010). The maize SBP-box transcription factor encoded by tasselsheath4 regulates bract development and the establishment of meristem boundaries. *Development*, 137(8), 1243–1250.

Crisp, P. A., Ganguly, D., Eichten, S. R., Borevitz, J. O., & Pogson, B. J. (2016). Reconsidering plant memory: Intersections between stress recovery, RNA turnover, and epigenetics. *Science Advances*, 2(2), e1501340.

Cuerda-Gil, D., & Slotkin, R. K. (2016). Non-canonical RNA-directed DNA methylation. *Nature Plants*, 2(11), 1–8.

Dalakouras, A., & Vlachostergios, D. (2021). Epigenetic approaches to crop breeding: Current status and perspectives. *Journal of Experimental Botany*, 72(15), 5356–5371.

de Alba, A. E. M., Elvira-Matelot, E., & Vaucheret, H. (2013). Gene silencing in plants: A diversity of pathways. *Biochimica et Biophysica Acta (BBA) – Gene Regulatory Mechanisms*, 1829(12), 1300–1308.

Deng, X., Song, X., Wei, L., Liu, C., & Cao, X. (2016). Epigenetic regulation and epigenomic landscape in rice. *National Science Review*, 3(3), 309–327.

Diallo, A. O., Ali-Benali, M. A., Badawi, M., Houde, M., & Sarhan, F. (2012). Expression of vernalization responsive genes in wheat is associated with histone H3 trimethylation. *Molecular Genetics and Genomics*, 287, 575–590.

Ding, Y., Chen, Z., & Zhu, C. (2011). Microarray-based analysis of cadmium-responsive microRNAs in rice (*Oryza sativa*). *Journal of Experimental Botany*, 62(10), 3563–3573.

Ding, Y., Fromm, M., & Avramova, Z. (2012). Multiple exposures to drought'train'transcriptional responses in *Arabidopsis*. *Nature Communications*, 3(1), 740.

Dinkar, V., Pandey, S., Kumar, A., Shiv, A., Lal, D., Bharati, A., Joshi, A., Adhikari, S., Singh, A., & Pandey, P. K. (2024). Epigenetic regulations under plant stress: A cereals perspective. *Environmental and Experimental Botany*, 220, 105688.

Dukowic-Schulze, S., Sundararajan, A., Ramaraj, T., Kianian, S., Pawlowski, W. P., Mudge, J., & Chen, C. (2016). Novel meiotic miRNAs and indications for a role of phasiRNAs in meiosis. *Frontiers in Plant Science*, 7, 762.

Ejaz, M., & von Korff, M. (2017). The genetic control of reproductive development under high ambient temperature. *Plant Physiology*, 173(1), 294–306.

Farlik, M., Sheffield, N. C., Nuzzo, A., Datlinger, P., Schönegger, A., Klughammer, J., & Bock, C. (2015). Single-cell DNA methylome sequencing and bioinformatic inference of epigenomic cell-state dynamics. *Cell Reports*, 10(8), 1386–1397.

Fei, Q., Yang, L., Liang, W., Zhang, D., & Meyers, B. C. (2016). Dynamic changes of small RNAs in rice spikelet development reveal specialized reproductive phasiRNA pathways. *Journal of Experimental Botany*, 67(21), 6037–6049.

Fieldes, M., Schaeffer, S., Krech, M., & Brown, J. (2005). DNA hypomethylation in 5-azacytidine-induced early-flowering lines of flax. *Theoretical and Applied Genetics*, 111, 136–149.

Folsom, J. J., Begcy, K., Hao, X., Wang, D., & Walia, H. (2014). Rice fertilization-independent endosperm1 regulates seed size under heat stress by controlling early endosperm development. *Plant Physiology*, 165(1), 238–248.

Forestan, C., Farinati, S., Rouster, J., Lassagne, H., Lauria, M., Dal Ferro, N., & Varotto, S. (2018). Control of maize vegetative and reproductive development, fertility, and rRNAs silencing by HISTONE DEACETYLASE 108. *Genetics*, 208(4), 1443–1466.

Gupta, C., & Salgotra, R. K. (2022). Epigenetics and its role in effecting agronomical traits. *Frontiers in Plant Science*, 13, 925688.

Hasan, M. A., Ahmed, J. U., Hossain, T., Mian, M. A. K., & Haque, M. M. (2013). Evaluation of the physiological quality of wheat seed as influenced by high parent plant growth temperature. *Journal of Crop Science and Biotechnology*, 16, 69–74.

Houston, K., McKim, S. M., Comadran, J., Bonar, N., Druka, I., Uzrek, N., Cirillo, E., Guzy-Wrobelska, J., Collins, N. C., & Halpin, C. (2013). Variation in the interaction between alleles of HvAPETALA2 and microRNA172 determines the density of grains on the barley inflorescence. *Proceedings of the National Academy of Sciences*, 110(41), 16675–16680.

Hu, L., Li, N., Xu, C., Zhong, S., Lin, X., Yang, J., Zhou, T., Yuliang, A., Wu, Y., & Chen, Y.-R. (2014). Mutation of a major CG methylase in rice causes genome-wide hypomethylation, dysregulated genome expression, and seedling lethality. *Proceedings of the National Academy of Sciences*, 111(29), 10642–10647.

Huang, J., Li, Z., & Zhao, D. (2016). Deregulation of the Os miR160 target gene OsARF18 causes growth and developmental defects with an alteration of auxin signaling in rice. *Scientific Reports*, 6(1), 29938.

Jin, X., Fu, Z., Lv, P., Peng, Q., Ding, D., Li, W., & Tang, J. (2015). Identification and characterization of microRNAs during maize grain filling. *PloS One*, 10(5), e0125800.

Kakoulidou, I., Avramidou, E. V., Baránek, M., Brunel-Muguet, S., Farrona, S., Johannes, F., Kaiserli, E.,

Lieberman-Lazarovich, M., Martinelli, F., & Mladenov, V. (2021). Epigenetics for crop improvement in times of global change. *Biology*, 10(8), 766.

Kapazoglou, A., Droscu, V., Argiriou, A., & Tsaftaris, A. S. (2013). The study of a barley epigenetic regulator, HvDME, in seed development and under drought. *BMC Plant Biology*, 13, 1–16.

Karan, R., DeLeon, T., Biradar, H., & Subudhi, P. K. (2012). Salt stress induced variation in DNA methylation pattern and its influence on gene expression in contrasting rice genotypes. *Plos One*, 7(6):e40203.

Kou, H., Li, Y., Song, X., Ou, X., Xing, S., Ma, J., Von Wettstein, D., & Liu, B. (2011). Heritable alteration in DNA methylation induced by nitrogen-deficiency stress accompanies enhanced tolerance by progenies to the stress in rice (*Oryza sativa* L.). *Journal of Plant Physiology*, 168(14), 1685–1693.

Kumar, S., Beena, A., Awana, M., & Singh, A. (2017). Physiological, biochemical, epigenetic and molecular analyses of wheat (Triticum aestivum) genotypes with contrasting salt tolerance. *Frontiers in Plant Science*, 8, 1151.

Kumar, S., & Singh, A. (2016). Epigenetic regulation of abiotic stress tolerance in plants. *Advances in Plants & Agriculture Research*, 5(5), 00179.

Latzel, V., Rendina González, A. P., & Rosenthal, J. (2016). Epigenetic memory as a basis for intelligent behavior in clonal plants. *Frontiers in Plant Science*, 7, 1354.

Li, J., Guo, G., Guo, W., Guo, G., Tong, D., Ni, Z., Sun, Q., & Yao, Y. (2012). miRNA164-directed cleavage of ZmNAC1 confers lateral root development in maize (*Zea mays* L.). *BMC Plant Biology*, 12, 1–14.

Li, J., Wang, Z., Peng, H., & Liu, Z. (2014). A MITE insertion into the 3′-UTR regulates the transcription of TaHSP16. 9 in common wheat. *The Crop Journal*, 2(6), 381–387.

Li, S., Lin, Y.-C. J., Wang, P., Zhang, B., Li, M., Chen, S., Shi, R., Tunlaya-Anukit, S., Liu, X., & Wang, Z. (2019). The AREB1 transcription factor influences histone acetylation to regulate drought responses and tolerance in *Populus trichocarpa*. *The Plant Cell*, 31(3), 663–686.

Li, X., Wang, X., He, K., Ma, Y., Su, N., He, H., Stolc, V., Tongprasit, W., Jin, W., & Jiang, J. (2008). High-resolution mapping of epigenetic modifications of the rice genome uncovers interplay between DNA methylation, histone methylation, and gene expression. *The Plant Cell*, 20(2), 259–276.

Liu, Q., Zhang, Y.-C., Wang, C.-Y., Luo, Y.-C., Huang, Q.-J., Chen, S.-Y., Zhou, H., Qu, L.-H., & Chen, Y.-Q. (2009). Expression analysis of phytohormone-regulated microRNAs in rice, implying their regulation roles in plant hormone signaling. *FEBS Letters*, 583(4), 723–728.

Liu, X., & Liu, X. (2022). PRC2, chromatin regulation, and human disease: Insights from molecular structure and function. *Frontiers in Oncology*, 12, 894585.

Lv, D.-K., Bai, X., Li, Y., Ding, X.-D., Ge, Y., Cai, H., Ji, W., Wu, N., & Zhu, Y.-M. (2010). Profiling of cold-stress-responsive miRNAs in rice by microarrays. *Gene*, 459(1–2), 39–47.

Moritoh, S., Eun, C. H., Ono, A., Asao, H., Okano, Y., Yamaguchi, K., Shimatani, Z., Koizumi, A., & Terada, R. (2012). Targeted disruption of an orthologue of DOMAINS REARRANGED METHYLASE 2, OsDRM2, impairs the growth of rice plants by abnormal DNA methylation. *The Plant Journal*, 71(1), 85–98.

Nobuta, K., Venu, R., Lu, C., Beló, A., Vemaraju, K., Kulkarni, K., Wang, W., Pillay, M., Green, P. J., & Wang, G.-l. (2007). An expression atlas of rice mRNAs and small RNAs. *Nature Biotechnology*, 25(4), 473–477.

Onufriev, A. V., & Schiessel, H. (2019). The nucleosome: From structure to function through physics. *Current Opinion in Structural Biology*, 56, 119–130.

Ou, X., Zhang, Y., Xu, C., Lin, X., Zang, Q., Zhuang, T., Jiang, L., Von Wettstein, D., & Liu, B. (2012). Transgenerational inheritance of modified DNA methylation patterns and enhanced tolerance induced by heavy metal stress in rice (*Oryza sativa* L.). *Plos One*, 7(9):e41143.

Papaefthimiou, D., Likotrafiti, E., Kapazoglou, A., Bladenopoulos, K., & Tsaftaris, A. (2010). Epigenetic chromatin modifiers in barley: III. Isolation and characterization of the barley GNAT-MYST family of histone acetyltransferases and responses to exogenous ABA. *Plant Physiology and Biochemistry*, 48(2–3), 98–107.

Rossi, V., Locatelli, S., Varotto, S., Donn, G., Pirona, R., Henderson, D. A., Hartings, H., & Motto, M. (2007). Maize histone deacetylase hda101 is involved in plant development, gene transcription, and sequence-specific modulation of histone modification of genes and repeats. *The Plant Cell*, 19(4), 1145–1162.

Sakai, Y., Suriyasak, C., Inoue, M., Hamaoka, N., & Ishibashi, Y. (2022). Heat stress during grain filling regulates seed germination through alterations of DNA methylation in barley (*Hordeum vulgare* L.). *Plant Molecular Biology*, 110(4), 325–332.

Singh, M., Singh, S., Randhawa, H., & Singh, J. (2013). Polymorphic homoeolog of key gene of RdDM pathway, ARGONAUTE4_9 class is associated with pre-harvest sprouting in wheat (*Triticum aestivum* L.). *PLoS One*, 8(10), e77009.

Song, X. J., Kuroha, T., Ayano, M., Furuta, T., Nagai, K., Komeda, N., Segami, S., Miura, K., Ogawa, D., & Kamura, T. (2015). Rare allele of a previously unidentified histone H4 acetyltransferase enhances grain weight, yield, and plant biomass in rice. *Proceedings of the National Academy of Sciences*, 112(1), 76–81.

Sun, C., Fang, J., Zhao, T., Xu, B., Zhang, F., Liu, L., Tang, J., Zhang, G., Deng, X., & Chen, F. (2012). The histone methyltransferase SDG724 mediates H3K36me2/3 deposition at MADS50 and RFT1 and promotes flowering in rice. *The Plant Cell*, 24(8), 3235–3247.

Surdonja, K., Eggert, K., Hajirezaei, M.-R., Harshavardhan, V. T., Seiler, C., Von Wirén, N., Sreenivasulu, N., & Kuhlmann, M. (2017). Increase of DNA methylation at the HvCKX2. 1 promoter by terminal drought stress in barley. *Epigenomes*, 1(2), 9.

Tang, X-M., Tao, X., Wang, Y., Ma, D-W., Li, D., Yang, H., & Ma, X-R. (2014). Analysis of DNA methylation of perennial ryegrass under drought using the methylation-sensitive amplification polymorphism (MSAP) technique. *Molecular Genetics and Genomics*, 289 (6), 1075–1084

Tonosaki, K., & Kinoshita, T. (2015). Possible roles for polycomb repressive complex 2 in cereal endosperm. *Frontiers in Plant Science*, 6, 144.

Tsukada, Y.-I., Fang, J., Erdjument-Bromage, H., Warren, M. E., Borchers, C. H., Tempst, P., & Zhang, Y. (2006). Histone demethylation by a family of JmjC domain-containing proteins. *Nature*, 439(7078), 811–816.

Ueda, M., & Seki, M. (2020). Histone modifications form epigenetic regulatory networks to regulate abiotic stress response. *Plant Physiology*, 182(1), 15–26.

Wang, B., Cheng, D., Chen, Z., Zhang, M., Zhang, G., Jiang, M., & Tan, M. (2019). Bioinformatic exploration of the targets of xylem sap miRNAs in maize under cadmium stress. *International Journal of Molecular Sciences*, 20(6), 1474.

Wang, H., Feng, Q., Zhang, M., Yang, C., Sha, W., & Liu, B. (2010). Alteration of DNA methylation level and pattern in sorghum (*Sorghum bicolor* L.) pure-lines and inter-line F1 hybrids following low-dose laser irradiation. *Journal of Photochemistry and Photobiology B: Biology*, 99(3), 150–153.

Wang, L., Sun, S., Jin, J., Fu, D., Yang, X., Weng, X., Xu, C., Li, X., Xiao, J., & Zhang, Q. (2015a). Coordinated regulation of vegetative and reproductive branching in rice. *Proceedings of the National Academy of Sciences*, 112(50), 15504–15509.

Wang, W.-S., Pan, Y.-J., Zhao, X.-Q., Dwivedi, D., Zhu, L.-H., Ali, J., Fu, B.-Y., & Li, Z.-K. (2011). Drought-induced site-specific DNA methylation and its association with drought tolerance in rice (*Oryza sativa* L.). *Journal of Experimental Botany*, 62(6), 1951–1960.

Wang, X., Cai, J., Liu, F., Dai, T., Cao, W., Wollenweber, B., & Jiang, D. (2014). Multiple heat priming enhances thermotolerance to a later high temperature stress via improving subcellular antioxidant activities in wheat seedlings. *Plant Physiology and Biochemistry*, 74, 185–192.

Wang, X., Vignjevic, M., Liu, F., Jacobsen, S., Jiang, D., & Wollenweber, B. (2015b). Drought priming at vegetative growth stages improves tolerance to drought and heat stresses occurring during grain filling in spring wheat. *Plant Growth Regulation*, 75, 677–687.

Wang, X., Xin, C., Cai, J., Zhou, Q., Dai, T., Cao, W., & Jiang, D. (2016). Heat priming induces trans-generational tolerance to high temperature stress in wheat. *Frontiers in Plant Science*, 7, 501.

Wang, X., Zhang, X., Chen, J., Wang, X., Cai, J., Zhou, Q., Dai, T., Cao, W., & Jiang, D. (2018). Parental drought-priming enhances tolerance to post-anthesis drought in offspring of wheat. *Frontiers in Plant Science*, 9, 261.

Wang, Y., Sun, F., Cao, H., Peng, H., Ni, Z., Sun, Q., & Yao, Y. (2012). TamiR159 directed wheat TaGAMYB cleavage and its involvement in anther development and heat response. *PloS One*, 7(11), e48445.

Wei, L., Zhang, D., Xiang, F., & Zhang, Z. (2009). Differentially expressed miRNAs potentially involved in the regulation of defense mechanism to drought stress in maize seedlings. *International Journal of Plant Sciences*, 170(8), 979–989.

Wu, K., Wang, S., Song, W., Zhang, J., Wang, Y., Liu, Q., Yu, J., Ye, Y., Li, S., & Chen, J. (2020). Enhanced sustainable green revolution yield via nitrogen-responsive chromatin modulation in rice. *Science*, 367(6478), eaaz2046.

Xing, L., Zhu, M., Zhang, M., Li, W., Jiang, H., Zou, J., Wang, L., & Xu, M. (2017). High-throughput sequencing of small RNA transcriptomes in maize kernel identifies miRNAs involved in embryo and endosperm development. *Genes*, 8(12), 385.

Xing, M.-Q., Zhang, Y.-J., Zhou, S.-R., Hu, W.-Y., Wu, X.-T., Ye, Y.-J., Wu, X.-X., Xiao, Y.-P., Li, X., & Xue, H.-W. (2015). Global analysis reveals the crucial roles of DNA methylation during rice seed development. *Plant Physiology*, 168(4), 1417–1432.

Xu, Y., Zhang, S., Lin, S., Guo, Y., Deng, W., Zhang, Y., & Xue, Y. (2016). WERAM: A database of writers, erasers and readers of histone acetylation and methylation in eukaryotes. *Nucleic Acids Research*, gkw1011.

Yan, K., Xu, H., Cao, W., & Chen, X. (2015). Salt priming improved salt tolerance in sweet sorghum by enhancing osmotic resistance and reducing root Na+ uptake. *Acta Physiologiae Plantarum*, 37, 1–10.

Yan, Y., Zhang, Y., Yang, K., Sun, Z., Fu, Y., Chen, X., & Fang, R. (2011). Small RNAs from MITE-derived stem-loop precursors regulate abscisic acid signaling and abiotic stress responses in rice. *The Plant Journal*, 65(5), 820–828.

Yang, H., Liu, X., Xin, M., Du, J., Hu, Z., Peng, H., Rossi, V., Sun, Q., Ni, Z., & Yao, Y. (2016). Genome-wide mapping of targets of maize histone deacetylase HDA101 reveals its function and regulatory mechanism during seed development. *The Plant Cell*, 28(3), 629–645.

Yu, Y., Zhen, S., Wang, S., Wang, Y., Cao, H., Zhang, Y., Li, J., & Yan, Y. (2016). Comparative transcriptome analysis of wheat embryo and endosperm responses to ABA and H 2 O 2 stresses during seed germination. *BMC Genomics*, 17, 1–18.

Zhai, J., Zhang, H., Arikit, S., Huang, K., Nan, G.-L., Walbot, V., & Meyers, B. C. (2015). Spatiotemporally dynamic, cell-type–dependent premeiotic and meiotic phasiRNAs in maize anthers. *Proceedings of the National Academy of Sciences*, 112(10), 3146–3151.

Zhang, H., Lang, Z., & Zhu, J.-K. (2018). Dynamics and function of DNA methylation in plants. *Nature Reviews Molecular Cell Biology*, 19(8), 489–506.

Zhang, M., An, P., Li, H., Wang, X., Zhou, J., Dong, P., Zhao, Y., Wang, Q., & Li, C. (2019). The miRNA-mediated post-transcriptional regulation of maize in response to high temperature. *International Journal of Molecular Sciences*, 20(7), 1754.

Zhang, Y.-C., Yu, Y., Wang, C.-Y., Li, Z.-Y., Liu, Q., Xu, J., Liao, J.-Y., Wang, X.-J., Qu, L.-H., & Chen, F. (2013). Overexpression of microRNA OsmiR397 improves rice yield by increasing grain size and promoting panicle branching. *Nature Biotechnology*, 31(9), 848.

Zhao, B., Liang, R., Ge, L., Li, W., Xiao, H., Lin, H., Ruan, K., & Jin, Y. (2007). Identification of drought-induced microRNAs in rice. *Biochemical and Biophysical Research Communications*, 354(2), 585–590.

Zhao, L., Wang, P., Hou, H., Zhang, H., Wang, Y., Yan, S., Huang, Y., Li, H., Tan, J., & Hu, A. (2014). Transcriptional regulation of cell cycle genes in response to abiotic stresses correlates with dynamic changes in histone modifications in maize. *PLoS One*, 9(8), e106070.

Zhao, M., Tai, H., Sun, S., Zhang, F., Xu, Y., & Li, W.-X. (2012). Cloning and characterization of maize miRNAs involved in responses to nitrogen deficiency. *PloS One*, 7(1), e29669.

Zhao, T., Zhan, Z., & Jiang, D. (2019). Histone modifications and their regulatory roles in plant development and environmental memory. *Journal of Genetics and Genomics*, 46(10), 467–476.

Zhao, X. Y., Hong, P., Wu, J. Y., Chen, X. B., Ye, X. G., Pan, Y. Y., Wang, J., & Zhang, X. S. (2016). The tae-miR408-mediated control of *TaTOC1* genes transcription is required for the regulation of heading time in wheat. *Plant Physiology*, 170(3), 1578–1594.

Zhao, Y., Xu, Z., Mo, Q., Zou, C., Li, W., Xu, Y., & Xie, C. (2013). Combined small RNA and degradome sequencing reveals novel miRNAs and their targets in response to low nitrate availability in maize. *Annals of Botany*, 112(3), 633–642.

Zheng, L., Zhang, X., Zhang, H., Gu, Y., Huang, X., Huang, H., Liu, H., Zhang, J., Hu, Y., & Li, Y. (2019). The miR164-dependent regulatory pathway in developing maize seed. *Molecular Genetics and Genomics*, 294, 501–517.

Zheng, X., Chen, L., Xia, H., Wei, H., Lou, Q., Li, M., Li, T., & Luo, L. (2017). Transgenerational epimutations induced by multi-generation drought imposition mediate rice plant's adaptation to drought condition. *Scientific Reports*, 7(1), 39843.

Zhou, L., Liu, Y., Liu, Z., Kong, D., Duan, M., & Luo, L. (2010). Genome-wide identification and analysis of drought-responsive microRNAs in *Oryza sativa*. *Journal of Experimental Botany*, 61(15), 4157–4168.

Zhou, M., Palanca, A. M. S., & Law, J. A. (2018). Locus-specific control of the de novo DNA methylation pathway in *Arabidopsis* by the CLASSY family. *Nature Genetics*, 50(6), 865–873.

Section III

Epigenetics in Plant Physiological Processes

11 Epigenetic Regulation of Photosynthesis in Plants

Hanuman Sharma, Renu Sharma, Apexa Pareek and Payal Lodha

11.1 INTRODUCTION

The sun is the sole energy source on Earth, and life on Earth is a gift from the sun. Every living being on Earth relies on the sun for energy either directly or indirectly. During photosynthesis, energy from the sun is stored by creating chemical bonds between carbon dioxide molecules that are present in the atmosphere. The organic compounds that are created because of this process are then utilized as a source of energy in different metabolic functions. These chemical molecules are continuously being passed on to other organisms, and they provide them with the energy that they have acquired from the sun.

Photosynthesis is the process that plants, algae and specific microbes use to convert light energy into chemical energy. It is a fundamental mechanism for these organisms, establishing the foundation of life on Earth. This process predominantly occurs in chloroplasts, where light energy is absorbed by pigments such as chlorophylls and carotenoids, facilitating the production of organic compounds like sugars (*Häder,* 2022; *Dai & Zhang,* 2019). Photosynthesis has two primary phases: the light reaction and the Calvin cycle. During the light reaction, chlorophyll absorbs light energy, resulting in the photolysis of water, which produces oxygen as a by-product, along with the synthesis of ATP and NADPH (*Häder,* 2022; *Sawicki & Chen,* 2020). The energy carriers are employed in the Calvin cycle to convert carbon dioxide into organic compounds, such as glucose, via a series of enzymatic processes. The efficacy of photosynthesis is affected by several parameters, including leaf structure, light intensity and environmental variables such as temperature and CO_2 concentration (*Fonseca-Pereira* et al., 2022; *Baslam* et al., 2020). Advancements in synthetic biology and genetic engineering seek to augment photosynthetic efficiency by optimizing aspects such as reengineering RuBisCO, integrating C4 pathways into C3 plants, and enhancing light usage efficiency (*Fonseca-Pereira* et al., 2022; *Baslam* et al., 2020). Furthermore, photosynthesis is essential for carbon capture, considerably impacting the global carbon cycle by sequestering over 100 billion tons of CO_2 each year (*Bose* et al., 2023; *Baslam* et al., 2020). The process is essential for food production and the synthesis of secondary metabolites with therapeutic potential (*Häder,* 2022). Climate change presents new obstacles, prompting continuous study aimed at comprehending and enhancing photosynthetic systems to provide sustainable food production and environmental stability (*Sharkey,* 2020).

11.1.1 Photosynthesis

Chlorophyll and pigments absorb light to produce ATP and NADPH for the Calvin cycle, which transforms carbon dioxide into organic molecules (*McKee & McKee,* 2020). Photosynthesis, being less effective than artificial solar technology, supports the global carbon cycle and affects climate (*Gloor* et al., 2022). Photosynthesis, which oxygenates the air, has helped complex life forms evolve (*Allakhverdiev* et al., 2019). Climate change requires improving photosynthesis, including light consumption efficiency and CO_2 concentrating mechanisms (*Fonseca-Pereira* et al., 2022; *Timppmann* et al., 2023) to meet the future food and energy demands. Photosynthesis generates medicinal secondary metabolites in addition to energy (*Häder,* 2022). Sustainable agriculture and renewable energy require photosynthetic efficiency research (*Fonseca-Pereira* et al., 2022; *Allakhverdiev,* 2019).

The Calvin cycle and light reactions are essential to photosynthesis, yet they perform different functions. The light reaction occurs in thylakoid membranes, where chlorophyll splits water molecules to release oxygen and generate ATP and NADPH. The Calvin cycle, which converts atmospheric CO_2 into organic molecules like glyceraldehyde-3-phosphate (GAP), using ATP and NADPH from light reactions, relies on these products (*Heldt & Piechulla,* 2005). Numerous enzymes, including RuBisCO, PRK and glyceraldehyde-3-phosphate dehydrogenase (GAPDH), control the Calvin cycle (*Kosztin & Schulten,* 2014). Redox state and nicotinamide dinucleotide availability affect these enzymes, with CP12 forming a complex with PRK and GAPDH to control activity (*McFarlane* et al., 2019). The Calvin cycle is efficiently maintained during daylight due to changes in the redox state mediated by the photosynthetic electron transport chain (*Champigny & Bismuth,* 1976). Additionally, photosystem I lectin affects Calvin cycle enzyme assembly on thylakoid membranes, enhancing RuBisCO activity during light-induced stroma alkalization (*Aleksidze* et al., 2002). A complex control links the Calvin cycle to light responses, increasing photosynthetic efficiency and carbon uptake in plants (*Wedel* et al., 1997).

DOI: 10.1201/9781003589914-14

Photosynthesis depends on nuclear and chloroplast genomes, which encode its components. In eukaryotes, the nuclear genome encodes the majority of the photosynthetic machinery, including structural proteins, enzymes and regulatory factors, due to the limited number of genes in the chloroplast genome (*Goldschmidt-Clermont*, 2013). This dual-genome system needs multilevel control for effective photosynthesis. Regulation occurs at transcriptional, post-transcriptional and post-translational levels, regulated by internal variables like redox status and circadian rhythms as well as environmental factors like light and abiotic stressors (*Hussain* et al., 2021). High-throughput genetic investigations in model organisms like *Chlamydomonas reinhardtii* reveal intricate regulatory networks in photosynthetic complex synthesis and control (*Kafri* et al., 2023) by identifying multiple genes. Environmental variables, including state transitions and qE quenching, affect photosynthetic efficiency by balancing energy distribution and dissipating surplus energy (*Minagawa*, 2017; *Minagawa* et al., 2015). Moreover, chloroplast thioredoxin systems regulate photosynthesis by controlling light and carbon fixation enzyme activity (*Nikkanen & Rintamäki,* 2019). Genetic engineering can improve photosynthetic efficiency and crop yields by manipulating regulatory processes, leading to over 40% yield increases in some cases (*Simkin* et al., 2019). Understanding and optimizing these regulatory processes is crucial in boosting photosynthetic efficiency, a primary goal of the current Green Revolution to boost agricultural output (*Xu*, 2001).

11.1.2 Epigenetics

Epigenetics involves various heritable changes in gene expression that occur without alterations to the DNA sequence, significantly influencing plant development, reproduction and reactions to environmental factors. The modifications encompass DNA methylation, histone modifications, and RNA interference, which together govern gene activity and chromatin structure, thus affecting plant phenotypes and their adaptation to environmental changes (*Plant Epigenomics,* 2022, 2023). In the realm of flowering and seed development, mechanisms like chromatin remodeling and DNA methylation play a crucial role. Important flowering regulators such as FLOWERING LOCUS C (*FLC*) and FLOWERING LOCUS T (*FT*) experience epigenetic changes in reaction to seasonal signals like photoperiod and vernalization, with the Polycomb repressive complex (PRC) significantly influencing these mechanisms (*Mahmood* et al., 2023). Furthermore, epigenetic modifications play a vital role in gametogenesis, embryogenesis and seed development, with DNA methylation patterns being especially significant for embryogenesis (*Mahmood* et al., 2023). Plants, due to their stationary nature, depend significantly on epigenetic regulation to adjust to abiotic stresses like temperature, water and salinity. Methylome DNA sequencing serves as a key technique for investigating these interactions (*Costa* et al., 2024). The transmission of epigenetic memory related to environmental signals occurs through mitosis and can occasionally extend to progeny, playing a role in phenotypic diversity and adaptation (Plant Epigenomics, 2023). Moreover, the diversity found in epigenetics provides extra avenues for variation in crop enhancement, with the possibility of utilizing epi-breeding to leverage these variations for the development of favorable traits in crops (*Jain* et al., 2021). In perennial plants such as almonds, epigenetic phenomena affect traits like dormancy and self-compatibility, underscoring the significance of epigenetics in natural adaptation and agricultural performance (*Fresnedo-Ramírez* et al., 2023). The domain of plant epigenetics is advancing swiftly, providing valuable insights into the intricate interactions between genetic and environmental influences on plant development and adaptation (*Shi* et al., 2024; *Kumari* et al., 2022).

11.1.3 Regulation of Photosynthesis

The regulation of photosynthesis in plants and photosynthetic bacteria takes place at various levels, encompassing genetic, transcriptional and post-transcriptional stages, which facilitates effective adaptation to environmental changes. The expression of photosynthetic genes at the genetic level is regulated by the interplay between nuclear and chloroplast genomes, with anterograde and retrograde signaling pathways facilitating the coordination of gene expression across these compartments. Anterograde signaling encompasses the roles of nuclear-encoded transcriptional and post-transcriptional regulators, whereas retrograde signaling employs photosynthetic mechanisms such as electron transport and redox signaling to affect nuclear gene expression (*Berry* et al., 2013). Light primarily drives transcriptional regulation, acting through photoreceptors like phytochrome to modulate gene expression. In cyanobacteria, the expression of genes in response to light is governed by cis-elements and trans-acting factors, where sigma factors are essential for transcriptional regulation (*Asayama,* 2006). In higher plants, the regulation of transcription is influenced by circadian rhythms, which align gene expression to enhance photosynthetic efficiency and reduce damage from intense light exposure (*Kloppstech,* 1997). The stability of mRNA and the rates of translation play a crucial role in the regulation of genes associated with photosynthesis after transcription; for example, the levels of mRNA are influenced by the rates of synthesis and decay, with mechanisms that either stabilize or destabilize mRNAs in reaction to environmental signals (*Klug,* 1995). Furthermore, redox signaling pathways, including those associated with plastoquinone, play a crucial role in regulating chloroplast gene transcription and post-translational modifications, thereby maintaining balanced photosystem activity (*Allen* et al., 2011). In C4 photosynthesis, regulation takes place at various levels, encompassing transcriptional and post-transcriptional processes, to achieve cell-specific protein accumulation (*Hibberd & Covshoff,* 2010). The regulation of photosynthesis is a complex, multilayered process that integrates various signals to maintain photosynthetic efficiency and adaptability under diverse environmental conditions.

11.1.3.1 Role of Epigenetic Mechanisms

Epigenetic mechanisms are essential in precisely regulating gene expression related to photosynthesis. This is achieved by altering chromatin structure and gene accessibility, which in turn affects how plants respond to environmental stimuli. DNA methylation, histone modifications and chromatin remodeling are essential epigenetic mechanisms that govern the regulation of photosynthetic genes, for example, in *Pinus radiata*, the expression of genes related to photosynthesis, such as *PrRBCA* and *PrRBCS*, is linked to histone modifications. This includes the removal of repressive marks like H3K9me3 and H3K27me3, as well as the acquisition of active marks like AcH4 during the maturation of needles, which improves photosynthetic ability (*Valledor* et al., 2015). In microalgae such as *Nannochloropsis oceanica*, histone modifications like H3K4me2 are associated with the regulation of carbon-concentrating mechanisms, crucial for photosynthesis in low CO_2 environment. The deletion of H3K4me2 methyltransferase in these algae leads to diminished photosynthetic efficiency, underscoring the significance of histone modifications in the regulation of photosynthesis (*Gong* et al., 2024). Moreover, modifications at the epigenetic level, including promoter methylation, can influence the expression of enzymes that play a role in carbon metabolism, thereby impacting both plant respiration and photosynthesis; for instance, the methylation of promoters for genes that encode enzymes involved in the tricarboxylic acid cycle can alter their activity in reaction to light, thus affecting the energy balance within photosynthesizing cells (*Fedorin* et al., 2024). Furthermore, the dynamic nature of the epigenome enables plants to adjust to environmental changes through the modulation of gene expression, a vital process for photosynthesis and responses to stress (*Banerjee & Roychoudhury,* 2018). The capacity for adaptability is demonstrated by plants, which can maintain specific epigenetic modifications as a form of stress memory, thereby improving their resilience to forthcoming environmental challenges (*Jin* et al., 2024; *Yung* et al., 2023). In summary, these epigenetic mechanisms establish a complex regulatory framework that precisely adjusts photosynthetic gene expression, allowing plants to enhance photosynthesis and respond to diverse environmental conditions.

11.1.3.2 Environmental Cues

Environmental factors like light, temperature and CO_2 levels play a crucial role in epigenetic regulation, subsequently impacting gene expression and development of plants. Light serves as a vital environmental cue that influences plant development via phytochrome-mediated mechanisms. Phytochromes, especially *PHYB1* and *PHYB2*, influence DNA methylation and chromatin remodeling, which affects fruit development and ripening by modifying the expression of essential transcription factors and biosynthetic genes in tomato (Bianchetti et al., 2022; Bianchetti et al., 2020). Temperature cues play a crucial role in processes such as thermomorphogenesis and heat stress tolerance in plants. The processes encompass epigenetic mechanisms, including histone modifications, notably H3K4 methylation, which are integral to temperature signaling pathways and stress memory (*Perrella* et al., 2022). Furthermore, epigenetic modifications such as DNA methylation and histone modifications play a vital role in regulating circadian clocks, which are shaped by environmental signals to uphold stable circadian rhythms (*Liu* et al., 2024). Alongside light and temperature, the levels of CO_2 and other non-living environmental factors such as salinity and water availability also induce epigenetic alterations. This phenomenon has been observed in several plant species, including *Arabidopsis thaliana* and *Oryza sativa*, where DNA methylation and histone modifications play crucial roles in responding to these stressors (*Costa* et al., 2024). The incorporation of these environmental signals via chromatin regulators enables plants to modify their developmental programs, thereby ensuring their survival and adaptability (*Wang & Sung*, 2024). Moreover, the involvement of non-coding RNAs, including lncRNAs, in the modulation of gene expression in reaction to environmental stresses such as temperature and salinity underscores the intricate nature of epigenetic regulation in plants (*Yung* et al., 2023). The results highlight the complex interplay between environmental signals and epigenetic processes, which govern how plants react to varying environmental conditions, ultimately influencing their growth, development and ability to withstand stress.

11.2 MOLECULAR MECHANISM OF EPIGENETIC REGULATION OF PHOTOSYNTHESIS

The intricate molecular mechanisms governing the epigenetic regulation of photosynthesis encompass a sophisticated interaction of histone modifications, DNA methylation and chromatin remodeling, all of which collectively shape gene expression and facilitate plant adaptation to environmental shifts. In microalgae, including *Nannochloropsis oceanica*, histone modifications such as H3K4me2 are essential for the regulation of carbon-concentrating mechanisms (CCMs) and photosynthesis, especially in low CO_2 environments. This modification influences the expression of crucial genes such as *NoHINT* and *NoPMA2*, which play a vital role in growth and photosynthetic efficiency (*Gong* et al., 2024). In plants, the regulation of gene expression through epigenetic mechanisms is crucial, as DNA methylation and histone modifications, including acetylation and methylation, play significant roles in the expression of genes related to photosynthesis during developmental stages such as needle maturation in conifers; for example, the expression of genes such as *PrRBCA* and *PrRBCS* correlates with alterations in histone modifications, notably the reduction of H3K9me3 and H3K27me3, alongside an elevation in AcH4, which are essential for achieving complete physiological competence (*Valledor* et al., 2015). Furthermore, the involvement of non-coding RNAs and chromatin remodelers in the repositioning of nucleosomes highlights the intricate nature of epigenetic regulation in plants, aiding in their responses to abiotic stresses and developmental signals (*Halder* et al., 2022). The conservation of these mechanisms across species underscores their essential function in the regulation of photosynthesis and plant development (*Valledor* et al., 2015;). Comprehending these epigenetic mechanisms offers valuable perspectives on improving plant productivity and stress resilience via specific modifications (*Bacova* et al., 2019).

11.2.1 Types of Epigenetic Regulation for Photosynthesis

The regulation of epigenetic factors is essential in influencing photosynthesis in both plants and algae, encompassing a range of mechanisms including DNA methylation, histone modifications and chromatin remodeling. The influence of these processes on gene expression is crucial for adapting to environmental changes and enhancing photosynthetic efficiency. The subsequent sections outline the various forms of epigenetic regulation related to photosynthesis, based on the analysis of the available literature.

11.2.1.1 DNA Methylation

DNA methylation serves as a crucial epigenetic mechanism influencing gene expression through the addition of methyl groups to cytosine residues within DNA. This alteration has the potential to either silence or activate genes, contingent upon the specific context in which it occurs. In pineapple, there are distinct differences in DNA methylation patterns between photosynthetic and non-photosynthetic tissues, particularly with a significant reduction of CHH methylation observed in non-photosynthetic tissues. The observed differential methylation is associated with the expression of genes related to the CAM pathway, including *beta-CA* and *PEPC*, which play a vital role in photosynthesis under arid conditions (*Shi* et al., 2024). In conifers such as *Pinus radiata*, alterations in DNA methylation correlate with the regulation of genes related to photosynthesis during the development of needles, highlighting its significance in developmental processes (*Valledor* et al., 2015).

11.2.1.2 Histone Modifications

Histone modifications, such as acetylation and methylation, are crucial in the regulation of chromatin structure and gene expression. In the microalga *Nannochloropsis oceanica*, specific histone modifications like H3K4me2 and H3K27ac are associated with the regulation of carbon-concentrating mechanisms (CCMs) in low CO_2 environment. The modifications affect the expression of essential photosynthetic genes and play a vital role in adapting to fluctuating environmental CO_2 levels (*Gong* et al., 2024). In *Pinus radiata*, histone acetylation (AcH4) and methylation (H3K9me3, H3K27me3) play a crucial role in regulating photosynthetic and stress response genes during needle maturation (Valledor et al., 2015).

11.2.1.3 Chromatin Remodeling

Chromatin remodeling entails the rearrangement of nucleosomes, influencing the accessibility of transcription factors to DNA and consequently modulating gene expression. In *Nannochloropsis oceanica*, the dynamic positioning of nucleosomes is linked to the regulation of CCM-related genes, underscoring the significance of chromatin remodeling in photosynthetic adaptation (*Gong* et al., 2024). The interaction between chromatin remodeling and histone modifications is observed in the regulation of photosynthetic genes in response to environmental factors, including light and nutrient availability (*Bacova* et al., 2019).

11.2.1.4 Mechanisms Involving RNA

Non-coding RNAs (ncRNAs) are increasingly recognized as significant regulators of epigenetic modifications, impacting gene expression via mechanisms like RNA-directed DNA methylation (RdDM). These RNA-based mechanisms play a crucial role in regulating photosynthetic genes through the modulation of chromatin structure and gene expression patterns (*Halder* et al., 2022).

Although these studies primarily concentrate on the positive regulation of photosynthesis via epigenetic mechanisms, it is crucial to acknowledge that epigenetic regulation may also result in the repression of photosynthetic genes, for example, in *Arabidopsis*, shade conditions can inhibit photosynthetic genes by interfering with the DNA binding of transcription factors, highlighting the intricate relationship between environmental factors and epigenetic regulation (*Kim* et al., 2003). This emphasizes the complex functions of epigenetic mechanisms in either facilitating or suppressing photosynthesis, influenced by the surrounding environment and the stage of development (Tables 11.1 –11.3).

11.2.2 Genes for Necessary Enzymes of Photosynthesis and Their Epigenetic Regulation

The *Calvin* cycle, an essential component of photosynthesis, encompasses various key enzymes that enable the transformation of carbon dioxide into organic compounds. Several enzymes are emphasized for their functions and regulations within the *Calvin* cycle. Fructose 1,6-bisphosphatase (FBPase) is identified as a soluble enzyme that transitions from an inactive to an active form via the reduction of disulfide groups. The process is enabled by the photochemical generation of reducing equivalents through the ferredoxin-thioredoxin system. The activation of FBPase

TABLE 11.1
Genes for leaf development and their epigenetic regulation

Gene/protein	Function	Epigenetic regulation	References
WUSCHEL (*WUS*)	Maintains the stem cell niche in the shoot apical meristem (SAM). Mutations lead to early termination of the shoot meristem and the absence of leaf primordia.	DNA methylation, histone modification, chromatin remodeling, and non-coding RNAs	*Cheng* et al. (2024)
KNOXI	Class I KNOTTED1-like homeobox proteins; maintain SAM and regulate compound leaf development. Misexpression leads to abnormal leaf shapes.	Histone modifications, specifically H3K4me3 and H3K9me2 marks	*De-la-Peña* et al. (2012)
ASYMMETRIC LEAVES 1 (AS1)	Represses *KNOXI* genes to influence leaf form and polarity. Mutations lead to altered leaf morphology.	Chromatin remodeling	*Matsumura* et al. (2016)
ASYMMETRIC LEAVES2 (AS2)	Works with AS1 to repress *KNOXI* genes and regulate adaxial-abaxial polarity.	Chromatin remodeling	*Matsumura* et al. (2016)
PIN-FORMED1 (PIN1)	Auxin efflux transporter; ensures proper auxin delivery for leaf primordia initiation.	Epigenetic regulation, specifically through histone deacetylation, affects the degradation of PIN-FORMED1 (PIN1) in *Arabidopsis*	*Nguyen* et al. (2013)
ARGOS and *SAUR*	Regulate auxin-mediated cell proliferation and expansion, influencing leaf size.	SAURs (SMALL AUXIN UP RNAs) are epigenetically regulated through the eviction of the auxin-induced histone variation H2A.Z	*Sun* et al. (2022)
TCP2, TCP3, TCP4	Regulate cell proliferation and leaf size; targeted by miR319.	miR319 is a kind of microRNA, a small non-coding RNA molecule that regulates gene expression	*Cao* et al. (2022)
miR396	Targets *GRF* genes to regulate leaf size and cell number. Overexpression reduces leaf size.	In *Arabidopsis thaliana*, miR396 adversely controls leaf size and vegetative phase transition	*Hou* et al. (2019)
GROWTH-REGULATING FACTOR (GRF)	Promotes cell proliferation and leaf growth. Mutations result in smaller leaves. Overexpression increases leaf size.	DNA methylation and histone modifications	*Qi* et al. (2023)
GRF-interacting factor *(GIF)*	Transcriptional co-activator; interacts with GRFs to promote cell proliferation.	GIF1 regulates gene expression by interacting with the SWI/SNF chromatin-remodeling complex, which affects DNA accessibility	*Li* et al. (2022)
DA1	Regulates cell proliferation duration, influencing organ size. Mutations lead to oversized leaves.	*DA1* expression is controlled by transcription factors and chromatin-remodeling enzymes that alter the epigenetic histone coding	*Chen* et al. (2021)
EOD1/BB	Activates DA1 peptidase activity to regulate organ growth.	DNA methylation and histone modification	*Lauria & Rossi* (2011)

is crucial for regulating CO_2 fixation in response to light. Sedoheptulose 1,7-bisphosphatase (SBPase) plays a role in the Calvin cycle. Like FBPase, SBPase is activated by light via a reduction mechanism. The activation of SBPase is essential for regulating the Calvin cycle, especially during the transition from dark to light conditions. Ribulose 5-phosphate kinase (Ru5P kinase) is an enzyme that exhibits activation in the presence of light. Recent studies indicate a significant activation of Ru5P kinase upon illumination, which is crucial for the role of the Calvin cycle in converting ribulose 5-phosphate into ribulose 1,5-bisphosphate, a substrate for CO_2 fixation. The light activation of NADP glyceraldehyde 3-phosphate dehydrogenase is minimal, whereas 3-phosphoglycerate kinase exhibits a moderate two-fold activation. These enzymes participate in the later phases of the *Calvin* cycle, facilitating the transformation of 3-phosphoglycerate into glyceraldehyde 3-phosphate, a precursor for glucose synthesis (*Wirtz* et al., 1982).

TABLE 11.2
Genes for chloroplast development and their epigenetic regulation

Category	Gene/protein	Function	Role in chloroplast development	References
Thylakoid assembly	*CURT1*	Membrane remodeling protein	Essential for the formation and structure of thylakoid membranes	*Sandoval-Ibañez* et al. (2021)
	VIPP1	Membrane remodeling protein	Crucial for proper thylakoid assembly	*Zeilfelder* and *Schroda* (2024)
Chloroplast division	FtsZ1, FtsZ2	Inner Z-ring formation	Facilitates contractile machinery for chloroplast division	*An* et al. (2024)
	PDV1, PDV2	Recruit DRP5B	Establishes outer contractile ring for chloroplast constriction	*Sun* et al. (2023)
	DRP5B	Dynamin protein	Completes division by membrane "pinching"	*Sun* et al. (2020)
Nuclear-encoded genes	Cp and PhANGs	Encode proteins for import, RNA processing, protein maturation and degradation	Regulate chlorophyll biosynthesis, metabolite transport and photosystem assembly	*Barajas-López* et al. (2013)
	TOC, TIC	Protein import complexes	Facilitate the import of nuclear-encoded proteins into chloroplasts	*Liu* et al. (2022)
	SIGMA factors (SIG)	Initiate transcription by plastid-encoded polymerases	Regulate transcription of chloroplast genes	*Li* et al. (2021)
Chloroplast genome	*PEP, NEP*	RNA polymerases	Facilitate transcription of chloroplast genes; essential for chloroplast gene expression	*Liebers* et al. (2018)
Retrograde signaling	*GUN1*	PPR family protein	Regulates retrograde signaling, mRNA turnover and tetrapyrrole synthesis under stress or developmental changes	*Tang* et al. (2024)
Transcription factors	*GLK1, GLK2*	GARP family transcription factors	Master regulators of chloroplast development genes; control cell-specific chloroplast biogenesis in C4 plants	*Ding* et al. (2023)
	GNC	GATA family transcription factor	Represses negative regulators of photosynthesis; positively influences chloroplast biogenesis	*Zubo* et al. (2018)
	CGA1 (GNL)	GATA family transcription factor	Activates chlorophyll biosynthesis genes; promotes chloroplast development in non-green tissues	*Xu* et al. (2017)
	HY5	Central light-responsive transcription factor	Promotes expression of photosynthesis-related genes in response to light	*Xu* et al. (2016)
Novel proteins	*BPG2, BPG3*	Novel chloroplast proteins	Regulate chloroplast development, rRNA accumulation and photosystem II electron transport	*Nakano* et al. (2010)
Light regulation	Phytochromes (*PHY*)	Red/far-red light receptors	Activates transcription factors (e.g., PIFs, *HY5*) for photosynthesis gene expression	*Griffin* et al. (2020)
	Cryptochromes (*CRY*)	Blue light receptors	Inhibits COP1 to positively regulate chloroplast-related genes	*Ma* et al. (2022)
	PIFs	Transcription factors inhibited by phytochromes	Repress chloroplast development genes in the dark	*Zhang* et al. (2021)
	COP1-SPA1	Ubiquitin-proteasome system components	Regulate degradation of phytochromes and transcription factors in dark conditions	*Herrera* (2017)
Hormonal regulation	*AHK2, AHK3, CRE1/AHK4*	Histidine kinases in cytokinin signaling	Regulate early CK-responsive genes involved in chloroplast function	*Cortleven* and *Schmülling* (2015)
	ARR1, ARR10, ARR12, CRF2	B-type response regulators and transcription factors	Positively regulate chlorophyll biosynthesis and chloroplast division in response to cytokinin	*Pogson* and *Albrecht* (2011)
	AUXIN RESPONSE FACTORS	Regulate genes in response to auxin	Influence chloroplast gene expression and development in roots and leaves	*Salazar-Iribe* & *De-la-Peña* (2020)
	DELLAs	Negative regulators of gibberellin signaling	Inhibit PIF proteins and promote photosynthesis-associated gene expression	*Cheminant* et al. (2011)

TABLE 11.3
Genes for photosynthetic regulation

Gene/protein name	Enzyme name	Function in photosynthesis	References
rbcL, rbcS	RuBisCO (ribulose-1,5-bisphosphate carboxylase/oxygenase)	Catalyzes CO_2 fixation in the Calvin cycle, converting RuBP and CO_2 into 3-phosphoglycerate (3-PGA).	*Xu* et al. (2023)
cbbM	Form II RuBisCO	Alternative RuBisCO form found in some bacteria, involved in CO_2 fixation.	*Romagnoli* and *Tabita* (2006)
Rca-α and Rca-β	RuBisCO activase (Rca)	Function in maintaining RuBisCo activation and providing thermo-protection during carbon fixation in C4 grasses.	*Kim* et al. (2021)
psaA, psaB	Photosystem I core proteins	These genes help to assemble the PSI complex.	*Chotewutmontri* et al. (2020)
psbA, psbD	Photosystem II core proteins (D1, D2)	Encode proteins involved in the chloroplast photosystem II complex.	*Shimmura* et al. (2017)
petA, petB, petD, PetG, PetL, PetN and PetM	Cytochrome b6f complex	Mediate electron transfer between photosystem II and photosystem I.	*Sandoval-Ibañez* et al. (2022)
atpH	ATP synthase subunits	Essential in photosynthesis by affecting the c-ring stoichiometry of ATP synthase in chloroplasts.	*Yamamoto* et al. (2023)
At1g15980 (Ndf1), At1g64770 (Ndf2), At3g16250 (Ndf4)	NADH dehydrogenase complex	Contributes to cyclic electron flow around photosystem I for ATP production.	*Takabayashi* et al. (2009)
ycf3, ycf4	Photosystem assembly factors	Assist in the assembly of photosystem I and II complexes.	*Khan* et al. (2022)
fbaA, fbaB	Fructose-1,6-bisphosphate aldolase	Functions in the Calvin cycle to convert fructose-1,6-bisphosphate to glyceraldehyde-3-phosphate and dihydroxyacetone phosphate.	*Assié* et al. (2020)
gapA	Glyceraldehyde-3-phosphate dehydrogenase	Plays a role in the Calvin cycle by converting 1,3-bisphosphoglycerate to glyceraldehyde-3-phosphate.	*Robbens* et al. (2007)

11.3 IMPACT OF ENVIRONMENT ON EPIGENETIC REGULATION OF PHOTOSYNTHESIS

The process of photosynthesis is intricate and shaped by numerous environmental factors, each significantly impacting the efficiency and rate of photosynthetic activity. Important elements encompass light intensity, temperature, CO_2 concentration, nutrient accessibility and abiotic stressors like drought and salinity. Grasping these influences is essential for enhancing plant productivity and adjusting to evolving environmental conditions (Figure 11.1).

11.3.1 Light Intensity

Light serves as a crucial catalyst in photosynthesis, supplying the necessary energy for transforming CO_2 and water into glucose and oxygen. Photosynthetic rates typically rise with increasing light intensity until they reach a certain threshold, after which they may level off or decline due to photoinhibition (*Taiz* et al., 2024; *Yamori* & *Yamori*, 2016). Recent studies investigated the impact of different light qualities on the photosynthetic processes of *Arabidopsis thaliana*, focusing on blue light (BL; 450 nm), amber light (AL; 595 nm) and red light (RL; 650 nm), with fluorescent light (FL; 400–700 nm) serving as a control. The process of photosynthesis, crucial for the growth of plants, is influenced by the quality and quantity of light. This, in turn, impacts the photosynthetic machinery, encompassing photosystem I (PSI), photosystem II (PSII) and associated electron transport carriers. The application of red light (RL) resulted in a notable increase in the net photosynthetic rate (Pn) across all the accessions of *A. thaliana.* The rise in Pn corresponded with improved leaf area development and biomass, suggesting that RL promotes effective energy uptake and plant adaptation. The study noted an increase in the chlorophyll a to b ratio under RL, indicating a proficient electron transfer system within the chloroplast membrane, which in turn enhances photosynthetic efficiency. Amber light (AL) demonstrated a more complex influence. While AL did not have a significant impact on photosynthetic activity compared to FL, it resulted in negative morphological characteristics, such as reduced leaf area or reduced growth and biomass in specific accessions. The persistent Pn under AL, even with diminished chlorophyll levels, suggests that AL could trigger unidentified mechanisms within

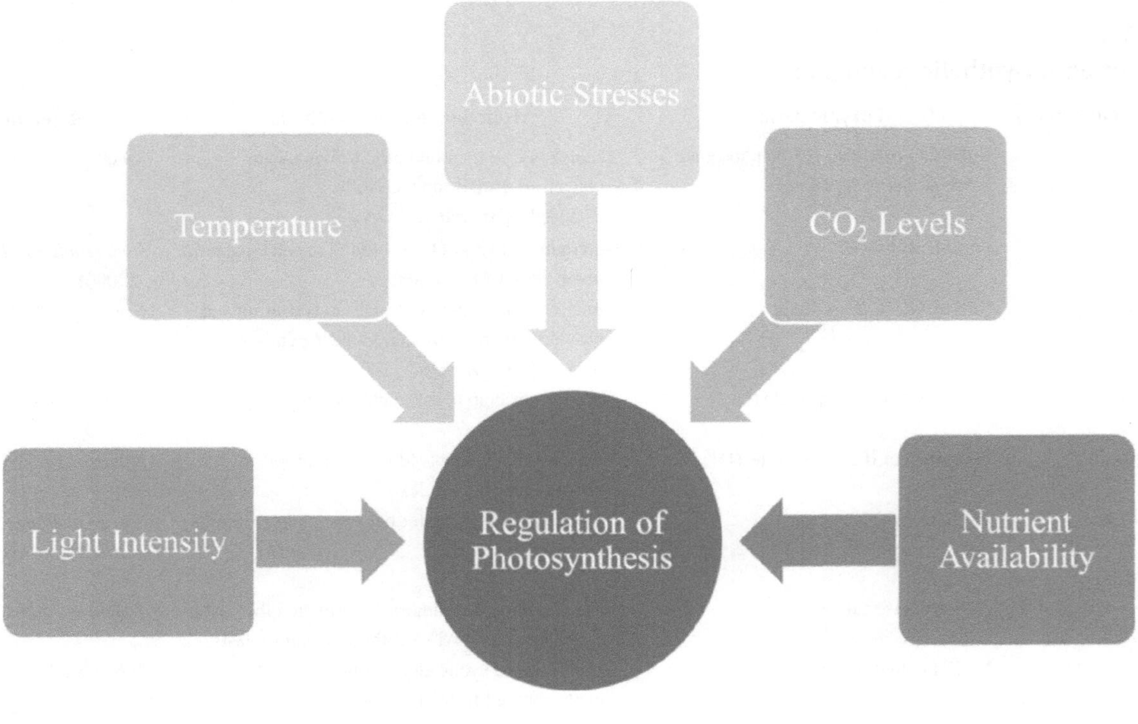

FIGURE 11.1 Regulators of the photosynthetic process.

the photosynthetic process. The study revealed changes in the expression of marker genes associated with photosynthetic light reactions, indicating potential protective strategies in reaction to light stress. BL was noted to increase the levels of protective pigments like anthocyanins, which help protect plants from stress caused by light exposure. However, the impact on biomass production differed among accessions, as some accessions showed decreased biomass despite increased leaf area growth. BL may initiate photoprotective mechanisms, leading to variations in pigment accumulation and biomass production across different accessions (*Yavari* et al., 2021). MET1-dependent CG methylation inhibits the expression of blue light receptor genes, *CRYPTOCHROME 1 (CRY1)* and *CRY2*, indicating that light signaling is affected by MET1-mediated DNA methylation (*Shim* et al., 2021).

The exposure to light leads to the demethylation of four heavily methylated cytosines within the maize *PEPC* promoter, which corresponds with an increase in *PEPC* expression in the leaves. The process of demethylation is uniquely observed in mesophyll (M) cells, while it is absent in bundle sheath (PBS) cells. The influence of light on methylation status is evident; however, cytosine methylation does not directly control the cell-specific expression of PEPC. This indicates a complex relationship among light, methylation and the regulation of gene expression (*Tolley* et al., 2012). Light triggers specific acetylation of histone H4 lysine 5 and histone H3 lysine 9 within the promoter and transcribed region of the C4-Pepc gene, showcasing a distinct pattern of epigenetic regulation. The acetylation process is completely reversible in the absence of light and does not rely on transcription, indicating that light has a direct effect on the activity of histone deacetylase (HDAC) (*Offermann* et al., 2008). The impact of illumination on NADP-malic enzyme (NADP-ME) activity is noteworthy, indicating a potential regulatory function of light in this process. Under light conditions, there was a significant increase in NADP-ME activity with different malate concentrations, suggesting feedback inhibition. Furthermore, the study emphasizes that the pattern of polymorphic gene expression could facilitate molecular modulation of NADP-ME in response to light and darkness, suggesting that light influences epigenetic regulation associated with this enzyme in C4 weed species (*Sarkar* et al., 2020). Light plays a crucial role in the initiation of *rbcL* mRNA translation, which is responsible for encoding the primary subunit of RuBisCo, within barley chloroplasts. This activation predominantly takes place during the elongation phase of translation, rather than at the initiation stage. Although translation initiation complexes form under both dark and light conditions, the transition from initiation to elongation is hindered in the absence of light. The removal of inhibition through illumination allows ribosomes to progress into the elongation phase, thereby promoting protein synthesis (*Kim & Mullet*, 2003). Plants have evolved strategies to enhance light absorption in low-light environment while safeguarding themselves from potential harm caused by excessive light, including dynamic photoprotection and non-photochemical quenching (*Yamori & Yamori*, 2016; *Kaiser* et al., 2015).

11.3.2 Temperature

The influence of temperature on enzymatic activities is significant, especially regarding the processes within the

Calvin–Benson cycle related to photosynthesis. Optimal temperature promotes enzyme activity, whereas extreme temperature can hinder photosynthesis by impacting enzyme stability and function (*Murchie* et al., 2022). Elevated temperature results in a significant increase in the expression of *DRM* genes, such as *DRM1*, in potato genotypes that exhibit heat tolerance. Increased temperature could potentially boost the activity of DNA methyltransferases and demethylases, thereby impacting epigenetic regulation. The pathways of active DNA demethylation and RNA-directed DNA methylation play a role in potentially enhancing the expression of positive regulators of tuberization during heat stress, which in turn affects the overall tuberization process in potato (*Dutta* et al., 2022). Heat stress influences epigenetic regulation in *Arabidopsis*, specifically noting that extended exposure to heat can result in the alleviation of silencing at loci within constitutive heterochromatin (*Popova* et al., 2013). Elevated temperature notably affects the epigenetic control of numerous genes, particularly those associated with light harvesting and components of the photosystem. High temperature stress resulted in enhanced DNA demethylation and elevated levels of histone modifications (H3K9ac and H3K4me3), potentially leading to the activation of gene expression. Transcription factors like HY5, HYH and PIF3 played significant roles in responding to temperature stresses, underscoring their involvement in the regulation of chlorophyll-binding and photosystem gene expression (*Yang* et al., 2022). The influence of temperature on the epigenetic control of HDA19/HD1 is a complex phenomenon that encompasses multiple mechanisms of chromatin alteration and gene expression management. In *Arabidopsis*, histone deacetylase *HDA19*, together with *HDA9* and *HDA15*, is essential for plant response to increased ambient temperature. Mutations in *hda19* lead to the increased expression of stress response genes at both 20°C and 27°C, suggesting its function in inhibiting these genes under standard conditions (Shen et al., 2019). The impact of high-temperature stress on DNA methylation patterns in *Pyropia haitanensis* could potentially alter the expression of various genes, particularly those associated with photosynthesis. The analysis revealed that the methylated regions of the sequenced fragments play a role in photosynthesis, indicating that changes in DNA methylation may influence the expression of photosystem I subunits. This epigenetic regulation could improve the ability of the organism to adapt to high-temperature stress by influencing the expression of genes related to stress responses in photosynthesis (*Yu* et al., 2018). Thermal acclimation enables plants to modify their photosynthetic systems in response to temperature variations, ensuring sustained photosynthetic efficiency over a spectrum of temperature (*Cousins* et al., 2014).

11.3.3 CO_2 Levels

Increased levels of CO_2 can boost photosynthetic rates by providing more CO_2 for carboxylation reactions. This is especially advantageous in scenarios where CO_2 acts as a limiting factor (*Taiz* et al., 2024). Increased CO_2 concentration exerts a complex influence on photosynthesis, affecting a range of physiological and biochemical processes within plants. Increased CO_2 concentration typically enhances photosynthetic rates by improving carbon fixation efficiency. This phenomenon is exemplified in *Ormosia hosiei*, where elevated CO_2 levels resulted in a higher maximum assimilation rate and stimulated both carbon and nitrogen metabolism, ultimately leading to greater biomass accumulation (*Wang* et al., 2024). Subtropical tree species exhibited enhanced leaf area-based photosynthetic rates and biomass under elevated CO_2 conditions, with evergreen species demonstrating a more significant response compared to their deciduous counterparts (*Fu* et al., 2024). Nonetheless, the advantages of increased CO_2 levels on photosynthesis can be influenced by environmental conditions like nitrogen availability and water supply; for example, in *Picea abies*, the positive effects of increased CO_2 on photosynthesis were diminished in scenarios of limited nitrogen and water, underscoring the critical role of these resources in optimizing the photosynthetic advantages of elevated CO_2 (*Ofori-Amanfo* et al., 2023). Furthermore, increased levels of CO_2 can alleviate the photoinhibition of photosystem I in fluctuating light environment, as demonstrated in *Arabidopsis thaliana*, by promoting a swift electron sink facilitated by CO_2 fixation (*Tan* et al., 2021). While there are beneficial outcomes, increased CO_2 levels can result in partial closure of stomata, leading to a decrease in water loss that outweighs any limitations on photosynthesis, ultimately improving water use efficiency (*Wolfe & Erickson*, 2022). Nonetheless, this stomatal response may result in reduced transpiration and heightened sensible heat, which could diminish precipitation and contribute to local warming, ultimately adversely affecting photosynthesis via climate feedback mechanisms (*Zhu* et al., 2017). Additionally, increased levels of CO_2 can enhance the temperature optimum for photosynthesis, as shown in studies involving winter wheat and rice, enabling these crops to sustain higher photosynthetic rates across a wider temperature spectrum (*Lv* et al., 2022). The increase in CO_2 typically boosts photosynthesis and biomass production; however, its impacts are intricate and can be affected by interactions with various environmental factors. This complexity requires a detailed comprehension of its function in plant physiology and ecosystem dynamics (*Wang* et al., 2022). Nonetheless, the reaction to elevated CO_2 levels can be influenced by additional elements like nutrient availability and water stress, which could restrict the possible enhancements in photosynthesis (*Morales* et al., 2020).

11.3.4 Nutrient Availability

Plants require 17 essential nutrients for optimal growth, categorized into macronutrients, mesonutrients and micronutrients. Epigenetic mechanisms are increasingly acknowledged for their crucial role in the way plants react to abiotic stresses. The impact of stress leads to the formation of cellular memory, influenced by DNA methylation, histone modifications and signaling proteins. Changes in DNA methylation due to stress can be inherited by future generations, playing a role in stress resilience and promoting diversity in breeding programs. The regulation of DNA methylation is a dynamic process that responds to environmental changes, facilitating stress adaptation in plants. The ability of DNA methylation to be reversible enables swift alterations to genomic DNA, presenting opportunities for innovative approaches in plant breeding. Investigating the responses of plant DNA methylation to nutrient stresses can facilitate the development of plant varieties that exhibit greater stress resistance, thereby advancing crop breeding initiatives.

Deficiencies in nitrogen and phosphorus present considerable obstacles to food production, particularly in scenarios with limited fertilization. On the other hand, excessive application of nitrogen and phosphorus fertilizers can result in environmental contamination. The prevalence of iron and zinc deficiencies necessitates the development of crop varieties that are abundant in these essential nutrients, aiming to combat the widespread nutritional gaps in human diets. The intricate nature of nutrient requirements in plants is further underscored by essential elements such as sulfur, potassium, calcium and magnesium. DNA methylation in plants plays a crucial role in adapting to changes in nutrient availability and ensuring nutrient balance.

RNA-dependent RNA polymerase 2 (RDR2) plays a role in biomass accumulation during nitrogen deficiency in *Arabidopsis thaliana*, suggesting a connection between RdDM and the regulation of nitrogen deficiency (*Meyer* et al., 2019). Alterations in shoot growth-associated single nucleotide polymorphisms (SNPs) and methylation patterns within gene regions subjected to low nitrogen treatment in *Arabidopsis thaliana* highlight the significance of epigenetic regulation in nitrogen-use efficiency (*Kuhlmann* et al., 2020). Low nitrogen levels induce the expression of methylases such as MET1, DRM1 and DRM2, indicating that these enzymes play a significant role in the response to nitrogen stress (*Fan* et al., 2022). In the context of phosphate starvation, alterations in DNA methylation (mC) were primarily observed in transposable elements (TEs), particularly those located near highly expressed stress-related genes (*Secco* et al., 2015). *Yong-Villalobos* et al. (2015) showed that a phosphorus deficiency led to extensive methylation changes throughout the genes in *Arabidopsis thaliana*, which were associated with variations in gene expression levels. Research has revealed a variety of differentially methylated regions (DMRs) when comparing low-phosphate and phosphate-replete conditions, suggesting that DNA methylation patterns play a significant role in the response to phosphorus stress. Reduced levels of Zn lead to decreased DNA methylation and inhibit the enzymes that play a role in maintaining DNA methylation (*Fan* et al., 2022). A deficiency in Zn results in a significant reduction of small RNAs linked to DNA methylation. Iron deficiency leads to extensive hypermethylation in plants, particularly within the CHH context (*Mager* et al., 2018). Genes activated by iron deficiency exhibit a positive correlation with 24 nt siRNAs, indicating that siRNAs are essential in guiding methylation alterations (*Sun* et al., 2021). The deficiency of sulfur induces a significant reaction in the sulfate transporter genes *SULTR1* and *SULTR2*, which show varying levels of methylation in mutated plants when compared to the wild type (*Huang* et al., 2016).

11.3.5 Abiotic Stresses (Drought and Salinity)

Drought stress diminishes photosynthesis mainly by causing stomatal closure, which restricts CO_2 absorption, as well as through non-stomatal factors that influence the photosynthetic machinery (*Morales* et al., 2020). The effects of drought on photosynthetic efficiency in plants are substantial, influencing both stomatal and non-stomatal factors. Drought in wheat leads to a decrease in the maximum functional intensity of the photosynthetic apparatus and slows gas exchange rates during the transition from dark to light. This suggests damage to mesophyll cells rather than an inhibition of stomatal conductance (*Kedruk* et al., 2024). In watermelon, drought stress similarly results in significant decreases in CO_2 assimilation, stomatal conductance and water use efficiency. However, short-term drought does not have a drastic impact on fruit productivity (*Silva* et al., 2024). Drought stress in cowpea leads to a linear decrease in photosynthesis and chlorophyll fluorescence, accompanied by a rise in intrinsic water-use efficiency. This indicates that both stomatal and non-stomatal limitations contribute to the reduction in photosynthesis (*Singh & Reddy,* 2011). Conversely, the cross-tolerance induced by soil salinity in tomato improves photosynthetic efficiency during subsequent drought stress by preserving non-photochemical quenching and cyclic electron flow, thereby safeguarding the photosynthetic apparatus (*Yang* et al., 2020). In Mediterranean plants such as rosemary and lavender, the reductions in photosynthetic parameters due to drought are linked to the downregulation of electron transport instead of photodamage, with complete recovery noted following rainfall (*Nogués & Alegre,* 2002). Furthermore, utilizing kartolin-4 in wheat has the potential to alleviate photosynthetic inhibition caused by drought, indicating a promising avenue for improving drought resilience via biochemical strategies (*Monakhova & Chernyad'ev*, 2002). In summary, drought presents a common challenge to photosynthetic efficiency; however, the degree and mechanisms of its impact differ among species and environmental contexts. Some plants display adaptive traits that enhance their resilience to water scarcity (*Ahmad* et al., 2023; *Wittmann & Pfanz*, 2008).

Salinity stress results in ionic imbalances and osmotic stress, adversely affecting photosynthesis by disrupting cellular homeostasis and enzyme activities. Salinity stress has a profound effect on photosynthetic genes in plants, resulting in a series of physiological and molecular disturbances. Elevated salt levels in the soil significantly impact the photosynthetic system by hindering water absorption, disturbing ion balance, and causing osmotic stress, all of which collectively diminish photosynthesis and lower crop yield (*Ahmed* et al., 2021). Salinity stress leads to detrimental effects on the photosystem II complex, decreases chlorophyll levels, and modifies chloroplast ultrastructure, all of which are essential elements of the photosynthetic machinery (*Zahra* et al., 2022; *Ahmed* et al., 2021). The stress contributes to the generation of reactive oxygen species (ROS), which further harms cellular structures and reduces photosynthetic efficiency (*Ahmed* et al., 2021). At the molecular level, salinity stress triggers the expression of numerous transcription factors and genes associated with photosynthesis, including those that code for photosystem proteins and enzymes that play a role in the *Calvin* cycle (*Ahmed* et al., 2021; *ElSayed* et al., 2022), for example, in rapeseed, the use of polyamines such as spermidine has been demonstrated to improve photosynthetic capacity by upregulating genes associated with the *Calvin* cycle and enhancing antioxidant defense mechanisms, which helps to alleviate the negative impacts of salinity (*ElSayed* et al., 2022). In tomato, salinity stress diminishes photosynthetic capacity while simultaneously increasing the expression of carotenoid biosynthesis genes. This may lead to improved fruit quality, even though it adversely affects overall plant growth (*Leiva-Ampuero* et al., 2020). Moreover, investigations into genetic and proteomic aspects have revealed mechanisms of salt tolerance that include the regulation of ion balance and the production of osmoprotectants, essential for sustaining photosynthetic function in a saline environment (*Balasubramaniam* et al., 2023).

The findings emphasize the intricate relationship between salinity stress and the regulation of photosynthetic genes, pointing to the possibilities for genetic and biotechnological strategies aimed at improving salt tolerance in crops (*Zahra* et al., 2022; *Feng* et al., 2020). Grasping these mechanisms is crucial for formulating strategies that enhance plant resilience to salinity, thereby safeguarding agricultural productivity in regions susceptible to salinity (*Trejo-Téllez*, 2023; *Roy,* 2023). Plants utilize a range of acclimation strategies, including osmotic adjustment and the activation of stress-responsive pathways, to alleviate the impacts of stress (*Murchie* et al., 2022).

The factors outlined are essential in shaping photosynthesis, yet it is crucial to acknowledge the dynamic and interconnected nature of these influences. Environmental conditions frequently vary, requiring plants to constantly modify their photosynthetic processes to ensure optimal efficiency. The dynamic modulation of photosynthesis represents a field of active inquiry, holding considerable implications for agriculture and ecosystem management amid the challenges posed by climate change (*Kaiser* et al., 2015; *Teskey* et al., 1995).

11.4 APPLICATIONS OF EPIGENETIC REGULATION ON PHOTOSYNTHESIS

Adaptability is essential for plant survival and productivity in diverse environmental conditions. The integration of epigenetic regulation with traditional breeding and biotechnological approaches presents a promising avenue for developing crop varieties that are better equipped to handle the stress associated with climate change. By understanding and manipulating the epigenetic mechanisms that control photosynthesis, scientists can enhance plant resilience, optimize growth and improve crop yield.

11.4.1 Improving Crop Yield and Productivity

A promising approach to improve crop yield via epigenetic regulation involves the targeted manipulation of genes associated with carbon fixation, including RuBisCO and phosphoenolpyruvate carboxylase (PEPC) (*Varotto* et al., 2020). By employing techniques like CRISPR-Cas9 to edit epigenetic markers associated with these genes, it is possible to potentially improve their expression under various environmental stresses, leading to increased photosynthetic efficiency (*Qi* et al., 2023). Furthermore, research indicates that small RNAs are essential in facilitating these epigenetic modifications, enabling plants to swiftly adjust to varying conditions, which may be critical in the context of climate change (*Hewezi,* 2018). The convergence of epigenetics and metabolic engineering presents novel avenues for fostering sustainable agricultural practices. It highlights the importance of a holistic strategy that merges genetic modification with conventional breeding techniques to enhance crop resilience and productivity.

11.4.2 Enhancing Stress Tolerance

Improving stress tolerance via epigenetic regulation not only boosts photosynthetic efficiency but also prepares crops to endure the negative impacts of climate change, including drought and heat stress. Recent findings have shown that epigenetic modifications can enable a form of "stress memory", allowing plants to respond more effectively to repeated environmental challenges (*Jin* et al., 2024). For example, altering histone acetylation patterns has demonstrated the ability to improve thermotolerance in multiple species, enabling them to sustain metabolic functions under extreme temperature conditions (*Dar* et al., 2022). Furthermore, combining these epigenetic approaches with conventional farming methods could produce crops that are not only more productive but also demand fewer resources, thereby supporting sustainable agriculture in a time characterized by increasing food needs and environmental challenges.

11.4.3 Climate-Resilient Agriculture

Combining epigenetic regulation with cutting-edge agricultural practices has the potential to drive major progress in the creation of climate-resilient crop varieties; for instance, utilizing genome editing technologies such as CRISPR-Cas9 enables accurate alterations of epigenetic traits and supports the incorporation of advantageous alleles from wild relatives into cultivated species, thereby improving genetic diversity and adaptability (*Jin* et al., 2024). Furthermore, recent investigations have shown that grasping the intricate interactions among various forms of epigenetic modifications – like DNA methylation and histone acetylation – can enhance breeding strategies focused on boosting stress resilience in crops (*Brunel-Muguet* et al., 2022). By concentrating on these diverse strategies, it is possible to develop a new generation of crops that can flourish in the face of increasingly unpredictable environmental conditions, ultimately safeguarding food security while reducing resource consumption.

11.4.4 Epigenetic Breeding Programs

With the increasing demand for sustainable agricultural practices, it is essential for epigenetic breeding programs to tackle the challenge of preserving genetic diversity among crop species. Incorporating epigenetic markers can enhance yield and stress tolerance while preserving beneficial traits from heirloom varieties or wild relatives (*Duarte-Aké* et al., 2023). This method enhances the ability of crops to adjust to various environmental conditions and minimizes dependence on chemical inputs. Moreover, recent findings indicate that the integration of conventional breeding techniques with cutting-edge genomic technologies may enhance the identification of crucial epigenetic regulators linked to favorable phenotypic traits, thus speeding up the creation of new cultivars that are both resilient and resource-efficient (*Brunel-Muguet* et al., 2022). By adopting this comprehensive approach, agricultural systems can enhance biodiversity while maintaining productivity, thereby establishing a stronger food security framework in response to climate variability.

11.4.5 Sustainable Agriculture

The influence of epigenetic regulation goes beyond the enhancement of specific crops, incorporating wider ecological factors that are essential for sustainable agricultural practices. For example, understanding the role of epigenetic mechanisms in shaping plant interactions with soil microbiomes can improve nutrient absorption and strengthen resistance to pathogens, ultimately fostering healthier ecosystems (*Brunel-Muguet* et al., 2022). Recent studies indicate that specific epigenetic modifications can influence root architecture, resulting in better symbiotic interactions with beneficial microbes, thereby enhancing overall plant health and productivity (*Jin* et al., 2024). This comprehensive perspective emphasizes the significance of incorporating epigenetics into breeding initiatives and underscores the need for agroecological methods that utilize these biological interactions. This approach ultimately promotes a more sustainable agricultural system that can adapt to environmental shifts while safeguarding food security.

11.4.6 Understanding Evolution and Adaptation

Deciphering the complex interplay between epigenetic regulation and plant evolution is essential for formulating approaches that improve crop adaptability. Recent studies indicate that epigenetic modifications can enable swift evolutionary responses to environmental stressors, allowing plants to rapidly adjust their phenotypes without requiring alterations in the underlying genetic code (*Jin* et al., 2024). This phenomenon emphasizes the importance of epigenetics in influencing plant resilience and illustrates how these mechanisms can be utilized to develop crops that can flourish under climate-induced challenges. Additionally, investigating the function of epigenetic memory – where past stress experiences shape the future reactions – may reveal innovative pathways for enhancing agricultural methods, especially in areas confronting severe weather conditions (*Brunel-Muguet* et al., 2022). By integrating insights from both evolutionary biology and epigenetics, it is possible to develop more sophisticated breeding programs that ensure long-term sustainability and food security amidst ongoing climatic challenges.

11.5 CONCLUSION

The regulation of epigenetic mechanisms is crucial in influencing photosynthesis, allowing plants to adapt to environmental changes effectively while enhancing energy conversion efficiency. Plants meticulously regulate the expression of genes linked to photosynthetic activity through various mechanisms, including DNA methylation, histone modifications, chromatin remodeling and non-coding RNAs, thereby achieving a harmonious balance between growth, development and stress adaptation. These modifications enable plants to control essential enzymes, maximize light usage and boost carbon uptake, leading to an overall enhancement in photosynthetic efficiency. Environmental factors like light intensity, temperature fluctuations, CO_2 availability and nutrient accessibility play a crucial role in shaping epigenetic modifications, which in turn affect photosynthetic efficiency and plant resilience. Epigenetic memory enables plants to preserve adaptive responses through generations, which can lead to significant long-term evolutionary benefits. Exploring the complex interactions between epigenetic processes and environmental factors offers important perspectives on how plants adapt and thrive in the face of shifting climate conditions. Recent advancements in epigenetic studies, alongside innovative genetic engineering tools such as CRISPR-Cas9, present exciting opportunities to boost crop yield, enhance stress resilience and create agricultural systems that can withstand climate challenges. By leveraging epigenetic modifications, it is possible to manipulate regulatory pathways to optimize photosynthetic efficiency, leading to higher productivity even under suboptimal growing conditions. Furthermore, epigenetic breeding strategies present significant opportunities for enhancing crop traits without modifying the fundamental genetic framework, thereby maintaining biodiversity while boosting agricultural productivity. Given the ongoing global challenges like food security, climate change and environmental stressors that jeopardize agricultural sustainability, it is crucial to incorporate epigenetic regulation into plant breeding and crop management strategies. A comprehensive exploration of epigenetic mechanisms governing photosynthesis has the potential to foster groundbreaking approaches for sustainable agriculture, promoting enduring resilience and effectiveness in crop production. In future, integrating various fields such as epigenetics, plant physiology and biotechnology will be essential for harnessing the complete capabilities of epigenetic regulation to enhance photosynthesis and ensure global food security.

11.6 FUTURE PERSPECTIVES

With the ongoing shifts in global climate and the rising demands of agriculture, understanding the mechanisms of epigenetic regulation in photosynthesis is important to gain significant importance. Future investigations should concentrate on elucidating the specific molecular mechanisms through which epigenetic modifications influence photosynthetic efficiency in response to different environmental conditions. A comprehensive exploration of the roles that DNA methylation, histone modifications and non-coding RNAs play in regulating photosynthetic gene expression will open innovative pathways for improving plant productivity and resilience.

A promising avenue is the advancement of strategies for crop improvement based on epigenetics. In contrast to conventional genetic modification, epigenetic modifications maintain the integrity of the DNA sequence, presenting a more sustainable and socially acceptable method for enhancing crop characteristics. Recent developments in epigenome editing technologies, particularly those utilizing CRISPR-Cas9 for epigenetic alterations, provide accurate regulation of gene expression. This capability allows for the meticulous adjustment of photosynthetic processes, aimed at improving carbon uptake, optimizing water usage and increasing resilience to stress.

The investigation into epigenetic memory in plants offers promising avenues for the production of climate-resilient crops. By understanding how plants retain epigenetic changes in response to environmental stressors, it is possible to harness these mechanisms to breed crops that are better equipped to withstand drought, extreme temperatures, and nutrient deficiencies. Exploring the inheritance of epigenetic modifications across generations is essential for incorporating epigenetics into sustainable crop enhancement initiatives.

Furthermore, the combination of epigenetic studies with synthetic biology and computational modeling has the potential to significantly enhance our understanding of plant adaptation and productivity. Innovative machine learning algorithms have the potential to forecast epigenetic changes linked to peak photosynthetic efficiency, paving the way for precise strategies in crop breeding. In a similar way, synthetic biology methods may facilitate the creation of artificial regulatory networks aimed at improving photosynthetic efficiency tailored to environmental conditions.

Future investigations should delve into the influence of epigenetics on plant–microbe interactions, especially concerning soil microbiomes. Exploring the impact of epigenetic modifications on symbiotic interactions with beneficial microbes may enhance nutrient absorption, bolster disease resistance and promote overall plant vitality, thereby advancing sustainable agricultural methods.

As we progress, collaboration across various fields such as plant biology, genetics, agronomy and data science will be crucial in realizing the complete potential of epigenetic regulation in photosynthesis. Integrating epigenetics into contemporary agricultural strategies allows for the development of high-yielding, climate-resilient crops that enhance global food security and reduce environmental impact. The ongoing investigation into epigenetic mechanisms in photosynthesis promises to enhance our understanding of plant science while also transforming sustainable agriculture for the generations to come.

REFERENCES

Ahmad, N., Irfan, A., Ahmad, H. R., Salma, H., Tahir, M. N., Tamimi, S. A., Sajid, Z. A., Liaquat, G., Nadeem, M., Ali, M. U., Abbasi, G. H., & Ayub, M. A. (2023). *Impact of Changing Abiotic Environment on Photosynthetic Adaptation in Plants.* https://doi.org/10.1007/978-3-031-43729-8_14.

Ahmed, K. B. M., Singh, S., Sadiq, Y., Khan, M. M. A., Uddin, M., Naeem, M., & Aftab, T. (2021). *Photosynthetic and Cellular Responses in Plants under Saline Conditions.* https://doi.org/10.1016/B978-0-323-90943-3.00007-9.

Aleksidze, G. Ya., Litvinov, A. I., & Vyskrebentseva, E. I. (2002). The model of Calvin cycle enzyme organization on thylakoid membranes with the involvement of the photosystem I lectin. *Russian Journal of Plant Physiology.* https://doi.org/10.1023/A:1013780732193.

Allakhverdiev, S. I., Subramanyam, R., & Tomo, T. (2019). International conference on "photosynthesis and hydrogen energy research for sustainability-2017." *Photosynthesis Research.* https://doi.org/10.1007/S11120-018-0596-7.

Allen, J. F., Santabarbara, S., Allen, C. A., & Puthiyaveetil, S. (2011). Discrete redox signaling pathways regulate photosynthetic light-harvesting and chloroplast gene transcription. *PLOS ONE.* https://doi.org/10.1371/JOURNAL.PONE.0026372.

An, J., Wang, L., Hong, C., & Gao, H. (2024). Evolution and functional differentiation of the C-terminal motifs of FtsZs during plant evolution. *Molecular Biology and Evolution.* https://doi.org/10.1093/molbev/msae145.

Asayama, M. (2006). The regulatory system for light-responsive gene expression in photosynthesizing bacteria: Cis-elements and trans-acting factors in transcription and post-transcription. *Bioscience, Biotechnology, and Biochemistry.* https://doi.org/10.1271/BBB.70.565.

Assié, A., Leisch, N., Meier, D. V., Meier, D. V., Gruber-Vodicka, H. R., Tegetmeyer, H. E., Tegetmeyer, H. E., Meyerdierks,

A., Kleiner, M., Hinzke, T., Hinzke, T., Joye, S. B., Saxton, M. A., Saxton, M. A., Dubilier, N., Dubilier, N., Petersen, J. M., & Petersen, J. M. (2020). Horizontal acquisition of a patchwork Calvin cycle by symbiotic and free-living Campylobacterota (formerly Epsilonproteobacteria). *The ISME Journal*, 14(1), 104–122. https://doi.org/10.1038/S41396-019-0508-7.

Bacova, R., Klejdus, B., Ryant, P., Cernei, N., Adam, V. and Huska, D., (2019). The effects of 5-azacytidine and cadmium on global 5-methylcytosine content and secondary metabolites in the freshwater microalgae *Chlamydomonas reinhardtii and Scenedesmus quadricauda. Journal of Phycology*, 55(2), pp.329–342. https://doi.org/10.1111/jpy.12819

Balasubramaniam, T., Shen, G., Esmaeili, N., & Zhang, Y. (2023). Plants' response mechanisms to salinity stress. *Plants*. https://doi.org/10.3390/plants12122253.

Banerjee, A., & Roychoudhury, A. (2018). The gymnastics of epigenomics in rice. *Plant Cell Reports*, 37, 25–49.

Barajas-López, J. de D., Kremnev, D., Shaikhali, J., Piñas-Fernández, A., & Strand, Å. (2013). PAPP5 is involved in the tetrapyrrole-mediated plastid signaling during chloroplast development. *PLOS ONE*, 8(3). https://doi.org/10.1371/JOURNAL.PONE.0060305.

Baslam, M., Mitsui, T., Hodges, M., Priesack, E., Herritt, M. T., Aranjuelo, I., & Sanz-Sáez, Á. (2020). Photosynthesis in a changing global climate: Scaling up and scaling down in crops. *Frontiers in Plant Science*. https://doi.org/10.3389/FPLS.2020.00882.

Berry, J. O., Yerramsetty, P., Zielinski, A. M., & Mure, C. M. (2013). Photosynthetic gene expression in higher plants. *Photosynthesis Research*. https://doi.org/10.1007/S11120-013-9880-8.

Bianchetti, R. E., Bellora, N., Haro, L. A. de, Zuccarelli, R., Rosado, D., Freschi, L., Rossi, M., & Bermúdez, L. (2022). Phytochrome-mediated light perception affects fruit development and ripening through epigenetic mechanisms. *Frontiers in Plant Science*. https://doi.org/10.3389/fpls.2022.870974.

Bianchetti, R. E., Bellora, N., La, de H., Zuccarelli, R., Rosado, D., Freschi, L., Rossi, M., Bermúdez, L., & Bermúdez, L. (2020). The regulatory effect of light over fruit development and ripening is mediated by epigenetic mechanisms. *bioRxiv*. https://doi.org/10.1101/2020.07.30.227447.

Bose, A. (2024). Bacterial photobiohybrids and photosynthesis: Optimizing energy harvesting with bacterial-semiconductor hybrids. *Open Access Government*. https://doi.org/10.56367/oag-042-11124.

Bose, A., O'Shea, R., De, S. K., & Murphy, J. D. (2023). *Carbon Capture by Photosynthesis of Plants*. https://doi.org/10.1016/b978-0-323-93940-9.00052-9.

Brunel-Muguet, S., Vetukuri, R. R., & Testillano, P. S. (2022). Epigenetics for crop adaptation to climate change. *Physiologia Plantarum*. https://doi.org/10.1111/ppl.13835.

Cao, B., Wang, H., Bai, J., Wang, X., Li, X., Zhang, Y., Yang, S., He, Y., & Ju, X. (2022). miR319-Regulated TCP3 modulates silique development associated with seed shattering in Brassicaceae. *Cells*, 11(19), 3096. https://doi.org/10.3390/cells11193096.

Champigny, M.-L., & Bismuth, E. (1976). Role of photosynthetic electron transfer in light activation of Calvin cycle enzymes. *Physiologia Plantarum*. https://doi.org/10.1111/J.1399-3054.1976.TB05034.X.

Cheminant, S., Wild, M., Bouvier, F., Pelletier, S., Renou, J.-P., Erhardt, M., Hayes, S., Terry, M. J., Genschik, P., & Achard, P. (2011). DELLAs regulate chlorophyll and carotenoid biosynthesis to prevent photooxidative damage during seedling deetiolation in Arabidopsis. *The Plant Cell*, 23(5), 1849–1860. https://doi.org/10.1105/TPC.111.085233.

Chen, Y., Inzé, D., & Vanhaeren, H. (2021). Post-translational modifications regulate the activity of the growth-restricting protease DA1. *Journal of Experimental Botany*, 72(9), 3352–3366. https://doi.org/10.1093/JXB/ERAB062.

Cheng, Y., Zhou, Y., & Wang, M. (2024). Targeted gene regulation through epigenome editing in plants. *Current Opinion in Plant Biology*, 80, 102552. https://doi.org/10.1016/j.pbi.2024.102552.

Chotewutmontri, P., Williams-Carrier, R., & Barkan, A. (2020). Exploring the link between photosystem II assembly and translation of the chloroplast psbA mRNA. *Plants*, 9(2), 152. https://doi.org/10.3390/PLANTS9020152.

Cortleven, A., & Schmülling, T. (2015). Regulation of chloroplast development and function by cytokinin. *Journal of Experimental Botany*, 66(16), 4999–5013. https://doi.org/10.1093/JXB/ERV132.

Costa, G. S. da, Cerqueira, A. F., Brito, C. R. de, Mielke, M. S., & Gaiotto, F. A. (2024). Epigenetics regulation in responses to abiotic factors in plant species: A systematic review. *Plants*. https://doi.org/10.3390/plants13152082.

Cousins, A. B., Johnson, M. P., & Leakey, A. D. B. (2014). Photosynthesis and the environment. *Photosynthesis Research*. https://doi.org/10.1007/S11120-013-9958-3.

Xu, D.Q., (2001). Progress in photosynthesis research: From molecular mechanisms to green revolution. *Acta Phytophysiologica Sinica*, *27*(2), pp.97–108.

Dai, W., & Zhang, Q. (2019). *Photosynthesis and Respiration in Woody Plants*. https://doi.org/10.1201/9780429190476-3.

Dar, F. A., Mushtaq, N., Saleem, S. M., Rehman, R. U., Dar, T. H., & Hakeem, K. R. (2022). Role of epigenetics in modulating phenotypic plasticity against abiotic stresses in plants. *International Journal of Genomics*. https://doi.org/10.1155/2022/1092894.

De-la-Peña, C., Nic-Can, G. I., Ojeda, G., Herrera-Herrera, J. L., López-Torres, A., Wrobel, K., & Robert-Díaz, M. L. (2012). KNOX1 is expressed and epigenetically regulated during in vitro conditions in *Agave* spp. *BMC Plant Biology*, 12(1), 203. https://doi.org/10.1186/1471-2229-12-203.

Ding, Q., Jia, L., Jiang, X., Wang, M.-X., Wang, Y. M., Jiang, H., Yu, F., & Yan, C. (2023). Genomic survey and expression analysis of GLKs in watermelon (*Citrullus lanatus*). *Horticulturae*, 9(2), 229. https://doi.org/10.3390/horticulturae9020229.

Duarte-Aké, F., Us-Camas, R., & De-la-Peña, C. (2023). Epigenetic regulation in heterosis and environmental stress: The challenge of producing hybrid epigenomes to face climate change. *Epigenomes*. https://doi.org/10.3390/epigenomes7030014.

Dutta, M., Raturi, V., Gahlaut, V., Kumar, A., Sharma, P., Verma, V., Gupta, V. K., Sood, S., & Zinta, G. (2022). The interplay of DNA methyltransferases and demethylases with tuberization genes in potato (*Solanum tuberosum* L.) genotypes under high temperature. *Frontiers in Plant Science*, 13. https://doi.org/10.3389/fpls.2022.933740.

ElSayed, A. I., Mohamed, A. H., Rafudeen, M. S., Omar, A. A., Awad, M. F., & Mansour, E. H. (2022). Polyamines mitigate the destructive impacts of salinity stress by enhancing photosynthetic capacity, antioxidant defense system, and upregulation of Calvin cycle-related genes in rapeseed (*Brassica napus* L.). *Saudi Journal of Biological Sciences*. https://doi.org/10.1016/j.sjbs.2022.02.053.

Fan, X., Peng, L., & Zhang, Y. (2022). Plant DNA methylation responds to nutrient stress. *Genes*, 13(6), 992. https://doi.org/10.3390/genes13060992.

Fedorin, D. N., Eprintsev, A. T., & Igamberdiev, A. U. (2024). The role of promoter methylation of the genes encoding the enzymes metabolizing di- and tricarboxylic acids in the regulation of plant respiration by light. *Journal of Plant Physiology*. https://doi.org/10.1016/j.jplph.2024.154195.

Feng, Y., Cui, J., Zhou, T., Liu, Y., Yue, C.-P., Huang, J., & Hua, Y.-P. (2020). Comprehensive dissection into morpho-physiologic responses, ionomic homeostasis, and transcriptomic profiling reveals the systematic resistance of allotetraploid rapeseed to salinity. *Plant Biology*. https://doi.org/10.1186/S12870-020-02734-4.

Fonseca-Pereira, P. da, Siqueira, J. A., Monteiro-Batista, R. de C., Vaz, M. G. M. V., Nunes-Nesi, A., & Araújo, W. L. (2022). Using synthetic biology to improve photosynthesis for sustainable food production. *Journal of Biotechnology*. https://doi.org/10.1016/j.jbiotec.2022.09.010.

Fresnedo-Ramírez, J., Anderson, E. S., D'Amico-Willman, K. M., & Gradziel, T. M. (2023). A review of plant epigenetics through the lens of almond. *The Plant Genome*. https://doi.org/10.1002/tpg2.20367.

Fu, P., Zhang, Y., Qi, J.-H., Zhang, Y.-J., Hao, G.-Y., & Finnegan, P. M. (2024). The impact of elevated CO_2 concentration on photosynthesis, growth, and hydraulics of evergreen and deciduous tree seedlings from a subtropical forest in Southwest China. *Agricultural and Forest Meteorology*. https://doi.org/10.1016/j.agrformet.2024.110021.

Gloor, E., Brienen, R. J. W., & Galbraith, D. (2022). *Photosynthesis in Action: The Global View*. https://doi.org/10.1016/b978-0-12-823781-6.00002-2.

Goldschmidt-Clermont, M. (2013). *Photosynthesis, Genetics of*. https://doi.org/10.1016/B978-0-12-374984-0.01159-1.

Gong, Y., Wang, Q., Wei, L., Wang, L., Lv, N., Du, X., Shen, C., Xin, Y., Sun, L., & Xu, J. (2024). The dynamic epigenomic landscape of carbon-concentrating mechanisms in the model industrial oleaginous microalga *Nannochloropsis oceanica*. https://doi.org/10.1101/2024.09.06.611585.

Griffin, J. H. C., Prado, K., Sutton, P., & Toledo-Ortiz, G. (2020). Coordinating light responses between the nucleus and the chloroplast, a role for plant cryptochromes and phytochromes. *Physiologia Plantarum*, 169(4), 515–528. https://doi.org/10.1111/PPL.13148.

Häder, D.-P. (2022). Photosynthesis in plants and algae. *Anticancer Research*. https://doi.org/10.21873/anticanres.16012.

Halder, K., Chaudhuri, A., Abdin, M.Z., Majee, M. and Datta, A., (2022). Chromatin-based transcriptional reprogramming in plants under abiotic stresses. *Plants*, *11*(11), p.1449. https://doi.org/10.3390/plants11111449

Heldt, H.-W., & Piechulla, B. (2005). The Calvin cycle catalyzes photosynthetic CO_2 assimilation. https://doi.org/10.1016/B978-012088391-2/50007-9.

Herrera, N. M. (2017). Analysis of COP1/SPA signaling events in plant developmental responses. https://kups.ub.uni-koeln.de/7957/.

Hewezi, T. (2018). Editorial: Epigenetic regulation of plant development and stress responses. *Plant Cell Reports*. https://doi.org/10.1007/S00299-017-2233-X.

Hibberd, J. M., & Covshoff, S. (2010). The regulation of gene expression required for C4 photosynthesis. *Annual Review of Plant Biology*. https://doi.org/10.1146/ANNUREV-ARPLANT-042809-112238.

Hou, N., Cao, Y., Li, F., Yuan, W., Bian, H., Wang, J., Zhu, M., & Han, N. (2019). Epigenetic regulation of miR396 expression by SWR1-C and the effect of miR396 on leaf growth and developmental phase transition in *Arabidopsis*. *Journal of Experimental Botany*, 70(19), 5217–5229. https://doi.org/10.1093/JXB/ERZ285.

Huang, X.-Y., Chao, D.-Y., Koprivova, A., Danku, J., Wirtz, M., Müller, S., Sandoval, F. J., Bauwe, H., Roje, S., Dilkes, B., Hell, R., Kopriva, S., & Salt, D. E. (2016). Nuclear localised MORE SULPHUR ACCUMULATION1 epigenetically regulates sulphur homeostasis in *Arabidopsis thaliana*. *PLOS Genetics*, 12(9), e1006298. https://doi.org/10.1371/journal.pgen.1006298.

Hussain, S., Ulhassan, Z., Brestic, M., Zivcak, M., Zhou, W., Allakhverdiev, S. I., Yang, X., Safdar, M. E., Yang, W., & Liu, W. (2021). Photosynthesis research under climate change. *Photosynthesis Research*. https://doi.org/10.1007/S11120-021-00861-Z.

Jain, N., Taak, Y., Choudhary, R., Yadav, S., Saini, N., Vasudev, S., & Yadava, D. K. (2021). *Advances and Prospects of Epigenetics in Plants*. https://doi.org/10.1016/B978-0-323-85652-2.00013-0.

Jin, Q., Chachar, M., Ali, A., Chachar, Z., Zhang, P., Riaz, A., Ahmed, N., & Chachar, S. (2024). Epigenetic regulation for heat stress adaptation in plants: New horizons for crop improvement under climate change. *Agronomy*. https://doi.org/10.3390/agronomy14092105.

Kafri, M., Patena, W., Wang, L., Gomer, G., Ergun, S. L., Sirkejyan, A. K., Goh, A., Wilson, A. T., Gavrilenko, S. E., Breker, M., Roichman, A., McWhite, C. D., Rabinowitz, J. D., Cross, F. R., Wühr, M., & Jonikas, M. C. (2023). Systematic identification and characterization of genes in the regulation and biogenesis of photosynthetic machinery. *Cell*. https://doi.org/10.1016/j.cell.2023.11.007.

Kaiser, E., Morales, A., Harbinson, J., Kromdijk, J., Heuvelink, E., & Marcelis, L. F. M. (2015). Dynamic photosynthesis in different environmental conditions. *Journal of Experimental Botany*. https://doi.org/10.1093/JXB/ERU406.

Kedruk, A. S., Kiriziy, D. A., Стасик, O. O., Sokolovska-Sergiienko, O. G., & Tarasiuk, M. V. (2024). Effects of drought on photosynthetic induction in leaves of different wheat genotypes under dark-to-light transition. *Regulatory Mechanisms in Biosystems*. https://doi.org/10.15421/022471.

Khan, M. S., Riaz, R. I., Majid, M. S., Mehmood, K., Mustafa, G., & Joyia, F. A. (2022). The tobacco chloroplast *YCF4* gene is essential for transcriptional gene regulation and plant photoautotrophic growth. *Frontiers in Plant Science*, 13. https://doi.org/10.3389/fpls.2022.1014236.

Kim, J., & Mullet, J. E. (2003). A mechanism for light-induced translation of the rbcL mRNA encoding the large subunit of Ribulose-1,5-bisphosphate carboxylase in barley chloroplasts. *Plant and Cell Physiology*, 44(5), 491–499. https://doi.org/10.1093/PCP/PCG061.

Kim, S. Y., Slattery, R. A., & Ort, D. R. (2021). A role for differential Rubisco activase isoform expression in C4 bioenergy grasses at high temperature. *Gcb Bioenergy*, 13(1), 211–223. https://doi.org/10.1111/GCBB.12768.

Kloppstech, K. (1997). Light regulation of photosynthetic genes. *Physiologia Plantarum*. https://doi.org/10.1111/J.1399-3054.1997.TB00001.X.

Klug, G. (1995). *Post-Transcriptional Control of Photosynthesis Gene Expression*. https://doi.org/10.1007/0-306-47954-0_59.

Kosztin, I., & Schulten, K. (2014). *Structure, Function, and Quantum Dynamics of Pigment-Protein Complexes*. https://doi.org/10.1017/CBO9780511863189.007.

Kuhlmann, M., Meyer, R. C., Jia, Z., Klose, D., Krieg, L.-M., Von Wirén, N., & Altmann, T. (2020). Epigenetic variation at a genomic locus affecting biomass accumulation under low nitrogen in *Arabidopsis thaliana*. *Agronomy*, 10(5), 636. https://doi.org/10.3390/agronomy10050636.

Kumari, P., Khan, S. I., Wani, I. A., Gupta, R., Verma, S., Alam, P., & Alaklabi, A. (2022). Unravelling the role of epigenetic modifications in development and reproduction of angiosperms: A critical appraisal. *Frontiers in Genetics*. https://doi.org/10.3389/fgene.2022.819941.

Lauria, M., & Rossi, V. (2011). Epigenetic control of gene regulation in plants. *Biochimica et Biophysica Acta*, 1809(8), 369–378. https://doi.org/10.1016/J.BBAGRM.2011.03.002.

Leiva-Ampuero, A., Leiva-Ampuero, A., Agurto, M., Agurto, M., Agurto, M., Matus, J. T., Hoppe, G., Hoppe, G., Huidobro, C., Huidobro, C., Inostroza-Blancheteau, C., Reyes-Díaz, M., Stange, C., Canessa, P., Canessa, P., Vega, A., & Vega, A. (2020). Salinity impairs photosynthetic capacity and enhances carotenoid-related gene expression and biosynthesis in tomato (*Solanum lycopersicum* L. cv. Micro-Tom). *PeerJ*. https://doi.org/10.7717/PEERJ.9742.

Li, C., Wang, J., Hu, Z., Yuanyan, X., Huang, Q., Yu, T., Yi, H., Lu, Y., Wang, J., & Cao, M. (2021). A valine residue deletion in ZmSig2A, a sigma factor, accounts for a revertible leaf-color mutation in maize. *Crop Journal*. https://doi.org/10.1016/J.CJ.2021.01.005.

Li, M., Zheng, Y., Cui, D., Du, Y., Zhang, D., Sun, W., Du, H., & Zhang, Z. (2022). GIF1 controls ear inflorescence architecture and floral development by regulating key genes in hormone biosynthesis and meristem determinacy in maize. *BMC Plant Biology*, 22(1). https://doi.org/10.1186/s12870-022-03517-9.

Liebers, M., Chevalier, F., Blanvillain, R., & Pfannschmidt, T. (2018). PAP genes are tissue- and cell-specific markers of chloroplast development. *Planta*, 248(3), 629–646. https://doi.org/10.1007/S00425-018-2924-8.

Liu, H., Li, A., Rochaix, J.-D., & Liu, Z. (2022). Architecture of chloroplast TOC–TIC translocon supercomplex. *Visual Education*, 615(7951), 349–357. https://doi.org/10.1038/s41586-023-05744-y.

Liu, X., Duan, Z., Yu, M., & Liu, X. (2024). Epigenetic control of circadian clocks by environmental signals. *Trends in Cell Biology*. https://doi.org/10.1016/j.tcb.2024.02.005.

Lv, C., Huang, Y., Sun, W., Yu, L., & Hu, Z. (2022). Elevated [CO_2] raises the temperature optimum of photosynthesis and thus promotes net photosynthesis of winter wheat and rice. *Physiologia Plantarum*. https://doi.org/10.1111/ppl.13757.

Ma, L., Jia, H., Shen, A. L., Ding, J., Wang, X., Wang, J., Wan, J., Yan, J., Zhang, D., Dong, X., & Yin, P. (2022). Two determinants influence CRY2 photobody formation and function. *Plant Biotechnology Journal*, 21(3), 460–462. https://doi.org/10.1111/pbi.13978.

Mager, S., Schönberger, B., & Ludewig, U. (2018). The transcriptome of zinc deficient maize roots and its relationship to DNA methylation loss. *BMC Plant Biology*, 18(1), 372. https://doi.org/10.1186/s12870-018-1603-z.

Mahmood, T., He, S., Abdullah, M., Sajjad, M., Jia, Y., Ahmar, S., Fu, G., Chen, B., & Du, X. (2023). Epigenetic insight into floral transition and seed development in plants. *Plant Science*. https://doi.org/10.1016/j.plantsci.2023.111926.

Matsumura, Y., Ohbayashi, I., Takahashi, H., Kojima, S., Ishibashi, N., Keta, S., Nakagawa, A., Hayashi, R., Sáez-Vásquez, J., Echeverria, M., Echeverria, M., Sugiyama, M., Nakamura, K., Machida, C., & Machida, Y. (2016). A genetic link between epigenetic repressor AS1-AS2 and a putative small subunit processome in leaf polarity establishment of *Arabidopsis*. *Biology Open*, 5(7), 942–954. https://doi.org/10.1242/BIO.019109.

McFarlane, C., Shah, N. R., Kabasakal, B. V., Echeverria, B., Cotton, C. A. R., Bubeck, D., & Murray, J. W. (2019). Structural basis of light-induced redox regulation in the Calvin-Benson cycle in cyanobacteria. *Proceedings of the National Academy of Sciences of the United States of America*. https://doi.org/10.1073/PNAS.1906722116.

McKee, T., & McKee, J. R. (2020). *Photosynthesis*. https://doi.org/10.1093/hesc/9780190847685.003.0013.

Meyer, R. C., Gryczka, C., Neitsch, C., Müller, M., Bräutigam, A., Schlereth, A., Schön, H., Weigelt-Fischer, K., & Altmann, T. (2019). Genetic diversity for nitrogen uses efficiency in *Arabidopsis thaliana*. *Planta*, 250(1), 41–57. https://doi.org/10.1007/s00425-019-03140-3.

Minagawa, J. (2017). *Chlamydomonas: Bioenergetic Pathways—Regulation of Photosynthesis*. https://doi.org/10.1007/978-3-319-66365-4_5.

Minagawa, J., Minagawa, J., Tokutsu, R., & Tokutsu, R. (2015). Dynamic regulation of photosynthesis in *Chlamydomonas reinhardtii*. *Plant Journal*. https://doi.org/10.1111/TPJ.12805.

Monakhova, O. F., & Chernyad'ev, I. I. (2002). Protective role of Kartolin-4 in wheat plants exposed to soil drought. *Applied Biochemistry and Microbiology*. https://doi.org/10.1023/A:1016243424428.

Morales, F., Ancín, M., Fakhet, D., González-Torralba, J., Gámez, A. L., Seminario, A., Soba, D., Mariem, S. B., Garriga, M., & Aranjuelo, I. (2020). *Photosynthetic Metabolism under Stressful Growth Conditions as a Bases for Crop Breeding and Yield Improvement*. https://doi.org/10.3390/PLANTS9010088.

Murchie, E. H., McAusland, L., & Burgess, A. J. (2022). *Abiotic Stress, Acclimation, and Adaptation in Carbon Fixation Processes*. https://doi.org/10.1016/b978-0-12-823781-6.00011-3.

Nakano, T., Asami, T., Yamagami, A., Shimada, S., & Matsui, M. (2010). *Transformed Plants or Algae with Highly Expressed Chloroplast Protein BPG2*. https://patents.google.com/patent/US8569580B2/en.

Nguyen, H., Kim, J. H., Jeong, C. Y., Hong, S. W., & Lee, H. (2013). Inhibition of histone deacetylation alters *Arabidopsis* root growth in response to auxin via PIN1 degradation. *Plant Cell Reports*, 32(10), 1625–1636. https://doi.org/10.1007/S00299-013-1474-6.

Nikkanen, L., & Rintamäki, E. (2019). Chloroplast thioredoxin systems dynamically regulate photosynthesis in plants. *Biochemical Journal*. https://doi.org/10.1042/BCJ20180707.

Nogués, S., & Alegre, L. (2002). An increase in water deficit has no impact on the photosynthetic capacity of field-grown Mediterranean plants. *Functional Plant Biology*. https://doi.org/10.1071/PP01117.

Offermann, S., Dreesen, B., Horst, I., Danker, T., Jaskiewicz, M., & Peterhänsel, C. (2008). Developmental and environmental signals induce distinct histone acetylation profiles on distal and proximal promoter elements of the C4-Pepc gene in maize. *Genetics*, 179(4), 1891–1901. https://doi.org/10.1534/GENETICS.108.087411.

Ofori-Amanfo, K. K., Klem, K., Veselá, B., Holub, P., Agyei, T., Juran, S., Grace, J., Markl, M., & Urban, O. (2023). The effect of elevated CO_2 on photosynthesis is modulated by nitrogen supply and reduced water availability in Picea abies. *Tree Physiology*. https://doi.org/10.1093/treephys/tpad024.

Perrella, G., Bäurle, I., & Zanten, M. van. (2022). Epigenetic regulation of thermomorphogenesis and heat stress tolerance. *New Phytologist*. https://doi.org/10.1111/nph.17970.

Plant Epigenomics. (2022). https://doi.org/10.1079/9781789247534.0007.

Plant Epigenomics. (2023). https://doi.org/10.1016/b978-0-323-91909-8.00007-4

Pogson, B. J., & Albrecht, V. (2011). Genetic dissection of chloroplast biogenesis and development: An overview. *Plant Physiology*, 155(4), 1545–1551. https://doi.org/10.1104/PP.110.170365.

Popova, O., Dinh, H. Q., Aufsatz, W., & Jonak, C. (2013). The RdDM pathway is required for basal heat tolerance in *Arabidopsis*. *Molecular Plant*, 6(2), 396–410. https://doi.org/10.1093/MP/SST023.

Qi, M., Zhao, Z., & Nixon, P. J. (2023). The photosynthetic electron transport chain of oxygenic photosynthesis. *Bioelectricity*. https://doi.org/10.1089/bioe.2023.0003.

Qi, Q., Hu, B., Jiang, W., Wang, Y., Yan, J., Ma, F., Guan, Q., & Xu, J. (2023). Advances in plant epigenome editing research and its application in plants. *International Journal of Molecular Sciences*, 24(4), 3442. https://doi.org/10.3390/ijms24043442.

Robbens, S., Petersen, J., Brinkmann, H., Rouzé, P., & Van de Peer, Y. (2007). Unique regulation of the Calvin cycle in the ultrasmall green alga *Ostreococcus*. *Journal of Molecular Evolution*, 64(5), 601–604. https://doi.org/10.1007/S00239-006-0159-Y.

Romagnoli, S., & Tabita, F. R. (2006). A novel three-protein two-component system provides a regulatory twist on an established circuit to modulate expression of the cbbI region of *Rhodopseudomonas palustris* CGA010. *Journal of Bacteriology*, 188(8), 2780–2791. https://doi.org/10.1128/JB.188.8.2780-2791.2006.

Roy, S. K. B. (2023). *Fundamental Response and Salinity Tolerance in Brassicaceae Plants*. https://doi.org/10.26434/chemrxiv-2023-0xhnx.

Salazar-Iribe, A., & De-la-Peña, C. (2020). Auxins, the hidden player in chloroplast development. *Plant Cell Reports*, 39(12), 1595–1608. https://doi.org/10.1007/S00299-020-02596-Y.

Sandoval-Ibañez, O., Rolo, D., Ghandour, R., Hertle, A., Armarego-Marriott, T., Sampathkumar, A., Zoschke, R., & Bock, R. (2022). De-etiolation-induced protein 1 (DEIP1) mediates assembly of the cytochrome b6f complex in *Arabidopsis*. *Nature Communications*, 13(1). https://doi.org/10.1038/s41467-022-31758-7.

Sandoval-Ibañez, O., Sandoval-Ibañez, O., Sharma, A., Bykowski, M., Borràs-Gas, G., Behrendorff, J. B. Y. H., Mellor, S. B., Qvortrup, K., Verdonk, J. C., Bock, R., Kowalewska, Ł., & Pribil, M. (2021). Curvature thylakoid 1 proteins modulate prolamellar body morphology and promote organized thylakoid biogenesis in *Arabidopsis thaliana*. *Proceedings of the National Academy of Sciences of the United States of America*, 118(42). https://doi.org/10.1073/PNAS.2113934118.

Sarkar, B., De, A. K., Saha, I., Ghosh, A., Dolui, D., & Adak, M. K. (2020). Modalities of NADP-malic enzyme activities under light and darkness indicate its regulation with reference to C4 weed. *Plant Science Today*, 7(4). https://doi.org/10.14719/PST.2020.7.4.754.

Sawicki, A., & Chen, M. (2020). *Molecular Mechanism of Photosynthesis Driven by Red-Shifted Chlorophylls*. https://doi.org/10.1007/978-981-15-3110-1_1.

Secco, D., Wang, C., Shou, H., Schultz, M. D., Chiarenza, S., Nussaume, L., Ecker, J. R., Whelan, J., & Lister, R. (2015). Stress induced gene expression drives transient DNA methylation changes at adjacent repetitive elements. *eLife*, 4, e09343. https://doi.org/10.7554/eLife.09343.

Sharkey, T. D. (2020). *Emerging Research in Plant Photosynthesis*. https://doi.org/10.1042/ETLS20200035.

Shen, Y., Lei, T., Cui, X., Liu, X., Zhou, S., Zheng, Y., Zheng, Y., Guérard, F., Issakidis-Bourguet, E., & Zhou, D.-X. (2019). *Arabidopsis* histone deacetylase HDA15 directly represses plant response to elevated ambient temperature. *Plant Journal*, 100(5), 991–1006. https://doi.org/10.1111/TPJ.14492.

Shi, M., Wei, Z., Zhang, P., Guan, C., Chachar, S., & Zhang, J. (2024). Epigenetic modifications and breeding applications in horticultural plants. *Horticulturae*. https://doi.org/10.3390/horticulturae10020143.

Shim, S., Lee, H. G., & Seo, P. J. (2021). MET1-dependent DNA methylation represses light signaling and influences plant regeneration in arabidopsis. *Molecules and Cells*, 44(10), 746–757. https://doi.org/10.14348/MOLCELLS.2021.0160.

Shimmura, S., Nozoe, M., Kitora, S., Kin, S., Matsutani, S., Ishizaki, Y., Nakahira, Y., Nakahira, Y., & Shiina, T. (2017). Comparative analysis of chloroplast psbD promoters in terrestrial plants. *Frontiers in Plant Science*, 8, 1186. https://doi.org/10.3389/FPLS.2017.01186.

Silva, D. M. R., Barros, A. C., Silva, R. B., Galdino, W. O., Souza, J. W. G. de, Marques, I. C., Sousa, J. I. de, Lira, V. da S., Melo, A. F., Abreu, L. da S. de, Júnior, E. de O. A., Barbosa, L. do N. S., Neto, A. L. dos S., Santos, V. R. dos, Júnior, F. G. B. F. F., Vargens, F. N., Luz, J. H. S. da, Ono, E. O., & Rodrigues, J. D. (2024). Impact of photosynthetic efficiency on watermelon cultivation in the face of drought. *Agronomy*. https://doi.org/10.3390/agronomy14050950.

Simkin, A. J., López-Calcagno, P. E., & Raines, C. A. (2019). Feeding the world: Improving photosynthetic efficiency for sustainable crop production. *Journal of Experimental Botany*. https://doi.org/10.1093/JXB/ERY445.

Singh, S. K., & Reddy, K. R. (2011). Regulation of photosynthesis, fluorescence, stomatal conductance and water-use efficiency of cowpea (*Vigna unguiculata* [L.] Walp.) under drought. *Journal of Photochemistry and Photobiology B-Biology*. https://doi.org/10.1016/J.JPHOTOBIOL.2011.07.001.

Sun, A.-Q., Yin, C., Ma, M., Zhou, Y.-N., Zheng, X., Tu, X., & Fang, Y. (2022). Feedback regulation of auxin signaling through the transcription of H2A.Z and deposition of H2A.Z to SMALL AUXIN UP RNAs in *Arabidopsis*. *New Phytologist*, 236(5), 1721–1733. https://doi.org/10.1111/nph.18440.

Sun, B., Zhang, Q., Yuan, H., Gao, W., Han, B., & Zhang, M. (2020). PDV1 and PDV2 Differentially affect remodeling and assembly of the chloroplast DRP5B ring. *Plant Physiology*, 182(4), 1966–1978. https://doi.org/10.1104/PP.19.01490.

Sun, Q., Cao, X., Liu, Z., An, C., Hu, J., Wang, Y., Qiao, M. X., Gao, T., Cheng, W., Zhang, Y., Feng, Y., & Gao, H. (2023). Structural and functional insights into the chloroplast division site regulators PARC6 and PDV1 in the intermembrane space. *Proceedings of the National Academy of Sciences of the United States of America*, 120(5), e2215575120. https://doi.org/10.1073/pnas.2215575120.

Sun, S., Zhu, J., Guo, R., Whelan, J., & Shou, H. (2021). DNA methylation is involved in acclimation to iron-deficiency in rice (Oryza *sativa*). *The Plant Journal*, 107(3), 727–739. https://doi.org/10.1111/tpj.15318.

Taiz, L., Møller, I. M., Murphy, A., & Peer, W. A. (2024). *Photosynthesis: Physiological and Ecological Considerations*. https://doi.org/10.1093/hesc/9780197614167.003.0009.

Takabayashi, A., Ishikawa, N., Obayashi, T., Obayashi, T., Ishida, S., Obokata, J., Endo, T., & Sato, F. (2009). Three novel subunits of Arabidopsis chloroplast NAD(P)H dehydrogenase identified by bioinformatic and reverse genetic approaches. *Plant Journal*, 57(2), 207–219. https://doi.org/10.1111/J.1365-313X.2008.03680.X.

Tan, S.-L., Huang, X., Li, W.-Q., Zhang, S.-B., & Huang, W. (2021). Elevated CO_2 concentration alters photosynthetic performances under fluctuating light in *Arabidopsis thaliana*. *Cells*. https://doi.org/10.3390/CELLS10092329.

Tang, Q., Xu, D., Lenzen, B., Brachmann, A., Yapa, M. M., Doroodian, P., Schmitz-Linneweber, C., Masuda, T., Hua, Z., Leister, D., & Kleine, T. (2024). GENOMES UNCOUPLED PROTEIN1 binds to plastid RNAs and promotes their maturation. *Plant Communications*, 101069. https://doi.org/10.1016/j.xplc.2024.101069.

Teskey, R. O., Sheriff, D. W., Hollinger, D. Y., & Thomas, R. B. (1995). *External and Internal Factors Regulating Photosynthesis*. https://doi.org/10.1016/B978-0-08-092591-2.50009-1.

Timpmann, K., Rätsep, M., & Freiberg, A. (2023). Enhancing solar spectrum utilization in photosynthesis: Exploring exciton and site energy shifts as key mechanisms. *Dental Science Reports*. https://doi.org/10.1038/s41598-023-49729-3.

Tolley, B. J., Woodfield, H., Wanchana, S., Bruskiewich, R., & Hibberd, J. M. (2012). Light-regulated and cell-specific methylation of the maize *PEPC* promoter. *Journal of Experimental Botany*, 63(3), 1381–1390. https://doi.org/10.1093/JXB/ERR367.

Trejo-Téllez, L. I. (2023). Salinity stress tolerance in plants. *Plants*. https://doi.org/10.3390/plants12203520.

Valledor, L., Pascual, J., Meijón, M., Escandón, M., & Cañal, M. J. (2015). Conserved epigenetic mechanisms could play a key role in regulation of photosynthesis and development-related genes during needle development of *Pinus radiata*. *PLOS ONE*. https://doi.org/10.1371/JOURNAL.PONE.0126405.

Varotto, S., Tani, E., Abraham, E. M., Krugman, T., Kapazoglou, A., Melzer, R., Radanović, A., & Miladinović, D. (2020). Epigenetics: Possible applications in climate-smart crop breeding. *Journal of Experimental Botany*. https://doi.org/10.1093/JXB/ERAA188.

Wang, M., Wang, M., Yu, D., & Wei, X. (2024). *Elevated CO2 Concentration Enhance Carbon and Nitrogen Metabolism and Biomass Accumulation of Ormosia hosiei*. https://doi.org/10.1016/j.plaphy.2024.108725.

Wang, W., & Sung, S. (2024). Chromatin sensing: Integration of environmental signals to reprogram plant development through chromatin regulators. *Journal of Experimental Botany*. https://doi.org/10.1093/jxb/erae086.

Wang, Z., Wang, C., & Liu, S. (2022). Elevated CO2 alleviates adverse effects of drought on plant water relations and photosynthesis: A global meta-analysis. *Journal of Ecology*. https://doi.org/10.1111/1365-2745.13988.

Wedel, N., Soll, J., & Paap, B. K. (1997). CP12 provides a new mode of light regulation of Calvin cycle activity in higher plants. *Proceedings of the National Academy of Sciences of the United States of America*. https://doi.org/10.1073/PNAS.94.19.10479.

Wirtz, W., Stitt, M., & Heldt, H. W. (1982). Light activation of calvin cycle enzymes as measured in pea leaves. *FEBS Letters*, 142(2), 223–226. https://doi.org/10.1016/0014-5793(82)80139-7.

Wittmann, C., & Pfanz, H. (2008). Antitranspirant functions of stem periderms and their influence on corticular photosynthesis under drought stress. *Trees-Structure and Function*. https://doi.org/10.1007/S00468-007-0194-3.

Wolfe, D. W., & Erickson, J. D. (2022). *Carbon Dioxide Effects on Plants*. https://doi.org/10.1201/9781315136967-8.

Xu, S., Meng, L., & Bao, Y. (2023). Molecular evolution of the *rbcS* multiple gene family in *Oryza punctata*. *Journal of Systematics and Evolution*. https://doi.org/10.1111/jse.13039.

Xu, X., Chi, W., Sun, X., Feng, P., Guo, H., Li, J., Lin, R., Lu, C., Wang, H., Leister, D., & Zhang, L. (2016). Convergence of light and chloroplast signals for de-etiolation through ABI4–HY5 and COP1. *Nature Plants*, 2(6), 16066. https://doi.org/10.1038/NPLANTS.2016.66.

Xu, Z., Casaretto, J. A., Bi, Y.-M., & Rothstein, S. J. (2017). Genome-wide binding analysis of AtGNC and AtCGA1 demonstrates their cross-regulation and common and specific functions. *Plant Direct*, 1(4). https://doi.org/10.1002/PLD3.16.

Yamamoto, H., Cheuk, A., Shearman, J., Nixon, P. J., Meier, T., & Shikanai, T. (2023). Impact of engineering the ATP synthase rotor ring on photosynthesis in tobacco chloroplasts. *Plant Physiology*, 192(2), 1221–1233. https://doi.org/10.1093/plphys/kiad043.

Yamori, W., & Yamori, W. (2016). Photosynthetic response to fluctuating environments and photoprotective strategies under abiotic stress. *Journal of Plant Research*. https://doi.org/10.1007/S10265-016-0816-1.

Yang, C., An, B., Wang, H.-P., Albaqami, M., Abou-Elwafa, S. F., Xu, L., & Xu, Y. (2022). Comparative transcriptomic, and epigenetic analyses reveal conserved and divergent regulatory pathways in barley response to temperature stresses. *Physiologia Plantarum*, 174(4), e13727. https://doi.org/10.1111/ppl.13727.

Yang, X., Li, Y., Li, Y., Chen, H., Chen, H., Huang, J., Huang, J., Zhang, Y., Zhang, Y., Qi, M., Qi, M., Liu, Y., Liu, Y., Li, T., & Li, T. (2020). Photosynthetic response mechanism of soil salinity-induced cross-tolerance to subsequent drought stress in tomato plants. https://doi.org/10.3390/PLANTS9030363.

Yavari, N., Tripathi, R. K., Wu, B. S., MacPherson, S., Singh, J., & Lefsrud, M. (2021). The effect of light quality on plant physiology, photosynthetic, and stress response in *Arabidopsis thaliana* leaves. *PLoS One*, *16*(3). https://doi.org/10.1371/JOURNAL.PONE.0247380.

Yong-Villalobos, L., González-Morales, S. I., Wrobel, K., Gutiérrez-Alanis, D., Cervantes-Peréz, S. A., Hayano-Kanashiro, C., Cropeza-Aburto, A., Cruz-Ramírez, A., Martínez, O., & Herrera-Estrella, L. (2015). Methylome analysis reveals an important role for epigenetic changes in the regulation of the *Arabidopsis* response to phosphate starvation. *Proceedings of the National Academy of Sciences*, 112(52). https://doi.org/10.1073/pnas.1522301112.

Yu, C., Xu, K., Wang, W., Xu, Y., Ji, D., Chen, C., & Xie, C. (2018). Detection of changes in DNA methylation patterns in *Pyropia haitanensis* under high-temperature stress using a methylation-sensitive amplified polymorphism assay. *Journal of Applied Phycology*, 30(3), 2091–2100. https://doi.org/10.1007/S10811-017-1363-4.

Yung, W.-S., Chan, T.-F., Kong, F., & Lam, H.-M. (2023). The Plant Genome special section: Epigenome and epitranscriptome in plant–environment interactions. *The Plant Genome*. https://doi.org/10.1002/tpg2.20404.

Zahra, N., Hinai, M. S. A., Hafeez, M., Rehman, A., Wahid, A., Siddique, K. H. M., & Farooq, M. (2022). Regulation of photosynthesis under salt stress and associated tolerance mechanisms. *Plant Physiology and Biochemistry*. https://doi.org/10.1016/j.plaphy.2022.03.003.

Zeilfelder, S., & Schroda, M. (2024). Das ESCRT-III-Homolog VIPP1 vermittelt Thylakoid-Biogenese und -Erhaltung. *Biospektrum*. https://doi.org/10.1007/s12268-024-2143-8.

Zhang, X., Xiong, L.-G., Luo, Y., Wen, B., Wang, K., Liu, Z., Huang, J., & Li, J. (2021). Identification, molecular characteristic, and expression analysis of PIFs related to chlorophyll metabolism in tea plant (*Camellia sinensis*). *International Journal of Molecular Sciences*, 22(20), 10949. https://doi.org/10.3390/IJMS222010949.

Zhu, P., Zhuang, Q., Ciais, P., Welp, L. R., Li, W., & Xin, Q. (2017). Elevated atmospheric CO_2 negatively impacts photosynthesis through radiative forcing and physiology-mediated climate feedback. *Geophysical Research Letters*. https://doi.org/10.1002/2016GL071733.

Zubo, Y. O., Blakley, I. C., Franco-Zorrilla, J. M., Yamburenko, M. V., Solano, R., Kieber, J. J., Loraine, A. E., & Schaller, G. E. (2018). Coordination of chloroplast development through the action of the GNC and GLK transcription factor families. *Plant Physiology*, 178(1), 130–147. https://doi.org/10.1104/PP.18.00414.

12 Nitrogen and Sulfur Metabolism and Their Epigenetic Regulation in Plants

Poonam Jodha, Gourav Chouhan, Lata Sharma, Gyan Singh Shekhawat and Suman Parihar

12.1 INTRODUCTION

Various elements in the soil act as important nutrients for plants. Nitrogen (N) and sulfur (S), as macronutrients, are essential for plant growth and development, contributing significantly to numerous physiological and metabolic functions. Amino acids such as methionine and cysteine, along with coenzymes and secondary metabolites, all require sulfur, whereas nitrogen is essential for nucleotides, amino acids and chlorophyll. The coordinated assimilation and metabolism of these elements are crucial for plant productivity, stress adaptation and overall health. For plant growth, development and production, nitrogen (N) is an essential macronutrient (Kant, 2018); therefore, it is also used as a crucial component in many growth media for *in vitro* production of plants (Chouhan et al., 2023; Parihar, 2017; 2018; 2020; 2023; Parihar & Dwivedi 2019; 2020). Nitrogen assimilation in plants is an intricate and vital procedure that involves the uptake and conversion of inorganic nitrogen sources, such as nitrate and ammonium, into organic forms. (Wang et al., 2012; Krapp et al., 2014; Weber & Burow, 2018). Several enzymes, such as glutamine synthetase (GS), glutamate synthase (GOGAT), nitrate reductase (NR) and nitrite reductase (NiR), play essential roles in nitrogen assimilation. Through the GS/GOGAT cycle, inorganic nitrogen is converted into organic nitrogen, which is crucial for plant growth and development. Additionally, N assimilation is tightly regulated by various metabolic pathways and enzymes that help maintain nitrogen homeostasis within the plant (Crawford et al., 2000; Sánchez-Rodrıguez et al., 2011; Iqbal et al., 2020; Lee et al., 2020).

For plants, sulfur is a vital nutrient, predominantly absorbed from the soil as inorganic sulfate. This sulfate is assimilated through the reductive sulfate assimilation pathway, ultimately being incorporated into S-containing compounds such as cysteine, methionine and glutathione, which are vital for plant growth, detoxification and stress response. The process of sulfate absorption and assimilation is controlled through a network of transporters, enzymes and transcription factors (Leustek & Saito, 1999; Leustek et al., 2000). The sulfate transporters SULTR1.1 and SULTR1.2 are essential for the efficient uptake of sulfate, while other transporters, such as SULTR3 and SULTR4.2, are involved in the storage and distribution of sulfate within the plant.

Agricultural practices often rely on the heavy use of fertilizers, particularly nitrogen (N) fertilizers, to enhance crop production; however, only 30–50% of the nitrogen fertilizers that are applied might be absorbed and integrated into the crops (Raun & Johnson, 1999). Therefore, overuse of fertilizer leads to significant environmental damage in addition to increased production input. A viable solution to this issue is improving nitrogen use efficiency (NUE) in crops, which refers to the amount of grain or biomass produced per unit of nitrogen applied (Li et al., 2017). Recently, epigenetics has gained attention as an important factor in regulating plant processes. Epigenetic changes, such as DNA methylation, histone modifications and non-coding RNAs, influence how genes related to nutrient uptake, transport and assimilation are controlled at both the transcriptional and post-transcriptional levels. These reversible modifications allow plants to dynamically adapt to environmental stimuli, such as nutrient availability and abiotic stresses, without altering their underlying genetic code. The integration of genetic, physiological and epigenetic approaches to improve nutrient efficiency and crop health is essential for optimizing agricultural productivity, reducing input costs and mitigating environmental impacts. Additionally, the optimization of the nitrogen-to-sulfur (N:S) ratio in fertilizers has practical implications for improving crop yield and quality. A balanced N:S ratio enhances nitrogen use efficiency, improves protein synthesis and supports overall plant health, which is essential for sustainable agricultural practices.

12.2 EPIGENETICS IN PLANTS: AN OVERVIEW

In plants, epigenetics involves inherited modifications in gene expression that occur without changes to the underlying DNA sequence. In general, "epigenetics" refers to a group of heritable molecular processes that do not involve changes to the DNA sequence itself. Instead, these processes encompass various protein complexes and regulatory mechanisms that influence gene activity and expression (Bender, 2002; Banerjee & Roychoudhury, 2017). These alterations are essential for plant growth, stress response and environmental change adaptation. Gene-to-environment or gene-to-gene interactions result in epigenetic modifications in the pattern of genomic expression, including the turning on and off of particular genes, which impact cell functionality

DOI: 10.1201/9781003589914-15

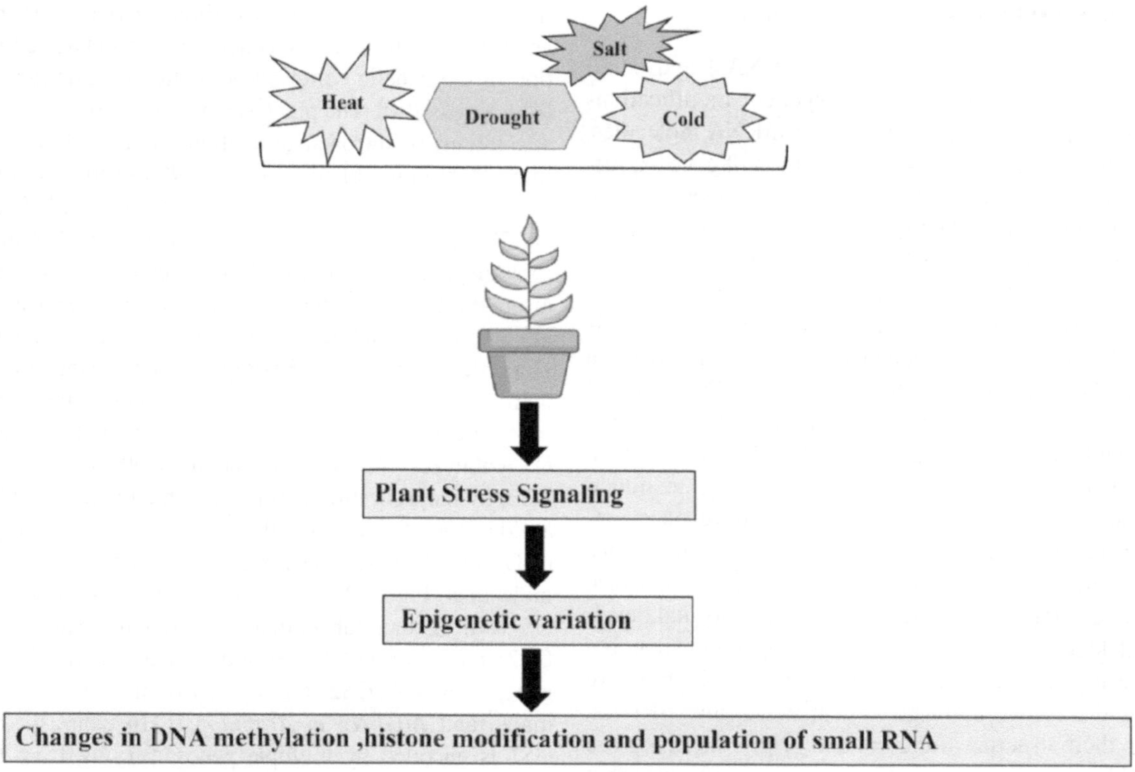

FIGURE 12.1 Epigenetic mechanism in plants.

without altering the DNA sequence. The key mechanisms of epigenetic regulation in plants comprise DNA methylation, RNA-mediated gene silencing and histone modification (Figure 12.1).

12.3 DNA METHYLATION

DNA methylation refers to the attachment of methyl groups to cytosine bases at the 5′ position, which can repress gene activity without altering the DNA sequence (Singh et al., 2019). There are two types of cytosine base methylation seen in plant species: symmetric (CG and CHG) and asymmetric (SHH) (Henderson & Jacobsen, 2007). When methylated cytosine is present in both complementary strands of DNA, this is known as symmetric methylation; when it occurs in just one strand, this is known as asymmetric methylation (Wambui et al., 2020). The DNA methyltransferases Dnmt1, Dnmt2 and Dnmt3 are conserved in all eukaryotes and have homologs in plants (Li & Zhang, 2014). In addition to being the plant homolog of mammalian Dnmt1, DNA METHYLTRANSFERASE 1 (MET1) may also be involved in CG de novo methylation. Similar to DNMT2 in mammals, plants also have DNMT2 homologs that are active transfer RNA (tRNA) methylases (Goll et al., 2006).

12.4 HISTONE MODIFICATION

Histone modification, a widely studied epigenetic marker, was first discovered by Vincent Allfrey in the early 1960s. These modifications involve covalent post-translational changes to histone proteins, such as acetylation, phosphorylation, ubiquitination, methylation and others. Numerous histone variations and histone-modifying enzymes are found in plants, and these are essential for the epigenetic control of plants in response to various environmental stimuli. The state of chromatin packing affects gene expression. The tail of the N-terminus of the highly conserved globular proteins that make up the histone family is located on the surface of the nucleosome octamer and is available for chemical modification. In eukaryotes, histones are the key proteins involved in organizing chromosomal DNA, comprising two copies each of the subunits H2A, H2B, H3 and H4. The residues within these subunits undergo multiple covalent modifications. The nucleosome is formed by wrapping each histone around 146 bp of DNA. Because of the wide variety of covalent histone modifications, researchers have even proposed a histone code that can identify the chromatin state of genes and chromosomal segments, and hence the transcriptional status (Turner, 2000; Jenuwein & Allis, 2001). Histone modifications are categorized into various types depending on the functional group attached to the amino acids in the N-terminal tail of the histone. According to Samantara et al. (2021), the alteration is both reversible and irreversible; that is, the action of the corresponding reversible enzymes might cause the epigenetic mark to be static or unmarked. AtSDG8 encodes a histone methyltransferase that affects plant growth and development, including flowering, shoot branching, touch sensitivity, pigment production, pathogen defense, carbon and light responses.

12.5 RNA-MEDIATED EPIGENETICS

Small RNAs, such as microRNAs (miRNAs) and small interfering RNAs (siRNAs), direct epigenetic modifications by targeting specific DNA sequences or mRNA molecules. RNA-mediated epigenetics is the regulation of gene expression and inheritance of traits via RNA molecules, without altering the underlying DNA sequences. This mechanism involves various types of RNA and epigenetic processes, playing critical roles in development, cellular differentiation and responses to environmental changes. RdDM is the term for the control of gene expression by methylation of certain DNA regions mediated by short RNA. RdDM is a well-known mechanism in plants that aids in genome defense and gene silencing. These small RNAs, which range in length from 21 to 24 nucleotides (nt), are found throughout the eukaryotic kingdoms and have occasionally been demonstrated to regulate common homologous pathways. Molecular markers such as retrotransposons, retroviruses, irregularly expressed genes and normal developmental loci are valuable for identifying gene silencing targets (Rapp & Wendel, 2005). In plants, small RNAs are categorized into two main groups: miRNAs and siRNAs, based on their structure, function and origin. miRNAs are short RNA sequences encoded by eukaryotic genomes that act as significant regulators of plant development; they were first identified in *Caenorhabditis elegans* (Lee et al., 1993).

12.6 NITROGEN METABOLISM AND EPIGENETICS

Proteins, phospholipids, nucleic acids and other secondary metabolites all require nitrogen for production. Ammonium (NH_4^+) or nitrate (NO_3^-) can be absorbed and used by plants as their main source of nitrogen. Protein synthesis, amino acid biosynthesis, nitrate assimilation, ammonium assimilation and nitrate absorption and transport within the plant comprise the linear process of nitrate utilization. Thus, nitrogen metabolism operates through a highly intricate network at both cellular and whole-plant levels, involving branching and merging pathways. This system interacts closely with carbon flow, pH regulation and transport of ions and assimilates across various plant tissues, ensuring coordinated nutrient assimilation and metabolic balance (Stitt et al., 2002).

12.6.1 Nitrogen Assimilation Pathways

The two primary sources of inorganic nitrogen in plants are nitrate and ammonium. After nitrate is taken up from the root, nitrate reductase (NR) in the cytoplasm first converts it to nitrite. Nitrite reductase (NiR) then transports the nitrite into plastids and converts it to ammonium. Nitrite reductase and nitrate reductase enzymes have been identified in various plants, with varying numbers reported for different species. Specifically, NiR is found as a single copy in *Arabidopsis*, three copies in rice, three copies in maize and two copies in wheat. Similarly, *NR* has been reported in two copies in *Arabidopsis*, three in rice, four in maize and three in wheat. These observations emphasize the variation in the genetic composition related to nitrogen metabolism across different plant species, reflecting their unique adaptations to nitrogen utilization. Besides, NR is also an important source of nitric oxide (NO) synthesis, a key signaling molecule in plants (Khator et al., 2024; Sharma et al., 2024; 2025). NO may inhibit or activate NR in tomato roots depending on nitrate levels, yet its role in regulating N uptake under fluctuating nitrate remains unclear (Sun et al., 2015). The glutamate synthase cycle, also known as the glutamine synthetase (GS) or glutamine-2-oxoglutarate aminotransferase (GOGAT) cycle, is essential for converting inorganic nitrogen into organic forms in plants. In this cycle, GS catalyzes the ATP-dependent synthesis of glutamine (Gln) by incorporating ammonium into glutamate (Glu). The amide group from Gln is then transferred to 2-oxoglutarate (2-OG) by GOGAT, resulting in the production of two Glu molecules (Figure 12.2). Higher plants possess two types of GS with distinct subcellular localizations: plastid-localized GS2 and cytosolic GS1. Plastid GS2 is encoded by a single gene, such as *OsGS2* in rice, *TaGS2* in wheat, *ZmGLN2* in maize and *AtGLN2* in *Arabidopsis.* In contrast, cytosolic GS1 is encoded by multiple genes, varying from three to five depending on the species, including *AtGLN1.1–1.5* in *Arabidopsis*, *OsGS1.1–1.3* in rice, *TaGS1.1–1.3* in wheat and *ZmGLN1.1–1.5* in maize. This distribution and genetic diversity highlight the complexity and adaptability of nitrogen metabolism in plants. Additionally, there are two types of glutamine-2-oxoglutarate aminotransferase (GOGAT) enzymes in plants, distinguished by their reductants: ferredoxin (Fd)-GOGAT and nicotinamide adenine dinucleotide (NADH)-GOGAT. Fd-GOGAT uses ferredoxin as a reductant and is encoded by genes such as *OsFd-GOGAT1* in rice, *TaFd-GOGAT1* in wheat, *ZmFd-GOGAT1* in maize and *AtGLU1* and *AtGLU2* in *Arabidopsis.* In contrast, NADH-GOGAT utilizes NADH as a reductant and is encoded by genes like *AtGLT* in *Arabidopsis*, *ZmNADHGOGAT1* in maize, *OsNADH-GOGAT1* and *OsNADH-GOGAT2* in rice and *TaNADH-GOGAT1* in wheat. Both types of GOGAT are localized in plastids and are typically encoded by one or two genes, emphasizing their integral role in nitrogen assimilation in plants. Moreover, nitrogen assimilation involves several key enzymes that facilitate the conversion of nitrogen into usable forms for plants. Asparagine synthetase (ASN) plays a central role by transferring the amide group from glutamine (Gln) or ammonium (NH_4^+) to aspartate (Asp), producing asparagine (Asn) and glutamate (Glu). Another enzyme, glutamate dehydrogenase (GDH), is involved in two processes: it adds ammonia to 2-oxoglutarate (2-OG) to produce Glu, and it removes the amine group from Glu, releasing ammonia and regenerating 2-OG. Additionally, carbamoyl phosphate synthetase (CPSase) uses bicarbonate (HCO_3) and ammonium, or the amide group from Gln, to synthesize carbamoyl phosphate (CP) and Glu. These enzymes work together to ensure that nitrogen is effectively integrated into the plant metabolic processes (Liu et al., 2022b).

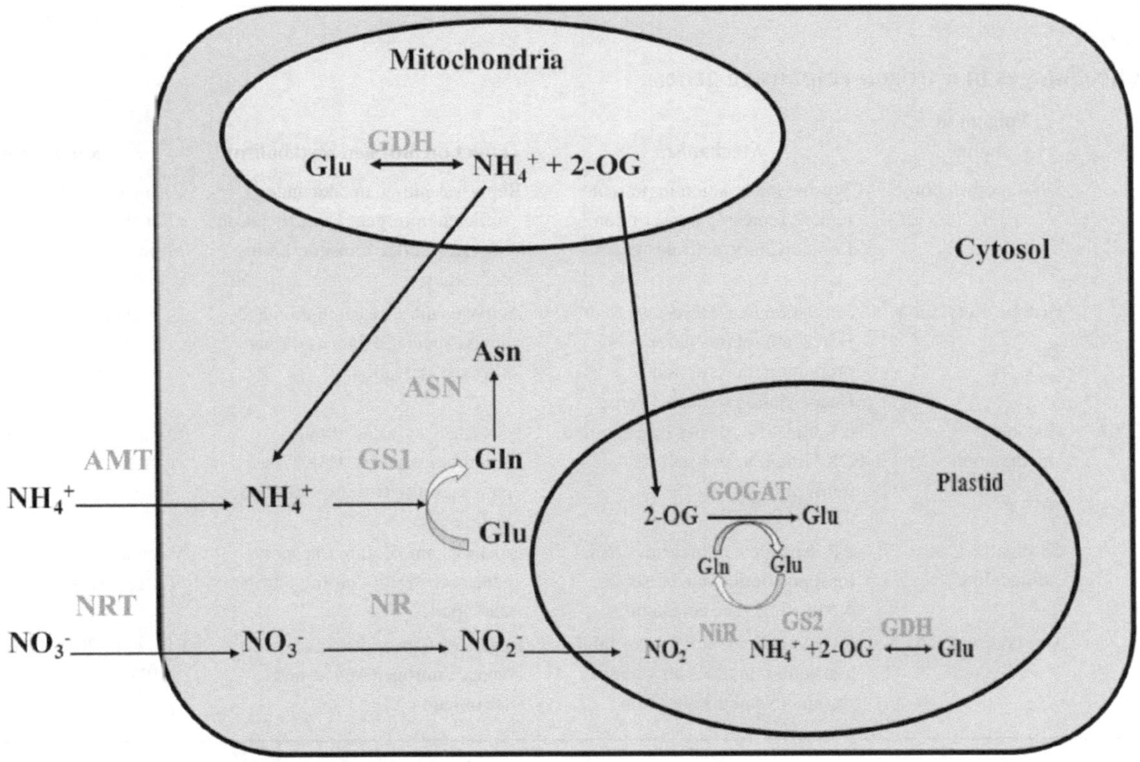

FIGURE 12.2 Nitrogen assimilation pathway in plants.

Plant nitrogen metabolism encompasses several interconnected processes critical for growth and development. The remobilization and storage of nitrate in various plant parts, along with de novo absorption of ammonium, play fundamental roles. Ammonium released during photorespiration is efficiently recycled (Hirel & Lea, 2001), while nitrogen is allocated to diverse branches of amino acid biosynthesis (Morot-Gaudry et al., 2001). Amino acids serve multiple functions, including protein synthesis, export, storage in vacuoles and redirection into secondary metabolic pathways that produce alkaloids, tetrapyrroles and phenylpropanoids (Heldt & Piechulla, 2021). To prevent alkalinization, malate acts as a counter-anion (Martinioa & Rentsch, 1994). In the GOGAT pathway, 2-oxoglutarate functions as the main ammonium acceptor (Heldt & Piechulla, 2021), while phosphorylation intermediates and various organic acids provide essential carbon precursors for amino acid synthesis (Morot-Gaudry et al., 2001). Additionally, the reduction of nitrate to ammonium relies on reducing equivalents generated through photosynthesis or carbohydrate breakdown (Kaiser et al., 1999; Foyer et al., 2001). These processes collectively highlight the intricate coordination involved in plant nitrogen metabolism.

12.6.2 Epigenetic Regulation of Nitrogen-Responsive Genes

Nitrogen-responsive genes play a vital role in controlling the absorption, assimilation and signaling of nitrogen in plants. Nitrate transporters (NRTs), ammonium transporters (AMTs), NIN-like proteins (NLPs) and the GS/GOGAT pathway enzymes are important genes in this process. Plants have a unique ability to sense nutrient availability quickly, triggering a series of reactions to preserve nutritional homeostasis (Vidal et al., 2020). Low- and high-affinity nitrate transporters help plants take up nitrate from the soil; for example, NRT1.1 from the NRT1 family acts as both a sensor and a transporter, while NRT2.1 from the NRT2 family is a high-affinity nitrate transporter that works effectively when nitrate levels are low (Muños et al., 2004; Gojon et al., 2011). Nitrate uptake and assimilation genes are activated by NIN LIKE PROTEIN 7 (NLP7) (Liu et al., 2022a). N causes speedy alterations in plant transcriptional responses by means of signal "messengers" like transcription factors (TFs) and calcium; these changes frequently occur during nitrate exposure (Varala et al., 2018). These reactions encompass a broad variety of activities, such as root development, nitrate uptake and plant architecture (O'Brien et al., 2016; Wang et al., 2018; Han et al., 2020). By adjusting gene expression without changing DNA sequences, epigenetics enables plants to quickly respond to nitrogen availability. In dicots, nitrogen absorption is regulated by histone modifications, such as the addition of methyl groups to specific sites on histone H3 (H3K27me3, H3K36me3) (Table 12.1). In *Arabidopsis*, nitrogen treatment activates the SDG8 enzyme in root cells, which modifies H3K36me3 to control genes involved in energy metabolism, including photosynthesis and nitrogen metabolism genes (Li et al., 2015). SDG8 also helps regulate the expression of key nitrogen-related genes like *NRT2.1*, *NRT1.5* and

TABLE 12.1
Epigenetic changes in nitrogen-responsive genes

Gene	Epigenetic change	Mechanism	Effect on nitrogen metabolism	References
NRT1.1	DNA methylation	Cytosine methylation in promoter regions represses transcription. Demethylation activates genes.	Represses nitrate uptake under sufficient nitrogen. Demethylation activates nitrate transport during deficiency.	Zhang et al. (2018); Sánchez-Vicente and Lorenzo (2021)
NLP7, NRT2.1	Histone acetylation	Acetylation (e.g. H3K9ac, H3K27ac) relaxes chromatin, promoting transcription. Deacetylation represses genes.	Activates nitrate assimilation and uptake regulating genes during nitrogen starvation.	Berr et al. (2011); Liu et al. (2022a)
NIA1, NRT1.1	Histone methylation	H2K34me3: Activates transcription. H2K27me3: Represses transcription.	H3K4me3 activates nitrate assimilation genes. H3K27me3 represses nitrate transporter under sufficiency.	Kou et al. (2011); li et al. (2015)
AMT1.1	Chromatin remodeling	ATP-dependent complexes crucial for repositioning nucleosomes, modulating gene accessibility.	Enhances ammonium transport gene accessibility during nitrogen starvation.	Yu et al. (2016); Wang et al. (2018); Yu et al. (2018)
miR444	Non-coding RNAs	miRNAs degrade mRNAs or inhibit translation. lncRNA modulate chromatin and transcription.	Regulates transcription factors to balance nitrogen uptake and assimilation.	Liu et al. (2019); Fukuda et al. (2019)

GLN1;4 in response to nitrogen (Li et al., 2020). In monocots, nitrogen uptake is influenced by chromatin remodeling. The *CHB101* gene encodes a component of the SWI/SNF chromatin remodeling complex, which facilitates ATP-dependent rearrangement of chromatin to support nitrogen absorption (Yu et al., 2016; Yu et al., 2018). DNA methylation often suppresses gene transcription. N deficit in maize was shown to reduce the overall methylated CG content in root from 12.0 to 26.6%, and hypomethylation was present in 95% of the DMRs under low N and control settings (Mager and Ludewig, 2018). Recently, *Arabidopsis* was used to analyze long non-coding RNAs (lncRNAs) in response to a 12-nutrient deficit (Fukuda et al., 2019). It has been observed that certain nutrient shortages only cause the expression of new lncRNAs. The possible targets of these novel lncRNAs were detected by the use of binding predictions and RNA-seq data; lncRNAs-RNA regulated pairings were detected for genes linked to P, K and N deficiencies. Specifically, it has been shown that the lncRNA TRANS-ACTING siRNA 3 (TAS3) may play a role in controlling *NRT2.4*. Nitrogen deprivation substantially induces the high-affinity nitrate transporter *NRT2.4* (Kiba et al., 2012).

12.7 SULFUR METABOLISM AND ITS EPIGENETIC REGULATION

Sulfur, one of the six macronutrients needed by plants, is present in several metabolites as well as the amino acids methionine and cysteine. It may be reasonable to assume that sulfur assimilation has received less attention given that sulfur is only 3–5% as abundant in plants as nitrogen. Sulfur requirements are even higher than phosphorus requirements for certain plants, such as oil crops, legumes and forages, in order to achieve optimum production and quality. Nature contains a lot of sulfur; for instance, the oceans contain 28 mM sulfate. The primary form of inorganic sulfur that is used by both aquatic and terrestrial species as a source is oxyanion sulfate. However, reduced sulfur is present in most metabolites as organic sulfide. As a result, organisms must expend energy on reducing the abundant sulfate. Therefore, sulfate assimilation is not a common mechanism. According to Patron et al. (2008), most parasite bacteria and metazoans do not reduce sulfate; however, certain bacterial and archaeal species specialize in oxidizing reduced forms of sulfur for energy production. Essential components of plant primary metabolism include sulfate uptake and absorption. Reduced sulfur is found in plants, which are used as food for both humans and animals.

12.7.1 Sulfur Assimilation

Higher plants primarily obtain sulfur in the form of inorganic sulfate from the soil. High-affinity sulfate transporters actively take up sulfate from the surrounding environment into the root symplast, and the reductive sulfate assimilation pathway reduces it into cysteine (Hell, 1997). Sulfate is frequently said to be transferred to the shoot from the root after absorption, where it undergoes reduction in the chloroplast before being assimilated into various organic sulfur compounds. Plants absorb sulfur from the soil as inorganic sulfate. Roots can also reduce sulfur and transport it to shoots, primarily as methionine and to a lesser extent as cysteine and glutathione (Pate, 1965). All enzymes required for sulfate assimilation are present in the roots, with sulfate reduction likely occurring in the plastids (Heiss et al., 1999; Lappartient et al., 1999; Lee & Leustek, 1999; Yonekura-Sakakibara et al., 2000). The expression and activity of these enzymes, such as APS reductase and ATP sulfurylase, are influenced by sulfate availability (Lappartient et al., 1999). In *Arabidopsis thaliana*, the two high-affinity sulfate transporters, SULTR1;1 and SULTR1;2, play key roles in root sulfate uptake (Takahashi et al., 2000; Yoshimoto et al., 2002; Rouached et al., 2008).

Once sulfate enters into root cells, it is either transported into plastids by SULTR3 sulfate transporters for incorporation into organic sulfur compounds or stored in vacuoles (Cao et al., 2013; Chen et al., 2015). In the sulfur assimilation pathway, ATP sulfurylase (ATPS) catalyzes the conversion of sulfate into adenosine 5′-phosphosulfate (APS) (Figure 12.3). APS then follows two primary pathways: it is either reduced to sulfite for further assimilation or phosphorylated to form 3′-phosphoadenosine 5′-phosphosulfate (PAPS) (Takahashi et al., 2011). APS kinase (APK) and APS reductase (APR) catalyze these two processes, respectively. In many sulfation reactions, PAPS supplies an activated sulfate, and sulfite reductase (SiR) further reduces sulfite to sulfide. Sulfide reacts with O-acetylserine (OAS) in a condensation process catalyzed by the enzyme O-acetylserine (thiol) lyase (OAS-TL), resulting in the synthesis of the sulfur-containing amino acid cysteine (Cys) (Figure 12.3). This Cys can serve as a precursor for protein synthesis or participate in the biosynthesis of other essential molecules such as glutathione (GSH) and methionine (Met). These two compounds can be utilized in the production of several S-containing derivatives, including phytochelatins and glucosinolates, which are crucial for plants to detoxify heavy metals and reduce biotic stress, correspondingly (Cobbett, 2000; Halkier & Gershenzon, 2006).

12.7.2 Epigenetic Regulation of Sulfur-Responsive Genes

Significant progress has been made in identifying the enzymes that regulate sulfur homeostasis at the genetic level in plants, along with the transporters involved in sulfate uptake. Regulation at both the transcriptional and post-transcriptional levels encompasses the control of sulfate uptake and distribution, sulfur assimilation and the recycling of sulfur-containing molecules. The control of sulfate absorption and distribution has been linked to a variety of trans-acting factors and cis-elements. SLIM1 (SULFUR LIMITATION 1) is an essential transcription factor regulating sulfur metabolism in plants. It controls sulfate uptake in roots by activating SULTR1;1 and SULTR1;2 and facilitates sulfate release from vacuoles by regulating SULTR4;2 (Maruyama-Nakashita et al., 2006). Several cis-elements, such as the UPE-box in the *UP9C* gene (Wawrzynska et al., 2010) and SURE elements in genes like *SULTR1;1* (Maruyama-Nakashita et al., 2005) and *sdi-1* (Howarth et al., 2009), respond to sulfur deficiency. Although SLIM1 does not directly bind to SURE elements, it interacts with the UPE-box to regulate sulfur-responsive genes like *LSU, APR* and *SULTR2;1*, ensuring efficient sulfur utilization (Wawrzynska et al., 2010; Wawrzynska & Sirko, 2016).

Sulfur homeostasis in plants is regulated by a multifaceted process that includes epigenetic regulation, which modifies the expression of gene response to S availability through histone modification, DNA methylation and non-coding RNAs (Koprivova & Kopriva, 2014). DNA methylation, a critical epigenetic marker, is reduced at the promoters of sulfur-responsive genes, such as APS reductase (APR3) and sulfate transporters (SULTR1;2), signaling their transcription when sulfur is deficient. miR395 plays a critical role in regulating sulfur uptake and assimilation by targeting and cleaving three isoforms of ATP sulfurylase (ATPS1, ATPS3 and ATPS4), which is a key enzyme in sulfate assimilation, as well as the low-affinity sulfate transporter SULTR2;1 (Kawashima et al., 2009; Kawashima et al., 2011; Matthewman et al., 2012) (Table 12.2). When sulfur level is low, the phloem partner cells in the root and shoot vascular tissue exhibit a high activation of miR395. The accumulation of miRNA395 in the presence of S deficit is regulated by sulfur limitation 1 (SLIM1), a transcription factor belonging to the ethylene insensitive 3 like (EIL) family. Additional studies have shown that the induction of miR395 accelerates sulfate translocation from roots to shoots (Liang et al., 2010; Kawashima et al., 2011).

12.7.3 Cross-Talk between Nitrogen and Sulfur

The cross-talk between nitrogen (N) and sulfur (S) in plants is a critical area of research, given their roles as essential macronutrients. Both elements are intricately linked in metabolic and physiological processes. Nitrogen is a fundamental element required for the synthesis of amino acids, nucleotides and chlorophyll, while sulfur is essential for producing cysteine, methionine and various coenzymes. The growth, development and stress responses of plants are significantly influenced by their interaction. Both nitrogen and sulfur are found in proteins, and a lack of either will significantly impair plant growth and protein synthesis. The imbalance in nitrate and sulfate ion supply disrupts the pools of these ions and their intermediate metabolites. Plants respond to these disturbances by regulating gene expression and possibly enzyme activity, highlighting the interconnectedness and coordination between nitrogen and sulfur assimilation pathways.

The absorption of sulfate (SO_4^{2-}) and nitrate (NO_3^-), the primary sources of N and S, respectively, is closely regulated. Adequate nitrogen levels promote the production of S-containing amino acids like methionine and cysteine. Conversely, sulfur deficiency reduces nitrate reductase activity, leading to impaired nitrogen assimilation (Kopriva et al., 2019). N and S metabolism are interdependent. The assimilation of sulfate depends on nitrogen availability since cysteine and methionine synthesis requires nitrogen. Additionally, glutathione (GSH), a major antioxidant, is

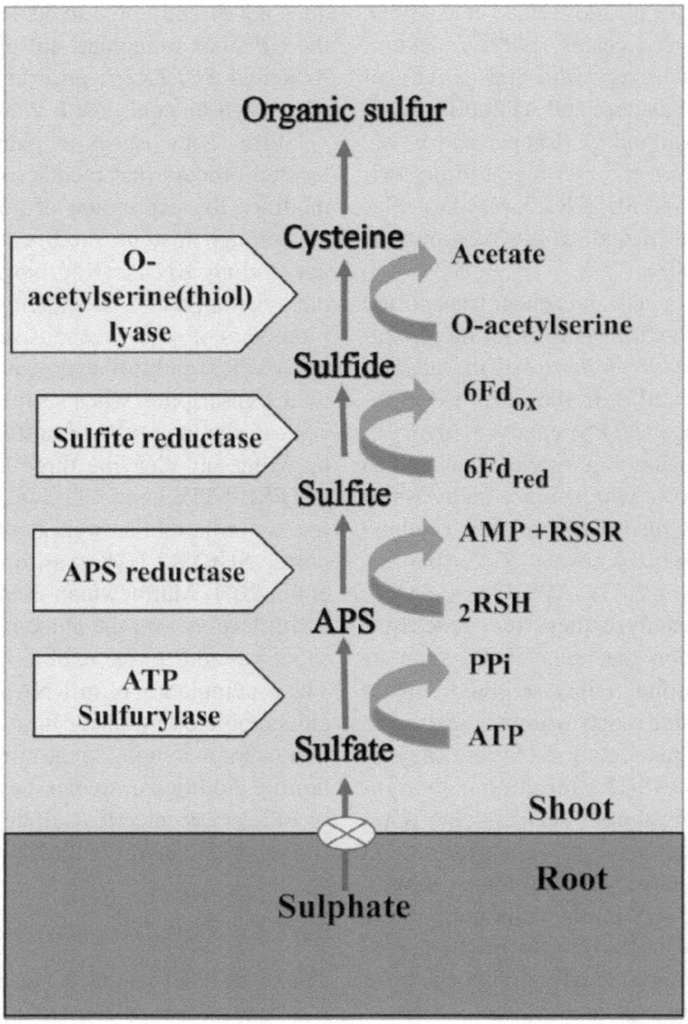

FIGURE 12.3 Sulfur assimilation pathway in plants.

TABLE 12.2
Epigenetic changes in sulfur-responsive genes

Gene	Function	Epigenetic control mechanism	Response type	References
SULTR1;2	High affinity sulfate uptake	miR395 repression under sufficiency	Sulfur uptake	Takahashi (2019); Rahikainen (2024)
APR1-3	Reduction of APS to sulfite	Histone acetylation	Sulfur assimilation	Rahikainen (2024)
SLIM1	Transcription factors for sulfur genes	Interaction with chromatin remodelers	Sulfur starvation response	Matthewman et al. (2012); Rahikainen (2024)
miR395	Targets *APS* and *SULTR* genes	Tissue-specific expression	Balance sulfur metabolism	Liang et al. (2010); Kawashima et al. (2011); Rahikainen (2024)
OAS-TL	Cysteine biosynthesis	Reduced DNA methylation	Organic sulfur incorporation	Hell (1997); Leustek & Saito (1999); Rahikainen (2024)

synthesized from cysteine and glutamate, demonstrating the interconnectedness of their pathways (Armstrong, 2006). In stress conditions, such as salt or heavy metal toxicity, the cross-talk becomes more evident. Sulfur compounds like GSH and phytochelatins detoxify reactive oxygen species and heavy metals, while nitrogen supports the synthesis of stress-related proteins The balance of N and S is crucial for effective stress mitigation (Barry et al., 2015). Cross-talk at the molecular level involves the regulation of key genes in N and S pathways; for example, the *SLIM1* transcription factor regulates sulfur-responsive genes in coordination with nitrogen signaling pathways. This indicates shared regulatory networks between these nutrients (Maruyama-Nakashita et al., 2015). Optimizing the N:S ratio in fertilizers enhances crop yield and quality. A sufficient supply of sulfur, for example, increases the efficiency of nitrogen usage, which benefits protein synthesis and the general health of plants (Zhao et al., 1999).

12.8 CONCLUSION AND FUTURE PERSPECTIVES

In conclusion, nitrogen (N) and sulfur (S) are the fundamental macronutrients that are crucial for plant growth, development and productivity. Their roles in essential physiological and metabolic processes, such as amino acid synthesis, coenzyme production and secondary metabolite formation, make them indispensable for healthy plant function. However, challenges related to the efficient uptake, assimilation and metabolism of nitrogen, especially in the context of fertilizer use, require focused efforts to improve nitrogen use efficiency (NUE) and sulfur management in crops. Excessive nitrogen fertilizer use has significant environmental consequences, making it imperative to explore sustainable practices that optimize nutrient uptake and minimize waste.

Advances in comprehension of the genetic, physiological and epigenetic mechanisms governing the uptake, assimilation and metabolism of nitrogen and sulfur can provide pathways to improve nutrient efficiency in plants. Epigenetic regulation, including DNA methylation, histone modifications and RNA-mediated gene silencing, presents promising avenues for modulating nutrient-responsive gene expression without altering the genetic code, offering adaptive mechanisms that help plants respond to fluctuating nutrient availability and environmental stress. The interaction between nitrogen and sulfur metabolism is essential for optimal plant health and stress resilience, and their cross-talk is pivotal in regulating nutrient homeostasis. Sulfur compounds like glutathione play critical roles in mitigating oxidative stress and heavy metal toxicity, illustrating the interdependence between these nutrients in stress response. Understanding how these elements influence each other at the molecular level will contribute to developing crops with better nutrient use efficiency, improved protein synthesis and enhanced resilience to environmental challenges. Future research into optimizing the nitrogen-to-sulfur (N:S) ratio in fertilizers, exploring epigenetic mechanisms and leveraging biotechnological approaches can transform agricultural practices. By developing crops that efficiently utilize both nitrogen and sulfur, it will be possible to reduce fertilizer dependency, enhance crop yield and support sustainable food production systems. Unraveling the complexities of nitrogen and sulfur assimilation and their genetic and epigenetic regulation holds the key to advancing global food security and promoting environmentally sustainable agriculture.

ACKNOWLEDGMENTS

The authors gratefully acknowledge UGC, New Delhi, for providing financial assistance through the CAS program.

REFERENCES

Armstrong, A. F., Logan, D. C., Tobin, A. K., O'toole, P., & Atkin, O. K. (2006). Heterogeneity of plant mitochondrial responses underpinning respiratory acclimation to the cold in *Arabidopsis thaliana* leaves. *Plant, Cell & Environment*, 29, 940–949.

Banerjee, A., & Roychoudhury, A. (2017). Epigenetic regulation during salinity and drought stress in plants: Histone modifications and DNA methylation. *Plant Gene*, 11, 199–204.

Barry, K. M., Janos, D. P., Nichols, S., & Bowman, D. M. (2015). *Eucalyptus obliqua* seedling growth in organic vs. mineral soil horizons. *Frontiers in Plant Science*, 6, 97.

Bender, J. (2002). Plant epigenetics. *Current Biology*, 12, R412–R414.

Berr, A., Shafiq, S., & Shen, W. H. (2011). Histone modifications in transcriptional activation during plant development. *Biochimica et Biophysica Acta-Gene Regulatory Mechanisms*, 1809, 567–576.

Cao, M. J., Wang, Z., Wirtz, M., Hell, R., Oliver, D. J., & Xiang, C. B. (2013). SULTR 3;1 is a chloroplast-localized sulfate transporter in *Arabidopsis thaliana*. *Plant Journal*, 73, 607–616.

Chen, C. Y., Wu, K., & Schmidt, W. (2015). The histone deacetylase HDA19 controls root cell elongation and modulates a subset of phosphate starvation responses in *Arabidopsis*. *Scientific Reports*, 5, 15708.

Chouhan, G., Solanki, N., Shekhawat, G. S., & Parihar, S. (2023). *Convolvulus prostratus* Forssk.: a memory boosting herb. *Plant Science Today*, 10, 86–93.

Cobbett, C. S. (2000). Phytochelatin biosynthesis and function in heavy-metal detoxification. *Current Opinion in Plant Biology*, 3, 211–216.

Crawford, N. M., Kahn, M. L., Leustek, T., & Long, S. R. (2000). Nitrogen and sulfur. In B. Buchanan, W. Gruissem, & R. Jones (Eds.), *Biochemistry and Molecular Biology of Plants* (pp. 786–849). UK: American Society of Plant Biologist, Wiley and Sons.

Foyer, C. H., Ferrario-Méry, S., & Noctor, G. (2001). Interactions between carbon and nitrogen metabolism. In P. J. Lea & J. F. Morot-Gaudry (Eds.), *Plant Nitrogen* (pp. 237–254). Berlin: Springer.

Fukuda, M., Nishida, S., Kakei, Y., Shimada, Y., & Fujiwara, T. (2019). Genome-wide analysis of long intergenic noncoding RNAs responding to low-nutrient conditions in *Arabidopsis thaliana*: Possible involvement of trans-acting siRNA3 in response to low nitrogen. *PCP*, 60, 1961–1973.

Gojon, A., Krouk, G., Perrine-Walker, F., & Laugier, E. (2011). Nitrate transceptor(s) in plants. *Journal of Experimental Botany*, 62, 2299–2308.

Goll, M. G., Kirpekar, F., Maggert, K. A., Yoder, J. A., Hsieh, C. L., Zhang, X., Golic, K. G., Jacobsen, S. E., & Bestor, T. H. (2006). Methylation of $tRNA^{Asp}$ by the DNA methyltransferase homolog Dnmt2. *Science*, 311, 395–398.

Halkier, B. A., & Gershenzon, J. (2006). Biology and biochemistry of glucosinolates. *Annual Review of Plant Biology*, 57, 303–333.

Han, X., Wu, K., Fu, X., & Liu, Q. (2020). Improving coordination of plant growth and nitrogen metabolism for sustainable agriculture. *aBiotech*, 1, 255–275.

Heiss, S., Schäfer, H. J., Haag-Kerwer, A., & Rausch, T. (1999). Cloning sulfur assimilation genes of *Brassica juncea* L.: Cadmium differentially affects the expression of a putative low-affinity sulfate transporter and isoforms of ATP sulfurylase and APS reductase. *Plant Molecular Biology*, 39, 847–857.

Heldt, H. W., & Piechulla, B. (2021). *Plant Biochemistry*. London: Academic Press, Elsevier.

Hell, R. (1997). Molecular physiology of plant sulfur metabolism. *Planta*, 202, 138–148.

Henderson, I. R., & Jacobsen, S. E. (2007). Epigenetic inheritance in plants. *Nature*, 447, 418–424.

Hirel, B., & Lea, P. J. (2001). Ammonia assimilation. In P. J. Lea & J. F. Morot-Gaudry (Eds.), *Plant Nitrogen* (pp. 79–99). Berlin: Springer.

Howarth, J. R., Parmar, S., Barraclough, P. B., & Hawkesford, M. J. (2009). A sulphur deficiency-induced gene, *sdi1*, involved in the utilization of stored sulphate pools under sulphur-limiting conditions has potential as a diagnostic indicator of sulphur nutritional status. *Plant Biotechnology Journal*, 7, 200–209.

Iqbal, A., Qiang, D., Alamzeb, M., Xiangru, W., Huiping, G., Hengheng, Z., Nianchang, P., Xiling, Z., & Meizhen, S. (2020). Untangling the molecular mechanisms and functions of nitrate to improve nitrogen use efficiency. *Journal of the Science of Food and Agriculture*, 100(3), 904–914.

Jenuwein, T., & Allis, C. D. (2001). Translating the histone code. *Science*, 293, 1074–1080.

Kaiser, W. M., Weiner, H., & Huber, S. C. (1999). Nitrate reductase in higher plants: A case study for transduction of environmental stimuli into control of catalytic activity. *Physiologia Plantarum*, 105, 384–389.

Kant, S. (2018). Understanding nitrate uptake, signaling and remobilisation for improving plant nitrogen use efficiency. *Seminars in Cell & Developmental Biology*, 74, 89–96.

Kawashima, C. G., Matthewman, C. A., Huang, S., Lee, B. R., Yoshimoto, N., Koprivova, A., Rubio-Somoza, R., Todesco, M., Rathjen, T., Saito, K., Takahashi, H., Dalmay, T., & Kopriva, S. (2011). Interplay of SLIM1 and miR395 in the regulation of sulfate assimilation in *Arabidopsis*. *Plant Journal*, 66, 863–876.

Kawashima, C. G., Yoshimoto, N., Maruyama-Nakashita, A., Tsuchiya, Y. N., Saito, K., Takahashi, H., & Dalmay, T. (2009). Sulphur starvation induces the expression of microRNA-395 and one of its target genes but in different cell types. *Plant Journal*, 57, 313–321.

Khator, K., Parihar, S., Jasik, J., & Shekhawat, G. S. (2024). Nitric oxide in plants: an insight on redox activity and responses toward abiotic stress signaling. *Plant Signaling & Behavior*, 19, 2298053.

Kiba, T., Feria-Bourrellier, A. B., Lafouge, F., Lezhneva, L., Boutet-Mercey, S., Orsel, M., Bréhaut, V., Miller, A., Daniel-Vedele, F., Sakakibara, H., & Krapp, A. (2012). The *Arabidopsis* nitrate transporter NRT2.4 plays a double role in roots and shoots of nitrogen-starved plants. *Plant Cell*, 24, 245–258.

Kopriva, S., Malagoli, M., & Takahashi, H. (2019). Sulfur nutrition: Impacts on plant development, metabolism, and stress responses. *Journal of Experimental Botany*, 70, 4069–4073.

Koprivova, A., & Kopriva, S. (2014). Molecular mechanisms of regulation of sulfate assimilation: First steps on a long road. *Frontiers in Plant Science*, 5, 589.

Kou, H. P., Li, Y., Song, X. X., Ou, X. F., Xing, S. C., Ma, J., Von Wettstein, D., & Liu, B. (2011). Heritable alteration in DNA methylation induced by nitrogen-deficiency stress accompanies enhanced tolerance by progenies to the stress in rice (*Oryza sativa* L.). *Journal of Plant Physiology*, 168, 1685–1693.

Krapp, A., David, L. C., Chardin, C., Girin, T., Marmagne, A., Leprince, A. S., Chaillou, S., Ferrario-Méry, S., Meyer, C., & Daniel-Vedele, F. (2014). Nitrate transport and signalling in *Arabidopsis*. *Journal of Experimental Botany*, 65, 789–798.

Lappartient, A. G., Vidmar, J. J., Leustek, T., Glass, A. D., & Touraine, B. (1999). Inter-organ signaling in plants: Regulation of ATP sulfurylase and sulfate transporter genes expression in roots mediated by phloem-translocated compound. *Plant Journal*, 18, 89–95.

Lee, R. C., Feinbaum, R. L., & Ambros, V. (1993). The *C. elegans* heterochronic gene lin-4 encodes small RNAs with antisense complementarity to lin-14. *Cell*, 75, 843–854.

Lee, S., & Leustek, T. (1999). The effect of cadmium on sulfate assimilation enzymes in *Brassica juncea*. *Plant Science*, 141, 201–207.

Lee, S., Marmagne, A., Park, J., Fabien, C., Yim, Y., Kim, S., Kim, T. H., Lim, P. O., Masclaux-Daubresse, C., & Nam, H. G. (2020). Concurrent activation of *OsAMT1; 2* and *OsGOGAT1* in rice leads to enhanced nitrogen use efficiency under nitrogen limitation. *Plant Journal*, 103, 7–20.

Leustek, T., Martin, M. N., Bick, J. A., & Davies, J. P. (2000). Pathways and regulation of sulfur metabolism revealed through molecular and genetic studies. *Annual Review of Plant Biology*, 51, 141–165.

Leustek, T., & Saito, K. (1999). Sulfate transport and assimilation in plants. *Plant Physiology*, 120, 637–644.

Li, E., & Zhang, Y. (2014). DNA methylation in mammals. *CSH Perspectives in Biology*, 6, a019133.

Li, H., Hu, B., & Chu, C. (2017). Nitrogen use efficiency in crops: Lessons from *Arabidopsis* and rice. *Journal of Experimental Botany*, 68, 2477–2488.

Li, Y., Brooks, M., Yeoh-Wang, J., McCoy, R. M., Rock, T. M., Pasquino, A., Moon, C. I., Patrick, R. M., Tanurdzic, M., & Ruffel, S. (2020). SDG8-mediated histone methylation and RNA processing function in the response to nitrate signaling. *Plant Physiology*, 182, 215–227.

Li, Y., Mukherjee, I., Thum, K. E., Tanurdzic, M., Katari, M. S., Obertello, M., Edwards, M. B., McCombie, W. R., Martienssen, R. A., & Coruzzi, G. M. (2015). The histone methyltransferase SDG8 mediates the epigenetic modification of light and carbon responsive genes in plants. *Genome Biology*, 16, 1–15.

Liang, G., Yang, F., & Yu, D. (2010). MicroRNA395 mediates regulation of sulfate accumulation and allocation in *Arabidopsis thaliana*. *Plant Journal*, 62, 1046–1057.

Liu, K. H., Liu, M., Lin, Z., Wang, Z. F., Chen, B., Liu, C., Guo, A., Konishi, M., Yanagisawa, S., Wagner, G., & Sheen, J. (2022a). NIN-like protein 7 transcription factor is a plant nitrate sensor. *Science*, 377, 1419–1425.

Liu, X., Hu, B., & Chu, C. (2022b). Nitrogen assimilation in plants: Current status and future prospects. *Journal of Genetics and Genomics*, 49, 394–404.

Liu, X., Liu, S., Wang, R., Chen, X., Fan, Z., Wu, B., & Zhou, T. (2019). Analyses of MiRNA functions in maize using a newly developed ZMBJ-CMV-2bN81-STTM vector. *Frontiers in Plant Science*, 10, 1277.

Mager, S., & Ludewig, U. (2018). Massive loss of DNA methylation in nitrogen-, but not in phosphorus-deficient *Zea mays* roots is poorly correlated with gene expression differences. *Frontiers in Plant Science*, 9, 497.

Martinoia, E., & Rentsch, D. (1994). Malate compartmentation - responses to a complex metabolism. *Annual Review of Plant Physiology and Plant Moleculat Biology*, 45, 1040–2519.

Maruyama-Nakashita, A., Nakamura, Y., Tohge, T., Saito, K., & Takahashi, H. (2006). *Arabidopsis* SLIM1 is a central transcriptional regulator of plant sulfur response and metabolism. *Plant Cell*, 18, 3235–3251.

Maruyama-Nakashita, A., Nakamura, Y., Watanabe-Takahashi, A., Inoue, E., Yamaya, T., & Takahashi, H. (2005). Identification of a novel cis-acting element conferring sulfur deficiency response in *Arabidopsis* roots. *Plant Journal*, 42, 305–314.

Maruyama-Nakashita, A., Watanabe-Takahashi, A., Inoue, E., Yamaya, T., Saito, K., & Takahashi, H. (2015). Sulfur-responsive elements in the 3′-nontranscribed intergenic region are essential for the induction of sulfate transporter 2;1 gene expression in *Arabidopsis* roots under sulfur deficiency. *Plant Cell*, 27, 1279–1296.

Matthewman, C. A., Kawashima, C. G., Húska, D., Csorba, T., Dalmay, T., & Kopriva, S. (2012). miR395 is a general component of the sulfate assimilation regulatory network in *Arabidopsis*. *FEBS Letters*, 586, 3242–3248.

Morot-Gaudry, J. F., Job, D., & Lea, P. J. (2001). Amino acid metabolism. In P. J. Lea & J. F. Morot-Gaudry (Eds.), *Plant Nitrogen* (pp. 167–211). Berlin: Springer.

Muños, S., Cazettes, C., Fizames, C., Gaymard, F., Tillard, P., Lepetit, M., Lejay, L., Gojon, A. (2004). Transcript profiling in the chl1-5 mutant of *Arabidopsis* reveals a role of the nitrate transporter NRT1.1 in the regulation of another nitrate transporter, NRT2.1. *Plant Cell*, 16, 2433–2447.

O'Brien, J. A., Vega, A., Bouguyon, E., Krouk, G., Gojon, A., Coruzzi, G., & Gutiérrez, R. A. (2016). Nitrate transport, sensing, and responses in plants. *Molecular Plant*, 9, 837–856.

Parihar, S. (2017). In vitro conservation protocol of *Ceropegia bulbosa*: An important medicinal and threatened plant species of Western Rajasthan. *Plant Science Today*, 4, 21-26.

Parihar, S. (2018). In vitro biochemical characterization of *Caralluma edulis* (Edgew.) Benth. & amp; Hook. f. and *Caralluma adscendens* (Roxb.) R. Br.: Medicinally potent Indian plant species. *Vegetos*, 31, 142–146.

Parihar, S. (2020). Protein profling of regenerative and non regenerative callus cultures of *Glossonema varians*: a rare, endemic and edible plant of Indian Thar Desert. *Vegetos*, 33, 385–389.

Parihar, S. (2023). In vitro regeneration of *Caralluma stalagmifera* var. *stalagmifera* through LCT and ex vitro rooting: A cost effective approach for conservation of succulents. *Vegetos, 36,* 1535–1545.

Parihar, S., & Dwivedi N. K. (2019). Comprehensive analysis of liquid and semisolid culture system for in vitro propagation and conservation of *Caralluma edulis*: an appetite suppressant medicinal succulent of the indian thar desert. *Plant Cell Biotechnology & Molecular Biology,* 20, 1020–1031.

Parihar S., & Dwivedi, N. K. (2020). A note on an important edible, rare and the famine food plant of Indian Thar Desert: *Glossonema varians* (Stocks) Benth. ex Hook. f. *Genetic Resources & Crop Evolution,* 67, 1929–1934.

Pate, J. S. (1965). Roots as organs of assimilation of sulfate. *Science*, 149, 547–548.

Patron, N. J., Durnford, D. G., & Kopriva, S. (2008). Sulfate assimilation in eukaryotes: Fusions, relocations and lateral transfers. *BMC Evolutionary Biology*, 8, 1–14.

Rahikainen, M. (2024). Epigenetic regulation of glucosinolate biosynthesis sees the light of day. *Plant Physiology*, 196, 689–690.

Rapp, R. A., & Wendel, J. F. (2005). Epigenetics and plant evolution. *New Phytologist*, 168, 81–91.

Raun, W. R., & Johnson, G. V. (1999). Improving nitrogen use efficiency for cereal production. *Agronomy Journal*, 91, 357–363.

Rouached, H., Wirtz, M., Alary, R., Hell, R., Arpat, A. B., Davidian, J. C., Fourcroy, P., & Berthomieu, P. (2008). Differential regulation of the expression of two high-affinity sulfate transporters, SULTR1.1 and SULTR1.2, in *Arabidopsis*. *Plant Physiology*, 147, 897–911.

Samantara, K., Shiv, A., de Sousa, L. L., Sandhu, K. S., Priyadarshini, P., & Mohapatra, S. R. (2021). A comprehensive review on epigenetic mechanisms and application of epigenetic modifications for crop improvement. *Environmental and Experimental Botany*, 188, 104479.

Sánchez-Rodríguez, E., del Mar Rubio-Wilhelmi, M., Ríos, J. J., Blasco, B., Rosales, M. Á, Melgarejo, R., Romero, L., & Ruiz, J. M. (2011). Ammonia production and assimilation: Its importance as a tolerance mechanism during moderate water deficit in tomato plants. *Journal of Plant Physiology*, 168, 816–823.

Sánchez-Vicente, I., & Lorenzo, O. (2021). Nitric oxide regulation of temperature acclimation: A molecular genetic perspective. *Journal of Experimental Botany*, 72, 5789–5794.

Sharma, L., Parihar, S., Seth, C. S. & Shekhawat, G. S. (2025). Abiotic stress tolerance: Role of Nitric Oxide signalling in developing climate-resilient plants. In R. K. Kalia & R. Pathak (Eds.), *Tree Biology and Biotechnology* (pp. 303–316). Singapore: Springer.

Sharma, L., Parihar, S., & Shekhawat, G. S. (2024). Nitric Oxide: A key bioactive regulator of plant tolerance mechanism under metal induced oxidative stress. In M. Faizan & S. Hayat (Eds.), *Plant Growth Regulators: Resilience for Sustainable Agriculture* (pp. 41–54). Singapore: Springer.

Singh, J., Mishra, V., Wang, F., Huang, H. Y., & Pikaard, C. S. (2019). Reaction mechanisms of Pol IV, RDR2, and DCL3 drive RNA channeling in the siRNA-directed DNA methylation pathway. *Molecular Cell*, 75, 576–589.

Stitt, M., Müller, C., Matt, P., Gibon, Y., Carillo, P., Morcuende, R., Scheible, W. R., & Krapp, A. (2002). Steps towards an integrated view of nitrogen metabolism. *Journal of Experimental Botany*, 53, 959–970.

Sun, H., Li, J., Song, W., Tao, J., Huang, S., Chen, S., & Zhang, Y. (2015). Nitric oxide generated by nitrate reductase increases nitrogen uptake capacity by inducing lateral root formation and inorganic nitrogen uptake under partial nitrate nutrition in rice. *Journal of Experimental Botany*, 66, 2449–2459.

Takahashi, H. (2019). Sulfate transport systems in plants: Functional diversity and molecular mechanism underlying regulatory coordination. *Journal of Experimental Botany*, 70, 4075–4087.

Takahashi, H., Kopriva, S., Giordano, M., Saito, K., & Hell, R. (2011). Sulfur assimilation in photosynthetic organisms: Molecular functions and regulations of transporters and assimilatory enzymes. *Annual Review of Plant Biology*, 62, 157–184.

Takahashi, H., Watanabe-Takahashi, A., Smith, F. W., Blake-Kalff, M., Hawkesford, M. J., & Saito, K. (2000). The roles of three functional sulphate transporters involved in uptake and translocation of sulphate in *Arabidopsis thaliana*. *Plant Journal*, 23, 171–182.

Turner, B. M. (2000). Histone acetylation and an epigenetic code. *Bioessays*, 22, 836–845.

Varala, K., Marshall-Colón, A., Cirrone, J., Brooks, M. D., Pasquino, A. V., Léran, S., Mittal, S., Rock, T. M., Edwards, M. B., Kim, G. J., & Ruffel, S. (2018). Temporal transcriptional logic of dynamic regulatory networks underlying nitrogen signaling and use in plants. *Proceedings of the National Academy of Sciences of the United States of America*, 115, 6494–6499.

Vidal, E. A., Alvarez, J. M., Araus, V., Riveras, E., Brooks, M. D., Krouk, G., Ruffel, S., Lejay, L., Crawford, N. M., Coruzzi, G. M., & Gutiérrez, R. A. (2020). Nitrate in 2020: Thirty years from transport to signaling networks. *Plant Cell*, 32, 2094–2119.

Wambui Mbichi, R., Wang, Q. F., & Wan, T. (2020). RNA directed DNA methylation and seed plant genome evolution. *Plant Cell Reports*, 39, 983–996.

Wang, Y. Y., Cheng, Y. H., Chen, K. E., & Tsay, Y. F. (2018). Nitrate transport, signaling, and use efficiency. *Annual Review of Plant Biology*, 69, 85–122.

Wang, Y. Y., Hsu, P. K., & Tsay, Y. F. (2012). Uptake, allocation and signaling of nitrate. *Trends in Plant Science*, 17, 458–467.

Wawrzyńska, A., Lewandowska, M., & Sirko, A. (2010). *Nicotiana tabacum* EIL2 directly regulates expression of at least one tobacco gene induced by sulphur starvation. *Journal of Experimental Botany*, 61, 889–900.

Wawrzyńska, A., & Sirko, A. (2016). EIN3 interferes with the sulfur deficiency signaling in *Arabidopsis thaliana* through direct interaction with the SLIM1 transcription factor. *Plant Science*, 253, 50–57.

Weber, K., & Burow, M. (2018). Nitrogen–essential macronutrient and signal controlling flowering time. *Physiologia Plantarum*, 162, 251–260.

Yonekura-Sakakibara, K., Onda, Y., Ashikari, T., Tanaka, Y., Kusumi, T., & Hase, T. (2000). Analysis of reductant supply systems for ferredoxin-dependent sulfite reductase in photosynthetic and nonphotosynthetic organs of maize. *Plant Physiology*, 122, 887–894.

Yoshimoto, N., Takahashi, H., Smith, F. W., Yamaya, T., & Saito, K. (2002). Two distinct high-affinity sulfate transporters with different inducibilities mediate uptake of sulfate in *Arabidopsis* roots. *Plant Journal*, 29, 465–473.

Yu, X., Jiang, L., Wu, R., Meng, X., Zhang, A., Li, N., Xia, Q., Qi, X., Pang, J., Xu, Z. Y., & Liu, B. (2016). The core subunit of a chromatin-remodeling complex, ZmCHB101, plays essential roles in maize growth and development. *Scientific Reports*, 6, 38504.

Yu, X., Meng, X., Liu, Y., Li, N., Zhang, A., Wang, T. J., Jiang, L., Pang, J., Zhao, X., Qi, X., & Zhang, M. (2018). The chromatin remodeler ZmCHB101 impacts expression of osmotic stress-responsive genes in maize. *Plant Molecular Biology*, 97, 451–465.

Zhang, H., Lang, Z., & Zhu, J. K. (2018). Dynamics and function of DNA methylation in plants. *Nature Reviews Molecular Cell Biology*, 19, 489–506.

Zhao, F. J., Salmon, S. E., Withers, P. J. A., Monaghan, J. M., Evans, E. J., Shewry, P. R., & McGrath, S. P. (1999). Variation in the breadmaking quality and rheological properties of wheat in relation to sulphur nutrition under field conditions. *Journal of Cereal Science*, 30, 19–31.

13 Epigenetic Regulation of Nitrogen and Sulfur Metabolism in Plants

Sumali Sucharita, Ankur Singh and Aryadeep Roychoudhury

13.1 INTRODUCTION

The word "epigenetics" refers to change in stable chromatin state that takes place during mitosis or meiosis and alter phenotypic traits and gene expression without changing the DNA sequence. Epigenetic factors include a wide variety of biochemical processes, including DNA methylation, histone modifications and small or long non-coding RNAs, which regulate chromatin structure and gene expression. A distinctive feature of gene silencing, viz. DNA cytosine methylation, is crucial for maintaining the integrity of the plant genome (Law & Jacobsen, 2010). Both normal metabolic processes such as respiration, photosynthesis, and photorespiration and stress responses in plant cells generate reactive oxygen species (ROS), which affect metabolic flux and the epigenetic process of gene regulation (Shen et al., 2016) (Figure 13.1).

Recent studies have contributed to elucidating the signal transduction pathways underlying abiotic stress responses and have shown the significance of epigenetic factors in plant responses to abiotic stimuli (Sahu et al., 2013; Kim et al., 2015). An example of how epigenetic regulation affects plant responses to their surroundings is vernalization, the process by which plants encounter cold temperature during winter in order to bloom in the spring (Zhao et al., 2018; Luo & He, 2020). Epigenetic mechanisms regulate stress-responsive genes both transcriptionally and post-transcriptionally, changing their chromatin status. Histone modifications and stable or heritable DNA methylation are examples of epigenetic processes that may help improve the memory of stress. Stresses can cause hypo- or hypermethylation of DNA, which can alter how genes are expressed. *ZmMI1* expression in maize roots under cold stress has been linked to a decrease in methylation in the nucleosome core DNA (Chinnusamy & Zhu, 2009).

Nucleosome core complex histones experience a number of post-translational changes. Additionally, different genes encode variants of each histone. The transcriptional state and level of expression of a gene are determined by its chromatin structure, which is largely determined by the combinations of histone variants and post-translational modifications known as the "histone code". Acetylation, specific phosphorylation and ubiquitination are histone modifications that promote transcription, whereas biotinylation and sumoylation suppress gene expression (Sridhar et al., 2007; Zhang et al., 2007; Nathan et al., 2006; Camporeale et al., 2007).

Plants generate a large number of low-molecular-weight organic compounds. The scientific community has classified these substances into three main categories based on their presumed roles: hormones, which regulate metabolism and organismal functions; secondary metabolites, which coordinate interactions between plants and the external environment; and primary metabolites, which are directly essential to crop growth and development (Erb & Kliebenstein, 2020). Plant growth and development depend directly on primary metabolites, which are extremely essential (Fernie & Pichersky, 2015). Secondary metabolites, which include major groups like terpenes, phenolics and nitrogen-containing compounds, are usually associated with one specific lineage and help plants to survive under biotic and abiotic stress conditions (Hartmann, 2007). Metabolites can affect gene expression by modulating epigenetic pathways and chromatin-modifying enzymes; for example, acetyl-CoA is a metabolite that serves as a cofactor for histone acetyltransferases (HATs), which are involved in epigenetic regulation. HAT enzymes use acetyl-CoA to acetylate lysine residues of histones to establish the active epigenetic state of chromatin in eukaryotic cells. Small plant gene families encode HATs and histone deacetylases (HDACs), which control histone acetylation. These genes are important for developmental processes, environmental responses and epigenetic regulation (Servet et al., 2010; Chen & Zhou, 2013; Ma et al., 2013; Grandperret et al., 2014). Metabolites that serve as cofactors or substrates for histone acetylation and deacetylation have an especially substantial impact on the activities of HATs and HDACs.

According to Liu et al. (2010), Chen et al. (2011) and Thorstensen et al. (2011), histone methylation and gene expression are significantly altered during developmental stages and in response to adverse conditions. This involves the actions of particular histone methyltransferases (HMTs) and demethylases (HDMs). On the other hand, new research suggests that certain HMTs play a special role in controlling the expression of metabolism-related genes in plants. According to Law and Jacobsen (2010) and Jones (2012), DNA cytosine methylation is required for transcriptional gene silencing of repetitive sequences and transposable elements in plants.

Sulfur, an essential nutrient for all living organisms, is found in the amino acids, cysteine and methionine, as

DOI: 10.1201/9781003589914-16

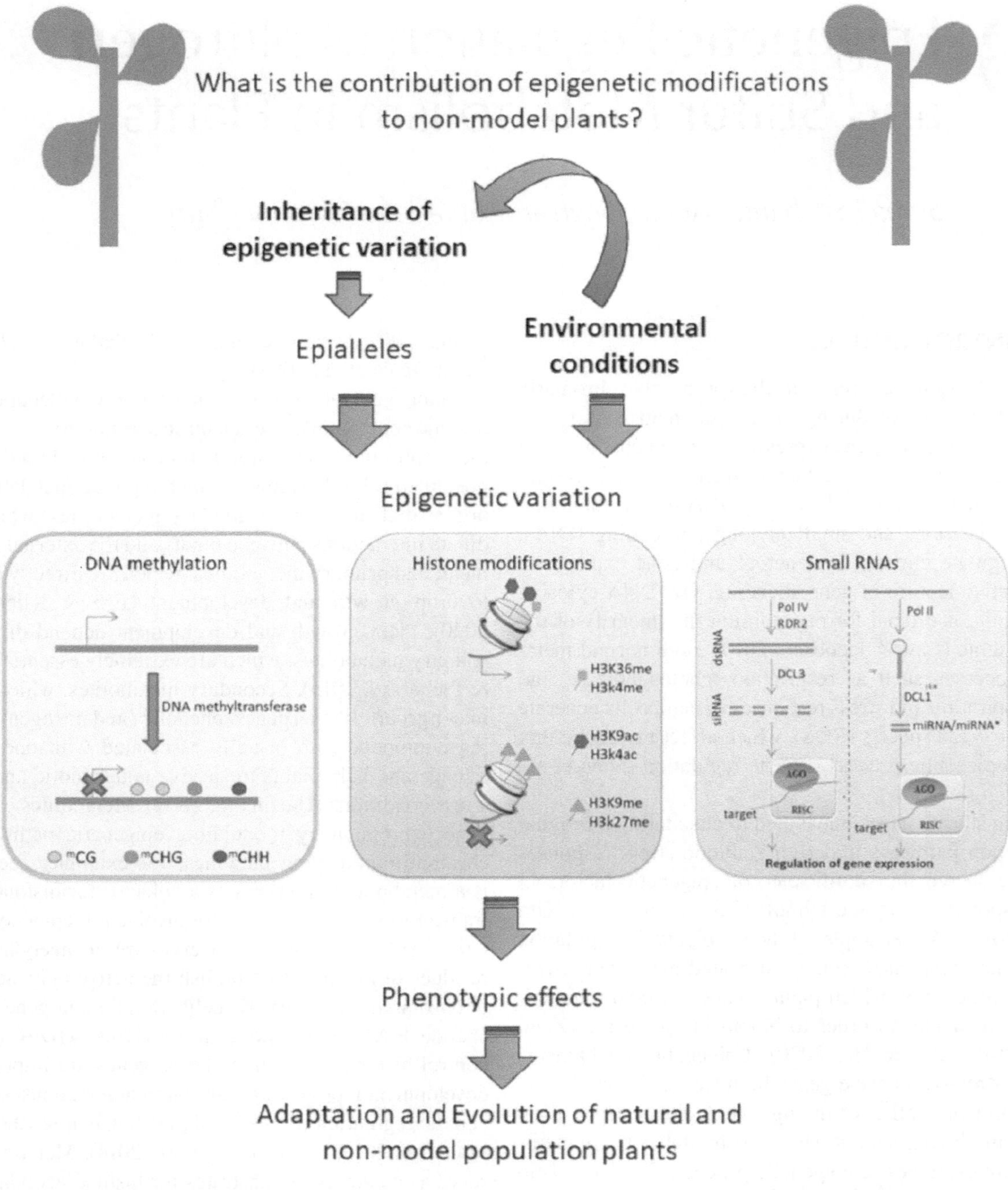

FIGURE 13.1 Epigenetic variation can facilitate the adaptation and evolution of non-model plants. Source: Taken from Thiebaut et al. (2019). This is an open-access article distributed under the terms of the Creative Commons Attribution License (CC BY).

well as in a variety of cofactors and prosthetic groups such as thiamine, Fe-S-center, S-adenosylmethionine and several primary and secondary metabolites. Taking up inorganic sulfate from the soil, plants can convert it to sulfide and then incorporate it into bioorganic compounds. In the sulfate assimilation pathway, ATP sulfurylase (ATPS) first converts sulfate to adenosine 5-phosphosulfate (APS) (Koprivova & Kopriva, 2014). Plants can use the organic sulfur that is mineralized into SO_4^{2-} as organic matter decomposes. In addition to organic matter, the soil contains a variety of minerals that contain different forms of sulfur. A number of biological processes such as energy production, photoprotection, photosynthesis and metabolic reactions rely on sulfur-containing compounds such as Fe-S clusters with proteins (Jogawat et al., 2021; Yadav et al., 2021; Bashir et al., 2011). Young leaf chloroplasts are the main organelles where SO_4^{2-} is assimilated for cysteine and methionine synthesis. Furthermore, sulfur compounds play a direct or indirect role in signaling, metabolism and managing biotic and abiotic stress responses in plants (Aarabi et al., 2020; Takahashi et al., 2011).

Sulfate is taken up from the soil and distributed within plant cells by a number of membrane transporters. A proton facilitates transport across the plasma membrane by maintaining a gradient that is maintained by proton ATPase. Their entry is part of the symport H^+ mechanism, along with sulfate. Due to the electrical gradient between the cytoplasm and vacuole sap, sulfate is transported across the tonoplast membrane. If H^+/SO_4^{2-} is present in the inner membrane of the chloroplast, it could help move SO_4^{2-} into chloroplasts (Li et al., 2020; Agrawal, 2003).

Nitrogen (N) ranks third among all the macronutrients required for plants, after carbon and oxygen. N plays an important role in all phases of plant development and has become the main component of proteins, nucleic acids, phospholipids, chlorophyll, hormones, vitamins and alkaloids (O'Brien et al., 2016; Chen et al., 2021; Jiang et al., 2024). According to Liu et al. (2023), low N stress may inhibit biomass accumulation, reduce photosynthesis, increase leaf senescence and cause irreversible damage to plants.

13.2 NITROGEN METABOLISM IN PLANTS

The physiological process that regulates nitrogen metabolism in plants includes nitrate reduction, ammonium assimilation in roots and leaves, nitrate or ammonium transport through cell membranes in roots and N_2 fixation within nodules for legumes (Gastal et al., 2015). The supply of nitrogenous compounds in the soil serves as the primary source of nitrogen for green plants. Nitrogenous compounds found in soil can be either organic or inorganic. Organic nitrogen compounds, such as urea, are generated when microorganisms ammonify and nitrify the remains of plants and animals, and they are incorporated into the soil humus. The nitrates, nitrites and ammonia compounds are inorganic nitrogenous substances found in soil. Since nitrate is the most crucial of these for plants, it must be brought back to maintain efficiency because, like other anions, it leaches out (Kochhar & Gujral, 2020).

13.2.1 Nitrate Reduction

Nitrate reductase (NR) and nitrite reductase (NiR) are two enzymes that control the reduction of nitrate (NO_3^-) to ammonium (NH_4^+). NR catalyzes the two-electron reduction of NO_3^- to nitrite (NO_2^-), while NiR transfers six electrons to convert NO_2^- to NH_4^+. Despite being spatially distinct between the cytoplasm (for NO_3^- reduction) and plastids/chloroplasts (for NO_2^- reduction), NO_3^- reduction can occur in both shoots and roots. Two identical subunits make up the cytosolic enzyme NR which catalyzes the reduction of NO_3^- into NO_2^-. Each subunit takes part in the transfer of electrons from NADH/NADPH to NO_3^- by means of three cofactors: a heme (bound to a domain resembling the family of cytochromes), a molybdenum cofactor and a flavin adenine dinucleotide. The NR enzyme is considered to be cytosolic. The three prosthetic groups of the enzyme, FAD, hemoglobin and a molybdenum cofactor, can be found in the three domains that make up the NR protein. They are used in that order to transport electrons from NADPH to nitrate. The NO_2^- produced by the reduction of NO_3^- is carried to the chloroplast, where the NiR enzyme subsequently reduces it to NH_4^+. NiR is found in the chloroplasts of leaves and the proplastids of roots.

13.2.2 Ammonium Assimilation

In plant tissues, NH_4^+ is generated through lignin synthesis, photorespiration, senescence-induced N remobilization, NO_3^- reduction and N_2 fixation in legumes. The integration process of NH_4^+ in the plastid/chloroplast is mainly carried out by the glutamine synthetase/glutamine-2-oxoglutarate aminotransferase pathway (GS/GOGAT pathway). The cytosol and plastids include plant GS in a number of different forms of the enzyme . Cytosolic GS is essential for both leaf senescence and the integration of NH_4^+ and the recycling of NH_4^+ that comes from different metabolic processes. There are two main GS isoforms in higher plants: cytosolic GS1 and plastid GS2. According to gene expression studies, the *GLN1* gene encodes *GS1*, while the *GS2* gene is encoded by a single nuclear gene, *GLN2*. The *GLN2* genes play a role in both the assimilation of NH_4^+ generated by photorespiration and the incorporation of NH_4^+ originating from NO_3^- reduction. GS catalyzes the conversion of NH_4^+ into glutamine. Following the reaction between glutamine and 2-oxoglutarate (2-OG), a by-product of the tricarboxylic acid (TCA) cycle, the GOGAT catalyzes the formation of two glutamate molecules. In plants, glutamate synthase comes in two distinct isoforms: Fd-GOGAT and NADH-GOGAT. Fd-GOGAT is mainly found in leaf chloroplasts, specifically in the phloem companion cells-sieve element complex in the leaf vein. It takes electrons from the reduced ferredoxin (from PSI). Non-photosynthetic tissues, including etiolated leaf tissues, roots and companion cells, contain NADH-GOGAT, which uses NADH (from respiration) as the electron donor.

13.2.3 Nitrate Transport

Once NO_3^- enters root cells, it can spread towards a number of cell layers and is found in different tissues. For long-distance movement through the transpiration stream, NO_3^- is loaded from the stelar cells into xylem vessels during NO_3^- shoot transportation. In plants, the petiole is a crucial organ for distributing NO_3^- to the leaves when it is transported to the aerial parts.

13.2.4 Ammonium Transport

NH_4^+ can be carried throughout the plant by apoplastic and symplastic pathways after its uptake by the roots and accumulated in root vacuoles or transferred to aerial parts. The concentration of NH_4^+ within xylem sap is in the millimolar range, indicating that large volumes of NH_4^+ are transferred from the roots to the shoots (Figure 13.2).

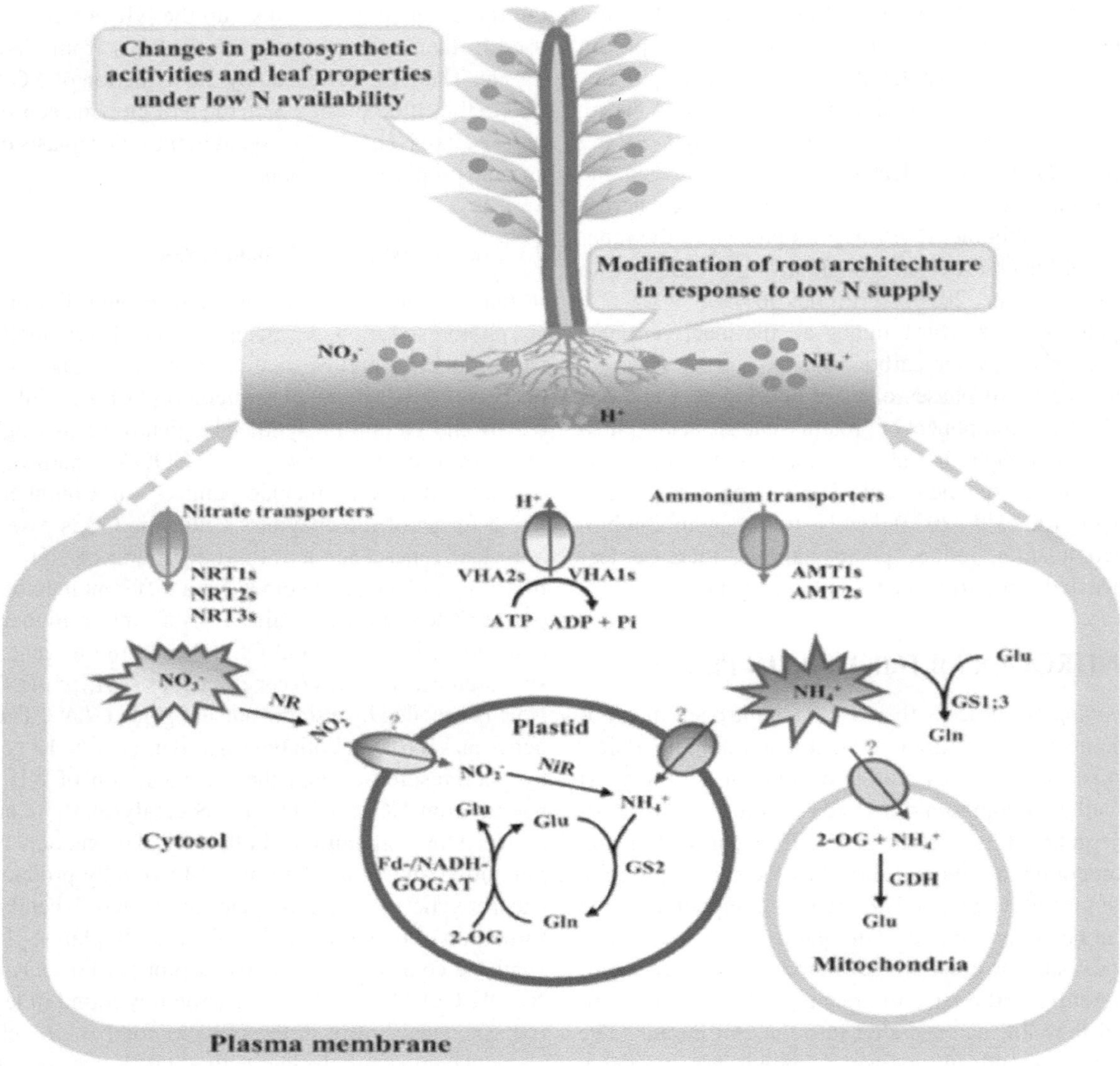

FIGURE 13.2 A schematic model of plant N metabolism. Through the use of plasma membrane H^+ ATPases (VHAs) in conjunction with ammonium (AMTs) and nitrate (NRTs) transporters, respectively, NH_4^+ and NO_3^- enter the cytosol during the uptake process. NH_4^+ and NO_3^- can be moved to leaves or other plant parts after being absorbed in the roots. The cytosolic nitrate reductase (NR) and the plastidic/chloroplastic nitrite reductase (NiR) transform NO_3^- into NH_4^+ during the assimilation process. Glutamate synthetase (GS) isoenzymes in the cytosol or plastid can then catalyze the conversion of NH_4^+ to glutamine (Gln). Fd- or NADH-dependent glutamate synthase (Fd/NADH-GOGAT) can further convert the Gln in the plastid with 2-oxoglutarate (2-OG) to glutamate (Glu). Moreover, glutamate dehydrogenase (GDH) in the mitochondrion can convert NH_4^+ to Glu. Source: Taken from Luo et al. (2013). This is an Open Access article distributed under the terms of the Creative Commons Attribution License (http://creativecommons.org/licenses/by/3.0/).

13.2.5 N_2 Fixation

Nitrogenase enzyme fixes atmospheric nitrogen through the symbiotic relationship between legume plant root and soil bacteria, or rhizobia, which includes the genera *Rhizobium*, *Bradyrhizobium*, *Mesorhizobium* and *Sinorhizobium* (Graham & Vance, 2003). This symbiotic relationship generates nodules, a newly differentiated special organ. Gram-negative alpha proteobacteria are the most prevalent microbial species that associate (endo-symbiotic interaction) with legumes of the Fabaceae (Papilionaceae) family in both plant and rhizobia symbiotic systems (Schultze & Kondorosi, 1998; Oldroyd & Downie, 2008; Desbrosses & Stougaard, 2011). Nitrogen fixation is also widely known for actinomycetes that associate with a wide variety of actinorhizal plants, such as *Frankia* sp. and *Parasponia* species (family Cannabaceae) (Santi et al., 2013).

13.3 REGULATION OF NITROGEN METABOLISM

13.3.1 Regulation at the Transcriptional Level

The mRNA involved in N assimilation and intake could be modified to regulate N metabolism. Because of adverse feedback caused by the N status of plants, the transcript

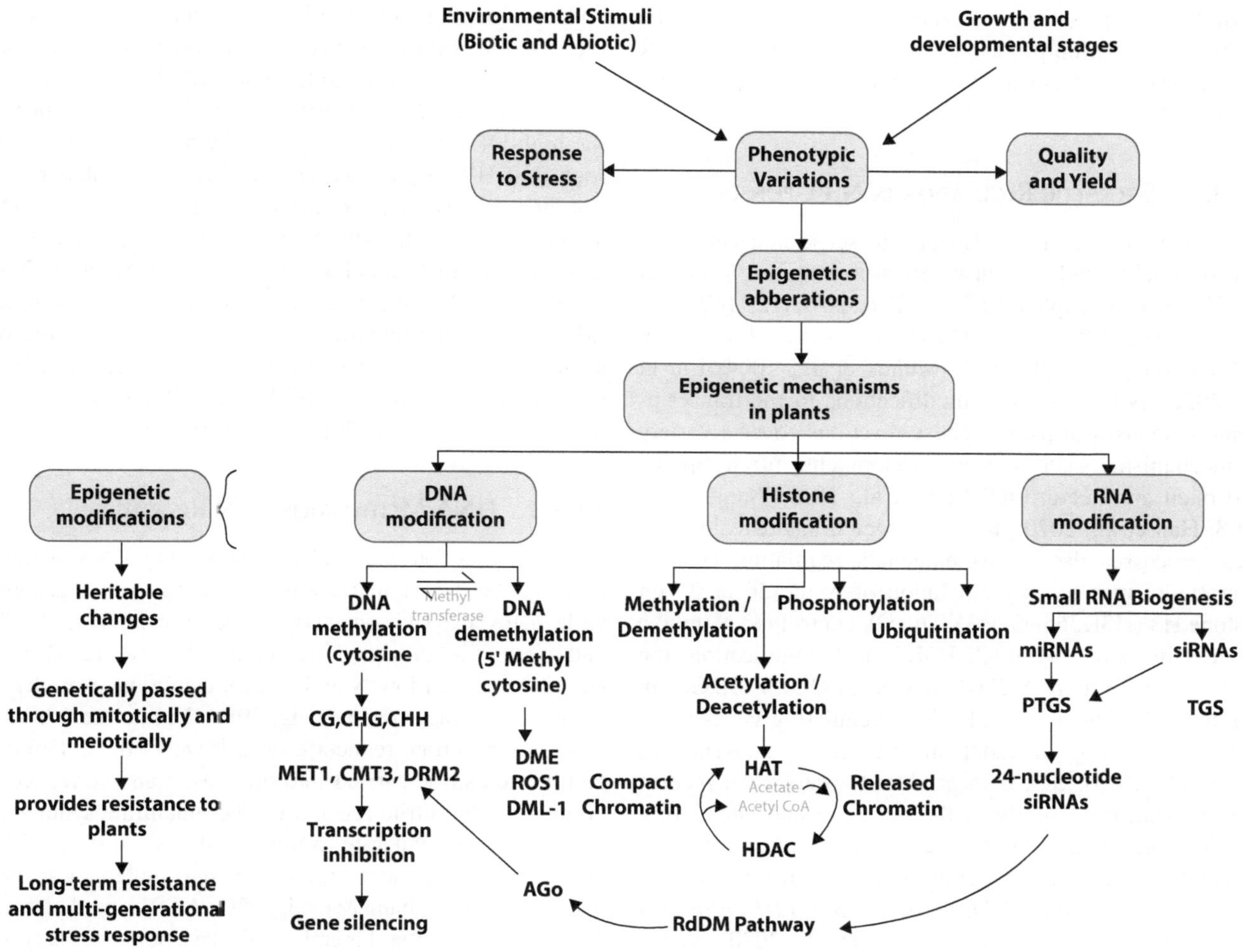

FIGURE 13.3 An overview of epigenetic modifications during plant growth and development. Source: Taken from Abdulraheem et al. (2024). This article is an open access article distributed under the terms and conditions of the Creative Commons Attribution (CC BY) license (https://creativecommons.org/licenses/by/4.0/).

level is upregulated when the N status is low and downregulated when the N status is high (Cerezo et al., 2001; Gazzarrini et al., 1999). Genes involved in the low affinity transport of NO_3^- were shown to have distinct regulatory features for *NRT1.1*, *NRT1.2* and *NRT1.5* based on transcriptional regulation. While *NRT1.2* is constitutively expressed both before and after NO_3^- exposure (Huang et al., 1999), *NRT1.1* mRNA levels are quickly induced by NO_3^- (Tsay et al., 1993). *NRT1.5* mRNA levels are induced by NO_3^- significantly but more slowly than *NRT1.1*, and additionally, a rise in pH and lower potassium level suppressed its expression (Lin et al., 2008). Regulation of genes by NO_2^- demonstrated that NO_3^-- and NO_2^--regulated genes largely overlap (Wang et al., 2007). According to Lejay et al. (2003), photosynthesis plays an essential role for controlling intake and incorporation of nitrogen since it ensures that N uptake correlates with the carbon status. The diurnal variation of the cycle of N uptake and decline is a key characteristic that can be linked with sugar formation via photosynthesis and its delivery to the roots (Delhon et al., 1996; Gastal & Saugier, 1989). Genes encoding NO_3^- transporters (Lejay et al., 1999; Matt et al., 2001; Ono et al., 2000) and NH_4^+ transporters (Gazzarrini et al., 1999; Lejay et al., 2003; Von Wire´n et al., 2000), as well as NR and NiR (Vincentz et al., 1993), are generally linked to seasonal variations in N uptake and assimilation. Furthermore, the reduction in the expression of both *AtNRT1.1* and *AtNRT2.1* during dark periods could be inhibited by providing a source of sucrose, suggesting that these genes are inducible by sugar and that sucrose could imitate the effects of light in their expression (Lejay et al., 1999; Lejay et al., 2003) (Figure 13.3).

13.3.2 Regulation at the Post-Transcriptional Level

According to Vidal and Gutierrez (2008), post-transcriptional control of mRNA translation and stability by microRNA has been demonstrated to be crucial for N response. Wirth et al. (2007) found that NRT2.1 proteins are relatively stable in response to light, sucrose or N treatments. NH_4^+ transporters have also been shown to be subject to post-translational regulation, which enables quick uptake. For post-translational regulation of N metabolism, NR regulation in higher plants is necessary. According to de Cires et al. (1993) and Kaiser and Huber (2001), the carbon-assimilation products are regulatory factors of NR activity and are associated with the post-translational regulation of NR.

13.3.3 Epigenetic Regulation in N Responses

Plants have the amazing capacity to sense nutrient availability quickly and initiate a series of reactions (Vidal et al., 2020). Transcripts for *Nitrate Transporter 1.1* (*NRT1.1*) or *NIN LIKE PROTEIN 7 (NLP7)* can sense N, a crucial nutrient (Gojon et al., 2011; Muños et al., 2004; Liu et al., 2022). N causes quick modifications to the transcriptional responses of plants. These reactions cover a variety of mechanisms, such as root development, nitrate uptake and plant architecture (O'Brien et al., 2016; Wang et al., 2018; Han et al., 2020). Recent research has revealed that these processes also involve epigenetic regulation. Histone alterations, such as tri-methylation of lysine 36 or 27 on histone H3 (H3K36me3, H3K27me3), control the N uptake process in dicots. Through H3K36me3 modification, the H3K36me3 "writer" SETDOMAIN GROUP8 (SDG8) in *Arabidopsis* plays a crucial role in regulating genes associated with energy metabolism, including photosynthesis (Li et al., 2015). SDG8 is produced when roots receive N treatment, particularly in the pericycle and lateral root cap (Gifford et al., 2008). According to recent research, SDG8 regulates the reprogramming of N-triggered gene expression, including *GLUTAMINE SYNTHETASE 1;4* (*GLN1;4*), *NRT2.1* and *NRT1.5* (Li et al., 2020). When treated with elevated N levels, *sdg8* mutants demonstrated reduced nitrate acquisition compared to wild-type plants (Li et al., 2020). In addition to root development, N supply has an epigenetic impact on plant architecture, especially on cultivation in monocots. N fertilizer causes alterations in the genome-wide H3K27me3 pattern in rice through *NITROGEN-MEDIATED TILLER GROWTH RESPONSE 5* (*NGR5*). By facilitating the recruitment of PRC2, this mechanism promotes the repression of branching-inhibitory genes through H3K27me3 modification, including *Dwarf14* (*D14*) and *SQUAMOSA-promoter binding protein-like 14* (*SPL14*) (Wu et al., 2020).

13.3.4 Histone Modifications in N Responses

Histone modification processes include methylation, phosphorylation, acetylation, ubiquitination and other covalent post-translational changes to histone proteins. The multiple functions of AtSDG8 include flowering, pigment synthesis, shoot branching, pathogen defense, touch response and carbon and light responses. It also encodes a histone methyl transferase (Cazzonelli et al., 2009, 2014; Dong et al., 2008; Kim et al., 2005; Lee et al., 2016; Li et al., 2015). When compared to N-sufficient conditions, N deficiency increased H3K36me3 levels in 143 genes in wild-type plants, while H3K36me3 levels decreased in 53 genes; in the *sdg8-5* mutants, these numbers dropped to 3 and 23. Furthermore, in *sdg8-5* mutants, N responses of increased lateral root growth and chlorophyll content were also reduced. According to Ho et al. (2009), Li et al. (2007) and Wang et al. (2020) AtNRT2.1 is the main component of the high-affinity transport system for nitrate uptake in the root. ZmCHB101 plays a role in leaf development, dehydration and abscisic acid response in maize. It also encodes a core subunit of the SWI/SNF-type ATP-dependent chromatin remodeling complex (Yu et al., 2016; Yu et al., 2018; Yu et al., 2019). Rice organ size and plant architecture are influenced by the integration of auxin and brassinosteroid signaling pathways by OsNGR5, which encodes a transcription factor with the APETALA2 (AP2) domain (Aya et al., 2014; Hirano et al., 2017; Qiao et al., 2017).

13.3.5 DNA Methylation in N Responses

Gene transcription is typically suppressed by DNA methylation. N deficiency in maize was found to reduce the genome-wide methylated CG content in the root from 26.6% to 12.0%, and over 95% of the Differentially Methylated Regions (DMRs) between low N and control conditions were hypomethylated (Mager & Ludewig, 2018). Nitrate breaks down to nitrite by nitrate reductase once it enters the cytoplasm, and nitrite can be carried into the plastid and converted to ammonium by nitrite reductase. The glutamine synthetase/glutamine-2-oxoglutarate aminotransferase (GS/GOGAT) cycle then incorporates ammonium into glutamate to produce glutamine (Zhang & Chu, 2020). A DNA methyl transferase that is essential for embryonic development is encoded by AtCMT3 and whose activity may be increased by AtSIZ1 sumoylation (Kim et al., 2015; Pillot et al., 2010).

13.4 SULFATE AND SULFUR AMINO ACID TRANSPORT

One of the most important macronutrients for plant growth and development is sulfur (S), which is also a necessary component for the physiological activity of plants. Plants absorb S as inorganic sulfate from the rhizosphere. Two root-specific high-affinity sulfate transporters SULTR1;1 and SULTR1;2 are primarily responsible for this process in *Arabidopsis thaliana* (Takahashi et al., 2000; Yoshimoto et al., 2002; Rouached et al., 2008). After passage into root cells, sulfate is either transferred into vacuoles for storage or into plastids by SULTR3 sulfate transporters where it is absorbed into organic S compounds (Cao et al., 2013; Chen et al., 2019). Many plant species have been found to contain the most prevalent sulfur metabolites, namely sulfate, cysteine and glutathione (Rennenberg et al., 2007). The primary source of sulfur for growth is sulfate that is absorbed by the roots; however, it must first be reduced to sulfur in order to undergo additional metabolism. The breakdown of sulfate to sulfide and the resulting formation into cysteine primarily occurs in the shoot and chloroplast, although all sulfate reduction enzymes are present in root plastids (De Kok et al., 2005).

The initial form or reduced sulfur donor of the majority of other organic sulfur compounds found in plants is cysteine (De Kok et al., 2005). Sulfate, the primary source of plant sulfur, gets into the plant through the root hair and spreads through the cortex and gathers actively in the primary endodermis. It enters the xylem vessels and travels through the transpiration stream to the leaves where it is released into the mesophyll cells. The primary site of sulfate reduction and the transformation of sulfur into organic form is the chloroplasts. Either cells grown in liquid media or isolated plant parts (roots, leaves or leaf disks) have been used to study short-distance transport. Out of several investigations into how SO_4^{2-} enters plant cells, a broad pathway can be drawn. (i) Transferring SO_4^{2-} into the cell requires specific transport proteins or permeases. (ii) Movement is dependent on energy since SO_4^{2-} accumulates opposite to an electrochemical gradient (Thompson et al., 1986).

13.4.1 Uptake and Assimilation of Sulfate

One of the main regulatory sites of sulfur assimilation appears to be the roots, which tightly regulate the uptake of sulfur and its transfer to the shoot. Specific sulfate transporter proteins absorb, transport and distribute sulfate at the subcellular level. The metabolic by-products of sulfur assimilation, such as glutathione, cysteine, and O-acetylserine, or sulfate itself, serve as signals for the expression of the sulfate transporters involved and for regulating the rate at which the root absorbs sulfate and transports it to the shoot (Hawkesford, 2000; Hawkesford et al., 2003a, 2003b; Buchner et al., 2004). Although all sulfate-reducing enzymes are found in root plastids, sulfate reduction primarily occurs in leaf chloroplasts. There are three steps involved in reducing sulfate to sulfide. Before being reduced to sulfite, sulfate must be activated to adenosine 5'-phosphosulfate (APS). ATP sulfurylase catalyzes the activation of sulfate, and its affinity for sulfate is relatively low (approximately 1 mM). The concentration of sulfate in situ in the chloroplast is probably one of the limiting/regulatory steps in sulfur reduction (Stulen & De Kok, 1993). APS reductase then catalyzes the reduction of APS to sulfite, with glutathione most likely acting as the reductant (Leustek & Saito, 1999; Kopriva & Koprivova, 2003). It is believed that APS reductase, with its low activity and rapid turnover rate, serves as a primary regulatory point in the sulfate reduction pathway (Brunold, 1990, 1993; Leustek & Saito, 1999; Kopriva & Koprivova, 2003; Saito, 2003). Sulfite reductase reduces sulfite with high affinity using ferredoxin as a reductant, and O-acetylserine(thiol)lyase catalyzes the incorporation of the resulting sulfide into cysteine using O-acetylserine as the substrate. Serine acetyltransferase catalyzes the synthesis of O-acetylserine and it is linked to O-acetylserine(thiol)lyase as part of an enzyme complex called cysteine synthase (Droux et al., 1998; Hell, 2003). The process by which sulfur and nitrogen are directly coupled in plants is via the production of cysteine (Brunold, 1990, 1993).

13.4.2 Regulation of Sulfur Homeostasis by Small RNAs

In order to identify and validate the function of sulfur-responsive miRNAs, several studies have been conducted. According to Li et al. (2016), these sulfur-responsive miRNAs typically target various transcription factors implicated in the auxin signaling pathway and response to stress. They also control the uptake, transport and assimilation of sulfate in plants. Five new miRNAs and 27 conserved miRNAs that express in *Arabidopsis* under SO_2 stress were found by deep sequencing (Li et al., 2016). In contrast to control, the SO_2-treated samples showed a seven-fold decrease in the expression of the novel miRNAs such as miR14, miR20 and miR43 and an upregulation of miR66 and miR67 by more than six-fold. Liang and Yu (2010) demonstrated that when miR395 levels were too high in *Arabidopsis*, it stopped the target genes *Sul2;1*, *APS1*, *APS3* and *APS4* from being expressed. Compared to the wild type, transgenic *Brassica napus* plants inducing miR395 build up more biomass and sulfur in Cd-treated plants. Three ATP sulfurylase genes (*ATPS1*, *ATPS3* and *ATPS4*) as well as a low-affinity sulfate transporter *SULTR2;1* in *Arabidopsis* were predicted to be targeted by miR395 (Jones-Rhoades & Bartel, 2004). Strong suppression of the generation of these four gene transcripts and an increase in the sulfate level in shoots is caused by the expression of the miR395-encoding gene.

13.5 EPIGENETIC REGULATION IN SULFUR METABOLISM

13.5.1 DNA Methylation

One of the most well-studied epigenetic modifications is DNA methylation, which is crucial for controlling gene expression, transposon silencing and imprinting. In plants, DNA methylation takes place in three distinct sequence contexts: CG, CHG and CHH (where H is A, C or T). DNA methylation can be defined as the transfer of a methyl group from S-adenosyl methionine (SAM) to the 5′ position of cytosine to form 5-methylcytosine (Huang et al., 2019). Since SAM, the universal methyl donor for DNA methylation, is produced from cysteine, which is the first organic sulfur compound in the primary sulfate assimilation pathway, it has been suggested that sulfate deficiency impacts genome-wide DNA methylation in plants. The biosynthesis of SAM is strictly regulated, and the amount of sulfate present influences the SAM concentration. When DNA methylation occurs, a specific DNA methyl transferase transfers the methyl group of SAM to cytosine, resulting in the synthesis of a molecule of S-adenosylhomocysteine (SAH). Modifications in global DNA methylation may also result from the depreciation of SAM biosynthesis by itself (Huang et al., 2019).

13.5.2 Histone Modifications

Histone modifications include methylation, acetylation, phosphorylation, ubiquitination and many other post-translational covalent changes on the amino-terminal tails of the core histones (Kouzarides, 2007; Liu et al., 2010; Bannister & Kouzarides, 2011). Gene expression is influenced by histone modifications, which change the accessibility of DNA to the transcriptional machinery. Histone methylation has a more complex impact on gene expression than histone acetylation and phosphorylation, which are typically linked to transcriptional activation (Berger, 2007). Gene activation is caused by H3K4me3 and H3K36me3, whereas H3K27me3 is linked to gene repression. A putative transcription factor, AL6, which functions as a histone mark reader and contains a plant homeodomain (PHD), can recognize and bind the H3K4me3 mark (Lee et al., 2009). Numerous genes involved in sulfate uptake and assimilation in *Arabidopsis* have histone methylations and acetylations such as H3K27me3 (Zhang et al., 2007), H3K4me3 and H3K36me3 (Luo et al., 2013), H3K23ac and H4K16ac (Lu et al., 2015) and H3K9ac (Zhou et al., 2010). It follows that histone modifications might be involved in preserving sulfur homeostasis.

13.6 CONCLUSION

Plant cells must coordinate epigenetics with metabolism and redox in order to quickly modify gene expression and metabolism in response to shifting environmental conditions. There are several ways in which metabolism, redox and epigenetics interact. In addition to chromatin modifiers controlling the activities of metabolic enzymes, chromatin states aid in defining the regulation of gene expression related to metabolism or redox, while key metabolites or redox-active molecules may directly influence the regulation of chromatin modifications. Furthermore, since biotic and abiotic stressors have a significant impact on plant metabolism and redox states, it will be crucial to investigate whether and how the alterations coordinate epigenetic systems to affect gene expression for plant adaptation and tolerance. Stress-responsive gene expression and plant development under stress have been demonstrated to be regulated by stress-induced alterations in histone variants, histone N-tail modifications and DNA methylation. Response to transient chromatin modification is mediated by acclimation. It has been demonstrated that many elements of the abiotic stress response are regulated by epigenetics and that treatments for abiotic stress either activate or repress the levels of genetic markers. Certain long non-coding RNAs may also be involved in different stress responses in addition to DNA methylation, histone modifications, chromatin modeling complexes and histone variants. The dynamic shifts in epigenetic markers on genes that respond to stress make their accessible or inaccessible chromatin status, which in turn controls the transcriptional or post-transcriptional expression of genes that respond to stress. There is growing evidence that epigenetic regulation plays a significant role in regulating plant responses to nutrient stressors and preserving nutrient homeostasis. Although the role of DNA methylation and histone modifications in the regulation of nitrogen and sulfur homeostasis is still poorly understood, miRNA-mediated gene silencing is well known to play a role in the regulation of sulfate uptake and adaptation.

REFERENCES

Aarabi, F., Naake, T., Fernie, A. R., & Hoefgen, R. (2020). Coordinating sulfur pools under sulfate deprivation. *Trends in Plant Science*, 25, 1227–1239.

Abdulraheem, M. I., Xiong, Y., Moshood, A. Y., Cadenas-Pliego, G., Zhang, H., & Hu, J. (2024). Mechanisms of plant epigenetic regulation in response to plant stress: Recent discoveries and implications. *Plants,* 13, 163.

Agrawal, M. (2003). Plant responses to atmospheric sulphur. In Y. P. Abrol & A. Ahmad (Eds.), *Sulphur in Plants* (pp. 279–293). Dordrecht: Springer.

Aya, K., Hobo, T., Sato-Izawa, K., Ueguchi-Tanaka, M., Kitano, H., & Matsuoka, M. (2014). A novel AP2-type transcription factor, SMALL ORGAN SIZE1, controls organ size downstream of an auxin signalling pathway. *Plant and Cell Physiology*, 55, 897–912.

Bannister, A. J., & Kouzarides, T. (2011). Regulation of chromatin by histone modifications. *Cell Research*, 21, 381–395.

Bashir, K., Ishimaru, Y., Shimo, H., Nagasaka, S., Fujimoto, M., Takanashi, H., Tsutsumi, N., An, G., Nakanishi, H., & Nishizawa, N. K. (2011). The rice mitochondrial iron transporter is essential for plant growth. *Nature Communications*, 2, 322.

Berger, S. L. (2007). The complex language of chromatin regulation during transcription. *Nature*, 447, 407–412.

Brunold, C. (1990). Reduction of sulfate to sulfide. In H. Rennenberg, C. Brunold, L. J. De Kok, & I. Stulen (Eds.), *Sulfur Nutrition and Sulfur Assimilation in Higher Plants: Fundamental, Environmental and Agricultural Aspects* (pp. 13–31). The Hague: SPB Academic Publishing.

Brunold, C. (1993). Regulatory interactions between sulfate and nitrate assimilation. In L. J. De Kok, I. Stulen, H. Rennenberg, C. Brunold, & W. Rauser (Eds.), *Sulfur Nutrition and Sulfur Assimilation in Higher Plants: Regulatory, Agricultural and Environmental Aspects* (pp. 125–138). The Hague: SPB Academic Publishing.

Buchner, P., Stuiver, C. E. E., Westerman, S., Wirtz, M., Hell, R., Hawkesford, M. J., & De Kok, L. J. (2004). Regulation of sulfate uptake and expression of sulfate transporter genes in *Brassica oleracea* as affected by atmospheric H_2S and pedospheric sulfate nutrition. *Plant Physiology*, 136, 3396–3408.

Camporeale, G., Oommen, A. M., Griffin, J. B., Sarath, G., & Zempleni, J. (2007). K12-biotinylated histone H4 marks heterochromatin in human lipoblastoma cells. *The Journal of Nutritional Biochemistry*, 18, 760–768.

Cao, M. J., Wang, Z., Wirtz, M., Hell, R., Oliver, D. J., & Xiang, C. B. (2013). SULTR 3; 1 is a chloroplast-localized sulfate transporter in *Arabidopsis thaliana. The Plant Journal*, 73, 607–616.

Cazzonelli, C. I., Cuttriss, A. J., Cossetto, S. B., Pye, W., Crisp, P., Whelan, J., Finnegan, E. J., Turnbull, C., & Pogson, B. J. (2009). Regulation of carotenoid composition and shoot branching in *Arabidopsis* by a chromatin modifying histone methyl transferase, *SDG8. The Plant Cell*, 21, 39–53.

Cazzonelli, C. I., Nisar, N., Roberts, A. C., Murray, K. D., Borevitz, J. O., & Pogson, B. J. (2014). A chromatin modifying enzyme, *SDG8,* is involved in morphological, gene expression, and epigenetic responses to mechanical stimulation. *Frontiers in Plant Science*, 5, 533.

Cerezo, M., Tillard, P., Filleur, S., Munos, S., Daniel-Vedele, F., & Gojon, A. (2001). Major alterations of the regulation of root NO_3- uptake are associated with the mutation of *Nrt2. 1* and *Nrt2. 2* genes in *Arabidopsis*. *Plant Physiology*, 127, 262–271.

Chen, H., Zhang, Q., Wang, X., Zhang, J., Ismail, A. M., & Zhang, Z. (2021). Nitrogen form-mediated ethylene signal regulates root-to-shoot K^+ translocation via NRT1. 5. *Plant, Cell & Environment*, 44, 3806–3818.

Chen, X., & Zhou, D. X. (2013). Rice epigenomics and epigenetics: Challenges and opportunities. *Current Opinion in Plant Biology*, 16, 164–169.

Chen, X., Hu, Y., & Zhou, D. X. (2011). Epigenetic gene regulation by plant Jumonji group of histone demethylase. *Biochimica et Biophysica Acta (BBA)-Gene Regulatory Mechanisms*, 1809, 421–426.

Chen, Z., Zhao, P. X., Miao, Z. Q., Qi, G. F., Wang, Z., Yuan, Y., Ahmad, N., Cao, M. J., Hell, R., Wirtz, M., & Xiang, C. B. (2019). SULTR3s function in chloroplast sulfate uptake and affect ABA biosynthesis and the stress response. *Plant Physiology*, 180, 593–604.

Chinnusamy, V., & Zhu, J. K. (2009). Epigenetic regulation of stress responses in plants. *Current Opinion in Plant Biology*, 12, 133–139.

de Cires, A., de la Torre, A., Delgado, B., & Lara, C. (1993). Role of light and CO_2 fixation in the control of nitrate-reductase activity in barley leaves. *Planta*, 190, 277–283.

De Kok, L. J., Castro, A., Durenkamp, M., Koralewska, A., Posthumus, F. S., Stuiver, C. E. E., Yang, L., & Stulen, I. (2005). Pathways of plant sulfur uptake and metabolism-an overview. *Landbauforschung Völkenrode Special Issue*, 283, 5–13.

Delhon, P., Gojon, A., Tillard, P., & Passama, L. (1996). Diurnal regulation of NO_3- uptake in soybean plants IV. Dependence on current photosynthesis and sugar availability to the roots. *Journal of Experimental Botany*, 47, 893–900.

Desbrosses, G. J., & Stougaard, J. (2011). Root nodulation: A paradigm for how plant-microbe symbiosis influences host developmental pathways. *Cell Host & Microbe*, 10, 348–358.

Dong, G., Ma, D. P., & Li, J. (2008). The histone methyltransferase *SDG8* regulates shoot branching in *Arabidopsis*. *Biochemical and Biophysical Research Communications*, 373, 659–664.

Droux, M., Ruffet, M. L., Douce, R., & Job, D. (1998). Interactions between serine acetyltransferase and O-acetylserine (thiol) lyase in higher plants: Structural and kinetic properties of the free and bound enzymes. *European Journal of Biochemistry*, 255, 235–245.

Erb, M., & Kliebenstein, D. J. (2020). Plant secondary metabolites as defences, regulators, and primary metabolites: The blurred functional trichotomy. *Plant Physiology*, 184, 39–52.

Fernie, A. R., & Pichersky, E. (2015). Focus issue on metabolism: Metabolites, metabolites everywhere. *Plant Physiology*, 169, 1421–1423.

Gastal, F., & Saugier, B. (1989). Relationships between nitrogen uptake and carbon assimilation in whole plants of tall fescue. *Plant, Cell & Environment*, 12, 407–418.

Gastal, F., Lemaire, G., Durand, J. L., & Louarn, G. (2015). Quantifying crop responses to nitrogen and avenues to improve nitrogen-use efficiency. In D. Calderini & V. O. Sadras (Eds.), *Crop Physiology* (pp. 161–206). Oxford: Oxford Academic Press.

Gazzarrini, S., Lejay, L., Gojon, A., Ninnemann, O., Frommer, W. B., & von Wirén, N. (1999). Three functional transporters for constitutive, diurnally regulated, and starvation-induced uptake of ammonium into *Arabidopsis* roots. *The Plant Cell*, 11, 937–947.

Gifford, M. L., Dean, A., Gutierrez, R. A., Coruzzi, G. M., & Birnbaum, K. D. (2008). Cell-specific nitrogen responses mediate developmental plasticity. *Proceedings of the National Academy of Sciences USA*, 105, 803–808.

Gojon, A., Krouk, G., Perrine-Walker, F., & Laugier, E. (2011). Nitrate transceptor (s) in plants. *Journal of Experimental Botany*, 62, 2299–2308.

Graham, P. H., & Vance, C. P. (2003). Legumes: Importance and constraints to greater use. *Plant Physiology*, 131, 872–877.

Grandperret, V., Nicolas-Frances, V., Wendehenne, D., & Bourque, S. (2014). Type-II histone deacetylases: Elusive plant nuclear signal transducers. *Plant, Cell & Environment*, 37, 1259–1269.

Han, X., Wu, K., Fu, X., & Liu, Q. (2020). Improving coordination of plant growth and nitrogen metabolism for sustainable agriculture. *a biotech*, 1, 255–275.

Hartmann, T. (2007). From waste products to eco chemicals: Fifty years' research of plant secondary metabolism. *Phytochemistry*, 68, 2831–2846.

Hawkesford, M. J., & Wray, J. L. (2000). Molecular genetics of sulphate assimilation. *Advances in Botanical Research*, 33, 159–223.

Hawkesford, M. J., Buchner, P., Hopkins, L., & Howarth, J. R. (2003a). The plant sulfate transporter family: Specialized functions, integration with whole plant nutrition. In J.-C. Davidian, D. Grill, L. J. De Kok, I. Stulen, M. J. Hawkesford, E. Schnug, & H. Rennenberg (Eds.), *Sulfur Transport and Assimilation in Plants: Regulation, Interaction and Signalling* (pp. 1–10). Leiden: Backhuys Publishers.

Hawkesford, M. J., Buchner, P., Hopkins, L., & Howarth, J. R. (2003b). Sulphate uptake and transport. In Y. P. Abrol & A. Ahmad (Eds.), *Sulphur in Plants* (pp. 71–86). Dordrecht: Kluwer Academic Publishers.

Hell, R. (2003). Metabolic regulation of cysteine synthesis and sulfur assimilation. The plant sulfate transporter family: Specialized functions, integration with whole plant nutrition. In J.-C. Davidian, D. Grill, L. J. De Kok, I. Stulen, M. J. Hawkesford, E. Schnug, & H. Rennenberg (Eds.), *Sulfur Transport and Assimilation in Plants: Regulation, Interaction and Signaling* (pp. 21–31). Leiden: Backhuys Publisher.

Hirano, K., Yoshida, H., Aya, K., Kawamura, M., Hayashi, M., Hobo, T., Sato-Izawa, K., Kitano, H., Ueguchi-Tanaka, M., & Matsuoka, M. (2017). SMALL ORGAN SIZE 1 and SMALL ORGAN SIZE 2/DWARF AND LOW-TILLERING form a complex to integrate auxin and *brassinosteroid* signalling in rice. *Molecular Plant*, 10, 590–604.

Ho, C. H., Lin, S. H., Hu, H. C., & Tsay, Y. F. (2009). CHL1 functions as a nitrate sensor in plants. *Cell*, 138, 1184–1194.

Huang, N. C., Liu, K. H., Lo, H. J., & Tsay, Y. F. (1999). Cloning and functional characterization of an *Arabidopsis* nitrate transporter gene that encodes a constitutive component of low affinity uptake. *Plant Cell*, 11, 1381–1392.

Huang, X. Y., Li, M., Luo, R., Zhao, F. J., & Salt, D. E. (2019). Epigenetic regulation of sulfur homeostasis in plants. *Journal of Experimental Botany*, 70, 4171–4182.

Jiang, H., Qi, C. H., Gao, H. N., Feng, Z. Q., Wu, Y. T., Xu, X. X., Cui, J. Y., Wang, X. F., Lv, Y. H., Gao, W. S., Jiang, Y. M., You, C. X., & Li, Y. Y. (2024). *MdBT2* regulates nitrogen-mediated cuticular wax biosynthesis via a MdMYB106-MdCER2L1 signalling pathway in apple. *Nature Plants*, 10, 131–144.

Jogawat, A., Yadav, B., Chhaya, & Narayan, O. P. (2021). Metal transporters in organelles and their roles in heavy metal transportation and sequestration mechanisms in plants. *Physiologia Plantarum*, 173, 259–275.

Jones, P. A. (2012). Functions of DNA methylation: Islands, start sites, gene bodies and beyond. *Nature Reviews Genetics*, 13, 484–492.

Jones-Rhoades, M. W., & Bartel, D. P. (2004). Computational identification of plant microRNAs and their targets, including a stress-induced miRNA. *Molecular Cell*, 14, 787–799.

Kaiser, W. M., & Huber, S. C. (2001). Post-translational regulation of nitrate reductase: Mechanism, physiological relevance and environmental triggers. *Journal of Experimental Botany*, 52, 1981–1989.

Kim, D. Y., Han, Y. J., Kim, S. I., Song, J. T., & Seo, H. S. (2015). *Arabidopsis* CMT3 activity is positively regulated by AtSIZ1-mediated sumoylation. *Plant Science*, 239, 209–215.

Kim, J. M., Sasaki, T., Ueda, M., Sako, K., & Seki, M. (2015). Chromatin changes in response to drought, salinity, heat, and cold stresses in plants. *Frontiers in Plant Science*, 6, 114.

Kim, S. Y., He, Y., Jacob, Y., Noh, Y. S., Michaels, S., & Amasino, R. (2005). Establishment of the vernalization-responsive, winter-annual habit in *Arabidopsis* requires a putative histone H3 methyl transferase. *The Plant Cell*, 17, 3301–3310.

Kochharr, S. L., & Gujral, S. K. (2020). Nitrogen metabolism. In *Plant Physiology: Theory and Applications* (pp. 313–351). Cambridge University.

Kopriva, S., & Koprivova, A. (2003). Sulphate assimilation: A pathway which likes to surprise. In Y. P. Abrol & A. Ahmad (Eds.), *Sulphur in Plants* (pp. 87–112). Dordrecht: Kluwer Academic Publishers.

Koprivova, A., & Kopriva, S. (2014). Molecular mechanisms of regulation of sulfate assimilation: First steps on a long road. *Frontiers in Plant Sciences*, 5, 589.

Kouzarides, T. (2007). Chromatin modifications and their function. *Cell*, 128, 693–705.

Law, J. A., & Jacobsen, S. E. (2010). Establishing, maintaining and modifying DNA methylation patterns in plants and animals. *Nature Reviews Genetics*, 11, 204–220.

Lee, S., Fu, F., Xu, S., Lee, S. Y., Yun, D. J., & Mengiste, T. (2016). Global regulation of plant immunity by histone lysine methyl transferases. *The Plant Cell*, 28, 1640–1661.

Lee, W. Y., Lee, D., Chung, W. I., & Kwon, C. S. (2009). *Arabidopsis* ING and Alfin1-like protein families localize to the nucleus and bind to *H3K4me3/2* via plant homeodomain fingers. *The Plant Journal*, 58, 511–524.

Lejay, L., Gansel, X., Cerezo, M., Tillard, P., Müller, C., Krapp, A., von Wirén, N., Daniel-Vedele, F., & Gojon, A. (2003). Regulation of root ion transporters by photosynthesis: Functional importance and relation with hexokinase. *The Plant Cell*, 15, 2218–2232.

Lejay, L., Tillard, P., Lepetit, M., Olive, F. D., Filleur, S., Daniel-Vedele, F., & Gojon, A. (1999). Molecular and functional regulation of two NO_3--uptake systems by N-and C-status of *Arabidopsis* plants. *The Plant Journal*, 18, 509–519.

Leustek, T., & Saito, K. (1999). Sulfate transport and assimilation in plants. *Plant Physiology*, 120, 637–643.

Li, H., Wang, Y., Chen, M., Xiao, P., Hu, C., Zeng, Z., Wang, C., Wang, J., & Hu, Z. (2016). Genome-wide long non-coding RNA screening, identification and characterization in a model microorganism *Chlamydomonas reinhardtii*. *Scientific Reports*, 6, 34109.

Li, Q., Gao, Y., & Yang, A. (2020). Sulfur homeostasis in plants. *International Journal of Molecular Sciences*, 21, 8926.

Li, W., Wang, Y., Okamoto, M., Crawford, N. M., Siddiqi, M. Y., & Glass, A. D. (2007). Dissection of the *AtNRT2. 1: AtNRT2. 2* inducible high-affinity nitrate transporter gene cluster. *Plant Physiology*, 143, 425–433.

Li, Y., Brooks, M., Yeoh-Wang, J., McCoy, R. M., Rock, T. M., Pasquino, A., Moon, C. I., Patrick, R. M., Tanurdzic, M., Ruffel, S., Widhalm, J. R., McCombie, W. R., & Coruzzi, G. M. (2020). SDG8-mediated histone methylation and RNA processing function in the response to nitrate signalling. *Plant Physiology*, 182, 215–227.

Li, Y., Mukherjee, I., Thum, K. E., Tanurdzic, M., Katari, M. S., Obertello, M., Edwards, M. B., McCombie, W. R., Martienssen, R. A., & Coruzzi, G. M. (2015). The histone methyl transferase SDG8 mediates the epigenetic modification of light and carbon responsive genes in plants. *Genome Biology*, 16, 1–15.

Liang, G., Yang, F., & Yu, D. (2010). MicroRNA395 mediates regulation of sulfate accumulation and allocation in *Arabidopsis thaliana*. *The Plant Journal*, 62, 1046–1057.

Lin, S. H., Kuo, H. F., Canivenc, G., Lin, C. S., Lepetit, M., Hsu, P. K., Tillard, P., Lin, H. L., Wang, Y. Y., Tsai, C. B., Gojon, A., & Tsay, Y. F. (2008). Mutation of the *Arabidopsis* NRT1. 5 nitrate transporter causes defective root-to-shoot nitrate transport. *The Plant Cell*, 20, 2514–2528.

Liu, C., Lu, F., Cui, X., & Cao, X. (2010). Histone methylation in higher plants. *Annual Review of Plant Biology*, 61, 395–420.

Liu, K. H., Liu, M., Lin, Z., Wang, Z. F., Chen, B., Liu, C., Guo, A., Konishi, M., Yanagisawa, S., Wagner, G., & Sheen, J. (2022). NIN-like protein 7 transcription factor is a plant nitrate sensor. *Science*, 377, 1419–1425.

Liu, Z., Ma, Z., Li, J., Bian, N., Guo, Z., Guo, J., Zhao, S., Chu, B., Niu, C., & Ma, F. (2023). Interfering small ubiquitin modifiers (SUMO) exhibits apple's enhanced tolerance to nitrogen deficiency. *Fruit Research*, 3, 24.

Lu, L., Chen, X., Sanders, D., Qian, S., & Zhong, X. (2015). High-resolution mapping of H4K16 and H3K23 acetylation reveals conserved and unique distribution patterns in *Arabidopsis* and rice. *Epigenetics*, 10, 1044–1053.

Luo, C., Sidote, D. J., Zhang, Y., Kerstetter, R. A., Michael, T. P., & Lam, E. (2013). Integrative analysis of chromatin states in *Arabidopsis* identified potential regulatory mechanisms for natural antisense transcript production. *The Plant Journal*, 73, 77–90.

Luo, J., Li, H., Liu, T., Polle, A., Peng, C., & Luo, Z. B. (2013). Nitrogen metabolism of two contrasting poplar species during acclimation to limiting nitrogen availability. *Journal of Experimental Botany*, 64, 4207–4224. https://doi.org/10.1093/jxb/ert234.

Luo, X., & He, Y. (2020). Experiencing winter for spring flowering: A molecular epigenetic perspective on vernalisation. *Journal of Integrative Plant Biology*, 62, 104–117.

Ma, X., Lv, S., Zhang, C., & Yang, C. (2013). Histone deacetylases and their functions in plants. *Plant Cell Reports*, 32, 465–478.

Mager, S., & Ludewig, U. (2018). Massive loss of DNA methylation in nitrogen-, but not in phosphorus-deficient *Zea mays* roots is poorly correlated with gene expression differences. *Frontiers in Plant Science*, 9, 497.

Matt, P., Geiger, M., Walch-Liu, P., Engels, C., Krapp, A., & Stitt, M. (2001). Elevated carbon dioxide increases nitrate uptake and nitrate reductase activity when tobacco is growing on nitrate, but increases ammonium uptake and inhibits nitrate reductase activity when tobacco is growing on ammonium nitrate. *Plant, Cell & Environment*, 24, 1119–1137.

Muños, S., Cazettes, C., Fizames, C., Gaymard, F., Tillard, P., Lepetit, M., Lejay, L., & Gojon, A. (2004). Transcript profiling in the chl1-5 mutant of *Arabidopsis* reveals a role of the nitrate transporter *NRT1. 1* in the regulation of another nitrate transporter, *NRT2. 1*. *The Plant Cell*, 16, 2433–2447.

Nathan, D., Ingvarsdottir, K., Sterner, D. E., Bylebyl, G. R., Dokmanovic, M., Dorsey, J. A., Whelan, K. A., Krsmanovic, M., Lane, W. S., Meluh, P. B., Johnson, E. S., & Berger, S. L. (2006). Histone sumoylation is a negative regulator in *Saccharomyces cerevisiae* and shows dynamic interplay with positive-acting histone modifications. *Genes & Development*, 20, 966–976.

O'Brien, J. A., Vega, A., Bouguyon, E., Krouk, G., Gojon, A., Coruzzi, G., & Gutiérrez, R. A. (2016). Nitrate transport, sensing, and responses in plants. *Molecular Plant*, 9, 837–856.

Oldroyd, G. E., & Downie, J. A. (2008). Coordinating nodule morphogenesis with rhizobial infection in legumes. *Annual Review of Plant Biology*, 59, 519–546.

Ono, F., Frommer, W. B., & von Wirén, N. (2000). Coordinated diurnal regulation of low-and high-affinity nitrate transporters in tomato. *Plant Biology*, 2, 17–23.

Pillot, M., Baroux, C., Vazquez, M. A., Autran, D., Leblanc, O., Vielle-Calzada, J. P., Grossniklaus, U., & Grimanelli, D. (2010). Embryo and endosperm inherit distinct chromatin and transcriptional states from the female gametes in *Arabidopsis*. *The Plant Cell*, 22, 307–320.

Qiao, S., Sun, S., Wang, L., Wu, Z., Li, C., Li, X., Wang, T., Leng, L., Tian, W., Lu, T., & Wang, X. (2017). The RLA1/SMOS1 transcription factor functions with OsBZR1 to regulate brassinosteroid signalling and rice architecture. *The Plant Cell*, 29, 292–309.

Rennenberg, H., Herschbach, C., Haberer, K., & Kopriva, S. (2007). Sulfur metabolism in plants: Are trees different? *Plant Biology*, 9, 620–637.

Rouached, H., Wirtz, M., Alary, R., Hell, R., Arpat, A. B., Davidian, J. C., Fourcroy, P., & Berthomieu, P. (2008). Differential regulation of the expression of two high-affinity sulfate transporters, *SULTR1. 1* and *SULTR1. 2* in *Arabidopsis*. *Plant Physiology*, 147, 897–911.

Sahu, P. P., Pandey, G., Sharma, N., Puranik, S., Muthamilarasan, M., & Prasad, M. (2013). Epigenetic mechanisms of plant stress responses and adaptation. *Plant Cell Reports*, 32, 1151–1159.

Saito, K. (2003). Molecular and metabolic regulation of sulfur assimilation: Initial approach by the postgenomics strategy. In J.-C. Davidian, D. Grill, L. J. De Kok, I. Stulen, M. J. Hawkesford, E. Schnug, & H. Rennenberg (Eds.), *Sulfur Transport and Assimilation in Plants: Regulation, Interaction and Signaling* (pp. 11–20). Leiden: Backhuys Publishers.

Santi, C., Bogusz, D., & Franche, C. (2013). Biological nitrogen fixation in non-legume plants. *Annals of Botany*, 111, 743–767.

Schultze, M., & Kondorosi, A. (1998). Regulation of symbiotic root nodule development. *Annual Review of Genetics*, 32, 33–57.

Servet, C., e Silva, N. C., & Zhou, D. X. (2010). Histone acetyltransferase AtGCN5/HAG1 is a versatile regulator of developmental and inducible gene expression in *Arabidopsis*. *Molecular Plant*, 3, 670–677.

Shen, Y., Issakidis-Bourguet, E., & Zhou, D. X. (2016). Perspectives on the interactions between metabolism, redox, and epigenetics in plants. *Journal of Experimental Botany*, 67, 5291–5300.

Sridhar, V. V., Kapoor, A., Zhang, K., Zhu, J., Zhou, T., Hasegawa, P. M., Bressan, R. A., & Zhu, J. K. (2007). Control of DNA methylation and heterochromatic silencing by histone H2B deubiquitinating. *Nature*, 447, 735–738.

Stulen, I., & De Kok, L. J. (1993). Whole plant regulation of sulfur metabolism. In L. J. De Kok, I. Stulen, H. Rennenberg, C. Brunold, & W. E. Rauser (Eds.), *Sulfur Nutrition and Sulfur Assimilation in Higher Plants: Regulatory, Agricultural and Environmental Aspects* (pp. 77–91). The Hague: SPB Academic Publishing.

Takahashi, H., Kopriva, S., Giordano, M., Saito, K., & Hell, R. (2011). Sulfur assimilation in photosynthetic organisms: Molecular functions and regulations of transporters and assimilatory enzymes. *Annual Review of Plant Biology*, 62, 157–184.

Takahashi, H., Watanabe-Takahashi, A., Smith, F. W., Blake-Kalff, M., Hawkesford, M. J., & Saito, K. (2000). The roles of three functional sulphate transporters involved in uptake and translocation of sulphate in *Arabidopsis thaliana*. *The Plant Journal*, 23, 171–182.

Thiebaut, F., Hemerly, A. S., & Ferreira, P. C. G. (2019). A role for epigenetic regulation in the adaptation and stress responses of non-model plants. *Frontiers in Plant Science*, 10, 246. https://doi.org/10.3389/fpls.2019.00246.

Thompson, J. F., Smith, I. K., & Madison, J. T. (1986). Sulfur metabolism in plants. *Sulfur in Agriculture*, 27, 57–121.

Thorstensen, T., Grini, P. E., & Aalen, R. B. (2011). SET domain proteins in plant development. *Biochimica et Biophysica Acta (BBA)-Gene Regulatory Mechanisms*, 1809, 407–420.

Tsay, Y. F., Schroeder, J. I., Feldmann, K. A., & Crawford, N. M. (1993). The herbicide sensitivity gene CHL1 of *Arabidopsis* encodes a nitrate-inducible nitrate transporter. *Cell*, 72, 705–713.

Vidal, E. A., & Gutiérrez, R. A. (2008). A systems view of nitrogen nutrient and metabolite responses in *Arabidopsis*. *Current Opinion in Plant Biology*, 11, 521–529.

Vidal, E. A., Alvarez, J. M., Araus, V., Riveras, E., Brooks, M. D., Krouk, G., Ruffel, S., Lejay, L., Crawford, N. M., Coruzzi, G. M., & Gutiérrez, R. A. (2020). Nitrate in 2020: Thirty years from transport to signalling networks. *The Plant Cell*, 32, 2094–2119.

Vincentz, M., Moureaux, T., Leydecker, M. T., Vaucheret, H., & Caboche, M. (1993). Regulation of nitrate and nitrite reductase expression in *Nicotiana plumbaginifolia* leaves by nitrogen and carbon metabolites. *The Plant Journal*, 3, 315–324.

von Wirén, N., Lauter, F. R., Ninnemann, O., Gillissen, B., Walch-Liu, P., Engels, C., Jost, W., & Frommer, W. B. (2000). Differential regulation of three functional ammonium transporter genes by nitrogen in root hairs and by light in leaves of tomato. *The Plant Journal*, 21, 167–175.

Wang, R. C., Xing, X. J., & Crawford, N. (2007). Nitrite acts as a transcriptome signal at micromolar concentrations in *Arabidopsis* roots. *Plant Physiology*, 145, 1735–1745.

Wang, W., Hu, B., Li, A., & Chu, C. (2020). NRT1. 1s in plants: Functions beyond nitrate transport. *Journal of Experimental Botany*, 71, 4373–4379.

Wang, Y. Y., Cheng, Y. H., Chen, K. E., & Tsay, Y. F. (2018). Nitrate transport, signalling, and use efficiency. *Annual Review of Plant Biology*, 69, 85–122.

Wirth, J., Chopin, F., Santoni, V., Viennois, G., Tillard, P., Krapp, A., Lejay, L., Daniel-Vedele, F., & Gojon, A. (2007). Regulation of root nitrate uptake at the NRT2. 1 protein level in *Arabidopsis thaliana. Journal of Biological Chemistry*, 282, 23541–23552.

Wu, K., Wang, S., Song, W., Zhang, J., Wang, Y., Liu, Q., Yu, J., Ye, Y., Li, S., Chen, J., Zhao, Y., Wang, J., Wu, X., Wang, M., Zhang, Y., Liu, B., Wu, Y., Harberd, N. P., & Fu, X. (2020). Enhanced sustainable green revolution yield via nitrogen-responsive chromatin modulation in rice. *Science*, 367, eaaz2046.

Yadav, B., Jogawat, A., Lal, S. K., Lakra, N., Mehta, S., Shabek, N., & Narayan, O. P. (2021). Plant mineral transport systems and the potential for crop improvement. *Planta*, 253, 1–30.

Yoshimoto, N., Takahashi, H., Smith, F. W., Yamaya, T., & Saito, K. (2002). Two distinct high-affinity sulfate transporters with different inducibilities mediate uptake of sulfate in *Arabidopsis* roots. *The Plant Journal*, 29, 465–473.

Yu, X., Jiang, L., Wu, R., Meng, X., Zhang, A., Li, N., Xia, Q., Qi, X., Pang, J., Xu, Z. Y., & Liu, B. (2016). The core subunit of a chromatin-remodeling complex, *ZmCHB101,* plays essential roles in maize growth and development. *Scientific Reports*, 6, 38504.

Yu, X., Meng, X., Liu, Y., Li, N., Zhang, A., Wang, T. J., Jiang, L., Pang, J., Zhao, X., Qi, X., Zhang, M., Wang, S., Liu, B., & Xu, Z. Y. (2018). The chromatin remodeler *ZmCHB101* impacts expression of osmotic stress-responsive genes in maize. *Plant Molecular Biology*, 97, 451–465.

Yu, X., Meng, X., Liu, Y., Wang, X., Wang, T. J., Zhang, A., Li, N., Qi, X., Liu, B., & Xu, Z. Y. (2019). The chromatin remodeler *ZmCHB101* impacts alternative splicing contexts in response to osmotic stress. *Plant Cell Reports,* 38, 131–145.

Zhang, K., Sridhar, V. V., Zhu, J., Kapoor, A., & Zhu, J. K. (2007). Distinctive core histone post-translational modification patterns in *Arabidopsis thaliana. PLoS One*, 2, e1210.

Zhang, X., Clarenz, O., Cokus, S., Bernatavichute, Y. V., Pellegrini, M., Goodrich, J., & Jacobsen, S. E. (2007). Whole-genome analysis of histone H3 lysine 27 trimethylation in *Arabidopsis. PLoS Biology*, 5, e129.

Zhang, Z., & Chu, C. (2020). Nitrogen-use divergence between indica and japonica rice: Variation at nitrate assimilation. *Molecular Plant*, 13, 6–7.

Zhao, X., Li, J., Lian, B., Gu, H., Li, Y., & Qi, Y. (2018). Global identification of *Arabidopsis* lncRNAs reveals the regulation of *MAF4* by a natural antisense RNA. *Nature Communications*, 9, 5056.

Zhou, J., Wang, X., He, K., Charron, J. B., Elling, A. A., & Deng, X. W. (2010). Genome-wide profiling of histone H3 lysine 9 acetylation and demethylation in *Arabidopsis* reveals correlation between multiple histone marks and gene expression. *Plant Molecular Biology*, 72, 585–595.

14 Epigenetic Regulation of Plant Response to Mineral Nutrient Availability

Recent Updates

Aman Singh, Haritha Si, Shelvy S, Archana Sahani and Lekshmy Sathee

14.1 INTRODUCTION

According to estimated figures worldwide, the total population by 2050 will reach 9.5 billion (Pardey et al., 2014). This continued increase in population increases the global food demand. Therefore, this high demand for food requires high agricultural productivity, crop yield and productivity. However, for sessile nature, plants face unfavorable environmental conditions. Plant stress is a significant issue that affects plant development, growth and productivity. Additionally, it produces unfavorable environmental conditions that interfere with regular physiological functions and hinder plant survival. To cope with these fluctuating environmental conditions, plants must adapt to fine-tuned responses. For development and reproduction, plants require a set of macro- and micro-mineral nutrients in addition to light, water and carbon dioxide. Mineral nutrient availability varies greatly depending on a number of factors, including the soil pH, element concentration, soil horizon and number of rain events. Plants have evolved strategies to address these challenges (Séré & Martin, 2020). The two primary adaptation methods that plants employ to address variations in nutrient availability are the regulation of root architecture and metabolic adjustment. Reprogramming gene expression is the most common method of these adaptive mechanisms (Shahzad & Amtmann, 2017). The transcriptional regulation of plants in response to nutrient availability and the factors that respond to particular nutrient-related stimuli, such as excess or deficit, have been extensively studied in recent years. One important way that plants react to and adjust to stress is through epigenetic regulation. The epigenetic mechanisms govern plant abiotic stress responses, especially DNA methylation, histone methylation/acetylation, chromatin remodeling and various metabolites (Ali et al., 2022).

The efficient use of nutrients is the cornerstone of plant productivity and growth and has immediate consequences on global food security and agricultural sustainability. Mineral nutrients are invaluable to plant life, with indispensable functions in numerous physiological, biochemical and molecular processes. Soil availability of these nutrients is naturally low or out of proportion due to climatic factors, agricultural practices and pollution. These deficiencies, excesses, toxicities and nutrient limitations are significant abiotic stressors with wide-ranging consequences on plant growth, productivity and tolerance. These stresses affect plants physiologically and even at the molecular level by causing epigenomic changes. An epigenomic change represents the modifications to nuclear DNA, histones and changes in the generation of regulatory non-coding RNAs that cause gene expression changes. The epigenetic modification of chromatin changes the manner how organisms respond to both endogenous and external stimuli. Symmetric CG, CHG methylation and asymmetric CHH methylation patterns are widely observed in plants (Meng et al., 2018). DNA methyltransferase 1 (MET1) is involved in CG methylation (Aufsatz et al., 2004); domains rearranged methylase 2 (DRM2) and chromomethylase 2 (CMT2) are responsible for CHH methylation and chromomethylase 3 (CMT3) for CHG methylation (Sathee et al., 2025). A detailed understanding of how epigenetic machinery works can be directed towards improving crop yield and sustainability of agriculture. Nutrient availability directly impacts epigenetic mechanisms like DNA methylation, histone modifications and non-coding RNAs (ncRNAs). Deficiency or toxicity of these nutrients triggers complex molecular responses that affect chromatin accessibility and gene expression. The key objective of this chapter is to discuss how the availability of mineral nutrients, both in limiting and excess amounts, influences the plant epigenome and gene expression. It combines recent advances in understanding the molecular and epigenetic responses to nutrient stress, emphasizing their importance in adaptation to a changing environment (Figure 14.1). Emerging evidence highlights the interdisciplinary significance of nutrient-induced epigenetic regulation, ranging from plant biology and soil science to agricultural biotechnology.

DOI: 10.1201/9781003589914-17

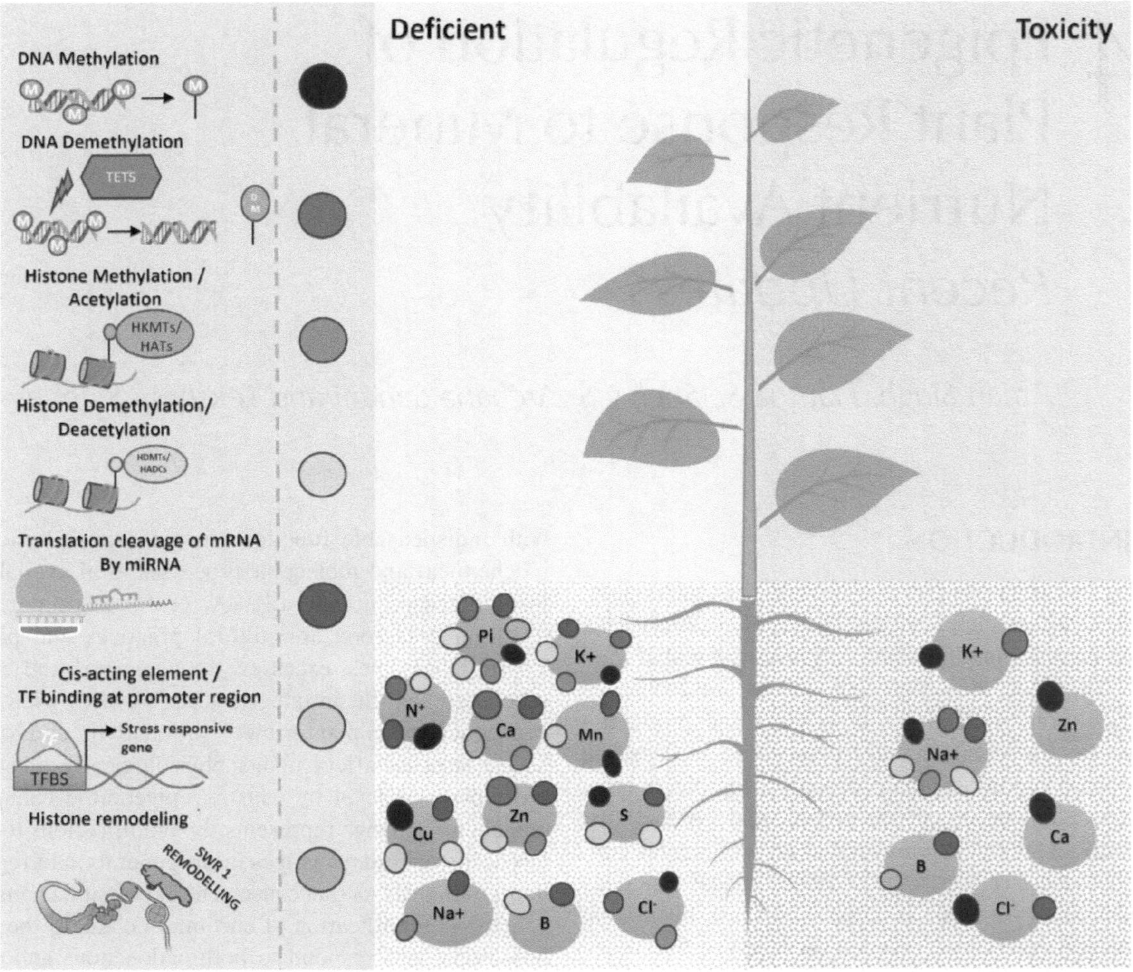

FIGURE 14.1 Epigenetic regulatory mechanisms that control plant responses to mineral nutrient deficiency and toxicity.

14.2 EPIGENOMICS OF PLANT NITROGEN (N) RESPONSE

Epigenetic regulation of N-response by miRNAs (Nischal et al., 2012; Islam et al., 2022) and by chromatin changes (Li et al., 2021) has been reported in several studies (Séré & Martin, 2020). Epigenetic regulation via G4 sequences or G-quadruplexes has also recently been reported to be involved in the epigenetic regulation of nitrogen use efficiency (NUE) (Sharma et al., 2023). Histone methyltransferase AtSDG8 (Wang et al., 2014; Li et al., 2015; Dong et al., 2008; Cazzonelli et al., 2009) plays pleiotropic role in plant growth, development and response to resources. AtSDG8 maintains the H3K36me3 levels of genomic regions encoding nitrate or nitrite reductases and asparagine synthetase (Li et al., 2015). H3K36me3 levels of approximately 140 genes were increased by N deficiency, while H3K36me3 levels showed a decreasing trend in over 50 genes, with a pronounced reduction observed in *sdg8-5* mutants (Wang et al., 2014).

AtNRT2.1, a high-affinity nitrate transporter, shows an abundance of the transcription repression marker H3K27me3 under high N conditions (Widiez et al., 2011). OsD14 (Zhang et al., 2017; Chen et al., 2015; Guo et al., 2016), OsNGR5 and AOsSPL14 regulate N-dependent tillering response. Deficiency of N decreases CG methylation from 26.6% to 12.0% in maize (Mager & Ludewig, 2018).

Recent research provides exciting information on epigenome changes under different N treatments; for example, it was discovered that high nitrogen insensitive 9 (HNI9) inhibits *NRT2.1* expression in roots. HNI9 encodes the IWS1 (interact with STP6) protein, which functions as a mediator of methylation and is a component of the RNA pol II complex. Under high N conditions, H3K27me3 levels were elevated in NRT2.1 chromatin in an HNI9-dependent manner, specifically at the cis-elements in the promoter region of *NRT2.1*, resulting in its repression. Through transgenerational inheritance of heavy cytosine methylation in rice, it has been discovered that the offspring of nitrogen-deficient plants are more tolerant to nitrogen deficiency. Recently, it has been reported that the parent seed nitrogen determines DNA methylation and significantly decreases the expression of *OsNAR2.1*, leading to a heritable phenotype of N deficiency. To summarize, epigenetics research, particularly in nitrogen response, is an emerging

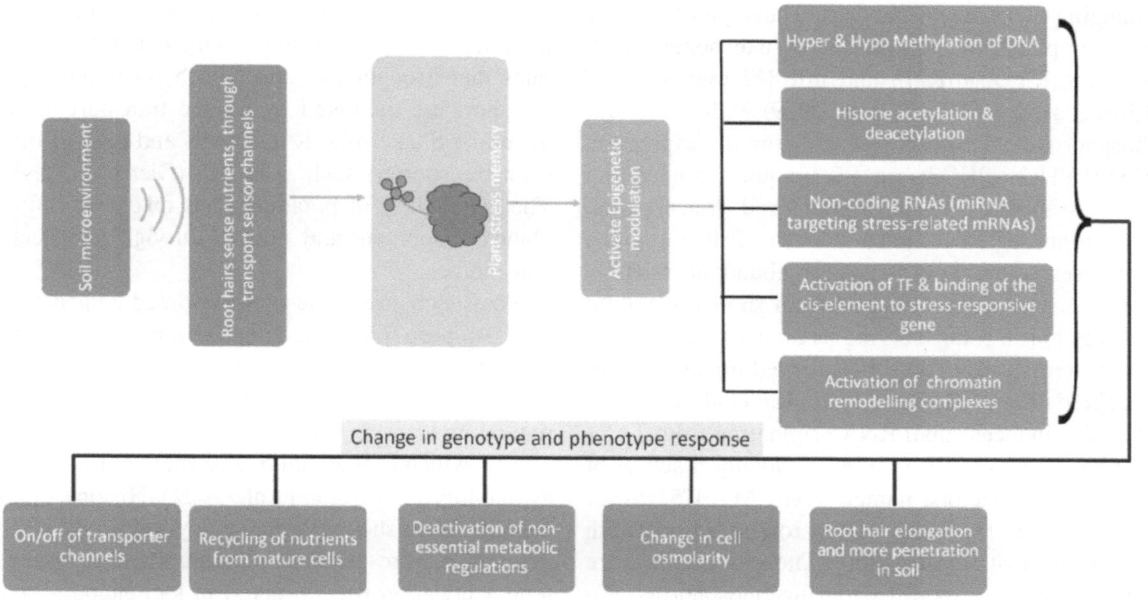

FIGURE 14.2 Nutrient stress signaling and epigenetic changes that coordinate plant phenotypic adaptive response towards mineral nutrient availability.

and rapidly evolving domain. By deciphering the epigenetic mechanisms that influence gene expression and plant adaptation to nitrogen availability, this field holds great potential for advancing sustainable agriculture and improving crop productivity in the face of changing environmental conditions (Figure 14.2).

14.3 MODULATION OF EPIGENOME BY PHOSPHORUS (P) STRESS

Plants require phosphorus to maintain energy metabolism, membrane dynamics, DNA structure and regulation, and many other biological processes. In soil, phosphorus is readily available in both organic and inorganic forms; plants assimilate it primarily in inorganic form (Pi), namely as $H_2PO_4^-$ and HPO_4^{2-}, but it has a very low diffusion rate (0.13 mm/day). Furthermore, it is easily fixed by the soil with additional nutrients (Khan et al., 2023). Thus, Pi is the second most significant limiting macronutrient for plant growth. The effective acquisition and utilization of available phosphorus from the soil is essential for plant growth, development and production. Plants have an intriguing process known as epigenetic modulation to mitigate the Pi deficit. This approach enables plants to activate particular molecules that improve Pi acquisition without modifying their fundamental genetic structure.

In plants, under phosphorus starvation conditions, the early response is triggered by genetic regulators like cis-acting elements, transcription factors and riboswitches. The MYB transcription factors PHOSPHATE STARVATION RESPONSE 1 (PHR1) and PHOSPHORUS RESPONSE 1 (PSR1) are key activators of numerous phosphate starvation response (*PSR*) genes (Rubio et al., 2001), as the PHR1 binding sequence (P1BS) is a conserved cis-element located in the promoter regions of many Pi-starvation-inducible genes, facilitating their transcriptional regulation. Chromatin-level regulation occurs by modulating the expression of *PSR* genes during Pi deficiency. The SWR1 chromatin remodeling complex, which replaces the H2A/H2B dimer of the nucleosome with the H2A.Z histone variant, provides a binding site for the transcription factor for mediating transcriptional response (Luo et al., 2020). Under Pi-deficient conditions, nuclear actin-related protein APR6, which is essential for the deposition of H2A.Z at several (*PSR*) genes, modulates chromatin level regulation, influencing the transcription of the (*PSR*) gene set (Smith et al., 2010). Despite H2A.Z seemingly exerting a detrimental effect on *PSR* gene expression, the underlying mechanisms remain ambiguous. Under typical circumstances, H2A.Z-containing nucleosomes establish a localized chromatin configuration at target loci that hinders gene activation, potentially by obstructing the attachment of transcriptional activators and/or facilitating the binding of transcriptional repressors (Smith et al., 2010).

DNA methylation governs transposable elements, repetitive sequences, genomic imprinting and gene silencing in plants. It occurs in three sequence settings (CG, CHG and CHH) and disrupts essential processes. Reduced Pi levels in *Arabidopsis* seedlings enhance 5-mC. When phosphorus levels are high, MET1 remains in a silenced methylation state. When phosphorus levels are low, PHR1 activates phosphorus starvation response (*PSR*) genes, which include MET1 (Yong-Villalobos et al., 2015). There are P1BS-binding sites in the upstream regions of other DNA methyltransferase genes, like *DMR1, CMT2, CMT3* and the demethylase *ROS1*. This means that PHR1 controls the expression of these genes. The control of phosphate transport, lipid remodeling and recycling is linked to these changing methylation patterns. These processes are necessary for plants to maintain phosphate homeostasis. Families of miR399, miR778 and miR827 regulate *PSR* genes (Bari et al., 2006; Hsieh et al., 2009). MiR399 modulates phosphorus homeostasis by regulating the expression of PHOSPHATE2 (PHO2), an E2 ubiquitin-conjugating enzyme. It stimulates Pi absorption and translocation under conditions of Pi scarcity, whereas PHO2 inhibits these processes when exogenous Pi is abundant. miR399 functions as a long-distance signal from shoots to inhibit *PHO2* expression in roots, serving as an early response to phosphorus deprivation. The shoot-derived miR399 mediates the degradation of PHO2 in roots (Lin et al., 2008). Pi deficiency also induces small RNA originating from TAS4 and miR828. TAS4-siR81(–) targets explicitly a subset of MYB transcription factors, namely PAP1/MYB75, PAP2/MYB90 and MYB113, which play a role in anthocyanin biosynthesis (Borevitz et al., 2000). The root architecture system is significantly modified during phosphorus deficiency, leading to diminished primary root growth and enhanced density and length of lateral roots. These modifications facilitate improved "topsoil foraging", enabling the extraction of Pi from the upper soil (Lynch & Brown, 2008). ALFIN-LIKE 6 (AL6) is part of a limited group of nuclear-localized plant homeodomain (PHD)-containing putative transcription factors (PHD finger). PHD fingers function as "readers" of histone post-translational modifications, specifically binding to H3K4me3 (trimethylated lysine 4 of histone 3) and H3K4me2 (Peña et al., 2006). Under phosphorus starvation, H3K4me3 is present at the promoter-proximal nucleosomes of the active *ETC1* gene, facilitating the interaction of AL6 with H3K4me3 (lysine 4-methylated histone H3) via its PHD domain, thereby positively influencing subsequent transcriptional processes. Thus, ETC1 may bind to and enhance the expression of NPC4, SQD2 and PS2 to facilitate root hair extension under phosphorus deprivation. Along with histone deacetylase, HDA19 influences genes that encode SPX (SYG1/Pho81/XPR) domain-containing proteins and genes associated with membrane lipid remodeling, a critical response to phosphorus shortage that enhances free phosphorus levels in plants (Chen et al., 2015).

14.4 EPIGENETIC REGULATION UNDER POTASSIUM (K) STRESS

Potassium regulates enzyme activation, osmoregulation and photosynthesis; calcium acts as a secondary messenger in stress signaling and maintains the structure of the cell wall. Zinc has an irreplaceable role in enzymatic catalysis, redox homeostasis and auxin biosynthesis. Potassium is a major macronutrient in many physiological and biochemical processes in plants, including enzyme activation, photosynthesis, protein synthesis, osmoregulation, etc. (Marschner, 2011). Potassium is the most common cation in plant cells and plays a primary role in ionic balance, turgor pressure and water use efficiency. Potassium further regulates the stomatal aperture, which dictates transpiration rates and drought tolerance. Both potassium uptake and transport are mediated by unique transport mechanisms, including Shaker-like K^+ channels and high-affinity potassium transporters such as HAK5 (Gierth & Mäser, 2007). The availability of potassium has extensive influences on plant development and growth through its effects on the epigenome.

Potassium stress causes widespread epigenetic changes that support plant acclimatization to nutrient stress. Research has shown that potassium deficiency induces hypomethylation of promoter regions of genes involved in potassium transport and stress response, such as OsHAK5 of rice, which is essential in absorbing potassium from low availability (Uchiyama et al., 2022). Histone modification has also been shown to play a key role in regulating gene expression in response to potassium stress. Increased histone acetylation was observed in loci encoding potassium transporters, which is consistent with enhanced transcriptional activity (Li et al., 2015). These changes in chromatin structure increase the accessibility of target genes to the transcription machinery, enabling plants to increase potassium uptake and utilization.

Recent research has highlighted the significant role of microRNAs (miRNAs), a type of non-coding RNA (ncRNA), in regulating potassium-related gene networks. For example, miR399, which has been classically associated with phosphate homeostasis, has also been implicated in potassium regulation by targeting genes associated with nutrient transport (He et al., 2018). These ncRNAs represent gene expression fine-tuners that enable plants to adjust their uptake of nutrients and energy investment by fluctuating potassium levels. Excessive potassium can also disrupt cellular homeostasis and induce toxicity, although the molecular and epigenetic mechanisms of this process are poorly understood. A study by Laanen et al. (2021) revealed that potassium toxicity in *Arabidopsis thaliana* leads to hypermethylation of stress response genes, suppressing their expression and further enhancing oxidative damage. Furthermore, histone modifications such as repressive methylation marks have been linked to altered gene expression under potassium toxicity. These epigenetic responses highlight the bifurcated function of potassium in regulating plant adaptation to both deficit and excess.

14.5 MODULATION OF EPIGENOME BY CALCIUM (Ca) AVAILABILITY

Calcium is a key secondary macronutrient, a structural constituent and a signaling molecule in plants. It is important in stabilizing cell walls and regulating enzymes, and it is a mediator in plant response signal transduction pathways to the environment (Ranty et al., 2016). Calcium homeostasis is maintained through a network of calcium channels, transporters and binding proteins, such as calmodulins, that regulate its intracellular concentration (Dayod et al., 2010). The fact that the signaling function of calcium in plants is dynamic underlines the importance of calcium in epigenome regulation. Calcium deficiency leads to drastic epigenetic changes that induce gene expression in calcium transport and stress adaptation. DNA hypomethylation at the promoters of calcium transporter genes has been reported, allowing higher expression to mitigate nutrient scarcity (Ding et al., 2022), for example, the upregulation of CAX1 and ACA4, the two significant calcium transporters, has been associated with hypomethylation and the induction of transcriptional activity. In addition to DNA methylation, histone modifications play a key role in calcium deficiency responses. Increased histone acetylation occurs in stress-related genes, inducing chromatin relaxation and transcriptional activation (Berr et al., 2010). Calcium limitation also influences the activity of ncRNAs, specifically miRNAs that target calcium-related genes; for example, miR319 regulates calcium signaling pathways during stress (Lu et al., 2023). These molecules play a dual role by fine-tuning gene expression and maintaining calcium homeostasis under limiting conditions.

Calcium toxicity is less common but also occurs in calcareous soils or as a reaction to overfertilization with calcium. Excessive calcium disrupts ion homeostasis, leading to toxicity symptoms such as chlorosis and necrosis. Experiments have shown that calcium toxicity triggers epigenetic modifications, including hypermethylation of stress-responsive genes and histone deacetylation, which repress gene expression and intensify cellular damage (Fu et al., 2013). Calcium signaling also involves epigenetic processes by calcium-dependent protein kinases (CDPKs), which phosphorylate histone-modifying enzymes and DNA methyltransferases, linking calcium availability to chromatin dynamics (Dekomah et al., 2022).

14.6 MODULATION OF EPIGENOME BY ZINC (Zn) AVAILABILITY

Zinc is a micronutrient critical in many physiological and biochemical processes in plants, including enzyme activity, protein synthesis and hormone regulation. It is also a critical structural component of various enzymes, such as superoxide dismutase and carbonic anhydrase, which control oxidative stress and metabolic regulation (Broadley et al., 2007). Specific transporters, such as ZIP and CDF family proteins, mediate its transport and uptake, which regulate its intracellular distribution and homeostasis (Guerinot, 2000).

Zinc deficiency induces widespread epigenetic modifications, enabling plants to acclimate to nutrient stress. DNA hypomethylation in stress-related genes has been observed under zinc-limiting conditions, leading to their activation and elevated expression of zinc transporters (Azimi et al., 2022). Histone acetylation also increases in zinc-uptake loci, facilitating chromatin accessibility and transcriptional activation (Broadley et al., 2007). miRNAs also play a role in zinc homeostasis, targeting genes involved in nutrient transport and signaling pathways (Islam et al., 2022). Zinc toxicity, which is most commonly caused by industrial pollution or excessive application of fertilizers, is a severe threat to plant health. High zinc induces hypermethylation of stress-responsive genes, suppressing their transcription and stress adaptation (Mazurek et al., 2024). Histone modifications, including the addition of repressive marks such as H3K9me2, actively participate in gene-silencing processes under zinc toxicity (Shafiq et al., 2019). Recent studies have also revealed the crosstalk between DNA methylation and histone modification in the zinc toxicity response, with a complex regulatory network that modulates gene expression and stress acclimation (Wei et al., 2023).

14.7 MODULATION OF EPIGENOME BY SULFUR (S) AVAILABILITY

Sulfur (S), which is taken by the plants primarily as inorganic sulfate (SO_4^{2-}) and other forms such as sulfite (SO_3^{2-}), sulfur dioxide (SO_2) and cysteine, was determined to be an essential element for plants in 1865. Sulfur is a crucial component of various proteins, cofactors, prosthetic groups, peptides, lipids, polysaccharides and many secondary metabolites such as glucosinolates and allins (Kopriva et al., 2015). Sulfur is required in large amounts for high quality and yield of plants belonging to the families Cruciferae and Liliaceae; e.g., oilseed rape crops require about 20–30 kg of S ha^{-1} (Scherer et al., 2001). Sulfur uptake from the soil is accomplished through high-affinity transporters such as sulfate transporter 1;1 (SULTR1;1) and sulfate transporter 1;2 (SULTR1;2) and through low-affinity transporters such as SULTR2;1, SULTR2;2 and SULTR3;5 (Islam et al., 2022).

lncRNAs in response to S deficiency have been studied in microalgae *Chlamydomonas reinhardtii*, but their functions are yet to be explored in plants (Li et al., 2016). Various differentially expressed miRNAs were found through computational approaches in response to S starvation, among which miR164c and miR395 were upregulated, while miR391 and miR845a were downregulated. miR395 targets ATP sulfurylases such as APS1, APS3 and APS4 and low-affinity S transporters such as SULTR2;1 and is extensively associated with S uptake, translocation and homeostasis (Liang et al., 2015). The target genes of miR395 show different correlation and expression levels in leaves and roots depending upon the duration of S starvation; for instance, APS4 drastically reduces in roots and shoots under S starvation due to high induction of miR395. Similarly, APS1 and APS3 levels deplete in shoots, whereas they slightly increase or maintain consistent expression in roots (Liang et al., 2010; Jagadeeswaran et al., 2014). Despite higher induction of miR395 in roots, SULTR2;1 is highly expressed due to their different spatial expression in phloem companion cells and xylem parenchyma cells, respectively. It is also due to SULFUR LIMITATION 1 (SLIM1), a central TF involved in glucosinolate degradation by regulation of high-affinity sulfate transporters SULTR1;1 and SULTR1;2 under limited S conditions, which might contribute to a complex regulation of miR395 in roots (Kawashima et al., 2011).

DNA methylation, which plays an important role in epigenetic regulation, requires methyl group donors. Poly-Glu-5-methyl-tetrahydrofolate (5-CH3-THF-Glun), a common form of folate, acts as a methyl donor. It plays a crucial role in the synthesis of methionine (Met) via the methylation of homocysteine (Hcy) through a folate-dependent Met synthase (Friso et al., 2002. Sulfur is crucial for the biosynthesis of the methyl group donor, S-adenosyl methionine (SAM). The condensation of S-containing amino acid cysteine (Cys) and O-phosphohomoserine (OPH) leads to the synthesis of cystathionine (Cyst), which later gets converted to Hcy and then Met, which serves as the donor of $-CH_3$ for SAM via Met adenosyltransferase (MAT) (Hesse & Hoefgen, 2003). SAM, being the methyl group donor for DNA methyltransferases, breaks down to S-adenosyl-Hcy (SAH), a feedback inhibitor, and thus it is rapidly broken down to Hcy and adenosine by SAH hydrolase (SAHH), thus completing the SAM cycle. SAHH is encoded by HOMOLOGY DEPENDENT GENE SILENCING1 (HOG1), and its mutation has led to poor DNA and histone methylation (Ouyang et al., 2012). Mutation in Folyl Polyglutamate Synthetase 1 (FPGS1) impairs the conversion of 5-CH3-THF-Glu1 to 5-CH3-THF-Glun, leading to elevated levels of SAH and reduced methylation potential in *Arabidopsis* (Zhou et al., 2013). Similarly, mutation/inhibition in methylenetetrahydrofolate dehydrogenase (MTHFD1), required for the conversion of 5,10-CH2-THF to THF, and dihydrofolate reductase (DHFR), involved in DHF to THF conversion, also led to a decrease in folate and SAM pool and caused genome-wide DNA hypomethylation and reduced H3K9 dimethylation levels in *Arabidopsis* (Groth et al., 2016; Loizaeu et al., 2008; Zhang et al., 2024) Thus, the alterations in the SAM cycle and SAM:SAH ratio, besides affecting the genome-wide DNA methylation patterns, also modify histone methylation (Meng et al., 2018). Under a deprived state, SAM levels decrease, and bisulfite sequencing studies revealed that cytosine methylation levels are altered in shoots and roots of *Arabidopsis*. Mutation in MORE SULFUR ACCUMULATION 1 (MSA1), a nuclear-localized gene characterized under the SHM protein family involved in SAM biosynthesis, leads to hypomethylated DNA and histones, causing a constitutive S deficiency response due to increased expression of high-affinity S transporters such as SULTR1;1, SULTR1;2, and S assimilation gene such as *APR3*. Under S-deprived conditions, cytosines in the promoters are demethylated, thus activating the gene expression by releasing the transcriptional repressor/binding of the transcriptional activator. For instance, in the *msa1-1* mutant, the flanking sequence of the sulfur-responsive element (SURE) found in the promoter region of *SULTR1;1* is hypomethylated in both S deficient and sufficient conditions, contributing to elevated S uptake under S deficient conditions (Huang et al., 2016).

14.8 EPIGENETIC MODULATION IN RESPONSE TO BORON (B) AVAILABILITY

Boron (B) was ascertained as an essential element in 1923 by Katherine Warington, whose studies in *Vicia faba* revealed that the presence of B in the form of boric acid (H_3BO_3) was necessary for proper growth and survival of the plant (Warington, 1923). Despite its early discovery as an essential microelement, its proper function and mechanism in plants remained unclear until the 1990s. In 1994, chemical and enzymatic studies in *Cucurbita pepo* L. and cultured *Nicotiana tabacum* cells showed the presence of B in pectin associated with cell walls owing to their role in cell wall division and elongation in plants (Hu & Brown, 1994).

The first symptom of B deficiency occurs at the root and leaf tips, floral buds and developing organs due to its high requirement for meristematic cells (Rerkasem, 1996). B deficiency alters the ion fluxes of H^+, K^+, PO_4^{3-} and Ca^{2+} and reduces H^+-ATPase activity, leading to the leakage of ions and adversely impacting the membrane permeability. Thus, B plays an important role in maintaining the electrochemical potential gradient across the plasma membrane (Robertson et al., 1974; Serrano, 1989; Goldbach et al., 1990; Schon et al., 1990; Barr et al., 1993). B-deficient plants accumulate large amounts of phenolic compounds, which act as a substrate for producing reactive oxygen species (ROS) and pave the way for oxidative damage to plasma membranes (Camacho-Cristobal et al., 2002). Besides these, B deficiency impairs several metabolic processes, which include carbohydrate metabolism, sugar transport and indole acetic acid (IAA) metabolism (Pollard et al., 1977). Long non-coding RNAs (lncRNAs) are capable of cis- or trans-regulation of target gene expression either through direct binding to transcripts or modulation of the transcription factors (TFs) (Wang et al., 2008). Under B deficiency in *Poncirus trifoliata* L. Raf, many TFs have been upregulated, such as WRKY, MYB, bHLH and bZIP, which might be modulated via lncRNAs (Zhou et al., 2019). Several miRNAs showed altered expression levels under B stress in *Citrus sinensis* leaves and roots (Lu et al., 2014; Lu et al., 2015). The expression of miR159, miR164, miR6260 and miR396 is downregulated in citrus leaves and enhances the levels of MYB, NAC, WRKY domain proteins and GRF TFs, respectively, as an adaptation to the stress response. To attenuate the overall plant growth under B deficiency conditions, miR160, miR393 and miR3946 targeting the auxin response factors (ARF10/16/17), auxin receptors (TIR1, AFB1, AFB2 and AFB3) and SAUR-like auxin-responsive protein family, respectively, are upregulated (Lu et al., 2015). miR408, which has been associated with abiotic stress response in *Arabidopsis* (Ma et al., 2015), drought stress tolerance in chickpea (Hajyzadeh et al., 2015), *Medicago truncula*ta (Trindade et al., 2010) and *Pisum sativum* (Jovanovic et al., 2014) is also associated with micronutrient homeostasis such as Zn, Mn, Cu and B (Patel et al., 2017). miR408 targets plantacyanin, laccases and Cu-SOD, and its downregulation during B deficiency alleviates oxidative stress by cell wall modifications and antioxidant activity. Repression of CRTISO (targeted by miR6025) and enhanced levels of PEPC (targeted by miR7539) cause loss of carotenoids and accumulation of photoassimilates under B stress, which is necessary for plant survival (Patel et al., 2017). miR474 targets proline dehydrogenase (PDH) and raises proline levels in the roots, which are not only an osmoprotectant but also involved in cell wall cross-linking. For proper growth and survival of the plant under stress conditions, miR157 (targeting SPL proteins), miR830 (targeting Kinesin motor-related protein, RanBP1), miR843 (targeting LRR) and miR5023 (targeting FKBP-type peptidyl-prolyl cis-trans isomerase family protein) were downregulated in citrus roots, which were responsible for increased flavonol content, vesicle/organelle transport, defense mechanism and lateral root formation, respectively. Under B stress conditions, root respiration is decreased as carbon reserves have to be maintained at optimum levels. Thus, energy-consuming pathways such as β-oxidation pathways are reduced by upregulation of miR821, which targets enoyl-CoA hydratase/isomerase. Nutrient homeostasis is preserved under B stress conditions by a reduction in the expression levels of miR5266 and miR395, which targets ammonium transporter AMT1;1 and sulfate transporter SULTR2;1 and improves the uptake of external ammonium and sulfate via roots, respectively (Lu et al., 2014).

Boron has a narrow optimum range in plants and, thus, attains toxic levels at ease (Mengel & Kirkby, 2001). B accumulation due to toxicity leads to leaf cupping, retarded shoot and root growth, burns in the tips and margins of mature leaves and less CO_2 fixation. B is phloem-mobile in some species such as *Prunus*, *Malus* and *Pyrus* sp. and leads to internal necrosis of fruits and stem dieback. Higher B levels in legumes such as soybeans affect ureide metabolism (Nable et al., 1997). Excess B levels are alleviated in plants through the regulation of B efflux transporters such as BOR1 and BOR2 in leaves and roots of wheat and barley (Hayes & Reid, 2004; Reid, 2007; Reid et al., 2009) and also via a multifunctional aquaporin NIP, an efflux transporter in barley roots (Schnurbusch et al., 2010). The epigenetic mechanisms responsible for B toxicity tolerance were initiated through high-throughput sequencing studies in citrus and barley. Since B is capable of forming esters with –OH compounds, contributing to cell wall cross-linkage and assembly, alterations of cell wall properties might restrict the uptake of excess B. Regulation of miR397a which targets LAC4 and LAC17, secondary cell wall biosynthesis genes, improved B toxicity tolerance in woody citrus (Huang et al., 2016). miR408 expression, previously known to be involved with Cu homeostasis, was downregulated under high B levels in barley. It potentially targeted AFG1 and heterotrimeric G proteins involved in signal transduction pathways in plants (Ozhuner et al., 2013). B at higher concentrations induced jasmonic acid (JA) and ethylene-mediated miRNAs in *Arabidopsis*, similar to the biotic stress response. miR319 and miR172 induced higher expression levels of MYB3 and TOE1 in wheat, and TOE1 showed significant differences in their expression levels in sensitive and tolerant wheat cultivars, ascertaining their role in B toxicity tolerance. Specific lncRNAs can mimic the transcripts, also known as endogenous target mimicry (ITM), which are targeted by miRNAs and degrade those miRNAs; for instance, TCONS_00032652 (which mimics tubby protein-like transcript targeted by miR5049) and TCONS_00043651 [which mimics phosphate transporter 2 (PHO2) and putative ubiquitin-conjugating enzyme (UBC) transcript targeted by miR399] were upregulated under excess B exposure in barley, which decoys and degrades their specific miRNAs through a complex interaction of miRNA–lncRNA–target transcript module (Unver & Tombuloglu, 2020). Excess B levels cause double-stranded breaks (DSBs) due to the hyperacetylation of histones and association of the chromatin remodeling complex called BRAHMA (BRM), a SWItch/Sucrose Non-Fermentable subunit with the acetylated histones. BRM comprises a bromodomain that recognizes acetylated Lys residues and opens the chromatin, making it vulnerable to DSBs. Studies in *Arabidopsis* mutants that are hypersensitive to excess boron revealed that the 19S regulatory particle (RP), viz. RPN2a, RPN8a, RPT2a, and RPT5a, along with the 20S core particle (CP), which form the 26S proteasome (26SP), are required for B toxicity tolerance. The 26S proteasome degrades BRM and SNL5, a component of the SIN3 complex that includes class I HDACs and prevents DSB (Sakamoto et al., 2018). Condensin II requires histone deacetylation for its effective chromatin loading and prevention of DSBs under B toxicity (Sakamoto et al., 2011). Histone protein H4 showed a drastic increase in excess B levels in *Citrus sinensis* roots, preventing DNA damage. EF1A is considered the eukaryotic equivalent of Ef-Tu and performs chaperone activity in *Arabidopsis* (Kunze et al., 2004). It is methylated at specific sites by the EF1A lysine methyltransferase 2 gene in *Gypsophila perfoliata* L. in response to B toxicity tolerance as B forms cis-diols easily, and methylation of such genes prevents the interruption of B in their transcription/translation (Akbudak et al., 2024). B-induced P deficiency has been reported in wheat (Singh et al., 1990) and tomato (Kaya et al., 2009), and upregulation of HDA19 in *Citrus sinensis* roots may be attributed to stress adaptive response to maintain P homeostasis (Guo et al., 2016).

14.9 EPIGENETIC MODULATION IN RESPONSE TO IRON (Fe) AVAILABILITY

Iron is an essential micronutrient for DNA synthesis, respiration, and photosynthesis. It is essential for synthesizing chlorophyll and acts as a cofactor for several enzymes (Rout & Sahoo, 2015). Despite its importance, Fe availability in soil often fluctuates between deficiency and toxicity due to its low solubility under aerobic conditions and excessive presence in waterlogged or acidic soils. Iron deficiency reduces chlorophyll production and pollen fertility, lowering crop quality and output. Fe shortage causes major epigenetic alterations, such as DNA methylation and histone modifications, that influence the expression of genes related to Fe acquisition and transport.

Transcription factors such as FIT (FER-LIKE IRON DEFICIENCY INDUCED TRANSCRIPTION FACTOR) and BASIC-HELIX-LOOP-HELIX (bHLH) family proteins are important for maintaining iron homeostasis because they control genes such as *IRON-REGULATED TRANSPORTER 1 (IRT1) and FERRIC REDUCTION OXIDASE 2 (FRO2)* (Xing et al., 2015). Iron reduction and transport are facilitated by the upregulation of IRT1 (iron-regulated transporter) and FRO2 (ferric reductase oxidase) in Fe-deficient environments (Connolly et al., 2003). Epigenetic processes control these responses; for example, restrictive histone modifications such as H4R3me2 and H3K27me3 suppress the expression of genes related to iron homeostasis in an iron-sufficient environment (Sere & Martin, 2020). SKB1 (SHK1 BINDING PROTEIN 1) symmetrically dimethylates H4R3, and polycomb repressive complex 2 (PRC2) deposits H3K27me3, which suppresses the expression of *FIT, IRT1 and FRO2* (Fan et al., 2014; Xing et al., 2015). These repressive marks are removed upon iron starvation, activating these critical genes. Additional histone changes, including H3K4 trimethylation (H3K4me3) and acetylation, activate iron uptake and storage genes; for example, H3K4me3 promotes the transcription of the *GRF11 (GENERAL REGULATORY FACTOR 11) and ferritin* genes, whereas GENERAL CONTROL NONDEPRESSIBLE 5 (GCN5)-mediated histone acetylation increases the expression of transporters such as AtFRD3 (FERRIC REDUCTASE DEFECTIVE 3) (Su et al., 2022). DNA methylation, particularly in the CHH context, contributes to iron deficiency responses, as observed in rice and barley, by influencing important transcription factors (OsIRO2 and OsbHLH156) (Liang et al., 2020; Sun et al., 2021). Moreover, DNA methylation plays an essential role in regulating Fe homeostasis. According to previous studies, hypomethylation in the promoter regions of Fe-responsive genes is caused by Fe deprivation, guaranteeing the continuous activation of these genes. Non-coding RNAs, such as long non-coding RNAs and microRNAs, target transcripts or alter chromatin accessibility to further fine-tune the Fe shortage response. According to Halliwell and Gutteridge (1986), excessive Fe availability, frequently present in acidic or wet soil, causes the Fenton reaction to produce too many reactive oxygen species (ROS). Plants respond to oxidative stress by activating antioxidant defense genes such as CAT (catalase) and APX (ascorbate peroxidase). Under conditions of Fe toxicity, epigenetic variations such as H3K4me3 (a mark related to active transcription) increase at the promoters of these genes. Interestingly, Fe toxicity also induces changes in chromatin architecture. Furthermore, RNA-directed DNA methylation (RdDM) in the CHH context plays an important role in regulating Fe deficiency responses (Sun et al., 2021; Wang et al., 2020), for example, the methylation of transposable elements (TEs) next to Fe-responsive genes increases their expression during deficit. In rice, long-distance mobile small RNAs (23–24 nt) are recruited to root and shoot tissues to modulate gene expression, highlighting the systemic nature of epigenetic regulation under Fe stress (Sun et al., 2021).

14.10 EPIGENETIC CHANGES MODULATED BY COPPER (Cu) AVAILABILITY

Copper is one of the most important cofactors that the different plant organelles require to carry out many physiological processes. Plant cells require copper to sustain vital functions, including photosynthesis, detoxifying reactive oxygen species, cell wall lignification and hormone perception. In soil, Cu exists in two ionic forms: the reduced cuprous ion, Cu (I), and the Cu oxide ion, Cu (II). Cu^+ is active in the redox state, mostly bound to protein, and is also used to channel electrons in chloroplasts and mitochondria. Cu^{2+} ion uptake occurs in solution form (Shabbir et al., 2020).

Copper uptake occurs through membrane-embedded copper transporters (COPTs/CTRs). Under Cu deficiency, in *Arabidopsis*, SQUAMOSA PROMOTER-BINDING PROTEIN-LIKE (SPL) regulates gene expression to maintain plant growth and cellular metabolism (Schulten et al., 2022). SPL7 activity enhances a high-affinity copper absorption system by regulating the optimal redistribution of copper to necessary copper proteins through particular miRNAs. SPL7 directly interacts with Cu response elements (CuREs) bearing the GTAC motif in the promoter regions of copper-responsive genes. It upregulates the expression of Cu uptake genes like *copper(II) ion reductases FERRIC REDUCTION OXIDASE 4 (FRO4), FRO5* and high-affinity copper transporters (COPT1, COPT2 and COPT6) (Jung et al., 2012; Puig, 2014). SPL7 activates the copper-responsive microRNA, viz., miR398, which regulates the transcript level of *COPPER/ZINC (Cu–Zn) SUPEROXIDE DISMUTASE, CSD1* and *CSD2* in *Arabidopsis*. SOD eliminates the harmful super radicals to generate H_2O_2 and O_2. PAA1/HM6 are Cu transporters in the chloroplast membrane, while PAA2/HMA8 are in the thylakoids (Rae et al., 2001; Abdel-Ghany et al., 2005). Cu/ZnSOD necessitates a copper chaperone known as Copper Chaperone for SOD (CCS) (Chu et al., 2005). CCS has an amino-terminal that is a metalloprotein-binding domain for copper, a central target-binding region for Cu/ZnSOD, and a carboxy-terminal domain containing cysteine residues (Arnesano et al., 2004). The CCS protein facilitates the transport of copper to its target through direct protein contact and contributes to superoxide dismutase activation by accelerating the creation of a disulfide bond (Arnesano et al., 2004). In the absence of optimal Cu levels, SPL7 induces miR398 (Dugas & Bartel, 2008), which targets the degraded mRNA of *CSD1* and *CSD2* genes and also switches to iron superoxide dismutase (FeSOD) so that little available Cu can be transferred to plastocyanin to carry out ETS. Under high copper toxicity, the transcript of miR398 is reduced. miR397 and miR408 in *Populus trichocarpa* and *Vitis vinifera* target to reduce the activity of laccase, a copper-containing oxidase, which facilitates the production of lignin from lignin monomers in plants and enhances the lignification process under less Cu concentration (Abdel-Ghany & Pilon, 2008).

Oxidative stress is a multidimensional physiological occurrence that arises along with biotic and abiotic stress. High copper ion levels within the cell build up ROS (Drazkiewicz et al., 2004). Under oxidative stress, plants safeguard the response of their internal organelle to a novel aspect of epi transcriptomic RNA regulation. The N6-methyladenosine m6A regulatory machinery is built post-transcriptionally by a conserved group of proteins at the consensus sequence RRACH (Kramer et al., 2020; Niu et al., 2013). Several proteins are involved in adding, removing and recognizing m6A, classified into writers, erasers and readers. Proteins include MTA, MTB, ALKBH9B and ECT2 (Yue et al., 2019). The m6A writer complex identifies the consensus motif RRACH, but not all RRACH sequences are modified. ALKBH9B and ALKBH10B in *Arabidopsis thaliana* serve as m6A erasers, oxidizing m6A methylation from single-stranded RNA molecules (Wei et al., 2018). YTH domain proteins function as m6A readers, such as ECT2. It is prevalent in the 3′ UTRs of target genes and is crucial for modulating 3′ UTR processing in the nucleus and RNA stability in the cytoplasm (Niu et al., 2013). ECT3 and ECT4 can bind to specific loci on m6A-modified cellular RNAs in the cytoplasm. Global m6A profiling of 20-day-old *Arabidopsis* seedlings exposed to copper oxidative stress showed transcripts exhibiting increased m6A, indicating a favorable correlation between m6A enrichment and transcript stability. Conversely, transcripts with no alterations in m6A levels typically do not demonstrate variations in abundance (Wei et al., 2018); however, the mechanism remains unknown.

14.11 EPIGENETIC MODULATION IN RESPONSE TO MOLYBDENUM (Mo) AVAILABILITY

Molybdenum (Mo) is required in trace amounts for plant growth. It is an essential component of enzymes involved in redox reactions such as nitrate reductase (NR), xanthine dehydrogenase (XDH), aldehyde oxidase (AO) and sulfite oxidase (SO) (Kaiser et al., 2005). Mo becomes biologically active inside the plant along with pyranopterin, forming a molybdenum cofactor (MoCo), which later becomes a part of the redox enzymes (Mendel et al., 2012). Although Mo, in the form of molybdate (MoO_4^{2-}), is found in sufficient amounts in soil, it is often sequestered by Fe and Al in acidic soil (Zimmer et al., 1999). Mo is necessary for N uptake in symbiotic plants, and its deficiency impairs the reduction of N to NO_3^- (Stout et al., 1948). It is also necessary to synthesize phytohormones such as abscisic acid (ABA) and auxin (Kaiser et al., 2005).

Unlike other micronutrients whose deficiency symptoms occur in young leaves, Mo deficiency symptoms occur throughout the plant as it is readily translocated (Gupta et al., 1997). Since, MoO_4^{2-} is similar to SO_4^{2-}, the uptake of Mo is facilitated by Mo-specific transporters such as MOT1 localized in the endo-membrane (Tejada-Jimenez et al., 2007) and MOT2 localized in the vacuole (Gasber et al., 2011) falling under sulfate transporter family. Mutation in the MOT1 transporter and low Mo supply in *Arabidopsis* led to increased expression of nitrate assimilation genes such as *NR*, *NiR*, *PHO1;H1*, involved in phosphate homeostasis; Cys, glutathione (GSH), adenosine 5′-phosphosulfate reductase (APR) and sulfite reductase (SiR), involved in S assimilation pathway; and S transporter genes such as *SULTR3;5* and *SULTR1;2*, thus proving the importance of Mo in nitrogen, carbon, sulfur and phosphorus metabolism (Ide et al., 2011). Mo also crosses pathways with Fe and Cu as MoCo biosynthesis requires Cu (Kuper et al., 2004), and Mo-dependent redox enzymes such as XDH, AO and NR need Fe in the form of heme (Bittner, 2014). miR1119 targets the molybdenum cofactor sulfurize family proteins as a stress response and was first computationally predicted in barley (Colaiacovo et al., 2010). ABA3, a molybdenum cofactor sulfurase, is responsible for activating AO and XDH by adding S and is quintessential for the induction of these enzymes under drought stress (Bittner et al., 2001).

Molybdenum toxicity is rarely observed in the field, and excess anthocyanin accumulation is observed under such conditions (Hansch et al., 2009). A recent study by Yang et al. (2020) reported that treating the soybean–rhizobia symbiotic system with various Mo nanoparticles in high concentration exhibited epigenetic toxicity based on dosage and time by altering the whole-genome methylation levels, which affect gene expression in soybean plants in response to Mo stress.

14.12 EPIGENETIC MODULATION IN RESPONSE TO MANGANESE (Mn) AVAILABILITY

Manganese (Mn) is another critical element for physiological and metabolic functions of plants. Mn, which functions as a cofactor for approximately 35 enzymes (Mousavi et al., 2011), is essential for photosynthesis, particularly as a component of the oxygen-evolving complex (OEC) in photosystem II, where it aids in the water-splitting reaction. It also helps with nitrogen metabolism, lignin and flavonoid secondary metabolite production, and stress responses by reducing oxidative damage (Lidon et al., 2004). However, Mn deficiency and toxicity can cause significant changes in plant growth and development, particularly in the epigenome, altering gene expression and metabolic activities.

Mn deficiency is prevalent in high-pH soil, such as calcareous and podzolic soils, where its bioavailability is reduced due to poor solubility (Pedas et al., 2005) (Hebbern et al., 2009; Hebbern et al., 2009). Deficiency affects several physiological processes, including photosynthesis and lignin biosynthesis, resulting in impaired water use efficiency, decreased chlorophyll content and increased susceptibility to pathogens (Alejandro et al., 2020). At the cellular level, Mn deficiency alters the chloroplast ultrastructure, reduces the number of functional photosystem II centers and decreases the number of chloroplasts in leaves (Pedas et al., 2005). Epigenetic changes, specifically DNA methylation, play a key role in regulating the expression of OsMTP11 in response to heavy metal stress in rice. Under Cd, Zn, Ni and Mn treatments, reduced DNA methylation levels were observed in the promoter region of *OsMTP11*, which was correlated with its induced expression. This finding suggests that demethylation of the promoter facilitates transcriptional activation of *OsMTP11*, enabling the plant to increase its capacity for metal tolerance. Thus, DNA methylation is an epigenetic mechanism modulating *OsMTP11* expression, linking environmental metal stress to adaptive gene regulation (Zhang & Liu, 2017). Jing et al. (2022) demonstrated that Mn deficiency leads to the differential expression of genes associated with DNA methylation and demethylation, suggesting an adaptive strategy to optimize metabolic functions under nutrient-limited conditions; for example, deficiency-specific methylation patterns were detected in pokeweed, an Mn hyperaccumulator, indicating adaptive responses to Mn limitation. Reactive oxygen species (ROS) produced under stress affect this process by modifying the expression of DNA methylases and demethylases. Pretreatment with NADPH oxidase inhibitors reduces ROS levels, altering the epigenetic landscape under Mn stress (Jing et al., 2022). These findings underscore the interconnectedness of Mn availability, oxidative stress and epigenomic modifications. Non-coding RNAs, particularly microRNAs such as miR399, miR395, and miR826, target genes such as *PHOSPHATE2 (PHO2)* to regulate Mn-phosphate interactions, indirectly influencing Mn homeostasis (Gong et al., 2016). Mn toxicity, on the other hand, occurs in acidic soil where too much Mn is available, leading to phytotoxic effects such as reduced photosynthesis, decreased enzyme activity and nutritional imbalances (Santos et al., 2017). According to Mousavi et al. (2011), oxidative stress and the production of ROS are the causes of toxicity, which manifest as chlorosis, necrosis and reduced growth. In their study of Mn-tolerant and Mn-sensitive *Chenopodium ambrosioides* populations, Ding et al. (2023) reported that Mn poisoning results in global changes in DNA methylation. Gene expression was downregulated in Mn-tolerant genotypes, especially in antioxidant defense and metal transport genes. The observed behavioral differences between tolerant and sensitive genotypes have been explained by discovering that essential genes are regulated by differentially methylated loci (DMLs). Similarly, *Physcomitrella patens* was shown to exhibit Mn-induced DNA hypomethylation when exposed to Mn nanoparticles (MnONPs), where high Mn concentration led to ROS-mediated genotoxicity and epigenetic alterations (Ghosh et al., 2019). The hypomethylation of DNA under MnONP exposure highlights the potential of epigenetic biomarkers for assessing Mn toxicity in plants. The transport and homeostasis of Mn are tightly regulated processes influenced by epigenetic modifications. NRAMP1, a key Mn transporter, dynamically cycles between the plasma membrane and endosomal compartments depending on Mn availability (Castaings et al., 2021). While Mn deficiency enhances NRAMP1 activity at the plasma membrane, toxicity triggers its internalization to prevent excess Mn uptake. Epigenetic regulation of NRAMP1 transcription and protein trafficking offers insights into adaptive mechanisms of plants to cope with the fluctuating Mn levels.

14.13 EPIGENETICS CHANGES MODULATED BY CHLORINE (Cl)

Chlorine has multiple functions in plants, acting as an osmotically active solute in vacuoles to regulate turgor and osmoregulation. It stabilizes the membrane potential, maintains intracellular pH gradients and contributes to electrical excitability (Geilfus, 2018; Li et al., 2017). Chlorine is critical in stomatal regulation, photosynthesis and enzymatic activities. For example, Cl^- is vital for the opening and closing of stomata, influencing water use efficiency and photosynthetic rates (Gong et al., 2011).

Cl deficiency is rare but can occur in sandy soil with high leaching rates. Cl deficiency leads to epigenetic modifications that influence the expression of genes involved in Cl transport and signaling; for example, changes in histone acetylation and methylation, which regulate the transcription of Cl transporter genes such as chloride channel (CLC) family members, have been observed in response to Cl deficiency.

Conversely, excessive Cl^- accumulation, often due to saline conditions, results in Cl toxicity, which can impair cellular functions and lead to oxidative stress. Certain desert plants, such as *Pugionium cornutum*, can remarkably tolerate high chlorine levels. These plants utilize chloride ions to increase growth under saline and drought conditions, indicating species-specific responses to chlorine dynamics (Liu & Chu, 2023). Epigenetic changes, including altered DNA methylation and histone modification patterns, affect the plant response to Cl toxicity. These changes regulate the expression of stress-responsive genes, enabling plants to adapt to high Cl^- levels and minimize cellular damage.

14.14 EPIGENETICS CHANGES MODULATED BY SODIUM

The role of sodium in plants is to maintain osmotic equilibrium and turgor pressure. Increased sodium levels in the soil cause salt stress, causing reduced crop yield. Plant epigenome machinery is highly active during salt stress, leading to genome-wide DNA hypermethylation of transposable elements with cytosine methylation (Han et al., 2020). OsHKT1;5 encodes a sodium transporter that prevents sodium accumulation and eliminates sodium ions from leaves. Under salinity stress, it is directly regulated by OsBAG4, which is a member of the BAG domain chaperone family (Harris et al., 2018). OsMYB106, a MYB transcription factor, and OsSUVH7, a DNA methylation reader, interact with OsBAG4 through its DNA binding domain (Yin et al., 2020). OsMYB106 interacts with the cis-regulatory elements of OsHKT1;5, while OsSUVH7 relates to the methylated miniature inverted-repeat transposable element (MITE). *HKT*s are crucial genes in wheat and *Arabidopsis*, governing the expression of *TaHKT2;1* and *TaHKT2;3* in both shoot and root tissues (Kumar et al., 2017). Salt stress induces genotype- and tissue-specific elevations in DNA methylation, resulting in the downregulation of *TaHKT2;1* and *TaHKT2;3* expressions, augmenting salt tolerance. Excessive methylation of the *HKT1* promoter may suppress transcription in both leaves and roots, but non-CG methylation may facilitate precise regulation of *HKT1* expression in leaves (Ashapkin et al., 2020). Histone modification tends to regulate gene expression during salt stress, facilitating activation or suppression of salt-responsive genes. In rice, the accumulation of AGO2 proteins at the BIG GRAINS3 (BG3) locus results in increased BG3 expression, which is accomplished due to elevated H3K4me3 levels and reduced H3K27me3 levels (Yin et al., 2020). BG3, which encodes a purine permease, plays a role in cytokinin mobilization from shoot to root during salt stress, promoting tolerance.

Histone demethylases erase the histone methylation mark Jumonji C (jmjC) demethylates via a ferrous ion [Fe(II)] and an α-ketoglutaric acid (α-KG)-dependent oxidative process. The knockout of *jmj15* altered H3K4me3 levels of several genes associated with salt and water deprivation stress under salt-stressed conditions (Shen et al., 2022). WRKY46 modulates lateral root growth in response to salt stress by modulating ABA signaling and auxin homeostasis. WRKY70 regulates the osmotic stress response by modulating the stomatal aperture. JMJ15 directly interacts with and eliminates the H3K4me3 modification from the promoter and coding regions of WRKY46 and WRKY70, mediating the repression of these *WRKY* genes and enhancing salt tolerance in plants (Shen et al., 2022). Under salt-stress conditions, SDG721, a SET DOMAIN GROUP protein with H3K4 methyltransferase activity, binds to and modifies the H3K4 mark in the promoter and coding regions of *OsHKT1;5*, thereby modulating its expression levels, which stimulates the expression of the *AtHKT1* gene in *Arabidopsis* (Liu et al., 2021; Liu et al., 2022). Epigenetic changes during salt stress include other regulations like histone acetylation, histone variations and non-coding RNAs (Zhang et al., 2024).

14.15 CONCLUSIONS

Nutrient availability associated epigenetic changes are recently being resolved in crop plants. Nutrient supply controls DNA methylation, histone modification and ncRNA abundance, enabling plants to adapt to nutrient toxicity and deficiencies. Understanding epigenomics over the past several years has elucidated the molecular mechanisms of such adaptation responses, with encouraging implications for developing nutrient use efficiency and crop tolerance. Exploiting this knowledge through biotechnological applications can help mitigate the pressing issues of global food security and environmental sustainability. Epigenetic variation is an aftermath of both environment and epigenetic processes (Figures 14.1 and 14.2). Changes in chromatin methylation and siRNA production cause the origin of epialleles. DNA methylation at the cytosine in CG, CHG and CHH nucleotide sequences is common. In plants, the domains rearranged methyltransferase 2 (DRM2) drives the methylation process. DNA methylation status determines the stress acclimatization capability of plants. Plants also use the RdDM pathway, under the control of siRNAs produced from dsRNA precursors (Zhang & Zhu, 2011). As the natural diversity in epigenetic changes is rare, targeted epigenome editing can be attempted in plants. The importance of epigenomics to improve nutrient use efficiency is huge. Deciphering the epigenetic regulation of nutrients will aid in improving crop nutrient response and nutrient use efficiency (NtUE). Moreover, the development of epigenome biomarkers may help in predicting NtUE of the genotype under consideration.

REFERENCES

Abdel-Ghany, S. E., & Pilon, M. (2008). MicroRNA-mediated systemic down-regulation of copper protein expression in response to low copper availability in *Arabidopsis*. *The Journal of Biological Chemistry*, 283(23), 15932. https://doi.org/10.1074/JBC.M801406200.

Abdel-Ghany, S. E., Müller-Moulé, P., Niyogi, K. K., Pilon, M., & Shikanai, T. (2005). Two P-type ATPases are required for copper delivery in *Arabidopsis thaliana* chloroplasts. *The Plant Cell*, 17(4), 1233. https://doi.org/10.1105/TPC.104.030452.

Akbudak, M. A., Cirik, N., Erdeger, S. N., Filiz, E., Dogu, S., & Bor, M. (2024). GpEF1A: A novel lysine methyltransferase gene from *Gypsophila perfoliata* L. involved in boron homeostasis. *Plant Biology*, 26, 727–734.

Alejandro, S., Höller, S., Meier, B., & Peiter, E. (2020). Manganese in plants: From acquisition to subcellular allocation. *Frontiers in Plant Science*, 11, 300. https://doi.org/10.3389/FPLS.2020.00300.

Ali, S., Khan, N., & Tang, Y. (2022). Epigenetic marks for mitigating abiotic stresses in plants. *Journal of Plant Physiology*, 275. https://doi.org/10.1016/J.JPLPH.2022.153740.

Arnesano, F., Banci, L., Bertini, I., Martinelli, M., Furukawa, Y., & O'Halloran, T. V. (2004). The unusually stable quaternary structure of human Cu,Zn-Superoxide sismutase 1 is controlled by both metal occupancy and disulfide status.

Journal of Biological Chemistry, 279(46), 47998–47993. https://doi.org/10.1074/JBC.M406021200.

Ashapkin, V. V., Kutueva, L. I., Aleksandrushkina, N. I., & Vanyushin, B. F. (2020). Epigenetic mechanisms of plant adaptation to biotic and abiotic stresses. *Mdpi.Com*, 21(20), 1–32. https://doi.org/10.3390/ijms21207457.

Aufsatz, W., Mette, M., Matzke, A., & Matzke, M. (2004). The role of MET1 in RNA-directed de novo and maintenance methylation of CG dinucleotides. *Plant Molecular Biology*, 54, 793–804.

Azimi, Z., Isa, M. R., Khan, J., Wang, S. M., & Ismail, Z. (2022). Association of zinc level with DNA methylation and its consequences: A systematic review. *Heliyon*, 8(10). https://doi.org/10.1016/j.heliyon.2022.e10815.

Bari, R., Pant, B. D., Stitt, M., & Scheible, W.-R. (2006). PHO2, MicroRNA399, and PHR1 define a phosphate-signaling pathway in plants. *Plant Physiology*, 141(3), 988–999. https://doi.org/10.1104/PP.106.079707.

Barr, R., Böttger, M., & Crane, F. L. (1993). The effect of boron on plasma membrane electron transport and associated proton secretion by cultured carrot cells. *Biochemistry and Molecular Biology International*, 31(1), 31–39.

Berr, A., McCallum, E. J., Ménard, R., Meyer, D., Fuchs, J., Dong, A., & Shen, W. H. (2010). Arabidopsis SET DOMAIN GROUP2 is required for H3K4 trimethylation and is crucial for both sporophyte and gametophyte development. *Plant Cell*, 22(10). https://doi.org/10.1105/tpc.110.079962.

Bittner, F. (2014). Molybdenum metabolism in plants and crosstalk to iron. *Frontiers in Plant Science*, 5, 28.

Bittner, F., Oreb, M., & Mendel, R. R. (2001). ABA3 is a molybdenum cofactor sulfurase required for activation of aldehyde oxidase and xanthine dehydrogenase in *Arabidopsis thaliana. Journal of Biological Chemistry*, 276(44), 40381–40384.

Borevitz, J. O., Xia, Y., Blount, J., Dixon, R. A., & Lamb, C. (2000). Activation tagging identifies a conserved MYB regulator of phenylpropanoid biosynthesis. *The Plant Cell*, 12(12), 2383–2393. https://doi.org/10.1105/TPC.12.12.2383.

Broadley, M. R., White, P. J., Hammond, J. P., Zelko, I., & Lux, A. (2007). Zinc in plants. *New Phytologist*, 173(4), 677–702. https://doi.org/10.1111/J.1469-8137.2007.01996.X.

Camacho-Cristóbal, J. J., Anzellotti, D., & González-Fontes, A. (2002). Changes in phenolic metabolism of tobacco plants during short-term boron deficiency. *Plant Physiology and Biochemistry*, 40(12), 997–1002.

Castaings, L., Alcon, C., Kosuth, T., Correia, D., & Curie, C. (2021). Manganese triggers phosphorylation-mediated endocytosis of the *Arabidopsis* metal transporter NRAMP1. *The Plant Journal: For Cell and Molecular Biology*, 106(5), 1328–1337. https://doi.org/10.1111/TPJ.15239.

Cazzonelli, C. I., Hou, X., Alagoz, Y., Rivers, J., Dhami, N., Lee, J., Marri, S., & Pogson, B. J. (2020). A cis-carotene derived apocarotenoid regulates etioplast and chloroplast development. *Elife*. 2020;9:e45310. Published 2020 Jan 31. doi:10.7554/eLife.45310

Chen, C.-Y., Wu, K., & Schmidt, W. (2015). The histone deacetylase HDA19 controls root cell elongation and modulates a subset of phosphate starvation responses in *Arabidopsis*. https://doi.org/10.1038/srep15708.

Chu, C. C., Lee, W. C., Guo, W. Y., Pan, S. M., Chen, L. J., Li, H. M., & Jinn, T. L. (2005). A copper chaperone for superoxide dismutase that confers three types of copper/zinc superoxide dismutase activity in *Arabidopsis. Plant Physiology*, 139(1), 425–436. https://doi.org/10.1104/PP.105.065284.

Colaiacovo, M., Subacchi, A., Bagnaresi, P., Lamontanara, A., Cattivelli, L., & Faccioli, P. (2010). A computational-based update on microRNAs and their targets in barley (*Hordeum vulgare* L.). *BMC Genomics*, 11, 1–15.

Connolly, E. L., Campbell, N. H., Grotz, N., Prichard, C. L., & Guerinot, M. L. (2003). Overexpression of the FRO2 ferric chelate reductase confers tolerance to growth on low iron and uncovers posttranscriptional control. *Plant Physiology*, 133(3), 1102–1110. https://doi.org/10.1104/PP.103.025122.

Dayod, M., Tyerman, S. D., Leigh, R. A., & Gilliham, M. (2010). Calcium storage in plants and the implications for calcium biofortification. In *Protoplasma* (Vol. 247, Issue 3). https://doi.org/10.1007/s00709-010-0182-0.

Dekomah, S. D., Bi, Z., Dormatey, R., Wang, Y., Haider, F. U., Sun, C., Yao, P., & Bai, J. (2022). The role of CDPKs in plant development, nutrient and stress signaling. In *Frontiers in Genetics* (Vol. 13). https://doi.org/10.3389/fgene.2022.996203.

Ding, G., Cao, L., Zhou, J., Li, Z., Lai, Y., Liu, K., Luo, Y., Bai, L., Wang, X., Wang, T., Wang, R., Yang, G., & Sun, S. (2022). DNA methylation correlates with the expression of drought-responsive genes and drought resistance in rice. *Agronomy*, 12(6). https://doi.org/10.3390/agronomy12061445.

Ding, S., Zhang, H., Zhou, C., Bao, Y., Xu, X., Chen, Y., Shen, Z., & Chen, C. (2023). Transcriptomic, epigenomic and physiological comparisons reveal key factors for different manganese tolerances in three *Chenopodium ambrosioides* L. populations. *Plant Physiology and Biochemistry*, 201, 107883. https://doi.org/10.1016/J.PLAPHY.2023.107883.

Dong G., Ma D. P., Li J. (2008). The histone methyltransferase SDG8 regulates shoot branching in *Arabidopsis. Biochemical and Biophysical Research Communications.* Sep 5;373(4):659–64. doi: 10.1016/j.bbrc.2008.06.096. Epub 2008 Jul 3. PMID: 18602372

Drązkiewicz, M., Skórzyńska-Polit, E., & Krupa, Z. (2004). Copper-induced oxidative stress and antioxidant defence in *Arabidopsis thaliana. Biometals: An International Journal on the Role of Metal Ions in Biology, Biochemistry, and Medicine*, 17(4), 379–387. https://doi.org/10.1023/B:BIOM.0000029417.18154.22.

Dugas, D. V., & Bartel, B. (2008). Sucrose induction of *Arabidopsis* MiR398 represses two Cu/Zn superoxide dismutases. *Plant Molecular Biology*, 67(4), 403–417. https://doi.org/10.1007/S11103-008-9329-1.

Fan, H., Zhang, Z., Wang, N., Cui, Y., Sun, H., Liu, Y., Wu, H., Zheng, S., Bao, S., & Ling, H. Q. (2014). SKB1/PRMT5-mediated histone H4R3 dimethylation of Ib subgroup bHLH genes negatively regulates iron homeostasis in *Arabidopsis thaliana. The Plant Journal*, 77(2), 209–221. https://doi.org/10.1111/TPJ.12380.

Friso, S., Choi, S. W., Girelli, D., Mason, J. B., Dolnikowski, G. G., Bagley, P. J., Olivieri, O., Jacques, P.F., Rosenberg, I.H., Corrocher, R., & Selhub, J. (2002). A common mutation in the 5, 10-methylenetetrahydrofolate reductase gene affects genomic DNA methylation through an interaction with folate status. *Proceedings of the National Academy of Sciences*, 99(8), 5606–5611.

Fu, Y. L., Zhang, G. Bin, Lv, X. F., Guan, Y., Yi, H. Y., & Gong, J. M. (2013). *Arabidopsis* histone methylase CAU1/PRMT5/SKB1 acts as an epigenetic suppressor of the calcium signaling gene cas to mediate stomatal closure in response to extracellular calcium. *Plant Cell*, 25(8). https://doi.org/10.1105/tpc.113.113886.

Gasber, A., Klaumann, S., Trentmann, O., Trampczynska, A., Clemens, S., Schneider, S., et al. (2011). Identification of an *Arabidopsis* solute carrier critical for intracellular transport and inter-organ allocation of molybdate. *Plant Biology (Stuttg)*, 13, 710–718. https://doi.org/10.1111/j.1438-8677.2011.00448.x.

Geilfus, C. M. (2018). Chloride: From nutrient to toxicant. *Plant and Cell Physiology*, 59(5), 877–886. https://doi.org/10.1093/PCP/PCY071.

Ghosh, I., Sadhu, A., Moriyasu, Y., Bandyopadhyay, M., & Mukherjee, A. (2019). Manganese oxide nanoparticles induce genotoxicity and DNA hypomethylation in the moss *Physcomitrella patens*. *Mutation Research – Genetic Toxicology and Environmental Mutagenesis*, 842, 146–157. https://doi.org/10.1016/J.MRGENTOX.2018.12.006.

Gierth, M., & Mäser, P. (2007). Potassium transporters in plants – Involvement in K^+ acquisition, redistribution and homeostasis. In *FEBS Letters* (Vol. 581, Issue 12). https://doi.org/10.1016/j.febslet.2007.03.035.

Goldbach, H. E., Hartmann, D., & Rötzer, T. (1990). Boron is required for the stimulation of the ferricyanide-induced proton release by auxins in suspension-cultured cells of *Daucus carota* and *Lycopersicon esculentum*. *Physiologia Plantarum*, 80(1), 114–118.

Gong, H., Blackmore, D., Clingeleffer, P., Sykes, S., Jha, D., Tester, M., & Walker, R. (2011). Contrast in chloride exclusion between two grapevine genotypes and its variation in their hybrid progeny. *Journal of Experimental Botany*, 62(3), 989–999. https://doi.org/10.1093/JXB/ERQ326.

Gong, J., Li, D., Li, H., Zhou, H., & Xu, J. (2016). Identification of manganese-responsive microRNAs in *Arabidopsis* by small RNA sequencing. *Original Paper Czech Journal of Genetics and Plant Breeding*, 55(2), 76–82. https://doi.org/10.17221/57/2018-CJGPB.

Groth, M., Moissiard, G., Wirtz, M., Wang, H., Garcia-Salinas, C., Ramos-Parra, P. A., ... & Jacobsen, S. E. (2016). MTHFD1 controls DNA methylation in *Arabidopsis*. *Nature Communications*, 7(1), 11640.

Guerinot, M. L. (2000). The ZIP family of metal transporters. In *Biochimica et Biophysica Acta - Biomembranes* (Vol. 1465, Issues 1–2). https://doi.org/10.1016/S0005-2736(00)00138-3.

Guo, P., Qi, Y. P., Yang, L. T., Ye, X., Huang, J. H., & Chen, L. S. (2016). Long-term boron-excess-induced alterations of gene profiles in roots of two citrus species differing in boron-tolerance revealed by cDNA-AFLP. *Frontiers in Plant Science*, 7, 898.

Gupta, U. C. (1997). Symptoms of molybdenum deficiency and toxicity in crops. *Molybdenum in Agriculture*, 2, 160–170.

Hajyzadeh, M., Turktas, M., Khawar, K. M., & Unver, T. (2015). miR408 overexpression causes increased drought tolerance in chickpea. *Gene*, 555(2), 186–193.

Halliwell, B., & Gutteridge, J. M. C. (1986). Oxygen free radicals and iron in relation to biology and medicine: Some problems and concepts. *Archives of Biochemistry and Biophysics*, 246(2), 501–514. https://doi.org/10.1016/0003-9861(86)90305-X.

Han, B., Xu, W., Ahmed, N., Yu, A., Wang, Z., & Liu, A. (2020). Changes and associations of genomic transcription and histone methylation with salt stress in castor bean. *Plant and Cell Physiology*, 61(6), 1120–1133. https://doi.org/10.1093/pcp/pcaa037.

Hänsch, R., & Mendel, R. R. (2009). Physiological functions of mineral micronutrients (Cu, Zn, Mn, Fe, Ni, Mo, B, cl). *Current Opinion in Plant Biology*, 12(3), 259–266.

Harris, C. J., Scheibe, M., Wongpalee, S. P., Liu, W., Cornett, E. M., Vaughan, R. M., Li, X., et al. (2018). A DNA methylation reader complex that enhances gene transcription. *Science*, 362(6419), 1182–1186. https://doi.org/10.1126/SCIENCE.AAR7854.

Hayes, J. E., & Reid, R. J. (2004). Boron tolerance in barley is mediated by efflux of boron from the roots. *Plant Physiology*, 136(2), 3376–3382.

Hebbern, C. A., Laursen, K. H., Ladegaard, A. H., Schmidt, S. B., Pedas, P., Bruhn, D., Schjoerring, J. K., Wulfsohn, D., & Husted, S. (2009). Latent manganese deficiency increases transpiration in barley (*Hordeum vulgare*). *Physiologia Plantarum*, 135(3), 307–316. https://doi.org/10.1111/J.1399-3054.2008.01188.X.

Hesse, H., & Hoefgen, R. (2003). Molecular aspects of methionine biosynthesis. *Trends in Plant Science*, 8(6), 259–262.

Hsieh, L. C., Lin, S. I., Chieh Shih, A. C., Chen, J. W., Lin, W. Y., Tseng, C. Y., Li, W. H., & Chiou, T. J. (2009). Uncovering small RNA-mediated responses to phosphate deficiency in *Arabidopsis* by deep sequencing. *Plant Physiology*, 151(4), 2120. https://doi.org/10.1104/PP.109.147280.

Hu, H., & Brown, P. H. (1994). Localization of boron in cell walls of squash and tobacco and its association with pectin (evidence for a structural role of boron in the cell wall). *Plant Physiology*, 105(2), 681–689.

Huang, J. H., Qi, Y. P., Wen, S. X., Guo, P., Chen, X. M., & Chen, L. S. (2016). Illumina microRNA profiles reveal the involvement of miR397a in *Citrus* adaptation to long-term boron toxicity via modulating secondary cell-wall biosynthesis. *Scientific Reports*, 6(1), 22900.

Ide, Y., Kusano, M., Oikawa, A., Fukushima, A., Tomatsu, H., Saito, K., et al. (2011). Effects of molybdenum deficiency and defects in molybdate transporter MOT1 on transcript accumulation and nitrogen/sulphur metabolism in *Arabidopsis thaliana*. *Journal of Experimental Botany*, 62, 1483–1497. https://doi.org/10.1093/jxb/erq345.

Islam, W., Tauqeer, A., Waheed, A., & Zeng, F. (2022). MicroRNA mediated plant responses to nutrient stress. In *International Journal of Molecular Sciences* (Vol. 23, Issue 5). https://doi.org/10.3390/ijms23052562.

Jagadeeswaran, G., Li, Y. F., & Sunkar, R. (2014). Redox signaling mediates the expression of a sulfate-deprivation-inducible micro RNA 395 in Arabidopsis. *The Plant Journal*, 77(1), 85–96.

Jing, M., Zhang, H., Wei, M., Tang, Y., Xia, Y., Chen, Y., Shen, Z., & Chen, C. (2022). Reactive oxygen species partly mediate DNA methylation in responses to different heavy metals in pokeweed. *Frontiers in Plant Science*, 13, 845108. https://doi.org/10.3389/FPLS.2022.845108/BIBTEX.

Jovanović, Ž., Stanisavljević, N., Mikić, A., Radović, S., & Maksimović, V. (2014). Water deficit down-regulates miR398 and miR408 in pea (*Pisum sativum* L.). *Plant Physiology and Biochemistry*, 83, 26–31.

Jung, H. Il, Gayomba, S. R., Rutzke, M. A., Craft, E., Kochian, L. V., & Vatamaniuk, O. K. (2012). COPT6 is a plasma membrane transporter that functions in copper homeostasis in *Arabidopsis* and is a novel target of SQUAMOSA promoter-binding protein-like 7. *The Journal of Biological Chemistry*, 287(40), 33252–33267. https://doi.org/10.1074/JBC.M112.397810.

Kaiser, B. N., Gridley, K. L., Ngaire Brady, J., Phillips, T., & Tyerman, S. D. (2005). The role of molybdenum in agricultural plant production. *Annals of Botany*, 96(5), 745–754.

Kawashima, C. G., Matthewman, C. A., Huang, S., Lee, B. R., Yoshimoto, N., Koprivova, A., ... & Kopriva, S. (2011). Interplay of SLIM1 and miR395 in the regulation of sulfate assimilation in *Arabidopsis. The Plant Journal*, 66(5), 863–876.

Kaya, C., Tuna, A. L., Dikilitas, M., Ashraf, M., Koskeroglu, S., & Guneri, M. (2009). Supplementary phosphorus can alleviate boron toxicity in tomato. *Scientia Horticulturae*, 121(3), 284–288. https://doi.org/10.1016/j.scienta.2009.02.011

Kayıhan, D.S., Kayıhan, C., Çiftçi, Y.Ö (2016). Excess boron responsive regulations of antioxidative mechanism at physio-biochemical and molecular levels in *Arabidopsis thaliana. Plant Physiology and Biochemistry.* 109:337–345. doi: 10.1016/j.plaphy.2016.10.016. Epub 2016 Oct 18. PMID: 27794275.

Khan, F., Siddique, A. B., Shabala, S., Zhou, M., & Zhao, C. (2023). Phosphorus plays key roles in regulating plants' physiological responses to abiotic stresses. *Plants*, 12(15), 2861. https://doi.org/10.3390/PLANTS12152861.

Kopriva, S. (2015). Plant sulfur nutrition: From Sachs to big data. *Plant Signaling & Behavior*, 10(9), e1055436.

Kosegarten, H., Appel, T. (2001). Boron. In: Mengel, K., Kirkby, E.A., Kosegarten, H., Appel, T. (eds) *Principles of Plant Nutrition.* Springer, Dordrecht. https://doi.org/10.1007/978-94-010-1009-2_18

Kramer, M. C., Janssen, K. A., Palos, K., Nelson, A. D. L., Vandivier, L. E., Garcia, B. A., Lyons, E., Beilstein, M. A., & Gregory, B. D. (2020). N6-Methyladenosine and RNA secondary structure affect transcript stability and protein abundance during systemic salt stress in *Arabidopsis. Plant Direct*, 4(7), e00239. https://doi.org/10.1002/PLD3.239.

Kumar, S., Beena, A. S., Awana, M., & Singh, A. (2017). Salt-induced tissue-specific cytosine methylation downregulates expression of HKT genes in contrasting wheat (*Triticum Aestivum* L.) genotypes. *DNA and Cell Biology*, 36(4), 283–294. https://doi.org/10.1089/DNA.2016.3505.

Kunze, G., Zipfel, C., Robatzek, S., Niehaus, K., Boller, T., & Felix, G. (2004). The N terminus of bacterial elongation factor Tu elicits innate immunity in *Arabidopsis* plants. *The Plant Cell*, 16(12), 3496–3507.

Kuper, J., Llamas, A., Hecht, H. J., Mendel, R. R., & Schwarz, G. (2004). Structure of the molybdopterin-bound Cnx1G domain links molybdenum and copper metabolism. *Nature*, 430(7001), 803–806.

Laanen, P., Saenen, E., Mysara, M., Van de Walle, J., Van Hees, M., Nauts, R., Van Nieuwerburgh, F., Voorspoels, S., Jacobs, G., Cuypers, A., & Horemans, N. (2021). Changes in DNA methylation in *Arabidopsis thaliana* plants exposed over multiple generations to gamma radiation. *Frontiers in Plant Science*, 12. https://doi.org/10.3389/fpls.2021.611783.

Li A., Hu B., Chu C. (2021). Epigenetic regulation of nitrogen and phosphorus responses in plants. *Journal of Plant Physiology.* 2021 Mar–Apr;258–259:153363. doi: 10.1016/j.jplph.2021.153363. Epub 2021 Jan 10. PMID: 33508741.

Li, B., Tester, M., & Gilliham, M. (2017). Chloride on the move. *Trends in Plant Science*, 22(3), 236–248. https://doi.org/10.1016/J.TPLANTS.2016.12.004.

Li, C., Chen, C., Gao, L., Yang, S., Nguyen, V., Shi, X., Siminovitch, K., Kohalmi, S. E., Huang, S., Wu, K., Chen, X., & Cui, Y. (2015). The Arabidopsis SWI2/SNF2 chromatin remodeler BRAHMA regulates polycomb function during vegetative development and directly activates the flowering repressor gene SVP. *PLoS Genetics*, 11(1). https://doi.org/10.1371/journal.pgen.1004944.

Li, H., Wang, Y., Chen, M., Xiao, P., Hu, C., Zeng, Z., ... & Hu, Z. (2016). Genome-wide long non-coding RNA screening, identification and characterization in a model microorganism *Chlamydomonas reinhardtii. Scientific Reports*, 6(1), 34109.

Li, Y., Mukherjee, I., Thum, K. E., Tanurdzic, M., Katari, M. S., Obertello, M., Edwards, M. B., McCombie, W. R., Martienssen, R. A., & Coruzzi, G. M. (2015). The histone methyltransferase SDG8 mediates the epigenetic modification of light and carbon responsive genes in plants. *Genome Biology*, *16*(1), 79. https://doi.org/10.1186/s13059-015-0640-2

Liang, G., Ai, Q., & Yu, D. (2015). Uncovering miRNAs involved in crosstalk between nutrient deficiencies in *Arabidopsis. Scientific Reports*, 5(1), 11813.

Liang, G., Yang, F., & Yu, D. (2010). MicroRNA395 mediates regulation of sulfate accumulation and allocation in *Arabidopsis thaliana. The Plant Journal*, 62(6), 1046–1057.

Liang, G., Zhang, H., Li, Y., Pu, M., Yang, Y., Li, C., Lu, C., Xu, P., & Yu, D. (2020). Oryza sativa FER-LIKE FE DEFICIENCY-INDUCED TRANSCRIPTION FACTOR (OsFIT/OsbHLH156) interacts with OsIRO2 to regulate iron homeostasis. *Journal of Integrative Plant Biology*, 62(5), 668–689. https://doi.org/10.1111/jipb.12933.

Lidon, F. C., Barreiro, M. G., & Ramalho, J. C. (2004). Manganese accumulation in rice: Implications for photosynthetic functioning. *Journal of Plant Physiology*, 161(11), 1235–1244. https://doi.org/10.1016/J.JPLPH.2004.02.003.

Lin, S. I., Chiang, S. F., Lin, W. Y., Chen, J. W., Tseng, C. T., Wu, P. C., & Chiou, T. Z. (2008). Regulatory network of MicroRNA399 and PHO2 by systemic signaling. *Plant Physiology*, 147(2), 732. https://doi.org/10.1104/PP.108.116269.

Liu, Y., Wang, J., Liu, B., & Xu, Z. Y. (2022). Dynamic regulation of DNA methylation and histone modifications in response to abiotic stresses in plants. *Journal of Integrative Plant Biology*, 64(12), 2252–2274. https://doi.org/10.1111/jipb.13368.

Liu, Y., & Chu, C. (2023). Improving maize seed protein content and nitrogen-use efficiency by a teosinte asparagine synthetase. *Molecular Plant*, 16(3), 497–499. https://doi.org/10.1016/J.MOLP.2022.12.001.

Liu, Y., Chen, X., Xue, S., Quan, T., Cui, D., Han, L., Cong, W., et al. (2021). SET DOMAIN GROUP 721 protein functions in saline–alkaline stress tolerance in the model rice variety kitaake. *Plant Biotechnology Journal*, 19(12), 2576–2588. https://doi.org/10.1111/PBI.13683.

Loizeau, K., De Brouwer, V., Gambonnet, B., Yu, A., Renou, J. P., Van Der Straeten, D., ... & Ravanel, S. (2008). A genome-wide and metabolic analysis determined the adaptive response of *Arabidopsis* cells to folate depletion induced by methotrexate. *Plant Physiology*, 148(4), 2083–2095.

Lu, H., Chen, L., Du, M., Lu, H., Liu, J., Ye, S., Tao, B., Li, R., Zhao, L., Wen, J., Yi, B., Tu, J., Fu, T., & Shen, J. (2023). miR319 and its target TCP4 involved in plant architecture regulation in *Brassica napus. Plant Science*, 326. https://doi.org/10.1016/j.plantsci.2022.111531.

Lu, Y. B., Qi, Y. P., Yang, L. T., Guo, P., Li, Y., & Chen, L. S. (2015). Boron-deficiency-responsive microRNAs and their targets in *Citrus sinensis* leaves. *BMC Plant Biology*, 15, 1–15.

Lu, Y. B., Yang, L. T., Qi, Y. P., Li, Y., Li, Z., Chen, Y. B., ... & Chen, L. S. (2014). Identification of boron-deficiency-responsive microRNAs in *Citrus sinensis* roots by Illumina sequencing. *BMC Plant Biology*, 14, 1–16.

Luo, Y.-X., Hou, X.-M., Zhang, C.-J., Tan, L.-M., Shao, C.-R., Lin, R.-N., Su, Y.-N., et al. (2020). A plant-specific SWR1 chromatin-remodeling complex couples histone H2A.Z deposition with nucleosome sliding. *The EMBO Journal*, 39(7). https://doi.org/10.15252/EMBJ.2019102008/SUPPL_FILE/EMBJ2019102008-SUP-0009-DATASETEV7.XLSX.

Lynch, J. P., & Brown, K. M. (2008). Root strategies for phosphorus acquisition (pp. 83–116). https://doi.org/10.1007/978-1-4020-8435-5_5.

Ma, C., Burd, S., & Lers, A. (2015). mi R 408 is involved in abiotic stress responses in *Arabidopsis*. *The Plant Journal*, 84(1), 169–187.

Mager, S., and Ludewig, U. (2018). Massive loss of DNA methylation in nitrogen-, but not in phosphorus-deficient *Zea mays* roots is poorly correlated with gene expression differences. *Frontiers in Plant Science* 9 (April). https://doi.org/10.3389/fpls.2018.00497.

Marschner, P. (2011). *Marschner's Mineral Nutrition of Higher Plants: Third Edition*. https://doi.org/10.1016/C2009-0-63043-9.

Mendel, R. R., & Kruse, T. (2012). Cell biology of molybdenum in plants and humans. *Biochimica et Biophysica Acta (BBA)-Molecular Cell Research*, 1823(9), 1568–1579.

Meng, J., Wang, L., Wang, J., Zhao, X., Cheng, J., Yu, W., ... & Gong, Z. (2018). METHIONINE ADENOSYLTRANSFERASE4 mediates DNA and histone methylation. *Plant Physiology*, 177(2), 652–670.

Mousavi, S., Shahsavari, M., & Rezaei, M. (2011). A general overview on manganese (Mn) importance for crops production. https://www.cabidigitallibrary.org/doi/full/10.5555/20113381121.

Nable, R. O., Banuelos, G. S., & Paull, J. G. (1997), Boron toxicity. *Plant Soil*, 193, 181–198.

Nischal, L., Mohsin, M., Khan, I., Kardam, H., Wadhwa, A., Abrol, Y. P., Iqbal, M., & Ahmad, A. (2012). Identification and comparative analysis of MicroRNAs associated with low-N tolerance in rice genotypes. *PloS one*. Accessed November 4, 2025. https://doi.org/10.1371/journal.pone.0050261.

Niu, Y., Zhao, X., Wu, Y. S., Li, M. M., Wang, X. J., & Yang, Y. G. (2013). N6-Methyl-Adenosine (M6A) in RNA: An old modification with a novel epigenetic function. *Genomics, Proteomics & Bioinformatics*, 11(1), 8–17. https://doi.org/10.1016/J.GPB.2012.12.002.

Ouyang, B., Fei, Z., Joung, J. G., Kolenovsky, A., Koh, C., Nowak, J., ... & Tsang, E. W. (2012). Transcriptome profiling and methyl homeostasis of an *Arabidopsis* mutant deficient in S-adenosylhomocysteine hydrolase1 (SAHH1). *Plant Molecular Biology*, 79, 315–331.

Ozhuner, E., Eldem, V., Ipek, A., Okay, S., Sakcali, S., Zhang, B., ... & Unver, T. (2013). Boron stress responsive microRNAs and their targets in barley. *PloS One*, 8(3), e59543.

Pardey, P. G., Beddow, J. M., Hurley, T. M., Beatty, T. K. M., & Eidman, V. R. (2014). A bounds analysis of world food futures: Global agriculture through to 2050. *Australian Journal of Agricultural and Resource Economics*, 58(4), 571–589. https://doi.org/10.1111/1467-8489.12072.

Patel, P., Yadav, K., & Ganapathi, T. R. (2017). Small and hungry: microRNAs in micronutrient homeostasis of plants. *MicroRNA*, 6(1), 22–41.

Pedas, P., Hebbern, C. A., Schjoerring, J. K., Holm, P. E., & Husted, S. (2005). Differential capacity for high-affinity manganese uptake contributes to differences between barley genotypes in tolerance to low manganese availability. *Plant Physiology*, 139(3), 1411–1420. https://doi.org/10.1104/PP.105.067561.

Peña, P. V., Davrazou, F., Shi, X., Walter, K. L., Verkhusha, V. V., Gozani, O., Zhao, R., & Kutateladze, T. G. (2006). Molecular mechanism of histone H3K4me3 recognition by plant homeodomain of ING2. *Nature*, 442(7098), 100–103. https://doi.org/10.1038/NATURE04814.

Pollard, A. S., Parr, A. J., & Loughman, B. C. (1977). Boron in relation to membrane function in higher plants. *Journal of Experimental Botany*, 28(4), 831–841.

Puig, S. (2014). Function and regulation of the plant COPT family of high-affinity copper transport proteins. *Advances in Botany*, 2014(1), 476917. https://doi.org/10.1155/2014/476917.

Rae, T. D., Torres, A. S., Pufahl, R. A., & O'Halloran, T. V. (2001). Mechanism of Cu,Zn-superoxide dismutase activation by the human metallochaperone HCCS. *Journal of Biological Chemistry*, 276(7), 5166–5176. https://doi.org/10.1074/JBC.M008005200.

Ranty, B., Aldon, D., Cotelle, V., Galaud, J. P., Thuleau, P., & Mazars, C. (2016). Calcium sensors as key hubs in plant responses to biotic and abiotic stresses. *Frontiers in Plant Science*, 7(MAR2016). https://doi.org/10.3389/fpls.2016.00327.

Reid, R. (2007). Identification of boron transporter genes likely to be responsible for tolerance to boron toxicity in wheat and barley. *Plant and Cell Physiology*, 48(12), 1673–1678.

Reid, R., & Fitzpatrick, K. (2009). Influence of leaf tolerance mechanisms and rain on boron toxicity in barley and wheat. *Plant Physiology*, 151(1), 413–420.

Rerkasem, B. (1996). Boron and plant reproductive development. *Sterility in Wheat in Subtropical Asia: Extent, Causes and Solutions*, 72, 32–34.

Robertson, G. A., & Loughman, B. C. (1974). Reversible effects of boron on the absorption and incorporation of phosphate in *Vicia fabal*. *New Phytologist*, 73(2), 291–298.

Rout, G. R., & Sahoo, S. (2015). Role of iron in plant growth and metabolism. *Reviews in Agricultural Science*, 3, 1–24. https://doi.org/10.7831/RAS.3.1.

Rubio, V., Linhares, F., Solano, R., Martín, A. C., Iglesias, J., Leyva, A., & Paz-Ares, J. (2001). A conserved MYB transcription factor involved in phosphate starvation signaling both in vascular plants and in unicellular algae. *Genes & Development*, 15(16), 2122–2133. https://doi.org/10.1101/GAD.204401.

Sakamoto, T., Inui, Y. T., Uraguchi, S., Yoshizumi, T., Matsunaga, S., Mastui, M., ... & Fujiwara, T. (2011). Condensin II alleviates DNA damage and is essential for tolerance of boron overload stress in *Arabidopsis*. *The Plant Cell*, 23(9), 3533–3546.

Sakamoto, T., Tsujimoto-Inui, Y., Sotta, N., Hirakawa, T., Matsunaga, T. M., Fukao, Y., ... & Fujiwara, T. (2018). Proteasomal degradation of BRAHMA promotes Boron tolerance in *Arabidopsis*. *Nature Communications*, 9(1), 5285.

Santos, E. F., Kondo Santini, J. M., Paixão, A. P., Júnior, E. F., Lavres, J., Campos, M., & Reis, A. R. dos. (2017). Physiological highlights of manganese toxicity symptoms in soybean plants: Mn toxicity responses. *Plant Physiology and Biochemistry*, 113, 6–19. https://doi.org/10.1016/J.PLAPHY.2017.01.022.

Scherer, H. W. (2001). Sulphur in crop production. *European Journal of Agronomy*, 14(2), 81–111.

Schnurbusch, T., Hayes, J., Hrmova, M., Baumann, U., Ramesh, S. A., Tyerman, S. D., ... & Sutton, T. (2010). Boron toxicity tolerance in barley through reduced expression of the multifunctional aquaporin HvNIP2; 1. *Plant Physiology*, 153(4), 1706–1715.

Schon, M. K., Novacky, A., & Blevins, D. G. (1990). Boron induces hyperpolarization of sunflower root cell membranes and increases membrane permeability to K^+. *Plant Physiology*, 93(2), 566–571.

Schulten, A., Pietzenuk, B., Quintana, J., Scholle, M., Feil, R., Krause, M., Romera-Branchat, M., et al. (2022). Energy status-promoted growth and development of *Arabidopsis* require copper deficiency response transcriptional regulator SPL7. *The Plant Cell*, 34(10), 3873–3898. https://doi.org/10.1093/PLCELL/KOAC215.

Séré, D., & Martin, A. (2020). Epigenetic regulation: Another layer in plant nutrition. *Plant Signaling and Behavior*, 15(1). https://doi.org/10.1080/15592324.2019.1686236/FORMAT/EPUB.

Serrano, R. (1989). Structure and function of plasma membrane ATPase. *Annual Review of Plant Biology*, 40(1), 61–94.

Shabbir, Z., Sardar, A., Shabbir, A., Abbas, G., Shamshad, S., Khalid, S., Natasha, Murtaza, G., Dumat, C., & Shahid, M. (2020). Copper uptake, essentiality, toxicity, detoxification and risk assessment in soil-plant environment. *Chemosphere*, 259. https://doi.org/10.1016/J.CHEMOSPHERE.2020.127436.

Shafiq, S., Zeb, Q., Ali, A., Sajjad, Y., Nazir, R., Widemann, E., & Liu, L. (2019). Lead, Cadmium and Zinc phytotoxicity alter DNA methylation levels to confer heavy metal tolerance in Wheat. *International Journal of Molecular Sciences*, 20(19). https://doi.org/10.3390/ijms20194676.

Shahzad, Z., & Amtmann, A. (2017). Food for thought: How nutrients regulate root system architecture. *Current Opinion in Plant Biology*, 39, 80–87. https://doi.org/10.1016/J.PBI.2017.06.008.

Sharma, N., Madan, B., Khan, M. S., Sandhu, K. S., & Raghuram, N. (2023). Weighted gene co-expression network analysis of nitrogen (N)-responsive genes and the putative role of G-quadruplexes in N use efficiency (NUE) in rice. *Frontiers in Plant Science,* 14. https://doi.org/10.3389/fpls.2023.1135675.

Shen, Y., Chi, Y., Lu, S., Lu, H., & Shi, L. (2022). Involvement of JMJ15 in the dynamic change of genome-wide H3K4me3 in response to salt stress. *Frontiers in Plant Science*, 13(September), 1009723. https://doi.org/10.3389/FPLS.2022.1009723/BIBTEX.

Singh, J.P., Dahiya, D.J. & Narwal, R.P. Boron uptake and toxicity in wheat in relation to zinc supply. *Fertilizer Research* 24, 105–110 (1990). https://doi.org/10.1007/BF01073228

Smith, A. P., Jain, A., Deal, R. B., Nagarajan, V. K., Poling, M. D., Raghothama, K. G., & Meagher, R. B. (2010). Histone H2A.Z regulates the expression of several classes of phosphate starvation response genes but not as a transcriptional activator. *Plant Physiology*, 152(1), 217. https://doi.org/10.1104/PP.109.145532.

Stout, P. R., & Meagher, W. R. (1948). Studies of the molybdenum nutrition of plants with radioactive molybdenum. *Science*, 108(2809), 471–473.

Su, J., Yao, Z., Wu, Y., Lee, J., & Jeong, J. (2022). Minireview: Chromatin-based regulation of iron homeostasis in plants. *Frontiers in Plant Science*, 13, 959840. https://doi.org/10.3389/FPLS.2022.959840.

Sun, S., Zhu, J., Guo, R., Whelan, J., & Shou, H. (2021). DNA methylation is involved in acclimation to iron-deficiency in rice (*Oryza sativa*). *The Plant Journal*, 107(3), 727–739. https://doi.org/10.1111/TPJ.15318.

Tejada-Jimenez, M., Llamas, A., Sanz-Luque, E., Galvan, A., & Fernandez, E. (2007). A high-affinity molybdate transporter in eukaryotes. *Proceedings of the National Academy of Sciences of the United States of America*, 104, 20126–20130. https://doi.org/10.1073/pnas.0704646104.

Trindade, I., Capitão, C., Dalmay, T., Fevereiro, M. P., & Santos, D. M. D. (2010). miR398 and miR408 are up-regulated in response to water deficit in *Medicago truncatula*. *Planta*, 231, 705–716.

Uchiyama, M., Fudaki, R., Kobayashi, T., Adachi, Y., Ukai, Y., Yoshihara, T., & Shimada, H. (2022). Rice OsHAK5 is a major potassium transporter that functions in potassium uptake with high specificity but contributes less to cesium uptake. *Bioscience, Biotechnology, and Biochemistry*, 86(11), 1599–1604. https://doi.org/10.1093/BBB/ZBAC152.

Unver, T., & Tombuloglu, H. (2020). Barley long non-coding RNAs (lncRNA) responsive to excess boron. *Genomics*, 112(2), 1947–1955.

Wang, S., Li, L., Ying, Y., Wang, J., Shao, J. F., Yamaji, N., Whelan, J., Ma, J. F., & Shou, H. (2020). A transcription factor OsbHLH156 regulates Strategy II iron acquisition through localising IRO2 to the nucleus in rice. *New Phytologist*, 225(3), 1247–1260. https://doi.org/10.1111/NPH.16232.

Wang, X., Arai, S., Song, X., Reichart, D., Du, K., Pascual, G., ... & Kurokawa, R. (2008). Induced ncRNAs allosterically modify RNA-binding proteins in cis to inhibit transcription. *Nature*, 454(7200), 126–130.

Warington, K. (1923). The effect of boric acid and borax on the broad bean and certain other plants. *Annals of Botany*, 37(148), 629–672.

Wei, W., Lu, L., Bian, X. H., Li, Q. T., Han, J. Q., Tao, J. J., Yin, C. C., Lai, Y. C., Li, W., Bi, Y. D., Man, W. Q., Chen, S. Y., Zhang, J. S., & Zhang, W. K. (2023). Zinc-finger protein GmZF351 improves both salt and drought stress tolerance in soybean. *Journal of Integrative Plant Biology*, 65(7). https://doi.org/10.1111/jipb.13474.

Widiez T., El Kafafi el S., Girin T., Berr A., Ruffel S., Krouk G., Vayssières A., Shen W. H., Coruzzi G. M., Gojon A., Lepetit M. (2011). High nitrogen insensitive 9 (HNI9)-mediated systemic repression of root NO_3- uptake is associated with changes in histone methylation. *Proceedings of the National Academy of Sciences of the United States of America*. 2011 Aug 9;108(32):13329–34. doi: 10.1073/pnas.1017863108. Epub 2011 Jul 25. PMID: 21788519; PMCID: PMC3156160

Xing, J., Wang, T., & Ni, Z. (2015). Epigenetic regulation of iron homeostasis in *Arabidopsis*. *Plant Signaling & Behavior*, 10(12), e1064574. https://doi.org/10.1080/15592324.2015.1064574.

Yang, J., Song, Z., Ma, J., & Han, H. (2020). Toxicity of molybdenum-based nanomaterials on the soybean–rhizobia symbiotic system: Implications for nutrition. *ACS Applied Nano Materials*, 3(6), 5773–5782.

Yin, W., Xiao, Y., Niu, M., Meng, W., Li, L., Zhang, X., Liu, D., et al. 2020. ARGONAUTE2 enhances grain length and salt tolerance by activating BIG GRAIN3 to modulate cytokinin distribution in rice. *The Plant Cell*, 32(7), 2292–2306. https://doi.org/10.1105/TPC.19.00542.

Yong-Villalobos, L., González-Morales, S. I., Wrobel, K., Gutiérrez-Alanis, D., Alan Cervantes-Peréz, S., Hayano-Kanashiro, C., Oropeza-Aburto, A., Cruz-Ramírez, A., Martínez, O., & Herrera-Estrellaa, L. (2015). Methylome analysis reveals an important role for epigenetic changes in the regulation of the *Arabidopsis* response to phosphate starvation. *Proceedings of the National Academy of Sciences of the United States of America*, 112(52), E7293–7302. https://doi.org/10.1073/PNAS.1522301112/SUPPL_FILE/PNAS.1522301112.SAPP.PDF.

Yue, H., Nie, X., Yan, Z., & Weining, S. (2019). N6-Methyladenosine regulatory machinery in plants: composition, function and evolution. *Plant Biotechnology Journal*, 17(7), 1194–1208. https://doi.org/10.1111/PBI.13149.

Zhang, D., Zhang, D., Zhang, Y., Li, G., Sun, D., Zhou, B., & Li, J. (2024). Insights into the epigenetic basis of plant salt tolerance. *International Journal of Molecular Sciences*, 25(21), 11698. https://doi.org/10.3390/IJMS252111698.

Zhang, M., & Liu, B. (2017). Identification of a rice metal tolerance protein OsMTP11 as a manganese transporter. *PLOS One*, 12(4), e0174987. https://doi.org/10.1371/JOURNAL.PONE.0174987.

Zhang, H., Zhu J.K (2011). RNA-directed DNA methylation. *Current Opinion in Plant Biology*. 2011;14(2):142–147. doi:10.1016/j.pbi.2011.02.003

Zhou, G. F., Zhang, L. P., Li, B. X., Sheng, O., Wei, Q. J., Yao, F. X., ... & Liu, G. D. (2019). Genome-wide identification of long non-coding RNA in trifoliate orange (*Poncirus trifoliata* (L.) Raf) leaves in response to boron deficiency. *International Journal of Molecular Sciences*, 20(21), 5419.

Zhou, H. R., Zhang, F. F., Ma, Z. Y., Huang, H. W., Jiang, L., Cai, T., ... & He, X. J. (2013). Folate polyglutamylation is involved in chromatin silencing by maintaining global DNA methylation and histone H3K9 dimethylation in *Arabidopsis. The Plant Cell*, 25(7), 2545–2559.

Zimmer, W., & Mendel, R. (1999). Molybdenum metabolism in plants. *Plant Biology*, 1(2), 160–168.

15 Epigenetic Regulation of Photoperiodic Flowering in Plants

Parama Majumdar and Suman Ganger

15.1 INTRODUCTION

Plants are sedentary organisms. For them to sustain in nature while surviving all the harsh as well as favorable conditions, it becomes imperative to have a continuous and meticulous interaction with the environment. Thus, we observe an extremely intricate system in plants to predict, adapt and respond to the environment, gauging multiple stimuli. Light is one such stimulus that acts as a primary regulator for growth and development of plants. This regulatory control is facilitated by the electromagnetic (EM) wavelengths ranging from ultraviolet (UV) to far-red areas (300–800 nm), which can function independently or in tandem and synergistically or antagonistically towards plant growth and development (Kotkar & Giri, 2020; Mirouze & Paszkowski, 2011).

These stimuli perceived from the external environment drive some vital transitions in the plant life cycle. The transition from the vegetative to the reproductive stage is one such critical process. In angiospermic plants, the onset of flowering is coordinated by a complex set of environmental cues that integrate with endogenous signals to trigger gene action. This action, in turn, is propelled through multiple hormone-driven cascades. Thus, the regulation of hormones, signaling and homeostasis are all vital checkpoints of this transition process. Working alone or in tandem, phytohormones promote flowering by such epigenetic regulation (Campos-Rivero et al., 2017a).

Comprehension of the function of epigenetic regulation in response to environmental stimuli has advanced significantly in the past few years. As the molecular mechanisms underlying chromatin folding and silencing are being understood, it is now possible to understand and predict new phenotypes (Thiebaut et al., 2019). It has also been demonstrated in certain instances that epigenetic changes can be permanently passed down to the next generations. Despite this, very little is known about how native plants respond to environmental changes in their natural habitat because the great bulk of research has been done using model plants, especially *Arabidopsis* (Hauser et al., 2011).

Chromatin compaction has been found to be affected by plant hormones, such as gibberellins, jasmonic acid, abscisic acid (ABA) and auxins that mediate their effect by DNA methylation and histone posttranslational modifications. Thus, these epigenetic changes have been implied to coordinate hormone signaling to navigate the onset of flowering (Achard et al., 2006). Studies have identified the importance of interactions among ethylene (ET), indole-3-acetic acid (IAA), cytokinins (Cks) and abscisic acid (ABA) for floral induction in *A. thaliana* (Campos-Rivero et al., 2017a; Domagalska et al., 2010a; Matsoukas, 2014).

15.2 PLAYERS IN FLORAL INDUCTION

A well-orchestrated confluence of six major pathways regulates the cascade of flowering response in angiosperms. The process is synchronized to the seasonal changes in day length and temperature by photoperiod and vernalization pathways; daily growth conditions are estimated by the ambient temperature pathway, and the age, autonomous and gibberellic acid pathways that act in a genetically programmed manner. Figure 15.1 summarizes the convergence of these six pathways.

15.2.1 Photoperiod Pathway and the Circadian Clock

The photoperiodic pathway represents one of the most conserved and extensively researched mechanisms of floral induction. It functions by transmitting information gathered from light stimuli and photoperiodic timing, thereby initiating the process of flowering. The flowering process in *Arabidopsis* is facilitated by long-day (LD) conditions and inhibited by the shorter days of winter. This response is regulated by the photoperiodic pathway, which operates within the leaves through a signaling cascade including GIGANTEA (GI) and the transcriptional regulator CONSTANS (CO) (Kardailsky et al., 1999; Kobayashi et al., 1999). CO functions as a transcription factor, which allows it to bind to DNA and modulate the expression of various genes. The levels of this protein are meticulously regulated throughout the day, with CO protein accumulating during long daylight hours in *Arabidopsis*, reaching its highest concentration at dusk. This daily regulation is essential for the process of photoperiodic flowering. CO serves as an upstream regulator of *FT*, primarily facilitating the activation of the *FT* gene expression. The FT (FLOWERING LOCUS T) protein is synthesized in the leaves and subsequently transported via the phloem to the shoot apex. FT functions as a mobile signal, often referred to as "florigen".

DOI: 10.1201/9781003589914-18

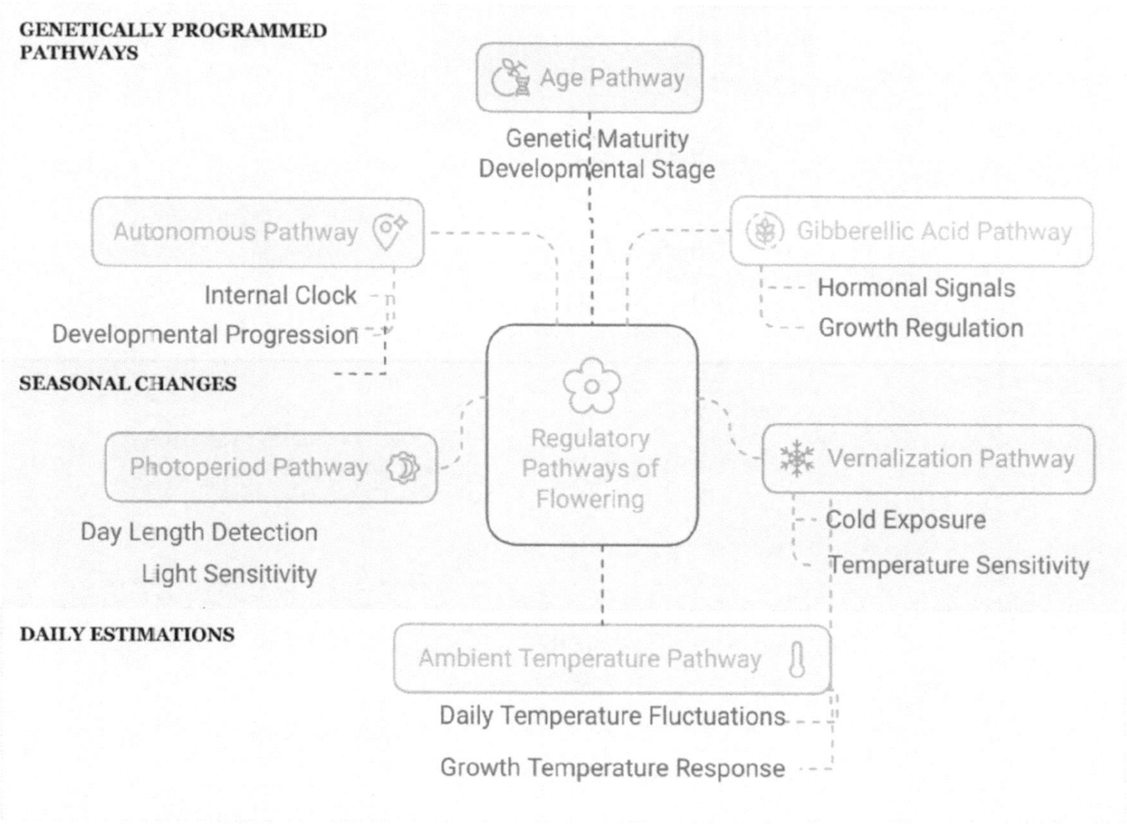

FIGURE 15.1 Regulatory pathways of flowering in angiosperms.

Upon arrival at the shoot apex, it initiates the transition from vegetative growth to reproductive growth, leading to flowering. FT also interacts with other proteins in the shoot apex to create a complex that activates the expression of genes critical for flower development. Figure 15.2 outlines the interrelationship of CO and FT and the flow of events that trigger floral development.

CO enhances flowering by triggering the transcription of the *FT* and *TWIN SISTER OF FT* (*TSF*) genes (Yamaguchi et al., 2005). Under long-day conditions, light enhances the interaction between the GI and a set of F-box ubiquitin ligases, which includes the FLAVIN-BINDING KELCH REPEAT F-BOX 1 (FKF1) protein. These interactions result in the stabilization of the F-box proteins, thereby promoting the degradation of various transcriptional repressors of CONSTANS (CO), notably a series of DOF transcription factors referred to as CYCLING DOF FACTORs (CDFs). CDFs play a role in inhibiting flowering by diminishing CO expression in the leaves. At the post-transcriptional level, CO is marked for degradation in the absence of light by the ubiquitin ligase CONSTITUTIVE PHOTOMORPHOGENIC 1 (COP1) and in the morning through a pathway activated by the photoreceptor phytochrome B (PHYB) (Nagano et al., 2020).

The overlaying mechanisms of transcriptional and post-transcriptional regulation ensure that CO activates the transcription of *FT* and *TSF* solely during prolonged daylight conditions. The expression and subsequent translation of FT occur within the leaves, after which the small FT protein is conveyed through the phloem to the meristem, initiating the floral development process. It is also possible for TSF to move through the phloem (Abe et al., 2005; Yamaguchi et al., 2005).

The circadian clock functions as a time-keeping system with a 24-hour cycle. This clock influences diurnal gene expression patterns in approximately 30% of the genes in *Arabidopsis*. It consists of three interconnected feedback loops. The primary loop includes the partially redundant transcription factors, viz., CIRCADIAN CLOCK ASSOCIATED 1 (CCA1) and LATE ELONGATED HYPOCOTYL (LHY), which inhibit the transcription of TIMING OF CAB 1 (TOC1). TOC1, which encodes a pseudoresponse regulator (PPR), promotes the transcription of LHY and CCA1. Similar feedback loops are present in the morning and evening, involving PRR7/PRR9 and GI, respectively (Wang et al., 2017). Numerous mutations that disrupt the circadian clock also affect flowering time; for instance, the circadian clock ensures that CO transcription reaches its peak late in the day, with GI amplifying this peak under long-day conditions. In genotypes where CO transcription occurs earlier in the day, the activation of CO by light is no longer confined to long days, resulting in FT transcription and flowering occurring earlier even under short-day (SD) conditions (Dunlap, 1996; Schaffer et al., 1998).

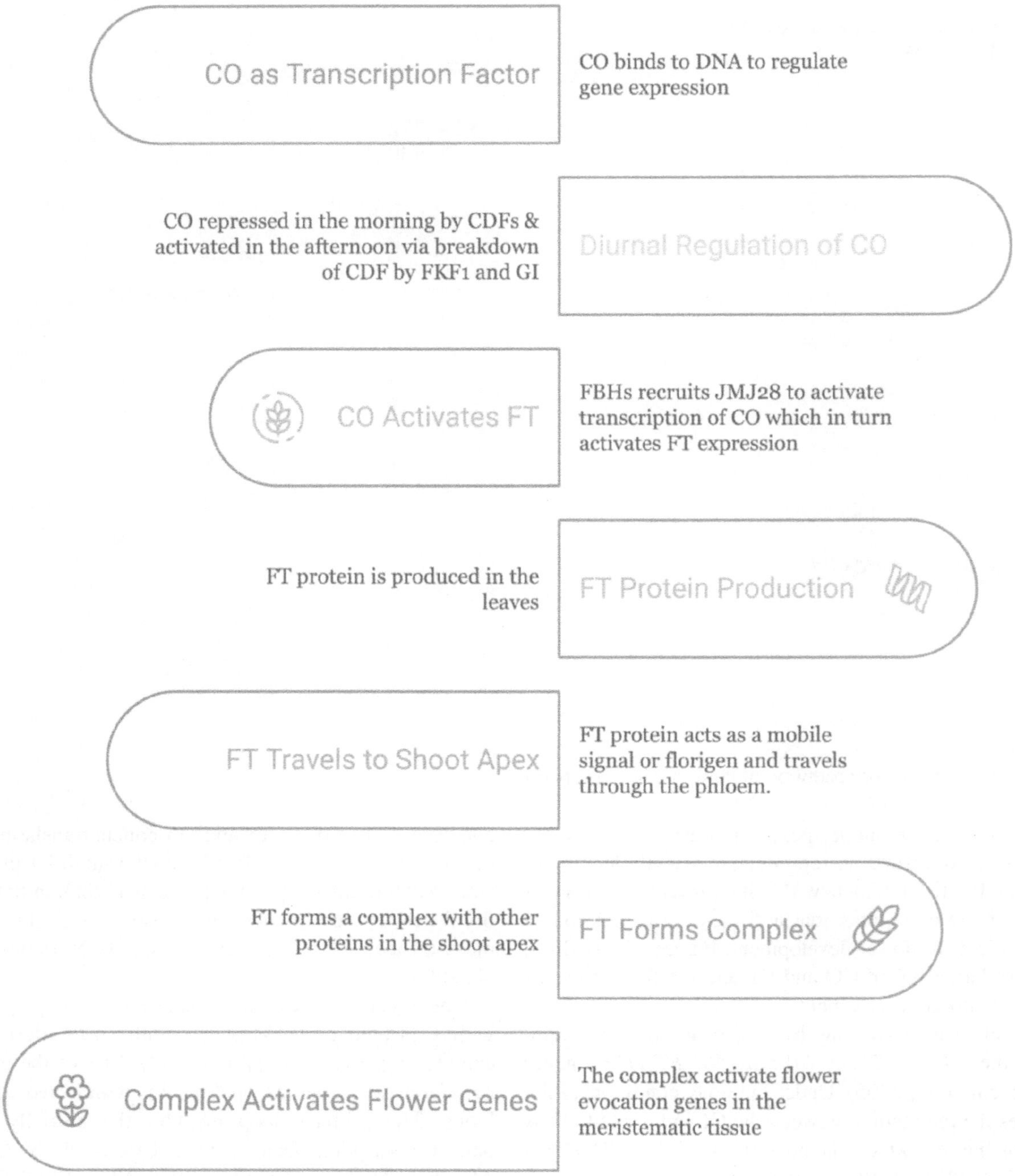

FIGURE 15.2 Regulation of FT via CO leading to transition from vegetative to reproductive development.

15.2.2 Vernalization Pathway

The pathway of vernalization promotes flowering by silencing *FLC* in reaction to extended periods of low temperature. FLOWERING LOCUS C (FLC) is a transcription factor belonging to the MADS-box family, functioning as a strong inhibitor of flowering. The activity of this factor varies across different *Arabidopsis* varieties and accounts for a significant portion of the differences in flowering time seen among these plants. In specific *Arabidopsis* varieties, exposure to cold induces antisense RNA transcription at the *FLC* locus, resulting in the synthesis of COOLAIR RNA, which subsequently silences *FLC* transcription. The pathways that regulate the floral repressor FLOWERING LOCUS C (FLC) are well-characterized (Matsoukas, 2014; Michaels, 2009).

15.2.3 Gibberellin Pathway

Gibberellin (GA) functions as a growth regulator that promotes flowering in *Arabidopsis*. The enzyme GIBBERELLIN 20 OXIDASE (GA20ox) is essential for GA biosynthesis, as it catalyzes several oxidation steps involving different precursors. Mutations that either impair this biosynthetic pathway or increase the degradation of GA lead to delayed flowering, particularly in short days. Before the onset of floral induction, the levels of bioactive GA (GA4) increase within the meristem (Fornara et al., 2010; Kinoshita et al., 2020; Singh & Roychoudhury, 2022). It is likely that GA4 is transported from other regions of the plant to the meristem, as mRNAs for GA20ox and other GA-biosynthetic enzymes do not significantly accumulate at this site. Nevertheless, these mRNAs do show increased levels in the petiole in response to changes in day length, indicating that the petioles may serve as a site for GA biosynthesis during the process of floral induction. The GA 2-oxidase (GA2ox) enzyme functions to deactivate GA through hydroxylation (Domagalska et al., 2010a; Mutasa-Göttgens & Hedden, 2009).

15.2.4 Ambient Temperature Pathway

Early flowering is exhibited by *Arabidopsis* plants when cultivated at elevated temperature, such as 23°C, compared to lower temperature like 16°C. SHORT VEGETATIVE PHASE (SVP), which is a MADS box transcription factor, has been found to be an integral part of this process, as *svp* mutants do not respond to variations in ambient temperature and bloom early in both temperature settings. At sub-ambient temperature, SVP inhibits the transcription of *FT*, thereby delaying flowering; however, the amount of *FT* mRNA rises significantly at higher temperature (Blümel et al., 2015; Hyun et al., 2016; Liu et al., 2009).

15.2.5 Effect of Age

Age of the plant acts as a significant genetically programmed switch. At maturity, the concentration of SQUAMOSA PROMOTER BINDING LIKE (SPL) transcription factors increases in plants. These SPLs promote flowering by activating the expression of several other transcription factors, such as LEAFY (LFY), FRUITFULL (FUL) and SOC1. The function of SPL proteins is suppressed by the microRNA miR-156, which is present in higher amounts in younger plants and decreases as the plant matures (Achard et al., 2004).

15.2.6 Meristem Responses

The process of floral induction involves a transformation of the shoot apical meristem, shifting from a vegetative meristem that facilitates leaf formation to an inflorescence meristem that is specifically focused on the development of flowers. This transition is characterized by an increase in height and a more domed shape of the meristem. These morphological alterations are accompanied by significant alterations in gene expression, notably the upregulation of *SOC1*, which encodes a MADS box transcription factor (Lee et al., 2008; Liu et al., 2008). The activation of SOC1 occurs swiftly when the plant transition occurs from short to long days, although it is also responsive to various other floral induction pathways; for instance, SOC1 is activated by gibberellins (GA) and the aging process, while its expression is inhibited by ambient temperature and vernalization pathways through direct repression by FLC and SVP. The activation of SOC1 over long-day conditions requires the presence of FT, which directly communicates with the meristem from the leaf. FT is essential for the initiation of flower development, as it interacts with the basic leucine zipper (bZIP) transcription factor FD and enhances the transcription of the MADS box factor APETALA1 (AP1) (Aukerman & Sakai, 2003; Madrid et al., 2021).

The activation of SOC1 triggers additional changes in gene expression within the meristem as it reshapes into an inflorescence meristem. This includes the expression of various transcription factors such as LEAFY (LFY), SPLs and AGAMOUS-LIKE 24 (AGL24). Within the inflorescence meristem, TFL1 (TERMINAL FLOWER 1), a protein similar to PEPB (phosphatidylethanolamine binding protein) that shares considerable sequence similarity with FT, inhibits the upregulation of *AP1* and *LFY* mRNA. This mechanism confines the expression of *AP1* and *LFY* to cells that are committed to developing into floral primordia, thereby ensuring the preservation of a reservoir of undifferentiated cells that support the indeterminate growth of the inflorescence meristem.

The action of all the pathways is brought together by the so-called *floral pathway integrator* (*FPI*) genes, which encompass FLOWERING LOCUS T (FT) (Kardailsky et al., 1999; Kobayashi et al., 1999), SUPPRESSOR OF OVEREXPRESSION OF CONSTANS 1 (SOC1) (Yoo et al., 2005) and AGAMOUS-LIKE24 (Lee et al., 2008; Liu et al., 2008). FLOWERING LOCUS C has also been extensively studied as a transcription factor that contains a MADS-domain and is identified as a key regulator of the flowering process in *Arabidopsis*. These integrators influence the *floral meristem identity (FMI)* genes *LEAFY (LFY)* (Lee et al., 2008), *FRUITFUL (FUL)* (Melzer et al., 2008) and *APETALA1 (AP1)* (Yamaguchi et al., 2005), leading to the initiation of flowering. Conversely, the pathways that facilitate floral induction also modulate the expression of floral repressors or translocatable florigen antagonists, referred to as antiflorigens (Matsoukas et al., 2012). Among these, the pathways that govern the floral repressor FLOWERING LOCUS C (FLC) are the most extensively studied (Michaels, 2009).

15.3 EPIGENETIC MECHANISMS UNDERLYING FLOWERING

The exploration of epigenetics in plants boasts a substantial and extensive history, beginning with early observations of non-Mendelian gene behaviors and advancing to pivotal findings regarding chromatin-modifying proteins and RNAs that facilitate gene silencing (Hauser et al., 2011). Plants exhibit exceptional proficiency in epigenetic regulation. They employ all the principal epigenetic mechanisms identified in eukaryotes, with the pathways involved being more intricately developed than in other biological groups. In plant genomes, DNA methylation is observed in CG, CHG and CHH sequence contexts (wherein H may be A, T or C), with patterns that indicate a balance among the enzymatic activities responsible for the installation, maintenance or removal of methylation. Similar to other eukaryotic organisms, histone-modifying enzymes play a crucial role in shaping epigenetic states in plants. These enzymes are encoded by relatively large gene families, which facilitate both diversified and overlapping functions (Feng et al., 2010; Zilberman et al., 2006).

Hormones, whether functioning independently or concomitantly, can facilitate the process of flowering through epigenetic regulation. Various plant hormones, including gibberellins, jasmonic acid, abscisic acid and auxins, significantly influence chromatin structure by affecting DNA methylation and posttranslational modifications of histones (Pikaard & Scheid, 2014). This suggests that epigenetic regulation may be integral to the flowering process, as mediated by hormonal activity. Furthermore, while miRNAs have traditionally been considered to operate separately from DNA methylation and histone modifications, their potential interactions with hormone signaling pathways—such as those involving auxins, gibberellins, ethylene, jasmonic acid and salicylic acid—should not be overlooked in the context of flowering regulation (Campos-Rivero et al., 2017a). Genetic screenings conducted in the model organism *Arabidopsis* have proven particularly fruitful, leading to the identification of over 130 epigenetic regulators to date (Pikaard & Scheid, 2014).

15.3.1 DNA Methylation

Methylation represents a fundamental aspect of epigenetic mechanisms. DNA methylation serves as a primary mechanism in the epigenetic regulation of various processes, including the response to environmental stimuli and the orchestration of embryonic development in plants. Among the various forms of methylation, DNA methylation, characterized by the covalent modification of the fifth cytidine site, is the most extensively researched epigenetic mechanism in plants. The three contexts of DNA methylation—CHH, CG and CHG (where H denotes a base, either A, T or C)—are critical for regulating gene expression in a temporal and spatial manner, as well as for condition-dependent phenotypic traits, such as variations in flower symmetry and phenotypic plasticity across different biological processes (Cubas et al., 1999; Kvist et al., 2018; Banerjee & Roychoudhury, 2018; Zhang et al., 2013). The significance of methylation in the pathways of vernalization and photoperiod has been progressively uncovered in recent years.

In *Arabidopsis*, the presence of far-red and blue light can promote flowering, whereas red light may delay this process. The primary regulators involved in the flowering pathway include CO, FT, CRYPTOCHROME 2 (CRY2), FHA (Forkhead-associated), GIGANTEA and FWA (FLOWERING WAGENINGEN), all of which are activated by photoperiod (Martinez-Zapater & Somerville, 1990; Searle & Coupland, 2004). Additionally, in *Arabidopsis*, external factors such as light (including photoperiod and light quality) and temperature (through vernalization) significantly affect the crucial transition to flowering, operating through the CO-FT and FLC pathways, respectively (Amasino, 2010a; Isabel Ortiz-Marchena et al., 2014).

Plants possess a distinct family of cytosine methyltransferases characterized by the presence of a chromodomain that interacts with methylated histones. The enzyme CHROMOMETHYLASE 3 (CMT3) plays a crucial role in the maintenance of CHG methylation. The chromodomain of CMT3 specifically binds to histone H3 that is dimethylated at lysine 9 (H3K9me2). This CHG methylation creates a binding site for the SET- AND RING-ASSOCIATED (SRA) domain of the H3K9 methyltransferase SUVH4, which subsequently leads to the dimethylation of H3K9. Consequently, CMT3 and SUVH4 form a self-reinforcing loop, wherein the marks of repressive DNA methylation and histone modification mutually specify each other, thereby sustaining an epigenetic state (Law et al., 2011).

Selection of photoperiod insensitive cultivars of wheat (*Triticum aestivum*) played a crucial role in "Green Revolution", as it enhanced the adaptability of the cultivars across a broader geographical spectrum. *Photoperiod-B1a* (*Ppd-B1a*) was one of the principal genes responsible for conferring photoperiod insensitivity in "Green Revolution" varieties, significantly contributing to the enhancement of wheat yield. An investigation into the mechanisms that govern the photoperiod insensitivity of Ppd-B1a alleles was conducted from an epigenetic standpoint, leading to the identification of two distinct methylation haplotypes of Ppd-B1. It was found that the haplotype with higher methylation (haplotype a) correlated with an increased number of gene copies and elevated expression levels of the *Ppd-B1* gene, resulting in earlier heading and enhanced photoperiod insensitivity (Sun et al., 2014).

DNA methylation plays a crucial role in the regulation of photoperiodic flowering in plants. MBD is a protein that identifies methylated DNA. In *Arabidopsis thaliana,* 13 putative genes encoding methyl-cytosine-binding domain (AtMBD) proteins have been identified and classified into different subgroups (Berg et al., 2003; Scebba et al., 2003). AtMBD5 and AtMBD7 proteins co-localize in vivo and demonstrate binding to the DECREASE IN DNA METHYLATION 1 (DDM1) protein in vitro (Zemach et al., 2008). Studies on loss of function mutations indicate that mutations in AtMBD9 result in a pleiotropic phenotype characterized by reduced histone acetylation and increased DNA methylation at the FLC locus, ultimately leading to an early flowering phenotype (Yaish et al., 2009).

The mutation of AtMBD8 postpones the flowering of *Arabidopsis* induced by photoperiods. The gene *AtMBD8*, encoding an *Arabidopsis* protein with a methyl-CpG-binding domain, and its mutant lines *atmbd8-1* and *2* were studied extensively; it was found that the *atmbd8-1* mutant plants are late flowering in both long- and short-day conditions and also correlated with an increase in the number of rosette leaves. The expression level of *FLC* did not show significant differences between the mutant and wild-type plants at any of the stages examined. This indicated that the *FT* genes downstream of FLC, specifically FT and SOC1, which are key promoters of flowering and are influenced directly or indirectly by CONSTANS (CO), were impacted. CONSTANS plays a crucial role in the perception of photoperiodic variations; mutations in *CO* lead to a late flowering phenotype exclusively under long-day (LD) conditions. The *atmbd8-1* mutants exhibited delayed flowering in both short- and long-day conditions, suggesting that *atmbd8-1* does not affect CO, a finding that was corroborated by quantitative PCR (qPCR) analysis. Consequently, it was concluded that AtMBD8 functions downstream of or in parallel to CO. The study also proposed that AtMBD8 positively regulates *FT* and indirectly influences SOC1 (Berg et al., 2003; Stangeland et al., 2009).

The mutation in *AtMBD9* leads to an early flowering phenotype, which is in sharp contrast to the late flowering phenotype observed in *atmbd8-1*. AtMBD9 has been demonstrated to regulate both shoot branching and flowering time (FlT). In *atmbd9* mutant plants, which exhibit earlier flowering compared to the wild type (wt), there is a down-regulation of flowering inhibitor, FLOWERING LOCUS C (FLC). The attenuation of FLC transcription in *atmbd9* mutants was linked to a notable decline in the acetylation levels of histones H3 and H4 within the FLC chromatin region (Peng et al., 2006; Scebba et al., 2003).

In a comprehensive study of the epigenetic and physiological impacts of gibberellin inhibitors on the floral transition of *Rhododendron*, it was observed that the application of plant growth regulators (PGRs) resulted in markedly elevated levels of 5-mdC in comparison to the control group during the final phase of floral bud development. The process of DNA methylation during the floral transition, along with the endogenous levels of polyamines (PAs), gibberellins and cytokinins, were examined following the addition of daminozide, paclobutrazol and chlormequat chloride, all of which come under the gibberellin synthesis inhibitors. The application of these gibberellin biosynthesis inhibitors led to notable alterations in the levels of polyamines, gibberellins and cytokinins, as well as in the overall DNA methylation status during the floral transition. A gradual increase in DNA methylation was noted, reaching its highest levels just before the flowering stage. Additionally, these changes in plant growth regulators and DNA methylation were linked to flower development. A positive relationship was also identified between flowering potential and the concentrations of dihydrozeatin riboside and GA9. As a result, DNA methylation, along with the concentrations of polyamines, gibberellins and cytokinins, was determined to be a significant predictive factor for floral capacity in *Rhododendron* spp. (Meijón et al., 2011).

In another study, cytidine analogue 5-azacytidine (azaC) was used to test the effect of methylation on photoperiodic flowering. 5-azacytidine (azaC) is a cytidine analogue that cannot be methylated at the fifth position of the ring. The incorporation of azaC in place of cytidine at the time of DNA replication leads to the hypomethylation of the DNA, and it is thus used as an epimutagen. azaC, when applied to germinating flax (*Linum usitatissimum* L.) seeds, was found to induce a number of early flowering lines (Fieldes & Amyot, 2011). Seeds of red- and green-leaved *Perilla frutescens* were subjected to azaC treatment and cultivated under long-day conditions, which are otherwise noninductive. Both varieties exhibited flowering induction as a result of azaC application. Additionally, the treatment of shoot apical meristems with azaC also prompted flowering. Consequently, it was suggested that DNA demethylation may play a role in the regulatory mechanism governing photoperiodic flowering (Takeno, 2010).

Photoperiodic flowering encompasses two primary processes: flower induction, where the stimulus for flowering originates in the leaves, and flower evocation, during which the shoot apical meristem reacts to the flowering signal conveyed from the leaves. Should the azaC treatment have prompted flower evocation, it would bypass the initial flower induction stage and lead to flowering quite early. However, the flowering triggered by azaC occurred later than that initiated by short-day conditions in control plants. This indicated that the process triggered by azaC was not flower evocation, but rather flower induction (Kondo et al., 2006).

A comprehensive genome-wide analysis was conducted to determine whether the DNA methylation state was influenced by inductive photoperiodic conditions, utilizing methylation sensitive-amplified fragment length polymorphism (MS-AFLP). Genomic DNA was extracted from the cotyledons of *Pharbitis nil* (syn. *Ipomoea nil*) subjected to short-day (SD) treatment and those from long-day (LD) control conditions following the SD treatment. The DNA was then digested with the restriction enzymes *Msp*I and *Hpa*II. The presence of methylated CCGG sequences would result in further fragmentation by these enzymes if demethylation occurred. Consequently, MS-AFLP is capable of identifying variations in the methylation state. DNA is hypermethylated under LD conditions and hypomethylated under SD conditions; thus, fragments exclusive to the LD sample indicate those that were demethylated due to the SD treatment. The observed patterns varied according to the photoperiodic conditions, indicating that the DNA methylation state was indeed modified by these conditions (Iwase et al., 2010; Kondo et al., 2010; Sherman & Talbert, 2011).

15.3.2 RNA Methylation

RNA methylation, akin to DNA methylation, plays a crucial role in the development of plants. Enzymatic covalent modifications of RNA serve as a crucial epigenetic mechanism that intricately and flexibly govern various cellular functions in eukaryotic organisms. Among these modifications, methylation at the N6 position of adenosine (m6A) is one of the most prevalent, found extensively in messenger RNA (mRNA), small nuclear RNA (snRNA), long noncoding RNA (lncRNA) and ribosomal RNA (rRNA). The presence of m6A on these RNA species influences RNA structure and alters protein–RNA interactions, thereby affecting RNA metabolism and a range of signaling pathways (Luo et al., 2022). N6-methyladenosine (m6A) and 5-methylcytidine (m5C) are the most commonly observed mRNA modifications in eukaryotic organisms. The adenosine bases in RNA can undergo temporary modifications through the addition of a methyl group, resulting in the formation of N6-methyladenosine (m6A). Recent research has revealed the significance of methylation in the processes of vernalization and photoperiod. Eukaryotic messenger RNA is subjected to N6-methyladenosine (m6A) modification, which constitutes approximately 50% of the posttranscriptional modifications of mRNA. These modifications are a reversible process, facilitated by methylases, commonly referred to as "writers", including mRNA ADENOSINE METHYLASE (MTA), METHYLTRANSFERASE B (MTB), FKBP12 INTERACTING PROTEIN 37 (FIP37), etc. These enzymes assemble posttranscriptionally on mRNAs that contain motifs such as RRACH. Demethylation enzymes, referred to as "erasers", including ALKBH9B and ALKBH10B, facilitate the removal of m6A modifications from adenine residues in mRNA. The "readers", predominantly belonging to the YTH family, identify these m6A-modified bases, thereby influencing the regulation of mRNA expression. Genes that possess m6A modifications typically exhibit enhanced expression and translation efficiency in comparison to their unmodified equivalents (Kierzek & Kierzek, 2003; Miao et al., n.d.; Yue et al., 2019).

Numerous studies have reported that in *Arabidopsis*, m6A methylation and m6A-mediated mRNA modifications participate in diverse developmental processes in plants, including embryogenesis, fruit ripening and the complex genetic mechanisms regulating flowering time (Duan et al., 2018a). In order to understand the role of m6A methylation of RNA, upland cotton cultivar TM-1 (photoperiod-insensitive) was studied along with the wild upland cotton yucatanense (photoperiod-sensitive). The differences in m6A alteration were detected in the genes associated with the photoperiod flowering pathway. A number of genes related to photoperiodic flowering exhibited higher levels of m6A modifications in yucatanense, indicating a correlation between m6A modification and photoperiod sensitivity (Yue et al., 2019).

The interplay between m6A modifications and mRNA expression is intricate. In certain cases, m6A modifications may decrease the stability of specific mRNA transcripts. The interaction with the m6A methyltransferase FIP37 enables CRY1 (cryptochrome 1) to modulate m6A modifications of genes associated with photomorphogenesis, such as PIF3, PIF4 and PIF5, resulting in a more rapid degradation of their corresponding transcripts. On the other hand, m6A modifications can also promote mRNA stability. Research has shown that in *Arabidopsis*, m6A modifications enhance mRNA stability by directly preventing localized ribonucleolytic cleavage of transcripts, leading to a greater abundance and improved translational efficiency of mRNAs that contain m6A compared to those that do not (Anderson et al., 2018; Yang et al., 2023).

A genetic screening aimed at identifying suppressors of late flowering in transgenic *Arabidopsis* plants that overexpress the miP1, a microprotein has led to the discovery of a novel allele of the FIONA1 (FIO1) m6A-methyltransferase. FIONA1 is functional as a U6 m6A methyltransferase in *Arabidopsis* (Wang et al., 2022). To investigate the early flowering phenotype observed in *fio1* mutant plants, a comprehensive methodology was employed, that included mRNA sequencing, nanopore direct RNA sequencing and meRIP sequencing to identify differentially expressed transcripts and differentially methylated RNAs. It was indicated that FIO1 is the long-sought methyltransferase responsible for the 3′-end methylation of the *FLOWERING LOCUS C (FLC)* transcript. Additionally, the genetic and biochemical analyses imply that 3′-methylation plays a role in stabilizing *FLC* mRNAs, while non-methylated FLC is susceptible to rapid degradation. In a genetic screening aimed at identifying early flowering mutants, it was discovered that the gene responsible for encoding the FIONA1 methyltransferase functions as a flowering repressor. An analysis of the alterations in the transcriptome and m6A-RNA-methylome indicated a significant number of differentially expressed genes and methylated RNAs. Notably, this includes the mRNA of the floral repressor FLOWERING LOCUS C, which exhibits a loss of m6A methylation in its 3′ untranslated region, leading to the destabilization of the FLC transcript (Sun et al., 2022; Zaccara et al., 2019).

15.3.3 Histone Modification

The N-terminal tails of histone proteins undergo various posttranslational modifications, including methylation, acetylation, phosphorylation, SUMOylation and ubiquitination. These modifications can modify the electronic charge and structural configuration of the histone tails, consequently affecting chromatin dynamics and gene expression (Banerjee & Roychoudhury, 2017; Steinbach, 2019).

15.3.3.1 Methylation

Histone methylation is an important and reversible post-translational modification. The primary site of methylation is on lysine (K) and arginine (R) residues located on the tails of histones H3 and H4. This reaction is facilitated by histone methyltransferases, with a possibility of three states of existence: mono-, di- and trimethylation (i.e. me1, me2 and me3, respectively). This process significantly influences a range of essential biological functions (Kornberg & Lorch, 1999; Luger & Hansen, 2005). The methylation of H3 lysine 9 (H3-K9) and H4-K20 has been found to be involved in the formation of heterochromatin; methylation of H3-K9 and H3-K27 has been linked to gene silencing and K4, K36 and K79 within histone H3 have been found to be associated with the positive regulation of gene expression (Martin & Zhang, 2005). TRITHORAX (TRX) and SET DOMAIN GROUP (SDG) are responsible for most of the histone methylations in *Arabidopsis,* mostly catalyzing H3K4me3, H3K36me3, H3K27me1, etc. (Liu et al., 2023). SDG8 has been found to affect the level of H3K36me3 in FT chromatin to influence flowering time (Shafiq et al., 2014; Shahbazian & Grunstein, 2007). Further, Jumonji domain-containing proteins, JMJ27, JMJ28 and JMJ29, have been recognized as demethylases for H3K9me1/2 that have been found to revoke the repression of gene expression (Berr et al., 2010; Hung et al., 2020). JMJ28 has the ability to reverse the repression of H3K9me2 at the *CO* promoter, thereby facilitating the activation of the *CO* expression profile. This process is aided by FLOWERING BHLHs (FBHs), which interact with the E-box cis-elements present in the *CO* promoter, leading to a significant increase in *CO* levels and promoting early flowering. In contrast, JMJ27 functions to inhibit flowering by repressing the transcription of *CO*. Additionally, the Morf-related genes, *MRG1* and *MRG2*, have been identified as readers of H3K36me3 and play a redundant role in the regulation of photoperiodic flowering (Dutta et al., 2017; Hung et al., 2020; Ito et al., 2012a).

15.3.3.2 Acetylation

Histone acetylation is another pivotal modification to regulate chromatin states and gene expression. Acetylation catalyzed by histone acetyltransferases (HATs) leads to the reduction of positive charge of lysine residues, subsequently avoiding the integration of histone tails with negatively charged DNA, thus opening corresponding DNA. Histone acetylation is generally considered a hallmark of gene activation. Deacetylation catalyzed by histone deacetylases (HDACs) can remove the acetyl groups from histones and usually function as repressors of gene transcription (Shahbazian & Grunstein, 2007).

An elevation in histone H3 tail modifications was observed in light conditions as opposed to dark conditions, with significant alterations noted in K9/K14ac, K4me2, K4me3, K27me3 and K36me3. These modifications represent the cumulative effects of all histone H3 changes, linked to both the upregulation and downregulation of gene expression in response to light exposure (Amasino, 2010b; Jan et al., 2011). Red- and far-red-light-driven responses through PhyA in *Arabidopsis* are brought about through an augmentation in acetylation of H3K9/14 (histone H3 lysine 9/14) and H3K27 and trimethylation of H3K4 at PhyA-controlled transcription sites. Likewise, PhyB action is documented to have possible links to HISTONE DEACETYLASE6 (HDA6) and METHYLTRANSFERASE 1 (MET1), which impact heterochromatin condensation levels (Jan et al., 2011).

In *Arabidopsis*, 12 histone deacetylase (HDACs) members have been identified, among which HDA5, HDA6, HDA9, HDA19 and HD2C have been observed to be associated with the flowering process. In *hda9* mutants, the acetylation of H3K9 and H3K27, along with the binding of RNA polymerase II to the floral activator AGL19, leads to the overexpression of AGL19, which subsequently elevates the levels of *FT* expression, leading to early flowering under short-day conditions. Furthermore, HDA9 interacts with the transcription factor ELONGATED HYPOCOTYL 5 (HY5), and this HDA9–HY5 complex interacts with the promoter of *COL5*, enhancing the deacetylation of H3K9 and H3K27, which in turn represses *COL5* transcription, ultimately inhibiting *FT* expression (Chu et al., 2022a; Yang et al., 2018; Yu et al., 2011).

15.3.3.3 Phosphorylation

Histone phosphorylation often occurs at serine (S) and threonine (T) sites of H3 and H2A tails, leading to chromosome condensation, gene activation or silencing, as well as DNA repair. Phosphorylation of H2A has been first reported in plants to regulate early seeding growth (Green et al., 1990; Huang et al., 2021). *Arabidopsis thaliana* MUT9P-LIKE-KINASE (MLK4) phosphorylates histone H2A on serine 95, a plant-specific modification in the histone core domain. Mutations in MLK4 resulted in delayed flowering under long-day conditions, while no significant phenotype was observed under short-day conditions. MLK4 has been shown to interact with CIRCADIAN CLOCK ASSOCIATED1 (CCA1), facilitating its binding to the promoter of *GIGANTEA* (*GI*). The loss of MLK4 function caused a delay in flowering by reducing the phosphorylation of H2A serine 95, which was accompanied by a decrease in the accumulation of H2A.Z and the acetylation of H4 at GI, ultimately leading to a reduction in GI expression. This finding shed light on the role of H2A serine 95 phosphorylation in promoting flowering time and indicates that the modulation of this phosphorylation by MLK4 is essential for the regulation of flowering time (Su et al., 2017).

15.3.3.4 SUMOylation

Small ubiquitin-like modifier (SUMO) is a polypeptide consisting of approximately 100 amino acids and is structurally similar to ubiquitin (Ub). The process of histone SUMOylation is orchestrated through a series of enzyme-dependent reactions that occur in multiple steps. This process involves SUMO E1 activating enzymes, E2 conjugating enzymes and E3 ligases. The E1 enzyme initiates the activation of SUMO proteins in an ATP-dependent manner, resulting in the formation of the E1–SUMO intermediate. Following this activation, SUMO is transferred to the E2 conjugating enzyme UBC9 via a transesterification reaction. Ultimately, the E3 ligases, specifically PIAS (protein inhibitor of activated STAT), facilitate the transfer of SUMO from UBC9 to designated lysine residues on the histone, which may include H4K5, K8, K12, K16, K20 and H2AK126 (Augustine & Vierstra, 2018; Park et al., 2011). Mutations in the SUMO protease ESD4 and its interacting partner NUA, which serves as a nuclear pore anchor, lead to an early onset of flowering regardless of day length. In the *esd4* mutant, the transcript levels of the floral repressor *FLC*

(*Flowering Locus C*) were diminished, while the expression of flowering-time genes that are typically inhibited by FLC was enhanced, thereby facilitating a quicker transition from vegetative growth to flowering (Murtas et al., 2003; Reeves et al., 2002; Xianfeng et al., 2007). Sumoylation and desumoylation play significant roles in abscisic acid (ABA) signaling within plants (Miura et al., 2009). The mutant of the SUMO E3 ligase, known as *siz1*, demonstrated early flowering under short-day conditions in *Arabidopsis*. The sumoylation of FLD by SIZ1 inhibits FLD activity, which in turn enhances the acetylation of histone 4 in FLC, leading to increased expression of FLC and a subsequent reduction in the expression of floral genes. In alignment with the phenotype observed in *siz1* mutants, the *sum1sum2* knockdown mutants also exhibited early flowering under short-day conditions. Conversely, the absence of SUM3 resulted in late flowering, while the overexpression of SUM3 induced early flowering (Park et al., 2011; van den Burg et al., 2010).

15.3.3.5 Polycomb-Group Proteins and Interacting Components

Polycomb-group (PcG) proteins are a family of proteins that primarily function as transcriptional repressors, interacting with each other to form large multi-protein complexes known as polycomb repressive complexes (PRC1 and PRC2), which work together to silence gene expression by modifying chromatin structure through histone methylation, primarily at lysine 27 on histone H3 (H3K27me3). The fundamental components of polycomb repressive complexes (PRCs), namely EMBRYONIC FLOWER 2 (EMF2) and VERNALIZATION 2 (VRN2), collaborate in various processes including seedling establishment, repression of FLOWERING LOCUS C (FLC) and the suppression of seed coat development prior to fertilization. Notably, only EMF2, rather than VRN2, demonstrates a synergistic interaction with SET DOMAIN GROUP 8 in the context of embryonic repression (Gendall et al., 2001; Jiang et al., 2008; Tang et al., 2012). In *Arabidopsis*, the presence of both the RING-finger domain and a newly characterized domain known as RAWUL (RING-finger and WD-associated ubiquitin-like) has been documented. Five homologs of RING-finger proteins have been identified: two AtRING1 proteins (AtRING1A and AtRING1B) and three AtBMI1 proteins (AtBMI1A, AtBMI1B and AtBMI1C). These proteins engage in mutual interactions and exhibit significant redundancy in their functions. AtRING1A and AtRING1B, along with AtBMI1A and AtBMI1B, are capable of catalyzing H2AK121ub in vitro, while AtBMI1A and AtBMI1B are essential for H2Aub in vivo. A specific subset of PRC2-target genes, including those involved in embryo maturation and *KNOX* genes, relies on AtRING1/AtBMI1-mediated H2Aub for their repression. Additionally, AtRING1 and AtBMI1 play a role in influencing flowering time (Bratzel et al., 2010; Sanchez-Pulido et al., 2008; Xu & Shen, 2008).

H3K27me3 chromatin in plants is characterized by a high presence of the canonical histone variant H3.1, while it shows a reduced presence of the transcription-related variant H3.3, as well as active modifications such as H3K4me3, H3K36me3 and the acetylation of lysine on histone H3 (H3Kac). Alongside this predominant H3K27me3 chromatin state, there are instances where H3K27me3 coexists with active histone modifications at specific loci. Notably, H3K27me3 and H3K4me3 are found to colocalize in certain regions, resulting in the formation of what is referred to as bivalent chromatin. Research on individual genes has also identified bivalent chromatin in plants, particularly in studies focused on the promoters of genes such as *FLC, FLOWERING LOCUS T* (*FT*), *KNUCKLES* (*KNU*) and *SPOROCYTELESS/NOZZLE* (*SPL/NZZ*). There is a pressing need for genome-wide mapping of bivalent chromatin in plants to assess the extent of this chromatin state. Furthermore, the flowering repressors FLC and FLOWERING LOCUS M (FLM) are known to interact with EMF1, facilitating the recruitment of a putative PRC1-like complex that establishes H3K27me3 and represses FT (Bratzel et al., 2010; Roudier et al., 2011; Sequeira-Mendes et al., 2014).

15.3.4 Chromatin Formation or Chromatin Remodeling

ATP-dependent chromatin-remodeling enzymes have the ability to modify the interaction between DNA and histone octamers, thereby influencing the accessibility of transcription factors and other DNA-binding proteins to the DNA. The SNF2 (sucrose non-fermenting 2) family of chromatin-remodeling proteins is classified within the SF2 group of DNA helicases, characterized by a helicase motif that comprises seven conserved blocks. These SNF2 proteins facilitate DNA translocation by harnessing energy from ATP hydrolysis, which enables them to mediate alterations in chromatin structure to enhance DNA accessibility (Clapier & Cairns, 2009). SWI/SNF (switch defective/sucrose nonfermentable), ISWI (imitation switch), CHD (chromodomain helicase DNA-binding), INO80 (inositol requiring 80) and SWR1 (SWI2/SNF2-related 1) subfamily SNF2 proteins represent the most extensively researched chromatin remodelers, typically operating within multi-subunit chromatin-remodeling complexes (Becker & Hörz, 2002; Clapier & Cairns, 2009). *Arabidopsis* possesses four SWI/SNF remodelers: BRM (BRAHMA), SYD (SPLAYED), MINU1 (MINUSCULE 1) and MINU2 (MINUSCULE 2). Notably, only BRM features a putative bromodomain located in its C-terminal region. These SWI/SNF remodelers play a crucial role in the regulation of essential processes related to plant growth and development, including the establishment and maintenance of meristems, cell differentiation, organ initiation, flower morphogenesis and the regulation of flowering time (Song et al., 2021). Regulatory networks involving transcription factors and chromatin remodelers play a crucial role in determining flowering time and flower development by integrating various internal and external signals (Wils & Kaufmann, 2017). These networks include pathways that respond to photoperiod, vernalization and temperature, which are essential for detecting long days, cold winters and ambient temperature. Additionally, they encompass pathways that react to internal factors, such as the age pathway and the gibberellin (GA) signaling pathway (Achard et al., 2004).

The core subunit components of the *Arabidopsis* SWR1 chromatin remodeling complex, which comprise PHOTOPERIOD-INDEPENDENT EARLY FLOWERING 1 (PIE1), ACTIN RELATED PROTEIN 6 (ARP6) and SWR1 COMPLEX 6 (SWC6)/SERRATED LEAVES AND EARLY FLOWERING (SEF), have been demonstrated to be crucial in governing the appropriate growth and development of various plant organs (Choi et al., 2016). Recent findings have shown that SWR1 plays a crucial role in regulating gene expression by creating distinct nucleosome structures. Specifically, it establishes a low accessibility configuration at the first nucleosome upstream of the transcription start site (TSS) and a high accessibility configuration at the first nucleosome downstream of the TSS (Dai et al., 2017; March-Díaz et al., 2007; Song et al., 2021). The deposition of H2A.Z at the transcription start site (TSS) of FLC by SWR1 is essential for the activation of FLC mediated by FRI. Evidence to support this assertion is that the functional impairment of SWR1 hinders the deposition of H2A.Z on FLC chromatin, which in turn leads to diminished expression and subsequent preponement of flowering (Choi et al., 2016). Additionally, research has indicated that mutations in *PIE1* markedly reduce FLC transcript levels, causing a transition from the winter-annual growth pattern to the accelerated flowering summer-annual growth pattern (Noh & Amasino, 2003).

A vital regulator of flowering time in *Arabidopsis* is the FT-encoded florigen protein (FT), which is activated by CONSTANS (CO) during extended daylight to encourage flowering (Corbesier et al., 2007). FT moves from the phloem to the shoot apical meristem (SAM), where it engages with the bZIP transcription factor FD to trigger the expression of floral identity genes, particularly LEAFY (LFY) and APETALA1 (AP1). This mechanism leads to the formation of the floral primordium, which eventually develops into three types of lateral floral organs: sepals, petals and stamens. LFY aids in the recruitment of SYD and BRM ATPases, which enhance the expression of the floral homeotic regulator AGAMOUS (AG) by removing H3K27me3 modifications from its second intron. The SWR1 complex is also involved in the regulation of *FT* expression. Any loss of function of SWR1 leads to temperature-insensitive activation of FT, resulting in early flowering (Wu et al., 2012). PICKLE (PKL), which is a member of the CHD3 subfamily II of ATP-dependent chromatin remodeling factors, has the ability to bind to both naked DNA and mononucleosomes (Ho et al., 2013). In *Arabidopsis*, PKL facilitates flowering by forming a physical interaction with CO, which assists in their binding to FT chromatin. This interaction increases H3 acetylation levels and enhances FT transcription in response to photoperiod. Furthermore, PKL can interact with ATX1, which helps to mitigate the repression of FT by polycomb group (PcG) proteins (Jing, Guo, & Lin, 2019; Jing, Guo, Zha, et al., 2019).

15.3.5 RNA Silencing

In long-day plants such as *Arabidopsis thaliana*, the expression of FT is known to be activated by the photoperiod pathway output, particularly during the dusk period of extended daylight. Recent studies have also highlighted the significance of FT silencing at various times throughout the day, which allows *Arabidopsis* to flower exclusively in response to long-day conditions. A plant-specific chromatin-silencing complex, referred to as EMBRYONIC FLOWER 1 (EMF1), plays a crucial role in repressing FT expression at times other than dusk during long days, thereby preventing its inappropriate temporal expression from interfering with the floral induction process associated with long days in *Arabidopsis* (He, 2015). The chromatin silencing of FT expression in *Arabidopsis* is mediated by the phloem EMF1c-PcG during long daylight periods. EMF1c-PcG is responsible for silencing FT expression at various times, excluding dusk, and this silencing is lifted only when the CO concentration attains a specific threshold close to dusk. CO plays a crucial role in interrupting the binding of EMF1-PcG to FT chromatin. The interrupted binding is represented by a dotted line, while solid lines with bars indicate repression or disruption (Wang, Gu, et al., 2014). The components of EMF1c, such as EMF1, LHP1, JMJ14 and BMI1s, are highly conserved across angiosperms. This conservation implies that the silencing of FT (or its orthologs) mediated by EMF1c may also be preserved in other plants that are sensitive to day length (Aubert et al., 2001a; Gaudin et al., 2001; Zhou & Ma, 2008).

In the short-day plant *Oryza sativa* (rice), the expression of an FT ortholog known as Hd3a, which is crucial for the induction of flowering in rice, is initiated at dawn during short-day conditions (8 hours of light and 16 hours of darkness) (Tamaki et al., 2007a). The phenomenon arises from the coordinated activity of a rice ortholog of the *Arabidopsis* CO and a B-type response regulator identified as Ehd1 (Kojima et al., 2002; Song et al., 2015).

The identification and functional analysis of long noncoding RNAs (lncRNAs) have highlighted the significance of the noncoding transcriptome in the survival and overall fitness of organisms. LncRNAs have emerged as important regulators in various biological processes, including development, disease and differentiation in animals. These molecules can be derived from intergenic or intronic regions, or from the complementary strand of coding genes, functioning as natural antisense transcripts. Functional investigations have uncovered mechanisms through which lncRNAs exert their effects, including chromatin remodeling, alterations in nuclear organization, RNA processing, RNA stability, translational regulation, assembly of protein complexes and the localization of proteins within the cell

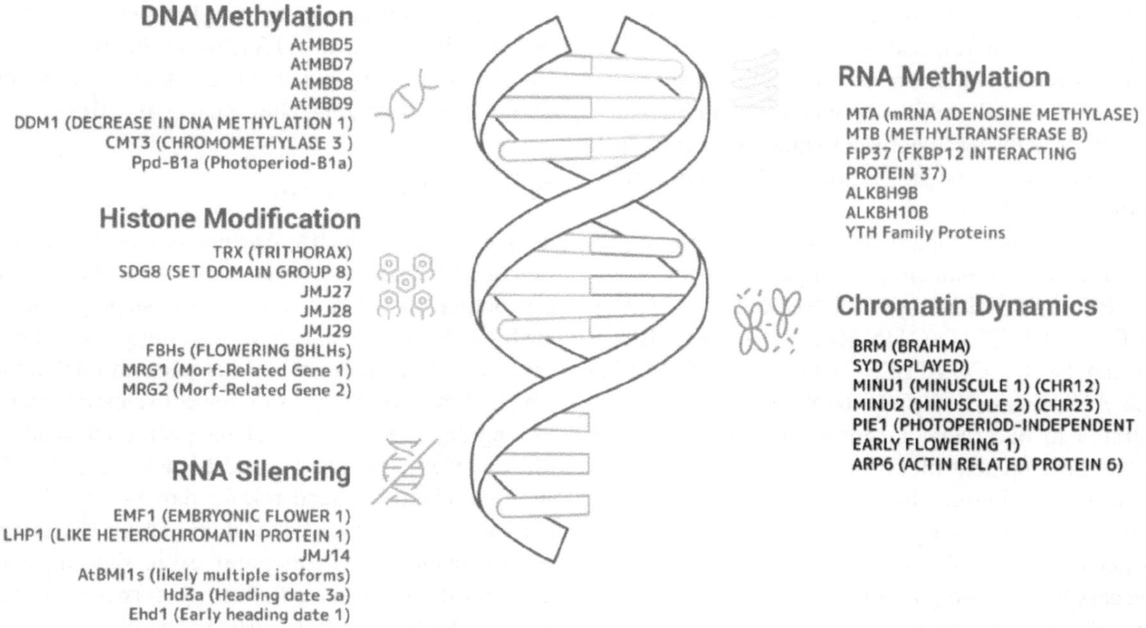

FIGURE 15.3 Epigenetic mechanisms in the foreground of flowering transition.

(Bardou et al., 2014; Ha & Kim, 2014; Heo & Sung, 2011; Rinn & Guttman, 2014). These diverse functions enable lncRNAs to engage in both long-term and dynamic biological processes, exemplified by light-responsive lnc-NATs in *Arabidopsis*. In this plant, the circadian clock is governed by several interconnected transcriptional loops, with chromatin remodeling playing a crucial role in establishing robust circadian rhythms (Hemmes et al., 2012; Malapeira et al., 2012).

In the realm of plant biology, the identification of long noncoding RNAs (lncRNAs) has progressed further with increasing evidence regarding their expression patterns—specific to tissue types, environmental conditions and developmental stages—indicating that they play significant biological roles. In *Arabidopsis*, lncRNAs contribute to the flowering response by modulating the transcription of FLOWERING LOCUS C (FLC) (Ariel et al., 2015; Bazin & Bailey-Serres, 2015). HIDDEN TREASURE 1 (HID1) plays a moderate role in regulating the expression of the PHYTOCHROME-INTERACTING FACTOR 3 (PIF3) transcription factor. Additionally, the AUXIN REGULATED PROMOTER LOOP (APOLO) influences the expression of PINOID (PID) by modulating the dynamics of chromosome loops, which in turn impacts auxin signaling (Ariel et al., 2014; Wang, Fan, et al., 2014).

The long noncoding RNA CDF5 LONG NONCODING RNA (FLORE), which is expressed in an antisense orientation to the *CYCLING DOF FACTOR 5* (*CDF5*) transcript, has been recognized as a primary mechanism for gene silencing in *Arabidopsis*. Additionally, the circadian oscillation of FLORE exhibits an antiphasic relationship with CDF5. By altering the levels of the FLORE transcript through T-DNA insertion mutagenesis or tissue-specific overexpression, an anti-parallel behavior has been noted, indicating a reciprocal inhibitory interaction. Furthermore, FLORE has been shown to operate not only in cis, influencing CDF5, but also in trans by modulating other CDFs, including CDF1 and CDF3. The manipulation of transcript levels for both *Arabidopsis* CYCLING DOF FACTOR 5 (CDF5) and CDF5 LONG NONCODING RNA (FLORE) has implications for flowering under long-day (LD) conditions. CDF proteins were found to directly suppress the transcription of CO and FT, thereby delaying flowering. A thorough analysis of the flowering time phenotype was conducted for all *CDF5* and *FLORE* mutants and transgenic lines cultivated under various photoperiods. Notably, a reduction in *CDF5* levels was observed in *cdf5-5'utr* mutants, which resulted in a slightly earlier flowering phenotype (Foo et al., 2016; Hemmes et al., 2012; Henriques et al., 2017; Malapeira et al., 2012). Figure 15.3 summarizes the primary genes working under different epigenetic modifiers, and the Table 15.1 enlists these genes with their function and putative effect on flowering.

15.4 ACTION OF PLANT GROWTH REGULATORS

Plant hormones are essential regulators of growth, development and environmental response and recently recognized as key players in modulating epigenetic mechanisms that control gene expression and phenotypic adaptability. Numerous studies underscore the significance of interactions among various plant hormones, including ethylene (ET), indole-3-acetic acid (IAA), cytokinins (Cks) and abscisic acid (ABA), in the process of flower induction in *Arabidopsis thaliana* (Domagalska et al., 2010b; Matsoukas, 2014). It is well established that hormonal signaling plays a crucial role in flowering by regulating key

TABLE 15.1
Summary of epigenetic modifiers, gene/proteins and their mechanism of action

Epigenetic modifiers	Gene/protein	Function	Effect on flowering	Species	References
DNA methylation	AtMBD5	Methyl-binding domain protein 5	Binds to DDM1 to regulate flowering time	*Arabidopsis thaliana*	Zemach et al. (2008)
	AtMBD7	Methyl-binding domain protein 7	Binds to DDM1 along with AtMBD5 to regulate flowering time	*Arabidopsis thaliana*	Zemach et al. (2008)
	AtMBD8	Methyl-binding domain protein 8	Positively regulates FT and indirectly influences SOC1	*Arabidopsis thaliana*	Berg et al. (2003)
	AtMBD9	Methyl-binding domain protein 9	Regulates flowering time	*Arabidopsis thaliana*	Peng et al. (2006); Scebba et al. (2003)
	DDM1 (DECREASE IN DNA METHYLATION 1)	Maintains DNA methylation, particularly at CG and CHG sequences	Regulates flowering time	*Arabidopsis thaliana*	Zemach et al. (2008)
	CMT3 (CHROMOMETHYLASE 3)	Cytosine methyltransferases maintains methylation, particularly CHG sequences	Activation of histone modifiers H3K9 methyltransferase	*Arabidopsis thaliana*	Law et al. (2011)
	Ppd-B1a (Photoperiod-B1a)	Confers photoperiod insensitivity when hypermethylated	Higher expression leads to earlier heading and enhanced photoperiod insensitivity	*Triticum aestivum*	H. Sun et al. (2014)
RNA methylation	MTA (mRNA ADENOSINE METHYLASE)	"Writer" enzyme: adds m6A (methyl group) to adenosine in RNA	Affects flowering time by modifying RNA expression of flowering-related genes	*Arabidopsis thaliana*	Chen et al. (2010a); Duan et al. (2018b)
	MTB (METHYLTRANSFERASE B)	"Writer" enzyme: adds m6A (methyl group) to adenosine in RNA	Affects flowering time by modifying RNA expression of flowering-related genes	*Arabidopsis thaliana*	Chen et al. (2010a); Duan et al. (2018b)
	FIP37 (FKBP12 INTERACTING PROTEIN 37)	"Writer" enzyme: adds m6A (methyl group) to adenosine in RNA	Affects flowering time by modifying RNA expression of flowering-related genes	*Arabidopsis thaliana*	Chen et al. (2010a); Duan et al. (2018b)
	ALKBH9B	"Eraser" enzyme: removes m6A (methyl group) from adenosine in RNA	Affects flowering time by modifying RNA expression of flowering-related genes	*Arabidopsis thaliana*	Chu et al. (2022b)
	ALKBH10B	"Eraser" enzyme: removes m6A (methyl group) from adenosine in RNA	Affects flowering time by modifying RNA expression of flowering-related genes	*Arabidopsis thaliana*	Chu et al. (2022b)
	YTH family proteins	"Reader" proteins: recognize and bind to m6A-modified RNA	Affect flowering time by influencing mRNA expression of genes involved in flowering pathways	*Arabidopsis thaliana*	Chu et al. (2022b)
RNA silencing	EMF1 (EMBRYONIC FLOWER 1)	Component of EMF1c-PcG complex; represses *FT* expression	Prevents flowering under non-inductive conditions (e.g., wrong time of day)	*Arabidopsis thaliana*	Aubert et al. (2001b)
	LHP1 (LIKE HETEROCHROMATIN PROTEIN 1)	Component of EMF1c–PcG complex; involved in chromatin silencing	Contributes to *FT* repression	*Arabidopsis thaliana*	Gaudin et al. (2001)
	JMJ14	Component of EMF1c–PcG complex; likely histone demethylase	Contributes to *FT* repression	*Arabidopsis thaliana*	Corbesier et al. (2007)
	AtBMI1s (likely multiple isoforms)	Component of EMF1c–PcG complex; polycomb group protein involved in gene silencing	Contributes to *FT* repression	*Arabidopsis thaliana*	Li et al. (2011)
	Hd3a (heading date 3a)	FT ortholog; promotes flowering	Essential for short-day flowering in rice	*Oryza sativa*	Tamaki et al. (2007b)
	Ehd1 (early heading date 1)	B-type response regulator	Involved in regulating Hd3a expression in rice	*Oryza sativa*	Kojima et al. (2002)

(Continued)

TABLE 15.1 CONTINUED

Epigenetic modifiers	Gene/protein	Function	Effect on flowering	Species	References
Chromosome remodeling	BRM (BRAHMA)	Chromatin remodeling; influences DNA accessibility	Regulates flowering time	*Arabidopsis thaliana*	Archacki et al. (2013a)
	SYD (SPLAYED)	Chromatin remodeling; influences DNA accessibility	Regulates flowering time	*Arabidopsis thaliana*	Vagner et al. (2001)
	MINU1 (MINUSCULE 1) (CHR12)	Chromatin remodeling; influences DNA accessibility	Regulates flowering time	*Arabidopsis thaliana*	Sang et al. (2012)
	MINU2 (MINUSCULE 2) (CHR23)	Chromatin remodeling; influences DNA accessibility	Regulates flowering time	*Arabidopsis thaliana*	Sang et al. (2012)
	PIE1 (PHOTOPERIOD-INDEPENDENT EARLY FLOWERING 1)	Component of SWR1 complex; regulates gene expression via nucleosome structure	Crucial for growth and development, including flowering	*Arabidopsis thaliana*	Noh & Amasino (2003)
	ARP6 (ACTIN RELATED PROTEIN 6)	Component of SWR1 complex; regulates gene expression via nucleosome structure	Crucial for growth and development, including flowering	*Arabidopsis thaliana*	Deal et al. (2005)
Histone modification	TRX (TRITHORAX)	Histone methyltransferase	Promotes flowering (generally)	*Oryza sativa*	Choi SC et al. (2014)
	SDG8 (SET DOMAIN GROUP 8)	Histone methyltransferase	Influences flowering time	*Arabidopsis thaliana*	Dong et al. (2008)
	JMJ27	Histone demethylase	Releases gene repression	*Arabidopsis thaliana*	Dutta et al. (2017)
	JMJ28	Histone demethylase	Activates CO expression, promotes early flowering	*Arabidopsis thaliana*	Hung et al. (2020)
	JMJ29	Histone demethylase	Releases gene repression	*Arabidopsis thaliana*	Hung et al. (2020)
	FBHs (FLOWERING BHLHs)	Transcription factor	Increases CO levels, promotes early flowering	*Arabidopsis thaliana*	Ito et al. (2012)
	MRG1 (Morf-Related Gene 1)	Reader of histone modifications	Regulates photoperiodic flowering	*Arabidopsis thaliana*	Bu et al. (2014)
	MRG2 (Morf-Related Gene 2)	Reader of histone modifications	Regulates photoperiodic flowering (redundant role with MRG1)	*Arabidopsis thaliana*	Bu et al. (2014)

genes such as *FLC, CO* and *FT*, for example, salicylic acid (SA) has been shown to influence the transcription of *CO, FLC, FT* and *SOC1* (Martínez et al., 2004). Notably, research has revealed that the late flowering phenotype observed in SA-deficient plants is associated with a 2.0–3.0-fold increase in *FLC* expression, alongside a reduction in FT levels compared to wild-type plants under both long-day (LD) and short-day (SD) conditions. Additionally, the expression dynamics of these genes are linked to chromatin modifications. The regulation of *FLC* and *FT* expression in *Arabidopsis* is governed by epigenetic mechanisms. Several studies have emphasized the role of the polycomb repressive complex 2 (PRC2) in silencing the floral repressor FLC locus through the histone modification H3K27me3 and the decrease of H3K4me2 levels at this locus. The *FLD* gene encodes a histone demethylase that facilitates the enzymatic demethylation of H3K4me2 and promotes the deacetylation of histone H4 within the FLC chromatin (Berr et al., 2015; Blümel et al., 2015; Kang et al., 2015; Martínez et al., 2004). Furthermore, the sumoylation and desumoylation activities of FLD have been identified as regulators of histone acetylation and deacetylation (Liu et al., 2007). Epigenetic mechanisms that modulate transcription through alterations in chromatin structure, influenced by DNA methylation, histone posttranslational modifications and chromatin remodeling, have been documented in the context of flowering (Liu et al., 2007; Song et al., 2021; Sun et al., 2014) and hormonal regulation (Campos-Rivero et al., 2017b; Yamamuro et al., 2015).

Several epigenetic mechanisms influence the transcription of genes involved in auxin biosynthesis. Comprehensive whole-genome tiling arrays reveal that H3K27me3 plays a repressive role over extensive genomic regions associated with auxin metabolism, including the *YUCCA (YUC), CYTOCHROME P450 (CYP) and TRYPTOPHAN AMINOTRANSFERASE 1/TRYPTOPHAN AMINOTRANSFERASE-RELATED (TAA1/TAR)* genes (Solanki & Shukla, 2023). This modification is primarily linked to the polycomb group pathway, which epigenetically suppresses gene expression. There are two polycomb repressive complexes (PRCs); PRC2 functions as a histone writing complex that deposits H3K27me3, which is subsequently recognized by PRC1. This process leads to the addition of ubiquitin on H2A, marking gene repression and further repression of the locus by additional mechanisms (Baile et al., 2022; Bieluszewski et al., 2021). Another significant epigenetic factor is the AUXIN-REGULATED PROMOTER LOOP (APOLO), which plays a crucial role in regulating both auxin biosynthesis and transport. APOLO connects LHP1 with DNA methylation through its interaction with the methylcytosine-binding protein VARIANT IN METHYLATION 1 (VIM1). This complex is capable of facilitating both RNA-directed DNA methylation (RdDM) and polycomb-mediated repression of the YUC2 locus. Additionally, YUC2 has been shown to be targeted by small RNAs, including siRNAs and miRNAs, in both direct and indirect manners (Ariel et al., 2020; Feng & Lu, 2017). In contrast to polycomb repression, the demethylation of H3K27me3 can be achieved by JUMONJI DOMAIN-CONTAINING (JMJ) proteins, converting it to H3K27me2 or further (Cui et al., 2016). Likewise, histone acetylation plays a role in altering plant chromatin structure and gene expression. The levels of histone acetylation are modulated by HISTONE ACETYLTRANSFERASES (HATs) and HISTONE DEACETYLASES (HDACs), which have been shown to influence the expression of related genes.

The GA signaling pathway comprises three primary elements: DELLA proteins, the GA-INSENSITIVE DWARF1 (GID1) receptor and the F-box protein GA-INSENSITIVE DWARF2/SLEEPY1 (GID2/SLY1). In the absence of gibberellins (GAs), DELLA proteins act to inhibit responses mediated by GAs. However, when GAs are present, they bind to the GID1 receptor, which facilitates their degradation via the proteasomal pathway (Gomi et al., 2004; Yan et al., 2021). Additionally, DELLA proteins engage with various components of chromatin-remodeling complexes, such as BRM, SWI3C and PKL, leading to either direct or indirect modifications of chromatin structure. The relationship between epigenetic factors and components of GA signaling is exemplified by the chromatin-remodeling protein PKL (Archacki et al., 2013b; Zhang et al., 2023). Furthermore, DELLA proteins also form direct interactions with histone H2A, which is crucial for the transcriptional regulation dependent on DELLA (Huang et al., 2023).

Recent studies indicate a significant role of histone posttranslational modifications in the regulation of abscisic acid (ABA) responses. It has been observed that both ABA and stress conditions enhance the presence of gene activation markers, specifically histone H3K9K14 acetylation and H3K4 trimethylation, while simultaneously reducing the gene repression marker, H3K9 dimethylation, associated with ABA and abiotic stress-responsive genes (Chen et al., 2010b). In addition to histone acetylation, histone methylation and chromatin-remodeling complexes, such as the SWI3B component of SWI/SNF, have been identified as essential for the activation of ABA-responsive genes and the pathways dependent on ABA (Benhamed et al., 2006; Ding et al., 2011). Moreover, the effects of ABA are influenced by microRNAs (miRNAs), and several miRNAs are also regulated by ABA treatment. Notably, the ABA-inducible Gma-miRNA172c from soybean has been shown to enhance tolerance to drought and salt stress. Transgenic *Arabidopsis* plants that overexpress Gma-miRNA172c exhibit increased responsiveness to ABA, characterized by an early flowering phenotype and elevated expression of both ABA signaling and flowering-promoting genes (Li et al., 2016).

The role of salicylic acid (SA) in the regulation of flowering has been recognized for many years. The application of exogenous SA to wild-type *Arabidopsis* plants resulted in a reduction of *FLC* transcript levels, while UV-C light exposure led to SA accumulation that activated *FT* expression. Interestingly, although SA seems to inhibit *FLC* expression, this mechanism is not critical for the late flowering phenotype observed in SA-deficient plants (Martínez et al., 2004).

Genetic analyses of the interactions between these genes and SA during photoperiod conditions revealed that exogenous SA can reverse the late flowering phenotype in the *co-1* mutant, but this effect is not observed in the *soc1* mutant under long-day conditions. This evidence implies that SA regulates flowering through interactions with the photoperiod-dependent pathway via a distinct CO-independent mechanism (Campos-Rivero et al., 2017a).

15.5 INHERITANCE OF EPIGENETIC CHARACTERS

Epigenetic modifications, such as DNA methylation, can be altered, leading to an epigenetic response. For transgenerational epigenetic inheritance to occur, it is essential that these modifications can be passed on to offspring. Consequently, it can be inferred that epigenetic marks may be transmitted through both mitosis and, in some cases, meiosis. The variation in methylation patterns of the same gene across different plant species is referred to as epialleles. These epialleles vary in the quantity or arrangement of methylated nucleotides within specific gene sequences, and it is crucial to understand that distinct epialleles can lead to different heritable phenotypes in subsequent generations (Hauser et al., 2011). It is also important to acknowledge that plants can perceive environmental conditions during their vegetative growth, which may induce epigenetic changes within a cell lineage that can subsequently give rise to a germline. Research utilizing the model organism *Arabidopsis* has demonstrated that stress-induced transgenerational responses are contingent upon alterations in DNA methylation (Paszkowski & Grossniklaus, 2011).

While epigenetic states tend to be stable, they can revert at certain frequencies, indicating that transgenerational epigenetic inheritance (TEI) possesses a stochastic element that must be taken into account. This inheritance involves the transference of traits that are not dictated by the DNA sequence, specifically meiotic epigenetic inheritance. In plants, the germ line undergoes differentiation later in development, as gametes are produced by haploid gametophytes. Therefore, TEI necessitates the maintenance of epigenetic states across numerous mitotic DNA replication cycles, meiosis and the post-meiotic mitoses that occur during gametophyte development. The mechanisms that ensure the continuity of epialleles throughout these various developmental stages remain inadequately understood, with the notable exception of cytosine methylation within a CG sequence context. The replication of CG methylation patterns is facilitated by the maintenance METHYLTRANSFERASE1 (MET1), which utilizes post-replicative, hemimethylated DNA to transfer methylation to the newly synthesized strand (Saze et al., 2003).

While it has been recently proposed that MET1 is downregulated in female gametes, it remains uncertain whether this observation accurately represents MET1 activity or the degradation of the associated complex fusion protein. Furthermore, the apparent reduction of MET1 activity in gametes appears to contradict genetic findings (Saze et al., 2003). Consequently, in the context of plants, there is currently insufficient evidence to support the notion of transgenerational resetting of CG methylation, a phenomenon that has been observed in mammals.

15.6 CONCLUSION

The transition from vegetative to flowering phase is a pivotal shift in the life cycle of angiosperms, necessitating a complex interaction of environmental factors, hormonal signals and detailed genetic regulation. As discussed in this review, this transition involves a meticulously coordinated series of events, with multiple pathways interacting to contribute essential elements to the overall mechanism. Although our comprehension of these processes has advanced considerably, especially through research on model organisms such as *Arabidopsis*, there remains a significant gap in our understanding of how these mechanisms function in wild plants in their natural environment. The photoperiodic pathway illustrates the capacity of the plant to detect and respond to variations in day length, utilizing the circadian clock and key components such as GI, CO and FT to translate this information into a specific developmental outcome. Epigenetic modifications serve as a means to connect environmental factors, such as light duration and temperature, with internal developmental signals. These complex interactions underscore the intricate regulatory network that controls flowering. Continued investigation into the specific functions and relationships of these hormones, along with a more profound understanding of the epigenetic processes involved, is expected to yield important insights into the complexities of flowering regulation (Deng et al., 2011; Wang & Sung, 2024). Epigenetic research provides significant opportunities for enhancing crop development by allowing precise adjustments to gene expression without modifying the DNA sequence itself. This approach can enhance stress resilience, increase yields and improve adaptability to changing environmental conditions. With the global population on the rise, there is a pressing need for climate-resilient crops that can effectively adjust to shifting climates. As a result, developing high-yield crop varieties with excellent agronomic traits has become a significant challenge in agricultural research. These innovations are vital for increasing yield and improving resistance to adverse conditions, ultimately benefiting farmers over time. Most agronomic traits are quantitative and influenced by complex genetic mechanisms, making crop improvement efforts more challenging. Epigenetics refers to heritable changes in gene expression that occur through mechanisms like DNA methylation, histone modifications or RNA interference, rather than alterations to the DNA sequence itself. These epigenetic changes influence gene expression by altering chromatin structure, which is essential for plant growth and determines how plants react to both external and internal stimuli. Additionally, epigenetic modifications, along with DNA sequence variations, support breeding efforts by offering valuable markers. They also take into account epigenome diversity to better predict plant performance and enhance crop yields. The manipulation of photoperiodic flowering through epigenetic modifications presents substantial opportunities for agricultural advancement. A comprehensive understanding of the genes and pathways governing flowering time in relation to day length facilitates the development of crops specifically designed for various latitudes and growing seasons. This is especially vital for expanding the geographical distribution of crops and adapting them to evolving climatic conditions. The discovery of genes such as *Ppd-B1a* in wheat, which confers photoperiod insensitivity, has empowered breeders to produce varieties that perform well across a broader range of latitudes. Moreover, investigating the epigenetic mechanisms that lie in the backdrop of these adaptations, including DNA methylation and histone modification, can pave the way for innovative breeding approaches. Selection of epigenetic markers linked to advantageous flowering times may allow breeders to expedite the development of crops that are resilient to climate change and capable of higher yields. Additionally, understanding how environmental factors affect epigenetic states can further lead to the development of new agricultural practices aimed at optimizing flowering time and enhancing productivity. Immense scope of research lies ahead to fully comprehend the intricate interactions among genetic, epigenetic and environmental factors that influence flowering time in crops, and this understanding will be critical for the sustainable intensification of agriculture and for securing global food supply in light of ongoing environmental challenges.

REFERENCES

Abe, M., Kobayashi, Y., Yamamoto, S., Daimon, Y., Yamaguchi, A., Ikeda, Y., Ichinoki, H., Notaguchi, M., Goto, K., & Araki, T. (2005). FD, a bZIP protein mediating signals from the floral pathway integrator FT at the shoot apex. *Science*, 309(5737), 1052–1056. https://doi.org/10.1126/SCIENCE.1115983/SUPPL_FILE/ABE.SOM.PDF.

Achard, P., Cheng, H., De Grauwe, L., Decat, J., Schoutteten, H., Moritz, T., Van Der Straeten, D., Peng, J., & Harberd, N. P. (2006). Integration of plant responses to environmentally activated phytohormonal signals. *Science*, 311(5757), 91–94. https://doi.org/10.1126/SCIENCE.1118642.

Achard, P., Herr, A., Baulcombe, D. C., & Harberd, N. P. (2004). Modulation of floral development by a gibberellin-regulated microRNA. *Development*, 131(14), 3357–3365. https://doi.org/10.1242/DEV.01206.

Amasino, R. (2010a). Seasonal and developmental timing of flowering. *The Plant Journal*, 61(6), 1001–1013. https://doi.org/10.1111/J.1365-313X.2010.04148.X.

Amasino, R. (2010b). Seasonal and developmental timing of flowering. *Plant Journal*, 61(6), 1001–1013. https://doi.org/10.1111/J.1365-313X.2010.04148.X.

Anderson, S. J., Kramer, M. C., Gosai, S. J., Yu, X., Vandivier, L. E., Nelson, A. D. L., Anderson, Z. D., Beilstein, M. A., Fray, R. G., Lyons, E., & Gregory, B. D. (2018). N6-methyladenosine inhibits local ribonucleolytic cleavage to stabilize mRNAs in *Arabidopsis*. *Cell Reports*, 25(5), 1146–1157.e3. https://doi.org/10.1016/j.celrep.2018.10.020.

Archacki, R., Buszewicz, D., Sarnowski, T. J., Sarnowska, E., Rolicka, A. T., Tohge, T., Fernie, A. R., Jikumaru, Y., Kotlinski, M., Iwanicka-Nowicka, R., Kalisiak, K., Patryn, J., Halibart-Puzio, J., Kamiya, Y., Davis, S. J., Koblowska, M. K., & Jerzmanowski, A. (2013a). BRAHMA ATPase of the SWI/SNF chromatin remodeling complex acts as a positive regulator of Gibberellin-mediated responses in *Arabidopsis*. *PLoS One*, 8(3), e58588. https://doi.org/10.1371/JOURNAL.PONE.0058588.

Archacki, R., Buszewicz, D., Sarnowski, T. J., Sarnowska, E., Rolicka, A. T., Tohge, T., Fernie, A. R., Jikumaru, Y., Kotlinski, M., Iwanicka-Nowicka, R., Kalisiak, K., Patryn, J., Halibart-Puzio, J., Kamiya, Y., Davis, S. J., Koblowska, M. K., & Jerzmanowski, A. (2013b). BRAHMA ATPase of the SWI/SNF chromatin remodeling complex acts as a positive regulator of Gibberellin-mediated responses in *Arabidopsis*. *PLoS One*, 8(3), e58588. https://doi.org/10.1371/JOURNAL.PONE.0058588.

Ariel, F., Jegu, T., Latrasse, D., Romero-Barrios, N., Christ, A., Benhamed, M., & Crespi, M. (2014). Noncoding transcription by alternative RNA polymerases dynamically regulates an auxin-driven chromatin loop. *Molecular Cell*, 55(3), 383–396. https://doi.org/10.1016/J.MOLCEL.2014.06.011/ASSET/A5B22950-6FF3-4820-8A90-4DC9AD2AEBD1/MAIN.ASSETS/GR6_LRG.JPG.

Ariel, F., Lucero, L., Christ, A., Mammarella, M. F., Jegu, T., Veluchamy, A., Mariappan, K., Latrasse, D., Blein, T., Liu, C., Benhamed, M., & Crespi, M. (2020). R-Loop mediated trans action of the APOLO long noncoding RNA. *Molecular Cell*, 77(5), 1055–1065.e4. https://doi.org/10.1016/J.MOLCEL.2019.12.015/ATTACHMENT/49C79015-E721-4DC8-8888-19E799FD3A14/MMC10.PDF.

Ariel, F., Romero-Barrios, N., Jégu, T., Benhamed, M., & Crespi, M. (2015). Battles and hijacks: Noncoding transcription in plants. *Trends in Plant Science*, 20(6), 362–371. https://doi.org/10.1016/J.TPLANTS.2015.03.003/ASSET/7DB1E544-BB62-4841-9443-727FC6971F04/MAIN.ASSETS/GR3.SML.

Aubert, D., Chen, L., Moon, Y.-H., Martin, D., Castle, L. A., Yang, C.-H., & Sung, Z. R. (2001a). EMF1, a novel protein involved in the control of shoot architecture and flowering in *Arabidopsis*. *The Plant Cell*, 13(8), 1865–1875. https://doi.org/10.1105/TPC.010094.

Aubert, D., Chen, L., Moon, Y.-H., Martin, D., Castle, L. A., Yang, C.-H., & Sung, Z. R. (2001b). EMF1, a novel protein involved in the control of shoot architecture and flowering in *Arabidopsis*. *The Plant Cell*, 13(8), 1865–1875. https://doi.org/10.1105/TPC.010094.

Augustine, R. C., & Vierstra, R. D. (2018). SUMOylation: Re-wiring the plant nucleus during stress and development. *Current Opinion in Plant Biology*, 45, 143–154. https://doi.org/10.1016/J.PBI.2018.06.006.

Aukerman, M. J., & Sakai, H. (2003). Regulation of flowering time and floral organ identity by a MicroRNA and its APETALA2-Like target genes. *Plant Cell*, 15(11), 2730–2741. https://doi.org/10.1105/TPC.016238.

Baile, F., Gómez-Zambrano, Á., & Calonje, M. (2022). Roles of Polycomb complexes in regulating gene expression and chromatin structure in plants. *Plant Communications*, 3(1), 100267. https://doi.org/10.1016/J.XPLC.2021.100267/ASSET/AB46CE7D-3BE3-4EFB-80C3-AEE1A6EFAFAF/MAIN.ASSETS/GR2.JPG.

Banerjee, A., & Roychoudhury, A. (2017). Epigenetic regulation during salinity and drought stress in plants: Histone modifications and DNA methylation. *Plant Gene*, 11, 199–204.

Banerjee, A., & Roychoudhury, A. (2018). The gymnastics of epigenomics in rice. *Plant Cell Reports*, 37, 25–49.

Bardou, F., Ariel, F., Simpson, C. G., Romero-Barrios, N., Laporte, P., Balzergue, S., Brown, J. W. S., & Crespi, M. (2014). Long noncoding RNA modulates alternative splicing regulators in *Arabidopsis*. *Developmental Cell*, 30(2), 166–176. https://doi.org/10.1016/j.devcel.2014.06.017.

Bazin, J., & Bailey-Serres, J. (2015). Emerging roles of long noncoding RNA in root developmental plasticity and regulation of phosphate homeostasis. *Frontiers in Plant Science*, 6(June), 144084. https://doi.org/10.3389/FPLS.2015.00400/BIBTEX.

Becker, P. B., & Hörz, W. (2002). Atp-dependent nucleosome remodeling. *Annual Review of Biochemistry*, 71, 247–273. https://doi.org/10.1146/ANNUREV.BIOCHEM.71.110601.135400/CITE/REFWORKS.

Benhamed, M., Bertrand, C., Servet, C., & Zhou, D. X. (2006). *Arabidopsis* GCN5, HD1, and TAF1/HAF2 interact to regulate histone acetylation required for light-responsive gene expression. *The Plant Cell*, 18(11), 2893–2903. https://doi.org/10.1105/TPC.106.043489.

Berg, A., Meza, T. J., Mahić, M., Thorstensen, T., Kristiansen, K., & Aalen, R. B. (2003). Ten members of the Arabidopsis gene family encoding methyl-CpG-binding domain proteins are transcriptionally active and at least one, AtMBD11, is crucial for normal development. *Nucleic Acids Research*, 31(18), 5291–5304. https://doi.org/10.1093/NAR/GKG735.

Berr, A., McCallum, E. J., Ménard, R., Meyer, D., Fuchs, J., Dong, A., & Shen, W. H. (2010). Arabidopsis SET DOMAIN GROUP2 is required for H3K4 trimethylation and is crucial for both sporophyte and gametophyte development. *The Plant Cell*, 22(10), 3232–3248. https://doi.org/10.1105/TPC.110.079962.

Berr, A., Shafiq, S., Pinon, V., Dong, A., & Shen, W. H. (2015). The trxG family histone methyltransferase SET DOMAIN GROUP 26 promotes flowering via a distinctive genetic pathway. *The Plant Journal: For Cell and Molecular Biology*, 81(2), 316–328. https://doi.org/10.1111/TPJ.12729.

Bieluszewski, T., Xiao, J., Yang, Y., & Wagner, D. (2021). PRC2 activity, recruitment, and silencing: A comparative perspective. *Trends in Plant Science Advances*, 26(11), 1186–1198. https://doi.org/10.1016/j.tplants.2021.06.006.

Blümel, M., Dally, N., & Jung, C. (2015). Flowering time regulation in crops — what did we learn from *Arabidopsis*? *Current Opinion in Biotechnology*, 32, 121–129. https://doi.org/10.1016/J.COPBIO.2014.11.023.

Bratzel, F., López-Torrejón, G., Koch, M., Del Pozo, J. C., & Calonje, M. (2010). Keeping cell identity in arabidopsis requires PRC1 RING-finger homologs that catalyze H2A monoubiquitination. *Current Biology*, 20(20), 1853–1859. https://doi.org/10.1016/j.cub.2010.09.046.

Bu, Z., Yu, Y., Li, Z., Liu, Y., Jiang, W., Huang, Y., & Dong, A. W. (2014). Regulation of *Arabidopsis* flowering by the histone mark readers MRG1/2 via interaction with CONSTANS to modulate FT expression. *PLOS Genetics*, 10(9), e1004617. https://doi.org/10.1371/JOURNAL.PGEN.1004617.

Campos-Rivero, G., Osorio-Montalvo, P., Sánchez-Borges, R., Us-Camas, R., Duarte-Aké, F., & De-la-Peña, C. (2017a). Plant hormone signaling in flowering: An epigenetic point of view. *Journal of Plant Physiology*, 214, 16–27. https://doi.org/10.1016/J.JPLPH.2017.03.018.

Campos-Rivero, G., Osorio-Montalvo, P., Sánchez-Borges, R., Us-Camas, R., Duarte-Aké, F., & De-la-Peña, C. (2017b). Plant hormone signaling in flowering: An epigenetic point of view. *Journal of Plant Physiology*, 214, 16–27. https://doi.org/10.1016/J.JPLPH.2017.03.018.

Chen, L. T., Luo, M., Wang, Y. Y., & Wu, K. (2010a). Involvement of *Arabidopsis* histone deacetylase HDA6 in ABA and salt stress response. *Journal of Experimental Botany*, 61(12), 3345–3353. https://doi.org/10.1093/JXB/ERQ154.

Chen, L. T., Luo, M., Wang, Y. Y., & Wu, K. (2010b). Involvement of *Arabidopsis* histone deacetylase HDA6 in ABA and salt stress response. *Journal of Experimental Botany*, 61(12), 3345–3353. https://doi.org/10.1093/JXB/ERQ154.

Choi, K., Kim, J., Müller, S. Y., Oh, M., Underwood, C., Henderson, I., & Lee, I. (2016). Regulation of MicroRNA-mediated developmental changes by the SWR1 chromatin remodeling complex. *Plant Physiology*, 171(2), 1128–1143. https://doi.org/10.1104/PP.16.00332.

Chu, L., Yang, C., Zhuang, F., Gao, Y., & Luo, M. (2022a). The HDA9-HY5 module epigenetically regulates flowering time in *Arabidopsis thaliana*. *Journal of Cellular Physiology*, 237(7), 2961–2968. https://doi.org/10.1002/JCP.30761.

Chu, L., Yang, C., Zhuang, F., Gao, Y., & Luo, M. (2022b). The HDA9-HY5 module epigenetically regulates flowering time in *Arabidopsis thaliana*. *Journal of Cellular Physiology*, 237(7), 2961–2968. https://doi.org/10.1002/JCP.30761.

Clapier, C. R., & Cairns, B. R. (2009). The biology of chromatin remodeling complexes. *Annual Review of Biochemistry*, 78, 273–304. https://doi.org/10.1146/ANNUREV.BIOCHEM.77.062706.153223/1.

Corbesier, L., Vincent, C., Jang, S., Fornara, F., Fan, Q., Searle, I., Giakountis, A., Farrona, S., Gissot, L., Turnbull, C., & Coupland, G. (2007). FT protein movement contributes to long-distance signaling in floral induction of *Arabidopsis*. *Science*, 316(5827), 1030–1033. https://doi.org/10.1126/SCIENCE.1141752/SUPPL_FILE/CORBESIER.SOM.PDF.

Cubas, P., Vincent, C., & Coen, E. (1999). An epigenetic mutation responsible for natural variation in floral symmetry. *Nature*, 401(6749), 157–161. https://doi.org/10.1038/43657.

Cui, X., Lu, F., Qiu, Q., Zhou, B., Gu, L., Zhang, S., Kang, Y., Cui, X., Ma, X., Yao, Q., Ma, J., Zhang, X., & Cao, X. (2016). REF6 recognizes a specific DNA sequence to demethylate H3K27me3 and regulate organ boundary formation in *Arabidopsis*. *Nature Genetics*, 48(6), 694–699. https://doi.org/10.1038/ng.3556.

Dai, X., Bai, Y., Zhao, L., Dou, X., Liu, Y., Wang, L., Li, Y., Li, W., Hui, Y., Huang, X., Wang, Z., & Qin, Y. (2017). H2A.Z represses gene expression by modulating promoter nucleosome structure and enhancer histone modifications in *Arabidopsis*. *Molecular Plant*, 10(10), 1274–1292. https://doi.org/10.1016/J.MOLP.2017.09.007/ATTACHMENT/22913D02-DBB7-4B3B-9AB5-8E5C4381EFF8/MMC2.XLSX.

Deal, R. B., Kandasamy, M. K., McKinney, E. C., & Meagher, R. B. (2005). The nuclear actin-related protein ARP6 is a pleiotropic developmental regulator required for the maintenance of FLOWERING LOCUS C expression and repression of flowering in *Arabidopsis*. *The Plant Cell*, 17(10), 2633–2646. https://doi.org/10.1105/TPC.105.035196.

Deng, W., Ying, H., Helliwell, C. A., Taylor, J. M., Peacock, W. J., & Dennis, E. S. (2011). FLOWERING LOCUS C (FLC) regulates development pathways throughout the life cycle of *Arabidopsis*. *Proceedings of the National Academy of Sciences of the United States of America*, 108(16), 6680–6685. https://doi.org/10.1073/PNAS.1103175108/-/DCSUPPLEMENTAL.

Ding, Y., Avramova, Z., & Fromm, M. (2011). The *Arabidopsis* trithorax-like factor ATX1 functions in dehydration stress responses via ABA-dependent and ABA-independent pathways. *The Plant Journal*, 66(5), 735–744. https://doi.org/10.1111/J.1365-313X.2011.04534.X.

Domagalska, M. A., Sarnowska, E., Nagy, F., & Davis, S. J. (2010a). Genetic analyses of interactions among gibberellin, abscisic acid, and brassinosteroids in the control of flowering time in *Arabidopsis thaliana*. *PLoS One*, 5(11), e14012. https://doi.org/10.1371/JOURNAL.PONE.0014012.

Domagalska, M. A., Sarnowska, E., Nagy, F., & Davis, S. J. (2010b). Genetic analyses of interactions among gibberellin, abscisic acid, and brassinosteroids in the control of flowering time in *Arabidopsis thaliana*. *PLoS One*, 5(11), e14012. https://doi.org/10.1371/JOURNAL.PONE.0014012.

Dong, G., Ma, D. P., & Li, J. (2008). The histone methyltransferase SDG8 regulates shoot branching in *Arabidopsis*. *Biochemical and Biophysical Research Communications*, 373(4), 659–664. https://doi.org/10.1016/J.BBRC.2008.06.096.

Duan, H. C., Wei, L. H., Zhang, C., Wang, Y., Chen, L., Lu, Z., Chen, P. R., He, C., & Jia, G. (2018a). ALKBH10B is an RNA N 6-methyladenosine demethylase affecting *Arabidopsis* floral transition. *The Plant Cell*, 29(12), 2995–3011. https://doi.org/10.1105/TPC.16.00912.

Duan, H. C., Wei, L. H., Zhang, C., Wang, Y., Chen, L., Lu, Z., Chen, P. R., He, C., & Jia, G. (2018b). ALKBH10B is an RNA N 6-methyladenosine demethylase affecting *Arabidopsis* floral transition. *The Plant Cell*, 29(12), 2995–3011. https://doi.org/10.1105/TPC.16.00912.

Dunlap, J. C. (1996). Genetic and molecular analysis of circadian rhythms. *Annual Review of Genetics*, 30, 579–601. https://doi.org/10.1146/ANNUREV.GENET.30.1.579.

Dutta, A., Choudhary, P., Caruana, J., & Raina, R. (2017). JMJ27, an *Arabidopsis* H3K9 histone demethylase, modulates defense against *Pseudomonas syringae* and flowering time. *The Plant Journal*, 91(6), 1015–1028. https://doi.org/10.1111/TPJ.13623.

Feng, J., & Lu, J. (2017). LHP1 could act as an activator and a repressor of transcription in plants. *Frontiers in Plant Science*, 8, 255451. https://doi.org/10.3389/FPLS.2017.02041/BIBTEX.

Feng, S., Cokus, S. J., Zhang, X., Chen, P. Y., Bostick, M., Goll, M. G., Hetzel, J., Jain, J., Strauss, S. H., Halpern, M. E., Ukomadu, C., Sadler, K. C., Pradhan, S., Pellegrini, M., & Jacobsen, S. E. (2010). Conservation and divergence of methylation patterning in plants and animals. *Proceedings of the National Academy of Sciences of the United States of America*, 107(19), 8689–8694. https://doi.org/10.1073/PNAS.1002720107/SUPPL_FILE/PNAS.201002720SI.PDF.

Fieldes, M. A., & Amyot, L. M. (2011). Evaluating the potential of using 5-azacytidine as an epimutagen. *Canadian Journal of Botany*, 77(11), 1617–1622. https://doi.org/10.1139/B99-136.

Foo, M., Somers, D. E., & Kim, P. J. (2016). Kernel architecture of the genetic circuitry of the *Arabidopsis* circadian system. *PLOS Computational Biology*, 12(2), e1004748. https://doi.org/10.1371/JOURNAL.PCBI.1004748.

Fornara, F., de Montaigu, A., & Coupland, G. (2010). SnapShot: Control of flowering in Arabidopsis. *Cell*, 141(3), 550–550.e2. https://doi.org/10.1016/J.CELL.2010.04.024.

Gaudin, V., Libault, M., Pouteau, S., Juul, T., Zhao, G., Lefebvre, D., & Grandjean, O. (2001). Mutations in LIKE HETEROCHROMATIN PROTEIN 1 affect flowering time and plant architecture in *Arabidopsis*. *Development*, 128(23), 4847–4858. https://doi.org/10.1242/DEV.128.23.4847.

Gendall, A. R., Levy, Y. Y., Wilson, A., & Dean, C. (2001). The VERNALIZATION 2 gene mediates the epigenetic regulation of vernalization in *Arabidopsis*. *Cell*, 107(4), 525–535. https://doi.org/10.1016/S0092-8674(01)00573-6/ASSET/473FFF24-44CE-43C0-9956-EFBF62D2EDEF/MAIN.ASSETS/GR6.JPG.

Gomi, K., Sasaki, A., Itoh, H., Ueguchi-Tanaka, M., Ashikari, M., Kitano, H., & Matsuoka, M. (2004). GID2, an F-box subunit of the SCF E3 complex, specifically interacts with phosphorylated SLR1 protein and regulates the gibberellin-dependent degradation of SLR1 in rice. *The Plant Journal*, 37(4), 626–634. https://doi.org/10.1111/J.1365-313X.2003.01990.X.

Green, G. R., Gustavsen, L. C., & Poccia, D. L. (1990). Phosphorylation of plant H2A histones. *Plant Physiology*, 93(3), 1241–1245. https://doi.org/10.1104/PP.93.3.1241.

Ha, M., & Kim, V. N. (2014). Regulation of microRNA biogenesis. *Nature Reviews Molecular Cell Biology*, 15(8), 509–524. https://doi.org/10.1038/nrm3838.

Hauser, M. T., Aufsatz, W., Jonak, C., & Luschnig, C. (2011). Transgenerational epigenetic inheritance in plants. *Biochimica et Biophysica Acta (BBA) – Gene Regulatory Mechanisms*, 1809(8), 459–468. https://doi.org/10.1016/J.BBAGRM.2011.03.007.

He, Y. (2015). Enabling photoperiodic control of flowering by timely chromatin silencing of the florigen gene. *Nucleus*, 6(3), 179–182. https://doi.org/10.1080/19491034.2015.1038000.

Hemmes, H., Henriques, R., Jang, I. C., Kim, S., & Chua, N. H. (2012). Circadian clock regulates dynamic chromatin modifications associated with *Arabidopsis* CCA1/LHY and TOC1 transcriptional rhythms. *Plant and Cell Physiology*, 53(12), 2016–2029. https://doi.org/10.1093/PCP/PCS148.

Henriques, R., Wang, H., Liu, J., Boix, M., Huang, L. F., & Chua, N. H. (2017). The antiphasic regulatory module comprising CDF5 and its antisense RNA FLORE links the circadian clock to photoperiodic flowering. *New Phytologist*, 216(3), 854–867. https://doi.org/10.1111/NPH.14703.

Heo, J. B., & Sung, S. (2011). Vernalization-mediated epigenetic silencing by a long intronic noncoding RNA. *Science*, 331(6013), 76–79. https://doi.org/10.1126/SCIENCE.1197349/SUPPL_FILE/HEO.SOM.PDF.

Ho, K. K., Zhang, H., Golden, B. L., & Ogas, J. (2013). PICKLE is a CHD subfamily II ATP-dependent chromatin remodeling factor. *Biochimica et Biophysica Acta (BBA) – Gene Regulatory Mechanisms*, 1829(2), 199–210. https://doi.org/10.1016/J.BBAGRM.2012.10.011.

Huang, T., Zhang, H., Zhou, Y., Su, Y., Zheng, H., & Ding, Y. (2021). Phosphorylation of histone H2A at serine 95 is essential for flowering time and development in *Arabidopsis*. *Frontiers in Plant Science*, 12, 761008. https://doi.org/10.3389/FPLS.2021.761008/BIBTEX.

Huang, X., Tian, H., Park, J., Oh, D. H., Hu, J., Zentella, R., Qiao, H., Dassanayake, M., & Sun, T. P. (2023). The master growth regulator DELLA binding to histone H2A is essential for DELLA-mediated global transcription regulation. *Nature Plants*, 9(8), 1291–1305. https://doi.org/10.1038/s41477-023-01477-y.

Hung, F. Y., Chen, J. H., Feng, Y. R., Lai, Y. C., Yang, S., & Wu, K. (2020). *Arabidopsis* JMJ29 is involved in trichome development by regulating the core trichome initiation gene GLABRA3. *The Plant Journal*, 103(5), 1735–1743. https://doi.org/10.1111/TPJ.14858.

Hyun, Y., Richter, R., Vincent, C., Martinez-Gallegos, R., Porri, A., & Coupland, G. (2016). Multi-layered regulation of SPL15 and cooperation with SOC1 integrate endogenous flowering pathways at the *Arabidopsis* shoot meristem. *Developmental Cell*, 37(3), 254–266. https://doi.org/10.1016/J.DEVCEL.2016.04.001.

Isabel Ortiz-Marchena, M., Albi, T., Lucas-Reina, E., Said, F. E., Romero-Campero, F. J., Cano, B., Teresa Ruiz, M., Romero, J. M., & Valverde, F. (2014). Photoperiodic control of carbon distribution during the floral transition in *Arabidopsis*. *The Plant Cell*, 26(2), 565–584. https://doi.org/10.1105/TPC.114.122721.

Ito, S., Song, Y. H., Josephson-Day, A. R., Miller, R. J., Breton, G., Olmstead, R. G., & Imaizumi, T. (2012). FLOWERING BHLH transcriptional activators control expression of the photoperiodic flowering regulator CONSTANS in *Arabidopsis*. *Proceedings of the National Academy of Sciences of the United States of America*, 109(9), 3582–3587. https://doi.org/10.1073/PNAS.1118876109/SUPPL_FILE/PNAS.201118876SI.PDF.

Iwase, Y., Shiraya, T., & Takeno, K. (2010). Flowering and dwarfism induced by DNA demethylation in *Pharbitis nil*. *Physiologia Plantarum*, 139(1), 118–127. https://doi.org/10.1111/J.1399-3054.2009.01345.X.

Jan, I. C., Chun, P. J., Hemmes, H., Jung, C., & Chua, N. H. (2011). Rapid and reversible light-mediated chromatin modifications of *Arabidopsis* phytochrome A locus. *The Plant Cell*, 23(2), 459. https://doi.org/10.1105/TPC.110.080481.

Jiang, D., Wang, Y., Wang, Y., & He, Y. (2008). Repression of FLOWERING LOCUS C and FLOWERING LOCUS T by the *Arabidopsis* polycomb repressive complex 2 components. *PLoS One*, 3(10), e3404. https://doi.org/10.1371/JOURNAL.PONE.0003404.

Jing, Y., Guo, Q., & Lin, R. (2019). The chromatin-remodeling factor PICKLE antagonizes polycomb repression of FT to promote flowering. *Plant Physiology*, 181(2), 656–668. https://doi.org/10.1104/PP.19.00596.

Jing, Y., Guo, Q., Zha, P., & Lin, R. (2019). The chromatin-remodelling factor PICKLE interacts with CONSTANS to promote flowering in *Arabidopsis*. *Plant, Cell & Environment*, 42(8), 2495–2507. https://doi.org/10.1111/PCE.13557.

Kang, M. J., Jin, H. S., Noh, Y. S., & Noh, B. (2015). Repression of flowering under a noninductive photoperiod by the HDA9-AGL19-FT module in *Arabidopsis*. *The New Phytologist*, 206(1), 281–294. https://doi.org/10.1111/NPH.13161.

Kardailsky, I., Shukla, V. K., Ahn, J. H., Dagenais, N., Christensen, S. K., Nguyen, J. T., Chory, J., Harrison, M. J., & Weigel, D. (1999). Activation tagging of the floral inducer FT. *Science*, 286(5446), 1962–1965. https://doi.org/10.1126/SCIENCE.286.5446.1962/SUPPL_FILE/1044618S3_THUMB.GIF.

Kierzek, E., & Kierzek, R. (2003). The thermodynamic stability of RNA duplexes and hairpins containing N6-alkyladenosines and 2-methylthio-N6-alkyladenosines. *Nucleic Acids Research*, 31(15), 4472–4480. https://doi.org/10.1093/NAR/GKG633.

Kinoshita, A., Vayssières, A., Richter, R., Sang, Q., Roggen, A., Van Driel, A. D., Smith, R. S., & Coupland, G. (2020). Regulation of shoot meristem shape by photoperiodic signaling and phytohormones during floral induction of *Arabidopsis*. *ELife*, 9, 1–29. https://doi.org/10.7554/ELIFE.60661.

Kobayashi, Y., Kaya, H., Goto, K., Iwabuchi, M., & Araki, T. (1999). A pair of related genes with antagonistic roles in mediating flowering signals. *Science*, 286(5446), 1960–1962. https://doi.org/10.1126/SCIENCE.286.5446.1960.

Kojima, S., Takahashi, Y., Kobayashi, Y., Monna, L., Sasaki, T., Araki, T., & Yano, M. (2002). Hd3a, a rice ortholog of the *Arabidopsis FT* gene, promotes transition to flowering downstream of Hd1 under short-day conditions. *Plant and Cell Physiology*, 43(10), 1096–1105. https://doi.org/10.1093/PCP/PCF156.

Kondo, H., Ozaki, H., Itoh, K., Kato, A., & Takeno, K. (2006). Flowering induced by 5-azacytidine, a DNA demethylating reagent in a short-day plant, *Perilla frutescens* var. crispa. *Physiologia Plantarum*, 127(1), 130–137. https://doi.org/10.1111/J.1399-3054.2005.00635.X.

Kondo, H., Shiraya, T., Wada, K. C., & Takeno, K. (2010). Induction of flowering by DNA demethylation in *Perilla frutescens* and *Silene armeria*: Heritability of 5-azacytidine-induced effects and alteration of the DNA methylation state by photoperiodic conditions. *Plant Science*, 178(3), 321–326. https://doi.org/10.1016/J.PLANTSCI.2010.01.012.

Kornberg, R. D., & Lorch, Y. (1999). Twenty-five years of the nucleosome, fundamental particle of the eukaryote chromosome. *Cell*, 98(3), 285–294. https://doi.org/10.1016/s0092-8674(00)81958-3.

Kotkar, H., & Giri, A. (2020). Plant epigenetics and the 'intelligent' priming system to combat biotic stress. *Epigenetics of the Immune System*, 16, 25–38. https://doi.org/10.1016/B978-0-12-817964-2.00002-2.

Kvist, J., Gonçalves Athanàsio, C., Shams Solari, O., Brown, J. B., Colbourne, J. K., Pfrender, M. E., & Mirbahai, L. (2018). Pattern of DNA methylation in *Daphnia*: Evolutionary perspective. *Genome Biology and Evolution*, 10(8), 1988–2007. https://doi.org/10.1093/GBE/EVY155.

Law, J. A., Vashisht, A. A., Wohlschlegel, J. A., & Jacobsen, S. E. (2011). SHH1, a homeodomain protein required for DNA methylation, as well as RDR2, RDM4, and chromatin remodeling factors, associate with RNA polymerase IV. *PLOS Genetics*, 7(7), e1002195. https://doi.org/10.1371/JOURNAL.PGEN.1002195.

Lee, J., Oh, M., Park, H., & Lee, I. (2008). SOC1 translocated to the nucleus by interaction with AGL24 directly regulates LEAFY. *The Plant Journal*, 55(5), 832–843. https://doi.org/10.1111/J.1365-313X.2008.03552.X.

Li, W., Wang, T., Zhang, Y., & Li, Y. (2016). Overexpression of soybean miR172c confers tolerance to water deficit and salt stress, but increases ABA sensitivity in transgenic *Arabidopsis thaliana*. *Journal of Experimental Botany*, 67(1), 175–194. https://doi.org/10.1093/JXB/ERV450.

Liu, C., Chen, H., Er, H. L., Soo, H. M., Kumar, P. P., Han, J. H., Liou, Y. C., & Yu, H. (2008). Direct interaction of AGL24 and SOC1 integrates flowering signals in *Arabidopsis*. *Development*, 135(8), 1481–1491. https://doi.org/10.1242/DEV.020255.

Liu, C., Xi, W., Shen, L., Tan, C., & Yu, H. (2009). Regulation of floral patterning by flowering time genes. *Developmental Cell*, 16(5), 711–722. https://doi.org/10.1016/J.DEVCEL.2009.03.011.

Liu, F., Quesada, V., Crevillén, P., Bäurle, I., Swiezewski, S., & Dean, C. (2007). The *Arabidopsis* RNA-binding protein FCA requires a lysine-specific demethylase 1 homolog to downregulate FLC. *Molecular Cell*, 28(3), 398–407. https://doi.org/10.1016/J.MOLCEL.2007.10.018.

Liu, S., He, M., Lin, X., & Kong, F. (2023). Epigenetic regulation of photoperiodic flowering in plants. *The Plant Genome*, 16(4), e20320. https://doi.org/10.1002/TPG2.20320.

Luger, K., & Hansen, J. C. (2005). Nucleosome and chromatin fiber dynamics. *Current Opinion in Structural Biology*, 15(2), 188–196. https://doi.org/10.1016/j.sbi.2005.03.006.

Luo, Q., Mo, J., Chen, H., Hu, Z., Wang, B., Wu, J., Liang, Z., Xie, W., Du, K., Peng, M., Li, Y., Li, T., Zhang, Y., Shi, X., Shen, W. H., Shi, Y., Dong, A., Wang, H., & Ma, J. (2022). Structural insights into molecular mechanism for N6-adenosine methylation by MT-A70 family methyltransferase METTL4. *Nature Communications*, 13(1), 1–11. https://doi.org/10.1038/s41467-022-33277-x.

Madrid, E., Chandler, J. W., & Coupland, G. (2021). Gene regulatory networks controlled by FLOWERING LOCUS C that confer variation in seasonal flowering and life history. *Journal of Experimental Botany*, 72(1), 4–14. https://doi.org/10.1093/JXB/ERAA216.

Malapeira, J., Khaitova, L. C., & Mas, P. (2012). Ordered changes in histone modifications at the core of the Arabidopsis circadian clock. *Proceedings of the National Academy of Sciences of the United States of America*, 109(52), 21540–21545. https://doi.org/10.1073/PNAS.1217022110/SUPPL_FILE/PNAS.201217022SI.PDF.

March-Díaz, R., García-Domínguez, M., Florencio, F. J., & Reyes, J. C. (2007). SEF, a new protein required for flowering repression in *Arabidopsis*, interacts with PIE1 and ARP6. *Plant Physiology*, 143(2), 893–901. https://doi.org/10.1104/PP.106.092270.

Martin, C., & Zhang, Y. (2005). The diverse functions of histone lysine methylation. *Nature Reviews Molecular Cell Biology*, 6(11), 838–849. https://doi.org/10.1038/nrm1761.

Martínez, C., Pons, E., Prats, G., & León, J. (2004). Salicylic acid regulates flowering time and links defence responses and reproductive development. *The Plant Journal: For Cell and Molecular Biology*, 37(2), 209–217. https://doi.org/10.1046/J.1365-313X.2003.01954.X.

Martinez-Zapater, J. M., & Somerville, C. R. (1990). Effect of light quality and vernalization on late-flowering mutants of *Arabidopsis thaliana*. *Plant Physiology*, 92(3), 770–776. https://doi.org/10.1104/PP.92.3.770.

Matsoukas, I. G. (2014). Interplay between sugar and hormone signaling pathways modulate floral signal transduction. *Frontiers in Genetics*, 5(Aug), 218. https://doi.org/10.3389/FGENE.2014.00218.

Matsoukas, I. G., Massiah, A. J., & Thomas, B. (2012). Florigenic and antiflorigenic signaling in plants. *Plant and Cell Physiology*, 53(11), 1827–1842. https://doi.org/10.1093/PCP/PCS130.

Meijón, M., Jesús Cañal, M., Valledor, L., Rodríguez, R., & Feito, I. (2011). Epigenetic and physiological effects of gibberellin inhibitors and chemical pruners on the floral transition of azalea. *Physiologia Plantarum*, 141(3), 276–288. https://doi.org/10.1111/J.1399-3054.2010.01430.X.

Melzer, S., Lens, F., Gennen, J., Vanneste, S., Rohde, A., & Beeckman, T. (2008). Flowering-time genes modulate meristem determinacy and growth form in *Arabidopsis thaliana*. *Nature Genetics*, 40(12), 1489–1492. https://doi.org/10.1038/ng.253.

Miao, Z., Zhang, T., Xie, B., Qi, Y., & Ma, C. (n.d.). *Evolutionary Implications of the RNA N 6-Methyladenosine Methylome in Plants*. https://doi.org/10.1093/molbev/msab299.

Michaels, S. D. (2009). Flowering time regulation produces much fruit. *Current Opinion in Plant Biology*, 12(1), 75–80. https://doi.org/10.1016/J.PBI.2008.09.005.

Mirouze, M., & Paszkowski, J. (2011). Epigenetic contribution to stress adaptation in plants. *Current Opinion in Plant Biology*, 14(3), 267–274. https://doi.org/10.1016/J.PBI.2011.03.004.

Miura, K., Lee, J., Jin, J. B., Yoo, C. Y., Miura, T., & Hasegawa, P. M. (2009). Sumoylation of ABI5 by the Arabidopsis SUMO E3 ligase SIZ1 negatively regulates abscisic acid signaling. *Proceedings of the National Academy of Sciences of the United States of America*, 106(13), 5418–5423. https://doi.org/10.1073/PNAS.0811088106/SUPPL_FILE/0811088106SI.PDF.

Murtas, G., Reeves, P. H., Fu, Y. F., Bancroft, I., Dean, C., & Coupland, G. (2003). A nuclear protease required for flowering-time regulation in Arabidopsis reduces the abundance of SMALL UBIQUITIN-RELATED MODIFIER conjugates. *The Plant Cell*, 15(10), 2308–2319. https://doi.org/10.1105/TPC.015487.

Mutasa-Göttgens, E., & Hedden, P. (2009). Gibberellin as a factor in floral regulatory networks. *Journal of Experimental Botany*, 60(7), 1979–1989. https://doi.org/10.1093/JXB/ERP040.

Nagano, S., Guan, K., Shenkutie, S. M., Feiler, C., Weiss, M., Kraskov, A., Buhrke, D., Hildebrandt, P., & Hughes, J. (2020). Structural insights into photoactivation and signalling in plant phytochromes. *Nature Plants*, 6(5), 581–588. https://doi.org/10.1038/S41477-020-0638-Y.

Noh, Y. S., & Amasino, R. M. (2003). PIE1, an ISWI family gene, is required for FLC activation and floral repression in Arabidopsis. *The Plant Cell*, 15(7), 1671–1682. https://doi.org/10.1105/TPC.012161.

Park, H. J., Kim, W. Y., Park, H. C., Lee, S. Y., Bohnert, H. J., & Yun, D. J. (2011a). SUMO and SUMOylation in plants. *Molecules and Cells*, 32(4), 305–316. https://doi.org/10.1007/S10059-011-0122-7/METRICS.

Paszkowski, J., & Grossniklaus, U. (2011). Selected aspects of transgenerational epigenetic inheritance and resetting in plants. *Current Opinion in Plant Biology*, 14(2), 195–203. https://doi.org/10.1016/J.PBI.2011.01.002.

Peng, M., Cui, Y., Bi, Y. M., & Rothstein, S. J. (2006). AtMBD9: A protein with a methyl-CpG-binding domain regulates flowering time and shoot branching in *Arabidopsis*. *The Plant Journal: For Cell and Molecular Biology*, 46(2), 282–296. https://doi.org/10.1111/J.1365-313X.2006.02691.X.

Pikaard, C. S., & Scheid, O. M. (2014). Epigenetic regulation in plants. *Cold Spring Harbor Perspectives in Biology*, 6(12), a019315. https://doi.org/10.1101/CSHPERSPECT.A019315.

Reeves, P. H., Murtas, G., Dash, S., & Coupland, G. (2002). early in short days 4, a mutation in *Arabidopsis* that causes early flowering and reduces the mRNA abundance of the floral repressor FLC. *Development*, 129(23), 5349–5361. https://doi.org/10.1242/DEV.00113.

Rinn, J., & Guttman, M. (2014). RNA and dynamic nuclear organization. *Science*, 345(6202), 1240–1241. https://doi.org/10.1126/SCIENCE.1252966.

Roudier, F., Ahmed, I., Bérard, C., Sarazin, A., Mary-Huard, T., Cortijo, S., Bouyer, D., Caillieux, E., Duvernois-Berthet, E., Al-Shikhley, L., Giraut, L., Desprás, B., Drevensek, S., Barneche, F., Dérozier, S., Brunaud, V., Aubourg, S., Schnittger, A., Bowler, C., ... Colot, V. (2011). Integrative epigenomic mapping defines four main chromatin states in *Arabidopsis*. *EMBO Journal*, 30(10), 1928–1938. https://doi.org/10.1038/EMBOJ.2011.103/SUPPL_FILE/EMBJ2011103.REVIEWER_COMMENTS.PDF.

Sanchez-Pulido, L., Devos, D., Sung, Z. R., & Calonje, M. (2008). RAWUL: A new ubiquitin-like domain in PRC1 Ring finger proteins that unveils putative plant and worm PRC1 orthologs. *BMC Genomics*, 9(1), 1–11. https://doi.org/10.1186/1471-2164-9-308/FIGURES/4.

Saze, H., Scheid, O. M., & Paszkowski, J. (2003). Maintenance of CpG methylation is essential for epigenetic inheritance during plant gametogenesis. *Nature Genetics*, 34(1), 65–69. https://doi.org/10.1038/ng1138.

Scebba, F., Bernacchia, G., De Bastiani, M., Evangelista, M., Cantoni, R. M., Cella, R., Locci, M. T., & Pitto, L. (2003). *Arabidopsis* MBD proteins show different binding specificities and nuclear localization. *Plant Molecular Biology*, 53(5), 755–771. https://doi.org/10.1023/B:PLAN.0000019118.56822.a9.

Schaffer, R., Ramsay, N., Samach, A., Corden, S., Putterill, J., Carré, I. A., & Coupland, G. (1998). The late elongated hypocotyl mutation of Arabidopsis disrupts circadian rhythms and the photoperiodic control of flowering. *Cell*, 93(7), 1219–1229. https://doi.org/10.1016/S0092-8674(00)81465-8.

Searle, I., & Coupland, G. (2004). Induction of flowering by seasonal changes in photoperiod. *EMBO Journal*, 23(6), 1217–1222. https://doi.org/10.1038/SJ.EMBOJ.7600117/ASSET/388689F9-5C0A-4957-8E0F-6B834F72D201/ASSETS/GRAPHIC/EMBJ7600117-FIG-0002-M.JPG.

Sequeira-Mendes, J., Aragüez, I., Peiró, R., Mendez-Giraldez, R., Zhang, X., Jacobsen, S. E., Bastolla, U., & Gutierreza, C. (2014). The functional topography of the *Arabidopsis* genome is organized in a reduced number of linear motifs of chromatin states. *The Plant Cell*, 26(6), 2351–2366. https://doi.org/10.1105/TPC.114.124578.

Shafiq, S., Berr, A., & Shen, W. H. (2014). Combinatorial functions of diverse histone methylations in *Arabidopsis thaliana* flowering time regulation. *New Phytologist*, 201(1), 312–322. https://doi.org/10.1111/NPH.12493.

Shahbazian, M. D., & Grunstein, M. (2007). Functions of Site-Specific histone acetylation and deacetylation. *Annual Review of Biochemistry*, 76, 75–100. https://doi.org/10.1146/ANNUREV.BIOCHEM.76.052705.162114/CITE/REFWORKS.

Sherman, J. D., & Talbert, L. E. (2011). Vernalization-induced changes of the DNA methylation pattern in winter wheat. *Genome*, 45(2), 253–260. https://doi.org/10.1139/G01-147.

Singh, A., & Roychoudhury, A. (2022) Mechanism of crosstalk between cytokinin and gibberellin. In T. Aftab (Ed.), *Auxins, Cytokinins and Gibberellins Signaling in Plants. Signaling and Communication in Plants* (pp. 77–90). Cham: Springer.

Solanki, M., & Shukla, L. I. (2023). Recent advances in auxin biosynthesis and homeostasis. *3 Biotech*, 13(9), 1–22. https://doi.org/10.1007/S13205-023-03709-6.

Song, Y. H., Shim, J. S., Kinmonth-Schultz, H. A., & Imaizumi, T. (2015). Photoperiodic flowering: Time measurement mechanisms in leaves. *Annual Review of Plant Biology*, 66, 441–464. https://doi.org/10.1146/ANNUREV-ARPLANT-043014-115555/CITE/REFWORKS.

Song, Z. T., Liu, J. X., & Han, J. J. (2021). Chromatin remodeling factors regulate environmental stress responses in plants. *Journal of Integrative Plant Biology*, 63(3), 438–450. https://doi.org/10.1111/JIPB.13064.

Stangeland, B., Maryann Rosenhave, E., Winge, P., Berg, A., Amundsen, S. S., Karabeg, M., Mandal, A., Bones, A. M., Grini, P. E., & Aalen, R. B. (2009). AtMBD8 is involved in control of flowering time in the C24 ecotype of *Arabidopsis thaliana*. *Physiologia Plantarum*, 136(1), 110–126. https://doi.org/10.1111/J.1399-3054.2009.01218.X.

Steinbach, Y. (2019). The *Arabidopsis thaliana* CONSTANS-LIKE 4 (COL4) – A modulator of flowering time. *Frontiers in Plant Science*, 10, 429068. https://doi.org/10.3389/FPLS.2019.00651/BIBTEX.

Su, Y., Wang, S., Zhang, F., Zheng, H., Liu, Y., Huang, T., & Ding, Y. (2017). Phosphorylation of histone H2A at serine 95: A plant-specific mark involved in flowering time regulation and H2A.Z deposition. *The Plant Cell*, 29(9), 2197–2213. https://doi.org/10.1105/TPC.17.00266.

Sun, B., Bhati, K. K., Song, P., Edwards, A., Petri, L., Kruusvee, V., Blaakmeer, A., Dolde, U., Rodrigues, V., Straub, D., Yang, J., Jia, G., & Wenkel, S. (2022). FIONA1-mediated methylation of the 3'UTR of FLC affects FLC transcript levels and flowering in *Arabidopsis*. *PLOS Genetics*, 18(9), e1010386. https://doi.org/10.1371/JOURNAL.PGEN.1010386.

Sun, C., Chen, D., Fang, J., Wang, P., Deng, X., & Chu, C. (2014). Understanding the genetic and epigenetic architecture in complex network of rice flowering pathways. *Protein and Cell*, 5(12), 889–898. https://doi.org/10.1007/S13238-014-0068-6/TABLES/2.

Sun, H., Guo, Z., Gao, L., Zhao, G., Zhang, W., Zhou, R., Wu, Y., Wang, H., An, H., & Jia, J. (2014). DNA methylation pattern of Photoperiod-B1 is associated with photoperiod insensitivity in wheat (*Triticum aestivum*). *New Phytologist*, 204(3), 682–692. https://doi.org/10.1111/NPH.12948.

Takeno, K. (2010). Epigenetic regulation of photoperiodic flowering. *Plant Signaling & Behavior*, 5(7), 788–791. https://doi.org/10.4161/PSB.5.7.11766.

Tamaki, S., Matsuo, S., Hann, L. W., Yokoi, S., & Shimamoto, K. (2007a). Hd3a protein is a mobile flowering signal in rice. *Science*, 316(5827), 1033–1036. https://doi.org/10.1126/SCIENCE.1141753/SUPPL_FILE/TAMAKI.SOM.PDF.

Tamaki, S., Matsuo, S., Hann, L. W., Yokoi, S., & Shimamoto, K. (2007b). Hd3a protein is a mobile flowering signal in rice. *Science*, 316(5827), 1033–1036. https://doi.org/10.1126/SCIENCE.1141753/SUPPL_FILE/TAMAKI.SOM.PDF.

Tang, X., Lim, M. H., Pelletier, J., Tang, M., Nguyen, V., Keller, W. A., Tsang, E. W. T., Wang, A., Rothstein, S. J., Harada, J. J., & Cui, Y. (2012). Synergistic repression of the embryonic programme by SET DOMAIN GROUP 8 and EMBRYONIC FLOWER 2 in *Arabidopsis* seedlings. *Journal of Experimental Botany*, 63(3), 1391–1404. https://doi.org/10.1093/JXB/ERR383.

Thiebaut, F., Hemerly, A. S., & Ferreira, P. C. G. (2019). A role for epigenetic regulation in the adaptation and stress responses of non-model plants. *Frontiers in Plant Science*, 10, 413504. https://doi.org/10.3389/FPLS.2019.00246/BIBTEX.

van den Burg, H. A., Kini, R. K., Schuurink, R. C., & Takken, F. L. W. (2010). *Arabidopsis* small ubiquitin-like modifier paralogs have distinct functions in development and defense. *The Plant Cell*, 22(6), 1998–2016. https://doi.org/10.1105/TPC.109.070961.

Wang, C., Yang, J., Song, P., Zhang, W., Lu, Q., Yu, Q., & Jia, G. (2022). FIONA1 is an RNA N 6-methyladenosine methyltransferase affecting *Arabidopsis* photomorphogenesis and flowering. *Genome Biology*, 23(1), 1–29. https://doi.org/10.1186/S13059-022-02612-2/FIGURES/6.

Wang, R., Xu, S., Wang, N., Xia, B., Jiang, Y., & Wang, R. (2017). Transcriptome analysis of secondary metabolism pathway, transcription factors, and transporters in response to methyl jasmonate in *Lycoris aurea. Frontiers in Plant Science*, 7. https://doi.org/10.3389/fpls.2016.01971.

Wang, W., & Sung, S. (2024). Chromatin sensing: Integration of environmental signals to reprogram plant development through chromatin regulators. *Journal of Experimental Botany*, 75(14), 4332–4345. https://doi.org/10.1093/JXB/ERAE086.

Wang, Y., Fan, X., Lin, F., He, G., Terzaghi, W., Zhu, D., & Deng, X. W. (2014). *Arabidopsis* noncoding RNA mediates control of photomorphogenesis by red light. *Proceedings of the National Academy of Sciences of the United States of America*, 111(28) 10359–10364. https://doi.org/10.1073/PNAS.1409457111/SUPPL_FILE/PNAS.201409457SI.PDF.

Wang, Y., Gu, X., Yuan, W., Schmitz, R. J., & He, Y. (2014). Photoperiodic control of the floral transition through a distinct polycomb repressive complex. *Developmental Cell*, 28(6), 727–736. https://doi.org/10.1016/j.devcel.2014.01.029.

Wils, C. R., & Kaufmann, K. (2017). Gene-regulatory networks controlling inflorescence and flower development in *Arabidopsis thaliana. Biochimica et Biophysica Acta (BBA) – Gene Regulatory Mechanisms*, 1860(1), 95–105. https://doi.org/10.1016/J.BBAGRM.2016.07.014.

Wu, M. F., Sang, Y., Bezhani, S., Yamaguchi, N., Han, S. K., Li, Z., Su, Y., Slewinski, T. L., & Wagner, D. (2012). SWI2/SNF2 chromatin remodeling ATPases overcome polycomb repression and control floral organ identity with the LEAFY and SEPALLATA3 transcription factors. *Proceedings of the National Academy of Sciences of the United States of America*, 109(9), 3576–3581. https://doi.org/10.1073/PNAS.1113409109/SUPPL_FILE/PNAS.201113409SI.PDF.

Xianfeng, M. X., Rose, A., Muthuswamy, S., Sun, Y. J., Venkatakrishnan, S., Zhao, Q., & Meier, I. (2007). NUCLEAR PORE ANCHOR, the *Arabidopsis* homolog of Tpr/Mlp1/Mlp2/Megator, is involved in mRNA export and SUMO homeostasis and affects diverse aspects of plant development. *The Plant Cell*, 19(5), 1537–1548. https://doi.org/10.1105/TPC.106.049239.

Xu, L., & Shen, W. H. (2008). Polycomb silencing of KNOX genes confines shoot stem cell niches in *Arabidopsis. Current Biology*, 18(24), 1966–1971. https://doi.org/10.1016/j.cub.2008.11.019.

Yaish, M. W. F., Peng, M., & Rothstein, S. J. (2009). AtMBD9 modulates *Arabidopsis* development through the dual epigenetic pathways of DNA methylation and histone acetylation. *The Plant Journal*, 59(1), 123–135. https://doi.org/10.1111/J.1365-313X.2009.03860.X.

Yamaguchi, A., Kobayashi, Y., Goto, K., Abe, M., & Araki, T. (2005). TWIN SISTER OF FT (TSF) acts as a floral pathway integrator redundantly with FT. *Plant and Cell Physiology*, 46(8), 1175–1189. https://doi.org/10.1093/PCP/PCI151.

Yamamuro, C., Zhu, J. K., & Yang, Z. (2015). Epigenetic modifications and plant hormone action. *Molecular Plant*, 9(1), 57. https://doi.org/10.1016/J.MOLP.2015.10.008.

Yan, B., Yang, Z., He, G., Jing, Y., Dong, H., Ju, L., Zhang, Y., Zhu, Y., Zhou, Y., & Sun, J. (2021). The blue light receptor CRY1 interacts with GID1 and DELLA proteins to repress gibberellin signaling and plant growth. *Plant Communications*, 2(6), 100245. https://doi.org/10.1016/J.XPLC.2021.100245/ATTACHMENT/9F69A789-BD12-495C-ACA4-EC031049B957/MMC2.PDF.

Yang, J., Li, L., Li, X., Zhong, M., Li, X., Qu, L., Zhang, H., Tang, D., Liu, X., He, C., & Zhao, X. (2023). The blue light receptor CRY1 interacts with FIP37 to promote N6-methyladenosine RNA modification and photomorphogenesis in Arabidopsis. *New Phytologist*, 237(3), 840–854. https://doi.org/10.1111/NPH.18583.

Yang, Z., Qian, S., Scheid, R. N., Lu, L., Chen, X., Liu, R., Du, X., Lv, X., Boersma, M. D., Scalf, M., Smith, L. M., Denu, J. M., Du, J., & Zhong, X. (2018). EBS is a bivalent histone reader that regulates floral phase transition in *Arabidopsis. Nature Genetics*, 50(9), 1247–1253. https://doi.org/10.1038/s41588-018-0187-8.

Yoo, S. K., Chung, K. S., Kim, J., Lee, J. H., Hong, S. M., Yoo, S. J., Yoo, S. Y., Lee, J. S., & Ahn, J. H. (2005). CONSTANS activates SUPPRESSOR OF OVEREXPRESSION OF CONSTANS 1 through FLOWERING LOCUS T to promote flowering in *Arabidopsis. Plant Physiology*, 139(2), 770–778. https://doi.org/10.1104/PP.105.066928.

Yu, C. W., Liu, X., Luo, M., Chen, C., Lin, X., Tian, G., Lu, Q., Cui, Y., & Wu, K. (2011). HISTONE DEACETYLASE6 interacts with FLOWERING LOCUS D and regulates flowering in *Arabidopsis. Plant Physiology*, 156(1), 173–184. https://doi.org/10.1104/PP.111.174417.

Yue, H., Nie, X., Yan, Z., & Weining, S. (2019). N6-methyladenosine regulatory machinery in plants: Composition, function and evolution. *Plant Biotechnology Journal*, 17(7), 1194–1208. https://doi.org/10.1111/PBI.13149.

Zaccara, S., Ries, R. J., & Jaffrey, S. R. (2019). Reading, writing and erasing mRNA methylation. *Nature Reviews Molecular Cell Biology*, 20(10), 608–624. https://doi.org/10.1038/s41580-019-0168-5.

Zemach, A., Gaspan, O., & Grafi, G. (2008). The three methyl-CpG-binding domains of AtMBD7 control its subnuclear localization and mobility. *Journal of Biological Chemistry*, 283(13), 8406–8411. https://doi.org/10.1074/jbc.M706221200.

Zhang, C., Jian, M., Li, W., Yao, X., Tan, C., Qian, Q., Hu, Y., Liu, X., & Hou, X. (2023). Gibberellin signaling modulates flowering via the DELLA–BRAHMA–NF-YC module in *Arabidopsis. The Plant Cell*, 35(9), 3470–3484. https://doi.org/10.1093/PLCELL/KOAD166.

Zhang, Y. Y., Fischer, M., Colot, V., & Bossdorf, O. (2013). Epigenetic variation creates potential for evolution of plant phenotypic plasticity. *New Phytologist*, 197(1), 314–322. https://doi.org/10.1111/NPH.12010.

Zhou, X., & Ma, H. (2008). Evolutionary history of histone demethylase families: Distinct evolutionary patterns suggest functional divergence. *BMC Evolutionary Biology*, 8(1), 1–16. https://doi.org/10.1186/1471-2148-8-294/FIGURES/7.

Zilberman, D., Gehring, M., Tran, R. K., Ballinger, T., & Henikoff, S. (2006). Genome-wide analysis of *Arabidopsis thaliana* DNA methylation uncovers an interdependence between methylation and transcription. *Nature Genetics*, 39(1), 61–69. https://doi.org/10.1038/ng1929.

16 Epigenetic Mechanisms of Plant Reproduction, Heterosis and Tissue Culture

Dilawar Ali, Asad Azeem and Sami Ul-Allah

16.1 INTRODUCTION

Epigenetics is the study of changes in organisms not due to DNA regulations but due to some other factors which play a crucial role in morphological and physiological characters (Guo et al., 2025). Epigenetics examines gene expression and causes changes in gene function without any change in DNA sequences, which may or may not be heritable and are essential for processes like plant reproduction, heterosis and tissue culture. DNA methylation, histone modifications and other epigenetic mechanisms influence plant reproduction, hybrid vigor and tissue culture (Nasir et al., 2024). It can be heritable or non-heritable, existing independently of the DNA sequence of an individual. Some epigenetic expressions help in identifying the role of proteins. Epigenetic regulations help in the development of hybrids and also cause inbreeding depressions. Moreover, various genes undergo epigenetic regulations and produce negative or positive effects. Along with the study of genetic mechanisms in hybrids, epigenetics also plays an important role in the development of hybrid vigor. In this sense, DNA methylation and histone modification help in the development of hybrids which show resistance to biotic and abiotic stress (Afreen et al., 2024). Many researches show that there is a solid connection between DNA methylation and heterosis in crop plants. Although there are a lot of molecular breeding techniques which can help in the development of hybrids, epigenetics can develop some novel offspring. This chapter delves into the major and interesting secrets of epigenetics and helps us to understand the micro-level study of inheritance and variations (Pagiatakis et al., 2021). Unlike mutation which cannot be controlled, epigenetics can be reversed. If any modification causes a harmful effect, it can be reversed. It may also pass from generation after generation.

16.2 MECHANISMS OF EPIGENETICS

Epigenetics has a wide range of mechanisms. Following are some key mechanisms of epigenetics (Figure 16.1).

16.2.1 DNA Methylation

DNA methylations modify a plant genotype, improve plant adaptability and increase plant yield (Sun et al., 2022). This regulation helps in masking the effect of susceptible genes and creates defensive approaches in plants against environmental stress, improving the fitness of plants for survival. Enzymatic addition of a methyl group to the cyclic carbon 5 of cytosine is the procedure of DNA methylation, representing a heritable epigenetic modification (Banerjee & Roychoudhury, 2017, 2018). This process can also reduce gene expression under stressful conditions, allowing plants to conserve energy and increase their chances of survival. While methylation is predominantly associated with host defense mechanisms in prokaryotes, it also plays various roles in eukaryotes, primarily regulating transposable elements within the genome (Akhter et al., 2021).

The extent of methylation varies significantly among organisms; for instance, methylated cytokines (5-mC) vary from kingdom to kingdom, and even within a kingdom, it ranges from 0 to 3% in insects, 10% in marine animals and 30% in plants (Chapelle & Silvestre, 2022). Variations in DNA methylation can occur in response to both internal and external signals. Stress-induced tissue level changes at developmental stages in DNA methylation have been documented in plants (Akhter et al., 2021; Azeem et al., 2025). A family of conserved DNA methyltransferases catalyzes the DNA methylation process, indicating that this alteration has an ancestral origin. DNA MTases are classified into different types: (a) maintenance methylases, which ensure consistent cytosine methylation throughout cell generations; (b) de novo methylases, which add methyl groups to cytosine that were previously unmethylated; and (c) domain-rearrangement methylases, which use RNA-directed DNA methylation to methylate homologous genes, especially under the guidance of short RNAs. Mutations in any of the ten or more DNA methyltransferases found in *Arabidopsis* can impair regular development or other traits of the plant. Demethylation can be done from DNA either through active or passive means. In passive mechanisms, the methyl group

DOI: 10.1201/9781003589914-19

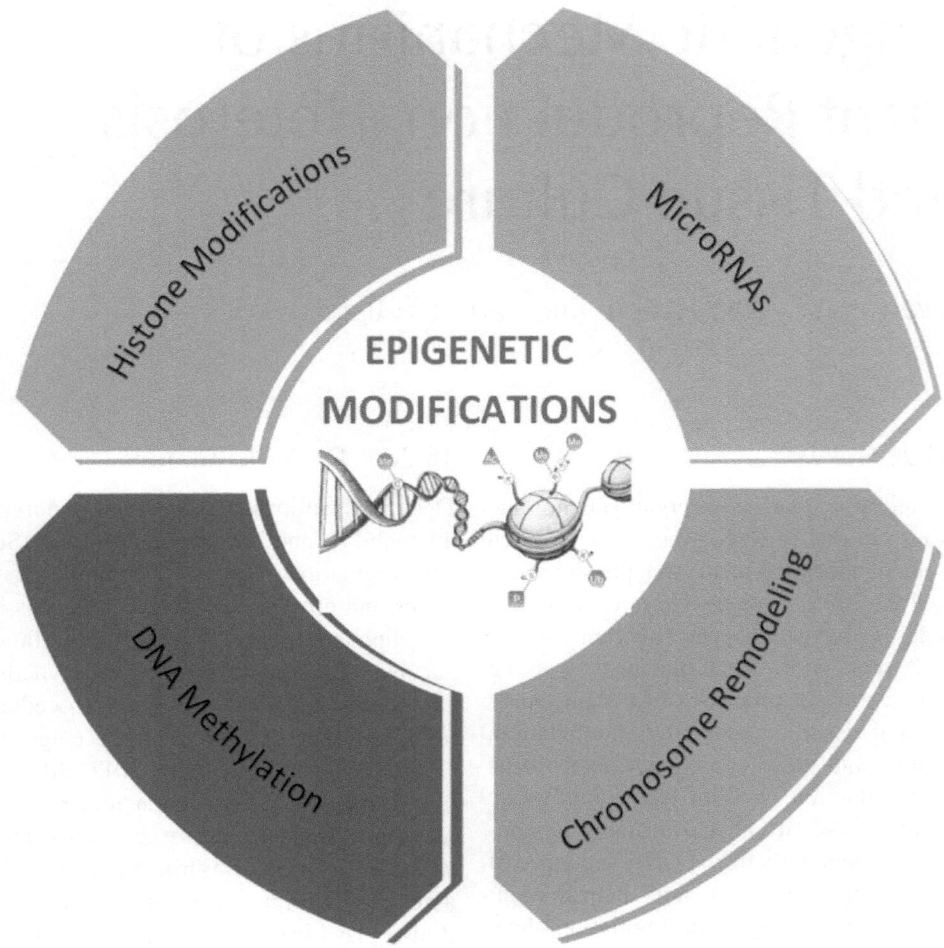

FIGURE 16.1 Epigenetic modification in plants.

is removed by replacing the methyl group with non-methyl cytosines during DNA replication (Acharjee et al., 2023). In contrast, active demethylation involves specific proteins that have been shown to demethylate DNA sequences in mammals and birds.

Two mutations in *Arabidopsis* have been identified: one is the gene effect on maternal expression, while the second induces transcriptional gene silencing (TGS) of transgenes and homologous loci. *DEMETER (DME1)* and *ROS1* genes are linked with the mutations and encode DNA glycosylases that remove 5-methylcytosine from DNA (Fei et al., 2021). The primary functions of methylation include: (i) serving as a heritable epigenetic mark that guides the developmental program of organisms, (ii) providing a defense mechanism against the activity of parasitic mobile elements, (iii) reducing transcriptional noise in organisms with many genes, and (iv) stabilizing the patterns of gene silencing to "memorize" gene activity.

16.2.2 Histone Modifications

Histone modifications serve as an important epigenetic mechanism that connects genetics and environmental factors to gene expression. Histones are fundamental proteins that help organize DNA (Zhang et al., 2021). These modifications have been recognized as key epigenetic regulators, with various alterations such as methylation, acetylation, phosphorylation, ribosylation, and ubiquitination occurring on conserved lysine residues in the amino-terminal tail domains. Typically, histone acetylation is associated with transcriptional regions, while in transcriptionally inactive euchromatic or heterochromatic areas, hypoacetylated histone proteins are present. Both active and inactive chromatin regions can be indicated by histone methylation; for example, gene action is linked with methylation of lysine 4 on histone H3 (H3-K4) and is primarily located at the promoters of active genes. In contrast, DNA silencing is marked by methylation of lysine 9 on the N terminus of histone H3 (H3-K9), commonly present at heterochromatic regions such as centromeres and telomeres. Methylation of lysine can exist in various forms: monomethyl, dimethyl or trimethyl. Additionally, histones may undergo other modifications, including phosphorylation, ribosylation, and ubiquitination, resulting in a complex array of potential modification combinations, often referred to as the "histone code." Cellular factors can recognize and interpret this code (Zhang et al., 2021).

Relationships between histone modifications and DNA methylation are very significant; for instance, H3-K9 methylation promotes DNA methylation and vice versa. Furthermore, histone deacetylases (HDACs), histone methyltransferases and methylcytosine-binding proteins are known to recruit DNA methyltransferases, although the initial triggers for recruiting these epigenetic modifiers to their specific targets remain unclear.

Multimeric protein complexes, also known as "cellular memory keys," are involved in the inheritance of gene expression through multiple rounds of cell division. These complexes were initially studied in *Drosophila*, with two primary groups identified: one is the polycomb group (PcG) and other is the trithorax group (trxG). While the PcG maintains homeotic genes in an inactive state, the trxG keeps them active. Similar complexes have been found in the kingdom Plantae, with the PcG group being the most thoroughly researched. The role of this complex has been documented in regulating the expression of homeotic genes that are involved in plant development (Shirasu et al., 2003). Although the PcG target genes in *Drosophila* and *Arabidopsis* do not share structural similarities, they perform similar functional functions in controlling developmental stages. A key component of the polycomb group complex in plants includes methyltransferases like the MEDEA protein in *Arabidopsis*, which particularly methylates H3-K27, resulting in silent chromatin (Santana et al., 2023).

16.2.3 RNA Interference

Gene silencing is brought about through RNA in which double stranded RNA (dsRNA) is used. The process involves developing small RNA (siRNA) through the Dicer enzyme. dsRNA is developed through transcription (Akbar et al., 2022). This mechanism was soon shown to induce gene silencing in other organisms, including plants (Akbar et al., 2022; Singh & Roychoudhury, 2023; Xu et al., 2024). During the transgenic approaches in tobacco, transcriptional gene silencing (TGS) was noted. The silenced transgenes exhibited DNA methylation (modification associated with gene silencing), which reverted when the complexes segregated in progeny. Posttranscriptional gene silencing (PTGS) was identified through two approaches. One involved testing antisense suppression. It is a potential method for selectively silencing plant genes. Antisense RNA is generated from a transgene in which the mRNA is complementary to the target gene. However, sense transgene RNAs, which should not base pair, often induce silencing unexpectedly. In another experiment aimed at enhancing floral coloration in petunia through the overexpression of a pigment synthesis transgene, partial or complete absence of color occurred due to mutual silencing of both genes, viz., transgene and gene present in the plant, later shown to operate at the posttranscriptional level. Similar to TGS, PTGS was frequently linked with DNA methylation of transgene sequences.

16.2.4 Chromatin Remodeling

Chromatin remodeling enables transcription factors to access the nucleosome octamer core, potentially resulting in irregular gene expression patterns. This process alters the structure of chromatin, making it more accessible for proteins that initiate transcription (Downs & Gasser, 2024). It helps in access to transcription factors and proteins. Euchromatin is the part of chromosomes in which transcription and translation occur and is loosely packed, while heterochromatin is the part of chromosomes where there is no transcription and translation and is tightly packed (Rhodes, 2024). In chromatin remodeling, ATP is the fueling agent that helps in the movement and reshaping of the nucleosomes. This method requires ATP, which helps in chromatin architecture. Chromatin remodeling can regulate or deactivate gene expression. Chromatin remodeling aberrations interfere with gene expression and cause various negative impacts on plants. Quantitative traits are sensitive to environmental changes, so chromatin remodeling can also be affected by the environment. Chromatin remodeling is innovative, helping to shape the gene and regulate the gene with its own interest.

16.2.5 Somaclonal Variation

Somaclonal variation refers to genetic or epigenetic changes that occur in plant cell and tissue culture. These variations can introduce desirable genetic diversity into a gene pool (Manchanda et al., 2024). Typically, somaclonal variation arises from heritable genetic alterations induced by tissue culture due to rearrangement of chromatin material. Any plants regenerated from these cells may be mistakenly considered as variants acquired during the culture process, even though they may have represented only a tiny fraction of the original explant. Recent findings indicate that somatically acquired epigenetic modifications can be stable through mitosis and heritable through meiosis, leading to a greater focus on DNA methylation as a source of somaclonal variation. Initial evidence for this connection came from studies on regenerants from crown-gall tumor lines, where alterations in T-DNA methylation correlated with phenotypic changes (Manchanda et al., 2024). Additional studies demonstrated demethylation and reactivation of silent Ac elements following tissue culture. Compelling evidence suggests that a significant portion of somaclonal variation might stem from diverse preexisting epigenetic states that remain stable in regenerated plants. It was shown that variation in antibiotic resistance among regenerated double transformants was due to the original transgene complex existing as an epigenetic mosaic with partial methylation. If similar findings are confirmed for endogenous plant genes, this could reveal substantial variability among individual somatic cells (Ibrahim et al., 2024).

Emerging technologies now allow for the analysis of individual somatic cells, which could be crucial, especially since some meristematic cells will eventually become gametes. It was also noted that both total methylation levels and specific site methylations change during the culture process, suggesting that methylation levels may decrease over time. Additionally, various studies indicate that the timing of development can influence methylation variations and patterns (Ibrahim et al., 2024).

16.3 EPIGENETIC INHERITANCE AND PLANT BREEDING

Epigenetic modifications – such as changes in chromosomal proteins and DNA methylation, often referred to as epimutations – have significant phenotypic effects on organisms. The study of epigenetic traits in plants has led geneticists to appreciate the importance of epigenetic inheritance systems in plant evolution, domestication and breeding (Cao & Chen, 2024). Over the past 50 years, the recognition of epigenetic inheritance in plants has grown, based on key findings that epigenetic phenotypes can arise from the following: (1) the activation, excision and translocation of transposable elements; (2) allelic interactions known as paramutations; (3) transgene silencing observed in transgenic plants since the late 1980s; and (4) epialleles of endogenous genes influencing floral induction, seed development and parental imprinting. Comprehensive reviews on these subjects are available, providing detailed insights. Focusing on plant breeding, the following sections will explore how epigenetic inheritance contributes to polymorphism and offers valuable variations for breeding purposes. Additionally, the role of DNA methylation in hybrid vigor and performance stability, as well as aspects of epigenetic transgene silencing in modern elite transgenic varieties, will also be discussed (Cao & Chen, 2024). Epigenetic inheritance helps the breeder in several key ways.

16.3.1 Generation of Variation

DNA methylation and histone modification help in creating the diversity in plants. This diversity helps in the selection of new traits which may be helpful for breeding.

16.3.2 Stability of Traits

Many epigenetic regulations are heritable, and they can be transferred from generation to generation. The stable nature of epigenetic changes helps breeders select desirable traits. It increases genetic diversity in plants.

16.3.3 Response to Environmental Stimuli

Epigenetics helps in the development of such types of plants that can tolerate environmental changes (Jin et al., 2024), for example, epigenetics that develop resistance to abiotic and biotic stress are useful in breeding programs. Breeders can exploit these mechanisms to develop a proper response to external stimuli.

16.3.4 Hybrid Vigor (Heterosis)

Hybrids are those offsprings that are superior to their parents. Epigenetics play a vital role in the development of plants that have superior traits such as yield, growth, resistance and many more (Cao & Chen, 2024).

16.3.5 Transgene Regulation

Epigenetic silencing is a technique that regulates the expression of foreign inserted genes, creating higher resistance to pests. Epigenetic approaches allow the breeder to develop transgenic plants.

16.3.6 Allelic Interactions

Paramutation, which creates interaction between alleles, causes heritable changes in gene expression; this is the phenomenon of epigenetics. It results in the creation of new phenotypes, which may be useful for trait improvement. With the help of epigenetics, plant breeders can widen the genetic bases of a plant, create diversity in plants, improve the crop for resistance and develop such types of plants that can perform in any environmental conditions. These techniques accelerate the breeding programs and approaches to the next level and improve strategies.

16.4 EPIGENETIC REGULATION OF PLANT REPRODUCTION

Epigenetic control is central to plant reproduction through the regulation of gene expression patterns that are vital for gametogenesis, fertilization, seed formation and transgenerational epigenetic inheritance (Liu et al., 2024). Together, DNA methylation, histone modifications, small RNAs and chromatin remodeling control reproductive outcomes by ensuring genomic stability, governing parent-of-origin effects and controlling gene activity at major developmental transitions. DNA methylation plays a crucial role in gametophyte and seed formation, where it is involved in ensuring accurate gene imprinting and silencing of transposable elements (TEs) for genome integrity. In male and female gametogenesis, characteristic patterns of DNA demethylation and remethylation coordinate the expression of important genes necessary for pollen viability, ovule development and embryo formation. In *Arabidopsis* and other model organisms, maternal allele demethylation in the endosperm, which is regulated by the DNA demethylase DEMETER (DME), is important for normal seed development through regulation of imprinted gene expression. Histone modification also plays a part in epigenetic control by affecting chromatin accessibility and transcriptional states (Liu et al., 2024). Histone acetylation, which is linked to active gene expression, is involved in gametophyte differentiation and pollen tube elongation, while histone methylation, especially H3K9me2 and H3K27me3, is involved in germline silencing and seed imprinting. Polycomb repressive complexes (PRCs) mediate histone methylation to control seed dormancy, germination and endosperm development. In addition, non-coding RNAs, including small interfering RNAs (siRNAs) and microRNAs (miRNAs), are crucial in plant reproduction as they regulate gene expression posttranscriptionally. siRNAs are implicated in RNA-directed DNA methylation (RdDM), a process important for the maintenance of genome stability in gametes and silencing of TE activation. In hybrid plants, differential expression of small RNAs can result in hybrid vigor or reproductive isolation caused by epigenetic incompatibilities (Abdulraheem et al., 2024). Chromatin remodeling is also responsible for plant reproduction at the core level by dynamically modifying nucleosome positions to control flowering time, meiosis and seed development genes. In flowering plants, epigenetic marks control flowering gene expression such as *FLC (FLOWERING LOCUS C)* to promote vegetative to reproductive transition properly. Environmental signals, for example, temperature and photoperiod, shape epigenetic changes, thus illustrating epigenetic control flexibility in plant reproduction. Epigenetic inheritance can also propagate some gained features to further generations, thereby impacting seed viability, dormancy and resilience against stress. Elucidation of epigenetic regulation of plant reproduction has important consequences for crop improvement, hybrid seed production and genetic improvement programs to improve reproductive efficiency and yield stability under variable environmental conditions (Abdulraheem et al., 2024).

16.5 HETEROSIS AND EPIGENETIC REGULATION

Heterosis refers to the advantage and superior performance of a hybrid compared to its parents and commercial varieties (Liao et al., 2025). Although the genetic underpinnings of heterosis have been well investigated, new studies emphasize the contribution of epigenetic regulation to the modulation of heterotic effects. Epigenetic pathways such as DNA methylation, histone modifications, small RNAs and chromatin remodeling are responsible for gene expression changes that affect hybrid performance. DNA methylation is essential in heterosis as it controls allele-specific expression, transposon silencing and genome stability in hybrids. The differential methylation patterns between hybrids and parental lines may result in changed gene expression, influencing growth and metabolic processes. Hybrids in certain instances have intermediate or new methylation patterns, which result in the activation or repression of important genes related to increased vigor (Liao et al., 2025). Histone modifications are also responsible for heterosis through the regulation of chromatin accessibility and transcriptional activity. Hybrids tend to have novel histone acetylation and methylation patterns that can activate or suppress gene expression, resulting in enhanced physiological traits. Non-coding RNAs, especially microRNAs (miRNAs) and small interfering RNAs (siRNAs), are important in epigenetic regulation of heterosis. miRNAs regulate gene expression after transcription by degrading or repressing mRNAs, thus modulating growth-related pathways. Differential miRNA expression in hybrids can lead to changed hormonal signaling, stress response and metabolic efficiency. Moreover, siRNAs play a role in RNA-directed DNA methylation (RdDM), which affects chromatin structure and transposon activity and is responsible for genome stability and hybrid vigor. Chromatin remodeling is also a significant epigenetic process implicated in heterosis. Chromatin-remodeling complexes in hybrids promote dynamic nucleosome positioning changes, providing access to growth, development and stress tolerance-related genes. These changes lead to gene expression patterns that are optimized and contribute to better hybrid traits. Epigenetic reprogramming of hybrids depends on environmental influences, developmental processes and genomic interactions between the parental alleles. Heterotic effects in some instances are sustained over generations, implying transgenerational epigenetic inheritance. Elucidating the role of epigenetic regulation in heterosis opens new avenues for improving crops and hybrid breeding programs. Through control of DNA methylation, histone modification and small RNA pathways, scientists can improve hybrid performance, enhance stress tolerance and maximize yield potential in crops of economic value. Combining epigenetic knowledge with conventional breeding strategies has the potential to create more efficient and sustainable farming practices (Paril et al., 2024).

16.6 EPIGENETIC MECHANISMS IN PLANT TISSUE CULTURE

Plant tissue culture refers to the regeneration of plants from somatic cells. Epigenetic regulation is important in plant tissue culture as it regulates gene expression without changing the sequence of DNA, affecting somatic embryogenesis, organogenesis, stress adaptation and clonal fidelity. DNA methylation, which is one of the major epigenetic modifications, is the process of adding methyl groups to cytosine residues within CpG, CHG and CHH context sequences (Pasternak & Steinmacher, 2024). This alteration is essential in gene silencing and activation at different stages of in vitro culture development. Excessive DNA methylation may result in diminished embryogenic capacity, whereas demethylation is reported to result in increased cell dedifferentiation, callus induction and somatic embryo development. Epigenetic drugs such as 5-azacytidine, an inhibitor of DNA methylation, have been found to enhance regeneration efficiency in some plant species. In addition to DNA methylation, histone modifications are crucial in chromatin dynamics and gene expression control (Zhang & Kiarasi, 2024). Transcriptional activation is facilitated by histone acetylation through the action of histone acetyltransferases (HATs), while histone deacetylation through histone deacetylases (HDACs) leads to gene repression and lowered organogenic capacity. Histone methylation can be both permissive and inhibitory of gene expression, with H3K4me3 being an activating mark and H3K9me3 and H3K27me3 being associated with transcriptional silencing. Non-coding RNAs (ncRNAs), especially microRNAs (miRNAs) and small interfering RNAs (siRNAs), play the role of other regulatory components in epigenetic regulation during plant tissue culture. miRNAs control gene expression at the posttranscriptional level through the degradation or translational suppression of particular mRNAs (Pasternak & Steinmacher, 2024), thus affecting developmental processes and stress responses. siRNAs are involved in RNA-directed DNA methylation (RdDM), a mechanism of chromatin state modification and silencing of transposons and repetitive elements that are essential for genome stability in regenerated plants. In addition, chromatin remodeling complexes, including SWI/SNF, CHD and ISWI, dynamically regulate nucleosome positioning to enable the activation of genes responsible for cellular reprogramming and totipotency acquisition in plant cells. These epigenetic changes are crucial for controlling somaclonal variation, a widespread phenomenon in plant tissue culture that occurs as a result of epigenetic instability and adds to genetic diversity but can also result in unwanted characteristics.

Epigenetic reprogramming is extremely sensitive to external factors like plant growth regulators (PGRs), stress and the composition of culture media. Auxins and cytokinins have an effect on DNA methylation and histone modification, hence on morphogenic potential. Environmental stress like osmotic stress, temperature stress and oxidative stress that occur during in vitro culture are capable of inducing epigenetic changes that result in stress memory, allowing plants to adapt more effectively to changing conditions. Moreover, epigenetic changes can be reversed, i.e., with the help of proper treatments like exposing them to certain hormones or epigenetic modifiers, gene expression profiles can be restored to increase regeneration efficacy and genetic stability. Knowledge of these epigenetic regulatory pathways is invaluable in terms of enhancing the effectiveness of plant tissue culture methods, clonal stability in micropropagation and employing epigenetic markers for crop improvement. Through the control of DNA methylation, histone modification, and small RNA pathways, scientists can design methods to increase the efficiency of plant regeneration, reduce unwanted somaclonal variation and enhance stress tolerance in tissue-cultured plants (Zhang & Kiarasi, 2024).

16.7 CONCLUSION

Epigenetic processes are central to plant reproduction, heterosis and tissue culture. The modifications like DNA methylation, histone modification, small RNA pathways and chromatin remodeling control important biological processes like gametogenesis, hybrid vigor and cellular reprogramming during in vitro culture. During plant reproduction, epigenetic control guarantees correct genome stability, parental imprinting and seed formation, affecting fertility and transgenerational inheritance. The specific regulation of DNA methylation and histone modifications dictates the activation or suppression of genes required for successful reproduction and seed viability. Likewise, heterosis is controlled by epigenetic elements that regulate gene expression patterns in hybrids, leading to increased growth, yield and stress tolerance. Epigenetic reprogramming in hybrids results in specific gene expression profiles that maximize physiological and metabolic functions, thereby enhancing hybrid performance. Epigenetic changes regulate somatic embryogenesis, organogenesis and stress tolerance in plant tissue culture through the regulation of cell differentiation and acquisition of totipotency. The use of epigenetic modifications may be crucial in developing climate-resilient crop cultivars.

REFERENCES

Abdulraheem, M. I., Xiong, Y., Moshood, A. Y., Cadenas-Pliego, G., Zhang, H., & Hu, J. (2024). Mechanisms of plant epigenetic regulation in response to plant stress: Recent discoveries and implications. *Plants*, 13(2), 163.

Acharjee, S., Chauhan, S., Pal, R., & Tomar, R. S. (2023). Mechanisms of DNA methylation and histone modifications. *Progress in Molecular Biology and Translational Science*, 197, 51–92.

Afreen, U., Mukhopadhyay, K., & Kumar, M. (2024). DNA methylation in wheat: Current understanding and future potential for enhancing biotic and abiotic stress tolerance. *Physiology and Molecular Biology of Plants*, 30(12):1921-1933

Akbar, S., Wei, Y., & Zhang, M.-Q. (2022). RNA interference: Promising approach to combat plant viruses. *International Journal of Molecular Sciences*, 23(10), 5312.

Akhter, Z., Bi, Z., Ali, K., Sun, C., Fiaz, S., Haider, F. U., & Bai, J. (2021). In response to abiotic stress, DNA methylation confers epigenetic changes in plants. *Plants*, 10(6), 1096.

Azeem, A., Hafeez, A., & Ul-Allah, S. (2025). Molecular breeding for crop plants: Improvement and practices. (Ed.) Azamel Husen. In *Agricultural Crop Improvement* (pp. 64–74). CRC Press.

Banerjee, A., & Roychoudhury, A. (2017). Epigenetic regulation during salinity and drought stress in plants: Histone modifications and DNA methylation. *Plant Gene*, 11, 199–204.

Banerjee, A., & Roychoudhury, A. (2018). The gymnastics of epigenomics in rice. *Plant Cell Reports*, 37, 25–49.

Cao, S. and Chen, Z. J., (2024). Transgenerational epigenetic inheritance during plant evolution and breeding. *Trends in Plant Science*, 29(11), pp.1203–1223.

Chapelle, V., & Silvestre, F. (2022). Population epigenetics: The extent of DNA methylation variation in wild animal populations. *Epigenomes*, 6(4), 31.

Downs, J. A., & Gasser, S. M. (2024). Chromatin remodeling and spatial concerns in DNA double-strand break repair. *Current Opinion in Cell Biology*, 90, 102405.

Fei, Y., Pyott, D. E., & Molnar, A. (2021). Temperature modulates virus-induced transcriptional gene silencing via secondary small RNAs. *New Phytologist*, 232(1), 356–371.

Guo, Y.-W., Liu, Y., Huang, P.-C., Rong, M., Wei, W., Xu, Y.-H., & Wei, J.-H. (2025). Adaptive changes and genetic mechanisms in organisms under controlled conditions: A review. *International Journal of Molecular Sciences*, 26(5), 2130.

Ibrahim, M. S. D., Hartati, R. S., Rubiyo, R., Reflinur, R., Purwito, A., & Sudarsono, S. (2024). Exploring indirect somatic embryogenesis and somaclonal variation for propagation of three Coffea arabica L. cultivars. *Chilean Journal of Agricultural Research*, 84(1), 15–27.

Jin, Q., Chachar, M., Ali, A., Chachar, Z., Zhang, P., Riaz, A., Ahmed, N., & Chachar, S. (2024). Epigenetic regulation for heat stress adaptation in plants: New horizons for crop improvement under climate change. *Agronomy*, 14(9), 2105.

Liao, X., Tang, M., Lei, W., Lin, H., Jin, S., Wang, P., & Ye, N. (2025). Epigenetic mechanisms of heterosis in tea plants: Insights into m6A RNA methylation and miRNA-mediated regulation. *Horticultural Plant Journal*. https://doi.org/10.1016/j.hpj.2024.11.005

Liu, J., Ke, M., Sun, Y., Niu, S., Zhang, W., & Li, Y. (2024). Epigenetic regulation and epigenetic memory resetting during plant rejuvenation. *Journal of Experimental Botany*, 75(3), 733–745.

Manchanda, P., Sharma, D., Kaur, G., Kaur, H., & Vanshika. (2024). Exploring the significance of somaclonal variations in horticultural crops. *Molecular Biotechnology*, 67(6), pp.2185–2203.

Nasir, L., Yimam, T., & Guadie, D. (2024). Review on epigenetic modification mechanisms and their application for the genetic improvement of plants. *SINET: Ethiopian Journal of Science*, 47(3), 258–286.

Pagiatakis, C., Musolino, E., Gornati, R., Bernardini, G., & Papait, R. (2021). Epigenetics of aging and disease: A brief overview. *Aging Clinical and Experimental Research*, 33, 737–745.

Paril, J., Reif, J., Fournier-Level, A., & Pourkheirandish, M. (2024). Heterosis in crop improvement. *The Plant Journal*, 117(1), 23–32.

Pasternak, T. P., & Steinmacher, D. (2024). Plant growth regulation in cell and tissue culture in vitro. *Plants*, 13(2), 327.

Rhodes, D. (2024). To slide or not to slide: Key role of the hexasome in chromatin remodeling revealed. *Nature Structural & Molecular Biology*, 31(5), 742–746.

Santana, D. A., Smith, M. d. A. C., & Chen, E. S. (2023). Histone modifications in Alzheimer's disease. *Genes*, 14(2), 347.

Shirasu, K., & Schulze-Lefert, P. (2003). Complex formation, promiscuity and multi-functionality: Protein interactions in disease-resistance pathways. *Trends in Plant Science*, 8(6), 252–258.

Singh, A., & Roychoudhury, A. (2023). RNA Interference (RNAi) technology: An effective tool in plant breeding. In A. Raina, M. R. Wani, R. A. Laskar, N. Tomlekova, & S. Khan (Eds.), *Advanced Crop Improvement, Volume 1* (pp. 309–320). Springer Nature Switzerland AG.

Sun, M., Yang, Z., Liu, L., & Duan, L. (2022). DNA methylation in plant responses and adaption to abiotic stresses. *International Journal of Molecular Sciences*, 23(13), 6910.

Xu, Q.-Q., Shang, F., Feng, S.-Y., Xie, Q.-P., Zhang, W., Wang, Z.-G., & Wang, J.-J. (2024). Design the fusion double-strand RNAs to control two global sap-sucking pests. *Pesticide Biochemistry and Physiology*, 205, 106114.

Zhang, S., & Kiarasi, F. (2024). Therapeutic effects of resveratrol on epigenetic mechanisms in age-related diseases: A comprehensive review. *Phytotherapy Research*, 38(5), 2347–2360.

Zhang, Y., Sun, Z., Jia, J., Du, T., Zhang, N., Tang, Y., Fang, Y. and Fang, D. (2021). Overview of histone modification. In: Fang, D., Han, J. (eds) *Histone Mutations and Cancer. Advances in Experimental Medicine and Biology*, 1283. Springer, Singapore. https://doi.org/10.1007/978-981-15-8104-5_1

17 Epigenetic Regulation in the Signaling of Phytohormones and Plant Growth Regulators

Uma S, Nisha Sam Nirmala, Ashwini R and Sivarama Krishna Lakkaboyana

17.1 INTRODUCTION

Conrad Waddington, a British developmental biologist, introduced the term "epigenetics" by combining "epigenesis" and "genetics" (Kinoshita et al., 2012). Earlier it was understood that the term epigenetics meant the study of the genes that play a role in the formation of a phenotype. Over time, the concept of epigenetics expanded to include the study of heritable modifications in gene function that occur independently of changes to the DNA sequence itself. Waddington's experiments on Drosophila embryos with environmental factors supported his "Genetic assimilation" theory that the genetic makeup of the organism changed over time to guarantee that the development proceeded in a particular manner despite typical environmental fluctuations (Raju et al., 2023). Epigenetics offers one of the reasons why cells and organisms with the same DNA can exhibit such significant phenotypic variations.

The regulation of plant hormone signaling, essential for plant growth and development, is significantly influenced by epigenetic mechanisms that facilitate responses to environmental cues. Epigenetic mechanisms, including DNA methylation, histone modification and small RNA-mediated gene regulation, interact with plant hormone signaling pathways to fine-tune gene expression and regulate various physiological processes. The complex crosstalk between epigenetic mechanisms and plant hormone signaling allows plants to reprogram their responses to changing environment, retain epigenetic memories and manifest phenotypic plasticity. Studies have demonstrated that epigenetic factors are integral to the signaling processes of several plant hormones, including auxin, gibberellin, cytokinin, abscisic acid and ethylene, highlighting the importance of epigenetics in plant hormone biology (Yamamuro et al., 2016). Understanding the epigenetic regulation of plant hormone signaling is essential for unraveling the molecular mechanisms underlying plant development and environmental responses and has significant implications for improving crop yield, disease resistance and stress tolerance.

17.2 HISTONE MODIFICATIONS IN PHYTOHORMONE SIGNALING

Histone modifications are key epigenetic mechanism that plays a crucial role in regulating phytohormone signaling. Chromatin is formed when DNA wraps around histone proteins, which provide a structural framework, and modifications to these histones can either relax or compact chromatin structure, thereby influencing gene expression. Regarding phytohormone signaling, histone modifications such as methylation, acetylation, phosphorylation and ubiquitination have been shown to regulate the expression of phytohormone-responsive genes (Huijia et al., 2022). Histone modifications, such as the trimethylation of histone H3 lysine 4 (H3K4me3), are linked to the active transcription of auxin-responsive genes. In contrast, histone deacetylation plays a role in silencing the gibberellin-responsive genes.

The regulation of histone modifications is a dynamic process controlled by various enzymes, such as histone methyltransferases, histone acetyltransferases, histone deacetylases and histone phosphorylases, which work together to modify histone proteins (Banerjee & Roychoudhury, 2017, 2018). These enzymes are often recruited to specific genomic regions by phytohormone signaling pathways, where they modify histones to regulate gene expression (Wang et al., 2020a), for instance, the *Arabidopsis* HMT (*Homocysteine methyltransferase*), SET DOMAIN GROUP 1 (SDG1), has been shown to interact with the auxin response factor, ARF1, to regulate auxin-responsive gene expression. Similarly, the HDAC (*Histone deacetylase*), HDA19, has been implicated in the regulation of abscisic acid (ABA)-responsive gene expression. The complex interplay between histone modifications and phytohormone signaling pathways highlights the importance of epigenetic regulation in modulating plant hormone responses.

17.2.1 Histone Methylation and Acetylation

Histone-modifying enzymes that affect chromatin structure and gene expression include histone acetyltransferases (HATs), histone deacetylases (HDACs) and histone methyltransferases (HMTs), which are essential for the epigenetic regulation of plants.

The following are the key aspects of histone modification in plants.

DOI: 10.1201/9781003589914-20

17.2.1.1 Histone Acetylation

Histone acetyltransferase (*HAT*) genes, including *HAC1* and *HAC5*, have the ability to acetylate histones and control phytohormone signaling. Gene expression is activated and chromatin structure is relaxed when acetyl groups are added to histone tails. Genes that react to phytohormones, including auxin, gibberellin and ethylene, as well as those involved in phytohormone biosynthesis, like the auxin biosynthesis gene, *YUCCA*, are activated by histone acetylation. The connection between auxin and ethylene signaling is one example of how histone acetylation affects the crosstalk among several phytohormone signaling pathways (Rando, 2012). Additionally, it affects growth and development by regulating the expression of genes involved in cell proliferation, expansion and specialization; these enzymes play a crucial role in shaping cellular processes. Specific enzymes involved in histone acetylation are:

1. Histone acetyltransferases (HATs): Enzymes that add acetyl groups to histone tails.
2. Histone deacetylases (HDACs): Enzymes that remove acetyl groups from histone tails.

Specific examples:

a. Auxin signaling: Histone acetylation can activate the expression of auxin-responsive genes, such as the Aux/IA family (AUX1), peptidyl-prolyl cis/trans isomerase (PIN1), auxin response factor (ARF1), and enhance auxin signaling (Figure 17.1).
b. Gibberellin signaling: Histone acetylation can regulate the expression of gibberellin-responsive genes, such as the GAI (*Gibberellic Acid Insensitive*) family (*GA20ox, GA3ox, GAI*), and modulate gibberellin signaling (Figure 17.2).
c. Ethylene signaling: Histone acetylation can activate the expression of ethylene-responsive genes (*ACS*), *1-aminocyclopropane-1-carboxylate oxidase* (*ACO*), *ethylene receptor 1 (ETR1)* such as the ethylene response factor (ERF) family, and enhance ethylene signaling.

Histone acetylation can recruit transcription factors, such as auxin response factor (ARF) and ERF, to phytohormone-responsive genes and activate their expression and remodel chromatin structure, making phytohormone-responsive genes more accessible to transcription factors and the transcriptional machinery. By controlling phytohormone signaling, modifications in histone acetylation can affect the growth and development of plants. It also affects plant stress responses, including drought and disease resistance, plant productivity and yield.

17.2.1.2 Histone Methylation

The addition of methyl groups to histone tails can either stimulate or silence gene expression, depending on the specific histone and methylation level. This process is critical in phytohormone signaling, where it influences gene expression and chromatin organization, ultimately shaping plant development and responses. There are three distinct sequences in which plant DNA methylation takes place: CG, CHG and CHH (H = A, C, T) (Fang et al., 2022). DNA methyltransferase MET1 maintains CG methylation, and DNA methyltransferase CMT3 maintains CHG methylation. CHH methylation must be maintained by the RNA-directed DNA methylation (RdDM) pathway and DNA methyltransferase CMT2.

Each generation precisely reproduces DNA methylation patterns. Here are some key effects of histone methylation on phytohormone signaling:

Histone methylation can activate or repress the expression of genes that respond to phytohormones, such as auxin, gibberellin and ethylene. By controlling the expression of genes involved in phytohormone signaling pathways, histone methylation can alter their function. Additionally, the methylation process affects the interactions between several phytohormone signaling pathways, including the

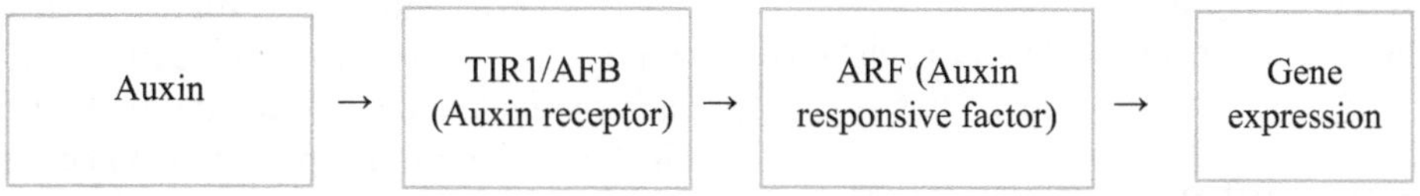

FIGURE 17.1 Auxin signaling pathway.

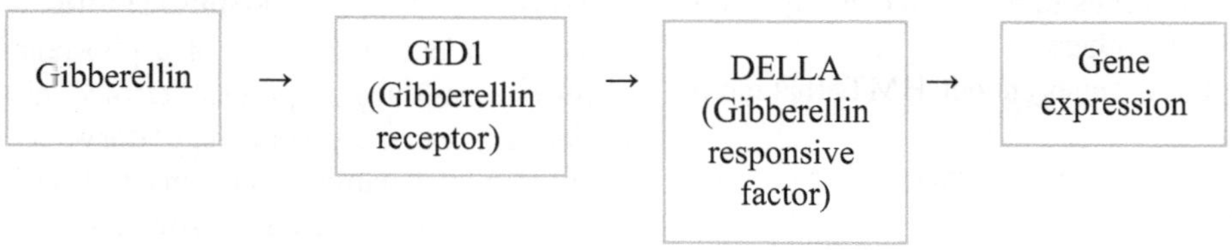

FIGURE 17.2 Gibberellin signaling pathway.

relationship between auxin and ethylene signaling and the genes involved in phytohormone production, including the *YUCCA* gene for auxin biosynthesis.

Specific enzymes involved in histone methylation are:

- Histone methyltransferases (HMTs): Enzymes that add methyl groups to histone tails.
- Histone demethylases (HDMs): Enzymes that remove methyl groups from histone tails.

By modulating the transcriptional landscape of genes orchestrating cellular specialization, expansion and proliferation, these enzymes exert profound influences on plant morphogenesis and ontogenesis, ultimately sculpting the architectural and functional attributes of plants.

a. Auxin signaling: Histone methylation can regulate the expression of auxin-responsive genes, such as the Aux/IAA (indole-3-acetic acid) family, and modulate auxin signaling.
b. Gibberellin signaling: Gibberellin signaling can be modulated by histone methylation, which regulates the expression of gibberellin-responsive genes, including those belonging to the GAI family, thereby influencing the gibberellin-mediated physiological processes.
c. Ethylene signaling: Ethylene signaling is subject to regulation by histone methylation, which controls the expression of ethylene-responsive genes, including those of the ERF family, thereby fine-tuning the ethylene-mediated responses in plants.

Histone methylation also plays a vital role in regulating phytohormone signaling by recruiting transcription factors like ARF (auxin response factor) and ERF (ethylene response factor) to specific genes and by remodeling chromatin structure to either facilitate or hinder access to the transcriptional machinery. This process is mediated by histone methyltransferases (HMTs), such as SET DOMAIN GROUP (SDG) proteins (Berr et al., 2009) which catalyze histone methylation and ultimately influence the expression of phytohormone-responsive genes.

In plants, several genes are responsible for histone methylation, including:

- *SET DOMAIN GROUP* (*SDG*) genes: These genes encode histone methyltransferases (HMTs) that catalyze the methylation of histones. There are several *SDG* genes in plants, including *SDG1, SDG2, SDG3* and others.
- *SET1* genes: These genes encode HMTs that methylate histones.
- *ASH1* genes: These genes encode HMTs that specifically methylate histone H3 lysine 36 (H3K36).
- *ATXR* genes: These genes encode HMTs that specifically methylate histone H3 lysine 4 (H3K4).
- *SUVH* genes: These genes encode HMTs that specifically methylate histone H3 lysine 9 (H3K9).

Changes in histone methylation can alter plant development and growth by regulating phytohormone signaling. Histone methylation can influence plant stress responses, such as drought and pathogen resistance, as well as plant yield and productivity by regulating phytohormone signaling.

Types of histone methylation
The main types of histone methylation in plants are:

- H3K4 methylation: Methylation of histone H3 lysine 4 (H3K4) is associated with active gene expression and is often found near the transcription start sites of active genes.
- H3K9 methylation: Methylation of histone H3 lysine 9 (H3K9) is associated with gene silencing and heterochromatin formation.
- H3K27 methylation: Methylation of histone H3 lysine 27 (H3K27) is associated with gene repression and is often found in developmental genes.
- H3K36 methylation: Methylation of histone H3 lysine 36 (H3K36) is associated with active gene expression and is often found in the body of active genes.
- H4K20 methylation: Methylation of histone H4 lysine 20 (H4K20) is associated with DNA damage response and genome stability.

17.2.2 Histone Phosphorylation and Ubiquitination

The histone core is comprised of an octameric complex, featuring paired molecules of histones H2A, H2B, H3 and H4.1. The amino-terminal extensions of all four histones, along with the carboxy-terminal extensions of H2A and H2B, collectively known as "histone tails", exhibit intrinsic disorder and project outward from the nucleosome (Eickbush et al., 1978). In contrast, the central regions of the histone proteins coalesce to form the globular core of nucleosomes, adopting a highly structured organization. Energetically controlling the chromatin structure is necessary to regulate DNA accessibility. Post-translational modifications such as phosphorylation and ubiquitination throw light on the nuclear process that modulates phytohormone signaling.

17.2.2.1 Histone Phosphorylation

The covalent attachment of a phosphate moiety to select histone residues, a process termed histone phosphorylation, serves as a pivotal regulatory mechanism governing chromatin dynamics and transcriptional responses in phytohormone-mediated signaling pathways. This reversible post-translational modification specifically targets serine, threonine, or tyrosine residues on histone proteins, thereby modulating chromatin architecture and gene expression.

a. Phosphorylation by kinases: Histone phosphorylation is catalyzed by kinases, which are enzymes that transfer phosphate groups from ATP to histone proteins. In phytohormone signaling, specific kinases are activated in response to hormone signals, leading to the phosphorylation of histones.
b. Recognition by phospho-binding domains: Phosphorylated histones are recognized by phospho-binding domains, such as 14-3-3 proteins or phospho-specific antibodies, which bind to the phosphorylated histones and recruit other proteins to the chromatin.
c. Chromatin remodeling: Histone phosphorylation can lead to chromatin remodeling, which involves the reorganization of nucleosomes and the relaxation of chromatin structure. This allows for the recruitment of transcriptional machinery and the activation of gene expression.
d. Transcriptional regulation: The phosphorylation of histones generates binding platforms for transcriptional regulators and other accessory proteins, facilitating the recruitment of these factors to specific genomic loci and thereby influencing the transcriptional landscape of phytohormone-responsive genes (Rossetto et al., 2012). For example, phosphorylated histone H2A (γ-H2A) can interact with the auxin response factor, ARF1, to regulate auxin-responsive gene expression.

Histone phosphorylation in phytohormone signaling:

- Auxin signaling: The kinase, *PINOID*, phosphorylates histone H2A, leading to the activation of auxin-responsive gene expression.
- Gibberellin signaling: The kinase, *GID1*, phosphorylates histone H3, leading to the activation of gibberellin-responsive gene expression.

17.2.2.2 Histone Ubiquitination

The process of histone ubiquitination entails the reversible conjugation of ubiquitin peptides to histone molecules, forming a covalent linkage that modulates chromatin function and regulates gene expression in response to various cellular signals. Ubiquitination can regulate chromatin structure and gene expression by targeting histones for degradation or by recruiting ubiquitin-binding proteins to the chromatin.

a. Ubiquitination by E3 ligases: Histone ubiquitination is catalyzed by E3 ligases, which are enzymes that recognize specific histone proteins and attach ubiquitin to them.
b. Recognition by ubiquitin-binding domains: Ubiquitinated histones are recognized by ubiquitin-binding domains, such as UIM (ubiquitin-interacting motif) or UBA (ubiquitin-associated) domains, which bind to the ubiquitinated histones and recruit other proteins to the chromatin.
c. Chromatin remodeling: Histone ubiquitination can lead to chromatin remodeling, which involves the reorganization of nucleosomes and the relaxation of chromatin structure.
d. Transcriptional regulation: The ubiquitination of histones creates a molecular scaffold that enables the recruitment of transcriptional regulators and associated proteins, thereby fine-tuning gene expression by facilitating or hindering the access of transcriptional machinery to specific genomic regions. For example, ubiquitinated histone H2B can interact with the transcription factor TAF1 (Transcription initiation factor 1) to regulate gene expression (Maraschin Fdos et al., 2009).
   - Auxin signaling: The E3 ligase, SCF(TIR1), ubiquitinates histone H2A, leading to the activation of auxin-responsive gene expression.
   - Gibberellin signaling: The E3 ligase, GID2, ubiquitinates histone H3, leading to the activation of gibberellin-responsive gene expression.

17.3 DNA METHYLATION AND DEMETHYLATION

DNA methylation is controlled by DNA methyltransferases (DNMTs) and DNA demethylases, which add or remove methyl groups from DNA (Moore et al., 2013).

The level of phytohormones in a plant can be influenced by DNA methylation, which can either activate or repress genes involved in phytohormone production. The expression of genes linked to phytohormones is impacted by DNA methylation, which also affects chromatin accessibility and structure. Variations in DNA methylation can impact signaling and phytohormone levels, which can alter the growth and development of plants and may affect stress responses and phytohormone signaling, which has an impact on the ability of a plant to withstand environmental stressors. Additionally, it affects the expression of genes related to the transport and distribution of phytohormones, which in turn affects the temporal and spatial distribution of phytohormones inside the plant. Thus, the alterations in plant shape and architecture are caused by DNA methylation, which plays a critical role in plant development.

17.3.1 Mechanism of Methylation

The mechanism of methylation in plant hormone regulation involves the following steps:

i. Recognition of target sequences: DMTs (DNA methyltransferases) recognize and bind to specific DNA sequences, often characterized by the presence of CG or CNG motifs. (CNG motif can be any of the following sequences i.e CAG, CCG, CTG and CGG), DMTs that recognize a "CNG" sequence are less specific than those recognizing only a "CG" (CpG) site, as the nucleotide at the middle position can vary. The flexibility at this "N" position allows the enzyme to target a broader set of sites in the genome.

ii. Methyl group transfer: The bound DMT enzyme transfers a methyl group from S-adenosylmethionine (SAM) to the target cytosine residue, resulting in the formation of 5-methylcytosine (5-mC).
iii. Maintenance of methylation: Once established, DNA methylation is maintained through the action of maintenance DMTs, which methylate newly synthesized DNA strands during replication.
iv. Regulation of gene expression: DNA methylation can regulate gene expression by:

Repressing transcription: Methylation of promoter regions can prevent the binding of transcription factors and RNA polymerase, leading to gene silencing.

Activating transcription: Methylation of specific sequences can also activate gene expression by recruiting chromatin-modifying complexes that facilitate transcription.

17.3.1.1 Types of DMTs

There are several types of DMTs involved in plant hormone regulation (Yamamuro et al., 2016) including:

i. MET1: Maintenance DMT that methylates CG sequences and is involved in the regulation of gene expression in response to plant hormones such as auxin and gibberellin.
ii. CMT3: Chromomethylase that methylates CNG sequences and is involved in the regulation of gene expression in response to environmental cues.
iii. DRM2: De novo DMT that methylates CG and CNG sequences and is involved in the regulation of gene expression in response to plant hormones such as auxin and cytokinin.

17.3.1.2 Role of Methylation in Plant Hormone Regulation

Methylation can regulate the expression of hormone signaling genes, such as receptors, transcription factors and downstream targets, which can influence the responsiveness of plants to hormones by regulating the expression of hormone-responsive genes. Methylation can also integrate hormone and environmental signals by regulating the expression of genes involved in stress responses and adaptation. The hormonal signaling is mediated as follows:

- Auxin signaling: Methylation of the *AUXIN RESPONSE FACTOR 1* (*ARF1*) gene promoter is required for auxin-induced gene expression.
- Gibberellin signaling: Methylation of the *GIBBERELLIN-INSENSITIVE* (*GAI*) gene promoter is required for gibberellin-induced gene expression.
- Cytokinin signaling: Methylation of the *CYTOKININ RESPONSE FACTOR 1* (*CRF1*) gene promoter is required for cytokinin-induced gene expression.

17.3.1.3 Mechanism of Demethylation

The mechanism of demethylation in plant hormone regulation involves the following steps:

1. Recognition of methylated DNA: DNA demethylases recognize and bind to methylated DNA sequences, specifically 5-mC residues.
2. Base excision repair (BER): The bound demethylase enzyme initiates a base excision repair (BER) pathway, which involves the removal of the methylated base (5-mC) from the DNA.
3. Glycosylase activity: The demethylase enzyme exhibits glycosylase activity, which breaks the N-glycosidic bond between the methylated base and the sugar–phosphate backbone.
4. AP endonuclease activity: The resulting abasic site is then cleaved by an AP endonuclease, which creates a nick in the DNA strand.
5. DNA repair synthesis: The nick is then repaired by DNA repair synthesis, which involves the incorporation of a new nucleotide to replace the removed methylated base.
6. Demethylation: The resulting DNA strand is now demethylated, and the gene is activated or repressed depending on the need for regulation.

17.3.1.4 Types of Demethylases

There are several types of demethylases involved in plant hormone regulation (Le et al., 2014) including:

1. REPRESSOR OF SILENCING 1 (ROS1): ROS1 is a DNA demethylase that plays a crucial role in the regulation of gene expression in response to plant hormones such as auxin and gibberellin.
2. DEMETER (DME): DME is a DNA demethylase that is involved in the regulation of gene expression during plant development and in response to environmental cues.
3. DEMETER-LIKE (DML): DML is a family of DNA demethylases that are involved in the regulation of gene expression in response to plant hormones and environmental cues.

Demethylation can activate gene expression by removing methyl groups from the promoter regions, allowing for the binding of transcription factors and the recruitment of RNA polymerase and repressive chromatin-modifying complexes.

Demethylation can regulate hormone signaling by modulating the expression of hormone-responsive genes as follows:

- Auxin signaling: Demethylation of the *AUXIN RESPONSE FACTOR 1* (*ARF1*) gene promoter is required for auxin-induced gene expression.
- Gibberellin signaling: Demethylation of the *GIBBERELLIN-INSENSITIVE* (*GAI*) gene promoter is required for gibberellin-induced gene expression.

17.4 SMALL RNA-MEDIATED GENE REGULATION

Small RNAs (sRNAs) play a crucial role in regulating gene expression in phytohormone signaling pathways (Guleria et al., 2011). There are two main types of sRNAs involved in phytohormone signaling: microRNAs (miRNAs) and small interfering RNAs (siRNAs).

siRNAs are made from double-stranded RNA (dsRNA) precursors, which are then converted into mature siRNAs by the enzyme DCL2 or DCL3. Mature siRNAs attach to specific target mRNAs by complementary base pairing, resulting in their destruction. The attached siRNA promotes the breakdown of the target mRNA, resulting in a drop in protein levels.

sRNAs can integrate hormone and environmental signals by regulating the expression of genes involved in stress responses and adaptation as follows:

- Auxin signaling: miR393 targets the auxin receptor gene *TIR1*, regulating auxin signaling and plant growth.
- Gibberellin signaling: miR159 targets the gibberellin-responsive gene *GAMYB*, regulating gibberellin-induced gene expression.
- Cytokinin signaling: siRNAs generated from the cytokinin-responsive gene *ARR4* regulate cytokinin-induced gene expression.

Accordingly, Dicer-like enzymes (DCLs) process precursor miRNAs and siRNAs into mature sRNAs. Argonaute proteins (AGOs) bind to mature sRNAs and guide them to target mRNAs, and RNA-dependent RNA polymerases (RDRs) generate dsRNA precursors for siRNA biogenesis.

17.4.1 MicroRNAs (miRNAs) and Their Targets

MicroRNAs (miRNAs) are small, non-coding RNA molecules that exert profound influence on the intricacies of gene regulation within phytohormone-mediated signaling networks (Yadav et al., 2024). Through high-affinity interactions with complementary mRNA sequences, miRNAs orchestrate the targeted dismantling or translational stalling of cognate transcripts, thereby recalibrating the phytohormone signaling paradigm and fine-tuning plant developmental responses.

Mechanism of miRNA-mediated gene regulation:

- miRNA biogenesis: miRNAs are transcribed from specific genes, called *MIR* genes, and processed into precursor miRNAs (pre-miRNAs) by the enzyme Dicer-like 1 (DCL1).
- miRNA processing: Pre-miRNAs are then processed into mature miRNAs by the enzyme DCL1 and other accessory proteins.
- miRNA binding to target mRNAs: Mature miRNAs bind to specific target mRNAs through complementary base pairing, leading to the repression of target gene expression.
- Target mRNA degradation or translational inhibition: The bound miRNA can either induce the degradation of the target mRNA or inhibit its translation, leading to a reduction in protein levels.

miRNAs target a wide range of genes involved in phytohormone signaling, including:

- Hormone receptors: miRNAs target hormone receptors, such as the auxin receptor TIR1, to regulate hormone signaling.
- Transcription factors: miRNAs target transcription factors, such as the auxin response factor ARF1, to regulate hormone-induced gene expression.
- Hormone biosynthesis genes: miRNAs target genes involved in hormone biosynthesis, such as the auxin biosynthesis gene *YUCCA*, to regulate hormone levels.
- Hormone signaling components: miRNAs target components of hormone signaling pathways, such as the auxin signaling component AUX/IAA, to regulate hormone signaling.

miRNAs and their specific targets in phytohormone signaling:

- miR393 and TIR1: miR393 targets the auxin receptor TIR1, regulating auxin signaling and plant growth.

- miR159 and GAMYB: miR159 targets the gibberellin-responsive gene *GAMYB*, regulating gibberellin-induced gene expression.
- miR160 and ARF10: miR160 targets the auxin response factor ARF10, regulating auxin-induced gene expression.
- miR167 and ARF8: miR167 targets the auxin response factor ARF8, regulating auxin-induced gene expression.

Further, Dicer-like 1 (DCL1) processes precursor miRNAs into mature miRNAs, Argonaute 1 (AGO1) binds to mature miRNAs and guides them to target mRNAs and microRNA-mediated gene silencing complex, which includes AGO1 and other proteins that are responsible for the repression of target gene expression.

17.4.2 Small Interfering RNAs (siRNAs) and Their Roles

Small interfering RNAs (siRNAs) are a class of small RNA molecules that play a crucial role in regulating gene expression in plants. siRNAs are involved in the RNA interference (RNAi) pathway, which is a conserved mechanism that regulates gene expression by degrading specific mRNA molecules.

Mechanism of siRNA-mediated gene regulation:

- siRNA biogenesis: siRNAs are generated from double-stranded RNA (dsRNA) precursors, which are processed into small RNA molecules by the enzyme Dicer-like 2 (DCL2) or DCL3.
- siRNA binding to target mRNAs: siRNAs attach to particular target mRNAs via complementary base pairing, resulting in the degradation of these target mRNAs and a subsequent decrease in protein levels.

Both auxin and gibberellin signaling have been demonstrated to be regulated by siRNAs, which target the auxin receptor gene *TIR1* and the gibberellin-responsive gene, respectively. It has been demonstrated that siRNAs and GAMYB (gibberellin- and abscisic acid-regulated MYB) control cytokinin signaling by focusing on the cytokinin-responsive gene *ARR4*. Furthermore, RNA-dependent RNA polymerase 6 (RDR6) produces dsRNA precursors for siRNA biogenesis. Dicer-like 2 (DCL2) and DCL3 convert dsRNA precursors into siRNAs, and the proteins Argonaute 1 (AGO1) and AGO4 attach to siRNAs and direct them to target mRNAs (Ye et al., 2016).

Consequently, siRNAs can target hormone signaling genes, including receptors, transcription factors and downstream targets, to regulate hormone signaling. They can influence the responsiveness of plants to hormones by regulating the expression of hormone-responsive genes. Additionally, siRNAs can integrate hormone and environmental signals by regulating the expression of genes involved in stress responses and adaptation.

17.5 CHROMATIN REMODELING AND PHYTOHORMONE SIGNALING

Phytohormones, such as auxin, gibberellin, cytokinin, abscisic acid and ethylene, play essential roles in plant growth and development, and their signaling pathways are intricately linked with chromatin remodeling. Further, chromatin remodeling complexes can be drawn to particular genomic areas by phytohormones, which can either open or close transcription factor access. By changing histone modifications including methylation, acetylation and phosphorylation, phytohormones can change the structure of chromatin. By altering chromatin structure, phytohormones can control access of transcription factors to particular genomic areas.

Chromatin remodeling is mediated by various protein complexes, including:

The SWI/SNF complex (SWItch/Sucrose Non-Fermentable), a molecular machine, fueled by ATP hydrolysis, dynamically reshapes chromatin topology, either granting or denying transcription factor access to specific genomic loci. Similarly, the (chromodomain helicase DNA-binding) family of CHD complex (Saha et al., 2024), another ATP-dependent chromatin remodeler, orchestrates precise adjustments to chromatin structure, thereby modulating transcriptional outcomes and influencing gene expression programs.

IN080 complex is involved in chromatin remodeling and DNA repair, and Nucleosome remodeling complex (NuRC) is involved in chromatin remodeling and transcriptional regulation.

- Auxin signaling: Auxin recruits the SWI/SNF complex to the *AUXIN RESPONSE FACTOR* (*ARF*) genes, allowing for their transcriptional activation.
- Gibberellin signaling: Gibberellin recruits the CHD complex to the *GIBBERELLIN-INSENSITIVE* (*GAI*) gene, allowing for its transcriptional activation.
- Cytokinin signaling: Cytokinin recruits the IN080 complex to the *CYTOKININ RESPONSE FACTOR* (*CRF*) genes, allowing for their transcriptional activation.

The intricate interplay between chromatin remodeling and phytohormone signaling is composed of a cohort of enzymatic regulators, including histone acetyltransferases (HATs), histone deacetylases (HDACs), histone methyltransferases (HMTs) and histone demethylases (HDMs).

These enzymes catalyze precise chromatin modifications, thereby reconfiguring chromatin topology and influencing gene expression.

Concurrently, a suite of chromatin remodeling complexes, encompassing SWI/SNF, CHD, IN080 and NuRC, execute dynamic rearrangements of chromatin structure, further calibrating the transcriptional landscape in response to phytohormone signals. Transcription factors such as ARF, GAI and CRF are critical for regulating gene expression in response to phytohormones.

17.6 EPIGENETIC REGULATION OF AUXIN, GIBBERELLIN AND OTHER PHYTOHORMONE SIGNALING PATHWAYS

Phytohormones, such as auxin, gibberellin, cytokinin, abscisic acid and ethylene, play essential roles in plant growth and development. The phytohormone signaling networks are subject to intricate epigenetic governance, wherein a triad of mechanisms – DNA methylation, histone tail modifications and non-coding RNA-mediated gene silencing – converge to fine-tune signal transduction, gene expression and, ultimately, plant developmental responses.

17.6.1 Auxin Signaling Pathway

Auxin is a key phytohormone involved in plant cell elongation, cell division and differentiation.

Components of auxin signaling pathway are Transport Inhibitor Resistant1/Auxin Signaling F-BOX (TIR1/AFB) proteins, Auxin/Indole-3-Acetic Acid (Aux/IAA) transcriptional coregulators and sequence-specific binding proteins called auxin response factors (ARFs) that promote stem cell maintenance, cellular reprogramming and plant responses to environmental stimuli (Wojcikowska et al., 2020).

The auxin signaling pathway is regulated by epigenetic mechanisms, including:

1. DNA methylation: DNA methylation plays a crucial role in regulating auxin signaling by silencing auxin-responsive genes.
2. Histone modification: Histone modification, such as histone acetylation and methylation, regulates auxin signaling by modifying chromatin structure. Histone acetylation is thought to stimulate gene transcription and encourage an open chromatin state. The Polycomb-group (PcG) and Trithorax-group (TrxG) proteins, respectively, catalyze the most common histone methylation marks, H3K27me3 and H3K4me3, which have been shown to play roles in controlling plant developmental plasticity.
3. Non-coding RNA-mediated gene regulation: Non-coding RNAs, such as microRNAs and siRNAs, regulate auxin signaling by targeting auxin-responsive genes. miR393 and miR160 modulate *TIR1, AFB1, AFB2, AFB3* and *ARF10, ARF16, ARF17* genes, respectively.

The intersection of non-coding RNAs and phytohormone signaling reveals a complex regulatory paradigm. MicroRNAs, such as miR160 and miR393, target key auxin-related genes, *ARF10* and *TIR1*, respectively, to modulate auxin-mediated gene expression and signaling. Long non-coding RNAs (lncRNAs) also participate in this regulatory network by influencing chromatin architecture, gene expression and microRNA activity.

The dynamic interplay between non-coding RNAs and phytohormone signaling enables plants to adapt to changing environmental conditions by fine-tuning hormone levels and signaling pathways. Moreover, this regulatory axis intersects with other epigenetic mechanisms, including DNA methylation and histone modification, to orchestrate plant development, growth and stress responses.

The intricate relationships between phytohormone signaling, non-coding RNAs and epigenetic mechanisms underscore the critical role of post-transcriptional regulation in plant hormone biology, revealing a rich and complex layer of gene regulation that shapes plant development and environmental responses.

17.6.2 Gibberellin Signaling Pathway

Gibberellin, a tetracyclic diterpenoid, is a phytohormone involved in plant growth and development, including seed germination, stem elongation and flower formation. The gibberellin signaling pathway is regulated by epigenetic mechanisms, including:

1. DNA methylation: DNA methylation regulates gibberellin signaling by silencing gibberellin-responsive genes.
2. Histone modification: Histone modification, such as histone acetylation and methylation, regulates gibberellin signaling by modifying chromatin structure.
3. Non-coding RNA-mediated gene regulation: Non-coding RNAs, such as microRNAs and siRNAs, regulate gibberellin signaling by targeting gibberellin-responsive genes.

The GA signaling pathway consists of three main components – DELLA proteins (DELLAs), GA-INSENSITIVE DWARF1 (GID1) and F-box protein GA-INSENSITIVE DWARF2/SLEEPY1. DELLAs prevent GA-mediated reactions when GA is not present. GAs bind to the GID1 receptor when they are present, which makes it easier for them to interact with DELLA repressors (Manoharlal et al., 2018). This encourages their breakdown via the proteasomal route, aided by the SCF complex that contains GID2/SLY1. The GA-dependent GID1–GID2/SLY1 module binds DELLAs and facilitates the proteasomal degradation of NITROGEN-MEDIATED TILLER GROWTH RESPONSE 5 (NGR5).

The plants exposed to GA3 showed an increase in chromatin de-condensation, a decrease in global DNA methylation, a decrease in DNA methyltransferases (NtDNMTs) activities along with a decrease in NtMET1 and NtCMT3 transcript levels, and a partial recovery of the phenotype and expression of the reporter transgene that was epigenetically silenced.

17.6.3 Other Phytohormone Signaling Pathways

Other phytohormone signaling pathways, including cytokinin, abscisic acid and ethylene, are also regulated by epigenetic mechanisms.

17.6.3.1 Cytokinin Signaling Pathway

Cytokinin is a phytohormone that plays a crucial role in plant cell division, differentiation and growth.

Cytokinin binds to its receptor, leading to the activation of CRF (cytokinin response factor) that regulates cytokinin-responsive gene expression. *CRF* is regulated by epigenetic mechanisms, DNA methylation and histone modification.

Regarding epigenetic regulation, cytokinins help the repressive histone mark H3K27me3 to be removed from the *WUS* locus during shoot regeneration in plants. Pluripotency is induced by histone acetylation and decreased H3K27me3 levels. PKL (ATP-dependent CHROMODOMAIN HELICASE DNA-BINDING 3 chromatin remodeling complex) promotes callogenesis, cell identity and the vegetative phase change via interaction with HDA9 and subsequent H3K27 deacetylation at miRNA156 (Rudolf et al., 2024). The auxin–cytokinin relationship is responsible for the first specification of the shoot meristem in callus, and ATXR2 controls cytokinin signaling and inhibits premature WUS activation to guarantee appropriate cell fate transition. Table 17.1 shows the significant transcription factors that are regulated during the epigenetic regulation in plants.

TABLE 17.1
Hormones and key transcription factors involved in plant epigenetic regulation

S. No	Hormone	Transcription factors/proteins
1	Cytokinin	Histidine kinases
		Histidine containing phosphotransmitters
		Response regulators (RRAs, RRBs and RRCs)
		TCP interactor containing ear motif protein 1 (TIE1)
		TCP interactor containing ear motif protein 2 (TIE2)
		Protein phosphatase with kelch-like domains 1 (PPKL1)
		WUSCHEL (WUS)
		Arabidopsis trithorax-related 2 (ATXR2)
2	Abscisic acid	Histone monoubiquitination1 (HUB1)
		CYP707A2
		NCED
		DOG1 (DELAY OF GERMINATION 1)
		ATS2
		PER1 (Period Circadian Regulator 1)
		VP1/ABI3-like (VAL protein)
		PIKLE (PKL)
		Histone deacetylases HDA6, HDA19
		ABA2
		AAO3
		ATS2
		FUS3
		FUSCA3
		LEC1 (LEAFY COTYLEDON1)
		NCED (9 cis-epoxycarotenoid dioxygenase)
		PER1 (peroxiredoxin family of antioxidants)
		UQ (ubiquitin)
		ZEP, zeaxanthin epoxidase
3	Ethylene	Ethylene Insensitive 3 (EIN3)
		EIN3 LIKE 1 (EIL1)
		FLC (Flowering Locus C)
		FLD (Flowering Locus D)

17.6.3.2 Abscisic Acid Signaling Pathway

Abscisic acid (ABA) is a phytohormone that plays a crucial role in plant stress responses, including drought, cold and salinity stress. The ABA signaling pathway is regulated by epigenetic mechanisms, including DNA methylation, histone modification and non-coding RNA-mediated gene regulation.

Through histone H2B monoubiquitination, histone monoubiquitination seems to control ABA levels in growing seeds. Histone deacetylases (HDACs)-dependent and PICKLE chromatin remodeling complexes (CRCs) suppress transcription factor networks required for seed maturation, while ABA either directly or indirectly stimulates the expression of these genes. ABI3 is a transcription factor that regulates ABA-responsive gene expression. ABI3 is regulated by epigenetic mechanisms, including DNA methylation and histone modification. ABA regulates stomatal response and stress-responsive gene expression through HDACs and HOS15-dependent histone deacetylation, as well as through the ATP-dependent SWITCH/SUCROSE NONFERMENTING CRC.

Non-coding RNAs, such as microRNAs (miRNAs) and small interfering RNAs (siRNAs), play a crucial role in regulating ABA signaling.

17.6.3.3 Ethylene Signaling Pathway

Ethylene is a phytohormone that plays a crucial role in plant growth and development, including fruit ripening, senescence and stress responses. The ethylene signaling pathway is regulated by epigenetic mechanisms.

Ethylene-responsive genes are often methylated, which silences their expression. Demethylation of these genes is required for ethylene-induced gene expression. Histone acetylation and methylation are associated with active transcription of ethylene-responsive genes, while histone deacetylation and demethylation are associated with repression of these genes. Non-coding RNAs, such as microRNAs (miRNAs) and small interfering RNAs (siRNAs), play a crucial role in regulating ethylene signaling. miRNAs and siRNAs target ethylene-responsive genes and regulate their expression. Ethylene Insensitive 2 (EIN2) is a transcription factor that regulates ethylene-responsive gene expression. Ethylene binds to its receptor, leading to the activation of EIN2. EIN2 then regulates the expression of ethylene-responsive genes through epigenetic mechanisms, including histone modification and DNA methylation.

Senescence induces DNA methylation and histone modification, leading to the activation of ethylene-responsive genes. In summary, epigenetic mechanisms, including DNA methylation, histone modification and non-coding RNA-mediated gene regulation, play crucial roles in regulating phytohormone signaling pathways, including auxin, gibberellin, cytokinin, abscisic acid and ethylene. Understanding the epigenetic regulation of phytohormone signaling pathways is essential for elucidating the complex regulatory networks that control plant growth and development.

17.6.3.4 Gene Silencing

Small RNAs, such as microRNAs (miRNAs) and small interfering RNAs (siRNAs), control the expression of genes by binding to specific mRNA regions and modifying their expression through post-transcriptional gene silencing. By altering the expression of genes involved in hormone production, signaling and response, non-coding RNA-mediated regulation contributes significantly to phytohormone signaling. Important participants in this process are microRNAs (miRNAs) and small interfering RNAs (siRNAs), which attach to particular messenger RNAs (mRNAs) and stop their translation, thus functioning as negative regulators of gene expression.

Proteins known as polycomb group (PcG) are essential for preserving gene silencing and controlling chromatin structure (Grossniklaus et al., 2014). By controlling the expression of genes involved in plant hormone production, signaling and response, they play a critical role in phytohormone signaling. PcG proteins can stop excessive cell growth and proliferation; PcG proteins like EMF2 and VRN2 can regulate auxin-responsive gene family members such as auxin/indole-3-acetic acid (Aux/IAA), Gretchen Hagen 3 (GH3) and short auxin-up RNA (SAUR). By suppressing the genes *GA20OX* and *GA3OX* involved in gibberellin production and signaling, PcG proteins can control gibberellin signaling. By controlling the expression of ethylene-responsive genes (*ERF1, ERF2*), PcG proteins can affect ethylene signaling.

17.7 CONCLUSION

The epigenetic landscape of plants is a complex and ever-changing entity, shaped by the interplay between genetic, environmental and hormonal factors. The complex interplay between epigenetic mechanisms, such as DNA methylation, histone modifications and non-coding RNAs, allows plants to fine-tune gene expression and adapt to changing conditions. As research continues to unravel the complexities of epigenetic mechanisms in plants, we can expect to uncover novel targets for crop improvement and sustainable agriculture practices.

REFERENCES

Banerjee, A., & Roychoudhury, A. (2017). Epigenetic regulation during salinity and drought stress in plants: Histone modifications and DNA methylation. *Plant Gene*, 11, 199–204.

Banerjee, A., & Roychoudhury, A. (2018). The gymnastics of epigenomics in rice. *Plant Cell Reports*, 37, 25–49.

Berr, A., Xu, L., Gao, J., Cognat, V., Steinmetz, A., Dong, A., & Shen, W. H. (2009). SET DOMAIN GROUP25 encodes a histone methyltransferase and is involved in FLOWERING LOCUS C activation and repression of flowering. *Plant Physiology*, 151(3), 1476–1485. https://doi.org/10.1104/pp.109.143941. Epub 2009 Sep 2. PMID: 19726574; PMCID: PMC2773071.

Eickbush, T. H., & Moudrianakis, E. N. (1978). The histone core complex: An octamer assembled by two sets of protein-protein interactions. *Biochemistry*, 17(23), 4955–4964. https://doi.org/10.1021/bi00616a016. PMID: 718868.

Fang, J., Jiang, J., Leichter, S. M., et al. (2022). Mechanistic basis for maintenance of CHG DNA methylation in plants. *Nature Communications*, 13, 3877. https://doi.org/10.1038/s41467-022-31627-3.

Grossniklaus, U., & Paro, R. (2014). Transcriptional silencing by polycomb-group proteins. *Cold Spring Harbor Perspectives in Biology*, 6(11), a019331. https://doi.org/10.1101/cshperspect.a019331. (Grossniklaus U, Paro R. Transcriptional silencing by polycomb-group proteins. Cold Spring Harb Perspect Biol. 2014 Nov 3;6(11):a019331. doi: 10.1101/cshperspect.a019331. PMID: 25367972; PMCID: PMC4413232.)

Guleria, P., Mahajan, M., Bhardwaj, J., & Yadav, S. K. (2011). Plant small RNAs: Biogenesis, mode of action and their roles in abiotic stresses. *Genomics Proteomics Bioinformatics*, 9(6), 183–199. https://doi.org/10.1016/S1672-0229(11)60022-3. PMID: 22289475; PMCID: PMC5054152.

Huijia, K., Tianyi, F., Jiabing, W., Yan, Z., & Wen-Hui, S. (2022). Histone modification and chromatin remodeling in plant response to pathogens. *Frontiers in Plant Science*, 13. ISSN=1664-462X, https://doi.org/10.3389/fpls.2022.986940.

Kinoshita, T., & Jacobsen, S. E. (2012). Opening the door to epigenetics in PCP. *Plant and Cell Physiology*, 53(5), 763–765. https://doi.org/10.1093/pcp/pcs061.

Le, T. N., Schumann, U., Smith, N. A., Tiwari, S., Au, P. C., Zhu, Q. H., Taylor, J. M., Kazan, K., Llewellyn, D. J., Zhang, R., Dennis, E. S., & Wang, M. B. (2014). DNA demethylases target promoter transposable elements to positively regulate stress responsive genes in *Arabidopsis*. *Genome Biology*, 15(9), 458. https://doi.org/10.1186/s13059-014-0458-3. PMID: 25228471; PMCID: PMC4189188.

Manoharlal, R., Saiprasad, G. V. S., Ullagaddi, C., et al. (2018). Gibberellin A_3 as an epigenetic determinant of global DNA hypo-methylation in tobacco. *Biologia Plantarum*, 62, 11–23. https://doi.org/10.1007/s10535-017-0738-3.

Maraschin Fdos, S., Memelink, J., & Offringa, R. (2009). Auxin-induced, SCF(TIR1)-mediated poly-ubiquitination marks AUX/IAA proteins for degradation. *Plant Journal*, 59(1), 100–109. https://doi.org/10.1111/j.1365-313X.2009.03854.x. Epub 2009 Feb 26. PMID: 19309453.

Moore, L., Le, T., & Fan, G. (2013). DNA methylation and its basic function. *Neuropsychopharmacology*, 38, 23–38. https://doi.org/10.1038/npp.2012.112.

Raju, A., Xue, B., & Leibler, S. (2023). A theoretical perspective on Waddington's genetic assimilation experiments. *Proceedings of the National Academy of Sciences of the United States of America*, 120(51), e2309760120. https://doi.org/10.1073/pnas.2309760120. Epub 2023 Dec 13. PMID: 38091287; PMCID: PMC10743363.

Rando, O. J. (2012). Combinatorial complexity in chromatin structure and function: Revisiting the histone code. *Current Opinion in Genetics and Development*, 22, 148–155. https://doi.org/10.1016/j.gde.2012.02.013.

Rossetto, D., Avvakumov, N., & Côté, J. (2012). Histone phosphorylation: A chromatin modification involved in diverse nuclear events. *Epigenetics*, 7(10), 1098–1108. https://doi.org/10.4161/epi.21975. Epub 2012 Sep 4. PMID: 22948226; PMCID: PMC3469451.

Rudolf, J., Tomovicova, L., Panzarova, K., Fajkus, J., Hejatko, J., & Skalak, J. (2024). Epigenetics and plant hormone dynamics: A functional and methodological perspective. *Journal of Experimental Botany*, 75(17), 5267–5294.

Saha, D., Animireddy, S., & Bartholomew, B. (2024). The SWI/SNF ATP-dependent chromatin remodeling complex in cell lineage priming and early development. *Biochemical Society Transactions*, 52(2), 603–616. https://doi.org/10.1042/BST20230416. PMID: 38572912; PMCID: PMC11088921.

Wang, L., Chen, H., Li, J., Shu, H., Zhang, X., Wang, Y., et al. (2020a). Effector gene silencing mediated by histone methylation underpins host adaptation in an oomycete plant pathogen. *Nucleic Acids Research*, 48, 1790–1799. https://doi.org/10.1093/nar/gkz1160.

Wojcikowska, B., Wojcik, A. M., & Gaj, M. D. (2020). Epigenetic regulation of auxin-induced somatic embryogenesis in plants. *International Journal of Molecular Sciences*, 21(7), 2307. https://doi.org/10.3390/ijms21072307. PMID: 32225116; PMCID: PMC7177879.

Yadav, A., Mathan, J., Dubey, A. K., & Singh, A. (2024). The emerging role of non-coding RNAs (ncRNAs) in plant growth, development, and stress response signaling. *Non-Coding RNA*, 10, 13. https://doi.org/10.3390/ncrna10010013.

Yamamuro, C., Zhu, J.-K., & Yang, Z. (2016). Epigenetic modifications and plant hormone action. *Molecular Plant*, 9(1), 57–70. ISSN 1674-2052. https://doi.org/10.1016/j.molp.2015.10.008.

Ye, R., Chen, Z., Lian, B., Jordan Rowley, M., Xia, N., Chai, J., Li, Y., He, X.-J., Wierzbicki, A.-J., & Qi, Y. (2016). A dicer-independent route for biogenesis of siRNAs that direct DNA methylation in *Arabidopsis*. *Molecular Cell*, 61(2), 222–235. ISSN 1097-2765, https://doi.org/10.1016/j.molcel.2015.11.015.(https://www.sciencedirect.com/science/article/pii/S1097276515009077).

18 Chromatin Modeling and Epigenetic Regulation in DNA Damage Repair in Plants

Sakshi Sharma, Ritika Dadhwal and Ritesh Banerjee

18.1 INTRODUCTION

Living organisms often experience damage to their cellular DNA, which can occur naturally or be brought on by a range of extrinsic genotoxins, such as chemical mutagens, ionizing radiation and UV light. The most common damage to DNA that occurs, which is frequently brought about by unavoidable errors in internal metabolism, is nucleotide modification, intra- or inter-strand crosslinking and phosphodiester bond breakage. Damaged DNA may have problems being correctly transcribed, duplicated or arranged if it is not repaired (Manova & Gruszka, 2015; Polyn et al., 2015). The cell may experience cytotoxic and genotoxic effects from the DNA damage. The chemical makeup of the lesion typically determines the true biological effects of these damaged products (Tuteja et al., 2009). Enzymatic activities such as nucleases, recombinases, polymerases, topoisomerases, kinases, ligases, glycosylases, helicases, phosphatases and demethylases that repair DNA chemically are implied by the term "DNA repair mechanisms" (Iyama & Wilson, 2013).

As plants are sessile, they must adapt to various environmental stresses, such as ozone, UV-B radiation, desiccation, rehydration, salt, extremes in temperature and toxins found in the air and soil, such as metals and metalloids. Other significant environmental agents that damage DNA include a number of chemical mutagens and crosslinking agents (such as cisplatin and mitomycin C), alkylating agents, aromatic chemicals, ionizing radiations, and bacterial and fungal toxins (Tuteja et al., 2009). Numerous repair pathways found in other eukaryotes are also found in plants; for example, UV-induced pyrimidine dimers (cyclobutane dimers and 6-4 photoproducts) can be eliminated by excision repair or photoreactivation, and higher plants have been shown to contain specific glycosylases that are particular to specific lesions. Plants are naturally exposed to solar radiation; therefore, the main DNA repair mechanism they require on an everyday basis is photoreactivation of UV-induced DNA damage. In the event that the plant genome experiences different kinds of DNA lesions, the two conventional excision repair techniques – base (BER) and nucleotide (NER) excision repair, which are frequently referred to as "dark repair" – are also accessible (Britt, 1996; Rastogi et al., 2010). In plants, the viability of using both traditional and molecular genetic methods for DNA repair has been demonstrated. Using the tiny dicot *Arabidopsis thaliana* as a model, major progress has been made in understanding DNA damage repair in higher plants (Hays, 2002). It has been determined that at least one *Arabidopsis* mutant is impaired in healing UV-induced lesions, while at least two radiation-sensitive mutants seem to be impaired in repairing double-strand breaks. Numerous genes associated with repair have been cloned from *Arabidopsis*, either by searching for homologs to known repair genes or by complementing repair-defective mutants from other species (Britt, 1996). Other than *Arabidopsis*, rice is the only higher plant with relatively well-characterized DNA repair mechanisms. These mechanisms include the identification and regulation of genes that regulate DNA repair and protection mechanisms, as well as the effects of various environmental and developmental variables on their activation and efficiency (Ueda & Nakamura, 2011). Some DNA repair processes are inherently error-prone, which can result in mutational changes that could be transmitted. However, some DNA repair processes are prone to errors, which enhances the genetic variability of the population and aids in the evolution of plant genomics (Schuermann et al., 2005).

The research on UV tolerance and DNA repair in higher plants is nonetheless in its early stages, despite recent advancements. Comprehending how higher plants repair their DNA is not only intriguing from a scientific perspective but could also play a major role in addressing future environmental and societal issues (Kimura & Sakaguchi, 2006). Numerous plant factors were identified using sequence homology-based analysis, in both model organisms and some species of crops. Even though there is not much knowledge about DNA repair mechanisms in plants, the present developments can be used to understand the difficulties in creating new kinds with distinctive traits (Spampinato, 2017). A general mechanism of epigenetic regulation in plants is shown in Figure 18.1.

DOI: 10.1201/9781003589914-21

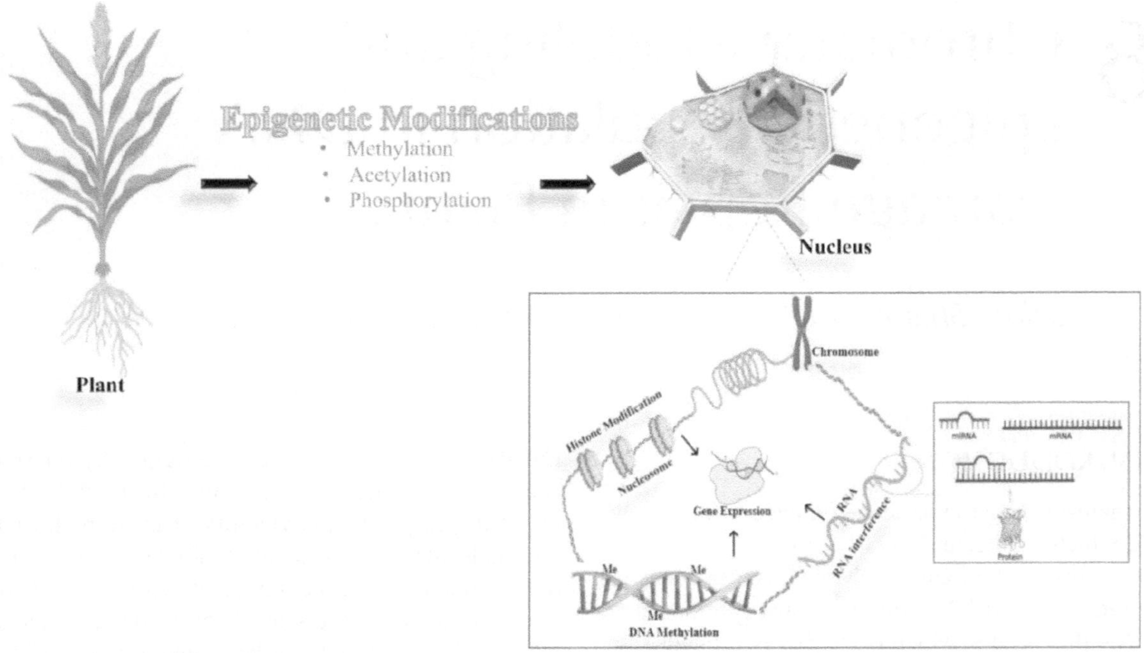

FIGURE 18.1 General mechanism of epigenetic regulation.

18.2 EPIGENETIC REGULATION FOR DNA DAMAGE RESPONSE IN PLANTS

The integrity of both the genome and epigenome is preserved during DNA damage repair (DDR). DNA damage compromises the integrity of the epigenome and causes DDR; however, DDR could be stopped by resetting the structures of the epigenome. Other epigenetic modifications, including DNA methylation, histone modification and RNA-assisted silencing, have been linked to genome maintenance in addition to chromatin remodeling and dynamics. These modifications also affect DNA damage signaling and repair pathways. Histone modifications, DNA methylation and variations in histone variants are collectively referred to as epigenetic regulation (Chinnusamy & Zhu, 2009). As of now, the field of epigenetics is the study of variations in gene expression patterns that are mitotically and/or meiotically heritable but do not result from changes in DNA sequence. Chromatin alterations and their roles in transcriptional silencing are usually the focus of epigenetic studies (Rajewsky et al., 2017). Three categories of genes that encode mediators, modifiers and epigenetic modulators are part of the suggested framework for epigenetic control. While epigenetic modifiers alter chromatin structure or DNA methylation, epigenetic moderators either activate or inhibit the epigenetic machinery, and epigenetic modifiers control the activity of epigenetic mediators. By altering the integrity through chromatin remodeling, histone modification, DNA methylation/demethylation and RNA silencing, epigenetic modifiers in DDR mediate DNA damage signaling and repair (Maréchal & Zou, 2013; Mailand & Bekker-Jensen, 2007).

Epigenetics is extremely important for plant development. They incorporate diverse environmental cues into distinct developmental or phenotypic responses. Thus, it is crucial to comprehend the mechanisms by which epigenetic alterations impact the way in which genotype is expressed as phenotype in plants. Numerous epigenetic phenomena have been identified in plants, including: (i) paramutation, which denotes inheritable shift in the expression status of an allele when it comes in contact with an allele that has the similar sequence but exhibits a distinct expression state; (ii) imprinting, which is defined as the selective expression of genes acquired by either the paternal or maternal parent; (iii) nucleolar dominance, which is the selective silencing of the ribosomal RNA genes acquired from one progenitor of a genetic hybrid; (iv) vernalization, which causes plants to flower in response to cold; (v) homology-dependent RNA-mediated methods, which have made significant advances in plant genetic engineering; (vi) RNA-mediated methylation of DNA, which results in downregulation of the gene; and (vii) RNA-mediated inactivation or degradation of mRNA (Rajewsky et al., 2017). Much like other higher species, plants use DNA methylation and histone modifications to start and control epigenetic regulation (Munshi et al., 2015). DNA methylation is essential for gene expression regulation, transposable element function, protection from foreign DNA and even inheritance of particular gene expression patterns (Boyko & Kovalchuk, 2008). The only epigenetic mark in DNA methylation is cytosine methylation. There are several other histone modifications, including acetylation, methylation, phosphorylation, ribosylation and ubiquitination of primarily H3 histones but also H4, H1 and H2A. These are the chromatin epigenetic markers that make up the "histone code". A common denominator for DNA methylation and histone changes is provided by RNA interference, which has rapidly entered the field of epigenetics (Tsaftaris et al., 2005). Epigenetic alterations are distinct to species, tissues, organelles and ages. They play a role in a number of activities, including genomic imprinting, transposon suppression and stress-induced defense responses (Yadav et al., 2018). Numerous components of the plant life cycle, including genomic integrity, transgene silencing, nucleosome organization, nucleolar dominance, paramutation, flowering and parent of origin-specific gene expression (imprinting), are significantly impacted by epigenetic alterations (Fujimoto et al., 2012). Histone modifications and DNA methylation are well-known examples of epigenetic changes (Fujimoto et al., 2012).

18.2.1 DNA Methylation

DNA methylation is the epigenetic phenomenon in plants that has been investigated the most (Ashapkin et al., 2020). Multiple interacting biological pathways establish and maintain DNA methylation, a crucial epigenetic alteration. Plants primarily exhibit cytosine methylation at a symmetrical CG dinucleotide position. It also occurs at asymmetric CHH sites (where H is A, C or T) and CHG, unlike animal sites. For correct cellular control during plant development, a dynamic interplay between methylation and demethylation, achieved by particular enzymes, is essential. "De novo" and "maintenance" DNA methyltransferases (MTases) carry out DNA methylation, which typically results in gene silencing; however, the inverse has sometimes been seen (Law & Jacobsen, 2010; Makarevich et al., 2008; Banerjee & Roychoudhury, 2017). In heterochromatic areas, including the centromeric and pericentromeric regions, which are primarily made up of transposons, DNA methylation is more abundant. To preserve genome integrity, the majority of transposons are stationary and are silenced by DNA methylation (Fujimoto et al., 2012). DNA methylation is an important constituent of epigenetic phenomena and controls gene expression during plant growth and development. The rate of DNA methylation differs in various species with 14% of methylated cytosine reported in *Arabidopsis thaliana*, 4% in *Mus musculus*, 2.3% in *Escherichia coli* and 0.03% in *Drosophila* (Yadav et al., 2018). The extent of DNA methylation in plant genomes is greater than in animal genomes. It can be symmetric, as in the case of CG, which is primarily regulated by DNA-METHYLTRANSFERASE-1 (MET1), or asymmetric, as in the case of CHG (where H is A, T or C), which is primarily regulated by CHROMOMETHYLASE 3 (CMT3), or asymmetric, as in the case of CHH DNA methylation, which is typically regulated by DOMAINS-REARRANGERSFERASEs (DRMs). The cytosine bases that make up 5-methylcytosine are methylated by DNA methyltransferases. Demethylating enzymes, such as DEMETER (DME), REPRESSOR OF SILENCING 1 (ROS1) and DEMETER-LIKE proteins (DML2/3) with DNA glycosylase domains, can eliminate the methyl groups in DNA. By changing the methylation state of promoters or coding regions, DNA methylation can modify the ability of genes to bind to transcriptional factors, hence controlling gene expression. Furthermore, DNA methylation can occur at repeats to silence those repeats and stabilize the heterochromatic structure (Miryeganeh, 2022). After treating DNA with sodium bisulfite, DNA methylation can be observed at high resolution with next-generation sequencing (Köhler & Springer, 2017).

18.2.2 Histone Modification

Research has demonstrated that along with DNA methylation, several kinds of histone modifications also alter the dynamic expression of genes during developmental stages or in response to environmental cues in plants. A chromatin structural component known as a nucleosome is formed when the DNA of eukaryote cells is wrapped around eight histone molecules, or an octamer. At certain genomic regions, nucleosomes have the ability to dismantle and reassemble in response to environmental and/or developmental stimuli (Miryeganeh, 2022). The H2A family of histones is the most varied and widely distributed in plants. The amino acids at the histone N-tails extending from the histone octamer can be changed by a variety of post-translational changes, including methylation, acetylation, H2B monoubiquitination and phosphorylation. These modifications alter chromatin structure and, consequently, the expression of the genes, as the interaction between DNA and histone and the accessibility of transcription factors change (Miryeganeh, 2022; Chinnusamy & Zhu, 2009; Banerjee & Roychoudhury, 2018). The *Arabidopsis* genome has 8 genes encoding histone H4 (HFO1–HFO8), 15 genes encoding histone H3 (HTR1–HTR15), 13 genes encoding histone H2A (HTA1–HTA13), 11 genes encoding histone H2B (HTB1–HTB11) and 3 genes encoding histone H1 (HON1–HON3). Among the 13 H2A-encoding genes, phylogenetic analysis could clearly assign 4 genes (HTA1/2/10/13) encoding canonical H2A proteins, 2 genes (HTA3/5) encoding H2A.X subtype variants, 3 genes (HTA8/9/11) encoding H2A.Z subtype variants and the other 4 genes (HTA4/6/7/12) encoding less well-categorized variants (March-Díaz & Reyes, 2009). The H2A.X variants possess a conserved C-terminal SQEF motif, where serine phosphorylation occurs at DNA damage sites, playing a crucial role in coordinating the DNA repair pathway (Van & Gasser, 2009; Zhu et al., 2012). The SWR1 ATP-dependent chromatin remodeling complex and histone chaperones NAP1 and Chz1 are involved in the exchange of H2A with H2A.Z in yeast (Luk et al., 2007). Within the various histone modifications, lysine 9 acetylation at histone H3 (H3K9ac) and lysine 4 trimethylation at histones H2 and H3 (H3K4me2/me3) are linked to stimulating gene transcription, whereas lysine 9 di-methylation at histone H9 (H3K9me2) and lysine 27 trimethylation at histone H2 and H3 (H3K27me2/me3) marks are implicated in suppressing gene transcription. Histone methyltransferases, histone demethylases, histone acetyltransferases (HATs) and histone deacetylases (HDAs or HDs) are the primary histone modifiers (Miryeganeh, 2022).

The patterns and functions of histone modification are also consistent throughout the history of eukaryotes. However, it is apparent that plants and mammals use different mechanisms for histone marks to control DNA methylation. According to structural analyses, ATRX-DNMT3-DNMT3L (ADD) domain of DNMT3L recruits DNMT3A to particular genomic areas and binds unmethylated histone 3 lysine 4 (H3K4) preferentially. DRM3 does not communicate with DRM2 and lacks the ADD domain. Although DNA methylation is linked to H3K4 demethylases, the role of H3K4 in DNA methylation and how it works in plants have not been established. It was recently found that histone 3 lysine 27 trimethylation (H3K27me3) may help avoid ectopic DNA methylation at cryptic sites in *Arabidopsis*. H3K27me3 has been linked to elevated DNA methylation in breast cancer cells, although more research is required to determine the functional importance of this linkage. The role of H3K36me3 in DNA methylation in plants is mainly unknown; however, it has recently been linked to directing DNMT3B to gene-body areas in mammals. However, recent research has revealed a variety of interactions between DNA methylation and histones. It is crucial for future research to comprehend how histone changes affect DNA methylation in relation to developmental processes and different environmental circumstances (Zhong, 2016).

18.3 CHROMATIN REMODELING

Histone composition can be changed by chromatin remodeling complexes (remodelers), which slide, evict or unwrap nucleosomes using ATP hydrolysis. By bringing remodelers to the damaged site, the nucleosome occupancy of the surrounding area is altered, ensuring that the repair factors are accessible. Furthermore, chromatin remodeling is necessary to control DNA strand resection, resolve intermediate structures during homologous recombination and control chromatin mobility to improve the efficiency of homologous recombination, as demonstrated in the yeast *Saccharomyces cerevisiae* (Gospodinov & Herceg, 2013; Seeber et al., 2013). Within the family of SWR1-like chromatin remodelers are the complexes SWR1 (Swi2/Snf2-RELATED1) and INO80 (INOSITOL-REQUIRING MUTANT80). They both have a conserved function in installing and removing the H2A.Z variation at transcriptional start sites, which regulates transcription across the genome. Genome stability is impacted by H2A.Z deposition because replicative polymerases and SWR-C activity work together to lower mutation rates (Van et al., 2015). The second major class of remodelers is Snf2-like complexes; however, it is less clear how they relate to DNA repair. Similar subunits can be found in *Arabidopsis*, including important chromatin-modifying (HDA1 and HDA6) and chromatin-interacting (PKL, PKR2 and FVE/MULTICOPY SUPPRESSOR OF IRA1 4 [MSI4]) components, although crucial functions in the regulation of development (cell identity, flowering period and various pleiotropic effects) are widely known (Donà & Mittelsten, 2015; Goodarzi et al., 2011).

18.4 DNA DAMAGE AND REPAIR PATHWAYS IN PLANTS

Living things are constantly in contact with substances that could compromise the integrity of their cells. This covers not only their genomes but also structural and enzymatic elements like lipids and proteins (Biedermann et al., 2011). Even though the genome of plants is extremely stable, exposure to various substances that break DNA, such as radiation, can still cause damage to the DNA. Mutations caused by structural or chemical damage to DNA can have either detrimental or beneficial (occasionally) effects on the organism (or its progeny). Base deletion, pyrimidine dimers, crosslinks, strand breakage and base modification, such as oxidation and alkylation, are examples of DNA damage (Tuteja et al., 2009; Biedermann et al., 2011). These genetically toxic agents can originate from their immediate environment and include chemicals or ionizing radiation, but DNA can also be harmed by reactive oxygen species (ROS), which are byproducts of everyday metabolism or result from insufficient defense against abiotic stressors. UV radiation can directly damage DNA by producing 6-4 photoproducts and cyclobutane pyrimidine dimer (CPD) photoproducts (Biedermann et al., 2011).

18.4.1 Major DNA Damage Repair Pathways in Plants

Further insights into the mechanisms of DNA repair have been obtained by recent research, and they will be discussed in the following sections: (1) photoreactivation, (2) nucleotide excision repair (NER), (3) base excision repair (BER), (4) mismatch repair (MMR) and (5) double-strand break repair (DSB repair) (Kimura & Sakaguchi, 2006; Tuteja et al., 2001; Manova & Gruszka, 2015).

18.4.1.1 Photoreactivation by Photolyases

By utilizing the energy of visible light, photolyases can repair UV-damaged DNA through a process known as photoreactivation. The thymine dimer structures of cyclobutane pyrimidine dimers (CPDs) and (6-4) photoproducts, which are both produced by UV irradiation, are promptly broken down by either CPD lyase or (6-4) phytolyase (Kimura & Sakaguchi, 2006). Particularly, photolyases bind to these DNA alterations and remove them by absorbing light with wavelength between 300 and 600 nm (Kimura et al., 2004). Photolyase genes have been identified in many plant species (Kimura & Sakaguchi, 2006). The primary mechanism in plants via which direct UV-induced DNA damage is repaired and CPDs and photoproducts are produced is dependent on the activity of photolyases. For photoreactivation, photolyases use two different types of chromophoric co-factors. The two-electron reduced version of FAD, FADH-, is one chromophore, and 7,8-dimethyl-8-hydroxy-5-deazariboflavin (8-HDF) or methenyltetrahydrofolate (MTHF) could be the other. MTHF or 8-HDF works as the light harvesting chromophores and transfers the energy to FADH- to lesion, triggering repair. An electron is transported from the excited FADH to CPD and (6-4) photoproducts upon direct binding by photolyases (Biedermann et al., 2011). Analysis of UV-sensitive mutants in *Arabidopsis* showed the presence of both CPD photolyase and (6-4) photolyase (Kimura & Sakaguchi, 2006). It has been shown that the constitutive production of CPD photolyases significantly increases UV tolerance in higher plants, but deletion of single members can result in enhanced UV sensitivity (Biedermann et al., 2011).

18.4.1.2 Nucleotide Excision Repair

NER facilitates repair of UV-induced lesions in a light-independent manner by two sub-pathways, the transcription-coupled repair (TCR) and the global genome repair (GGR), that process bulky DNA lesions along actively transcribed DNA strands or throughout the genome, respectively. By removing 24-32 nucleotide single-strand oligonucleotides from the damaged DNA strand, the NER mechanism eliminates DNA photolesions, followed by the restoration of an intact double helix by DNA repair synthesis. During this repair process, about 30 key proteins become mobilized. While TCR and GGR have different recognition steps, their excision repair procedures are similar as shown in Figure 18.2 (Molinier, 2017).

18.4.1.3 Base Excision Repair

Base excision repair involves using one of the numerous lesion-specific glycosylases to remove a single damaged base while maintaining the sugar–phosphate backbone of the DNA. An apurinic/apyrimidinic (AP) endonuclease or AP lyase then recognizes the resultant abasic sites and nicks the DNA backbone at the AP site. The damaged DNA is repaired to its original sequence through the coordinated actions of exonucleases, a repair polymerase and DNA ligase. Recent findings indicate that the repair polymerase, pol β, can excise the 5′-deoxyribose phosphate residue, which is produced by the combined activity of DNA glycosylases and class II AP endonucleases (Britt, 1996; Figure 18.3).

18.4.1.4 Mismatch Repair Mechanism

In addition to correcting mismatches, the MMR process also corrects short insertion/deletion loops (IDLs), which result from mis-incorporated or accidentally deleted nucleotides during DNA replication in the S phase (Hao, 2019). MutS complexes with DNA-binding capabilities identify lesions in the first stage of this process (Warren et al., 2007; Szurman-Zubrzycka et al., 2023). MutSα (MSH2/MSH6 heterodimer) and MutSβ (MSH2/MSH3 heterodimer) complexes are generally used to identify mismatches in eukaryotes. However, plants possess an additional complex involved in this step – MutSγ (MSH2/MSH7 heterodimer). MutSα identifies single-base mismatches (including oxidative and methylated mispairs) and very short IDLs, MutSβ identifies longer IDLs (with up to 16 unpaired nucleotides) and MutSγ detects a limited variety of single-base mismatches in plants. Additionally, MutSα can detect UV-induced lesions such as CPD and 6-4 PPs and start MMR to repair them; however, NER processes often repair these types of lesions. ATR (Ataxia Telangiectasia and Rad3)-related kinase can be activated by the MSH2 protein, which is a part of every known MutS complex implicated in MMR. In an ATP-dependent way, the MutS complexes recruit the MutLα complex to the lesion site. With its endonuclease activity, the MutLα complex nicks the daughter strand 5' to the mismatch, starting a repair response. In plants, the MutLα complex is a heterodimer of MLH1 and PMS1. MutLα recruits Exo1 to the nick. In the daughter strand, 5' → 3' excision is carried out by the exonuclease Exo1. To enable Exo1 to proceed with excision, the MutS must separate from the lesion site when the excision reaches the mismatch. It has been established that the *Arabidopsis* homolog of human and yeast DNA2-JHS1 (JING HE SHENG 1) has a role in meristem maintenance, cell cycle control and DNA repair in plants. Following appropriate DNA re-synthesis by DNA Pol δ with assistance from RFC, PCNA and RPA, the MMR pathway is completed, and the nick is ligated (Szurman-Zubrzycka et al., 2023).

18.4.1.5 DNA Double-Strand Break Repair

All organisms experience DNA damage, which can be brought on by either the exogenous environment or endogenous metabolism. Damage to DNA has a significant impact on the stability of chromosome structure and the maintenance of genomic information, which may result in cancer, cell death and abnormal development. In order to counteract the consequences of genotoxic stress, living organisms have evolved effective DNA repair systems. DNA damage known as double-strand breaks (DSBs) is commonly experienced during meiosis and DNA replication. In response to double-strand breaks, cells regulate the cell cycle, initiate DNA repair or trigger programmed cell death. In order to eliminate errors from DSBs, multicellular organisms have developed a DNA repair process (Yuan et al., 2014). Depending on the type of DNA damage and the cell cycle stage, the two most conserved repair mechanisms are non-homologous end joining (NHEJ) and homologous recombination (HR) (Fernandez et al., 2021), The successive organogenesis of plant meristems depends on the preservation of genomic information, and HR is always used to fix DSBs in mitotic cells (Yuan et al., 2014).

18.4.1.5.1 Homologous Recombination

All living things utilize HR, a DNA metabolic mechanism that offers high-fidelity, template-dependent repair or tolerance of complex DNA damages (DSBs, DNA gaps and DNA interstrand crosslinks) (Li & Heyer, 2008). In homologous recombination (HR), the broken DNA helix is unwound and the damaged strands are invaded to form a homologous DNA duplex molecule. This molecule could be a sister chromatid, homologous chromosome or viral genome with sequence similarity to the genomic DNA. HR repairs the DNA double-strand break by means of replication, using the homologous strand as a template. Therefore,

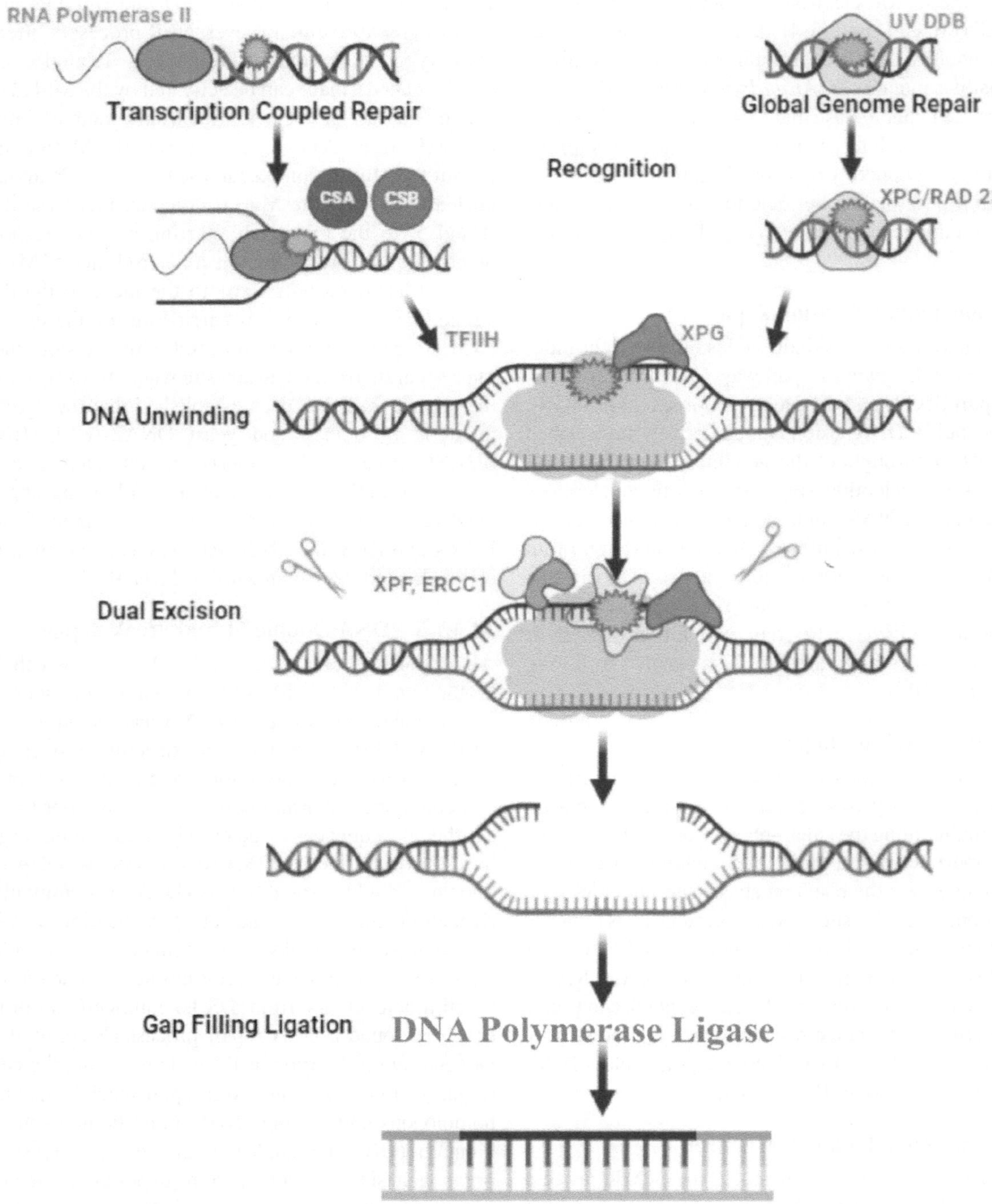

FIGURE 18.2 Nucleotide excision repair (NER). NER pathways could be categorized into two categories: global genome repair (GGR) and transcription-coupled repair (TCR). The identification of DNA damage, unwinding of double-stranded DNA nearby, excision of the damaged nucleotides, and filling of the gap by ligation and DNA synthesis comprise the NER pathway.

HR requires large areas of DNA homology in a different DNA duplex, and typically there is no genetic information loss (Featherstone & Jackson, 1999). Synthesis-dependent strand annealing and DSBRs are the two main HR pathways, which repair DSBs in DNA (Banerjee & Roy, 2021). Presynapsis, synapsis and postsynapsis are the three conceptual stages of homologous recombination.

- DSB ends are recognized and transformed into a single-stranded tail with a 3′-OH end during presynapsis.
- A D-loop is produced in synapsis by the Rad51-single-stranded DNA filament invading DNA strands. Following the D-loop intermediate, at least three alternative routes are proposed. After DNA synthesis, the invasive strand is released and annealed with the second end in SDSA (Synthesis-Dependent Strand Annealing), resulting in localized conversion without crossover. Multiple rounds of invasion, synthesis and disengagement may occur during this process. Break-induced replication (BIR) causes loss-of-heterozygosity (LOH) by

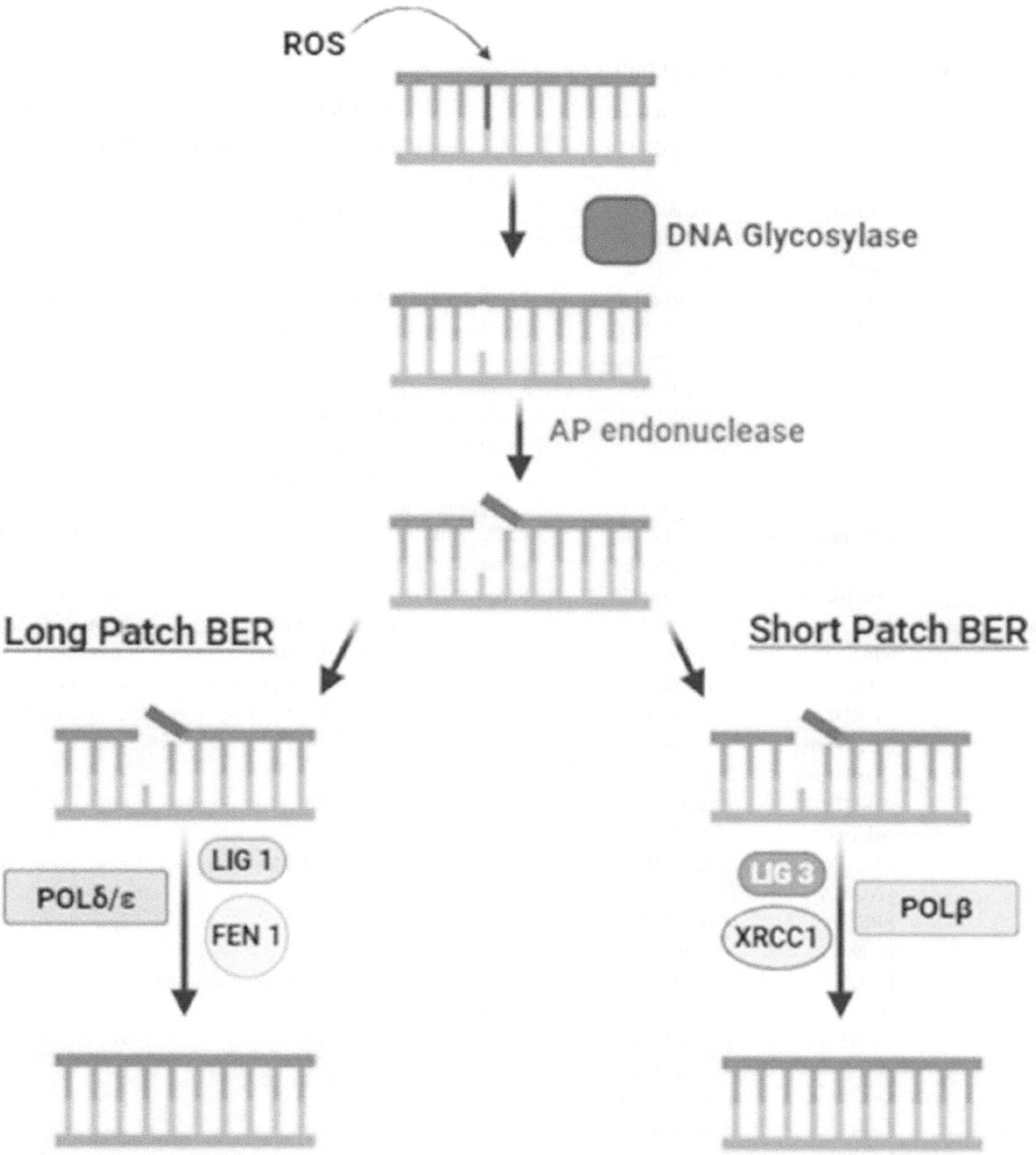

FIGURE 18.3 Schematic model for base excision repair.

assembling the D-loop into a complete replication fork and duplicating the entire distal portion of the chromosome.

- When both ends of the DSB are engaged, either by independent strand invasion or by second end capture, double Holliday junction development takes place. The junction can be processed by a resolvase, generating either non-crossover or crossover products or dissolved by BLM-mediated branch migration and TOPOIIIa-catalyzed dissolution of a hemicatenane, resulting exclusively in non-crossover products (Li & Heyer, 2008).

18.4.1.5.2 Nonhomologous End Joining (NHEJ)

NHEJ is the primary mechanism for DSB repair in the majority of higher eukaryotic cells (Bray & West, 2005). Instead of using a homologous sequence to repair DNA lesions, this DNA repair method rejoins the two broken ends of the DNA, which frequently results in errors or deletions that cause mutations (Gulfishan et al., 2023). The initial recognition of the DSB may be mediated by a complex of KU70 and KU80 (Bray & West, 2005). The genes Ku70, Ku80, XRCC4 and ligase IV are the key players in NHEJ in *Arabidopsis*. They have been found to have mutations that increase sensitivity to DSB-causing events. Broken ends of DNA are resected in NHEJ, a simply classified process. Where there are complementary nucleotides, junctions are formed through the rejoining of the two separate strands. The end portions are typically clipped, re-joined and the micro homologies are exposed at the junction site. Usually, the NHEJ mechanism results in deletions at the junction location due to DSB resection (Gulfishan et al., 2023) (Table 18.1).

TABLE 18.1
Plants as model organisms to study DNA repair

Repair	Gene	Acc. no.	Function	References
BER-related proteins/ genes	FPG	At1g52500	Bifunctional glycosylases (glycosylase and AP lyase activity)	Szurman-Zubrzycka et al. (2023)
	MBD4L	At3g07930	Monofunctional glycosylases	
	APE1L	At3g48425	AP endonucleases	
	APE2	At4g36050		
	ARP/APE1	AT2G41460	Apurinic/apyrimidinic (AP) endonuclease	Biedermann et al. (2011)
	OGG1	At1g21710	Lesion recognition and removal	Spampinato (2017)
	NTH2	At1g05900		
	XRCC1	At1g80420	Stimulation of phosphatase activity of ZDP (ZINC 4 FINGER DNA 3′-PHOSPHOESTERASE) and enhancement of DNA ligation by interaction with LIG1	Szurman-Zubrzycka et al. (2023)
Photoreactivation	CPD photolyase	AK111418	Removal of CPD	Kimura and Sakaguchi (2006)
	(6-4) photolyase	AP005008	Removal of (6-4) photoproduct	
	PHR1/UVR2	AT1G12370	Repair of CPDs	Biedermann et al. (2011)
Nucleotide excision repair	XPB	AK060447	DNA helicase, subunit of TFIIH	Kimura and Sakaguchi (2006)
	XPD	AK099724	DNA helicase, subunit of THIIH	
	AtCEN2	AT4G37010	Recognition of CPDs and 6-4 PPs, GGR	Biedermann et al. (2011)
	CDKD1;1	At1g73690	Module of TFIIH with kinase activity	Szurman-Zubrzycka et al. (2023)
	CDKD1;2	At1g66750		
	CDKD1;3	At1g18040		
NHEJ/HR	AtKU70	AT1G16970	Binding of DNA double-strand break ends	Riha et al. (2002)
	AtKU80	AT1G48050		
	AtRAD51/RAD51A	AT5G20850	DNA binding, mediation of interstrand pairing	Biedermann et al. (2011)
	SRS2	At4g25120	Resolution of the D-loop	Spampinato (2017)
	Rad9	At3g05480	DNA end processing	Szurman-Zubrzycka et al. (2023)
	RAD51	At5g20850	Recombinase initiating homology search and strand invasion	
MMR	MSH2	At3g18524	Recognition of mismatches	Szurman-Zubrzycka et al. (2023)
	MSH3	At4g25540		
	MSH6	At4g02070		
	MSH7	At3g24495		
	MSH1	AK110146	–	Kimura and Sakaguchi (2006)
	MutS homolog	AK067042	–	
	MSHα – MSH2–MSH6	–	Mismatch error and 1–2 bases of deletion/insertion loopout	de Medeiros et al. (2015)

18.5 ROLE OF CHROMATIN REMODELING IN DNA DAMAGE REPAIR

18.5.1 Chromatin Remodeling

Changes in the chromatin structure are necessary for transcription and replication to enable the binding of the relevant factors. There is mounting evidence that DNA repair works similarly in this situation (Dinant et al., 2008). Chromatin modification is crucial for nuclear functions that rely on DNA, including transcription, replication and repair (Morrison & Shen, 2005). Chromatin remodeling influences the accessibility of proteins to DNA in two ways: covalent histone alterations via post-translational modifications and displacement of histones or entire nucleosomes, either by removal or by sliding along the DNA (Dinant et al., 2008). Two key components used to carry out the remodeling process in eukaryotes are ATP-dependent chromatin remodeling complexes and histone-modifying enzymes (Liu et al., 2012). These complexes alter chromatin processes, such as creating DNA superhelical torsion, breaking DNA–histone bonds and moving nucleosomes, by using the energy of ATP hydrolysis. The existence of an SNF2-like DEAD/H(SF2) ATPase component in all ATP-dependent chromatin remodeling complexes places them in the SWI/SNF chromatin remodeling superfamily (Morrison & Shen, 2005).

Recent research has demonstrated the importance of ATP-dependent chromatin remodeling complexes for the

DDR (Lans et al., 2012). By using the energy from ATP hydrolysis, ATP-dependent chromatin remodeling complexes can slide or evict nucleosomes and catalyze the breaking of DNA–histone interactions. They also regulate nucleosome assembly and composition by exchanging histones and their variants in collaboration with histone chaperones. A catalytic component with an ATPase domain belonging to the SWI2/SNF2-family is essential to these complexes (Aydin et al., 2014). Chromatin remodelers are further classified into four groups, namely SWI/SNF, ISWI, CHD and INO80, based on their distinct ATPase-flanking domains (Liu et al., 2012), and have been demonstrated to mediate DNA repair and DDR signaling-related chromatin remodeling in *Arabidopsis thaliana* (Kim, 2019).

SWI/SNF: The SWI2/SNF2-RELATED 1 (SWR1) chromatin remodeling complex keeps the right expression of direct HS response genes (like HSP70) and catalyzes the substitution of nucleosome of its variant, H2A.Z, for canonical histone H2A (Song et al., 2021). SWI/SNF contributes to gene expression by promoting *de novo* DNA methylation and regulating microRNA (miRNA) processing and influences splicing. SWI/SNF activities are stimulus gated via complex subunit abundance, post-translational modifications, prion-like domains and the local chromatin context. Inhibition of the SWI/SNF component, SWI3B, by RNA interference increased the accumulation of DNA damage in plants. A chromosomal ATPase involved in double-strand break DNA repair, viz , structural maintenance complex 5 (SMC5), is recruited by SWI3B. UV-B exposure upregulates CHC101/SWBP73B in maize and *Arabidopsis*, preventing UV-B-induced DNA damage in maize (Bieluszewski et al., 2023).

ISWI: Imitation Switch (ISWI) belongs to the family of ATP-dependent chromatin remodeling factors, which use energy from ATP hydrolysis to move or reorganize nucleosomes. AtISWI forms complexes with AtDDT-domain proteins, which include DNA-binding homeobox and various transcription factors, many of which have distinct domain structures compared to DDT-domain proteins in yeast and animals. According to *in vivo* research, AtISWI has a role in the development of the uniformly distributed nucleosome organization pattern in gene bodies, linked to high gene transcription levels. Furthermore, meristem fate transition and organ formation are two of the several developmental processes in *A. thaliana* that are mediated by AtISWI and the AtDDT-domain protein RINGLET (RLT) (Li et al., 2017). The CHR11 and CHR17 proteins from *Arabidopsis* are examples of ISWI chromatin remodelers in plants. The nucleosome distribution patterns, which are linked to gene expression, are formed in part by the AtISWI proteins. Moreover, the role of the ISWI chromatin remodelers in plant DDR still remains underexplored (Kim, 2019).

CHD: *Arabidopsis* encodes four CHD family chromatin remodelers: CHROMATIN REMODELING 5 (CHR5), a CHD1 homolog and three CHD3 homologs – PICKLE (PKL), PKR and PKR2. These proteins regulate plant development and stress response by controlling gene expression. The *Arabidopsis* CHR5 has recently been shown to play additional roles in controlling the immunological response of plants and remodeling nucleosome occupancy. Nevertheless, there is no evidence that the CHD proteins contribute to plant DDR (Kim, 2019).

INO80 and SWR1: INO80 and SWR1 chromatin remodeling complexes (INO80-c and SWR1-c) are ATP-dependent regulators that influence histone variant H2A.Z incorporation into nucleosomes, a key process in eukaryotic gene regulation (Wang et al., 2019). The functions of plant INO80 have only been studied in *Arabidopsis*, and are shown to be required for somatic HR (Zhang et al., 2015). The *Arabidopsis* INO80 (AtINO80) has a dual role in the transcription and HR repair of DNA damage (Fritsch et al., 2004). The AtINO80-mediated chromatin remodeling is therefore crucial in genome stability maintenance and plant development (Zhang et al., 2015). The *Arabidopsis* SWR1 (AtSWR1) complex, composed of core subunits, PHOTOPERIOD-INDEPENDENT EARLY FLOWERING1 (PIE1), ARP6 and SWR1 COMPLEX SUBUNIT6 (SWC6), facilitates the exchange of H2A with H2A.Z in nucleosomes. Meiosis and somatic HR repair depend on the AtSWR1 complex. Furthermore, the AtSWR1 subunits and H2AZ might perform non-redundant roles in *Arabidopsis* gene regulation and plant immunology (Kim, 2019).

18.6 CONCLUSION

In conclusion, the maintenance of genomic integrity in plants is a complex yet vital process that requires robust DNA repair mechanisms, particularly due to their constant environmental challenges including pollutants, UV radiation and chemical mutagens. DNA damage, if left unrepaired, can lead to important genotoxic and cytotoxic consequences, with potential impacts on plant growth, development and evolution. Photoreactivation, excision repair and mismatch repair are some of the DNA repair processes found in plants that help to counteract damage to DNA from both endogenous and external sources. The evolutionary adaptation of these repair mechanisms, particularly in model plants like *Arabidopsis thaliana* and crop species like rice, highlights the interplay between DNA repair, environmental stress response and genetic diversity.

With alterations in DNA methylation, histone modifications and chromatin remodeling impacting genome stability, the importance of epigenetic control in the DNA damage response (DDR) is becoming more widely acknowledged. These epigenetic modifications are critical for plant development, particularly in response to environmental signals. The integration of epigenetic regulation and DNA repair mechanisms ensures the maintenance of genome integrity and the ability to adapt to changing environments. While significant progress has been made in unraveling DNA repair mechanisms in plants, there is still much to explore, especially in crop species. A deeper understanding of these pathways will not only contribute to fundamental biological knowledge but also offer practical applications in agriculture, possibly resulting in the creation of crop types that are more resilient to environmental stressors.

REFERENCES

Ashapkin, V. V., Kutueva, L. I., Aleksandrushkina, N. I., & Vanyushin B. F. (2020). Epigenetic mechanisms of plant adaptation to biotic and abiotic stresses. *International Journal of Molecular Sciences*, 21(20), 7457.

Aydin, Ö. Z., Vermeulen, W., & Lans, H. (2014). ISWI chromatin remodeling complexes in the DNA damage response. *Cell Cycle*, 13(19), 3016–3025.

Banerjee, A., & Roychoudhury, A. (2017). Epigenetic regulation during salinity and drought stress in plants: Histone modifications and DNA methylation. *Plant Gene*, 11, 199–204.

Banerjee, A., & Roychoudhury, A. (2018). The gymnastics of epigenomics in rice. *Plant Cell Reports*, 37, 25–49.

Banerjee, S., & Roy, S. (2021). An insight into understanding the coupling between homologous recombination mediated DNA repair and chromatin remodeling mechanisms in plant genome: An update. *Cell Cycle*, 20(18), 1760–1784.

Biedermann, S., Mooney, S., & Hellmann, H. (2011). Recognition and repair pathways of damaged DNA in higher plants. (Ed.) Chen, C.C. In *Selected Topics in DNA Repair* (pp. 201–236).

Bieluszewski, T., Prakash, S., Roulé, T., & Wagner, D. (2023). The role and activity of SWI/SNF chromatin remodelers. *Annual Review of Plant Biology*, 74(1), 139–163.

Boyko, A., & Kovalchuk, I. (2008). Epigenetic control of plant stress response. *Environmental and Molecular Mutagenesis*, 49(1), 61–72.

Bray, C. M., & West, C. E. (2005). DNA repair mechanisms in plants: Crucial sensors and effectors for the maintenance of genome integrity. *New Phytologist*, 168(3), 511–528.

Britt, A. B. (1996). DNA damage and repair in plants. *Annual Review of Plant Biology*, 47(1), 75–100.

Chinnusamy, V., & Zhu, J. K. (2009). Epigenetic regulation of stress responses in plants. *Current Opinion in Plant Biology*, 12(2), 133–139.

de Medeiros, N. M., de Medeiros, A. L., Silva, H. C., & Scortecci, K. C. (2015). Recent advances in plant DNA repair. Chen C. C. In *Advances in DNA Repair.*InTech, Croatia (pp. 3–42).

Donà M., Mittelsten Scheid O (2015). DNA damage repair in the context of plant chromatin. *Plant Physiology.* 2015 Aug 1;168(4):1206–18.

Dinant, C., Houtsmuller, A. B., & Vermeulen, W. (2008). Chromatin structure and DNA damage repair. *Epigenetics & Chromatin*, 1, 1–3.

Featherstone, C., & Jackson, S. P. (1999). DNA double-strand break repair. *Current Biology*, 9(20), R759–R761.

Fernandez, A., O'Leary, C., O'Byrne, K. J., Burgess, J., Richard, D. J., & Suraweera, A. (2021). Epigenetic mechanisms in DNA double strand break repair: A clinical review. *Frontiers in Molecular Biosciences*, 8, 685440.

Fritsch, O., Benvenuto, G., Bowler, C., Molinier, J., & Hohn, B. (2004). The INO80 protein controls homologous recombination in *Arabidopsis thaliana. Molecular Cell*, 16(3), 479–485.

Fujimoto, R., Sasaki, T., Ishikawa, R., Osabe, K., Kawanabe, T., & Dennis, E. S. (2012). Molecular mechanisms of epigenetic variation in plants. *International Journal of Molecular Sciences*, 13(8), 9900–9922.

Goodarzi A. A., Kurka T., Jeggo P.A. (2011). KAP-1 phosphorylation regulates CHD3 nucleosome remodeling during the DNA double-strand break response. *Nature Structural & Molecular Biology*, 18: 831–839

Gospodinov, A., & Herceg, Z. (2013). Chromatin structure in double strand break repair. *DNA Repair*, 12(10), 800–810.

Gulfishan, M., Bhat, T. A., & Jahan, A. (2023). DNA repair pathways and genes in plants. (Eds) Ahmad, T., Khalid, B., Hakeem, R. In *Biotechnologies and Genetics in Plant Mutation Breeding 2023* (pp. 109–129). Apple Academic Press, UK.

Hao, P. (2019). *Single Molecule Study of DNA Mismatch Repair.* North Carolina State University, USA.

Hays, J. B. (2002). *Arabidopsis thaliana*, a versatile model system for study of eukaryotic genome-maintenance functions. *DNA Repair*, 1(8), 579–600.

Iyama, T., & Wilson III, D. M. (2013). DNA repair mechanisms in dividing and non-dividing cells. *DNA Repair*, 12(8), 620–636.

Kim, J. H. (2019). Chromatin remodeling and epigenetic regulation in plant DNA damage repair. *International Journal of Molecular Sciences*, 20(17), 4093.

Kimura, S., & Sakaguchi, K. (2006). DNA repair in plants. *Chemical Reviews*, 106(2), 753–766.

Kimura, S., Tahira, Y., Ishibashi, T., Mori, Y., Mori, T., Hashimoto, J., & Sakaguchi, K. (2004). DNA repair in higher plants; photoreactivation is the major DNA repair pathway in non-proliferating cells while excision repair (nucleotide excision repair and base excision repair) is active in proliferating cells. *Nucleic Acids Research*, 32(9), 2760–2767.

Köhler, C., & Springer, N. (2017). Plant epigenomics—deciphering the mechanisms of epigenetic inheritance and plasticity in plants. *Genome Biology*, 18, 1–3.

Lans, H., Marteijn, J. A., & Vermeulen, W. (2012). ATP-dependent chromatin remodeling in the DNA-damage response. *Epigenetics & Chromatin*, 5, 1–4.

Law, J. A., & Jacobsen, S. E. (2010). Establishing, maintaining and modifying DNA methylation patterns in plants and animals. *Nature Reviews Genetics*, 11(3), 204–220.

Li, D., Liu, J., Liu, W., Li, G., Yang, Z., Qin, P., & Xu, L. (2017). The ISWI remodeler in plants: Protein complexes, biochemical functions, and developmental roles. *Chromosoma,* 126, 365–373.

Li, X., & Heyer, W. D. (2008). Homologous recombination in DNA repair and DNA damage tolerance. *Cell Research*, 18(1), 99–113.

Liu, B, Yip, R. K. H., & Zhou, Z. (2012). Chromatin remodeling, DNA damage repair and aging. *Current Genomics*, 13(7), 533–547.

Luk, E., Vu, N. D., Patteson, K., Mizuguchi, G., Wu, W. H., Ranjan, A., Backus, J., Sen, S., Lewis, M., Bai, Y., & Wu,

C. (2007). Chz1, a nuclear chaperone for histone H2AZ. *Molecular Cell*, 25(3), 357–368.

Mailand, N., Bekker-Jensen, S., Faustrup, H., Melander, F., Bartek, J., Lukas, C., & Lukas, J. (2007). RNF8 ubiquitylates histones at DNA double-strand breaks and promotes assembly of repair proteins. *Cell*, 131, 887–900.

Makarevich, G., Villar, C. B., Erilova, A., & Köhler, C. (2008). Mechanism of PHERES1 imprinting in *Arabidopsis. Journal of Cell Science*, 121(6), 906–912.

Manova, V., & Gruszka, D. (2015). DNA damage and repair in plants–from models to crops. *Frontiers in Plant Science*, 6, 885.

March-Díaz, R., & Reyes, J. C. (2009). The beauty of being a variant: H2A. Z and the SWR1 complex in plants. *Molecular Plant*, 2(4), 565–577.

Maréchal, A., & Zou, L. (2013). DNA damage sensing by the ATM and ATR kinases. *Cold Spring Harbor Perspectives in Biology*, 5(9), a012716.

Miryeganeh, M. (2022). Epigenetic mechanisms of senescence in plants. *Cells*, 11(2), 251.

Molinier, J. (2017). Genome and epigenome surveillance processes underlying UV exposure in plants. *Genes,* 8(11), 316.

Morrison, A. J., & Shen, X. (2005). DNA repair in the context of chromatin. *Cell Cycle, 4*(4), 568–571.

Munshi, A., Ahuja, Y. R., & Bahadur, B. (2015). Epigenetic mechanisms in plants: An overview. (Eds) Bahadur, B., Rajam, V. M., Sahijram, L., Krishnamurthy, K.V. In *Plant Biology and Biotechnology: Volume II: Plant Genomics and Biotechnology*. Springer, India (pp. 265–278).

Polyn, S., Willems, A., & De Veylder, L. (2015). Cell cycle entry, maintenance, and exit during plant development. *Current Opinion in Plant Biology*, 23, 1–7. https://doi.org/10.1016/j.pbi.2014.09.012.

Rajewsky, N., Jurga, S., & Barciszewski, J., (Eds.). (2017). *Plant Epigenetics.* Springer International Publishing.

Rastogi, R. P., Kumar, A., Tyagi, M. B., & Sinha, R. P. (2010). Molecular mechanisms of ultraviolet radiation-induced DNA damage and repair. *Journal of Nucleic Acids*, 592980. https://doi.org/10.4061/2010/592980.

Riha, K., Watson, J. M., Parkey, J., & Shippen, D. E. (2002). Telomere length deregulation and enhanced sensitivity to genotoxic stress in *Arabidopsis* mutants deficient in Ku70. *EMBO Journal*, 21, 2819–2826.

Schuermann, D., Molinier, J., Fritsch, O., & Hohn, B. (2005). The dual nature of homologous recombination in plants. *TRENDS in Genetics*, 21(3), 172–181.

Seeber, A., Hauer, M., & Gasser, S. M. (2013). Nucleosome remodelers in double-strand break repair. *Current Opinion in Genetics & Development*, 23(2), 174–184.

Song, Z. T., Liu, J. X., & Han, J. J. (2021). Chromatin remodeling factors regulate environmental stress responses in plants. *Journal of Integrative Plant Biology*, 63(3), 438–450.

Spampinato, C. P. (2017). Protecting DNA from errors and damage: An overview of DNA repair mechanisms in plants compared to mammals. *Cellular and Molecular Life Sciences*, 74, 1693–1709.

Szurman-Zubrzycka, M., Jędrzejek, P., & Szarejko, I. (2023). How do plants cope with DNA damage? A concise review on the DDR pathway in plants. *International Journal of Molecular Sciences*, 24(3), 2404.

Tsaftaris, A. S., Polidoros, A. N., Koumproglou, R. A., Tani, E. L., Kovacevic, N. I., & Abatzidou, E. L. (2005). Epigenetic mechanisms in plants and their implications in plant breeding. In *The Wake of the Double Helix: From the Green Revolution to the Gene Revolution* (pp. 157–171). Bologna: Avenue Media.

Tuteja, N., Ahmad, P., Panda, B. B., & Tuteja, R. (2009). Genotoxic stress in plants: Shedding light on DNA damage, repair and DNA repair helicases. *Mutation Research/Reviews in Mutation Research*, 681(2–3), 134–149.

Tuteja, N., Singh, M. B., Misra, M. K., Bhalla, P. L., & Tuteja, R. (2001). Molecular mechanisms of DNA damage and repair: Progress in plants. *Critical Reviews in Biochemistry and Molecular Biology*, 36(4), 337–397.

Ueda, T., & Nakamura, C. (2011). Ultraviolet-defense mechanisms in higher plants. *Biotechnology & Biotechnological Equipment*, 25(1), 2177–2182.

Van, C., Williams, J. S., Kunkel, T. A., & Peterson, C. L. (2015). Deposition of histone H2A. Z by the SWR-C remodeling enzyme prevents genome instability. *DNA Repair*, 25, 9–14.

Van Attikum H., Gasser S.M (2009). Crosstalk between histone modifications during the DNA damage response. *Trends in Cell Biology*. 2009 May 1;19(5):207–17.

Wang, J., Gao, S., Peng, X., Wu, K., & Yang, S. (2019). Roles of the INO80 and SWR1 chromatin remodeling complexes in plants. *International Journal of Molecular Sciences*, 20(18), 4591.

Warren, J. J., Pohlhaus, T. J., Changela, A., Iyer, R. R., Modrich, P. L., & Beese, L. S. (2007). Structure of the human MutSα DNA lesion recognition complex. *Molecular Cell*, 26(4), 579–592.

Yadav, C. B., Pandey, G., Muthamilarasan, M., & Prasad, M. (2018). Epigenetics and epigenomics of plants. *Plant Genetics and Molecular Biology*, 2018, 237–261.

Yuan, D., Lai, J., Xu, P., Zhang, S., Zhang, J., Li, C., Wang, Y., Du, J., Liu, Y., & Yang, C. (2014). AtMMS21 regulates DNA damage response and homologous recombination repair in *Arabidopsis. DNA Repair*, 21, 140–147.

Zhang, C., Cao, L., Rong, L., An, Z., Zhou, W., Ma, J., Shen, W. H., Zhu, Y., & Dong, A. (2015). The chromatin-remodeling factor At INO 80 plays crucial roles in genome stability maintenance and in plant development. *The Plant Journal*, 82(4), 655–668.

Zhong, X. (2016). Comparative epigenomics: A powerful tool to understand the evolution of DNA methylation. *New Phytologist*, 210(1), 76–80.

Zhu H., Guo F. J., Zhao W., Zhou J., Liu Y., Song F., Wang Y. (2012). ATF4 and IRE1α inhibit DNA repair protein DNA-dependent protein kinase 1 induced by heat shock. *Molecular and Cellular Biochemistry*, 371(1), 225–232.

Section IV

Epigenetics in Plant Stress Responses

19 MicroRNAs

Mastering Epigenetic Regulation in Plant Stress Responses

Subhasis Karmakar and Sabarinathan Selvaraj

19.1 INTRODUCTION

Plant stress responses are vital for survival and adaptation in dynamic and challenging environment. As sessile organisms, plants are exposed to a variety of stressors, which include biotic factors like pathogens and herbivory, as well as abiotic factors, viz. drought, salinity, extreme temperature and heavy metals, all of which can negatively impact their physiology. To overcome these challenges, plants have acquired complex and highly specialized mechanisms to detect, communicate and acclimatize to an adverse environment (Taria et al., 2022). A crucial adaptation of these mechanisms is their ability to recognize and distinguish between various stress cues (Volkov & Ranatunga, 2006). Plants depend on specialized sensors to detect environmental fluctuations, including alterations in humidity, light, temperature, and the presence of pathogenic organisms (Mittler et al., 2022). These signals trigger biochemical and physiological responses that help reduce damage and improve resilience. In reaction to biotic stress, plants activate defense strategies such as producing secondary metabolites and phytochemicals, strengthening cell walls and inducing localized programmed cell death at infection sites. They also emit aroma compounds to attract natural predators of herbivores (Clavijo McCormick et al., 2012). Alternatively, abiotic stress responses include strategies like stomatal closure to conserve water during drought, accumulation of osmotic regulators to maintain cell turgor, root adaptations for better water absorption, membrane stabilization under extreme temperature, production of heat shock proteins and activation of antioxidant systems to safeguard cellular components (Russo et al., 2022).

Epigenetics, or "beyond the gene", has become a crucial field in plant functional genomics, providing valuable insights into stress adaptation mechanisms. Regulatory processes like DNA methylation, histone modifications, chromatin remodeling and gene silencing using small RNAs are essential. Small RNAs, viz., miRNAs and siRNAs, play a vital role in regulating stress responses and advancements in genome sequencing have underscored their potential for enhancing crop improvement (Banerjee et al., 2017).

At the molecular level, stress responses are regulated by a wide range of transcription factors, diverse protein kinases and key signaling molecules (Muthusamy et al., 2021). Among these, microRNAs (miRNAs) serve as essential regulators that regulate gene expression patterns to improve stress adaptation. These small RNAs influence morphogenesis, development, agronomic traits and stress resilience by regulating gene expression post-transcriptionally. Decoding these regulatory networks is vital for tackling challenges such as agricultural sustainability and climate adaptation. Investigating the intricate mechanisms of plant stress adaptation, particularly the role of miRNAs, paves the way for increasing crop tolerance, promoting sustainable agriculture and reducing the effects of environmental stress on ecosystems.

19.2 BIOGENESIS OF miRNA

miRNA genes are primarily located in intergenic regions (IGRs) but can also be found within introns, either in sense or antisense orientations. They are transcribed in the nucleus by RNA polymerase II into precursors (pre-miRNAs) that form long hairpin structures, typically ranging from 70 to several hundreds of nucleotides (Axtell et al., 2011). These precursors are processed into a hairpin-like structure and cleaved by Dicer-like (DCL) enzymes to produce mature miRNAs. DCL1 interacts with chromatin at *MIR* gene loci, while regulatory proteins regulate miRNA transcription by interacting with the processing components (Wang et al., 2019). Plant-derived miRNAs, with near-perfect complementarity to their mRNA targets, induce mRNA silencing or cleavage. After processing, miRNAs are transported to the cytoplasm by HASTY (HST) and incorporated into the RNA-induced silencing complex (RISC), where they mediate the cleavage of mRNA or translation inhibition (Park et al., 2005). Furthermore, some miRNAs, referred to as mitrons, are sourced from the introns of host genes. Although mitrons are widespread in animals, only a small number have been identified in rice and *Arabidopsis*, with their precursors found in introns of unidentified protein-coding genes.

DOI: 10.1201/9781003589914-23

19.3 BIOTIC AND ABIOTIC CHALLENGES IN GROWTH AND SURVIVAL

Stressors in plants pose major threats to growth and productivity, affecting ecosystems and agriculture. Understanding these challenges is central to devising approaches to alleviate their impact and improve global food safety and environmental sustainability. Abiotic stressors, such as temperature extremes, drought and salinity disrupt plant growth and yield. Frost damages tissues, excessive heat accelerates water loss and drought causes wilting, stunted growth and even plant death (Volkov & Ranatunga, 2006). Salinity leads to osmotic stress and ion toxicity, while extreme pH levels restrict nutrient availability. UV radiation damages tissues, insufficient light reduces photosynthesis and pollutants, including heavy metals and chemicals, impairing plant metabolism (Ahanger et al., 2017). Pathogens cause diseases, while herbivores inflict damage and spread infections. Competition for sunlight, water and nutrients further imposes stress on plants and invasive species or allelopathic interactions disrupt ecosystems and suppress native plant growth. Climate change intensifies these stressors, introducing new pests and diseases to previously unaffected regions (Hu et al., 2022).

Despite these challenges, plants have developed advanced morphological, physiological and biochemical strategies. However, the rising frequency and severity of stressors driven by climate change and human activities demand innovative, sustainable solutions to protect plant health and ensure agricultural resilience (Singh et al., 2021).

19.4 PLANT STRESS-RESPONSIVE REGULATORY PATHWAYS

Complex systems enable plants to adapt to diverse biotic and abiotic stress, forming the foundation of their stress response strategies. These responses are triggered through the detection of stress signals via specialized receptors, leading to signaling cascades involving proteins such as protein kinases and essential transcription factors. ABA (abscisic acid) plays a pivotal role in mediating responses to drought and salinity stress. Stress-induced gene expression regulates key processes like osmotic adjustment and antioxidant defense, enabling plants to endure hostile conditions (Lamers et al., 2020). In biotic stress responses, phytohormones like jasmonic acid (JA) and salicylic acid (SA) trigger defense mechanisms, such as cell wall strengthening and the synthesis of defense-related compounds (Zhang et al., 2023). Epigenetic changes, including DNA methylation and histone modifications, play a role in stress response, enabling plants to adapt more efficiently to repeated stress events (Chinnusamy et al., 2014). Comprehending these pathways is vital for strengthening crop resilience and solving global challenges like environmental shifts and agricultural sustainability.

19.5 EPIGENETIC MECHANISMS IN PLANT STRESS ADAPTATION

Epigenetics deals with the study of DNA and chromatin modifications that result in heritable alterations in gene expression, without any adjustment to the underlying DNA sequence. These reversible modifications, which are often linked to gene activation or suppression, are pivotal in plant growth and stress-responsive phenotypic flexibility (Gallusci et al., 2017). Epigenetic mechanisms, including miRNAs/siRNAs, histone modifications and DNA methylation, are vital for regulating stress responses and can be passed down through generations. Exploring these processes provides essential insights for developing stress-resilient crops and establishing epigenetics as a key focus in plant research.

19.5.1 Histone Modifications

Various post-translational modifications occur, creating a "histone code" that regulates chromatin structure and gene expression in the N-terminal regions of histones. Modifications like acetylation, phosphorylation and ubiquitination promote transcription, whereas biotinylation and sumoylation suppress it (Banerjee & Roychoudhury, 2017, 2018), for example, H3K4 trimethylation promotes transcription, whereas H3K9 and H3K27 dimethylation are linked to repression (Sridhar et al., 2007). Stress-induced gene regulation is tightly connected to these modifications. In tomatoes, drought stress induces the H1-S histone variant, which reduces stomatal conductivity (Scippa et al., 2004). Submergence stress in rice induces H3K4 trimethylation of H3K4 and H3 acetylation at genes such as *ADH1* and *PDC*, leading to enhancement in expression (Tsuji et al., 2006). In *Arabidopsis*, histone deacetylases (*HDA6* and *HDA19*) play critical roles during stress responses by removing acetyl groups (Zhou et al., 2005), while stress and hormones regulate HDAC activity in rice (Fu et al., 2007). Drought stress is linked to epigenetic changes such as H3K4 trimethylation and H3K9 acetylation (Kim et al., 2008). Additionally, the histone acetyltransferase *GCN5* promotes the induction of stress-responsive genes in *Arabidopsis* (Stockinger et al., 2001).

19.5.2 DNA Methylation

Cytosine methylation on DNA, including both asymmetric and symmetric types, is related to suppressive chromatin states and gene silencing. DRM1 and DRM2 facilitate de novo methylation, while MET1 and CMT3 sustain symmetric methylation (Henderson & Jacobsen, 2007). Recent studies suggest that proteins like MET1 and CMT3 can contribute to de novo methylation, whereas DRM1 and DRM2 may play roles in maintaining symmetric methylation (Lister et al., 2008). Stress-driven modifications in DNA methylation can significantly affect gene expression; for instance, cold stress triggers hypomethylation and demethylation found in maize and tobacco, respectively (Choi & Sano, 2007; Steward et al., 2002). Conversely, drought stress has been shown to induce hypermethylation, as observed in pea (Labra et al., 2002) and *Mesembryanthemum crystallinum* (Dyachenko et al., 2006). Environmental stress can induce the activation of transposable elements via DNA demethylation, as demonstrated in *Antirrhinum majus* under cold stress (Hashida et al., 2006). Histone modifications also affect DNA methylation; for instance, mutations in *HDA6* and *HDA101* are associated with altered DNA methylation, resulting in the reactivation of previously silenced genes (Earley et al., 2006).

19.5.3 RNA-Directed DNA Methylation

Genetic studies in *Arabidopsis* mutants reveal that small interfering RNAs (siRNAs) play an intricate role in RNA-directed DNA methylation (RdDM) and contribute to the methylation of more than one-third of methylated loci (Pontes et al., 2006). ROS1, a DNA glycosylase, actively demethylates DNA to counterbalance RdDM, while ROS3 facilitates sequence-specific demethylation (Gong et al., 2002).

Abiotic stress, particularly temperature extremes, significantly influences small RNA regulation, with low-temperature promoting gene silencing and high temperature inhibiting or delaying it (Tsuji et al., 2006). In *Arabidopsis*, stress-regulated siRNAs, such as SRO5-P5CDH nat-siRNA, enhance salt stress tolerance by repressing proline degradation (Borsani et al., 2005). Moreover, abiotic stress together with abscisic acid (ABA) modulates the activity of *RdDM*-related genes, highlighting a connection between siRNAs, histone modifications and DNA methylation under stress conditions.

19.6 EXPLORING THE REGULATORY FUNCTIONS OF SMALL RNAS AND miRNAS IN PLANT STRESS TOLERANCE

Small RNAs (sRNAs), especially microRNAs, are key regulators of gene activity in plants, playing a vital role in adaptation to environmental changes (Wang et al., 2023). Typically spanning 20–24 nucleotides, miRNAs originate from precursor molecules and guide the RNA-induced silencing complex (RISC) toward specific mRNAs, initiating their degradation or preventing translation (Adhikari et al., 2023). They are critical for fundamental developmental processes, ensuring proper plant structure and function. Under environmental stress, miRNAs control the regulation of stress-related genes, boosting the capacity of plants to adapt and endure challenging conditions (Banerjee et al., 2016; Li et al., 2017). Furthermore, miRNAs modulate plant immunity by controlling genes involved in antimicrobial compound synthesis and pathogenesis-related (PR) protein production, equipping plants to defend against pathogens and herbivores. Similarly, small interfering RNAs (siRNAs) facilitate genome stability by directing RNA-dependent DNA methylation (RdDM), which inhibits the activity of transposable elements and redundant sequences (Šečić et al., 2021).

miRNAs also impact plant hormone pathways by regulating genes associated with hormone biosynthesis, signaling and regulation, thus coordinating stress adaptations and developmental processes (Banerjee et al., 2017). The diversity of miRNAs across plant species contributes to unique adaptations, with evolutionary processes shaping distinct miRNA families and their functions. In agriculture, miRNAs offer significant potential for improving stress tolerance, yield and nutritional quality (Chakraborty & Roychoudhury, 2023). By targeting specific miRNAs through genetic engineering, researchers are exploring promising strategies for crop improvement (Wang et al., 2023). Their central role in plant biology underscores their importance in advancing agricultural productivity and ecological sustainability.

19.7 HARNESSING miRNAS FOR PLANT RESILIENCE TO ABIOTIC STRESS

A regulatory network linking miRNAs to plant stress responses illustrates how miRNAs regulate stress-related genes and interact with signaling pathways, transcription factors, protein kinases and epigenetic modifications to enhance stress tolerance and resilience (Figure 19.1).

19.7.1 Role in Low-Temperature Stress

Temperature, light, water and minerals are crucial for plant development. However, abiotic stress can severely impair plant growth and reduce crop yield by triggering oxidative stress (Kerchev et al., 2020). miRNAs are crucial in alleviating stress by modulating the activity of the target genes. In response to cold stress, specific miRNAs help plants adapt by modulating critical pathways; for instance, in maize, miR160 and miR319 regulate processes related to cold stress responses, such as circadian rhythm and hormone signaling. Similarly, in potato, stu-miR530, stu-miR167b and stu-miR156d are involved in cold-stress-related pathways (Yan et al., 2021). In rice, miR1320 and miR156 enhance cold tolerance by targeting cold-responsive pathways (Huo et al., 2022). miR169 also plays a role in enhancing cold tolerance by alteration in auxin biosynthesis (Aslam et al., 2020). These findings highlight the critical role of miRNAs in plant stress responses, offering valuable insights into their participation in enhancing plant adaptation and resilience to abiotic stress conditions.

19.7.2 Role in Heat Stress Response

Heat stress, exacerbated by global warming, presents a major threat to plant growth and agricultural productivity. It interferes with vital physiological processes like photosynthesis, respiration and metabolism (Begum et al., 2022). miRNAs are essential in alleviating heat stress by

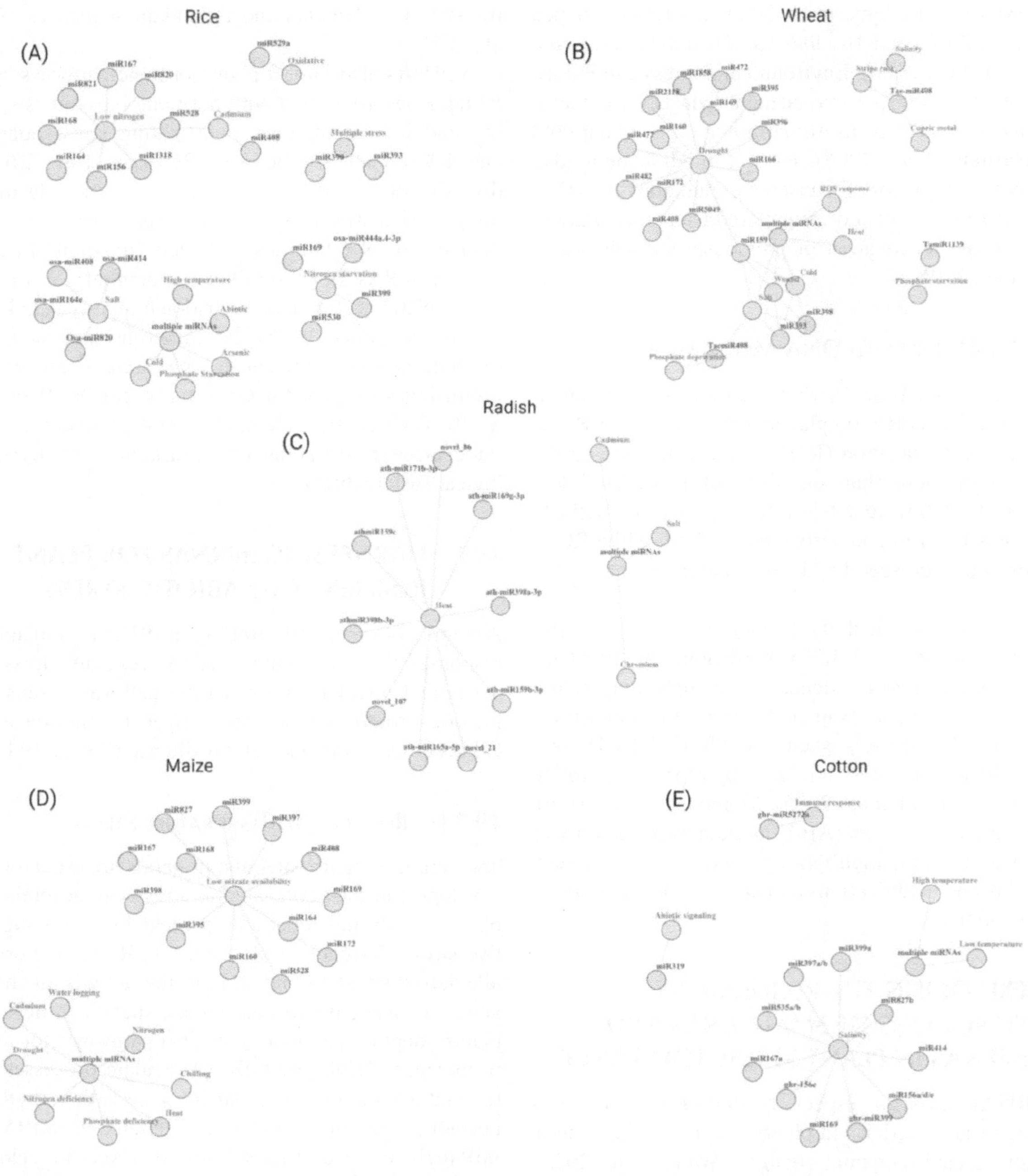

FIGURE 19.1 Regulatory network of miRNAs in stress responses, showing how they control the key stress-related genes and interact with signaling pathways, transcription factors, protein kinases and epigenetic mechanisms to enhance plant resilience and stress tolerance.

controlling the activity of the target genes. In maize, the miR160, miR169, miR159 and miR156 families exhibit varying levels of expression response to heat stress, with miR156 being significantly upregulated (Zhang et al., 2019). miRNAs such as miR156 and miR160 improve heat tolerance in diverse plant species, for example, *Arabidopsis* and cotton (Chen et al., 2020; Stief et al., 2014). Overexpression of miR160 improves thermotolerance in *Arabidopsis* (Lin et al., 2018), while miR156 improves heat stress tolerance in alfalfa (Arshad et al., 2020). In wheat, miRNAs such as miR159c and miR165b have been associated with terminal heat tolerance (Sihag et al., 2021). Additionally, species-specific miRNAs like miR824 and miR858 in *Arabidopsis* play roles in heat stress regulation by modulating transcription factors and reactive oxygen species (ROS) metabolism (Zhang et al., 2019). These findings underscore the diverse and critical roles of miRNAs in heat stress adaptation, offering potential avenues for enhancing crop resilience under rising temperatures.

19.7.3 Role in Drought Stress

Drought is a major limitation to plant development and crop yield, significantly endangering global food security

(Fahlgren et al., 2007). Plants cope with drought by employing a range of physiological and biochemical adaptations (Zhu, 2016). miRNAs have been recognized as crucial regulators of drought stress responses, with evidence from studies on several plant species (Das & Mondal, 2021; Ferdous et al., 2017). Phytohormones, especially abscisic acid (ABA), are crucial for drought tolerance by controlling stomatal behavior and interacting with other hormones such as salicylic acid and auxins. miRNA-target regulated pathways like miR159-MYB and miR169-NFYA function in ABA-dependent pathways, while miR156-SPL and miR393-TIR1 are involved in ABA-independent mechanisms (Singroha et al., 2021). In *Arabidopsis*, miR165/166 has been shown to mediate drought tolerance through ABA-dependent pathways, whereas miR169c and miR398c negatively regulate drought tolerance by targeting transcription factors (Zhou et al., 2020). On the other hand, miR172c and miR394 positively influence drought tolerance (Li et al., 2016). Barley and cardamom display genotype-specific miRNA expression under drought conditions; for example, miR169b is upregulated in drought-tolerant barley, while miR159 and miR156 exhibit differential expression in cardamom (Anjali et al., 2019; Ferdous et al., 2017). In maize, miRNAs like miR398 and miR402 regulate drought response pathways that are dependent on ABA, while miR160-ARF10 and miR528-POD influence auxin and ROS pathways, respectively (Das & Mondal, 2021). In wheat, drought-responsive miRNAs such as tae-miR9657b-5p and ata-miR396c-5p contribute to improved stress resilience by safeguarding the photosynthetic apparatus and preserving redox homeostasis (Yang et al., 2022). Overall, miRNAs play a vital role in regulating plant responses to drought through both ABA-dependent and ABA-independent pathways. Their expression is shaped by environmental conditions, genotypic variation and developmental stages (Ferdous et al., 2015).

19.7.4 Role in Salt Stress

Salinity is a vital environmental cue that negatively impacts plant growth, progression and global harvest yield (Wani et al., 2020). Salinity stress triggers osmotic imbalance, ion toxicity and oxidative damage, which hinder water uptake and plant growth (İbrahimova et al., 2021). Enhancing salt tolerance in crops is therefore critical, with miRNAs emerging as key post-transcriptional regulators in salt stress responses (Sun et al., 2019). In *Arabidopsis*, miR397 has been shown to reduce lignin content and salt sensitivity by targeting *LACCASE* genes under NaCl stress (Nguyen et al., 2020). Similarly, miR393 regulates salt stress via RACK1A-mediated ABA signaling, balancing auxin and ABA pathways (Denver & Ullah, 2019). In maize, salt-tolerant lines display distinct miRNA expression profiles compared to salt-sensitive lines (Ding et al., 2009). In other crops, miR156 plays a pivotal role in modulating ion homeostasis and enhancing both biomass and salt resistance in alfalfa (Arshad et al., 2017). In apple, the miR156-SPL module controls salt resistance by activating MdWRKY100 (Ma et al., 2021). miR319 enhances salinity resistance and ethylene synthesis in switchgrass by repressing *PvTCP* genes (Liu et al., 2019). The miR390-TAS3-ARF signaling pathway plays a role in salinity adaptation in Jerusalem artichoke by regulating *ARF* genes (Wen et al., 2020), while miR167 modulates ARF6 in *Tamarix chinensis* under salt stress (Ye et al., 2020).

Antioxidant systems also contribute to salt tolerance; for example, miR172 and IDS1 regulate ROS homeostasis in cereals (Cheng et al., 2021), and miR414c influences ROS metabolism in cotton by repressing *GhFSD1* (Wang et al., 2019). Targeting miRNA-regulated genes, including *ARF, IDS, AFB, SPL* and *TCP*, presents useful strategies for improvement in salt tolerance in crops, contributing to the development of resilient agricultural systems.

19.8 ROLE OF miRNA IN BIOTIC STRESS RESPONSE

In rice, miRNA expression fluctuates during microbial infections, underscoring their vital role in plant defense mechanisms. For example, the pathogen *Magnaporthe oryzae*, which causes rice blast disease, triggers the upregulation of miR319, leading to the inhibition of its target gene, *OsTCP21* (Zhang et al., 2018). Similarly, overexpressing *osa-miR398b* enhances resistance against rice blast (Yang et al., 2012). Additionally, limiting nitrogen availability has been found to strengthen resistance to bacterial blight caused by *Xanthomonas oryzae*, indicating that nutrient levels play a significant role in shaping plant–pathogen interactions (Tran et al., 2018). In wheat, fungal infections caused by *Blumeria graminis* f. sp. tritici *(Bgt)* lead to extensive reprogramming of miRNA expression, with several miRNAs being upregulated and others downregulated during the infection process (Sun et al., 2014). Similarly, infection by *Fusarium culmorum*, a significant pathogen of wheat and barley, notably induces the upregulation of miRNAs such as ath-miR2933, gma-miR5783 and gma-miR171 (Xin et al., 2010). Resistance to stripe rust in wheat, a disease that thrives in cooler climates, has been linked to several miRNAs, including miR1138, miR138, miR167, miR171, miR444, and miR1129. These miRNAs play crucial roles in regulating key stress response pathways (Gupta & Huang, 2014). Furthermore, abiotic stress-responsive miRNAs, such as Ta-miR113, Ta-miR1122, Ta-miR1117 and Ta-miR1113, have been identified for their participation in improving stress tolerance (Li et al., 2015). In coffee plants, miRNAs deploy a vital role in defense against leaf rust caused by *Hemileia vastatrix* (Rebijith et al., 2013), for instance, car-miR428d and car-miR2118e target the genes encoding disease-resistant proteins with nucleotide-binding ARC domains, enhancing disease resistance (Devi et al., 2016). Similarly, in tea plants, miRNAs like novel-miR1 and csn-miR426 have been associated with improved disease resistance (Jeyaraj et al., 2020). Additionally, other miRNAs, such as csn-miR156 and csn-miR319, are linked

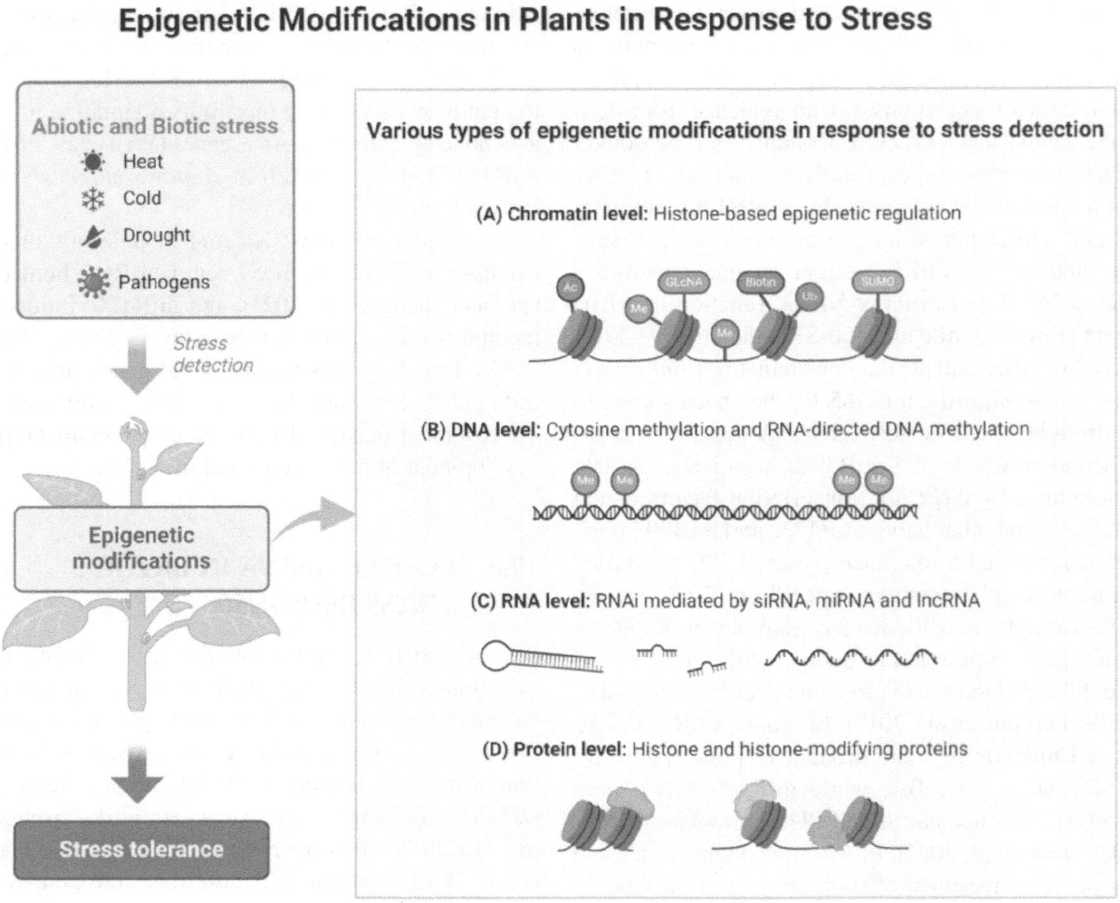

FIGURE 19.2 Epigenetic modifications in plants, including DNA methylation, histone changes and small RNA-mediated silencing, regulate adaptation and defense against environmental changes and pathogen attacks.

to responses against biotic stress in tea (Jeyaraj et al., 2017). Furthermore, Zhou et al. (2013) recognized 14 novel miRNAs targeting 51 mRNAs associated with redox processes, stress tolerance and nutrient transport, underscoring the complex regulatory networks governed by miRNAs during plant–pathogen interactions.

19.9 EPIGENETIC MODIFICATIONS AND miRNA-MEDIATED STRESS REGULATION IN PLANTS

Epigenetic modifications are crucial in miRNA-mediated stress responses, allowing plants to adapt to environmental challenges and maintain stress memory. In response to stress, miRNAs frequently target repressors of stress pathways, forming feedback loops that enhance adaptive mechanisms (Sanz-Carbonell et al., 2019). DNA methylation and histone modifications, which drive enduring changes to improve stress tolerance, are also regulated by miRNAs. Additionally, the interaction between various small RNAs, such as siRNAs, lncRNAs and miRNAs, forms intricate regulatory networks that precisely control gene activity in response to stress (Tiwari & Rajam, 2022). These interactions, combined with epigenetic modifications, strengthen the plant capacity to respond effectively to environmental challenges (Figure 19.2). Stress memory and priming are key outcomes of miRNA-mediated epigenetic regulation. Stress-induced miRNAs trigger epigenetic changes that enable plants to retain memory of past stress events, leading to quicker and more efficient responses to repeated stress (Singh et al., 2021). Moreover, miRNAs can influence protein stability and activity indirectly, through the regulation of enzymes that control post-translational modifications. A better understanding of how miRNAs regulate epigenetic changes, interact with hormones and create feedback loops is crucial for improving plant stress resilience (Table 19.1). This knowledge can be applied to develop stress-tolerant crops, an essential step in tackling the challenges brought about by climate change and the increasing demand for sustainable agriculture (Sanz-Carbonell et al., 2019).

19.10 INTERPLAY IN PLANT STRESS RESPONSES: RESOURCE ALLOCATION AND EPIGENETIC REGULATION

Interactions between stress-responsive pathways are crucial for how plants manage their limited resources when faced with multiple stressors. When faced with multiple stresses,

TABLE 19.1
A brief overview of small RNA (sRNA)-induced DNA methylation, its target genes and their established roles in plant development and stress tolerance

Plants	Small RNA-mediated methylation	Example	Methylated loci	Proteins participating in DNA methylation	DNA methylation pattern	Predicted and confirmed functions of sRNAs in regulating plant responses to stress and development	References
Zea mays	Phased small interfering RNAs (phasiRNAs) are generated from PHAS loci transcripts that are cleaved by miRNAs	miR2118, miR2275	PHAS loci	AGO104 or AGO18b	Cis-directed DNA methylation at the originating PHAS loci	Promote the processing of long non-coding precursor transcripts to produce phased small RNAs (phasiRNAs), which are essential for regulating gene expression	Dukowic-Schulze et al. (2016)
Arabidopsis	miRNA-guided easiRNAs derived from transposons	miR156, miR159, miR169	Transposon transcripts (e.g., VANDAL, ATHILA, etc.)	AGO1	Disrupting the RDR2–DCL3 complex, thereby inhibiting RdDM mediated by 24-nt siRNAs	miR156: regulates developmental transitions; miR159: targets MYB and TCP transcription factors during plant development; miR169: involved in nitrogen and stress responses	Creasey et al. (2014)
Arabidopsis	ta-siRNAs derived from miRNA-cleaved *TAS* gene transcripts	miR173, miR390	TAS1/2/3	AGO4/6 and Pol V	Cis-directed DNA methylation at *TAS* loci near miRNA cleavage sites	Directs TAS ncRNA to produce tasiRNAs	Wu et al. (2012)
Oryza sativa	siRNAs originating from miRNA loci within introns	miR1863, miR437	Intron in the locus LOC_Os12g09290.1	AGO4	Initiate cis-directed DNA methylation	Methylates both the miRNA region and sequences beyond the stem loop, suppressing the target gene expression	Chen et al.(2011)
Arabidopsis	siRNAs from miRNA loci	miR2328, miR2381	At4g16580, At5g08490	AGO4	Trans-directed RdDM at target loci, decreasing the accumulation of target gene transcripts	Methylates approximately 80 nt upstream and downstream of complementary sites on target genes via the NRPD1-dependent pathway	Chellappan et al. (2010)
Arabidopsis	Canonical miRNA	miR165, miR166	PHB and PHV	ND	Induces DNA methylation downstream of the miRNA binding site	miR165 targets class III HD-ZIP transcription factors, while miR166 influences Squamosa promoter-binding protein-like factors, both essential for stress response mechanisms	Bao et al. (2004)

plants must distribute resources such as nutrients, energy, and water across various response mechanisms (Aftab & Roychoudhury, 2021). Epigenetic modifications play a crucial role in this process. These changes can regulate genes involved in various stress responses and even produce transgenerational effects, boosting the plant's adaptability over time. This interaction also extends beyond individual plants, influencing relationships with herbivores and pathogens, as well as broader ecological functions like nutrient cycling and ecosystem health. This often leads to trade-offs, where prioritizing one response may reduce the ability of plants to effectively manage others (Aftab & Roychoudhury, 2021), for example, in *Arabidopsis thaliana*, miR395 modulates sulfate assimilation by targeting ATP sulfurylases, enzymes involved in the conversion of sulfate to adenosine 5′-phosphosulfate (Ai et al., 2016). This intricate regulatory system allows plants to process signals from different stress factors, prioritize their responses and efficiently allocate resources. Improving our understanding of these mechanisms is crucial for boosting stress resistance in crops and ensuring the resilience of ecosystems in the face of environmental challenges.

19.11 EPIGENETIC AND POST-TRANSCRIPTIONAL ROLES OF miRNAS IN STRESS TOLERANCE

Epigenetic regulation is crucial for incorporating miRNAs into stress response networks, affecting DNA methylation and histone modifications that lead to enduring changes in gene expression and stress retention (Yao et al., 2019). miRNAs also control gene modulation after transcription by interacting with target mRNAs, influencing their stability and translation efficiency, thereby enabling precise regulation of protein levels under stress conditions (Sharma et al., 2021). These functions enhance plant growth, developmental processes, stress resilience and defense mechanisms by adding complexity to gene networks, supported by interactions between different signaling pathways (Zhang et al., 2022). miRNA-driven stress tolerance functions through complex signaling pathways. Plants sense environmental stressors like drought or salinity via receptors that trigger hormonal responses, especially involving ABA. miRNAs regulate these pathways by targeting genes involved in hormone synthesis and signaling (Villani et al., 2023). Furthermore, stress-responsive miRNAs regulate transcription factors (TFs), which serve as key hubs in stress signaling pathways, controlling the expression of target genes involved in stress responses (Achkar, 2016). Together, these mechanisms enhance plant resilience and adaptability to stressful conditions (Zhang et al., 2022).

Stress-responsive miRNAs react to environmental signals by targeting genes like TFs, stress-related proteins and signaling components. This regulation, involving mRNA degradation or inhibition of translation, precisely modulates gene expression patterns (Villani et al., 2023). miRNA-driven stress pathways frequently intersect with other signaling networks, forming crosstalk that connects stress and defense responses (Pagano et al., 2021).

Calcium ions act as secondary messengers, where stress-induced fluctuations in Ca^{2+} concentration initiate the activation of transcription factors, protein kinases and other key regulatory proteins (Ketehouli et al., 2022). Reactive oxygen species (ROS), including hydrogen peroxide, also function as signaling molecules, influencing gene expression and activating downstream response pathways (Peng et al., 2023). Phosphorylation events, driven by protein kinases and phosphatases, are essential for regulating protein activity and controlling gene expression (Chakradhar et al., 2019).

Epigenetic changes, including DNA methylation and histone modifications, frequently accompany miRNA-mediated stress responses, strengthening stress memory and long-term gene expression regulation. Feedback loops further refine these pathways, ensuring accurate stress responses while avoiding excess activation (Shi et al., 2023). These complex signaling networks, involving miRNA regulation, hormonal signaling, crosstalk, calcium and ROS signaling, kinase and phosphatase activity, and epigenetic changes, enhance the adaptability and resilience of plants to environmental stressors (Ketehouli et al., 2022).

19.12 INTEGRATION OF miRNAS AND TRANSCRIPTION FACTORS IN STRESS SIGNALING PATHWAYS

Transcription factors (TFs) serve as crucial intermediaries in miRNA-regulated stress reactions, helping plants adapt to environmental stressors (Zhang, 2015). Stress-responsive miRNAs typically target transcription factors (TFs), leading to their downregulation through mRNA degradation or blocking translation, thus regulating stress response genes (Ahmed et al., 2019). Key TF families, viz., MYB, AP2/ERF, NAC, WRKY and bZIP, are delimited by specific miRNAs, such as miR159, which marks MYB33 and MYB101, the key players in drought and ABA signaling pathways (Zhang et al., 2022). TFs targeted by miRNAs control downstream stress-related genes, influencing antioxidant defense, osmotic regulation and hormone signaling (Sanz-Carbonell et al., 2019). This hierarchical control places miRNAs as central coordinators, directing the regulation of genes in response to stress. Additionally, miRNAs regulate hormone signaling by targeting TFs linked to hormone-responsive genes, thus integrating stress signaling pathways (Arora et al., 2013). Feedback loops further enhance rapid stress responses, while miRNA-mediated TF regulation can drive epigenetic changes, creating stress memory. Thus, miRNAs and TFs form a regulatory axis critical for plant adaptability and resilience to stress, providing insights for improving crop stress tolerance (Ahmed et al., 2019).

19.13 BIOTECHNOLOGICAL ADVANCES IN HARNESSING miRNA FOR STRESS-TOLERANT CROP BREEDING

MicroRNAs (miRNAs) have become valuable tools in biotechnology for boosting stress resistance in crops. These small regulatory RNAs are crucial in altering the activity of genes involved in stress responses, holding significant potential for crop improvement. The first step in harnessing miRNAs for crop enhancement involves identifying those that are stress-responsive through differential expression studies under different stress conditions, followed by functional analysis to uncover their roles and target genes (Basso et al., 2019). To manipulate miRNA expression, biotechnological strategies include the use of miRNA mimics to overexpress specific miRNAs and antagomirs (inhibitors) to downregulate them. These strategies enable targeted modulation of miRNA activity, improving the plant capacity to tolerate stress (Chaudhary et al., 2021). Transgenic techniques enable targeted overexpression or silencing of stress-responsive miRNAs, thereby improving plant resilience by modifying the regulatory networks (Figure 19.3). Additionally, RNA interference (RNAi) can be used to downregulate miRNAs that hinder stress tolerance by delivering dsRNA (double-stranded RNA) molecules (Wang et al., 2019). Another approach involves the design of artificial miRNAs (amiRNAs) that target specific stress-related genes, providing a precise and adaptable tool for crop improvement (Kotowska-Zimmer et al., 2021). Genome editing techniques like CRISPR-Cas9 offer another promising method for directly manipulating miRNA loci, enabling precise control over miRNA expression to bolster crop stress tolerance (Gaj et al., 2016). Integrating multiple stress-related miRNAs or genetic factors through stackable traits can further boost crop tolerance to both drought and pathogens by leveraging the synergistic effects of miRNAs. Marker-assisted selection (MAS) aids breeding programs by using molecular markers associated with stress-responsive miRNAs to identify and select plants with enhanced stress resilience (Zhou & Luo, 2013). miRNA-modified crops undergo extensive field testing to assess their performance and safety before commercialization, ensuring that farmers can access stress-resilient varieties (Castro et al., 2021). Given their significant potential to boost stress resistance, miRNAs could play a crucial role in ensuring long-term food availability and supporting agricultural resilience amid climate change. Furthermore, thorough risk evaluations and regulatory assessments are vital for their broad implementation (Chaudhary et al., 2021).

19.14 EMERGING CHALLENGES AND FUTURE PERSPECTIVES

The use of microRNAs (miRNAs) to enhance plant stress resilience offers significant potential but also presents several important challenges. A major concern is the risk of off-target effects, where miRNA manipulation could unintentionally influence non-target genes and biological pathways, potentially disrupting plant physiology. Additionally, the environmental impact, including potential unintended effects on non-target species and ecosystems, must be carefully assessed. Unlike conventional genetically modified organisms (GMOs), regulatory frameworks based on miRNAs are still in their infancy, complicating the approval process. Additionally, developing efficient delivery systems for precise miRNA manipulation continues to be a significant challenge. Another complexity lies in understanding how miRNAs interact with the complex combination of multiple stresses (Jiang et al., 2022). miRNA-driven approaches present considerable promise for precision agriculture by enabling the precise adjustment of stress-responsive miRNAs, customized for specific crops and environmental conditions. Future research will aim to develop customized solutions, including stacking stress-responsive traits and designing synthetic miRNAs to enhance stress tolerance. Technologies such as CRISPR-Cas9, along with other gene-editing tools, offer fine-tuned regulation of miRNA expression. By combining multi-omics approaches, researchers can gain deeper insights into miRNA-mediated stress responses, enabling the creation of more effective strategies. Rigorous environmental monitoring will be crucial to ensure the safety of miRNA-modified crops, while ethical and equitable considerations should steer the application of these technologies (Jiang et al., 2022). miRNA-driven strategies offer significant potential for improving crop stress resilience and advancing more sustainable agricultural practices. Despite the challenges, ongoing research and collaboration among scientists, regulators and key stakeholders will be crucial to fully realize the potential of miRNA technology in securing global food security (Post et al., 2020).

19.15 CONCLUSION

In summary, miRNAs are crucial in regulating plant responses to diverse stressors by influencing the genes and pathways involved in stress adaptation. Present in numerous plant species, stress-responsive miRNAs enable plants to manage challenges such as drought, salinity, heat and pathogen attacks. Their capacity to influence multiple genes facilitates an integrated response to environmental pressures, highlighting their potential to develop stress-resilient crops. With the escalating impacts of climate change, leveraging miRNA-based approaches offers promising opportunities to enhance sustainable agriculture and bolster food security. These technologies can help develop crops that are more resilient to stress, improving productivity while decreasing reliance on chemical inputs and fostering environmentally sustainable farming practices. By increasing resilience in crops, particularly in regions vulnerable to extreme weather conditions, miRNA applications can significantly benefit global communities. Nonetheless, their safe and effective implementation requires robust ethical considerations, thorough environmental assessments

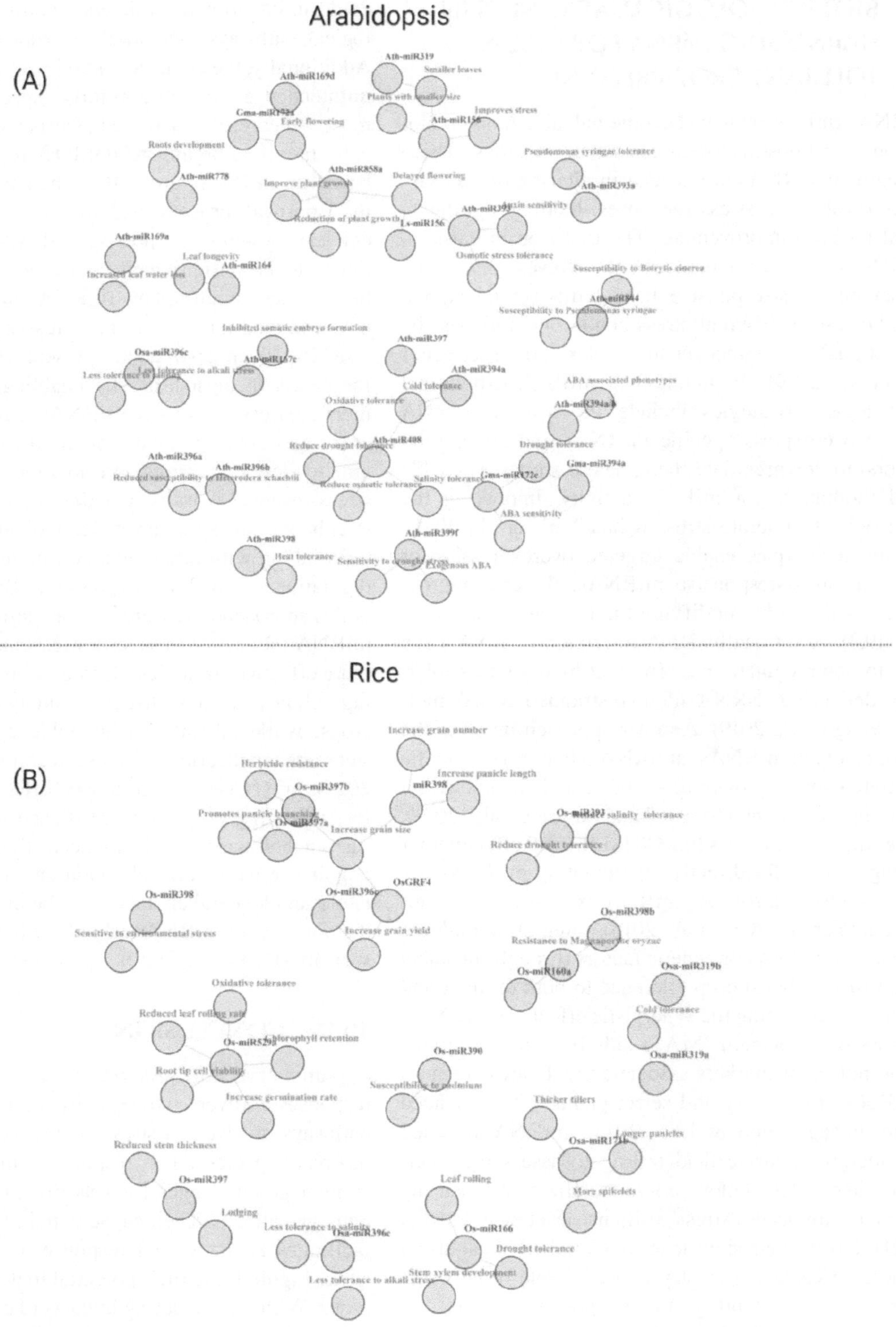

FIGURE 19.3 Biotechnological advances in harnessing miRNA for developing stress-tolerant crops. The figure illustrates the role of engineered miRNAs in modulating gene expression to enhance plant resilience against abiotic and biotic stresses (a special emphasis on *Arabidopsis* and rice). Key approaches include miRNA overexpression, target mimics and genome editing strategies for precise regulation of stress-responsive pathways.

and well-structured regulatory policies. Harnessing the potential of miRNAs provides a pathway to address critical agricultural challenges, safeguarding food supplies and preserving environmental health for future generations.

REFERENCES

Achkar, N. P., Cambiagno, D. A., & Manavella, P. A. (2016). miRNA biogenesis: a dynamic pathway. *Trends in Plant Science*, 21(12), 1034–1044.

Adhikari, A., et al. (2023). microRNAomic profiling of maize root reveals multifaceted mechanisms to cope with Cr (VI) stress. *Plant Physiology and Biochemistry*, 198, 107693.

Aftab, T., & Roychoudhury, A. (2021). Crosstalk among plant growth regulators and signaling molecules during biotic and abiotic stresses: Molecular responses and signaling pathways. *Plant Cell Reports*, 40, 2017–2019.

Ahanger, M. A., Akram, N. A., Ashraf, M., Alyemeni, M. N., Wijaya, L., & Ahmad, P. (2017). Plant responses to environmental stresses – From gene to biotechnology. *AoB Plants*, 9, plx025.

Ahmed, W., Xia, Y., Zhang, H., Li, R., Bai, G., Siddique, K. H. M., & Guo, P. (2019). Identification of conserved and novel miRNAs responsive to heat stress in flowering Chinese cabbage using high-throughput sequencing. *Scientific Reports*, 9, 1–11.

Ai, Q., Liang, G., Zhang, H., & Yu, D. (2016). Control of sulfate concentration by miR395-targeted APS genes in *Arabidopsis thaliana. Plant Diversity*, 38, 92–100.

Anjali, N., Nadiya, F., Thomas, J., & Sabu, K. K. (2019). Identification and characterization of drought responsive microRNAs and their target genes in cardamom (*Elettaria cardamomum* Maton). *Plant Growth Regulation*, 87, 201–216.

Arora, S., Rana, R., Chhabra, A., Jaiswal, A., & Rani, V. (2013). MiRNA-transcription factor interactions: A combinatorial regulation of gene expression. *Molecular Genetics and Genomics*, 288, 77–87.

Arshad, M., Gruber, M. Y., Wall, K., & Hannoufa, A. (2017). An insight into microRNA156 role in salinity stress responses of alfalfa. *Frontiers in Plant Science*, 8, 1–15.

Arshad, M., Puri, A., Simkovich, A. J., Renaud, J., Gruber, M. Y., Marsolais, F., & Hannoufa, A. (2020). Label-free quantitative proteomic analysis of alfalfa in response to microRNA156 under high temperature. *BMC Genomics*, 21, 1–18.

Aslam, M., Sugita, K., Qin, Y., & Rahman, A. (2020). Aux/iaa14 regulates microrna-mediated cold stress response in *Arabidopsis* roots. *International Journal of Molecular Sciences*, 21, 1–25.

Axtell, M. J., Westholm, J. O., & Lai, E. C. (2011). Vive la différence: Biogenesis and evolution of microRNAs in plants and animals. *Genome Biology*, 12, 1–13.

Banerjee, A., & Roychoudhury, A. (2017). Epigenetic regulation during salinity and drought stress in plants: Histone modifications and DNA methylation. *Plant Gene*, 11, 199–204.

Banerjee, A., & Roychoudhury, A. (2018). The gymnastics of epigenomics in rice. *Plant Cell Reports*, 37, 25–49.

Banerjee, A., Roychoudhury, A., & Krishnamoorthi, S. (2016). Emerging techniques to decipher microRNAs (miRNAs) and their regulatory role in conferring abiotic stress tolerance of plants. *Plant Biotechnology Reports*, 10(4), 185–205.

Banerjee, S., Sirohi, A., Ansari, A. A., & Gill, S. S. (2017). Role of small RNAs in abiotic stress responses in plants. *Plant Gene*, 11, 180–189.

Bao, N., Lye, K., & Barton, M. K. (2004). MicroRNA binding sites in Arabidopsis class III HD-ZIP mRNAs are required for methylation of the template chromosome. *Developmental Cell*, 7, 653–662.

Basso, M. F., Ferreira, P. C. G., Kobayashi, A. K., Harmon, F. G., Nepomuceno, A. L., Molinari, H. B. C., & Grossi-de-Sa, M. F. (2019). MicroRNAs and new biotechnological tools for its modulation and improving stress tolerance in plants. *Plant Biotechnology Journal*, 17, 1482–1500.

Begum, Y., et al. (2022). Regulatory role of microRNAs (miRNAs) in the recent development of abiotic stress tolerance of plants. *Gene*, 821, 146283.

Borsani, O., Zhu, J., Verslues, P. E., Sunkar, R., & Zhu, J. K. (2005). Endogenous siRNAs derived from a pair of natural cis-antisense transcripts regulate salt tolerance in *Arabidopsis. Cell*, 123, 1279–1291.

Castro, B., Citterico, M., Kimura, S., Stevens, D. M., Wrzaczek, M., & Coaker, G. (2021). Stress-induced reactive oxygen species compartmentalization, perception and signalling. *Nature Plants*, 7, 403–412.

Chakraborty, S., & Roychoudhury, A. (2023). MicroRNA-mediated regulation of heavy metal stress in plants. In D. Pandita & A. Pandita (Eds.), *Plant MicroRNAs and Stress Response* (pp. 183–197). Boca Raton: CRC Press.

Chakradhar, T., Reddy, R. A., & Chandrasekhar, T. (2019). Protein kinases and phosphatases in stress transduction: Role in crop improvement. In *Plant Signaling Molecules* (pp. 533–547). Woodhead Publishing.

Chaudhary, S., Grover, A., & Sharma, P. C. (2021). MicroRNAs: Potential targets for developing stress-tolerant crops. *Life*, 11, 1–21.

Chellappan, P., Xia, J., Zhou, X., Gao, S., Zhang, X., Coutino, G., et al. (2010). siRNAs from miRNA sites mediate DNA methylation of target genes. *Nucleic Acids Research*, 38, 6883–6894.

Chen, D., Meng, Y., Yuan, C., Bai, L., Huang, D., Lv, S., et al. (2011). Plant siRNAs from introns mediate DNA methylation of host genes. *RNA*, 17, 1012–1024.

Chen, J., Pan, A., He, S., Su, P., Yuan, X., Zhu, S., & Liu, Z. (2020). Different microRNA families involved in regulating high temperature stress response during cotton (*Gossypium hirsutum* l.) anther development. *International Journal of Molecular Sciences*, 21, 1–16.

Cheng, X., He, Q., Tang, S., Wang, Haoran, Zhang, X., Lv, M., et al. (2021). The miR172/IDS1 signaling module confers salt tolerance through maintaining ROS homeostasis in cereal crops. *New Phytologist*, 230, 1017–1033.

Chinnusamy, V., Dalal, M., & Zhu, J. K. (2013). Epigenetic regulation of abiotic stress responses in plants. In M. A. Jenks & P. M. Hasegawa (Eds.), *Plant Abiotic Stress* (2nd ed., pp. 203–229). Hoboken, New Jersey: John Wiley & Sons.

Choi, C. S., & Sano, H. (2007). Abiotic-stress induces demethylation and transcriptional activation of a gene encoding a glycerophosphodiesterase-like protein in tobacco plants. *Molecular Genetics and Genomics*, 277, 589–600.

Clavijo McCormick, A., Unsicker, S. B., & Gershenzon, J. (2012). The specificity of herbivore-induced plant volatiles in attracting herbivore enemies. *Trends in Plant Science*, 17, 303–310.

Creasey, K. M., Zhai, J., Borges, F., Van Ex, F., Regulski, M., Meyers, B. C., & Martienssen, R. A. (2014). MiRNAs trigger widespread epigenetically activated siRNAs from transposons in *Arabidopsis*. *Nature*, 508, 411–415.

Das, R., & Mondal, S. K. (2021). Plant miRNAs: Biogenesis and its functional validation to combat drought stress with special focus on maize. *Plant Gene*, 27, 100294.

Denver, J. B., & Ullah, H. (2019). miR393s regulate salt stress response pathway in Arabidopsis thaliana through scaffold protein RACK1A mediated ABA signaling pathways. *Plant Signaling & Behavior*, 14, 1–7.

Devi, K. J., Chakraborty, S., Deb, B., & Rajwanshi, R. (2016). Computational identification and functional annotation of microRNAs and their targets from expressed sequence tags (ESTs) and genome survey sequences (GSSs) of coffee (*Coffea arabica* L.). *Plant Gene*, 6, 30–42.

Ding, D., et al. (2009). Differential expression of miRNAs in response to salt stress in maize roots. *Annals of Botany*, 103, 29–38.

Dukowic-Schulze, S., Sundararajan, A., Ramaraj, T., Kianian, S., Pawlowski, W. P., Mudge, J., & Chen, C. (2016). Novel meiotic miRNAs and indications for a role of phasiRNAs in meiosis. *Frontiers in Plant Science*, 7, 1–18.

Dyachenko, O. V., Zakharchenko, N. S., Shevchuk, T. V., Bohnert, H. J., Cushman, J. C., & Buryanov, Y. I. (2006). Effect of hypermethylation of CCWGG sequences in DNA of *Mesembryanthemum crystallinum* plants on their adaptation to salt stress. *Biochemistry*, 71, 461–465.

Earley, K., Lawrence, R. J., Pontes, O., Reuther, R., Enciso, A. J., Silva, M., et al. (2006). Erasure of histone acetylation by *Arabidopsis* HDA6 mediates large-scale gene silencing in nucleolar dominance. *Genes & Development*, 20, 1283–1293.

Fahlgren, N., Howell, M. D., Kasschau, K. D., Chapman, E. J., Sullivan, C. M., Cumbie, J. S., et al. (2007). High-throughput sequencing of *Arabidopsis* microRNAs: Evidence for frequent birth and death of MIRNA genes. *PLoS One*, 2, e219.

Ferdous, J., Hussain, S. S., & Shi, B. J. (2015). Role of microRNAs in plant drought tolerance. *Plant Biotechnology Journal*, 13, 293–305.

Ferdous, J., Sanchez-Ferrero, J. C., Langridge, P., Milne, L., Chowdhury, J., Brien, C., & Tricker, P. J. (2017). Differential expression of microRNAs and potential targets under drought stress in barley. *Plant, Cell & Environment*, 40, 11–24.

Fu, W., Wu, K., & Duan, J. (2007). Sequence and expression analysis of histone deacetylases in rice. *Biochemical and Biophysical Research Communications*, 356, 843–850.

Gaj, T., Sirk, S. J., Shui, S., & Liu, J. (2016). Genome-editing technologies: Principles and applications. *Cold Spring Harbor Perspectives in Biology*, 8, a023754.

Gallusci, P., Dai, Z., Génard, M., Gauffretau, A., Leblanc-Fournier, N., Richard-Molard, C., et al. (2017). Epigenetics for plant improvement: Current knowledge and modeling avenues. *Trends in Plant Science*, 22, 610–623.

Gong, Z., Morales-Ruiz, T., Ariza, R. R., Roldán-Arjona, T., David, L., & Zhu, J. K. (2002). ROS1, a repressor of transcriptional gene silencing in *Arabidopsis*, encodes a DNA glycosylase/lyase. *Cell*, 111, 803–814.

Gupta, B., & Huang, B. (2014). Mechanism of salinity tolerance in plants: Physiological, biochemical, and molecular characterization. *International Journal of Genomics*, 2014, 701596.

Hashida, S. N., Uchiyama, T., Martin, C., Kishima, Y., Sano, Y., & Mikami, T. (2006). The temperature-dependent change in methylation of the *Antiirhinum* transposon Tam3 is controlled by the activity of its transposase. *Plant Cell*, 18, 104–118.

Henderson, I. R., & Jacobsen, S. E. (2007). Epigenetic inheritance in plants. *Nature*, 447, 418–424.

Hu, Z., Lin, S., Wang, H., & Dai, J. (2022). Seasonal variations of cold hardiness and dormancy depth in five temperate woody plants in China. *Frontiers in Forests and Global Change*, 5, 1–16.

Huo, C., Zhang, B., & Wang, R. (2022). Research progress on plant noncoding RNAs in response to low-temperature stress. *Plant Signaling & Behavior*, 17, 2004035.

İbrahimova, U., Kumari, P., Yadav, S., Rastogi, A., Antala, M., Suleymanova, Z., et al. (2021). Progress in understanding salt stress response in plants using biotechnological tools. *Journal of Biotechnology*, 329, 180–191.

Jeyaraj, A., Elango, T., Li, X., & Guo, G. (2020). Utilization of microRNAs and their regulatory functions for improving biotic stress tolerance in tea plant [*Camellia sinensis* (L.) O. Kuntze]. *RNA Biology*, 17, 1365–1382.

Jeyaraj, A., Liu, S., Zhang, X., Zhang, R., Shangguan, M., & Wei, C. (2017). Genome-wide identification of microRNAs responsive to Ectropis oblique feeding in tea plant (*Camellia sinensis* L.). *Scientific Reports*, 7, 1–16.

Jiang, X., Zhang, W., Fernie, A. R., & Wen, W. (2022). Combining novel technologies with interdisciplinary basic research to enhance horticultural crops. *Plant Journal*, 109, 35–46.

Kerchev, P., van der Meer, T., Sujeeth, N., Verlee, A., Stevens, C. V., Van Breusegem, F., & Gechev, T. (2020). Molecular priming as an approach to induce tolerance against abiotic and oxidative stresses in crop plants. *Biotechnology Advances*, 40, 107503.

Ketehouli, T., et al. (2022). Overview of the roles of calcium sensors in plants' response to osmotic stress signalling. *Functional Plant Biology*, 49, 588–589.

Kim, J. M., To, T. K., Ishida, J., Morosawa, T., Kawashima, M., Matsui, A., et al. (2008). Alterations of lysine modifications on the histone H3 N-tail under drought stress conditions in *Arabidopsis thaliana*. *Plant & Cell Physiology*, 49, 1580–1588.

Kotowska-Zimmer, A., Pewinska, M., & Olejniczak, M. (2021). Artificial miRNAs as therapeutic tools: Challenges and opportunities. *Wiley Interdisciplinary Reviews: RNA*, 12, 1–33.

Labra, M., Ghiani, A., Citterio, S., Sgorbati, S., Sala, F., Vannini, C., et al. (2002). Analysis of cytosine methylation pattern in response to water deficit in pea root tips. *Plant Biology*, 4, 694–699.

Lamers, J., Der Meer, T. Van, & Testerink, C. (2020). How plants sense and respond to stressful environments. *Plant Physiology*, 182, 1624–1635.

Li, et al. (2015). GmMIR160A, a class of soybean microRNA gene, negatively regulates progress of leaf senescence. *Guihaia*, 35, 84–91.

Li, S., Castillo-González, C., Yu, B., & Zhang, X. (2017). The functions of plant small RNAs in development and in stress responses. *Plant Journal*, 90, 654–670.

Li, W., Wang, T., Zhang, Y., & Li, Y. (2016). Overexpression of soybean miR172c confers tolerance to water deficit and salt stress, but increases ABA sensitivity in transgenic *Arabidopsis thaliana*. *Journal of Experimental Botany*, 67, 175–194.

Lin, J. S., Kuo, C. C., Yang, I. C., Tsai, W. A., Shen, Y. H., Lin, C. C., et al. (2018). MicroRNA160 modulates plant development and heat shock protein gene expression to mediate heat tolerance in *Arabidopsis*. *Frontiers in Plant Science*, 9, 1–16.

Lister, R., O'Malley, R. C., Tonti-Filippini, J., Gregory, B. D., Berry, C. C., Millar, A. H., & Ecker, J. R. (2008). Highly integrated single-base resolution maps of the epigenome in *Arabidopsis*. *Cell*, 133, 523–536.

Liu, Y., Li, D., Yan, J., Wang, K., Luo, H., & Zhang, W. (2019). MiR319 mediated salt tolerance by ethylene. *Plant Biotechnology Journal*, 17, 2370–2383.

Ma, Y., Xue, H., Zhang, F., Jiang, Q., Yang, S., Yue, P., et al. (2021). The miR156/SPL module regulates apple salt stress tolerance by activating MdWRKY100 expression. *Plant Biotechnology Journal*, 19, 311–323.

Mittler, R., et al. (2022). Reactive oxygen species signalling in plant stress responses. *Nature Reviews Molecular Cell Biology*, 23, 663–679.

Muthusamy, M., Kim, J. H., Kim, J. A., & Lee, S. I. (2021). Plant rna binding proteins as critical modulators in drought, high salinity, heat, and cold stress responses: An updated overview. *International Journal of Molecular Sciences*, 22, 6731.

Nguyen, D. Q., Brown, C. W., Pegler, J. L., Eamens, A. L., & Grof, C. P. L. (2020). Molecular manipulation of microRNA397 abundance influences the development and salt stress response of *Arabidopsis thaliana*. *International Journal of Molecular Sciences*, 21, 1–24.

Pagano, L., Rossi, R., Paesano, L., Marmiroli, N., & Marmiroli, M. (2021). miRNA regulation and stress adaptation in plants. *Environmental and Experimental Botany*, 184, 104369.

Park, Y. M., Wu, G., Gonzalez-Sulser, A., Vaucheret, H., & Poethig, R. S. (2005). Nuclear processing and export of microRNAs in Arabidopsis. *Proceedings of the National Academy of Sciences of the United States of America*, 102, 3691–3696.

Peng, H.-X., Chai, F., Chen, K. H., Huang, Y. X., Wei, G. J., Yuan, H., et al. (2023). Reactive oxygen species-mediated mitophagy and cell apoptosis are involved in the toxicity of aluminum chloride exposure in GC-2spd. *Biological Trace Element Research*, 202, 2616–2629.

Pontes, O., Li, C. F., Nunes, P. C., Haag, J., Ream, T., Vitins, A., et al. (2006). The *Arabidopsis* chromatin-modifying nuclear siRNA pathway involves a nucleolar RNA processing center. *Cell*, 126, 79–92.

Post, M. J., Levenberg, S., Kaplan, D. L., Genovese, N., Fu, J., Bryant, C. J., et al. (2020). Scientific, sustainability and regulatory challenges of cultured meat. *Nature Food*, 1, 403–415.

Rebijith, K. B., Asokan, R., Ranjitha, H. H., Krishna, V., & Nirmalbabu, K. (2013). In silico mining of novel microRNAs from coffee (*Coffea arabica*) using expressed sequence tags. *Journal of Horticultural Science and Biotechnology*, 88, 325–337.

Russo, A., Pollastri, S., Ruocco, M., Monti, M. M., & Loreto, F. (2022). Volatile organic compounds in the interaction between plants and beneficial microorganisms. *Journal of Plant Interactions*, 17, 840–852.

Sanz-Carbonell, A., Marques, M. C., Bustamante, A., Fares, M. A., Rodrigo, G., & Gomez, G. (2019). Inferring the regulatory network of the miRNA-mediated response to biotic and abiotic stress in melon. *BMC Plant Biology*, 19, 1–17.

Scippa, G. S., Di Michele, M., Onelli, E., Patrignani, G., Chiatante, D., & Bray, E. A. (2004). The histone-like protein H1-S and the response of tomato leaves to water deficit. *Journal of Experimental Botany*, 55, 99–109.

Šečić, E., Kogel, K. H., & Ladera-Carmona, M. J. (2021). Biotic stress-associated microRNA families in plants. *Journal of Plant Physiology*, 263, 153451.

Sharma, R., Upadhyay, S., Bhattacharya, S., & Singh, A. (2021). Abiotic stress-responsive miRNA and transcription factor-mediated gene regulatory network in *Oryza sativa*: Construction and structural measure study. *Frontiers in Genetics*, 12, 1–13.

Shi, X., et al. (2023). The mechanism of Ca^{2+} signal transduction in plants responding to abiotic stresses. *Environmental and Experimental Botany*, 216, 105514.

Sihag, P., Sagwal, V., Kumar, A., Balyan, P., Mir, R. R., Dhankher, O. P., & Kumar, U. (2021). Discovery of miRNAs and development of heat-responsive miRNA-SSR markers for characterization of wheat germplasm for terminal heat tolerance breeding. *Frontiers in Genetics*, 12, 1–12.

Singh, P., Dutta, P., & Chakrabarty, D. (2021). miRNAs play critical roles in response to abiotic stress by modulating cross-talk of phytohormone signaling. *Plant Cell Reports*, 40, 1617–1630.

Singroha, G., Sharma, P., & Sunkur, R. (2021). Current status of microRNA-mediated regulation of drought stress responses in cereals. *Physiologia Plantarum*, 172, 1808–1821.

Sridhar, V. V., Kapoor, A., Zhang, K., Zhu, J., Zhou, T., Hasegawa, P. M., et al. (2007). Control of DNA methylation and heterochromatic silencing by histone H2B deubiquitination. *Nature*, 447, 735–738.

Steward, N., Ito, M., Yamaguchi, Y., Koizumi, N., & Sano, H. (2002). Periodic DNA methylation in maize nucleosomes and demethylation by environmental stress. *Journal of Biological Chemistry*, 277, 37741–37746.

Stief, A., Altmann, S., Hoffmann, K., Pant, B. D., Scheible, W. R., & Bäurle, I. (2014). *Arabidopsis* miR156 regulates tolerance to recurring environmental stress through SPL transcription factors. *Plant Cell*, 26, 1792–1807.

Stockinger, E. J., Mao, Y., Regier, M. K., Triezenberg, S. J., & Thomashow, M. F. (2001). Transcriptional adaptor and histone acetyltransferase proteins in *Arabidopsis* and their interactions with CBF1, a transcriptional activator involved in cold-regulated gene expression. *Nucleic Acids Research*, 29, 1524–1533.

Sun, F., Guo, G., Du, J., Guo, W., Peng, H., Ni, Z., et al. (2014). Whole-genome discovery of miRNAs and their targets in wheat (*Triticum aestivum* L.). *BMC Plant Biology*, 14, 1–17.

Sun, X., Lin, L., & Sui, N. (2019). Regulation mechanism of microRNA in plant response to abiotic stress and breeding. *Molecular Biology Reports*, 46, 1447–1457.

Taria, S., Kumar, M., Alam, B., Kumar, S., Kumar, S., Roy, S., Kumar, S., & Rane, J. (2022). Abiotic stress and plant response: Adaptive mechanisms of plants against multiple stresses. In G. Santoyo, A. Kumar, M. Aamir, & S. Uthandi (Eds.), *Mitigation of Plant Abiotic Stress by Microorganisms* (pp. 1–17). Cambridge: Academic Press.

Tiwari, R., & Rajam, M. V. (2022). RNA- and miRNA-interference to enhance abiotic stress tolerance in plants. *Journal of Plant Biochemistry and Biotechnology*, 31, 689–704.

Tran, T. T., Pérez-Quintero, A. L., Wonni, I., Carpenter, S. C. D., Yu, Y., Wang, L., et al. (2018). Functional analysis of African *Xanthomonas oryzae* pv. oryzae TALomes reveals a new susceptibility gene in bacterial leaf blight of rice. *PLoS Pathogens*, 14, 1–25.

Tsuji, H., Saika, H., Tsutsumi, N., Hirai, A., & Nakazono, M. (2006). Dynamic and reversible changes in histone H3-Lys4 methylation and H3 acetylation occurring at submergence-inducible genes in rice. *Plant & Cell Physiology*, 47, 995–1003.

Villani, V., Di Marco, G., Iacovelli, F., Pietrucci, D., Canini, A., & Gismondi, A. (2023). Profile and potential bioactivity of the miRNome and metabolome expressed in *Malva sylvestris* L. leaf and flower. *BMC Plant Biology*, 23, 1–21.

Volkov, A. G., & Ranatunga, D. R. A. (2006). Plants as environmental biosensors. *Plant Signaling & Behavior*, 1, 105–115.

Wang, Y., et al. (2023). Superstar microRNA, miR156, involved in plant biological processes and stress response: A review. *Scientia Horticulturae*, 316, 112010.

Wang, J., Mei, J., & Ren, G. (2019). Plant microRNAs: Biogenesis, homeostasis, and degradation. *Frontiers in Plant Science*, 10, 1–12.

Wang, W., Liu, D., Chen, D., Cheng, Y., Zhang, X., Song, L., et al. (2019). MicroRNA414c affects salt tolerance of cotton by regulating reactive oxygen species metabolism under salinity stress. *RNA Biology*, 16, 362–375.

Wani, S. H., Kumar, V., Khare, T., Guddimalli, R., Parveda, M., Solymosi, K., et al. (2020). Engineering salinity tolerance in plants: Progress and prospects. *Planta*, 251, 1–29.

Wen, F. L., Yue, Y., He, T. F., Gao, X. M., Zhou, Z. S., & Long, X. H. (2020). Identification of miR390-TAS3-ARF pathway in response to salt stress in *Helianthus tuberosus* L. *Gene*, 738, 144460.

Wu, L., Mao, L., & Qi, Y. (2012). Roles of DICER-LIKE and ARGONAUTE proteins in TAS-derived small interfering RNA-triggered DNA methylation. *Plant Physiology*, 160, 990–999.

Xin, M., Wang, Y., Yao, Y., Xie, C., Peng, H., Ni, Z., & Sun, Q. (2010). Diverse set of microRNAs are responsive to powdery mildew infection and heat stress in wheat *(Triticum aestivum L.)*. *BMC Plant Biology*, 10.

Yan, C., Zhang, N., Wang, Q., Fu, Y., Wang, F., Su, Y., et al. (2021). The effect of low temperature stress on the leaves and MicroRNA expression of potato seedlings. *Frontiers in Ecology and Evolution*, 9, 1–16.

Yang, G., Pan, Y., Zhao, Q., Huang, J., Pan, W., Cui, L., et al. (2022). Genome-wide identification and characterization of RNA/DNA differences associated with *Fusarium graminearum* infection in wheat. *International Journal of Molecular Sciences*, 23, 7982.

Yang, G. D., Yan, K., Wu, B. J., Wang, Y. H., Gao, Y. X., & Zheng, C. C. (2012). Genomewide analysis of intronic microRNAs in rice and *Arabidopsis*. *Journal of Genetics*, 91, 313–324.

Yao, Q., Chen, Y., & Zhou, X. (2019). The roles of microRNAs in epigenetic regulation. *Current Opinion in Chemical Biology*, 51, 11–17.

Ye, Y., Wang, J., Wang, W., & Xu, L. A. (2020). ARF family identification in *Tamarix chinensis* reveals the salt responsive expression of TcARF6 targeted by miR167. *PeerJ*, 2020, 1–15.

Zhang, B. (2015). MicroRNA: A new target for improving plant tolerance to abiotic stress. *Journal of Experimental Botany*, 66, 1749–1761.

Zhang, F., Yang, J., Zhang, N., Wu, J., & Si, H. (2022). Roles of microRNAs in abiotic stress response and characteristics regulation of plant. *Frontiers in Plant Science*, 13.

Zhang, M., An, P., Li, H., Wang, X., Zhou, J., Dong, P., et al. (2019). The miRNA-mediated post-transcriptional regulation of maize in response to high temperature. *International Journal of Molecular Sciences*, 20.

Zhang, X., Bao, Y., Shan, D., Wang, Z., Song, X., Wang, Z., et al. (2018). *Magnaporthe oryzae* induces the expression of a microrna to suppress the immune response in rice. *Plant Physiology*, 177, 352–368.

Zhang, Y., Xu, J., Li Ruofan, Ge, Y., Li, Y., & Li, R. (2023). Plants' response to abiotic stress: Mechanisms and strategies. *International Journal of Molecular Sciences*, 24, 1–17.

Zhou, C., Zhang, L., Duan, J., Miki, B., & Wu, K. (2005). Histone Deacetylase19 is involved in jasmonic acid and ethylene signaling of pathogen response in *Arabidopsis*. *Plant Cell*, 17, 1196–1204.

Zhou M., Luo H. (2013) MicroRNA-mediated gene regulation: Potential applications for plant genetic engineering. *Plant Molecular Biology*, 83, 59–75. https://doi.org/10.1007/s11103-013-0089-1

Zhou, Y., Liu, W., Li, X., Sun, D., Xu, K., Feng, C., et al. (2020). Integration of sRNA, degradome, transcriptome analysis and functional investigation reveals gma-miR398c negatively regulates drought tolerance via GmCSDs and GmCCS in transgenic *Arabidopsis* and soybean. *BMC Plant Biology*, 20, 1–19.

Zhu, J. K. (2016). Abiotic stress signaling and responses in plants. *Cell*, 167(2), 313–324.

20 Crop Improvement through Epigenetic Technologies

Case Studies

Ashwini R, Nisha Sam Nirmala, Uma S, Sivarama Krishna Lakkaboyana and Pamarthi Satyanarayan

20.1 INTRODUCTION

Epigenetics was defined by Arthur Riggs and his colleagues as the "study of heritable changes in gene function that occurs either mitotically or meiotically without altering the DNA sequences or influencing gene" (Riggs et al., 1996; Riggs & Porter, 1996; Russo et al., 1996). Before reaching this definition, epigenetics has undergone several debates and experimentations. The quest for finding the involvement of DNA, genes and other related components began a long time ago. Epigenetics is coherently linked with evolution and development, and the embryologists took sides in preformationism and epigenesis when the development of organisms was studied. In preformationism, it was believed that there were preformed elements that enlarge during growth and development, whereas in epigenesis, it was believed that there exists a complex developmental plan involving chemical reactions and soluble components that play an essential role in growth and development. Subsequently, studies on varying phenotypic characters in *Drosophila* caused by changes in chromosomes themselves rather than genes, discovery of transposable elements, and further studies revealed that though DNA can be the primary source of genetic information that transfers characters from one generation to another, there are other factors and proteins involved (Felsenfeld, 2014).

Over the past 50 years, the basic ideology of epigenetics has undergone several modifications with respect to the increased knowledge obtained in molecular mechanisms and their functions (Felsenfeld, 2014). Initially, "epigenetics" was denoted as "epigenesis" by the physician and physiologist William Harvey as the development with increased complexity from homogeneous material in the egg. In the year 1942, the term "epigenetics" was introduced, establishing its relationship with "epigenesis" by embryologist Conrad Waddington. He defined that "epigenetics is a complex developmental process that lies between genotype and phenotype" (Waddington, 1942). During this period, epigenetics was more related to gene expression. Later studies showed that gene expression and regulation may not be attributed to epigenetics. When the non-specific inhibition of gene expression was identified in eukaryotes by histone complexes other than genes, the concept of epigenetics was concluded to be not related to genes alone (Wooland, 1987; Morange, 2009). Several studies on chromatin network, histone proteins and DNA methylation showed that changes in these components may exhibit different phenotypes even when there is no alteration in the genetic makeup of the organism, and also, they have the ability to be inherited from one generation to another. In the present day, epigenetics refers to the study of heritable changes in gene expression that occur due to changes in the chromosomes rather than the DNA sequence itself (Berger et al., 2009).

20.2 MECHANISM OF EPIGENETICS

Epigenetic modifications were found to be obtained through mechanisms such as DNA methylation, histone modification and modification in non-coding RNA (Figure 20.1).

20.3 EPIGENETICS IN CROP IMPROVEMENT

The increase in world population along with the rising climate change and fluctuations due to global warming has created a challenge in global production of food crops (Takeda & Matsuoka, 2008). In addition to global warming, the biotic stress such as pathogens, pests and weeds has also caused increased concerns in crop production (Oerke, 2006). The need for mitigating these biotic and abiotic stressors for enhancing the quality and quantity of food production has been of utmost importance to scientists and the government for several decades. Crop improvement was intended to be achieved by plant physiology, genetics and biotechnology (Lombardo et al., 2016). Ongoing and past research efforts have focused on these areas to enhance crop yield and improve agricultural productivity (El-Sharkawy, 2006; Chen et al., 2024). Though these methods can bring about successful beneficial changes in crops, they involve modifying the genetic makeup of the

DOI: 10.1201/9781003589914-24

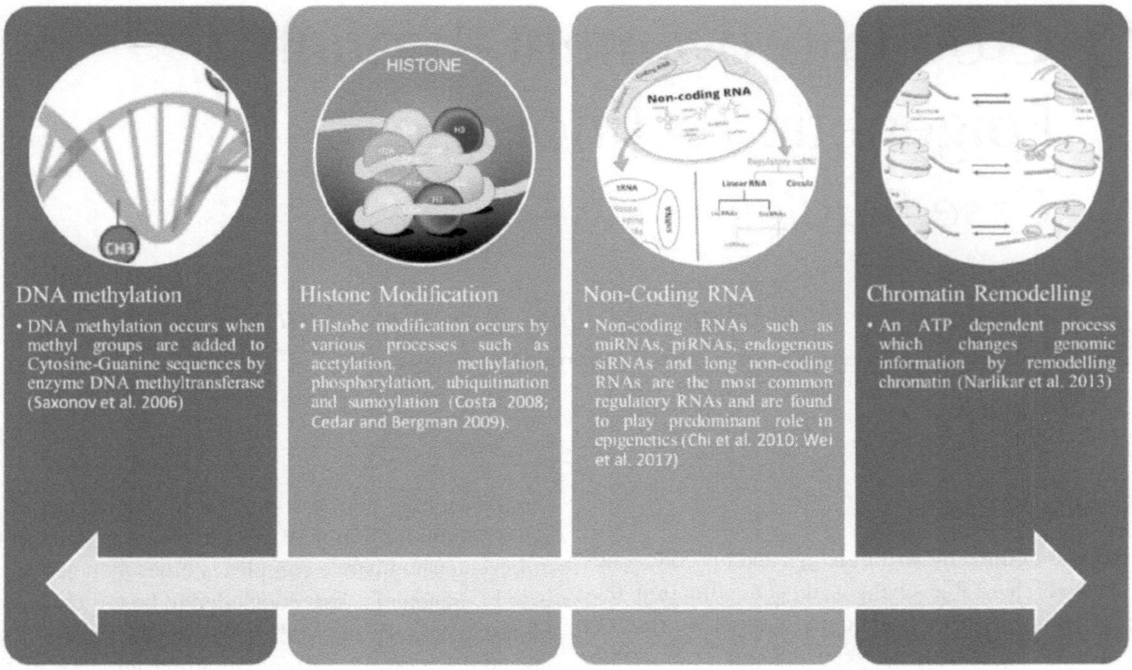

FIGURE 20.1 Mechanism of epigenetic modifications (Phillips & Shaw, 2008; Almatroudi, 2022).

plant. Epigenetics, which brings about inheritable phenotypic changes by altering the components associated with DNA without changing the genetic makeup, is considered more beneficial and advantageous over the other methods for better crop improvement (Merce et al., 2020).

Plants are exposed to various ecological factors. Their growth and development are adversely affected by environmental conditions such as day length, light, temperature and water availability (Xiao et al., 2017). These climatic features are sensed by plants and activate transcription factors, which in turn activate gene expression required for plant survival and growth. This helps the plant to adapt to various environmental challenges (Mirouze & Paszkowski, 2011). The concept of adapting to climate concerns was utilized by scientists to bring out new essential traits in plants by epigenetics. As these epigenetic traits are reversible and can be inherited from one generation to another, they are considered the most advantageous method for developing desired characters in plants (Iwasaki & Paszkowski, 2014; Singh et al., 2019b). By the process of DNA methylation, histone modification, non-coding RNA modification and chromosome remodeling, plants resistant to abiotic stress such as drought (Forestan et al., 2018; Benoit et al., 2019), heat (Wang et al., 2016a; Hossain et al., 2017), cold (Hebrard et al., 2016; Tang et al., 2018), salinity (Zhu et al., 2015), biotic stress such as bacterial and fungal pathogens (Choi et al., 2012; Kong et al., 2020), plants with enhanced growth (Virdi et al., 2015), altered photoperiodic flowering (Song et al., 2017) and sex determination (Akagi et al., 2016) were developed. These crops were developed based on various studies and knowledge gained after performing several epigenetic modifications in plants, including the most studied model plant *Arabidopsis thaliana*. Further experiments are being carried out to bring out essential traits in plants, aiming to improve crop production sustainability in the face of increased environmental challenges and a growing global population. The present chapter focuses on various epigenetic case studies in plants, highlighting strategies for developing crops with enhanced productivity and quality.

20.4 EARLIER STUDIES ON EPIGENETICS IN CROP IMPROVEMENT

The role of epigenetics in plants was studied for a long time and initial studies were primarily aimed at understanding the phenomena of environmental changes in plants. During the recent past and at present, the knowledge gained from previous studies has been applied to improve crop production through epigenetics. Most of the epigenetic studies were performed in the model plant *Arabidopsis* and other food crops such as maize and wheat. Later, studies were conducted in other plants as well. Table 20.1 describes some of the early studies of epigenetics performed in plants, especially *Arabidopsis*, maize and wheat.

20.5 CASE STUDIES ON THE ROLE OF EPIGENETICS IN CROP IMPROVEMENT

Several studies have been performed on epigenetic modifications for improving crop productivity by protecting them from biotic or abiotic stress and by enhancing desired characters. The case studies that highlight the role of epigenetic modification for various desirable trait expressions in plants are discussed as follows:

TABLE 20.1
Earlier epigenetic regulatory studies in plants

S. No.	Epigenetic study performed	Identification from the study	References
***Arabidopsis*:**			
1.	DNA methylation on transposable elements and the role of siRNA in methylation	Methylation is lost frequently and siRNA or miRNA does not influence methylation. It is influenced by DNA methyltransferase MET1. Methylation does not affect gene expression unless it occurs in the promoter or regulatory region	Vaughn et al. (2007)
2.	Role of siRNA in DNA methylation and heterochromatinization of hAT element of Landsberg *erecta* (Ler) and Columbia (Col) of *Arabidopsis*	siRNA-directed methylation can cause variations in closely related species	Zhai JiXian et al. (2008)
3.	Adaptive epigenetic memory when exposed to different temperature	Adaptive response was observed in plants treated with mild heat even in F3 generations, proving adaptive epigenetic memory	Whittle et al. (2009)
4.	Utilizing Chip-Chip and Chip-seq techniques for identifying the binding sites of epigenetic regulators	Complex network of epigenetic regulators was found to control the expression of genes in higher plants	Luo et al. (2009)
5.	Focuses on the two-step process that facilitates the transmission of epigenetic traits across generations	Role of DNA methylation and chromatin remodeling was established in the study. The study also revealed that some epigenetic modifications will revert back when the alterations in methylation and chromatin remodeling occur	Blevins et al. (2014)
6.	Change in DNA methylation 1-2 (ddm1-2)-derived epigenetic recombinant inbred lines (epiRILs), resulting in altered growth and morphology when exposed to neutral and saline environment	Epigenetic mechanisms are crucial in enabling plants to adapt and respond to stress conditions, influencing their growth patterns, morphological changes and ability to exhibit plasticity	Kooke et al. (2015)
7.	Expression of germ-line-specific repressors and activators	Repressive epigenetic marks such as H3K9me2 and H3K27me3 were found in somatic cells, which differentiates them from somatic and germ-line cells. This control of gene expression was identified to be controlled by epigenetic modifications such as DNA methylation and histone modification.	Jung et al. (2015)
8.	Epigenetic regulation of gene responsiveness	Histone variant H2A.Z and trimethylation of H3lysine4 (H3K4me3) were found to regulate gene responsiveness, with the latter found to be increased with respect to stress. It was also found to play a role in epigenetic memory	To and Kim (2014)
9.	Preconditioning epigenetic response to high vapor pressure deficit	High vapor pressure deficit (vpd) at two gene loci was found to be associated with leaf stomatal frequency. Preconditioned parents and offsprings showed resistance to drought over two generations. This is attributed to the inability of gene *cmt3* to maintain the asymmetric context methylation	Tricker et al. (2013)
***Maize*:**			
10.	DNA methylation in maize and its response to environmental stress	The periodic DNA methylation observed in the plant is regulated by epigenetic modifications such as DNA methylation, histone modification and chromatin remodeling with respect to environmental stress	Steward et al. (2002)
11.	Unstable factor for orange 1 (Ufo1) for bringing out epigenetic changes in pericarp color 1 of maize	Among the hypermethylated six genes of P1, *P1-wr* was found to have reduced methylations due to Ufo1 developing pigmentation in the kernels. These changes were also found not to be inherited and are shown to cause only transient changes in DNA methylation of particular loci	Chopra et al. (2003)
12.	Epigenetic modification of the pericarp color (P1) allele in tissue-specific mosaicism	Hypermethylation and hypomethylation of varying regions in p1-mm (pericarp mosaic) resulted in different types of color patterns	Robbins et al. (2009)
13.	Role of epigenetic reprogramming during vegetative phase change	Cytosine methylation of RNA silencing pathways and histone modification play a major role in the transition from vegetative phase	Li et al. (2010)
14.	Heritable epigenetic modifications in maize	Cytosine methylations that are inherited were observed	Eichten et al. (2011)
15.	The role of chromatin modifications in various zones of the root meristem	The root zone-specific chromatin distribution was observed when histone modification was analyzed	Yan et al. (2014)

(*Continued*)

TABLE 20.1 CONTINUED

Wheat:			
16.	Epigenetic and genetic modification in Class E MADS box gene in wheat hexaploid	In the wheat variety WLHS1, the homeolog WLHS1-A was found to have structural modification, WLHS1-B was found to be silenced predominantly by cytosine methylation and only WLHS1-D was active	Shitsukawa et al. (2007)
17.	DNA methylation changes due to salt stress in Dekang 961 and Lumai 15	Demethylation and hypomethylation of CCGG sites were observed in the cultivars owing to their varied adaptation to salt stress	Zhong et al. (2009)
18.	Expression divergence of three TaEXPA1 in hexaploid wheat due to epigenetic modifications	*TaEXPA1* was found to be silenced in roots due to an increased level of H3K9 dimethylation, decreased levels of H3K4 trimethylation and H3K9 acetylation. Reactivation of *TaEXPA1-A* and *TaEXPA1-D* expression in leaves was correlated with increased levels of H3K4 trimethylation and H3K9 acetylation, and decreased levels of H3K9 dimethylation in their promoters and a higher level of cytosine methylation was detected in the promoter region of *TaEXPA1-B*, which may contribute to its silencing in leaves after the formation of hexaploid	Hu Zhao Rong et al. (2013)
19.	Role of silencing in nucleolar organizing regions in deactivation of A and D in wheat	CHG, CHH methylation, increased levels of H3K27me3 and H3K9me2 lead to the asymmetric epigenetic changes in the nucleolar organizing regions of A and D, leading to their disappearance in wheat	
20.	The role of genetic and epigenetic modifications in the BBAA component of wheat during its evolution to hexaploidy	DNA methylations and genetic modifications were found to be observed during their evolutionary process	Liu et al. (2015)
Other crops:			
21.	Epigenetic modification of *Petunia hybrida* by the *A1* gene from maize	Varying expression pattern and color were observed in the flowers of *Petunia hybrida* based on the number of integrated copies of *A1* maize cDNA. In addition, methylation of 35S RNA promoter and its influence on the *A1* gene was found to control the expression pattern	Linn et al. (1990)
22.	The role of DNA methylation in the salt- and drought-tolerant rice cultivars IR-64 (stress sensitive), Nagina 22 (drought tolerant) and Pokkali (salt tolerant)	Variations in DNA methylations, differentially methylated regions and abundance of small RNA in hypermethylated regions in the rice cultivars were identified for their salt and drought tolerance or sensitivity	Garg et al. (2015)

20.5.1 Epigenetic Regulation for Drought Tolerance in Maize

Drought was identified to be one of the predominant factors limiting plant growth and productivity. The adverse climate change and reduced precipitation are the main factors for frequent onset of drought in various places around the world (Mishra & Singh, 2010; Lobell et al., 2011). Drought is classified into four types as meteorological, agricultural, hydrological and socio-economic, with each being interlinked to the other (Mishra & Singh, 2010; Zhang et al., 2017). The reduction in precipitation leading to reduction in water required for plant growth results in drought episodes. These reduced precipitation and drought have a negative impact on the growth and development of plants (Barnabas et al., 2008; Praba et al., 2009). This in turn affects the water table and water cycle. Finally, all these factors combined have their impact on human life. Drought was found to have adverse consequences for the human population and also the economy, thereby threatening global food security (Christie et al., 2013). Plants that are exposed to drought conditions may either have very little water resources or lose water by transpiration. The severity of drought varies with factors like precipitation, nature of the soil and its water-holding capacity, and loss of water through evapotranspiration (Anjum et al., 2011). The response of plants to drought also varies based on the stage of growth and environmental factors (Demirevska et al., 2009). Several measures are put forward and carried out to maintain hydrology regimes for increasing crop productivity (Bodner et al., 2015).

Environmental stress due to water deficit was identified to bring about epigenetic changes in plants such as citrus and *Medicago* (Neves et al., 2017; Ventouris et al., 2020). In tomato, epigenetic modification was observed during self-grafting, whereas in *Linum usitatissimum*, alternative splicing was identified as a result of epigenetic modification in response to drought stress (Fuentes-Merlos et al., 2023; Wang et al., 2023). In addition, studies on epigenetic modifications by DNA methylation and histone modification in plants such as *Arabidopsis* and *Medicago* revealed that they respond to this epigenetic modification and develop drought tolerance (Khan et al., 2020; Nishio et al., 2024; Zi et al., 2024). Epigenetic modifications by ABA-mediated root morphology were observed in *Arabidopsis*, which is induced by the microbe *Pseudomonas argentinensis* SA190 (Alwutayd et al., 2023). Different modifications identified in maize for exhibiting tolerance to drought are shown in Figure 20.2.

DNA methylation is the predominantly observed epigenetic modification (Banerjee & Roychoudhury, 2017). In the inbred maize cultivar W22, 18 differentially methylated fragments (DMF) were identified by amplified methylated polymorphism (AMP). Among the 18 DMFs, DMF 4, DMF 10 and DMF 13 located in lncRNA and DMF 18 located in Nudix hydroxylase 2 were found to be demethylated when exposed to drought stress. However, the DMF 3, which is located upstream of protein-dehydration-induced 19 homolog 3, was found to be methylated under drought stress. These methylation patterns suggested that DMF 4, 10, 13 and 18 are hypomethylated and DMF 3 is hypermethylated when the wheat cultivar W22 was exposed to drought conditions. The location of these DMFs in the above-mentioned sequences (lncRNA, Nudix hydroxylase 2 and protein-dehydration-induced 19 homolog 3) was reported by previous studies to have resistance towards abiotic stress in plants. The methylation changes of these sequences in wheat confirm that these regions can be used as regulators to develop drought tolerance in wheat (Sallam et al., 2019).

Wang et al. (2021) studied the role of DNA methylation in inbred drought-tolerant varieties of wheat AC7643_DT and RIL208_DT along with drought-sensitive varieties of wheat AC7729/TZSRW_DS and RIL64_DS, showing transposable elements (TEs) as important regulatory factors for exhibiting drought tolerance. It was identified that DNA methylation was higher in the gypsy region of TEs, DNA transposons, intergenic regions and TE-related genes. In addition, DNA methylation was observed in parental lines and also in derived Recombination Inbred Lines, indicating the inheritance and involvement of recombination. The identified DNA methylations were found to cause differences in alternative spicing, siRNA and regulation of Differentially Expressesd Genes (DEGs) for expressing or silencing gene clusters with respect to drought. Apart

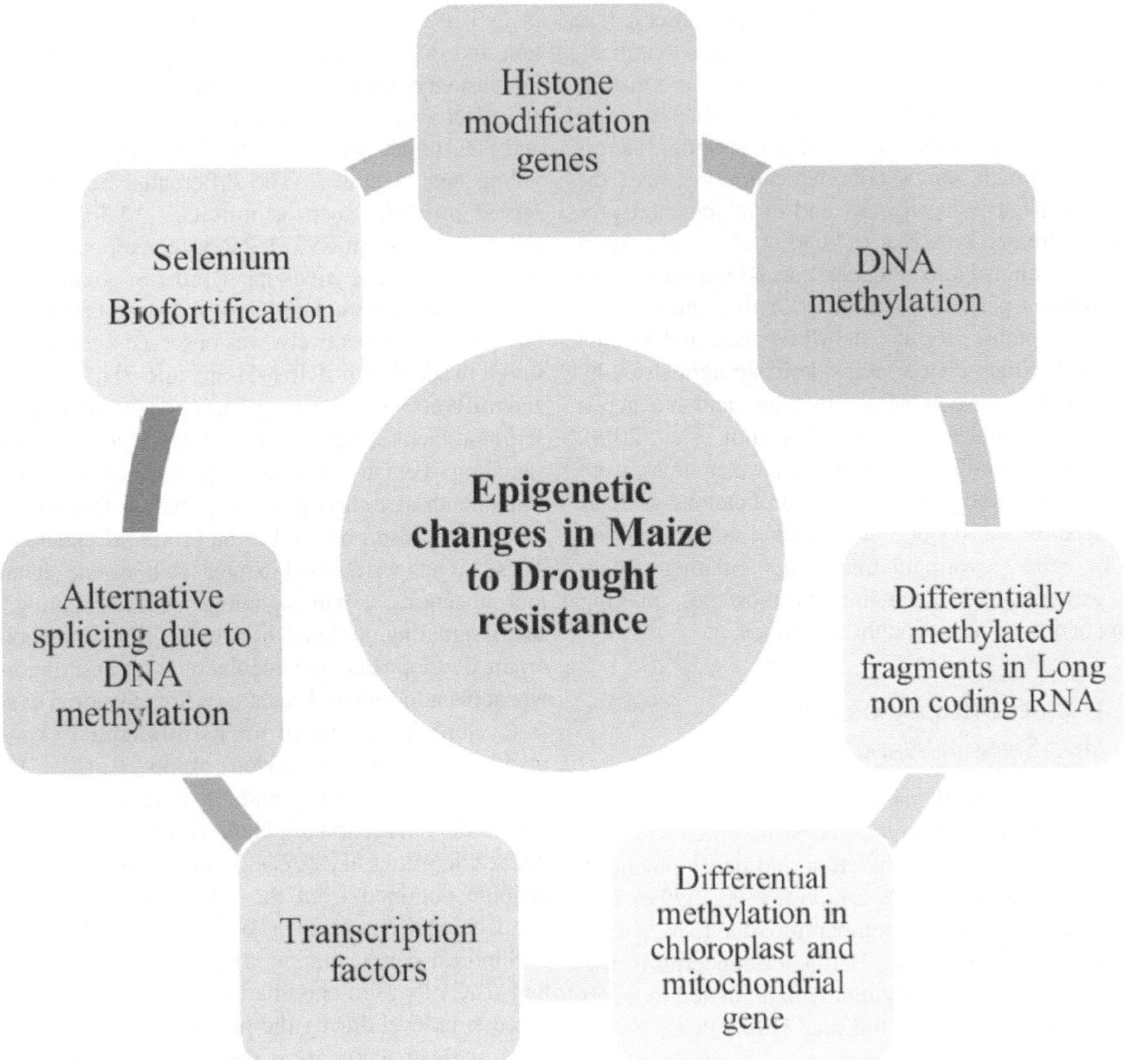

FIGURE 20.2 Epigenetic modifications identified in maize for developing tolerance to drought.

from nuclear DNA, methylation in organellar DNA such as chloroplasts and mitochondria could also make the plants drought tolerant. Differential methylation was observed in photosystems I and II of chloroplasts and the *nad4* gene of mitochondria. Upregulation of genes such as *ps3*, *rsp2A*, *ccmFC*, *atpl* along with some uncharacterized genes in mitochondria and *psbA*, *psbD*, *psbc*, *psaA*, *atpA*, *ndhk*, *ndhB1* and *B2* genes in chloroplasts was observed in water-stressed plants, which was attributed to hypomethylation. The methylation changes in turn led to the upregulation of the above-mentioned genes, helping the plants survive in a water-stressed environment (Bhanu et al., 2020).

In addition to DNA methylation, histone modifications were also found to be effective in bringing drought-tolerant characteristics in maize. Ten histone modifications were observed in maize with increased production of ZmJMJ2 in drought-tolerant cultivar ND476 than the drought-sensitive ZX978 cultivar. Similarly, ZmSDG5, ZmSDG24 and ZmJMJ4 were found to be induced in drought-tolerant variety C7-2, whereas in drought-sensitive variety Lv28 and H082183, ZmJMJ5 was upregulated under both moderate and severe drought treatment. Increased expression of *ZmSDG1* and decreased expression of *ZmHDA2* were observed in H082183 after moderate and severe drought treatment, respectively. This shows that differentially expressed histone modifications are responsible for drought tolerance in maize (Zheng et al., 2021). Further, a hub of transcription factors was observed, which shows communication between the drought stress triggered pathways and developmental processes under stressed conditions (Miao et al., 2017). Apart from these, selenium biofortification caused changes in the DNA methylation patterns, especially for the genes encoding phytoene synthase, sorbitol dehydrogenase and alcohol dehydrogenase, which play a vital role in drought stress by regulating the leaf carotenoids, osmolytes, and biochemical and environmental adaptations (Bocchini et al., 2018). Overall, environmental factors, especially water stress, were identified to bring about epigenetic modifications such as DNA methylation and histone modification in transcription elements by either downregulating or upregulating certain genes of genomic and organellar chromosomes, making them adapt and survive in drought conditions.

20.5.2 Epigenetic Regulation for Heat Stress in Wheat

Heat stress is defined as the rise in temperature beyond the tolerable threshold level. It is the period in which irreversible damage is caused in the function and development of the plant (Peet & Willits, 1998; Siddique et al., 1999). The continuous increase in Earth's temperature due to increased emissions of greenhouse gases is raising concerns in the world. The temperature of air globally is predicted to reach an increase of 1.8°C to 4°C by the year 2100 (IPCC, 2007). The rising temperature causes changes in the growing season and regional distribution of agricultural crops. This alters the threshold temperature, making the plants grow, mature and produce flowers, fruits and vegetables faster than the actual expected period (Porter, 2005). The threshold level varies for different plants. Plants exposed to higher temperature undergo adverse modifications at their molecular levels, which include cellular injury due to the production of more Reactive Oxygen Species (ROS), altered or loss of membrane integrity, changes in protein formation, protein aggregation, up or down regulation of certain genes and also altered cell division (Smertenko et al., 1997; Howarth, 2005; Hasanuzzaman et al., 2012; Hasanuzzaman et al., 2013a). With respect to heat stress, plants produce enzymes that neutralize ROS, along with stress-related proteins especially the heat shock proteins (HSPs), and also carry out other mechanisms to protect themselves from heat stress (Hasanuzzaman et al., 2013b). However, these combating mechanisms to heat stress are not efficient in all plants.

Among the crops that are highly affected by heat stress, wheat is worth mentioning since it has a low threshold level to temperature (Stone & Nicolas, 1994). Being an important food crop globally, heat stress on wheat affects global food security (Tripathi et al., 2016). Studies on epigenetic modification in wheat suggest that the plant undergoes changes mainly in its miRNA and histone proteins that alter the upregulation and downregulation of genes providing heat tolerance. For instance, PARE (Pare Analysis of RNA Ends) analysis of miRNAs in wheat cultivar showed that certain miRNAs and their targets were conserved and play a vital role in providing heat tolerance. The differential expression of conserved miRNAs such as miR166–12.1-3p, miR167–14-3p, miR528–1.2-5p, miR9772–2.2-5p corresponding to the targets Homeobox leucine-zipper protein, auxin response factor, superoxide dismutase [Cu-Zn], L-ascorbate oxidase and F box proteins, respectively, was observed with respect to heat stress. In addition, miR169–51-3p, miR171–2-3p, miR2020b-1 and miR9662–1.1-3p corresponding to the target nuclear transcription factor Y subunit, Floral development drought stress signaling, Pentatricopeptide repeat-containing protein and Mitochondrial transcription termination factor-like, respectively, were also observed to be produced under heat stress. These targets were found to alter the hormone signaling, leaf morphogenesis, auxin signaling, ABA signaling, drought stress signaling, antioxidant activity, RNA metabolism, and organ development and regulation of transcription for the wheat plant to survive heat stress (Ravichandran et al., 2019).

Cyclophilin is known for its differential expression in plants with respect to various abiotic stresses. It controls transcription, signaling and plant development and also aids in the interaction of plants with pathogens (Singh et al., 2020; Olejnik et al., 2021). *In silico* analysis of the wheat genome obtained from the International Wheat Genome Sequencing Consortium (IWGSC) showed over 83 wheat cyclophilin genes, among which 58 TaCYPs were identified. TaCYPs were classified into five groups based on the expression level during the process of acclimation. During the process of acclimation, the genes from group 1 to group 4 were upregulated with only some genes downregulated. Comparing genes from group 1 with group 2, the latter showed the highest expression of *TaCYP* genes in the acclimation stage, which were stable in the initial stages of the heat exposure to lethal levels. Group 3 genes showed stable expression for 4.5 h of lethal exposure, whereas group 4 genes showed a decline in their expression after 2 h of lethal level exposure, with exceptions such as *TaCYP46-1-7B*, *TaCYP46-2-7A* and *TaCYP46-3-7D* whose expression remained unaffected, and *TaCYP48-1-2D*, *TaCYP48-2-2A* and *TaCYP48-3-2B* whose expression was higher. Unlike other *CYP* gene groups, those belonging to group 5 were not affected by any of the heat stress regimes. The presence of heat shock elements (HSEs) was identified to be present in 32 cyclophilin regions, which were responsible for plant development under heat stress (Singh et al., 2019a).

An increased production of histone acetyltransferase of wheat, viz., TaHAG1, was observed in transgenic wheat. The increased *TaHAG1* expression was found to interact with NAC domain-containing protein TaNACL, which in turn activated *TaG1* and *TaPSBR1* that enhanced thermotolerance in wheat (Lin et al., 2022).

Exposure of plants to a moderate level of heat stress followed by severe heat stress is called heat priming (Bäurle, 2016; Liu et al., 2021). This heat priming strategy is used to identify the epigenetic modifications and their consequences in plants. Heat priming in the bread wheat Gladius and Paragon showed that the heat-tolerant variety Gladius could withstand heat stress, whereas deleterious effects were observed in the heat-sensitive variety Paragon (Mendanha et al., 2018). Further, primed plants with or without recovery followed by heat stress were observed to have better biomass, photosynthetic rate, stomatal conductance and protein content. In addition, genes corresponding to the production of antioxidants such as CAT, APX and SOD were found to be higher in plants not primed than in the primed plants, indicating the coping mechanism exhibited by non-primed plants to protect themselves from heat stress. Heat-responsive genes such as *sHSP1, HSP70-1, HSP90, HSP26, F3-1, TaFBR1, TaFBR2, TaFBR3, GS1, GS2, DREB, MAPK3, MAPK6, RBOH, LRRK, LysM* and *TaCRK3* were also found to show higher expression in non-primed plants, which indicated the responsiveness to heat stress (Khanzada et al., 2024). The gene *Lysine specific histone demethylase 1* (*LSD1*) responsible for histone demethylation was observed to have epigenetic modification in heat-stressed plants. Further, protein modifications pertaining to photosynthesis, energy production, protein destination and storage were also observed in the subsequent generation, indicating epigenetic memory and inheritance (Wang et al., 2016b). These studies suggested that epigenetic modifications can bring about desired characteristics for better crop production in wheat.

20.5.3 Epigenetic Regulation for Salt Tolerance in Rice

Salinization of soil is considered one of the major threats to the environment and agriculture (Hazell & Wood, 2008). It has been estimated that around 20% of cultivable land is affected by salinization (Qadir et al., 2014; Arora, 2019). Increased salinization is attributed to climate change, environmental deterioration factors, improper use of water resources such as recycling water containing a large number of salt and poor agronomic practices, such as high use of fertilizers (Ziska et al., 2012; Park et al., 2016; Arora, 2019; Kumar & Sharma, 2020). Salt stress has a negative impact on plant growth, especially during the period of germination and the seedling stage. Salinity causes the accumulation of salts, especially Na in the roots thereby affecting growth. In addition, salt stress results in hormonal imbalance, low availability of nutrients, osmotic stress, oxidative stress, ion toxicity and increased susceptibility to pathogens (Zhu, 2002; Atta et al., 2022a; Atta et al., 2022b; Hasanuzzaman & Fujita, 2022; Lu et al., 2023). All these result in secondary effects such as improper growth due to reduced photosynthesis, changes in physiology and reduced cell division and expansion (Van Zelm et al., 2020). In order to combat the problems associated with salt stress, plants carry out several mechanisms, which are shown in Figure 20.3.

Despite these combating mechanisms, plants suffer from salt stress, which in turn affects productivity. Epigenetic modifications were identified to bring about desired changes in plants with respect to salt stress. The growth and productivity of rice are adversely affected by salt stress, and approaches to increase tolerance of rice to salt stress are essential to meet the global food demand (Liu et al., 2022). The most commonly found epigenetic phenomena in rice plants to develop salt tolerance are DNA methylation. Differentially methylated cytosines and differentially expressed genes were observed in the promoter and coding regions. Hypermethylation in the CHH context was observed in the Pokkali rice variety (salt tolerant), thereby increasing gene expression. In addition, the accumulation of dmC-SNPs was observed in the stressed varieties indicating the role of epialleles in abiotic stress response (Rajkumar et al., 2020). DNA methylation was identified to provoke a feedback mechanism in the tetraploid rice cultivars with respect to salt stress. In tetraploid rice, CHH hypomethylation was observed which in turn activates jasmonic acid (JA)-related genes increasing their production in response to salt stress. This higher JA-Ile content enhances JA signaling cascade, thus lowering the sodium uptake. Under non-stress conditions, CHH methylation represses the neighboring transcriptional elements (TEs). This feedback regulation of hypermethylation and methylation with respect to salt stress was identified to be beneficial for rice cultivars, especially the tetraploids (Wang et al., 2021).

When comparing the effect of DNA methylation with respect to salt stress in different rice cultivars, viz., IR-29 (salt sensitive) and FL478 (salt tolerant), demethylation was observed in the roots and seedling stage of IR-29 helping the plant to overcome salt stress. This demethylation was observed in low quantities in the salt-tolerant variety FL478. The methylated fragments were found homologous to GYF

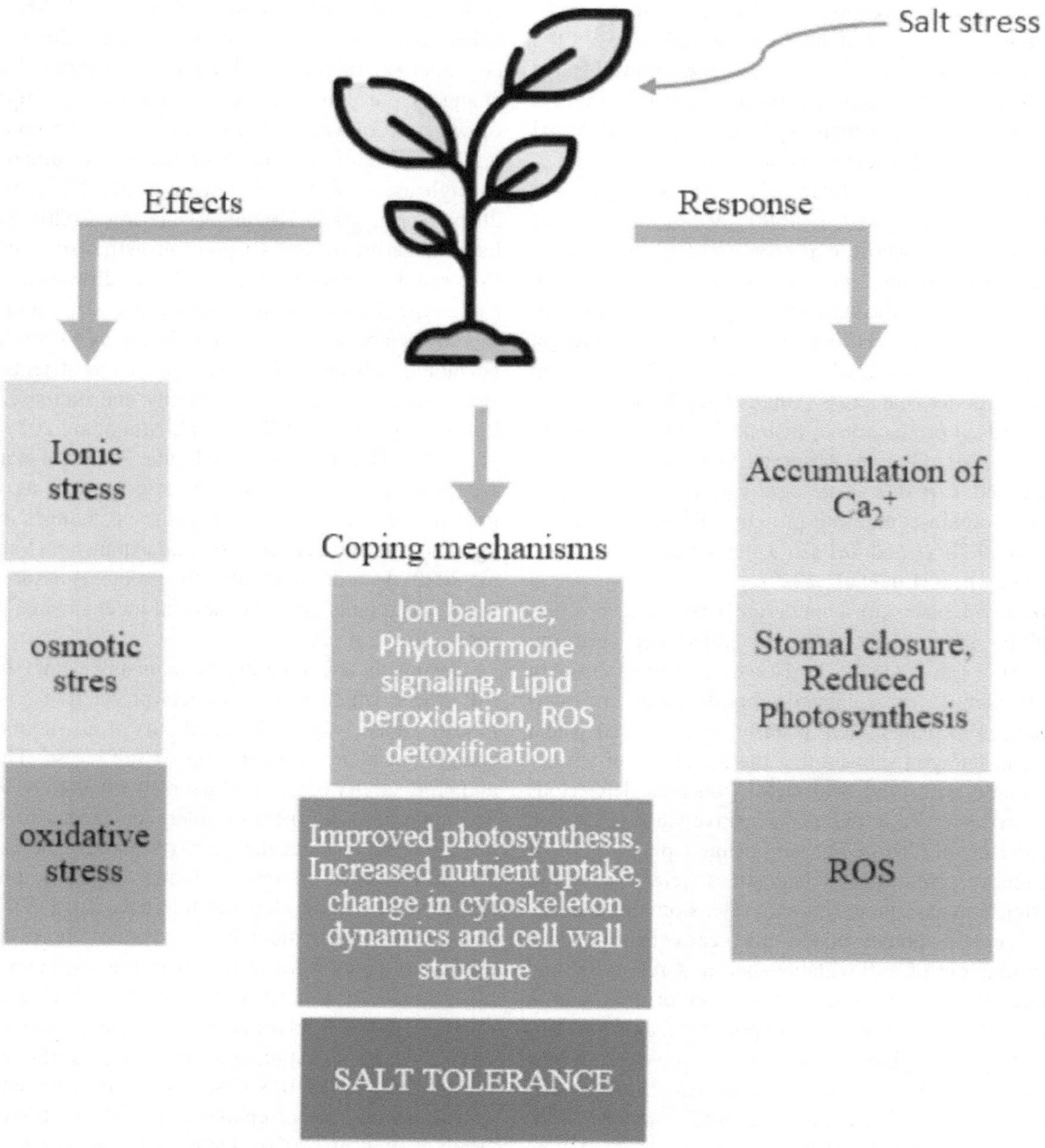

FIGURE 20.3 Effects, response and combating mechanisms in plants with respect to salt stress (Zhao et al., 2021; Atta et al., 2023).

domain-containing proteins, WRKY transcription factors, zinc finger domains, histone-lysine *N*-methyltransferases, cytochrome b6-f complex subunit 4, Ser/Thr phosphatase family of proteins, NB-ARC domain-containing proteins, photosystem I assembly protein, retrotransposon, NADPH quinone oxidoreductase, genes encoding hydrolases, cytochrome subunits, transcription factors, kinases and genes involved in redox regulation, such as acetyl-CoA, peroxidases and ABC transporters. The remaining three genes were identified as putative proteins. This concludes that methylation affects a wide variety of genes in IR-29 and FL478, helping them to reduce salt stress (Wang et al., 2011). Salt stress-induced DNA methylation in the introgression line IL177-103 was found to increase relative water content, ABA content, ROS scavenging enzyme activity with lowered Na^+ concentration in shoots than the parent IR-64. Such DNA methylation was observed to change genotypes predominantly in roots of salt tolerant introgression line IL177-103 than the parent IR-64, thus providing epigenetic markers for stress tolerance (Wang et al., 2015).

WRKY transcription factor family is identified as one of the most important regulators of stress tolerance. Among them, WRKY50 and WRKY72 are known to be the regulators of salt stress. A total of 72 methylated cytosines corresponding to WRKY50 and 83 methylated cytosines corresponding to WRKY72 in the rice cultivars showed that WRKY50 was downregulated and WRKY72 was upregulated with respect to salt stress (Celik et al., 2019). On studying methylome flexibility, relaxation of DNA methylation was observed in the salt-tolerant variety Pokkali.

Demethylation in the Pokkali variety was identified due to the increased activity of DNA demethylases under salt stress, whereas in the salt-sensitive cultivar IR-29, the activity of DNA demethylases and methyltransferase was attributed to the lower plasticity of demethylation. This confirms that the salt-tolerant variety can alter its methylation levels with respect to salt stress, whereas the salt-sensitive variety could not alter methylation when compared with the salt-tolerant variety (Ferreira et al., 2015).

miRNA was identified to regulate gene expression under salt stress. Investigation of miRNA osa-miR393a and its target gene *TIR1* was performed in the salt-tolerant and salt-sensitive variety FL478 and IR-29, respectively. A time-dependent expression pattern was observed for osa-miR393a and *TIR1* in roots and shoots. In addition, downregulation of osa-miR393a and upregulation of *TIR1* were observed in the rice cultivar FL478. This is attributed to the methylation in the promoter of the miRNA on exposure to salt stress (Ganie et al., 2016). In addition to miRNAs, histone deacetylases play a crucial role in salt stress. A member of the eHDAC RPD3/HDA1 family, HDA710 was studied in rice cultivars. The rice cultivar with knocked out *hda710* showed a positive correlation with stress response with increased expression of ABA- and salt-stress-responsive genes such as *OsLEA3*, *OsABI5*, *OsbZIP72* and *OsNHX*. This is attributed to the increased H4 acetylation levels in the corresponding gene promoter regions indicating the role of HDA710 in the regulation of gene expression with respect to salt stress (Ullah et al., 2021). Similarly, histone deacetylase OsHDA706 was found to enhance salt tolerance in rice cultivars by regulating the expression of gene encoding clade A protein phosphatase 2C, viz., OsPP2C49 by deacetylation of H4K5 and H4K8 in histone H4 (Liu et al., 2023). These studies reveal that epigenetic modifications such as DNA methylation, histone modification and miRNA methylation enhance salt stress in rice cultivars.

20.5.4 Epigenetic Regulation of Crop Yield in Soybean

Soybean is called the "king of beans" and is considered one of the dominant sources of vegetable oil (Sugiyama et al., 2015; Voora et al., 2024). Nine essential amino acids were identified to be present with protein ranging from 36% to 40%, making it a rich source of plant-based protein (Banaszkiewicz, 2011). They are also rich in isoflavones, which have potential health benefits against heart disease and cancer (Messina et al., 2022). Being an important agricultural crop worldwide, increasing its productivity is essential for the global market. However, the increasing abiotic and biotic stresses due to various reasons may reduce its global productivity, which can be enhanced by epigenetic regulations.

Crop yield in soybean can be improved by enhancing flower production, increasing seed size, increasing isoflavone content and growth. DNA methylation study in symptomatic and asymptomatic plants of floral malady showed 392 methylation-sensitive amplified polymorphisms (MSAP) sites in symptomatic plants and 402 MSAP sites in asymptomatic plants. Differential methylations such as hemi-methylation, full methylation, hyper-methylation and no methylation were observed in some regions of the observed MSAP sites. In addition, the genotype JS-335 was found to have higher epigenetic modifications in the symptomatic variety, and polymorphisms associated with DNA methylations were observed to be associated with floral malady (Kad et al., 2022). The seed size was improved by altering the DNA demethylase GmDMEa in the cultivar Dongnong Soybean, which is known for its small bean size. The study revealed that the cultivar with edited GmDMEa produced larger seeds without altering the nutritional value or seed architecture (Wang et al., 2024).

The seed quality can be improved by increasing the isoflavone content, which is one of the main nutrients with medicinal value in soybeans. DNA cytosine methylation in the regions of *IFS1* and *IFS2* genes responsible for isoflavone content was observed in the isoflavone-rich cultivar NRC37 and low isoflavone cultivar NRC7. In addition, the accumulation of isoflavone was found to be higher in NRC37 after 65 days of flowering than in NRC7. This indicates that DNA methylation plays a crucial role in the isoflavone content of soybean cultivars, and epigenetic modifications could bring about beneficial changes (Gupta et al., 2019). The overall growth and yield of the plant can be improved by several epigenetic regulations; for instance, the cultivar containing RNAi suppression of the plant-specific gene *MutS Homolog 1* (*MSH1*) showed beneficial developmental changes and abiotic stress responses. The variety with MSH1 showed increased pods per plant, seed weight and maturity time. These characters were also observed in subsequent generations, indicating the epigenetic memory (Raju et al., 2018). In a study by Mirakhorli et al. (2022), nitric oxide was used as an epigenetic regulator to enhance the growth and productivity of soybean. It was observed that nitric oxide enhanced the expression of genes such as *HSF34*, *R2R3MYB*, *WRKY1* and *MAPK1,* caused upregulation of histone deacetylases, CAT1, and enhanced production of photosynthetic pigments. In addition, it also increased the concentration of soluble phenols and proline in leaves, thereby exerting positive regulation in the overall growth and nutritional content of the plant. These studies prove that epigenetic regulation has a positive impact on the crop growth and yield of soybean.

20.5.5 Epigenetic Regulation of Biotic Stress in Tomato

Plants are prone to pathogenic infections in addition to being exposed to abiotic stress such as heat, drought, salinity and cold (Jones & Dangl, 2006). The developing environmental fluctuations are increasing the incidence of biotic stress in plants. Tomato is one of the fruits or vegetables most commonly used in plant research (The tomato genome consortium, 2012; Gao et al., 2019; Zhou et al., 2022). Tomato is also prone to several diseases caused by bacteria, viruses and fungi. Histone epigenetic marks were identified in tomato against the infectious pathogen *Botrytis cinerea.* H3K4 methylation increased the expression of DES (divinyl ethyl synthase), LoxD (lipoxygenase D), DOX1 (alpha dioxygenase 1), PR2 (pathogenesis-related protein 2), WRKY53 and WRKY33. These histone modifications function as epigenetic marks with respect to pathogenic infection by *Botrytis cinerea* (Crespo-Salvador et al., 2018). Similarly, histone marks in intronic and exonic regions showed that increased H3K9ac mark along with early genes such as *SlyDES, SlyDOX1* and *SlyLoxD* encoding oxylipin-pathway enzymes, and *SlyWRKY75* encoding for a transcriptional regulator of hormonal signaling. The mRNA was found to be higher, whereas the miRNA was found to be reduced in plants with infection by *Botrytis cinerea.* The intronic region with increased H3K4 methylation in the miRNA binding site upon *Botrytis cinerea* infection was identified, and the gene encoding Sly-miR1127-3p was located, which was responsible for reduced H3K4 methylation in plants infected with *Botrytis cinerea* (Crespo-Salvador et al., 2020).

Epigenetic regulation for the viral disease, tomato leaf curl New Delhi virus infection and Tomato yellow leaf curl virus infection has also been studied. In the tolerant cultivar, resistance to Tomato leaf curl New Delhi virus infection led to the increased accumulation of viral IR specific nucleotides siRNA and AC1-specific siRNA. Combination of DNA methylation and RNA degradation by siRNA was identified to confer resistance against the pathogen (Sahu et al., 2014). In tomato yellow leaf curl virus infection, DNA methylation induced the differential gene expression, especially in the miRNA and phasiRNA conferring resistance to the disease (Romero-Rodríguez et al., 2023). In tomato spotted wilt virus infection, epigenetic regulation of auxin response factor 8 and the miR167a triggered resistance of plants against the pathogen. The study showed that pathogens downregulate ARF8, disrupting auxin production, paving the way for their multiplication and survival. Modification in this gene thereby enhances auxin production that can make plants resistant to the pathogen and subsequent infection. Here, auxin serves as an epigenetic regulator for plant resistance against the pathogen (Werghi et al., 2021).

In addition to auxin, essential oils were also found to serve as epigenetic modulators against pathogenic infection. In a study by Soudani et al. (2022), absinthium essential oil (AEO) was found to show resistance against *Fusarium oxysporum* sp. *oxysporum radicis lycopersici* (Fol). Symptoms associated with the disease such as weight loss, necrosis and chlorosis were found to be reduced in AEO-treated seedlings. This has been attributed to the production of vanillic acid, coumarin, lycopene, oleamide and an unknown compound by the AEO. In addition, DNA methylation was induced in the *StNRPD2* gene of RNA polymerase IV and V by the direct RNA-directed DNA methylation. These epigenetic changes were identified to bring about resistance to the disease caused by *Fusarium oxysporum* in the tomato seedlings and even in the later stages. These studies confirm the role of epigenesis in reducing biotic stress in plants.

20.6 CONCLUSION

Epigenetic modifications include DNA methylation, histone modification and chromatin remodeling. With respect to improving crop productivity in plants, epigenetic regulation such as DNA methylation, histone methylation and miRNA modification was found to play a significant role. With the advent of gene editing technologies such as CRISPR-Cas9, available genomic information and advanced bioinformatic tools, the study of epigenetic modification is increasing. The epigenetic regulation is reversible, and studies on the time period that a plant takes to reverse the changes and inheritance pattern could help improve crop productivity. Though several studies are performed in epigenetic regulations, it is essential to know more about the epigenetic modifications to produce plants tolerant to abiotic stress such as drought, salinity and heats as well as biotic stress. Epigenetic regulation of plants can be a promising mechanism to improve crop productivity and meet the global food demands.

REFERENCES

Akagi, T., Henry, I. M., Kawai, T., Comai, L., & Tao, R. (2016). Epigenetic regulation of the sex determination gene MeGI in polyploid persimmon. *The Plant Cell*, 28(12), 2905–2915.

Almatroudi, A. (2022). Non-coding RNAs in tuberculosis epidemiology: Platforms and approaches for investigating the genome's dark matter. *International Journal of Molecular Sciences*, 23(8), 4430.

Alwutayd, K. M., Rawat, A. A., Sheikh, A. H., Almeida-Trapp, M., Veluchamy, A., Jalal, R., .Karampelias M., Froehlich K., Alzaed W., Tabassum N., Schley T. R., Schäffner A. R., Daur I., Saad M. M.,& Hirt, H. (2023). Microbe-induced drought tolerance by ABA-mediated root architecture and epigenetic reprogramming. *EMBO Reports*, 24(8), e56754.

Anjum, S. A., Wang, L. C., Farooq, M., Hussain, M., Xue, L. L., & Zou, C. M. (2011). Brassinolide application improves the drought tolerance in maize through modulation of enzymatic antioxidants and leaf gas exchange. *Journal of Agronomy and Crop Science*, 197(3), 177–185.

Arora, N. K. (2019). Impact of climate change on agriculture production and its sustainable solutions. *Environmental Sustainability*, 2(2), 95–96.

Atta, K., Adhikary, S., Mondal, S., Mukherjee, S., Pal, A., Mondal, S., Jana, K.,& Biswas, B. (2022b). A review on stress physiology and breeding potential of an underutilized, multipurpose legume: Rice bean (*Vigna umbellata*). In *Developing Climate Resilient Grain and Forage Legumes* (pp. 235–253).

Atta, K., Mondal, S., Gorai, S., Singh, A. P., Kumari, A., Ghosh, T., Roy, A., Hembram, S., Gaikwad, D. J., Mondal, S., Bhattacharya, S., Jha, U. C., & Jespersen, D. (2023). Impacts of salinity stress on crop plants: Improving salt tolerance through genetic and molecular dissection. *Frontiers in Plant Science*, 14, 1241736.

Atta, K., Sen, J., Chettri, P., & Pal, A. K. (2022a). Antioxidant responses of ricebean [*Vigna umbellata* (Thunb.) Ohwi and Ohashi] seedling under iso-osmotic potential of salinity and drought stress. *Legume Research-An International Journal*, 45(4), 429–434.

Banaszkiewicz, T. (2011). Nutritional value of soybean meal. *Soybean and Nutrition*, 2011, 1–20.

Banerjee, A., & Roychoudhury, A. (2017). Epigenetic regulation during salinity and drought stress in plants: Histone modifications and DNA methylation. *Plant Gene*, 11, 199–204.

Barnabás, B., Jäger, K., & Fehér, A. (2008). The effect of drought and heat stress on reproductive processes in cereals. *Plant, Cell & Environment*, 31(1), 11–38.

Bäurle, I. (2016). Plant heat adaptation: Priming in response to heat stress. *F1000Research*, 5, F1000-Faculty.

Benoit, M., Drost, H. G., Catoni, M., Gouil, Q., Lopez-Gomollon, S., Baulcombe, D. & Paszkowski, J. (2019). Environmental and epigenetic regulation of Rider retrotransposons in tomato. *PLoS Genetics*, 15(9), e1008370.

Berger, S. L., Kouzarides, T., Shiekhattar, R., & Shilatifard, A. (2009). An operational definition of epigenetics. *Genes & Development*, 23(7), 781–783.

Bhanu, B. D., Ulaganathan, K., & Shanker, A. K. (2020). Water stress responsive differential methylation of organellar genomes of *Zea mays* Z59. *American Journal of Plant Sciences*, 11(7), 1077.

Blevins, T., Pontvianne, F., Cocklin, R., Podicheti, R., Chandrasekhara, C., Yerneni, S., ... & Pikaard, C. S. (2014). A two-step process for epigenetic inheritance in *Arabidopsis*. *Molecular Cell*, 54(1), 30–42.

Bocchini, M., D'Amato, R., Ciancaleoni, S., Fontanella, M. C., Palmerini, C. A., Beone, G. M., ... & Businelli, D. (2018). Soil selenium (Se) biofortification changes the physiological, biochemical and epigenetic responses to water stress in *Zea mays* L. by inducing a higher drought tolerance. *Frontiers in Plant Science*, 9, 389.

Bodner, G., Nakhforoosh, A., & Kaul, H. P. (2015). Management of crop water under drought: A review. *Agronomy for Sustainable Development*, 35, 401–442.

Cedar, H., & Bergman, Y. (2009). Linking DNA methylation and histone modification: Patterns and paradigms. *Nature Reviews Genetics* 10(5), 295–304.

Çelik, Ö., Meriç, S., Ayan, A., & Atak, Ç. (2019). Epigenetic analysis of WRKY transcription factor genes in salt stressed rice (*Oryza sativa* L.) plants. *Environmental and Experimental Botany*, 159, 121–131.

Chen, Y., Zhang, B., & Annis, G. B. (2024). Plant biotechnology and genetics for sustainable agriculture and global food security. *Frontiers in Plant Science*, 15, 1479632.

Chi, P., Allis, C. D., & Wang, G. G. (2010). Covalent histone modifications-miswritten, misinterpreted and mis-erased in human cancers. *Nature Reviews Cancer*, 10(7), 457–469.

Choi, S. M., Song, H. R., Han, S. K., Han, M., Kim, C. Y., Park, J., Lee, Y. H., Jeon, J. S., Noh, Y. S.,& Noh, B. (2012). HDA19 is required for the repression of salicylic acid biosynthesis and salicylic acid-mediated defense responses in *Arabidopsis*. *The Plant Journal*, 71(1), 135–146.

Chopra, S., Cocciolone, S. M., Bushman, S., Sangar, V., McMullen, M. D., & Peterson, T. (2003). The maize unstable factor for orange1 is a dominant epigenetic modifier of a tissue specifically silent allele of pericarp color1. *Genetics*, 163(3), 1135–1146.

Christie, I., Fernandes, E., Messerli, H., & Twining-Ward, L. (2013). Tourism in Africa: Harnessing tourism for growth and improved livelihoods *(English)*. Washington DC ; World Bank. http://documents.worldbank.org/curated/en/723511468102894381

Costa, F. F. (2008). Non-coding RNAs, epigenetics and complexity. *Gene*, 410(1), 9–17.

Crespo-Salvador, Ó., Escamilla-Aguilar, M., López-Cruz, J., López-Rodas, G., & González-Bosch, C. (2018). Determination of histone epigenetic marks in *Arabidopsis* and tomato genes in the early response to *Botrytis cinerea*. *Plant Cell Reports*, 37, 153–166.

Crespo-Salvador, Ó., Sánchez-Giménez, L., López-Galiano, M. J., Fernández-Crespo, E., Scalschi, L., García-Robles, I., ... & González-Bosch, C. (2020). The histone marks signature in exonic and intronic regions is relevant in early response of tomato genes to *Botrytis cinerea* and in miRNA regulation. *Plants*, 9(3), 300.

Demirevska, K., Zasheva, D., Dimitrov, R., Simova-Stoilova, L., Stamenova, M., & Feller, U. (2009). Drought stress effects on Rubisco in wheat: Changes in the Rubisco large subunit. *Acta Physiologiae Plantarum*, 31, 1129–1138.

Eichten, S. R., Swanson-Wagner, R. A., Schnable, J. C., Waters, A. J., Hermanson, P. J., Liu, S., ... & Springer, N. M. (2011). Heritable epigenetic variation among maize inbreds. *PLoS Genetics*, 7(11), e1002372.

El-Sharkawy, M. A. (2006). Utility of basic research in plant/crop physiology in relation to crop improvement: A review and a personal account. *Brazilian Journal of Plant Physiology*, 18, 419–446.

Felsenfeld, G. (2014). A brief history of epigenetics. *Cold Spring Harbor Perspectives in Biology*, 6(1), a018200.

Ferreira, L. J., Azevedo, V., Maroco, J., Oliveira, M. M., & Santos, A. P. (2015). Salt tolerant and sensitive rice varieties display differential methylome flexibility under salt stress. *PLoS One*, 10(5), e0124060.

Forestan, C., Farinati, S., Rouster, J., Lassagne, H., Lauria, M., Dal Ferro, N., & Varotto, S. (2018). Control of maize vegetative and reproductive development, fertility, and rRNAs silencing by HISTONE DEACETYLASE 108. *Genetics*, 208(4), 1443–1466.

Fuentes-Merlos, M. I., Bamba, M., Sato, S., & Higashitani, A. (2023). Self-grafting-induced epigenetic changes leading to drought stress tolerance in tomato plants. *DNA Research*, 30(4), dsad016.

Ganie, S. A., Dey, N., & Mondal, T. K. (2016). Promoter methylation regulates the abundance of osa-miR393a in contrasting rice genotypes under salinity stress. *Functional & Integrative Genomics*, 16, 1–11.

Gao, L., Gonda, I., Sun, H., Ma, Q., Bao, K., Tieman, D. M., Burzynski-Chang E. A., Fish T. L., Stromberg K. A., Sacks G. L., Thannhauser T. W., Foolad M. R., Diez M. J., Blanca J., Canizares J., Xu Y., van der Knaap E., Huang S., Klee H. J., Giovannoni J. J.,& Fei, Z. (2019). The tomato pan-genome uncovers new genes and a rare allele regulating fruit flavor. *Nature Genetics*, 51(6), 1044–1051.

Garg, R., Narayana Chevala, V. V. S., Shankar, R., & Jain, M. (2015). Divergent DNA methylation patterns associated with gene expression in rice cultivars with contrasting drought and salinity stress response. *Scientific Reports*, 5(1), 14922.

Guo Xiang, G. X., & Han FangPu, H. F. (2014). Asymmetric epigenetic modification and elimination of rDNA sequences by polyploidization in wheat.*The Plant Cell*, *26*(11), 4311–4327

Gupta, O. P., Dahuja, A., Sachdev, A., Jain, P. K., Kumari, S., & Praveen, S. (2019). Cytosine methylation of isoflavone synthase gene in the genic region positively regulates its expression and isoflavone biosynthesis in soybean seeds. *DNA and Cell Biology*, 38(6), 510–520.

Hasanuzzaman, M., & Fujita, M. (2022). Plant responses and tolerance to salt stress: Physiological and molecular interventions. *International Journal of Molecular Sciences*, 23(9), 4810.

Hasanuzzaman, M., Hossain, M. A., da Silva, J. A. T., & Fujita, M. (2012). Plant response and tolerance to abiotic oxidative stress: Antioxidant defense is a key factor. In *Crop Stress and its Management: Perspectives and Strategies* (pp. 261–315).Dordrecht: Springer Netherlands.

Hasanuzzaman, M., Nahar, K., & Fujita, M. (2013a). Extreme temperature responses, oxidative stress and antioxidant defense in plants. In *Abiotic Stress: Plant Responses and Applications in Agriculture* (p. 169).IntechOpen, London, 169–205

Hasanuzzaman, M., Nahar, K., Alam, M. M., Roychowdhury, R., & Fujita, M. (2013b). Physiological, biochemical, and molecular mechanisms of heat stress tolerance in plants. *International Journal of Molecular Sciences*, 14(5), 9643–9684.

Hazell, P., & Wood, S. (2008). Drivers of change in global agriculture. *Philosophical Transactions of the Royal Society B: Biological Sciences*, 363(1491), 495–515.

Hébrard, C., Peterson, D. G., Willems, G., Delaunay, A., Jesson, B., Lefèbvre, M., Barnes, S. & Maury, S. (2016). Epigenomics and bolting tolerance in sugar beet genotypes. *Journal of Experimental Botany*, *67*(1), 207–225.

Hossain, M. S., Kawakatsu, T., Kim, K. D., Zhang, N., Nguyen, C. T., Khan, S. M., Batek, J. M., Joshi, T., Schmutz, J., Grimwood, J.,& Stacey, G. (2017). Divergent cytosine DNA methylation patterns in single-cell, soybean root hairs. *New Phytologist*, 214(2), 808–819.

Howarth, C. J. (2005). Genetic improvements of tolerance to high temperature. In *Abiotic Stresses: Plant Resistance through Breeding and Molecular Approaches* (pp. 277–300). New York: Howarth Press Inc.

https://biomodal.com/blog/how-histone-modifications-impact-gene-regulation/.

Hu, Z., Han, Z., Song, N., Chai, L., Yao, Y., Peng, H., Ni, Z., & Sun, Q. (2013). Epigenetic modification contributes to the expression divergence of three TaEXPA1 homoeologs in hexaploid wheat (*Triticum aestivum*). *New Phytologist*, 197(4), 1344–1352.

Intergovernmental Panel on Climate Change (IPCC). (2007). Climate change 2007–The physical science basis. In *Contribution of Working Group I to the Fourth Assessment Report of the Intergovernmental Panel on Climate Change*. Cambridge: Cambridge University Press.

Iwasaki, M., & Paszkowski, J. (2014). Epigenetic memory in plants. *The EMBO Journal*, 33(18), 1987–1998.

Jones, J. D., & Dangl, J. L. (2006). The plant immune system. *Nature*, 444(7117), 323–329.

Jung, C. H., O'Brien, M., Singh, M. B., & Bhalla, P. L. (2015). Epigenetic landscape of germline specific genes in the sporophyte cells of *Arabidopsis thaliana*. *Frontiers in Plant Science*, 6, 328.

Kad, S. K., VishwanathraoJadhav, P., Thakur, P. P., Kale, P. B., Magar, S. G., Moharil, M. P., ... & Dani, R. G. (2022). MSAP marker based epigenetic regulation in symptomatic and asymptomatic floral malady of soybean (*Glycine max*). *Brazilian Archives of Biology and Technology*, 65, e22210115.

Khan, I. U., Ali, A., Khan, H. A., Baek, D., Park, J., Lim, C. J., Zareen, S., Jan, M., Lee, S. Y., Pardo, J. M., & Yun, D. J. (2020). PWR/HDA9/ABI4 complex epigenetically regulates ABA dependent drought stress tolerance in *Arabidopsis*. *Frontiers in Plant Science*, 11, 623.

Khanzada, A., Wang, X., Malko, M. M., Wu, Y., Samo, A., & Dong, J. (2024). Response of the persistence of heat stress tolerance in winter wheat seedling to heat priming at early growth stages. *Plant Stress*, 11, 100323.

Kong, L., Zhi, P., Liu, J., Li, H., Zhang, X., Xu, J., Zhou, J., Wang, X., & Chang, C. (2020). Epigenetic activation of Enoyl-CoA Reductase by an acetyltransferase complex triggers wheat wax biosynthesis. *Plant Physiology*, 183(3), 1250–1267.

Kooke, R., Johannes, F., Wardenaar, R., Becker, F., Etcheverry, M., Colot, V., Vreugdenhil, D.,& Keurentjes, J. J. (2015). Epigenetic basis of morphological variation and phenotypic plasticity in *Arabidopsis thaliana*. *The Plant Cell*, 27(2), 337–348.

Kumar, P., & Sharma, P. K. (2020). Soil salinity and food security in India. *Frontiers in Sustainable Food Systems*, 4, 533781.

Li, H., Freeling, M., & Lisch, D. (2010). Epigenetic reprogramming during vegetative phase change in maize. *Proceedings of the National Academy of Sciences*, 107(51), 22184–22189.

Lin, J., Song, N., Liu, D., Liu, X., Chu, W., Li, J., Chang, S., Liu, Z., Chen, Y., Yang, Q., & Hu, Z. (2022). Histone acetyltransferase TaHAG1 interacts with TaNACL to promote heat stress tolerance in wheat. *Plant Biotechnology Journal*, 20(9), 1645.

Linn, F., Heidmann, I., Saedler, H., & Meyer, P. (1990). Epigenetic changes in the expression of the maize A1 gene in *Petunia hybrida*: Role of numbers of integrated gene copies and state of methylation. *Molecular & General Genetics: MGG*, 222(2–3), 329–336.

Liu, B., Zhang, L., Rusalepp, L., Kaurilind, E., Sulaiman, H. Y., Püssa, T., & Niinemets, Ü. (2021). Heat priming improved heat tolerance of photosynthesis, enhanced terpenoid and benzenoid emission and phenolics accumulation in *Achillea millefolium*. *Plant, Cell & Environment*, 44(7), 2365–2385.

Liu, C., Mao, B., Yuan, D., Chu, C., & Duan, M. (2022). Salt tolerance in rice: Physiological responses and molecular mechanisms. *The Crop Journal*, 10(1), 13–25.

Liu, C., Yang, X., Zhang, H., Wang, X., Zhang, Z., Bian, Y., ... & Liu, B. (2015). Genetic and epigenetic modifications to the BBAA component of common wheat during its evolutionary history at the hexaploid level. *Plant Molecular Biology*, 88(1–2), 53–64.

Liu, K., Chen, J., Sun, S., Chen, X., Zhao, X., Hu, Y., Qi, G., Li, X., Xu, B., Miao, J.,& Gong, Z. (2023). Histone deacetylase OsHDA706 increases salt tolerance via H4K5/K8 deacetylation of OsPP2C49 in rice. *Journal of Integrative Plant Biology*, 65(6), 1394–1407.

Lobell, D. B., Schlenker, W., & Costa-Roberts, J. (2011). Climate trends and global crop production since 1980. *Science*, 333(6042), 616–620.

Lombardo, L., Coppola, G., & Zelasco, S. (2016). New technologies for insect-resistant and herbicide-tolerant plants. *Trends in Biotechnology*, 34(1), 49–57.

Lu, K., Guo, Z., Di, S., Lu, Y., Muhammad, I. A. R., Rong, C., Ding, Y., Li, W.,& Ding, C. (2023). OsMFT1 inhibits seed germination by modulating abscisic acid signaling and gibberellin biosynthesis under salt stress in rice. *Plant and Cell Physiology*, 64(6), 674–685.

Luo, C., Durgin, B. G., Watanabe, N., & Lam, E. (2009). Defining the functional network of epigenetic regulators in *Arabidopsis thaliana*. *Molecular Plant*, 2(4), 661–674.

Mendanha, T., Rosenqvist, E., Hyldgaard, B., & Ottosen, C. O. (2018). Heat priming effects on anthesis heat stress in wheat cultivars (*Triticum aestivum* L.) with contrasting tolerance to heat stress. *Plant Physiology and Biochemistry*, 132, 213–221.

Mercé, C., Bayer, P. E., Tay Fernandez, C., Batley, J., & Edwards, D. (2020). Induced methylation in plants as a crop improvement tool: Progress and perspectives. *Agronomy*, 10(10), 1484.

Messina, M., Duncan, A., Messina, V., Lynch, H., Kiel, J., & Erdman Jr, J. W. (2022). The health effects of soy: A reference guide for health professionals. *Frontiers in Nutrition*, 9, 970364.

Miao, Z., Han, Z., Zhang, T., Chen, S., & Ma, C. (2017). A systems approach to a spatio-temporal understanding of the drought stress response in maize. *Scientific Reports*, 7.

Mirakhorli, T., Oraghi Ardebili, Z., Ladan-Moghadam, A., & Danaee, E. (2022). Nitric oxide improved growth and yield in soybean (*Glycine max*) by mediating physiological, anatomical, and transcriptional modifications. *Journal of Plant Growth Regulation*, 41(3), 1331–1343.

Mirouze, M., & Paszkowski, J. (2011). Epigenetic contribution to stress adaptation in plants. *Current Opinion in Plant Biology*, 14(3), 267–274.

Mishra, A. K., & Singh, V. P. (2010). A review of drought concepts. *Journal of Hydrology*, 391(1–2), 202–216.

Morange, M. (2009). What history tells us XVI. A third pillar for molecular biology: Molecular embryology. *Journal of Biosciences*, 34(1), 17–20.

Narlikar, G. J., Sundaramoorthy, R., & Owen-Hughes, T. (2013). Mechanisms and functions of ATP-dependent chromatin-remodeling enzymes. *Cell*, 154(3), 490–503.

Neves, D. M., Almeida, L. A. D. H., Santana-Vieira, D. D. S., Freschi, L., Ferreira, C. F., Soares Filho, W. D. S., ... & Gesteira, A. D. S. (2017). Recurrent water deficit causes epigenetic and hormonal changes in citrus plants. *Scientific Reports*, 7(1), 13684.

Nishio, H., Kawakatsu, T., & Yamaguchi, N. (2024). Beyond heat waves: Unlocking epigenetic heat stress memory in *Arabidopsis*. *Plant Physiology*, 194(4), 1934–1951.

Oerke, E. C. (2006). Crop losses to pests. *The Journal of Agricultural Science*, 144(1), 31–43.

Olejnik, P., Mądrzak, C. J., & Nuc, K. (2021). Cyclophilins and their functions in abiotic stress and plant–microbe interactions. *Biomolecules*, 11(9), 1390.

Park, H. J., Kim, W. Y., & Yun, D. J. (2016). A new insight of salt stress signalingin plant. *Molecules and Cells*, 39(6), 447–459.

Peet, M. M., & Willits, D. H. (1998). The effect of night temperature on greenhouse grown tomato yields in warm climates. *Agricultural and Forest Meteorology*, 92(3), 191–202.

Phillips, T., & Shaw, K. (2008). Chromatin remodeling in eukaryotes. *Nature Education*, 1(1), 209.

Porter, J. R. (2005). Rising temperatures are likely to reduce crop yields. *Nature*, 436(7048), 174–174.

Praba, M. L., Cairns, J. E., Babu, R. C., & Lafitte, H. R. (2009). Identification of physiological traits underlying cultivar differences in drought tolerance in rice and wheat. *Journal of Agronomy and Crop Science*, 195(1), 30–46.

Qadir, M., Quillérou, E., Nangia, V., Murtaza, G., Singh, M., Thomas, R. J., Drechsel, P., & Noble, A. D. (2014, November). Economics of salt-induced land degradation and restoration. *Natural Resources Forum*, 38(4), 282–295.

Rajkumar, M. S., Shankar, R., Garg, R., & Jain, M. (2020). Bisulphite sequencing reveals dynamic DNA methylation under desiccation and salinity stresses in rice cultivars. *Genomics*, 112(5), 3537–3548.

Raju, S. K. K., Shao, M. R., Sanchez, R., Xu, Y. Z., Sandhu, A., Graef, G., & Mackenzie, S. (2018). An epigenetic breeding system in soybean for increased yield and stability. *Plant Biotechnology Journal*, 16(11), 1836–1847.

Ravichandran, S., Ragupathy, R., Edwards, T., Domaratzki, M., & Cloutier, S. (2019). MicroRNA-guided regulation of heat stress response in wheat. *BMC Genomics*, 20, 1–16.

Riggs, A. D., Martienssen, R. A., & Russo, V. E. A. (1996). Introduction. In V. E. A. Russo, et al. (Ed.), *Epigenetic Mechanisms of Gene Regulation* (pp. 1–4). New York: Cold Spring Harbor Laboratory Press.

Riggs, A. D., & Porter, T. N. (1996). Overview of epigenetic mechanisms. *Cold Spring Harbor Monograph Series*, 32, 29–46.

Robbins, M. L., Wang, P., Sekhon, R. S., & Chopra, S. (2009). Gene structure induced epigenetic modifications of pericarp color1 alleles of maize result in tissue-specific mosaicism. *PLoS One*, 4(12), e8231.

Romero-Rodríguez, B., Petek, M., Jiao, C., Križnik, M., Zagorščak, M., Fei, Z., ... & Castillo, A. G. (2023). Transcriptional and epigenetic changes during tomato yellow leaf curl virus infection in tomato. *BMC Plant Biology*, 23(1), 651.

Russo, V. E., Martienssen, R. A., & Riggs, A. D. (Eds.). (1996). *Epigenetic Mechanisms of Gene Regulation* (pp. xii+-692). Cold Spring Harbor Laboratory Press, U.S.

Sahu, P. P., Sharma, N., Puranik, S., & Prasad, M. (2014). Post-transcriptional and epigenetic arms of RNA silencing: A defense machinery of naturally tolerant tomato plant against Tomato leaf curl New Delhi virus. *Plant Molecular Biology Reporter*, 32, 1015–1029.

Sallam, N., Moussa, M., Yacout, M., & El-Seedy, A. (2019). Differential DNA methylation under drought stress in maize. *International Journal of Current Microbiology and Applied Sciences*, 8(8), 2527–2543.

Saxonov, S., Berg, P., & Brutlag, D. L. (2006). A genome-wide analysis of CpG dinucleotides in the human genome distinguishes two distinct classes of promoters. *Proceedings of the National Academy of Sciences*, 103(5), 1412–1417.

Shitsukawa, N., Tahira, C., Kassai, K. I., Hirabayashi, C., Shimizu, T., Takumi, S., Mochida, K., Kawaura, K., Ogihara, Y., & Murai, K. (2007). Genetic and epigenetic alteration among three homoeologous genes of a class E MADS box gene in hexaploid wheat. *The Plant Cell*, 19(6), 1723–1737.

Siddique, K. H. M., Loss, S. P., Regan, K. L., & Jettner, R. L. (1999). Adaptation and seed yield of cool season grain legumes in Mediterranean environments of south-western Australia. *Australian Journal of Agricultural Research*, 50(3), 375–388.

Singh, H., Kaur, K., Singh, M., Kaur, G., & Singh, P. (2020). Plant cyclophilins: Multifaceted proteins with versatile roles. *Frontiers in Plant Science*, 11, 585212.

Singh, H., Kaur, K., Singh, S., Kaur, P., & Singh, P. (2019a). Genome-wide analysis of cyclophilin gene family in wheat and identification of heat stress responsive members. *Plant Gene*, 19, 100197.

Singh, J., Mishra, V., Wang, F., Huang, H. Y., & Pikaard, C. S. (2019b). Reaction mechanisms of Pol IV, RDR2, and DCL3

drive RNA channeling in the siRNA-directed DNA methylation pathway. *Molecular Cell*, 75(3), 576–589.
Smertenko, A., Dráber, P., Viklický, V., & Opatrný, Z. (1997). Heat stress affects the organization of microtubules and cell division in *Nicotiana tabacum* cells. *Plant, Cell & Environment*, 20(12), 1534–1542.
Song, Q., Zhang, T., Stelly, D. M., & Chen, Z. J. (2017). Epigenomic and functional analyses reveal roles of epialleles in the loss of photoperiod sensitivity during domestication of allotetraploid cottons. *Genome Biology*, 18, 1–14.
Soudani, S., Poza-Carrión, C., De la Cruz Gómez, N., González-Coloma, A., Andrés, M. F., & Berrocal-Lobo, M. (2022). Essential oils prime epigenetic and metabolomic changes in tomato defense against *Fusarium oxysporum*. *Frontiers in Plant Science*, 13, 804104.
Steward, N., Ito, M., Yamaguchi, Y., Koizumi, N., & Sano, H. (2002). Periodic DNA methylation in maize nucleosomes and demethylation by environmental stress. *Journal of Biological Chemistry*, 277(40), 37741–37746.
Stone, P. J., & Nicolas, M. E. (1994). Wheat cultivars vary widely in their responses of grain yield and quality to short periods of post-anthesis heat stress. *Functional Plant Biology*, 21(6), 887–900.
Sugiyama, A., Ueda, Y., Takase, H., & Yazaki, K. (2015). Do soybeans select specific species of *Bradyrhizobium* during growth? *Communicative & Integrative Biology*, 8(1), e992734.
Takeda, S., & Matsuoka, M. (2008). Genetic approaches to crop improvement: Responding to environmental and population changes. *Nature Reviews Genetics*, 9(6), 444–457.
Tang, X., Liu, G., Zhou, J., Ren, Q., You, Q., Tian, L., Xin, X., Zhong, Z., Liu, B., Zheng, X., & Zhang, Y. (2018). A large-scale whole-genome sequencing analysis reveals highly specific genome editing by both Cas9 and Cpf1 (Cas12a) nucleases in rice. *Genome Biology*, 19, 1–13.
The Tomato Genome Consortium. (2012). The tomato genome sequence provides insights into fleshy fruit evolution. *Nature*, 485, 635–641.
To, T. K., & Kim, J. M. (2014). Epigenetic regulation of gene responsiveness in *Arabidopsis*. *Frontiers in Plant Science*, 4, 548.
Tricker, P. J., Rodríguez López, C. M., Hadley, P., Wagstaff, C., & Wilkinson, M. J. (2013). Pre-conditioning the epigenetic response to high vapor pressure deficit increases the drought tolerance of *Arabidopsis thaliana*. *Plant Signaling & Behavior*, 8(10), e25974.
Tripathi, A., Tripathi, D. K., Chauhan, D. K., Kumar, N., & Singh, G. S. (2016). Paradigms of climate change impacts on some major food sources of the world: A review on current knowledge and future prospects. *Agriculture, Ecosystems & Environment*, 216, 356–373.
Ullah, F., Xu, Q., Zhao, Y., & Zhou, D. X. (2021). Histone deacetylase HDA710 controls salt tolerance by regulating ABA signaling in rice. *Journal of Integrative Plant Biology*, 63(3), 451–467.
Van Zelm, E., Zhang, Y., & Testerink, C. (2020). Salt tolerance mechanisms of plants. *Annual Review of Plant Biology*, 71(1), 403–433.
Vaughn, M. W., Tanurdžić, M., Lippman, Z., Jiang, H., Carrasquillo, R., Rabinowicz, P. D., ... & Martienssen, R. A. (2007). Epigenetic natural variation in *Arabidopsis thaliana*. *PLoS Biology*, 5(7), e174.
Ventouris, Y. E., Tani, E., Avramidou, E. V., Abraham, E. M., Chorianopoulou, S. N., Vlachostergios, D. N., ... & Kapazoglou, A. (2020). Recurrent water deficit and epigenetic memory in *Medicago sativa* L. varieties. *Applied Sciences*, 10(9), 3110.
Virdi, K. S., Laurie, J. D., Xu, Y. Z., Yu, J., Shao, M. R., Sanchez, R., Kundariya, H., Wang, D., Riethoven, J. J. M., Wamboldt, Y., & Mackenzie, S. A. (2015). *Arabidopsis* MSH1 mutation alters the epigenome and produces heritable changes in plant growth. *Nature Communications*, 6(1), 6386.
Voora, V., Bermúdez, J., Le, H., Larrea, C., & Luna, E. (2024). Soybean prices and sustainability. *IISD Market Report*, 2.
Waddington, C. H. (1942). The epigenotype. *Endeavour*, 1, 18–20.
Wang, X., Xin, C., Cai, J., Zhou, Q., Dai, T., Cao, W., & Jiang, D. (2016b). Heat priming induces Trans-generational tolerance to high temperature stress in wheat. *Frontiers in Plant Science*, 7.
Wang, L., Cao, S., Wang, P., Lu, K., Song, Q., Zhao, F. J., & Chen, Z. J. (2021). DNA hypomethylation in tetraploid rice potentiates stress-responsive gene expression for salt tolerance. *Proceedings of the National Academy of Sciences*, 118(13), e2023981118.
Wang, L., Wang, L., Tan, M., Wang, L., Zhao, W., You, J., Wang, L., Yan, X.,& Wang, W. (2023). The pattern of alternative splicing and DNA methylation alteration and their interaction in linseed (*Linum usitatissimum* L.) response to repeated drought stresses. *Biological Research*, 56(1), 12.
Wang, Q., Xu, J., Pu, X., Lv, H., Liu, Y., Ma, H., Wu, F., Wang, Q., Feng, X., Liu, T., & Lu, Y. (2021). Maize DNA methylation in response to drought stress is involved in target gene expression and alternative splicing. *International Journal of Molecular Sciences*, 22(15), 8285.
Wang, W., Huang, F., Qin, Q., Zhao, X., Li, Z., & Fu, B. (2015). Comparative analysis of DNA methylation changes in two rice genotypes under salt stress and subsequent recovery. *Biochemical and Biophysical Research Communications*, 465(4), 790–796.
Wang, W., Zhang, T., Liu, C., Liu, C., Jiang, Z., Zhang, Z., Ali, S., Li, Z., Wang, J., Sun, S., & Xie, L. (2024). A DNA demethylase reduces seed size by decreasing the DNA methylation of AT-rich transposable elements in soybean. *Communications Biology*, 7(1), 613.
Wang, W., Zhao, X., Pan, Y., Zhu, L., Fu, B., & Li, Z. (2011). DNA methylation changes detected by methylation-sensitive amplified polymorphism in two contrasting rice genotypes under salt stress. *Journal of Genetics and Genomics*, 38(9), 419–424.
Wang, Y., Li, H., Sun, Q., & Yao, Y. (2016a). Characterization of small RNAs derived from tRNAs, rRNAs and snoRNAs and their response to heat stress in wheat seedlings. *PLoS One*, 11(3), e0150933.
Wei, J. W., Huang, K., Yang, C., & Kang, C. S. (2017). Non-coding RNAs as regulators in epigenetics. *Oncology Reports*, 37(1), 3–9.
Werghi, S., Herrero, F. A., Fakhfakh, H., & Gorsane, F. (2021). Auxin drives tomato spotted wilt virus (TSWV) resistance through epigenetic regulation of auxin response factor ARF8 expression in tomato. *Gene*, 804, 145905.
Whittle, C. A., Otto, S. P., Johnston, M. O., & Krochko, J. E. (2009). Adaptive epigenetic memory of ancestral temperature regime in *Arabidopsis thaliana*. *Botany*, 87(6), 650–657.
Woodland, H. R. (1987). Gene Activity in Early Development: By Davidson, E.H.. Orlando, Florida: Academic Press.(1986). 670 pp. 305-306

Xiao, J., Jin, R., & Wagner, D. (2017). Developmental transitions: Integrating environmental cues with hormonal signaling in the chromatin landscape in plants. *Genome Biology*, 18, 1–14.

Yan, S., Zhang, Q., Li, Y., Huang, Y., Zhao, L., Tan, J., He, S., & Li, L. (2014). Comparison of chromatin epigenetic modification patterns among root meristem, elongation and maturation zones in maize (*Zea mays* L.). *Cytogenetic and Genome Research*, 143(1–3), 179–188.

Zhai, J., Liu, J., Liu, B., Li, P., Meyers, B. C., Chen, X. M., & Cao, X. (2008). Small RNA-directed epigenetic natural variation in *Arabidopsis thaliana*. *PLoS Genetics*, 4(4), e1000056.

Zhang, L., Jiao, W., Zhang, H., Huang, C., & Tong, Q. (2017). Studying drought phenomena in the Continental United States in 2011 and 2012 using various drought indices. *Remote Sensing of Environment*, 190, 96–106.

Zhao, S., Zhang, Q., Liu, M., Zhou, H., Ma, C., & Wang, P. (2021). Regulation of plant responses to salt stress. *International Journal of Molecular Sciences*, 22(9), 4609.

Zheng, L., Ma, S., Shen, D., Fu, H., Wang, Y., Liu, Y., Shah, K., Yue, C., & Huang, J. (2021). Genome-wide identification of Gramineae histone modification genes and their potential roles in regulating wheat and maize growth and stress responses. *BMC Plant Biology*, 21, 1–18.

Zhong, L., Xu, Y. H., & Wang, J. B. (2009). DNA-methylation changes induced by salt stress in wheat *Triticum aestivum*. *African Journal of Biotechnology*, 8(22).

Zhou, Y., Zhang, Z., Bao, Z., Li, H., Lyu, Y., Zan, Y., Wu, Y., Cheng, L., Fang, Y., Wu, K., & Huang, S. (2022). Graph pangenome captures missing heritability and empowers tomato breeding. *Nature*, 606(7914), 527–534.

Zhu, J. K. (2002). Salt and drought stress signal transduction in plants. *Annual Review of Plant Biology*, 53(1), 247–273.

Zhu, N., Cheng, S., Liu, X., Du, H., Dai, M., Zhou, D. X., Yang, W., & Zhao, Y. (2015). The R2R3-type MYB gene OsMYB91 has a function in coordinating plant growth and salt stress tolerance in rice. *Plant Science*, 236, 146–156.

Zi, N., Ren, W., Guo, H., Yuan, F., Liu, Y., & Fry, E. (2024). DNA methylation participates in drought stress memory and response to drought in *Medicago ruthenica*. *Genes*, 15(10), 1286.

Ziska, L. H., Bunce, J. A., Shimono, H., Gealy, D. R., Baker, J. T., Newton, P. C., Reynolds, M.P., & Wilson, L. T. (2012). Food security and climate change: On the potential to adapt global crop production by active selection to rising atmospheric carbon dioxide. *Proceedings of the Royal Society B: Biological Sciences*, 279(1745), 4097–4105.

21 Epigenetic Regulation in Plant–Microbe Interaction and Plant Immunity

Bijendra Kumar Singh, Ashish Chandran and Satyajit Singh

21.1 INTRODUCTION

Plants must initiate quick and efficient defense responses in situ since they are sessile and photoautotrophic, fending off ongoing attacks from a variety of diseases. For plants to maximize ecological fitness, defense must be perfectly synchronized with growth and development. They engage in cooperative or competitive interactions with bacteria, nematodes, fungi, oomycetes, insects and other plants on a regular basis. Because they are sessile organisms, plants are usually vulnerable to a wide range of dangerous bacteria. By linking pathogen detection to prompt and effective defense activation, pathogen effectors drive plant mechanisms to develop in a way that minimizes fitness costs throughout the dynamic plant–pathogen interactions. All accessible plant tissues are colonized by bacteria, fungi, protists, nematodes and viruses that make up the plant microbiome. The surrounding microbiota interacts with the host plant in intricate and dynamic ways. These interactions can increase plant resistance to environmental stressors and are greatly impacted by the surroundings (Michmizos & Hilioti, 2019; Trivedi et al., 2020; Regmi et al., 2024). Since plants are sessile and must adapt to their local environment, flexibility is crucial to their existence. Plants may swiftly adapt to their environment and improve their fitness through interactions between the microbiota and epigenetic regulators. Phenotypic changes ascribed to mutualists and their transmission to progeny are commonly referred to as epigenetic variation. However, this kind of plasticity is more like an interspecies induction of changes mediated by epigenetics rather than epigenetics-induced changes based solely on epigenetic heritable pathways (Chiquito-Contreras et al., 2024; Vílchez et al., 2024). Epigenetics is the study of mechanisms that change gene expression in response to stimuli that occur independently of changes in the underlying DNA sequence. DNA methylation, chromatin remodeling proteins, histone modifications and noncoding RNAs (ncRNA) that silence DNA are some of these processes that have been elucidated. In plants, the development of plant diseases is regulated by important epigenetic mechanisms such as DNA methylation and histone modifications. Epigenetic pathways are also used by the pathogens to boost virulence during infection. It has been proposed that plant pathogens use epigenetic modifications to their avirulence effector genes as switches to avoid host recognition (Ramos-Cruz et al., 2021; Wen et al., 2023).

Plant defense is triggered by a variety of immunogenic cues. Immunogenic microbial patterns that are relatively constant across taxonomic groups are known as microbe-associated molecular patterns (MAMPs). Similar to peptidoglycan and the bacterial flagellin epitope flg22, plants use lytic enzymes to aid in the release of MAMP epitopes from invading pathogens. In addition to poisons and enzymes that break down plant cell walls, diseases can damage plant tissues by releasing extracellular ATP, nicotinamide adenine dinucleotide (NAD) and fragments of plant cell walls. These plant-derived chemicals function as damage-associated molecular patterns (DAMPs), which may trigger immunological responses. Furthermore, plants actively track alterations in the integrity of their cell walls, which likely helps identify diseases (Tang et al., 2017; Buscaill et al., 2019; Aerts et al., 2022). As a co-evolutionary reaction, plants have evolved the ability to recognize deadly diseases early in an attack and build an appropriate defense response. One well-known example of this kind of co-evolved plant defense is "effector-triggered immunity" (ETI), which enables plants to detect the presence or activity of certain pathogen effectors. Although it needs *resistance* (*R*) genes in the host plant, ETI can render some virulent diseases avirulent (Zhang et al., 2019; Ahmad et al., 2024). The relationship between ribonucleic acid (RNA) and genetic information is complex. The discovery of protein production mechanisms indicates that RNA is tied to genetic information in at least three separate ways. It is capable of directly transmitting messenger RNA (mRNA). It is capable of both structural and enzymatic roles, just as ribosomal RNA (rRNA). As an alternative, it can serve as transfer RNA (tRNA) at the junction of the protein sequence and genetic information. Recent studies have shown that DNA sequences from intronic, intergenic or protein-coding genes may be used to create ncRNAs. Based on their regulatory roles, ncRNAs in plants may be categorized into two major types. Because housekeeping ncRNAs are necessary for fundamental biological processes in life, the composition is generally constant. Regulatory ncRNAs have distinct manifestations in space and time due to variations in their size, shape and accumulation patterns (Wierzbicki et al., 2021; Hu et al., 2024). A physiological state known as "priming" is induced in the host plant by beneficial microbial

DOI: 10.1201/9781003589914-25

colonization. After priming is triggered, plants show stronger and faster defensive responses against the future invasion of pathogens, which is a common feature of systemic resistance created by beneficial bacteria. Plants can recognize the beneficial bacteria by recognizing the common microbial compounds they produce, such as ligands, flagellin, lipopolysaccharides (LPS), exopolysaccharides and chitin oligosaccharides (Manoharan et al., 2023). The chapter deals with the epigenetic regulations due to plant–microbe interactions affecting the plant immunity.

It covers the importance and impacts of epigenetic modifications on plant–microbe interactions, domains involved in the plant immunological transcriptome regulation, immune signaling and interactions of epigenetic pathways and significance of noncoding RNAs in the regulation of epigenomics. Moreover, the beneficial microbe-induced epigenetic modifications that lead to disease resistance and crop improvement have also been discussed in the chapter.

21.2 IMPACT OF EPIGENETIC MODIFICATIONS ON PLANT–MICROBE INTERACTIONS

The method of gene expression known as epigenetic modification is brought about by modifications to the structure and functions of chromatin without changing the DNA sequence. Since DNA in eukaryotes (plants, animals and fungi) is compact and encircled by histone proteins, epigenetic alteration has been accomplished by histone modifications and DNA cytosine methylation (Ramos-Cruz et al., 2021; Regmi et al., 2024). Histone tail post-translational modifications, methylation–demethylation and acetylation–deacetylation, for instance, affect chromatin structure and have the potential to initiate and deactivate gene expression. Since bacteria lack histones, adenine and cytosine methylation is the primary epigenetic alteration that influences gene transcription while simultaneously shielding DNA from cleavage by bacterial restriction-change systems, which attack foreign DNA. In both bacteria and eukaryotes, cytosine methylation in post-replicative nuclear DNA is a conserved kind of epigenetic alteration that suppresses transposon activity and controls gene expression (Zhang et al., 2020; Wen et al., 2023). Epigenetics describes processes that occur independently of modifications in the underlying DNA sequence and alter gene expression in response to stimuli. Among these mechanisms, which have been clarified, are DNA methylation, chromatin remodeling proteins and histone modifications, and noncoding RNAs (ncRNAs) that silence DNA (Gómez-Díaz et al., 2012; Liang et al., 2020; Ramos-Cruz et al., 2021).

Histone changes and DNA methylation are significant epigenetic processes that control the growth of plant diseases, much like in plants. There is increasing evidence that infections also employ epigenetic pathways to increase virulence during infection. It has been suggested that epigenetic changes to avirulence effector genes in plant pathogens act as switches to prevent host identification (Gómez-Díaz et al., 2012; Jeon et al., 2015). Histone deacetylation appears to be involved in controlling the biosynthesis of secondary metabolites, such as phytohormones, pigments and mycotoxins, which are crucial for the virulence of the phytopathogenic fungus *Fusarium fujikuroi*, the cause of the bakanae disease in rice (Studt et al., 2013; Vílchez et al., 2024).

Plants must adapt to local environmental circumstances due to their sessile nature, which means that plasticity is crucial to their survival. Interactions between the microbiota and epigenetics enable plants to quickly adapt to their surroundings and so enhance their fitness. Epigenetic variation is frequently used to describe phenotypic alterations attributed to mutualists and their transfer to offspring. However, rather than being epigenetics-induced changes based only on epigenetic heritable pathways, this type of plasticity is more akin to an interspecies induction of changes mediated by epigenetics. To completely comprehend the consequences for evolutionary biology, more research into the processes behind this occurrence is necessary (Chiquito-Contreras et al., 2024; Regmi et al., 2024). In addition to bilateral interactions, it is intriguing to consider if the host epigenetic configuration affects or responds to the broader state of plant–microbe associations. High-throughput sequencing has made it possible to evaluate the holobiont – the host and all its related microorganisms – both quantitatively and qualitatively (Wibowo et al., 2018; Ramos-Cruz et al., 2021). Notable changes are observed in the makeup of microbiomes linked to plants when specific plant epigenetic regulatory systems are triggered or inhibited. These processes can influence the recruitment of beneficial microbes from the environment to fight environmental challenges and the timing of plant defense systems following pathogen infection. Notably, several of these systems have been shown to predict strong reactions to comparable circumstances in the future and prime responses for interactions. Furthermore, by passing on alteration patterns, these systems may be passed down to the next generations.

In a similar way, microbes use certain epigenetic regulatory systems to communicate with plants (Chen et al., 2022; Kaushal et al., 2021). Numerous studies in the scientific literature have revealed changes in host physiology, morphology and behavior brought by the pathogens. The alterations that have been demonstrated to be the consequence of the manipulative approach of the pathogens to maximize their survival and transmission are arguably the most intriguing instances of these modifications (Lefèvre et al., 2009; Gómez-Díaz et al., 2012). The RNA silencing pathway, a highly advanced mechanism based on post-transcriptional gene silencing (PTGS), protects plants against a variety of viruses by targeting transcripts from both DNA and RNA viruses. Viral transcription can result in the production of double-stranded RNAs (dsRNAs), which are then broken down into 24 nucleotide (nt) small interfering RNAs (siRNAs) by the DCL2 and DCL4 proteins, preventing their translation. The RdDM (RNA-directed DNA methylation)

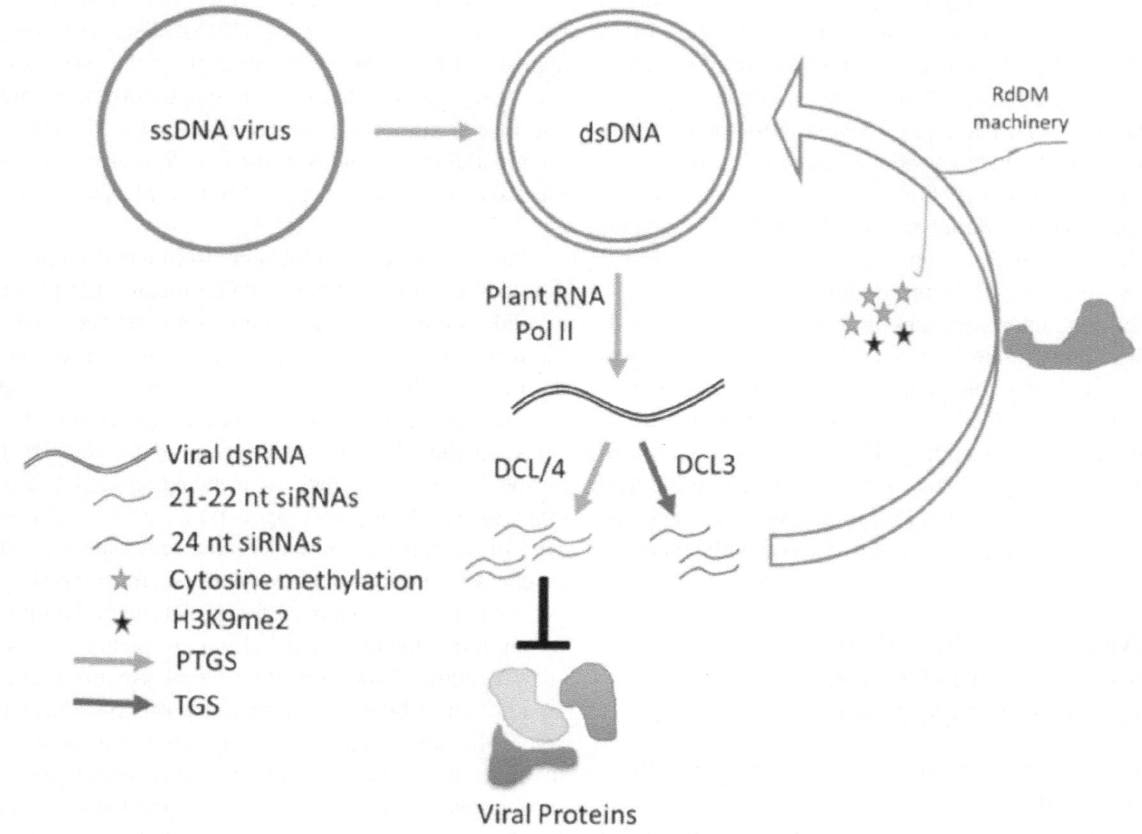

FIGURE 21.1 Plants use transcriptional gene silencing (TGS) and post-transcriptional gene silencing (PTGS) as an effective protection mechanism against viruses.

machinery may be directed to the viral genome, where repressive DNA methylation and histone methylation marks are placed, suppressing its expression. DCL3 can also cleave such dsRNAs into 21–22 nt siRNAs and charge them into AGO4 proteins as shown in Figure 21.1. Members of the Geminiviridae family use their host replication machinery to replicate their single-stranded circular DNA genome. A viral double-stranded DNA chain is created during replication, and this chain is linked to eukaryotic histones to form tiny structures that resemble chromosomes. It is interesting to note that plants also employ transcriptional gene silencing (TGS) to protect themselves from viral assaults. When a virus replicates, they use their DNA and histone methylation machinery to stop it. Upon infection, *Arabidopsis* mutants lacking DNA methylation also exhibit decreased H3K9me2 levels in the viral genome and increased vulnerability to geminiviruses (Raja et al., 2008; Wieczorek & Obrępalska-Stęplowska, 2015; Ramirez-Prado et al., 2018).

21.3 DOMAINS INVOLVED IN THE REGULATION OF PLANT IMMUNOLOGICAL TRANSCRIPTOME

Numerous microorganisms, such as bacteria, fungi, viruses and oomycetes, are naturally found in terrestrial plants. In contrast to commensal or beneficial bacteria, pathogenic microbes can cause tissue damage, disrupt physiological functions, stop development and deprive their hosts of nutrition by producing poisons, enzymes that break down cell wall and virulence proteins. Being able to differentiate infections from commensals or beneficial microorganisms and to effectively activate defenses during pathogen assault is a significant challenge for terrestrial plants (Dou & Zhou, 2012; Teixeira et al., 2019). Higher plants can detect a wide range of immunogenic signals linked to pathogen infections because they have a vast number of cell-surface and intracellular immune receptors. Conceptually similar to animal toll-like receptors, cell-surface immunological receptors are made up of receptor-like proteins (RLPs) and receptor kinases (RKs). Nucleotide-binding, leucine-rich repeat (NLR) receptors are intracellular immune receptors found in plants and mammals. In plants, NLRs specifically detect the more variable pathogen effector proteins that are delivered into the plant cell, while cell-surface immune receptors, also known as pattern recognition receptors (PRRs), oversee perceiving patterns of immunogenic molecules derived from microbes or the host (Tang et al., 2017; Kourelis & Van Der Hoorn, 2018; Van de Weyer et al., 2019).

Several types of immunogenic signals trigger plant defenses. Microbe-associated molecular patterns (MAMPs) are immunogenic microbial patterns that are comparatively conserved within taxonomic groupings. As with the bacterial flagellin epitope flg22 and peptidoglycan, plants employ lytic enzymes to help release MAMP epitopes from invasive pathogens. Furthermore, infections can harm plant tissues by releasing nicotinamide adenine dinucleotide (NAD), extracellular ATP and plant cell wall fragments using poisons and enzymes that break down plant cell wall. These compounds generated from plants have the potential to activate immune responses by acting as damage-associated molecular patterns (DAMPs). Additionally, plants actively monitor changing cell wall integrity, which probably aids in disease detection (Liu et al., 2014; Bacete et al., 2018; Buscaill et al., 2019). To encourage parasitism, pathogenic microorganisms release effector proteins into plant cells or extracellular spaces. However, when the host plant develops appropriate immune receptors, host–pathogen co-evolution may make these pathogen effectors immunogenic. PRRs sense MAMPs, DAMPs and immunogenic apoplastic effectors, while NLRs perceive intracellular effectors. As a result, immune receptors perceive various danger signals during diseases (Cui et al., 2015; Kourelis & Van Der Hoorn, 2018). To maximize ecological fitness, plants must also carefully coordinate defense with growth and development. Pattern-triggered immunity (PTI) and effector-triggered immunity (ETI) are the two tiers of the innate immune system that plants use to fight against infections. Pathogen- or microbe-associated molecular patterns (PAMPs/MAMPs) and endogenous damage-associated molecular patterns (DAMPs) activate plasma membrane (PM) resident pattern recognition receptors (PRRs), which in turn cause PTI. Localized programmed cell death is frequently associated with ETI, which is triggered by pathogen-derived effectors through intracellular nucleotide-binding domain leucine-rich repeat-containing receptors (NLRs) (Macho & Zipfel, 2014; Zebell & Dong, 2015).

The following families comprise the primary transcription factor families found in plants that control genes susceptible to infections, viz., APETALA2/ethylene-responsive element binding factors family (AP2/ERF); basic leucine zipper containing domain proteins (bZIP); myelocytomatosis-related proteins (MYC); myeloblastosis-related proteins (MYB); and, more recently, the no apical meristem (NAM), *Arabidopsis* transcription activation factor (ATAF) and cup-shaped cotyledon (CUC), also known as the Nuclear Localization Sequence (NAC) family. Every family of transcription factors has a unique binding domain, such as helix turn helix, zinc finger or bZIP. The distinctions between these domains are important characteristics that make one family apart from another. These domains bind to DNA cis-elements linked to the response to a particular environmental stress set (Van Verk et al., 2009; Pérez-Rodríguez et al., 2010). The nuclear lamina (NL), a thick fibrous protein meshwork situated underneath the inner nuclear membrane (INM), is made up of lamin and lamin-like proteins as well as other proteins connected to the nucleoskeleton. NL components are crucial for many cellular functions, such as cellular signaling, gene transcription, epigenetic regulation, cell cycle progression and cell differentiation, in addition to controlling nuclear structure and dynamics. In plants and animals, the main components of the nucleoskeleton are nuclear matrix constituent proteins (NMCPs) and lamins, respectively. With a lengthy C-terminal tail domain that contains an NLS sequence and a short N-terminal head on either side of the central coiled-coil rod domain, lamins and NMCPs have a similar domain layout (Burke & Stewart, 2013; Pawar & Kutay, 2021). Mutations in CRWN1 cause jasmonic acid (JA) to accumulate in *Arabidopsis*, upregulate the expression of genes sensitive to JA and increase resistance to the necrotrophic disease *Botrytis cinerea*. A complex function for CRWN1 in controlling plant responses to various pathogens is suggested by the observation that *crwn1* mutants also exhibit impaired PTI but unaltered baseline resistance to the hemibiotrophic bacterial pathogen *Pseudomonas syringae*. The nucleus is one of the most important downstream convergence sites for immunological signaling, despite not being situated at the forefront of host–microbe interactions. Coordination of immune activation and amplitude in plants is closely linked to NE-associated core molecular machinery and processes, which include nuclear transport receptors, structurally and functionally conserved nucleoporins, and nucleoskeletal proteins (Jarad et al., 2020; Blunt et al., 2020; Fang & Gu, 2021). The bacterial elongation factor Tu (EF-Tu) receptor EFR, which recognizes an 18-amino-acid peptide (elf18) of EF-Tu, and the bacterial flagellin receptor FLS2, which recognizes a conserved 22-amino acid peptide (flg22) of flagellin, are two examples of the well-characterized *Arabidopsis* PRRs. Leucine-rich repeat receptor-like kinases (LRR-RLKs) include FLS2 and EFR. When these PRRs detect the appropriate MAMPs, they instantly heterodimerize with the LRR-RLK family co-receptor BAK1 to start immunological signaling. Following MAMP sensing, the receptor-like cytoplasmic kinase (RLCK) BIK1 and its homolog PBL1 constitutively bind to FLS2/EFR and BAK1, get quickly phosphorylated and are released from the receptor complexes. The PM-resident NADPH oxidase, viz., RBOHD is directly phosphorylated by BIK1, and this, together with calcium signaling-mediated RBOHD regulation, causes the temporary generation of reactive oxygen species (ROS) (Böhm, et al., 2014; Macho and Zipfel, 2014; Li et al., 2016). With the help of general transcription factors, the multi-subunit enzyme RNA polymerase II (RNAPII) catalyzes the transcription of eukaryotic mRNA by binding specifically to the promoters of target genes. By interacting with gene-specific transcription factors that bind to cis-elements in promoters, the mediator complex acts as a transcription cofactor to control RNAPII activities at every stage of transcription cycles and enhances transcriptional specificity. A few other co-activators or co-repressors, including those that control histone modification and chromatin remodeling, also control the initiation and elongation of mRNA transcripts. Highly complex regulatory systems regulate the transcriptional responses to pathogen infections at several levels (Li et al., 2016).

Like cytokines in animal immunity, phytocytokines are immunogenic peptides that are actively produced by plants and are detected by PRRs or PRR-associated receptors. These peptides can alter immune responses. Despite our poor understanding of phytocytokines, protease-mediated maturation has been identified as a key mechanism controlling the apoplast peptide accumulation; for example, plant elicitor peptides (Peps) are well-known phytocytokines that activate defense via PEPR1 and PEPR2, and leucine-rich repeat RKs (LRR-RKs). Multiple calcium-dependent cysteine proteases called cytoplasmic type-II metacaspases (MCs) regulate the cleavage, maturation and release of Peps into the apoplast (Gust et al., 2017; Hander, 2019; Shen et al., 2019).

21.4 INTERACTION BETWEEN IMMUNE SIGNALING AND EPIGENETIC PATHWAYS

Most dangerous insects and microorganisms can be fended off by plants. Many plants, for example, respond to attacks by releasing substances that are harmful to the pathogens. The innate immune system of plants regulates these inducible defenses. When pathogenic microorganisms attack, "pathogen-associated molecular patterns" (PAMPs; also known as "microbe-associated molecular patterns" or MAMPs) usually set off innate immune responses. The plant is shielded from most potentially dangerous microorganisms by the ensuing "PAMP-triggered immunity" (PTI). A small percentage of virulent plant infections, however, have developed defense-suppressing effector molecules that deregulate the signaling pathways governing plant innate immunity to suppress PTI. Plants have developed the capacity to identify virulent diseases early in an attack and mount a suitable defense response as a co-evolutionary response. "Effector-triggered immunity" (ETI), a well-known illustration of this type of co-evolved plant defense, allows the plant to identify the presence or activity of certain pathogen effectors. ETI can make some virulent infections avirulent, but it requires *resistance* (*R*) genes in the host plant. Avirulent strains are always under selection pressure to evolve novel effectors that are either unknown or capable of suppressing ETI, even if ETI offers complete protection against certain pathovars of biotrophic diseases. As a result, single *R* genes may not usually provide long-term disease protection (Jones & Dangl, 2006; Boyd et al., 2013; Pastor et al., 2013). After recovering from biotic stress, plants may develop higher levels of resistance. Priming, which gives the plant an improved defensive capability that mediates a quicker and/or stronger immunological response upon subsequent assaults by pests and diseases, is usually the basis for this induced resistance (IR). In addition to developing during recovery from biotic stress, non-pathogenic microorganisms or chemical stimuli can also cause priming. Chemical priming stimuli can include endogenous stress signaling substances like β-aminobutyric acid (BABA), salicylic acid (SA), or jasmonic acid (JA), or microbe-derived MAMPs like chitin. Furthermore, xenobiotic substances like (R)-β-homoserine (RBH) and benzothiadiazole (BTH) can partially replicate the effects of biological priming stimuli (Pastor et al., 2013; Mauch-Mani et al., 2017; Yassin et al., 2021). Priming is a type of phenotypic plasticity that is theoretically comparable to acquired immunity in vertebrates, even though it depends on distinct processes, whereas innate immunity is genetically encoded into the plant DNA. Compared to naïve plants that have previously been exposed to priming stimuli, primed plants are more sensitive to fending off invaders and can induce PTI-related defenses more quickly and powerfully. Although priming can last throughout the life cycle of plants and, in some situations, be passed on to subsequent generations to provide protection against the same disease to which the parent plants had been exposed, priming is reversible due to its ecological costs. Priming includes a geographical component in addition to these temporal alterations in the plant immune system; long-distance (systemic) defensive signals frequently cause it to grow in plant portions that are distant from the original assault locations. In certain instances, volatile organic chemicals (VOCs) can even spread priming to the neighboring plants (Luna Diez et al., 2012; Wilkinson et al., 2019; López Sánchez et al., 2021). In response to invasive infections, hormone signal transduction pathways may interact antagonistically or cooperatively. The two main hormonal routes involved in plant defense are salicylic acid (SA) and JA. Plant defense systems are adaptable and may be adapted to the specific traits of the pathogens. While the JA/ET pathway aids in defense against infections by herbivores or necrotrophic diseases, the SA pathway is mainly triggered in response to biotrophic pathogens. To conserve energy and resources, just one of these routes is usually triggered. Other hormones play a supportive role in plant defense, even though JA, ET and SA are the main hormones involved (Koornneef & Pieterse, 2008; Tsuda et al., 2008). ABA, auxins, gibberellins, cytokinins and brassinosteroids are some of these hormones. While JA and ET are essential defense hormones in response to herbivores and necrotrophic pathogens, SA plays a crucial role in the battle against biotrophic pathogens. Studies in mutant plants have indicated that PRRs trigger complicated defense signaling cascades. Pathogen-associated molecular patterns (PAMPs) like fungal chitin and bacterial flagellin are recognized by transmembrane pattern recognition receptors (PRRs) in pattern-triggered immunity (PTI), which triggers PTI on the cell surface. PRRs start intricate defense signaling cascades when they are activated. Innate immune receptors, which are expressed in all cells and identify invasion signals to cause PTI or ETI, are essential to plant immunity (Pieterse et al., 2009; Zhai et al., 2019; Movahedi et al., 2022). Innate immune receptors, which are expressed in all cells and identify invasion signals to cause PTI or ETI, are essential to plant immunity. Plants employ receptor barriers as an intrinsic defense mechanism to stop pathogen invasion. Innate immune receptors, which are expressed in all cells and detect invasion signals, are necessary for microbial or modified host molecules to produce PTI or ETI. Plants employ receptor barriers as an intrinsic defense mechanism to stop pathogen invasion. PRRs in the plasma membrane identify microbial or modified host molecules, which triggers PTI. PRRs recognize and bind to danger-associated molecular patterns, microbe-associated molecular patterns (MAMPs) and damaged endogenous molecules (Wan et al., 2019; Zhai et al., 2019; Albert et al., 2020).

21.5 SIGNIFICANCE OF NONCODING RNAS IN THE REGULATION OF EPIGENOMICS

Genetic information and ribonucleic acids (RNAs) have a convoluted interaction. There are at least three different ways that RNA is related to genetic information, as demonstrated by the discovery of protein biosynthesis processes. It can transmit genetic information directly as messenger RNA (mRNA). Like ribosomal RNA (rRNA), it can carry out structural and enzymatic functions. Alternatively, it can function as transfer RNA (tRNA) at the interface between genetic information and protein sequence (Rheinberger, 2004; Wierzbicki et al., 2021). The term "noncoding RNA" (ncRNA), which is nonspecific and negatively inflected, is used to refer to all other RNA types, even though mRNA was not the first RNA to be found. As the ncRNA library grew, ncRNAs were viewed as the next frontier for discovery rather than just molecular fossils. Many crucial processes in this contemporary RNA environment were known to be ncRNA-driven (Wierzbicki et al., 2021; Numan et al., 2024). According to recent research, ncRNAs can be produced from DNA sequences found in intronic, intergenic or protein-coding genes. In plants, ncRNAs may be subdivided into two main groups based on their regulatory functions. The composition is largely consistent since housekeeping ncRNAs are essential for basic biological functions in life. Because of their differences in size, shape and accumulation patterns, regulatory ncRNAs exhibit spatially and temporally unique expressions (Axtell, 2013; Bhat et al., 2020; Hu et al., 2024). While the production of siRNA and tsRNA in plants is poorly understood, that of several regulatory ncRNAs has been well documented. While the synthesis of siRNA and tsRNA in plants is still poorly understood, that of several regulatory ncRNAs has been well documented to date. Additional research has demonstrated that ncRNAs contribute to the preservation of homeostasis in plants through interactions with other biomolecules and microbes, which are crucial for plant growth, development, differentiation and reproduction (Achkar et al., 2016; Wang et al., 2017; Chao et al., 2022).

Large endogenous ncRNAs with a range of regulatory potentials have now been found in a number of plant species. Regulatory ncRNAs can be further divided into tiny RNAs (18–30 nt), medium-sized ncRNAs and large RNAs based on their average size. Large endogenous ncRNAs with a range of regulatory potentials have now been found in several plant species. Regulatory ncRNAs can be further classified as short RNAs (18–30 nt), medium-sized ncRNAs (31–200 nt) and lncRNAs (>200 nt) based on their average size. By controlling the expression of target genes, the regulatory small RNAs (sRNAs), specifically siRNA and miRNA, play crucial roles with regard to how the organism responds to environmental changes or stress (Liu et al., 2015; D'Ario et al., 2017; Liu & Zhong, 2024). Through ncRNA-associated interactions with DNA, RNA, protein and microorganisms, respectively, recent research has shown that ncRNAs contribute to the maintenance of plant homeostasis and have significant effects on plant growth, development, differentiation and reproduction. These ncRNAs usually impact biological processes and change the destiny of cells by interacting with genes or gene products (such as proteins and different RNAs) in the cytoplasm and nucleus. Our knowledge of the terrain of the dynamic intermolecular networks within plant cells will thus be advanced by more discussions of the interactions between ncRNAs and other biological macromolecules (Fernandes et al., 2019; Li & Liu, 2019; Shi et al., 2020). When pathogenic microbes attack plants, they use plasma membrane-associated pattern recognition receptors (PRRs) to identify pathogen-associated chemicals and initiate immunotherapy. Certain microbes cause illnesses by secreting small ncRNAs as effector molecules, which block the signal transduction of plant PTIs (Jones et al., 2016). Furthermore, plants primarily use extracellular vesicles to deliver short noncoding RNAs (ncRNAs) into pathogens in order to inhibit genes linked to virulence. In response to infection by harmful microbes, plants have developed two mechanisms for controlling small ncRNAs in addition to the innate immune system. The first is that endogenous sRNAs from plants have a role in immune response control; for instance, *Arabidopsis* miR863-3p successively silences both positive and negative plant immunity regulators to fine-tune plant immune responses during infection (Katiyar-Agarwal et al., 2006; Jones et al., 2016; Niu DongDong et al., 2016). Through the formation of different ribonucleoprotein (RNP) complexes with proteins, ncRNAs accomplish and control a variety of activities; for instance, snRNPs, which are made up of proteins and snRNAs, can control both alternative and canonical splicing. Additionally, nucleolar RNP particles post-transcriptionally modify many nucleotides in pre-rRNA, pre-snRNA and pre-tRNA. By interacting with AGO family proteins during the RNA interference pathway, small noncoding RNAs (ncRNAs) like siRNA and miRNA can also affect regulation of gene expression (Lafontaine, 2015; Miyoshi et al., 2016; Hegge et al., 2019). Congener and isotypic ncRNA interactions can affect translation and epigenetic changes, among other biological processes. miRNA response elements (MRE), including circRNAs, lncRNAs and eRNAs, function as competing endogenous RNAs (ceRNAs) with profound effects on post-transcriptional gene regulation in a range of physiological and pathological processes (Salmena et al., 2011; Zhang et al., 2017). RNA interference (RNAi) is the collective term for small RNAs, which include microRNAs (miRNAs), small interfering RNAs (siRNAs) and piwi-interacting RNAs (piRNAs). These noncoding RNAs have a length of about 20–30 nucleotides and play crucial roles in controlling biological processes like development, reproduction and stress responses. DICER (known for its role as a riboendonuclease in the small RNA pathway) or DICER-like (DCL) proteins produce small RNAs, which are then loaded onto RNAi effector proteins called Argonautes (AGOs) to control target mRNA expression by either transcriptional or translational suppression. RNA polymerase II first transcribes plant miRNA precursors, which have poorly base-paired hairpin loop structures. DCL endonuclease subsequently cleaves the precursors to create miRNA/miRNA double-stranded RNA. An active RNA-induced silencing complex (RISC) is formed when one of the two strands of the double strand – a guide strand (mature miRNA) and a passenger strand (miRNA) – bind to AGOs. SiRNAs are produced from precisely paired double-stranded RNA (dsRNA) precursors, as opposed to miRNAs. These dsRNA precursors are produced by cellular RNA-dependent RNA polymerase (RDR) or antisense transcription (Katiyar et al., 2010; Rogers & Chen, 2013). Several RNA-binding proteins have been reported to be involved in the plant immunity, some of which have been mentioned in Table 21.1.

TABLE 21.1
RNA-binding proteins involved in plant immunity

RNA-binding proteins	Properties	Mode of action	Plant origin	References
tcI14	Have SR-rich repeat region RNP1 and RNP2	Performs alternative splicing	Tobacco	Petitot et al. (1997)
PvPRP1	Have role in mRNA destabilization	RNA-binding activity inversely correlates with *PvPRP1* mRNA levels	*Arabidopsis*	Zhang and Mehdy (1994)
PRP-BP	Binds with U-rich regions	The *PvPRP1* mRNA-binding protein play role in elicitor-induced destabilization of *PvPRP7* mRNA	Bean	Zhang and Mehdy (1994)
RPP13		Hypersensitive response and ETI, as well as the overlap between genes activated by Nuclear Hormone Receptors (NHR) and one for host resistance	Wheat	Rezaei et al. (2022)
GaPR10	Having K-A-X-E-X-Y-L domain and a P-loop	Performs ribonuclease activity	Cotton	Zhou et al. (2002)
GRP7	Possess RNA Recognition Motif (RRM), glycine-rich motif, RNP-binding domain	Pathogenic defense, mRNA stability and control	*Arabidopsis*	Staiger et al. (2003)
MOS2	Possess G-patch and KOW motif	Involved in RNA binding	*Arabidopsis*	Zhang et al. (2005)
DCL2	Possess double-stranded RNA-binding motif, double-stranded RNA-binding domain, helicase super family c-terminal domain	Production of ta-siRNAs involved in RNA interference, defense response to a virus and maintenance of DNA methylation	*Arabidopsis*	Ellendorff et al. (2009)
DCL4	Having ribonuclease III C-terminal domain, PAZ domain, double-stranded RNA-binding domain	DNA methylation maintenance, RNA interference-related lsiRNA, siRNA and ta-siRNA synthesis, and virus-induced gene silencing	*Arabidopsis*	Ellendorff et al. (2009)
MOS11	Having homologous to protein CIP29 of human RNA binding	Involved in mRNA export	*Arabidopsis*	Germain et al. (2010)
eIF4E	Having eukaryotic initiation factor 4E homologous domain	Translational initiation, components of five cap-binding complex (CBC), response to virus	*Arabidopsis*	Piron et al. (2010)
ALFIN-LIKE 6 (AL6)		ChEP-P is ideally suited to derive comprehensive understanding of plant chromatin-related activities	*Arabidopsis*	Vélez-Bermúdez and Schmidt (2021)
CPR5		Involved in RNA polyadenylation factor cleavage and polyadenylation specificity factor co-ordinately functions downstream of CPR5	*Arabidopsis*	Peng et al. (2022)

21.6 BENEFICIAL MICROBE-INDUCED EPIGENETIC MODIFICATIONS FOR DISEASE RESISTANCE

Any organism that lives in this interconnected natural ecosystem must protect itself or face extinction. Defense is a survival strategy. Every living organism has at least one defense mechanism, and plants are no exception. Plants must fight against and counteract the acts of their enemies (pathogens) in a variety of ways, including via production of secondary metabolites called phytoalexins in order to maintain their health, vigor and survival (Khare et al., 2017). For plants to survive, they must be able to protect themselves from invasions or infections by defending themselves both locally and systemically. These defense signals, which result in several defense reactions against the invader and other unrelated pathogenic species, are set off by pathogens at the infection site (Pieterse et al., 2014). Since viruses exploit the host cell machinery to infect others, they are definitely hard to regulate. Globally, viral infections constitute a significant barrier to sustainable agricultural production. The knowledge of viral pathogenicity and host responses in antiviral resistance has been enhanced by recently established methods in plant virology on RNA silencing, such as virus-induced gene silencing, large-scale genomic analysis and epigenetic analysis. *R* gene-mediated resistance, recessive resistance and antiviral RNA silencing are examples of plant antiviral activity (Maule et al., 2007; Akhter et al., 2019; Jones, 2021).

Induced systemic resistance (ISR) and systemic acquired resistance (SAR) are two types of plant systemic resistance that are caused by pathogenic and non-pathogenic microorganisms, respectively. Beneficial microbial colonization causes the plant host to enter a physiological condition known as "priming." As a typical characteristic of systemic resistance established by beneficial bacteria, plants exhibit greater and quicker defense responses against the subsequent invasion of pathogens after "priming" is activated (Yu et al., 2022). By identifying the common microbial molecules they produce, such as flagellin, lipopolysaccharides (LPS), exopolysaccharides and chitin oligosaccharides, plants can detect the beneficial microorganisms. The receptor proteins initiate a signal cascade that includes oxidative burst, Ca^{2+} influx, MAPK activation and hormone signaling activation by binding with ligands, enlisting co-receptors and forming complexes to phosphorylate the downstream substrates (Zipfel et al., 2017; Bazin et al., 2020). Numerous eubacteria have high conservation of the N-terminal portion of flagellin, which includes the 22-amino acid epitope flg22. Beneficial microorganisms like *Bacillus subtilis* and *Burkholderia phytofirmans* have flagellin that their host plants can recognize. The first receptor known to recognize flagellin from PGPR is FLAGELLIN-SENSING2 (FLS2). FLS2 and its co-receptors, BRI1-associated kinase (BAK1) and BAK1-LIKE1 (BKK1) heterodimerize in response to flg22 perception. This phosphorylates their interacting receptor-like cytoplasmic kinase *Botrytis*-induced kinase1 (BIK1), which initiates the PTI signaling (Zipfel, 2014; Deng et al., 2019; Yu et al., 2022). *Arabidopsis thaliana* BAK1 operated as a positive regulator in signaling, as evident from *bak1* mutants that displayed aberrant PTI responses but normal flagellin binding. When flagellin is detected, BIK1 gets phosphorylated. This causes the FLS2/BAK1 complex to become trans-phosphorylated, which transmits flagellin signaling and initiates intracellular signaling cascades. The *bik1* mutant exhibits the same deficiencies as the *bak1* mutant in flagellin-mediated responses to the invasion of non-pathogenic microorganisms, suggesting that BIK1 plays a crucial role in MAMP signal transduction and systemic resistance. Additionally, rhizobia and Arbuscular Mycorrhizal Fungi (AMF) use signals from chitin-derived oligosaccharides to establish symbiosis with the host. A shared symbiotic route that regulates both arbuscular mycorrhizal symbioses and the rhizobia-legume to create mycorrhization and nodulation is activated by nod factors, such as acylated lipochitooligosaccharides that are provided by rhizobia and recognized by LysM receptor-like kinases (Gough & Cullimore, 2011; Zipfel et al., 2017).

Through the JA or ET signaling pathway, several important transcription factors are engaged in the network that regulates ISR. Responses to interactions between microorganisms and plants are linked to WRKY transcription factors. When *Arabidopsis thaliana* plants interact with the beneficial fungus *T. atroviride*, there is a time-dependent differential expression of *WRKY* genes. It was obvious that positive regulators in the JA-mediated pathway, including AtWRKY8 and AtWRKY33, would be expressed more than the *WRKY* genes controlled by the SA pathway (Sáenz-Mata et al., 2014). Through the JA and SA signaling pathways, respectively, WRKY11 and WRKY70 regulated the ISR triggered by *Bacillus cereus* strain AR156 in *Arabidopsis*. The transcription factors known as MYB family proteins control how plants develop and react to biotic and abiotic stress. By functioning upstream of ethylene in the signaling pathway, MYB72 was necessary for the first signaling stages of beneficial microbe-mediated ISR and was triggered following colonization of *Pseudomonas fluorescens* WCS417r. Beneficial microbe-triggered ISR required the basic helix-loop-helix (bHLH) transcription factor MYC2, whereas pathogens targeted its activity by suppressing host immunity through effectors (Dubos et al., 2010; Jiang et al., 2016). A transcription factor that controls the expression of pathogen response genes and stops the course of disease development is encoded by ethylene response factor 1 (ERF1). Both JA and ET have the ability to quickly and cooperatively activate ERF1 expression. WRKY11 and WRKY70 were implicated in the control of *B. cereus* strain AR156-triggered ISR in *Arabidopsis*, via the JA and SA signaling pathways, respectively. Short RNAs are essential for inherent defenses against bacteria, fungi and viruses. The first miRNA to be linked to PTI is miR393, which is triggered by flg22 to silence its receptors and hence suppress auxin signaling. There is growing evidence that plant microRNAs cause ETI by targeting conserved regions of genes that encode NB-LRR. Pretreatment with *B. cereus* AR156 inhibits the miR825/miR825 pair, which targets toll-interleukin-like receptor NB-LRR (TNLs)-type *resistance* (*R*) genes, and initiates ISR signaling (Niu et al., 2016; Nie et al., 2019). Moreover, through the suppression of miR472 and the activation of coiled-coil NB-LRR-mediated basal immunity in *Arabidopsis*, *B. cereus* AR156 initiates ISR against *P. syringae* pv. tomato DC3000. By suppressing *Arabidopsis*-specific miR846 expression, *Bacillus amyloliquefaciens* FZB42 inoculation creates systemic resistance through a signaling route that is reliant on JA. In response to FZB42 and the loss-of-function mutant FZB42 △sfp △alss (deficient in inducing ISR), maize was shown to exhibit differential expression of 217 unique miRNAs and 146 recognized miRNAs. Among these, four miRNAs that were particularly downregulated in FZB42 therapy (zma-miR169a-5p, zma-miR169c-5p, zma-miR169i-5p and zma-miR395b-5p) were found as potential ISR-associated miRNAs (Xie et al., 2018; Jiang et al., 2020).

21.7 CONCLUSION AND FUTURE PERSPECTIVES

We are beginning to understand the complex and intimate relationship between the plant microbiome and effects on plant productivity and fitness. Plants do not have specialized mobile immune cells as mammals do. Every plant cell is thought to be able to initiate a successful immune response. Plants, on the other hand, have many cell and organ types that may experience varying degrees of infection severity and frequency. Plant gene expression is often profoundly and dynamically reprogrammed during pathogen invasion, which is essential for initiating strong and efficient host defense responses. Our knowledge of the genome-wide transcriptional landscape driving plant defense responses has significantly increased because of recent research, which has also shown the involvement of some pathway-specific transcription factors in controlling the expression of plant immune genes. Furthermore, the precise roles of general transcriptional machinery in controlling primary immune gene expression are becoming clearer, as are the routes from pathogen sensing to gene activation (Tsuda and Somssich, 2015; Li et al., 2016; Ye et al., 2024). It is now evident from extensive and productive research conducted over the past few decades that chromatin dynamics and epigenetic modifications are essential components of signaling pathways that connect transcription factors, downstream gene regulation and transcription machinery for signaling output. Plants can detect microbial diseases due to specific patterns of identification, which sets off a defense mechanism. The importance of epigenetic changes in plant immunity is supported by recent studies, especially when it comes to the establishment of immunogenic memory. Furthermore, there are several emerging technologies that might greatly advance our comprehension of these systems in the future. Future studies will be necessary to identify the precise epigenetic changes influencing plant immunity that are passed on to progeny and how the selection pressure of coexisting pathogens affects the stability of these changes.

REFERENCES

Achkar, N. P., Cambiagno, D. A., & Manavella, P. A. (2016). miRNA biogenesis: A dynamic pathway. *Trends in Plant Science*, 21(12), 1034–1044.

Aerts, N., Chhillar, H., Ding, P., & Van Wees, S. C. (2022). Transcriptional regulation of plant innate immunity. *Essays in Biochemistry*, 66(5), 607–620.

Ahmad, N., Hussain, H., Naeem, M., ur Rahman, S., Khan, K. A., Iqbal, B., & Umar, A. W. (2024). Metabolites-induced co-evolutionary warfare between plants, viruses, and their associated vectors: So close yet so far away. *Plant Science*, 346, 112165.

Akhter, M. S., Akanda, A. M., Kobayashi, K., Jain, R. K., & Mandal, B. (2019). Plant virus diseases and their management in Bangladesh. *Crop Protection*, 118, 57–65.

Albert, I., Hua, C., Nürnberger, T., Pruitt, R. N., & Zhang, L. (2020). Surface sensor systems in plant immunity. *Plant Physiology*, 182(4), 1582–1596.

Axtell, M. J. (2013). Classification and comparison of small RNAs from plants. *Annual Review of Plant Biology*, 64(1), 137–159.

Bacete, L., Melida, H., Miedes, E., & Molina, A. (2018). Plant cell wall-mediated immunity: Cell wall changes trigger disease resistance responses. *The Plant Journal*, 93(4), 614–636.

Bazin, J., Mariappan, K., Jiang, Y., Blein, T., Voelz, R., Crespi, M., & Hirt, H. (2020). Role of MPK4 in pathogen-associated molecular pattern-triggered alternative splicing in *Arabidopsis*. *PLoS Pathogens*, 16(4), e1008401.

Bhat, S. S., Bielewicz, D., Gulanicz, T., Bodi, Z., Yu, X., Anderson, S. J., ... & Szweykowska-Kulinska, Z. (2020). mRNA adenosine methylase (MTA) deposits m6A on pri-miRNAs to modulate miRNA biogenesis in *Arabidopsis thaliana*. *Proceedings of the National Academy of Sciences*, 117(35), 21785–21795.

Blunt, E. L., Shandler, J. A., Hughes, E. J., Sussman, H., Christopherson, R. C., & Richards, E. J. (2020). Coordination of NMCP1-and NMCP2-class proteins within the plant nucleoskeleton. *Molecular Biology of the Cell*, 31(26), 2948–2958.

Böhm, H., Albert, I., Fan, L., Reinhard, A., & Nürnberger, T. (2014). Immune receptor complexes at the plant cell surface. *Current Opinion in Plant Biology*, 20, 47–54.

Boyd, L. A., Ridout, C., O'Sullivan, D. M., Leach, J. E., & Leung, H. (2013). Plant–pathogen interactions: Disease resistance in modern agriculture. *Trends in Genetics*, 29(4), 233–240.

Burke, B., & Stewart, C. L. (2013). The nuclear lamins: Flexibility in function. *Nature Reviews Molecular Cell Biology*, 14(1), 13–24.

Buscaill, P., Chandrasekar, B., Sanguankiattichai, N., Kourelis, J., Kaschani, F., Thomas, E. L., ... & Van Der Hoorn, R. A. (2019). Glycosidase and glycan polymorphism control hydrolytic release of immunogenic flagellin peptides. *Science*, 364(6436), eaav0748.

Chao, H., Hu, Y., Zhao, L., Xin, S., Ni, Q., Zhang, P., & Chen, M. (2022). Biogenesis, functions, interactions, and resources of non-coding RNAs in plants. *International Journal of Molecular Sciences*, 23(7), 3695.

Chen, C., Wang, M., Zhu, J., Tang, Y., Zhang, H., Zhao, Q., ... & Shen, Z. (2022). Long-term effect of epigenetic modification in plant–microbe interactions: Modification of DNA methylation induced by plant growth-promoting bacteria mediates promotion process. *Microbiome*, 10(1), 36.

Chiquito-Contreras, C. J., Meza-Menchaca, T., Guzmán-López, O., Vásquez, E. C., & Ricaño-Rodríguez, J. (2024). Molecular insights into plant–microbe interactions: A comprehensive review of key mechanisms. *Frontiers in Bioscience-Elite*, 16(1), 9.

Cui, H., Tsuda, K., & Parker, J. E. (2015). Effector-triggered immunity: From pathogen perception to robust defense. *Annual Review of Plant Biology*, 66(1), 487–511.

D'Ario, M., Griffiths-Jones, S., & Kim, M. (2017). Small RNAs: Big impact on plant development. *Trends in Plant Science*, 22(12), 1056–1068.

Deng, Y., Chen, H., Li, C., Xu, J., Qi, Q., Xu, Y., ... & Sun, M. (2019). Endophyte *Bacillus subtilis* evade plant defense by producing lantibiotic subtilomycin to mask self-produced flagellin. *Communications Biology*, 2(1), 368.

Dou, D., & Zhou, J. M. (2012). Phytopathogen effectors subverting host immunity: Different foes, similar battleground. *Cell Host & Microbe*, 12(4), 484–495.

Dubos, C., Stracke, R., Grotewold, E., Weisshaar, B., Martin, C., & Lepiniec, L. (2010). MYB transcription factors in *Arabidopsis. Trends in Plant Science*, 15(10), 573–581.

Ellendorff, U., Fradin, E. F., De Jonge, R., & Thomma, B. P. (2009). RNA silencing is required for *Arabidopsis* defence against *Verticillium* wilt disease. *Journal of Experimental Botany*, 60(2), 591–602.

Fang, Y., & Gu, Y. (2021). Regulation of plant immunity by nuclear membrane-associated mechanisms. *Frontiers in Immunology*, 12, 771065.

Fernandes, J. C., Acuña, S. M., Aoki, J. I., Floeter-Winter, L. M., & Muxel, S. M. (2019). Long non-coding RNAs in the regulation of gene expression: Physiology and disease. *Non-coding RNA*, 5(1), 17.

Germain, H., Qu, N., Cheng, Y. T., Lee, E., Huang, Y., Dong, O. X., ... & Li, X. (2010). MOS11: A new component in the mRNA export pathway. *PLoS Genetics*, 6(12), e1001250.

Gómez-Díaz, E., Jorda, M., Peinado, M. A., & Rivero, A. (2012). Epigenetics of host–pathogen interactions: The road ahead and the road behind. *PLoS Pathogens*, 8(11), e1003007.

Gough, C., & Cullimore, J. (2011). Lipo-chitooligosaccharide signaling in endosymbiotic plant-microbe interactions. *Molecular Plant-Microbe Interactions*, 24(8), 867–878.

Gust, A. A., Pruitt, R., & Nürnberger, T. (2017). Sensing danger: Key to activating plant immunity. *Trends in Plant Science*, 22(9), 779–791.

Hander, T., Fernández-Fernández, A. D., Kumpf, R. P., Willems, P., Schatowitz, H., Rombaut, D., et al. (2019). Damage on plants activates Ca^{2+}-dependent metacaspases for release of immunomodulatory peptides. *Science*, 22(363), 6433.

Hegge, J. W., Swarts, D. C., Chandradoss, S. D., Cui, T. J., Kneppers, J., Jinek, M., ... & Van Der Oost, J. (2019). DNA-guided DNA cleavage at moderate temperatures by *Clostridium butyricum* Argonaute. *Nucleic Acids Research*, 47(11), 5809–5821.

Hu, J., Xu, T., & Kang, H. (2024). Crosstalk between RNA m6A modification and epigenetic factors in plant gene regulation. *Plant Communications*, 5(10).

Jarad, M., Mariappan, K., Almeida-Trapp, M., Mette, M. F., Mithöfer, A., Rayapuram, N., & Hirt, H. (2020). The lamin-like LITTLE NUCLEI 1 (LINC1) regulates pattern-triggered immunity and jasmonic acid signaling. *Frontiers in Plant Science*, 10, 1639.

Jeon, J., Choi, J., Lee, G. W., Park, S. Y., Huh, A., Dean, R. A., & Lee, Y. H. (2015). Genome-wide profiling of DNA methylation provides insights into epigenetic regulation of fungal development in a plant pathogenic fungus, *Magnaporthe oryzae. Scientific Reports*, 5(1), 8567.

Jiang, C. H., Huang, Z. Y., Xie, P., Gu, C., Li, K., Wang, D. C., ... & Guo, J. H. (2016). Transcription factors WRKY70 and WRKY11 served as regulators in rhizobacterium *Bacillus cereus* AR156-induced systemic resistance to *Pseudomonas syringae* pv. tomato DC3000 in *Arabidopsis. Journal of Experimental Botany*, 67(1), 157–174.

Jiang, C., Fan, Z., Li, Z., Niu, D., Li, Y., Zheng, M., ... & Guo, J. (2020). Bacillus cereus AR156 triggers induced systemic resistance against *Pseudomonas syringae* pv. tomato DC3000 by suppressing miR472 and activating CNLs-mediated basal immunity in *Arabidopsis. Molecular Plant Pathology*, 21(6), 854–870.

Jones, J. D., & Dangl, J. L. (2006). The plant immune system. *Nature*, 444(7117), 323–329.

Jones, J. D., Vance, R. E., & Dangl, J. L. (2016). Intracellular innate immune surveillance devices in plants and animals. *Science*, 354(6316), aaf6395.

Jones, R. A. (2021). Global plant virus disease pandemics and epidemics. *Plants*, 10(2), 233.

Katiyar-Agarwal, S., & Jin, H. (2010). Role of small RNAs in host-microbe interactions. *Annual Review of Phytopathology*, 48(1), 225–246.

Katiyar-Agarwal, S., Morgan, R., Dahlbeck, D., Borsani, O., Villegas Jr, A., Zhu, J. K., ... & Jin, H. (2006). A pathogen-inducible endogenous siRNA in plant immunity. *Proceedings of the National Academy of Sciences*, 103(47), 18002–18007.

Kaushal, R., Peng, L., Singh, S. K., Zhang, M., Zhang, X., Vílchez, J. I., ... & Zhang, H. (2021). Dicer-like proteins influence *Arabidopsis* root microbiota independent of RNA-directed DNA methylation. *Microbiome*, 9, 1–20.

Khare, D., Choi, H., Huh, S. U., Bassin, B., Kim, J., Martinoia, E., ... & Lee, Y. (2017). *Arabidopsis* ABCG34 contributes to defense against necrotrophic pathogens by mediating the secretion of camalexin. *Proceedings of the National Academy of Sciences*, 114(28), E5712–E5720.

Koornneef, A., & Pieterse, C. M. (2008). Cross talk in defense signaling. *Plant Physiology*, 146(3), 839–844.

Kourelis, J., & Van Der Hoorn, R. A. (2018). Defended to the nines: 25 years of resistance gene cloning identifies nine mechanisms for R protein function. *The Plant Cell*, 30(2), 285–299.

Lafontaine, D. L. (2015). Noncoding RNAs in eukaryotic ribosome biogenesis and function. *Nature Structural & Molecular Biology*, 22(1), 11–19.

Lefèvre, T., Lebarbenchon, C., Gauthier-Clerc, M., Missé, D., Poulin, R., & Thomas, F. (2009). The ecological significance of manipulative parasites. *Trends in Ecology & Evolution*, 24(1), 41–48.

Li, B., Meng, X., Shan, L., & He, P. (2016). Transcriptional regulation of pattern-triggered immunity in plants. *Cell Host & Microbe*, 19(5), 641–650.

Li, J., & Liu, C. (2019). Coding or noncoding, the converging concepts of RNAs. *Frontiers in Genetics*, 10, 496.

Liang, Z., Riaz, A., Chachar, S., Ding, Y., Du, H., & Gu, X. (2020). Epigenetic modifications of mRNA and DNA in plants. *Molecular Plant*, 13(1), 14–30.

Liu, J., & Zhong, X. (2024). Epiallelic variation of non-coding RNA genes and their phenotypic consequences. *Nature Communications*, 15(1), 1375.

Liu, X., Grabherr, H. M., Willmann, R., Kolb, D., Brunner, F., Bertsche, U., ... & Gust, A. A. (2014). Host-induced bacterial cell wall decomposition mediates pattern-triggered immunity in *Arabidopsis. Elife*, 3, e01990.

Liu, X., Hao, L., Li, D., Zhu, L., & Hu, S. (2015). Long non-coding RNAs and their biological roles in plants. *Genomics, Proteomics and Bioinformatics*, 13(3), 137–147.

López Sánchez, A., Pascual-Pardo, D., Furci, L., Roberts, M. R., & Ton, J. (2021). Costs and benefits of transgenerational induced resistance in *Arabidopsis. Frontiers in Plant Science*, 12, 644999.

Luna, E., Bruce, T. J., Roberts, M. R., Flors, V., & Ton, J. (2012). Next-generation systemic acquired resistance. *Plant Physiology*, 158(2), 844–853.

Macho, A. P., & Zipfel, C. (2014). Plant PRRs and the activation of innate immune signaling. *Molecular Cell*, 54(2), 263–272.

Manoharan, B., Narayanasamy, S., Joshi, J. B., Jegadeesan, S., Qi, S., Dai, Z., ... & Uthandi, S. (2023). Molecular events and defence mechanism against biotic stress induced by bio-priming of beneficial microbes. In *Microbial Biocontrol: Molecular Perspective in Plant Disease Management* (pp. 61–87). Singapore: Springer Nature Singapore.

Mauch-Mani, B., Baccelli, I., Luna, E., & Flors, V. (2017). Defense priming: An adaptive part of induced resistance. *Annual Review of Plant Biology*, 68(1), 485–512.

Maule, A. J., Caranta, C., & Boulton, M. I. (2007). Sources of natural resistance to plant viruses: Status and prospects. *Molecular Plant Pathology*, 8(2), 223–231.

Michmizos, D., & Hilioti, Z. (2019). A roadmap towards a functional paradigm for learning & memory in plants. *Journal of Plant Physiology*, 232, 209–215.

Miyoshi, T., Ito, K., Murakami, R., & Uchiumi, T. (2016). Structural basis for the recognition of guide RNA and target DNA heteroduplex by Argonaute. *Nature Communications*, 7(1), 11846.

Movahedi, A., Aghaei-Dargiri, S., Barati, B., Kadkhodaei, S., Wei, H., Sangari, S., ... & Xu, C. (2022). Plant immunity is regulated by biological, genetic, and epigenetic factors. *Agronomy*, 12(11), 2790.

Nie, P., Chen, C., Yin, Q., Jiang, C., Guo, J., Zhao, H., & Niu, D. (2019). Function of miR825 and miR825* as negative regulators in Bacillus cereus AR156-elicited systemic resistance to *Botrytis cinerea* in *Arabidopsis thaliana. International Journal of Molecular Sciences*, 20(20), 5032.

Niu, D., Lii, Y. E., Chellappan, P., Lei, L., Peralta, K., Jiang, C., Guo, J., Coaker, G., & Jin, H. (2016). miRNA863-3p sequentially targets negative immune regulator ARLPKs and positive regulator SERRATE upon bacterial infection. *Nature Communications*, 7(1), 11324.

Niu, D., Xia, J., Jiang, C., Qi, B., Ling, X., Lin, S., ... & Zhao, H. (2016). Bacillus cereus AR156 primes induced systemic resistance by suppressing miR825/825* and activating defense-related genes in *Arabidopsis. Journal of Integrative Plant Biology*, 58(4), 426–439.

Numan, M., Sun, Y., & Li, G. (2024). Exploring the emerging role of long non-coding RNAs (lncRNAs) in plant biology: Functions, mechanisms of action, and future directions. *Plant Physiology and Biochemistry*, 108797.

Pastor, V., Luna, E., Mauch-Mani, B., Ton, J., & Flors, V. (2013). Primed plants do not forget. *Environmental and Experimental Botany*, 94, 46–56.

Pawar, S., & Kutay, U. (2021). The diverse cellular functions of inner nuclear membrane proteins. *Cold Spring Harbor Perspectives in Biology*, 13(9), a040477.

Peng, S., Guo, D., Guo, Y., Zhao, H., Mei, J., Han, Y., ... & Wang, S. (2022). CONSTITUTIVE EXPRESSER OF PATHOGENESIS-RELATED GENES 5 is an RNA-binding protein controlling plant immunity via an RNA processing complex. *The Plant Cell*, 34(5), 1724–1744.

Pérez-Rodríguez, P., Riano-Pachon, D. M., Corrêa, L. G. G., Rensing, S. A., Kersten, B., & Mueller-Roeber, B. (2010). PlnTFDB: Updated content and new features of the plant transcription factor database. *Nucleic Acids Research*, 38(suppl_1), D822–D827.

Petitot, A. S., Blein, J. P., Pugin, A., & Suty, L. (1997). Cloning of two plant cDNAs encoding a β-type proteasome subunit and a transformer-2-like SR-related protein: Early induction of the corresponding genes in tobacco cells treated with cryptogein. *Plant Molecular Biology*, 35, 261–269.

Pieterse, C. M., Leon-Reyes, A., Van der Ent, S., & Van Wees, S. C. (2009). Networking by small-molecule hormones in plant immunity. *Nature Chemical Biology*, 5(5), 308–316.

Pieterse, C. M., Zamioudis, C., Berendsen, R. L., Weller, D. M., Van Wees, S. C., & Bakker, P. A. (2014). Induced systemic resistance by beneficial microbes. *Annual Review of Phytopathology*, 52(1), 347–375.

Piron, F., Nicolaï, M., Minoïa, S., Piednoir, E., Moretti, A., Salgues, A., ... & Bendahmane, A. (2010). An induced mutation in tomato eIF4E leads to immunity to two potyviruses. *PLoS One*, 5(6), e11313.

Raja, P., Sanville, B. C., Buchmann, R. C., & Bisaro, D. M. (2008). Viral genome methylation as an epigenetic defense against geminiviruses. *Journal of Virology*, 82(18), 8997–9007.

Ramirez-Prado, J. S., Piquerez, S. J., Bendahmane, A., Hirt, H., Raynaud, C., & Benhamed, M. (2018). Modify the histone to win the battle: Chromatin dynamics in plant–pathogen interactions. *Frontiers in Plant Science*, 9, 355.

Ramos-Cruz, D., Troyee, A. N., & Becker, C. (2021). Epigenetics in plant organismic interactions. *Current Opinion in Plant Biology*, 61, 102060.

Regmi, R., Sharma, A., & Qi, Y. (2024). Epigenetic regulations during plant-microbe interactions. In: Witzany, G. (eds) *Epigenetics in Biological Communication* (pp. 399–419). Cham: Springer Nature Switzerland.

Rezaei, A., Mahdian, S., Hashemi-Petroudi, S. H., Goodwin, P. H., Babaeizad, V., & Rahimian, H. (2022). Nonhost resistance EST profiling of wheat interacting with *Blumeria graminis* f. sp. hordei identifies genes for durable resistance to powdery mildew. *European Journal of Plant Pathology*, 162(4), 793–806.

Rheinberger, H. J. (2004). A history of protein biosynthesis and ribosome research. *Protein Synthesis and Ribosome Structure: Translating the Genome*, 1–51.

Rogers, K., & Chen, X. (2013). Biogenesis, turnover, and mode of action of plant microRNAs. *The Plant Cell*, 25(7), 2383–2399.

Sáenz-Mata, J., Salazar-Badillo, F. B., & Jiménez-Bremont, J. F. (2014). Transcriptional regulation of *Arabidopsis thaliana* WRKY genes under interaction with beneficial fungus *Trichoderma atroviride. Acta Physiologiae Plantarum*, 36, 1085–1093.

Salmena, L., Poliseno, L., Tay, Y., Kats, L., & Pandolfi, P. P. (2011). A ceRNA hypothesis: The Rosetta Stone of a hidden RNA language? *Cell*, 146, 353–358.

Shen, W., Liu, J., & Li, J. F. (2019). Type-II metacaspases mediate the processing of plant elicitor peptides in *Arabidopsis*. *Molecular Plant*, 12(11), 1524–1533.

Shi, T., Guo, D., Xu, H., Su, G., Chen, J., Zhao, Z., ... & Lu, W. (2020). HOTAIRM1, an enhancer lncRNA, promotes glioma proliferation by regulating long-range chromatin interactions within HOXA cluster genes. *Molecular Biology Reports*, 47, 2723–2733.

Staiger, D., Zecca, L., Kirk, D. A. W., Apel, K., & Eckstein, L. (2003). The circadian clock regulated RNA-binding protein AtGRP7 autoregulates its expression by influencing alternative splicing of its own pre-mRNA. *The Plant Journal*, 33(2), 361–371.

Studt, L., Schmidt, F. J., Jahn, L., Sieber, C. M. K., Connolly, L. R., Niehaus, E. M., ... & Tudzynski, B. (2013). Two histone deacetylases, FfHda1 and FfHda2, are important for *Fusarium fujikuroi* secondary metabolism and virulence. *Applied and Environmental Microbiology*, 79(24), 7719–7734.

Tang, D., Wang, G., & Zhou, J. M. (2017). Receptor kinases in plant-pathogen interactions: More than pattern recognition. *The Plant Cell*, 29(4), 618–637.

Teixeira, P. J. P., Colaianni, N. R., Fitzpatrick, C. R., & Dangl, J. L. (2019). Beyond pathogens: Microbiota interactions with the plant immune system. *Current Opinion in Microbiology*, 49, 7–17.

Trivedi, P., Leach, J. E., Tringe, S. G., Sa, T., & Singh, B. K. (2020). Plant–microbiome interactions: From community assembly to plant health. *Nature Reviews Microbiology*, 18(11), 607–621.

Tsuda, K., & Somssich, I. E. (2015). Transcriptional networks in plant immunity. *New Phytologist*, 206(3), 932–947.

Tsuda, K., Sato, M., Glazebrook, J., Cohen, J. D., & Katagiri, F. (2008). Interplay between MAMP-triggered and SA-mediated defense responses. *The Plant Journal*, 53(5), 763–775.

Van de Weyer, A. L., Monteiro, F., Furzer, O. J., Nishimura, M. T., Cevik, V., Witek, K., ... & Bemm, F. (2019). A species-wide inventory of NLR genes and alleles in *Arabidopsis thaliana*. *Cell*, 178(5), 1260–1272.

Van Verk, M. C., Gatz, C., & Linthorst, H. J. (2009). Transcriptional regulation of plant defense responses. *Advances in Botanical Research*, 51, 397–438.

Vélez-Bermúdez, I. C., & Schmidt, W. (2021). Chromatin enrichment for proteomics in plants (ChEP-P) implicates the histone reader ALFIN-LIKE 6 in jasmonate signalling. *BMC Genomics*, 22, 1–21.

Vílchez, J. I., Varotto, S., & Jung, H. W. (2024). Epigenetic regulation behind plant-microbe interactions. *Frontiers in Plant Science*, 15, 1385356.

Wan, W. L., Fröhlich, K., Pruitt, R. N., Nuernberger, T., & Zhang, L. (2019). Plant cell surface immune receptor complex signaling. *Current Opinion in Plant Biology*, 50, 18–28.

Wang, H. L. V., & Chekanova, J. A. (2017). Long noncoding RNAs in plants. *Long Non Coding RNA Biology*, 133–154.

Wen, H. G., Zhao, J. H., Zhang, B. S., Gao, F., Wu, X. M., Yan, Y. S., ... & Guo, H. S. (2023). Microbe-induced gene silencing boosts crop protection against soil-borne fungal pathogens. *Nature Plants*, 9(9), 1409–1418.

Wibowo, A., Becker, C., Durr, J., Price, J., Spaepen, S., Hilton, S., ... & Gutierrez-Marcos, J. (2018). Partial maintenance of organ-specific epigenetic marks during plant asexual reproduction leads to heritable phenotypic variation. *Proceedings of the National Academy of Sciences*, 115(39), E9145–E9152.

Wieczorek, P., & Obrępalska-Stęplowska, A. (2015). Suppress to survive – implication of plant viruses in PTGS. *Plant Molecular Biology Reporter*, 33, 335–346.

Wierzbicki, A. T., Blevins, T., & Swiezewski, S. (2021). Long noncoding RNAs in plants. *Annual Review of Plant Biology*, 72(1), 245–271.

Wilkinson, S. W., Magerøy, M. H., López Sánchez, A., Smith, L. M., Furci, L., Cotton, T. A., ... & Ton, J. (2019). Surviving in a hostile world: Plant strategies to resist pests and diseases. *Annual Review of Phytopathology*, 57(1), 505–529.

Xie, S., Jiang, H., Ding, T., Xu, Q., Chai, W., & Cheng, B. (2018). *Bacillus amyloliquefaciens* FZB42 represses plant miR846 to induce systemic resistance via a jasmonic acid-dependent signalling pathway. *Molecular Plant Pathology*, 19(7), 1612–1623.

Yassin, M., Ton, J., Rolfe, S. A., Valentine, T. A., Cromey, M., Holden, N., & Newton, A. C. (2021). The rise, fall and resurrection of chemical-induced resistance agents. *Pest Management Science*, 77(9), 3900–3909.

Ye, D., Zhang, S., Gao, X., Li, X., Jin, X., Shi, M., ... & Zhou, W. (2024). Mining of disease–resistance genes in *Crocus sativus* based on transcriptome sequencing. *Frontiers in Genetics*, 15, 1349626.

Yu, Y., Gui, Y., Li, Z., Jiang, C., Guo, J., & Niu, D. (2022). Induced systemic resistance for improving plant immunity by beneficial microbes. *Plants*, 11(3), 386.

Zebell, S. G., & Dong, X. (2015). Cell-cycle regulators and cell death in immunity. *Cell Host & Microbe*, 18(4), 402–407.

Zhai, K., Deng, Y., Liang, D. I., Tang, J., Liu, J., Yan, B., ... & He, Z. (2019). RRM transcription factors interact with NLRs and regulate broad-spectrum blast resistance in rice. *Molecular Cell*, 74(5), 996–1009.

Zhang, H., Roberts, D. N., & Cairns, B. R. (2005). Genome-wide dynamics of Htz1, a histone H2A variant that poises repressed/basal promoters for activation through histone loss. *Cell*, 123(2), 219–231.

Zhang, M., da Silva, J. A. T., Yu, Z., Wang, H., Si, C., Zhao, C., ... & Duan, J. (2020). Identification of histone deacetylase genes in *Dendrobium officinale* and their expression profiles under phytohormone and abiotic stress treatments. *PeerJ*, 8, e10482.

Zhang, P., Meng, X., Chen, H., Liu, Y., Xue, J., Zhou, Y., & Chen, M. (2017). PlantCircNet: A database for plant circRNA–miRNA–mRNA regulatory networks. *Database*, 2017, bax089.

Zhang, R., Zheng, F., Wei, S., Zhang, S., Li, G., Cao, P., & Zhao, S. (2019). Evolution of disease defense genes and their regulators in plants. *International Journal of Molecular Sciences*, 20(2), 335.

Zhang, S., & Mehdy, M. C. (1994). Binding of a 50-kD protein to a U-rich sequence in an mRNA encoding a proline-rich protein that is destabilized by fungal elicitor. *The Plant Cell*, 6(1), 135–145.

Zhou, X. J., Lu, S., Xu, Y. H., Wang, J. W., & Chen, X. Y. (2002). A cotton cDNA (GaPR-10) encoding a pathogenesis-related 10 protein with in vitro ribonuclease activity. *Plant Science*, 162(4), 629–636.

Zipfel, C. (2014). Plant pattern-recognition receptors. *Trends in Immunology*, 35(7), 345–351.

Zipfel, C., & Oldroyd, G. E. (2017). Plant signalling in symbiosis and immunity. *Nature*, 543(7645), 328–336.

22 Epigenetic Regulation in Plant Tolerance to Salinity and Drought Stress

Faiza Munir, Shahab, Qurat-Ul Ain-Ali,
Nimra Ghafoor, Ramsha Tariq and Alvina Gul

22.1 INTRODUCTION

As plants are sessile organisms, many factors such as salinity and drought affect crop production. Epigenetic regulation, which alters plant genes, is one of the main strategies that plants use to adapt to challenging environment. Epigenetic inheritance is a phenomenon whereby changes elicited by abiotic stress affect gene expression patterns and can be inherited by new cells during vegetative growth and by offspring. The sophisticated regulation of genes and their functions has occurred through layers of epigenetic information like DNA methylation, non-coding RNAs and histones, enabling plants to cope with unfavorable conditions like drought and salinity (Zhao et al., 2019). Being highly reversible, these modifications enable plants to adapt and respond quickly to stress stimuli, consequently modifying their physiological and molecular mechanisms. In other words, epigenetic regulation is essential for maintaining the cellular stress response and integrating environmental signals with the genome of the plant. These responses vary depending on the plant species and kind of stress, which points to the fact that plant stress tolerance is not a simple process (Zhang et al., 2018). Epigenetics comprises heritable changes in chromatin functions where no alteration in the DNA sequence takes place. All these processes are significant throughout the life cycles of plants as well as animals (Duan et al., 2018).

Drought and salinity severely constrain plant growth and development. These stresses induce intricate physiological, molecular and epigenetic reactions in a plant to enhance their endurance (Luo & He, 2020). Concerning such abiotic stress, various coping strategies have been developed to tackle stress in plants, including physiological changes brought about by regulation via epigenetics (Cheong, 2021). Epigenetic alterations take place in all forms of life to regulate their development and interaction with their surroundings. Especially, plants require intervention as these changes can prompt realization of environmental stimuli or change in plant growth and developmental pattern. Understanding these changes is currently easier because there are several genome sequences of various species including *Arabidopsis*, rice, poplar, maize and moss (Hemenway & Gehring, 2023; Lu & Hyun, 2021).

The method of epigenetic control that reorganizes chromatin structure in order to finalize gene expression on certain genomic sites is still actively developing (Banerjee & Roychoudhury, 2017). This chapter envisages novel knowledge on epigenetic changes in plants under salinity and drought stress, with reference to the ways in which these changes translate to gene regulation and physiological plant responses. It also highlights how epigenetic memory helps plants to "remember" past stress, thus preparing themselves well for any stress in the future. Altogether, the results of these investigations reveal useful opportunities for developing stress-tolerant crops that could offer sustainable solutions to many challenges facing agriculture due to climate change as well as resource scarcity.

22.2 AGRICULTURE AND FOOD SECURITY AFFECTED BY SALT AND DROUGHT STRESS

In the context of agriculture, abiotic stress is regarded as a non-biological factor that can negatively impact plant growth and, consequently, compromise yields of crops (Rajput et al., 2022). It is broadly expected that abiotic and biotic factors like drought and salinity present a significant threat to global food security (Ortas et al., 2021). Irrigation and salinity are major limitations to crop production and are expected to occur more often in future. Global soil salinity is predicted to increase by 2050 and involve half of the cultivated lands mainly in semi-arid and arid regions (Angon et al., 2022). Both salt and drought stress influence the growth and development of plants, alteration in enzymatic activity, crop yield and its productivity in a given field (Roy & Chowdhury, 2021).

Salt accumulation hinders water uptake and nutrient availability for plants in nearly 20% of irrigated lands globally (Hailu & Mehari, 2021). Excessive salt levels hinder water absorption and nutrient uptake, leading to plant stress and reduced yield. The use of salt-tolerant cultivars and effective irrigation practices can help mitigate production

DOI: 10.1201/9781003589914-26

losses due to salinity (Egamberdieva et al., 2019). In drought-prone and semi-arid regions, agricultural production declines by as much as 60%, leading to food insecurity and widening social and economic disparities. Drought impacts agricultural productivity and livelihoods more than any other natural disaster. Prolonged water shortages caused by drought stress can lead to a 15–50% reduction in crop yield (Lottering et al., 2021). Agronomy is facing a major challenge due to salt stress with some crop species more sensitive to salinity than others. These include rice, wheat, barley, soybean, potato, tomato, citrus fruits, cotton, sugar beet and maize (Selvaraj et al., 2021). These crops are sensitive to salt stress and often exhibit growth inhibition, lower yield and poor quality, which is not healthy for the regional and global economy in the context of food security (Zhu et al., 2023). The rising tendency in the frequency of drought will increase future yield risks, especially in rice, soybean, wheat and maize (Leng & Hall, 2019). Thus, food security is expected to be even more threatened by salinity and drought stress caused by the changing climate globally (Mehrabi et al., 2022). It is therefore crucial that appropriate methods be employed to enhance plant resistance to soil salinity and drought stress in order to guarantee sustainable agriculture and food security in saline and drought-affected regions (Dhankher et al., 2018).

22.2.1 Role of Epigenetics in Enhancing Stress Tolerance in Crops

Epigenetic regulation shows a beneficial function in the ability of plants to respond and recover from stress. Non-coding RNAs, DNA methylation and histone modifications have been studied for their involvement in plant stress response. Stress responses are important as they allow the plant to reduce stress-associated damages (Abdulraheem et al., 2024). Epigenetic inheritance occurs in two forms in the plant kingdom: getting messages not contained in the genes to the next generations and maintaining epigenetic changes from one generation to the next, being different in the next generation (Lloyd & Lister, 2022). It is crucial to know which aspects and factors make each type of epigenetic inheritance different and how it works. This area of study is particularly interesting as stress can lead to an activation of signaling pathways that strengthen stress gene responses. The information derived from this research can be applied in formulating ways and means of enhancing food productivity, crop yield and quality, and increased stress tolerance of crops, which will have positive impacts on the sectors of agriculture and the environment (Akhter et al., 2021; Balazova et al., 2022).

Epigenetic regulation is crucial during growth and development, as well as in flowering, fruit ripening and protection against pathogens. Epigenetic regulation also enhances the plant's ability to tolerate stress factors including modulation in temperature, salinity and nutrient status (Duarte-Aké et al., 2023). Genomic repatterning or modulating epigenetic networks is highly challenging to improve crop yield, growth, quality and production. These advancements aid sustainable agriculture where epigenetic elements such as small RNA, DNA methylation and histone modifications mediate key agronomic genes in crop plants. These processes affect gene transcription and translation of growth regulators, control morphogenesis and seedling development, germination, fruit setting, and are also directly responsible for crop yield and quality (Ashapkin et al., 2020). Moreover, the epigenetic changes may be durable and passed down to offsprings, so that can pass stress experiences from generation to generation (Sudan et al., 2018).

22.2.2 Epigenetic Mechanisms in Plants under Salinity and Drought Stress

In response to stress, plants display several epigenetic mechanisms of changing gene expression, which enable the plants to survive under adverse conditions. They include histone modification, small RNA-induced gene silencing, DNA methylation and chromatin change. Enhancing genomic stability and controlling gene expression are also critical because every machinery has its own distinct function (Banerjee & Roychoudhury, 2017). Whenever the plant is affected by abiotic stresses including temperature, salinity, drought and flooding stress, it leads to epigenetic changes. These include histone modifications and DNA methylation, which influence the expression of genes as well as small non-coding RNA molecules known as microRNAs (miRNAs) that are involved in stress responses. In the DNA methylation process, methyl groups (CH_3) are added to DNA, thereby inactivating or promoting certain stress-inducible genes; for instance, Polycomb repressive complex 2 (PRC2) and S-adenosylmethionine (SAM) catalyze gene methylation, which in turn modifies the chromatin structure to allow or restrict gene access. Furthermore, they modulate gene expression at the post-transcriptional level by binding to target mRNAs. These dynamic epigenetic modifications alter the expression levels of stress-responsive genes in plants, improve physiological function, and promote adaptation and tolerance. Altogether, these mechanisms can allow plants to maintain their survival and growth under the negative effects of the environment and keep them active and strong (Rehman et al., 2020).

22.2.2.1 DNA Methylation

DNA methylation is a complicated process that involves several cofactors and enzymes. It begins when a DNA methyltransferase enzyme binds to a cytosine–guanine (CpG) dinucleotide (Jogam et al., 2022). Most of the DNA methylation observed universally takes place at the carbon-5 position of cytosine and yields 5-methylcytosine (5mC). This alteration substitutes the structure of chromatin and suppresses the transcription of genes located on the chromosome. DNA methylation is another epigenetic modification that is essential in controlling the expression of genes under plant stress and developmental processes. It influences various plant characteristics, such as leaf shape, degrees of stress tolerance and disease susceptibility (Zhang et al., 2023).

High salinity causes changes in ionic and osmotic regulation, protein synthesis, energy production, photosynthesis and lipid metabolism (Muchate et al., 2016). Recent findings show that DNA methylation contributes significantly to the altered gene expression patterns under salinity stress (Al-Harrasi et al., 2018). Increased salinity was observed to lead to alteration in the DNA methylation pattern throughout the genome. Analysis on alfalfa (*Medicago* spp.) revealed that salinity stress results in the increase of the methylome and alfalfa seedlings that have been treated with 5-AzaC (a DNA methylation inhibitor) showing lesser salt tolerance (Al-Lawati et al., 2016). When plant is exposed to salt stress, it leads to the expression of genes, viz., *chalcone isomerase* (*CHI*), *chalcone synthase* (*CHS*), *flavonol synthase* (*FLS*), *anthocyanidin synthase* (*ANS*), *dihydroflavonol 4-reductase* (*DFR*) and the antioxidant genes such as *glutathione peroxidase* (*GPx*). The altered expression of these genes affects the methylation status in transgenic *Arabidopsis* (Bharti et al., 2015). Drought stress is quite complex in plants, and DNA methylation is an important regulatory system for gene expression. Analysis of the degree of methylation of individual cytosines in the genome under drought stress shows that the level of methylation in stressed plants is higher than that of control plants; for example, *Morus alba* elevated methylation level by 8.64% compared with control, and *Populus trichocarpa* by 2.29% for methylation levels (Li et al., 2020). Likewise in maize, suppression of *ZmNAC111* under drought stress is via DNA methylation of its promoter region, with the result being increased drought stress susceptibility. The enzymes that are involved in the process of maintaining the DNA methylation status in the plant genome come under two categories. During drought stress conditions, cytosine-5-methyl transferases and cytosine-5-demethylases play a significant role. Some facts about enhanced expressions of 5mC methyltransferase and demethylase enzymes under drought stress are reported in plant species including *Malus domestica* (Xu et al., 2018), *Solanum melongena* (Moglia et al., 2019) and *Solanum lycopersicum* (Kumar et al., 2016).

22.2.2.2 Histone Modifications

Histone modifications are important to plant stress including salt and water stress because they involve chromatin and gene regulation. These modifications include the addition of the methyl, acetyl, phosphoryl or ubiquitin chemical groups to histones, which are proteins that surround the DNA. Histone modification enables plants to regulate the chromatin density, thereby either repressing or initiating gene activity, which is linked to stress-related genes (Yung et al., 2021).

Drought, salinity, temperature stress and high metal toxicity cause signaling cascades in plants and alter histone modifications. Research has revealed that drought stress enhances the degree of histone H3 acetylation, and this has impacts on the stress-responsive genes in *Arabidopsis thaliana* (Nunez-Vazquez et al., 2022). This acetylation implicates stress-responsive genes and confirms that histone modifications may control the efficiency of the ability of the organism to cope with water-deficit conditions. Similarly, the earlier exploration of salt stress revealed that changes in histone methylation affected the genes involved in ion regulation and osmotic stability (Ueda & Seki, 2020). This is due to alterations in gene expression through histone acetylation, with studies showing that HDA6 histone deacetylase-dependent stress-responsive genes are necessary for drought tolerance in *Arabidopsis thaliana* (Zhao et al., 2019). It has been proposed that the MAPK (mitogen-activated protein kinase) cascade is involved in the phosphorylation of histone, with an impact on gene regulation during stress (Mozgova et al., 2019). The present data suggest that histone modifications, including acetylation, methylation, ubiquitination, sumoylation and ADP-ribosylation, can regulate gene expression and chromatin organization in plants. These modifications are stress-responsive, but the exact roles of these modifications in plant stress response are not well understood (Wang & Qiao, 2020).

22.2.2.3 Non-coding RNAs

In addition to mRNAs, there are small interfering RNAs (siRNAs) and non-coding RNAs called microRNAs (miRNAs) that regulate gene expression at both the transcriptional and translational levels. These short, non-coding RNA molecules program the RNA-induced silencing complex (RISC) to either inhibit translation or degrade target mRNAs. The miRNAs are processed from internal hairpin loop precursors, whereas siRNAs originate from external double-stranded RNA (dsRNA) or long internal dsRNA precursors (Jagadhesan et al., 2022; Xie & Yu, 2015). Plant utilizes miRNAs in response to abiotic stresses such as heat and drought stress. The miR159 has been reported to regulate drought stress tolerance genes in *Arabidopsis thaliana* (Wang & Qiao, 2020). In *Arabidopsis*, not only flowering time, but also fruits ripen through regulation of Squamosa promoter-binding protein (SPL) transcription factors by miR156 (Begum, 2022). In this way, miR156/SPLs modulate histone changes at the flowering locus, so that timely flowering can take place. Small RNAs can potentially come directly from the transposable elements or stress-responsive genes triggered during stressful conditions (Zhan & Meyers, 2023). Stress-responsive genes are turned on and off by small RNA molecules that control plant stress responses and help plants grow in harsh conditions; for example, miR398 is induced under conditions of oxidative stress and the transcripts of the *copper superoxide dismutase* (*CSD*) gene in *Arabidopsis thaliana* (Singroha et al., 2021). Plant produces ROS in chloroplasts and peroxisomes when subjected to drought stress. Thus, to counteract the toxic impact of ROS on the cells, plants synthesize antioxidant enzymes like catalase, superoxide dismutase, peroxidase, ascorbate peroxidase and glutathione reductase enzymes (He et al., 2023; Zhou et al., 2019). Notably, some of the plant miRNAs such as miR528 and miR398 are known to be involved in the regulation of oxidative stress networks (Qiu et al., 2018; Yang et al., 2022).

22.2.2.4 Chromatin Remodeling

Chromatin remodeling complexes in plants are known to be essential in determining and modulating the epigenome mainly under stress (Tonosaki et al., 2022). These complexes are involved in the process of altering chromatin structure that in turn involves changes in DNA accessibility to both the transcription machinery and regulatory proteins (Pandey et al., 2016). Chromatin remodeling complexes help plant stress by changing structure that activates or inhibits stress-responsive genes required to signal survival and modulate gene expression (Wang & Qiao, 2020). Chromatin remodeling complexes are significant for epigenetic mechanisms as these complexes facilitate rapid and reversible changes in gene expression to stress in plants (Pecinka et al., 2020). This dynamic regulation of expression facilitates an appropriate response to nearly any stress without the need to rewrite the genetic code. Apart from playing a role in the formation of chromatin remodeling complexes, they also cooperate with other epigenetic functions such as DNA methylation and those that involve small RNA. These layers and organization of epigenetic regulation and their integration are important for stress response and plant resistance. Furthermore, new studies suggest that environmental cues control the business and terminal predestination of chromatin remodeling complexes and organize a connection between epigenetic modifications and stress-responsive gene expression (Kim, 2021; Roca Paixão et al., 2019).

22.3 ROLE OF EPIGENETICS IN ABIOTIC STRESS RESPONSES

Recent analysis has revealed that stress conditions result in epigenetic changes within the plants. Such changes allow temporary variation of gene activity and several vital physiological functions in plants, including osmotic balance and water conservation. These modifications enable transient changes in gene expression and enable several critical physiological processes, including osmotic balance and water retention in plants (Chang et al., 2020).

22.3.1 Ion Transport and Homeostasis

More precisely, for salt stress, DNA methylation changes regulate the genes associated with membrane transport, heavy metal translocation and organic acid secretion as a response to stress in plants. By genotypic analysis and tissue-specific study, authors realized that the density of cytosine methylation on high-affinity potassium transporter genes *TaHKT2;1* and *TaHKT2;3* increases during NaCl treatment. These genes were downregulated in the wheat genotype Kharchia-65, and it has a positive impact to provide better salt tolerance (Kumar et al., 2017b).

In plants, the Ca^{2+} Calcineurin B-Like Protein (CBL), CBL Interacting Protein Kinase (CIPK) complex, plays a critical role in maintaining cell ion homeostasis (Zhu, 2016). *Salt overly sensitive genes* (*SOS*) activated under conditions such as excess magnesium (Mg^{2+}), low potassium (K^+), high sodium (Na^+) and high pH elevate intracellular calcium (Ca^{2+}) signaling to activate proteins like Na^+/H^+ antiporter – SOS1, potassium channel – AKT 1, magnesium transporter and H^+ ATPase. In this study, we documented that HKT1, a High-Affinity K^+ Channel 1 that regulates the activity of Na^+ influx, and the SOS pathway are both involved in modulating plant salinity tolerance. In *Arabidopsis*, there is a putative siRNA target region, approximately 2.6 kb upstream of the *HKT1* gene start codon that is almost completely methylated at every cytosine residue in wild-type plants (Baek et al., 2011).

Histone deacetylases have multiple activities related to plant salt stress tolerance. It has been identified that Class II histone deacetylases including *HDA5*, *HDA14*, *HDA15* and *HDA18* promote salt tolerance in *Arabidopsis*. On the other hand, Class I histone deacetylase HDA19 is detrimental to salt tolerance as it reduces histone acetylation and represses transcription of the abscisic acid (ABA) receptor genes; pyrabactin resistance 1 like (PYR/PYL) proteins including PYL4, PYL5 and PYL6 (Ueda et al., 2017).

High-throughput evaluation of DNA methylation has been reported to be affected by drought in rice, maize, wheat, barley and cotton, and drought-affected methylation loci in *Vigna radiata* (Naderi et al., 2024; Zhao et al., 2022); for instance, in maize roots, drought stress increases global DNA methylation level with the methylome of drought-tolerant inbred lines being less affected in comparison with the methylome of drought-sensitive inbred lines (Wang et al., 2021). Drought stress also produces a similar number of methylations and demethylations in barley leaves but a higher number of methylations in barley roots. This implies that regulation of the methylome may be organ-specific and may be important for drought stress tolerance (Chwialkowska et al., 2016).

22.3.1.1 Role of Ion Transporters and Channels in Salinity Tolerance

Ion transporters, especially those regulating sodium (Na^+) and potassium (K^+), are extensively studied in plants to unravel the changes in expression pattern of genes under salt stress (Qiu et al., 2004). Plant transporters, such as HKT1;1 and NHX1, function to either exclude Na^+ or store it to maintain the correct balance of ions under high saline conditions (Wang et al., 2022). The Salt Overly Sensitive (SOS1) transporter is a particular example of such a protein; it belongs to the Na^+/H^+ antiporter family that works as a transport pump removing toxic amounts of Na^+ from the cytoplasm (Zhao et al., 2021). Knowledge of these transporters can go a long way to increasing crop salt tolerance either by breeding optimization or by genetic manipulation of Na^+ management systems. In maize, uptake of Na^+ from salt stress hinders cell growth and causes cellular injury and growth inhibition (Ganotra et al., 2022). Moreover, it becomes difficult for the plant to uptake nutrients such as K^+, Mg^{2+} and Ca^{2+} ions that are essential in maintaining the health of the cells. Na^+ accumulates in the older leaves, and after a while, they become poisonous to the plant (Lam et al., 2022). Amending the plant with the presence of calcium silicate (Ca_2SiO_4) together with salt increases the ability of the plant to conserve higher levels of K^+, Ca^{2+} and Mg^{2+} to relieve some of the stress (Mahmood et al., 2024). This can be because Ca^{2+} aids the membranes to firm up and control the ions, while silicon appears to assist the plant in retaining K^+, thereby giving it a better standing in a salty environment (Li & Yang, 2023).

22.3.1.2 Sodium Channels and Transporters

To survive saline environments, plants need to maintain a high K^+/Na^+ ratio (Liang et al., 2019). When salt levels rise, Na^+ floods into the cells, disrupting growth. Several channels, like non-selective cation channels, glutamate receptors and cyclic nucleotide-gated channels, help control the influx of Na^+ entering the roots (Seifikalhor et al., 2019). To prevent Na^+ accumulation in the shoots, high-affinity potassium transporters move it back down to the roots. Plants also have a backup plan: they store extra Na^+ in vacuoles using Na^+/H^+ exchangers and use SOS1 transporters to push Na^+ out of the cells (Gayatri et al., 2021).

22.3.1.3 Potassium Channel

Potassium is vital for plant growth, influencing yield or stress tolerance and key processes like protein production and carbon dioxide uptake (Sardans & Peñuelas, 2021). While soil K^+ levels can be low (0.1–1 mM), plants keep K^+ at a much higher level inside their cells, around 100 mM. However, during salinity stress, Na^+ competes with K^+, making it harder for plants to absorb enough K+ (Abdulraheem et al., 2024). This shortage can lead to oxidative stress and stunted growth (Shohan et al., 2019). To adapt, plants fine-tune their K^+ channels and transporters, which belong to several families like HAK, HKT and shaker-like channels (Pottosin & Dobrovinskaya, 2014). These aid in the uptake of K^+ by plants and the exclusion of Na^+. Inward rectifying channels transport K^+ into the roots through channels as well as the outward channels to properly regulate ion throughput depending on signals such as ROS (Chen et al., 2023). When K^+ levels are low, auxin response factor 2 increases K^+ acquisition and when K^+ concentration is high, NHX1 and NHX2 transport the K^+ to regulate turgor pressure and cell volume (Mulet et al., 2023). Modification in expression of genes encoding K^+ transporters can enhance salt tolerance in crops (Joshi et al., 2022).

22.3.1.4 Salt Overly Sensitive Pathway and Epigenetic Control

Histone methylation is an epigenetic mechanism that regulates the Salt Overly Sensitive (SOS) pathway critical for sodium efflux control in salinity-stressed plants (Liu et al., 2023). Histone modifications can affect expression levels of SOS pathway genes under salt stress (Duscha et al., 2020). Specifically, histone H3 methylation at lysine 4 around the *SOS1* and *SOS2* promoters enhances their transcriptional activation and increases accessibility in response to salt stress (Li & Yang, 2023).

22.3.2 Hormonal Signaling in Stress Responses

One of the most exciting features of plant biology is transgenerational epigenetic transmission and the ability for stress responses that can be passed on from one parental generation to another, not by changing the DNA sequence (Shah et al., 2023). This would mean that plants that survive drought may actually pass on a propensity to be more resilient to similar experiences in the future (Myers et al., 2023); for instance, DNA methylation patterns were altered in drought-exposed *Arabidopsis thaliana* studies that were maintained during the susceptible generation but not further passed to subsequent generations. On chromosome 6, short arm lies approximately 155 kb with 13 exons spanning the *FKBP5* gene (Fichman et al., 2022). Its expression is induced by several receptors, including the glucocorticoid receptor (GR), androgen receptor and progesterone receptor (Fichman et al., 2022). Induction through GR implies that the receptor binds to Glycogen Response Elements (GREs) that are spread out in a region of more than 100 kb. These GREs are located outside the *FKBP5* gene and mainly in defined introns of the gene (Sharma & Ahammed, 2024). This activation is further supported by the three-dimensional chromatin looping, which remodels the distal enhancer elements to the transcription start site (TSS) of the *FKBP5* allowing interaction with RNA polymerase II. Interestingly, the *FKBP5* locus shows the epigenetic feature of open chromatin, such as specific DNA methylation and histone acetylation, among healthy mature plants (Manh et al., 2023). Such changes indicate that within specific tissues, chromatin in the region of the *FKBP5* gene may be primed for activation regardless of the action of steroid hormones. ABA levels have been found to increase to significant levels in salt-stressed plants and reached their highest concentration at six hours after the imposition of salt stress. For a number of genes among the differentially expressed genes involved in ABA production, the RNA-seq study pointed at cytochrome P450 707A (*CYP707A*) among other genes (Pan et al., 2024). It shows best expression at one hour and then slowly decreases. However, other genes such as the Nine-Cis-Epoxycarotenoid Dioxygenase (*NCED*) and Abscisic Aldehyde Oxidase 3 (*AAO3*) show a gradual rise in expression approximately three hours after salt stress has been imposed (Minnoye et al., 2021). This suggests that while *NCED* and *AAO3* allow sustained synthesis of ABA under prolonged stress, *CYP707A* is probably implicated in the temporary increase of ABA. The relation between these genes may play a role in controlling the level of ABA (Chen et al., 2020). Of the very limited number of upward controllers of salt, it has been demonstrated that the increase of ABRE-Binding Factor (ABF) is enhanced by the ABA signal transduction. In addition, several genes that are involved in regulating ABA, including Abscisic Acid Deficient 2 (*ABA2*), *CYP707A*, *NCED*, Violaxanthin De-epoxidase (*VDE*), Zeaxanthin Epoxidase (*ZEP*) and *AAO3*, activate *ABF* to perform stress-related physiological responses (Li et al., 2023).

22.3.2.1 Epigenetic Regulation of Hormone-Responsive Genes during Drought and Salinity

Several hormonal stimuli regulate the diverse responses of plants subject to drought stress. Forms of epigenetic changes include small RNA, histones and DNA methylation (Yamamuro et al., 2016). They are required for changing the signal transduction for hormones (Dar et al., 2022). Figure 22.1 demonstrates the link between how drought perception by abscisic acid takes place and how there are changes to the DNA and histone modification of plant stress response genes. This figure illustrates that ABA-mediated drought perception leads to specific changes in plant stress-responsive genes, which can be termed as epigenetic regulation such as DNA methylation and histone modifications. Epigenetic processes assist plants in growing more robust to the shocks by the occurrence of heritable changes. A detailed depiction of the ABA signaling pathway is available on the upper side of the flowchart starting with the ABA receptor PYR/PYL/RCAR. These receptors counteract the activity of endogenous proteins known as type 2C protein phosphatases (PP2Cs), which are inhibitors of ABA signaling when bound by ABA. This inhibition activates the main positive regulators and the secondary messengers, Snf1-related protein kinases 2 (SnRK2s) (Rudolf et al., 2024). It regulates the genes that code for stress response and metabolic networks needed by the plant to cope with drought (Sallam & Moussa, 2021). A model in the upper part of the diagram of epigenetic regulation and ABA signaling pathway crosstalk in the stress response genes under drought stress is shown in Figure 22.1.

Histone modification, DNA methylation and chromatin remodeling mechanisms play significant roles in plant salt stress tolerance (Banerjee & Roychoudhury, 2017). Histone deacetylases also contribute to this balance, affecting abscisic acid signaling and stress-responsive gene regulation either as salinity protectors or exacerbators (Bharti et al., 2015). Another source is histone methylation that, in addition to the above-mentioned mechanisms, offers the so-called stress memory, by which plants can maintain certain adaptive signals that could have originally responded to prior salinity stress (Karan et al., 2012).

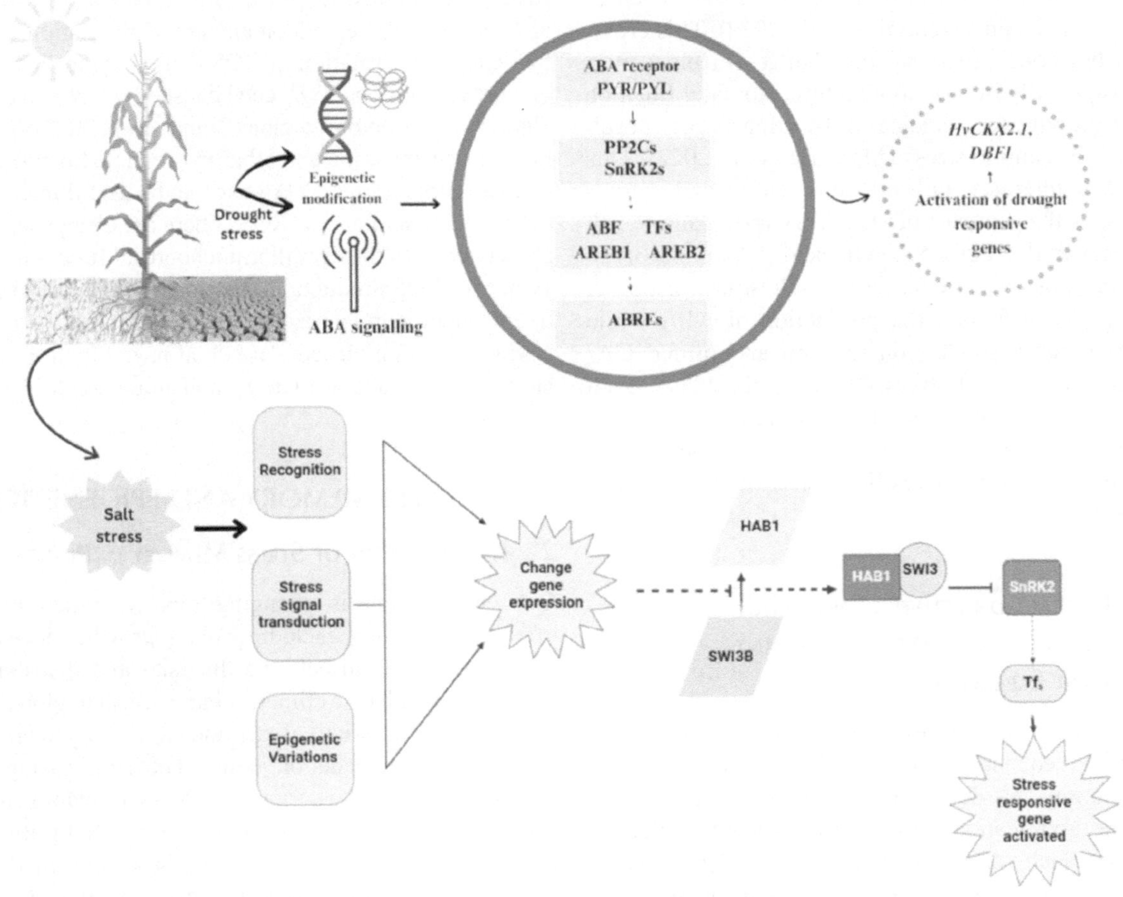

FIGURE 22.1 Epigenetic and signaling pathways in plant stress responses to drought and salt stress.

DNA methylation changes the stress-responsive genes both positively and negatively throughout their regulation (Garg et al., 2015). The reduction in the level of DNA methylation at specific loci during salinity stress results in the activation of antioxidant and metabolic genes (Banerjee et al., 2016). Most of these methylation changes are tissue-specific and differ across plant genotypes, linking gene expression with chromatin and small RNA functions (Saez et al., 2008). Chromatin remodeling also contributes to stress adaptation from remodeling chromatin, so that these specific genes may be expressed. Remodeling complexes modulate the genes and proteins that form a signaling pathway, viz., the abscisic acid regulators (Yuan et al., 2013). Overall, these mixed epigenetic effectors allow plants to fine-tune their behavior to salinity–stress conditions and achieve progressive activation/repression of their genes to increase survival and tolerance (Kim et al., 2015).

22.3.2.2 Convergence of Hormonal Signals and Epigenetic Changes in Stress Adaptation

Plant hormones function together and play a role in growth regulation and stress responses for salt stress. With some hormones having inhibitory effects, others enhance stress tolerance. The regulation of genes that are responsible for the synthesis and breakdown of hormones changes the levels of hormones within the plant tissues, such as leaves, roots and stems (Jeremias et al., 2022). This crosstalk with hormones also impacts changes in plant morphology, especially in root architecture, stomatal function and shoot growth; for instance, ABA impacts root development under osmotic stress (Majnarić et al., 2022). This is because it interacts with ethylene, auxin and cytokinin. It causes the increase of rice ABA levels due to salt stress (Khan et al., 2020). Salicylic acid (SA) and ABA in combination positively regulate the salt stress response; conversely, SA decreases the production of ABA while forcing gibberellin production to keep an optimal GA/ABA ratio through salt stress (Khan et al., 2020). Such a fine-tuned hormonal crosstalk at different plant organs allows cereals and other plants to adjust their physiology and further increase their resilience to salt stress (Garai et al., 2021).

22.3.3 Interaction between Epigenetic Modifications and Antioxidant Gene Expression

Changes in a single gene expression without changes in the nucleotide sequence of DNA are known as epigenetics. These alterations are due to alterations in DNA methylation, histone modifications and non-coding RNAs including microRNAs. Such cellular switching mechanisms control numerous biological functions and stress-inducible pathways through turning on or off the specific genetic information processing (Abdulraheem et al., 2024).

22.3.3.1 Antioxidant Gene Epigenetic Activation

As part of the antioxidant reaction, the ones that help counter oxidative damage such as glutathione peroxidase, catalase and superoxide dismutase may undergo epigenetic changes and have their activity heightened in the event of stress. The worst thing is that under oxidative stress, such genes can become hypomethylated – they obtain less methylation, turn out to be transcribed with greater intensity, and thus reinforce the antioxidant protection (Kowluru, 2023). The best studied histone modification associated with antioxidant gene activation is histone acetylation, which leads to chromatin relaxation and gene expression. In addition to the regulation of the expression of many antioxidant proteins via its binding to the promoter region of antioxidant genes, the transcription factor NRF2 has a critical role in the response to oxidative stress (Bouyahya et al., 2024). Recent evidence suggests that miRNAs are epigenetic master regulators of antioxidant genes; for instance, ***miR-144*** has been proved to target the mRNA of NRF2 and downregulate the specific mRNA, thereby resulting in the decrease in the transcription of antioxidant genes (Ramakrishnan et al., 2022).

22.3.3.2 Epigenetic Control of Reactive Oxygen Species Generation

NADPH Oxidase (NOX) Enzymes: They mostly lead to ROS generation. Changes in the NOX expression as well as NOX activity depend on epigenetic regulation. One way by which the formation of ROS can be controlled includes downregulating the *NOX* gene caused by DNA hypermethylation of its promoter region (Chung et al., 2022). Similarly, histone deacetylation and other irreversible histone changes that can suppress the *NOX* gene can be useful in preventing the production of excess ROS. There are also genes that are associated with the overall mitochondrial function and necessary for ROS production; they are also subject to changes in epigenetic regulation. Sirtuins (SIRTs) are the family of NAD^+-dependent deacetylases that have functions in regulating the production of ROS and mitochondrial function (Guo et al., 2023).

22.4 STRESS MEMORY AND EPIGENETICS

22.4.1 Concept of Stress Memory in Plants

Our environment is changing rapidly, which adversely affects agriculture, including plant growth, development and yield, which affects the diversity and distribution of plants. These adverse climatic changes lead to global warming, which is a serious issue, and potentially increase the abiotic stress incidents on plants. These increasing abiotic stresses have posed a serious challenge to plant health and overall cultivation (Lagiotis et al., 2023). Naturally, plants have a mechanism to cope with these unwanted abiotic stresses; hence, they reprogram their developmental regulation that favors mechanisms to cope with stresses. These responses are tissue-specific and depend on the ontogenetic stage (Gray & Brady, 2016). It is intimated in several studies that both biotic and abiotic stresses can prompt genomic reprogramming by calibrating the gene expression at different levels in plants (Etalo et al., 2013). Plants naturally have evolved such complex and efficient regulatory mechanisms to cope with the damage inflicted upon them as a result of a harsh environment (Atkinson & Urwin, 2012). Stress memory in plants is important in many ways, being pivotal in adapting to such adverse environmental challenges that our world is facing. This stress memory, which embraces the changes occurring in plants in response to stresses, could be physiological or molecular changes allowing the plants to retain the memories of past stresses (Siddique et al., 2024).

22.4.1.1 Drought Stress Memory

Stress mechanism is an indispensable process in plants that increases their endurance to these intermittent stresses like drought. *Arabidopsis thaliana* exhibits substantial stress memory for drought. Abscisic acid maintains the stomatal conductance and transpiration rate (Fleta-Soriano & Munné-Bosch, 2016). Stress memory of drought in *Brassica napus* plants has shown enhanced seedling strength to cope with stress conditions and increased growth under consecutive drought conditions. It is one of the examples showing the transgenerational effect and how it improves the durability of progeny (Hatzig et al., 2018). During subsequent drought conditions in *Saccharum officinarum*, a quicker recovery of CO_2 assimilation was noted as well as an increased carboxylation efficiency. This well-suited example of memory in sugarcane allows the plants to respond more efficiently to consecutive drought conditions (Kambona et al., 2023).

22.4.1.2 Salinity Stress Memory

Arabidopsis thaliana is shown to have the capability of retaining salinity stress memory across generations. Plants with salinity tolerance have offspring with even more increased salt stress tolerance, giving the idea that epigenetic modifications linked to stress exposure can be transferred through generations (Ma et al., 2020). Former salt stress in *Triticum aestivum* cultivars has been shown to have increased plant growth and yield under frequent salt stress conditions These plants have shown enhanced photosynthetic efficiency, which is important for maintaining plant productivity in saline environment (Dos Santos et al., 2022). When *Oryza sativa* plants are exposed to high salt stress, they start to have changes in ion transport mechanisms and osmotic balance. These changes in plants contribute to increased salt tolerance in the later stages of stress, which gives a clear example of how stress memory works in plants (Chen et al., 2013).

22.4.2 Epigenetic Basis of Stress Memory

22.4.2.1 Case Studies of Stress Memory in Crops

Stress memory is significant in comprehending the mechanisms of how plants cope with subsequent stresses. Some case studies have shown stress memory in different crops. *Arabidopsis thaliana,* for instance, has shown physiological changes enhancing its resilience to recurrent drought conditions. One example shows that stomata remain closed when plants are in the recovery phase during drought conditions. This phenomenon helps the plants to conserve water for themselves in case they face water deficiency in the future. In *Zea mays* and *Craterostigma* plantagineum, there is a pre-exposure period in which the plant is shown to have a higher relative water content during recurrent dehydration periods. This explains well the ability of plants to preserve stress memory (Sun et al., 2022). In case of *Solanum tuberosum* tubers, exposure to priming by mild dehydration treatment produces more tubers in comparison to high drought conditions. It provides the example of how important an initial stress is to plants to enhance their productivity (Ramírez et al., 2015).

22.4.3 Transgenerational Epigenetic Inheritance

The transmission of epigenetic marks and modifications from one generation to future generations without any alterations in the DNA sequence is transgenerational epigenetic inheritance. It allows the transfer of acquired traits and also affects gene expression and certain morphological changes in future generations. The complete understanding of such mechanisms is very important in learning how plants remember the past stresses and how they cope with future stress conditions (Hauser et al., 2011).

22.4.3.1 Examples of Transgenerational Epigenetic Inheritance

The *B1* gene in maize is involved in anthocyanin production, which exhibits paramutation. It showed that when plants with high expression of the *B1* allele are crossed with the low expression B' allele, the offspring show reduced pigmentation, which is due to epigenetic silencing of the *B1* gene (Giacopelli & Hollick, 2015). Stress tolerance is heritable and changes the DNA methylation levels accordingly (Nguyen et al., 2022).

22.4.4 Implications for Breeding Programs Focused on Stress Resilience

As the climate has been changing adversely, it requires the development of such crop varieties that can withstand the increasing challenges of abiotic stresses like drought and salinity as well as extreme heat and cold temperature. Breeding programs with prime focus on stress resilience are important for ensuring food security as well as agricultural sustainability. The prime aim of breeding programs should be the exploration and utilization of genetic diversity in crops. These genetic resources contain the traits necessary for resilience to stresses (Marone et al., 2021), for instance, N22, a rice landrace, for drought tolerance is an important genetic donor (Binodh et al., 2023). Breeding programs should be directed to create more varieties with increased tolerance to multiple stresses. It requires a detailed understanding of plant responses to combined stresses (Slama et al., 2015). Several techniques have allowed breeders to identify and choose the desired traits. DNA-based molecular markers allow more precision with selection of genotypes. This allows the breeding process to succeed more rapidly and effectively (Jonah et al., 2011). Understanding how plants remember the past stresses is important in adjusting their physiological responses. The initial exposure of plants to stress helps them to prime responses more efficiently to the future stresses (Sadhukhan et al., 2022). Breeding programs require successful collaborations between private and public institutions and research organizations (Domínguez, 2022).

22.4.5 Molecular Markers for Stress Memory

Identifying biomarkers for stress memory is important for carrying out breeding programs for developing resilient crops. Some types of molecular markers used for stress memory in plants against abiotic stresses include DNA methylation patterns, histone modifications and non-coding RNAs (Niculescu et al., 2020). The biochemical markers include the expression levels of heat shock proteins (HSPs) produced in response to stresses (Goswami et al., 2010). A plant undergoing stress conditions triggers the accumulation of ROS, which makes it an effective indicator for studying the stress level (Sharma et al., 2022). Abscisic acid levels before stress exposure and after stress exposure make it a good indicator of the stress memory of plants (Wahab et al., 2022).

Several technological advancements have encouraged the identification of biomarkers. Next-generation sequencing (NGS) allows detailed analysis of genome-wide epigenetic modifications, helping the identification of markers related to stress memory (Trippa et al., 2024). Transcriptomic approaches involve RNA sequencing that helps in analyzing the changes in gene expression profiling related to stress memory (Satrio et al., 2024).

22.5 ADVANCES IN EPIGENETIC RESEARCH AND TECHNIQUES

22.5.1 Emerging Discoveries in Epigenetic Modifications under Abiotic Stress

Epigenetic regulation is a promising emerging field that has gained significant interest recently because of its enormous potential to enhance plant yield, boost disease resistance and increase plant adaptability. Over the last five to ten years, important discoveries have been made that have enhanced the understanding of epigenetic modifications in plants under abiotic stress conditions. Studies have shown that DNA methylation and histone acetylation activity can be manipulated to cause changes in the structure of DNA that can ultimately suppress or activate gene expression (Sun et al., 2022).

It has been established that drought stress initiates change in the pattern of DNA methylation in stress-responsive genes. As a result, gene expression also increases imparting resilience to plants against drought stress. Similarly, other stresses including heat and salt also cause epigenetic changes in stress-responsive genes of plants. According to Surdonja et al. (2017), following exposure to drought stress, the DM levels in the promoter regions of the *HvCKX 2.1* gene were elevated, resulting in the quick emergence of shoots due to the high level of cytokinins. Furthermore, epigenetic studies in rice showed that drought-induced mutations are stable and are passed on to future generations (Lu et al., 2022). The transcript levels of the stress-related genes involved in metabolic processes and regulations changed in rice post DM modification (Feng et al., 2016).

Another important discovery in epigenetic modification linked with abiotic stress is the finding of stress memory in plants (Asensi-Fabado et al., 2017). Stress memory has been studied in *Arabidopsis thaliana*, and the findings revealed that exposure to drought stress led to several alterations in DM patterns of stress-responsive genes even after the stress was removed. Hormones play a central role in plant stress responses, and modern research has shown that epigenetic modifications of hormones are involved in the environmental adaptation of plants to abiotic stress. ABA is an extremely important plant hormone involved in stress responses in plants, and studies have unfolded that DNA methylation and histone modification regulate ABA signaling and thus play a vital function in ABA-regulated stress responses (Zhang et al., 2022).

Plants utilize secondary metabolites as defense tools for surviving harsh environmental conditions. These secondary metabolites include alkaloids, phenolic compounds and osmoprotectant molecules that are encoded by stress-regulating genes. Current research has proved that epigenetic regulation is also involved in the biosynthesis of these compounds (Tran et al., 2023). Furthermore, a study revealed that histone modifications and DNA methylation alter gene expression to increase the synthesis of anthocyanins. This particular phenolic compound protects plants from osmotic and oxidative stress (Zhang et al., 2018).

22.5.2 State-of-the-Art Tools and Techniques for Studying Plant Epigenetics

Understanding epigenetic mechanisms requires various advanced tools that can examine variations in gene expression, DNA modifications, chromatin structure and histone acetylation without disturbing the underlying genetic sequence. However, sophisticated high-throughput technologies have made studying epigenetic processes easier and understanding their functions in stress responses.

22.5.2.1 Chip-Seq (Chromatin Immunoprecipitation Sequencing)

Chip-Seq is a powerful tool for studying epigenetic modifications that was invented by the integration of chromatin immunoprecipitation with next-generation sequencing. This technique is mainly used to determine the binding sites of proteins or histone proteins in DNA. In Chip-Seq technique, first the DNA–protein complexes are precipitated with antibodies that are specific to the protein of interest. The immunoprecipitated DNA is then isolated, purified and sequenced. The sequence reads are used to map binding sites or identify histone modification patterns on the plant genome. Chip-Seq offers high resolution, greater coverage and accurate quantification of data for studying epigenetic changes (Nakato & Sakata, 2021; Park, 2009).

22.5.2.2 Methylation-Sensitive Restriction Enzymes (MSREs)

Methylation-sensitive restriction enzyme analysis is used to study cytosine methylation in CpG residues. Restriction enzymes such as *Sma* I, *Acc* II, *Not* I, *Acc* II and *Pma*C I lack the ability to digest methylated cytosine residues present in methylated DNA. Yet, these restriction enzymes efficiently digest unmethylated DNA, which is later amplified using PCR. Thus, amplicons of methylated DNA are detected while the unmethylated DNA remains undigested. However, complete digestion of DNA is necessary for PCR and accurate generation of data. The lack of availability of CpG site-specific MSREs and the ability of MSRE to cut multiple CpG sites are the drawbacks of this analysis (Subedi et al., 2020).

22.5.2.3 Bisulfite Sequencing Techniques for the Study of Epigenetics: Genome-Wide Bisulfite Sequencing

This technique is used to study cytosine methylation patterns in the target sequence. Bisulfite treatment converts unmethylated cytosine in uracil while the methylated cytosine remains intact; thus, comparison of treated sequences with the reference sample helps to identify the methylated cytosine. Genome-wide bisulfite sequencing is a powerful technique as it helps to locate the methylation areas at a single base pair difference (Grehl et al., 2020).

22.5.2.4 Methyl-Sensitive Amplification Polymorphism (MSAP)

MSAP is a molecular technique that identifies variation in DNA methylation with the help of two isoschizomers, *Hpa*II and *Msp*I. The *Msp*I and *Hpa*II enzymes have the same restriction digestion sequence (5'-CCGG-3') but possess different sensitivity to methylation. *Hpa*II enzymes cannot digest methylated cytosine while the *Msp*I enzyme can cut methylated cytosine at its restriction sites. Thus, the *Hpa*II enzyme is called methylation sensitive and the *Msp*I enzyme is called methylation insensitive. Afterwards, the digested fragments are ligated to an adaptor and amplified using fluorescent tagged primers. The digested amplicons are sequenced and compared to other accessible sequences in the database to study DNA methylation patterns in the target sample. This technique does not sequence a specific probe and the presence or absence of the amplified DNA depicts the methylation pattern of the target loci. This technique is useful to compare methylation patterns in plants under different developmental stages or stress conditions (Sharma et al., 2021).

22.5.2.5 Chromatin-Immunoprecipitation PCR (Chop-PCR)

Integrating partial digestion of MSREs with PCR amplification, the Chop-PCR technique locates DNA methylation or histone modifications in the genome at a target locus. MSRE digestion shields the digestion of cytosine at the target sites; thus, no digestion occurs, whereas MSREs cleave the unmethylated sites. Hence, PCR products are formed at the methylation site while no amplicon is formed at the unmethylated site (Dasgupta & Chaudhuri, 2019).

22.5.3 CRISPR and Epigenome Editing for Stress Tolerance

CRISPR system is a revolutionary genome editing technique that utilizes guide RNA (gRNA) to modify target sequences. After binding gRNA with the target sequence, Cas proteins cleave the DNA to allow modifications, i.e., insertions or deletions at the target site using the DNA-repair mechanism of the host cell (Mali et al., 2013). CRISPR-based epigenome editing enhances the tolerance in plants to abiotic stress by targeting the epigenetic marks that control the stress-regulating genes. In this technique, the CRISPR-Cas9 is fused with specific epigenetic modifiers like methyltransferase or histone acetyltransferase. The fusion of Cas9 with an epigenetic modifier then allows histone modification, chromatin structure modification or DNA methylation without disturbing the DNA sequence (Fadul et al., 2023). Roca Paixão et al. (2019) improved drought tolerance of *Arabidopsis thaliana* by fusing *Arabidopsis histone acetyltransferase* (*HAT1*) gene with Cas9 protein. AtHAT1 triggers chromatin relaxation and allows exposure of DNA to transcriptional machinery, thus initiating gene expression activity. HAT1: CRISPR Cas9 construct results in the histone acetylation of the promoter of abscisic acid-responsive element-binding protein 1 (AREB1), a key drought-responsive gene. The expression of *AREB1* gene was enhanced and the transgenic *Arabidopsis* plants showed enhanced tolerance to drought stress as compared to the control plants (de Melo et al., 2020). Similarly, the fusion of Cas9 with effector domains like KRAB/SID (Kruppel-associated box/Silencing Domain) can increase the transcriptional repression of genes that are negative regulators of stress tolerance in plants. Hence, CRISPR-based epigenome editing is a smart, powerful and accurate approach to enhance the abiotic stress ability of plants (Miglani et al., 2020).

22.6 PRACTICAL APPLICATIONS IN CROP IMPROVEMENT

22.6.1 Epigenetic Breeding for Stress Tolerance

22.6.1.1 Strategies for Integrating Epigenetic Insights into Crop Breeding Programs for Drought and Salinity Tolerance

The merger of epigenetics with breeding programs for drought and salt stress resistance in plants is a promising technique to enhance the resilience of crops without modifying the genetic sequence. As epigenetic modifications such as non-coding RNAs, DNA methylation and histone modifications regulate the expression of stress-responsive genes, epigenetics should be incorporated into plant breeding programs to enhance the resilience of crops in response to environmental stresses. As epigenetic variations are heritable, plant breeders can generate stress-tolerant varieties by exploiting natural phenotypic variation in the crop plants. Studies showed that exposure to drought stress caused several epimutations in rice (Lu et al., 2020). These drought-induced epimutations were stable and passed on to successive generations, suggesting that epimutations are heritable, so that epigenetics can be exploited in plant breeding (Zheng et al., 2017). Epialleles play an important role in the heritable transfer of modifications caused by epigenetic modifications and result in the formation of epigenetic recombinant inbred lines (epiRILS). These breeding lines are similar to traditional breeding lines and show heritable epigenetic modification for various traits of agricultural importance (Dalakouras & Vlachostergios, 2021).

22.6.1.2 Marker-Assisted Selection in Epigenome

Epialleles act as markers in epigenetics as they contribute to phenotypic variations in plants, either they occur naturally in the genome or can be artificially induced. Epialleles are reliable epi-markers if they are stable, heritable, and can be used in epi-breeding. Furthermore, epigenome-wide association can be used as a marker to understand phenotypic variation better. After gathering epigenome profiles, epigenome-wide association is performed to know the epigenetic variation that leads to phenotypic variation. Differentially methylated regions (DMRs) are another epi-marker that can be used to select a specific desirable agronomic trait. DMRs associated with metabolism, stress tolerance, fruit ripening and fruit quality have been recently discovered (Kakoulidou et al., 2021; Rajkumar et al., 2020). High-throughput sequencing showed that under salt stress, the tolerant rice plants exhibited various DMR regions in the epigenome while the control samples lacked DMRs. Furthermore, as DMRs affect the chromatin structures of genomic regions, they also influenced the functions of other genes. Thus, these epi-markers can be used to identify new key players involved in the regulation of stress responses and new stress-responsive genes with potential application in crop improvement (Turcotte et al., 2022).

22.6.2 Future Prospects for Epigenetic-Based Crop Improvement

22.6.2.1 Potential of Epigenetic Engineering to Enhance Crop Resilience to Climate Change

Epigenetic engineering has significant potential to improve crop resistance to stresses in adverse environmental conditions. Researchers can develop stress-tolerant crops by elucidating epigenetic stress-responsive regulatory networks; for example, histone modifications or DM patterns can be manipulated to induce drought, salinity and biotic stress tolerance (Kakoulidou et al., 2021). Likewise, identification of stress-responsive ncRNAs and their possible targets could open avenues in breeding programs and genetic engineering aimed at the improvement of stress tolerance of crops. As epigenetic modifications are inherited to several generations, transgenerational stress tolerance of plants can be achieved through epigenetics. Genetic engineers can improve the stress tolerance of crops by introducing genes that encode enzymes involved in epigenetic modifications (Table 22.1) (Chang et al., 2020). Overexpression of genes encoding methyltransferase has improved stress tolerance in many important crops. Similarly, high levels of DM have been linked to improved drought tolerance in maize

TABLE 22.1
Improvement of stress tolerance in crops by epigenetics

Specie name	Epigenetic modification	Stress tolerance	References
Cannabis sativa	Methylation of *COR* gene	Cold acclimatization	Mayer et al. (2015)
Hordeum vulgare	Acetylation of H3K27	Water logging	Wang et al. (2018)
Triticum aestivum	Acetylation of TaHKT1;4	Salt stress tolerance	Kumar et al. (2017a)
Zea mays	Hypermethylation	Drought tolerance	Wang et al. (2021)
Solanum lycoperiscum	Non-coding RNAs	Salt tolerance	Çakır et al. (2021)
Arabidopsis thaliana	DNA methylation	Drought tolerance	Ashapkin et al. (2020)
Gossypium hirsutum	Histone deacetylation	Drought tolerance	Zhang et al. (2020)
Vitis vinifera	Chromatin modeling	Drought, heat tolerance	Kapazoglou et al. (2021)
Oryza sativa	DNA methylation	Drought tolerance	Varotto et al. (2020)
Nicotiana tabacum	DNA methylation	Salt tolerance	Yamagishi and Kikuta (2021)
Cajanus cajan	DNA methylation	Drought tolerance	Sinha et al. (2020)
Vicia faba	DNA methylation	Drought tolerance	Abid et al. (2017)
Oryza sativa	DNA methylation	Drought and salt tolerance	Rajkumar et al. (2020)
Zea mays	DNA methylation	Heat stress tolerance	Qian et al. (2019)

and rice. Conclusively, by understanding histone modifications, DM variations and regulatory RNAs, researchers can develop stress-tolerant crops that will help to achieve food security and overcome environmental adversities (Fan et al., 2024).

22.6.2.2 Opportunities and Challenges in Translating Epigenetic Research to Field Applications

Understanding epigenetic regulations is important for the achievement of sustainable agriculture, crops and the environment. Epigenetic regulation helps to identify key pathways involved in stress regulation and the mechanism to manipulate those pathways to achieve desired traits in plants. The knowledge of epigenetics can also be used to design and adopt efficient management strategies that will help farmers make decisions about crop management practices. Thus, epigenetics plays a key role in enhancing stress tolerance, precision agriculture and conserving biodiversity of plants (Li et al., 2021).

Translating epigenetic research in field or practical applications to develop stress-tolerant plants faces several challenges. The mechanism of epigenetic regulation is complex and the influence of epigenetic marks on gene expression is not yet completely understood, which also makes it difficult to interpret epigenetic data (Varotto et al., 2020). Moreover, interpreting epigenetic data requires sophisticated tools that are extremely costly. The lack of standardized tools and procedures makes it difficult to implement epigenetic research in the field (EL Sabagh et al., 2022; Manasa et al., 2022). Different protocols yield different results, making it difficult to make consistent guidelines for crop breeding through epigenetics. Thus, these drawbacks prevent the research of epigenetics from being implemented in fields. However, these shortcomings can be overcome through continuous efforts, advanced technology, consistent collaboration among agricultural scientists and breeders along with an effective regulatory policy (Manasa et al., 2022).

22.7 CONCLUSION

Epigenetics is the study of heritable changes in gene expression linked to specific phenotypes without the alteration of the underlying DNA sequence. The major epigenetic modifications include DNA methylation, histone modifications, chromatin remodeling and non-coding RNA. Epigenetic regulation of plants in response to stress is a vast area of research full of potential for future applications in crop improvement. Epigenetic modifications perform a major role in modulating the response of many stress-responsive genes; thus, they can affect plant responses to environmental stresses. A better understanding of the mechanism of epigenetic modifications will help researchers improve agricultural productivity through the production of stress-tolerant crops. Recent studies have revealed that the integration of epigenetics with CRISPR and other genetic engineering tools has successfully generated drought-tolerant crops. One of the most unique and significant aspects of epigenetic regulation of plants in response to environmental stress is transgenerational inheritance. This depicts that epigenetic modifications are inheritable, which means that drought or salt stress in a plant can impart stress tolerance in successive generations. This makes epigenetic modifications extremely important in plant breeding for the generation of epigenetic recombinant breeding lines. Thus, epigenetics offers diverse, flexible and precise methods of manipulating gene expression. However, there is a need to standardize tools to measure and interpret epigenetic data and discover epigenomes of more crop plants to generate diverse stress-tolerant crops.

REFERENCES

Abdulraheem, M. I., Xiong, Y., Moshood, A. Y., Cadenas-Pliego, G., Zhang, H., & Hu, J. (2024). Mechanisms of plant epigenetic regulation in response to plant stress: Recent discoveries and Implications. *Plants*, 13(2), 163. https://www.mdpi.com/2223-7747/13/2/163

Abid, G., Mingeot, D., Muhovski, Y., Mergeai, G., Aouida, M., Abdelkarim, S., Aroua, I., El Ayed, M., M'hamdi, M., Sassi, K. & Jebara, M. (2017). Analysis of DNA methylation patterns associated with drought stress response in faba bean (*Vicia faba* L.) using methylation-sensitive amplification polymorphism (MSAP). *Environmental and Experimental Botany*, 142, 34–44.

Akhter, Z., Bi, Z., Ali, K., Sun, C., Fiaz, S., Haider, F. U., & Bai, J. (2021). In response to abiotic stress, DNA methylation confers epigenetic changes in plants. *Plants*, 10(6), 1096.

Al-Harrasi, I., Al-Yahyai, R., & Yaish, M. W. (2018). Differential DNA methylation and transcription profiles in date palm roots exposed to salinity. *PLoS One*, 13(1), e0191492.

Al-Lawati, A., Al-Bahry, S., Victor, R., Al-Lawati, A., & Yaish, M. (2016). Salt stress alters DNA methylation levels in alfalfa (*Medicago* spp). *Genet Mol Res*, 15(1), 15018299.

Angon, P., Tahjib-Ul-Arif, M., Samin, S., Habiba, U., Hossain, M., & Brestic, M. (2022). How do plants respond to combined drought and salinity stress? – A systematic review. *Plants*, 11, 2884.

Asensi-Fabado, M.-A., Amtmann, A., & Perrella, G. (2017). Plant responses to abiotic stress: The chromatin context of transcriptional regulation. *Biochimica et Biophysica Acta (BBA)-Gene Regulatory Mechanisms*, 1860(1), 106–122.

Ashapkin, V. V., Kutueva, L. I., Aleksandrushkina, N. I., & Vanyushin, B. F. (2020). Epigenetic mechanisms of plant adaptation to biotic and abiotic stresses. *International Journal of Molecular Sciences*, 21(20), 7457.

Atkinson, N. J., & Urwin, P. E. (2012). The interaction of plant biotic and abiotic stresses: From genes to the field. *Journal of Experimental Botany*, 63(10), 3523–3543.

Baek, D., Jiang, J., Chung, J.-S., Wang, B., Chen, J., Xin, Z., & Shi, H. (2011). Regulated *AtHKT1* gene expression by a distal enhancer element and DNA methylation in the promoter plays an important role in salt tolerance. *Plant and Cell Physiology*, 52(1), 149–161.

Balazova, E., Balazova, A., & Oblozinsky, M. (2022). Epigenetic modifications in plants-impact on phospholipid signaling and secondary metabolism. *Chemické Listy*, 116(7), 416–422.

Banerjee, A., & Roychoudhury, A. (2017). Epigenetic regulation during salinity and drought stress in plants: Histone modifications and DNA methylation. *Plant Gene*, 11, 199–204. https://doi.org/https://doi.org/10.1016/j.plgene.2017.05.011.

Banerjee, A., Roychoudhury, A., & Krishnamoorthi, B. S. (2016). Emerging techniques to decipher microRNAs (miRNAs) and their regulatory role in conferring abiotic stress tolerance of plants. *Plant Biotechnology Reports*, 10, 185–205.

Begum, Y. (2022). Regulatory role of microRNAs (miRNAs) in the recent development of abiotic stress tolerance of plants. *Gene*, 821, 146283.

Bharti, Mahajan, M., Vishwakarma, A. K., Bhardwaj, J., & Yadav, S. K. (2015). *AtROS1* overexpression provides evidence for epigenetic regulation of genes encoding enzymes of flavonoid biosynthesis and antioxidant pathways during salt stress in transgenic tobacco. *Journal of Experimental Botany*, 66(19), 5959–5969. https://doi.org/10.1093/jxb/erv304.

Binodh, A. K., Kathiresan, P. K., Thankappan, S., & Senthil, A. (2023). Acclimatization of non-cultivated rice landraces to early moisture stress mediated by enzymatic antioxidants and osmolyte accumulation. *Biocatalysis and Agricultural Biotechnology*, 47, 102623.

Bouyahya, A., Bakrim, S., Aboulaghras, S., El Kadri, K., Aanniz, T., Khalid, A., Abdalla, A.N., Abdallah, A.A., Ardianto, C.,Ming, L. C., & El Omari, N. (2024). Bioactive compounds from nature: Antioxidants targeting cellular transformation in response to epigenetic perturbations induced by oxidative stress. *Biomedicine & Pharmacotherapy*, 174, 116432.

Çakır, Ö., Arıkan, B., Karpuz, B., & Turgut-Kara, N. (2021). Expression analysis of miRNAs and their targets related to salt stress in *Solanum lycopersicum* H-2274. *Biotechnology & Biotechnological Equipment*, 35(1), 275–282.

Chang, Y. N., Zhu, C., Jiang, J., Zhang, H., Zhu, J. K., & Duan, C. G. (2020). Epigenetic regulation in plant abiotic stress responses. *Journal of Integrative Plant Biology*, 62(5), 563–580.

Chen, J., Song, Y., Zhang, H., & Zhang, D. (2013). Genome-wide analysis of gene expression in response to drought stress in *Populus simonii*. *Plant Molecular Biology Reporter*, 31, 946–962.

Chen, K., Li, G. J., Bressan, R. A., Song, C. P., Zhu, J. K., & Zhao, Y. (2020). Abscisic acid dynamics, signaling, and functions in plants. *Journal of Integrative Plant Biology*, 62(1), 25–54.

Chen, X., Feng, Y., Quinn, R. J., Pountney, D. L., Richardson, D. R., Mellick, G. D., & Ma, L. (2023). Potassium channels in Parkinson's disease: Potential roles in its pathogenesis and innovative molecular targets for treatment. *Pharmacological Reviews*, 75(4), 758–788.

Cheong, J. J. (2021). Genetic and epigenetic changes in plants in response to abiotic stress. *Genes*, *12*(10), 1603.

Chung, S., Kwon, C., & Lee, J.-H. (2022). Epigenetic control of abiotic stress signaling in plants. *Genes & Genomics*, 44(3), 267–278..

Chwialkowska, K., Nowakowska, U., Mroziewicz, A., Szarejko, I., & Kwasniewski, M. (2016). Water-deficiency conditions differently modulate the methylome of roots and leaves in barley (*Hordeum vulgare* L.). *Journal of Experimental Botany*, 67(4), 1109–1121.

Dalakouras, A., & Vlachostergios, D. (2021). Epigenetic approaches to crop breeding: Current status and perspectives. *Journal of Experimental Botany*, 72(15), 5356–5371.

Dar, F. A., Mushtaq, N. U., Saleem, S., Rehman, R. U., Dar, T. U. H., & Hakeem, K. R. (2022). Role of epigenetics in modulating phenotypic plasticity against abiotic stresses in plants. *International Journal of Genomics*, 2022(1), 1092894.

Dasgupta, P., & Chaudhuri, S. (2019). Analysis of DNA methylation profile in plants by Chop-PCR. In *Plant innate immunity: Methods and protocols* (pp. 79-90). New York, NY: Springer New York. doi: 10.1007/978-1-4939-9458-8_9. PMID: 31041765.

de Melo, B. P., Lourenço-Tessutti, I. T., Paixão, J. F. R., Noriega, D. D., Silva, M. C. M., de Almeida-Engler, J., Fontes, E.P.B., & Grossi-de-Sa, M. F., (2020). Transcriptional modulation of AREB-1 by CRISPRa improves plant physiological performance under severe water deficit. *Scientific Reports*, 10(1), 16231.

Dhankher, O. P., & Foyer, C. H. (2018). Climate resilient crops for improving global food security and safety. *Plant, Cell, & Environment*, 41, 877–884.

Domínguez, C. (2022). *Public-private Partnerships in the Development and Delivery of Crop Variety Innovations: Lessons from the MasAgro Consortium for Hybrid Maize in Mexico*. Reading: University of Reading. (Doctoral dissertation, University of Reading).

Dos Santos, T. B., Ribas, A. F., de Souza, S. G. H., Budzinski, I. G. F., & Domingues, D. S. (2022). Physiological responses to drought, salinity, and heat stress in plants: A review. *Stresses*, 2(1), 113–135.

Duan, C.-G., Zhu, J.-K., & Cao, X. (2018). Retrospective and perspective of plant epigenetics in China. *Journal of Genetics and Genomics*, 45(11), 621–638.

Duarte-Aké, F., Us-Camas, R., & De-la-Peña, C. (2023). Epigenetic regulation in heterosis and environmental stress: The challenge of producing hybrid epigenomes to face climate change. *Epigenomes*, 7(3), 14.

Duscha, K., Martins Rodrigues, C., Müller, M., Wartenberg, R., Fliegel, L., Deitmer, J. W., Jung M., Zimmermann R & Neuhaus, H. E. (2020). 14-3-3 proteins and other candidates form protein-protein interactions with the cytosolic C-terminal end of *SOS1* affecting its transport activity. *International Journal of Molecular Sciences*, *21*(9), 3334.

Egamberdieva, D., Wirth, S., Bellingrath-Kimura, S. D., Mishra, J., & Arora, N. K. (2019). Salt-tolerant plant growth promoting rhizobacteria for enhancing crop productivity of saline soils. *Frontiers in Microbiology*, 10, 2791.

EL Sabagh, A., Islam, M. S., Hossain, A., Iqbal, M. A., Mubeen, M., Waleed, M., Reginato, M., Battaglia, M., Ahmed, S., Rehman, A., Arif, M., (2022). Phytohormones as growth regulators during abiotic stress tolerance in plants. *Frontiers in Agronomy*, *4*, 765068.

Etalo, D. W., Stulemeijer, I. J., Peter van Esse, H., de Vos, R. C., Bouwmeester, H. J., & Joosten, M. H. (2013). System-wide hypersensitive response-associated transcriptome and metabolome reprogramming in tomato. *Plant Physiology*, 162(3), 1599–1617.

Fadul, S. M., Arshad, A., & Mehmood, R. (2023). CRISPR-based epigenome editing: Mechanisms and applications. *Epigenomics*, 15(21), 1137–1155.

Fan, Y., Sun, C., Yan, K., Li, P., Hein, I., Gilroy, E. M., Kear, P., Bi, Z., Yao, P., & Liu, Z. (2024). Recent advances in studies of genomic DNA methylation and its involvement in regulating drought stress response in crops. *Plants*, 13(10), 1400.

Feng, S. J., Liu, X. S., Tao, H., Tan, S. K., Chu, S. S., Oono, Y. Zhang, X.D., Chen, J., & Yang, Z. M. (2016). Variation of DNA methylation patterns associated with gene expression in rice (*Oryza sativa*) exposed to cadmium. *Plant, Cell & Environment*, 39(12), 2629–2649.

Fichman, Y., Zandalinas, S. I., Peck, S., Luan, S., & Mittler, R. (2022). HPCA1 is required for systemic reactive oxygen species and calcium cell-to-cell signaling and plant acclimation to stress. *The Plant Cell*, 34(11), 4453–4471.

Fleta-Soriano, E., & Munné-Bosch, S. (2016). Stress memory and the inevitable effects of drought: A physiological perspective. *Frontiers in Plant Science*, 7, 143.

Ganotra, J., Sharma, B., Biswal, B., Bhardwaj, D., & Tuteja, N. (2022). Emerging role of small GTPases and their interactome in plants to combat abiotic and biotic stress. *Protoplasma*, *260*(4), 1007–1029.. https://doi.org/10.1007/s00709-022-01830-6.

Garai, S., Citu, Singla-Pareek, S. L., Sopory, S. K., Kaur, C., & Yadav, G. (2021). Complex networks of prion-like proteins reveal cross talk between stress and memory pathways in plants. *Frontiers in Plant Science*, 12, 707286.

Garg, R., Narayana Chevala, V., Shankar, R., & Jain, M. (2015). Divergent DNA methylation patterns associated with gene expression in rice cultivars with contrasting drought and salinity stress response. *Scientific Reports*, 5, 14922. https://doi.org/10.1038/srep14922.

Gayatri, V., Sergey, S., Anne-Aliénor, V., Gopalasamudram Neelakantan, H., Suji, S., Shalini, P., Sellamuthu, G., Harikrishnan, M., Kumari, K., Shabala, L., &Zhong-Hua, C. (2021). To exclude or to accumulate? Revealing the role of the sodium HKT1;5 transporter in plant adaptive responses to varying soil salinity. *Plant Physiology and Biochemistry*, 169, 333–342. https://doi.org/https://doi.org/10.1016/j.plaphy.2021.11.030.

Giacopelli, B. J., & Hollick, J. B. (2015). Trans-homolog interactions facilitating paramutation in maize. *Plant Physiology*, 168(4), 1226–1236.

Goswami, A., Banerjee, R., & Raha, S. (2010). Mechanisms of plant adaptation/memory in rice seedlings under arsenic and heat stress: Expression of heat-shock protein gene *HSP70*. *AoB Plants*, 2010, plq023.

Gray, S. B., & Brady, S. M. (2016). Plant developmental responses to climate change. *Developmental Biology*, 419(1), 64–77.

Grehl, C., Wagner, M, Lemnian, I., Glaser, B., & Grosse, I. (2020). Performance of mapping approaches for whole-genome bisulfite sequencing data in crop plants. *Frontiers in Plant Science*, 11, 176.

Guo, Z., Ma, X., Zhang, R. X., & Yan, H. (2023). Oxidative stress, epigenetic regulation and pathological processes of lens epithelial cells underlying diabetic cataract. *Advances in Ophthalmology Practice and Research*, 3(4), 180–186.

Hailu, B., & Mehari, H. (2021). Impacts of soil salinity/sodicity on soil-water relations and plant growth in dry land areas: A review. *Journal of Natural Sciences Research*, 12, 1–10.

Hatzig, S. V., Nuppenau, J.-N., Snowdon, R. J., & Schießl, S. V. (2018). Drought stress has transgenerational effects on seeds and seedlings in winter oilseed rape (*Brassica napus* L.). *BMC Plant Biology*, 18, 1–13.

Hauser, M.-T., Aufsatz, W., Jonak, C., & Luschnig, C. (2011). Transgenerational epigenetic inheritance in plants. *Biochimica et Biophysica Acta (BBA)-Gene Regulatory Mechanisms*, 1809(8), 459–468.

He, M.-Y., Ren, T. X., Jin, Z. D., Deng, L., Liu, H. J., Cheng, Y. Y., Li, Z. Y., Liu, X. X., Yang, Y., &Chang, H. (2023). Precise analysis of potassium isotopic composition in plant materials by multi-collector inductively coupled plasma mass spectrometry. *Spectrochimica Acta Part B: Atomic Spectroscopy*, 209, 106781.

He, M. Y., Ren, T. X., Jin, Z. D., Deng, L., Liu, H. J., Cheng, Y. Y., Li, Z. Y., Liu, X.X., Yang, Y., & Chang, H. (2023). Precise analysis of potassium isotopic composition in plant materials by multi-collector inductively coupled plasma mass spectrometry. *Spectrochimica Acta Part B: Atomic Spectroscopy*, *209*, 106781

Hemenway, E. A., & Gehring, M. (2023). Epigenetic regulation during plant development and the capacity for epigenetic memory. *Annual Review of Plant Biology*, 74(1), 87–109.

Jagadhesan, B., Das, S., Singh, D., Jha, S. K., Durgesh, K., & Sathee, L. (2022). Micro RNA mediated regulation of nutrient response in plants: The case of nitrogen. *Plant Physiology Reports*, 27(3), 345–357.

Jeremias, G., Gonçalves, F. J., Asselman, J., & Pereira, J. L. (2022). Epigenetic regulation: The cross-talk among development, adaptive strategies, and microevolutionary change. In *Epigenetics, Development, Ecology and Evolution* (pp. 5–33). Cham: Springer International Publishing.

Jogam, P., Sandhya, D., Alok, A., Peddaboina, V., Allini, V. R., & Zhang, B. (2022). A review on CRISPR/Cas-based epigenetic regulation in plants. *International Journal of Biological Macromolecules*, 219, 1261–1271.

Jonah, P., Bello, L., Lucky, O., Midau, A., & Moruppa, S. (2011). The importance of molecular markers in plant breeding programmes. *Global Journal of Science Frontier Research*, 11(5), 5–12.

Joshi, S., Nath, J., Singh, A. K., Pareek, A., & Joshi, R. (2022). Ion transporters and their regulatory signal transduction mechanisms for salinity tolerance in plants. *Physiologia Plantarum*, *174*(3), e13702.

Kakoulidou, I., Avramidou, E. V., Baránek, M., Brunel-Muguet, S., Farrona, S., Johannes, F., Kaiserli, E., Lieberman-Lazarovich, M., Martinelli, F., Mladenov, V., Testillano, P. S., Vassileva, V., & Maury, S. (2021). Epigenetics for crop improvement in times of global change. Biology, 10(8), 766.

Kambona, C. M., Koua, P. A., Léon, J., & Ballvora, A. (2023). Intergenerational and transgenerational effects of drought stress on winter wheat (*Triticum aestivum* L.). *Physiologia Plantarum*, *175*(4), e13951.

Kapazoglou, A., Tani, E., Avramidou, E. V., Abraham, E. M., Gerakari, M., Megariti, S., Doupis, G., & Doulis, A. G. (2021). Epigenetic changes and transcriptional reprogramming upon woody plant grafting for crop sustainability in a changing environment. *Frontiers in Plant Science*, *11*, 613004.

Karan, R., DeLeon, T., Biradar, H., & Subudhi, P. K. (2012). Salt stress induced variation in DNA methylation pattern and its influence on gene expression in contrasting rice genotypes. *PLoS One*, 7(6), e40203. https://doi.org/10.1371/journal.pone.0040203.

Khan, N., Bano, A., Ali, S., & Babar, M. A. (2020). Crosstalk amongst phytohormones from planta and PGPR under biotic and abiotic stresses. *Plant Growth Regulation*, 90(2), 189–203. https://doi.org/10.1007/s10725-020-00571-x.

Kim, J.-H. (2021). Multifaceted chromatin structure and transcription changes in plant stress response. *International Journal of Molecular Sciences*, 22(4), 2013.

Kim, J. M., Sasaki, T., Ueda, M., Sako, K., & Seki, M. (2015). Chromatin changes in response to drought, salinity, heat, and cold stresses in plants. *Frontiers Plant Science*, 6, 114. https://doi.org/10.3389/fpls.2015.00114.

Kowluru, R. A. (2023). Cross talks between oxidative stress, inflammation and epigenetics in diabetic retinopathy. *Cells*, 12(2), 300. https://www.mdpi.com/2073-4409/12/2/300.

Kumar, R., Chauhan, P. K., & Khurana, A. (2016). Identification and expression profiling of DNA methyltransferases during development and stress conditions in Solanaceae. *Functional & Integrative Genomics*, 16, 513–528.

Kumar, S., Beena, A., Awana, M., & Singh, A. (2017a). Physiological, biochemical, epigenetic and molecular analyses of wheat (*Triticum aestivum*) genotypes with contrasting salt tolerance. *Frontiers in Plant Science*, 8, 1151. https://doi.org/10.3389/fpls.2017.01151.

Kumar, S., Beena, A. S., Awana, M., & Singh, A. (2017b). Salt-induced tissue-specific cytosine methylation downregulates expression of *HKT* genes in contrasting wheat (*Triticum aestivum* L.) genotypes. *DNA and Cell Biology*, 36(4), 283–294.

Lagiotis, G., Madesis, P., & Stavridou, E. (2023). Echoes of a stressful past: Abiotic stress memory in crop plants towards enhanced adaptation. *Agriculture*, 13(11), 2090.

Lam, Y., Winch, P. J., Nizame, F. A., Broaddus-Shea, E. T., Harun, M. G. D., & Surkan, P. J. (2022). Salinity and food security in southwest coastal Bangladesh: Impacts on household food production and strategies for adaptation. *Food Security*, *14*(1), 229–248.

Leng, G., & Hall, J. (2019). Crop yield sensitivity of global major agricultural countries to droughts and the projected changes in the future. *Science of the Total Environment*, 654, 811–821.

Li, J., & Yang, Y. (2023). How do plants maintain pH and ion homeostasis under saline-alkali stress? [Review]. *Frontiers in Plant Science*, *14*, 1217193. https://doi.org/10.3389/fpls.2023.1217193.

Li, R., Hu, F., Li, B., Zhang, Y., Chen, M., Fan, T., & Wang, T. (2020). Whole genome bisulfite sequencing methylome analysis of mulberry (*Morus alba*) reveals epigenome modifications in response to drought stress. *Scientific Reports*, 10(1), 8013.

Li, S., He, X., Gao, Y., Zhou, C., Chiang, V. L., & Li, W. (2021). Histone acetylation changes in plant response to drought stress. *Genes*, 12(9), 1409.

Li, W., Deng, M., Wang, S., Wang, C., Guo, M., Song, Y., Guo, J., Yan, J., Ma, F., Guan, Q., & Xu, J. (2023). HISTONE DEACETYLASE 6 interaction with ABSCISIC ACID-INSENSITIVE 5 decreases apple drought tolerance. *Plant Physiology*, *193*(4), 2711–2733

Liang, Q., Wang, Q., Zhang, Q., Wei, J., Lim, SX., Zhu, R., Hu, J., Wei, W., Lee, C., Sow, C., Zhang, W., & Wee, ATS. High-performance, room temperature, ultra-broadband photodetectors based on air-stable PdSe2. *Advanced Materials*, *31*(24), 1807609.

Liu, X., Jiang, W., Li, Y., Nie, H., Cui, L., Li, R., Tan, L., Peng, L., Li, C., &Luo, J. (2023). FERONIA coordinates plant growth and salt tolerance via the phosphorylation of phyB. *Nature Plants*, 9(4), 645–660.

Lloyd, J. P., & Lister, R. (2022). Epigenome plasticity in plants. *Nature Reviews Genetics*, 23(1), 55–68.

Lottering, S., Mafongoya, P., & Lottering, R. (2021). Drought and its impacts on small-scale farmers in sub-Saharan Africa: A review. *South African Geographical Journal*, 103(3), 319–341.

Lu, L., Zhai, X., Li, X., Wang, S., Zhang, L., Wang, L.,Jin, X., Liang, L., Deng, Z., & Li, Z. (2022). Met1-specific motifs conserved in OTUB subfamily of green plants enable rice OTUB1 to hydrolyse Met1 ubiquitin chains. *Nature Communications*, 13(1), 4672.

Lu, X., & Hyun, T. K. (2021). The role of epigenetic modifications in plant responses to stress. *Botanica Serbica*, 45(1), 3–12.

Lu, Y., Zhou, D.-X., & Zhao, Y. (2020). Understanding epigenomics based on the rice model. *Theoretical and Applied Genetics*, 133, 1345–1363.

Liu, X., Jiang, W., Li, Y., Nie, H., Cui, L., Li, R., Tan, L., Peng, L., Li, C., & Zhao, C. (2023). FERONIA coordinates plant growth and salt tolerance via the phosphorylation of phyB. *Nature Plants*, *9*(4), 645–660.

Luo, X., & He, Y. (2020). Experiencing winter for spring flowering: A molecular epigenetic perspective on vernalization. *Journal of Integrative Plant Biology*, 62(1), 104–117.

Mahmood, M. Z., Odeibat, H. A., Ahmad, R., Gatasheh, M. K., Shahzad, M., & Abbasi, A. M. (2024). Low apoplastic Na^+ and intracellular ionic homeostasis confer salinity tolerance upon Ca2SiO4 chemigation in Zea mays L. under salt stress. *Frontiers in Plant Science*, 14, 1268750. https://doi.org/10.3389/fpls.2023.1268750.

Majnarić, L. T., Bosnić, Z., Štefanić, M., & Wittlinger, T. (2022). Cross-talk between the cytokine IL–37 and thyroid hormones in modulating chronic inflammation associated with target organ damage in age-related metabolic and vascular conditions. *International Journal of Molecular Sciences*, 23(12), 6456.

Ma, Y., Dias, M. C., & Freitas, H. (2020). Drought and salinity stress responses and microbe-induced tolerance in plants. *Frontiers in Plant Science*, *11*, 591911.

Mali, P., Yang, L., Esvelt, K. M., Aach, J., Guell, M., DiCarlo, J. E., Norville J.E., & Church, G. M. (2013). RNA-guided human genome engineering via Cas9. *Science*, *339*(6121), 823–826.

Manasa S, L., Panigrahy, M., Panigrahi, K. C., & Rout, G. R. (2022). Overview of cold stress regulation in plants. *The Botanical Review*, 88(3), 359–387.

Manh, M. B., Ost, C., Peiter, E., Hause, B., Krupinska, K., & Humbeck, K. (2023). WHIRLY1 acts upstream of ABA-related reprogramming of drought-induced gene expression in barley and affects stress-related histone modifications. *International Journal of Molecular Sciences*, 24(7), 6326.

Marone, D., Russo, M. A., Mores, A., Ficco, D. B., Laidò, G., Mastrangelo, A. M., & Borrelli, G. M. (2021). Importance of landraces in cereal breeding for stress tolerance. *Plants*, 10(7), 1267.

Mayer, B. F., Ali-Benali, M. A., Demone, J., Bertrand, A., & Charron, J. B. (2015). Cold acclimation induces distinctive changes in the chromatin state and transcript levels of COR genes in *Cannabis sativa* varieties with contrasting cold acclimation capacities. *Physiologia Plantarum*, 155(3), 281–295.

Mehrabi, Z., Delzeit, R., Ignaciuk, A., Levers, C., Braich, G., Bajaj, K., Amo-Aidoo, A., Anderson, W., Balgah, R.A., Benton, T. G. & Chari, M.M.(2022). Research priorities for global food security under extreme events. *One Earth*, 5(7), 756–766.

Miglani, G. S., Kaur, A., & Kaur, L. (2020). Plant gene expression control using genome-and epigenome-editing technologies. *Journal of Crop Improvement*, 34(1), 1–63.

Minnoye, L., Marinov, G. K., Krausgruber, T., Pan, L., Marand, A. P., Secchia, S., Greenleaf, W.J., Furlong, E.E., Zhao, K.,Schmitz, R. J. & Bock, C. (2021). Chromatin accessibility profiling methods. *Nature Reviews Methods Primers*, 1(1), 10.

Moglia, A., Gianoglio, S., Acquadro, A., Valentino, D., Milani, A. M., Lanteri, S., & Comino, C. (2019). Identification of DNA methyltransferases and demethylases in *Solanum melongena* L., and their transcription dynamics during fruit development and after salt and drought stresses. *PLoS One*, 14(10), e0223581.

Mozgova, I., Mikulski, P., Pecinka, A., & Farrona, S. (2019). Epigenetic mechanisms of abiotic stress response and memory in plants. *Epigenetics in Plants of Agronomic Importance: Fundamentals and Applications: Transcriptional Regulation and Chromatin Remodelling in Plants*, 1–64.

Muchate, N. S., Nikalje, G. C., Rajurkar, N. S., Suprasanna, P., & Nikam, T. D. (2016). Plant salt stress: Adaptive responses, tolerance mechanism and bioengineering for salt tolerance. *The Botanical Review*, 82, 371–406.

Mulet, J. M., Porcel, R., & Yenush, L. (2023). Modulation of potassium transport to increase abiotic stress tolerance in plants. *Journal of Experimental Botany*, 74(19), 5989–6005.

Myers , R. J., Fichman, Y., Zandalinas, S. I., & Mittler, R. (2023). Jasmonic acid and salicylic acid modulate systemic reactive oxygen species signaling during stress responses. *Plant Physiology*, 191(2), 862–873.

Naderi, S., Maali-Amiri, R., Sadeghi, L., & Hamidi, A. (2024). Physio-biochemical and DNA methylation analysis of the defense response network of wheat to drought stress. *Plant Physiology and Biochemistry*, 209, 108516.

Nakato, R., & Sakata, T. (2021). Methods for ChIP-seq analysis: A practical workflow and advanced applications. *Methods*, 187, 44–53.

Nguyen, N. H., Vu, N. T., & Cheong, J.-J. (2022). Transcriptional stress memory and transgenerational inheritance of drought tolerance in plants. *International Journal of Molecular Sciences*, 23(21), 12918.

Niculescu, A., Le-Niculescu, H., Roseberry, K., Wang, S., Hart, J., Kaur, A., Robertson, H., Jones, T., Strasburger, A., Williams, A., Kurian, S.M., Lamb, B., Shekhar, A., Lahiri, D.K., & Saykin, A.J. (2020). Blood biomarkers for memory: Toward early detection of risk for Alzheimer disease, pharmacogenomics, and repurposed drugs. *Molecular Psychiatry*, 25(8), 1651–1672.

Nunez-Vazquez, R., Desvoyes, B., & Gutierrez, C. (2022). Histone variants and modifications during abiotic stress response. *Frontiers in Plant Science*, 13, 984702.

Ortas, I., Rafique, M., & Çekiç, F. (2021). Do Mycorrhizal Fungi enable plants to cope with abiotic stresses by overcoming the detrimental effects of salinity and improving drought tolerance?. In *Symbiotic Soil Microorganisms. Soil Biology*, 60, 391–428. Cham: Springer International Publishing.

Pan, W., Yu-Hang, H., Liao-Yuan, Z., Yong, M., Zhen-Fu, M., Tian-Ru, W., Zhang, Z.L., Fang, J. C., Liu, X. Q., You, Q., & Zhang, Y.. (2024). Experimental study of the influencing factors and mechanisms of the pressure-reduction and augmented injection effect by nanoparticles in ultra-low permeability reservoirs. *Petroleum Science*, 21(3), 1915–1927. https://doi.org/https://doi.org/10.1016/j.petsci.2023.11.024.

Pandey, G., Sharma, N., Pankaj Sahu, P., & Prasad, M. (2016). Chromatin-based epigenetic regulation of plant abiotic stress response. *Current Genomics*, 17(6), 490–498.

Park, P. J. (2009). ChIP–seq: Advantages and challenges of a maturing technology. *Nature Reviews Genetics*, 10(10), 669–680.

Pecinka, A., Chevalier, C., Colas, I., Kalantidis, K., Varotto, S., Krugman, T., Michailidis, C., Vallés, M. P., Muñoz, A. & Pradillo, M. (2020). Chromatin dynamics during interphase and cell division: Similarities and differences between model and crop plants. *Journal of Experimental Botany*, 71(17), 5205–5222.

Pottosin, I., & Dobrovinskaya, O. (2014). Non-selective cation channels in plasma and vacuolar membranes and their contribution to K+ transport. *Journal of Plant Physiology*, 171(9), 732–742.

Qian, Y., Hu, W., Liao, J., Zhang, J., & Ren, Q. (2019). The Dynamics of DNA methylation in the maize (*Zea mays* L.) inbred line B73 response to heat stress at the seedling stage. *Biochemical and Biophysical Research Communications*, 512(4), 742–749.

Qiu, Q.-S., Guo, Y., Quintero, F. J., Pardo, J. M., Schumaker, K. S., & Zhu, J.-K. (2004). Regulation of vacuolar Na^+/H^+ exchange in *Arabidopsis thaliana* by the salt-overly-sensitive (SOS) pathway. *Journal of Biological Chemistry*, 279(1), 207–215.

Qiu, Z., He, Y., Zhang, Y., Guo, J., & Wang, L. (2018). Characterization of miRNAs and their target genes in He-Ne laser pretreated wheat seedlings exposed to drought stress. *Ecotoxicology and Environmental Safety*, 164, 611–617.

Rajkumar, M. S., Shankar, R., Garg, R., & Jain, M. (2020). Bisulphite sequencing reveals dynamic DNA methylation under desiccation and salinity stresses in rice cultivars. *Genomics*, 112(5), 3537–3548.

Rajput, S., Sengupta, P., Kohli, I., Varma, A., Singh, P. K., & Joshi, N. C. (2022). Role of *Piriformospora indica* in inducing soil microbial communities and drought stress tolerance in plants. In *New and Future Developments in Microbial Biotechnology and Bioengineering* (pp. 93–110). Elsevier.

Ramakrishnan, M., Papolu, P. K., Satish, L., Vinod, K. K., Wei, Q., Sharma, A., Emamverdian, A., Zou, L.H., &Zhou, M. (2022). Redox status of the plant cell determines epigenetic modifications under abiotic stress conditions and during developmental processes. *Journal of Advanced Research*, 42, 99–116.

Ramírez, D., Rolando, J., Yactayo, W., Monneveux, P., Mares, V., & Quiroz, R. (2015). Improving potato drought tolerance through the induction of long-term water stress memory. *Plant Science*, 238, 26–32.

Rehman, M., Tanti, B. (2020). Understanding epigenetic modifications in response to abiotic stresses in plants. *Biocatalysis and Agricultural Biotechnology*, 27, 101673.

Roca Paixão, J. F., Gillet, F.-X., Ribeiro, T. P., Bournaud, C., Lourenço-Tessutti, I. T., Noriega, D. D., Melo, B.P., de Almeida-Engler, J., & Grossi-de-Sa, M. F. (2019). Improved drought stress tolerance in *Arabidopsis* by CRISPR/dCas9 fusion with a Histone AcetylTransferase. *Scientific Reports*, 9(1), 8080.

Roy, S., & Chowdhury, N. (2020). *Salt Stress in Plants and Amelioration Strategies: A Critical Review.* IntechOpen.

Rudolf, J., Tomovicova, L., Panzarova, K., Fajkus, J., Hejatko, J., & Skalak, J. (2024). Epigenetics and plant hormone dynamics: A functional and methodological perspective. *Journal of Experimental Botany*, *75*(17), 5267–5294. .

Sadhukhan, A., Prasad, S. S., Mitra, J., Siddiqui, N., Sahoo, L., Kobayashi, Y., & Koyama, H. (2022). How do plants remember drought? *Planta*, 256(1), 7.

Saez, A., Rodrigues, A., Santiago, J., Rubio, S., & Rodriguez, P. L. (2008). HAB1-SWI3B interaction reveals a link between abscisic acid signaling and putative SWI/SNF chromatin-remodeling complexes in *Arabidopsis. Plant Cell*, 20(11), 2972–2988. https://doi.org/10.1105/tpc.107.056705.

Sallam, N., & Moussa, M. (2021). DNA methylation changes stimulated by drought stress in ABA-deficient maize mutant vp10. *Plant Physiology and Biochemistry*, 160, 218–224.

Sardans, J., & Peñuelas, J. (2021). Potassium control of plant functions: Ecological and agricultural implications. *Plants*, 10(2), 419.

Satrio, R.D., Fendiyanto, M.H., Miftahudin, M. (2024). Tools and techniques used at global scale through genomics, transcriptomics, proteomics, and etabolomics to investigate plant stress responses at the molecular level. In *Molecular Dynamics of Plant Stress and Its Management* (pp. 555–607). Singapore: Springer Nature Singapore.

Seifikalhor, M., Aliniaeifard, S., Shomali, A., Azad, N., Hassani, B., Lastochkina, O., & Li, T. (2019). Calcium signaling and salt tolerance are diversely entwined in plants. *Plant Signaling & Behavior*, 14(11), 1665455.

Selvaraj, A., Gautam, J., Verma, S., Verma, G., & Jain, S. (2021). Life cycle sustainability assessment of crops in India. *Current Research in Environmental Sustainability*, 3, 100074.

Shah, S. H., Islam, S., Mohammad, F., & Siddiqui, M. H. (2023). Gibberellic acid: A versatile regulator of plant growth, development and stress responses. *Journal of Plant Growth Regulation*, 42(12), 7352–7373.

Sharma, A., & Ahammed, G. J. (2024). *Melatonin in Plants: Role in Plant Growth, Development, and Stress Response*. Singapore: Springer.

Sharma, M., Kumar, P., Verma, V., Sharma, R., Bhargava, B., & Irfan, M. (2022). Understanding plant stress memory response for abiotic stress resilience: Molecular insights and prospects. *Plant Physiology and Biochemistry*, 179, 10–24.

Sharma, S., Kaur, R., Kumar, K., Kumar, D., & Solanke, A. K. U. (2021). Genetic variability in *Rubus ellipticus* collections assessed by morphological traits and EST-SSR markers. *Journal of Plant Biochemistry and Biotechnology*, 30, 37–55.

Siddique, A. B., Parveen, S., Rahman, M. Z., & Rahman, J. (2024). Revisiting plant stress memory: Mechanisms and contribution to stress adaptation. *Physiology and Molecular Biology of Plants*, 30(2), 349–367.

Singroha, G., Sharma, P., & Sunkur, R. (2021). Current status of microRNA-mediated regulation of drought stress responses in cereals. *Physiologia Plantarum*, 172(3), 1808–1821.

Shohan, M. U. S., Sinha, S., Nabila, F. H., Dastidar, S. G., & Seraj, Z. I. (2019). HKT1; 5 transporter gene expression and association of amino acid substitutions with salt tolerance across rice genotypes. *Frontiers in Plant Science*, *10*, 1420.

Sinha, P., Singh, V. K., Saxena, R. K., Kale, S. M., Li, Y., Garg, V., Meifang, T., Khan, A. W., Kim, K.D., & Chitikineni, A. (2020). Genome-wide analysis of epigenetic and transcriptional changes associated with heterosis in pigeonpea. *Plant Biotechnology Journal*, 18(8), 1697–1710.

Slama, I., Abdelly, C., Bouchereau, A., Flowers, T., & Savouré, A. (2015). Diversity, distribution and roles of osmoprotective compounds accumulated in halophytes under abiotic stress. *Annals of Botany*, 115(3), 433–447.

Subedi, U., Ozga, J. A., Chen, G., Foroud, N. A., & Singer, S. D. (2020). CRISPR/Cas-mediated genome editing for the improvement of oilseed crop productivity. *Critical Reviews in Plant Sciences*, 39(3), 195–221.

Sudan, J., Raina, M., & Singh, R. (2018). Plant epigenetic mechanisms: Role in abiotic stress and their generational heritability. *3 Biotech*, 8, 1–12.

Sun, M., Yang, Z., Liu, L., & Duan, L. (2022). DNA methylation in plant responses and adaption to abiotic stresses. *International Journal of Molecular Sciences*, 23(13), 6910.

Surdonja, K., Eggert, K., Hajirezaei, M.-R., Harshavardhan, V. T., Seiler, C., Von Wirén, N., Sreenivasulu, N., &Kuhlmann, M. (2017). Increase of DNA methylation at the HvCKX2. 1 promoter by terminal drought stress in barley. *Epigenomes*, 1(2), 9.

Tonosaki, K., Fujimoto, R., Dennis, E. S., Raboy, V., & Osabe, K. (2022). Will epigenetics be a key player in crop breeding?. *Frontiers in Plant Science*, 13, 958350.

Tran, T. L. C., Callahan, D. L., Islam, M. T., Wang, Y., Arioli, T., & Cahill, D. (2023). Comparative metabolomic profiling of *Arabidopsis thaliana* roots and leaves reveals complex response mechanisms induced by a seaweed extract. *Frontiers in Plant Science*, 14, 1114172.

Trippa, D., Scalenghe, R., Basso, M. F., Panno, S., Davino, S., Morone, C.,Giovino, A., Oufensou, S., Luchi, N., Yousefi, S. & Martinelli, F. (2024). Next-generation methods for early disease detection in crops. *Pest Management Science*, 80(2), 245–261.

Turcotte, H., Hooker, J., Samanfar, B., & Parent, J.-S. (2022). Can epigenetics guide the production of better adapted cultivars? *Agronomy*, 12(4), 838.

Ueda, M., Matsui, A., Tanaka, M., Nakamura, T., Abe, T., Sako, K.,Sasaki, T., Kim, J. M., Ito, A., Nishino, N., & Shimada, H.(2017). The distinct roles of class I and II RPD3-like histone deacetylases in salinity stress response. *Plant Physiology*, 175(4), 1760–1773.

Ueda, M., & Seki, M. (2020). Histone modifications form epigenetic regulatory networks to regulate abiotic stress response. *Plant Physiology*, 182(1), 15–26.

Varotto, S., Tani, E., Abraham, E., Krugman, T., Kapazoglou, A., Melzer, R., Radanović, A., & Miladinović, D. (2020). Epigenetics: Possible applications in climate-smart crop breeding. *Journal of Experimental Botany*, 71(17), 5223–5236.

Wahab, A., Abdi, G., Saleem, M. H., Ali, B., Ullah, S., Shah, W., Mumtaz, S., Yasin, G., Muresan, C. C., & Marc, R. A. (2022). Plants' physio-biochemical and phyto-hormonal responses to alleviate the adverse effects of drought stress: A comprehensive review. *Plants*, 11(13), 1620.

Wang, C.-F., Han, G.-L., Yang, Z.-R., Li, Y.-X., & Wang, B.-S. (2022). Plant salinity sensors: Current understanding and future directions. *Frontiers in Plant Science*, 13, 859224.

Wang, L., & Qiao, H. (2020). Chromatin regulation in plant hormone and plant stress responses. *Current Opinion in Plant Biology*, 57, 164–170.

Wang, Q., Liu, C., Dong, Q., Huang, D., Li, C., Li, P., & Ma, F. (2018). Genome-Wide Identification and Analysis of Apple NITRATE TRANSPORTER 1/PEPTIDE TRANSPORTER Family (NPF) Genes Reveals MdNPF6.5 Confers High Capacity for Nitrogen Uptake under Low-Nitrogen Conditions. *International Journal of Molecular Sciences*, 19(9), 2761. https://www.mdpi.com/1422-0067/19/9/2761.

Wang, Q., Xu, J., Pu, X., Lv, H., Liu, Y., Ma, H., Wu, F., Wang, Q., Feng, X.,Liu, T., & Tang, Q. (2021). Maize DNA methylation in response to drought stress is involved in target gene expression and alternative splicing. *International Journal of Molecular Sciences*, 22(15), 8285.

Xie, M., & Yu, B. (2015). siRNA-directed DNA methylation in plants. *Current Genomics*, 16(1), 23–31.

Xu, J., Zhou, S., Gong, X., Song, Y., van Nocker, S., Ma, F., & Guan, Q. (2018). Single-base methylome analysis reveals dynamic epigenomic differences associated with water deficit in apple. *Plant Biotechnology Journal*, 16(2), 672–687.

Yamagishi, K., & Kikuta, Y. (2021). Nucleoside derivatives of 5-methylcytosine suppress 5-azacytidine-induced reactivation of a silent transgene in suspension-cultured tobacco cells. *Plant Biotechnology*, 38(1), 173–178.

Yamamuro, C., Zhu, J.-K., & Yang, Z. (2016). Epigenetic modifications and plant hormone action. *Molecular Plant*, 9(1), 57–70.

Yang, Y., Huang, J., Sun, Q., Wang, J., Huang, L., Fu, S.,Qin, S., Xie, X., Ge, S., Li, X. & Cheng, Z. . (2022). microRNAs: Key players in plant response to metal toxicity. *International Journal of Molecular Sciences*, 23(15), 8642.

Yuan, L., Liu, X., Luo, M., Yang, S., & Wu, K. (2013). Involvement of histone modifications in plant abiotic stress responses. *Journal of Integrative Plant Biology*, 55(10), 892–901. https://doi.org/10.1111/jipb.12060.

Yung, W. S., Li, M. W., Sze, C. C., Wang, Q., & Lam, H. M. (2021). Histone modifications and chromatin remodelling in plants in response to salt stress. *Physiologia Plantarum*, 173(4), 1495–1513.

Zhan, J., & Meyers, B. C. (2023). Plant small RNAs: Their biogenesis, regulatory roles, and functions. *Annual Review of Plant Biology*, 74(1), 21–51.

Zhang, F., Yang, J., Zhang, N., Wu, J., & Si, H. (2022). Roles of microRNAs in abiotic stress response and characteristics regulation of plant. *Frontiers in Plant Science*, 13, 919243.

Zhang, H., Lang, Z., & Zhu, J.-K. (2018). Dynamics and function of DNA methylation in plants. *Nature Reviews Molecular Cell Biology*, 19(8), 489–506.

Zhang, H., Zhang, X., & Xiao, J. (2023). Epigenetic regulation of nitrogen signaling and adaptation in plants. *Plants*, 12(14), 2725.

Zhang, J.-B., He, S.-P., Luo, J.-W., Wang, X.-P., Li, D.-D., & Li, X.-B. (2020). A histone deacetylase, *GhHDT4D*, is positively involved in cotton response to drought stress. *Plant Molecular Biology*, 104(1), 67–79. https://doi.org/10.1007/s11103-020-01024-9.

Zhao, P., Ma, B., Cai, C., & Xu, J. (2022). Transcriptome and methylome changes in two contrasting mungbean genotypes in response to drought stress. *BMC Genomics*, 23(1), 80.

Zhao, S., Zhang, Q., Liu, M., Zhou, H., Ma, C., & Wang, P. (2021). Regulation of plant responses to salt stress. *International Journal of Molecular Sciences*, 22(9), 4609.

Zhao, T., Zhan, Z., & Jiang, D. (2019). Histone modifications and their regulatory roles in plant development and environmental memory. *Journal of Genetics and Genomics*, 46(10), 467–476.

Zheng, X., Chen, L., Xia, H., Wei, H., Lou, Q., Li, M., Li, T., & Luo, L. (2017). Transgenerational epimutations induced by multi-generation drought imposition mediate rice plant's adaptation to drought condition. *Scientific Reports*, 7(1), 39843.

Zhou, R., Kong, L., Yu, X., Ottosen, C.-O., Zhao, T., Jiang, F., & Wu, Z. (2019). Oxidative damage and antioxidant mechanism in tomatoes responding to drought and heat stress. *Acta Physiologiae Plantarum,* 41, 1–11.

Zhu, H., Wen, T., Sun, M., Ali, I., Sheteiwy, M. S., Wahab, A., Tan, W., Wen, C., He, X., & Wang, X. (2023). Enhancing rice yield and nitrogen utilization efficiency through optimal planting density and reduced nitrogen rates. *Agronomy*, 13(5), 1387.

23 Epigenetic Regulation in Plant Tolerance to Heat Stress and Thermomorphogenesis

Louis Noble, M. S. Amritha and Jos T. Puthur

23.1 INTRODUCTION

Plants develop various mechanisms to cope with many environmental pressures, including heat stress during their development and growth (Jablonka & Lamb, 2002). In order to achieve agro-sustainability, it is essential to know how heat stress tolerance is dealt with at the molecular level in the context of rising global temperature (Rapp & Wendel, 2005). It has been shown that epigenetic regulation is a key mechanism in dealing with abiotic stress like heat stress, as it alters gene expression in the absence of alteration of the underlying DNA sequence of any gene (Eichten et al., 2014). DNA methylation, histone modification and chromatin remodeling are some of the ways via which plants deal with stress by changing their transcriptional circuitry and keeping their genomic stability (Gao et al., 2014, He et al., 2022).

DNA methylation can either positively or negatively affect the regulation of the transcription of genes that respond to heat stress, depending on where it is in the genome (Zhang et al., 2021). Hypermethylation of gene promoters may inhibit some unnecessary attributes that are expensive to maintain during stress, whereas the hypomethylation of certain defense genes triggers key metabolic responses, including those of heat shock proteins (Feng et al., 2022). Indeed, some works indicate that plants can use specific patterns of DNA methylation to induce a phenomenon called "heat stress memory", allowing them to tolerate high temperature in the future (Li et al., 2023). There is further evidence to highlight how useful it is for the plants to possess this kind of methylation "memory." Moreover, adaptation to heat stress involves various histone modifications, including but not limited to acetylation and methylation, for instance, histone acetyltransferases like GCN5 (GENERAL CONTROL NON-REPRESSED PROTEIN 5) target specific loci of heat-responsive genes so that they enhance gene expression (Wang et al., 2024). Stress causes the loss of methylation markers like H3K27me3, which restrict gene activity (Kim et al., 2015). These strong and quick changes in transcriptional activities, especially when cells are under severe stress, are made possible by chromatin modifiers and histone variants (He et al., 2022). By remodeling chromatin, this basic change in its structure makes it possible to control gene accessibility. Heat stress moves nucleosomes with the help of chromatin remodelers that need ATP. This makes stress-related loci available for the efficient transcription of protective genes like heat shock proteins (Zhao et al., 2020). This process not only facilitates gene expression but also silences noncritical genes to conserve resources during stress (Kumar & Wigge, 2010). Other chromatin-interacting epigenetic regulators are also involved in the post-stress recovery by maintaining genome integrity (Zheng et al., 2024).

In precise terms, epigenetics controls more than just the stress response. It also controls thermomorphogenesis, which is when changes in development and shape caused by heat are controlled by epigenetics. High temperature modulates hypocotyl elongation and flowering time via histone modification and DNA methylation, illustrating the complex interconnection between environmental signals and epigenetic control (Han et al., 2024, Zhou et al., 2024). These mechanisms are crucial for plants to modify growth patterns in response to varying temperature. Stress memory is also epigenetically regulated, whereby plants maintain specific epigenetic marks to increase responsiveness to repeated heat stress. According to Pratx et al. (2023), stress-responsive loci keep histone marks like H3K4me3 during recovery, which makes the transcriptional response to heat stronger later on. This adaptive trait not only reflects the evolutionary importance of epigenetics but also presents potential for crop improvement.

As climate change intensifies, we can leverage our understanding of epigenetic mechanisms to develop new crops capable of thriving in elevated temperature. It is believed that incorporating the epigenetic component into traditional breeding or biotechnology will improve crop yield in an unfavorable environment (Lämke & Bäurle, 2017). This chapter investigates various aspects of epigenetic regulation of plant heat stress responses and thermomorphogenesis, as well as explaining how these processes work, their relationships and their impacts on sustainable agriculture.

DOI: 10.1201/9781003589914-27

23.2 EPIGENETICS IN PLANTS: MECHANISM AND FUNCTION

Epigenetics is defined as the study of heritable variations in gene expression that do not involve changes to the DNA sequence (Jablonka & Lamb, 2002; Banerjee & Roychoudhury, 2017). These changes are capable of functioning as molecular switches that regulate gene accessibility to transcriptional machinery, which allows for dynamic responses by plants to environmental stimuli (Rapp & Wendel, 2005; Banerjee & Roychoudhury, 2018a). Important changes in DNA methylation, modifications of histones, and variation in histone variants and nucleosome positioning are the fundamental components of epigenetics (Eichten et al., 2014) (Figure 23.1). All these events ensure the desired gene expression in plants to achieve both short- and long-term tolerance to heat stress (Gao et al., 2014, He et al., 2022).

Epigenetic reprogramming in response to heat stress triggers the rapid activation of protective genes in plants; for example, *Arabidopsis thaliana* is capable of rapidly changing DNA methylation patterns and levels of histone acetylation in response to enhanced temperature, which results in the expression of HSPs and other defense genes, such as ABA-inducible genes, that enhance stress responses rapidly (Lämke & Bäurle, 2017, Peirats et al., 2016). Epigenetics helps to ensure genome stability by silencing transposable elements that can be activated under stress conditions (Liu et al., 2021). Maintaining the expression of stress-responsive genes while safeguarding the structural genome integrity explains the significance of epigenetic regulation in plants during heat stress (Lämke & Bäurle, 2017). Research on agricultural crops like wheat and maize has reinforced this notion by showing that these mechanisms are conserved in different species of plants, and thus, their importance is universal (Kim et al., 2015).

23.3 HEAT STRESS-INDUCED CHROMATIN MODIFIERS AND CHROMATIN REMODELING

Chromatin modifiers are specialized proteins that regulate chromatin structure, which facilitates or restricts gene expression. These modifiers include DMTs (DNA methyltransferases) and DDMs (DNA demethylases) for DNA methylation and HATs (histone acetyltransferases), HDACs (histone deacetylases), HMTs (histone methyltransferases) and HDMs (histone demethylases) for histone modifications (Lämke & Bäurle, 2017). Additionally, there are reports on ATP-dependent chromatin remodelers, which enable the repositioning of nucleosomes with the help of ATP hydrolysis to ensure the accessibility of stress-responsive genes (Zhao et al., 2020). SWI/SNF complex is an extensively studied chromatin remodeling complex, especially during heat stress, which can activate heat shock transcription factors (HSFs) that modulate HSP expression (Suter & Widmer, 2013). In case of barley (*Hordeum vulgare*), they demonstrated that chromatin modifiers like SDG8 (SET DOMAIN GROUP 8), a histone methyltransferase, can add methyl groups to H3K4 and activate heat stress-responsive pathways, enhancing thermotolerance (He et al., 2021). Chromatin modifiers coordinate with transcriptional factors and interact with co-activators to form complexes that can direct the epigenetic changes to specific loci. RdDM pathways (RNA-directed DNA methylation) are a similar gene regulatory mechanism during heat stress (Lämke and Bäurle 2017). Analyzing these relationships could help understand the molecular events resulting in heat stress response in plants (He et al., 2021).

Chromatin remodeling involves dynamic changes to chromatin structure, allowing or restricting access to genes. It is an important step in the activation of the heat-responsive genes. While under heat stress, chromatin in the vicinity of stress-related genes undergoes structural modification from heterochromatin to euchromatin; this is done through ATP hydrolysis as reported by Zhao et al. (2020). This chromatin restructuring enables the binding of transcriptional factors to specific gene loci to facilitate the transcription of HSPs and other protective proteins. Research within *Arabidopsis* suggests the existence of chromatin remodeling complexes like CHR (CHROMATIN REMODELING FACTOR), which play a vital role in heat stress adaptation (Han et al., 2024). DNA damage repair also entails chromatin remodeling via SWI2/SNF2–related 1 chromatin remodeling complex (SWR1-C), which is responsible for heat stress-induced DNA repair (Fritsch et al., 2004, Shaked et al., 2006, Rosa et al., 2013). SWR1-mediated DNA damage repair is accomplished in two distinct ways: (1) remodeling of chromatin and (2) histone tail modifications (Van Attikum & Gasser, 2005). Likewise, various plants under high temperature exhibit changes in nucleosome positioning around certain stress-responsive genes, which shows that this mechanism is evolutionarily conserved (Zheng et al., 2024). A further example of this evolutionary mechanism in plants is in the report stating that *A. thaliana* effectively utilizes H2A.Z (a histone variant) in controlling its thermal-sensory response (Kumar & Wigge, 2010). Plants were more sensitive to heat stress when they failed to incorporate this histone variant in the nucleosome. In addition to activation, chromatin remodeling leads to silencing, subsequently leading to the possibility of conserving energy resource availability during times of stress. This reveals how chromatin remodeling is an important aspect in the regulation of stress-responsive genes.

23.4 DNA METHYLATION CHANGES AND HEAT STRESS RESPONSES

As an epigenetic mechanism, DNA methylation has an important role in the activation and suppression of genes that respond to heat stress in plants (Figure 23.2). In general, DNA methylation at the promoters of genes prevents

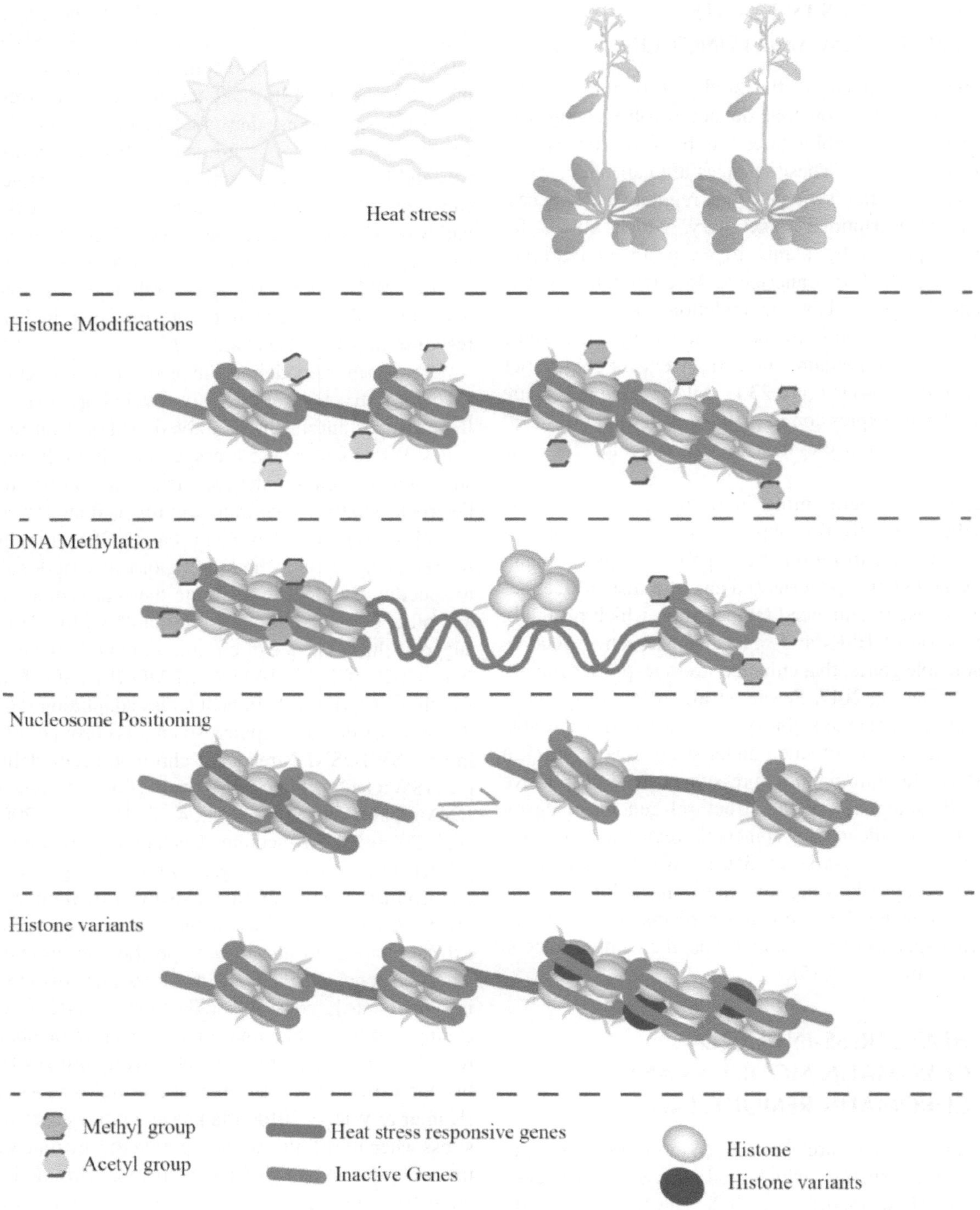

FIGURE 23.1 Schematic representation of epigenetic modifications induced by heat stress, facilitating the activation of stress-responsive genes to enhance stress tolerance and thermomorphogenesis.

transcription by blocking the action of transcription factors (Zhang et al., 2021). In comparison, genes that code for heat shock proteins and other stress-responsive proteins may be expressed in the presence of methylation-free promoter regions (Feng et al., 2022). Modifications to DNA methylation patterns are important in regulating gene expression at the cellular level, which is necessary for plants to efficiently respond to heat stress.

Heat stress can lead to hypermethylation or hypomethylation at targeted genomic regions. Genes responsible for growth and development can be silenced by hypermethylation so that plants can effectively work for stress response activities by conserving energy and resources (Li et al., 2023), whereas genes encoding antioxidant enzymes and molecular chaperones will be activated via hypomethylation (Wang et al., 2024). Added together, altered DNA

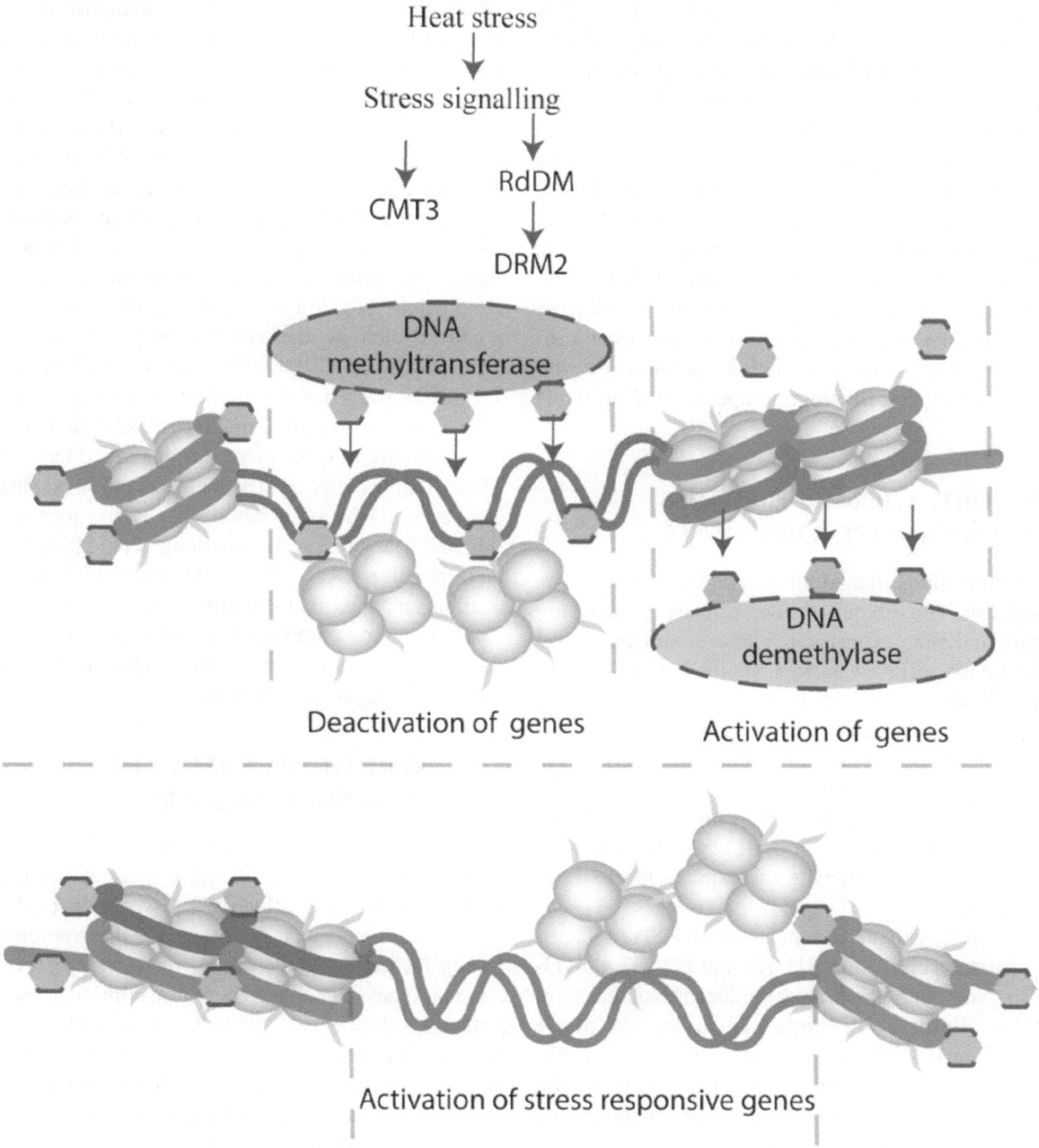

FIGURE 23.2 Heat stress modulates DNA methylation dynamics by activating DNA methyltransferases for gene silencing and DNA demethylases for gene activation. This regulation leads to the activation of heat stress-responsive genes, enhancing stress tolerance.

methylation patterns are responsible for the advanced heat stress memory and provide better tolerance to plants in future heat stress events. Cytosine methylation changes were evaluated using a comparison study on heat stress-tolerant and heat stress-sensitive varieties of *Brassica napus* L. (Gao et al., 2014). Interestingly, it reveals that more demethylation takes place in heat-tolerant genotypes and methylation in heat-sensitive genotypes, which in turn indicates the number of genes subjected to activation and deactivation in plants of different tolerant levels.

In *Arabidopsis thaliana*, heat stress greatly alters the patterns of DNA methylation, which in turn alters the expression of many genes. The most notable DNA region in that regard is the retrotransposon *ONSEN* that is actively expressed under heat stress conditions. Its expression is controlled with the help of DNA methyltransferase CHROMOMETHYLASE3 (CMT3). CMT3 mutations result in high CHH methylation of *ONSEN*, suggesting that this methyltransferase is responsible for transposon activation during heat conditions (Matsunaga et al., 2012). Furthermore, the heat stress-induced gene *AtSN1*, a SINE-type retrotransposon, has been shown to be more methylated at elevated temperature, and this suppression leads to lower expression. This methylation shift exhibited a strong correlation with the components of RNA-directed DNA methylation (RdDM), such as DRM2 (DOMAINS

REARRANGED METHYLTRANSFERASE2), NRPD1 (NUCLEAR RNA POLYMERASE D1) and NRPE1 (NUCLEAR RNA POLYMERASE E1), emphasizing the significant role of RdDM in modulating transcription of the *AtSN1* gene under heat stress (Ito et al., 2011; Matsunaga et al., 2012). Additionally, further research on *Arabidopsis* indicated that heat stress induces DNA demethylation, especially after exposure to normal temperature. Stress-associated genes, such as heat shock proteins and some translation-regulating proteins for which the DNA methylation changes are maintained, are believed to be inducing the state of memory. This means that some species or varieties actively demethylate to counter stress and cope with environmental changes, like high temperature (Korotko et al., 2021).

23.5 MODIFICATION OF HISTONES AND HEAT STRESS RESPONSES

Histone modifications, including acetylation, methylation, phosphorylation and ubiquitination, influence chromatin architecture and gene expression. These modifications are influenced by two primary factors: (1) the type of modification and (2) the position of modification. In addition to DNA methylation, histone acetylation also serves as a modification technique, especially apparent when *Gossypium hirsutum* is subjected to heat stress (He et al., 2022). The chromatin modification targets H3K4me2 (methylation mark) and H4K5ac (acetylation mark), which have been proposed to particularly regulate the expression of stress-responsive genes.

In *Arabidopsis*, enhanced expression of HSPs is facilitated by histone acetylation of H3K9 and H3K27, providing cellular protection against heat-induced damage (Kim et al., 2015). Similarly, in crops like maize, the active transcription mark H3K4, associated with histone methylation, is high at heat-responsive loci. Conversely, the repressive marks, such as H3K27me3, are absent during stress-induced active gene modification. In *Arabidopsis thaliana*, the histone acetyltransferase GCN5 (GENERAL CONTROL NON-REPRESSED PROTEIN 5) facilitates the response to heat stress by modulating the acetylation of histone proteins. GCN5 facilitates the acetylation by adding acetyl functional groups to the lysine residues 9 and 14 in histone H3 (H3K9ac and H3K14ac) during the transcription of heat stress-responsive genes such as *HSFA3* (HEAT SHOCK TRANSCRIPTION FACTOR A3) and *UVH6* (UV HYPERSENSITIVE 6), thereby enhancing thermotolerance (Wang et al., 2024). Test evidence indicates that GCN5 loss-of-function mutants have diminished levels of H3K9ac and H3K14ac at genes associated with suboptimal expression and impaired heat tolerance.

Wu et al. (2022) conducted a genome-wide examination of roses (*Rosa* spp.) and found many genes responsive to heat stress, including those for histone acetylation, such as histone acetyltransferases (HATs) and histone deacetylases (HDACs). This study specifically demonstrated that certain *HAT* genes (*HAG1, HAG4, HAC2*) were increased under high-temperature conditions, indicating that as the temperature-sensitive genes were active, the amount of histone acetylation appeared to increase. However, under heat stress, some *HDAC* (*HDT1, HDA19, RPD3*) genes showed downregulation, which may indicate a decrease in their activity in histone deacetylation. These findings demonstrate how heat stress alters histone acetylation and gene expression differently in roses than under typical growth circumstances. Additionally, during recovery studies, gene products such as the H3K4 methyltransferases SDG25 (SET DOMAIN GROUP 25) and ATX1 (ARABIDOPSIS TRITHORAX1) are essential for sustaining the expression of these genes. Mutation of genes *SDG25* and ATX1 leads to an increase in DNA methylation and reduces H3K4me3. These findings imply that a balance between histone modifications and DNA methylation is necessary for their expression and heat stress response (Song et al., 2021). Chromatin modifiers like HATs, HDACs, HMTs and HDMs are modification enzymes that coordinate these processes (Figure 23.3). Signaling pathways frequently control their activity, demonstrating how biochemical and epigenetic networks interact during stress reactions.

23.6 ROLE OF EPIGENETICS IN HEAT STRESS MEMORY

Plants employ epigenetic alterations such as DNA methylation and histone modification to establish heat stress memory, facilitating effective stress responses. H3K4me3 in *Arabidopsis* plants is maintained at the promoter of heat-responsive genes, which during recovery preserve epigenetic marks and hence enhance transcriptional responses at post-stress (Pratx et al., 2023). The pattern of histone alterations during recovery from heat stress serves as evidence of heat stress memory markers. Moreover, the researchers observed that environmental factors had tangible effects on epigenetic markers. HDA6 (HISTONE DEACETYLASE 6) is another crucial component in histone acetylation-mediated heat stress memory. In *Arabidopsis*, HDA6 interacts with the RdDM system and represses the expression of heat-responsive genes. The mutants exhibited significant abnormalities due to the absence of functional HDA6, indicating the pronounced role of histone modification during heat stress reactions. The findings demonstrate that HDA6 is crucial for the silencing of specific transposable elements and aids in the memory of heat stress, hence facilitating the activation of heat stress-responsive genes (Ramakrishnan et al., 2022).

Recent research has identified several genes whose methylation patterns are altered by heat stress, resulting in their expression and the phenomenon known as heat stress memory. In *Fragaria vesca* (woodland strawberry), the gene associated with the cytochrome P450 family 87 (*CYP87*) exhibited differential methylation in the CHH

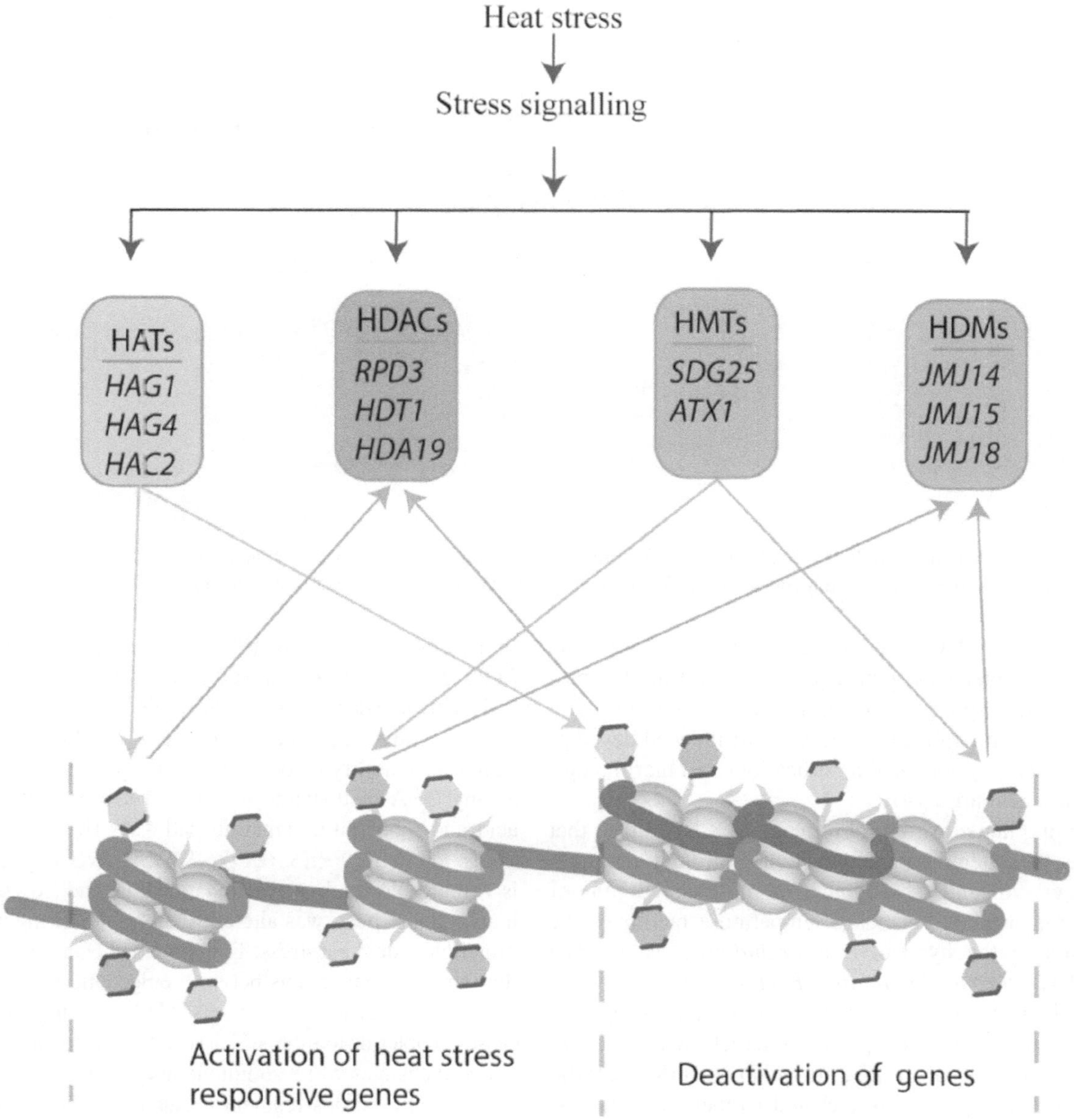

FIGURE 23.3 Heat stress regulated histone modifications by activating histone acetyltransferases (HATs), histone deacetylases (HDACs), histone methyltransferases (HMTs) and histone demethylases (HDMs), leading to the activation or repression of heat stress-responsive genes to enhance stress tolerance.

context inside its promoter region following heat stress exposure. This modified methylation was linked to altered gene expression and may thus be regarded as an adaptive response of the plant to thermal stress. These findings underscore the significance of DNA methylation modifications in the preservation and expression of genes during and following heat stress, particularly in the regulation of the memory of previous stress conditions (Lopez et al., 2024).

The expression of certain genes and the placement of nucleosomes are crucial in the development of stress memory. The mutation of the *FGT1* (*FORGETTER 1*) gene in *Arabidopsis* led to the elimination of heat stress memory, highlighting the importance of nucleosome remodeling by *FGT1* with regard to the plant ability to manage memory stress (Brzezinka et al., 2016). The repetitive DNA sequences termed transposable elements, commonly silenced under normal conditions, exhibit transcriptional activity during stress conditions; for example, the ONSEN retrotransposon is activated in response to thermal stress conditions. This phenomenon is significantly inhibited by small RNAs linked to RNA-mediated DNA methylation, suggesting that specific alterations in DNA methylation patterns allow plants to retain stress memory (Ito et al., 2011, Matsunaga et al., 2012).

23.7 EPIGENETICS AND THERMOMORPHOGENESIS

The term "thermomorphogenesis" describes how plants adapt to temperature variations in terms of morphology,

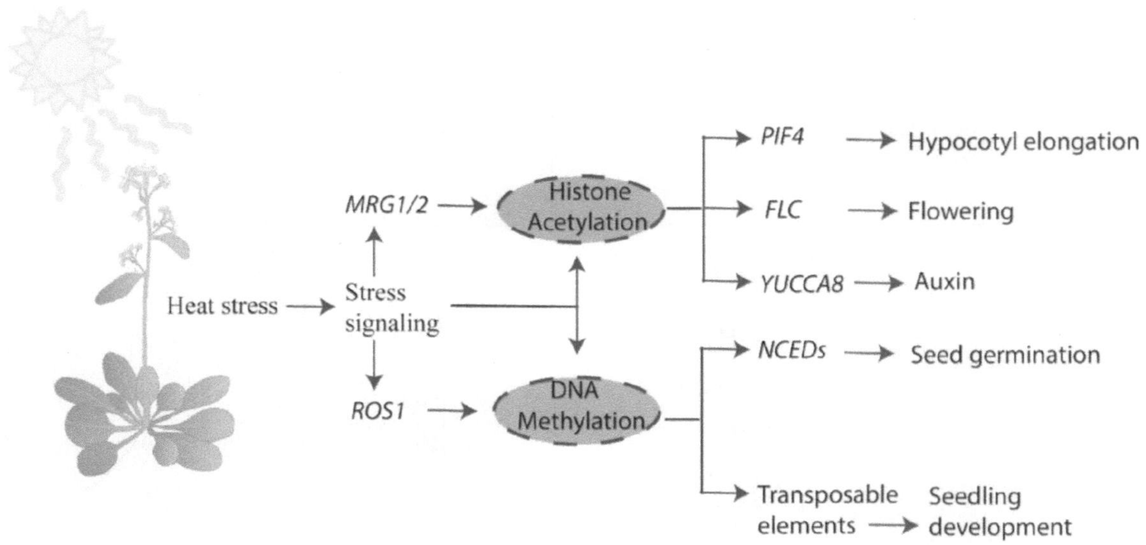

FIGURE 23.4 Heat stress-induced epigenetic modifications in genes regulating hypocotyl elongation, variation in flowering time, auxin biosynthesis, seed germination and seedling development, facilitating effective thermomorphogenesis.

physiology and development (Zhou et al., 2024). The function of epigenetic pathways in generating thermomorphogenic responses is examined in this section, with particular attention given to temperature-induced developmental shifts, physiological modifications and morphological changes (Figure 23.4).

Certain plants exhibit physiological adaptations that enhance their temperature sensitivity through the augmented growth and development of particular organs or tissues; for instance, increased temperature facilitates the elongation of the hypocotyl in *Arabidopsis* plants. The *PIF4* (*Phytochrome Interacting Factor 4*) gene governs this mechanism, and its expression is subject to epigenetic regulation. The PIF4 protein interacts with *MRG1/2* (*MORF-related gene 1 and 2*) to enhance H4K5 acetylation on genes responsive to elevated temperature, such as *YUCCA8*, a crucial component in the auxin biosynthetic pathway (Zhou et al., 2024). Besides physical alterations, certain genetic modifications are recognized to facilitate the plant ability to implement essential physiological changes, such as hormone alteration. Histone modification and DNA methylation have been demonstrated to influence the biosynthesis and function of auxins, gibberellins and ethylene (Zioutopoulou et al., 2021). Research indicates that HDA9 (Histone deacetylase 9) is essential for the activation of *YUCCA8*, as it eliminates the H2A.Z histone variant from *YUCCA8*-associated nucleosomes at elevated temperature (Van Der Woude et al., 2019). Metabolic reprogramming occurs, characterized by the overproduction of HSPs and antioxidants in response to temperature stress. These reactions are epigenetically regulated through chromatin remodeling and long noncoding RNAs (Shahbaz 2024).

Studies are available on how organisms may develop and evolve because of epigenetic alterations resulting from heat stress exposure. Temperature significantly influences the timing of specific developmental stages, such as flowering in multiple species. In *Arabidopsis*, the *FLT* (*FLOWERING LOCUS T*) gene, a crucial regulator of flowering, is epigenetically activated due to alterations in CG, CHG and CHH methylation within the promoters (Han et al., 2024). Temperature signals can modify seed dormancy and germination by influencing DNA methylation in genes that control the abscisic acid and gibberellin signaling pathways (Malabarba et al., 2021). *ROS1* (*REPRESSOR OF SILENCING 1*) gene, which is responsible for hypermethylation that induces seed dormancy under heat, was altered, thereby impairing seed germination under heat stress. Thermomorphogenesis illustrates the intricate connections between epigenetic regulation and temperature variations in plants. Understanding these processes enables us to more efficiently recognize and cope with climatic changes while simultaneously exploring techniques for cultivating crops resilient to such modifications. While considerable progress has been made in this area (Table 23.1), increased focus is required on the several epigenetic pathways governing thermomorphogenesis, their targeted responses and potential agricultural implications (Banerjee & Roychoudhury, 2018b).

23.8 CONCLUSION

Epigenetic control is a primary way of comprehending how plants manage heat stress and thermomorphogenesis. The enhancement of stress tolerance results from the synergistic effects of DNA methylation, histone modification and chromatin remodeling. These activities not only facilitate averting immediate threats but also implant stress memory, thereby preparing plants for future occurrences of heat stress and mitigating the negative effects. The capacity of epigenetics to enhance the flexibility of plants to cope with more challenging environmental conditions is crucial. The regulatory mechanisms and biochemical processes associated with memory are inadequately defined. A deeper level

TABLE 23.1
Heat stress-induced epigenetic changes for thermotolerance and thermomorphogenesis

Plant	Heat stress	Epigenetic change	Targets	Thermotolerance/ thermomorphogenesis	References
Gossypium hirsutum	42°C (3–6 h) and 42°C (1–10 days)	H3K4me2, H4K5ac and DNA hypomethylation	*HSFA1a* *HSFA2* *ERF1A* *HXK1*	Enhanced thermotolerance	He et al. (2022)
Arabidopsis thaliana	4°C for 1 h followed by 37°C for 24 h	DNA methylation	Transposable elements	Maintain genome integrity and thermotolerance	Liu et al. (2021)
Arabidopsis thaliana	12°C for 48 h followed by 27°C for 3 h	Alteration in H2A.Z nucleosome occupancy	*HSP70*	Thermosensory response for thermotolerance	Kumar and Wigge (2010)
Arabidopsis thaliana	42°C for 24 h	DNA demethylation	*HSP90* *HSF1*	Enhanced thermotolerance	Korotko et al. (2021)
Arabidopsis thaliana	35°C for 5 days	H3K4me3 histone methylation	*SDG25* *ATX1*	Enables short-term heat stress memory	Song et al. (2021)
Rosa chinensis	35°C for 50 days	Histone acetylation and deacetylation	*RPD3* *HDT1* *HDA19*	Enhanced thermotolerance	Wu et al. (2022)
Arabidopsis thaliana	37°C for 60 min, 22°C for 90 min and 44°C for 45 min	Alteration in turnover of histone H3.3 and maintaining H3k4 methylation	*HSP70* *HSP18* *HSA32*	Induce heat stress memory	Pratx et al. (2023)
Arabidopsis thaliana	44°C for 45 min	Nucleosome remodeling	*FGT1*	Maintaining heat stress memory	Brzezinka et al. (2016)
Fragaria vesca	37°C for 7 days	DNA methylation	*HSFB19* *HSFA9a* *HSFB3a*	Maintaining heat stress memory	Lopez et al. (2024)
Arabidopsis thaliana	28°C for 3 days	H4k5 acetylation	*MRG1/2* *PIF2*	Hypocotyl elongation	Zhou et al. (2024)
Arabidopsis thaliana	27°C for 4 days	H2A.Z histone variant maintenance by HDAC9	*YUCCA8*	Hypocotyl elongation	Van Der Woude et al. (2019)
Arabidopsis thaliana	10°C and 16°C during flowering stages	DNA methylation	*FLC*	Variation in flowering time	Han et al. (2024)
Arabidopsis thaliana	27°C during germination	DNA methylation	*ROS1*	Seed germination and seedling development	Malabarba et al. (2021)

of understanding in this area can be beneficial for the creation of innovative approaches focused on augmenting crop resilience through targeted breeding programs or epigenetic engineering. An important illustration of this process is thermomorphogenesis, which demonstrates how plants modify their structure in response to environmental stimuli via epigenetic regulation of temperature-induced developmental alterations. A comprehensive understanding of the epigenetic regulation of flowering, seed germination and hypocotyl elongation could benefit plant development, surpassing current predictions in response to climate change.

23.9 FUTURE PERSPECTIVES

Further investigations should focus on identifying the epigenetic markers for heat stress tolerance in various crops. Research contrasting heat-tolerant and heat-sensitive genotypes could reveal novel paths for the selection of stress-resistant cultivars. The utilization of multi-omics, encompassing transcriptomics, epigenomics, proteomics and advanced bioinformatics, is essential for elucidating plant response to heat stress. Biotechnological methods, including CRISPR-mediated epigenome editing, can significantly contribute towards customizing stress-resistant genes in crops. Furthermore, collaborations among researchers in academia, industry and policy sectors will facilitate the development of crop varieties with climatic adaptations to improve global food security. Ultimately, the application of epigenetic knowledge in agriculture will yield innovative strategies to address climate change while fostering a resilient and sustainable agricultural ecosystem.

DECLARATION OF COMPETING INTEREST

The authors declare that they have no known competing financial interests or personal relationships that could have appeared to influence the work reported in this paper.

ACKNOWLEDGMENTS

NL greatly acknowledges the financial assistance provided by the University Grant Commission (UGC), Government of India, in the form of SRF. MSA greatly acknowledges the financial assistance provided by the Council of Scientific and Industrial Research (CSIR), Government of India, in the form of SRF. NL, MSA and JTP express sincere thanks to the Department of Science and Technology (DST), Government of India, for granting funds under the Fund for Improvement of S&T Infrastructure (FIST) programme (SR/FST/LSI-532/2012) and DST-Science and Engineering Research Board (DST-SERB) Government of India for granting funds under the SURE grant (SUR/2022/002538-G).

REFERENCES

Attikum, H. V., & Gasser, S. M. (2005). ATP-dependent chromatin remodeling and DNA double-strand break repair. *Cell Cycle*, 4(8), 1011–1014.

Banerjee, A., & Roychoudhury, A. (2017). Epigenetic regulation during salinity and drought stress in plants: Histone modifications and DNA methylation. *Plant Gene*, 11, 199–204.

Banerjee, A., & Roychoudhury, A. (2018a). The gymnastics of epigenomics in rice. *Plant Cell Reports*, 37, 25–49.

Banerjee, A., & Roychoudhury, A. (2018b). Small heat shock proteins: Structural assembly and functional responses against heat stress in plants. In P. Ahmad, M. A. Ahanger, V. P. Singh, D. K. Tripathi, P. Alam, & M. N. Alyemeni (Eds.), *Plant Metabolites and Regulation Under Environmental Stress* (pp. 367–376). Amsterdam: Elsevier (Academic Press).

Brzezinka, K., Altmann, S., Czesnick, H., Nicolas, P., Gorka, M., Benke, E., Tina, K., Jahne, F., Graf, A., Kappel, C, & Bäurle, I. (2016). *Arabidopsis* FORGETTER1 mediates stress-induced chromatin memory through nucleosome remodeling. *elife*, 5, e17061.

Eichten, S. R., Schmitz, R. J., & Springer, N. M. (2014). Epigenetics: Beyond chromatin modifications and complex genetic regulation. *Plant Physiology*, 165(3), 933–947.

Feng, S., Zhang, Y., Li, X., & Chen, Q. (2022). DNA methylation and demethylation in plant abiotic stress responses. *Plant Signaling & Behavior*, 17(1), 2021018.

Fritsch, O., Benvenuto, G., Bowler, C., Molinier, J., & Hohn, B. (2004). The INO80 protein controls homologous recombination in *Arabidopsis thaliana*. *Molecular Cell*, 16(3), 479–485.

Gao, G., Li, J., Li, H., Li, F., Xu, K., Yan, G., ... & Wu, X. (2014). Comparison of the heat stress induced variations in DNA methylation between heat-tolerant and heat-sensitive rapeseed seedlings. *Breeding Science*, 64(2), 125–133.

Han, Y., Liu, L., Lei, M., Liu, W., Si, H., Ji, Y., ... & Zan, Y. (2024). Divergent flowering time responses to increasing temperatures are associated with transcriptome plasticity and epigenetic modification differences at *FLC* promoter region of *Arabidopsis thaliana*. *Molecular Ecology*, e17544.

He, K., Cao, X., & Deng, X. (2021). Histone methylation in epigenetic regulation and temperature responses. *Current Opinion in Plant Biology*, 61, 102001.

He, S., Zhang, Y., Wang, J., Wang, Y., Ji, F., Sun, L., ... & Hao, F. (2022). H3K4me2, H4K5ac and DNA methylation function in short-and long-term heat stress responses through affecting the expression of the stress-related genes in *G. hirsutum*. *Environmental and Experimental Botany*, 194, 104699.

Ito, H., Gaubert, H., Bucher, E., Mirouze, M., Vaillant, I., & Paszkowski, J. (2011). An siRNA pathway prevents transgenerational retrotransposition in plants subjected to stress. *Nature*, 472(7341), 115–119.

Jablonka, E., & Lamb, M. J. (2002). The changing concept of epigenetics. *Annals of the New York Academy of Sciences*, 981(1), 82–96.

Kim, J. M., Kim, K., Punj, V., Liang, G., Ulmer, T. S., Lu, W., & An, W. (2015). Linker histone H1. 2 establishes chromatin compaction and gene silencing through recognition of H3K27me3. *Scientific Reports*, 5(1), 16714.

Korotko, U., Chwiałkowska, K., Sańko-Sawczenko, I., & Kwasniewski, M. (2021). DNA demethylation in response to heat stress in *Arabidopsis thaliana*. *International Journal of Molecular Sciences*, 22(4), 1555.

Kumar, S. V., & Wigge, P. A. (2010). H2A. Z-containing nucleosomes mediate the thermosensory response in *Arabidopsis*. *Cell*, 140(1), 136–147.

Lämke, J., & Bäurle, I. (2017). Epigenetic and chromatin-based mechanisms in environmental stress adaptation and stress memory in plants. *Genome Biology*, 18(1), 124.

Li, X., Wang, X., Zhang, Y., & Chen, Q. (2023). Epigenetic regulation of plant responses to heat stress. *Journal of Integrative Plant Biology*, 65(1), 1–18.

Liu, S., de Jonge, J., Trejo-Arellano, M. S., Santos-González, J., Köhler, C., & Hennig, L. (2021). Role of H1 and DNA methylation in selective regulation of transposable elements during heat stress. *New Phytologist*, 229(4), 2238–2250.

López, M. E., Denoyes, B., & Bucher, E. (2024). Epigenomic and transcriptomic persistence of heat stress memory in strawberry (*Fragaria vesca*). *BMC Plant Biology*, 24(1), 405.

Malabarba, J., Windels, D., Xu, W., & Verdier, J. (2021). Regulation of DNA (de) methylation positively impacts seed germination during seed development under heat stress. *Genes*, 12(3), 457.

Matsunaga, W., Kobayashi, A., Kato, A., & Ito, H. (2012). The effects of heat induction and the siRNA biogenesis pathway on the transgenerational transposition of ONSEN, a copia-like retrotransposon in *Arabidopsis thaliana*. *Plant and Cell Physiology*, 53(5), 824–833.

Peirats-Llobet, M., Han, S. K., Gonzalez-Guzman, M., Jeong, C. W., Rodriguez, L., Belda-Palazon, B., ... & Rodriguez, P. L. (2016). A direct link between abscisic acid sensing and the chromatin-remodeling ATPase BRAHMA via core ABA signaling pathway components. *Molecular Plant*, 9(1), 136–147.

Pratx, L., Wendering, P., Kappel, C., Nikoloski, Z., & Bäurle, I. (2023). Histone retention preserves epigenetic marks during heat stress-induced transcriptional memory in plants. *The EMBO Journal*, 42(24), e113595.

Ramakrishnan, M., Zhang, Z., Mullasseri, S., Kalendar, R., Ahmad, Z., Sharma, A., ... & Wei, Q. (2022). Epigenetic stress memory: A new approach to study cold and heat stress responses in plants. *Frontiers in Plant Science*, 13, 1075279.

Rapp, R. A., & Wendel, J. F. (2005). Epigenetics and plant evolution. *New Phytologist*, 168(1), 81–91.

Rosa, M., Von Harder, M., Aiese Cigliano, R., Schlögelhofer, P., & Mittelsten Scheid, O. (2013). The *Arabidopsis* SWR1 chromatin-remodeling complex is important for DNA repair, somatic recombination, and meiosis. *The Plant Cell*, 25(6), 1990–2001.

Shahbaz, M. (2024). Heat and wheat: Adaptation strategies with respect to heat shock proteins and antioxidant potential; An era of climate change. *International Journal of Biological Macromolecules*, 256, 128379.

Shaked, H., Avivi-Ragolsky, N., & Levy, A. A. (2006). Involvement of the *Arabidopsis* SWI2/SNF2 chromatin remodeling gene family in DNA damage response and recombination. *Genetics*, 173(2), 985–994.

Song, Z. T., Zhang, L. L , Han, J. J., Zhou, M., & Liu, J. X. (2021). Histone H3K4 methyltransferases SDG25 and ATX1 maintain heat-stress gene expression during recovery in *Arabidopsis. The Plant Journal*, 105(5), 1326–1338.

Suter, L., & Widmer, A. (2013). Environmental heat and salt stress induce transgenerational phenotypic changes in *Arabidopsis thaliana. PloS One*, 8(4), e60364.

Van Der Woude, L. C., Perrella, G., Snoek, B. L., Van Hoogdalem, M., Novák, O., Van Verk, M. C., ... & Van Zanten, M. (2019). HISTONE DEACETYLASE 9 stimulates auxin-dependent thermomorphogenesis in *Arabidopsis thaliana* by mediating H2A. Z depletion. *Proceedings of the National Academy of Sciences*, 116(50), 25343–25354.

Wang, F., Li, C. H., Liu, Y., He, L. F., Li, P., Guo, J. X., ... & Guo, Y. D. (2024). Plant responses to abiotic stress regulated by histone acetylation. *Frontiers in Plant Science*, 15, 1404977.

Wang, X., Li, X., & Chen, Q. (2024). The role of DNA methylation in plant heat stress tolerance: A review. *Plant Science*, 332, 111318.

Wu, Q., Huang, Q., Guan, H., Zhang, X., Bao, M., Bendahmane, M., & Wu, Q. (2022). Comprehensive genome-wide analysis of histone acetylation genes in roses and expression analyses in response to heat stress. *Genes*, 13(6), 980.

Zhang, Y., Li, X., & Chen, Q. (2021). DNA methylation and demethylation in plant abiotic stress responses. *Plant Signaling & Behavior*, 17(1), 2021018.

Zhao, J., Lu, Z., Wang, L., & Jin, B. (2020). Plant responses to heat stress: Physiology, transcription, noncoding RNAs, and epigenetics. *International Journal of Molecular Sciences*, 22(1), 117.

Zheng, S., Zhao, W., Liu, Z., Geng, Z., Li, Q., Liu, B., ... & Bai, J. (2024). Establishment and maintenance of heat-stress memory in plants. *International Journal of Molecular Sciences*, 25(16), 8976.

Zhou, N., Li, C., Xie, W., Liang, N., Wang, J., Wang, B., ... & Dong, A. (2024). Histone methylation readers MRG1/2 interact with PIF4 to promote thermomorphogenesis in *Arabidopsis. Cell Reports*, 43(2), 1–20.

Zioutopoulou, A., Patitaki, E., Xu, T., & Kaiserli, E. (2021). The epigenetic mechanisms underlying thermomorphogenesis and heat stress responses in *Arabidopsis. Plants*, 10(11), 2439.

24 Cold Stress Tolerance and Epigenetic Memory in Plants

Priya Kushwaha, Rahul Verma and Amit Kumar Singh

24.1 INTRODUCTION

Change in environmental conditions causes environmental stress, which has a detrimental influence on plant growth and crop yield, causing irreversible damage or mortality if the plant continues under stressed conditions for an extended period. Plants, usually being stationary organisms, inevitably encounter various environmental stressors throughout their lives. These stressors can be classified as abiotic, which include edaphic factors like drought, salinity and heavy metal toxicity, as well as atmospheric factors such as heat, cold, high light and ultraviolet (UV) radiation (1). Among these, cold stress is considered a significant environmental factor that limits crop yield and agricultural development, particularly in hilly regions (2). Non-freezing low temperature can cause chilling injuries that lead to various physiological issues in plants. These include damage to photosynthesis, chlorosis, uncontrolled apoptosis, loss of membrane fluidity, and ultimately, wilting (100). Cold stress can be divided into two categories based on plant sensitivity: freezing stress, which occurs at temperature below 0°C, and chilling stress, which occurs between 0°C and 15°C (3, 99).

Plants in temperate climates have a greater ability to tolerate cold compared to tropical and subtropical climates. This increased cold tolerance is due to a process known as cold acclimatization, which involves various molecular and biochemical changes (4). A key floral regulatory mechanism "vernalization" helps to prevent flowers from blooming too early in the fall or winter. Following extended exposure to cold temperature, vernalization progressively enhances readiness to bloom, and this process varies depending on the species (5). Vernalization acts as a physiological "memory response" that is linked to epigenetic regulation, as demonstrated by research on the model plant *Arabidopsis thaliana* (L.) (6).

Extreme temperature caused by climate change significantly impact plant growth by influencing gene expression. When plants encounter stressful conditions, they depend on their stress memory, which includes transcriptional, somatic, intergenerational and transgenerational memory, to improve their chances of survival (7). Unlike humans and animals, plants do not possess brains, so their stress memory operates differently. This memory is crucial for gene regulation as plants detect and respond to cold and heat stress, and it is regulated by epigenetic modifications. These modifications help facilitate the necessary molecular and physiological changes. Moreover, stress memories influenced by epigenetics can be passed down to subsequent generations (8).

The memory mechanism involves a signaling protein and salicylic acid (9). During this process, the transcription of genes responsive to salicylic acid is activated. Changes in the chromatin of these target genes occur in response to elevated levels of salicylic acid; for instance, the promoter regions of stress-related genes show increased acetylation of H3 and H4 histones, as well as methylation of H3K4 when stimulated by salicylic acid (10). However, chromatin modifications are dynamic and can alter in response to stress (11). Consequently, variations in histone modifications can either enhance or suppress the expression of stress memory genes (12). Additionally, changes in the *Flowering Locus C* (*FLC*) gene associated with H3K27me3 help maintain cold stress memory (13). This indicates that chromatin accessibility and structure are critical for gene expression and are often associated with various epigenetic codes. These codes encompass DNA methylation, histone variations, histone post-translational modifications (PTMs) and certain non-coding RNAs (14). Furthermore, the inheritance of stress memory occurs through histone modifications and the replacement of histone variants (15).

24.2 ADAPTATIONS TO COLD STRESS IN PLANTS

24.2.1 Morphological Adaptations

Under low-temperature conditions, cells experience dehydration, resulting in osmotic stress that impacts membrane integrity and permeability. Morphologically, plants tend to grow shorter, display reduced leaf expansion and quantity, and show thickened epidermis. Additionally, cold temperature can make the plasma membrane more rigid by altering the types and compositions of membrane lipids (16). By increasing lipid unsaturation and altering lipid class, composition and lipid-to-protein ratio, the plasma membrane maintains its structural and functional integrity. The desaturation of fatty acids, along with the proportions of phospholipids, galactolipids, sphingolipids and sterols, plays a significant role in membrane fluidity. Many plants, including herbaceous and woody species, as well

DOI: 10.1201/9781003589914-28

as dicotyledonous and monocotyledonous varieties, such as oats, rye, mulberries, orchard grass and *Arabidopsis*, have been observed to respond to cold stress by exhibiting higher levels of specific phospholipids (phosphatidylcholine and phosphatidylethanolamine) while showing a decrease in sterols and cerebrosides in their plasma membranes (17).

A 10% drop in phosphatidylcholine was observed in maize at low temperature, suggesting that a key reason for regulating the lipid content of the membrane is to mitigate low-temperature stress (18). Additionally, the expression of various enzymes involved in lipid pathways increases the proportion of unsaturated fatty acids; for example, genetic studies indicate that fatty acid desaturase (FAD) likely plays a role in controlling membrane fluidity, which contributes to cold tolerance in *Tetrahymena thermophila* (19).

24.2.2 Physiological Adaptations

Cold or freezing stress damages the integrity of the chloroplast membrane, which leads to a decrease in the effectiveness of photosystems I and II, as well as a reduction in intercellular CO_2 levels and the amounts of chlorophyll-a and chlorophyll-b. All these factors contribute to a decline in net photosynthesis (20). To counteract, plants produce accessory pigments such as carotenoids, violaxanthin, antheraxanthin and zeaxanthin to help maintain photosynthetic efficiency. Additionally, they increase the activity of enzymes associated with the Calvin cycle and sugar metabolism, and they boost the synthesis of D1 protein, which is essential for maintaining the efficiency of photosystem II (PSII) (21). Cold stress also alters the morphology and architecture of stomata. Studies on *Stevia* and other plants have shown that cold stress can lead to an increase in stomatal size (22). This change in stomatal shape may significantly reduce water use efficiency and lower intracellular CO_2 content.

24.2.3 Biochemical Adaptations

Plants produce and accumulate various cryoprotectants, including soluble sugars and specific amino acids such as proline, glycine, alanine and serine, to mitigate biochemical stress caused by cold temperature (23). Additionally, secondary metabolites, betaines and polyamines also play a role in cold tolerance. To survive cold stress, plants utilize sugars like sucrose, glucose, fructose, galactose, raffinose and trehalose. These sugars help control osmotic potential, inhibit the formation of ice crystals, scavenge reactive oxygen species and enhance membrane stability. Cold-adapted plants, such as *Arabidopsis thaliana*, chickpeas, barley and winter wheat, have been shown to accumulate proline (24). Proline is considered a biomarker for cold tolerance because it scavenges reactive oxygen species, stabilizes membranes and proteins, maintains osmotic potential and regulates the expression of genes associated with osmotic stress.

PAs play a crucial role in enhancing osmolyte accumulation, regulating gene expression, stabilizing membranes and maintaining redox homeostasis. They also promote seed germination, improve fruit quality and protect the photosynthetic machinery from cold or freezing stress (25). When there is an excessive buildup of reactive oxygen species (ROS), a higher quantity of antioxidants is needed to counteract their harmful effects. As a result, secondary metabolites are produced, including phenols, flavonoids, coumarins, catechins, tocopherols, derivatives of cinnamic acid, lignins and polyfunctional organic acids (Figure 24.1) (26).

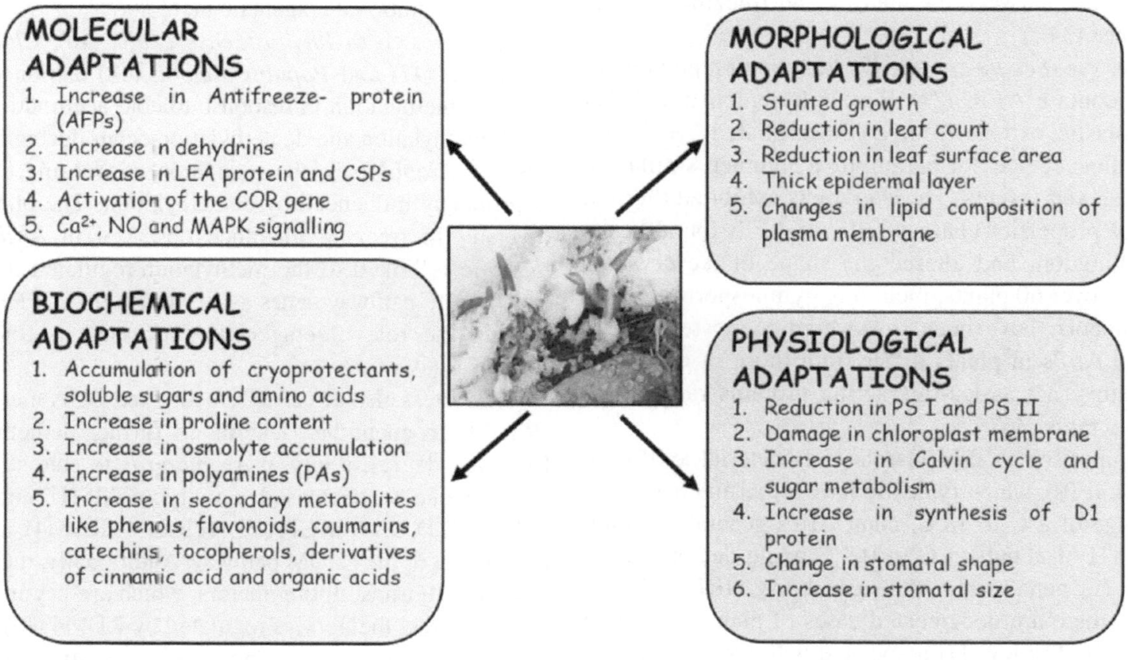

FIGURE 24.1 Adaptations of plants to cold stress by morphological, physiological, biochemical and molecular means.

24.2.4 Molecular Adaptations

Plants produce stress-responsive proteins at the molecular level, including antifreeze proteins (AFPs), HSPs, CSPs, LEA and dehydrins. These proteins aid in protein folding and refolding, prevent the denaturing of globular macromolecules, facilitate cellular protein transport and inhibit ice crystal formation.

Cold-hardy plants produce a specific type of protein known as AFPs to survive freezing temperature. These proteins lower the freezing point of the cytosol, prevent the growth of ice crystals in the apoplast and reduce freezing damage. The initial discovery of AFPs was made by De Vries and Wohlschlag (1969), who found a plasma protein that could decrease the freezing point of blood in Arctic fish (27). Some of these proteins were glycosylated and are referred to as antifreeze glycoproteins (AFGPs), while a smaller number were purely proteinaceous and categorized as AFPs. Collectively, these proteins are known as AF(G)Ps. Since their initial identification in marine teleosts, AF(G)Ps have also been found in bacteria (28), molds fungi (29), sea-ice diatoms (30), plants (31) and snow algae (32).

Adaptive conflict and environmental pressures are believed to have driven the development of antifreeze (AF) genes through processes such as intra-gene and whole-gene duplication, sequence divergence of specific genes like sialic acid synthase, trypsinogen and C-type lectin, as well as from non-coding DNA sequences. Due to their unique structure, AFPs can attach to small ice crystals, preventing their growth (33). Additionally, AFPs function in a non-colligative manner. They possess two distinct characteristics: thermal hysteresis (TH) and ice recrystallization inhibition (IRI). IRI refers to the ability to prevent the formation of ice crystals, while TH is the difference between the melting and freezing point of a substance (34).

Winter rye (*Secale cereale* L.) was the first plant identified to contain AFPs (31). Research demonstrated that the apoplastic extract from cold-acclimated rye leaves could influence ice formation in a manner similar to that of fish and insects. These extracts exhibited thermal hysteresis properties characteristic of AFPs found in the animal kingdom and altered the shape of ice crystals. Currently, over 60 plants, including gymnosperms, dicots and monocots, have been discovered to possess AFPs (35). The AFPs in plants differ from those in mammals because they act as ice-interacting proteins rather than cryoprotectants (36).

In comparison to fish, insects and bacterial antifreeze proteins (AFPs), which typically have a thermal hysteresis (TH) range of 2°C to 13°C, plant AFPs generally exhibit a lower TH of about 0.1°C to 0.5°C while demonstrating a strong ice penalty reduction capability (IRI) (37, 38). However, there are documented cases of plant AFPs with a higher TH. The low TH associated with more controlled ice crystal development, combined with a high IRI, allows AFPs to remain effective even at very low concentration. Thus, AFPs with low TH and high IRI seem to be an evolutionary adaptation in plants.

24.3 EPIGENETIC REGULATION OF COLD STRESS IN PLANTS

24.3.1 DNA Methylation

DNA methylation involves the addition of a methyl group to the fifth carbon position of a cytosine ring. It plays a crucial role in maintaining genome stability, regulating genomic imprinting (39) and silencing transposable elements (TE) (40). Both methylation and demethylation are known to disrupt the regulation of multiple genes in response to environmental cues (41). This disruption influences gene expression by either recruiting proteins that repress genes or inhibiting the binding of DNA transcription factors (42). Environmental factors such as temperature, photoperiod, UV radiation, water availability and salinity can alter the methylation patterns in the coding regions of certain stress-sensitive genes, thus controlling their expression (43).

Numerous studies indicate that the transcriptional response to cold stress is regulated by DNA methylation, specifically by two modifications: 5-methylcytosine (5mC) and N6-methyladenine (6mA) (44). While 5mC has been extensively researched under cold stress, 6mA – a relatively novel epigenetic marker – has not received as much attention. 6mA is widely present in the genomes of rice and *Arabidopsis* and is linked to stress responses and gene expression; however, its role in stress memory remains unclear (45). Therefore, this chapter focus solely on 5mC. Plants can respond to cold stress in two ways: they can either transmit DNA methylation changes induced by cold stress within the same generation or pass these changes on as transgenerational epigenetic memory.

Species such as *Phyllostachys edulis* (46), *Chorispora bungeana* (47) and *Populus simonii* (48) exhibit changes in DNA methylation in response to cold acclimation. Both DNA methylation and demethylation occur during this process. In *Arabidopsis,* the regulation of freezing tolerance is primarily influenced by demethylation. The phenotypic diversity in freezing tolerance observed in *Arabidopsis thaliana* is linked to the methylation-regulated transcription of CBF pathway genes associated with AtICE1, underscoring the role of epigenetic processes in cold stress responses (49).

Cold stress also affects RNA-directed DNA methylation (RdDM) in cucumber, leading to further demethylation. However, the sensitivity of *Arabidopsis* to cold stress was reduced due to decreased activity of MSH1 and RNA-DIRECTED DNA METHYLATION 4 (RDM4), a crucial component of the RdDM pathway. Additionally, the expression of CBF transcription factors, which are key in managing cold stress memory, is regulated by RDM4 (13).

In vernalized plants, exposure to cold stress results in reduced levels of 5mC and an increase in the expression of

Casein Kinase II A-Subunit (CKA2) and B-Subunit (CKB4). This increase enhances CK2 activity while decreasing the expression of the clock gene Circadian Clock Associated 1 (CCA1), which plays a vital role in photoperiod awareness (50). Vernalization also affects non-CG methylation, which can be maintained through mitosis, and induces DNA demethylation that is not preserved in plants (45).

A study conducted on the cold-sensitive crop *Oryza sativa* revealed changes in the methylation patterns. Pan et al. (51) analyzed individuals from two different genotypes – one sensitive to cold stress and the other tolerant – exposing them to cold stress conditions at 4°C for two days. Following the treatment, they assessed the methylation patterns using AFLP-MS analysis. Only the roots of the cold-tolerant genotype exhibited increased DNA methylation during cold stress. Furthermore, it was observed that the cold-induced alterations in DNA methylation could not be reversed even after the plants recovered. The methylation rate of seedlings that were grown at normal temperature before the cold stress treatment showed slight recovery, but it never returned to its original level (51). This response has also been noted in *Gossypium hirsutum*, suggesting that the loci with altered DNA methylation may be mitotically inherited and are involved in cold stress responses (52).

Mitochondrial malate dehydrogenase (BramMDH1) exhibited increased expression in *Brassica rapa* due to cold-induced DNA demethylation. Its overexpression enhanced heat tolerance in *Arabidopsis*, but this effect was not observed in *Brassica rapa*. This suggests that while DNA demethylation is sufficient to boost cold tolerance in *Brassica rapa*, it does not improve heat tolerance (53, 101).

In *Rosa hybrida*, the expression of RhAG was reduced due to cold-induced DNA methylation, particularly in the CHH context. This heightened suppression of RhAG resulted in an increased number of petals formed in response to cold temperature (54). Additionally, cold tolerance is regulated by a decrease in DNA and histone methylation – specifically H3K9me2 – and an increase in histone acetylation (H3K9ac) within heterochromatic regions. This indicates that modifications to both DNA and histones are essential for effective cold acclimation (55). In *Hevea brasiliensis*, the expression of HbICE1 and HbCBF2 is linked to cold tolerance as well as DNA demethylation (Table 24.1) (56).

24.3.2 Histone Modification

A nucleosome is a structural unit of eukaryotic DNA, consisting of proteins called histones (64). It features an octameric core made up of two copies of each of the four core histone proteins: H2A, H2B, H3 and H4. This core encircles approximately 150 base pairs of DNA, forming the nucleosome. Histones play crucial structural roles and are essential for transitioning between active and inactive chromatin

TABLE 24.1
Mechanism of DNA methylation in various plants

Plant species	Regulation	References
Arachis hypogaea	Elevated transcript levels of *C5-MTase* and *DME*-like genes highlight their significant role in the DNA methylation alterations that respond to cold stress	Wang et al. (57)
Zea mays	Demethylation is vital for controlling hormones, photosynthesis and the activation of transposons, emphasizing its critical importance in cold stress tolerance	Shan et al. (58)
	ZmMI1 comprises a segment of a coding region for a potential protein along with a retrotransposon-like sequence, clearly indicating a genome-wide demethylation in root tissues	Steward et al. (41)
Gossypium hirsutum	Abiotic stress-related gene methylation is regulated. GhTPS is demethylated and shows increased expression under cold treatment, whereas GhCLSD, GhARK and GhARM are methylated, leading to reduced expression in cold environment	Fan et al. (59)
Arabidopsis thaliana	All members of the CBF pathway correlate negatively with *AtICE1* gene methylation levels	Xie et al. (60)
	The long-term cold exposure in cold-sensitive *msh1* mutants leads to significant hypomethylation in the asymmetric CHH motif	Kenchanmane-Raju et al. (61)
Brassica rapa	Promoter demethylation of *BramMDH1, BraKAT2, BraSHM4* and *Bra4CL2* decisively enhances their transcriptional activities	Liu et al. (53)
Oryza sativa	Hyperacetylation of histone H3 activates cold-induced OsDREB1b transcriptionally	Roy et al. (62)
Lolium perenne	LpCYP72A161, a CG-methylated gene, showed a flexible shift of gene body methylation in exon 1, and the methylation level was inversely linked with transcript abundance	Dai et al. (63)
Chorispora bungeana	Alterations in DNA methylation across the genome	Song et al. (47)
Populus simonii	The induction of genes Establishment and maintenance of DNA methylation is associated with PsiCMT3.1, PsiCMT3.2, PsiDRM2, PsiMET1.1, PsiDNMT2, PsiROS1, PsiDME1 and PsiDME3	Song et al. (48)

states. They facilitate access for proteins and enzymes that bind to specific DNA sequences, thereby allowing gene expression (65).

Various modifications to histones, such as acetylation, methylation, ubiquitination and phosphorylation of the N- and C-terminal tails, as well as changes to the histone cores themselves, can lead to alterations in histone modifications (66). Research has shown that di-methylated and tri-methylated histone H3 at lysine 27 (H3K27me2/3) marks genes that are repressed or silent, which are crucial for plant development and stress responses (67).

Studies indicate that during cold stress, histone modifications, including methylation and acetylation, change and affect various biological pathways, including those related to cold-responsive genes that can be genetically manipulated (68). When plants were returned to normal conditions, the cold-dependent alternative splicing observed in the *JMJC5* gene of *Medicago truncatula* was reversed. This suggests that cold-dependent alternative splicing is associated with the epigenetic regulation of cold stress (69).

The cold stress response is significantly influenced by genomic regions containing active cis-acting elements, identified through chromatin research as DNase I hypersensitive sites (DHSs) in potato. These cold stress-induced DHSs, which are enhanced in genic regions, regulate the expression of cold-responsive genes linked to bivalent histone modifications, specifically H3K4me3 and H3K27me3 (70).

In a study involving *Arabidopsis thaliana hda6* mutants, it was observed that plants lacking *hda6* function had reduced survival rates under low-temperature stress compared to control samples. Unlike the *DREB/CBF* and *RESPONSIVE TO DESICCATION* (*RD*) genes, which showed increased transcript accumulation starting 2 h after exposure, control samples demonstrated upregulation of HDA6 only after prolonged exposure (72 h) to low temperature. As cold stress intensified, there was a progressive increase in H3K27me3 levels in *Arabidopsis*, corresponding with the expression of two cold-responsive genes, *COR15A* and *ATGOLS3*. After returning to normal conditions, plants that had been exposed to cold maintained lower levels of H3K27me3 in *COR15A* and *ATGOLS3*, along with possible histone replacement. However, there was no increase in the expression of these two genes, suggesting that while H3K27me3 did not inhibit gene activation, it could be inherited through cell division, serving as a marker for cold stress memory (71).

Plants that overexpress the *HD2D* gene take longer to bloom, yet they demonstrate increased resilience to environmental stressors like cold temperature (72). Data suggest that in certain situations, negative regulation is essential for the induction of specific proteins, even when potential genes for histone deacetylases (such as HDA, HD2C and HD2D) are activated in response to freezing conditions, low temperature and acclimatization. In rice plants, a response to cold and stress tolerance develops as a result of chromatin remodeling in the promoter region and the area upstream of the *OsDREB1b* gene. This process is facilitated by dynamic changes in the acetylation marks on histone H3 (62). Additionally, it has been suggested that the modulation of fatty acid biosynthesis – crucial for maintaining membrane integrity during low-temperature exposure – is correlated with high acetylation levels of histones H3 and H4 in the promoters of ω-3 *fatty acid desaturase* (*FAD*) genes in banana fruit (73). This indicates that histone modifications play a significant role in cold stress responses in plants.

Gene expression is often enhanced by histone acetylation. The histone acetyltransferase HAC1 is crucial for establishing stress memory and improving cold memory. Moreover, histone H3 acetylation promotes the activation of *cold-responsive* (*COR*) genes during cold stress (74). Cold exposure also enhances histone deacetylation. In *Zea mays* L., histone deacetylases (HDACs) are upregulated, leading to the deacetylation of lysine residues on the histone subunits H3 and H4 following cold acclimatization. As a result, the expression of the cold-responsive genes *ZmDREB1* and *ZmCOR413* decreases when HDAC activity is suppressed (Table 24.2) (75).

24.3.3 Non-coding RNAs

Highly condensed DNA, known as heterochromatin, not only suppresses target gene expression but also helps maintain genomic integrity. Small non-coding RNAs play a significant role in controlling protein-coding transcription and influencing epigenetic gene silencing pathways (81). These small RNAs are partly responsible for the dynamic changes in gene expression that occur due to cold stress. Similarly, short RNA molecules can alter chromatin structure through RNA interference (RNAi) pathways, often resulting in the formation of heterochromatin and the subsequent silencing of target genes, which regulates various biological processes (82). In plants, the two main types of small regulatory RNAs, categorized by their biosynthesis and function, are microRNA (miRNA) and small interfering RNA (siRNA) (83).

It is expected that various cold-responsive microRNAs will target DNA methylation genes in *Prunus persica* (84). Bioinformatic analyses have shown that cold-sensitive miRNA genes may be involved in multiple signaling pathways; for instance, miR169a contains both abscisic acid response elements (ABRE) and low-temperature response elements (LTRE), whereas miR169g contains only ABRE and miR169h lacks both. This indicates that the regulation of this gene family could involve ABA-dependent pathways, non-ABA-dependent pathways or a combination of both (85). Thiebaut et al. (2012) provided empirical evidence in their research on sugarcane, demonstrating that both cold and ABA treatments positively regulate miR319 (86). Additionally, miRNA search revealed the presence of conserved miRNA sequences (miR169e, miR172b and miR397) in the genome of *Brachypodium distachyon*. These sequences are associated with responses to low-temperature stress in *Arabidopsis thaliana*, rice and Chinese cottonwood, and they are also regulated by cold conditions in *B. distachyon* (Table 24.3) (87).

TABLE 24.2
Histone modifications in various plants

Plant species	Regulation	References
Hordeum vulgare	bPb-2778, comparable to *DDM1* gene sequences, is located on chromosome 3H. Generates an ATP-dependent DNA helicase that can modify chromatin	Fiust and Rapacz (76)
Chorispora bungeana	Removing histone H3 from the promoter and middle-coding region enhances *CbADH1* expression	Liu et al. (53)
Arabidopsis thaliana	Identified and characterized the interaction between HOS15 and histone H4. High levels of acetylated histone H4 in hos15 indicate vulnerability to subzero temperature	Zhu et al. (77)
	Cold exposure reduced H3K27me3 levels in promoter regions but not in coding areas of cold-responsive *COR15A* and *ATGOLS3* genes	Kwon et al. (71)
	The *HDA6 histone deacetylase* gene is necessary for cold acclimation and freezing tolerance. In contrast to cold-treated wild-type plants, cold-treated HDA6 exhibited decreased freezing tolerance	To et al. (78)
	H3K27me3-dependent chromatin remodeling activates the CBF3-mediated pathway	Yang et al. (79)
Solanum tuberosum	Genes with the bivalent H3K4me3 and H3K27me3 marks are upregulated in stress response and downregulated in developmental stages	Zeng et al. (70)
Medicago truncatula	MtJMJC5 cold-dependent alternative splicing is probably a species- or genus-specific method of RNA-level gene expression control	Shen et al. (69)
Zea mays L.	Reduced acetylation of H3K9, H4K5 and H44 leads to the activation of cold-induced ZmDREB1. HDA6, a histone deacetylase, favorably regulates the expression of cold-induced ZmDREB1	Hu et al. (75)
	A decrease in DNA methylation and H3K9me2 and a rise in H3K9ac are linked to the transcriptional activation of TR-1, knob tandem-repetitive sequences	Hu et al. (80)
	Acetylation of histones H3K9 and H4K5 at gene promoters affects cell cycle, regulates expression in response to stress and inhibits seedling growth and development	Zhao et al. (2021)
Cannabis sativa	During cold acclimatization, some varieties with more effective genes involved in de novo DNA methylation showed increased transcript levels, as well as alterations in H3K9ac, H3K27me3 and methylcytosine levels	Mayer et al. (2015)

TABLE 24.3
Non-coding RNA regulation in different plants

Plant species	Regulation	References
Arabidopsis thaliana	miRNAs and sRNAs from cis- and trans-*NAT* gene pairs, as well as sRNAs produced from lncRNAs regulate gene expression under cold acclimation	Tiwari et al. (88)
Saccharum officinarum	At 4°C, miR319 expression was increased, but in cold-tolerant cultivars, the effects were delayed	Thiebaut et al. (86)
Brachypodium distachyon	Determine the conserved miRNA sequences in *Arabidopsis* that react to low-temperature stress	Zhang et al. (87)

24.4 EPIGENETIC STRESS MEMORY

After a period of stress, most stress-induced epigenetic changes typically return to their original levels. However, some of these changes can reactivate during sexual reproduction or embryonic development, while others may remain stable and be inherited as epigenetic stress memory across generations through both mitotic and meiotic cell divisions (89). This type of epigenetic memory does not involve changes to the DNA nucleotide sequences but relies on specific modifications to chromatin, influencing gene activation and related processes in response to various stressors over different time scales.

There are three forms of epigenetic memory: cellular, transcriptional and transgenerational. These forms are dynamic and can be inherited through meiosis or mitosis, triggered by previous stress exposure. Somatic memory in cells is a heritable trait that can be passed down through mitosis. This type of memory is influenced by tissue-specific transcriptional programs and transcription factors (TFs) established during early development. Additionally, transcriptional memory involves mitotically heritable changes that are passed down across generations in response to stressful events. In contrast, meiotically heritable changes in transgenerational memory may impact the behavior of offspring (89).

Somatic stress memory typically lasts only a few days or weeks within a generation, for instance, the H3K27me3 modification of the *FLC* gene preserves vernalization

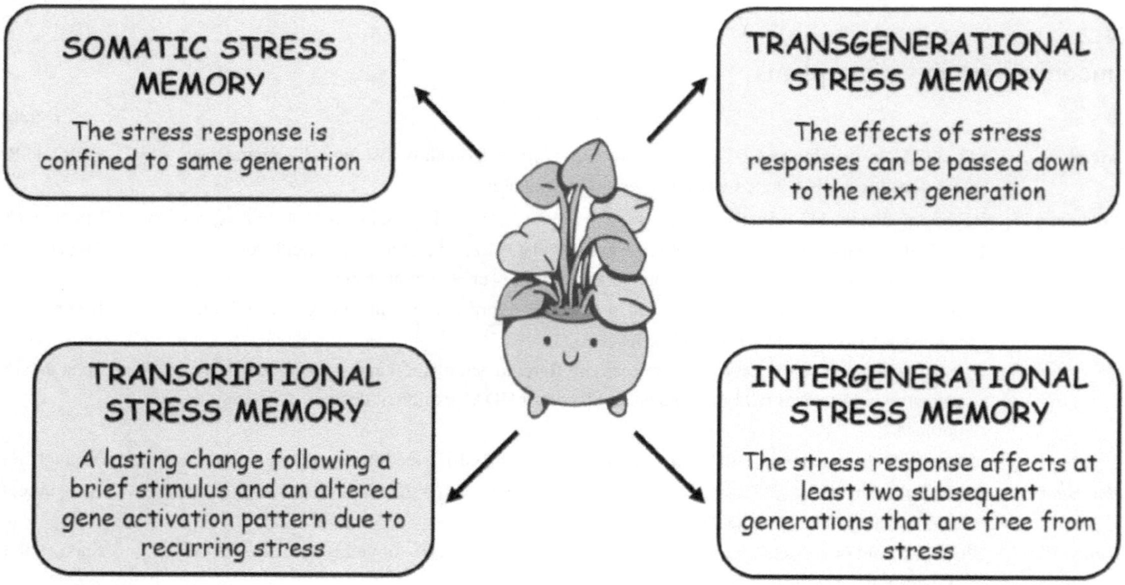

FIGURE 24.2 Different types of stress memory in plants.

memory for a few weeks (90). Both PRC1 and PRC2 are essential for maintaining the activity of somatic memory genes by facilitating histone methylation (specifically H3K27me), which serves as a repressive marker during mitosis. This modification is crucial for maintaining and transmitting active chromatin states. In subsequent generations, where epigenetic modifications can occur autonomously, transgenerational memory can last for a longer duration (91). The inheritance of stress memory across generations is significantly influenced by the genome-wide distribution of DNA methylation and regulatory mechanisms related to chromatin alterations (92). Intergenerational memory can endure for at least two generations without exposure to stress (93).

Intergenerational memory can be reversed after a stress-free generation, indicating that this memory is influenced by stress. Most of this memory is inherited from the mother, while paternal inheritance is suppressed by the DEMETER (DME) DNA demethylase but can be restored in *dme* mutants. The regulation of intergenerational memory involves RdDM-mediated demethylation (94). In primed plants, histones H3K4me2 and H3K4me3 play a role in controlling acquired transcriptional memory, while hypermethylation of these histone marks helps maintain transcriptional memory during heat stress (95, 96).

Arabidopsis thaliana demonstrated enhanced freezing tolerance after undergoing cold treatment, indicating that it has an initial experience memory (97). Consequently, some research suggests that, following an initial stress event, certain cold response genes may acquire activation marks that lead to a primed state. In contrast, other genes may gain repressive marks that inhibit or limit their reactivation during subsequent stress exposure. This results in a differential cold response when the plant encounters stress for a second time (74).

The transcription regulation of the cold-sensitive genes *COR15A* and *ATGOLS3* is associated with the repressive mark of facultative heterochromatin, specifically the trimethylation of histone H3 on lysine 27 (H3K27me3). Following a period of transcriptional repression, H3K27me3 levels in the promoter regions of *COR15A* and *ATGOLS3* progressively decline over three days. This decline suggests that cell division may influence the levels of H3K27me3 to some degree. Additionally, the duration of exposure to cold plays a crucial role in memory acquisition (71). *In vitro* studies show that *A. thaliana* seedlings exposed to *night time* cold temperature exhibit a steady improvement in freezing resistance, provided that the intermittent room temperature remains below 32 h. Seedlings that undergo extended cold treatment demonstrate increased freezing tolerance, and this tolerance further increases in response to stress three days later, indicating the presence of a cold stress memory (Figure 24.2) (98).

24.5 CONCLUSION

For optimal growth, plants require appropriate temperature, and their responses to stress are complex processes that vary by species. Thousands of genes and dynamic epigenetic processes, influenced by the type of stress, regulate how plants react. Although the mechanisms in crop plants are not yet fully understood, the epigenetic processes in *Arabidopsis* have been well characterized. These processes transmit epigenetic stress memories during both mitotic and meiotic cell divisions. Epigenetic stress memory fine-tunes the genetic relationships within plants. Additionally, histone changes and controlling epigenetic modifications play a significant role in this process, for example, stress-related modifications like H3K9me2 and H3K27me3 are linked to chromatin architecture and DNA methylation, which are crucial for establishing memory under stress conditions.

Two climatic scenarios can help predict how plants might respond in terms of fitness when faced with the loss of epigenetic memory: (1) milder conditions resulting from a gradual increase in temperature and (2) harsher conditions caused by more frequent extreme temperature events. In milder circumstances, such as prolonged high temperature, plants from colder climates may be more likely to survive and produce more flowers, which enhances their reproduction and dispersal. Since critical environmental conditions are absent in this scenario, the loss of epigenetic memory may not negatively impact plant fitness. Conversely, during extreme cold events, there is a trade-off between the expression of cold resistance genes and overall plant fitness. Plants that retain their epigenetic memory might demonstrate greater fitness during these extreme cold spells; however, this advantage may diminish under warmer temperature.

REFERENCES

1. Surekha, C., Aruna, L., Hossain, M. A., Wani, S. H., & Neelapu, N. R. R. (2015). Present status and future prospects of transgenic approaches for salt tolerance in plants/crop plants. In S. H. Wani & M. A. Hossain (Eds.), *Managing Salt Tolerance in Plants: Molecular and Genomic Perspectives* (pp. 329–352). New York, NY: CRC Press.
2. Sanghera, G. S., Wani, S. H., Hussain, W., & Singh, N. B. (2011). Engineering cold stress tolerance in crop plants. *Current Genomics*, 12, 30–43. https://doi.org/10.2174/138920211794520178.
3. Wani, S. H., Sah, S. K., Sanghera, G., Hussain, W., & Singh, N. B. (2016). Genetic engineering for cold stress tolerance in crop plants. In Atta-ur-Rahman (Ed.), *Advances in Genome Science* (Vol. 4, pp. 173–201). London, UK: Bentham Science.
4. Wani, S. H., Singh, N. B., Haribhushan, A., & Mir, J. I. (2013). Compatible solute engineering in plants for abiotic stress tolerance-Role of glycine betaine. *Current Genetics*, 14, 157–165. https://doi.org/10.2174/1389202911314030001.
5. Yamaguchi-Shinozaki, K., & Shinozaki, K. (2006). Transcriptional regulatory networks in cellular responses and tolerance to dehydration and cold stresses. *Annual Review of Plant Biology*, 57, 781–803. https://doi.org/10.1146/annurev.arplant.57.032905.105444.
6. Song, J., Angel, A., Howard, M., & Dean, C. (2012). Vernalization—a cold induced epigenetic switch. *Journal of Cell Science*, 125, 3723–3731. https://doi.org/10.1242/jcs.084764.
7. Gallusci, P., Agius, D. R., Moschou, P. N., Dobranszki, J., Kaiserli, E., & Martinelli, F. (2022). Deep inside the epigenetic memories of stressed plants. *Trends in Plant Science*, 1–12. https://doi.org/10.1016/j.tplants.2022.09.004.
8. Sharma, M., Kumar, P., Verma, V., Sharma, R., Bhargava, B., & Irfan, M. (2022). Understanding plant stress memory response for abiotic stress resilience: Molecular insights and prospects. *Plant Physiology and Biochemistry*, 179, 10–24. https://doi.org/10.1016/ j.plaphy.2022.03.004.
9. Wu, Y., Zhang, D., Chu, J. Y., Boyle, P., Wang, Y., Brindle, I. D., De Luca, V. & Despres, C., (2012). The *Arabidopsis* NPR1 protein is a receptor for the plant defense hormone salicylic acid. *Cell Reports*, 1, 639–647. https://doi.org/10.1016/j.celrep.2012.05.008.
10. Butterbrodt, T., Thurow, C., & Gatz, C. (2006). Chromatin immunoprecipitation analysis of the tobacco PR-1a- and the truncated CaMV 35S promoter reveals differences in salicylic acid-dependent TGA factor binding and histone acetylation. *Plant Molecular Biology*, 61, 665–674. https://doi.org/10.1007/s11103-006-0039-2.
11. Bhadouriya, S. L., Mehrotra, S., Basantani, M. K., Loake, G. J., & Mehrotra, R. (2021). Role of chromatin architecture in plant stress responses: An update. *Frontiers in Plant Science*, 11, 603380–603380. https://doi.org/10.3389/fpls.2020.603380.
12. Liu, H., Able, A. J., & Able, J. A. (2022). Priming crops for the future: Rewiring stress memory. *Trends in Plant Science*, 27, 699–716. https://doi.org/10.1016/j.tplants.2021.11.015.
13. Xie, W., Tang, Q., Yan, F., & Tao, Z. (2021). Transcriptional memory and response to adverse temperatures in plants. *Journal of Zhejiang University: Science B*, 22, 791–804. https://doi.org/10.1631/jzus.B2100287.
14. Joseph, F. M., & Young, N. L. (2022). Histone variant-specific post translational modifications. *Seminars in Cell & Developmental Biology*, 135, 73–84. https://doi.org/10.1016/ j.semcdb.2022.02.012.
15. Borg, M., Jiang, D., & Berger, F. (2021). Histone variants take center stage in shaping the epigenome. *Current Opinion in Plant Biology*, 61, 101991. https://doi.org/10.1016/ j.pbi.2020.101991.
16. Wang, B., Chai, H., Zhong, Y., Shen, Y., Yang, W., Chen, J., et al. (2020). The DEAD-box RNA helicase SHI2 functions in repression of salt-inducible genes and regulation of cold-inducible gene splicing. *Journal of Experimental Botany*, 71(4), 1598–1613. https://doi.org/10.1093/jxb/erz523.
17. Uemura, M., Tominaga, Y., Nakagawara, C., Shigematsu, S., Minami, A., & Kawamura, Y. (2006). Responses of the plasma membrane to low temperatures. *Physiologia Plantarum*, 126(1), 81–89. https://doi.org/10.1111/j.1399-3054.2005.00594.x.
18. Zhao, X., Wei, Y., Zhang, J., Yang, L., Liu, X., Zhang, H., et al. (2021). Membrane lipids' metabolism and transcriptional regulation in maize roots under cold stress. *Frontiers in Plant Science*, 12, 639132. https://doi.org/10.3389/fpls.2021.639132.
19. Granel, M. L. S., Cánepa, C., Cid, N. G., Navarro, J. C., Monroig, Ó., Verstraeten, S. V., et al. (2019). Gene identification and functional characterization of a Δ12 fatty acid desaturase in Tetrahymena thermophila and its influence in homeoviscous adaptation to low temperature. *Biochimica et Biophysica Acta – Molecular and Cell Biology of Lipids*, 1864 (11), 1644–1655. https://doi.org/10.1016/j.bbalip.2019.08.003.
20. Gao, X., Dong, J., Rasouli, F., Pouya, A. K., Tahir, A. T., & Kang, J. (2022). Transcriptome analysis provides new insights into plants responses under phosphate starvation in association with chilling stress. *BMC Plant Biology*, 22(1), 26–14. https://doi.org/10.1186/s12870-021-03381-z.
21. Wu, P., Xiao, C., Cui, J., Hao, B., Zhang, W., Yang, Z., et al. (2021). Nitric oxide and its interaction with hydrogen peroxide enhance plant tolerance to low temperatures by improving the efficiency of the calvin cycle and the ascorbate–glutathione cycle in cucumber seedlings. *Journal of Plant Growth Regulation*, 40(6), 2390–2408. https://doi.org/10.1007/s00344-020-10242-w.

22. Hajihashemi, S., Noedoost, F., Geuns, J., Djalovic, I., & Siddique, K. H. (2018). Effect of cold stress on photosynthetic traits, carbohydrates, morphology, and anatomy in nine cultivars of *Stevia rebaudiana. Frontiers in Plant Science*, 2018, 1430. https://doi.org/10. 3389/fpls.2018.01430.
23. Ramazan, S., Nazir, I., Yousuf, W., John, R., & Allakhverdiev, S. (2022). Environmental stress tolerance in maize (*Zea mays*): Role of polyamine metabolism. *Functional Plant Biology*. https://doi.org/10.1071/FP21324.
24. Primo-Capella, A., Martínez-Cuenca, M. R., Gil-Muñoz, F., & Forner-Giner, M. A. (2021). Physiological characterization and proline route genes quantification under long-term cold stress in *Carrizo citrange. Scientia Horticulturae*, 276, 109744. https://doi.org/10. 1016/j.scienta.2020.109744.
25. Alcázar, R., Bueno, M., & Tiburcio, A. F. (2020). Polyamines: Small amines with large effects on plant abiotic stress tolerance. *Cells*, 9 (11), 2373. https://doi.org/10.3390/ cells9112373.
26. Shkryl, Y., Yugay, Y., Avramenko, T., Grigorchuk, V., Gorpenchenko, T., Grischenko, O., et al. (2021). CRISPR/Cas9-Mediated knockout of HOS1 reveals its role in the regulation of secondary metabolism in *Arabidopsis thaliana. Plants*, 10(1), 104. https://doi.org/10.3390/plants10010104.
27. DeVries, A. L., & Wohlschlag, D. E. (1969). Freezing resistance in some Antarctic fishes. *Science*, 163(3871), 1073–1075. https://doi.org/10.1126/science.163.3871.1073.
28. Singh, P., Hanada, Y., Singh, S. M., & Tsuda, S. (2014). Antifreeze protein activity in Arctic cryoconite bacteria. *FEMS Microbiology Letters*, 351(1), 14–22. https://doi.org/10. 1111/1574-6968.12345.
29. Hoshino, T., Kiriaki, M., Ohgiya, S., Fujiwara, M., Kondo, H., Nishimiya, Y., et al. (2003). Antifreeze proteins from snow mold fungi. *Canadian Journal of Botany*, 81(12), 1175–1181. https://doi.org/10.1139/b03-116.
30. Gwak, I. G., Kim, H. J., Kang, S. H., & Jin, E. (2010). Antifreeze protein in Antarctic marine diatom, *Chaetoceros neogracile. Marine Biotechnology*, 12(6), 630–639. https://doi.org/10.1007/s10126-009-9250-x.
31. Griffith, M., Paul, A., Daniel, S. C., Yang, W-C., Moffatt, B. A., & Hon, W. C. (1992). Antifreeze protein produced endogenously in winter rye leaves. *Plant Physiology*, 100(2), 593–596. https://doi.org/10.1104/pp.100.2.593.
32. Leya, T. (2013). Polyextremophiles. In *Snow Algae: Adaptation Strategies to Survive on Snow and Ice* (pp. 401–423). Dordrecht: Springer. https://doi.org/10.1007/978-94-007 6488-0_17.
33. Carvajal-Rondanelli, P. A., Marshall, S. H., & Guzman, F. (2011). Antifreeze glycoprotein agents: Structural requirements for activity. *Journal of the Science of Food and Agriculture*, 91(14), 2507–2510. https://doi.org/10.1002/jsfa.4473.
34. Maddah, M., Shahabi, M., & Peyvandi, K. (2021). How does dc AFP, a plant antifreeze protein, control ice inhibition through the kelvin effect? *Industrial & Engineering Chemistry Research*, 60(50), 18230–18242. https://doi.org/10.1021/acs.iecr.1c02559.
35. Yu, H., Zheng, H., Liu, Y., Yang, Q., Li, W., Zhang, Y., et al. (2021). Antifreeze protein from *Ammopiptanthus nanus* functions in temperature-stress through domain A. *Scientific Reports*, 11(1), 8458–8510. https://doi.org/10.1038/s41598-021-88021-0.
36. Griffith, M., Lumb, C., Wiseman, S. B., Wisniewski, M., Johnson, R. W., & Marangoni, A. G. (2005). Antifreeze proteins modify the freezing process in planta. *Plant Physiology*, 138(1), 330–340. https://doi.org/10.1104/pp.104.058628.
37. Gupta, R., & Deswal, R. (2014). Antifreeze proteins enable plants to survive in freezing conditions. *Journal of Biosciences*, 39(5), 931–944. https://doi.org/10.1007/s12038-014-9468-2.
38. Zhang, D. Q., Liu, B., Feng, D. R., He, Y. M., & Wang, J. F. (2004). Expression, purification, and antifreeze activity of carrot antifreeze protein and its mutants. *Protein Expression and Purification*, 35(2), 257–263. https://doi.org/10.1016/j.pep.2004.01.019.
39. Bartels, A., Han, Q., Nair, P., Stacey, L., Gaynier, H., Mosley, M., Huang, Q. Q., Pearson, J. K., Hsieh, T., Charles, A. Y., & Xiao, W. (2018). Dynamic DNA methylation in plant growth and development. *International Journal of Molecular Sciences*, 19(7), 2144. https://doi.org/10.3390/ijms19072144.
40. Law, J. A., & Jacobsen, S. E. (2010). Establishing, maintaining and modifying DNA methylation patterns in plants and animals. *Nature Reviews Genetics*, 11(3), 204–220.
41. Steward, N., Ito, M., Yamaguchi, Y., Koizumi, N., & Sano, H. (2002). Periodic DNA methylation in maize nucleosomes and demethylation by environmental stress. *Journal of Biological Chemistry*, 277(40), 37741–37746. https://doi.org/10.1074/jbc.M204050200.
42. Moore, L. D., Le, T., & Fan, G. (2013). DNA methylation and its basic function. *Neuropsychopharmacology*, 38(1), 23–38. https://doi.org/10.1038/npp.2012.112.
43. Sudan, J., Raina, M., & Singh, R. (2018). Plant epigenetic mechanisms: Role in abiotic stress and their generational heritability. *3 Biotech*, 8(3), 172. https://doi.org/10.1007/s13205-018-1202-6.
44. Verma, N., Giri, S. K., Singh, G., Gill, R., & Kumar, A. (2022). Epigenetic regulation of heat and cold stress responses in crop plants. *Plant Gene*, 29, 100351. https://doi.org/10.1016/j.plgene.2022.100351.
45. Liu, J., & He, Z. (2020). Small DNA methylation, big player in plant abiotic stress responses and memory. *Frontiers in Plant Science*, 11. https://doi.org/10.3389/fpls.2020.595603.
46. Ding, Y., Zou, L.-H., Wu, J., Ramakrishnan, M., Gao, Y., Zhao, L., et al. (2022). The pattern of DNA methylation alteration, and its association with the expression changes of non-coding RNAs and mRNAs in moso bamboo under abiotic stress. *Plant Science*, 325, 111451. https://doi.org/10.1016/j.plantsci.2022.111451.
47. Song, Y., Liu, L., Feng, Y., Wei, Y., Yue, X., He, W., et al. (2015). Chilling- and freezing-induced alterations in cytosine methylation and its association with the cold tolerance of an alpine subnival plant, *Chorispora bungeana. PLoS One*, 10, e0135485. https://doi.org/10.1371/journal.pone.0135485.
48. Song, Y., Ci, D., Tian, M., & Zhang, D. (2016). Stable methylation of a non coding RNA gene regulates gene expression in response to abiotic stress in *Populus simonii. Journal of Experimental Botany*, 67, 1477–1492. https://doi.org/10.1093/jxb/erv543.
49. Foust, C. M., Preite, V., Schrey, A. W., Alvarez, M., Robertson, M. H., Verho even, K. J. F., & Richards, C. L. (2016). Genetic and epigenetic differences associated with environmental gradients in replicate populations of two salt marsh perennials. *Molecular Ecology*, 25(8), 1639–1652. https://doi.org/10.1111/mec.13522.

50. Duan, W., Zhang, H., Zhang, B., Wu, X., Shao, S., Li, Y., et al. (2017). Role of vernalization-mediated demethylation in the floral transition of *Brassica rapa*. *Planta*, 245, 227–233. https://doi.org/10.1007/s00425-016-2622-3.
51. Pan, J., Wang, W., Zhao, X., Zhu, L., Fu, B., & Li, Z. (2011). DNA methylation alterations of rice in response to cold stress. *Plant Omics Journal*, 4(7), 364–369.
52. Deng, J., Kou, S., Zhang, C., Zou, Q., Li, P., et al. (2018). DNA methylation and plant stress responses. *Journal of Plant Physiology and Pathology*, 6, 4. https://doi.org/10.4172/2329-955X.1000182.
53. Liu, T., Li, Y., Duan, W., Huang, F., & Hou, X. (2017). Cold acclimation alters DNA methylation patterns and confers tolerance to heat and increases growth rate in *Brassica rapa*. *Journal of Experimental Botany*, 68, 1213–1224. https://doi.org/10.1093/jxb/erw496.
54. Ma, N., Chen, W., Fan, T., Tian, Y., Zhang, S., Zeng, D., et al. (2015). Low temperature-induced DNA hypermethylation attenuates expression of RhAG, an AGAMOUS homolog, and increases petal number in rose (*Rosa hybrida*). *BMC Plant Biology*, 15, 237. https://doi.org/10.1186/s12870-015-0623-1.
55. Kim, J.-M., Sasaki, T., Ueda, M., Sako, K., & Seki, M. (2015). Chromatin changes in response to drought, salinity, heat, and cold stresses in plants. *Frontiers in Plant Science*, 6. https://doi.org/10.3389/fpls.2015.00114.
56. Tang, X., Wang, Q., Yuan, H., & Huang, X. (2018). Chilling-induced DNA demethylation is associated with the cold tolerance of *Hevea brasiliensis*. *BMC Plant Biology*, 18, 70. https://doi.org/10.1186/s12870-018-1276-7.
57. Wang, H., Wang, H., Shao, H., & Tang, X. (2016). Recent advances in utilizing transcription factors to improve plant abiotic stress tolerance by transgenic technology. *Frontiers in Plant Science*, 7, 67. https://doi.org/10. 3389/fpls.2016.00067.
58. Shan, X., Wang, X., Yang, G., Wu, Y., Su, S., Li, S., Liu, H., & Yuan, Y. (2013). Analysis of the DNA methylation of maize (*Zea mays* L.) in response to cold stress based on methylation-sensitive amplified polymorphisms. *Journal of Plant Biology*, 56(1), 32–38. https://doi.org/ 10.1007/s12374-012-0251-3.
59. Fan, H. H., Wei, J., Li, T. C., Li, Z. P., Guo, N., Cai, Y. P., & Lin, Y. (2013). DNA methylation alterations of upland cotton (*Gossypium hirsutum*) in response to cold stress. *Acta Physiologiae Plantarum*, 35(8), 2445–2453. https://doi.org/10.1007/s11738-013-1278-x.
60. Xie, H., Sun, Y., Cheng, B., Xue, S., Cheng, D., Liu, L., & Qiang, S. (2019). Variation in ICE1 methylation primarily determines phenotypic variation in freezing tolerance in *Arabidopsis thaliana*. *Plant and Cell Physiology*, 60(1), 152–165.
61. Kenchanmane, R. S. K., Shao, M. R., Wamboldt, Y., & Mackenzie, S. (2018). Epigenomic plasticity of *Arabidopsis* msh1 mutants under prolonged cold stress. *Plant Direct*, 2(8), e00079. https://doi.org/10.1002/pld3.79.
62. Roy, D., Paul, A., Roy, A., Ghosh, R., Ganguly, P., & Chaudhuri, S. (2014). Differential acetylation of histone H3 at the regulatory region of *OsDREB1b* promoter facilitates chromatin remodelling and transcription activation during cold stress. *PLoS One*, 9(6), e100343. https://doi.org/10.1371/journal.pone.0100343.
63. Dai, Y., Mao, P., Tao, X., Wang, Y., Wei, C. M., & Ma, X. R. (2017). Flexible shift on gene body methylation and transcription of LpCYP72A161 exposed to temperature stress in perennial ryegrass. *Environmental and Experimental Botany*, 143, 29–37. https://doi.org/10.1016/j.envexpbot.2017.08.002.
64. Deal, R. B., & Henikoff, S. (2011). Histone variants and modifications in plant gene regulation. *Current Opinion in Plant Biology*, 14(2), 116–122. https:// doi.org/10.1016/j.pbi.2010.11.005.
65. Park, J., Lim, C. J., Shen, M., Park, H. J., Cha, J. Y., Iniesto, E., & Yun, D. J. (2018). Epigenetic switch from repressive to permissive chromatin in response to cold stress. *Proceedings of the National Academy of Sciences of the United States of America*, 115(23), 5400–5409. https://doi.org/10.1073/pnas.1721241115.
66. Berr, A., Shafiq, S., & Shen, W. H. (2011). Histone modifications in transcriptional activation during plant development. *Biochimica et Biophysica Acta*, 1809(10), 567–576. https://doi.org/10.1016/j.bbagrm.2011.07.001.
67. Li, T., Chen, X., Zhong, X., Zhao, Y., Liu, X., Zhou, S., Cheng, S., & Zhou, D. X. (2013). Jumonji C domain protein JMJ705-mediated removal of histone H3 lysine 27 trimethylation is involved in defense related gene activation in rice. *Plant Cell*, 25, 4725–4736. https://doi.org/10.1105/tpc.113.118802.
68. Kidokoro, S., Shinozaki, K., & Yamaguchi-Shinozaki, K. (2022). Transcriptional regulatory network of plant cold-stress responses. *Trends in Plant Science*, 27, 922–935. https://doi.org/10.1016/j.tplants.2022.01.008.
69. Shen, Y., Wu, X., Liu, D., Song, S., Liu, D., & Wang, H. (2016). Cold-dependent alternative splicing of a Jumonji c domain-containing gene MtJMJC5 in *Medicago truncatula*. *Biochemical and Biophysical Research Communications*, 474, 271–276. https://doi.org/10.1016/j.bbrc.2016.04.062.
70. Zeng, Z., Zhang, W., Marand, A. P., Zhu, B., Buell, C. R., & Jiang, J. (2019). Cold stress induces enhanced chromatin accessibility and bivalent histone modifications H3K4me3 and H3K27me3 of active genes in potato. *Genome Biology*, 20, 123. https://doi.org/10.1186/s13059-019-1731-2.
71. Kwon, C. S., Lee, D., Choi, G., & Chung, W. I. (2009). Histone occupancy dependent and-independent removal of H3K27 trimethylation at cold-responsive genes in *Arabidopsis*. *The Plant Journal*, 60, 112–121. https://doi.org/10.1111/j.1365-313X.2009.03938.x.
72. Han, Z., Yu, H., Zhao, Z., Hunter, D., Luo, X., Duan, J., & Tian, L. (2016). AtHD2D gene plays a role in plant growth, development, and response to abiotic stresses in *Arabidopsis thaliana*. *Frontiers in Plant Science*, 7, 310. https://doi.org/10.3389/fpls.2016.00310.
73. Song, C., Yang, Y., Yang, T., Ba, L., Zhang, H., Han, Y., Xiao, Y., Shan, W., Kuang, J., Chen, J., & Lu, W. (2019). MaMYB4 recruits histone deacetylase MaHDA2 and modulates the expression of ω-3 fatty acid desaturase genes during cold stress response in banana fruit. *Plant and Cell Physiology*, 60(11), 2410–2422. https://doi.org/10.1093/pcp/pcz142.
74. Friedrich, T., Faivre, L., Bäurle, I., & Schubert, D. (2019). Chromatin-based mechanisms of temperature memory in plants. *Plant, Cell & Environment*, 42, 762–770. https://doi.org/10.1111/pce.13373.
75. Hu, Y., Zhang, L., Zhao, L., Li, J., He, S., Zhou, K., et al. (2011). Trichostatin a selectively suppresses the cold-induced transcription of the ZmDREB1 gene in maize. *PLoS One*, 6, e22132. https://doi.org/10.1371/journal.pone.0022132.

76. Fiust, A., & Rapacz, M. (2020). Downregulation of three novel candidate genes is important for freezing tolerance of field and laboratory cold acclimated barley. *Journal of Plant Physiology*, 244, 153049. https://doi.org/10.1016/j.jplph.2019.153049.
77. Zhu, J., Jeong, J. C., Zhu, Y., Sokolchik, I., Miyazaki, S., Zhu, J. K., Hasegawa, P. M., Bohnert, H. J., Shi, H., Yun, D. J., & Bressan, R. A. (2008). Involvement of *Arabidopsis* HOS15 in histone deacetylation and cold tolerance. *Proceedings of the National Academy of Sciences of the United States of America*, 105(12), 4945–4950. https://doi.org/10.1073/pnas.0801029105.
78. To, T. K., Nakaminami, K., Kim, J. M., Morosawa, T., Ishida, J., Tanaka, M., Yokoyama, S., Shinozaki, K., & Seki, M. (2011). *Arabidopsis* HDA6 is required for freezing tolerance. *Biochemical and Biophysical Research Communications*, 406(3), 414–419. https://doi.org/10.1016/j.bbrc.2011.02.058.
79. Yang, R., Hong, Y., Ren, Z., Tang, K., Zhang, H., Zhu, J. K., & Zhao, C. (2019). A role for PICKLE in the regulation of cold and salt stress tolerance in *Arabidopsis. Frontiers in Plant Science*, 10, 900.
80. Hu, Y., Zhang, L., He, S., Huang, M., Tan, J., Zhao, L., Yan, S., Li, H., Zhou, K., Liang, Y., & Li, L. (2012). Cold stress selectively unsilences tandem repeats in heterochromatin associated with accumulation of H3K9ac. *Plant, Cell & Environment*, 35(12), 2130–2142. https://doi.org/10.1111/j.1365-3040.2012.02541.x.
81. Djupedal, I., & Ekwall, K. (2009). Epigenetics: Heterochromatin meets RNAi. *Cell Research*, 19(3), 282–295. https://doi.org/10.1038/cr.2009.13.
82. Holoch, D., & Moazed, D. (2015). RNA-mediated epigenetic regulation of gene expression. *Nature Reviews Genetics*, 16, 71–84. https://doi.org/10.1038/nrg3863.
83. Meister, G. (2013) Argonaute proteins: Functional insights and emerging roles. *Nature Reviews Genetics*, 14(7), 447–459. https://doi.org/10.1038/nrg3462.
84. Barakat, A., Sriram, A., Park, J., Zhebentyayeva, T., Main, D., & Abbott, A. (2012). Genome wide identification of chilling responsive micro RNAs in *Prunus persica. BMC Genomics*, 13, 481.
85. Zhou, X., Wang, G., Sutoh, K., Zhu, J. K., & Zhang, W. (2008). Identification of cold-inducible microRNAs in plants by transcriptome analysis. *Biochimica et Biophysica Acta*, 1779(11), 780–788. https://doi.org/10. 1016/j.bbagrm.2008.04.005.
86. Thiebaut, F., Rojas, C. A., Almeida, K. L., Grativol, C., Domiciano, G. C., Lamb, C. R., de Almeida, J., Hemerly, A., & Ferreira, P. (2012). Regulation of miR319 during cold stress in sugarcane. *Plant, Cell & Environment*, 35(3), 502–512. https://doi.org/10.1111/j.1365-3040.2011.02430.x.
87. Zhang, J., Xu, Y., Huan, Q., & Chong, K. (2009). Deep sequencing of Brachy podium small RNAs at the global genome level identifies micro RNAs involved in cold stress response. *BMC Genomics*, 10, 449. https://doi.org/10.1186/1471-2164-10-449.
88. Tiwari, B., Habermann, K., Arif, M. A., Weil, H. L., Garcia-Molina, A., Kleine, T., Mühlhaus, T., & Frank, W. (2020). Identification of small RNAs during cold acclimation in *Arabidopsis thaliana. BMC Plant Biology*, 20(1), 1–25. https://doi.org/10.1186/s12870-020-02511-3.
89. D'urso, A., & Brickner, J. H. (2014). Mechanisms of epigenetic memory. *Trends in Genetics*, 30, 230–236. https://doi.org/10.1016/j.tig.2014.04.004.
90. Yang, H., Berry, S., Olsson, T. S. G., Hartley, M., Howard, M., & Dean, C. (2017). Distinct phases of polycomb silencing to hold epigenetic memory of cold in *Arabidopsis. Science*, 357, 1142–1145. https://doi.org/10.1126/science.aan1121.
91. Shanker, A. K., Bhanu, D., & Maheswari, M. (2020). Epigenetics and transgenerational memory in plants under heat stress. *Plant Physiology Reports*, 25, 583–593. https://doi.org/10.1007/s40502-020-00557-x.
92. Cong, W., Miao, Y., Xu, L., Zhang, Y., Yuan, C., Wang, J., et al. (2019). Transgenerational memory of gene expression changes induced by heavy metal stress in rice (*Oryza sativa* l.). *BMC Plant Biology*, 19, 282. https://doi.org/10.1186/s12870-019 1887-7.
93. Oberkofler, V., Pratx, L., & Bäurle, I. (2021). Epigenetic regulation of abiotic stress memory: Maintaining the good things while they last. *Current Opinion in Plant Biology*, 61, 102007. https://doi.org/10.1016/j.pbi.2021.102007.
94. Wibowo, A., Becker, C., Marconi, G., Durr, J., Price, J., Hagmann, J., et al. (2016). Hyperosmotic stress memory in *Arabidopsis* is mediated by distinct epigenetically labile sites in the genome and is restricted in the male germline by DNA glycosylase activity. *eLife*, 5, e13546. https://doi.org/10.7554/eLife.13546.044.
95. Jaskiewicz, M., Conrath, U., & Peterhänsel, C. (2011). Chromatin modification acts as a memory for systemic acquired resistance in the plant stress response. *EMBO Reports*, 12, 50–55. https://doi.org/10.1038/embor.2010.186.
96. Lämke, J., Brzezinka, K., Altmann, S., & Bäurle, I. (2016). A hit-and-run heat shock factor governs sustained histone methylation and transcriptional stress memory. *The EMBO Journal*, 35, 162–175. https://doi.org/10.15252/embj.201592593.
97. Zuther, E., Schaarschmidt, S., Fischer, A., Erban, A., Pagter, M., Mubeen, U., et al. (2019). Molecular signatures associated with increased freezing tolerance due to low temperature memory in *Arabidopsis. Plant, Cell & Environment*, 42, 854–873. https://doi.org/10.1111/pce.13502.
98. Leuendorf, J. E., Frank, M., & Schmülling, T. (2020). Acclimation, priming and memory in the response of *Arabidopsis thaliana* seed lings to cold stress. *Scientific Reports*, 10, 689. https://doi.org/10.1038/s41598-019-56797-x.
99. Mukhopadhyay, J., & Roychoudhury, A. (2018). Cold-induced injuries and signaling responses in plants. In S. H. Wani & V. Herath (Eds.), *Cold Tolerance in Plants Physiological, Molecular and Genetic Perspectives* (pp. 1–35). Switzerland: Springer Nature Switzerland.
100. Banerjee, A., & Roychoudhury, A. (2019). Cold stress and photosynthesis. In P. Ahmad, M. A. Ahanger, M. N. Alyemeni, & P. Alam (Eds.), *Photosynthesis, Productivity, and Environmental Stress* (pp. 27–37). Hoboken: John Wiley & Sons Ltd.
101. Banerjee, A., Wani, S. H., & Roychoudhury, A. (2017). Epigenetic control of plant cold responses. *Frontiers in Plant Science*, 8, 1643.
102. Mayer, B.F., Ali-Benali, M.A., Demone, J., Bertrand, A. & Charron, J.B., 2015. Cold acclimation induces distinctive changes in the chromatin state and transcript level of COR genes in Cannabis sativa varities with contrasting cold acclimation capacities. *Physiologia Plantarum*, 155(3), pp.281–295.

25 Epigenetic Responses to Heavy Metal Stress in Plants

Priya Kushwaha, Rahul Verma and Amit Kumar Singh

25.1 INTRODUCTION

The term "epigenetics" was initially used in developmental studies, but its definition has since changed. Epigenetics now refers to heritable changes in gene expression that occur without altering the DNA sequence (15). Several epigenetic factors – such as DNA cytosine methylation, histone modifications and small RNAs (sRNAs) – influence gene expression in response to various stimuli (16). In plants, epigenetic inheritance occurs in two ways: first, by conveying information across generations that is not contained in the DNA, and second, by maintaining epigenetic changes within an individual, even after being reset between generations. Plants can address environmental pressure and various stresses through different techniques, including regulating gene expression, undergoing epigenetic alterations and introducing heritable genetic mutations (17).

Abiotic and biotic stress factors can restrict plant biomass and growth. Both types of stress negatively impact plant growth, quality and production. Damage inflicted on plants by pests and various pathogens is categorized as biotic stress. In contrast, abiotic stressors include organic pollutants, heavy metals, salinity, extreme temperature (both high and low) and drought conditions.

Potentially toxic elements (PTEs), commonly known as heavy metals, can have harmful biological effects on living organisms when present at high levels (1). Although these elements naturally occur in metalliferous rocks and soil, their distribution in the environment is increasing due to activities such as mining, industrial processes, use of fertilizers and the land application of household and treated wastewater (2). PTE contamination is being recognized worldwide as a serious environmental issue, particularly in agricultural areas, due to its effects on crop production, the entry of heavy metals into the food chain and the potential harm to human health. Such activities can cause metals to leach into groundwater or accumulate on the soil surface (2, 3).

One alarming characteristic of heavy metals is that they are non-biodegradable and cannot be naturally removed from the environment. While some metals remain immobile at their source, others are mobile and can be absorbed by plants through diffusion, endocytosis or metal transporters (4). PTEs are categorized into two groups: essential elements (e.g., copper, iron, nickel, zinc) that play a vital role in growth and development, and non-essential elements (e.g., arsenic, cadmium, mercury), which can be toxic even at low concentration (58, 59). Zinc, copper and nickel are essential micronutrient cofactors for enzymes and are required in trace amounts. However, pesticides containing harmful metals, such as cadmium and lead, pose risks when their concentrations exceed safe levels (5).

When PTE concentration surpasses toxicity thresholds, they disrupt protein structures, replace essential metals in biomolecules or as cofactors, interfere with functional groups and generate reactive oxygen species (ROS) (6). Consequently, plants growing in metal-polluted environment may show abnormal metal deposition in various tissues and organs, suffer from metabolic disorders and experience changes in physiological and biochemical processes, all of which can reduce growth and biomass yield (7). Metal exposure can lead to decreased chlorophyll levels, altered chloroplast ultrastructure and inhibited photosynthetic enzymes, resulting in lower photosynthetic efficiency. Additionally, plants exposed to heavy metals typically show altered mineral compositions (8, 9).

Numerous cellular pathways influence metal toxicity, with the formation of ROS being the most commonly documented process leading to oxidative stress in plants (9). Metals may bind to protein functional groups, altering protein structure and affecting protein-folding processes, potentially resulting in protein aggregation. Generally, proteins are seen as primary targets of heavy metals through mechanisms such as complexation or the displacement of essential ions in metalloproteins. Heavy metals and metalloids can compete with free ions that serve as cofactors or ligands for enzymes, which can have detrimental effects (10). Metal stress can also trigger genotoxic effects, such as a lower mitotic index and increased chromosomal abnormalities, including stickiness, bridges, fragmentation and the production of micronuclei (11).

Nonetheless, plants have developed defense mechanisms against elevated external PTE concentration in naturally metalliferous environment or those already contaminated. Many species employ avoidance strategies to prevent hazardous metals from entering plant tissues. These methods include physical barriers like thick cuticles and cell walls, selective transport across the plasma membrane and the release of root exudates to modify metal availability. If PTEs breach these defenses, plants respond by developing

DOI: 10.1201/9781003589914-29

tolerance mechanisms, primarily through repairing damage and avoiding vital cellular interactions with metals (12).

Metal detoxification strategies include forming chelating agents, such as organic acids, nicotinamide and glutathione, as well as compartmentalizing metals away from the cytoplasm. Plant tolerance often relies on activating redox detoxification and signaling pathways, which involve hormones like jasmonic acid, ethylene and salicylic acid to coordinate defense responses and mitigate cellular damage (6, 13). Angiosperms and pteridophytes exhibit unique hypertolerance and hyperaccumulation traits. A small subset of metallophyte or pseudo-metallophyte species, known as hyperaccumulator plants, can absorb, transport and accumulate PTEs in their aboveground tissues to very high concentrations without displaying overt signs of toxicity (14, 63).

Epigenetic changes play a crucial role in how plants respond to stress and regulate gene expression. These changes modify chromatin structure through histone modifications and the action of non-coding RNA molecules. By altering the structure of DNA or proteins, these modifications can impact gene function. Importantly, plants can transmit "stress memories" to future generations due to these lasting changes (16, 64).

Together with genetic variants linked to the evolution of tolerance and accumulation traits, epigenetic processes effectively enable plants to adapt to their specific ecosystems. Epigenetic alterations, which modify DNA function without changing the nucleotide sequence, include cytosine methylation and post-translational modification of lysine and arginine residues in histone proteins. These modifications can influence transcription patterns by affecting DNA condensation, accessibility to regulatory proteins and components of the transcriptional machinery. The ability of plants to adapt to harsh conditions through epigenetic control is notable, as it can target both heritable traits and temporary responses. Plant adaptability to various habitats, including interactions with biotic components, contributes to its integration into an adaptive network. Factors such as stress type, genotype, tissue and organism significantly determine stress-induced methylation patterns, which, in turn, affect the regulation of stress-responsive genes. In plants, DNA methylation usually occurs at cytosine residues in CG contexts (known as CG methylation), but it can also occur in other sequence contexts, including CHG and CHH, where H can be A, C or T. These changes in DNA methylation (DMs) and histone modifications (HMs) can influence gene expression by inhibiting the binding of transcription factors or altering DNA accessibility for regulatory proteins.

Research has shown that certain enzymes, such as DNA methyltransferase DRM2, play a critical role in stress-induced epigenetic changes. DRM2 is essential for the modifications that occur in response to stress. Additionally, histone acetyltransferases and histone deacetylases have been associated with histone modifications that respond to stress (23). Plants respond to stress in two primary ways: through short-term acclimatization and long-term alterations that influence future responses. They can retain and retrieve information from previous stimuli, which imprints future stress events. Plants may retain memories of past stress exposure and pass them on to future generations through intricate epigenetic pathways.

In addition to epigenetic mechanisms, plants can also respond to stress via epitranscriptomic processes. While epigenetics focuses on chemical modifications of DNA, epitranscriptomics refers to covalent modifications of RNA (18). The methylation of adenosine at position N6 (m6A) is the most prevalent and extensively studied ribonucleotide alteration, found in 70–100% of mRNA transcripts, depending on the plant species and organelle (19). Furthermore, m6A can appear at various locations on a transcript (20). Studies on mutants defective in RNA adenosine methylation or demethylation machinery have shown that m6A is essential for plant growth and stress response (21).

Epitranscriptomic alterations are considered to be a quicker response to environmental changes, while epigenetic changes typically contribute to long-term adaptation. Metal pollution in the environment can negatively impact agricultural production and food safety. However, there is a lack of research on the epigenetic and epitranscriptomic modifications caused by these contaminants. Non-essential metals such as cadmium (Cd), lead (Pb), aluminum (Al) and mercury (Hg) are highly toxic even in small amounts, posing significant risks. Metal poisoning in plants can result in reduced root development, smaller leaf areas, root swelling and chlorosis (yellowing of leaves) (8). To improve agricultural output, growth, quality and overall productivity, it is essential to manipulate these epigenetic processes effectively. Epigenetic systems influence important agricultural traits, such as growth rate, seed germination and fruit development through mechanisms like DNA methylation, histone modifications and small RNAs. These epigenetic processes play a crucial role in sustainable agriculture by directly affecting agricultural production, yield and quality (Figure 25.1).

25.2 DNA METHYLATION OF SOME HEAVY METALS

25.2.1 Lead

Lead poisoning can have several harmful effects on plants, disrupting their water balance, reducing seed germination and stunting overall development. It can cause damage to cell membranes, lead to the closure of stomata and induce oxidative stress (58). Additionally, lead toxicity inhibits various enzymes essential for processes like photosynthesis, nitrogen metabolism and nucleic acid metabolism. Key enzymes affected by lead include 'δ-aminolevulinic acid dehydratase (ALAD), ribulose-1,5-bisphosphate carboxylase/oxygenase (Rubisco), phosphoenolpyruvate carboxylase (PEPC), glutamine synthase, nitrate reductase, deoxyribonucleases and ribonucleases (23).

Research examining the effects of cadmium (Cd), zinc (Zn) and lead (Pb) on DNA methylation in transporter gene promoters of maize has shown that lead exposure increases

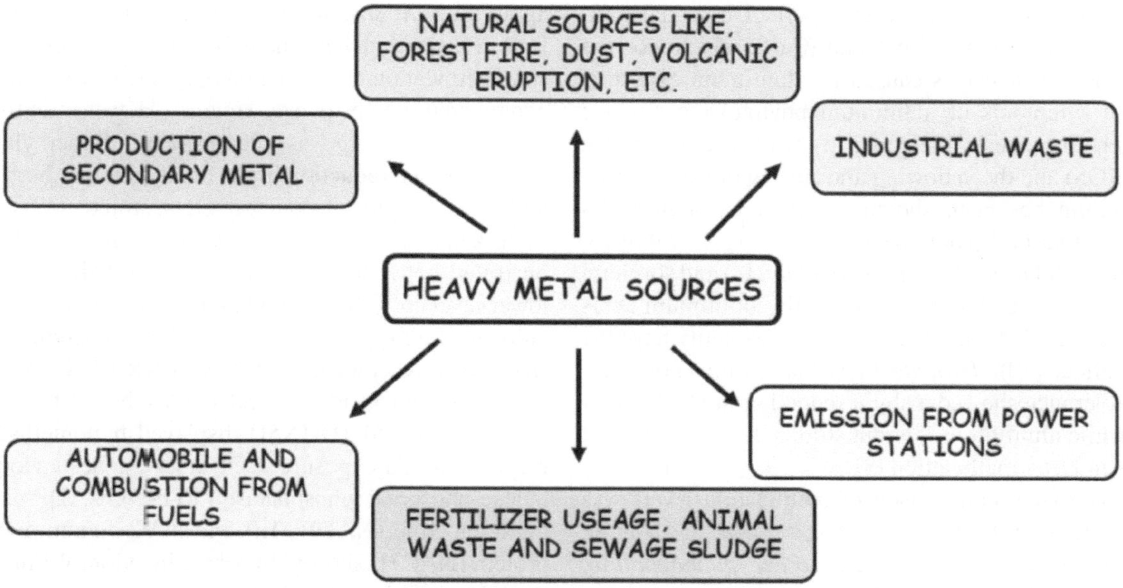

FIGURE 25.1 Different sources of heavy metals.

methylation levels of the *ZIP2* and *ZIP8* promoters (24). It also reduces the expression of several histone deacetylase genes, including *ZmHDlb, ZmHDA102, ZmHDA110, ZmHD2a, ZmHD2b, ZmHD2c* and *ZmHDA106*. Furthermore, lead has been shown to affect the methylation and expression of transporter genes in radish (*Raphanus sativus* L.). Whole-genome bisulfite sequencing (WGBS) identified 11,930 differentially methylated regions (DMRs) when comparing lead-treated plants to a control group. The most common response observed was hypermethylation (25).

In terms of the specific methylation contexts, the CG context exhibited the highest methylation rate at approximately 49%, followed by CHH at about 29% and CHG at around 26%. This methylation influenced gene expression in various ways: 76 hypermethylated regions showed downregulation, while 38 hypomethylated regions also exhibited downregulation. In contrast, 72 hyper-DMRs and 42 hypo-DMRs were activated. Significant alterations in the methylation and expression levels of numerous genes associated with transcription factors and transporters were observed following lead treatment. Specifically, transporter genes such as *RsABCF5, RsABCG14* and *RsZIP11*, along with the RsWRKY46 transcription factor, displayed increased methylation levels and reduced expression. Conversely, genes encoding MATE (Multidrug and Toxic Compound Extrusion) efflux proteins, heavy metal transport proteins (HMT) and ZFP (Zinc Finger Protein) transcription factors showed decreased levels of 5-methylcytosine (5mC) and increased expression.

25.2.2 Arsenic

Arsenic is found in the Earth's crust in various oxidative forms, with arsenate [As(V)] and arsenite [As(III)] being the most common and toxic to plants. As(III) is often more dangerous than As(V) because it can bind to sulfhydryl groups in proteins, altering their structure and function. As(V) also negatively affects plant metabolism by competing with phosphate in cellular processes and can sometimes even replace it. Importantly, arsenic can accumulate in the edible parts of plants, posing significant risks to human health (61). The contamination of water and crops with arsenic is a serious concern in certain regions of the world, affecting an estimated 2 lakh million people. Long-term exposure to high levels of this metalloid has been linked to various types of cancer, including skin, liver, urinary tract and blood cancers (26).

The Cretan brake fern (*Pteris cretica* L.), known as a hyperaccumulator, exhibits reduced DNA methylation, a change observed only in older fronds. In contrast, younger fronds maintain their methylation state regardless of the arsenic (As) concentration (27). Evidence suggests that arsenic uptake, similar to metal uptake, is regulated by DNA methylation. In experiments, the overexpression of VIM1 (Variant in Methylation/Orthrus), a methyl-cytosine-binding protein, resulted in greater arsenic tolerance and lower levels of this metalloid in *Arabidopsis thaliana*. Conversely, VIM1-defective mutants accumulated more arsenic and displayed more severe poisoning symptoms compared to wild-type plants. VIM1 plays a role in mediating the DNA methylation of transporter gene promoters, leading to reduced expression and uptake of arsenic (28).

25.2.3 Aluminium

Aluminium (Al) is the most abundant metal in the Earth's crust and is considered the primary phytotoxic agent in acidic soil due to its high bioavailability at low pH levels. Aluminium mainly affects the plant root system. When plants are exposed to this metal, several responses occur: the cell wall thickens, cell division is halted and root development is delayed. These changes ultimately reduce the plant ability to absorb water and nutrients. The inhibitory effects of aluminium also impact the transport of essential ions, such as potassium (K^+) and calcium (Ca^{2+}), which worsens the nutritional deficiencies caused by aluminium exposure. Additional symptoms of aluminium phytotoxicity include reduced photosynthesis, increased levels of reactive oxygen species (ROS) and the yellowing and curling of leaves (29).

Aluminium has been shown to affect DNA methylation levels in several plant species, including *Arabidopsis*, wheat, maize, tobacco (*Nicotiana tabacum* L.) and sorghum (*Sorghum bicolor* L. Moench). Generally, aluminium tends to demethylate DNA; however, it can occasionally have the opposite effect (30). *Triticale* lines that exhibit poor aluminium tolerance showed a slight reduction in DNA methylation, while aluminium-tolerant strains displayed modest increases in DNA methylation (31).

A genome-wide comparison of methylation levels was conducted between farmed rice (*Oryza sativa*) and a more aluminium-tolerant wild species, *Oryza glumaepatula*. This analysis identified 4,633 differentially methylated regions (DMRs) in *O. sativa*, of which 38% exhibited hypermethylation, while 62% showed hypomethylation. In contrast, among the DMRs in *O. glumaepatula*, 28% were hypomethylated and 72% were hypermethylated. Additionally, aluminium-responsive genes were associated with differentially expressed genes (DEGs) in *O. sativa* and with 37 genes in *O. glumaepatula* (32).

Barley plants (*Hordeum vulgare* L.) respond to aluminium stress by demethylating a retrotransposon located in the upstream sequence of the *HvAACT1* gene (aluminium-activated citrate transporter 1). This process enhances the expression of *HvAACT1* in the roots and promotes the detoxification of aluminium externally (33).

25.2.4 Cadmium

Cadmium is highly mobile, making it easily absorbed by plants and allowing it to transfer throughout their organs. Exposure to this metal disrupts the balance of water and minerals, lowers chlorophyll levels and reduces photosynthetic efficiency. Additionally, cadmium induces oxidative stress, leading to increased lipid peroxidation, as well as the oxidation of proteins and nucleic acids (60). The presence of elevated micronuclei and chromosomal abnormalities indicates its harmful effects on genetic material. Other symptoms of cadmium exposure in plants include stunted growth, reduced seed germination and structural changes at the microscopic level (34).

The type of plant, the amount of metal administered and the duration of treatment all influence the degree and pattern of DNA methylation in plants exposed to cadmium (Cd). Results from microsatellite amplification polymorphism (MSAP)-PCR analysis revealed that while there were no significant changes in the internal cytosines, cadmium treatment increased the methylation of external cytosines in Arabidopsis (*Arabidopsis thaliana* L.) (35). Additionally, MSAP findings indicated that both foliar and soil applications of Cd elevated DNA methylation in thorn apple (*Datura stramonium* L.) plants (36). Cadmium and lead (Pb) caused hemimethylation, which refers to cytosine methylation occurring on only one DNA strand, in the endangered species *Isoetes sinensis*. However, in the metal-treated plants, there was a decrease in full methylation levels, meaning a reduction in the methylation of both internal and external cytosines on both DNA strands (37).

In kenaf (*Hibiscus cannabinus* L.), the F1 hybrid demonstrated better tolerance to cadmium (Cd) and exhibited lower levels of DNA methylation compared to the parental cultivars Fuhong 992 and P3A. Genes encoding leucine-rich repeat receptor-like kinases (LRR-RLK), NRT1/PTR family protein (NPF2.7) and DEAD-box ATP-dependent RNA helicase 51 (DHX51) displayed hypomethylation in response to Cd exposure. Conversely, genes encoding trehalose-phosphate phosphatase (TTP-D), NADP-dependent malic enzyme (NADP-ME) and NAC-domain-containing protein (NAC71) showed hypermethylation. While NPF2.7, DHX51, TTP-D and NADP-ME were upregulated, LRR-RLK and NAC71 were downregulated (38). In contrast, Cd treatment had no impact on DNA methylation levels in soybean (*Glycine max* L.), as assessed using the enzyme-linked immunosorbent assay (ELISA) technique, both immediately after the treatment and seven days later during the recovery phase.

Whole-genome bisulfite sequencing (WGBS) of rice (*Oryza sativa* L.) identified over 2,400 differentially methylated regions (DMRs). A significant finding was the hypermethylation of regions upstream of various genes (39). Functional analysis indicated that the majority of the genes associated with these methylation changes were categorized under "cellular processes", "metabolic processes" and "stimulus-response".

Additionally, the methyl-sens comet assay demonstrated that cadmium (Cd) exposure in *Noccaea caerulescens* stimulated DNA methylation, resulting in increased expression of the *MET1* gene. In contrast, *Arabidopsis thaliana* displayed an increased expression of *DRM2*, a decrease in *MET1* expression and a reduction in overall DNA methylation. The increased DNA methylation in the hyperaccumulator may enhance metal tolerance and provide DNA protection.

Changes in the expression or activity of methylation and demethylation enzymes can influence DNA methylation patterns. In the seagrass species *Zostera marina*, exposure to cadmium (Cd) has been shown to increase the expression of the methyltransferases, viz., CMT3 and DRM2 while simultaneously decreasing the expression of the demethylase ROS1. This shift in enzyme expression may lead to an increase in DNA 5-methylcytosine (5mC) levels (40).

Two populations of *Arabidopsis helleri*, I16 from Italy and PL22 from Poland, demonstrated different responses in the expression of genes that regulate DNA methylation following cadmium exposure (41). The PL22 population showed increased expression of several DNA methylation-related genes, including *MET1, WIM1* and *DRM2*. Additionally, genes associated with histone modifications exhibited heightened expression, such as those encoding the histone-lysine N methyltransferase KRYPTONITE (KYP), histone H3 lysine 9 demethylases (INCREASE IN BONSAI METHYLATION 1, IBM1) and histone deacetylase 8 (HDA8). These changes were accompanied by chromatin condensation. In contrast, the I16 population exhibited a lower expression level of all these genes.

Changes in the methylation state and expression level of transporter genes may be necessary for increased cadmium (Cd) uptake. In maize (*Zea mays* L.), cadmium exposure elevated the methylation level of transporter genes *IRT1*, *ZIP1*, *ZIP2* and *ZIP6* and significantly increased the expression of several methyltransferase genes, including *Met1*, *Met2a*, *Met2b*, *Met3a*, *Met3b*, *Met3c* and *Met* (24).

Comparing the responses of metal-tolerant and metal-sensitive cultivars of wheat (*Triticum aestivum* L.) to cadmium (Cd), lead (Pb) and zinc (Zn), it was observed that the metal-tolerant cultivar had a lower methylation level in the promoter regions of many transporter genes. This reduction in methylation was associated with higher expression levels of the transporter genes *TaABCC2*, *TaABCC3*, *TaABCC4* and *TaHMA2*. Consequently, decreased methylation of transporter genes may have facilitated metal sequestration and efflux in vacuoles, ultimately enhancing plant tolerance to these metals (Table 25.1) (42).

TABLE 25.1
Effects of different heavy metals in DNA methylation in different plants

Heavy metal	Treatment	Plant species	Effect on plant	References
Arsenic (As)	100 and 250 mg/kg of soil; 122 days	*Pteris cretica* L.	Reduction in global methylation in mature frond and no change in methylation of younger fronds	Zemanova et al. (27)
	35 μM; 21 days	*A. thaliana* L.	Overexpression of VIM1 causes increased methylation in transporter gene promoters and reduced As accumulation	Bahmani et al. (28)
Aluminium (Al)	15, 20 and 30 ppm; pH 4.5, 24 h	*Triticale*	Higher methylation in Al-tolerant line and lower in Al-sensitive lines	Niedziela (31)
Cadmium (Cd)	40 μM for 7 days	*A. thaliana* L.	Increased DNA methylation, particularly in CHH and CHG contexts, resulting in altered expression of methylation-related genes. Reduce the expression of ROS1, DML2 and DML3 demethylases	Fan et al. (43)
	5 μM in *A. thaliana*, 50 μM in *Noccaea caerulescens* 7 days	*A. thaliana* L., *Noccaea caerulescens* L.	Increased DNA methylation is linked with greater DNA damage in *Arabidopsis*. In *N. caerulescens*, less DNA methylation is associated with reduced DNA damage	Galati et al. (44)
	0.89 and 8.9 μM; 6 days	*Zostera marina*	Increase in expression of *CMT3* and *DRM2* gene and reduced expression of *ROS1*	Greco et al. (40)
	5 μM for 5 days (*A. thaliana*); 50 μM for 14 days (*Arabidopsis halleri*)	*A. thaliana* L., *A. helleri* L.	Reduced expression of MET1 and DRM2 expressing genes in *A. thaliana*; enhanced expression of MET1 and DRM2 in *A. helleri*; population PL22 is associated with enhanced CpG methylation. Reduced expression of MET1 and DRM2 in *A. halleri* population I16, associated with decreased CpG methylation	Galati et al. (41)
	80 μM for 4 days	*Oryza sativa* L.	Higher methylation in CHH site and hypomethylation in CHG and CG site	Feng et al. (39)
	75, 150 and 225 ppm 2 months	*Datura stramonium* L.	Increase in hypermethylation and decrease in hypomethylation	Shirkhani et al. (36)
Cadmium (Cd) and lead (Pb)	Cd: 100, 250 and 500 mg/l. Pb: 500, 1000 and 2000 mg/l; 10 days	*Isoetes sinensis* L.	Increase in CCGG hemi-methylation and decrease in methylation	Ding et al., 2018 (37)
	100 μM for 2 days	*Zea mays* L.	Amended methylation level of the promoter regions of the IRT and ZIP transporters	Shafiq et al. (24)
	100 μM for 2 days	*Triticum aestivum* L.	Metal tolerance reduces methylation of promoter regions of transporter genes via modulating the expression of related genes	Shafiq et al. (42)
	Cd 15 mg/l, Pb 200 mg/l, Cd + Pb (15 + 200 mg/l)	*Amaranthus cruentus* L.	Elevated expression of *CMT1-3*, *DRM2*, *MET1* and *DML2* gene families	Lancıkova et al. (45)
Lead (Pb)	200 mg/l	*Raphanus sativus* L.	Hypermethylation leads to alterations in methylation levels and expression patterns of genes producing transporters and transcription factors	Tang et al. (25)

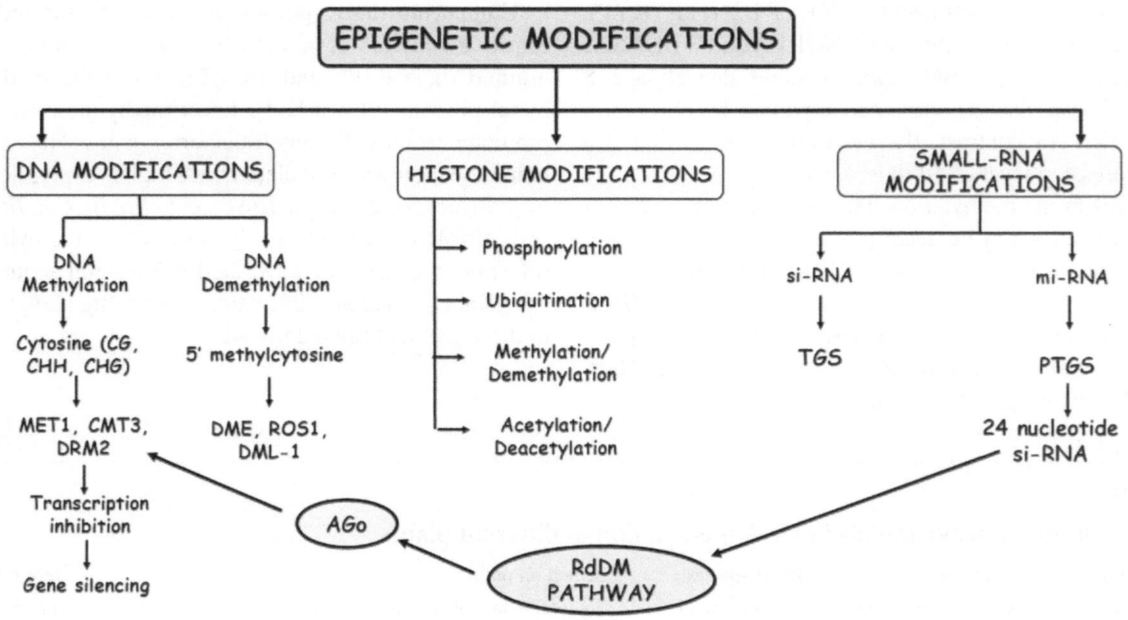

FIGURE 25.2 Types of epigenetic modification due to heavy metal stress.

25.3 HISTONE MODIFICATIONS

Histone expression is influenced by exposure to non-essential metals at various levels, affecting both transcript and protein levels. For instance, exposure to cadmium (Cd) led to an accumulation of the H2A histone protein in soybean cell suspensions. In barley, a metal-sensitive line exhibited a decrease in the number of transcripts encoding the H2A, H2B, H3 and H4 histones in response to Cd, while a metal-tolerant line showed no changes (46). Immunofluorescence analysis revealed alterations in histone modifications in the root meristem cells of broad bean (*Vicia faba*) (47). Plants treated with cadmium (Cd) exhibited a significant reduction in the demethylation of histone 3 at lysine 79 (H3K79Me2), a modification associated with replication initiation events. This decrease was observed in all three stages of the cell cycle – G1, S and G2 – when compared to the control group.

Metals also influence the expression of genes related to histone modifications; for example, cotton (*Gossypium* L.) and maize showed increased expression of genes involved in histone acetylation and deacetylation in response to cadmium (Cd) and lead (Pb), respectively. Interestingly, different lines of the same plant species can display varying expression patterns (24, 48). In the *A. halleri* PL22 population, treatment with Cd increased the abundance of transcripts for the *KYP*, *IBM1* and *HDA8* genes, whereas it decreased transcript levels in the I16 line (41). Exposure to cadmium (Cd) was found to reduce the phosphorylation of histone H3 at threonine 45 (H3T45Ph), a modification important for DNA replication. Additionally, in plants treated with cadmium, there was an increase in histone H3 acetylation on lysine 56 (H3K56Ac) during the S phase of the cell cycle.

Mutant plants may exhibit changes in histone expression following exposure to metal stress. One significant group of histone demethylases includes proteins known as Jumonji-C domain-containing proteins (JMJ). In tomato plants treated with mercury (Hg), lead (Pb) and copper (Cu), there was an increase in the expression of the *SlJMJ524* gene (49). Overexpressing this gene in *Arabidopsis* modulated plant tolerance to cadmium (Cd). Young seedlings displayed lower tolerance, while mature plants exhibited increased tolerance.

Additionally, the overexpression of the *SlJMJ524* gene resulted in heightened expression of genes involved in metal absorption, such as *ZIP1*, *IRT1*, *COPT2* and *NRAMP1*, as well as transport genes like *ABCC1*, *PDR1*, *GSH1* and *GSH2*. However, the expression of other genes associated with transport and detoxification, particularly those encoding glutathione reductase (GR1) and phytochelatin synthase (PCS1 and PCS2), was reduced (Figure 25.2).

25.4 MICRO-RNAs IN HEAVY METAL TOLERANCE

A significant number of heavy metal-responsive microRNAs (miRNAs) have been identified in various metal hyperaccumulators, including *Sedum alfredii, Oryza sativa, Zea mays, Brassica juncea, Medicago truncatula* and *Hordeum vulgare* (50). Advanced genomic technologies such as next-generation sequencing and miRNA microarrays have made these discoveries possible; for instance, in a study involving radish, 81 known and 72 novel miRNAs were identified through Solexa sequencing, with 54 known and 16 novel miRNAs showing differential expression in response to chromium (Cr) stress (51). Similarly, small RNA sequencing revealed 18 known and 148 novel miRNAs that were differentially expressed in *Brassica parachinensis* cultivars with low and high cadmium (Cd) accumulation under Cd stress (52).

Many of the target genes for metal-responsive miRNAs are transcription factors, indicating that miRNAs play a vital role in the transcriptional regulatory network that governs how plants respond to heavy metal stress (62). The analysis of metal-regulated miRNA targets has also identified genes involved in various processes related to heavy metal responses in plants. These processes include metal uptake and transport, sulfate allocation and assimilation, protein folding, the antioxidant system, phytohormone signaling, miRNA biogenesis and feedback regulation (53).

Recent research has highlighted the role of cadmium-responsive microRNA, miR166, through findings from qRT-PCR and miRNA microarray experiments. It was observed that the expression of miR166 was significantly suppressed in rice seedlings exposed to cadmium (Cd) (54). In contrast, transgenic rice plants that overexpressed miR166 showed a decrease in oxidative damage caused by Cd. More importantly, these transgenic plants demonstrated reduced Cd accumulation and transfer from roots to shoots in rice (55).

A key player in this process is OsHMA2, a zinc/cadmium transporter found in rice roots that facilitates the root-to-shoot transfer of Cd (56). Mutations in *OsHMA2* have been shown to reduce Cd translocation (57). In the 35S:MIR166 plants, *OsHMA2* expression was downregulated, which aligns with the observed decrease in root-to-shoot Cd transfer in transgenic rice plants (55). Therefore, the reduced translocation of Cd in plants overexpressing miR166 may be partially attributed to the impact on *OsHMA2* expression.

25.5 CONCLUSION

Understanding the epigenetic mechanisms that enable plants to endure various stresses could enhance crop resilience and improve agricultural productivity. However, studies on how heavy metal stress affects epigenetic changes in plants are relatively scarce and have primarily focused on a limited number of species. Many of these studies employed less sensitive techniques, such as microsatellite amplification polymorphism (MSAP), high-performance liquid chromatography (HPLC) and immunolabeling, rather than more advanced methods like bisulfite sequencing and single-base resolution methylome analysis. As a result, the findings in the existing literature regarding epigenetic modifications may be overestimated. The high cost of next-generation sequencing methods limits their use in ecological research. Many studies provide evidence that metalloids and non-essential metals can affect the DNA methylation state of cells. However, various factors influence the specific patterns of DNA methylation changes observed, including the type of plant, the metal used, its concentration and the duration of treatment. Notable differences have been observed between metal-sensitive and metal-tolerant lines, such as in rice and triticale, as well as between F1 hybrids and their parent plants. Although the impact of metals on histone modifications has not been extensively studied, important metals have been shown to influence the transcriptional and post-transcriptional expression of genes associated with histone biosynthesis and modifications. Metals also affect the pattern of histone modifications, such as methylation and acetylation.

MicroRNAs (miRNAs) can mediate DNA methylation, and in various plant species – such as *Arabidopsis thaliana*, *Phaseolus vulgaris*, *Medicago truncatula*, *Brassica napus*, *Oryza sativa* and *Nicotiana tabacum* – metals have been shown to cause the upregulation or downregulation of numerous miRNAs. miRNAs play important roles in signaling, gene regulation, detoxification and control of metal absorption and chelation under metal stress. However, it remains unclear how these RNA molecules contribute to metal-dependent methylation. Further research is needed in this area, especially considering that miRNAs may also influence DNA methylation in response to other stress conditions. Epigenetic changes are vital in regulating gene expression, influencing how plants respond to various stressors such as drought, heat, salinity and disease. Research on the epigenetic regulation of plants in response to stress is crucial and holds significant promise for future agricultural applications.

REFERENCES

1. Pourret, O., & Hursthouse, A. (2019). It's time to replace the term 'heavy metals' with 'potentially toxic elements' when reporting environmental research. *International Journal of Environmental Research and Public Health*, 16, 4446.
2. Rai, P. K., Lee, S. S., Zhang, M., Tsang, Y. F., & Kim, K. H. (2019). Heavy metals in food crops: Health risks, fate, mechanisms, and management. Environment International, 125, 365–385.
3. Srivastava, V., Sarkar, A., Singh, S., Singh, P., De Araujo, A. S., & Singh, R. P. (2017). Agroecological responses of heavy metal pollution with special emphasis on soil health and plant performances. *Frontiers in Environmental Science*, 5, 64.
4. Burakova, E. A., Dyachkova, T. P., Rukhov, A. V., Tugolukov, E. N., Galunin, E. V., Tkachev, A. G., & Ali, I. (2018). Novel and economic method of carbon nanotubes synthesis on a nickel magnesium oxide catalyst using microwave radiation. *Journal of Molecular Liquids*, 253, 340–346.
5. Ali, I., Alothman, Z. A., & Alwarthan, A. (2017). Supra molecular mechanism of the removal of 17-β-estradiol endocrine disturbing pollutant from water on functionalized iron nano particles. *Journal of Molecular Liquids*, 241, 123–129.
6. Furini, A. (2012). *Plants and Heavy Metals*. Dordrecht, The Netherlands: Springer.
7. Viehweger, K. (2014). How plants cope with heavy metals. *Botanical Studies*, 55, 35.
8. Riyazuddin, R., Nisha, N., Ejaz, B., Khan, M. I. R., Kumar, M., Ramteke, P. W., & Gupta, R., (2021). Comprehensive review on the heavy metal toxicity and sequestration in plants. *Biomolecules*, 12, 43. https://doi.org/10.3390/biom1201004.
9. Noor, I., Sohail, H., Sun, J., Nawaz, M. A., Li, G., Hasanuzzaman, M., et al. (2022). Heavy metal and metalloid

toxicity in horticultural plants: Tolerance mechanism and remediation strategies. *Chemosphere*, 303, 135196. https://doi.org/10.1016/j.chemosphere.2022.135196.
10. Rahman, Z., & Singh, V. P. (2019). The relative impact of toxic heavy metals (THMs) (arsenic (As), cadmium (Cd), chromium (Cr)(VI), mercury (Hg), and lead (Pb)) on the total environment: An overview. *Environmental Monitoring and Assessment*, 191(7), 419. https://doi.org/10.1007/s10661-019-7528-7.
11. Dutta, S., Mitra, M., Agarwal, P., Mahapatra, K., De, S., Sett, U., et al. (2018). Oxidative and genotoxic damages in plants in response to heavy metal stress and maintenance of genome stability. *Plant Signaling & Behavior*, 13, 8. https://doi.org/10.1080/ 15592324.2018.1460048.
12. Thakur, M., Praveen, S., Divte, P. R., Mitra, R., Kumar, M., Gupta, C. K., Kalidindi, U., Bansal, R., Roy, S., Anand, A., et al. (2022). Metal tolerance in plants: Molecular and physicochemical interface determines the 'not so heavy effect' of heavy metals. *Chemosphere*, 287, 131957.
13. Shahid, M., Natasha, Khalid, S., Abbas, G., Niazi, N. K., Murtaza, B., Rashid, M. I., & Bibi, I. (2019). Redox mechanisms and plant tolerance under heavy metal stress: Genes and regulatory networks. In G. Sablok (Ed.), *Plant Metallomics and Functional Omics: A System-Wide Perspective* (pp. 71–105). Cham, Switzerland: Springer.
14. VanderEnt, A., Baker, A. J. M., Reeves, R. D., Pollard, A. J., & Schat, H. (2013). Hyperaccumulators of metal and metalloid trace elements: Facts and fiction. Plant and Soil, 362, 319–334.
15. Iwasaki, M., & Paszkowski, J. (2014). Epigenetic memory in plants. *EMBO Journal*, 33, 1987–1998. https://doi.org/10.15252/embj.201488883.
16. Bilichak, A., Ilnytskyy, Y., Woycicki, R., Kepeshchuk, N., Fogen, D., & Kovalchuk, I. (2015). The elucidation of stress memory inheritance in *Brassica rapa* plants. *Frontiers in Plant Science*, 6. https://doi.org/10.3389/fpls.2015.00005.
17. Galviz, Y. C. F., Ribeiro, R. V., & Souza, G. M. (2020). Yes, plants do have memory. *Theoretical and Experimental Plant Physiology*, 32, 195–202. https://doi.org/10.1007/s40626-020-00181-y.
18. Saletore, Y., Chen-Kiang, S., & Mason, C. M. (2013). Novel RNA regulatory mechanisms revealed in the epitranscriptome. *RNA Biology*, 10(3), 342–346. https://doi.org/10.4161/rna.23812.
19. Wang, Z., Tang, K., Zhang, D., Wan, Y., Wen, Y., Lu, Q., et al. (2017). High throughput m6A-seq reveals RNA m6A methylation patterns in the chloroplast and mitochondria transcriptomes of *Arabidopsis thaliana*. *PLoS One*, 12(11), e0185612. https://doi.org/10.1371/journal.pone.0185612.
20. Li, Y., Wang, X., Li, C., Hu, S., Yu, J., & Song, S. (2014). Transcriptome-wide N6 methyladenosine profiling of rice callus and leaf reveals the presence of tissue-specific competitors involved in selective mRNA modification. *RNA Biology*, 11(9), 1180–1188. https://doi.org/10.4161/rna.36281.
21. Tang, J., Chen, S., & Jia, G. (2023). Detection, regulation, and functions of RNA N6-methyladenosine modification in plants. *Plant Communications*, 4, 100546. https://doi.org/10.1016/j.xplc.2023.100546.
22. Tang, W. Y., & Ho, S. M. (2007). Epigenetic reprogramming and imprinting in origins of disease. *Reviews in Endocrine and Metabolic Disorders*, 8, 173–182.
23. Zulfiqar, U., Farooq, M., Hussain, S., Maqsood, M., Hussain, M., Ishfaq, M., et al. (2019). Lead toxicity in plants: Impacts and remediation. *Journal of Environmental Management*, 250, 109557. https://doi.org/10.1016/j.jenvman.2019.109557.
24. Shafiq, S., Ali, A., Sajjad, Y., Zeb, Q., Shahzad, M., Khan, A. R., et al. (2020). The interplay between toxic and essential metals for their uptake and translocation is likely governed by DNA methylation and histone deacetylation in maize. *International Journal of Molecular Sciences*, 21, 6959. https://doi.org/10.3390/ijms21186959.
25. Tang, M., Xu, L., Wang, Y., Dong, J., Zhang, X., Wang, K., et al. (2021). Melatonin induced DNA demethylation of metal transporters and antioxidant genes alleviates lead stress in radish plants. *Horticulture Research*, 8(1), 124. https://doi.org/10.1038/s41438-021-00561-8.
26. Genchi, G., Lauria, G., Catalano, A., Carocci, A., & Sinicropi, M. S. (2022). Arsenic: A review on a great health issue worldwide. *Applied Sciences*, 12, 6184. https://doi.org/10.3390/app12126184.
27. Zemanova, V., Popov, M., Pavlıkova, D., Kotrba, P., Ceska, J., & Pavlık, M. (2020). Effect of arsenic stress on 5-methylcytosine, photosynthetic parameters and nutrient content in arsenic hyperaccumulator *Pteris cretica* (L.) var. Albolineata. *BMC Plant Biology*, 130(2020), 20. https://doi.org/10.1186/s12870-020-2325-6.
28. Bahmani, R., Modareszadeh, R., Kim, D., & Hwang, S. (2023). Methylcytosine binding protein VIM1 decreases As(III) accumulation by epigenetic downregulation of the As(III) importer NIP3;1 in *Arabidopsis*. *Journal of Hazardous Materials*, 441, 129987. https://doi.org/10.1016/j.jhazmat.2022.129987.
29. Liu, H., Zhu, R., Shu, K., Lv, W., Wang, S., & Wang, C. (2022). Aluminum stress signaling, response, and adaptive mechanisms in plants. *Plant Signaling & Behavior*, 17(1), 2057060. https://doi.org/10.1080/15592324.2022.2057060.
30. Gallo-Franco, J. J., Sosa, C. C., Ghneim-Herrera, T., & Quimbaya, M. (2020). Epigenetic control of plant response to heavy metal stress: A new view on aluminium tolerance. *Frontiers in Plant Science*, 11. https://doi.org/10.3389/fpls.2020.602625.
31. Niedziela, A. (2018). The influence of Al^{3+} on DNA methylation and sequence changes in the triticale (× Triticosecale Wittmack) genome. *Journal of Applied Genetics*, 59, 405–417. https://doi.org/10.1007/s13353-018-0459-0.
32. Gallo-Franco, J. J., Ghneim-Herrera, T., Tobar-Tosse, F., Romero, M., Chaura, J., & Quimbaya, M. (2022). Whole-genome DNA methylation patterns of *Oryza sativa* (L.) and *Oryza glumaepatula* (Steud) genotypes associated with aluminium response. *Plant Direct*, 6(8), e430. https://doi.org/10.1002/pld3.430.
33. Kashino-Fujii, M., Yokosho, K., Yamaji, N., Yamane, M., Saisho, D., Sato, K., et al. (2018). Retrotransposon insertion and DNA methylation regulate aluminum tolerance in European barley accessions. *Plant Physiology*, 178, 716–727. https://doi.org/10.1104/pp.18.00651.
34. Zulfiqar, U., Ayub, A., Hussain, S., Waraich, E. A., El-Esawi, M. A., Ishfaq, M., et al. (2021). Cadmium toxicity in plants: Recent progress on morpho-physiological effects and remediation strategies. *Journal of Soil Science and Plant Nutrition*, 22, 212–269. https://doi.org/10.1007/s42729-021-00645-3.
35. Wang, H., He, L., Song, J., Cui, W., & Zhang, Y. (2016). Cadmium-induced genomic instability in *Arabidopsis*: Molecular toxicological biomarkers for early diagnosis of cadmium stress. *Chemosphere*, 150, 258–265. https://doi.org/10.1016/j.chemosphere.2016.02.042.

36. Shirkhani, Z., Chehregani Rad, A., & Mohsenzadeh, F. (2022). Effects of soil and foliar cadmium application on morphological, physiological, genetic and epigenetic characteristics of *Datura stramonium* L. *International Journal of Environmental Science and Technology*, 19, 9067–9082. https://doi.org/10.1007/s13762-022-04128-5.
37. Ding, G. H., Guo, D. D., Guan, Y., Chi, C.-Y., & Liu, B.-D. (2018). Changes of DNA methylation of *Isoetes sinensis* under Pb and Cd stress. *Environmental Science and Pollution Research*, 26, 3428–3435. https://doi.org/10.1007/s11356-018-3864-3.
38. Luo, D., Lu, H., Wang, C., Mubeen, S., Cao, S., Yue, J., et al. (2023). Physiological and DNA methylation analysis provides epigenetic insights into kenaf cadmium tolerance heterosis. *Plant Science*, 331, 111663. https://doi.org/10.1016/j.plantsci.
39. Feng, S. J., Liu, X. S., Tao, H., Tan, S. K., Chu, S. S., & Oono, Y. (2016). Variation of DNA methylation pattern associated with gene expression in rice (*Oryza sativa*) exposed to cadmium. *Plant, Cell & Environment*, 39, 2629–2649. https://doi.org/10.1111/pce.12793.
40. Greco, M., Saez, C. A., Contreras, R. A., Rodrıguez-Rojas, F., Bitonti, M. B., & Brown, M. T. (2019). Cadmium and/or copper excess induce interdependent metal accumulation, DNA methylation, induction of metal chelators and antioxidant defences in the seagrass *Zostera marina*. *Chemosphere*, 224, 111–119. https://doi.org/10.1016/j.chemosphere.2019.02.123.
41. Galati, S., Gullì, M., Giannelli, G., Furini, A., DalCorso, G., Fragni, R., et al. (2021). Heavy metals modulate DNA compaction and methylation at CpG sites in the metal hyperaccumulator *Arabidopsis halleri*. *Environmental and Molecular Mutagenesis*, 62, 133–142. https://doi.org/10.1002/em.22421.
42. Shafiq, S., Zeb, Q., Ali, A., Sajjad, Y., Nazir, R., Widemann, E., et al. (2019). Lead, cadmium and zinc phytotoxicity alter DNA methylation levels to confer heavy metal tolerance in wheat. *International Journal of Molecular Sciences*, 20, 4676. https://doi.org/10.3390/ijms20194676.
43. Fan, S. K., Ye, J. Y. Zhang, L. L., Chen, H. S., Zhang, H. H., Zhu, Y. X., et al. (2020). Inhibition of DNA demethylation enhances plant tolerance to cadmium toxicity by improving iron nutrition. *Plant, Cell & Environment*, 43, 275–291. https://doi.org/10.1111/pce.13670.
44. Galati, S., DalCorso, G., Furini, A., Fragni, R., Maccari, C., Mozzoni, P., et al. (2023). DNA methylation is enhanced during Cd hyperaccumulation in *Noccaea caerulescens* ecoptype Ganges. *Environmental Science and Pollution Research*, 30, 26178–26190. https://doi.org/10.1007/s11356-022 23983-w.
45. Lancıkova, V., Kacırova, J., & Hricova, A. (2023). Identification and gene expression analysis of cytosine-5 DNA methyltransferases and demethylase genes in *Amaranthus cruentus* L. under heavy metal stress. *Frontiers Plant Science*, 13. https://doi.org/10.3389/fpls.2022.1092067.
46. Cao, F., Chen, F., Sun, H., Zhang, G., Chen, Z.-H., & Wu, F. (2014). Genome-wide transcriptome and functional analysis of two contrasting genotypes reveals key genes for cadmium tolerance in barley. *BMC Genomics*, 15, 611. https://doi.org/10.1186/1471-2164 15-611.
47. Zabka, A., Gocek, N., Winnicki, K., Szczeblewski, P., Laskowski, T., & Polit, J. T. (2021). Changes in epigenetic patterns related to DNA replication in *Vicia faba* root meristem cells under cadmium-induced stress conditions. *Cells*, 10, 3409. https://doi.org/10.3390/cells10123409.
48. Imran, M., Shafiq, S., Naeem, M. K., Widemann, E., Munir, M. Z., Jensen, K. B., et al. (2020). Histone deacetylase (HDAC) gene family in allotetraploid cotton and its diploid progenitors: in silico identification, molecular characterization, and gene expression analysis under multiple abiotic stresses, DNA damage and phytohormone treatments. *International Journal of Molecular Sciences*, 21, 321. https://doi.org/10.3390/ijms21010321.
49. Li, Q., Sun, W., Chen, C., Dong, D., Cao, Y., Dong, Y., et al. (2022). Overexpression of histone demethylase gene SlJMJ524 from tomato confers Cd tolerance by regulating metal transport-related protein genes and flavonoid content in *Arabidopsis*. *Plant Science*, 318, 111205. https://doi.org/10.1016/j.plantsci.2022.111205.
50. Gao, J., Luo, M., Peng, H., Chen, F. B., & Li, W. B. (2019). Characterization of cadmium responsive MicroRNAs and their target genes in maize (*Zea mays*) roots. *BMC Molecular Biology*, 20, 14.
51. Liu, W., Xu, L., Wang, Y., Shen, H., Zhu, X. W., Zhang, K. Y., Chen, Y. L., Yu, R. G., Limera, C., & Liu, L. W. (2015). Transcriptome-wide analysis of chromium-stress responsive microRNAs to explore miRNA-mediated regulatory networks in radish (*Raphanus sativus* L.). *Scientific Reports*, 5, 14024.
52. Zhou, Q., Yang, Y. C., Shen, C., He, C. T., Yuan, J. G., Yang, Z. Y. (2017). Comparative analysis between low- and high-cadmium accumulating cultivars of *Brassica parachinensis* to identify difference of cadmium-induced microRNA and their targets. *Plant Soil*, 420(1), pp.223–237.
53. Noman, A., & Aqeel, M. (2017). miRNA-based heavy metal homeostasis and plant growth. *Environmental Science and Pollution Research*, 24, 10068.
54. Ding, Y. F., Chen, Z., & Zhu, C. (2011). Microarray-based analysis of cadmium-responsive microRNAs in rice (*Oryza sativa*). *Journal of Experimental Botany*, 62, 3563–3573.
55. Ding, Y. F., Gong, S. H., Wang, Y., Wang, F. J,, Bao, H. X. G., Sun, J. W., Cai, C., Yi, K. K., Chen, Z. X., & Zhu, C. (2018). MicroRNA166 modulates cadmium tolerance and accumulation in rice. *Plant Physiology*, 177, 1–13.
56. Takahashi, R., Ishimaru, Y., Shimo, H., Ogo, Y., Senoura, T., Nishizawa, N. K., & Nakanishi, H. (2012). The OsHMA2 transporter is involved in root-to-shoot translocation of Zn and Cd in rice. *Plant, Cell & Environment*, 35, 1948–1957.
57. Satoh–Nagasawa, N., Mori, M., Nakazawa, N., Kawamoto, T., Nagato, Y., Sakurai, K., Takahashi, H., Watanabe, A., & Akagi, H. (2012). Mutations in rice (*Oryza sativa*) heavy metal ATPase 2 (OsHMA2) restrict the translocation of zinc and cadmium. *Plant and Cell Physiology*, 53, 213–224.
58. Roychoudhury, A., & Chakraborty, S. (2020). Cellular and molecular phytotoxicity of lead and mercury. In M. Faisal et al. (Eds.), *Cellular and Molecular Phytotoxicity of Heavy Metals* (pp. 373–387). Cham, Switzerland: Springer Nature Switzerland.
59. Samanta, S., & Roychoudhury, A. (2021). Heavy metal toxicity in plants: Physiological and molecular adaptations. In T. Aftab & K. R. Hakeem (Eds.), *Heavy Metal Toxicity in Plants* (pp. 1–9). Boca Raton: CRC Press Taylor & Francis Group.
60. Datta, K., & Roychoudhury, A. (2022). Cadmium-induced oxidative stress and remediation in plants. In M. Naeem, T. Aftab, A. A. Ansari, S. Gill, & A. Macovei (Eds.),

Hazardous and Trace Materials in Soil and Plants (pp. 247–261). Amsterdam: Academic Press (Elsevier).

61. Upadhyaya, G., & Roychoudhury, A. (2022). Arsenic-toxicity and tolerance: Phytochelatin-mediated detoxification and genetic engineering-based remediation. In N. K. Niazi, I. Bibi, & T. Aftab (Eds.), *Global Arsenic Hazard Ecotoxicology and Remediation* (pp. 481–508). Cham: Springer Nature Switzerland.
62. Chakraborty, S., & Roychoudhury, A. (2023). MicroRNA-mediated regulation of heavy metal stress in plants. In D. Pandita & A. Pandita (Eds.), *Plant MicroRNAs and Stress Response* (pp. 183–197). Boca Raton: CRC Press (Taylor and Francis Group).
63. Roychoudhury, A., Pradhan, S., Chaudhuri, B., & Das, K. (2012). Phytoremediation of toxic metals and the involvement of *Brassica* species. In N. A. Anjum, M. E. Pereira, I. Ahmad, A. C. Duarte, S. Umar, & N. A. Khan (Eds.), *Phytotechnologies: Remediation of Environmental Contaminants* (pp. 219–251). Boca Raton: CRC Press, Taylor and Francis Group, Boca Raton.
64. Banerjee, A., & Roychoudhury, A. (2017). Epigenetic regulation during salinity and drought stress in plants: Histone modifications and DNA methylation. *Plant Gene*, 11, 199–204.

26 Epigenetic Memory of Ultraviolet and Light Stress in Plants
Mechanism, Crosstalk and Application

Aryadeep Roychoudhury, Swayambhik Mukherjee and Abhinaba Banerjee

26.1 INTRODUCTION

Sunlight serves both as an energy source and a crucial environmental cue for plants, allowing them to regulate their growth and development (Müller-Xing et al., 2014). A wide range of developmental processes and adaptations, including germination, de-etiolation, shadow avoidance, stomatal development, circadian rhythm and flowering, are regulated by solar signals in higher plants, such as the model plant *Arabidopsis thaliana* (Arabidopsis) (Müller-Xing et al., 2014). High light, and in particular its inherent ultraviolet (UV) component, can cause stress reactions in plants in addition to regulating growth and development (which is now referred to as "normal" photomorphogenesis) (Müller-Xing et al., 2014; Banerjee & Roychoudhury, 2016). About 9% of the light emitted by the sun constitutes UV radiation, which is classified into three categories by the ISO 21348 standard: UV-C (200–280 nm), UV-B (280–315 nm) and UV-A (315–400 nm) (del Socorro Sánchez Correa et al., 2023; Gill et al., 2015). Exposure duration, fluence rate and whether or not plants have been acclimated to light before, altogether influence the kind of stress response that light causes, for instance, UV-B-induced light stress results in DNA damage, reactive oxygen species (ROS) generation, decreased pathogen resistance and altered cellular functions (Jenkins, 2009; Li et al., 2013). Because of the metabolic disruption and impaired photosynthetic electron transport, or because of higher activity of membrane-localized NADPH-oxidases and peroxidases, UV-B radiation encourages the production of ROS (Jenkins, 2009; Hideg et al., 2012).

Although the term epigenetics has been used in a variety of contexts, it is currently most commonly used to refer to mitotically or meiotically heritable modifications in gene function that are not brought about by variations in DNA sequence and that frequently exhibit non-Mendelian characteristics like reversibility. Generally, epigenetic effects result from chromatin modifications, of which many types have been reported, such as covalent modifications in histone residues, incorporation of histone variations, DNA methylation (DNAme), histone methylation, histone acetylation and functioning of small RNAs (Chen et al., 2010; Banerjee & Roychoudhury, 2017). In plants, the histone H3 variation H3.3 is linked to active histone modifications such as RNA polymerase II (RNA Pol II) and H3K4me3 marks, as well as expressed genes, while H3.1 is linked to repressive marks (Stroud et al., 2012). DNAme adds a methyl group to the fifth carbon of cytosine nucleotide of DNA (Bird, 2002). CpG methylation, CHG (H standing for A, C or T) and CHH (an asymmetrical site) methylations are the three primary patterns seen in plant DNAme (Chen et al., 2010). Lysine acetylations (H3K_ac) and methylations (H3K_me) account for the majority of histone H3 modifications. While DNAme is typically linked to gene repression and H3K_ac to transcription activation, the impact of H3K_me varies depending on the particular residue and the quantity of methyl groups (me, me2 or me3) (Müller-Xing et al., 2014). Target gene is often upregulated by H3K4me1, H3K4me2, H3K4me3, H3K36me2 and H3K36me3, whereas H3K9me2, H3K9me3 and H3K27me3 downregulate the expression (Benhamed et al., 2008 ; Zhou, 2009; Chen et al., 2010).

Ding et al. (2012) researched on drought stress responses in leaf cells that have primarily stopped dividing. They defined "transcriptional memory" as information that remains after the plant has recovered from the initial stress and that affects subsequent transcriptional responses to stressors encountered later. Although some scientists prefer to use the term more strictly for heritable modifications alone, the term "epigenetics" is now used to refer to all memory phenomena, based on chromatin modification in accordance with the definition given by Bird (2007). There are three classes of epigenetic stress memory (Chinnusamy & Zhu, 2009): (1) "short-term stress memory" (also known as "transcriptional memory"; Ding et al., 2012), which are persistent but non-heritable changes in histone variants, histone modification and/or DNAme that affect subsequent transcription responses; (2) "long-term stress memory" (Müller-Xing et al., 2014), which are heritable during mitosis; and (3) "transgenerational stress memory" (Müller-Xing et al., 2014), which is heritable during both meiosis and mitosis.

DOI: 10.1201/9781003589914-30

26.2 UV/LIGHT STRESS IN PLANTS

26.2.1 DNA Damage and Stress Response

Double-strand breaks are among the DNA damages caused by UV exposure (Ries et al., 2000). Additionally, when DNA absorbs UV-B rays, covalent bonds between neighboring pyrimidines can form cyclobutane pyrimidine dimers (CPDs), which obstruct gene transcription and DNA replication (Jansen et al., 1998). Chromatin alterations and remodeling are part of evolutionary conserved DNA repair pathways that fix a variety of DNA lesions, including UV-induced damage (Schmidt & Jackson, 2013). This may connect epigenetic memory with UV. Light-dependent photolyases rectify pyrimidine dimerization and reinstate the original structure of DNA (Jansen et al., 1998). According to Landry et al. (1997), the *Arabidopsis uvr2-1* mutant, which has a mutation in the type II CPD photolyase PHR1, exhibits hypersensitivity to UV-B and a loss of CPD photorepair. This suggests that a functioning DNA repair system is essential to preserve genome integrity and thereby improve tolerance to light-induced stress. High levels of UV-B radiation cause photolyase PHR1 and AtRAD1, a putative marker for recombination repair, to be substantially expressed. They also cause a higher frequency of somatic homologous DNA recombination in *Arabidopsis* (Ries et al., 2000). It is noteworthy that the frequency of homologous recombination was higher in the offspring of UV-B-exposed *uvr2-1* plants, which do not have photoreactivation repair of CPDs, than in the parent population. This suggests that the recombination process triggered by UV-B may have an impact on the genome stability of subsequent generations (Ries et al., 2000).

The *DECREASE IN DNA METHYLATION1 (DDM1)* mutant primarily loses DNAme, especially at heterochromatin, and H3K9me2 is mostly replaced by H3K4me2. DDM1 encodes an ATP-dependent SWI2/SNF2 chromatin remodeling factor that is necessary for normal patterns of genomic DNAme in *Arabidopsis* (Gendrel et al., 2002). According to Questa et al. (2013), DDM1 may be implicated in UV-B-induced DNA repair because DDM1 transcription is suppressed by UV-B radiation; *ddm1* mutants exhibit higher levels of DNA repair enzymes and are more sensitive to UV-B radiation than those of wild-type plants. Different UV response ecotypes were screened using transcriptional profiling of high-altitude maize landraces. Several gene clusters associated with chromatin remodeling were found in the UV-B-responsive transcripts, suggesting that multiple chromatin factors may be involved in UV-B tolerance (Casati et al., 2006). According to Campi et al. (2011), maize lines that had their putative chromatin remodeling or histone modification genes, *NFC102/NFC4* or *SDG102/SDG26*, knocked down by RNA interference (RNAi), revealed increased DNA damage and hypersensitivity to UV-B. Since UV-B can promote histone H3 and H4 acetylation in maize (Casati et al., 2008), researchers also looked at how histone acetylation affected DNA repair in *Arabidopsis*. In an investigation on the function of the histone acetyltransferases HAM1 and HAM2 in DNA damage repair, the *ham1* mutant showed greater DNA damage and more CPDs, indicating that at least HAM1 is crucial for UV-B-induced DNA repair in *Arabidopsis* (Campi et al., 2011). The important histone H3/H4 chaperone ANTI-SILENCING FUNCTION1 (ASF1) interacts with histone acetyltransferases, which in turn connect with E2F transcription factors to participate in a number of DNA repair and cell cycle regulation activities (Lario et al., 2013). UV-B stimulated the expression of ASF1A and ASF1B, and RNAi knockdown lines of both the genes displayed heightened sensitivity to UV-B in comparison to plants of the wild type. This suggests that ASF1A and ASF1B are related to DNA repair, following UV-B irradiation and are controlled by cell cycle progression (Lario et al., 2013). Chromatin remodeling, histone acetylation and photolyase-dependent photorepair are critical processes during UV-B-induced DNA repair, indicating that chromatin alterations also play a role in DNA repair in plants. If the chromatin alterations brought upon by DNA repair leave any trace of the UV source of the DNA damage, more research will be required to come to a concrete inference.

26.2.2 Oxidative Stress and Response

Excess UV irradiation in plants causes oxidative stress via the generation of ROS, which in turn cause enhanced protein and lipid oxidation (Rajneesh et al., 2017). Table 26.1 shows the damage that ROS cause to biomolecules, including DNA, lipids and proteins. The exposure to UV-B radiation can also activate UV-B chromophores, including phenolic compounds, aromatic amino acids and NADH [nicotinamide adenine dinucleotide-hydrogen (reduced)], which combine with molecular oxygen to generate singlet oxygen and superoxide anion (Rajneesh et al., 2017). ROS are partially excited or reduced forms of atmospheric oxygen that can harm cells (Rajneesh et al., 2017). They are produced indirectly by UV radiation and include $O_2^{\bullet-}$, $OH^{\bullet}$, H_2O_2 and $^1O_2$ (Halliwell & Gutteridge, 2015). Under UV-B stress, $^1O_2$ is produced in a relatively smaller amount than $O_2^{\bullet-}$, which is the predominant form of ROS (Hideg et al., 2002; Barta et al., 2004).

There is ample evidence of the alterations in secondary metabolism that occur in plants of all taxa when exposed to UV light. UV-B light, for example, increased type I flavone synthase activity in the moss *Pohlia nutans*, thereby enhancing flavone production (Wang et al., 2020). *Taxus cuspidata* accumulated flavonoids and toxoids in response to UV-B exposure (3 W/m^2) (Jiao et al., 2022). Additionally, following seven days of exposure to UV-A, the flavonoid content of *Scutellaria baicalensis* reached its maximum concentration (41.86 mg g^{-1}) (Miao et al., 2007). *Pisum sativum* leaves exposed to UV-B radiation had higher levels of trigonelline and nicotinamide; this is known as nicotinamide induction, which is a process caused by oxidative stress (Berglund et al., 1996). A study comparing two

TABLE 26.1
Damage caused by ROS to DNA, lipids and proteins

Oxidative damage to DNA	Oxidative damage to lipids	Oxidative damage to proteins	References
ROS can compromise the genomic integrity of plants by causing single-strand and double-strand breaks in DNA	ROS can damage membrane lipids by targeting polyunsaturated fatty acids, initiating lipid peroxidation and jeopardizing membrane integrity	Modification of amino acid residues leads to altered protein structure and function. This includes oxidation, hydroxylation and carbonylation of side chains	Vanderauwera et al. (2011), Sharma et al. (2012) and Krajčovičová-Kudláčková et al. (2006)
Mutations may result from ROS-induced changes to DNA bases, such as the synthesis of 8-oxo-2'-deoxyguanosine	Reactive aldehydes like malondialdehyde are produced by lipid peroxidation and have the potential to further damage the cellular constituents	Oxidative modifications can lead to the formation of protein aggregates, which jeopardize cellular functions	Tripathi et al. (2023), Chmielowska-BÄ et al. (2015) and Juan et al. (2021)
ROS have the ability to remove DNA bases, forming abasic sites that interfere with transcription and replication	The physical characteristics of membranes are changed by oxidative damage, which impacts their permeability and fluidity	Proteins become more prone to proteolytic degradation when they are oxidatively damaged	Tripathi et al. (2023) Sharma et al. (2012) and Xiong et al. (2006)
Replication and repair processes can be hampered by the ability of ROS to create covalent interactions between proteins and DNA	Lipid peroxidation can propagate oxidative stress by generating additional ROS, thereby amplifying cellular damage	Metabolic processes are affected when active sites of enzymes are oxidized by ROS through reduced or suppressed catalytic activity	Vanderauwera et al. (2011), Chmielowska-BÄ et al. (2015) and Juan et al. (2021)

distinct ecotypes of the *Paubrasilia echinata* revealed that UV-B radiation reduced photochemical efficiency, biomass accumulation, CO_2 assimilation and stem growth in the shade-tolerant ecotype, while increasing flavonoids, lignin and antioxidant qualities in the sun-tolerant ecotype, but decreasing cell respiration (Cuzzuol et al., 2020). After 13 days of exposure to UV-B rays, *Nymphoides humboldtiana* exhibited increased antioxidant activity and the formation of flavonoids such as epicatechin, quercetin, phloroglucinol, chlorogenic acid and ferulic acid (Nocchi et al., 2020). The manufacture of flavonoids, especially flavone C-glycosides, which are metabolites found in the most metabolically active cells, was enhanced in *Colobanthus quitensis* when exposed to UV-B light (Contreras et al., 2018). Under the influence of UV radiation, *Melissa officinalis* displayed alterations in glycolysis and phenylpropanoid pathway with varying recovery timeframes (Kim et al., 2018).

Numerous studies have also been conducted on the effects of UV radiation exposure on plants used in traditional medicine, for instance, exposure to UV-B inhibits the development and secondary metabolism of *Morus alba*, a plant used in traditional Chinese medicine (Li et al., 2022). Prolonged exposure to UV-B radiation increases the production of flavonoids, which are advantageous as active ingredients in medicinal products (Zhao et al., 2020). Under UV-B radiation, two distinct species of the Chinese herb, *Astragalus*, altered the production of secondary metabolites; *A. membranaceus* generated more derivatives of hydroxycinnamic acid, whereas *Astragalus mongholicus* accumulated myricitrin and isoflavones, demonstrating varying levels of tolerance to UV-B stress (Liu et al., 2020). In Asian nations, *Lunarera japonica* blossoms are utilized as a medicinal herb. *L. japonica* raises the concentrations of secondary metabolites like isochlorogenic acid, secoxyloganin and secologanic acids, as well as oxidative pentose phosphate when exposed to UV light (Zhu et al., 2017).

These studies demonstrate how UV light has compelled plants to alter their metabolism, especially the secondary metabolism, in order to improve their ecological success rate. However, this also has significant ramifications for commercially valuable plants (del Socorro Sánchez Correa et al., 2023).

26.2.3 Impact on Photosynthesis and Plant Growth

Since photosynthesis is a light-dependent process, UV radiation almost certainly has an impact on it. Numerous studies have documented the damage that UV light causes to particular locations inside the photosynthetic machinery of green plants (Liu et al., 2022). A large portion of the damage is brought about by the increased generation of ROS, which are implicated in UV-induced reactions as both harmful and signaling agents in regular cellular functions. ROS can have adverse effects on proteins that comprise the light-harvesting complexes of the photosystems or those present in the protein complexes that house the electron carriers of photosynthesis. Their buildup has even been shown to destroy ribulose bisphosphate carboxylase/oxygenase (del Socorro Sánchez Correa et al., 2023), which lowers atmospheric carbon fixation and plant biomass. The oxidation of fatty acids in the membranes is another significant damage caused by ROS. When combined with peroxidation and photooxidation due to UV light, this damages the essential integrity of the thylakoid membrane in the chloroplast, changing the way the membrane-embedded photosynthetic complexes are organized and lowering their photosynthetic capacity (Swarna et al., 2012). UV light damages the plant proteins as well; in fact, it damages the enzymes that produce pigments like chlorophylls, which is one of the effects on photosynthesis (Ranjbarfordoei et al., 2011). Furthermore, UV light breaks down pigments, particularly carotenoids and chlorophyll b. As a result, exposure to this kind of radiation can lead to an imbalance in the quantity of pigments, which leads to changes in the photosynthetic apparatus, as was recently observed in maize. Fluorescence and chlorophyll concentration dropped in a number of maize lines following 19 days of UV exposure, but in varying amounts and in a line-dependent fashion (Jovanic et al., 2022). A number of components of photosystem II are UV-sensitive. Since many of these proteins are components of electron transport centers and this complex is made up of pigments and proteins, their modification or degradation impacts the electron transport chain of photosynthesis, lowering its levels under UV radiation stress (del Socorro Sánchez Correa et al., 2023). Proteins D1 and D2, located in the reaction core of photosystem II, are particularly vulnerable to even low UV radiation intensities (1 μmol m^{-2} s^{-1}). The breakdown of proteins D1 and D2 contributes to the modification of manganese (Mn) oxidizing group of water, which collectively results in the loss of functionality of the reaction center and inhibition of electron transport (Kataria et al., 2014). Investigating downstream transporters is challenging when damage occurs at the start of the electron transport chain of photosynthesis; however, some reports show that prolonged exposure to UV light alters the ratio of photosystems II and I because of a decrease in absorption at 700 nm, which is absorbed by photosystem I (Renger et al., 1986).

26.3 MECHANISMS OF EPIGENETIC REGULATION IN UV/LIGHT STRESS

26.3.1 DNA Methylation and Demethylation

DNA methylation, leading to the formation of 5-methyl cytosine (5-mC), is a significant epigenetic marker. The dynamic result of methylation and demethylation processes or DNA methylation patterns is necessary for both the control of gene activity and long-term silencing of transposable elements (TEs) (Law & Jacobsen, 2010). *Artemisia annua* was found to have a decrease in DNA methylation at seven potential transcription factor binding sites in response to UV-B exposure (Pandey & Pandey-Rai, 2015). One important gene that regulates the manufacture of artemisinin is DOUBLE BOND REDUCTASE 2 (DBR2) (Pandey & Pandey-Rai, 2015). By causing DNA demethylation in the *DBR2* promoter regions, which include more WRKY transcription factor binding sites, UV-B treatment increased the expression of the *DBR2* gene (Pandey & Pandey-Rai, 2015). More recently, UV-B-induced flavonoid production in *Artemisia annua* was shown to involve DNA methylation (Pandey et al., 2019). The entire genome of *Artemisia annua* was found to be hypomethylated as a result of UV-B therapy. According to Pandey et al. (2019), UV-B radiation specifically prompted the demethylation of the *AaPAL1* promoter region, which released the binding sites of many transcription factors and enhanced the expression of MYB factors, including MYB1, MYC and WRKY.

When maize was exposed to UV-A/B radiation, minor alterations in DNA methylation were seen (Eichten & Springer, 2015). DNA damage and UV responsiveness are also determined by the structure and organization of the genome-epigenome. According to reports, 5-mC next to pyrimidine has a higher UV-B absorbance than unmethylated cytosines, making it more likely to produce pyrimidine dimers (Rochette et al., 2009). Given that DNA methylation is localized in particular genomic locations, it is possible that 5-mC-enriched regions, such as constitutive heterochromatin, could be hot spots of photodamage, affecting the flexibility and stability of the genome. This finding is supported by evidence showing a strong positive connection between UV-induced mutations in *Arabidopsis thaliana* and DNA methylation patterns (Willing et al., 2016).

26.3.2 Histone Modifications

Epigenetic modifications against UV stress have been shown to provide plant "memory" (Müller-Xing et al., 2014). Additionally, UV-B stress causes the production of ROS, which impairs metabolism, reduces resistance to infections and changes cellular functions (Jenkins, 2009; Hideg et al., 2012; Li et al., 2013). Different light conditions caused noticeable differences in the histone modifications that were seen at particular loci in *Arabidopsis* seedlings (Sharma et al., 2023). The study found that changes in the H3K9ac status, which is also impacted by the amount and quality of light, may result from transitions between dark and light situations (Sharma et al., 2023). Remarkably, HY5, DET1 and COP1, the important regulators of UV signaling and photomorphogenesis, seem to play a role in regulating H3K9ac levels (Sharma et al., 2023). Notably, in mutant variants of these three proteins, the transcription of particular genes was strongly correlated with their H3K9ac status (Sharma et al., 2023). Furthermore, in wheat and *Arabidopsis*, exposure to UV-B enhanced the H3K9/K14 acetylation of the *ELIP1 (Early Light-Inducible Protein 1)* promoter (Sharma et al., 2023). Through genome methylation and homologous frequency, it was found that

Arabidopsis plants exposed to UV-C inherited stress tolerance to untreated genotypes (Boyko et al., 2010).

Heritable variations in the response of *Arabidopsis* to UV-B exposure were reported by Lang-Mladek et al. (2010), which had an impact on the epigenetic regulation of a reporter gene. These abnormalities were linked to changes in chromatin structure and adjustments in histone H3 acetylation. In maize seedlings exposed to UV-C and gamma irradiation treatments, chromatin changes were noted, and a direct correlation between germination rate, resistance to UV-C exposure and the prevalence of chromosomal abnormalities has been demonstrated (Sharma et al., 2023). In *Arabidopsis* seedlings exposed to gamma rays, researchers observed histone changes associated with the genes involved in DNA damage repair (Kim et al., 2019). The transcription of genes involved in DNA damage repair increased after gamma irradiation. Investigations have shown that histone H3K4me3 at the transcription start sites is linked to the increased expression of certain genes. This finding was arrived at after several chromatin modification investigations utilizing the standard chromatin immunoprecipitation (ChIP) method (Jung et al., 2019). DNA methylation analysis at various levels within the loci of At1g31280 (AtAgo2), At4g19130 (AtRPA1E), At5g20850 (AtRad51) and At5g24280 (AtGMI1) has been revealed by a number of investigations. MethylC-Seq (bisulfite conversion followed by sequencing) was used to undertake research on genomic DNA taken from soil-grown plants that were four weeks old. The results of these investigations led scientists to a noteworthy discovery. Following gamma irradiation exposure, there was a considerable increase in the expression of genes involved in DNA damage repair (Dong et al., 2020; Cruz et al., 2013; Nielsen et al., 2019). H3K9ac was found at the AtAgo2 locus, whereas H3K4me3 was seen at gamma ray-inducible loci after DNA damage repair. Interestingly, the study showed that gamma radiation-induced histone alterations are site-specific. Moreover, the activation of gamma radiation-induced genes at these particular loci appears to be more dependent on the modifications H3K4me3 and H3K9ac rather than on changes in DNA methylation (Yadav et al., 2017; Guidotti, 2020).

26.4 EPIGENETIC MEMORY OF UV/LIGHT STRESS

According to D'Urso and Brickner (2014), transgenerational memory is the term used to describe meiotically heritable alterations in organisms in response to events in earlier generations. According to Chandler and Walbot (1986), one of the traditional instances of transmission of epigenetic information over generations is the silencing of transposable elements (TEs) by DNAme, which is crucial for the integrity of plant genomes. In maize, quiet Mutator (Mu) TEs can be reactivated by UV-B radiation treatments (Walbot, 1992; Walbot, 1999). The two MuDR TE genes exhibit increased expression following UV-B treatment, which is also accompanied by decreased DNA and H3K9me2 methylation and increased histone H3 acetylation. These chromatin alterations may be the cause of transposon reactivation by UV-B in maize (Questa et al., 2010). The frequency of homologous recombination can be increased by plant responses to UV radiation. Offspring of parent plants that have previously been exposed to these abiotic stresses exhibit identical increases in homologous recombination, higher DNA methylation and improved stress tolerance (Boyko et al., 2010). There is still much to learn about the molecular processes behind transgenerational stress memory. One theory is that stress causes the production of ROS, phytohormones and secondary metabolites rather quickly. As the stress level rises, a long-lasting reaction that includes DNA methylation and histone changes may be passed on to offspring (Pastor et al., 2012). Transgenerational memory has been studied using UV light because high levels of UV radiation can harm DNA, cause genomic instability, induce homologous recombination and promote chromatin remodeling, especially DNA and histone alterations, which can be passed on to the offspring (Müller-Xing et al., 2014). Using UV-C and flagellin, Molinier et al. (2006) simulated environmental stress and examined the heritable recombination rates caused by stress. Even four generations after the stress treatment, they demonstrated an increase in homologous recombination using transgenic reporter lines that enable histochemical detection of homologous recombination (Molinier et al., 2006). Studies conducted by other groups under comparable UV-C stress conditions, however, revealed that genes involved in somatic homologous recombination, including RAD51, BRCA1, MIM1 and ATM, were upregulated in the parent generation under stress, but not in the next two progeny generations. This was linked to an increase in somatic homologous recombination in the parent generation, but not in the next two progeny generations (Pecinka et al., 2009). The activation of a transcriptionally silenced *GUS* reporter gene and multiple stress-related genes by UV-B treatment was linked to increased histone acetylation (H3K9ac1) in Arabidopsis (Lang-Mladek et al., 2010). This is in line with earlier findings on *Arabidopsis* (Cloix & Jenkins, 2008) and maize (Casati et al., 2008). However, the effects of UV-B stress on reporter gene activation were limited to a few cells in non-stressed progeny plants and did not last longer than two generations of non-stressed progeny. Additionally, seed aging counteracted the inheritance of the release of stress-induced gene silencing (Lang-Mladek et al., 2010). Though the effect was limited to the next generation, UV-C, salt, temperature and drought can cause changes in global DNA methylation and gene transcription (Boyko et al., 2010). According to Boyko et al. (2010), *dicer-like 2* (*dcl2*) and *dicer-like 3* (*dcl3*) mutants showed impaired somatic homologous recombination in response to stress. Interestingly, the offspring of *dcl4* mutant plants exposed to UV-C showed hypermethylation in their genomes, indicating that Dicer-dependent small RNA pathways play a role in transgenerational memory regarding environmental stress (Müller-Xing et al., 2014).

The processes underlying these impacts remain poorly known, aside from the controversy surrounding the generational transfer of environmental-triggered epigenetic characteristics. According to a recent study, a reduction in DNA methylation 1 (DDM1) and Morpheus' Molecule 1 (MOM1) acts redundantly to restore the stress-induced epigenetic state to its pre-stress state (Müller-Xing et al., 2014). This hinders the mitotic and meiotic inheritance of stress-induced epigenetic changes, potentially removing the foundation for transgenerational memory in plants (Iwasaki & Paszkowski, 2014).

26.5 INTERACTION OF UV/LIGHT STRESS WITH OTHER ENVIRONMENTAL STRESSORS

Plants often experience multiple environmental stresses simultaneously, and their responses are interconnected through complex regulatory networks. Epigenetic mechanisms play a crucial role in integrating these signals, allowing plants to adapt effectively. This section explores the crosstalk between UV/light stress and other environmental stressors, such as drought, salinity and temperature, as well as the role of phytohormones in coordinating these responses.

26.5.1 Crosstalk in Epigenetic Regulation

UV and light stress interact with other environmental stressors, leading to synergistic or antagonistic effects on plant physiology and gene expression; for example, drought and UV stress both impact stomatal regulation, water use efficiency and photosynthesis. Epigenetic modifications in drought-responsive genes, such as *DREB* and *NCED*, overlap with those observed under light stress, suggesting a shared regulatory mechanism (Kim et al., 2019). Similarly, salinity stress triggers ionic imbalance and ROS accumulation, which are also hallmarks of UV stress. Changes in DNA methylation under UV stress may prime plants for better salt tolerance by enhancing antioxidant defense pathways (Hideg et al., 2012). Temperature stress further complicates these interactions. High temperature exacerbates UV stress by increasing ROS accumulation and accelerating DNA damage, while cold stress-induced chromatin remodeling may interact with UV stress pathways (Hasanuzzaman et al., 2013; Zhu et al., 2017). These overlapping stress response pathways highlight the integrative role of epigenetic regulation in plant adaptation.

26.5.2 Role of Phytohormones

Phytohormones play a critical role in mediating plant responses to combined stresses. Abscisic acid (ABA), a key hormone in drought and salinity responses, is influenced by light stress. UV-B exposure enhances ABA biosynthesis, promoting stomatal closure and stress tolerance (Tossi et al., 2009). Auxins, which regulate growth and development, are also modulated by light stress. UV stress can epigenetically alter auxin-related genes, influencing root architecture and stress responses (Hectors et al., 2007). Additionally, salicylic acid (SA) and jasmonic acid (JA) are involved in UV-induced stress responses and plant defense. SA enhances DNA repair under UV stress, while JA modulates ROS signaling, highlighting the interconnected nature of stress responses (Shilpa et al., 2024). These hormonal pathways, regulated by epigenetic mechanisms, enable plants to fine-tune their responses to multiple stressors.

26.6 APPLICATION OF EPIGENETIC MODIFICATIONS UNDER UV/LIGHT STRESS

Epigenetic mechanisms offer a promising avenue for improving plant resilience to environmental stresses, including UV and light stress. Understanding how plants regulate stress responses at the epigenetic level can help in developing strategies to enhance crop productivity, stress tolerance and adaptability to changing climate.

26.6.1 Crop Improvement and Stress Tolerance

Epigenetic breeding strategies can be used to develop stress-resilient crop varieties. Unlike genetic modifications, epigenetic changes do not alter the DNA sequence, but regulate gene expression, making them a valuable tool for crop improvement (Springer & Schmitz, 2017). For example, *Arabidopsis* and tomato plants, exposed to UV-B stress, showed stable changes in DNA methylation that improved stress tolerance in subsequent generations (Lang-Mladek et al., 2010). Epigenome-wide association studies (EWAS) can further identify epigenetic markers linked to stress tolerance traits, guiding plant breeding programs for better stress adaptation (Richards et al., 2017). Epigenetic engineering,

TABLE 26.2
Key epigenetic modifications under UV and light stress

Modification type	Role in stress response	Example genes	References
DNA methylation	Silencing of stress-sensitive genes	Photorepair genes	Lang-Mladek et al. (2010)
Histone acetylation	Activation of stress-responsive genes	*ELIP* genes	Kim et al. (2019)
miRNA regulation	Post-transcriptional gene silencing	miR398, miR414	Sunkar et al. (2012)

particularly CRISPR-based tools like CRISPR-dCas9, allows for precise modification of epigenetic marks without altering the DNA sequence. CRISPR-based activation of stress-responsive genes through histone acetylation has been shown to improve light stress tolerance in plants (Lowder et al., 2015).

26.6.2 Sustainable Agriculture Practices

Stress priming through mild UV exposure can induce beneficial epigenetic modifications, preparing plants for future stress conditions. This approach can be applied in controlled environment, such as greenhouses, to improve crop performance in the field (Mauch-Mani et al., 2017). Additionally, natural compounds that modify epigenetic markers, such as flavonoids and plant hormones, can be used to induce beneficial stress responses in organic farming systems (Vriet et al., 2013). Chemical epigenetic modulators, such as histone deacetylase inhibitors, offer another avenue for enhancing stress tolerance. Treatment with these compounds has been shown to increase drought and UV stress tolerance in crops like rice and maize (Garg et al., 2015). These sustainable practices, combined with advances in epigenetic research, hold great promise for improving crop resilience and ensuring food security in a changing climate.

26.7 CONCLUSIONS AND FUTURE PERSPECTIVES

Epigenetic research has significantly advanced our understanding of plant responses to UV and light stress. This chapter highlights how epigenetic mechanisms, including DNA methylation and histone modifications, regulate stress responses and memory in plants. UV and light stress induce epigenetic modifications that influence gene expression and stress memory, allowing plants to adapt to changing environmental conditions. The crosstalk between UV/light stress and other environmental stressors, such as drought, salinity and temperature extremes underscores the integrative nature of plant stress regulation. Future research should focus on expanding our understanding of transgenerational epigenetic inheritance. While stress memory is well documented, more studies are needed to elucidate how epigenetic marks are passed to future generations and how their stability and reversibility can be optimized for crop improvement (Heard & Martienssen, 2014). Advanced technologies, such as single-cell epigenomics and CRISPR-based epigenome editing, offer exciting opportunities for precise analysis and manipulation of stress-related epigenetic marks (Lowder et al., 2015). Additionally, artificial intelligence (AI)-driven analysis can help predict how epigenetic modifications influence stress responses, accelerating the development of stress-resilient crops (Libbrecht & Noble, 2015). The translational potential of epigenetic memory studies is immense. Applying this knowledge in agriculture can lead to the development of climate-resilient crops and sustainable farming practices, for example, stress priming through mild UV exposure can be used to prepare crops for future stress conditions, while epigenetic biomarkers can aid in early stress detection and precision breeding (Springer & Schmitz, 2017). Integrating epigenetic approaches with traditional breeding and genetic engineering holds great promise for revolutionizing plant stress tolerance strategies and ensuring food security in a changing climate.

REFERENCES

Banerjee, A., & Roychoudhury, A. (2016). Plant responses to light stress: Oxidative damages, photoprotection, and role of phytohormones. In G. J. Ahammed & Yu, J.-Q. (Eds.), *Plant Hormones under Challenging Environmental Factors* (pp. 181–213). Springer Publishers.

Banerjee, A., & Roychoudhury, A. (2017). Epigenetic regulation during salinity and drought stress in plants: Histone modifications and DNA methylation. *Plant Gene*, 11, 199–204.

Barta, C., Kálai, T., Hideg, K., Vass, I., & Hideg, É. (2004). Differences in the ROS-generating efficacy of various ultraviolet wavelengths in detached spinach leaves. *Functional Plant Biology: FPB*, 31(1), 23–28. https://doi.org/10.1071/FP03170.

Benhamed, M., Martin-Magniette, M., Taconnat, L., Bitton, F., Servet, C., De Clercq, R., De Meyer, B., Buysschaert, C., Rombauts, S., Villarroel, R., Aubourg, S., Beynon, J., Bhalerao, R. P., Coupland, G., Gruissem, W., Menke, F. L., Weisshaar, B., Renou, J., Zhou, D., & Hilson, P. (2008). Genome-scale *Arabidopsis* promoter array identifies targets of the histone acetyltransferase GCN5. *The Plant Journal*, 56(3), 493–504. https://doi.org/10.1111/j.1365-313x.2008.03606.x.

Berglund, T., Kalbin, G., Strid, Å., Rydström, J., & Ohlsson, A. B. (1996). UV-B- and oxidative stress-induced increase in nicotinamide and trigonelline and inhibition of defensive metabolism induction by poly(ADP-ribose)polymerase inhibitor in plant tissue. *FEBS Letters*, 380(1–2), 188–193. https://doi.org/10.1016/0014-5793(96)00027-0.

Bird, A. (2002). DNA methylation patterns and epigenetic memory. *Genes & Development*, 16(1), 6–21. https://doi.org/10.1101/gad.947102.

Bird, A. (2007). Perceptions of epigenetics. *Nature*, 447(7143), 396–398. https://doi.org/10.1038/nature05913.

Boyko, A., Blevins, T., Yao, Y., Golubov, A., Bilichak, A., Ilnytskyy, Y., Hollunder, J., Meins, F., Jr, & Kovalchuk, I. (2010). Transgenerational adaptation of *Arabidopsis* to stress requires DNA methylation and the function of Dicer-like proteins. *PLoS ONE*, 5(3), e9514. https://doi.org/10.1371/journal.pone.0009514.

Campi, M., D'Andrea, L., Emiliani, J., & Casati, P. (2011). Participation of chromatin-remodeling proteins in the repair of ultraviolet-B-damaged DNA. *Plant Physiology*, 158(2), 981–995. https://doi.org/10.1104/pp.111.191452.

Casati, P., Campi, M., Chu, F., Suzuki, N., Maltby, D., Guan, S., Burlingame, A. L., & Walbot, V. (2008). Histone acetylation and chromatin remodeling are required for UV-B–dependent transcriptional activation of regulated genes in maize. *The Plant Cell*, 20(4), 827–842. https://doi.org/10.1105/tpc.107.056457.

Casati, P., Stapleton, A. E., Blum, J. E., & Walbot, V. (2006). Genome-wide analysis of high-altitude maize and gene knockdown stocks implicates chromatin remodeling proteins in response to UV-B. *The Plant Journal*, 46(4), 613–627. https://doi.org/10.1111/j.1365-313x.2006.02721.x.

Chandler, V. L., & Walbot, V. (1986). DNA modification of a maize transposable element correlates with loss of activity. *Proceedings of the National Academy of Sciences of the United States of America*, 83(6), 1767–1771. https://doi.org/10.1073/pnas.83.6.1767.

Chen, M., Lv, S., & Meng, Y. (2010). Epigenetic performers in plant. *Development Growth & Differentiation*, 52(6), 555–566. https://doi.org/10.1111/j.1440-169x.2010.01192.x.

Chinnusamy, V., & Zhu, J. (2009). Epigenetic regulation of stress responses in plants. *Current Opinion in Plant Biology*, 12(2), 133–139. https://doi.org/10.1016/j.pbi.2008.12.006.

Chmielowska-BÄK, J., Izbiańska, K., & Deckert, J. (2015). Products of lipid, protein and RNA oxidation as signals and regulators of gene expression in plants. *Frontiers in Plant Science*, 6. https://doi.org/10.3389/fpls.2015.00405.

Cloix, C., & Jenkins, G. I. (2008). Interaction of the *Arabidopsis* UV-B-specific signaling component UVR8 with chromatin. *Molecular Plant*, 1(1), 118–128. https://doi.org/10.1093/mp/ssm012.

Contreras, R. A., Pizarro, M., Köhler, H., Zamora, P., & Zúñiga, G. E. (2018). UV-B shock induces photoprotective flavonoids but not antioxidant activity in Antarctic *Colobanthus quitensis* (Kunth) Bartl. *Environmental and Experimental Botany*, 159, 179–190. https://doi.org/10.1016/j.envexpbot.2018.12.022.

Cruz, C., Della Rosa, M., Krueger, C., Gao, Q., Horkai, D., King, M., Field, L., & Houseley, J. (2018). Tri-methylation of histone H3 lysine 4 facilitates gene expression in ageing cells. *eLife*, 7, e34081. https://doi.org/10.7554/eLife.34081.

Cuzzuol, G. R. F., Gama, V. N., Zanetti, L. V., Werner, E. T., & Pezzopane, J. E. M. (2020). UV-B effects on growth, photosynthesis, total antioxidant potential and cell wall components of shade-tolerant and sun-tolerant ecotypes of *Paubrasilia echinata*. *Flora*, 271, 151679. https://doi.org/10.1016/j.flora.2020.151679.

del Socorro Sánchez Correa, M., el Rocío Reyero Saavedra, M., Estrella Parra, A., Nolasco Ontiveros, E., del Carmen Benítez Flores, J., Gerardo Ortiz Montiel, J., Eduardo Campos Contreras, J., López Urrutia, E., Guillermo Ávila Acevedo, J., Edith Jiménez Nopala, G., and Montserrat Espinosa González, A. (2023). Ultraviolet radiation and its effects on plants. Abiotic stress in plants: Adaptations to climate change. *IntechOpen*. https://doi.org/10.5772/intechopen.109474.

Ding, Y., Fromm, M., & Avramova, Z. (2012). Multiple exposures to drought 'train' transcriptional responses in *Arabidopsis*. *Nature Communications*, 3(1). https://doi.org/10.1038/ncomms1732.

Dong, W., Gao, T., Wang, Q., Chen, J., Lv, J., & Song, Y. (2020). Salinity stress induces epigenetic alterations to the promoter of MsMYB4 encoding a salt-induced MYB transcription factor. *Plant Physiology and Biochemistry: PPB*, 155, 709–715. https://doi.org/10.1016/j.plaphy.2020.08.015.

D'Urso, A., & Brickner, J. H. (2014). Mechanisms of epigenetic memory. *Trends in Genetics* 30(6), 230–236. https://doi.org/10.1016/j.tig.2014.04.004.

Eichten, S. R., & Springer, N. M. (2015). Minimal evidence for consistent changes in maize DNA methylation patterns following environmental stress. *Frontiers in Plant Science*, 6, 308. https://doi.org/10.3389/fpls.2015.00308.

Garg, R., Chevala, V. N., Shankar, R., & Jain, M. (2015). Divergent DNA methylation patterns associated with gene expression in rice cultivars with contrasting drought and salinity stress response. *Scientific Reports*, 5, 14922. https://doi.org/10.1038/srep14922.

Gendrel, A., Lippman, Z., Yordan, C., Colot, V., & Martienssen, R. A. (2002). Dependence of heterochromatic Histone H3 methylation patterns on the *Arabidopsis* gene DDM1. *Science*, 297(5588), 1871–1873. https://doi.org/10.1126/science.1074950.

Gill, S. S., Anjum, N. A., Gill, R., Jha, M., & Tuteja, N. (2015). DNA damage and repair in plants under ultraviolet and ionizing radiations. *The Scientific World Journal*, 2015, 250158. https://doi.org/10.1155/2015/250158.

Guidotti, N. (2020). Development of chemical biology tools to reveal the function and organization of the silenced chromatin state. https://doi.org/10.5075/epfl-thesis-7726.

Halliwell, B., & Gutteridge, J. M. C. (2015). *Free Radicals in Biology and Medicine* (5th ed.). New York: Oxford University Press. http://dx.doi.org/10.1093/acprof:oso/9780198717478.001.0001.

Hasanuzzaman, M., Nahar, K., Alam, M. M., Roychowdhury, R., & Fujita, M. (2013). Physiological, biochemical, and molecular mechanisms of heat stress tolerance in plants. *International Journal of Molecular Sciences*, 14(5), 9643–9684. https://doi.org/10.3390/ijms14059643.

Heard, E., & Martienssen, R. A. (2014). Transgenerational epigenetic inheritance: Myths and mechanisms. *Cell*, 157(1), 95–109. https://doi.org/10.1016/j.cell.2014.02.045.

Hectors, K., Prinsen, E., De Coen, W., Jansen, M. A. K., & Guisez, Y. (2007). *Arabidopsis thaliana* plants acclimated to low dose rates of ultraviolet B radiation show specific changes in morphology and gene expression in the absence of stress symptoms. *New Phytologist*, 175(2), 255–270. https://doi.org/10.1111/j.1469-8137.2007.02092.x.

Hideg, E., Barta, C., Kálai, T., Vass, I., Hideg, K., & Asada, K. (2002). Detection of singlet oxygen and superoxide with fluorescent sensors in leaves under stress by photoinhibition or UV radiation. *Plant & Cell Physiology*, 43(10), 1154–1164. https://doi.org/10.1093/pcp/pcf145.

Hideg, É., Jansen, M. A., & Strid, Å. (2012). UV-B exposure, ROS, and stress: Inseparable companions or loosely linked associates? *Trends in Plant Science*, 18(2), 107–115. https://doi.org/10.1016/j.tplants.2012.09.003.

Iwasaki, M., & Paszkowski, J. (2014). Identification of genes preventing transgenerational transmission of stress-induced epigenetic states. *Proceedings of the National Academy of Sciences of the United States of America*, 111(23), 8547–8552. https://doi.org/10.1073/pnas.1402275111.

Jansen, M. A., Gaba, V., & Greenberg, B. M. (1998). Higher plants and UV-B radiation: Balancing damage, repair and acclimation. *Trends in Plant Science*, 3(4), 131–135. https://doi.org/10.1016/s1360-1385(98)01215-1.

Jenkins, G. I. (2009). Signal transduction in responses to UV-B radiation. *Annual Review of Plant Biology*, 60(1), 407–431. https://doi.org/10.1146/annurev.arplant.59.032607.092953.

Jiao, J., Xu, X. J., Lu, Y., Liu, J., Fu, Y. J., Fu, J. X., & Gai, Q. Y. (2022). Identification of genes associated with biosynthesis of bioactive flavonoids and taxoids in *Taxus cuspidata* Sieb. Et Zucc. plantlets exposed to UV-B radiation. *Gene*, 823, 146384. https://doi.org/10.1016/j.gene.2022.146384.

Jovanić, B., Radenković, B., Despotović-Zrakić, M., Bogdanović, Z., & Barać, D. (2022). Effect of UV-B radiation on chlorophyll fluorescence, photosynthetic activity and relative chlorophyll content of five different corn hybrids. *Journal of Photochemistry and Photobiology*, 10, 100115. https://doi.org/10.1016/j.jpap.2022.100115.

Juan, C. A., Pérez de la Lastra, J. M., Plou, F. J., & Pérez-Lebeña, E. (2021). The chemistry of reactive oxygen species (ROS) revisited: Outlining their role in biological macromolecules (DNA, lipids and proteins) and induced pathologies. *International Journal of Molecular Sciences*, 22(9), 4642. https://doi.org/10.3390/ijms22094642.

Jung, I. J., Ahn, J. W., Jung, S., Hwang, J. E., Hong, M. J., Choi, H. I., & Kim, J. B. (2019). Overexpression of rice jacalin-related mannose-binding lectin (OsJAC1) enhances resistance to ionizing radiation in *Arabidopsis*. *BMC Plant Biology*, 19(1), 561. https://doi.org/10.1186/s12870-019-2056-8.

Kataria, S., Jajoo, A., & Guruprasad, K. N. (2014). Impact of increasing Ultraviolet-B (UV-B) radiation on photosynthetic processes. *Journal of Photochemistry and Photobiology B Biology*, 137, 55–66. https://doi.org/10.1016/j.jphotobiol.2014.02.004.

Kim, J. S., Lim, J. Y., Shin, H., Kim, B. G., Yoo, S. D., Kim, W. T., & Huh, J. H. (2019). ROS1-dependent DNA demethylation is required for ABA-inducible NIC3 expression. *Plant Physiology*, 179(4), 1810–1821. https://doi.org/10.1104/pp.18.01471.

Kim, S., Lee, H., & Kim, K. H. (2018). Metabolomic elucidation of recovery of *Melissa officinalis* from UV-B irradiation stress. *Industrial Crops and Products*, 121, 428–433. https://doi.org/10.1016/j.indcrop.2018.05.002.

Krajčovičová-Kudláčková, M., Dušinská, M., Valachovičová, M., Blažíček, P., & Pauková, V. (2006). Products of DNA, protein and lipid oxidative damage in relation to vitamin C plasma concentration. *Physiological Research*, 227–231. https://doi.org/10.33549/physiolres.930761.

Landry, L. G., Stapleton, A. E., Lim, J., Hoffman, P., Hays, J. B., Walbot, V., & Last, R. L. (1997). An *Arabidopsis* photolyase mutant is hypersensitive to ultraviolet-B radiation. *Proceedings of the National Academy of Sciences*, 94(1), 328–332. https://doi.org/10.1073/pnas.94.1.328.

Lang-Mladek, C., Popova, O., Kiok, K., Berlinger, M., Rakic, B., Aufsatz, W., Jonak, C., Hauser, M. T., & Luschnig, C. (2010). Transgenerational inheritance and resetting of stress-induced loss of epigenetic gene silencing in *Arabidopsis*. *Molecular Plant*, 3(3), 594–602. https://doi.org/10.1093/mp/ssq014.

Lario, L. D., Ramirez-Parra, E., Gutierrez, C., Spampinato, C. P., & Casati, P. (2013). ANTI-SILENCING FUNCTION1 proteins are involved in ultraviolet-induced DNA damage repair and are cell cycle regulated by E2F transcription factors in *Arabidopsis*. *Plant Physiology*, 162(2), 1164–1177. https://doi.org/10.1104/pp.112.212837.

Law, J. A., & Jacobsen, S. E. (2010). Establishing, maintaining and modifying DNA methylation patterns in plants and animals. *Nature reviews Genetics*, 11(3), 204–220. https://doi.org/10.1038/nrg2719.

Li, J., Yang, L., Jin, D., Nezames, C. D., Terzaghi, W., & Deng, X. W. (2013). UV-B-induced photomorphogenesis in *Arabidopsis*. *Protein & Cell*, 4(7), 485–492. https://doi.org/10.1007/s13238-013-3036-7.

Li, Y., Liu, S., Shawky, E., Tao, M., Liu, A., Sulaiman, K., Tian, J., & Zhu, W. (2022). SWATH-based quantitative proteomic analysis of *Morus alba* L. leaves after exposure to ultraviolet-B radiation and incubation in the dark. *Journal of Photochemistry and Photobiology B Biology*, 230, 112443. https://doi.org/10.1016/j.jphotobiol.2022.112443.

Libbrecht, M. W., & Noble, W. S. (2015). Machine learning applications in genetics and genomics. *Nature Reviews Genetics*, 16(6), 321–332. https://doi.org/10.1038/nrg3920.

Liu, L., Wang, X., & Chang, C. (2022). Toward a smart skin: Harnessing cuticle biosynthesis for crop adaptation to drought, salinity, temperature, and ultraviolet stress. *Frontiers in Plant Science*, 13. https://doi.org/10.3389/fpls.2022.961829.

Liu, Y., Liu, J., Wang, H., Wu, K., Guo, X., Mu, L., & Tang, Z. (2020). Comparison of the global metabolic responses to UV-B radiation between two medicinal *Astragalus* species: An integrated metabolomics strategy. *Environmental and Experimental Botany*, 176, 104094. https://doi.org/10.1016/j.envexpbot.2020.104094.

Lowder, L. G., Zhang, D., Baltes, N. J., Paul, J. W., III, Tang, X., Zheng, X., Voytas, D. F., Hsieh, T. F., Zhang, Y., & Qi, Y. (2015). A CRISPR/Cas9 toolbox for multiplexed plant genome editing and transcriptional regulation. *Plant Physiology*, 169(2), 971–985. https://doi.org/10.1104/pp.15.00636.

Mauch-Mani, B., Baccelli, I., Luna, E., & Flors, V. (2017). Defense priming: An adaptive part of induced resistance. *Annual Review of Plant Biology*, 68, 485–512. https://doi.org/10.1146/annurev-arplant-042916-041132.

Miao, N., Yun, C., Shi, Y., Gao, Y., Wu, S., Zhang, Z., Han, S., Wang, H., & Wang, W. (2007). Enhancement of flavonoid synthesis and antioxidant activity in *Scutellaria baicalensis* aerial parts by UV-A radiation. *Industrial Crops and Products*, 187, 115532. https://doi.org/10.1016/j.indcrop.2022.115532.

Molinier, J., Ries, G., Zipfel, C., & Hohn, B. (2006). Transgeneration memory of stress in plants. *Nature*, 442(7106), 1046–1049. https://doi.org/10.1038/nature05022.

Müller-Xing, R., Xing, Q., & Goodrich, J. (2014). Footprints of the sun: Memory of UV and light stress in plants. *Frontiers in Plant Science*, 5, 474. https://doi.org/10.3389/fpls.2014.00474.

Nielsen, M., Ard, R., Leng, X., Ivanov, M., Kindgren, P., Pelechano, V., & Marquardt, S. (2019). Transcription-driven chromatin repression of Intragenic transcription start sites. *PLoS Genetics*, 15(2), e1007969. https://doi.org/10.1371/journal.pgen.1007969.

Nocchi, N., Duarte, H. M., Pereira, R. C., Konno, T. U. P., & Soares, A. R. (2020). Effects of UV-B radiation on secondary metabolite production, antioxidant activity, photosynthesis and herbivory interactions in *Nymphoides humboldtiana* (Menyanthaceae). *Journal of Photochemistry and Photobiology B Biology*, 212, 112021. https://doi.org/10.1016/j.jphotobiol.2020.112021.

Pandey, N., & Pandey-Rai, S. (2015). Deciphering UV-B-induced variation in DNA methylation pattern and its influence on regulation of DBR2 expression in *Artemisia annua* L. *Planta*, 242(4), 869–879. https://doi.org/10.1007/s00425-015-2323-3.

Pandey, N., Goswami, N., Tripathi, D., Rai, K. K., Rai, S. K., Singh, S., & Pandey-Rai, S. (2019). Epigenetic control of UV-B-induced flavonoid accumulation in *Artemisia annua* L. *Planta*, 249(2), 497–514. https://doi.org/10.1007/s00425-018-3022-7.

Pastor, V., Luna, E., Mauch-Mani, B., Ton, J., & Flors, V. (2012). Primed plants do not forget. *Environmental and Experimental Botany*, 94, 46–56. https://doi.org/10.1016/j.envexpbot.2012.02.013.

Pecinka, A., Rosa, M., Schikora, A., Berlinger, M., Hirt, H., Luschnig, C., & Mittelsten Scheid, O. (2009). Transgenerational stress memory is not a general response in *Arabidopsis*. *PLoS ONE*, 4(4), e5202. https://doi.org/10.1371/journal.pone.0005202.

Qüesta, J. I., Fina, J. P., & Casati, P. (2013). DDM1 and ROS1 have a role in UV-B induced- and oxidative DNA damage in *A. thaliana*. *Frontiers in Plant Science*, 4. https://doi.org/10.3389/fpls.2013.00420.

Qüesta, J. I., Walbot, V., & Casati, P. (2010). Mutator transposon activation after UV-B involves chromatin remodeling. *Epigenetics*, 5(4), 352–363. https://doi.org/10.4161/epi.5.4.11751.

Rajneesh, Pathak, J., Chatterjee, A., Singh, S. P., & Sinha, R. P. (2017). Detection of reactive oxygen species (ROS) in cyanobacteria using the oxidant-sensing probe 2',7'-dichlorodihydrofluorescein diacetate (DCFH-DA). *Bio-protocol*, 7(17), e2545. https://doi.org/10.21769/BioProtoc.2545.

Ranjbarfordoei, A., Samson, R., & Van Damme, P. (2011). Photosynthesis performance in sweet almond [*Prunus dulcis* (Mill) D. Webb] exposed to supplemental UV-B radiation. *Photosynthetica*, 49(1). https://doi.org/10.1007/s11099-011-0017-z.

Renger, G., Voss, M., Gräber, P., & Schulze, A. (1986). Effect of UV irradiation on different partial reactions of the primary processes of photosynthesis. In: Worrest, R.C., Caldwell, M.M. (eds) *Stratospheric Ozone Reduction, Solar Ultraviolet Radiation and Plant Life*. NATO ASI Series, vol 8. Springer, Berlin, Heidelberg. (pp. 171–184). https://doi.org/10.1007/978-3-642-70090-3_13.

Richards, C. L., Alonso, C., Becker, C., Bossdorf, O., Bucher, E., Colomé-Tatché, M., Durka, W., Engelhardt, J., Gaspar, B., Gogol-Döring, A., Grosse, I., van Gurp, T. P., Heer, K., Kronholm, I., Lampei, C., Latzel, V., Mirouze, M., Opgenoorth, L., Paun, O., Prohaska, S. J., Rensing, S. A., Stadler, P. F., Trucchi, E., Ullrich, K., & Verhoeven, K. J. F. (2017). Ecological plant epigenetics: Evidence from model and non-model species, and the way forward. *Ecology Letters*, 20(12), 1576–1590. https://doi.org/10.1111/ele.12858.

Ries, G., Heller, W., Puchta, H., Sandermann, H., Seidlitz, H. K., & Hohn, B. (2000). Elevated UV-B radiation reduces genome stability in plants. *Nature*, 406(6791), 98–101. https://doi.org/10.1038/35017595.

Rochette, P. J., Lacoste, S., Therrien, J. P., Bastien, N., Brash, D. E., & Drouin, R. (2009). Influence of cytosine methylation on ultraviolet-induced cyclobutane pyrimidine dimer formation in genomic DNA. *Mutation Research*, 665(1–2), 7–13. https://doi.org/10.1016/j.mrfmmm.2009.02.008.

Schmidt, C. K., & Jackson, S. P. (2013). On your MARK, get SET(D2), go! H3K36ME3 primes DNA mismatch repair. *Cell*, 153(3), 513–515. https://doi.org/10.1016/j.cell.2013.04.018.

Sharma, M., Sidhu, A. K., Samota, M. K., Gupta, M., Koli, P., & Choudhary, M. (2023). Post-translational modifications in histones and their role in abiotic stress tolerance in plants. *Proteomes*, 11(4), 38. https://doi.org/10.3390/proteomes11040038.

Sharma, P., Jha, A. B., Dubey, R. S., & Pessarakli, M. (2012). Reactive oxygen species, oxidative damage, and antioxidative defense mechanism in plants under stressful conditions. *Journal of Botany*, 2012, 1–26. https://doi.org/10.1155/2012/217037.

Shilpa, A., Thakur, R., & Prasad, P. (2024). Epigenetic regulation of abiotic stress responses in plants. *Biochimica et Biophysica Acta (BBA) – General Subjects*, 1868(9), 130661. https://doi.org/10.1016/j.bbagen.2024.130661.

Springer, N. M., & Schmitz, R. J. (2017). Exploiting induced and natural epigenetic variation for crop improvement. *Nature Reviews Genetics*, 18(9), 563–575. https://doi.org/10.1038/nrg.2017.45.

Stroud, H., Otero, S., Desvoyes, B., Ramírez-Parra, E., Jacobsen, S. E., & Gutierrez, C. (2012). Genome-wide analysis of histone H3.1 and H3.3 variants in *Arabidopsis thaliana*. *Proceedings of the National Academy of Sciences*, 109(14), 5370–5375. https://doi.org/10.1073/pnas.1203145109.

Sunkar, R., Li, Y. F., & Jagadeeswaran, G. (2012). Functions of microRNAs in plant stress responses. *Trends in Plant Science*, 17(4), 196-203. https://doi.org/10.1016/j.tplants.2012.01.010.

Swarna, S., Lorente, C., Thomas, A. H., & Martin, C. B. (2012). Rate constants of quenching of the fluorescence of pterins by the iodide anion in aqueous solution. *Chemical Physics Letters*, 542, 62–65. https://doi.org/10.1016/j.cplett.2012.06.013.

Tossi, V., Lamattina, L., & Cassia, R. (2009). An increase in the concentration of abscisic acid is critical for nitric oxide-mediated plant adaptive responses to UV-B irradiation. *New Phytologist*, 181(4), 871–879. https://doi.org/10.1111/j.1469-8137.2008.02722.x.

Tripathi, D., Oldenburg, D. J., & Bendich, A. J. (2023). Oxidative and glycation damage to mitochondrial DNA and plastid DNA during plant development. *Antioxidants*, 12(4), 891. https://doi.org/10.3390/antiox12040891.

Vanderauwera, S., Suzuki, N., Miller, G., Van De Cotte, B., Morsa, S., Ravanat, J., Hegie, A., Triantaphylidès, C., Shulaev, V., Van Montagu, M. C. E., Van Breusegem, F., & Mittler, R. (2011). Extranuclear protection of chromosomal DNA from oxidative stress. *Proceedings of the National Academy of Sciences*, 108(4), 1711–1716. https://doi.org/10.1073/pnas.1018359108.

Vriet, C., Russinova, E., & Reuzeau, C. (2013). From squalene to brassinolide: The steroid metabolic and signaling pathways across the plant kingdom. *Molecular Plant*, 6(2), 888–920. https://doi.org/10.1093/mp/sst096.

Walbot V. (1992). Reactivation of Mutator transposable elements of maize by ultraviolet light. *Molecular & General Genetics* 234(3), 353–360. https://doi.org/10.1007/BF00538694.

Walbot V. (1999). UV-B damage amplified by transposons in maize. *Nature*, 397(6718), 398–399. https://doi.org/10.1038/17043.

Wang, H., Liu, S., Wang, T., Liu, H., Xu, X., Chen, K., & Zhang, P. (2020). The moss flavone synthase I positively regulates the tolerance of plants to drought stress and UV-B radiation. *Plant Science*, 298, 110591. https://doi.org/10.1016/j.plantsci.2020.110591.

Willing, E. M., Piofczyk, T., Albert, A., Winkler, J. B., Schneeberger, K., & Pecinka, A. (2016). UVR2 ensures transgenerational genome stability under simulated natural UV-B in *Arabidopsis thaliana*. *Nature Communications*, 7, 13522. https://doi.org/10.1038/ncomms13522.

Xiong, Y., Contento, A. L., Nguyen, P. Q., & Bassham, D. C. (2006). Degradation of oxidized proteins by autophagy during oxidative stress in *Arabidopsis*. *Plant Physiology*, 143(1), 291–299. https://doi.org/10.1104/pp.106.092106.

Yadav, R. K., Jablonowski, C. M., Fernandez, A. G., Lowe, B. R., Henry, R. A., Finkelstein, D., Barnum, K. J., Pidoux,

A. L., Kuo, Y. M , Huang, J., O'Connell, M. J., Andrews, A. J., Onar-Thomas, A., Allshire, R. C., & Partridge, J. F. (2017). Histone H3G34R mutation causes replication stress, homologous recombination defects and genomic instability in *S. pombe*. *eLife*, 6, e27406. https://doi.org/10.7554/eLife.27406.

Zhao, B., Wang, L., Pang, S., Jia, Z., Wang, L., Li, W., & Jin, B. (2020). UV-B promotes flavonoid synthesis in *Ginkgo biloba* leaves. *Industrial Crops and Products*, 151, 112483. https://doi.org/10.1016/j.indcrop.2020.112483.

Zhou, D. (2009). Regulatory mechanism of histone epigenetic modifications in plants. *Epigenetics*, 4(1), 15–18. https://doi.org/10.4161/epi.4.1.7371.

Zhu, W., Zheng, W., Hu, X., Xu, X., Zhang, L., & Tian, J. (2017). Variations of metabolites and proteome in *Lonicera japonica* Thunb. Buds and flowers under UV radiation. *Biochimica Et Biophysica Acta (BBA) – Proteins and Proteomics*, 1865(4), 404–413. https://doi.org/10.1016/j.bbapap.2017.01.004.

27 Ethical and Regulatory Considerations in Plant Epigenetic Regulation during Diverse Stress Conditions

Muhammad Younis, Hafiz Muhammad Nadeem, Jianxu Wang, Sana Khalid, Tasveer Zahra Tariq, Behzad Murtaza and Muhammad Shahid

27.1 INTRODUCTION

During the last two or three decades, the world has witnessed fast and unchecked urban and industrial expansions, partially due to a rapid increase in world population. It is expected that these technological advancements and the world population will continue expanding in the future, thereby inducing numerous environmental changes. For example, environmental contamination by different types of organic and inorganic pollutants is increasing with the passage of time due to their enhanced use in different industrial processes (Khalid et al., 2024). The world climate is changing rapidly due to different human activities. The manufacturing and application of different types of synthetic chemicals have increased in the agricultural sector. The release of wastewater and solid waste from industry and cities has further enhanced environmental pollution (Shahid et al., 2023). Moreover, the natural resources are either getting contaminated or depleted (soil, water, air) at a rapid pace. These environmental changes, deteriorations and contaminations can negatively affect plants, animals and humans because of their continuous exposure (Shahid et al., 2023; Ejaz et al., 2025).

In contrast to animals and humans, plants cannot avoid environmental stresses by moving from one place to another. Naturally, plants have to face all the stresses at their place of germination and growth. Nowadays, plants have to face a huge number of environmental stresses due to extreme climatic conditions and environmental contaminations (Natasha et al., 2022), for example, severe weather conditions (low or high temperatures), shortage of good quality irrigation water, variation in rainfall patterns, heavy metal contamination of soil, water and air, and numerous other unfavorable environmental conditions can severely impact plant growth and production (Shahid et al., 2021; Shahid et al., 2023). Hence, plants experience various stresses of varying intensity and duration throughout their life span. It is estimated that the majority of plant species may experience a growth reduction of more than 50% due to environmental stressors (Shiade et al., 2024). Overall, plant growth and productivity are greatly hampered under these unfavorable environmental conditions, which also impede growth and development.

To withstand these various types of environmental stresses and obtain cellular tolerance, plants possess various signaling pathways under varied environmental stresses (Natasha et al., 2020). Generally, plants respond to these environmental stresses by altering their transcriptome and metabolome through signaling systems (Natasha et al., 2019; Shilpa et al., 2024). In fact, plants are naturally furnished with a variety of adaptive and tolerance mechanisms to cope with environmental changes including contamination and severe conditions of temperature and water (Gaude et al., 2024). It is revealed that the majority of plant species experience growth loss of more than 50% as a result of environmental stressors (Shiade et al., 2024).

When plants get exposed to different types of environmental stresses, they have a natural ability to tolerate these stresses via various mechanisms and processes; for example, numerous recent studies have delineated the important role of epigenetic pathways in regulating and tolerating environmental stresses, both biotic and abiotic (Ma et al., 2024; Shilpa et al., 2024; Xiang et al., 2024). Complex systems that regulate the modulation of epigenetic changes in their genomic DNA have been developed by plants (Iqbal et al., 2024). In order to better execute tolerance against environmental stresses, plants need to understand, recognize and memorize their past stress experiences (Kotkar & Giri 2020). Epigenetic alterations can also incur in plants the identification of stress memory, which helps them to better tolerate such stresses in future exposures (Banerjee & Roychoudhury 2017; Banerjee & Roychoudhury 2018). Highlighting the crucial role of epigenetics in assisting plants in thriving and surviving under various environmental stresses such as drought, heat, salt, heavy metals and cold is the main goal of this book chapter.

27.2 ENVIRONMENTAL STRESSES IN PLANTS

According to the projection of the United Nations, human population may reach up to 9.7 billion by 2050. Currently, about 800 million people are experiencing "inhumane

DOI: 10.1201/9781003589914-31

levels" of hunger. In order to ensure global food security, farm production must increase by 60% till 2050 (Van Dijk et al., 2021). Moreover, ensuring nutritional security is a major challenge that current agricultural systems must deal with (Faizan et al., 2024). In fact, there are numerous environmental stresses that are capable of impeding plant growth and farm production. External factors that harmfully affect plant growth, yield and development are collectively termed as plant stresses. Various environmental stressors are experienced by plants, which decrease and restrict agricultural yield.

Abiotic stress and biotic stress are the two categories of environmental stressors that plants experience (Sharma et al., 2022; Tavakoli et al., 2025). Worldwide, abiotic stressors such as salinity, radiation, flood, heavy metals, drought and temperature extremes are responsible for the loss of plant growth. Moreover, living things such as fungi, bacteria, viruses, insects, weeds and arachnids can mediate biotic stress in plants. These different types of environmental stressors cause reduced photosynthesis, stunted growth and decreased production by upsetting several plant physiological and biochemical attributes (Penna et al., 2021; Sharma et al., 2022; Rafiq et al., 2024).

Plants are harmed by certain environmental stressors, which cause them to display numerous metabolic dysfunctions. Mild or short-term environmental stresses can help plants recover from injuries since their effects are transient, but severe environmental stresses can even cause agricultural plants to die by stopping seed development, flowering and causing senescence (Khalid et al., 2017; Penna et al., 2021). These plants will be regarded as being susceptible to environmental stresses. However, certain plant species can withstand severe environmental stresses (Khalid et al., 2017).

Environmental stresses are capable of inducing intricate molecular and cellular interactions in plants. These environmental stresses negatively impact several plant attributes and upset the delicate balance of physiological processes. Under these circumstances of environmental stresses, plants frequently experience oxidative stress due to enhanced generation of reactive oxygen species (ROS) (Das & Roychoudhury 2014; Shahid et al., 2014). These ROS are capable of interacting with cellular components and damaging them. Environmental stressors have complex impacts on plant production that are more than a decrease in yield. These negative impacts ripple across ecological equilibriums, economic stabilities and food security (Sharma et al., 2022). Food scarcity, economic instability and rising food prices can result from lower crop yield, particularly in regions where people rely mainly on agriculture. Hence, it is highly imperative to comprehend how environmental stresses affect plant growth, development and productivity. These problems associated with environmental stresses warrant abrupt attention and can be tackled by establishing stress-resilient plants (Dinkar et al., 2024). Abundant research has been carried out in this regard to tolerate environmental stresses and improve crop production.

Overall, plants are not able to avoid these environmental cues because they are sessile by nature. Plants respond to environmental stress in various ways, such as mediating alterations in gene expression, cellular metabolism, growth rates and crop yield (Shahid 2021). Establishing techniques to minimize the effects of environmental stresses while boosting plant species tolerance and resilience requires an in-depth understanding of the different mechanisms by which plants react to these environmental stresses. To combat these environmental stresses, plants are equipped with numerous defense/tolerance mechanisms. Generally, plants can detect the environmental stresses followed by the activation and development of the appropriate biological reactions (Shahid et al., 2017). These adaptation and tolerance strategies are accomplished via different types of signal transduction routes to convey impulses from sensors on the cytoplasm or cell surface to the transcriptional machinery in the nucleus (Shahid et al., 2018). Various transcriptional modifications ultimately mediate plant stress tolerance. In order to detect environmental stress and produce the proper biochemical and physiological responses, the signaling pathway serves as a bridge and is highly important. Plants can avoid and tolerate environmental stresses by adapting to specific tolerance/defense mechanisms, even though they lack the adaptive immune system. Despite tremendous advancements in this field, there exists a gap regarding the interaction between epigenetic mechanisms taking place under different types of environmental stresses in plants (Rehman & Tanti 2020).

27.3 EPIGENETIC MODIFICATIONS

It is well established that plants are equipped with a great deal of phenotypic flexibility. In the majority of the cases, environmental adaptations are the most likely explanation for epigenetic variability (Penna et al., 2021; Verma et al., 2022; Iqbal et al., 2024). The probability that the epigenetic changes would be inherited by succeeding plant generations governs their capacity to swiftly adapt to environmental stresses (Kotkar & Giri 2020). Numerous studies have reported that plants use epigenetic memory as a crucial adaptation/tolerance mechanism to environmental stresses, especially under extreme biotic and abiotic stresses (Chen et al., 2024; Farahani et al., 2024; Guarino et al., 2024). This skill mainly explains why plants are capable of surviving harsh environmental changes (Ali et al., 2022). The epigenetic alterations in plants under various environmental stresses occur throughout their lifespan during different stages of growth and development (Barozai &Aziz 2018).

Epigenetic inheritance is the process by which environmental stresses alter gene expression, which is then passed on to newly generated cells during their vegetative growth and even to the progeny (Banerjee et al., 2017; Yajnik et al., 2024). According to the definition of epigenetics, a complex web of developmental events interlinks genotype and phenotype with one another. The study of processes and substances that sustain different gene activity levels within the same DNA sequence is called epigenetics. These pathways are capable of altering the chromatin state of genes by changing the transcriptional and post-transcriptional gene expression involved in tolerating environmental stress. Therefore, understanding the dynamic interactions between epigenetic alterations and plant responses to environmental stresses can lead to the development of stress-resistant plants (Penna et al., 2021; Iqbal et al., 2024). By integrating the latest understandings and data of epigenetics and genetics, it is possible to establish potential plants with beneficial genetic traits and stress tolerance potential (Mirouze and Paszkowski 2011; Magar et al., 2024). These advancements can be of great importance in deriving high plant yield under different types of environmental stresses.

27.4 EPIGENETIC MECHANISMS IN PLANTS

Dynamic alterations to DNA, RNA and histone proteins, along with regulations to chromatin structure and conformation, are the key factors of epigenetic mechanisms (Chinnusamy & Zhu 2009; Kotkar & Giri 2020; Gaude et al., 2024). Altogether, these changes impact gene expression without altering DNA sequences. Moreover, by modifying chromatin modifications, mRNA quantity and translation, non-coding RNAs govern gene expression (Figure 27.1). The stress-sensitive genes, genetic makeup and epigenetic markers play a crucial role in regulating environmental stresses in plants (Dinkar et al., 2024). Many reports are available on genetic regulation directly linked with stress-responsive genes. In fact, plants are naturally equipped with sensing and signaling processes to distinguish and react to different types of environmental stresses (Magar et al., 2024). Nowadays, it is well established that the epigenetic alterations in plants mediate innovative ways for tailoring plant reactions to tolerating and mitigating environmental stresses (Iqbal et al., 2024). Epigenetic alterations not only govern gene regulations under environmental stresses but also mediate a stress memory to tolerate potential future stresses (Magar et al., 2024). Reports reveal that despite lacking a nervous system, an efficient molecular network in plants helps them to recognize, memorize, select and impart decisions with respect to environmental stress type (Gallusci et al., 2023). In fact, epigenetic alterations in DNA and associated histone proteins can mediate short- and long-term memories in a crop or impart transgenerational impacts (Magar et al., 2024). In this way, plants are capable of transmitting their experiences to the next generations via epigenetic alterations that make up an efficient genetic inheritance (Gallusci et al., 2023).

Addition of 5-methylcytosine (5mC) is the most predominant type of DNA methylation in plants (Goud et al., 2024). It can take place in all cytosine sequence contexts. When concentrated at gene promoters or transposable elements, DNA methylation generally subdues transcription (Zhang et al., 2018). However, rarely, transcription is aided by DNA methylation (Ma et al., 2024). The fundamental chromatin subunit, nucleosomes, is formed when histone proteins bind to DNA. Numerous post-translational variations on histones, including phosphorylation, methylation and sumoylation, can modify chromatin activity. In this way, they influence gene transcription and define the chromatin functions (Millán-Zambrano et al., 2022).

Chromatin accessibility is dynamic and reacts to environmental stresses and endogenous stimuli. The alterations in RNA are considered another type of epigenetic regulation besides chromatin modifications. It is revealed that cellular RNAs can exhibit hundreds of distinctive modifications to date (Cappannini et al., 2024). N6-methyladenosine (m6A) and 5-methylcytosine (5mC), two of the most prevalent and thoroughly researched internal alterations on mRNA, have an impact on nearly every facet of mRNA metabolism. Recently, Aoyagi Blue and Satake (2022) revealed that BRU1/TSK/MGO3 causes chromatin modification and mediates the maintenance of genome integrity, meristem and chromatin inheritance state. SDE3 induces RNA silencing and is involved in antiviral defense via post-transcriptional gene silencing (Aoyagi Blue and Satake 2022). They reported differences in diverse epigenetic regulatory genes with enhanced copy numbers in long-lived trees (Gaude et al., 2024). Li et al. (2024) reported the role of ROXY19 in

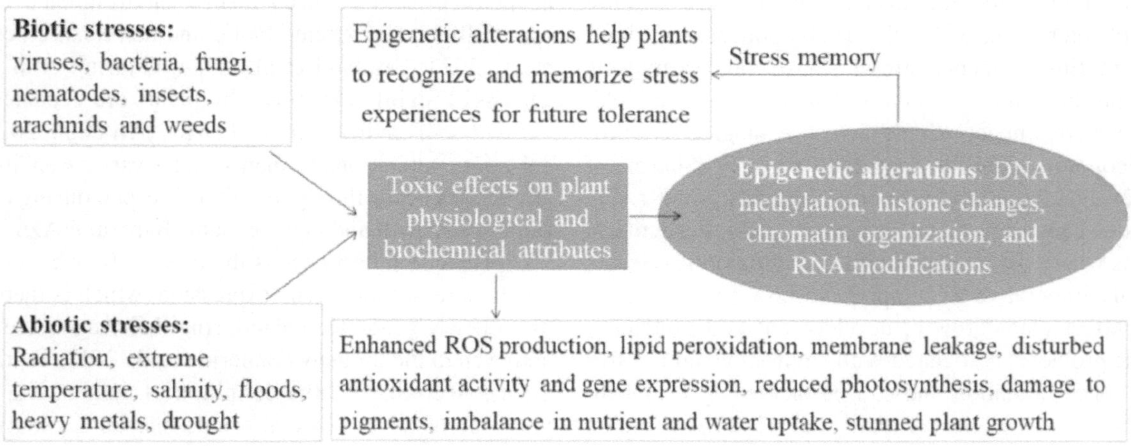

FIGURE 27.1 Possible epigenetic regulation of plants under biotic and abiotic stresses.

Arabidopsis under xenobiotic chemicals by silencing tolerance genes. Xu et al. (2021) reported the role of lysine acetylation for epigenetic regulations and chromatin functions.

Reports also revealed that cell redox status governs the enzymes and transcription factors associated with epigenetic alterations (Ramakrishnan et al., 2022). However, the unbalanced generation of reactive redox radicals such as ROS keeps alerting the cell redox status, mainly via oxidative stress. In this way, this cell redox status mediates epigenetic changes to alleviate the effect of environmental stresses and maintain plant yield (Ramakrishnan et al., 2022). The latest developments also revealed the role of different plant amendments, for example, Ahmad et al. (2023) delineated the role of melatonin in epigenetic mechanisms. They reported that melatonin can mediate epigenetic alterations and gene expression via methylation of DNA, alterations of histone and regulations of miRNA. In this way, melatonin can play a key role in tolerating environmental stresses by altering plant epigenetics (Ahmad et al., 2023).

27.5 EPIGENETIC REGULATION OF PLANTS UNDER ENVIRONMENTAL STRESS

27.5.1 Heat Stress

The main causes of enhanced global temperature are human activities such as greenhouse gas emissions and the burning of fossil fuels. Moreover, rapid industrialization, unbalanced urbanization and technological advancements are also contributing to climate change while regularly ignoring environmental sustainability (Foyer & Kranner 2023; Nadeem et al., 2024). The variations in temperature are the most obvious aspect of climate change. Therefore, the harms to crop output associated with these changes are inevitable.

Each plant has a range of ideal temperature where it can reach its full genetic potential (Haider et al., 2021). Heat stress is defined as temperature above their normal ranges. It is reported that a temperature rise of 1°C can decrease the production of *Triticum aestivum*, *Oryza sativa*, *Zea mays* and *Glycine max*, respectively, by 6%, 3%, 7% and 3% (Zhao et al., 2017). Temperature for heat stress varied for different species. Temperatures above 30°C, 29°C and 32°C during reproduction reduce yield, respectively, for *Glycine max*, *Zea mays* and *Gossypium herbaceum* (Gao et al., 2022). Heat stress is frequently caused by exposure to high temperatures above 34°C for *Oryza sativa* and above 32°C for *Triticum aestivum* (Gao et al., 2022). Heat stress in *Arabidopsis* is caused by brief exposure to temperature above 37°C.

The plant growth and development can be significantly impacted under high temperature. It is revealed that under high temperature, plants experience decreased photosynthetic activity, reduced production, protein damage, stunted growth and membrane fluidity (Kumari et al., 2024). Depending on the plant type and the intensity and duration of the stress, plant production may decrease from 10% to 40% or even more. High temperature mainly decreases crop production and seed development while hastening plant flowering and aging. Crop yield is predicted to decrease by 3% to 8% for every degree Celsius rise.

According to the data recorded, 2023 was the warmest year, with average temperatures 1.48°C higher than pre-industrial levels (Soares & Fidalgo 2024). The increased temperature and altered rainfall patterns have compromised food security and impacted water security. Therefore, it is becoming highly difficult to sustain the rapidly growing population under a changing climate. According to some predictions, the agricultural system needs to triple its output by 2030 to meet the growing food demand worldwide (Foyer & Kranner 2023).

Plants have developed acclimatization, adaptation and selective pressure to deal with biotic and abiotic stressors (Foyer & Kranner 2023). Plants are generally equipped with genetic and epigenetically controlled defenses against environmental stresses including heat stress (Ma et al., 2024; Shilpa et al., 2024). These mechanisms include upregulating stress-responsive genes and DNA methylation and histone changes (Verma et al., 2022). Under high temperature, chromatin regulators are capable of maintaining the speed of these gene networks. Furthermore, the significance and participation of chromatin regulators and histone alterations in plant stress memory have been demonstrated by genetic and biochemical understanding of plant responses to environmental stresses (Haider et al., 2021; Sharma et al., 2022).

Recently, it has been demonstrated that one of the most effective and efficient ways to govern plant survival against a variety of abiotic challenges, including heat stress, is through epigenetic regulators (Verma et al., 2022) (Table 27.1). Histone alterations and nucleosome occupancy are linked to heat stress memory at the somatic level, which is partially mediated by both chromatin-based and epigenetic modifications (Haider et al., 2021). The epigenetic control of stress-responsive gene expression is primarily arbitrated by histone acetyltransferase and histone deacetylases (HDACs) 3, 4, 5 and 6. Through the epigenetic modifications of gene expression, a number of HDACs have been demonstrated to regulate plant detoxification to high temperature, for example, HDA9 and HDA19 contribute positively to the plant response to high temperature. Furthermore, it was demonstrated that HD2B and HD2C react with ARGONAUTE 4 to enhance DNA methylation and sustain heterochromatin stability when exposed to high temperature (Chen et al., 2024). After being exposed to mild or non-lethal heat stress, many plants develop heat stress memory. This allows the plants to react more quickly to severe recurrent heat stress events, increasing their thermotolerance to even very high-temperature stress (Gao et al., 2022).

Plant growth and development can be severely inhibited and even killed by extremely high temperature that severely disturb cellular homeostasis. The architecture of chromatin is drastically altered by heat stress. Bao et al. (2025) delineated that transgenic *Solanum lycopersicum* showed

TABLE 27.1
Plant epigenetic modifications under different environmental stresses

Stress type	Stress effect	Plant	Epigenetic changes	Effect of epigenetic changes	References
Salt stress	WRKYs regulate tolerance against stresses	*Oryza sativa*	WRKY50 and WRKY72	The regulation of WRKY50 and WRKY72 varies under salinity stress. Results revealed that 83 and 72 methylated cytosines correlated, respectively, to WRKY72 and WRKY50	Çelik et al. (2019)
Heat stress	High temperature decreased the protein lysine acetylation	*Oryza sativa*	HDA714 protein	The enhanced expression of HDA714 improved *Oryza sativa* tolerance to high temperature. Overall, HDA714 deacetylates cellular proteins, controls metabolic activities and confers heat tolerance in *Oryza sativa*	Chen et al. (2024)
Heat stress	–	*Oryza sativa*	HDA714	*Oryza sativa* cytoplasmic histone deacetylase HDA714 controls translational activities and ribosomal protein lysine acetylation	Xu et al. (2021)
Chromium stress	Chromium exposure mediated significant changes in the gene expression profile, genomic DNA methylation profile and in gene expression	*Arabidopsis thaliana*	DNA methylation	F1 plants exposed to Cr stress modified several genes in their methylation status to maintain plant defenses. RNA processing, photorespiration, protein translation, DNA transcription and plant resilience were modulated by epigenetic mechanisms	Guarino et al. (2024)
Xenobiotic chemicals	Xenobiotic stress induces ROXY19 expression in *Arabidopsis*	*Arabidopsis thaliana*	Gene repressions, DNA methylations and histone deacetylations	ROXY19-TPL module induced DNA methylation-mediated gene repression and histone deacetylation. Methylation status was not modified in ROXY19OE	Li et al. (2024)
Cadmium	Stress caused disruption and decondensation of 45S rDNA chromatin	*Brassica rapa*	DNA methylation	Cd enhanced the levels of H4K5ac, H3K9ac and H3K9me2 in 45S rDNA. Individual site-specific hypomethylation activated 45 S rDNA transcription	Xiang et al. (2024)
Cadmium	The phytotoxicity of Cd resulted in a decrease in shoot biomass and caused DNA methylation	*Thymus vulgaris*	DNA hypomethylation	The application of nitric oxide and ascorbic acid increased threefold the unmethylation percentages	Farahani et al. (2024)
Drought stress	A multi-omics analysis delineated the response of *Camellia sinensis* under drought stress at various regulatory levels	*Camellia sinensis*	Transcriptomes, DNA methylomes and phosphoproteomes	Drought stress caused variations in genes responsive to phytohormone signaling and epigenetic modifications	Yue et al. (2023)
–	–	85 plant species	BRU1/TSK/MGO3	BRU1/TSK/MGO3 caused chromatin modifications. SDE3 induced RNA silencing and post-transcriptional gene silencing	Aoyagi Blue and Satake (2022)
Salinity	Salinity stress affected membrane stability, mitotic index and increased chromosomal abnormalities	*Vicia faba*	DNA methylation	DNA methyltransferases genes (*MT1 and MT2*) were upregulated eightfold in salinity stressed shoot	
Heat stress	Heat stress causes MdALKBH1A overexpression	*Solanum lycopersicum*	RNA demethylase gene, demethylase gene MdALKBH1A	Overexpressing MdALKBH1A improved plasma membrane stability	Bao et al. (2025)

(*Continued*)

TABLE 27.1 CONTINUED

Stress type	Stress effect	Plant	Epigenetic changes	Effect of epigenetic changes	References
Fixed-carbon starvation	Heat stress causes MdALKBH1A overexpression	*Solanum lycopersicum*	RNA demethylase gene, demethylase gene MdALKBH1A	Overexpressing MdALKBH1A increased photosynthetic rates, enhanced plasma membrane stability and decreased autophagic activity	Bao et al. (2025)
Salinity	Salinity-induced OsULT1 expression at transcription regulators	*Oryza sativa*	ULTRAPETALA1 (OsULT1)	Chromatin modifications governed the expressions of H3K27me3 and H3K4me3. OsULT1 was bound to AP2/ERF under salinity stress	Chakrabarty et al. (2024)
Mercury	Hg exposure induces oxidative stress and growth reduction	*Oryza sativa*	OsMET1–2: DNA methyltransferase-coding gene	Mercury-resistant line (RHg) in *Oryza sativa* delineated enhanced tolerance under Hg stress. There was amelioration in oxidative stress with less Hg phytouptake and aerial transfer. 34 mercury resistance-related genes were identified to be involved in phytohormone signaling	Cong et al. (2024)

an enhanced overexpression of MdALKBH1A under high temperature. They proposed that this sensitivity of *Solanum lycopersicum* to heat stress was correlated to lipid peroxidation and low antioxidant activity and membrane stability (Bao et al., 2025). Chen et al. (2024) reported the role of HDA714 to tolerate heat stress in *Oryza sativa*. Under high-temperature treatment, overexpression of HDA714 in *Oryza sativa* mediated deacetylation of metabolic enzymes. They revealed that HDA714 reacts to high temperature in *Oryza sativa* to govern metabolic activity, deacetylate cellular protein and stimulate stress granule development (Chen et al., 2024).

Crop species like rice and tomato have also been shown to exhibit 3D chromatin organization modifications brought on by heat stress (Ma et al., 2024). In response to heat stress, a heat stress transcription factor called HSFA1a helps in temporarily constructing enhancer–promoter chromatin loops in tomato, which in turn promotes the expression of genes that are sensitive to heat stress (Huang et al., 2023). Changes in chromatin accessibility are correlated with chromatin architectural changes brought on by heat stress in rice.

27.5.2 Cold Stress

Chilling (0–15°C) and freezing (<0°C) are examples of low temperature stress. While freezing results in the creation of ice crystals, which can permanently harm cellular function, chilling stress usually prevents plant growth and development. According to data from several agencies, cold periods in April 2024 seriously damaged European produce and vineyards. Significant financial losses resulted in 2024, which is regarded as one of the worst in the past century.

The CBF-COR signaling system controls plant responses to cold stress (Mukhopadhyay and Roychoudhury 2018; Shi et al., 2018). Membrane breakdown, electrolyte leakage and decreased enzyme activity are all consequences of cold stress, which alters cellular structures and functions (Nian et al., 2024). Based on the crop and the intensity of the stress episode, yield losses from cold stress might range from 10% to 40% or more. Plant physiology is also impacted by extremely low temperature, which include uncontrolled apoptosis, chlorosis, wilting, loss of membrane fluidity and photosynthesis impairments (Dev et al., 2025).

For reproduction to be successful, the developmental shift from the growth stage to the reproductive stage needs to coincide with temperature fluctuations. In order to set seeds under ideal temperature circumstances, many overwintering plants coordinate their flowering period with seasonal temperature variations during the physiological process (Gao et al., 2022). In one manner or another, and to varied degrees, plants adjust to cold. These processes are significantly influenced by epigenetic factors, primarily through different chromatin alterations (Bulgakov et al., 2024).

27.5.3 Drought Stress

Drought stress is a consequence of soil water deficiency, one of the negative impacts of climate change. A molecular approach was used in numerous investigations to determine the mechanisms underlying drought tolerance. Worldwide, drought has resulted in significant economic losses and had a great effect on plant growth and development (Yue et al., 2023). Limited water availability during drought stress causes changes in plant hormone levels, decreased stomatal conductance and impaired photosynthesis. Stunted growth, decreased leaf area and damaged root architecture are the results of these consequences, which impair plant growth and development (Rabara et al., 2021; Babić et al., 2024). Therefore, it has been predicted that by 2050, dryness will affect plant productivity in more than half of arable soil. More than 10% of fertile land is becoming arid due to global warming, which is also causing more than 50% less crop production in Middle Eastern nations.

The length, intensity, genotype and developmental stage of stress all affect how the plants react to drought stress. However, it is clear that the main effects of drought stress include slowing down the rates of growth and cell division. These changes result in smaller leaves, shorter stems and weaker root systems. Furthermore, changes in the physiochemical makeup and movement of xylem sap in dry soil impair nutrient uptake (Mansoor & Chung 2024). The density of essential oil-secreting glands and leaf area usually have an inverse relationship during drought. As the leaf area decreases, the number of glands increases, improving the quantity and quality of essential oil (Emami Bistgani et al., 2024). Additionally, earlier studies have demonstrated that the genes involved in essential oil transfer play a crucial role in the operation of these glands, with their activity changing in response to drought stress. As a result, new terpene compounds with distinctive molecular structures and medicinal qualities are produced (Emami Bistgani et al., 2024).

When plants are exposed to drought stress, the internal tolerance mechanisms of plants get activated. In majority of the plants, there exists a unique network comprising translation, transcription and epigenetic alterations that take place in nearly all cellular organelles such as the vacuoles, mitochondria, chloroplasts, cell wall, membrane system and peroxisomes (Babić et al., 2024; Emami Bistgani et al., 2024). These processes have been reported to modulate the plant response to drought stress. It has been shown that the drought stress response depends critically on calcium, abscisic acid and ROS. In order to maintain the biochemical and physiological responses in different cellular compartments and promote drought stress tolerance, these processes activate a large number of genes involved in stress tolerance and transcription factor regulation (Zhang et al., 2025).

Physiological (stomata closure, decreased photosynthesis), morphological (root elongation, early maturation), molecular (signal transduction, gene expression changes) and biochemical (ABA production, proline accumulation) mechanisms are all used by plants to withstand drought (Oguz et al., 2022; Zhang et al., 2025). Various studies have reported that wheat, rice, cotton, barley, maize and mung bean are among the crop species whose DNA methylation dynamics have been shown to be impacted by drought (Naderi et al., 2024).

Drought stress causes JMJ27 to become more abundant, which further decreases H3K9me2 at GOLS2 and RD20, enabling drought stress to fully activate them (Wang et al., 2021). In *Oryza sativa*, the activation of genes involved in drought stress response is positively correlated with drought-induced chromatin accessibility (Rajkumar et al., 2024). OsNMCP1 may promote chromatin accessibility by preventing the SWI/SNF complex from silencing genes during drought stress (Yang et al., 2020). Yue et al. (2023) reported the role of transcriptome, gene expression, DNA methylome, phosphoproteome and proteome in *Camellia sinensis* under drought stress.

27.5.4 Salt Stress

Because a high concentration of salt in the soil changes plant growth and production by mediating ion toxicity, metabolic and oxidative damages, and osmotic stress, excessive salinity poses a serious environmental risk to crops. According to the United Nations Food and Agriculture Organization, soil salinization causes up to 1.5 million hectares of farmland to cease production annually, resulting in a US$31 million loss in agricultural productivity annually (Food and Agriculture Organization of the United Nations, 2023). Since osmotic and salt stress are closely related, they will be discussed simultaneously. An effective salt stress response is based on a deep reprogramming of gene expression brought on by salt stress.

Ion balance and water absorption are disturbed by salinity stress. Reduced nutrient uptake, stunted growth, chlorosis and eventually decrease in production are the results of salinity-induced osmotic damage, ion toxicity and oxidative stress (Joshi et al., 2025). Depending on variables including crop sensitivity, irrigation management and soil salinity, the decrease in crop production can range greatly, from 5% to 50% or more. Ionic stress and osmotic stress are the two main stresses that high salt directly causes. These stresses set off their corresponding signaling and activate a number of ABA-dependent and salt overly sensitive pathways (Roychoudhury et al. 2013).

Ion toxicity, nutritional imbalance, hyperosmotic stress and oxidative stress that result in cell death are all toxic impacts of salinity stress on plant cells (Hao et al., 2021; Zhang et al., 2025). Depending on its length and intensity, salinity stress can significantly disrupt normal metabolic and physiological processes (Hao et al., 2021). Osmotic stress mediated by elevated salt concentration in soil decreases the potential of the root system to uptake water from the soil. This, in turn, causes a rise in cytoplasmic Na^+ concentration, which ultimately prevents the absorption of other vital ions like K^+ and disrupts a variety of metabolic processes (Hao et al., 2021). Furthermore, osmotic and ionic stress can result in secondary stress in plants by building up ROS (Huang et al., 2022).

Under salt stress, *Arabidopsis* accrues around 45% more differentially CG methylation sites compared to control at the epigenetic level (Jiang et al., 2014). In the mangrove *Bruguiera gymnorhiza*, high salinity conditions similarly cause DNA to become hypermethylated (Miryeganeh et al., 2022). Salt causes distinct DNA methylation modifications in *Oryza sativa*, *Triticum aestivum* and *Cicer arietinum* (Gupta & Garg 2023). Chakrabarty et al. (2024) showed that overexpression of OsULT1 under salinity stress was controlled by chromatin modifications including H3K27me3 and H3K4me3. Moreover, under salinity stress, OsULT1 was linked to AP2/ERF core motif A/GCCGAC as shown by DNA binding analysis. The overexpression of OsULT1 under salinity stress mediated various developmental alterations with increased height of plants and a greater number of panicles and tillers. Salinity stress was alleviated by OsULT1OE with decreased ROS level, increased seed germination and high post-stress rescue. Çelik et al. (2019) reported that the WRKY transcription factor family can regulate tolerance against environmental stresses, especially salt stress. The regulation of WRKY50 and WRKY72 in *Oryza sativa* varies under salinity stress. Results revealed that 83 and 72 methylated cytosines correlated, respectively, with the regulation of WRKY72 and WRKY50 (Çelik et al., 2019).

27.5.5 Heavy Metal Stress

In the periodic table, 53 of 90 naturally reported elements are categorized as heavy metals since they have a density greater than 5 g/cm^3. Since they are necessary for different plant metabolisms, 17 of these heavy metals are categorized as micronutrients of biological significance. Even these micronutrients may be detrimental to the growth and survival of the plant at larger quantities (Shahid et al., 2014). Plants do not require any of the other elements, such as mercury (Hg), cadmium (Cd), silver (Ag), chromium (Cr), lead (Pb) and aluminium (Al), for their normal metabolic activities and therefore show various toxicity symptoms upon exposure to these heavy metals.

New developments have improved the discharge of several heavy metals into the environment. Concerns over the sustainability of food crops and human health have been raised by some concerning reports of heavy metal environmental contamination around the world in recent years (Natasha et al., 2021; Tian et al., 2025). Several studies have identified various sources of heavy metals in the ecosystem, including vehicle gas emissions, mining, sewage sludge, improper industrial waste disposal, air deposition, agrochemical use, farmyard manure, waste incineration and the extensive use of synthetic agrochemicals (Sarwar et al., 2020).

One of the main abiotic stressors on plants that prevents them from growing is heavy metals. Heavy metals can have a variety of harmful consequences in plants because of their toxicity and persistence in the environment. The four heavy metals that have a particularly negative impact on humans and plants include Pb, Hg, Cd and As (Pourrut et al., 2011; Shahid et al., 2017). Long-term environmental exposure to Pb and Cd has been linked to an increased risk of all cancer deaths (Pourrut et al., 2011; Shahid et al., 2017). They could accumulate in the environment for a long time. Heavy metals from soil can accumulate in plant systems and pollute the food chain, raising major concerns about the quality and safety of food. Heavy metals are extremely stable and difficult to decompose. Agricultural soil sometimes includes high quantities of elements that are detrimental to plants, such as Ni, Hg, Cd, Cr, As, Cu, Zn and Pb. Across practically all pollution levels, Cd, Cr, As, Hg and Pb are the most dangerous and harmful to plant life.

When exposed to different types of heavy metals, the main reaction within plants is the production of ROS. In fact, almost all heavy metals are capable of over-generating ROS via various mechanisms, which can lead to oxidative damages (Shahid et al., 2014). In fact, different forms of ROS, especially hydroxide, can penetrate the membrane and thereby cause oxidative damage. ROS-induced lipid peroxidation is a highly reported effect following heavy metal exposure to plants (Chukwu & Gulser 2025). Numerous studies have been conducted in the past to better understand how plants react to heavy metal stress. Many physiological, morphological, cellular, biochemical and molecular processes related to heavy metal stress in plants have been better understood as a result of these investigations (Natasha et al., 2020; Natasha et al., 2022). Plants can tolerate heavy metal toxicity by modifying their absorption and root–shoot transfer, chelation, activating antioxidants and activating ion transporters. Additionally, a number of chemical and biological techniques have been used to mitigate the harmful impacts of heavy metals from contaminated water and land resources (Saini et al., 2021; Samanta and Roychoudhury 2021).

Heavy metals affect the diversity of soil bacteria by causing DNA damage, impeding protein synthesis and cell division, and denaturing proteins. Plant growth and production as well as metabolic processes are impacted by the inhibition or promotion of the activities of certain antioxidants such as catalase and glutathione peroxidase, which results in changes in physiological and biochemical markers (Chukwu & Gulser 2025). Additionally, heavy metals affect the aboveground components of plants, which usually show up as morphological symptoms, including height and color. Plant dwarfing, leaf deformation and color alterations are common. The physiological and biochemical markers of the aboveground portions of plants under heavy metal stress are impacted by incomplete reductase activity, reduced pigment concentration and suppression of photosynthesis (Zhai et al., 2023). The recent developments in the fields of functional genomics and genome sequencing have improved our understanding of epigenetic regulations under environmental stresses. The reports dealing with recognizing epigenetic alterations/markers under environmental stresses have advanced the knowledge about stress tolerance, establishing methods to edit the epigenome, and integrating epigenetic and breeding data for generating stress-tolerant plants (Iqbal et al., 2024).

By negatively impacting enzyme activity, protein structure and membrane integrity, heavy metals can interfere with plant metabolism. In *Arabidopsis* and rice, Cd stress raises total DNA methylation levels. Similar to DNA methylation, RNA modifications such as m6A are altered by Cd stress (Ma et al., 2024). Farahani et al. (2024) reported epigenetic DNA methylation in *Zataria multiflora* under Cd treatment supplemented with ascorbic acid and nitric oxide. There occurred 3.0-times higher unmethylation by amendments compared to control. Moreover, the transcription of *CYP71D178* and *CYP71D180* genes was increased. Overall, the results revealed the role of ascorbic acid and nitric oxide towards Cd detoxification by epigenetic DNA methylations, phenylpropanoid metabolism, enhancing proline levels and improving antioxidant activities (Farahani et al., 2024).

Similarly, Guarino et al. (2024) reported that treating *Arabidopsis thaliana* with Cr caused alterations in the gene expressions and genomic DNA methylation profile. Results showed Cr-mediated alterations in gene methylation status. These findings revealed that *Arabidopsis* exposed to Cr maintain their epigenetic status over generations via DNA methylation. Recently, Xiang et al. (2024) revealed molecular mechanisms linking Cd toxicity, rDNA epigenetic alterations and nucleolus disintegration in *Brassica rapa*. They showed that exposure of *Brassica rapa* to varied Cd levels induced decondensation of 45S rDNA chromatin, nucleolus disruption and an enhancement in 5′ external transcribed spacer region transcription. They also revealed modifications in histone methylation and acetylation under Cd stress (Xiang et al., 2024). Moreover, the expression levels of H4K5ac, H3K9ac and H3K9me2 increased in promoter regions of the 45S rDNA. It was proposed that hypomethylation causes the induction of 45S rDNA transcriptions (Xiang et al., 2024). Nevertheless, more research is needed to determine how RNA modification dynamics affect reactions to heavy metal stress. Cong et al. (2024) assessed the Hg-resistant line (RHg) in *Oryza sativa* obtained by a DNA methyltransferase-coding gene, *OsMET1–2* heterozygous mutant. Results revealed amelioration in oxidative stress with less Hg phytouptake and aerial transfer in *Oryza sativa*. Moreover, 34 Hg resistance-related genes were identified to be involved in phytohormone signaling.

27.6 CONCLUSIONS

It is evident that plants will continue to face different types of environmental stresses. Even the number, intensity and duration of these environmental stresses may increase in the future, keeping in view the rapid increase in world population and allied environmental changes. Numerous recent studies have shown that environmental stresses are capable of inducing physiological, molecular and biochemical alterations in various plant attributes, thereby compromising agricultural production. Therefore, plants have to cope with and tolerate these environmental stresses to maintain their growth, yield and production, a highly imperative factor for sustaining a rapidly growing population. Recently, numerous studies have delineated the role of different types of genetic and epigenetic modifications during environmental stresses on plants. It is well established that plants can mediate different types of epigenetic alterations such as histone changes and DNA methylation. These epigenetic alterations help plants to withstand environmental stresses and maintain their growth and production. However, these epigenetic alterations are yet to be fully understood for different environmental stresses and for different plant species. Moreover, the role of various exogenous applications such as plant hormones and organic and inorganic amendments in epigenetic mechanisms is not fully explained.

ACKNOWLEDGMENT

Professor Dr Jianxu Wang acknowledges ANSO project "Funded by CAS-ANSO Fellowship. Grant No. CAS-ANSO-FS-2024-06". Prof. Wang thanks Guizhou Provincial Major Scientific and Technological Program ([2025]002).

REFERENCES

Ahmad, N., Naeem, M., Ali, H., Alabbosh, K. F., Hussain, H., Khan, I., Siddiqui, S. A., Khan, A. A., & Iqbal, B. (2023). From challenges to solutions: The impact of melatonin on abiotic stress synergies in horticultural plants via redox regulation and epigenetic signaling. *Scientia Horticulturae,* 321, 112369.

Ali, S., Khan, N., & Tang, Y. (2022). Epigenetic marks for mitigating abiotic stresses in plants. *Journal of Plant Physiology*, 275, 153740.

Aoyagi Blue, Y., & Satake, A. (2022). Analyses of gene copy number variation in diverse epigenetic regulatory gene families across plants: Increased copy numbers of BRUSHY1/TONSOKU/MGOUN3 (BRU1/TSK/MGO3) and SILENCING DEFECTIVE 3 (SDE3) in long-lived trees. *Plant Gene*, 32, 100384.

Babić, I., Šamec, D., Hižak, M., Huđ, A., Senko, H., Šangut, I. J., Mlinarić, S., Petek, M., Palijan, G., Kolić, N. U., & Petrić, I. (2024). Evaluating drought impact on white cabbage: Plant stress response and soil microbiome adaptation. *Plant Stress*, 14, 100683.

Banerjee, A., & Roychoudhury, A. (2017). Epigenetic regulation during salinity and drought stress in plants: Histone modifications and DNA methylation. *Plant Gene*, 11, 199–204.

Banerjee, A., & Roychoudhury, A. (2018). The gymnastics of epigenomics in rice. *Plant Cell Reports*, 37, 25–49.

Banerjee, A., Wani, S. H., & Roychoudhury, A. (2017). Epigenetic control of plant cold responses. *Frontiers in Plant Science*, 8, 1643.

Bao, R., Guo, T., Yang, Z., Feng, C., Wu, J., Fu, X., Hu, L., Liu, C., & Ma, F. (2025). Overexpression of the apple m6A demethylase gene MdALKBH1A regulates resistance to heat stress and fixed-carbon starvation. *Journal of Integrative Agriculture*, 24(4), 1489–1502.

Barozai, M. Y. K., & Aziz, A. N. (2018). Recent plant growth and stress management related significant advancements in epigenetics. *Annals of Agrarian Science*, 16, 416–421.

Bulgakov, V. P., Fialko, A. V., & Yugay, Y. A. (2024). Involvement of epigenetic factors in flavonoid accumulation during plant cold adaptation. *Plant Physiology and Biochemistry*, 216, 109096.

Cappannini, A., Ray, A., Purta, E., Mukherjee, S., Boccaletto, P., Moafinejad, S. N., Lechner, A., Barchet, C., Klaholz, B. P., & Stefaniak, F. (2024). MODOMICS: A database of RNA modifications and related information. 2023 update. *Nucleic Acids Research*, 52, D239–D244.

Çelik, Ö., Meriç, S., Ayan, A., & Atak, Ç. (2019). Epigenetic analysis of WRKY transcription factor genes in salt stressed rice (*Oryza sativa* L.) plants. *Environmental and Experimental Botany*, 159, 121–131.

Chakrabarty, J., Parveen, R., Datta, S., Ghosh, B., Roy, V., Ghosh, Z., & Chaudhuri, S. (2024). ULTRAPETALA 1 regulates the growth and development of rice plants to promote resilience to salinity stress. *Environmental and Experimental Botany*, 223, 105780.

Chen, Z., Xu, Q., Wang, J., Zhao, H., Yue, Y., Liu, B., Xiong, L., Zhao, Y., & Zhou, D.-X. (2024). A histone deacetylase confers plant tolerance to heat stress by controlling protein lysine deacetylation and stress granule formation in rice. *Cell Reports*, 43, 114642.

Chinnusamy, V., & Zhu, J.-K. (2009). Epigenetic regulation of stress responses in plants. *Current Opinion in Plant Biology*, 12, 133–139.

Chukwu, E. C., & Gulser, C. (2025). Morphological, physiological, and anatomical effects of heavy metals on soil and plant health and possible remediation technologies. *Soil Security*, 18, 100178.

Cong, W., Li, N., Miao, Y., Huang, Y., Zhao, W., Kang, Y., Zhang, B., Wang, J., Zhang, J., Lv, Y., Li, J., Zhang, J., Gong, L., Liu, B., & Ou, X. (2024). DNA hypomethylation-associated transcriptional rewiring enables resistance to heavy metal mercury (Hg) stress in rice. *Journal of Hazardous Materials*, 461, 132649.

Das, K., & Roychoudhury, A. (2014). Reactive oxygen species (ROS) and response of antioxidants as ROS-scavengers during environmental stress in plants. *Frontiers in Environmental Science*, December; 2, 53.

Dev, W., Sultana, F., Li, H., Hu, D., Peng, Z., He, S., Zhang, H., Waqas, M., Geng, X., & Du, X. (2025). Molecular mechanisms of cold stress response in cotton: Transcriptional reprogramming and genetic strategies for tolerance. *Plant Science*, 352, 112390.

Dinkar, V., Pandey, S., Kumar, A., Shiv, A., Lal, D., Bharati, A., Joshi, A., Adhikari, S., Aparna, Singh, A., Pandey, P. K., & Chandra, A. K. (2024). Epigenetic regulations under plant stress: A cereals perspective. *Environmental and Experimental Botany*, 220, 105688.

Ejaz, U., Khan, S. M., Shah, S. F. A., Khalid, N., Jehangir, S., Rizvi, Z. F., & Svenning, J.-C. (2025). Integrative data-driven analytics for assessing ecological and human health risks of soil heavy metal contamination. *Journal of Hazardous Materials Advances*, 17, 100596.

Emami Bistgani, Z., Barker, A. V., & Hashemi, M. (2024). Physiology of medicinal and aromatic plants under drought stress. *The Crop Journal*, 12, 330–339.

Faizan, M., Sharma, P., Sultan, H., Alam, P., Sehar, S., Rajput, V. D., & Hayat, S. (2024). Nano-priming: Improving plant nutrition to support the establishment of sustainable agriculture under heavy metal stress. *Plant Nano Biology*, 10, 100096.

Farahani, F., Iranbakhsh, A., Ebadi, M., Oraghi Ardebili, Z., & Haghighat, S. (2024). Nitric oxide and ascorbic acid confer cadmium (Cd) tolerance by improving plant terpenoid metabolism and epigenetically modifying DNA methylation. *Environmental Pollution*, 362, 124917.

Foyer, C. H., & Kranner, I. (2023). Plant adaptation to climate change. In *Vol 480* (pp 1865–1869). Portland Press Ltd. Biochem J. 480(22):1865–1869.

Gallusci, P., Agius, D. R., Moschou, P. N., Dobránszki, J., Kaiserli, E., & Martinelli, F. (2023). Deep inside the epigenetic memories of stressed plants. *Trends in Plant Science*, 28, 142–153.

Gao, Z., Zhou, Y., & He, Y. (2022). Molecular epigenetic mechanisms for the memory of temperature stresses in plants. *Journal of Genetics and Genomics*, 49, 991–1001.

Gaude, A. A., Siqueira, R. H., Botelho, S. B., & Jalmi, S. K. (2024). Epigenetic arsenal for stress mitigation in plants. *Biochimica et Biophysica Acta (BBA) – General Subjects*, 1868, 130620.

Goud, V. K., Goud, A. C., Ramassamy, S., Jayanthi, M., Medha, R., & Chandrashekar, L. (2024). Whole blood global DNA methylation [5-methylcytosine (5-mC)] levels in psoriatic patients before and after methotrexate treatment. *Gene Reports*, 37, 102083.

Guarino, F., Cicatelli, A., Nissim, W. G., Colzi, I., Gonnelli, C., Basso, M. F., Vergata, C., Contaldi, F., Martinelli, F., & Castiglione, S. (2024). Epigenetic changes induced by chronic and acute chromium stress treatments in *Arabidopsis thaliana* identified by the MSAP-Seq. *Chemosphere*, 362, 142642.

Gupta, K., & Garg, R. (2023). Unravelling differential DNA methylation patterns in genotype dependent manner under salinity stress response in chickpea. *International Journal of Molecular Sciences*, 24, 1863.

Haider, S., Iqbal, J., Shaukat, M., Naseer, S., & Mahmood, T. (2021). The epigenetic chromatin-based regulation of somatic heat stress memory in plants. *Plant Gene*, 27, 100318.

Hao, S., Wang, Y., Yan, Y., Liu, Y., Wang, J., & Chen, S. (2021). A review on plant responses to salt stress and their mechanisms of salt resistance. *Horticulturae*, 7, 132.

Huang, X., Tanveer, M., Min, Y., & Shabala, S. (2022). Melatonin as a regulator of plant ionic homeostasis: Implications for abiotic stress tolerance. *Journal of Experimental Botany*, 73, 5886–5902.

Huang, Y., An, J., Sircar, S., Bergis, C., Lopes, C. D., He, X., Da Costa, B., Tan, F.-Q., Bazin, J., & Antunez-Sanchez, J. (2023). HSFA1a modulates plant heat stress responses and alters the 3D chromatin organization of enhancer-promoter interactions. *Nature Communications*, 14, 469.

Iqbal, B., Ahmad, N., Li, G., Jalal, A., Khan, A. R., Zheng, X., Naeem, M., & Du, D. (2024). Unlocking plant resilience: Advanced epigenetic strategies against heavy metal and metalloid stress. *Plant Science*, 349, 112265.

Jiang, C., Mithani, A., Belfield, E. J., Mott, R., Hurst, L. D., & Harberd, N. P. (2014). Environmentally responsive genome-wide accumulation of de novo *Arabidopsis thaliana* mutations and epimutations. *Genome Research*, 24, 1821–1829.

Joshi, P. S., Singla Pareek, S. L., & Pareek, A. (2025). Shaping resilience: The critical role of plant response regulators in salinity stress. *Biochimica et Biophysica Acta (BBA) – General Subjects*, 1869, 130749.

Khalid, S., Shahid, M., Bibi, I., Nadeem, H. M., Younis, M., Natasha, N., Murtaza, B., & Niazi, N. K. (2024). Assessing the arsenic contents and associated risks in groundwater of Vehari and Lodhran districts, Pakistan. *Water*, 16, 3055.

Khalid, S., Shahid, M., Niazi, N. K., Murtaza, B., Bibi, I., & Dumat, C. (2017). A comparison of technologies for remediation of heavy metal contaminated soils. *Journal of Geochemical Exploration*, 182, 247–268.

Kotkar, H., & Giri, A. (2020). Chapter 2 - Plant epigenetics and the 'intelligent' priming system to combat biotic stress. In D. Kabelitz & J. Bhat (Eds.), *Epigenetics of the Immune System* (Vol 16, pp. 25–38). London: Academic Press.

Kumari, S., Basu, S., & Kumar, G. (2024). A systematic review on the implications of concurrent heat and drought stress in modulating floral development in plants. *Plant Science*, 349, 112248.

Li, N., Yu, P., Zeng, Y., Chen, J., Yang, W., Qin, G., Du, S., Han, X., & Huang, L.-J. (2024). Characterization of epigenetic modifications in a plant-specific glutaredoxin-mediated repression of stress-responsive gene expression. *Environmental and Experimental Botany,* 217, 105582.

Ma, L., Xing, L., Li, Z., & Jiang, D. (2024). Epigenetic control of plant abiotic stress responses. *Journal of Genetics and Genomics.* 52(2), 129–144.

Magar, N. D., Shah, P., Barbadikar, K. M., Bosamia, T. C., Madhav, M. S., Mangrauthia, S. K., Pandey, M. K., Sharma, S., Shanker, A. K., Neeraja, C. N., & Sundaram, R. M. (2024). Long non-coding RNA-mediated epigenetic response for abiotic stress tolerance in plants. *Plant Physiology and Biochemistry*, 206, 108165.

Mansoor, S., & Chung, Y. S. (2024). Functional phenotyping: Understanding the dynamic response of plants to drought stress. *Current Plant Biology*, 38, 100331.

Millán-Zambrano, G., Burton, A., Bannister, A. J., & Schneider, R. (2022). Histone post-translational modifications – cause and consequence of genome function. *Nature Reviews Genetics*, 23, 563–580.

Mirouze, M., & Paszkowski, J. (2011). Epigenetic contribution to stress adaptation in plants. *Current Opinion in Plant Biology*, 14, 267–274.

Miryeganeh, M., Marlétaz, F., Gavriouchkina, D., & Saze, H. (2022). De novo genome assembly and in natura epigenomics reveal salinity-induced DNA methylation in the mangrove tree *Bruguiera gymnorhiza. New Phytologist*, 233, 2094–2110.

Mukhopadhyay, J., & Roychoudhury, A. (2018). Cold-induced injuries and signaling responses in plants. In S. H. Wani & V. Herath (Eds.), *Cold Tolerance in Plants Physiological, Molecular and Genetic Perspectives* (pp. 1–35). Cham: Springer Nature Switzerland.

Nadeem, H., Ashraf, M., Rasool, G., Shflot, A. S., & Malik, M. Y. (2024). Thermophoretic convection in porous atmosphere due to boosting temperature of plume: Climate change effects. *Case Studies in Thermal Engineering*, 64, 105537.

Naderi, S., Maali-Amiri, R., Sadeghi, L., & Hamidi, A. (2024). Physio-biochemical and DNA methylation analysis of the defense response network of wheat to drought stress. *Plant Physiology and Biochemistry*, 209, 108516.

Natasha, Shahid, M., Khalid, S., Bibi, I., Bundschuh, J., Niazi, N. K., & Dumat, C. (2020). A critical review of mercury speciation, bioavailability, toxicity and detoxification in soil-plant environment: Ecotoxicology and health risk assessment. *Science of The Total Environment*, 711, 134749.

Natasha, Shahid, M., Khalid, S., Dumat, C., Pierart, A., & Niazi, N. K. (2019). Biogeochemistry of antimony in soil-plant system: Ecotoxicology and human health. *Applied Geochemistry*, 106, 45–59.

Natasha, N., Shahid, M., Bibi, I., Iqbal, J., Khalid, S., Murtaza, B., Bakhat, H. F., Farooq, A. B. U., Amjad, M., Hammad, H. M., & Niazi, N. K. (2022). Zinc in soil-plant-human system: A data-analysis review. *Science of the Total Environment*, 808, 152024.

Natasha, N., Shahid, M , Khalid, S., Niazi, N. K., Murtaza, B., Ahmad, N., Farooq, A., Zakir, A., Imran, M., & Abbas, G. (2021). Health risks of arsenic buildup in soil and food crops after wastewater irrigation. *Science of The Total Environment*, 772, 145266.

Nian, Y., Aslam, M. M., Wang, X., Gu, H., Li, W., & Shao, Y. (2024). The *CpCOR1* gene enhances cold tolerance and antioxidant activity of papaya fruit in response to postharvest chilling stress. *Postharvest Biology and Technology*, 218, 113154.

Oguz, M. C., Aycan, M., Oguz, E., Poyraz, I., & Yildiz, M. (2022). Drought stress tolerance in plants: Interplay of molecular, biochemical and physiological responses in important development stages. *Physiologia*, 2, 180–197.

Penna, S., Negi, P., & Kumar, V. (2021). Chapter 5 – Genomics and epigenetics of plant abiotic stress. In G. Fink (Ed.), *Stress: Genetics, Epigenetics and Genomics* (pp. 49–56). London: Academic Press.

Pourrut, B., Shahid, M., Dumat, C., Winterton, P., & Pinelli, E. (2011). Lead uptake, toxicity, and detoxification in plants. *Reviews of Environmental Contamination and Toxicology Volume*, 213, 113–136.

Rabara, R., Msanne, J., Basu, S., Ferrer, M., & Roychoudhury, A. (2021). Coping with inclement weather conditions due to high temperature and water deficit in rice: An insight from genetic and biochemical perspectives. *Physiologia Plantarum*, 172, 487–504.

Rafiq, M., Shahid, M., Bibi, I., Khalid, S., Tariq, T. Z., Al-Kahtani, A. A., ZA, A. L., Murtaza, B., & Niazi, N. K. (2024). Role of organic and inorganic amendments on physiological attributes of germinating pea seedlings under arsenic stress. *International Journal of Phytoremediation*, 26, 1243–1252.

Rajkumar, M. S., Tembhare, K., Garg, R., & Jain, M. (2024). Genome-wide mapping of DNase I hypersensitive sites revealed differential chromatin accessibility and regulatory DNA elements under drought stress in rice cultivars. *The Plant Journal*. 119(4), 2063–2079.

Ramakrishnan, M., Papolu, P. K., Satish, L., Vinod, K. K., Wei, Q., Sharma, A., Emamverdian, A., Zou, L.-H., & Zhou, M. (2022). Redox status of the plant cell determines epigenetic modifications under abiotic stress conditions and during developmental processes. *Journal of Advanced Research*, 42, 99–116.

Rehman, M., & Tanti, B. (2020). Understanding epigenetic modifications in response to abiotic stresses in plants. *Biocatalysis and Agricultural Biotechnology*, 27, 101673.

Roychoudhury, A., Paul, S., & Basu, S. (2013). Cross-talk between abscisic acid-dependent and abscisic acid-independent pathways during abiotic stress. *Plant Cell Reports*, 32(7), 985–1006.

Saini, S., Kaur, N., & Pati, P. K. (2021). Phytohormones: Key players in the modulation of heavy metal stress tolerance in plants. *Ecotoxicology and Environmental Safety*, 223, 112578.

Samanta, S., & Roychoudhury, A. (2021). Heavy metal toxicity in plants: Physiological and molecular adaptations. In T. Aftab & K. R. Hakeem (Eds.), *Heavy Metal Toxicity in Plants* (pp. 1–9). Boca Raton: CRC Press Taylor & Francis Group.

Sarwar, T., Shahid, M., Khalid, S., Shah, A. H., Ahmad, N., Naeem, M. A., ul Haq, Z., Murtaza, B., & Bakhat, H. F. (2020). Quantification and risk assessment of heavy metal build-up in soil–plant system after irrigation with untreated city wastewater in Vehari, Pakistan. *Environmental Geochemistry and Health*, 42, 4281–4297.

Shahid, M. (2021). Effect of soil amendments on trace element-mediated oxidative stress in plants: Meta-analysis and mechanistic interpretations. *Journal of Hazardous Materials*, *407*, 124881.

Shahid, M., Al-Kahtani, A. A., ZA, A. L., Bibi, I., Tariq, T. Z., Natasha, N., Khalid, S., Niazi, N. K., & Murtaza, B. (2023). Effect of freshwater and wastewater irrigation on buildup of toxic elements in soil and maize crop. *Environmental Monitoring and Assessment*, 195, 438.

Shahid, M., Dumat, C., Khalid, S., Niazi, N. K., & Antunes, P. M. (2017). Cadmium bioavailability, uptake, toxicity and detoxification in soil-plant system. *Reviews of Environmental Contamination and Toxicology*, 241, 73–137.

Shahid, M., Natasha, Dumat, C., Niazi, N. K., Xiong, T. T., Farooq, A. B. U., & Khalid, S. (2021). Ecotoxicology of heavy metal (loid)-enriched particulate matter: Foliar accumulation by plants and health impacts. *Reviews of Environmental Contamination and Toxicology*, 253, 65–113.

Shahid, M., Pinelli, E., & Dumat, C. (2018). Tracing trends in plant physiology and biochemistry: Need of databases from genetic to kingdom level. *Plant Physiology and Biochemistry*, 127, 630–635.

Shahid, M., Pourrut, B., Dumat, C., Nadeem, M., Aslam, M., & Pinelli, E. (2014). Heavy-metal-induced reactive oxygen species: Phytotoxicity and physicochemical changes in plants. *Reviews of Environmental Contamination and Toxicology*, 232, 1–44.

Sharma, S., Chauhan, A., Dobbal, S., & Kumar, R. (2022). Biology of plants coping stresses: Epigenetic modifications and genetic engineering. *South African Journal of Botany*, 144, 270–283.

Shi, Y., Ding, Y., & Yang, S. (2018). Molecular regulation of CBF signaling in cold acclimation. *Trends in Plant Science*, 23, 623–637.

Shiade, S. R. G., Zand-Silakhoor, A., Fathi, A., Rahimi, R., Minkina, T., Rajput, V. D., Zulfiqar, U., & Chaudhary, T. (2024). Plant metabolites and signaling pathways in response to biotic and abiotic stresses: Exploring bio stimulant applications. *Plant Stress*, 12, 100454.

Shilpa, Thakur, R., & Prasad, P. (2024). Epigenetic regulation of abiotic stress responses in plants. *Biochimica et Biophysica Acta (BBA) – General Subjects*, 1868, 130661.

Soares, C., & Fidalgo, F. (2024). Tackling plant responses towards abiotic stress combinations – A focus on redox regulation and epigenetic mechanisms. *Plant Stress*, 13, 100550.

Tavakoli, F., Hajiboland, R., Haeili, M., Sadeghzadeh, N., & Nikolic, M. (2025). Effect of elevated ammonium on biotic and abiotic stress defense responses and expression of related genes in cucumber (*Cucumis sativus* L.) plants. *Plant Physiology and Biochemistry*, 218, 109310.

Tian, W., Wang, J., Man, Y., Anderson, C. W., & Feng, X. (2025). Novel insights into HgO oxidation in rice leaf: Catalase functions and transcriptome responses. *Environmental Science & Technology*. 59(1).

Van Dijk, M., Morley, T., Rau, M. L., & Saghai, Y. (2021). A meta-analysis of projected global food demand and population at risk of hunger for the period 2010–2050. *Nature Food*, 2, 494–501.

Verma, N., Giri, S. K., Singh, G., Gill, R., & Kumar, A. (2022). Epigenetic regulation of heat and cold stress responses in crop plants. *Plant Gene*, 29, 100351.

Wang, Q., Liu, P., Jing, H., Zhou, X. F., Zhao, B., Li, Y., & Jin, J. B. (2021). JMJ27-mediated histone H3K9 demethylation positively regulates drought-stress responses in *Arabidopsis*. *New Phytologist*, 232, 221–236.

Xiang, Y., Zhang, M., Hu, Y., Wang, L., Xiao, X., Yin, F., Cao, X., Sui, M., & Yao, Y. (2024). Epigenetic modifications of 45S rDNA associates with the disruption of nucleolar organisation during Cd stress response in Pakchoi. *Ecotoxicology and Environmental Safety*, 270, 115859.

Xu, Q., Liu, Q., Chen, Z., Yue, Y., Liu, Y., Zhao, Y., & Zhou, D.-X. (2021). Histone deacetylases control lysine acetylation of ribosomal proteins in rice. *Nucleic Acids Research*, 49, 4613–4628.

Yajnik, K. N., Singh, I. K., & Singh, A. (2024). lncRNAs and epigenetics regulate plant's resilience against biotic stresses. *Plant Physiology and Biochemistry*, 214, 108892.

Yang, J., Chang, Y., Qin, Y., Chen, D., Zhu, T., Peng, K., Wang, H., Tang, N., Li, X., & Wang, Y. (2020). A lamin-like protein OsNMCP1 regulates drought resistance and root growth through chromatin accessibility modulation by interacting with a chromatin remodeller OsSWI3C in rice. *New Phytologist*, 227, 65–83.

Yue, C., Cao, H., Zhang, S., Shen, G., Wu, Z., Yuan, L., Luo, L., & Zeng, L. (2023). Multilayer omics landscape analyses reveal the regulatory responses of tea plants to drought stress. *International Journal of Biological Macromolecules*, 253, 126582.

Zhai, Y., Zhou, L., Qi, H., Gao, P., & Zhang, C. (2023). Application of visible/near-infrared spectroscopy and hyperspectral imaging with machine learning for high-throughput plant heavy metal stress phenotyping: A review. *Plant Phenomics*, 5, 0124.

Zhang, H., Lang, Z., & Zhu, J.-K. (2018). Dynamics and function of DNA methylation in plants. *Nature Reviews Molecular Cell Biology*, 19, 489–506.

Zhang, S., Zheng, D., Gao, Y., She, M., Wu, Z., Lu, Y., & Zhang, Z. (2025). The TIFY transcription factor ZmJAZ13 enhances plant tolerance to drought and salt stress by interacting with ZmbHLH161 and ZmA0A1D6GLB9. *Plant Science*, 352, 112388.

Zhao, C., Liu, B., Piao, S., Wang, X., Lobell, D. B., Huang, Y., Huang, M., Yao, Y., Bassu, S., & Ciais, P. (2017). Temperature increase reduces global yields of major crops in four independent estimates. *Proceedings of the National Academy of Sciences,* 114, 9326–9331.

28 Unraveling the Prospects of Epigenetic Memory
Insights into the Resilience against Abiotic Stressors

Dibakar Ghosh and Swarnendu Roy

28.1 INTRODUCTION

Abiotic stressors represent a significant challenge to global agricultural systems, causing substantial crop losses and threatening food security, particularly in the face of climate change, escalating the frequency and severity of these challenges (Shahzad et al., 2021). Addressing these issues is essential for achieving several United Nations Sustainable Development Goals (SDGs), particularly Goal 2 (Zero Hunger) and Goal 13 (Climate Action) (Atukunda et al., 2021). A compelling and innovative strategy to mitigate crop losses is the application of epigenetic memory, enhancing resilience against environmental stresses and helping to sustain yield while reducing reliance on agrochemicals (Flors et al., 2024). By integrating epigenetic principles into modern crop improvement strategies, we can make substantial strides toward sustainable agriculture and food security for a growing global population (Komen et al., 2020). The ability of plants to endure and adapt to abiotic stresses is vital for their survival, growth and agricultural productivity (Kumari et al., 2022). With climate change introducing heightened challenges such as drought, salinity, extreme temperature and heavy metal toxicity, there is an increasing scientific focus on understanding the mechanisms that drive plant resilience (Iqbal et al., 2024). While traditional breeding methods and genetic modification have been used to boost stress tolerance, recent research highlights the transformative potential of epigenetic memory in enhancing plant resilience to abiotic stressors (Jin et al., 2024).

Plants and humans have distinct methods for storing memories, yet they share similar epigenetic mechanisms. Although plants lack a nervous system, they can "remember" stress through epigenetic modifications such as DNA methylation and histone alterations, primarily occurring in their roots and leaves (Pissolato et al., 2024). The roots function as memory hubs, retaining stress responses that shape future growth, while the leaves absorb environmental cues, like drought conditions, triggering systemic responses. This memory enables plants to control gene expression and enhance resilience (De Bang et al., 2021). In contrast, humans store memories in the brain, relying on neuronal plasticity, synaptic strength and epigenetic regulation (Maity et al., 2022). Human neurons employ DNA methylation and histone modifications to stabilize long-term memory (Gallusci et al., 2023). Both systems also experience a form of "forgetting" over time, as epigenetic markers gradually diminish. Stress can impact memory in both plants and humans; for example, drought conditions prime plant roots, while chronic stress can impair human cognition (Echeverria et al., 2021). Despite their differences, the epigenetic memory found in plant roots and leaves, as well as in human neurons, illustrates a shared adaptive strategy for survival and learning (Liu et al., 2022).

Epigenetics encompasses heritable changes in gene expression that occur independently of DNA sequence alterations, involving mechanisms such as DNA methylation, histone modifications and non-coding RNAs. These processes fundamentally shape chromatin architecture and gene activity (Bure et al., 2022). Epigenetic memory specifically defines how previous environmental experiences inform future stress responses through stable epigenetic marks, allowing plants to react more swiftly and effectively to similar stressors, thereby promoting resilience across generations and within the lifetime of an individual plant (Fabrizio et al., 2019; Mertens et al., 2021). The growing acknowledgment of the importance of epigenetic memory in plant biology underscores its role in adapting to abiotic stress (Vaschetto, 2024). Empirical evidence shows that plants experiencing abiotic stress undergo epigenetic changes that endure even after stress exposure (Kakoulidou et al., 2021). This lasting epigenetic reprogramming creates a "priming" effect, enhancing anticipatory responses to future stress events (Deans, 2024), for instance, studies on *Arabidopsis* and several other model organisms demonstrate that drought-induced DNA methylation patterns can be reliably inherited across generations, significantly improving the tolerance of the progeny to water scarcity (Banerjee & Roychoudhury, 2018; Sun et al., 2021).

A classic instance of epigenetic memory is vernalization, where prolonged exposure to cold temperature prompts epigenetic modifications that initiate flowering, revealing

DOI: 10.1201/9781003589914-32

how plants harness epigenetic regulation to interpret environmental signals and optimize developmental processes (Kakoulidou et al., 2021). Similar responses occur with exposure to salinity, heat and oxidative stress, leading to changes in histone acetylation and methylation patterns that activate stress-responsive genes (Rashid et al., 2022). These epigenetic transformations not only enhance immediate stress responses but also establish an adaptive memory that can be passed on to future generations (Ambeskovic et al., 2019). Exploiting epigenetic memory to develop stress-resilient crops presents a unique opportunity for sustainable agricultural practices. Traditional breeding methods can be protracted and can yield unpredictable results due to complex genetic interactions. In contrast, epigenetic modifications can be swiftly activated in response to environmental cues without requiring direct genetic alterations, offering an effective strategy to enhance plant adaptability (Jin et al., 2024). Furthermore, a comprehensive exploration of the epigenetic landscape in crop species will illuminate the gene regulatory networks implicated in stress tolerance, enabling targeted interventions to improve agricultural productivity under challenging conditions (Ahmed et al., 2023). Despite the immense promise of epigenetic memory, several challenges remain to be addressed. The stability and heritability of epigenetic modifications in field environment are critical research areas (Amaral et al., 2020). While certain epigenetic markers demonstrate transgenerational stability, others may reset during reproductive cycles, potentially limiting their long-term effectiveness (Xavier et al., 2019). Additionally, the interactions between genetic and epigenetic factors in stress response mechanisms must be further examined to optimize the application of epigenetic modifications for crop improvement (Samantara et al., 2021). Advances in omics technologies, such as high-throughput sequencing and epigenome editing, provide powerful tools to decode the molecular foundations of epigenetic memory and its vital role in plant adaptation (Mahmood et al., 2022).

This chapter aims to explore the prospects of epigenetic memory in enhancing resilience against abiotic stressors. First, the fundamental mechanisms of epigenetic regulation have been discussed, and key findings on epigenetic memory in various plant species and their potential applications in agriculture have been highlighted. By unraveling the intricate relationship between epigenetic modifications and plant stress responses, we can pave the way for innovative strategies to bolster crop resilience and ensure global food security in the face of environmental challenges.

28.2 TYPES OF EPIGENETIC MEMORIES

Epigenetic memories can be categorized based on the longevity and the response time of the plants after priming. Plants can have somatic (or intra-generational) memories of stress that last for varying periods within the plant generation following the events of immediate stress response. However, intergenerational memories persist in the next generation of exposed progeny of plants, and transgenerational memory is transmitted to further generations even in the absence of stress (Srivastava et al., 2021).

Several molecular mechanisms contribute to the somatic memory of abiotic stress. This memory is associated with the persistence of stress-induced metabolites, the stalling of RNA polymerase II, which limits transcription, and the accumulation of proteins such as mitogen-activated protein kinases (MAPKs) (Mozgova et al., 2019). Additionally, it involves the stable maintenance of stress-induced changes in chromatin during mitosis (Mognato et al., 2021). Chromatin-based transcriptional memory can be classified into two types based on the transcription levels of stress response genes. Type I refers to the scenario where the transcriptional activity of stress-responsive genes continues, while Type II describes a situation in which the initial stress-induced transcription ceases. However, upon a second exposure to stress, a modified response can be triggered compared to infant plants (Bäurle, 2018). However, the somatic memory of abiotic stresses typically lasts only a few days to weeks (Oberkofler et al., 2021).

Key chromatin-based mechanisms involved in developing intra-generational epigenetic memory include nucleosome occupancy and remodeling, histone post-translational modifications (PTMs), DNA methylation and RNA interference. The molecular aspects of somatic memory are primarily studied in the short-lived annual plant *Arabidopsis thaliana*, although they are likely to be significant in long-lived perennial species as well (Auge et al., 2023). Evidence for this is seen in the persistence of DNA methylation changes observed in the shoot apical meristem (SAM) of *Populus* species grown under different water conditions (Sow et al., 2020). Furthermore, winter-dormant SAMs from trees grown in varying environment exhibit differentially methylated regions in genes related to abiotic stress responses, SAM organization and phytohormone signaling. This suggests that growth conditions during the vegetative phase can influence future organ development and overall tree performance (Sow et al., 2020). Notably, global DNA methylation changes occur during bud dormancy and breaking, influenced by DNA demethylases (Kumar et al. 2016), indicating the process of active reprogramming. However, it remains unclear whether environmentally induced epialleles (alleles that exhibit different epigenetic or chromatin states at a particular locus) can evade global DNA methylation reprogramming (Angers et al., 2010).

Contrarily, transgenerational stress memory can potentially be mediated by the transmission of structural variations in the genome, the inheritance of epialleles and maternal effects, which involve seed provisioning. This provisioning can vary based on environmental conditions during the growth of the maternal plant and includes different levels of resources such as mRNA, hormones, proteins, starch, lipids and other reserve molecules stored in the seed (Kaur et al., 2021). Separating maternal effects from the inheritance of epialleles poses challenges, leading to debates about the transgenerational inheritance of acquired epialleles as a form of environmental memory and its adaptive significance (Fallet, 2024). Nonetheless, when considering maternal effects and possible induced structural variations, intergenerational memory mechanisms can contribute to adaptive transgenerational plasticity and rapid environmental adaptation in plants (Auge et al., 2023). The transmission of acquired epialleles between generations is typically prevented by the active resetting of chromatin states during sexual reproduction (Bošković & Rando, 2018). Several features of plant development make them more susceptible to the transgenerational inheritance of acquired epialleles. Firstly, the germline forms late in development from stem cells within the shoot apical meristem (SAM), allowing exposure of somatic tissue to environmental factors that can affect the germline. However, it is important to recognize that mechanisms that restrict changes impacting genome and epigenome stability might be more stringent in germline stem cells than in vegetative tissue (Quadrana & Colot, 2016). Secondly, stress-induced epigenetic changes can be replicated and maintained during plant sexual reproduction. In plants, the reprogramming (specifically, a reduction) of DNA methylation mainly occurs in the companion cells, the vegetative cell nucleus in pollen and the central cell nucleus in the ovule, rather than in the nuclei (sperm cell and egg cell) that will fuse to form the zygote during fertilization (Han et al., 2019). Despite this, significant global epigenetic reprogramming does occur during gametogenesis, which is linked to processes like histone replacement and DNA demethylation (Kawashima & Berger, 2014). Even though epigenetic reprogramming during sexual reproduction constrains the transmission of acquired epialleles, there are more abundant examples of sexual transmission of epialleles in plants than in mammals, suggesting a greater potential for transgenerational epiallele inheritance in plants (Kumari et al., 2022). Lastly, it is important to note that plants have a remarkable capacity for vegetative reproduction, which may increase the likelihood of epiallele retention and later expression in sexually propagating individuals (Hung et al., 2018). On the other hand, intergenerational epigenetic memory allows plants to inherit stress tolerance without changes in DNA sequence. This memory is primarily mediated by DNA methylation, histone modifications and small RNAs, which regulate gene expression across generations (Nguyen et al., 2022). In *Arabidopsis thaliana*, parental drought exposure leads to DNA methylation changes, enhancing water-use efficiency and stomatal regulation in offspring (Tricker et al., 2013). Moreover, salinity stress modifies histone marks in genes like *SOS1* in rice, improving salt tolerance in progeny (Singroha et al., 2022).

28.3 MOLECULAR REGULATION OF ABIOTIC STRESS RESPONSE THROUGH EPIGENETIC MEMORY

Responses to stress lead to extensive modifications in chromatin organization and gene expression and may also influence genetic sequences (Kim, 2021). When organisms face stress, alterations occur at every level of chromatin structure. This encompasses changes in DNA methylation, nucleosome composition, occupancy, histone variants, post-translational modifications (PTMs) of histones and the overall organization of chromatin (Millán-Zambrano et al., 2022). It is difficult to distinguish the direct impacts of stress on chromatin structure and nuclear arrangement from its influences on gene expression, resulting in much of the information regarding the mechanisms behind stress-induced alterations being correlational (Aristizabal et al., 2020). Although the relationships between chromatin reorganization and responses to different types of stress are well characterized, how the chromatin structure modified by stress is inherited remains not fully understood. The molecular mechanisms accountable for the regulation of abiotic stress responses through epigenetic inheritance are discussed as follows.

28.3.1 Changes in Chromatin Structure

Chromatin regulation involves changes in its structure that affect how easily DNA can be accessed by transcription factors (Isbel et al., 2022). Accessible chromatin areas usually lack nucleosomes and often act as regulatory sites, including promoters and enhancers (Klemm et al., 2019). Chromatin accessibility can change due to internal signals from within the cell or external environmental factors. ATP-dependent chromatin remodelers help establish this accessibility (Banerjee & Roychoudhury 2018; Mansisidor & Risca, 2022). They manage several processes, such as assembling nucleosomes, moving them around, removing them and swapping out histone variants (Escobar et al., 2021). In addition to these changes, the three-dimensional (3D) organization of chromatin also affects accessibility and gene activity. This 3D organization includes different structures, such as chromatin loops and chromosome domains, as well as active and inactive compartments, and chromosome territories (Rowley & Corces, 2018), for instance, regulatory regions like enhancers and promoters can interact dynamically through long-range chromatin contacts, influencing gene transcription (Schoenfelder & Fraser, 2019).

28.3.2 Alterations in DNA Methylation and the Function of Transposable Element

In plants, DNA methylation often occurs as 5-methylcytosine (5mC) in different cytosine contexts. Normally, DNA methylation helps to turn off genes at transposable elements (TEs) and gene promoters (Deniz et al., 2019). However, in some cases, DNA methylation can help turn on gene expression. The process of adding DNA methylation involves the RNA-directed DNA methylation (RdDM) pathway. In this pathway, the enzyme DOMAINS REARRANGED METHYLASE 2 (DRM2) adds new methylation to DNA (Zhang & Zhu, 2024). After methylation is done, several enzymes maintain it. These include METHYLTRANSFERASE 1 (MET1), CHROMOMETHYLASE 2 (CMT2), CMT3 or DRM2, depending on the specific context of 5mC (Shaikh et al., 2022). DNA methylation can be removed in two ways. The first is passive dilution, which occurs when maintenance fails after DNA replication. The second is active demethylation, which is carried out by a group of enzymes called DNA glycosylases (Chang, 2023). In *Arabidopsis*, these enzymes include REPRESSOR OF SILENCING 1 (ROS1), DEMETER (DME), DEMETER-LIKE 2 (DML2) and DML3. These enzymes remove methylated cytosines directly, helping to demethylate DNA through a base excision repair pathway (Parrilla-Doblas et al., 2019).

28.3.3 Role of Small RNAs in Post-Transcriptional Regulation of Stress-Responsive Genes

In addition to chromatin modifications and DNA methylation, RNA modifications constitute another layer of epigenetic regulation. To date, researchers have identified hundreds of different modifications in cellular RNAs (Schaefer, 2021). Among the most prevalent and well-studied internal modifications of messenger RNA (mRNA) are N6-methyladenosine (m6A) and 5-methylcytosine (m5C) (Rengaraj et al., 2021). These modifications impact nearly every aspect of mRNA metabolism, including maturation, nuclear transport, cellular localization, stability and translation, thereby playing vital roles in plant development and responses to environmental changes (Shoaib et al., 2022). Similar to DNA and histone modifications, m6A and m5C are dynamically regulated; for instance, in plants, m6A is added by a multicomponent methyltransferase complex and FIONA1, while it is removed by ALPHA-KETOGLUTARATE-DEPENDENT DIOXYGENASE ALKB HOMOLOG5 (ALKBH5) ortholog proteins. Furthermore, it is recognized by a group of proteins that contain YTH domains (Ge et al., 2024; Tang et al., 2023).

28.3.4 Histone Modifications and Memory – DNA Interactions with Chromatin Remodelers

Histone proteins interact with DNA to form nucleosomes, the fundamental units of chromatin. Various post-translational modifications of histones – including acetylation, methylation, ubiquitination, sumoylation and phosphorylation – can significantly influence chromatin activity (Ramazi et al., 2020). These modifications play a crucial role in determining chromatin function and affecting gene transcription (Morgan & Shilatifard, 2020). They may directly alter chromatin properties or engage effector proteins that recognize specific histone modifications, facilitating downstream processes (Zhao et al., 2021), for example, histone acetylation is thought to promote transcription by weakening the interactions between DNA and histones (Verza et al., 2020). In contrast, the effects of histone methylation can vary based on the specific position and degree of methylation (mono-, di- or tri-methylation). The dimethylation of histone H3 at lysine 9 (H3K9me2), which is associated with heterochromatin, is linked to transcriptional silencing, while trimethylation at lysine 27 (H3K27me3), found in euchromatin, is also correlated with gene silencing (Acharjee et al., 2023). Conversely, trimethylation at lysine 4 (H3K4me3) and trimethylation at lysine 36 (H3K36me3) are modifications associated with active transcription (Lam et al., 2022). These histone modifications are dynamically regulated by specific enzymes, for instance, histone acetylation is performed by histone acetyltransferases and reversed by histone deacetylases (Narita et al., 2019). The addition of H3K4me3 is mediated by the evolutionarily conserved COMPASS complex, while H3K27me3 is established by the Polycomb Repressive Complex 2 (PRC2). Both complexes consist of several core structural subunits and histone methyltransferases (Douillet et al., 2020; Guo et al., 2021). The removal of histone methylation is carried out by histone demethylases, which include either Jumonji C (JmjC) domain-containing proteins or Lysine-Specific Demethylase 1 (LSD1)-like proteins (Perillo et al., 2020).

28.3.5 Histone Variants and Nucleosome Assembly

In addition to histone modifications, histone variants, i.e., sequence variations of histones, contribute to the diversity of nucleosomes. These variants not only differ in their sequences but also display distinct expression patterns and localizations within chromatin (Wong & Tremethick, 2024). The deposition and removal of these variants are orchestrated by specific histone chaperones and chromatin remodelers, which influence nucleosome properties and, consequently, affect chromatin activities and transcription (Clapier, 2021; Wu et al., 2023). The repressed genes are typically marked by the histone variant H2A.Z across the entire gene body, while active genes tend to be enriched with H3.3 (Foroozani et al., 2022). In *Arabidopsis*, the loss of either H2A.Z or H3.3 significantly affects the expression of genes that respond differently to various environmental and developmental cues. This underscores the vital roles of H2A.Z and H3.3 in regulating plant responsiveness (Cheng et al., 2024).

28.4 EPIGENETIC MEMORY OF PLANTS FOR ABIOTIC STRESS TOLERANCE

Plants are perpetually subjected to various environmental challenges that influence their growth, development and survival. Several abiotic stresses, such as drought, salinity, extreme temperature and nutrient deficiencies, threaten global agriculture and food security (Chaffai et al., 2024). Unlike animals, plants cannot escape adverse conditions; therefore, they depend on complex molecular mechanisms to perceive, respond to and adapt to stress (Kosakivska et al., 2021). A noteworthy aspect of plant responses to stress is the concept of epigenetic memory, which facilitates heritable changes in gene expression without necessitating alterations to the DNA sequence. This capability to retain stress-related memory confers an evolutionary advantage upon plants, enhancing their ability to respond effectively to recurrent stress events (Lagiotis et al., 2023). Even though environmental stresses frequently manifest in combination in natural settings, the majority of research conducted to date has focused predominantly on investigating individual stress types (Rojas-Carvajal et al., 2022). The following section comprehensively summarizes the principal abiotic stresses and the associated plant responses concerning memory (Table 28.1).

28.4.1 Epigenetics and Drought Stress

Drought is a significant factor affecting agricultural output, and its impact is expected to grow due to climate change. Research on drought memory in *Arabidopsis* reveals that prior desiccation treatments can alter gene expression and stress responses in subsequent drought (Sadhukhan et al., 2022). Studies in *Arabidopsis* identified genes responsive to drought that exhibit altered transcriptional and chromatin changes after predisposition to drought stress. Many of these genes are also involved in responses to other stresses like salt and temperature (Hussain et al., 2021). Furthermore, imbibed seeds treated with polyethylene glycol (PEG) showed improved survival rates post-rehydration, indicating lasting drought tolerance (Lyall et al., 2014). Drought causes various changes in crops, including reduced germination, biomass and physiological activity. Rice is particularly sensitive, with yield losses up to 92%. However, pre-treatment with mild drought enhances resistance to oxidative stress caused by subsequent drought (Wahab et al., 2022). Similarly, wheat seedlings acclimated by dehydration for a mild 9-day stress and a severe 11-day stress showed improved water retention and limited membrane damage when stressed further (Selote and Khanna-Chopra 2010). Similarly, in maize, exposure to repeated drought cycles resulted in better water content compared to controls, highlighting both conserved and species-specific responses in stress memory (Mozgova et al., 2019). Moreover, in potatoes, mild dehydration improved overall plant growth under stress but did not affect tuber yield under severe drought conditions (Obidiegwu et al., 2015).

The *Arabidopsis* demethylase JMJ27 limits H3K9me2 accumulation to prevent transcriptional silencing of drought tolerance regulators GOLS2 and RD20, with drought stress increasing JMJ27 levels (Wang et al., 2021). ATX1, an H3K4 methyltransferase, enhances H3K4me3 and *NCED3* expression, a key gene in ABA biosynthesis, under drought stress (Ding et al., 2011) (Figure 28.1a). In contrast, ATX4 and ATX5 positively regulate AHG3, a negative ABA signaling regulator (Liu et al., 2018). The H3K4 demethylase

TABLE 28.1
Priming agents and their mechanism of action in enhancing plant resilience against various abiotic stresses

Priming agent	Applied stress	Plant species	Observed mechanism	References
High temperature and intense light	Exposure to high temperature after 2 days of recovery	Tomato	Formation of secondary metabolites in chlorophyll pigments	Zhou et al. (2020)
Drought stress	Initial drought stress applied for several days post-germination, followed by a 50-day drought cycle after a 6-day recovery	Cowpea	Increased synthesis of IAA (Indole Acetic Acid) and ABA (abscisic acid)	Tankari et al. (2021)
Low-concentration of salt shock	Later exposure to high salinity	Faba bean	Enhanced antioxidative enzyme activities	Nasrallah and Atia (2022)
Polyethylene glycol (PEG)-mediated seed treatment	Salinity stress at a later growth phase	Tomato	Elevated photosynthetic activity and improved fruit quality	Habibi et al. (2023)
Salinity stress at the five-leaf stage	Water withheld during the reproductive phase	Rice	Greater tolerance at physiological, transcriptional and hormonal levels	Rossatto et al. (2023)
Hydrogen peroxide (H_2O_2) treatment of seed	Waterlogging stress at the V3 stage	Maize	Increased tolerance to waterlogging via stomatal regulation and enhanced photosynthesis	Wang and Hu (2022)
Zinc seed priming	Growth in Zn-deficient soil	Maize	Improved early development and grain yield	Tamindžić et al. (2021)
Selenium seed priming	Drought stress applied during multiple growth stages (leaf, flowering and seed-filling)	Quinoa	Enhanced biochemical and photosynthetic performance	Raza et al. (2024)

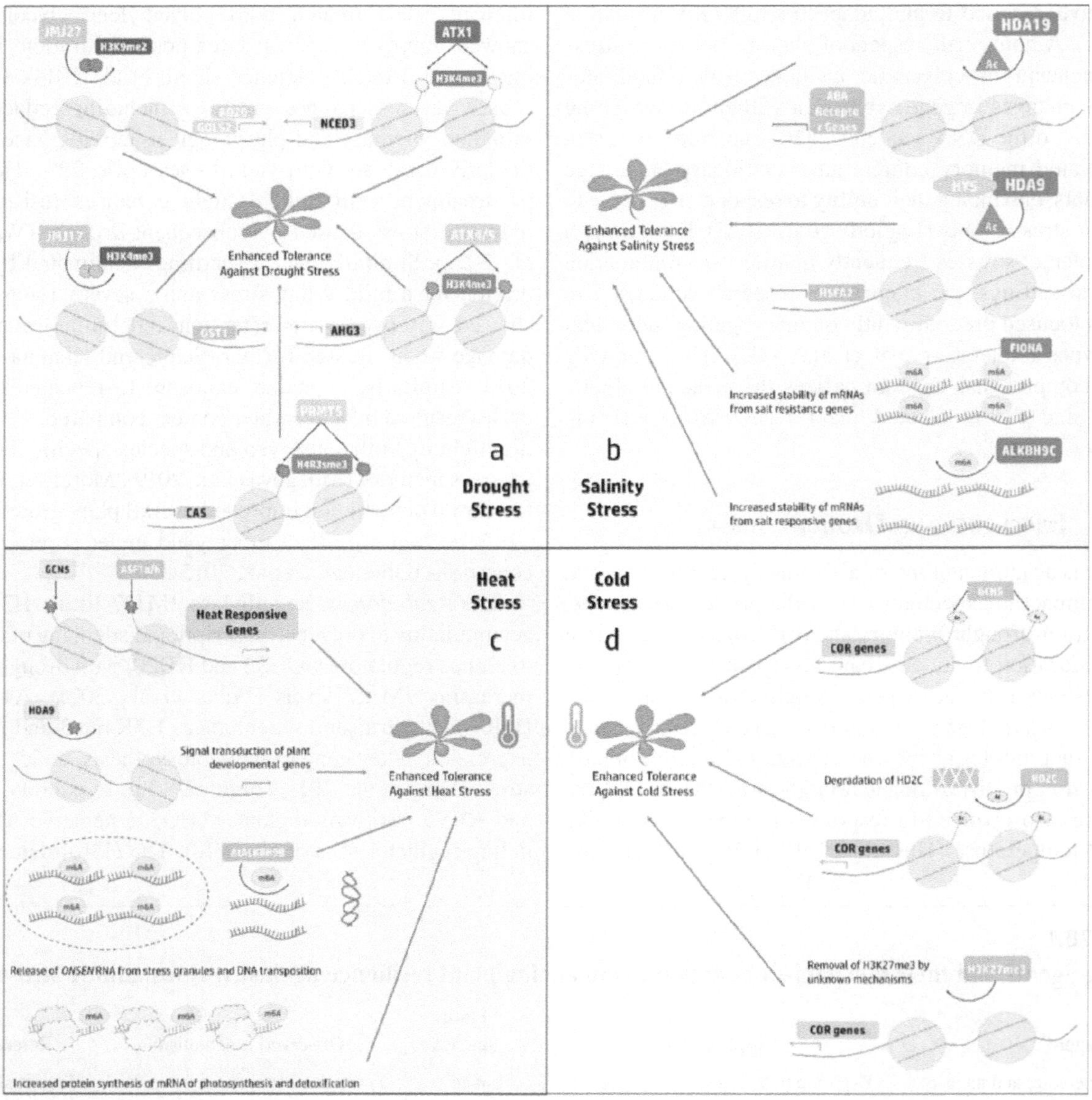

FIGURE 28.1 Molecular mechanism of plant response to various abiotic stresses through epigenetic control. (a) Histone lysine and arginine methylation affect gene transcription under drought stress, (b) histone acetylation-deacetylation affects gene transcription and m6A writing-erasing affects mRNA stability under salinity stress, (c) histone acetylation-deacetylation controls gene transcription and RNA modifications govern TE transposition and protein synthesis under heat stress and (d) control of histone acetylation and H3K27me3 helps activate *COR* genes under cold stress.

JMJ17 represses OST1, which is crucial for stomatal closure and drought tolerance (Huang et al., 2019). In rice, OsHUB2, an E3 ligase, activates ABA-related transcription factor OsbZIP46 under drought, while its repressor MODD reduces expression when stress subsides (Ma et al., 2019). The H3K36 methyltransferase SET DOMAIN activates key ABA biosynthesis genes *OsNCED3* and *OsNCED5*, promoting drought tolerance (Chen et al., 2021). Drought stress increases chromatin accessibility, promoting drought response gene expression, aided by NMCP1 and countered by the SWI/SNF complex (Rajkumar et al., 2024). Additionally, loss of OsCHR4 enhances wax deposition and drought tolerance by modulating gene expression (Guo et al., 2019). Overall, these studies highlight the significance of drought epigenetic memory in enhancing crop resilience to climate-induced stress.

28.4.2 Epigenetics, Osmotic and Salinity Stress

High salinity poses a significant threat to agricultural production, wreaking havoc on plants growing in both naturally saline soil and irrigated lands that suffer from high evaporation or poor water management (Sharma & Naik, 2024). This salt-induced osmotic stress drastically hampers plant growth by limiting water uptake, leading to stomatal closure and a decline in photosynthetic activity. Additionally, the ionic stress from excessive concentration of specific salts disrupts the balance of essential ions, hinders metabolic activity and jeopardizes the integrity of plasma membranes (Chaudhary et al., 2024). However, epigenetic memory can be a beacon of hope in tackling the consequences of salinity (Banerjee & Roychoudhury 2017). Priming plants with mild salt treatment has demonstrated remarkable potential to boost their resilience against subsequent salt stress. This approach enhances key physiological parameters and promotes plant vigor and fitness without imposing significant growth disadvantages (Carillo et al., 2011). Research has highlighted that *Arabidopsis* plants primed with low NaCl concentration exhibit reduced sodium accumulation in shoots, higher biomass and improved survival rates following subsequent stress exposure. Notably, this stress memory was found to last for at least ten days, showcasing an impressive linkage between salinity and drought resilience (Paul et al. 2023). Moreover, the accumulation of proline, a vital amino acid associated with metal chelation, antioxidation and signaling, increased significantly in salt-primed plants. This enhancement is linked to the transcription of the *P5CS1* gene (DELTA1-PYRROLINE-5-CARBOXYLATE SYNTHASE 1), which is critical for proline biosynthesis (Sellamuthu et al., 2024). This trend is not confined to *Arabidopsis* only; similar benefits have been observed in crops like wheat, where mild NaCl priming fosters tolerance to more intense salt stress, culminating in vibrant photosynthesis and better osmotic regulation (Huang et al., 2024). In rice, pre-treating seedlings with sublethal NaCl doses led to superior ion absorption and osmotic adjustment capabilities compared to non-primed plants (Abo-Shanab et al., 2025). The positive effects of salt priming are also evident in sorghum, maize, pea, potato and olive, with results varying according to cultivar types (Truşcă et al. 2023; Abdul Rahman et al., 2021). Particularly in potato, salt-sensitive cultivars experienced the most significant benefits, while olive showed varied enhancements across different plant organs (Levy et al., 2013).

Salt stress activates GCN5, which adds H3K9 and H3K14 acetylation to the *CTL1, PGX3* and *MYB54* genes, enhancing cell wall integrity and salt tolerance in *Arabidopsis* (Zheng et al., 2019). Similar histone acetyltransferase functions have been observed in wheat and soybean (Zheng et al., 2021; Feng et al., 2022). Class II histone deacetylases HDA5/14/15/18 promote salt resistance, while class I deacetylase HDA19 inhibits it by reducing acetylation at ABA receptor genes (Ueda et al., 2017). HY5 recruits HDA9 to suppress the salt resistance-related HSFA2 (Yang et al., 2023). The deacetylase SRT2 helps counteract the inhibitory effects of salt on seed germination by repressing VESICLE-ASSOCIATED MEMBRANE PROTEIN 714 (Tang et al., 2022). GmHDA13 in soybean collaborates with GmFVE to regulate H3K9 acetylation at salt-responsive genes, while salt stress induces the NF-Y subunit GmNFYA to enhance gene activation (Lu et al., 2021). In rice, the roles of HDA706 and HDA710 in regulating salt tolerance have been noted, with IDS1 repressing salt tolerant genes like *LEA1* and *SOS1* (Liu et al., 2023; Cheng et al., 2018). Meanwhile, OsHKT2;1 acts as a negative regulator, with salt stress promoting PSEUDO-RESPONSE REGULATOR 73 (OsPRR73) that inhibits OsHKT2;1 via HISTONE DEACETYLASE 10 (HDAC10) (Wei et al., 2021).

Dynamic RNA modifications also influence salt tolerance. The m6A methyltransferase FIONA1 enhances resistance in *Arabidopsis* by stabilizing the mRNA of salt-responsive genes (Amara et al., 2024; Cai et al., 2024). Other m6A components, including MTA and VIR, play vital roles in salt tolerance by destabilizing the negative regulators (Hu et al., 2021). Loss of the demethylase ALKBH9C results in increased m6A levels and reduced stability of salt-responsive genes, indicating varied impacts of m6A modification on mRNA stability (Amara et al., 2022) (Figure 28.1b). The evidence compels that salt priming is a transformative strategy that can revolutionize how we approach crop resilience in increasingly saline environment (Singh et al., 2021). By harnessing this technique, we can cultivate stronger, more adaptable plants that thrive amid challenging conditions, ultimately securing agricultural productivity for the future.

28.4.3 Epigenetics and Thermal Stress

Climate change models predict rising temperature, which could severely impact food security by impairing plant growth and reproduction. Plants have evolved stress memory mechanisms to cope with environmental challenges such as heat and cold stress (Hilker, 2019). These memory responses have been extensively studied in *Arabidopsis* and applied to agronomic crops to improve stress resilience. Heat stress memory in *Arabidopsis* has primarily been studied at the seedling stage, where it lasts up to three days but may decay within hours after recovery (Charng et al., 2023). Rice research has shown that the duration of heat stress memory varies across cultivars (Lawas et al., 2019). Studies in barley demonstrated that acute heat shock priming can also counteract the negative effects of salt stress on root elongation (Chakraborty & Dwivedi, 2021). Heat priming also enhances tolerance to heavy metal stress. In wheat, a heat shock improved seedling survival when exposed to iron and cadmium (Zang et al., 2017). Similarly, in wild tomato cell cultures, heat pre-treatment reduced cadmium-induced membrane damage (Muhammad et al., 2019). Similarly, rice seedlings showed reduced cadmium-induced chlorosis following a short-term heat pre-treatment (Kanu et al., 2017).

Under heat stress, normally silenced transposable elements (TEs) like the ONSEN retrotransposon in *Arabidopsis* become activated. This activation depends on heat stress transcription factors HSFA1 and HSFA2, while the RdDM pathway prevents overactivation (Cavrak et al., 2014). ONSEN integration occurs in regions rich in histone variant H2A.Z and H3K27me3, impacting gene expression and creating transcriptional novelties (Roquis et al., 2021) (Figure 28.1c). The CMT2 locus is linked to temperature seasonality in *Arabidopsis*; its mutation leads to reduced DNA CHH methylation and enhanced ONSEN activation under heat stress (Nozawa et al., 2022). Histone H1 also limits ONSEN activation, influencing plant survival under heat stress (Liu et al., 2021). DNA methylation levels decrease in various plants in response to heat (Hossain et al., 2017). GCN5 and ASF1a/b are crucial for heat stress tolerance, promoting gene activation (Hu et al., 2015). The deacetylase HD2C, induced by heat, prevents overactivation of heat-responsive genes (Buszewicz et al., 2016). PP2AB'b stabilizes HDA9, which represses genes related to signaling and development during heat stress (Niu et al., 2022). Additionally, sumoylation is essential for the rapid transcriptional response to heat, regulating the balance between stress response and development (Han et al., 2021).

Low temperature significantly affects crop productivity by impairing germination, chlorophyll production, organ expansion and reproductive development (Mukhopadhyay & Roychoudhury 2018). Cold stress memory has been widely studied in *Arabidopsis*, where long-term cold stress (LTC) but not short-term cold stress (STC) improves photosynthetic efficiency after a subsequent cold event (van Buer et al., 2016). Cold priming enhances tolerance in chilling-sensitive crops. In tomato, heat pre-treatment reduces post-harvest chilling injury symptoms such as aroma loss, oxidative stress and failure to ripen (Albornoz et al., 2019). In rice, prior exposure to moderate cold prevented cold-induced water uptake issues, leaf wilting and color bleaching (Ahamed et al., 2012). Maize cold priming protected the photosynthetic apparatus, especially in cold-resistant inbred lines. In wheat, cold priming led to increased soluble protein content, decreased water content and proline accumulation, with memory effects lasting up to 30 days, particularly in winter cultivars (Baier et al., 2019). In canola, cold priming improved survival after freezing exposure (Zhu et al., 2021). Like heat priming, cold priming can also enhance tolerance to other stress types. Mustard seedlings primed with cold exhibited improved resistance to salt and drought. In pea, cold priming mitigated oxidative damage caused by cadmium, copper and high light intensity (Sayed & Gadallah, 2019). Cold priming also increased heat stress resistance in several species, including wheat and canola (Ahmad et al., 2022).

In *Arabidopsis*, GCN5 is essential for freezing tolerance, as its loss leads to reduced activation of *COR* genes during cold treatment, indicating that GCN5-related histone acetylation aids COR activation (Vlachonasios et al., 2003) (Figure 28.1d). Similarly, histone hyperacetylation is linked to the cold-induced activation of OsDREB1b in rice (Roy et al., 2014). The cold response also involves histone deacetylation; *Arabidopsis* histone deacetylase HDA6 is upregulated by cold and is crucial for freezing tolerance (To et al., 2011). Cold triggers the degradation of HD2C, leading to increased H3 acetylation and COR activation (Park et al., 2018). The *COR* genes *COR15A* and *GALACTINOL SYNTHASE3* (*ATGOLS3*) are marked by repressive H3K27me3, which is gradually removed during cold treatment, even after returning to normal conditions (Kwon et al., 2009). This aligns with findings that H3K27me3 reprogramming lags behind transcriptional changes (Pan et al., 2023). The CHD3-type chromatin remodeler PICKLE (PKL) and the SWI/SNF complex subunit LEAF AND FLOWER RELATED (LFR) are both important for cold-regulated gene expression (Ma et al., 2023). Additionally, loss of the m6A methyltransferase MTA impacts *CBF* and *COR* gene expression, affecting freezing tolerance (Banerjee et al. 2017; Govindan et al., 2022). Therefore, both heat and cold stress memory play critical roles in plant adaptation to environmental challenges. Henceforth, these studies suggest that heat or cold pre-treatment can enhance tolerance not only to thermal but also to other stresses like heavy metals.

28.4.4 Epigenetics during UV Stress and Chemical Agents

UV-B radiation, a component of solar radiation, presents challenges to plants due to their sessile nature. UV-B stress varies by dosage and duration, i.e., acute (seconds to hours) versus chronic (hours to days) (Piccini, 2022; Hideg et al., 2013). Acute high-dose UV-B causes severe damage, leading to programmed cell death, whereas chronic low-dose exposure can trigger acclimation mechanisms (Hideg et al., 2013). *Arabidopsis* subjected to prolonged low-dose UV-B retained stress memory for up to nine days, exhibiting morphological changes such as reduced rosette size, shorter inflorescence height, increased flowering stems and enhanced axillary branching (Zhang et al., 2021). Despite reduced leaf area, photosynthetic efficiency remained stable due to increased pigment accumulation (Jin et al., 2016). Additionally, repeated short UV-B pulses led to elevated flavonoid levels, indicating their role in stress memory and acclimation (Höll et al., 2019). Low-dose UV-B effects have also been studied in crops, revealing benefits at morphological, physiological and metabolic levels. Crops like wheat, maize, rice, cabbage (*Brassica oleracea*), rapeseed (*Brassica napus*), and legumes such as mung bean (*Vigna radiata*) and soybean (*Glycine max*) exhibited improved germination, growth, pigment production and stress tolerance (Ebert, 2022). In maize, UV-B exposure enhanced germination rates, while increased pigment levels were observed in cabbage, kidney beans, soybeans and rice (Singh et al., 2016; Olsson et al., 1998; Yanqun et al., 2003).

Beyond UV-B, stress memory can be induced through chemical priming using natural and synthetic compounds, including amino acids, hormones, nutrients, pesticides and reactive oxygen-nitrogen-sulfur species (RONSS) (Hiiback, 2023; Antoniou et al., 2016). Chemical priming enhances resilience to various stressors, for example, the non-protein amino acid β-aminobutyric acid (BABA) increased *Arabidopsis* tolerance to salinity and drought by reducing wilting and water loss (Decsi et al., 2024). BABA also enhanced systemic acquired resistance (SAR) against pathogens, suggesting overlapping pathways for biotic and abiotic stress responses. Furthermore, melatonin pre-treatment improved plant growth metrics like fresh weight, root length and shoot length under cold stress and upregulated cold-responsive genes during recovery, mirroring transcriptomic patterns seen in mild cold pre-conditioning (Colombage et al., 2023; Bai et al., 2024). Similarly, spermine, a natural polyamine, mitigated chlorosis in *Arabidopsis* cotyledons under stress. Transgenic lines overexpressing spermine biosynthetic genes exhibited improved heat stress tolerance, whereas knock-out mutants showed hypersensitivity (Kohli et al., 2019). Chemical seed priming in crops has gained attention as a practical approach in improving stress resilience. Therefore, by pre-conditioning plants at early growth stages, priming treatments provide long-term benefits, enhancing their ability to withstand environmental stressors.

28.5 STRATEGIES FOR MODULATING EPIGENETIC MEMORY FOR ABIOTIC STRESS TOLERANCE

Plants possess the ability to adjust their genetic makeup in response to environmental stimuli. While most of these adjustments are temporary, some can persist for extended periods, effectively creating a type of memory within the plant. A well-known example of this phenomenon is vernalization in *Arabidopsis*. When these plants are exposed to cold temperature for a significant duration, they induce changes that promote blooming at a later time (Tun et al., 2021). Moreover, plants can learn from experiences when exposed to stressful environments. When subjected to stressful conditions, they can recall these experiences, allowing them to prepare more actively for future encounters. This process is commonly referred to as priming or acclimation (Charng et al., 2023). Although these memories are short-lived and not transmitted to the next generation, in certain cases, some memories are inheritable. For instance, specific alterations in DNA caused by salt stress can be passed on to offspring, enhancing their capacity to withstand stress. In the absence of continued stress, this memory tends to diminish over time in subsequent generations (Liu et al., 2022). Resetting this stress memory could enable plants to redirect their resources towards growth rather than solely maintaining the memory (Sharma et al., 2022).

Repeated drought exposure in *Arabidopsis* boosts their resistance to drought and strengthens their capability to activate genes, such as *RESPONSIVE TO DEHYDRATION 28B* (*RD28B*) and *RESPONSIVE TO ABA 18* (*RAB18*), that aid essential services in drought responses. This enhancement is due to the enduring presence of H3K4me3 and halted RNA Polymerase II at these genes, lasting for up to five days after the initial drought (Farkas & Dobránszki, 2024). A similar study revealed that some drought-related changes in H3K4me3 are retained in other genes during recovery following a drought episode (Ost et al., 2023). In *Arabidopsis*, mild salt stress was found to induce notable alterations in the H3K27me3 profile, which may persist for a minimum of 10 days following the cessation of stress exposure. This modification is associated with the subsequent responses of plants to stress (Sani et al., 2013). Furthermore, salt stress can result in a persistent rise of H3K4me3 at the *P5CS1* gene, which plays a crucial role in producing proline, a compound that assists plants in conferring tolerance to stress conditions. Increased levels of H3K4me3 at P5CS1 help plants produce more proline during subsequent salt stress events. This process depends on light and is regulated by a factor called HY5 (Yung et al., 2021).

H3K4me3 plays a crucial role in the memory associated with heat stress in plants. After an initial exposure to high temperature, plants were found to become better prepared to endure more intense heat stress in the future. In *Arabidopsis*, this memory typically lasted for several days following the first heat treatment (Nishio et al., 2024). Simultaneously, certain heat-responsive genes show heightened activity following the initial stress, with even greater activity during repeated heat events (Zhou et al., 2022). These genes often have higher levels of H3K4me3, regulated by specific transcription factors that respond to heat stress (Kappel et al., 2023). While two enzymes, ATX1 and SDG25, are known to assist with heat stress tolerance, they do not seem to influence heat stress memory (Pratx et al., 2024). However, the exact enzymes responsible for adding H3K4me3 at genes related to heat stress memory remain unidentified. Multiple mechanisms play significant roles in maintaining the memory of heat stress. The protein FORGETTER1 (FGT1) works in conjunction with chromatin remodelers, such as SWI/SNF (BRM) and ISWI (CHR11 and CHR17), to preserve transcriptional memory. This occurs by loosening the DNA structure of genes related to heat stress (Krzysztof et al., 2016). Furthermore, a protein named TONSOKU (TSK) is necessary for preserving these epigenetic modifications connected to heat stress memory (Wang et al., 2023). Also, a specific histone variant known as H3.3 supports active transcription and helps maintain the elevated levels of H3K4me3 (Lowe et al., 2019). The turnover rate of H3.3 at genes related to heat stress memory is slower compared to non-memory genes, which helps maintain the presence of H3K4me3. Furthermore, the TARGET OF RAPAMYCIN (TOR) kinase, which integrates signals related to nutrition, energy, growth and stress, encourages the accumulation of H3K4me3 at heat stress memory genes. TOR influences the expression of ATX1 through E2F transcription factors (Sharma et al., 2022). Since ATX1 is not essential for maintaining heat stress memory, TOR may influence this process through alternative mechanisms (Zheng et al., 2024).

28.6 CONCLUSION AND FUTURE PERSPECTIVES

Global environmental changes pose significant challenges to agricultural productivity and food security, primarily due to crop loss caused by abiotic stresses.

Recent research has concentrated on understanding how these stresses impact plant growth, development, yield and adaptive responses. Increasing global average temperature, exacerbated by climate change, has intensified the adverse effects of abiotic stresses, especially heat, drought, salinity and cold. These conditions disrupt plant growth and trigger rapid alterations in structural organization, photosynthetic efficiency and metabolic pathways, which encompass both primary and secondary processes, as well as vital functions like lipid and hormonal signaling. This review highlights recent advancements in the study of plant priming and the mechanisms underpinning somatic and transgenerational memory related to various abiotic stresses. The concept of stress memory – the capacity of plants to retain knowledge of prior stress exposure – holds significant promise for enhancing plant resilience in natural environment. Current evidence indicates that the memory of abiotic stress is managed through multiple mechanisms, including chromatin remodeling, DNA methylation patterns, retrotransposon activity and microRNA regulation, all of which play crucial roles in effective stress management in plants. Nonetheless, further research is essential to unravel the intricate molecular pathways and evolutionary ecology that govern these processes. Additionally, innovative genome-editing techniques, particularly CRISPR-Cas9, offer exciting opportunities to delve into the mechanisms of stress memory and their implications for sustainable agricultural practices.

REFERENCES

Abdul Rahman, N. S. N., Abdul Hamid, N. W., & Nadarajah, K. (2021). Effects of abiotic stress on soil microbiome. *International Journal of Molecular Sciences*, 22(16), 9036.

Abo-Shanab, W. A., Elshobary, M. E., Czubacka, A., & Diab, R. H. (2025). Improvement of salt tolerance in *Vicia faba* (L.) seedlings: A comprehensive investigation of the effects of exogenous calcium chloride. *BMC Plant Biology*, 25(1), 200.

Acharjee, S., Chauhan, S., Pal, R., & Tomar, R. S. (2023). Mechanisms of DNA methylation and histone modifications. *Progress in Molecular Biology and Translational Science*, 197, 51–92.

Ahamed, A., Murai-Hatano, M., Ishikawa-Sakurai, J., Hayashi, H., Kawamura, Y., & Uemura, M. (2012). Cold stress-induced acclimation in rice is mediated by root-specific aquaporins. *Plant and Cell Physiology*, 53(8), 1445–1456.

Ahmad, M., Waraich, E. A., Hussain, S., Ayyub, C. M., Ahmad, Z., & Zulfiqar, U. (2022). Improving heat stress tolerance in *Camelina sativa* and *Brassica napus* through thiourea seed priming. *Journal of Plant Growth Regulation*, 41(7), 2886–2902.

Ahmed, M., Tariq, H., Haroon, F., Malik, S., Dar, M., Solangi, F., & Aamir, S. (2023). Harnessing epigenetic modifications for targeted trait manipulation in Pakistani wheat. *International Journal of Precision Farming*, 1(1), 50–55.

Albornoz, K., Cantwell, M. I., Zhang, L., & Beckles, D. M. (2019). Integrative analysis of postharvest chilling injury in cherry tomato fruit reveals contrapuntal spatio-temporal responses to ripening and cold stress. *Scientific Reports*, 9(1), 2795.

Amara, U., Hu, J., Park, S. J., & Kang, H. (2024). ECT12, an YTH-domain protein, is a potential mRNA m6A reader that affects abiotic stress responses by modulating mRNA stability in *Arabidopsis*. *Plant Physiology and Biochemistry*, 206, 108255.

Amara, U., Shoaib, Y., & Kang, H. (2022). ALKBH9C, a potential RNA m6A demethylase, regulates the response of *Arabidopsis* to abiotic stresses and abscisic acid. *Plant, Cell & Environment*, 45(12), 3566–3581.

Amaral, J., Ribeyre, Z., Vigneaud, J., Sow, M. D., Fichot, R., Messier, C., Pinto, G., Nolet, P., & Maury, S. (2020). Advances and promises of epigenetics for forest trees. *Forests,* 11(9), 976.

Ambeskovic, M., Babenko, O., Ilnytskyy, Y., Kovalchuk, I., Kolb, B., & Metz, G. A. (2019). Ancestral stress alters lifetime mental health trajectories and cortical neuromorphology via epigenetic regulation. *Scientific Reports*, 9(1), 6389.

Angers, B., Castonguay, E., & Massicotte, R. (2010). Environmentally induced phenotypes and DNA methylation: How to deal with unpredictable conditions until the next generation and after. *Molecular Ecology*, 19(7), 1283–1295.

Antoniou, C., Savvides, A., Christou, A., & Fotopoulos, V. (2016). Unravelling chemical priming machinery in plants: The role of reactive oxygen–nitrogen–sulfur species in abiotic stress tolerance enhancement. *Current Opinion in Plant Biology*, 33, 101–107.

Aristizabal, M. J., Anreiter, I., Halldorsdottir, T., Odgers, C. L., McDade, T. W., Goldenberg, A., Mostafavi, S., Kobor, M. S., Binder, E. B., Sokolowski, M. B., & O'Donnell, K. J. (2020). Biological embedding of experience: A primer on epigenetics. *Proceedings of the National Academy of Sciences*, 117(38), 23261-23269.

Atukunda, P., Eide, W. B., Kardel, K. R., Iversen, P. O., & Westerberg, A. C. (2021). Unlocking the potential for achievement of the UN Sustainable Development Goal 2 – 'Zero Hunger' – in Africa: Targets, strategies, synergies and challenges. *Food & Nutrition Research*, 65, 10–29219.

Auge, G., Hankofer, V., Groth, M., Antoniou-Kourounioti, R., Ratikainen, I., & Lampei, C. (2023). Plant environmental memory: Implications, mechanisms and opportunities for plant scientists and beyond. *AoB Plants*, 15(4), plad032.

Bai, Y., Wang, X., Wang, X., Dai, Q., Zhan, X., Gong, H., & Guo, J. (2024). Effects of combined application of selenium and various plant hormones on the cold stress tolerance of tomato plants. *Plant and Soil*, 501(1), 471–489.

Baier, M., Bittner, A., Prescher, A., & van Buer, J. (2019). Preparing plants for improved cold tolerance by priming. *Plant, Cell & Environment*, 42(3), 782–800.

Banerjee, A., & Roychoudhury, A. (2017). Epigenetic regulation during salinity and drought stress in plants: Histone modifications and DNA methylation. *Plant Gene*, 11, 199–204.

Banerjee, A., & Roychoudhury, A. (2018). The gymnastics of epigenomics in rice. *Plant Cell Reports,* 37, 25–49.

Banerjee, A., Wani, S. H., & Roychoudhury, A. (2017). Epigenetic control of plant cold responses. *Frontiers in Plant Science*, 8, 1643.

Bäurle, I. (2018). Can't remember to forget you: Chromatin-based priming of somatic stress responses. *Seminars in Cell & Developmental Biology,* 83, 133–139.

Bošković, A., & Rando, O. J. (2018). Transgenerational epigenetic inheritance. *Annual Review of Genetics*, 52(1), 21–41.

Bure, I. V., Nemtsova, M. V., & Kuznetsova, E. B. (2022). Histone modifications and non-coding RNAs: Mutual epigenetic regulation and role in pathogenesis. *International Journal of Molecular Sciences*, 23(10), 5801.

Buszewicz, D., Archacki, R., Palusiński, A., Kotliński, M., Fogtman, A., Iwanicka-Nowicka, R., Sosnowska, K., Kuciński, J., Pupel, P., Olędzki, J., & Koblowska, M. K. (2016). HD2C histone deacetylase and a SWI/SNF chromatin remodelling complex interact and both are involved in mediating the heat stress response in *Arabidopsis*. *Plant, Cell & Environment*, 39(10), 2108–2122.

Cai, J., Hu, J., Xu, T., & Kang, H. (2024). FIONA1-mediated mRNA m6A methylation regulates the response of *Arabidopsis* to salt stress. *Plant, Cell & Environment*, 47(3), 900–912.

Carillo, P., Annunziata, M. G., Pontecorvo, G., Fuggi, A., & Woodrow, P. (2011). Salinity stress and salt tolerance. *Abiotic Stress in Plants-mechanisms and Adaptations*, 1, 21–38.

Cavrak, V. V., Lettner, N., Jamge, S., Kosarewicz, A., Bayer, L. M., & Mittelsten Scheid, O. (2014). How a retrotransposon exploits the plant's heat stress response for its activation. *PLoS Genetics*, 10(1), e1004115.

Chaffai, R., Ganesan, M., & Cherif, A. (2024). Abiotic stress in plants: Challenges and strategies for enhancing plant growth and development. In R. Chaffai, M. Ganesan, & A. Cherif (Eds.), *Plant Adaptation to Abiotic Stress: From Signaling Pathways and Microbiomes to Molecular Mechanisms* (pp. 1–30). Singapore: Springer Nature Singapore.

Chakraborty, P., & Dwivedi, P. (2021). Seed priming and its role in mitigating heat stress responses in crop plants. *Journal of Soil Science and Plant Nutrition*, 21(2), 1718–1734.

Chang, C. (2023). *New DNA Repair and Demethylation Functions in Uracil DNA Glycosylase Superfamily.* All Dissertations, 1, 3313.

Charng, Y. Y., Mitra, S., & Yu, S. J. (2023). Maintenance of abiotic stress memory in plants: Lessons learned from heat acclimation. *The Plant Cell*, 35(1), 187–200.

Chaudhary, M. T., Majeed, S., Rana, I. A., Ali, Z., Jia, Y., Du, X., Hinze, L., & Azhar, M. T. (2024). Impact of salinity stress on cotton and opportunities for improvement through conventional and biotechnological approaches. *BMC Plant Biology*, 24(1), 20.

Chen, K., Du, K., Shi, Y., Yin, L., Shen, W. H., Yu, Y., Liu, B., & Dong, A. (202[illegible]). H3K36 methyltransferase SDG708 enhances drought tolerance by promoting abscisic acid biosynthesis in rice. *The New Phytologist*, 230(5), 1967–1984.

Cheng, X., Zhang, S., Tao, W., Zhang, X., Liu, J., Sun, J., Zhang, H., Pu, L., Huang, R., & Chen, T. (2018). INDETERMINATE SPIKELET1 recruits histone deacetylase and a transcriptional repression complex to regulate rice salt tolerance. *Plant Physiology*, 178(2), 824–837.

Cheng, Y. J., Wang, J. W., & Ye, R. (2024). Histone dynamics responding to internal and external cues underlying plant development. *Plant Physiology*, 194(4), 1980–1997.

Clapier, C. R. (2021). Sophisticated conversations between chromatin and chromatin remodelers, and dissonances in cancer. *International Journal of Molecular Sciences*, 22(11), 5578.

Colombage, R., Singh, M. B., & Bhalla, P. L. (2023). Melatonin and abiotic stress tolerance in crop plants. *International Journal of Molecular Sciences*, 24(8), 7447.

De Bang, T. C., Husted, S., Laursen, K. H., Persson, D. P., & Schjoerring, J. K. (2021). The molecular–physiological functions of mineral macronutrients and their consequences for deficiency symptoms in plants. *New Phytologist*, 229(5), 2446–2469.

Deans, C. (2024). *Biological Prescience: The Role of Anticipation in Organismal Design*. Authorea Preprints.

Decsi, K., Ahmed, M., Rizk, R., Abdul-Hamid, D., Kovács, G. P., & Tóth, Z. (2024). Emerging trends in non-protein amino acids as potential priming agents: Implications for stress management strategies and unveiling their regulatory functions. *International Journal of Molecular Sciences*, 25(11), 6203.

Deniz, Ö., Frost, J. M., & Branco, M. R. (2019). Regulation of transposable elements by DNA modifications. *Nature Reviews Genetics*, 20(7), 417–431.

Ding, Y., Avramova, Z., & Fromm, M. (2011). The *Arabidopsis* trithorax-like factor ATX1 functions in dehydration stress responses via ABA-dependent and ABA-independent pathways. *The Plant Journal*, 66(5), 735–744.

Douillet, D., Sze, C. C., Ryan, C., Piunti, A., Shah, A. P., Ugarenko, M., Marshall, S. A., Rendleman, E. J., Zha, D., Helmin, K. A., & Shilatifard, A. (2020). Uncoupling histone H3K4 trimethylation from developmental gene expression via an equilibrium of COMPASS, Polycomb and DNA methylation. *Nature Genetics*, 52(6), 615–625.

Ebert, A. W. (2022). Sprouts and microgreens—novel food sources for healthy diets. *Plants*, 11(4), 571.

Echeverria, V., Echeverria, F., Barreto, G. E., Echeverría, J., & Mendoza, C. (2021). Estrogenic plants: To prevent neurodegeneration and memory loss and other symptoms in women after menopause. *Frontiers in Pharmacology*, 12, 644103.

Escobar, T. M., Loyola, A., & Reinberg, D. (2021). Parental nucleosome segregation and the inheritance of cellular identity. *Nature Reviews Genetics*, 22(6), 379–392.

Fabrizio, P., Garvis, S., & Palladino, F. (2019). Histone methylation and memory of environmental stress. *Cells*, 8(4), 339.

Fallet, M. (2024). Epigenetic inheritance. In G. Witzany (Ed.), *Epigenetics in Biological Communication* (pp. 87–130). Cham: Springer Nature Switzerland.

Farkas, D., & Dobránszki, J. (2024). Vegetal memory through the lens of transcriptomic changes–recent progress and future practical prospects for exploiting plant transcriptional memory. *Plant Signaling & Behavior*, 19(1), 2383515.

Feng, P., Sun, X., Liu, X., Li, Y., Sun, Q., Lu, H., Li, M., Ding, X., & Dong, Y. (2022). Epigenetic regulation of plant tolerance to salt stress by histone acetyltransferase GsMYST1 from wild soybean. *Frontiers in Plant Science*, 13, 860056.

Flors, V., Kyndt, T., Mauch-Mani, B., Pozo, M. J., Ryu, C. M., & Ton, J. (2024). Enabling sustainable crop protection with induced resistance in plants. *Frontiers in Science*, 2, 1407410.

Foroozani, M., Holder, D. H., & Deal, R. B. (2022). Histone variants in the specialization of plant chromatin. *Annual Review of Plant Biology*, 73(1), 149–172.

Gallusci, P., Agius, D. R., Moschou, P. N., Dobránszki, J., Kaiserli, E., & Martinelli, F. (2023). Deep inside the epigenetic memories of stressed plants. *Trends in Plant Science*, 28(2), 142–153.

Ge, L., Pan, F., Jia, M., Pott, D. M., He, H., Shan, H., Lozano-Durán, R., Wang, A., Zhou, X., & Li, F. (2024). RNA modifications in plant biotic interactions. *Plant Communications*, 6(2), 101232.

Govindan, G., Sharma, B., Li, Y. F., Armstrong, C. D., Merum, P., Rohila, J. S., Gregory, B. D., & Sunkar, R. (2022). mRNA N6-methyladenosine is critical for cold tolerance in *Arabidopsis*. *The Plant Journal*, 111(4), 1052–1068.

Guo, T., Wang, D., Fang, J., Zhao, J., Yuan, S., Xiao, L., & Li, X. (2019). Mutations in the rice OsCHR4 gene, encoding a CHD3 family chromatin remodeler, induce narrow and rolled leaves with increased cuticular wax. *International Journal of Molecular Sciences*, 20(10), 2567.

Guo, Y., Zhao, S., & Wang, G. G. (2021). Polycomb gene silencing mechanisms: PRC2 chromatin targeting, H3K27me3'readout', and phase separation-based compaction. *Trends in Genetics*, 37(6), 547–565.

Habibi, N., Aryan, S., Amin, M. W., Sanada, A., Terada, N., & Koshio, K. (2023). Potential benefits of seed priming under salt stress conditions on physiological, and biochemical attributes of micro-tom tomato plants. *Plants*, 12(11), 2187.

Han, D., Chen, C., Xia, S., Liu, J., Shu, J., Nguyen, V., Lai, J., Cui, Y., & Yang, C. (2021). Chromatin-associated SUMOylation controls the transcriptional switch between plant development and heat stress responses. *Plant Communications*, 2(1), 100091.

Han, Q., Bartels, A., Cheng, X., Meyer, A., An, Y. Q. C., Hsieh, T. F., & Xiao, W. (2019). Epigenetics regulates reproductive development in plants. *Plants*, 8(12), 564.

Hideg, É., Jansen, M. A., & Strid, Å. (2013). UV-B exposure, ROS, and stress: Inseparable companions or loosely linked associates?. *Trends in Plant Science*, 18(2), 107–115.

Hiiback, K. M. (2023). *Identification and Systemic Investigation of Novel Chemical Seed Priming Agents and Their Persistent Effect on Plant Phenotype and Metabolism* (Doctoral dissertation, University of Toronto (Canada)).

Hilker, M., & Schmülling, T. (2019). Stress priming, memory, and signalling in plants. *Plant, Cell & Environment*, 42(3), 753–761.

Höll, J., Lindner, S., Walter, H., Joshi, D., Poschet, G., Pfleger, S., Ziegler, T., Hell, R., Bogs, J., & Rausch, T. (2019). Impact of pulsed UV-B stress exposure on plant performance: How recovery periods stimulate secondary metabolism while reducing adaptive growth attenuation. *Plant, Cell & Environment*, 42(3), 801–814.

Hossain, M. S., Kawakatsu, T., Kim, K. D., Zhang, N., Nguyen, C. T., Khan, S. M., Batek, J. M., Joshi, T., Schmutz, J., Grimwood, J., & Stacey, G. (2017). Divergent cytosine DNA methylation patterns in single-cell, soybean root hairs. *New Phytologist*, 214(2), 808–819.

Hu, J., Cai, J., Park, S. J., Lee, K., Li, Y., Chen, Y., Yun, J. Y., Xu, T. & Kang, H. (2021). N 6-Methyladenosine mRNA methylation is important for salt stress tolerance in *Arabidopsis*. *The Plant Journal*, 106(6), 1759–1775.

Hu, Z., Song, N., Zheng, M., Liu, X., Liu, Z., Xing, J., Ma, J., Guo, W., Yao, Y., Peng, H., & Sun, Q. (2015). Histone acetyltransferase GCN 5 is essential for heat stress-responsive gene activation and thermotolerance in *Arabidopsis*. *The Plant Journal*, 84(6), 1178–1191.

Huang, R., Dai, M., Jiang, S., Guo, Z., & Shi, H. (2024). Chloroplast-localized PvBASS2 regulates salt tolerance in the C4 plant seashore paspalum. *The Plant Journal*, 119(6), 2782–2796.

Huang, S., Zhang, A., Jin, J. B., Zhao, B., Wang, T. J., Wu, Y., Wang, S., Liu, Y., Wang, J., Guo, P., & Xu, Z. Y. (2019). *Arabidopsis* histone H3K4 demethylase JMJ 17 functions in dehydration stress response. *New Phytologist*, 223(3), 1372–1387.

Hung, Y. H., Liu, F., Zhang, X. Q., Xiao, W., & Hsieh, T. F. (2018). Sexual and non-sexual reproduction: Inheritance and stability of epigenetic variations and consequences for breeding application. In M. Mirouze, E. Bucher, & P. Gallusci (Eds.), *Advances in Botanical Research* (Vol. 88, pp. 117–163). Academic Press.

Hussain, Q., Asim, M., Zhang, R., Khan, R., Farooq, S., & Wu, J. (2021). Transcription factors interact with ABA through gene expression and signaling pathways to mitigate drought and salinity stress. *Biomolecules*, 11(8), 1159.

Iqbal, B., Ahmad, N., Li, G., Jalal, A., Khan, A. R., Zheng, X., Naeem, M., & Du, D. (2024). Unlocking plant resilience: Advanced epigenetic strategies against heavy metal and metalloid stress. *Plant Science*, 349, 112265.

Isbel, L., Grand, R. S., & Schübeler, D. (2022). Generating specificity in genome regulation through transcription factor sensitivity to chromatin. *Nature Reviews Genetics*, 23(12), 728–740.

Jin, H., Li, M., Duan, S., Fu, M., Dong, X., Liu, B., Feng, D., Wang, J., & Wang, H. B. (2016). Optimization of light-harvesting pigment improves photosynthetic efficiency. *Plant Physiology*, 172(3), 1720–1731.

Jin, Q., Chachar, M., Ali, A., Chachar, Z., Zhang, P., Riaz, A., Ahmed, N., & Chachar, S. (2024). Epigenetic regulation for heat stress adaptation in plants: New horizons for crop improvement under climate change. *Agronomy*, 14(9), 2105.

Kakoulidou, I., Avramidou, E. V., Baránek, M., Brunel-Muguet, S., Farrona, S., Johannes, F., Kaiserli, E., Lieberman-Lazarovich, M., Martinelli, F., Mladenov, V., & Maury, S. (2021). Epigenetics for crop improvement in times of global change. *Biology*, 10(8), 766.

Kanu, A. S., Ashraf, U., Bangura, A., Yang, D. M., Ngaujah, A. S., & Tang, X. (2017). Cadmium (Cd) stress in rice; phytoavailability, toxic effects, and mitigation measures-a critical review. *IOSR-JESTFT*, 11, 7–23.

Kappel, C., Friedrich, T., Oberkofler, V., Jiang, L., Crawford, T., Lenhard, M., & Bäurle, I. (2023). Genomic and epigenomic determinants of heat stress-induced transcriptional memory in *Arabidopsis*. *Genome Biology*, 24(1), 129.

Kaur, M., Tak, Y., Bhatia, S., Asthir, B., Lorenzo, J. M., & Amarowicz, R. (2021). Crosstalk during the carbon–nitrogen cycle that interlinks the biosynthesis, mobilization and accumulation of seed storage reserves. *International Journal of Molecular Sciences*, 22(21), 12032.

Kawashima, T., & Berger, F. (2014). Epigenetic reprogramming in plant sexual reproduction. *Nature Reviews Genetics*, 15(9), 613–624.

Kim, J. H. (2021). Multifaceted chromatin structure and transcription changes in plant stress response. *International Journal of Molecular Sciences*, 22(4), 2013.

Klemm, S. L., Shipony, Z., & Greenleaf, W. J. (2019). Chromatin accessibility and the regulatory epigenome. *Nature Reviews Genetics*, 20(4), 207–220.

Kohli, S. K., Bali, S., Tejpal, R., Bhalla, V., Verma, V., Bhardwaj, R., Alqarawi, A. A., Abd_Allah, E. F., & Ahmad, P. (2019). In-situ localization and biochemical analysis of bio-molecules reveals Pb-stress amelioration in *Brassica juncea* L. by co-application of 24-Epibrassinolide and Salicylic Acid. *Scientific Reports*, 9(1), 3524.

Komen, J., Tripathi, L., Mkoko, B., Ofosu, D. O., Oloka, H., & Wangari, D. (2020). Biosafety regulatory reviews and leeway to operate: Case studies from sub-Sahara Africa. *Frontiers in Plant Science*, 11, 130.

Kosakivska, I. V., Babenko, L. M., Romanenko, K. O., Korotka, I. Y. & Potters, G. (2021). Molecular mechanisms of plant adaptive responses to heavy metals stress. *Cell Biology International*, 45(2), 258–272.

Krzysztof, B., Altmann, S., Hjördis, C., Nicolas, P., Michal, G., Benke, E., Tina, K., Felix, J., Graf, A., Kappel, C., & Isabel, B. (2016). *Arabidopsis* FORGETTER1 mediates stress-induced chromatin memory through nucleosome remodeling. *eLife*, 5, e17061.

Kumar, G., Rattan, U. K., & Singh, A. K. (2016). Chilling-mediated DNA methylation changes during dormancy and its release reveal the importance of epigenetic regulation during winter dormancy in apple (*Malus* x *domestica* Borkh.). *PLoS One*, 11(2), e0149934.

Kumari, P., Khan, S., Wani, I. A., Gupta, R., Verma, S., Alam, P., & Alaklabi, A. (2022). Unravelling the role of epigenetic modifications in development and reproduction of angiosperms: A critical appraisal. *Frontiers in Genetics*, 13, 819941.

Kumari, V. V., Banerjee, P., Verma, V. C., Sukumaran, S., Chandran, M. A. S., Gopinath, K. A., Venkatesh, G., Yadav, S. K., Singh, V. K., & Awasthi, N. K. (2022). Plant nutrition: An effective way to alleviate abiotic stress in agricultural crops. *International Journal of Molecular Sciences*, 23(15), 8519.

Kwon, C. S., Lee, D., Choi, G., & Chung, W. I. (2009). Histone occupancy-dependent and -independent removal of H3K27 trimethylation at cold-responsive genes in *Arabidopsis*. *The Plant Journal*, 60(1), 112–121.

Lagiotis, G., Madesis, P., & Stavridou, E. (2023). Echoes of a stressful past: Abiotic stress memory in crop plants towards enhanced adaptation. *Agriculture*, 13(11), 2090.

Lam, U. T. F., Tan, B. K. Y., Poh, J. J. X., & Chen, E. S. (2022). Structural and functional specificity of H3K36 methylation. *Epigenetics & Chromatin*, 15(1), 17.

Lawas, L. M. F., Li, X., Erban, A., Kopka, J., Jagadish, S. K., Zuther, E., & Hincha, D. K. (2019). Metabolic responses of rice cultivars with different tolerance to combined drought and heat stress under field conditions. *GigaScience*, 8(5), 50.

Levy, D., Coleman, W. K., & Veilleux, R. E. (2013). Adaptation of potato to water shortage: Irrigation management and enhancement of tolerance to drought and salinity. *American Journal of Potato Research*, 90, 186–206.

Liu, H., Able, A. J., & Able, J. A. (2022). Priming crops for the future: Rewiring stress memory. *Trends in Plant Science*, 27(7), 699–716.

Liu, H. C., Lämke, J., Lin, S. Y., Hung, M. J., Liu, K. M., Charng, Y. Y., & Bäurle, I. (2018). Distinct heat shock factors and chromatin modifications mediate the organ-autonomous transcriptional memory of heat stress. *The Plant Journal*, 95(3), 401–413.

Liu, K., Chen, J., Sun, S., Chen, X., Zhao, X., Hu, Y., Qi, G., Li, X., Xu, B., Miao, J., & Gong, Z. (2023). Histone deacetylase OsHDA706 increases salt tolerance via H4K5/K8 deacetylation of OsPP2C49 in rice. *Journal of Integrative Plant Biology*, 65(6), 1394–1407.

Liu, S., de Jonge, J., Trejo-Arellano, M. S., Santos-González, J., Köhler, C., & Hennig, L. (2021). Role of H1 and DNA methylation in selective regulation of transposable elements during heat stress. *New Phytologist*, 229(4), 2238–2250.

Lowe, B. R., Maxham, L. A., Hamey, J. J., Wilkins, M. R., & Partridge, J. F. (2019). Histone H3 mutations: An updated view of their role in chromatin deregulation and cancer. *Cancers*, 11(5), 660.

Lu, L., Wei, W., Tao, J. J., Lu, X., Bian, X. H., Hu, Y., Cheng, T., Yin, C. C., Zhang, W. K., Chen, S. Y., & Zhang, J. S. (2021). Nuclear factor Y subunit GmNFYA competes with GmHDA13 for interaction with GmFVE to positively regulate salt tolerance in soybean. *Plant Biotechnology Journal*, 19(11), 2362–2379.

Lyall, R., Ingle, R. A., & Illing, N. (2014). The window of desiccation tolerance shown by early-stage germinating seedlings remains open in the resurrection plant, *Xerophyta viscosa*. *PLoS One*, 9(3), e93093.

Ma, S., Tang, N., Li, X., Xie, Y., Xiang, D., Fu, J., Shen, J., Yang, J., Tu, H., Li, X., & Xiong, L. (2019). Reversible histone H2B monoubiquitination fine-tunes abscisic acid signaling and drought response in rice. *Molecular Plant*, 12(2), 263–277.

Ma, T., Wang, S., Sun, C., Tian, J., Guo, H., Cui, S., & Zhao, H. (2023). *Arabidopsis* LFR, a SWI/SNF complex component, interacts with ICE1 and activates ICE1 and CBF3 expression in cold acclimation. *Frontiers in Plant Science*, 14, 1097158.

Mahmood, U., Li, X., Fan, Y., Chang, W., Niu, Y., Li, J., Qu, C., & Lu, K. (2022). Multi-omics revolution to promote plant breeding efficiency. *Frontiers in Plant Science*, 13, 1062952.

Maity, S., Abbaspour, R., Nahabedian, D., & Connor, S. A. (2022). Norepinephrine, beyond the synapse: Coordinating epigenetic codes for memory. *International Journal of Molecular Sciences*, 23(17), 9916.

Mansisidor, A. R., & Risca, V. I. (2022). Chromatin accessibility: Methods, mechanisms, and biological insights. *Nucleus*, 13(1), 238–278.

Mertens, D., Boege, K., Kessler, A., Koricheva, J., Thaler, J. S., Whiteman, N. K., & Poelman, E. H. (2021). Predictability of biotic stress structures plant defence evolution. *Trends in Ecology & Evolution*, 36(5), 444–456.

Mognato, M., Burdak-Rothkamm, S., & Rothkamm, K. (2021). Interplay between DNA replication stress, chromatin dynamics and DNA-damage response for the maintenance of genome stability. *Mutation Research/Reviews in Mutation Research*, 787, 108346.

Morgan, M. A., & Shilatifard, A. (2020). Reevaluating the roles of histone-modifying enzymes and their associated chromatin modifications in transcriptional regulation. *Nature Genetics*, 52(12), 1271–1281.

Mozgova, I., Mikulski, P., Pecinka, A., & Farrona, S. (2019). Epigenetic mechanisms of abiotic stress response and memory in plants. In R. Alvarez-Venegas, C. De-la-Peña, & J. Casas-Mollano (Eds.), *Epigenetics in Plants of Agronomic Importance: Fundamentals and Applications: Transcriptional Regulation and Chromatin Remodelling in Plants* (pp. 1–64). Cham: Springer.

Muhammad, T., Zhang, J., Ma, Y., Li, Y., Zhang, F., Zhang, Y., & Liang, Y. (2019). Overexpression of a mitogen-activated protein kinase SlMAPK3 positively regulates tomato tolerance to cadmium and drought stress. *Molecules*, 24(3), 556.

Mukhopadhyay, J., & Roychoudhury, A. (2018). Cold-induced injuries and signaling responses in plants. In S. H. Wani & V. Herath (Ed.), *Cold Tolerance in Plants Physiological, Molecular and Genetic Perspectives* (pp. 1–35). Cham: Springer Nature Switzerland.

Narita, T., Weinert, B. T., & Choudhary, C. (2019). Functions and mechanisms of non-histone protein acetylation. *Nature reviews Molecular Cell Biology*, 20(3), 156–174.

Nasrallah, A. K., Atia, M. A., Abd El-Maksoud, R. M., Kord, M. A., & Fouad, A. S. (2022). Salt priming as a smart approach to mitigate salt stress in Faba bean (*Vicia faba* L.). *Plants*, 11(12), 1610.

Nguyen, N. H., Vu, N. T., & Cheong, J. J. (2022). Transcriptional stress memory and transgenerational inheritance of drought tolerance in plants. *International Journal of Molecular Sciences*, 23(21), 12918.

Nishio, H., Kawakatsu, T., & Yamaguchi, N. (2024). Beyond heat waves: Unlocking epigenetic heat stress memory in *Arabidopsis*. *Plant Physiology*, 194(4), 1934–1951.

Niu, Y., Bai, J., Liu, X., Zhang, H., Bao, J., Zhao, W., Hou, Y., Deng, X., Yang, C., Guo, L., & Zheng, S. (2022). HISTONE DEACETYLASE 9 transduces heat signal in plant cells. *Proceedings of the National Academy of Sciences*, 119(45), e2206846119.

Nozawa, K., Masuda, S., Saze, H., Ikeda, Y., Suzuki, T., Takagi, H., Tanaka, K., Ohama, N., Niu, X., Kato, A., & Ito, H. (2022). Epigenetic regulation of ecotype-specific expression of the heat-activated transposon ONSEN. *Frontiers in Plant Science*, 13, 899105.

Oberkofler, V., Pratx, L., & Bäurle, I. (2021). Epigenetic regulation of abiotic stress memory: Maintaining the good things while they last. *Current Opinion in Plant Biology*, 61, 102007.

Obidiegwu, J. E., Bryan, G. J., Jones, H. G., & Prashar, A. (2015). Coping with drought: Stress and adaptive responses in potato and perspectives for improvement. *Frontiers in Plant Science*, 6, 542.

Ost, C., Cao, H. X., Nguyen, T. L., Himmelbach, A., Mascher, M., Stein, N., & Humbeck, K. (2023). Drought-stress-related reprogramming of gene expression in barley involves differential histone modifications at ABA-related genes. *International Journal of Molecular Sciences*, 24(15), 12065.

Pan, J., Zhang, H., Zhan, Z., Zhao, T., & Jiang, D. (2023). A REF6-dependent H3K27me3-depleted state facilitates gene activation during germination in *Arabidopsis*. *Journal of Genetics and Genomics*, 50(3), 178–191.

Park, J., Lim, C. J., Shen, M., Park, H. J., Cha, J. Y., Iniesto, E., Rubio, V., Mengiste, T., Zhu, J. K., Bressan, R. A., & Lee, S. Y. (2018). Epigenetic switch from repressive to permissive chromatin in response to cold stress. *Proceedings of the National Academy of Sciences*, 115(23), 5400–5409.

Parrilla-Doblas, J. T., Roldán-Arjona, T., Ariza, R. R., & Córdoba-Cañero, D. (2019). Active DNA demethylation in plants. *International Journal of Molecular Sciences*, 20(19), 4683.

Paul, A., Mondal, S., Pal, A., Biswas, S., Chakraborty, K., Mazumder, A., Biswas, A. K., & Kundu, R. (2023). Seed priming with NaCl helps to improve tissue tolerance, potassium retention ability of plants, and protects the photosynthetic ability in two different legumes, chickpea and lentil, under salt stress. *Planta*, 257(6), 111.

Perillo, B., Tramontano, A., Pezone, A., & Migliaccio, A. (2020). LSD1: More than demethylation of histone lysine residues. *Experimental & Molecular Medicine*, 52(12), 1936–1947.

Piccini, C. (2022). Physiological, biochemical and metabolomic changes in two Italian cultivars of *Olea europaea* in response to UV-B radiations. Doctoral Dissertation, Università degli Studi di Siena.

Pissolato, M. D., Martins, T. S., Fajardo, Y. C., Souza, G. M., Machado, E. C., & Ribeiro, R. V. (2024). Stress memory in crops: What we have learned so far. *Theoretical and Experimental Plant Physiology*, 36(3), 535–565.

Pratx, L., Crawford, T., & Bäurle, I. (2024). Mechanisms of heat stress-induced transcriptional memory. *Current Opinion in Plant Biology*, 81, 102590.

Quadrana, L., & Colot, V. (2016). Plant transgenerational epigenetics. *Annual Review of Genetics*, 50(1), 467–491.

Rajkumar, M. S., Tembhare, K., Garg, R., & Jain, M. (2024). Genome-wide mapping of DNase I hypersensitive sites revealed differential chromatin accessibility and regulatory DNA elements under drought stress in rice cultivars. *The Plant Journal*, 119(4), 2063–2079.

Ramazi, S., Allahverdi, A., & Zahiri, J. (2020). Evaluation of post-translational modifications in histone proteins: A review on histone modification defects in developmental and neurological disorders. *Journal of Biosciences*, 45(1), 135.

Rashid, M. M., Vaishnav, A., Verma, R. K., Sharma, P., Suprasanna, P., & Gaur, R. K. (2022). Epigenetic regulation of salinity stress responses in cereals. *Molecular Biology Reports*, 49(1), 761–772.

Raza, M. A. S., Aslam, M. U., Valipour, M., Iqbal, R., Haider, I., Mustafa, A. E. Z. M., Elshikh, M. S., Ali, I., Roy, R., & Elshamly, A. M. (2024). Seed priming with selenium improves growth and yield of quinoa plants suffering drought. *Scientific Reports*, 14(1), 886.

Rengaraj, P., Obrdlík, A., Vukić, D., Varadarajan, N. M., Keegan, L. P., Vaňáčová, Š., & O'Connell, M. A. (2021). Interplays of different types of epitranscriptomic mRNA modifications. *RNA Biology*, 18(supl), 19–30.

Rojas-Carvajal, M., Sequeira-Cordero, A., & Brenes, J. C. (2022). The environmental enrichment model revisited: A translatable paradigm to study the stress of our modern lifestyle. *European Journal of Neuroscience*, 55, 2359–2392.

Roquis, D., Robertson, M., Yu, L., Thieme, M., Julkowska, M., & Bucher, E. (2021). Genomic impact of stress-induced transposable element mobility in *Arabidopsis*. *Nucleic Acids Research*, 49(18), 10431–10447.

Rossatto, T., Souza, G. M., do Amaral, M. N., Auler, P. A., Pérez-Alonso, M. M., Pollmann, S., & Braga, E. J. B. (2023). Cross-stress memory: Salt priming at vegetative growth stages improves tolerance to drought stress during grain-filling in rice plants. *Environmental and Experimental Botany*, 206, 105187.

Rowley, M. J., & Corces, V. G. (2018). Organizational principles of 3D genome architecture. *Nature Reviews Genetics*, 19(12), 789–800.

Roy, D., Paul, A., Roy, A., Ghosh, R., Ganguly, P., & Chaudhuri, S. (2014). Differential acetylation of histone H3 at the regulatory region of OsDREB1b promoter facilitates chromatin remodelling and transcription activation during cold stress. *PLoS One*, 9(6), e100343.

Sadhukhan, A., Prasad, S. S., Mitra, J., Siddiqui, N., Sahoo, L., Kobayashi, Y., & Koyama, H. (2022). How do plants remember drought?. *Planta*, 256(1), 7.

Samantara, K., Shiv, A., de Sousa, L. L., Sandhu, K. S., Priyadarshini, P., & Mohapatra, S. R. (2021). A comprehensive review on epigenetic mechanisms and application of epigenetic modifications for crop improvement. *Environmental and Experimental Botany*, 188, 104479.

Sani, E., Herzyk, P., Perrella, G., Colot, V., & Amtmann, A. (2013). Hyperosmotic priming of *Arabidopsis* seedlings establishes a long-term somatic memory accompanied by specific changes of the epigenome. *Genome Biology*, 14(6), 59.

Sayed, S., & Gadallah, M. (2019). Hydrogen peroxide supplementation alleviates the deleterious effects of cadmium on photosynthetic pigments and oxidative stress and improves growth, yield and pods quality of pea (*Pisum sativum* L.) plants. *Acta Physiologiae Plantarum*, 41, 1–12.

Schaefer, M. R. (2021). The regulation of RNA modification systems: The next frontier in epitranscriptomics?. *Genes*, 12(3), 345.

Schoenfelder, S., & Fraser, P. (2019). Long-range enhancer–promoter contacts in gene expression control. *Nature Reviews Genetics*, 20(8), 437–455.

Sellamuthu, G., Tarafdar, A., Jasrotia, R. S., Chaudhary, M., Vishwakarma, H., & Padaria, J. C. (2024). Introgression of Δ 1-pyrroline-5-carboxylate synthetase (PgP5CS) confers enhanced resistance to abiotic stresses in transgenic tobacco. *Transgenic Research*, 33(3), 131–147.

Selote, D. S., & Khanna-Chopra, R. (2010). Antioxidant response of wheat roots to drought acclimation. *Protoplasma*, 245, 153–163.

Shahzad, A., Ullah, S., Dar, A. A., Sardar, M. F., Mehmood, T., Tufail, M. A., Shakoor, A., & Haris, M. (2021). Nexus on climate change: Agriculture and possible solution to cope future climate change stresses. *Environmental Science and Pollution Research*, 28, 14211–14232.

Shaikh, A. A., Chachar, S., Chachar, M., Ahmed, N., Guan, C., & Zhang, P. (2022). Recent advances in DNA methylation and their potential breeding applications in plants. *Horticulturae*, 8(7), 562.

Sharma, L. K., & Naik, R. (2024). Saline Wetland ecosystems. In L. K. Sharma & R. Nair (Eds.), *Conservation of Saline Wetland Ecosystems: An Initiative towards UN Decade of Ecological Restoration* (pp. 33–75). Singapore: Springer Nature Singapore.

Sharma, M., Sharma, M., Jamsheer K, M., & Laxmi, A. (2022). A glucose–target of rapamycin signaling axis integrates environmental history of heat stress through maintenance of transcription-associated epigenetic memory in *Arabidopsis*. *Journal of Experimental Botany*, 73(20), 7083–7102.

Shoaib, Y., Usman, B., Kang, H., & Jung, K. H. (2022). Epitranscriptomics: An additional regulatory layer in plants' development and stress response. *Plants*, 11(8), 1033.

Singh, V. J., Vinod, K. K., Krishnan, S. G., & Singh, A. K. (2021). Rice adaptation to climate change: Opportunities and priorities in molecular breeding. In M. A. Hossain, L. Hassan, K. M. Ifterkharuddaula, A. Kumar, & R. Henry (Eds.), *Molecular Breeding for Rice Abiotic Stress Tolerance and Nutritional Quality* (pp. 1–25). Hoboken, NJ: Wiley Online Library.

Singh, V. P., Kumar, J., Singh, M., Singh, S., Prasad, S. M., Dwivedi, R., & Singh, M. P. V. V. B. (2016). Role of salicylic acid-seed priming in the regulation of chromium (VI) and UV-B toxicity in maize seedlings. *Plant Growth Regulation*, 78, 79–91.

Singroha, G., Kumar, S. Gupta, O. P., Singh, G. P., & Sharma, P. (2022). Uncovering the epigenetic marks involved in mediating salt stress tolerance in plants. *Frontiers in Genetics*, 13, 811732.

Sow, M. D., Le Gac, A. L., Fichot, R., Lanciano, S., Delaunay, A., Le Jan, I., Lesage-Descauses, M. C., Citerne, S., Caius, J., Brunaud, V., & Maury, S. (2020). Hypomethylated poplars show higher tolerance to water deficit and highlight a dual role for DNA methylation in shoot meristem: Regulation of stress response and genome integrity. *BioRxiv*, 2020–04.

Srivastava, A. K., Suresh Kumar, J., & Suprasanna, P. (2021). Seed 'primeomics': Plants memorize their germination under stress. *Biological Reviews*, 96(5), 1723–1743.

Sun, C., Ali, K., Yan, K., Fiaz, S., Dormatey, R., Bi, Z., & Bai, J. (2021). Exploration of epigenetics for improvement of drought and other stress resistance in crops: A review. *Plants*, 10(6), 1226.

Tamindžić, G., Ignjatov, M., Milošević, D., Nikolić, Z., Kravljanac, L. K., Jovičić, D., Dolijanović, Ž., & Savić, J. (2021). Seed priming with zinc improves field performance of maize hybrids grown on calcareous chernozem. *Italian Journal of Agronomy*, 16(3), 1795.

Tang, J., Chen, S., & Jia, G. (2023). Detection, regulation, and functions of RNA N6-methyladenosine modification in plants. *Plant Communications*, 4(3), 100–546.

Tang, W. S., Zhong, L., Ding, Q. Q., Dou, Y. N., Li, W. W., Xu, Z. S., Zhou, Y. B., Chen, J., Chen, M., & Ma, Y. Z. (2022). Histone deacetylase AtSRT2 regulates salt tolerance during seed germination via repression of vesicle-associated membrane protein 714 (VAMP714) in *Arabidopsis*. *New Phytologist*, 234(4), 1278–1293.

Tankari, M., Wang, C., Ma, H., Li, X., Li, L., Soothar, R. K., Cui, N., Zaman-Allah, M., Hao, W., Liu, F., & Wang, Y. (2021). Drought priming improved water status, photosynthesis and water productivity of cowpea during post-anthesis drought stress. *Agricultural Water Management*, 245, 106565.

To, T. K., Nakaminami, K., Kim, J. M., Morosawa, T., Ishida, J., Tanaka, M., Yokoyama, S., Shinozaki, K., & Seki, M. (2011). Arabidopsis HDA6 is required for freezing tolerance. *Biochemical and Biophysical Research Communications*, 406(3), 414–419.

Tricker, P. J., Rodríguez López, C. M., Hadley, P., Wagstaff, C., & Wilkinson, M. J. (2013). Pre-conditioning the epigenetic response to high vapor pressure deficit increases the drought tolerance of *Arabidopsis thaliana*. *Plant Signaling & Behavior*, 8(10), e25974.

Trușcă, M., Gâdea, Ș., Vidican, R., Stoian, V., Vâtcă, A., Balint, C., Stoian, V. A., Horvat, M., & Vâtcă, S. (2023). Exploring the research challenges and perspectives in ecophysiology of plants affected by salinity stress. *Agriculture*, 13(3), 734.

Tun, W., Yoon, J., Jeon, J. S., & An, G. (2021). Influence of climate change on flowering time. *Journal of Plant Biology*, 64(3), 193–203.

Ueda, M., Matsui, A., Tanaka, M., Nakamura, T., Abe, T., Sako, K., Sasaki, T., Kim, J. M., Ito, A., Nishino, N., & Seki, M. (2017). The distinct roles of class I and II RPD3-like histone deacetylases in salinity stress response. *Plant Physiology*, 175(4), 1760–1773.

van Buer, J., Cvetkovic, J., & Baier, M. (2016). Cold regulation of plastid ascorbate peroxidases serves as a priming hub controlling ROS signaling in *Arabidopsis thaliana*. *BMC Plant Biology*, 16, 1–20.

Vaschetto, L. M. (2024). The role of epigenetic variation in plant adaptation. In L. M. Vaschetto (Eds.), *Epigenetics in Crop Improvement: Safeguarding Food Security in an Ever-Changing Climate* (pp. 29–47). Cham: Springer Nature Switzerland.

Verza, F. A., Das, U., Fachin, A. L., Dimmock, J. R., & Marins, M. (2020). Roles of histone deacetylases and inhibitors in anticancer therapy. *Cancers*, 12(6), 1664.

Vlachonasios, K. E., Thomashow, M. F., & Triezenberg, S. J. (2003). Disruption mutations of ADA2b and GCN5 transcriptional adaptor genes dramatically affect *Arabidopsis* growth, development, and gene expression. *The Plant Cell*, 15(3), 626–638.

Wahab, A., Abdi, G., Saleem, M. H., Ali, B., Ullah, S., Shah, W., Mumtaz, S., Yasin, G., Muresan, C. C., & Marc, R. A. (2022). Plants' physio-biochemical and phyto-hormonal responses to alleviate the adverse effects of drought stress: A comprehensive review. *Plants*, 11(13), 1620.

Wang, S., Hu, J., Ren, B., Liu, P., Zhao, B., & Zhang, J. (2022). Effects of hydrogen peroxide priming on yield, photosynthetic capacity and chlorophyll fluorescence of waterlogged summer maize. *Frontiers in Plant Science*, 13, 1042920.

Wang, L., Xue, M., Zhang, H., Ma, L., & Jiang, D. (2023). TONSOKU is required for the maintenance of repressive chromatin modifications in *Arabidopsis*. *Cell Reports*, 42(7), 112738.

Wang, Q., Liu, P., Jing, H., Zhou, X. F., Zhao, B., Li, Y., & Jin, J. B. (2021). JMJ27-mediated histone H3K9 demethylation positively regulates drought-stress responses in *Arabidopsis*. *New Phytologist*, 232(1), 221–236.

Wei, H., Wang, X., He, Y., Xu, H., & Wang, L. (2021). Clock component OsPRR73 positively regulates rice salt tolerance by modulating OsHKT2; 1-mediated sodium homeostasis. *The EMBO Journal*, 40(3), e105086.

Wong, L. H., & Tremethick, D. J. (2024). Multifunctional histone variants in genome function. *Nature Reviews Genetics*, *26*(2), 82–104.

Wu, X., Zhang, X., Huang, B., Han, J., & Fang, H. (2023). Advances in biological functions and mechanisms of histone variants in plants. *Frontiers in Genetics*, 14, 1229782.

Xavier, M. J., Roman, S. D., Aitken, R. J., & Nixon, B. (2019). Transgenerational inheritance: How impacts to the epigenetic and genetic information of parents affect offspring health. *Human Reproduction Update*, 25(5), 519–541.

Yang, F., Sun, Y., Du, X., Chu, Z., Zhong, X., & Chen, X. (2023). Plant-specific histone deacetylases associate with ARGONAUTE4 to promote heterochromatin stabilization and plant heat tolerance. *New Phytologist*, 238(1), 252–269.

Yanqun, Z., Yuan, L., Haiyan, C., & Jianjun, C. (2003). Intraspecific differences in physiological response of 20 soybean cultivars to enhanced ultraviolet-B radiation under field conditions. *Environmental and Experimental Botany*, 50(1), 87–97.

Yung, W. S., Li, M. W., Sze, C. C., Wang, Q., & Lam, H. M. (2021). Histone modifications and chromatin remodelling in plants in response to salt stress. *Physiologia Plantarum*, 173(4), 1495–1513.

Zang, X., Geng, X., Wang, F., Liu, Z., Zhang, L., Zhao, Y., & Peng, H. (2017). Overexpression of wheat ferritin gene TaFER-5B enhances tolerance to heat stress and other abiotic stresses associated with the ROS scavenging. *BMC Plant Biology*, 17, 1–13.

Zhang, H., & Zhu, J. K. (2024). Epigenetic gene regulation in plants and its potential applications in crop improvement. *Nature Reviews Molecular Cell Biology*, *26*(1), 51–67.

Zhang, X., Li, C., Tie, D., Quan, J., Yue, M., & Liu, X. (2021). Epigenetic memory and growth responses of the clonal plant *Glechoma longituba* to parental recurrent UV-B stress. *Functional Plant Biology*, 48(8), 827–838.

Zhao, S., Allis, C. D., & Wang, G. G. (2021). The language of chromatin modification in human cancers. *Nature Reviews Cancer*, 21(7), 413–430.

Zheng, M., Lin, J., Liu, X., Chu, W., Li, J., Gao, Y., An, K., Song, W., Xin, M., Yao, Y., & Hu, Z. (2021). Histone acetyltransferase TaHAG1 acts as a crucial regulator to strengthen salt tolerance of hexaploid wheat. *Plant Physiology*, 186(4), 1951–1969.

Zheng, M., Liu, X., Lin, J., Liu, X., Wang, Z., Xin, M., Yao, Y., Peng, H., Zhou, D. X., Ni, Z., & Hu, Z. (2019). Histone acetyltransferase GCN 5 contributes to cell wall integrity and salt stress tolerance by altering the expression of cellulose synthesis genes. *The Plant Journal*, 97(3), 587–602.

Zheng, S., Zhao, W., Liu, Z., Geng, Z., Li, Q., Liu, B., Li, B., & Bai, J. (2024). Establishment and maintenance of heat-stress memory in plants. *International Journal of Molecular Sciences*, 25(16), 8976.

Zhou, R., Yu, X., Li, X., Dos Santos, T. M., Rosenqvist, E., & Ottosen, C. O. (2020). Combined high light and heat stress-induced complex response in tomato with better leaf cooling after heat priming. *Plant Physiology and Biochemistry*, 151, 1–9.

Zhou, Y., Xu, F., Shao, Y., & He, J. (2022). Regulatory mechanisms of heat stress response and thermomorphogenesis in plants. *Plants*, 11(24), 3410.

Zhu, Z. H., Sami, A., Xu, Q. Q., Wu, L. L., Zheng, W. Y., Chen, Z. P., Jin, X. Z., Zhang, H., Li, Y., Yu, Y., & Zhou, K. J. (2021). Effects of seed priming treatments on the germination and development of two rapeseed (*Brassica napus* L.) varieties under the co-influence of low temperature and drought. *PLoS One*, 16(9), e0257236.

29 Epigenetics

A Tool for Modifying Plants to Combat against Environmental Stress

Sachin Patil and Jagannath Patil

29.1 INTRODUCTION

Epigenetics is one of the prominent fields of genetics, completely interfering with classical information on the interaction between the hereditary material (DNA) and the phenotype (Russo et al., 1996). Epigenetic modifications are heritable modifications in the genetic material such as alterations in the pattern of DNA methylation and histone proteins. These mechanisms do not affect the DNA sequence but can reversibly modify the expression of genes. Plants experience several epigenetic modifications throughout their life span, directly affecting gene expression in the current and subsequent generations. The epigenetic changes in plants are involved in regulating several important biological phenomena such as the silencing of transgenes and transposons, the regulation of imprinting, growth and development (True et al., 1993). Various environmental factors influence epigenetic mechanisms in plants (Iwasaki et al., 2014). Plant response to environmental stresses is a complex phenomenon and involves several mechanisms activated and controlled by massive changes in gene expression and nuclear organization (Budak et al., 2015). The environmental stresses, viz., biotic or abiotic, influence the plant epigenetic mechanisms. Abiotic stresses include high or low temperature, excess rainfall, drought, high or low light intensity, nutrient deficiency, heavy metal toxicity and salinity, whereas biotic stresses include bacterial, viral, fungal and insect infestations (Law et al., 2010). Plants are exposed to environmental stresses due to unpredictable weather and climate change, affecting their growth, development and productivity. To combat the abiotic stresses, plants have evolved different approaches through epigenetics-mediated gene regulation to survive during the stresses (Kumar & Sing, 2016). The environmental stresses may lead to variations in gene expression and are transmitted to newly designed cells during their vegetative growth and even transferred to the offspring, known as epigenetic inheritance (Cheong, 2021). Recent advances in our understanding of the epigenetic mechanisms that contribute to abiotic stress tolerance in plants are truly exciting, and they give us hope for mitigating the impact of these stresses on various crop plants. Genetic engineering may be used to incorporate genes that encode specific enzymes for epigenetic modifications, viz., a gene encoding the enzyme DNA methyltransferases, when inserted in a plant, a methyl group is added to DNA that improves stress tolerance capacity in various plant species (Asensi-Fabado et al., 2017).

29.2 EPIGENETIC MECHANISMS AND MARKS FOR ABIOTIC STRESS

29.2.1 DNA Methylation

In plants, chromatin can be modified by using the DNA methylation mechanism that occurs in CG, CHG and CHH (H = A, T or C) contexts through distinct pathways. The plant enzymes, viz., methyltransferase 1 (MET1) and chromomethylase 3 (CMT3), are responsible for controlling the CG and CHG methylation, respectively. There are two pathways through which one can perform CHH methylation. The RNA-dependent DNA methylation pathway is carried out by the enzyme methyltransferase, which involves the biogenesis of small interfering RNAs that are targeted to corresponding genomic loci by ARGONAUTE (AGO) family members. The second pathway requires the enzyme chromomethylase 2 (CMT2) with a decrease in DNA Methylation 1 (DDM1) in histone H1-enriched chromatin regions (Zemach et al., 2013 and Zhang et al., 2018). Furthermore, DNA methylation may be removed by a family of bifunctional methyl-cytosine glycosylases-apurinic/apyrimidinic lyases through a base editing repair machinery. DNA methylation may affect gene expression, regulate imprinting, activate transposable elements and transposable element-associated genes, particularly in response to environmental cues (Penterman et al., 2007 and Law and Jacobsen, 2010).

29.2.2 Histone Protein Modification

Chromatin modifications stimulate the accessibility of DNA sequences, such as coding and regulatory DNA sequences, to the transcriptional machinery, helping in the regulation of gene expression (Li et al., 2008). This DNA accessibility for transcription is facilitated by core nucleosome aligning. The core nucleosome is composed of four types of histone proteins (H2A, H2B, H3, H4) in two copies

DOI: 10.1201/9781003539914-33

forming a histone octamer and about 146 base pairs of DNA in two turns. The unpackaged DNA in nucleosomes, with variable length, is reliant on the compaction level of nucleosomes (determined by the balance between the DNA self-repulsion and other factors) and is allied to the linker histone protein (H1). The interaction between histone proteins and DNA is influenced by reversible covalent post-translational modifications in the histone tails that protrude from the nucleosome core particle. The post-translational alterations dynamically control the level of chromatin condensation and DNA accessibility (Wei et al., 2017). The two well-known reversible plant histone modifications are histone acetylation/deacetylation and histone methylation/demethylation. Histone acetylation is correlated with transcriptional activation whereas histone deacetylation is correlated with transcriptional repression. The histone acetylation marks are H3K9ac (histone 3 lysine 9 acetylation), H3K14ac (histone 3 lysine 14 acetylation) and H3K36ac (histone 3 lysine 36 acetylation) associated with gene activation. The markers associated with histone deacetylation are histone deacetylases (HDACs) during transcriptional repression. However, histone methylation has a gene activating or repressing role depending on the site of modifications. The histone methylation markers H3K4 (histone H3 lysine 4) and H3K36 (histone 3 lysine 36 methylation) are related to gene expression, while H3K9 (histone 3 lysine 9 methylation) and H3K27 (histone 3 lysine 27 methylation) are related to gene repression and heterochromatin formation. In plants, both transposable elements and repetitive sequence-enriched heterochromatic regions are marked by histone H3 lysine 9 monomethylation and dimethylation (H3K9me1 and H3K9me2), which maintain the repressive transcriptional state. Heterochromatin regions are also associated with H3K27me1, while the trimethylated state of H3K27 (H3K27me3) exerts a repressive role in euchromatin regions. In particular, H3K27me3 is preferentially enriched in the entire transcribed region of inactive genes (Liu et al., 2010).

29.2.3 Epigenetics during Abiotic Stress

To cope with environmental stress, plants have developed genetic and epigenetic mechanisms that enable them to withstand single or combined stresses and their interactions. An experiment was carried out by Brzezinka et al. (2016) in *Arabidopsis* plants to identify genes that are specifically required for heat stress. The authors identified the gene *FORGETTER1* (*FGT1*) and revealed that the FGT1 protein binds directly to a specific class of heat-inducible genes that are relevant for heat stress memory. Their findings could lead to new approaches in crop breeding programs for enhancing the resistance to abiotic stress, as knowledge of stability and heritability features of epigenetic marks and epigenetic regulatory mechanisms are crucial for breeding applications. Further, various epigenetic mechanisms were identified from different crop plants exposed to different abiotic stresses (Table 29.1).

29.2.4 DNA Methylation in Drought Stress

Earth temperature is increasing yearly at an alarming rate and it is predicted that before the end of the twenty-first century, it will increase by 4–5°C (21). Unpredictable rainfall has influenced the precipitation rate and groundwater levels, and rising temperature produces enormous heat that creates drought conditions. Drought is an important abiotic stressor that causes a huge biodiversity loss or loss of crop production worldwide. DNA methylation is an important tool to mitigate such stress. Wang et al. (2011) conducted a detailed study to understand DNA methylation patterns to mitigate abiotic stress. They noted that genes adjacent to repetitive elements undergo epigenetic silencing. The plants develop antisilencing mechanisms like REPRESSOR OF SILENCING 1 (ROS1)-mediated DNA demethylation. They reported that ANTI-SILENCING 1 (ASI1) exhibits a bromo-adjacent homology domain and constitutes an RNA recognition motif-containing protein. Studies were also conducted on tree species to understand epigenetic regulations. In a study carried out on *Populus trichocarpa*, ISWI-(IMITATION SWITCH)-responsive element-binding protein 1 was reported to control H3K9ac and Pol II activity along with *PtrNAC* genes. Drought-responsive genes *PtrNAC006/007/120* increase H3K9ac and Pol II at promoters by employing histone acetyl-transferase factor ADA26-GCN5, which ultimately controls sensitivity to drought (Wassenegger, 1994; Huang & Zhu, 2014; Harris et al., 2018).

29.2.5 Histone Acetylation Mechanism in Drought Stress

Drought stress impacts the health of ecosystems because it has adverse effects on the growth, development, regulation and productivity of plants. However, to mitigate this adverse effect, plants have developed drought avoidance and resilience strategies for survival, such as increasing water absorption and conduction, reducing water loss and maintaining growth. Various studies have been carried out to understand how plants respond and regulate drought stress. Epigenetic marks are a group of regulators affecting drought response and resilience in plants through the modification of chromatin structure to control the transcription of pertinent genes. Histone acetylation is a ubiquitous epigenetic mark. The level of histone acetylation, which is regulated by histone acetyltransferases (HATs) and histone deacetylases (HDACs), determines whether the chromatin is open or closed, thereby controlling the access of DNA-binding proteins for transcriptional activation. Histone acetylation is mostly associated with a high rate of gene expression, and deacetylation is associated with the repression of the transcriptional process. Acetylation may lead to the production of chromatin structure that is easy for transcription. The HATs and HDACs are two vital enzymes that control transcription by catalyzing the removal and addition of an acetyl moiety, respectively. Histone acetylation

TABLE 29.1
Epigenetic mechanisms in crops exposed to different abiotic stresses

Crop name	Abiotic stress	Epigenetic mechanisms	References
Maize	Drought	Enhancement in H3K36me3, H3K9ac and H3K4me3	Xu et al. (2017)
		Modifications of H3K4me3 and H3K9ac dynamics	Forestan et al. (2020) and Varroto et al. (2020)
	Heat	H3K4me2 and H3K9ac alterations	Wang et al. (2011), Hou et al. (2019) and Varroto et al. (2020)
		Increased histone acetylation and decreased H3K9me3	
	Cold	Enhancement in H3K9ac and decrease in DNA methylation and H3K9me2	Steward et al. (2002), Hu et al. (2012) and Varroto et al. (2020)
		Reduction in histone acetylation in euchromatin-associated gene regions	
		DNA demethylation	
Wheat	Heat	Increased histone demethylation of the various genes	Wang et al. (2011) and Varroto et al. (2020)
	Salinity	Hypermethylation of cytosines at *HKT* genes	Zhong et al. (2010) and Kumar et al. (2017)
		5-mC depletion	Varroto et al. (2020)
Barley	Drought	Hc-siRNA-mediated hyper-methylation at CYTOKININ-OXIDASE 2.1 promoter	Surdonja et al. (2017) and Varroto et al. (2020)
		Increase in H3 and loss in H3K9me2	Temel et al. (2017) and Varroto et al. (2020)
		Accumulation of miR408 transcripts	Kantar et al. (2010) and Varroto et al. (2020)
Rice	Drought	Hypomethylation	Gayacharan and Joel (2013) and Varroto et al. (2020)
		Upregulation of miR408 expression	Mutum et al. (2013) and Varroto et al. (2020)
		Site-specific DNA methylation	Wang et al. (2011) and Varroto et al. (2020)
	Salinity	Demethylation at promoter region of *OsMYB91* gene and rapid histone modifications at *OsMYB9* locus	Zhu et al. (2015) and Varroto et al. (2020)
		DNA methylation	Karan et al. (2012), Ferreira et al. (2019) and Varroto et al. (2020)
Soybean	Drought	miR1514a modulation of a NAC transcription factor transcript	Sosa-Valencia et al. (2017) and Varroto et al. (2020)
		Upregulation of isomiRNAs	Kulcheski et al. (2011) and Varroto et al. (2020)
	Heat	Hypomethylation of cytosine	Hossain et al. (2017) and Varroto et al. (2020)
Pea	Drought	Hypermethylation of cytosine residues	Labra et al. (2002) and Varroto et al. (2020)
Chickpea	Drought	Accumulation of miR408 transcripts	Hajyzadeh et al. (2015) and Varroto et al. (2020)
	Drought + Salinity	Accumulation of miRNAs at root apex	Khandal et al. (2017) and Varroto et al. (2020)
Cowpea	Drought	Increase of *P5CS* transcripts and low expression of vun-miR5021 and vun-miR156b-3p	Shui et al. (2013) and Varroto et al. (2020)
Bean	Drought	Dicistronic arrangement of miR398a and miR2119	De la Rosa et al. (2019)and Varroto et al. (2020)
Faba bean	Drought	Increased DNA demethylation	Abid et al. (2017) and Varroto et al. (2020)
Alfalfa	Drought	Overexpression of miR156	Arshad et al. (2018) and Varroto et al. (2020)
Rapeseed	Heat	Increased DNA demethylation	Gao et al. (2014) and Varroto et al. (2020)
	Salinity	Increased DNA demethylation	Marconi et al. (2013) and Varroto et al. (2020)
Tomato	Drought	RNA-dependent DNA methylation	Benoit et al. (2019) and Varroto et al. (2020)
		Increased *Asr1* and *Asr2* expression due to demethylation of putative regulatory and transcribed regions	González et al. (2011)
	Cold	Increased DNA methylation	Zhang et al. (2016)

changes trigger transcriptional activations or repressions for drought stress response.

29.2.6 DNA Methylation and Salt Stress

Salt stress is also an important abiotic stress factor that affects the growth and development of plants. Usually, salt-tolerant plants exhibit improved DNA methylation as compared to salt-sensitive plants under stress environment. DNA methylation is an epigenetic modification that helps to mitigate the salt stress environment. There are various studies conducted on plant DNA methylation under salt stress conditions. Kinoshita and Seki (2014) reported that the increase in DNA methylation level of *Hordeum vulgare* leaves compared to roots indicated that salt-mediated DNA methylation is organ specific. Chen et al. (2019) reported that the methylation level in *Glycine max* roots was lower under salt stress compared to the plants grown in normal salt concentration.

29.3 CONCLUSION

Plants adopt well-adapted strategies to endure several environmental stresses. In this chapter, we have highlighted the different epigenetic mechanisms adopted by plants to tackle abiotic stresses. Epigenetic regulations like DNA methylation/demethylation and histone modification are the important mechanisms that endure several environmental stresses by upregulating or downregulating several genes. The regulation of such genes and the underlying epigenetic control can immensely help the scientific community in developing stress-tolerant plants.

ACKNOWLEDGMENT

The author, Dr Sachin Patil, is very much thankful to the Principal, Dr Ghali College, Gadhinglaj, for providing the facilities and constant support to write the book chapter.

REFERENCES

Abid G, Mingeot D, Muhovski Y, et al. (2017). Analysis of DNA methylation patterns associated with stress response in faba bean (*Vicia faba* L.) using methylation-sensitive amplification polymorphism (MSAP). *Environmental and Experimental Botany*, 142, 34–44.

Arshad M, Gruber MY, Hannoufa A. (2018). Transcriptome analysis of microRNA156 overexpression alfalfa roots under drought stress. *Scientific Reports*, 8, 9363.

Asensi-Fabado. M.-A., Amtmann, A., & Perrella, G. (2017). Plant responses to abiotic stress: The chromatin context of transcriptional regulation. *Biochimica et Biophysica Acta (BBA) – Gene Regulatory Mechanisms*, 1860, 106–122.

Benoit M, Drost HG, Catoni M, Gouil Q, Lopez-Gomollon S, Baulcombe D, Paszkowski J. (2019). Environmental and epigenetic regulation of Rider retrotransposons in tomato. *PloS Genetics*, 15, e1008370.

Brzezinka, K, Altmann, S, Czesnick, H, Nicolas, P, Gorka, M, Benke, Z, Kabelitz, T, Jähne, F, Graf, A, Kappel, C and Bäurle, I. (2016). Arabidopsis FORGETTER1 mediates stress-induced chromatin memory through nucleosome remodeling. *eLife* 5:e17061. https://doi.org/10.7554/eLife.17061

Budak H, Hussain B, Khan Z, Ozturk NZ and Ullah N (2015). From genetics to functional genomics: Improvement in drought signaling and tolerance in wheat. *Frontiers in Plant Science*, 6:1012. doi: 10.3389/fpls.2015.01012

Cheong, J. J. (2021). Genetic and epigenetic changes in plants in response to abiotic stress. *Genes*, 12, 1603. https://doi.org/10.3390/genes12101603.

Chin DC, Kumar RS, Suen CS, Chien CY, Hwang MJ, Hsu CH, Xuhan X, Lai ZX, Yeh KW (2019). Plant cytosolic ascorbate peroxidase with dual catalytic activity modulates abiotic stress tolerances. *iScience*, 16:31–49. https://doi.org/10.1016/j.isci.2019.05.014

De la Rosa C, Covarrubias AA, Reyes JL. (2019). A dicistronic precursor encoding miR398 and the legume-specific miR2119 coregulates CSD1 and ADH1 mRNAs in response to water deficit. *Plant, Cell & Environment*, 42, 133–144.

Ferreira D, Escudeiro A, Adega F and Chaves R (2019). DNA methylation patterns of a satellite non-coding sequence – FA-SAT in cancer cells: Its expression cannot be explained solely by DNA methylation. *Frontiers in Genetics*, 10:101. doi: 10.3389/fgene.2019.00101

Forestan, C., Farinati, S., Zambelli, F., Pavesi, G., Rossi, V., & Varotto, S. (2020). Epigenetic signatures of stress adaptation and flowering regulation in response to extended drought and recovery in Zea mays. *Plant, Cell & Environment*, *43*(1), 55–75.

Gao G, Li J, Li H, Li F, Xu K, Yan G, Chen B, Qiao J, Wu X. (2014). Comparison of the heat stress induced variations in DNA methylation between heat-tolerant and heat-sensitive rapeseed seedlings. *Breeding Science*, 64, 125–133.

Gayacharan and Joel, A.J. (2013). Epigenetic responses to drought stress in rice (Oryza sativa L.). *Physiology and Molecular Biology of Plants*, 19, 379–387, doi:10.1007/s12298-013-0176-4

González RM, Ricardi MM, Iusem ND. (2011). Atypical epigenetic mark in an atypical location: cytosine methylation at asymmetric (CNN) sites within the body of a non-repetitive tomato gene. *BMC Plant Biology*, 11, 94.

Hajyzadeh M, Turktas M, Khawar KM, Unver T. (2015). miR408 overexpression causes increased drought tolerance in chickpea. *Gene*, 555, 186–193.

Harris, C. J., Scheibe, M., Wongpalee, S. P., Liu, W., Cornett, E. M., Vaughan, R. M., et al. (2018). A DNA methylation reader complex that enhances gene transcription. *Science*, 362, 1182–1186. https://doi.org/10.1126/science.aar7854.

Hossain MS, Kawakatsu T, Kim KD, et al. (2017). Divergent cytosine DNA methylation patterns in single-cell, soybean root hairs. *New Phytologist*, 214, 808–819.

Hou, J., Lu, D., Mason, A.S., Li, B., Xiao, M., An, S. & Fu, D. (2019). Non-coding RNAs and transposable elements in plant genomes: emergence, regulatory mechanisms and roles in plant development and stress responses. *Planta*. 250, 23–40. https://doi.org/10.1007/s00425-019-03166-7

Hu, Y., Liu, D., Zhong, X., Zhang, C., Zhang, Q. & Zhou, D. (2012). CHD3 protein recognizes and regulates methylated histone H3 lysines 4 and 27 over a subset of targets in the rice genome. *Proceedings of the National Academy of Sciences USA*, 109 (15) 5773-5778, https://doi.org/10.1073/pnas.1203148109.

Huang, C. F., & Zhu, J. K. (2014). RNA splicing factors and RNA-directed DNA methylation. *Biology*, 3, 243–254. https://doi.org/10.3390/biology3020243.

Iwasaki, M., & Paszkowski, J. (2014). Epigenetic memory in plants. *The EMBO Journal*, 33(18), 1987–1998. https://doi.org/10.15252/embj.201488883.

Jon Penterman, Rie Uzawa, Robert L. Fischer. (2007). Genetic interactions between DNA demethylation and methylation in arabidopsis. *Plant Physiology.* 145 (4). 1549–1557, https://doi.org/10.1104/pp.107.107730

Kantar, M., Unver, T. and Budak, H. (2010) Regulation of barley miRNAs upon dehydration stress correlated with target gene expression. *Functional & Integrative Genomics*, 10, 493–507

Karan R, DeLeon T, Biradar H, Subudhi PK. (2012). Salt stress induced variation in DNA methylation pattern and its influence on gene expression in contrasting rice genotypes. *PLOS One*. 7(6):e40203. doi: 10.1371/journal.pone.0040203.

Kinoshita, T., and Seki, M. (2014). Epigenetic memory for stress response and adaptation in plants. *Plant and Cell Physiology*, 55 (11). 1859–1863.

Khandal H, Parween S, Roy R, Meena MK, Chattopadhyay D. (2017). MicroRNA profiling provides insights into post-transcriptional regulation of gene expression in chickpea root apex under salinity and water deficiency. *Scientific Reports*, 7, 4632.

Kumar, S., & Singh, A. (2016). Epigenetic regulation of abiotic stress tolerance in plants. *Advances in Plants & Agriculture Research,* 5(3), 00179. https://doi.org/10.15406/apar.2016.05.00179.

Kumar, S., Singh, A.K., and Mohapatra, T. (2017). Epigenetics: History, present status and future perspective. *Indian Journal of Genetics and Plant Breeding*, 77(4): 445–463. DOI: 10.5958/0975-6906.2017.00061.X

Kulcheski FR, De OLFV, Molina LG et al (2011). Identification of novel soybean microRNAs involved in abiotic and biotic stresses. *BMC Genomics,* 12:307. https://doi.org/10.1186/1471-2164-12-307

Labra M, Ghiani A, Citterio S, Sgorbati S, Sala F, Vannini C, Ruffini-Castiglione M, Bracale M. (2002). Analysis of cytosine methylation pattern in response to water deficit in pea root tips. *Plant Biology,* 4, 694–699.

Law, J. A., & Jacobsen, S. E. (2010). Establishing, maintaining and modifying DNA methylation patterns in plants and animals. *Nature Reviews Genetics*, 11(3), 204–220. https://doi.org/10.1038/nrg2719.

Li, X, Wang, X, He, K, Ma, Y, Su, N, He, H, Stolc, V, Tongprasit, W, Jin, W, Jiang, J, Terzaghi, W, Li, S, and Deng SW. (2008). High-resolution Mapping of epigenetic modifications of the rice genome uncovers interplay between DNA methylation, histone methylation, and gene expression. *The Plant Cell*, 20 (2). 259–276, https://doi.org/10.1105/tpc.107.056879

Liu, C, Lu, F, Cui, X, and Cao, X. (2010). Histone methylation in higher plants. *Annual Review of Plant Biology*, 61, 395–420. https://doi.org/10.1146/annurev.arplant.043008.091939

Marconi G, Pace R, Traini A, Raggi L, Lutts S, Chiusano M, Guiducci M, Falcinelli M, Benincasa P, Albertini E. (2013). Use of MSAP markers to analyse the effects of salt stress on DNA methylation in rapeseed (*Brassica napus* var. oleifera). *PLoS One*, 8, e75597.

Mutum, R.D., Balyan, S.C., Kansal, S., Agarwal, P., Kumar, S., Kumar, M., Raghuvanshi, S. (2013). Evolution of variety-specific regulatory schema for expression of osa-miR408 in indica rice varieties under drought stress. *The FEBS Journal*, 280, 1717–1730.

Russo, V. E. A., Martienssen, R. A., & Riggs, A. D. (1996). *Epigenetic Mechanisms of Gene Regulation.* Woodbury: Cold Spring Harbor Laboratory Press.

Shui XR, Chen ZW, Li JX. (2013). MicroRNA prediction and its function in regulating drought-related genes in cowpea. *Plant Science*, 210, 25–35.

Sosa-Valencia G, Palomar M, Covarrubias AA, Reyes JL. (2017). The legume miR1514a modulates a NAC transcription factor transcript to trigger phasiRNA formation in response to drought. *Journal of Experimental Botany,* 68, 2013–2026.

Steward, N., Ito, M., Yamaguchi, Y., Koizumi, N. and Sano, H. (2002). Periodic DNA methylation in maize nucleosomes and demethylation by environmental stress. *Journal of Biological Chemistry*, 277, 37741–37746.

Surdonja, K.; Eggert, K.; Hajirezaei, M.-R.; Harshavardhan, V.T.; Seiler, C.; Von Wirén, N.; Sreenivasulu, N.; Kuhlmann, M (2017). Increase of DNA methylation at the HvCKX2.1 promoter by terminal drought stress in barley. *Epigenomes* 2017, 1, 9. https://doi.org/10.3390/epigenomes1020009

Temel, A.; Janack, B.; Humbeck, K. Drought stress-related physiological changes and histone modifications in barley primary leaves at HSP17 Gene. *Agronomy* 2017, 7, 43. https://doi.org/10.3390/agronomy7020043

True, W. R., Rice, J., Eisen, S. A., Heath, A. C., Goldberg, J., Lyons, M. J., & Nowak, J. (1993). A twin study of genetic and environmental contributions to liability for posttraumatic stress symptoms. *Archives of General Psychiatry*, 50, 257–264. https://doi.org/10.1001/archpsyc.1993.01820160019002.

Varotto, S, Tani, E, Abraham, E, Krugman, T, Kapazoglou, A, Melzer, R, Radanović, R and Miladinović, R. (2020). Epigenetics: possible applications in climate-smart crop breeding. *Journal of Experimental Botany*, 71 (17). 5223–5236, https://doi.org/10.1093/jxb/eraa188

Wang, W. S., Pan, Y. J., Zhao, X. Q., Dwivedi, D., Zhu, L. H., Ali, J., Fu, B. Y., & Li, Z. K. (2011). Drought-induced site-specific DNA methylation and its association with drought tolerance in rice (*Oryza sativa* L.). *Journal of Experimental Botany*, 62, 1951–1960.

Wassenegger, M., Heimes, S., Riedel, L., & Sänger, H. L. (1994). RNA-directed de novo methylation of genomic sequences in plants. *Cell*, 76, 567–576. https://doi.org/10.1016/0092-8674(94)90119-8.

Wei, X, Song, X, Wei, L, Tang, S, Sun, J, Hu, P, Cao, X. (2017). An epiallele of rice AK1 affects photosynthetic capacity. *Journal of Integrative Plant Biology*, 59, 158–163. doi: 10.1111/jipb.12518

Zemach, M.Y. Kim, P. Silva, J.A. Rodrigues, B. Dotson, M.D. Brooks, & D. Zilberman (2010). Local DNA hypomethylation activates genes in rice endosperm, *Proceedings of the National Academy of Sciences*, U.S.A. 107 (43) 18729–18734, https://doi.org/10.1073/pnas.1009695107

Zhang B., Tieman D.M., Jiao C., Xu Y., Chen K., Fei Z., Giovannoni J.J., Klee H.J. (2016). Chilling-induced tomato flavor loss is associated with altered volatile synthesis and transient changes in DNA methylation. *Proceedings of the National Academy of Sciences, USA* 113, 12580–12585.

Zhang, H., Lang, Z. & Zhu, J.K. Dynamics and function of DNA methylation in plants. *Nature Reviews Molecular Cell Biology*, 19, 489–506 (2018). https://doi.org/10.1038/s41580-018-0016-z

Zhong L, D'Urso A, Toiber D, Sebastian C, Henry RE, et al. (2010). The histone deacetylase Sirt6 regulates glucose homeostasis via Hif1alpha. *Cell,* 140: 280–293.

Zhu N, Cheng S, Liu X, Du H, Dai M, Zhou DX, Yang W, Zhao Y. (2015). The R2R3-type MYB gene OsMYB91 gas a function in coordinating plant growth and salt stress tolerance in rice. *Plant Science*, 236, 146–156. doi:10.1016/j.plantsci.2015.03.023

30 Epigenetic Regulations in Plant–Environment Interactions

Nisha Sam Nirmala, Eyerusalem Goitom Kidane, Uma S, Ashwini Ravi and Sivarama Krishna Lakkaboyana

30.1 INTRODUCTION

Plants, as sessile organisms, show specialized adaptations to their specific habitats, perpetually facing many adversities, arising from abiotic stresses, such as inadequate water, salinity and extreme temperature, and biotic stresses, such as frequent pathogen attacks and herbivorous foraging. Unlike animals that can escape inhospitable surroundings, plants must rely on advanced molecular processes to sense, respond to and adjust to their habitats (Ashapkin et al., 2020). Conrad Waddington first used the term epigenetics in the 1940s. It describes heritable alterations in gene expression occurring without changes to the main DNA sequence (Waddington, 2011). This concept is a remarkably important mechanism underlying these flexible responses. Over many years, studies have definitively demonstrated epigenetic regulation to be important to plant biology; this importance allows plants to change their gene expression conforming to all ecological signals, along with the capacity to convey each of these flexible attributes to every descendant generation (He & Li, 2018; Kumar & Rani, 2023).

Epigenetic regulation in plants is thoroughly orchestrated through the cooperative function of many molecular routes; these include DNA methylation, histone alterations, coupled with multiple non-coding RNAs. DNA methylation, or the addition of a methyl group to many cytosine residues, represents a greatly scrutinized epigenetic marker (Zhang et al., 2006). Within plants, this action occurs in three sequential contexts: CG, CHG and in conjunction with CHH. The enzyme methyltransferase 1 (MET1) is responsible for maintaining CG methylation, while chromomethylase 3 (CMT3) is responsible for sustaining CHG methylation. Domains rearranged methyltransferase 2 (DRM2) diligently eases de novo methylation in all contexts, as it is skillfully directed by small interfering RNAs (siRNAs) through the RNA-directed DNA methylation (RdDM) pathway (Cao & Jacobsen, 2002; Matzke et al., 2014). Histone modifications like acetylation, methylation, phosphorylation and ubiquitination mediate chromatin structure and gene expression. Histone methyltransferases (HMTs) and histone acetyltransferases (HATs) powerfully trigger these particular changes, while histone demethylases (HDMs) and histone deacetylases (HDACs) aggressively oppose them (Zhang et al., 2009).

Non-coding RNAs, including all siRNAs and microRNAs (miRNAs), constitute another undeniably important component of the epigenetic machinery. Small RNAs have the ability to guide the silencing of specific genes; this can occur through RNA interference (RNAi) or, in addition, by directing DNA methylation to target loci (Lee & Carroll, 2018). The RdDM pathway is an important mechanism for establishing and maintaining DNA methylation at many transposable elements and repetitive sequences.

Plants use each of these epigenetic mechanisms in order to interact with their environment, along with allowing them to respond to as well as sense multiple external stimuli. Ecological stresses, such as drought, salinity, extreme temperature and pathogen attacks, can induce changes in DNA methylation and histone modifications, which result in changed gene expression patterns that improve stress tolerance (Chinnusamy & Zhu, 2009; Mirouze & Paszkowski, 2011). These epigenetic changes fine-tune each gene expression profile when plants face environmental challenges. As a result, plants can better survive and reproduce under many stressful conditions. Epigenetic memory is among the extremely engaging parts of plant epigenetics; it allows all plants to "remember" past ecological experiences, using every bit of what they know to respond far better to all recurring stresses. Stress priming is a clear example because if plants are first exposed to stress, they can respond better to stress later.

In summary, epigenetic mechanisms fundamentally allow plants to shift and change in response to multiple abiotic and biotic stresses, which means that they are key for important plant–environment interactions. Several mechanisms, including DNA methylation, histone modifications and non-coding RNAs, regulate gene expression and chromatin structure. Epigenetic memory also points out how plants can retain a certain amount of information on past stressful events, which allows them to respond with greater strength to every future difficulty. It is important to understand these epigenetic processes in order to develop strategies. These strategies can improve plant resilience in addition to crop yield as ecological conditions change.

DOI: 10.1201/9781003589914-34

30.2 MECHANISMS OF EPIGENETIC MODIFICATIONS IN PLANTS

Epigenetic modifications in plants are heritable changes in gene expression that do not change the underlying DNA sequence. These modifications play a crucial role in regulating gene expression and enabling plants to adapt to stressors and environmental changes. Plants utilize many epigenetic mechanisms to fine-tune gene expression patterns. These mechanisms include DNA methylation, histone modification and the activity of non-coding RNAs.

30.2.1 DNA Methylation

DNA methylation is a critical epigenetic modification in plants, involving the addition of a methyl group to cytosine residues. This modification primarily occurs at CpG dinucleotides but can also occur in CHG and CHH contexts, where H represents adenine (A), cytosine (C) or thymine (T) (Kawakatsu et al., 2017). Key mechanisms of DNA methylation involved in gene regulation include a dynamic process responsive to environmental stimuli and stress. Stress-induced methylation can suppress transcription factor binding to modulate gene expression. Different stressors induce unique methylation patterns based on plant genotype, tissue type and environmental conditions; for example, drought stress in *Arabidopsis thaliana* results in global hypomethylation of stress-responsive genes (Lee et al., 2023), Saline stress induces demethylation in salt-sensitive rice varieties more than in salt-tolerant ones (Ferreira et al., 2015). Enzymes involved include DNA methyltransferases like DRM2 that initiate and maintain methylation patterns under stress (Saeed et al., 2022), as well as demethylases that actively remove methyl groups to modulate stress responses dynamically (Luo et al., 2013).

Methylation plays important roles in gene regulation and phenotypic plasticity in gene silencing where it suppresses gene expression by reducing transcription factor accessibility at promoter regions (Kooke et al., 2015). Heritable phenotypic changes come along with modification in DNA methylation during enhanced disease resistance (Zhou et al., 2005) or altered flowering time (Shafiq et al., 2013). In case of stress adaptation, plants remember past "stress", enabling them to respond faster and in a more robust way to similar future challenges (Mirouze & Paszkowski, 2011). What makes this even better is that this epigenetic memory can also be inherited by offspring, aiding transgenerational stress tolerance. In addition, DNA methylation has implications in plant development, defense mechanisms against pathogens and environmental adaptation. It regulates critical traits like flowering time (Shafiq et al., 2013) and leaf morphology by repressing or activating specific genes (Wang et al., 2023). Pathogen infection leads to targeted changes in methylation, activating defense genes. Stress-induced methylation modifications are crucial for surviving adverse conditions like drought (Li et al., 2021), salinity (Qin et al., 2022) and extreme temperature (Duarte-Aké et al., 2023).

30.2.2 Histone Modifications and Chromosome Remodeling

In general, chromatin structures play a crucial role in regulating gene expression in plants as they control access to transcription factors as well as being a machinery for DNA (Kang et al., 2022). There are two main mechanisms that are used to modulate chromatin structure: chromatin remodeling and histone modification (Wang & Qiao, 2020). Histone modifications are post-translational changes made to histone proteins which influence the structure of chromatin and regulation of gene expression. These modifications specifically occur at the amino acid residues in histones where they have a crucial role in transcriptional regulation at the time of development and stress response (Ageeva-Kieferle et al., 2021). The major types of histone modification include acetylation, methylation and phosphorylation at distinct lysine, arginine, threonine, tyrosine and serine residues. Depending on the degree and site of modifications, they can either activate or repress gene expression; for example, increased acetylation of H3 histone in *Arabidopsis thaliana* during drought stress typically relaxes the chromatin structure and enhances transcription of genes involved in water conservation and stress tolerance (Kim et al., 2008), while methylation at H3K4 leads to gene silencing, which has a significant role in regulating responses to abiotic stress (Song et al., 2020).

Chromatin remodeling includes ATP-dependent modifications of nucleosome positioning that regulate DNA accessibility in replication and transcription (He et al., 2023). Key mechanisms in chromatin remodeling involve nucleosome sliding where remodeling complexes slide nucleosomes along DNA exposing regulatory regions for transcription, chromatin looping that brings large-scale structural changes that create loops, which bring distant regulatory elements closer to target genes (Stoddard et al., 2018) and histone exchange where different histone variants are incorporated or removed, altering nucleosome stability and chromatin accessibility (Rutowicz et al., 2015), for example: remodeling complexes such as SNF/SWI facilitate chromatin condensation for repression or opening for active transcription (Buszewicz et al., 2016).

30.2.3 RNA-Based Mechanisms

siRNAs and miRNA are short non-coding RNA molecules. They play a vital role in gene expression, post-transcriptional gene silencing, chromatin remodeling and enhancing plant stress response. siRNA originates from double-stranded RNA, which arises from endogenous transposons or exogenous sources like viruses (Dowen et al., 2012), whereas microRNAs originate from endogenous single-stranded RNA (ssRNA) precursors that fold into hairpin structures.

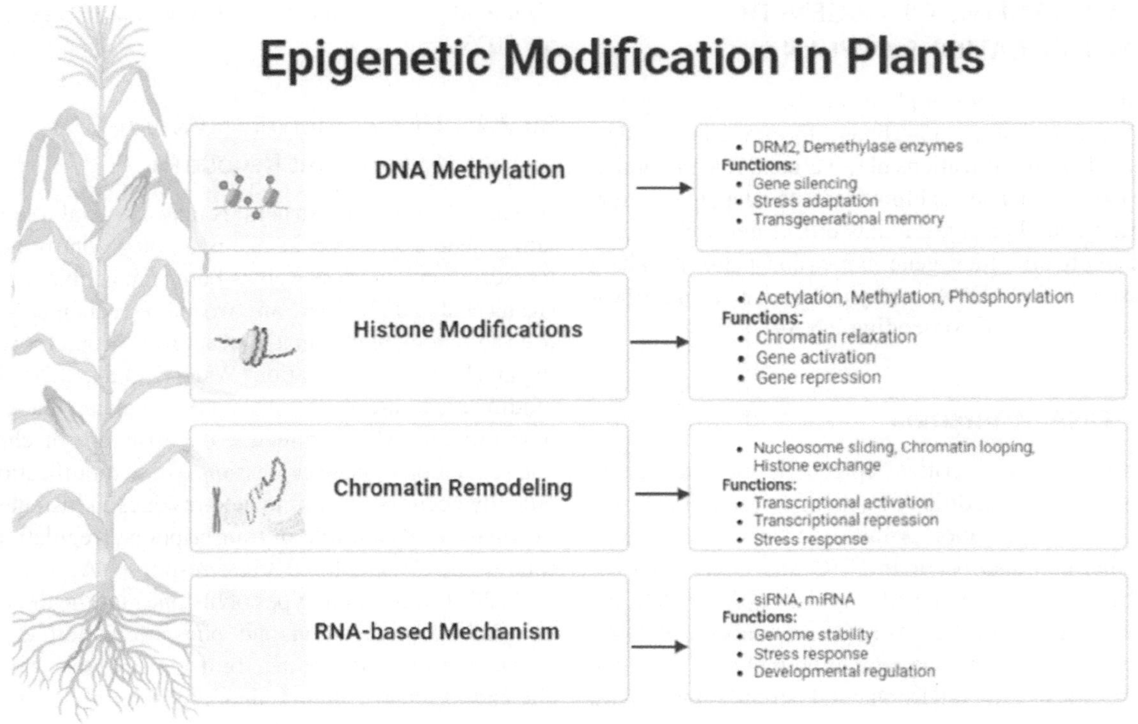

FIGURE 30.1 Types of epigenetic modifications in plants.

Dicer-like enzymes are responsible for the precursors to mature into miRNA (Sheng et al., 2018). They both guide RNA-induced silencing complexes (RISCs) to target complementary sequences for RNA degradation or translational inhibition (Xie & Yu, 2015). miRNAs are mainly involved in fine-tuning gene expression by targeting stress-responsive genes during stress responses, for example, miR159 modulates drought tolerance genes in *Arabidopsis thaliana*, enhancing water retention and survival under dehydration. miR398 regulates oxidative stress by targeting the transcripts of *copper/zinc superoxide dismutase* (*CSD*) genes, modulating reactive oxygen species levels (Singroha et al., 2021). siRNAs are key in maintaining genome stability by silencing transposable elements through RdDM (Panda & Slotkin, 2013) by guiding DNA methyltransferase DRM2 to specific genomic regions to modify chromatin structure and silence stress-related genes (Gallego-Bartolomé, 2020).

Small RNAs play a crucial role in regulating stress responses including gene silencing by targeting specific mRNA transcripts for degradation or suppression, which enables plants to conserve energy and resources (Park et al., 2017). siRNAs regulate genes involved in abiotic stress tolerance, such as salt, drought and heat stress. They also modulate different pathways responsible for oxidative stress management (Singroha et al., 2021), hormone signaling (Sah et al., 2016) and water conservation to ensure adaptive responses, for example, miR156 and miR528 demonstrate their role in balancing stress responses with developmental needs. miR156 regulates flowering time and stress responses by targeting squamosa promoter-binding-like (SPL) transcription factors (Begum, 2022). It modulates histone modifications at flowering loci, balancing stress and developmental needs. Small RNAs act synergistically with other epigenetic mechanisms such as DNA methylation and histone modifications to coordinate gene expression under stress. The RdDM pathway employs siRNAs to direct DNA methylation, silencing transposable elements and other genomic regions during stress (Panda & Slotkin, 2013). Figure 30.1 represents the different types of epigenetic modifications commonly occurring in plants.

Table 30.1 summarizes few examples of different epigenetic modifications in some of the plants and their implications.

30.3 PLANTS AND THEIR ENVIRONMENTAL INTERACTIONS

Plants are in continuous, interactive dynamics with their environment, which contains a huge number of abiotic and biotic features. Such interactions have profound impacts on their growth, reproductive processes and evolutionary paths. Since they are immobile, these plants must adapt to such critical environmental stressors as drought, nutrient scarcity or herbivory, thus further increasing the demands for their survival and adaptability, requiring complex physiological and genetic systems. In addition, other interactions of a plant with its environment include symbiotic relationships with pollinators and microbes, which greatly affect ecosystems and ecological communities. The intricate interactions between plants and their environment, focusing on three principal themes – responses of plants to stress, mutualistic relationships and dynamics of reproduction, are represented in Figure 30.2.

TABLE 30.1
Types of epigenetic modifications in different plant species and their implications

Plant species	Epigenetic modification	Mechanism	Key features	Functions	References
Arabidopsis thaliana	DNA methylation	Methylation of cytosine residues at CpG, CHG and CHH contexts; modulates stress responses and flowering time	Heritable changes in gene expression; stress-induced methylation patterns	Regulates flowering time, enhances stress tolerance and enables transgenerational memory	Chodavarapu et al. (2010)
Oryza sativa (Rice)	Histone modification	Acetylation of histone H3 under drought stress increases transcription of stress-responsive genes; methylation at H3K9 silences transposons	Reversible and site-specific modifications of histone tails	Controls chromatin structure and transcription; modulates stress and development responses	Fang et al. (2013)
Zea mays (Maize)	RNA-directed methylation	siRNAs guide DNA methylation to silence transposable elements and regulate stress-responsive genes under heat and drought conditions	Small RNA molecules that regulate gene silencing	Maintains genome stability and controls stress-responsive genes	Guo et al. (2021)
Arabidopsis thaliana	Chromatin remodeling	SWI/SNF complexes reposition nucleosomes to enhance or repress transcription of genes involved in abiotic stress tolerance and growth regulation	ATP-dependent restructuring of chromatin for DNA accessibility	Facilitates transcriptional activation or repression; regulates heat stress responses and development	Buszewicz et al. (2016)
Petunia axillaris and *Petunia inflata*	miRNA-mediated regulation	miR156 regulates flowering and stress responses by targeting SPL transcription factors, balancing developmental needs with stress adaptation	Targets mRNAs for degradation or translational repression	Fine-tunes gene expression; balances growth and stress adaptation	Zhou et al. (2021)
Glycine max (Soybean)	DNA methylation	Environmental stress, such as salinity, induces specific methylation patterns that modulate genes linked to salt tolerance	Environmentally responsive and reversible methylation patterns	Enhances salinity tolerance and regulates stress-responsive genes	Song et al. (2012)
Populus trichocarpa	Chromatin remodeling	Nucleosome positioning and chromatin loops regulate genes involved in growth and wood formation under varying environmental conditions	Dynamic restructuring of chromatin architecture	Modulates wood formation, growth and stress adaptation	Chen et al. (2019)
Medicago truncatula	siRNA-mediated silencing	siRNAs target genes involved in symbiotic nitrogen fixation, regulating root nodulation under nutrient stress	siRNAs derived from double-stranded RNA; highly sequence-specific	Controls nitrogen fixation and root development under stress	Floss et al. (2013)
Arabidopsis thaliana	miRNA-mediated regulation	miR398 modulates oxidative stress responses by targeting genes encoding copper/zinc superoxide dismutase, ensuring adaptive responses to high temperature	Regulates oxidative stress by targeting antioxidant enzyme genes	Enhances tolerance to oxidative stress and heat	Guan et al. (2013)

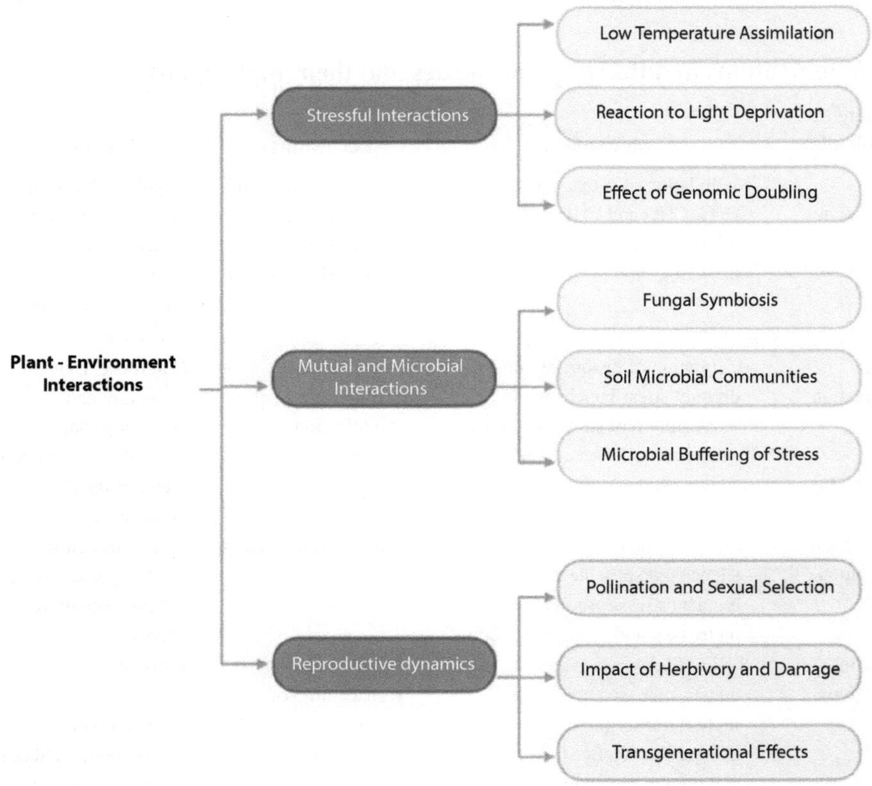

FIGURE 30.2 Plant–environment interactions.

30.3.1 Stressful Interactions

Plants are continually subjected to a broad spectrum of stressors from both abiotic and biotic sources. The stressful abiotic interactions presented are generally low-temperature acclimation, reduced light intensity responses and genome doubling as a process of evolutionary significance.

Low temperature causes damage to the cells and disrupts physiological processes of the plant, making it less viable. To cope with such stress, plants have developed a more complex system of increasing tolerance towards freezing temperature. In the model organism *Arabidopsis thaliana*, cold acclimation mainly operates through the C-repeat binding factor (CBF) pathway (Park et al., 2017). However, interestingly, the gene *CBF2* is found to be a critical component in this adaptation process. The mutation of the *CBF2* gene significantly affects the freezing tolerance ability of plants. Studies with CRISPR-Cas9 gene editing technologies have provided increased insights into the functional roles of CBF2 (Jia et al., 2016; Zhao et al., 2016). Targeted mutations in *CBF2* demonstrated alterations in freezing tolerance, indicating the role of the gene in cold tolerance. The results suggest that the CBF pathway has underlying complex regulatory networks governing comprehensive changes in gene expression as a response to cold stress. Understanding these pathways opens up possible avenues for the generation of frost-resistant crops, an extremely important area with regard to climate change and extreme weather events (Sanderson et al., 2019).

The ability of light to determine growth and development in plants helps them execute photosynthesis and photomorphogenesis. Survival under low light conditions requires physiological and developmental adjustment. One such response is known as the "block of greening" or BOG in *Arabidopsis thaliana*, where the seedlings grown in far-red light fail to synthesize chlorophyll upon transfer to white light in the absence of exogenous carbon sources such as sucrose. This response is mediated through phytochrome A, which is one of the types of photoreceptors that regulate the expression of genes for chlorophyll biosynthesis. Mutations in nuclear-encoded sigma factor genes like *SIG6* have been reported in relation to this process. SIG6 has been found to be integral in the regulation of plastid genes at the transcriptional level during chloroplast development. The *SIG6* gene mutations may actually disrupt the greening process and therefore suggest a role for the gene in responding to changing light conditions. Genetic information on light perception and response is a basis for understanding how plants acclimate to shaded environment or sudden changes in the availability of light (Alameldin et al., 2020).

The presence of multiple chromosome sets in plant species is one of the commonest features that act as an important evolutionary adaptation mechanism. In *Fragaria* (strawberries), polyploidy has been linked to dramatic changes in their morphological and physiological traits. Stomatal density and dimensions in polyploids have been found to be different and may, therefore, influence transpiration rates and photosynthetic efficiency. These changes may then turn out to be advantageous concerning water use efficiency and enable polyploid species to survive in such environment where the water supply is unreliable. In addition, genome duplication could lead to high genetic and phenotypic variation that can introduce new functions or options of genes conferring more and different types of resistance and tolerance toward stressful conditions. Phylogenetic analysis showed that polyploidization within *Fragaria* species is related to complex evolutionary histories implicated by hybridization events and genomic reorganizations. This results in phenotypic plasticity and allows the polyploid plants to grow within a range of ecological niches and tolerate multiple environmental stressors (Yang & Davis, 2017).

30.3.2 Mutualism and Microbial Interactions

Although the mutualistic interactions between plants and microbes play an essential role in nutrient acquisition-related stress tolerances as well as overall plant fitness, the complexity of plant–microbe–environment dynamics includes everything from symbioses of fungi and bacteria to the impact of soil microbial communities. The two most common types of microbial mutualists demonstrated to be vital for plant fitness are endophytic fungi/bacteria and nitrogen-fixing bacteria. These endophytes are within the plant tissues, where they pose no harm. The growth of the plant is also enhanced in a further process due to higher assimilation of nutrients, growth promoters and hormones. Nitrogen transfer by various endophytic fungi like *Metarhizium* and *Beauveria* has been found to improve nutritional status in the host plants (Liu-Xu et al., 2024). It is well documented that the endophytic fungi of *Theobroma cacao* (cacao) promote secondary metabolite production, thereby increasing the resistance and resilience of the host plant to pathogens (Baron & Rigobelo, 2021). Nitrogen-fixing nodules are formed in legumes as a result of symbiotic association with nitrogen-fixing bacteria, mainly rhizobia. In *Medicago sativa*, or alfalfa, studies have revealed that autotetraploid plants have greater mutualistic efficiency with rhizobia than their diploid siblings and may imply that polyploidy enhances the benefits that can be obtained from microbial interaction. More root biomass and nodule development in polyploid plants would naturally lead to enhanced nitrogen fixation and assimilation (Goh et al., 2013).

Microorganisms in soil serve as discriminatory agents. The key influences on plant traits, including flowering phenology, growth and stress responses, are driven by strong selective forces in soil microbial communities. Experimental manipulations of the soil microbiome in a species like *Ipomoea purpurea* (morning glory) showed that microbial communities can be powerful evolutionary agents modulating life history traits. Some soil bacteria could produce phytohormones or modify the levels of plant hormones, hence affecting plant growth and stress tolerance (Singh et al., 2022). Such results therefore highlight the dual role of soil microorganisms as a collaborator in ecological processes and catalysts in evolutionary change.

Microbial mutualisms are known to alleviate environmental stress, but this would depend on the nature of the stressor. Rhizosphere microorganisms help in plant resistance to abiotic stress through the production of phytohormones and mineral solubilization, nitrogen fixation, provision of defense against pathogens and others (Tiwari et al., 2024). However, protective effects of microbial mutualists are compromised by some stressors. Research on *Lemna minor* (duckweed) has shown that although helpful microbes can enhance the tolerance of the plants to some pollutants, high salt concentrations can suppress these protective effects. This implies an intricate relationship between plant species, microbial communities and environmental conditions (O'Brien et al., 2020). The complicated interactions between plants and their associated microbial communities are of significant interest to plant ecology and evolutionary sciences. Important insights into how plants adapt can be gained through these studies, as well as possible ways of increasing crop resistance and productivity by manipulating plant–microbe interactions.

30.3.3 Reproductive Dynamics and Environmental Stress

Plant reproductive dynamics are closely associated with environmental conditions such that environmental stress factors play a vital role in modulating reproductive strategy and reproductive performance. Understanding how plants alter their reproductive mechanisms to adapt to environmental stressors is critical in understanding plant evolution and in developing adaptive agricultural systems. Extensive research has shown tremendous insight into documenting environmental stressors that significantly impact the reproductive features of plants and consequently influence mechanisms and processes of sexual selection involved in pollination. Rising and falling temperature has resulted in disturbances in the development and viability of pollen, which are among the critical determinants of pollination success rates (Zinn et al., 2010). Drought affects floral traits; they vary in size, shape and nectar composition and therefore in attractiveness and responsiveness of visiting pollinators. Such changes can enhance the prospects for reproduction in stressful environmental conditions.

Herbivory causes stress, which triggers a host of reproductive responses in diverse plant species. In some cases, severe leaf loss has been linked to shifts in reproductive trade-offs, for example, extending or shortening the length of flowering, or increasing investment in reproductive parts, to compensate for tissue loss and guarantee reproductive success. Certain plant species can increase their reproductive effort following herbivory as a way of compensating for any fitness losses (Quiroz-Pacheco et al., 2020). Indeed, maternal herbivory may imprint transgenerationally on offspring for characteristics related to reproductive success or stress resilience.

The inheritance of stress-induced traits underlines long-term consequences of environmental interactions with plant reproduction. Transgenerational plasticity enables plants to 'prepare' their offspring when exposed to the same conditions, possibly at higher fitness costs, in an unpredictable environment. Exposure to drought can thus result in offspring with changes in reproductive strategy, for example, a changed flowering time or increased mechanisms related to drought tolerance. In light of these results, there is now evidence for variation in fitness as heritable across individuals. Thus, environmental factors could very likely make a contribution to adaptive changes in succeeding generations (Binder et al., 2024). Overall, it can be seen that environmental stresses have major implications for plant reproductive dynamics. The adaptive responses highlighted above in terms of immediate changes in phenotypic expressions as well as in transgenerational effects underscore the sensitivity and resilience of plant reproductive strategies in response to environmental stressors. Thus, further exploration of these dynamics is necessary for more enlightenment into the evolution of plants and how strategies can be developed to effectively and efficiently promote crop improvement with regard to climatic global changes.

30.4 EPIGENETICS FOR IMPROVED ENVIRONMENTAL RESILIENCE

Plant response to varying and dynamic environment decides their effective growth and yield. Extreme climatic conditions and weather parameters have continually threatened plant survival and production in recent years. The tactic epigenetic players of the plants are the crucial weapons that make the plants resilient warriors against the adverse environment. Understanding and introducing these epigenetic changes in plants using crop improvement strategies can lead us to the production of environment-resilient crops, meeting global agricultural demands. This section focuses on enlisting the epigenetic modifications reported in plants during the prominent environmental interactions/changes, restricting to various abiotic and biotic environmental influencers.

30.4.1 Modifications during Abiotic Stress

In the changing environment, major factors like the variations in temperature, water levels, salinity of the soil, nutrition, UV exposure and pollutants like heavy metals induce changes in the expression of stress-responsive genes of plants, brought about by the epigenetic variations. These variations are crucial in making the plants thrive in such harsh stressful conditions (Zhang et al., 2018) and are discussed below.

30.4.1.1 Temperature-Induced Variations

Among all of the epigenetic mechanisms regulating heat-responsive genes in plants, DNA methylation is the most widely studied. In *Arabidopsis*, it was found that a methyltransferase, chromomethylase 2 (CMT2), had seasonal correlation in its expression when *cmt2* mutants were shown to be highly tolerant to heat stress, suggesting a regulatory modification occurring as a plastic response for natural adaptation to changes in temperature (Shen et al., 2014). The study on RNA-mediated methylation of important DNA methylation genes (*MET1*, *CMT3 and DRM2*), two specific RNA polymerases, Pol IV and Pol V during heat stress revealed that there was a positive correlation between the upregulation of these modulators and methylation in *Arabidopsis* (Naydenov et al., 2015). Another major player during heat stress in plants, the heat shock proteins (HSPs), were observed to undergo epigenetic changes during heat stress in a number of cases. Acetylation and trimethylation were observed in the heat shock protein genes like *HSP18*, *HSP70*, *APX2* and *HSP22.0* during heat stress (Lämke et al., 2015). It was also observed that DNA methylation kept the *SDC* gene of *Arabidopsis* silenced; however, during heat stress, the gene was activated via a demethylation process in order to render tolerance to heat stress (Sanchez & Paszkowski, 2014). Silencing of genes as a result of heat stress is also reported in *Arabidopsis* by a study led by Popova et al. (2013). Mutants defective in methylation, histone modification, chromatin remodeling and/or siRNA-mediated silencing pathways showed that heat stress response is dependent on the RNA-dependent DNA methylation pathway. Expression of miRNA was also found to be changing during heat stress in different plants like *Betula* and *Poplar*. GCN5, a histone acetyltransferase, is crucial for heat tolerance in *Arabidopsis* via regulating the expression of different heat stress-responsive genes. It was found that the loss of function of this gene ended up making the plants vulnerable to heat changes, making them more susceptible to heat (Hu et al., 2015). It was further documented that the reduced tolerance to heat was due to the inadequate activation of genes encoding heat shock transcription factors, viz., HSFA2 and HSFA3, along with multiprotein bridging factor 1c (MBF1c) (Liu et al., 2015).

The C-repeat binding factor (CBF)-mediated cold stress-responsive mechanism was well studied in various plants. It is a well characterized pathway, supported by numerous studies. Stress induced by extremely low temperature (cold stress) was found to increase the transcription levels of CBF that can lead to upregulation of *cold-responsive (COR)* genes (Zhu, 2016). In *Arabidopsis*, *PKL*, a chromatin remodeling gene, was mutated and the resulting mutants were found to be highly sensitive to cold stress, showing the participation of this gene in the CBF-mediated cold stress response (Yang et al., 2019). It was further observed that there was significant downregulation in the transcription levels of *CBF3*, *COR15B* and *RD29A* in the *pkl* mutants during cold stress. The *COR* genes also showed higher methylation frequencies at H3K27me3 for which the *PKL* gene is an influencer (Carter et al., 2018). Thus, it was documented that the *PKL* gene through DNA methylation and H3K27me3 can affect the cold-tolerance memory in plants by regulating the expression of *COR* genes (Chang et al., 2019).

The *COR* genes were found to possess higher histone acetylation efficiently regulated by histone acetyltransferases (HAT) and histone deacetylases (HDAC) (Park et al., 2018). Some of the *COR* genes were shown to undergo histone acetylation in their promoter regions upon exposure to cold (Pavangadkar et al., 2010). A study in maize showed the reduction in DNA methylation and histone demethylation caused by the upregulation of HDACs led to the deacetylation of H3 and H4 (Ding et al., 2012). *ZmMI1*, a prominent stress-responsive gene, which is generally methylated, was observed to undergo demethylation during cold stress in maize, where there was a report of methylation almost genome-wide except this gene (Hu et al., 2012).

Histone deacetylation and cold tolerance were found to go hand in hand via a study in *Arabidopsis*. An interaction between HOS15, a protein involved in histone deacetylation, and histone deacetylase 2C (HD2C), which is a WD40 repeat-containing protein, affected *COR* gene expression via interactions at their promoter regions (Park et al., 2018). *hos15* mutants were found to be sensitive to cold, showing the role of cold-induced changes in HOS15 that were necessary to degrade HD2C, which can otherwise negatively regulate the expression of *COR* genes. Thus, HOS15 was considered a positive regulator of cold stress response in plants (Ashapkin et al., 2020).

30.4.1.2 Epigenetic Modifications during Drought Stress

There has been constant development in the understanding of abscisic acid (ABA) and drought stress tolerance in plants (Shinozaki & Yamaguchi-Shinozaki, 2006) and the involvement of epigenetic mechanisms in its regulation (Zhu et al., 2018). Expression of genes related to drought stress has a closer correlation with histone modifications (Kim et al., 2015). ABA has a predominant role in arresting the growth in seedlings and this was related to the changes in histone proteins (Ryu et al., 2014). Major regulators of the ABA signaling pathway, ABI3 and ABI5, were known to be repressed following the histone deacetylation of the ABI3 chromatin by a histone deacetylation complex, which can be suppressed during the early development of seedlings. A set of other ABA-inducible genes was observed to be regulated by another epigenetic mechanism, the ROS1-dependent DNA methylation (Kim et al., 2019). In comparison to the non-drought conditions, there was an increase in H3K4me3 and H3K9Ac paired with a reduced nucleosome density on the genes involved in the response to drought, supporting that the depth of epigenetic changes during drought stress is related to the severity of the stress. When the plants recovered from drought, there was an eventual decrease in H3K9Ac along with H3K4me3 and lower transcription of drought stress-related genes (Ashapkin et al., 2020). NCED3, an important enzyme in ABA synthesis, showed an increased accumulation of H3K4me3 during drought stress (Ding et al., 2011). A set of drought stress-responsive genes namely *RD29A, RD29B, RD22* and *RELATED TO AP2.4* (*RAP2.4*) was known to get upregulated via the H3K4me3 and H3K9Ac in their promoter regions and the increase is evident remarkably during the severity of the stress, as reported in *Arabidopsis* (Ramirez-Prado et al., 2019).

In *Arabidopsis*, the growth arrest mediated by ABA was shown to be controlled via demethylation of histones by JMJ30 and JMJ32 (Wu et al., 2019). The expression of JMJ30 was found to be activated by ABI3, which in turn demethylates the promoter region of *SnRK2.8 kinase*, activating its expression that is then involved in the activation of ABI3. In maize, it was shown that insertion of transposable elements, MITE in the promoter region of NAC along with H3K9me2 accumulation resulted in its repression (Mao et al., 2015), thus proving the role of histone demethylation in drought stress adaptation. Drought stress-responsive genes of *Arabidopsis* were found to be induced by the increased H3K4me3 and H3K9Ac on their promoter and acetylation of H3K23 and H3K27 in the gene (Kim et al., 2012). The transcription of various dehydration stress-responsive genes was induced by the trimethylation of lysine 4 of histone 3 by ATX1. Decreased germination and rapid transpiration were reported in *atx1* mutants leading to decreased drought tolerance, which also reported a decrease in a key enzyme of ABA biosynthesis (Ding et al., 2011).

In addition to histone modification, DNA methylation is another modification mechanism involved in drought response. Through the studies and findings in *Populus*, it was evident that changes in the methylation levels of several drought-responsive genes led to changes in their expression drastically (Liang et al., 2014). While genome-wide methylation levels of cytosines reported during drought stress correlated positively with the gene expression pattern, it was also seen that higher intensities of methylation led to silencing of the genes. It was shown from various studies that reduced as well as increased DNA methylation of certain transcription factors (TFs) were both correlated with drought induction in plants. Extensive methylation-mediated DNA remodeling was observed in *Poplar* during drought response, and the expression of genes was regulated by the methylation of the neighboring repetitive elements.

The wild apple tree *Malus pruniflora* was tolerant to drought in comparison to the *Malus domestica,* and it was found that the tolerance was the effect of lower methylation levels of the promoter and thus higher expression of the *DREB2A* gene, linked to heat and drought resistance (Sakuma et al., 2006). Drought-induced methylation was reported to be 70% in rice. While the epigenetic changes were reversed and seen at 29% after recovery of the plants, the methylation patterns remained the same. Significant levels of tissue-specific developmental variation in DNA methylation were observed, indicating the role of drought stress-induced DNA methylation patterns to be an important regulatory mechanism for plant response to drought stress (Wang et al., 2010). It was also reported that there were hypomethylation patterns in the susceptible cultivars of rice, whereas the resistant/tolerant varieties exhibited hypermethylation of the genes involved in drought-responsive stress (Gayacharan & Joel, 2013).

30.4.1.3 Epigenetic Regulations during Salt-Induced Stress

Salinity leads to a series of stressful conditions for the plants, including stress via osmotic pressure and related damages, sodium ion toxicity, etc. (Ashapkin et al., 2020). It is well proven that regulation of epigenetic modifications via DNA methylation is vital among the various mechanisms involving salt stress response. Genotype as well as organ-specific regulation of gene expression in rice plants was reported to show methylation of the promoter in both tolerant and sensitive cultivars. The differential expression of the *DRM2* gene correlated with hypomethylation and the gene was upregulated in salt-sensitive cultivars only, showing the role of DNA methylation in salt stress response in rice (Ferreira et al., 2015).

A transporter protein HKT1 is known to mediate sodium ion influx in plants, and it coordinates the salt overly sensitive (SOS) pathway leading to salt tolerance (Rus et al., 2001). In a wild phenotype of *Arabidopsis*, a putative small RNA target region, approximately 2.6 kb upstream of the start codon of the *HKT1* gene, was identified that showed extensive global methylation. While studying the mutant *rdr2*, there was a significant reduction in the methylation of CHG and CHH while CG methylation remained unchanged, proving the role of small RNA in the methylation of non-CG islands (Baek et al., 2010).

There is a correlation between the downregulation of *HKT* genes in wheat and the hypermethylation during salt stress. Differential expression levels of *HKT* genes were observed in different wheat cultivars exhibiting different levels of tolerance to salt stress that eventually regulate the epigenetic modifications at different levels (Kumar et al., 2017). Variations were observed in the levels of methylated cytosine in the gene *HKT2* in different tissues, in stress conditions and in different cultivars. With regard to salt stress, it increased significantly and led to the downregulation of *HKT2* genes. The link between differential expression of *HKT* genes and DNA methylation was further supported by the findings of Wang et al. in wheat (Wang et al., 2011).

A salt stress-induced transcription factor, MYB74, was found to have a highly methylated promoter under normal conditions, which usually helps in maintaining low levels of transcripts. During salt stress, hypomethylation of the gene along with low levels of an siRNA was reported, leading to an increased expression of MYB74 (Xu et al., 2015). Five siRNAs were believed to target a small promoter region of the gene and reduction in their accumulation correlated with salt stress, thus suggesting the role of these RNA molecules in the regulation of DNA methylation and also the expression of the *MYB74* gene. Variations in chromatin structure as brought about by the histone modification are also an important player in mediating tolerance to high salt stress. Salt tolerance conferring genes generally had a deposition of histone marks like H3K9K14Ac and H3K4me3 while there was a decrease in H3K9me2 and H3K27me3 depositions (Yolcu et al., 2016).

The role of histone deacetylases 2 (HD2) in salt stress response is documented by the studies conducted in *hd2c* mutants that showed higher sensitivity to NaCl during germination and also less tolerance to salt-induced stress. They also exhibited high sensitivity to ABA as the HD2C protein was shown to repress ABA1 and ABI2 via its interactions with HDA6 under normal conditions (Luo et al., 2011). Thus, the association of HD2C with HDA6 through modifications of histones for the regulation of gene expression is evident. Mutations in *HDA6* increased susceptibility to salt stress in *Arabidopsis* plants (Chen et al., 2010), further supporting its role in gene regulation. Reports on the crosstalk between histone acetylation and DNA methylation during stress responses have been shown in soybean where during high salinity stress, there was a combined activity of DNA methylation and histone modification in the regulation of certain stress-induced transcription factors (TFs) (Song et al., 2012).

30.4.1.4 Modifications during UV Stress

Sunlight serves as an essential environmental signal for the regulation of growth and development for plants. While considering the essentiality of sunlight, stress is induced in the plants due to high light intensity and UV. Various factors involving the exposure time, fluence rate and prior exposure can determine the type of stress response in plants (Muller-Xing et al., 2014). It was reported from the *Arabidopsis* plants exposed to UV-B stress that a silent reporter gene was being controlled by epigenetic modifications involving changes in the conformation of chromatin and acetylation of histone 3 (Lang-Mladek et al., 2010). There was a report from *Artemisia* on hypomethylation of DNA in response to UV stress. *DBR2*, a key regulatory gene for artemisinin synthesis, was found to be expressed in response to UV-B treatment, upon demethylation of its promoter region (Pandey & Pandey-Rai, 2015). The UV-B-induced synthesis of flavonoids in *Artemisia* was reported to be involved in DNA methylation. Promoter region demethylation of the *AaPAL1* gene upon UV-B treatment led to the release of several TF binding sites and was correlated with the increase in expression of MYB TFs (Pandey et al., 2019).

30.4.1.5 Epigenetic Modifications during Heavy Metal Toxicity

Various metabolic mechanisms including but not limited to enzyme activity, protein structure and integrity of membranes can be disturbed and disrupted by heavy metal-induced stress (Feng et al., 2023). Stress induced by cadmium (Cd) was found to increase the overall DNA methylation status in rice and *Arabidopsis* (Feng et al., 2016; Fan et al., 2020). Certain DNA demethylase genes are repressed during Cd exposure, leading to increased DNA methylation and tolerance to metal stress correlated with improved Fe nutrition. Wheat varieties resistant to a range of heavy metals (lead, cadmium and zinc) showed a higher activation of metal detoxification transporters via DNA hypomethylation of their promoter regions when compared with the less tolerant varieties (Shafiq et al., 2019). A similar metal detoxification transporter, OsZIP1, was observed to undergo demethylation at H3K9 during Cd stress in rice, which resulted in its overexpression, leading to tolerance to metals like zinc, copper and cadmium (Liu et al., 2019).

Contradictory to the above observations, hypermethylation at the genome level was reported in *Noccaea caerulescens* plants, which is a hyperaccumulator of nickel (Ni). The study performed a comparative analysis of DNA methylation levels during Ni stress in *Noccaea* and *Arabidopsis*, which found that there was less methylation in *Arabidopsis* plants susceptible to Ni, as supported by the differential expression of key DNA methylation genes like *MET1*, *DRM2* and *HDA8* in both species (Gullì et al., 2018). A study of methylation patterns was performed in wild tobacco plants during aluminium (Al) stress and showed that Al-induced stress promotes demethylation of the coding region of *NtGPDL* gene linked to the extracellular matrix and involved in Al stress responses Choi and Sano (2007). The *SAMS1* gene from *Andropogon virginicus* was found to confer resistance to Al stress upon transformation in *Arabidopsis*. Overexpression of this gene in transformants resulted in DNA and histone 3 methylation patterns upon exposure to Al, also reporting differential methylation patterns in different regions of the gene itself (Ezaki et al., 2016).

30.4.1.6 Epigenetic Changes during Nutrient Stress

In the modern agricultural era where heavy usage of fertilizers is leading to adverse effects in the environment as well as in the plant responses, it becomes essential to unravel the adaptive stress-responsive mechanisms in plants, which can aid towards sustainable agriculture. Epigenetic modification profiling in response to low N in wheat showed that there were cultivar-specific variations in H3K27ac and H3K27me3, leading to cultivar-specific adaptation strategies to N deficiency (Zhang et al., 2023). Low N tolerance in apple trees was found to be promoted by an m6A reader, MhYTP2, via interfering with the stability of the mRNA of *MdATG8A*, a key gene in autophagy regulation along with other bound mRNAs (Guo et al., 2023). Anthocyanin accumulation is known to be a hallmark response in plants responding to N deficiency. The HDA15 of *Arabidopsis* was found to repress genes for the biosynthesis and regulation of anthocyanins, thus reducing the excess accumulation of anthocyanins (Liao et al., 2022). It is evident that during N deficiency, there is a need for autophagy for N remobilization and reuse (Masclaux-Daubresse et al., 2017). The degradation of HY5, a protein essential for autophagy repression, was reported during N deficiency conditions leading to the activation of genes involved in autophagy, *ATG5* and *ATG8*, via blocking the histone 3 deacetylation by HDA9 (Yang et al., 2020).

Phosphate (Pi) starvation is found to induce extensive remodeling of DNA methylation patterns in *Arabidopsis* as well as in rice (Yong-Villalobos et al., 2015; Secco et al., 2015). DNA methylation under low Pi stress was found to be associated with altering the transcriptional levels of certain genes in soybean and tomato (Chu et al., 2020; Tian et al., 2021). The proper distribution of Pi in the roots and shoots of *Arabidopsis* was found to be regulated by a histone acetyltransferase, GCN5 under Pi deficiency conditions via the regulation of PHO2, a translocator protein achieved by the H3 acetylation and activation of the long non-coding RNA, At4. (Wang et al., 2019). During Pi deficiency, a complete remodeling of the root system architecture involving reduced root growth, increase in lateral root number and enhanced density of root hairs was found to be induced (Svistoonoff et al., 2007). Root architecture regulating genes, *LPR1* and *LPR2* of *Arabidopsis* undergo histone 3 deacetylation via a complex of HDC1 and HDA6 which gets recruited by the chromatin remodeler BRAHMA (BRM). During deficiency in Pi, there was degradation of BRM and HDC1, which led to the increased H3 acetylation, and thus the activation of LPR1 and LPR2 (Xu et al., 2020; Li et al., 2022).

30.4.2 Epigenetic Changes during Biotic Stress

30.4.2.1 Virus-Induced Modifications

Unlike animals, plants lack an extensive adaptive immune system strategy but have evolved a more complex RNA-based defense mechanism of regulating and protecting themselves against invading pathogens. As the viral transcription and replication occur inside the host cell, plants are believed to have evolved the ability to detect such exogenous nucleic acids of the virus as a potential threat. The control of viral virulence through post-transcriptional gene silencing (PTGS) was one among the first evidences for the involvement of epigenetic modulations in plant defense against biotic factors (Ramirez-Prado et al., 2018). It is a general perception that all of the epigenetic mechanisms involving RNA intermediates were initially part of the defense mechanisms against viral particles.

Plants recognize the double-stranded viral RNA molecules upon infection, and their degradation into siRNAs is marked by the endoribonucleases, DCL2 and DCL4. Genomes of single stranded RNA viruses get converted to double strands by RNA-dependent RNA polymerases for the DCL family endoribonucleases to act upon. Transcriptional gene silencing (TGS) is another mechanism by which there is permanent defense against plant viruses through RdDM, which takes place as Pol IV-dependent transcription for the viral genome, yielding 24-nucleotide siRNAs that get incorporated into AGO4, forming the siRNA–AGO4 complex that can act as controlling factors for viral DNA methylation via DRM2. MET1 and CMT2 help in the maintenance of the methylation pattern (Arora et al., 2022).

A m6A-specific methylation of the alfalfa mosaic virus (AMV) genome was found to control infection in *Arabidopsis* (Martínez-Pérez et al., 2017). An *Arabidopsis* protein, ALKBH9B (At2g17970) possessing demethylase activity removed m6A from ssRNA molecules and accumulated them in the siRNA bodies in the cytoplasm, in vitro. This activity showed the involvement of the protein in the silencing of mRNA and its decay. When ALKBH9B activity was suppressed, the m6A molecules accumulated in the AMV genome and impaired systemic invasion of the virus.

Changes in DNA methylation at the genomic regions of defense and stress-related genes were reported during tomato plant infection with yellow leaf curl sardinia virus (Mason et al., 2008). A study in watermelon plants infected with Cucumber green mottle mosaic virus identified numerous differentially methylated regions (DMRs), which led to the generation of whole-genome methylation profiles for watermelon (Zeng et al., 2020). While studying the pattern of DNA methylation in tobacco infected with the tobacco mosaic virus (TMV), it was found that there was both hyper- and hypo-methylation of certain regions as an outcome of TMV infection. Hypermethylation was observed at the global genome level and actin loci of the F1 progeny of infected plants, whereas the LRR-containing loci of the F1 progeny reported hypomethylation (Boyko et al., 2007).

30.4.2.2 Microbes

Pathogen-associated molecular pattern (PAMP)-triggered immunity (PTI) is the first layer of active defense against pathogens, which relies on the recognition of PAMPs or microbe-associated molecular patterns (MAMPs) by pattern-recognition receptors (PRRs). Following this, there is an activation of the immune responses in the host plant resulting in basal immunity. The pathogens on their part produce effector molecules to override the defense strategies of the plant and neutralize the PTI. As a counterattack, these effector molecules of the pathogen get recognized by plant disease resistance proteins, leading to a more potent immune response, the effector-triggered immunity (ETI). It is proven that both PTI and ETI involve massive gene expression changes that are highly regulated by epigenetic mechanisms like DNA methylation (Ramirez-Prado et al., 2018; Deleris et al., 2016).

In *Arabidopsis*, a series of defense mechanisms is activated in response to the bacterial virulence factors produced during infection by a bacterial pathogen *Pseudomonas syringae* pv. *tomato* (*Pst*)(Pst), including the upregulation of SA signaling. It was found that mutant plants defective in CG and non-CG methylation (*met1 & drm1, drm2, cmt3-ddc*) showed marked resistance against the pathogen in comparison to the wild types. In the mutants, there was an enormous increase in the expression of numerous genes, almost more than tenfold. Pathogen exposure led to changes in the expression levels of numerous pathogen-responsive genes as well. These data support the fact that changes in DNA methylation, as in the mutants, could be an important defense mechanism against *Pst* (Dowen et al., 2012). A Gene Ontology (GO) analysis was performed for differentially methylated regions (DMRs) induced by virulent *Pst* or SA and it was shown that genes involved in plant defense were enriched, whereas the avirulent bacteria induced DMRs that corresponded to genes involved in transcriptional regulation, while only fewer of the defense-related genes.

A well-characterized plant PRR, FLS2, which can sense the bacterial flagellin-derived peptide, flg22, leads to changes in the expression of multiple genes (Yu et al., 2013). A study involving flg22 treatment in plants showed that the reactivation of multiple transposable elements and RdDM targets leads to the inhibition of TGS. Upregulation of an early defense gene, FRK1 was found to be an effect of significant downregulation of *RdDM* genes. There was a progressive CHH-demethylation of the retroelements AtSN1 and LTR of ONSEN, leading to their activation, showing the role of demethylation in transcriptional activation. Mutants defective for RdDM surprisingly exhibited lower bacterial growth than in the wild types with a consistent PR1 expression. As proof for the involvement of plant NOD-like receptors (NLRs), which can trigger a hypersensitive response (HR) and PR1 induction, the *met1 nrpd2* double mutants showed both HR-like response and increased PR1 levels, also indicating the control of NLRs via RdDM.

Innate immune mechanisms of plants, though short-termed, can elicit the priming of inducible defense mechanisms that help a plant respond faster and stronger to recurring stimuli. Priming is induced by microbes and manifests as pathogen-induced systemic acquired resistance (SAR). Priming states can be either short-termed, disappearing within a few days, or long-lasting, being transmitted between generations. One of the priming mechanisms involving the SA-dependent immunity has been observed to be transgenerational, suggesting the epigenetic mechanisms behind its activation (Sánchez et al., 2016). *Arabidopsis* DNA methylation mutants, *nrpe1, drd1, ddm1, ros1* and *cmt3*, infected with the oomycete *Hyaloperonospora arabidopsidis* (*Hpa*) were used for the study of the involvement of DNA methylation in resistance against biotrophic pathogens (Sánchez et al., 2016). The levels of resistance exhibited by the mutants to the pathogen were in the increasing order of *ddm1, nrpe1, drd1* and *cmt3* and definitely well above the resistance exhibited by the wild-type plants, while *ros1*, the DNA-hypermethylated mutant, was more susceptible to infection than the wild types. The *ddm1* mutant showed the strongest level of resistance among all genotypes tested. Consistent with respective resistant phenotypes, *nrpe1* mutants had stronger induction of the SA-induced *PR1* gene, while *ros1* mutant plants showed decreased PR1, indicating that DNA hypomethylation primes SA-dependent defense against *Hpa*. The same mutants were also studied for the resistance mechanisms against the necrotrophic fungus *Plectosphaerella cucumerina* and *Alternaria brassicicola*, against which the basal resistance mechanism operates via the induction of jasmonic acid (JA)-mediated responses. In both cases, in the mutants of the RdDM pathway, there was repression of JA-inducible defense genes. Six days post-infection, it was observed that *ros1* mutants exhibited higher tolerance than the *nrpe1* mutants when compared to the wild-type plants, showing the role of DNA hypermethylation for increased resistance against necrotrophic fungi (Sánchez et al., 2016).

While analyzing the epigenetic recombinant in-bred lines (epiRIL) of *Arabidopsis* in connection with *Hpa*-resistant epigenomes, four epigenetic quantitative trait loci (epiQTLs) were identified, which accounted for most of the variation in disease resistance (Furci et al., 2019). While none of the epiRILs exhibited increased expression of SA-dependent PR1, there was increased callose deposition, which is an SA-independent pathogen-inducible defense strategy. Genome-wide priming of the defense genes was found to be associated with higher resistance in the *Hpa*-resistant epiRILs. A large set of genes involved in SA-dependent and SA-independent defensive responses showed more intense *Hpa*-induced expression in the epiRILs compared with wild-type plants. These observations suggest that the increased resistance exhibited in epiRILs is associated with priming response. It was also shown that though a larger proportion of defense-related genes were placed far apart rather than being located within the epiQTL borders, the four epiRILs showed a larger proportion of similar defense-related genes in spite of having different epiQTL combinations. This suggests that the *Hpa*-inducible genes might be getting trans-regulated via DNA methylation. A similar observation was made in the study involving resistance mechanisms of *Arabidopsis* to *Plasmodiophora brassicae*, where 16 epiQTLs were detected to influence the resistance mechanism. Six epiQTLs were mapped to be closer to the resistance genes and QTLs, proving that both allelic and epiallelic variations interact with the environment, leading to variable resistance to the pathogen (Liégard et al., 2018). Increased resistance to *Phytophthora infestans* in potato was associated with priming response via β-aminobutyric acid (BABA). There was upregulation of SA-responsive genes in the first unstressed generation of the BABA-primed parent plants upon infection with the pathogen (Meller et al., 2018). A bivalent histone mark configuration, H3K4me2 and H3K27me3, was observed in the SAR regulator genes *NPR1* (*Non-expressor of PR*) and *SNI1* (*Suppressor of NPR1 inducible*) during the early priming phase, enabling a readily switchable chromatin configuration state contributing to intergenerational stress memory via increased *PR1* and *PR2* gene transcripts. Histone acetylation was not in play for the normal maintenance of BABA-primed defense memory until the infection with *P. infestans*, after which HAT activity was mandated in both parents and progeny for the increased expression of *PR* genes. Upon treatment with BABA, there was a drastic downregulation of the pathogen gene, *Pitef1*, up to fourfold, compared to the unprimed inoculated plants, while the progeny exhibited 2.5-fold downregulation of the same gene (Kuźnicki et al., 2019). The expression of the plant genes *MET1*, *CMT3* and *DRM2* was analyzed after treatment with BABA and it was found that MET1 was unaffected while CMT3 and DRM2 were upregulated about 7.0-fold and 18.0-fold, respectively. Thus, priming with BABA was dependent on DNA methylation by CMT3 and DRM2. For the identification of HMTs regulating PTI responses in *Arabidopsis*, a reverse genetic screen was used. HMT encoding genes of *Arabidopsis* are termed as *SDG* (*SET Domain Group*) genes. Fungal infection in plants triggers *pep1*-mediated PTI response, which is regulated by SDG25 and SDG8. Ten mutant *SDG* genes responsive to infection were tested for fungal resistance to *Botrytis cinerea* and/or PTI. Increased sensitivity to *B. cinerea* infection was reported in *sdg8* and *sdg25* mutant plants before and after pretreatment with pep1, while displaying decreased pep1-triggered immunity compared to wild-type plants. Thus, it could be assumed that SDG8 and SDG25 contribute actively to pep1-triggered immunity against fungal infection. Significant upregulation of both *SDG25* and *SDG8* genes during *B. cinerea* infection and pep1 treatment shows the involvement of histone modification in the resistance against fungal infection (Lee et al., 2016). A similar study was conducted to analyze the role of *SDG25* and *SDG8* during *Pst* infection and showed that the same genes were also implicated in immunity against bacterial pathogens. Both the genes play roles in regulating SAR and SDG8 contributes to ETI response while SDG25 contributes to certain effectors. In the double-mutant analysis (*sgd8 sdg25*), almost 6,000 genes failed to be induced by the pathogen in comparison to the wild types, proving the role of HMTs as epigenetic transcription regulators. In addition to the downregulated genes, almost 5,000 genes were upregulated in the mutants in comparison to the wild-type plants, showing the role of SDGs in negatively regulating few transcripts during *B. cinerea* infection. H3K36 and H3K4 methylation is affected by SDG8, while H3K4 methylation is affected by SDG25, globally. In the *sdg8 sdg25* double mutants, H3K4me3, H3K36me2 and H3K36me3 levels were significantly diminished (Lee et al., 2016). *B. cinerea* infection in tomato showed increased upregulation of *DES*, *DOX1* and *LoxD* genes that encode key enzymes in the oxylipin pathway, and WRKY75 that encodes a stress-responsive TF. Concurrent to their activation, there was an increase in H3K4me3 and H3K9ac levels in all the pathogen-induced genes upon infection. The same set of genes was also increased in transcript levels in response to *Pst* DC3000 along with an increase in H3K4me3 and H3K9ac. *DES* and *DOX1* expression was significantly less than with *B. cinerea*, whereas WRKY75 showed a significant increase in both histone marks along the gene (Crespo-Salvador et al., 2020).

30.4.2.3 Pests

Induced changes in the methylation patterns were studied by WGBS analysis in the soybean cyst nematode-infected and healthy soybean roots. In the context of CG and CHG methylation, hypermethylated and hypomethylated regions were identified and hypomethylated regions were almost double in number as the hypermethylated regions in the infected roots in comparison to the uninfected roots, showing that hypomethylation was the prevalent trend in all sequence contexts. In total, 703 and 1,346 genes were found to be hyper- and hypo-methylated, respectively (Rambani et al., 2015).

The effects of genome-wide DNA methylation during SCN infection were studied in both resistant and susceptible soybean. The DNA methylomes of the highly susceptible and highly resistant lines showed so much variation in spite of high genetic similarity between the lines. Out of the DMRs identified between the susceptible and resistant lines, many were identified to be overlapping with protein-coding genes and TEs. In the gene bodies, CG and CHG-DMRs were found to be predominant. The gene promoter regions had a lesser distribution of CG-DMRs and the LTR retrotransposons had TE-associated DMRs. GO analysis of the DEGs showed significant enrichment for genes related to wounding and defense responses, membrane disassembly and intracellular signal transduction, indicating that DEGs between the two lines under non-infected conditions depict their differential capability to respond to the nematode infestation (Rambani et al., 2020).

The methylation of genome body and H3K9 was observed in *Arabidopsis* mutant screening, regulating the resistance against infestation by the pest, *Myzus persicae*, the green peach aphid. (Annacondia et al., 2021). It was shown that the *nrpd1* and *kyp* mutant plants are more resistant to aphids, but it is still not clear how the crosstalk between DNA and H3K9 methylation contributes to resistance against the aphid.

30.5 EPIGENETIC MEMORY FOR INCREASED DEFENSE AND ENVIRONMENTAL ADAPTATION

30.5.1 Priming Response to Environmental Challenges

Most of the environmental conditions/stresses encountered by the plants may occur more often in their life span. In such cases, plants tend to display hypersensitive responses to recurring stress, which leads to enhanced defense against most of the limiting factors of the environment (Avramova, 2015; Kinoshita & Seki, 2014; Martinez-Medina et al., 2016). These priming effects due to the initial exposure to stress help the plants to memorize their past interactions with the environment for better preparation during the recurrence of the event and have come to be termed as stress priming or memory (Ding et al., 2012; Lämke & Bäurle, 2017). Chromatin modifications that are environment-induced contribute primarily to plant environmental memory (He & Li, 2018), for example, several cycles of dehydration in *Arabidopsis* plants led to the ability of the plants to retain more water than those plants that were experiencing dehydration for the first time (Ding et al., 2012).

In response to priming, certain genes are expressed at similar levels during every exposure while others increase in expression significantly at repeated treatments and behave as stress-responsive and stress-memory genes, respectively (Chang et al., 2019). During the recovery period, there are two distinct epigenetic marks that are found in the memory genes – high levels of H3K4me3 and stalled form of RNA polymerase II – Ser5P PolII (phosphorylated at the serine 5th). These marks were found to increase dynamically only at stress exposure and decrease to basal levels during recovery in the stress-responsive genes, whereas in the memory genes, they persisted as long as the transcriptional memory lasted. RNA-Seq analysis of transcriptomes in *Arabidopsis* prior to and after the first and third exposure to dehydration stress led to the identification of a high diversity of memory-type responses (Ding et al., 2013). GO enrichment analysis identified the memory genes to be implicated in cold/heat acclimation, salt/ABA treatment and genes involved in signaling pathways, namely ABA, ethylene, auxin, gibberellic acid, JA and SA.

Study on *Arabidopsis* plants exposed to priming heat stress through a ChIP-qPCR study showed the heat stress-induced gene, *HSP22.0*, to be involved in stress memory (Lämke et al., 2015). After two days of exposure, *HSP22.0* transcripts were still increased while the transcripts of *HSP70* were brought to the basal level. H3K9ac marking was significantly enriched in both genes at 4 h after stress exposure, declined rapidly after 52 h in HSP70 while it still increased in *HSP22.0*. Another epigenetic marking, H3K4me3, was observed at HSP70, at threefold enrichment, 4 h after HS and then declined to basal levels for the next two days. In contrast, a 75-fold enrichment was observed for the marking in HSP22.0, which remained the same even at 52 h. With regard to H3K4me2, no changes were seen in HSP70, while HSP22.0 showed an enrichment at the later time points, 28 h and 52 h but absent at 4 h. These observations state that H3K4me3 and H3K4me2 marks at HSP22.0 might be representing transcriptional memory of the priming heat stress.

Using epiRILs generated from *ddm1* mutants crossing to wild-type plants, a screening of *Arabidopsis* plants was performed for the identification of *Hpa*-resistant *Arabidopsis*. The resistant lines did not show any retardation in growth and had a stronger induction of PR1 upon *Hpa* infection than the wild-type plants, revealing a priming of SA-inducible defenses. Transcriptome and DNA methylome analysis of the resistant lines revealed that there was genome-wide priming of defense-related genes, which was sufficient to quantitatively increase disease resistance and can be heritable (Furci et al., 2019). Therefore, DNA methylation epigenomic responses in previous generations could effectively contribute to transgenerational acquired resistance in *Arabidopsis*.

Priming of *Arabidopsis* PTI against *Pectobacterium carotovorum* by BABA treatment is mediated by H3K9K14ac and H3K4me2 (Po-Wen et al., 2012). It is also reported that treatment of common bean with BABA or 2,6-dichloroisonicotinic acid leads to the enhancement in H3K4me3 and H3K36me3 markings in defense-related genes at the promoter-exon regions (Martínez-Aguilar et al., 2016).

Treatment with BABA in potato showed priming effects as well as induced intergenerational resistance against oomycete *Phytophthora infestans*. This response is found to be mediated via epigenetic enrichment of H3K4me2 and H3K27me3 dynamics. Upon treatment with BABA, H3K4me2 was shown to be transiently induced in NPR1 and SNI1, leading to the activation of SA-responsive genes and showed enhanced markings of defense response genes *WRKY1*, *PR1* and *PR2* in primed plants and their progenies (Meller et al., 2018).

30.5.2 Transgenerational Memory of Adaptations

In addition to the memory during priming, there are ample reports of an epigenetic change being transmitted between generations. For an epigenetic mark to be stably inherited in multiple generations, there is a requirement for its passage through gametes, not being removed during ontogenesis (Lal et al., 2018). As the germline cells of the plants are developed from the meristematic cells in the later part of their life cycle, epigenetic changes that are acquired during the vegetative phase are transmitted to the progeny and the subsequent generations (Kinoshita & Seki, 2014). The term intergenerational memory refers to the detectable memory only in the first stress-free generation of the primed plants, possibly mediated by flower/seed development conditions, whereas if the memory is retained for at least two stress-free generations, it is referred to as transgenerational memory, which generally has an epigenetic basis (Lämke & Bäurle, 2017). Plants achieve long-term stable memory mainly through DNA methylation/demethylation, whereas other epigenetic mechanisms mainly contribute to short-term memory (Joseph & Shah, 2022).

Considering the DNA methylation context, CG and CHG are faithfully preserved in both paternal and maternal germlines, whereas CHH methylation gets eliminated in the paternal germline due to the loss of DRM2 activity that subsequently leads to the rescued activity of RdDM in male gametes. Thus, any of the CHH methylation restored in the embryo should possibly be from the maternal epigenetic contributions (Calarco et al., 2012). In plants, the gene body methylations are mostly limited to CG marks while the promoter regions and the gene borders are enriched by CHG and CHH marks (Cokus et al., 2008). As reported by Dalakouras et al. (2012), the faithful preservation of CG methylation and CHH methylation resetting in the germlines through de novo methylation driven by the female gametes suggests the coordination of transgenerational epigenetic stability by CG methylation.

Recently researchers have focused their attention on RNA-mediated transgenerational memory, and its induction in plants has received much attention due to the inherent physiological characteristics/ability of plants to transport RNAs through plasmodesmata and the vascular system throughout the plant (Uddin & Kim, 2013; Atkins et al., 2010). The large amount of complex RNA pool transmitted to the next generation via gametes has the potential to mediate transgenerational inheritance in many species (Riddle, 2014), for instance, the transport of siRNAs leads to the systemic spread of gene silencing transgenerationally and is capable of inducing epigenetic effects like paramutation (Shah, 2021).

The small RNA-mediated epigenetic changes, in most cases, ultimately contribute to the establishment of DNA methylation (RdDM), which is heritable (Bond & Baulcombe, 2013), for example, fungal elicitors induced the accumulation of a set of small RNAs involved in RdDM in *Arabidopsis* (Baldrich et al., 2014). Even though different possibilities of RNA-mediated transgenerational memory are postulated, there are only a few reports in this regard under biotic stress in plants. Histone modifications are also reported to play certain roles in intergenerational memory; for example, enhanced histone acetylation of SA-inducible gene promoters was noted in the progeny of *P. syringae* pv. *tomato* DC3000-inoculated *Arabidopsis*, which resulted in inherited SAR (Luna et al., 2011).

Upon caterpillar herbivory, JA treatment or mechanical damage, *Arabidopsis* and tomato plants exhibited transgenerational priming of JA-mediated defense responses (Rasmann et al., 2011). There was a decline in the growth of the corn earworm *Helicoverpa zea* on the progeny of treated tomato plants by about 40% when compared to the control plants. Feeding of the *Arabidopsis* progeny by white cabbage butterfly, *Pieris rapae* was diminished by 40% through parental feeding induced transgenerational memory. This memory of stress priming lasted for two generations and disappeared in the third generation in *Arabidopsis*. Mutants deficient in JA signaling and/or the biogenesis of siRNAs failed to exhibit the inherited resistance. Thus, the study has led to the suggestive conclusions that RdDM-induced changes in DNA methylation may serve as epigenetic memory marks, leading to the enhancement in defense-related mechanisms, transgenerationally.

By imposing drought on drought-tolerant and drought-sensitive rice varieties through 11 successive generations, two epimutation accumulation (epi-MA) lines were raised which were then studied for the role of epigenetic mechanisms in adaptation to the adverse environment in rice (Zheng et al., 2017). The DNA methylomes of 32 progeny plants were subjected to WGBS for observing the frequency and distribution of induced epimutations. It was observed that the F11 progeny of both lines exhibited higher tolerance to osmotic stress in comparison to their respective parental plants (F0). Upon analysis of the DNA methylomes of three stressed generations (F0, F10 and F11), drought-treated and normal plants in each of them, and a recovery generation F10R1, derived from stressed F10 by cultivation in well-watered conditions for one more generation, it was found that drought treatment induced similar patterns of epimutations in F0 and successive generations, depicting the long-term effect of drought on DNA methylation patterns. There was directionality observed in the methylation as shown by the clustering of the epimutations in the progeny. The successive sibling generations, F10R1 and F11, were more similar to F10 than to F0, indicating directionality in the inheritance of epimutations. However, the count of epimutations did not increase linearly with time. The number of epimutations between F10 and F11 through a single generation was unusually higher than the total number of mutations between F0 and F11, which might be due to the fact that many of the epimutations were not inherited stably between generations.

There are instances of other environmental stresses being primed through the biotic stress-induced epigenetic memory in plants, like in the case of infection with *P. syringae* pv. *maculicola* or treatment with BTH [benzo-(1,2,3)-thiadiazole-7-carbothioic acid *S*-methyl ester] in *Arabidopsis*. These factors elicited defense responses in the plants as well as modified promoters of several WRKY transcription factors by H3K4 methylation and H3 and H4 acetylation, which eventually primed the plants for a subsequent water infiltration stress (Jaskiewicz et al., 2010).

30.6 IMPLICATIONS OF PLANT–ENVIRONMENT EPIGENETIC REGULATIONS

Decoding the mechanisms of plant epigenetic regulation in response to environmental cues has prominent and significant implications for agriculture and crop improvement. They can be used potentially for the strategic development of enhanced stress tolerance and performance in plants, with globally changing growth conditions. Locus-specific epigenetic breeding technology has been made possible at a higher precision by the recent advancements in engineered DNA-binding domain development (Bilichak & Kovalchuk, 2016). Plants have the potential to decide what information to preserve or omit when exposed to stress. Plants can create epigenetic stress memories, either temporary or permanent, via epialleles (Sudan et al., 2018; Kumar & Rani, 2023). By unraveling the transgenerational epigenetic inheritance mechanisms, potential breeding strategies for the enhancement of stress tolerance in plants can be developed (Tonosaki et al., 2022; Oberkofler et al., 2021; Li et al., 2023).

Changes in the climatic conditions are largely unpredictable and immense, and have put agricultural production under intense stress due to its adverse impacts on the yield and quality of the crop. Genes encoding the enzymes involved in the identified epigenetic regulation can be used as tools for engineering the plants (Kumar & Rani, 2023). The genes encoding DMTs have been demonstrated in different plant species for improving tolerance against multiple environmental conditions, such as drought tolerance in rice and maize (Deleris et al., 2016; Kinoshita & Seki, 2014). Apart from genetic engineering, traditional breeding could also be exploited for the application of epigenetic regulatory mechanisms for improved crop varieties (Kumar, 2016).

Identification of specific molecular and regulatory pathways involved during the adaptation to environmental factors will be made possible by unraveling the epigenetic responses of a plant to its growing environment (Chang et al., 2019). The knowledge can then be applied for the management strategies to minimize adverse effects on crop quality and yield (Pathak et al., 2021).

Precision agricultural practices are the need of the day as we can see multiple varying parameters affecting plant growth and development. Study of specific genes and their regulatory mechanisms will pave the way for proposing a plant for their cultivation and harvest, given the input and probabilities towards the yield of the plant. Planning connected to irrigation, treatment strategies and certain site-specific growth strategies can be performed via gathering information on plant responses (Duarte-Aké et al., 2023).

Conservation strategies towards the critically endangered plant species for their economic value can be effectively designed by unfolding the ways in which the plant responds to its growing conditions and understanding the epigenetic regulatory network behind each response. Ecosystem restoration programs can be planned with the gathered information for the reintroduction of native plant species and reclamation of degraded/polluted lands (Li et al., 2021; Kakoulidou et al., 2021). Epigenetic modifications that involve the interactions between plants and the beneficial microbiome of the soil and the interactions between pollinators and other organisms of the food chain can be applied in the programs aiming at preserving and promoting biodiversity niches. A more sustainable and resilient agricultural system can be achieved by harnessing the knowledge on epigenetic regulations in plants, which can adapt effectively and efficiently to climate changes and environmental pressure (Kumar & Rani, 2023).

30.7 CONCLUSION

Understanding epigenetic changes and their mechanisms in complex plant–environment interactions sheds light on plant resilience and adaptation. This chapter provided a detailed report on the epigenetic markings involving various environmental factors, both biotic and abiotic, in model plants and a few of the crops. These epigenetic variations can be seen as a unique adaptation mechanism of plants with much evolutionary significance, as they can potentially impact and change gene expression patterns and can also be stably inherited in successive generations of the plants. The potential for transgenerational inheritance of stress memory makes plants better adapted to the recurring challenges. Such dynamic adjustments made to their gene expression profiles in response to the changing climatic features provide a survival advantage for plants and open up a promising crop breeding and improvement strategy that could be adopted by the scientific community for the generation of resilient crop systems. Translating the stated research findings on epigenetic regulations into practically applicable tools for crop improvement faces certain challenges, majorly due to the complexity of the plant–environment interactions and also in understanding the complex regulatory networking and crosstalk involved in the epigenetic changes. Protocol standardization along with the development of cost-effective large-scale epigenomic analysis techniques should facilitate the applicability of epigenomic research. Nevertheless, this research area focusing on epigenomics-based plant plasticity and adaptability to the ever-changing environment holds great promise for the development of a precision agricultural system but still demands technological advancements, collaborative, interdisciplinary approaches and active engagement and efforts from the research community.

REFERENCES

Ageeva-Kieferle, A., Georgii, E., Winkler, B., Ghirardo, A., Albert, A., Hüther, P., Mengel, A., Becker, C., Schnitzler, J., Durner, J., & Lindermayr, C. (2021). Nitric oxide coordinates growth, development, and stress response via histone modification and gene expression. *Plant Physiology*, 187(1), 336–360. https://doi.org/10.1093/plphys/kiab222.

Alameldin, H. F., Oh, S., Hernandez, A. P., & Montgomery, B. L. (2020). Nuclear-encoded sigma factor 6 (SIG 6) is involved in the block of greening response in *Arabidopsis thaliana*. *American Journal of Botany*, 107(2), 329–338. https://doi.org/10.1002/ajb2.1423.

Annacondia, M. L., Markovic, D., Reig-Valiente, J. L., Scaltsoyiannes, V., Pieterse, C. M. J., Ninkovic, V., Slotkin, R. K., & Martinez, G. (2021). Aphid feeding induces the relaxation of epigenetic control and the associated regulation of the defense response in *Arabidopsis*. *New Phytologist*, 230(3), 1185–1200. https://doi.org/10.1111/nph.17226.

Arora, H., Singh, R. K., Sharma, S., Sharma, N., Panchal, A., Das, T., Prasad, A., & Prasad, M. (2022). DNA methylation dynamics in response to abiotic and pathogen stress in plants. *Plant Cell Reports*, 41(10), 1931–1944. https://doi.org/10.1007/s00299-022-02901-x.

Ashapkin, V. V., Kutueva, L. I., Aleksandrushkina, N. I., & Vanyushin, B. F. (2020). Epigenetic mechanisms of plant adaptation to biotic and abiotic stresses. *International Journal of Molecular Sciences*, 21(20), 7457. https://doi.org/10.3390/ijms21207457.

Atkins, C. A., Smith, P. M. C., & Rodriguez-Medina, C. (2010). Macromolecules in phloem exudates—A review. *Protoplasma*, 248(1), 165–172. https://doi.org/10.1007/s00709-010-0236-3.

Avramova, Z. (2015). Transcriptional 'memory' of a stress: Transient chromatin and memory (epigenetic) marks at stress-response genes. *The Plant Journal*, 83(1), 149–159. https://doi.org/10.1111/tpj.12832.

Baek, D., Jiang, J., Chung, J., Wang, B., Chen, J., Xin, Z., & Shi, H. (2010). Regulated *AtHKT1* gene expression by a distal enhancer element and DNA methylation in the promoter plays an important role in salt tolerance. *Plant and Cell Physiology*, 52(1), 149–161. https://doi.org/10.1093/pcp/pcq182.

Baldrich, P., Kakar, K., Siré, C., Moreno, A. B., Berger, A., García-Chapa, M., López-Moya, J. J., Riechmann, J. L., & Segundo, B. S. (2014). Small RNA profiling reveals regulation of *Arabidopsis* miR168 and heterochromatic siRNA415 in response to fungal elicitors. *BMC Genomics*, 15(1). https://doi.org/10.1186/1471-2164-15-1083.

Baron, N. C., & Rigobelo, E. C. (2021). Endophytic fungi: A tool for plant growth promotion and sustainable agriculture. *Mycology&#58 an International Journal on Fungal Biology*, 13(1), 39–55. https://doi.org/10.1080/21501203.2021.1945699.

Begum, Y. (2022). *Regulatory Role of microRNAs (miRNAs) in the Recent Development of Abiotic Stress Tolerance of Plants*. https://www.semanticscholar.org/paper/Regulatory-role-of-microRNAs-(miRNAs)-in-the-recent-Begum/c1f49e6a954a4086e811c14ec033a9dc3fb85654.

Bilichak, A., & Kovalchuk, I. (2016). Transgenerational response to stress in plants and its application for breeding. *Journal of Experimental Botany*, 67(7), 2081–2092. https://doi.org/10.1093/jxb/erw066.

Binder, M., Zinger, E., Hadany, L., & Ohad, N. (2024). Transgenerational effects of stress on reproduction strategy in the mixed mating plant *Lamium amplexicaule*. *BMC Plant Biology*, 24(1). https://doi.org/10.1186/s12870-024-05458-x.

Bond, D. M., & Baulcombe, D. C. (2013). Small RNAs and heritable epigenetic variation in plants. *Trends in Cell Biology*, 24(2), 100–107. https://doi.org/10.1016/j.tcb.2013.08.001.

Boyko, A., Kathiria, P., Zemp, F. J., Yao, Y., Pogribny, I., & Kovalchuk, I. (2007). Transgenerational changes in the genome stability and methylation in pathogen-infected plants. *Nucleic Acids Research*, 35(5), 1714–1725. https://doi.org/10.1093/nar/gkm029.

Buszewicz, D., Archacki, R., Palusiński, A., Kotliński, M., Fogtman, A., Iwanicka-Nowicka, R., Sosnowska, K., Kuciński, J., Pupel, P., Olędzki, J., Dadlez, M., Misicka, A., Jerzmanowski, A., & Koblowska, M. K. (2016). HD2C histone deacetylase and a SWI/SNF chromatin remodelling complex interact and both are involved in mediating the heat stress response in *Arabidopsis*. *Plant Cell & Environment*, 39(10), 2108–2122. https://doi.org/10.1111/pce.12756.

Calarco, J. P., Borges, F., Donoghue, M. T., Van Ex, F., Jullien, P. E., Lopes, T., Gardner, R., Berger, F., Feijó, J. A., Becker, J. D., & Martienssen, R. A. (2012). Reprogramming of DNA methylation in pollen guides epigenetic inheritance via small RNA. *Cell*, 151(1), 194–205. https://doi.org/10.1016/j.cell.2012.09.001.

Cao, X., & Jacobsen, S. E. (2002). Role of the *Arabidopsis* DRM methyltransferases in de novo DNA methylation and gene silencing. *Current Biology*, 12(13), 1138–1144. https://doi.org/10.1016/s0960-9822(02)00925-9.

Carter, B., Bishop, B., Ho, K. K., Huang, R., Jia, W., Zhang, H., Pascuzzi, P. E., Deal, R. B., & Ogas, J. (2018). The chromatin remodelers PKL and PIE1 act in an epigenetic pathway that determines H3K27me3 homeostasis in *Arabidopsis*. *The Plant Cell*, 30(6), 1337–1352. https://doi.org/10.1105/tpc.17.00867.

Chang, Y., Zhu, C., Jiang, J., Zhang, H., Zhu, J., & Duan, C. (2019). Epigenetic regulation in plant abiotic stress responses. *Journal of Integrative Plant Biology*, 62(5), 563–580. https://doi.org/10.1111/jipb.12901.

Chen, H., Chen, L., Patel, K., Li, Y., Baulcombe, D. C., & Wu, S. (2010). 22-nucleotide RNAs trigger secondary siRNA biogenesis in plants. *Proceedings of the National Academy of Sciences*, 107(34), 15269–15274. https://doi.org/10.1073/pnas.1001738107.

Chen, H., Wang, J. P., Liu, H., Li, H., Lin, Y. J., Shi, R., Yang, C., Gao, J., Zhou, C., Li, Q., Sederoff, R. R., Li, W., & Chiang, V. L. (2019). Hierarchical transcription factor and chromatin binding network for wood formation in *Populus trichocarpa*. *The Plant Cell*, 31(3), 602–626. https://doi.org/10.1105/tpc.18.00620.

Chinnusamy, V., & Zhu, J. (2009). *Epigenetic Regulation of Stress Responses in Plants*. https://www.semanticscholar.org/paper/Epigenetic-regulation-of-stress-responses-in-Chinnusamy-Zhu/66de96057f0e4275975856f29299c4490c20b3a4.

Chodavarapu, R. K., Feng, S., Bernatavichute, Y. V., Chen, P. Y., Stroud, H., Yu, Y., Hetzel, J. A., Kuo, F., Kim, J., Cokus, S. J., Casero, D., Bernal, M., Huijser, P., Clark, A. T., Krämer, U., Merchant, S. S., Zhang, X., Jacobsen, S. E., & Pellegrini, M. (2010). Relationship between nucleosome positioning and DNA methylation. *Nature*, 466(7304), 388–392. https://doi.org/10.1038/nature09147.

Choi, C., & Sano, H. (2007). Abiotic-stress induces demethylation and transcriptional activation of a gene encoding a glycerophosphodiesterase-like protein in tobacco plants. *Molecular Genetics and Genomics*, 277(5), 589–600. https://doi.org/10.1007/s00438-007-0209-1.

Chu, S., Zhang, X., Yu, K., Lv, L., Sun, C., Liu, X., Zhang, J., Jiao, Y., & Zhang, D. (2020). Genome-wide analysis reveals dynamic epigenomic differences in soybean response to low-phosphorus stress. *International Journal of Molecular Sciences*, 21(18), 6817. https://doi.org/10.3390/ijms21186817.

Cokus, S. J., Feng, S., Zhang, X., Chen, Z., Merriman, B., Haudenschild, C. D., Pradhan, S., Nelson, S. F., Pellegrini, M., & Jacobsen, S. E. (2008). Shotgun bisulphite sequencing of the *Arabidopsis* genome reveals DNA methylation patterning. *Nature*, 452(7184), 215–219. https://doi.org/10.1038/nature06745.

Crespo-Salvador, Ó., Sánchez-Giménez, L., López-Galiano, M., Fernández-Crespo, E., Scalschi, L., García-Robles, I., Rausell, C., Real, M., & González-Bosch, C. (2020). The histone marks signature in exonic and intronic regions is relevant in early response of tomato genes to *Botrytis cinerea* and in miRNA regulation. *Plants*, 9(3), 300. https://doi.org/10.3390/plants9030300.

Dalakouras, A., Dadami, E., Zwiebel, M., Krczal, G., & Wassenegger, M. (2012). Transgenerational maintenance of transgene body CG but not CHG and CHH methylation. *Epigenetics*, 7(9), 1071–1078. https://doi.org/10.4161/epi.21644.

Deleris, A., Halter, T., & Navarro, L. (2016). DNA methylation and demethylation in plant immunity. *Annual Review of Phytopathology*, 54(1), 579–603. https://doi.org/10.1146/annurev-phyto-080615-100308.

Ding, B., Del Rosario Bellizzi, M., Ning, Y., Meyers, B. C., & Wang, G. (2012). HDT701, a histone H4 deacetylase, negatively regulates plant innate immunity by modulating histone H4 acetylation of defense-related genes in rice. *The Plant Cell*, 24(9), 3783–3794. https://doi.org/10.1105/tpc.112.101972.

Ding, Y., Avramova, Z., & Fromm, M. (2011). The *Arabidopsis* trithorax-like factor ATX1 functions in dehydration stress responses via ABA-dependent and ABA-independent pathways. *The Plant Journal*, 66(5), 735–744. https://doi.org/10.1111/j.1365-313x.2011.04534.x.

Ding, Y., Fromm, M., & Avramova, Z. (2012). Multiple exposures to drought 'train' transcriptional responses in *Arabidopsis*. *Nature Communications*, 3(1). https://doi.org/10.1038/ncomms1732.

Ding, Y., Liu, N., Virlouvet, L., Riethoven, J., Fromm, M., & Avramova, Z. (2013). Four distinct types of dehydration stress memory genes in *Arabidopsis thaliana*. *BMC Plant Biology*, 13(1). https://doi.org/10.1186/1471-2229-13-229.

Dowen, R. H., Pelizzola, M., Schmitz, R. J., Lister, R., Dowen, J. M., Nery, J. R., Dixon, J. E., & Ecker, J. R. (2012). Widespread dynamic DNA methylation in response to biotic stress. *Proceedings of the National Academy of Sciences*, 109(32). https://doi.org/10.1073/pnas.1209329109.

Duarte-Aké, F., Us-Camas, R., & De-La-Peña, C. (2023). Epigenetic regulation in heterosis and environmental stress: The challenge of producing hybrid epigenomes to face climate change. *Epigenomes*, 7(3), 14. https://doi.org/10.3390/epigenomes7030014.

Ezaki, B., Higashi, A., Nanba, N., & Nishiuchi, T. (2016). An *S-adenosyl Methionine Synthetase* (SAMS) Gene from *Andropogon virginicus* L. confers aluminum stress tolerance and facilitates epigenetic gene regulation in *Arabidopsis thaliana*. *Frontiers in Plant Science*, 7. https://doi.org/10.3389/fpls.2016.01627.

Fan, S. K., Ye, J. Y., Zhang, L. L., Chen, H. S., Zhang, H. H., Zhu, Y. X., Liu, X. X., & Jin, C. W. (2020). Inhibition of DNA demethylation enhances plant tolerance to cadmium toxicity by improving iron nutrition. *Plant Cell & Environment*, 43(1), 275–291. https://doi.org/10.1111/pce.13670.

Fang, H., Liu, X., Thorn, G., Duan, J., & Tian, L. (2013). Expression analysis of histone acetyltransferases in rice under drought stress. *Biochemical and Biophysical Research Communications*, 443(2), 400–405. https://doi.org/10.1016/j.bbrc.2013.11.102.

Feng, D., Wang, R., Sun, X., Liu, L., Liu, P., Tang, J., Zhang, C., & Liu, H. (2023). Heavy metal stress in plants: Ways to alleviate with exogenous substances. *The Science of the Total Environment*, 897, 165397. https://doi.org/10.1016/j.scitotenv.2023.165397.

Feng, S. J., Liu, X. S., Tao, H., Tan, S. K., Chu, S. S., Oono, Y., Zhang, X. D., Chen, J., & Yang, Z. M. (2016). Variation of DNA methylation patterns associated with gene expression in rice (*Oryza sativa*) exposed to cadmium. *Plant Cell & Environment*, 39(12), 2629–2649. https://doi.org/10.1111/pce.12793.

Ferreira, L. J., Azevedo, V., Maroco, J., Oliveira, M. M., & Santos, A. P. (2015). Salt tolerant and sensitive rice varieties display differential methylome flexibility under salt stress. *PLoS ONE*, 10(5), e0124060. https://doi.org/10.1371/journal.pone.0124060.

Floss, D. S., Schmitz, A. M., Starker, C. G., Gantt, J. S., & Harrison, M. J. (2013). Gene silencing in *Medicago truncatula* roots using RNAi. *Methods in Molecular Biology*, 163–177. https://doi.org/10.1007/978-1-62703-613-9_12.

Furci, L., Jain, R., Stassen, J., Berkowitz, O., Whelan, J., Roquis, D., Baillet, V., Colot, V., Johannes, F., & Ton, J. (2019). Identification and characterisation of hypomethylated DNA loci controlling quantitative resistance in *Arabidopsis*. *eLife*, 8. https://doi.org/10.7554/elife.40655.

Gallego-Bartolomé, J. (2020). DNA methylation in plants: Mechanisms and tools for targeted manipulation. *New Phytologist*, 227(1), 38–44. https://doi.org/10.1111/nph.16529.

Gayacharan, N., & Joel, A. J. (2013). Epigenetic responses to drought stress in rice (*Oryza sativa* L.). *Physiology and Molecular Biology of Plants*, 19(3), 379–387. https://doi.org/10.1007/s12298-013-0176-4.

Goh, C., Vallejos, D. F. V., Nicotra, A. B., & Mathesius, U. (2013). The impact of beneficial plant-associated microbes on plant phenotypic plasticity. *Journal of Chemical Ecology*, 39(7), 826–839. https://doi.org/10.1007/s10886-013-0326-8.

Guan, Q., Lu, X., Zeng, H., Zhang, Y., & Zhu, J. (2013). Heat stress induction of miR398 triggers a regulatory loop that is critical for thermotolerance in *Arabidopsis*. *The Plant Journal*, 74(5), 840–851. https://doi.org/10.1111/tpj.12169.

Gullì, M., Marchi, L., Fragni, R., Buschini, A., & Visioli, G. (2018). Epigenetic modifications preserve the hyperaccumulator *Noccaea caerulescens* from Ni geno-toxicity. *Environmental and Molecular Mutagenesis*, 59(6), 464–475. https://doi.org/10.1002/em.22191.

Guo, T., Yang, Z., Bao, R., Fu, X., Wang, N., Liu, C., & Ma, F. (2023). The m6A reader MhYTP2 regulates the stability of its target mRNAs contributing to low nitrogen tolerance in apple (*Malus domestica*). *Horticulture Research*, 10(6). https://doi.org/10.1093/hr/uhad094.

Guo, W., Wang, D., & Lisch, D. (2021). RNA-directed DNA methylation prevents rapid and heritable reversal of transposon silencing under heat stress in *Zea mays*. *PLoS Genetics*, 17(6), e1009326. https://doi.org/10.1371/journal.pgen.1009326.

He, M., Ren, T. X., Jin, Z. D., Deng, L., Liu, H. J., Cheng, Y. Y., Li, Z. Y., Liu, X. X., Yang, Y., & Chang, H. (2023). Precise analysis of potassium isotopic composition in plant materials by multi-collector inductively coupled plasma mass spectrometry. *Spectrochimica Acta Part B Atomic Spectroscopy*, 209, 106781. https://doi.org/10.1016/j.sab.2023.106781.

He, Y., & Li, Z. (2018). Epigenetic environmental memories in plants: Establishment, maintenance, and reprogramming. *Trends in Genetics*, 34(11), 856–866. https://doi.org/10.1016/j.tig.2018.07.006.

Hu, Y., Zhang, L., He, S., Huang, M., Tan, J., Zhao, L., Yan, S., Li, H., Zhou, K., Liang, Y., & Li, L. (2012). Cold stress selectively unsilences tandem repeats in heterochromatin associated with accumulation of H3K9ac. *Plant Cell & Environment*, 35(12), 2130–2142. https://doi.org/10.1111/j.1365-3040.2012.02541.x.

Hu, Z., Song, N., Zheng, M., Liu, X., Liu, Z., Xing, J., Ma, J., Guo, W., Yao, Y., Peng, H., Xin, M., Zhou, D., Ni, Z., & Sun, Q. (2015). Histone acetyltransferase GCN5 is essential for heat stress-responsive gene activation and thermotolerance in *Arabidopsis*. *The Plant Journal*, 84(6), 1178–1191. https://doi.org/10.1111/tpj 13076.

Jaskiewicz, M., Conrath, U., & Peterhänsel, C. (2010). Chromatin modification acts as a memory for systemic acquired resistance in the plant stress response. *EMBO Reports*, 12(1), 50–55. https://doi.org/10.1038/embor.2010.186.

Jia, Y., Ding, Y., Shi, Y., Zhang, X., Gong, Z., & Yang, S. (2016). The *cbfs* triple mutants reveal the essential functions of CBFs in cold acclimation and allow the definition of CBF regulons in *Arabidopsis*. *New Phytologist*, 212(2), 345–353. https://doi.org/10.1111/nph.14088.

Joseph, J. T., & Shah, J. M. (2022). Biotic stress-induced epigenetic changes and transgenerational memory in plants. *Biologia*, 77(8), 2007–2021. https://doi.org/10.1007/s11756-022-01053-3.

Kakoulidou, I., Avramidou, E. V., Baránek, M., Brunel-Muguet, S., Farrona, S., Johannes, F., Kaiserli, E., Lieberman-Lazarovich, M., Martinelli, F., Mladenov, V., Testillano, P. S., Vassileva, V., & Maury, S. (2021). Epigenetics for crop improvement in times of global change. *Biology*, 10(8), 766. https://doi.org/10.3390/biology10080766.

Kang, H., Fan, T., Wu, J., Zhu, Y., & Shen, W. (2022). Histone modification and chromatin remodeling in plant response to pathogens. *Frontiers in Plant Science*, 13. https://doi.org/10.3389/fpls.2022 986940.

Kawakatsu, T., Nery, J. R., Castanon, R., & Ecker, J. R. (2017). Dynamic DNA methylation reconfiguration during seed development and germination. *Genome Biology*, 18(1). https://doi.org/10.1186/s13059-017-1251-x.

Kim, J., Lim, J. Y., Shin, H., Kim, B., Yoo, S., Kim, W. T., & Huh, J. H. (2019). ROS1-dependent DNA demethylation is required for ABA-inducible *NIC3* expression. *Plant Physiology*, 179(4), 1810–1821. https://doi.org/10.1104/pp.18.01471.

Kim, J., Sasaki, T., Ueda, M., Sako, K., & Seki, M. (2015). Chromatin changes in response to drought, salinity, heat, and cold stresses in plants. *Frontiers in Plant Science*, 6. https://doi.org/10.3389/fpls.2015.00114.

Kim, J., To, T. K., Ishida, J., Matsui, A., Kimura, H., & Seki, M. (2012). Transition of chromatin status during the process of recovery from drought stress in *Arabidopsis thaliana*. *Plant and Cell Physiology*, 53(5), 847–856. https://doi.org/10.1093/pcp/pcs053.

Kim, J., To, T. K., Ishida, J., Morosawa, T., Kawashima, M., Matsui, A., Toyoda, T., Kimura, H., Shinozaki, K., & Seki, M. (2008). Alterations of lysine modifications on the histone H3 N-tail under drought stress conditions in *Arabidopsis thaliana*. *Plant and Cell Physiology*, 49(10), 1580–1588. https://doi.org/10.1093/pcp/pcn133.

Kinoshita, T., & Seki, M. (2014). Epigenetic memory for stress response and adaptation in plants. *Plant and Cell Physiology*, 55(11), 1859–1863. https://doi.org/10.1093/pcp/pcu125.

Kooke, R., Johannes, F., Wardenaar, R., Becker, F., Etcheverry, M., Colot, V., Vreugdenhil, D., & Keurentjes, J. J. (2015). Epigenetic basis of morphological variation and phenotypic plasticity in *Arabidopsis thaliana*. *The Plant Cell*, 27(2), 337–348. https://doi.org/10.1105/tpc.114.133025.

Kumar, M., & Rani, K. (2023). Epigenomics in stress tolerance of plants under the climate change. *Molecular Biology Reports*, 50(7), 6201–6216. https://doi.org/10.1007/s11033-023-08539-6.

Kumar, S. (2016). Epigenetic regulation of abiotic stress tolerance in plants. *Advances in Plants & Agriculture Research*, 5(3). https://doi.org/10.15406/apar.2016.05.00179.

Kumar, S., Beena, A. S., Awana, M., & Singh, A. (2017). Salt-iInduced tissue-specific cytosine methylation down-regulates expression of *HKT* genes in contrasting wheat (*Triticum aestivum* L.) genotypes. *DNA And Cell Biology*, 36(4), 283–294. https://doi.org/10.1089/dna.2016.3505.

Kuźnicki, D., Meller, B., Arasimowicz-Jelonek, M., Braszewska-Zalewska, A., Drozda, A., & Floryszak-Wieczorek, J. (2019). BABA-induced DNA methylome adjustment to intergenerational defense priming in potato to *Phytophthora infestans*. *Frontiers in Plant Science*, 10. https://doi.org/10.3389/fpls.2019.00650.

Lal, S. K., Kumar, S., Sheri, V., Mehta, S., Varakumar, P., Ram, B., Borphukan, B., James, D., Fartyal, D., & Reddy, M. K. (2018). *Seed Priming: An Emerging Technology to Impart Abiotic Stress Tolerance in Crop Plants*. https://www.semanticscholar.org/paper/Seed-Priming%3A-An-Emerging-Technology-to-Impart-in-Lal-Kumar/5865577fe8ba74acd2854d78276d4fa2e2097ee6.

Lämke, J., & Bäurle, I. (2017). Epigenetic and chromatin-based mechanisms in environmental stress adaptation and stress memory in plants. *Genome Biology*, 18(1). https://doi.org/10.1186/s13059-017-1263-6.

Lämke, J., Brzezinka, K., Altmann, S., & Bäurle, I. (2015). A hit-and-run heat shock factor governs sustained histone methylation and transcriptional stress memory. *The EMBO Journal*, 35(2), 162–175. https://doi.org/10.15252/embj.201592593.

Lang-Mladek, C., Popova, O., Kiok, K., Berlinger, M., Rakic, B., Aufsatz, W., Jonak, C., Hauser, M., & Luschnig, C. (2010).

Transgenerational inheritance and resetting of stress-induced loss of epigenetic gene silencing in *Arabidopsis*. *Molecular Plant*, 3(3), 594–602. https://doi.org/10.1093/mp/ssq014.

Lee, C. H., & Carroll, B. J. (2018). Evolution and diversification of small RNA pathways in flowering plants. *Plant and Cell Physiology*. https://doi.org/10.1093/pcp/pcy167.

Lee, S., Choi, J., Park, J., Hong, C. P., Choi, D., Han, S., Choi, K., Roh, T., Hwang, D., & Hwang, I. (2023). DDM1-mediated gene body DNA methylation is associated with inducible activation of defense-related genes in *Arabidopsis*. *Genome Biology*, 24(1). https://doi.org/10.1186/s13059-023-02952-7.

Lee, S., Fu, F., Xu, S., Lee, S. Y., Yun, D., & Mengiste, T. (2016). Global regulation of plant immunity by histone lysine methyl transferases. *The Plant Cell*, tpc.00012.2016. https://doi.org/10.1105/tpc.16.00012.

Li, S., He, X., Gao, Y., Zhou, C., Chiang, V. L., & Li, W. (2021). Histone acetylation changes in plant response to drought stress. *Genes*, 12(9), 1409. https://doi.org/10.3390/genes12091409.

Li, T., Zhang, R., Satheesh, V., Wang, P., Ma, G., Guo, J., An, G., & Lei, M. (2022). The chromatin remodeler BRAHMA recruits HISTONE DEACETYLASE6 to regulate root growth inhibition in response to phosphate starvation in *Arabidopsis*. *Journal of Integrative Plant Biology*, 64(12), 2314–2326. https://doi.org/10.1111/jipb.13345.

Li, Y., Mo, X., Xiong, J., Huang, K., Zheng, M., Jiang, Q., Su, G., Ou, Q., Pan, H., & Jiang, C. (2023). Deciphering the probiotic properties and safety assessment of a novel multi-stress-tolerant aromatic yeast *Pichia kudriavzevii* HJ2 from marine mangroves. *Food Bioscience*, 56, 103248. https://doi.org/10.1016/j.fbio.2023.103248.

Liang, D., Zhang, Z., Wu, H., Huang, C., Shuai, P., Ye, C., Tang, S., Wang, Y., Yang, L., Wang, J., Yin, W., & Xia, X. (2014). Single-base-resolution methylomes of *Populus trichocarpa* reveal the association between DNA methylation and drought stress. *BMC Genomic Data*, 15(S1). https://doi.org/10.1186/1471-2156-15-s1-s9.

Liao, H., Yang, C., & Hsieh, M. (2022). Nitrogen deficiency- and sucrose-induced anthocyanin biosynthesis is modulated by HISTONE DEACETYLASE15 in *Arabidopsis*. *Journal of Experimental Botany*, 73(11), 3726–3742. https://doi.org/10.1093/jxb/erac067.

Liégard, B., Baillet, V., Etcheverry, M., Joseph, E., Lariagon, C., Lemoine, J., Evrard, A., Colot, V., Gravot, A., Manzanares-Dauleux, M. J., & Jubault, M. (2018). Quantitative resistance to clubroot infection mediated by transgenerational epigenetic variation in *Arabidopsis*. *New Phytologist*, 222(1), 468–479. https://doi.org/10.1111/nph.15579.

Liu, J., Feng, L., Li, J., & He, Z. (2015). Genetic and epigenetic control of plant heat responses. *Frontiers in Plant Science*, 06. https://doi.org/10.3389/fpls.2015.00267.

Liu, X., Huang, J., Tang, Y., Zhang, S., & Lu, J. (2019). Topoisomerase I inhibitory and photocleavage activity by ruthenium complexes containing a new polyaza ligand. *Inorganic and Nano-Metal Chemistry*, 49(9), 283–290. https://doi.org/10.1080/24701556.2019.1661446.

Liu-Xu, L., González-Hernández, A. I., Camañes, G., Vicedo, B., Scalschi, L., & Llorens, E. (2024). Harnessing green helpers: Nitrogen-fixing bacteria and other beneficial microorganisms in plant–microbe interactions for sustainable agriculture. *Horticulturae*, 10(6), 621. https://doi.org/10.3390/horticulturae10060621.

Luna, E., Bruce, T. J., Roberts, M. R., Flors, V., & Ton, J. (2011). Next-generation systemic acquired resistance. *Plant Physiology*, 158(2), 844–853. https://doi.org/10.1104/pp.111.187468.

Luo, M., Hung, F., Yang, S., Liu, X., & Wu, K. (2013). Histone lysine demethylases and their functions in plants. *Plant Molecular Biology Reporter*, 32(2), 558–565. https://doi.org/10.1007/s11105-013-0673-1.

Luo, M., Liu, X., Singh, P., Cui, Y., Zimmerli, L., & Wu, K. (2011). Chromatin modifications and remodeling in plant abiotic stress responses. *Biochimica Et Biophysica Acta (BBA) - Gene Regulatory Mechanisms*, 1819(2), 129–136. https://doi.org/10.1016/j.bbagrm.2011.06.008.

Mao, H., Wang, H., Liu, S., Li, Z., Yang, X., Yan, J., Li, J., Tran, L. P., & Qin, F. (2015). A transposable element in a *NAC* gene is associated with drought tolerance in maize seedlings. *Nature Communications*, 6(1). https://doi.org/10.1038/ncomms9326.

Martínez-Aguilar, K., Ramírez-Carrasco, G., Hernández-Chávez, J. L., Barraza, A., & Alvarez-Venegas, R. (2016). Use of BABA and INA As activators of a primed state in the common bean (*Phaseolus vulgaris* L.). *Frontiers in Plant Science*, 7. https://doi.org/10.3389/fpls.2016.00653.

Martinez-Medina, A., Flors, V., Heil, M., Mauch-Mani, B., Pieterse, C. M., Pozo, M. J., Ton, J., Van Dam, N. M., & Conrath, U. (2016). Recognizing plant defense priming. *Trends in Plant Science*, 21(10), 818–822. https://doi.org/10.1016/j.tplants.2016.07.009.

Martínez-Pérez, M., Aparicio, F., López-Gresa, M. P., Bellés, J. M., Sánchez-Navarro, J. A., & Pallás, V. (2017). Arabidopsis m6A demethylase activity modulates viral infection of a plant virus and the m6A abundance in its genomic RNAs. *Proceedings of the National Academy of Sciences*, 114(40), 10755–10760. https://doi.org/10.1073/pnas.1703139114.

Masclaux-Daubresse, C., Chen, Q., & Havé, M. (2017). Regulation of nutrient recycling via autophagy. *Current Opinion in Plant Biology*, 39, 8–17. https://doi.org/10.1016/j.pbi.2017.05.001.

Mason, G., Noris, E., Lanteri, S., Acquadro, A., Accotto, G. P., & Portis, E. (2008). Potentiality of methylation-sensitive amplification polymorphism (MSAP) in identifying genes involved in tomato response to Tomato yellow leaf curl Sardinia virus. *Plant Molecular Biology Reporter*, 26(3), 156–173. https://doi.org/10.1007/s11105-008-0031-x.

Matzke, M. A., Kanno, T., & Matzke, A. J. (2014). RNA-directed DNA methylation: The evolution of a complex epigenetic pathway in flowering plants. *Annual Review of Plant Biology*, 66(1), 243–267. https://doi.org/10.1146/annurev-arplant-043014-114633.

Meller, B., Kuźnicki, D., Arasimowicz-Jelonek, M., Deckert, J., & Floryszak-Wieczorek, J. (2018). BABA-primed histone modifications in potato for intergenerational resistance to *Phytophthora infestans*. *Frontiers in Plant Science*, 9. https://doi.org/10.3389/fpls.2018.01228.

Mirouze, M., & Paszkowski, J. (2011). Epigenetic contribution to stress adaptation in plants. *Current Opinion in Plant Biology*, 14(3), 267–274. https://doi.org/10.1016/j.pbi.2011.03.004.

Muller-Xing, R., Xing, Q., & Goodrich, J. (2014). Footprints of the sun: Memory of UV and light stress in plants. *Frontiers in Plant Science*, 5. https://doi.org/10.3389/fpls.2014.00474.

Naydenov, M., Baev, V., Apostolova, E., Gospodinova, N., Sablok, G., Gozmanova, M., & Yahubyan, G. (2015). High-temperature effect on genes engaged in DNA methylation and affected by DNA methylation in *Arabidopsis*. *Plant Physiology and Biochemistry*, 87, 102–108. https://doi.org/10.1016/j.plaphy.2014.12.022.

Oberkofler, V., Pratx, L., & Bäurle, I. (2021). Epigenetic regulation of abiotic stress memory: Maintaining the good things while they last. *Current Opinion in Plant Biology*, 61, 102007. https://doi.org/10.1016/j.pbi.2021.102007.

O'Brien, A. M., Laurich, J., Lash, E., & Frederickson, M. E. (2020). Mutualistic outcomes across plant populations, microbes, and environments in the duckweed *Lemna minor*. *Microbial Ecology*, 80(2), 384–397. https://doi.org/10.1007/s00248-019-01452-1.

Panda, K., & Slotkin, R. K. (2013). Proposed mechanism for the initiation of transposable element silencing by the RDR6-directed DNA methylation pathway. *Plant Signaling & Behavior*, 8(8), e25206. https://doi.org/10.4161/psb.25206.

Pandey, N., Goswami, N., Tripathi, D., Rai, K. K., Rai, S. K., Singh, S., & Pandey-Rai, S. (2019). Epigenetic control of UV-B-induced flavonoid accumulation in *Artemisia annua* L. *Planta*, 249(2), 497–514. https://doi.org/10.1007/s00425-018-3022-7.

Pandey, N., & Pandey-Rai, S. (2015). Deciphering UV-B-induced variation in DNA methylation pattern and its influence on regulation of DBR2 expression in *Artemisia annua* L. *Planta*, 242(4), 869–879. https://doi.org/10.1007/s00425-015-2323-3.

Park, J., Dempewolf, E., Zhang, W., & Wang, Z. (2017). RNA-guided transcriptional activation via CRISPR/dCas9 mimics overexpression phenotypes in *Arabidopsis*. *PLoS ONE*, 12(6), e0179410. https://doi.org/10.1371/journal.pone.0179410.

Park, J., Lim, C. J., Shen, M., Park, H. J., Cha, J., Iniesto, E., Rubio, V., Mengiste, T., Zhu, J., Bressan, R. A., Lee, S. Y., Lee, B., Jin, J. B., Pardo, J. M., Kim, W., & Yun, D. (2018). Epigenetic switch from repressive to permissive chromatin in response to cold stress. *Proceedings of the National Academy of Sciences*, 115(23). https://doi.org/10.1073/pnas.1721241115.

Pathak, H., Kumar, M., Molla, K. A., & Chakraborty, K. (2021, August 12). *Abiotic Stresses in Rice Production: Impacts and Management* https://epubs.icar.org.in/ejournal/index.php/OIJR/article/view/113846/0.

Pavangadkar, K., Thomashow, M. F., & Triezenberg, S. J. (2010). Histone dynamics and roles of histone acetyltransferases during cold-induced gene regulation in *Arabidopsis*. *Plant Molecular Biology*, 74(1–2), 183–200. https://doi.org/10.1007/s11103-010-9665-9.

Popova, O. V., Dinh, H. Q., Aufsatz, W., & Jonak, C. (2013). The RDDM pathway is required for basal heat tolerance in *Arabidopsis*. *Molecular Plant*, 6(2), 396–410. https://doi.org/10.1093/mp/sst023.

Po-Wen, C., Singh, P., & Zimmerli, L. (2012). Priming of the *Arabidopsis* pattern-triggered immunity response upon infection by necrotrophic *Pectobacterium carotovorum* bacteria. *Molecular Plant Pathology*, 14(1), 58–70. https://doi.org/10.1111/j.1364-3703.2012.00827.x.

Qin, X., Zhang, K., Fan, Y., Fang, H., Nie, Y., & Wu, X. (2022). The bacterial MtrAB two-component system regulates the cell wall homeostasis responding to environmental alkaline stress. *Microbiology Spectrum*, 10(5). https://doi.org/10.1128/spectrum.02311-22.

Quiroz-Pacheco, E. N., Mora, F., Boege, K., Domínguez, C. A., & Del-Val, E. (2020). Effects of herbivory and its timing on reproductive success of a tropical deciduous tree. *Annals of Botany*, 126(5), 957–969. https://doi.org/10.1093/aob/mcaa117.

Rambani, A., Pantalone, V., Yang, S., Rice, J. H., Song, Q., Mazarei, M., Arelli, P. R., Meksem, K., Stewart, C. N., & Hewezi, T. (2020). Identification of introduced and stably inherited DNA methylation variants in soybean associated with soybean cyst nematode parasitism. *New Phytologist*, 227(1), 168–184. https://doi.org/10.1111/nph.16511.

Rambani, A., Rice, J. H., Liu, J., Lane, T., Ranjan, P., Mazarei, M., Pantalone, V., Stewart, C. N., Staton, M., & Hewezi, T. (2015). The methylome of soybean roots during the compatible interaction with the soybean cyst nematode. *Plant Physiology*, 168(4), 1364–1377. https://doi.org/10.1104/pp.15.00826.

Ramirez-Prado, J. S., Abulfaraj, A. A., Rayapuram, N., Benhamed, M., & Hirt, H. (2018). Plant immunity: From signaling to epigenetic control of defense. *Trends in Plant Science*, 23(9), 833–844. https://doi.org/10.1016/j.tplants.2018.06.004.

Ramirez-Prado, J. S., Latrasse, D., Rodriguez-Granados, N. Y., Huang, Y., Manza-Mianza, D., Brik-Chaouche, R., Jaouannet, M., Citerne, S., Bendahmane, A., Hirt, H., Raynaud, C., & Benhamed, M. (2019). The polycomb protein LHP1 regulates *Arabidopsis thaliana* stress responses through the repression of the MYC2-dependent branch of immunity. *The Plant Journal*, 100(6), 1118–1131. https://doi.org/10.1111/tpj.14502.

Rasmann, S., De Vos, M., Casteel, C. L., Tian, D., Halitschke, R., Sun, J. Y., Agrawal, A. A., Felton, G. W., & Jander, G. (2011). Herbivory in the previous generation primes plants for enhanced insect resistance. *Plant Physiology*, 158(2), 854–863. https://doi.org/10.1104/pp.111.187831.

Riddle, N. C. (2014). Heritable generational epigenetic effects through RNA. In *Elsevier eBooks* (pp. 105–119). https://doi.org/10.1016/b978-0-12-405944-3.00010-6.

Rus, A., Yokoi, S., Sharkhuu, A., Reddy, M., Lee, B., Matsumoto, T. K., Koiwa, H., Zhu, J., Bressan, R. A., & Hasegawa, P. M. (2001). AtHKT1 is a salt tolerance determinant that controls Na^+ entry into plant roots. *Proceedings of the National Academy of Sciences*, 98(24), 14150–14155. https://doi.org/10.1073/pnas.241501798.

Rutowicz, K., Puzio, M., Halibart-Puzio, J., Lirski, M., Kroteń, M. A., Kotliński, M., Kniżewski, Ł., Lange, B., Muszewska, A., Śniegowska-świerk, K., Kościelniak, J., Iwanicka-Nowicka, R., Żmuda, K., Buza, K., Janowiak, F., Jõesaar, I., Laskowska-Kaszub, K., Fogtman, A., Zielenkiewicz, P., Tiuryn J., Kollist H., Siedlecki P., Ginalski K., Świeżewski S., Koblowska M., Archacki R., Wilczyński B., Rapacz M., Jerzmanowski, A. (2015). A specialized histone H1 variant is required for adaptive responses to complex abiotic stress and related DNA methylation in *Arabidopsis*. *Plant Physiology*, 00493.2015. https://doi.org/10.1104/pp.15.00493.

Ryu, H., Cho, H., Bae, W., & Hwang, I. (2014). Control of early seedling development by BES1/TPL/HDA19-mediated epigenetic regulation of ABI3. *Nature Communications*, 5(1). https://doi.org/10.1038/ncomms5138.

Saeed, F., Chaudhry, U. K., Bakhsh, A., Raza, A., Saeed, Y., Bohra, A., & Varshney, R. K. (2022). Moving beyond DNA sequence to improve plant stress responses. *Frontiers in Genetics*, 13. https://doi.org/10.3389/fgene.2022.874648.

Sah, S. K., Reddy, K. R., & Li, J. (2016). Abscisic acid and abiotic stress tolerance in crop plants. *Frontiers in Plant Science*, 7. https://doi.org/10.3389/fpls.2016.00571.

Sakuma, Y., Maruyama, K., Osakabe, Y., Qin, F., Seki, M., Shinozaki, K., & Yamaguchi-Shinozaki, K. (2006). Functional analysis of an *Arabidopsis* transcription factor, DREB2A, involved in drought-responsive gene expression. *The Plant Cell*, 18(5), 1292–1309. https://doi.org/10.1105/tpc.105.035881.

Sánchez, A. L., Stassen, J. H., Furci, L., Smith, L. M., & Ton, J. (2016). The role of DNA (de)methylation in immune responsiveness of *Arabidopsis*. *The Plant Journal*, 88(3), 361–374. https://doi.org/10.1111/tpj.13252.

Sanchez, D. H., & Paszkowski, J. (2014). Heat-Induced release of epigenetic silencing reveals the concealed role of an imprinted plant gene. *PLoS Genetics*, 10(11), e1004806. https://doi.org/10.1371/journal.pgen.1004806.

Sanderson, B. J., Park, S., Jameel, M. I., Kraft, J. C., Thomashow, M. F., Schemske, D. W., & Oakley, C. G. (2019). Genetic and physiological mechanisms of freezing tolerance in locally adapted populations of a winter annual. *American Journal of Botany*, 107(2), 250–261. https://doi.org/10.1002/ajb2.1385.

Secco, D., Wang, C., Shou, H., Schultz, M. D., Chiarenza, S., Nussaume, L., Ecker, J. R., Whelan, J., & Lister, R. (2015). Stress induced gene expression drives transient DNA methylation changes at adjacent repetitive elements. *eLife*, 4. https://doi.org/10.7554/elife.09343.

Shafiq, S., Berr, A., & Shen, W. (2013). Combinatorial functions of diverse histone methylations in *Arabidopsis thaliana* flowering time regulation. *New Phytologist*, 201(1), 312–322. https://doi.org/10.1111/nph.12493.

Shafiq, S., Zeb, Q., Ali, A., Sajjad, Y., Nazir, R., Widemann, E., & Liu, L. (2019). Lead, cadmium and zinc phytotoxicity alter DNA methylation levels to confer heavy metal tolerance in wheat. *International Journal of Molecular Sciences*, 20(19), 4676. https://doi.org/10.3390/ijms20194676.

Shah, J. M. (2021). Epimutations and mutations, nurturing phenotypic diversity. *Genetica*, 150(3–4), 171–181. https://doi.org/10.1007/s10709-021-00124-8.

Shen, X., De Jonge, J., Forsberg, S. K. G., Pettersson, M. E., Sheng, Z., Hennig, L., & Carlborg, Ö. (2014). Natural CMT2 variation is associated with genome-wide methylation changes and temperature seasonality. *PLoS Genetics*, 10(12), e1004842. https://doi.org/10.1371/journal.pgen.1004842.

Sheng, P., Fields, C., Aadland, K., Wei, T., Kolaczkowski, O., Gu, T., Kolaczkowski, B., & Xie, M. (2018). Dicer cleaves 5′-extended microRNA precursors originating from RNA polymerase II transcription start sites. *Nucleic Acids Research*, 46(11), 5737–5752. https://doi.org/10.1093/nar/gky306.

Shinozaki, K., & Yamaguchi-Shinozaki, K. (2006). Gene networks involved in drought stress response and tolerance. *Journal of Experimental Botany*, 58(2), 221–227. https://doi.org/10.1093/jxb/erl164.

Singh, S. K., Wu, X., Shao, C., & Zhang, H. (2022). Microbial enhancement of plant nutrient acquisition. *Stress Biology*, 2(1). https://doi.org/10.1007/s44154-021-00027-w.

Singroha, G., Sharma, P., & Sunkur, R. (2021). Current status of microRNA-mediated regulation of drought stress responses in cereals. https://www.semanticscholar.org/paper/Current-status-of-microRNA-mediated-regulation-of-Singroha-Sharma/f8809f810bdfbb50c615014be6d6a27accdb666f.

Song, Y., Ji, D., Li, S., Wang, P., Li, Q., & Xiang, F. (2012). The dynamic changes of DNA methylation and histone modifications of salt responsive transcription factor genes in soybean. *PLoS ONE*, 7(7), e41274. https://doi.org/10.1371/journal.pone.0041274.

Song, Z., Zhang, L., Han, J., Zhou, M., & Liu, J. (2020). Histone H3K4 methyltransferases *SDG25* and *ATX1* maintain heat-stress gene expression during recovery in *Arabidopsis*. *The Plant Journal*, 105(5), 1326–1338. https://doi.org/10.1111/tpj.15114.

Stoddard, C. I., Feng, S., Campbell, M. G., Liu, W., Wang, H., Zhong, X., Bernatavichute, Y., Cheng, Y., Jacobsen, S. E., & Narlikar, G. J. (2018). A nucleosome bridging mechanism for activation of a maintenance DNA methyltransferase. *Molecular Cell*, 73(1), 73–83.e6. https://doi.org/10.1016/j.molcel.2018.10.006.

Sudan, J., Raina, M., & Singh, R. (2018). Plant epigenetic mechanisms: Role in abiotic stress and their generational heritability. *3 Biotech*, 8(3). https://doi.org/10.1007/s13205-018-1202-6.

Svistoonoff, S., Creff, A., Reymond, M., Sigoillot-Claude, C., Ricaud, L., Blanchet, A., Nussaume, L., & Desnos, T. (2007). Root tip contact with low-phosphate media reprograms plant root architecture. *Nature Genetics*, 39(6), 792–796. https://doi.org/10.1038/ng2041.

Tian, P., Lin, Z., Lin, D., Dong, S., Huang, J., & Huang, T. (2021). The pattern of DNA methylation alteration, and its association with the changes of gene expression and alternative splicing during phosphate starvation in tomato. *The Plant Journal*, 108(3), 841–858. https://doi.org/10.1111/tpj.15486.

Tiwari, P., Bose, S. K., Park, K., Dufossé, L., & Fouillaud, M. (2024). Plant-Microbe Interactions under the Extreme Habitats and Their Potential Applications.https://www.semanticscholar.org/paper/Plant-Microbe-Interactions-under-the-Extreme-and-Tiwari-Bose/fd26d638a8407f6b38ee7933a508224c119e55db.

Tonosaki, K., Fujimoto, R., Dennis, E. S., Raboy, V., & Osabe, K. (2022). Will epigenetics be a key player in crop breeding? *Frontiers in Plant Science*, 13. https://doi.org/10.3389/fpls.2022.958350.

Uddin, M. N., & Kim, J. (2013). Intercellular and systemic spread of RNA and RNAi in plants. *Wiley Interdisciplinary Reviews - RNA*, 4(3), 279–293. https://doi.org/10.1002/wrna.1160.

Waddington, C. H. (2011). The epigenotype. *International Journal of Epidemiology*, 41(1), 10–13. https://doi.org/10.1093/ije/dyr184.

Wang, L., & Qiao, H. (2020). Chromatin regulation in plant hormone and plant stress responses. *Current Opinion in Plant Biology*, 57, 164–170. https://doi.org/10.1016/j.pbi.2020.08.007.

Wang, L., Wang, L., Tan, M., Wang, L., Zhao, W., You, J., Wang, L., Yan, X., & Wang, W. (2023). The pattern of alternative splicing and DNA methylation alteration and their interaction in linseed (*Linum usitatissimum* L.) response to repeated drought stresses. *Biological Research*, 56(1). https://doi.org/10.1186/s40659-023-00424-7.

Wang, T., Xing, J., Liu, Z., Zheng, M., Yao, Y., Hu, Z., Peng, H., Xin, M., Zhou, D., & Ni, Z. (2019). Histone acetyltransferase GCN5-mediated regulation of long non-coding RNA At4 contributes to phosphate starvation response in *Arabidopsis*. *Journal of Experimental Botany*, 70(21), 6337–6348. https://doi.org/10.1093/jxb/erz359.

Wang, W., Pan, Y., Zhao, X., Dwivedi, D., Zhu, L., Ali, J., Fu, B., & Li, Z. (2010). Drought-induced site-specific DNA methylation and its association with drought tolerance in rice (*Oryza sativa* L.). *Journal of Experimental Botany*, 62(6), 1951–1960. https://doi.org/10.1093/jxb/erq391.

Wang, W., Zhao, X., Pan, Y., Zhu, L., Fu, B., & Li, Z. (2011). DNA methylation changes detected by methylation-sensitive amplified polymorphism in two contrasting rice genotypes under salt stress. *Journal of Genetics and Genomics/Journal of Genetics and Genomics*, 38(9), 419–424. https://doi.org/10.1016/j.jgg.2011.07.006.

Wu, J., Ichihashi, Y., Suzuki, T., Shibata, A., Shirasu, K., Yamaguchi, N., & Ito, T. (2019). Abscisic acid-dependent histone demethylation during postgermination growth arrest in *Arabidopsis*. *Plant Cell & Environment*, 42(7), 2198–2214. https://doi.org/10.1111/pce.13547.

Xie, M., & Yu, B. (2015). siRNA-directed DNA methylation in plants. *Current Genomics*, 16(1), 23–31. https://doi.org/10.2174/1389202915666141128002211.

Xu, J. M., Wang, Z. Q., Wang, J. Y., Li, P. F., Jin, J. F., Chen, W. W., Fan, W., Kochian, L. V., Zheng, S. J., & Yang, J. L. (2019). Low phosphate represses histone deacetylase complex1 to regulate root system architecture remodeling in *Arabidopsis*. *New Phytologist*, 225(4), 1732–1745. https://doi.org/10.1111/nph.16264.

Xu, R., Wang, Y., Zheng, H., Lu, W., Wu, C., Huang, J., Yan, K., Yang, G., & Zheng, C. (2015). Salt-induced transcription factorMYB74is regulated by the RNA-directed DNA methylation pathway in *Arabidopsis*. *Journal of Experimental Botany*, 66(19), 5997–6008. https://doi.org/10.1093/jxb/erv312.

Xu, W., Li, K., Li, S., Hou, Q., Zhang, Y., Liu, K., & Sun, Q. (2020). The R-loop atlas of *Arabidopsis* development and responses to environmental stimuli. *The Plant Cell*, 32(4), 888–903. https://doi.org/10.1105/tpc.19.00802

Yang, C., Shen, W., Yang, L., Sun, Y., Li, X., Lai, M., Wei, J., Wang, C., Xu, Y., Li, F., Liang, S., Yang, C., Zhong, S., Luo, M., & Gao, C. (2020). HY5-HDA9 module transcriptionally regulates plant autophagy in response to light-to-dark conversion and nitrogen starvation. *Molecular Plant*, 13(3), 515–531. https://doi.org/10.1016/j.molp.2020.02.011.

Yang, R., Hong, Y., Ren, Z., Tang, K., Zhang, H., Zhu, J., & Zhao, C. (2019). A role for PICKLE in the regulation of cold and salt stress tolerance in *Arabidopsis*. *Frontiers in Plant Science*, 10. https://doi.org/10.3389/fpls.2019.00900.

Yang, Y., & Davis, T. (2017). A new perspective on polyploid *Fragaria* (Strawberry) genome composition based on large-scale, multi-locus phylogenetic analysis. https://www.semanticscholar.org/paper/A-New-Perspective-on-Polyploid-Fragaria-Genome-on-Yang-Davis/5ba4339c006636615975fe786b6efd2d331f58ab.

Yolcu, S., Ozdemir, F., Güler, A., & Bor, M. (2016). Histone acetylation influences the transcriptional activation of POX in *Beta vulgaris* L. and *Beta maritima* L. under salt stress. *Plant Physiology and Biochemistry*, 100, 37–46. https://doi.org/10.1016/j.plaphy.2015.12.019.

Yong-Villalobos, L., González-Morales, S. I., Wrobel, K., Gutiérrez-Alanis, D., Cervantes-Peréz, S. A., Hayano-Kanashiro, C., Oropeza-Aburto, A., Cruz-Ramírez, A., Martínez, O., & Herrera-Estrella, L. (2015). Methylome analysis reveals an important role for epigenetic changes in the regulation of the *Arabidopsis* response to phosphate starvation. *Proceedings of the National Academy of Sciences*, 112(52). https://doi.org/10.1073/pnas.1522301112.

Yu, A., Lepère, G., Jay, F., Wang, J., Bapaume, L., Wang, Y., Abraham, A., Penterman, J., Fischer, R. L., Voinnet, O., & Navarro, L. (2013). Dynamics and biological relevance of DNA demethylation in *Arabidopsis* antibacterial defense. *Proceedings of the National Academy of Sciences*, 110(6), 2389–2394. https://doi.org/10.1073/pnas.1211757110.

Zeng, W., Huang, H., Lin, X., Zhu, C., Kosami, K., Huang, C., Zhang, H., Duan, C., Zhu, J., & Miki, D. (2020). Roles of DEMETER in regulating DNA methylation in vegetative tissues and pathogen resistance. *Journal of Integrative Plant Biology*, 63(4), 691–706. https://doi.org/10.1111/jipb.13037.

Zhang, H., Jin, Z., Cui, F., Zhao, L., Zhang, X., Chen, J., Zhang, J., Li, Y., Li, Y., Niu, Y., Zhang, W., Gao, C., Fu, X., Tong, Y., Wang, L., Ling, H., Li, J., & Xiao, J. (2023). Epigenetic modifications regulate cultivar-specific root development and metabolic adaptation to nitrogen availability in wheat. *Nature Communications*, 14(1). https://doi.org/10.1038/s41467-023-44003-6.

Zhang, H., Lang, Z., & Zhu, J. (2018). Dynamics and function of DNA methylation in plants. *Nature Reviews Molecular Cell Biology*, 19(8), 489–506. https://doi.org/10.1038/s41580-018-0016-z.

Zhang, X., Bernatavichute, Y. V., Cokus, S., Pellegrini, M., & Jacobsen, S. E. (2009). Genome-wide analysis of mono-, di- and trimethylation of histone H3 lysine 4 in *Arabidopsis thaliana*. *Genome Biology*, 10(6). https://doi.org/10.1186/gb-2009-10-6-r62.

Zhang, X., Yazaki, J., Sundaresan, A., Cokus, S., Chan, S. W., Chen, H., Henderson, I. R., Shinn, P., Pellegrini, M., Jacobsen, S. E., & Ecker, J. R. (2006). Genome-wide high-resolution mapping and functional analysis of DNA methylation in *Arabidopsis*. *Cell*, 126(6), 1189–1201. https://doi.org/10.1016/j.cell.2006.08.003.

Zhao, C., Zhang, Z., Xie, S., Si, T., Li, Y., & Zhu, J. (2016). Mutational evidence for the critical role of CBF transcription factors in cold acclimation in *Arabidopsis*. *Plant Physiology*, 171(4), 2744–2759. https://doi.org/10.1104/pp.16.00533.

Zheng, X., Chen, L., Xia, H., Wei, H., Lou, Q., Li, M., Li, T., & Luo, L. (2017). Transgenerational epimutations induced by multi-generation drought imposition mediate rice plant's adaptation to drought condition. *Scientific Reports*, 7(1). https://doi.org/10.1038/srep39843.

Zhou, C., Zhang, L., Duan, J., Miki, B., & Wu, K. (2005). Histone deacetylase19 is involved in jasmonic acid and ethylene signaling of pathogen response in *Arabidopsis*. *The Plant Cell*, 17(4), 1196–1204. https://doi.org/10.1105/tpc.104.028514.

Zhou, Q., Shi, J., Li, Z., Zhang, S., Zhang, S., Zhang, J., Bao, M., & Liu, G. (2021). miR156/157 targets SPLs to regulate flowering transition, plant architecture and flower organ size in *Petunia*. *Plant and Cell Physiology*, 62(5), 839–857. https://doi.org/10.1093/pcp/pcab041.

Zhu, H., Xie, W., Xu, D., Miki, D., Tang, K., Huang, C., & Zhu, J. (2018). DNA demethylase ROS1 negatively regulates the imprinting of DOGL4 and seed dormancy in *Arabidopsis thaliana*. *Proceedings of the National Academy of Sciences*, 115(42). https://doi.org/10.1073/pnas.1812847115.

Zhu, J. (2016). Abiotic stress signaling and responses in plants. *Cell*, 167(2), 313–324. https://doi.org/10.1016/j.cell.2016.08.029.

Zinn, K. E., Tunc-Ozdemir, M., & Harper, J. F. (2010). Temperature stress and plant sexual reproduction: Uncovering the weakest links. *Journal of Experimental Botany*, 61(7), 1959–1968. https://doi.org/10.1093/jxb/erq053.

Index

S

T

U

V

W